To the Shmu

Lecture Notes on Mathematical Modelling in the Life Sciences

Series Editors
Angela Stevens
Michael C. Mackey

For further volumes:
http://www.springer.com/series/10049

Lecture Notes on Mathematical Modelling in the Life Sciences

Paul C. Bressloff

Waves in Neural Media

From Single Neurons to Neural Fields

Springer

Paul C. Bressloff
University of Utah
Department of Mathematics
Salt Lake City, UT, USA

ISSN 2193-4789 ISSN 2193-4797 (electronic)
ISBN 978-1-4614-8865-1 ISBN 978-1-4614-8866-8 (eBook)
DOI 10.1007/978-1-4614-8866-8
Springer New York Heidelberg Dordrecht London

Library of Congress Control Number: 2013948234

Mathematics Subject Classification (2010): 35C07, 82C31, 92C20, 92C30, 92C37, 82C31

Printed on acid-free paper

Springer is part of Springer Science+Business Media (www.springer.com)

Preface

This is a book on mathematical neuroscience, in which the unifying theme is wavelike phenomena at multiple spatial and temporal scales. There are already several excellent books on applications of dynamical systems theory and bifurcation theory to neuroscience, but these are mainly concerned with mathematical models of neurons and networks that take the form of systems of ordinary differential equations (ODEs). Spatial aspects are usually restricted to standard partial differential equation (PDE) models of action potential propagation along axons and the spatial spread of voltage along dendrites (the cable equation). However, there are many other interesting spatiotemporal phenomena that are not usually addressed in any detail, at least within a single book. These include electrical and chemical waves along spiny dendrites, calcium waves in astrocytes, excitable and oscillatory waves in cortical and subcortical structures, binocular rivalry waves and ambiguous perception, oscillatory waves and swimming locomotion, cytoskeletal waves and neuritogenesis, spreading depression and migraine auras, the spread of prions and other neurodegenerative diseases within the brain, and epileptic waves in complex neural networks. All of these phenomena are linked by the notion of the time-dependent propagation or spread of some quantity in space, whether it is chemical concentration, a protein aggregate, voltage, or spiking activity of a population of neurons. They should be distinguished from another important class of spatial phenomena, namely, pattern-forming instabilities, which could be the subject of a future book!

In order to develop the themes of this book, I have endeavored to give a pedagogical introduction to the mathematical modeling of biological neurons and networks, as well as a toolbox of analytical techniques used to study wavelike phenomena, including the construction of traveling wave solutions of PDES and nonlocal equations, stability analysis, phase-plane analysis, singular perturbation theory, phase-reduction and averaging methods, and homogenization theory. A common thread throughout the book is an emphasis on stochasticity. In recent years, there has been an explosion of interest in the role of noise in biological systems, partly driven by the presence of low copy numbers in gene and biochemical networks. Noise is also ubiquitous in the brain, ranging from fluctuations in gene expression and the opening of ion-channel proteins to the Poisson-like spiking activity of individual cortical neurons, the presence of fluctuations during perceptual decision making and bistable perception, and the generation of spontaneous activity during resting brain states. Hence, the book contains an introduction to the theory of stochastic processes and how it can be used to study the effects of noise on wave propagation and wave initiation. In the latter case, one typically has to solve an escape problem, in which noise drives some quantity such as voltage, calcium concentration, or population spiking activity above a threshold. This results, for example, in the generation of a spontaneous action potential, a calcium spark, or a transition between cortical up and down states, respectively.

Before giving a more detailed description of the contents of this book, I thought that I would elaborate a little on my own approach to neuroscience. This is not to favor one approach over another—a diversity of methodological and conceptual frameworks is crucial in trying to understand such a complex system as the brain—but rather to highlight my own tastes and biases. Let me begin with a story (perhaps apocryphal). Several years ago a well-known mathematical biology journal was relaunched with a new editorial board. At the final organizational meeting of the editors in chief, after some good food and wine, a satisfied glow spread around the room as the list of associate editors was drawn up, covering the major areas of biology. However, someone then sheepishly pointed out that one area was missing—neuroscience! I suspect that neuroscience has a slightly complicated relationship with the rest of theoretical biology due to the ambiguous notion of "computation." One sense of the word, which applies to all areas of biology, refers to the high level of biological detail and computational complexity in a model, which means that it is usually less amenable to mathematical analysis than simpler reduced models. The focus tends to be on issues such as parameter searches and data fitting, sensitivity analysis, numerical convergence, and computational efficiency. It follows that one way to characterize a modeling approach is where it lies on the mathematical–computational biology axis. Within this context, the present book is on mathematical rather than computational neuroscience, since I consider reduced, mathematically tractable models of wavelike phenomena.

The other sense of the word "computation" refers to a functional interpretation of the brain as a biological computer that processes sensory inputs via a sequence of internal brain states to produce a motor output or a change in cognitive state. A common assumption is that if we had a complete understanding of the physiology of the heart at the molecular, cellular, and systems levels, then we would understand the function of the heart, whereas the same is not true of the brain. That is, what makes neuroscience so challenging, beyond the incredible complexity of the brain as a biological organ, is how one links physiological brain states to behavior and cognition. Consequently, a major drive in computational neuroscience (in the second sense) is to link physiological descriptions of neurons and networks with computation and information processing. Popular themes are spike trains and neural coding, statistical correlations, the role of feedback in predictive coding, learning algorithms, Bayesian statistics, and maximum entropy. The current book is not concerned with this form of computational neuroscience either, since I do not really discuss the possible functional role of wavelike neural phenomena, although the role of waves in development and disease is discussed at length. There have been some attempts to interpret the computational role of waves, based on the observation that cortical waves often occur between sensory processing events. However, such an observation is correlative at best and provides very little insight into the causal mechanisms of perception. Unfortunately, the same could be said of many of the overhyped results coming from functional magnetic resonance imaging (fMRI), where parts of the brain light up when someone is telling a joke or planning to commit a crime.

In the media lately, a number of distinguished physicists have had some harsh words to say about philosophy and its relevance to scientific questions. I personally disagree with such attitudes, partly due to my own interests in philosophy and also because I have found it useful to stand back occasionally in order to place my own work in a broader context. I think that this is particularly important in a field such as neuroscience, with all of its potential clinical, psychological, ethical, sociological, and cultural ramifications. Of course, reading philosophy does not make me a philosopher, so I proceed with caution. Nevertheless, I am a naturalist, in the sense that I do not believe the brain is made of different "stuff" or obeys different physical laws from planets, sushi, and hot tubs. On the other hand, I tend not to view the brain in functionalist terms, that is, as the hardware to the mind's software. I favor a more direct biological interpretation in terms of an evolved and adaptive system that enables an organism to carry out goal-directed behavior in a changing physical and social environment (although many people believe that this can be reformulated in terms of algorithms and computation). A related idea is that meaning or semantics cannot be generated from a purely syntactic process, since one needs to take into account the evolutionary and biographical history of an embodied mind acting within a societal and cultural context. Another reason I tend to steer clear of computational approaches is that I fear that using such a paradigm is in danger of imposing order, rationality, and engineering design principles on a rather messy and haphazard system. I tend to agree with the idea that the brain is a "kludge," in which modules that evolved to solve simpler problems have been thrown together suboptimally as the complexity of an organism has increased. I also feel that randomness plays a significant role in how we "think," which counters the idea of following a computer program or algorithm. I do not mean randomness in the sense of exploiting stochasticity to enhance performance (e.g., simulated annealing or random search), which itself could be algorithmic, but randomness as irrationality. People commit suicide, skateboard down Everest on a piano, and drink root beer (an alien concept to an Englishman).

The "elephant in the room" is the so-called hard problem of consciousness, that is, of explaining how and why we have qualia or phenomenal experiences. (This is a controversial issue, since some view qualia as an illusion based on misguided folk psychology or poor intuition pumps.) A related question is how one goes from objective brain states to subjective mental states. I would describe myself as a non-reductionist, in the sense that I believe mental properties supervene on (but are distinct from) physical properties and are an emergent feature of the physical brain. However, a systematic framework for understanding this emergence is currently beyond our grasp. An analogy is how one goes from microscopic descriptions of molecules to macroscopic descriptions of states of matter. The latter also supervene on the former, but in this case we do have a framework for crossing the divide, namely, statistical mechanics. Hence, I do not believe we will discover a "statistical mechanics" of the brain, at least not during the funding periods of the European "Human Brain Project" and the USA's "BRAIN Initiative!" An interesting alternative to emergentism is neutral monism, which posits that protean versions of mental properties inhere in all physical objects—a form of panpsychism.

Having elucidated what this book is not about and why, let me now turn to its contents. The book is divided into three parts: Neurons (Chaps. 1–4), Networks (Chaps. 5–8), and Development and disease (Chap. 9). Chapter 1 provides a detailed introduction to the working parts of a neuron, including conductance-based models of action potential generation, synaptic and dendritic processing, and ion channels. Two important mathematical topics are also introduced. First, the dynamics of a periodically forced neural oscillator is used to introduce phase-reduction and averaging methods, phase-resetting curves, and synchronization. These are later applied to the study of waves in oscillatory neural media (Chap. 5). Second, a detailed account of stochastic ion channels and membrane voltage fluctuations is given, which also provides background material on stochastic processes. A major theme is how to model and analyze stochastic hybrid systems, in which a continuous variable (e.g., voltage) couples to a discrete jump Markov process (e.g., number of open ion channels). Spontaneous action potential generation is formulated as a first passage time problem, which is solved using perturbation methods such as WKB and matched asymptotics. These methods are later used to analyze related problems such as the generation of calcium sparks (Chap. 4) and bistability in populations of spiking neurons (Chap. 6). Chapter 2 covers the classical problem of waves in one-dimensional excitable media, as exemplified by the FitzHugh–Nagumo model of action potential propagation along an axon. Standard methods for analyzing front and pulse solutions of PDEs are described, including phase-plane analysis, singular perturbation methods and slow–fast analysis, and Evans functions for wave stability. In addition, the problem of wave propagation failure in myelinated axons is considered, where an averaging method is used to determine the effects of spatial discreteness on wave speed. This method is later used to analyze wave propagation failure in inhomogeneous neural fields (Chap. 7). Finally, stochastic traveling waves are considered, where formal perturbation methods are used to show how to separate out fast fluctuations of the wave profile from the slow diffusive-like wandering of the wave.

Chapter 3 presents two different models of traveling waves along spiny dendrites: a spike–diffuse–spike model of propagating voltage spikes mediated by active dendritic spines and a reaction–diffusion model of Ca^{2+}/calmodulin-dependent protein kinase II (CaMKII) translocation waves. The former model introduces methods that are also used to analyze solitary waves propagating in spiking neural networks (Chap. 5). The latter model turns out to be identical in form to the diffusive susceptible-infected (SI) model of the spread of epidemics, which is a generalization of the scalar Fisher–KPP equation of population genetics. One characteristic feature of such equations is that they support traveling fronts propagating into an unstable steady state, in which the wave speed and longtime asymptotics are determined by the dynamics in the leading edge of the wave—the so-called pulled fronts. In particular, a sufficiently localized initial perturbation will asymptotically approach the traveling front solution that has the minimum possible wave speed. Hence, pulled fronts have very different properties from those considered in Chap. 2. Homogenization methods are also presented, which allow one to approximate the discrete distribution of spines by a smooth distribution. Chapter 4 gives a comprehensive review of calcium wave modeling, with an emphasis on their role in neuronal calcium signaling. Two models of intracellular waves are considered in some detail: a reaction–diffusion model of calcium dynamics and the fire–diffuse–fire model of calcium release. The latter is formally very similar to the spike–diffuse–spike model of spiny dendrites and is analyzed accordingly. Stochastic models of spontaneous calcium release (calcium puffs and sparks) are then analyzed using the stochastic methods introduced in Chap. 1. Finally, several models of intercellular calcium waves in astrocytes are presented. Traditionally, astrocytes were thought to be physiologically passive cells that only play a supporting role in the central nervous system by regulating and optimizing the environment within which neurons operate. However, there is an increasing amount of empirical data indicating that astrocytes play a more active role in modulating synaptic transmission and neuronal signal processing.

In the second part of the book, we consider a variety of neural network models that have been used to investigate the mechanisms and properties of wave propagation in cortex and other parts of the nervous system. There has been a rapid increase in the number of computational studies of network dynamics, which are based on biophysically detailed conductance-based models of synaptically (and possibly electrically) coupled neurons. These models provide considerable insights into the role of ionic currents, synaptic processing, and network structure on spatiotemporal dynamics, but they tend to be analytically intractable. This has motivated an alternative approach to network dynamics, involving simplified neuron models that hopefully capture important aspects of wave phenomena, while allowing a more concise mathematical treatment. In the case of oscillatory networks, such a simplification can be achieved by reducing a conductance-based neuron model to a phase model along the lines of Chap. 1. Alternatively, one can use a simplified spiking neuron model such as integrate-and-fire in order to investigate waves in excitable and oscillatory neural media. Both of these approaches are considered in Chap. 5, which also provides a summary of various wave phenomena in cortical and subcortical structures. Chapter 6 describes the construction of population-based rate models under the assumption that the spiking of individual neurons is unimportant. The issue of how stochasticity at the single-cell level manifests itself at the population level is discussed, introducing topics such as balanced networks, Poisson statistics, and asynchronous states. Stochastic methods are then used to analyze bistability in a stochastic population model. Finally, the transition from spatially structured neural networks to continuum neural fields is highlighted. The latter take the form of nonlocal integrodifferential equations, in which the integral kernel represents the distribution of synaptic connections.

Chapter 7 develops the theory of waves in excitable neural fields, where the fundamental network element is a local population of cells rather than a single neuron. It is shown how many of the PDE methods and results from the analysis of waves in reaction–diffusion equations considered in Part I can be extended to the nonlocal equations of neural field theory. First, the existence and stability of solitary traveling fronts and pulses in one-dimensional excitatory neural fields are considered. In the case of traveling pulses, it is necessary to include some form of local negative feedback mechanism such as synaptic depression or spike frequency adaptation. Two approaches to analyzing wave propagation failure in inhomogeneous neural media are then presented: one based on averaging methods and the other on interfacial dynamics. Finally, wave propagation in stochastic neural fields is analyzed and oscillatory waves in two-dimensional neural media are briefly discussed. In Chap. 8, neural field theory is used to model binocular rivalry waves. During binocular rivalry, visual perception switches back and forth between different images presented to the two eyes. The resulting fluctuations in perceptual dominance and suppression provide a basis for noninvasive studies of the human visual system and the identification of possible neural mechanisms underlying conscious visual awareness. Various psychophysical experiments have demonstrated that the switch between a dominant and suppressed visual percept propagates as a traveling front for each eye. In addition to considering the particular problem of binocular rivalry waves, the more general issue of how to develop neural field models of the functional architecture of primary visual cortex (V1) is discussed.

In the final part of the book, Chap. 9, a variety of topics regarding wavelike phenomena in the developing and diseased brain are presented. First, the possible role of calcium and retinal waves in early development is summarized. There is then a detailed description and analysis of cytoskeletal waves involved in neurite growth and cell polarization. This introduces another interesting phenomenon, namely, wave pinning. Three distinct examples of waves in the diseased brain are considered: spreading depression and migraine auras, epileptic waves, and the spread of neurodegenerative waves due to protein aggregation. In order to tackle the latter two phenomena, a review of complex network theory is included, covering topics such as small-world networks, scale-free networks, neuronal avalanches, branching processes, and epidemics on networks.

I end this introduction by noting that I sometimes have to deal with comments such as "neuroscience is pre-Newtonian" or "systems neuroscience lacks a Navier–Stokes equation." My reaction to such comments is that it is for these very reasons that neuroscience is so exciting. One does not have to be as smart as Isaac Newton to work at the frontiers of a twenty-first-century science and have the chance to make a significant contribution. Of course, it also means that the contents of this book could one day be irrelevant and pass into dust the way of my flesh. If this does happen dear reader, please do not ask for your money back!

Salt Lake City, UT, USA — Paul C. Bressloff

Acknowledgements

I would first like to thank my past graduate students: Pete Roper, Stefan Folias, Andy Oster, Berton Earnshaw, Will Nesse, Zack Kilpatrick, Jay Newby, Yi-Ming Lai, and Matt Webber. I have really enjoyed mentoring all of my students and collaborating on so many fun projects. There are a number of people who have inspired my own work in spatiotemporal aspects of mathematical neuroscience, either independently or in collaboration: Shun-Ichi Amari, Carson Chow, Jack Cowan, Steve Coombes, Bard Ermentrout, Olivier Faugeras, Marty Golubitsky, David Holcman, Jim Keener, Carlo Laing, John Rinzel, Jon Rubin, Dave Terman, Bill Troy, and Hugh Wilson. I would like to give a particular thanks to Jack Cowan, for his encouragement and guidance, and Steve Coombes for keeping the fire burning in the UK. I thank my parents Andy and Jackie for their unstinting love and support, Neil and his family for putting up with the crazy brother, and, last but not least, my wife Alessandra, who gave me the greatest gift of all, our son Luca. Ti amo e grazie mille per tutto.

Contents

Part I Neurons

Part I
Neurons

Chapter 1
Single Neuron Modeling

In this introductory chapter we describe conductance-based models of single neurons, which take into account action potential generation, the effects of external stimuli, synaptic and dendritic processing, and ion channel fluctuations. In the appendix we provide an informal introduction to stochastic calculus. In Chap. 2 we turn to the classical theory of waves in one-dimensional neural media, as exemplified by action potential propagation along axons. This will introduce some of the techniques used throughout the book, including phase-plane analysis, slow–fast systems, stability theory and Evans functions, averaging theory, and stochastic wave propagation. We then consider two examples of intracellular waves propagating along spiny dendrites (Chap. 3): the spike–diffuse–spike model of calcium-mediated spikes and a reaction–diffusion model of CaMKII translocation waves. A number of additional methods will be presented, including the analysis of waves in spiking models, homogenization theory, and the theory of pulled fronts. Finally, in Chap. 4 we consider both intracellular and intercellular calcium waves, emphasizing aspects that are specific to neurons. In particular, we highlight the important role of calcium signaling in astrocytes.

1.1 Conductance-Based Model of a Neuron

Cortical neurons typically consist of a cell body (or soma) where the nucleus containing DNA is located, a branching output structure known as the *axon* and a branching input structure known as the *dendritic tree*; see Fig. 1.1. Neurons mainly communicate with each other by sending electrical impulses or spikes (action potentials) along their axons. (Some neurons are also coupled diffusively via gap junctions [126].) These axons make contacts on the dendrites of other neurons via microscopic junctions known as *synapses*. The arrival of an electrical spike at a synaptic junction leads to the flow of electrical current along the dendritic tree of the stimulated neuron. If the total synaptic current from all of the activated synapses forces the electrical potential within the cell body to cross some threshold, then the neuron

P.C. Bressloff, *Waves in Neural Media: From Single Neurons to Neural Fields*,
Lecture Notes on Mathematical Modelling in the Life Sciences,
DOI 10.1007/978-1-4614-8866-8_1,

fires a spike. The standard biophysical model for describing the dynamics of a single neuron with somatic membrane potential v is based upon conservation of electric charge:

$$C\frac{dv}{dt} = -I_{\text{con}} + u + I_{\text{ext}}, \tag{1.1}$$

where C is the cell capacitance, I_{con} is the membrane current, u denotes the sum of synaptic currents entering the cell body, and I_{ext} describes any externally injected currents. Ions can diffuse in and out of the cell through ion specific channels embedded in the cell membrane. Ion pumps within the cell membrane maintain concentration gradients, such that there is a higher concentration of Na^+ and Ca^{2+} outside the cell and a higher concentration of K^+ inside the cell. The membrane current through a specific channel varies approximately linearly with changes in the potential v relative to some equilibrium or reversal potential, which is the potential at which there is a balance between the opposing effects of diffusion and electrical forces. Summing over all channel types, the total membrane current (flow of positive ions) leaving the cell through the cell membrane is

$$I_{\text{con}} = \sum_s g_s(v - V_s), \tag{1.2}$$

where g_s is the conductance due to channels of type s and V_s is the corresponding reversal potential. In the case of a channel selective to a single ion, V_s satisfies the Nernst equation

$$V_s = \frac{k_B T}{q} \ln\left(\frac{[outside]_s}{[inside]_s}\right), \tag{1.3}$$

where q is the charge of the ion, k_B is the Boltzmann constant, T is temperature (in degrees Kelvin), and $[outside]_s, [inside]_s$ denote the extracellular and intracellular concentrations of the given ion. Typical values for the common ion species are $V_K \approx -75\,\text{mV}$, $V_{\text{Na}} \approx 50\,\text{mV}$, $V_{\text{Ca}} \approx 150\,\text{mV}$, and $V_{\text{Cl}} \approx -60\,\text{mV}$ (which is close to the resting potential of the cell).

The generation and propagation of an action potential arises from nonlinearities associated with active membrane conductances. Recordings of the current flowing through single channels indicate that channels fluctuate rapidly between open and closed states in a stochastic fashion. Nevertheless, most models of a neuron use deterministic descriptions of conductance changes, under the assumption that there are a large number of approximately independent channels of each type. It then follows from the law of large numbers that the fraction of channels open at any given time is approximately equal to the probability that any one channel is in an open state. The conductance g_s for ion channels of type s is thus taken to be the product $g_s = \bar{g}_s P_s$ where $\bar{g}_s$ is equal to the density of channels in the membrane multiplied by the conductance of a single channel and P_s is the fraction of open channels. The voltage dependence of the probabilities P_s in the case of a delayed-rectifier K^+ current and a fast Na^+ current was originally obtained by Hodgkin and Huxley [279] as part of their Nobel Prize winning work on the generation of action potentials in the squid giant axon. The delayed-rectifier K^+ current is responsible for

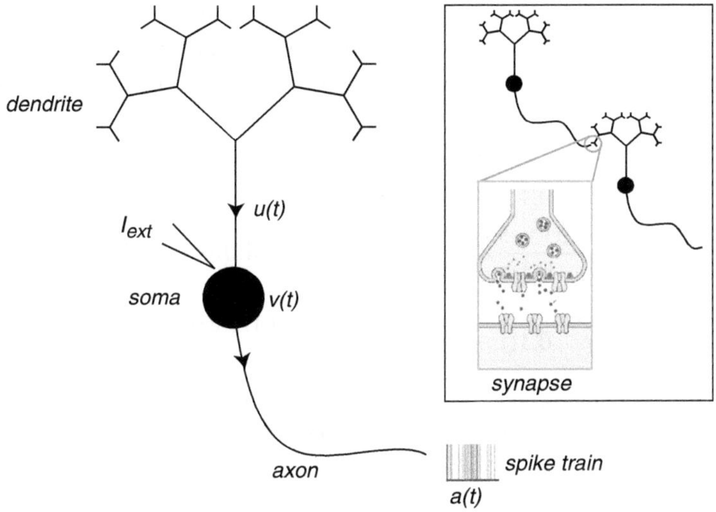

Fig. 1.1 Basic structure of a neuron. (Inset shows a synaptic connection from an upstream or presynaptic neuron and a downstream or postsynaptic neuron.) See text for details

terminating an action potential by repolarizing a neuron. One finds that opening of the K^+ channel requires structural changes in 4 identical and independent subunits so that $P_K = n^4$ where n is the probability that any one gate subunit has opened. In the case of the fast Na^+ current, which is responsible for the rapid depolarization of a cell leading to action potential generation, the probability of an open channel takes the form $P_{Na} = m^3 h$ where m^3 is the probability that an activating gate is open and h is the probability that an inactivating gate is open. Depolarization causes m to increase and h to decrease, whereas hyperpolarization has the opposite effect.

The dynamics of the gating variables m, n, h are usually formulated in terms of a simple kinetic scheme that describes voltage-dependent transitions of each gating subunit between open and closed states. More specifically, for each $X \in \{m, n, h\}$

$$\frac{dX}{dt} = \alpha_X(v)(1 - X) - \beta_X(v)X, \tag{1.4}$$

where $\alpha_X(v)$ is the rate of the transition *closed* $\rightarrow$ *open* and $\beta_X(v)$ is the rate of the reverse transition *open* $\rightarrow$ *closed*. Equation (1.4) can be rewritten in the alternative form

$$\tau_X(v)\frac{\mathrm{d}X}{\mathrm{d}t} = X_\infty(v) - X, \qquad \text{with } X \in \{m, n, h\}, \tag{1.5}$$

where

$$\tau_X(v) = \frac{1}{\alpha_X(v) + \beta_X(v)}, \qquad X_\infty(v) = \alpha_X(v)\tau_X(v).$$

It follows that the conductance variables m, n, and h approach the asymptotic values $m_\infty(v)$, $n_\infty(v)$, and $h_\infty(v)$ exponentially with time constants $\tau_m(v)$, $\tau_n(v)$, and $\tau_h(v)$, respectively. From basic thermodynamic arguments, the opening and closing rates are expected to be exponential functions of the voltage. Hodgkin and Huxley [279] fitted exponential-like functions to the experimental data obtained from the squid axon:

$$\alpha_m = \frac{0.1(v+40)}{1-\exp[-0.1(v+40)]} \qquad \alpha_h = 0.07\exp[-0.05(v+65)],$$
$$\alpha_n = \frac{0.01(v+55)}{1-\exp[-0.1(v+55)]} \qquad \beta_m = 4.0\exp[-0.556(v+65)],$$
$$\beta_h = \frac{1}{1+\exp[-0.1(v+35)]} \qquad \beta_n = 0.125\exp[-0.125(v+65)].$$

All potentials are measured in mV, all times in ms, and all currents in μA/cm^2. The corresponding asymptotic functions $X_\infty(v)$ and time constants $\tau_X(v)$ are plotted in Fig. 1.2.

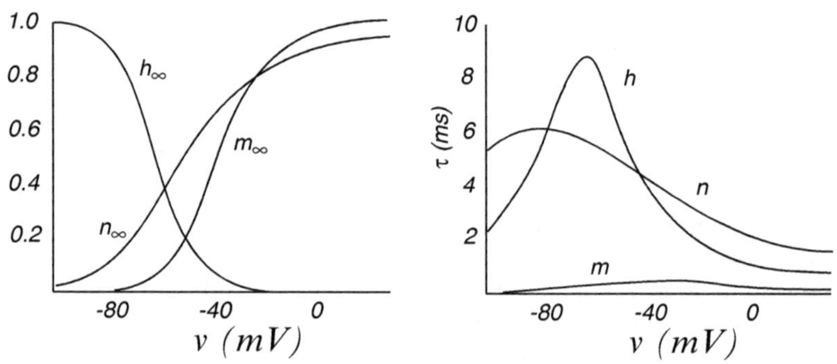

Fig. 1.2 Voltage-dependent steady-state levels of activation and inactivation (*left panel*) and voltage-dependent time constants (*right panel*) for the Hodgkin–Huxley model

We can now write down the Hodgkin–Huxley model for the generation of an action potential, which takes the membrane current to be the sum of a leakage current, a delayed-rectifier K^+ current, and a fast Na^+ current,

$$C\frac{dv}{dt} = f(v,m,n,h) + I_{\text{ext}}, \tag{1.6}$$

with

$$f(v,m,n,h) = -\bar{g}_{\text{Na}}m^3h(v-V_{\text{Na}}) - \bar{g}_{\text{K}}n^4(v-V_{\text{K}}) - \bar{g}_{\text{L}}(v-V_{\text{L}}). \tag{1.7}$$

The maximal conductances and reversal potentials used in the original model are $\bar{g}_{\text{L}} = 0.003\,\text{ms/mm}^2$, $\bar{g}_{\text{K}} = 0.36\,\text{mS/mm}^2$, $\bar{g}_{\text{Na}} = 1.2\,\text{mS/mm}^2$, $V_{\text{L}} = -54.387\,\text{mV}$, $V_{\text{K}} = -77\,\text{mV}$, and $V_{\text{Na}} = 50\,\text{mV}$. Note that the leakage current groups together various voltage-independent processes such as the currents carried by ion pumps that maintain the concentration gradients. The variables m,n,h evolve according to (1.4). The temporal evolution of the variables v,f,m,n,h during a single action potential is shown in Fig. 1.3. Injection of a depolarizing current induces a rapid increase in the m variable describing activation of the Na^+ current. Since the slower h variable is initially around 0.6, there is a large influx of Na^+ ions, producing a sharp downward spike in the membrane current and a rapid depolarization through positive feedback. However, the rise in the membrane potential causes the Na^+ conductance to inactivate by driving h towards zero. In addition, the depolarization activates the K^+ conductance, resulting in a subsequent hyperpolarization.

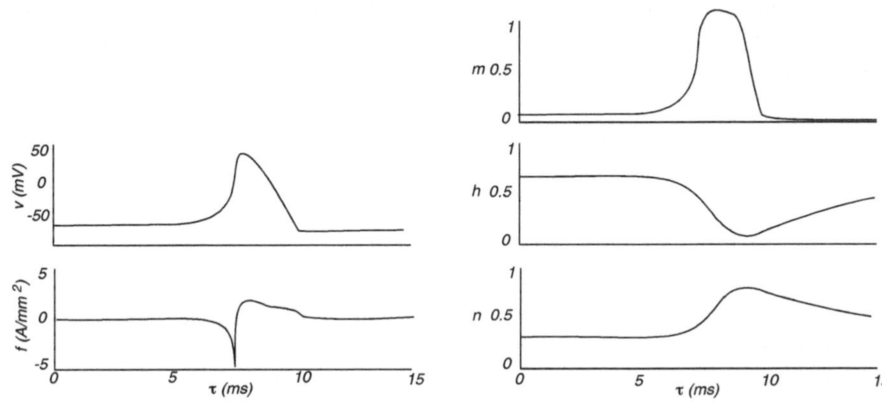

Fig. 1.3 The dynamics of v, f, n, m, h in the Hodgkin–Huxley model during the firing of an action potential induced by a current injection at $t = 5$ ms

Once a neuron has generated an action potential, it propagates as a traveling pulse along the axon of the neuron. In order to model a propagating action potential, it is necessary to combine the Hodgkin–Huxley conductance-based model given by (1.4) and (1.6) with a one-dimensional cable equation describing passive voltage changes along an axon or dendrite. The result is a partial differential equation of the form

$$C\frac{\partial v}{\partial t} = K\frac{\partial^2 v}{\partial x^2} + f(v, m, n, h), \tag{1.8}$$

where K is related to the membrane space constant of the cable; see Sect. 1.4, and $X = m, n, h$ evolve according to (1.4). Equation (1.8) is an example of a nonlinear reaction–diffusion equation used to model wave propagation in an excitable medium; one characteristic of an excitable medium is that it cannot support the passing of another wave until a certain amount of time has passed (known as the refractory period). A rigorous proof of the existence of traveling wave solutions of the spatially extended Hodgkin–Huxley equations has been developed [104, 267]. However, following standard treatments of waves in excitable media [242, 322, 444], we will develop the theory by considering the simpler FitzHugh–Nagumo model [192, 446]; see Chap. 2.

There is an ongoing debate about how best to characterize the output of a neuron, from either a dynamical systems or an information processing perspective. It is generally agreed that the detailed shape of an action potential is usually unimportant, so, at the fine temporal scale of a few milliseconds, the output of a neuron can be represented in terms of the times at which the neuron fires an action potential (or spike). Given the conductance-based model (1.1), one typically identifies a firing threshold κ such that if $v(t)$ crosses the threshold from below at time $t = T$, then the neuron fires an action potential. If T^m denotes the mth firing time of the neuron since $t = 0$, say, then we have the threshold condition

$$T^m = \inf\{t, t > T^{m-1} | v(t) = \kappa, \dot{v}(t) > 0\}. \tag{1.9}$$

As an alternative to a spike timing representation of neuronal output, one can consider a rate-based representation that is obtained by filtering the spike train with some causal integral kernel $\Gamma(t)$, $\Gamma(t) = 0$ for $t < 0$:

$$z(t) = \sum_m \Gamma(t - T^m) = \int_{-\infty}^{\infty} \Gamma(t - \tau) a(\tau) d\tau, \tag{1.10}$$

where

$$a(t) = \sum_m \delta(t - T^m). \tag{1.11}$$

For example, if $\Gamma(t) = T^{-1}$ for $0 \leq t < T$ and is zero otherwise, then $z(t)$ simply counts the number of spikes within the time interval $[t - T, t]$. In the special case of a regular spike train with $T^{m+1} - T^m = \Delta_0$ for all m, $z(t) = 1/\Delta_0$ in the limit $T \to \infty$.

Suppose, for the moment, that we ignore synaptic currents and consider what happens as the external input I_{ext} to a neuron is increased. Experimentally it is found that most cortical neurons switch from a resting state characterized by a low rate of (noise-driven) spontaneous firing to an active state characterized by either tonic (regular, repetitive) firing or bursting [131]. There has been considerable theoretical work on the transitions from resting to active states in conductance-based models based on bifurcation theory; see [173, 301] for excellent reviews. We will focus on tonic firing neurons, since these comprise the majority of cells in cortical networks. In the case of constant input $I_{\text{ext}} = I$, the firing rate $\bar{z}$ (mean number of spikes per second) of the neuron is typically found to be a nonlinear function of the input:

$$\bar{z} = F(I) \tag{1.12}$$

with the form of F depending on the nature of the bifurcation from the stable resting state to repetitive firing. A common bifurcation scenario in conductance-based models of cortical neurons is a saddle–node on an invariant circle [173, 301], which is classified as type I excitability. Close to the bifurcation point (see Fig. 1.4), we have

$$F(I) = F_0 \sqrt{I - I_c}, \tag{1.13}$$

where I_c is the critical current for onset of regular spiking. (Another common form of excitability is type II, in which the loss of stability of the resting state and the transition to repetitive firing occur via a Hopf bifurcation. There are also more exotic forms of excitability as detailed elsewhere [201, 423].) If one includes stochastic effects arising from synaptic and membrane noise, for example, then the effective mean firing rate becomes a smooth sigmoid-like function of injected current,

$$F(I) = \frac{F_0}{1 + \mathrm{e}^{-\eta(I-\kappa)}}, \tag{1.14}$$

where η is the gain and κ is the firing threshold. In the high-gain limit $\eta \to \infty$, this reduces to a Heaviside firing rate function

$$F(I) = F_0 H(I-\kappa) = \begin{cases} F_0 & \text{if} I > \kappa \\ 0 & \text{if} I < \kappa. \end{cases} \tag{1.15}$$

Yet another commonly used firing rate function is the piecewise linear function

$$F(I) = \begin{cases} 0, & I < \kappa, \\ \eta(u-\kappa), & \kappa < I < \kappa + \eta^{-1}, \\ 1, & I > \kappa + \eta^{-1}. \end{cases} \tag{1.16}$$

This preserves the hard threshold of the saddle–node on a limit cycle bifurcation but ensures that the firing rate saturates at high input currents.

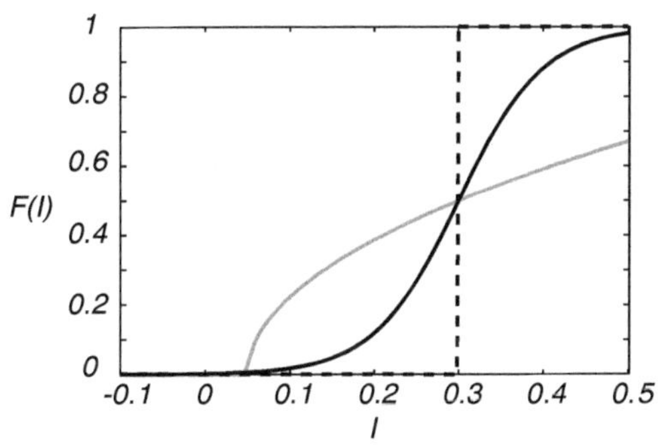

Fig. 1.4 Various forms of the nonlinear firing rate function $F(I)$. Sigmoid function (*black curve*) and Heaviside function (*dashed curve*) have a threshold $\kappa = 0.3$, whereas the square root function (*gray curve*) has a critical current $I_c = 0.05$

The issue of how to represent a single neuron becomes even more salient when considering synaptically coupled spiking networks (Chap. 5) and neural fields (Chap. 6). In order to make analytical progress, it is necessary either to consider a very simple model of a spiking neuron such as integrate-and-fire [323] (Sect. 5.3) or to carry out some form of reduction of a conductance-based model. In the case of weakly coupled neural oscillators, one can reduce the dynamics to a network of coupled phase oscillators; see Chap. 5. The basic ideas underlying so-called phase reduction methods can be understood by considering the simpler case of a single, periodically forced neural oscillator.

1.2 Periodically Forced Neural Oscillator

A conductance-based model of a neuron with constant input current can be formulated as an M-dimensional ($M \geq 2$) system of ODEs

$$\frac{d\mathbf{x}}{dt} = \mathbf{f}(\mathbf{x}), \qquad \mathbf{x} = (x_1, \ldots x_M). \tag{1.17}$$

Here x_1, say, represents the membrane potential of the neuron (treated as a point processor) and x_m, $m > 1$, represent various ionic channel gating variables. Suppose that the neuron has a stable periodic solution $\mathbf{x}(t) = \mathbf{x}(t + \Delta_0)$ where $\omega_0 = 2\pi/\Delta_0$ is the *natural frequency* of the oscillator. In *phase space* the solution is an isolated attractive trajectory called a *limit cycle*. The dynamics on the limit cycle can be described by a uniformly rotating phase such that

$$\frac{d\theta}{dt} = \omega_0, \tag{1.18}$$

and $\mathbf{x}(t) = \mathbf{g}(\theta(t))$ with $\mathbf{g}$ a 2π-periodic function. Note that the phase is *neutrally stable* with respect to perturbations along the limit cycle—this reflects invariance of an autonomous dynamical system with respect to time shifts. Now suppose that a small external periodic input is applied to the oscillator such that

$$\frac{d\mathbf{x}}{dt} = \mathbf{f}(\mathbf{x}) + \varepsilon\mathbf{p}(\mathbf{x},t), \tag{1.19}$$

where $\mathbf{p}(\mathbf{x},t) = \mathbf{p}(\mathbf{x},t+\Delta)$ and $\omega = 2\pi/\Delta$ is the forcing frequency. If the amplitude ε is sufficiently small and the cycle is stable, then deviations transverse to the limit cycle are small so that the main effect of the perturbation is to induce shifts in the phase. Therefore, we need to extend the definition of phase to a neighborhood of the limit cycle. This leads to the notion of an *isochrone* [221, 350, 679].

1.2.1 Isochrones and Phase-Resetting Curves

Suppose that we observe the unperturbed system stroboscopically at time intervals of length Δ_0. This leads to a Poincare mapping

$$\mathbf{x}(t) \to \mathbf{x}(t + \Delta_0) \equiv \mathscr{P}(\mathbf{x}(t)).$$

This mapping has all points on the limit cycle as fixed points. Choose a point $\mathbf{x}^*$ on the cycle and consider all points in the vicinity of $\mathbf{x}^*$ that are attracted to it under the action of $\mathscr{P}$. They form an $(M-1)$-dimensional hypersurface $\mathscr{I}$, called an isochrone, crossing the limit cycle at $\mathbf{x}^*$ (see Fig. 1.5). A unique isochrone can be drawn through each point on the limit cycle so we can parameterize the isochrones by the phase, $\mathscr{I} = \mathscr{I}(\theta)$. Finally, we extend the definition of phase by taking all points $\mathbf{x} \in \mathscr{I}(\theta)$ to have the same phase, $\Theta(\mathbf{x}) = \theta$, which then rotates at the natural frequency ω_0 (in the unperturbed case). Hence, for an unperturbed oscillator in the vicinity of the limit cycle, we have

$$\omega_0 = \frac{d\Theta(\mathbf{x})}{dt} = \sum_k \frac{\partial \Theta}{\partial x_k} \frac{dx_k}{dt} = \sum_k \frac{\partial \Theta}{\partial x_k} f_k(\mathbf{x}).$$

Now consider the perturbed system but with the unperturbed definition of the phase:

$$\frac{d\Theta(\mathbf{x})}{dt} = \sum_k \frac{\partial \Theta}{\partial x_k}(f_k(\mathbf{x}) + \varepsilon p_k(\mathbf{x},t)) = \omega_0 + \varepsilon \sum_k \frac{\partial \Theta}{\partial x_k} p_k(\mathbf{x},t).$$

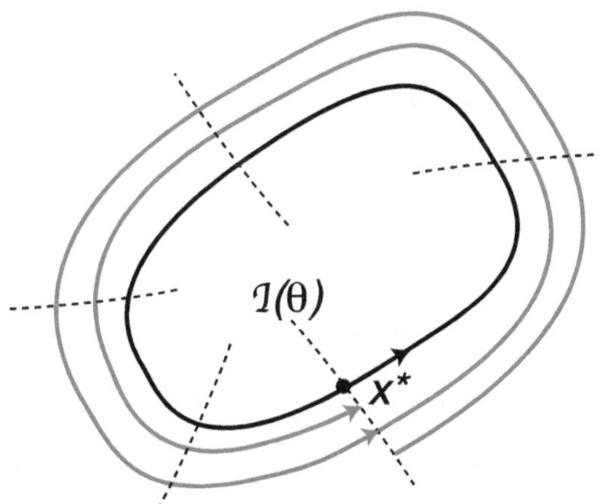

Fig. 1.5 Isochrones in the neighborhood of a stable limit cycle

To a first approximation we can neglect deviations of $\mathbf{x}$ from the limit cycle which we denote by $\mathbf{x}^*$:

$$\frac{d\Theta(\mathbf{x})}{dt} = \omega_0 + \varepsilon \sum_k \frac{\partial \Theta(\mathbf{x}^*)}{\partial x_k} p_k(\mathbf{x}^*,t).$$

Finally, since points on the limit cycle are in 1:1 correspondence with the phase θ, we obtain the closed phase equation

$$\frac{d\theta}{dt} = \omega_0 + \varepsilon Q(\theta,t), \tag{1.20}$$

where

$$Q(\theta,t) = \sum_k \frac{\partial \Theta(\mathbf{x}^*(\theta))}{\partial x_k} p_k(\mathbf{x}^*(\theta),t) \tag{1.21}$$

is a 2π-periodic function of θ and a Δ-periodic function of t.

Consider, as an example, the complex amplitude equation that arises for a limit cycle oscillator close to a Hopf bifurcation [248]:

$$\frac{dA}{dt} = (1+i\eta)A - (1+i\alpha)|A|^2 A, \quad A \in \mathbb{C}. \tag{1.22}$$

In polar coordinates $A = Re^{i\phi}$,

$$\frac{dR}{dt} = R(1-R^2), \quad \frac{d\phi}{dt} = \eta - \alpha R^2.$$

The solution for arbitrary initial data $R(0) = R_0$, $\theta(0) = \theta_0$ is

$$\begin{aligned} R(t) &= \left[1 + \frac{1-R_0^2}{R_0^2}\mathrm{e}^{-2t}\right]^{-1/2}, \\ \phi(t) &= \phi_0 + \omega_0 t - \frac{\alpha}{2}\log(R_0^2 + (1-R_0^2)\mathrm{e}^{-2t}), \end{aligned} \tag{1.23}$$

where $\omega_0 = \eta - \alpha$ is the natural frequency of the stable limit cycle at $R = 1$. Strobing the solution at times $t = n\Delta_0$, we see that

$$\lim_{n\to\infty} \phi(n\Delta_0) = \phi_0 - \alpha \ln R_0.$$

Hence, we can define a phase on the whole plane

$$\Theta(R,\phi) = \phi - \alpha \ln R. \tag{1.24}$$

It follows that the isochrones are logarithmic spirals with $\phi - \alpha \ln R = \text{constant}$. Now rewrite (1.22) in Cartesian coordinates

$$\begin{aligned} \frac{dx}{dt} &= x - \eta y - (x^2+y^2)(x - \alpha y) + \varepsilon \cos \omega t, \\ \frac{dy}{dt} &= y + \eta y - (x^2+y^2)(y + \alpha x), \end{aligned}$$

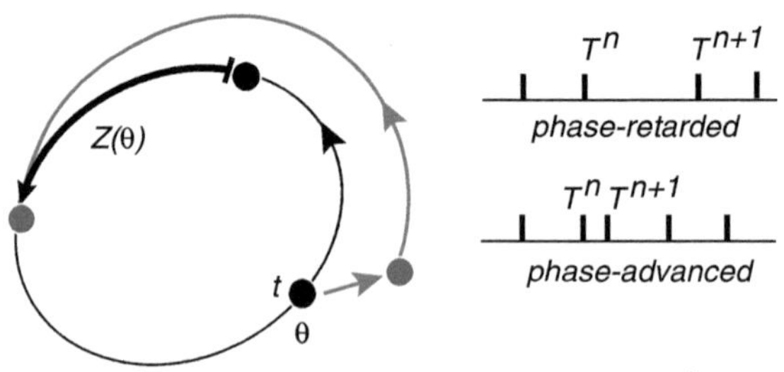

Fig. 1.6 Phase shift $Z(\theta)$ induced by a small perturbation of the membrane potential applied at time $t = 0$ when the phase of the limit cycle is θ. *Black* (*gray*) *dots* represent points on the unperturbed (perturbed) trajectory. The phase shift induces a corresponding shift in successive firing times

where we have added a periodic modulation in the x-direction. Rewrite the phase (1.24) as

$$\Theta = \tan^{-1}\frac{y}{x} - \frac{\alpha}{2}\log(x^2+y^2),$$

so that

$$\frac{\partial\Theta}{\partial x} = -\frac{y}{x^2+y^2} - \alpha\frac{x}{x^2+y^2}.$$

On the limit cycle $\mathbf{x}_0(\theta) = (\cos\theta, \sin\theta)$, we have

$$\frac{\partial\Theta(\mathbf{x}_0)}{\partial x} = -\sin\theta - \alpha\cos\theta.$$

It follows that the corresponding phase equation is

$$\frac{d\theta}{dt} = \omega_0 - \varepsilon(\alpha\cos\theta + \sin\theta)\cos\omega t.$$

The phase reduction method is particularly useful because the function $Q(\theta,t)$ can be related to an easily measurable property of a neural oscillator, namely, its *phase-resetting curve* (PRC), which we denote by the 2π-periodic function $Z(\theta)$. The PRC is found experimentally (or numerically) by perturbing the oscillator with a brief depolarizing voltage stimulus of size $\varepsilon\Delta V$ at different times in its cycle and measuring the resulting phase shift from the unperturbed system [221, 679]; see Fig. 1.6. Taking the coordinate x_1 as the membrane potential, it follows from (1.20) that

$$\frac{d\theta}{dt} = \omega_0 + \varepsilon\Delta x_1\frac{\partial\Theta(\mathbf{x}^*(\theta))}{\partial x_1}\delta(t-t_0). \tag{1.25}$$

Integrating this equation over a small interval around t_0, we see that the impulse induces a phase shift $\Delta\theta = (\varepsilon\Delta x_1)Z(\theta_0)$ where $Z(\theta) = \partial\Theta(\mathbf{x}^*(\theta))/\partial x_1$ and $\theta_0 = \theta(t_0)$. Thus comparing the phase at large times for the unperturbed and perturbed cases generates the PRC. Given the PRC $Z(\theta)$, the response of the neuron to a more general time-dependent voltage perturbation $\varepsilon P(t)$ is determined by the phase equation

$$\frac{d\theta}{dt} = \omega_0 + \varepsilon Z(\theta)P(t). \tag{1.26}$$

We can also express the PRC in terms of the firing times of a neuron (assuming fast reconvergence to the limit cycle). Suppose that there exists a well-defined threshold κ signaling the onset of fast somatic membrane depolarization and the subsequent firing of an action potential spike. Let T^n denote the nth firing time of the neuron as defined by (1.9). Since the membrane voltage $v(t) = x_1(\theta(t))$, the threshold corresponds to a particular phase of the limit cycle, which we choose to be $\theta = 0$. In the absence of perturbations, we have $\theta(t) = 2\pi t/\Delta_0$, so that the firing times are $T^n = n\Delta_0$ where Δ_0 is the natural period of oscillation. On the other hand, a small perturbation applied at the point θ on the limit cycle at time $t, T^n < t < T^{n+1}$, induces a phase shift that changes the next time of firing according to (see Fig. 1.6)

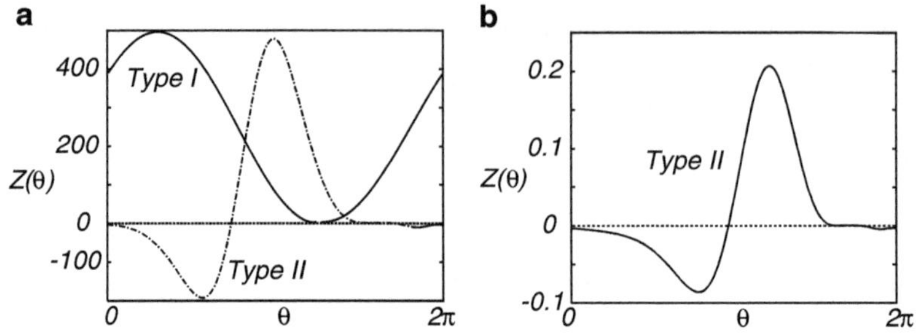

Fig. 1.7 (**a**) Morris–Lecar model showing two different response types. In both cases $v_k = -0.7$, $u_L = -0.5$, $u_{Ca} = 1$, $g_K = 2$, $g_L = 0.5$, $v_1 = -0.01$, $v_2 = 0.15$. For a type I response, $g_{Ca} = 1.33$, $v_3 = 0.1$, $v_4 = 0.145$, $\phi = 1/3$, and $I = 0.0695$. For a type II response, $g_{Ca} = 1.1$, $v_3 = 0$, $v_4 = 0.3$, $\phi = 0.2$, and $I = 0.25$. Responses have been scaled to the same ranges. (**b**) Hodgkin–Huxley model with external drive $I = 10$ showing type II phase-resetting curve

$$\frac{T^{n+1} - T^n}{\Delta_0} = 1 - \frac{(\varepsilon \Delta V) Z(\theta)}{2\pi}. \tag{1.27}$$

For certain types of neuron a depolarizing stimulus always advances the onset of the next spike, that is, the PRC is always positive, whereas for others the stimulus may also delay the next spike. Oscillators with a strictly positive PRC are called type I whereas those for which the PRC has a negative regime are called type II. A numerical example illustrating both types of PRC is shown in Fig. 1.7a for the Morris–Lecar model of a neuron, which was originally introduced to describe how under constant current injection barnacle muscle fibers respond with a host of oscillatory voltage waveforms [440]. It takes the form

$$\begin{aligned}\frac{dv}{dt} &= I - g_L(v - v_L) - g_K w(v - v_K) - g_{Ca} m_\infty(v)(v - v_{Ca}),\\ \frac{dw}{dt} &= \lambda(v)(w_\infty(v) - w),\end{aligned} \tag{1.28}$$

with

$$\begin{aligned}m_\infty(v) &= 0.5(1 + \tanh[(v - v_1)/v_2]),\\ w_\infty(v) &= 0.5(1 + \tanh[(v - v_3)/v_4]),\\ \lambda(v) &= \phi \cosh[(v - v_3)/(2v_4)].\end{aligned}$$

Here, g_L is the leakage conductance, g_K, g_{Ca} are potassium and calcium conductances, v_L, v_K, v_{Ca} are corresponding reversal potentials, $m_\infty(v)$, $w_\infty(v)$ are voltage-dependent gating functions, and $\lambda(v)$ is a voltage-dependent rate. The type II PRC for the Hodgkin–Huxley model is shown in Fig. 1.7b.

1.2.2 Phase-Locking and Synchronization

Now suppose that $Q(\theta, t)$ in (1.20) is expanded as a double Fourier series

$$Q(\theta,t)=\sum_{l,k}a_{l,k}\mathrm{e}^{ik\theta+il\omega t}.$$

Substitute for θ using the zero-order approximation $\theta=\omega_0 t+\theta_0$:

$$Q(\theta,t)=\sum_{l,k}a_{l,k}\mathrm{e}^{ik\theta_0+i(k\omega_0+l\omega)t}.$$

It follows that Q contains fast oscillating terms (compared to the time scale Δ_0/ε) together with slowly varying terms that satisfy the *resonance condition*

$$k\omega_0+l\omega\approx 0. \tag{1.29}$$

Only the latter will lead to large variations in the phase, so we can average the forcing term Q keeping only the resonant terms. The simplest case is $\omega\approx\omega_0$ for which the resonant terms satisfy $l=-k$ and

$$Q(\theta,t)\to\sum_{k}a_{-k,k}\mathrm{e}^{ik(\theta-\omega t)}=q(\theta-\omega t). \tag{1.30}$$

The phase equation then becomes

$$\frac{d\theta}{dt}=\omega_0+\varepsilon q(\theta-\omega t).$$

The phase difference between the oscillator and external drive, $\psi=\theta-\omega t$, then satisfies the equation

$$\frac{d\psi}{dt}=-\Delta\omega+\varepsilon q(\psi), \tag{1.31}$$

where $\Delta\omega=\omega-\omega_0$ is the degree of *frequency detuning*. Similarly, if $\omega\approx m\omega_0/n$, then

$$Q(\theta,t)\to\sum_{k}a_{-nj,mj}\mathrm{e}^{ij(m\theta-n\omega t)}=\hat{q}(m\theta-n\omega t), \tag{1.32}$$

and

$$\frac{d\psi}{dt}=m\omega_0-n\omega+\varepsilon m\hat{q}(\psi), \tag{1.33}$$

where $\psi=m\theta-n\omega t$.

The above is an example of an application of the averaging theorem [248]. Assuming that $\Delta\omega=\omega-\omega_0=\mathcal{O}(\varepsilon)$ and defining $\psi=\theta-\omega t$, we have

$$\frac{d\psi}{dt}=-\Delta\omega+\varepsilon Q(\psi+\omega t,t)=\mathcal{O}(\varepsilon). \tag{1.34}$$

Define

$$q(\psi)=\lim_{T\to\infty}\frac{1}{T}\int_0^T Q(\psi+\omega t,t)dt, \tag{1.35}$$

and consider the averaged equation

$$\frac{d\psi}{dt} = -\Delta\omega + \varepsilon q(\psi). \tag{1.36}$$

It is easy to establish that q only contains the resonant terms of Q as above. The averaging theorem ensures that there exists a change of variables that maps solutions of the full equation to those of the averaged equation to leading order in ε. The question then remains as to what extent solutions of the averaged equations are a good approximation to the solutions of the full equation. In general, one can only establish that a solution of the full equation is ε-close to a corresponding solution of the averaged system for times of $\mathcal{O}(\varepsilon^{-1})$. No such problem occurs however for hyperbolic fixed points corresponding to phase-locked states.

Suppose that the 2π-periodic function $q(\psi)$ has a unique maximum q_{max} and a unique minimum q_{min} in the interval $[0, 2\pi)$. We can then distinguish between two regimes [502]:

Synchronization regime: If the degree of detuning for a given drive amplitude is sufficiently small,

$$\varepsilon q_{min} < \Delta\omega < \varepsilon q_{\max},$$

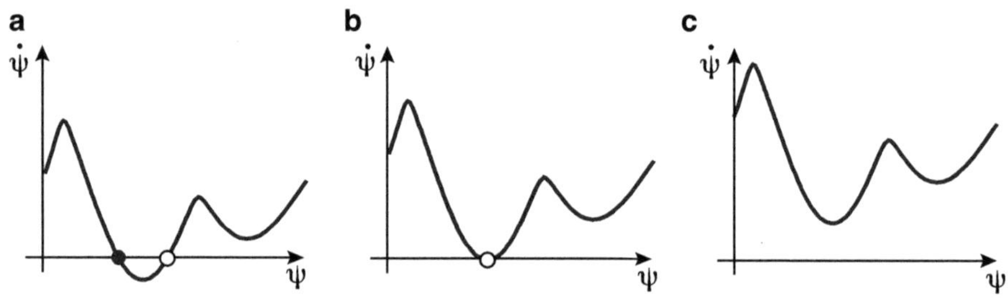

Fig. 1.8 Saddle–node bifurcation signaling a transition from a synchronized to a drifting state as the size of frequency detuning $|\Delta\omega|$ increases (**a**) Synchronization regime. (**b**) Saddle–node bifurcation. (**c**) Drift regime

then there exists at least one pair of stable/unstable fixed points (ψ_s, ψ_u). (This follows from the fact that $q(\psi)$ is 2π-periodic and continuous so it has to cross any horizontal line an even number of times.) The system evolves to the synchronized state

$$\theta(t) = \omega t + \psi_s,$$

in which the oscillator is *phase-locked* to the external drive and is *frequency entrained*. Note that the stability of a phase-locked state is determined by the sign of $q'(\psi)$ with $q'(\psi_s) < 0$ and $q'(\psi_u) > 0$ (see Fig. 1.8a).

Drift regime: As $|\Delta\omega|$ increases, it approaches one of the critical values $\varepsilon q_{min,max}$ where the two fixed points coalesce in a saddle–node bifurcation and phase-locking disappears; see Fig. 1.8b, c. Hence, if the degree of tuning is large, then $d\psi/dt$ never changes sign and the oscillation frequency differs from the drive frequency ω. The phase $\psi(t)$ rotates through 2π with period

$$T_\psi = \left| \int_0^{2\pi} \frac{d\psi}{\varepsilon q(\psi) - \Delta\omega} \right|. \tag{1.37}$$

The mean frequency of rotation is thus $\Omega = \omega + \Omega_\psi$ where $\Omega_\psi = 2\pi/T_\psi$ is known as the *beat frequency*. One is often interested in how the behavior varies in the $(\Delta\omega, \varepsilon)$-plane (see Fig. 1.9). First the boundary between the two regimes consists of the two straight lines $\Delta\omega = \varepsilon q_{max,min}$. Second, close to the boundary Ω_ψ has a characteristic form. Suppose, for example, that $\Delta\omega - \Delta\omega_{max}$ is small for fixed ε with $\Delta\omega_{max} = \varepsilon q_{max}$. The integral in (1.37) is then dominated by a small region around ψ_{max}. Expanding $q(\psi)$ as a Taylor series,

$$\begin{aligned} \Omega_\psi = \frac{2\pi}{T_\psi} &\approx 2\pi \left| \int_{-\infty}^{\infty} \frac{d\psi}{\varepsilon q''(\psi_{max})\psi^2 - (\Delta\omega - \Delta\omega_{max})} \right|^{-1} \\ &= \sqrt{\varepsilon |q''(\psi_{max})| (\Delta\omega - \Delta\omega_{max})}. \end{aligned} \tag{1.38}$$

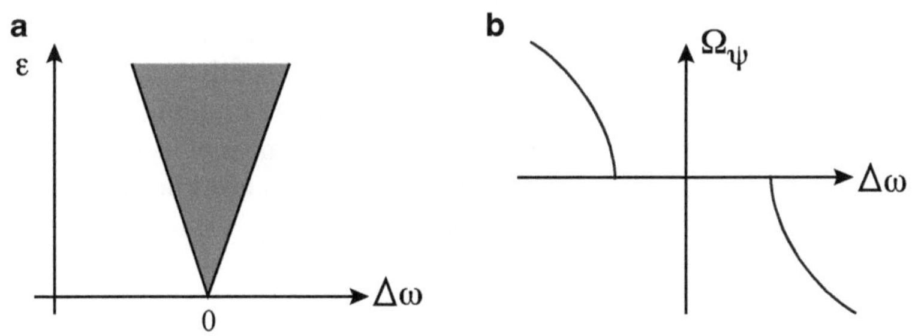

Fig. 1.9 (**a**) Synchronization regime (*shaded*) in $(\Delta\omega, \varepsilon)$-plane. (**b**) Variation of beat frequency with $\Delta\omega$ for fixed ε

1.3 Synaptic Processing

In the conductance-based model given by (1.6), we decomposed the total input current to the soma into an external part $I_{ext}(t)$ and a synaptic part $u(t)$. In this section, we consider the current generated at a single synapse and the sequence of events underlying conductance changes in the postsynaptic membrane due to the arrival of an action potential at the presynaptic terminal. We then show how these conductance changes can be modeled in terms of a kinetic scheme describing the opening and closing of ion channels in the postsynaptic membrane.

1.3.1 Excitatory and Inhibitory Synapses

The basic stages of synaptic processing induced by the arrival of an action potential at an axon terminal are shown in Fig. 1.10. (See [99] for a more detailed description.) An action potential arriving at the terminal of a presynaptic axon causes voltage-gated Ca^{2+} channels within an active zone to open. The influx of Ca^{2+} produces a high concentration of Ca^{2+} near the active zone [45, 195], which in turn causes vesicles containing neurotransmitter to fuse with the presynaptic cell membrane and release their contents into the synaptic cleft (a process known as exocytosis). The released neurotransmitter molecules then diffuse across the synaptic cleft and bind to specific receptors on the postsynaptic membrane. These receptors cause ion channels to open, thereby changing the membrane conductance and membrane potential of the postsynaptic cell. A single synaptic event due to the arrival of an action potential at time T induces a synaptic current of the form

$$I_{\text{syn}}(t) = g_{\text{syn}}(t-T)(V_{\text{syn}} - v(t)), \tag{1.39}$$

where v is the voltage of the postsynaptic neuron, V_{syn} is the synaptic reversal potential, and $g_{\text{syn}}(t)$ is the change in synaptic conductance with $g_{\text{syn}}(t) = 0$ for $t < 0$. The sign of V_{syn} relative to the resting potential V_{rest} (typically $V_{\text{rest}} \approx -65\,\text{mV}$) determines whether the synapse is excitatory ($V_{\text{syn}} > V_{\text{rest}}$) or inhibitory ($V_{\text{syn}} < V_{\text{rest}}$). For simplicity, it is often assumed that a neuron spends most of its time close to rest such that $V_{\text{syn}} - v \approx V_{\text{syn}} - V_{\text{rest}}$, with the factor $V_{\text{syn}} - V_{\text{rest}}$ absorbed into g_{syn}. One is then effectively taking the arrival of a spike as generating a synaptic current rather than a change in conductance.

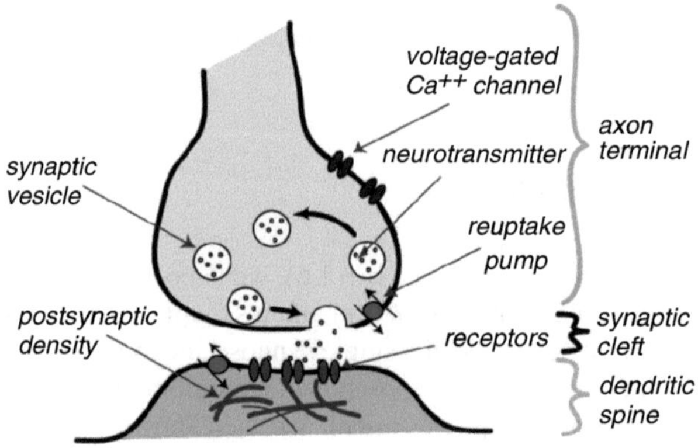

Fig. 1.10 Basic stages of synaptic processing shown for an excitatory synapse. See text for details

The predominant fast, excitatory neurotransmitter of the vertebrate central nervous system is the amino acid *glutamate*, whereas in the peripheral nervous system, it is *acetylcholine*. Glutamate-sensitive receptors in the postsynaptic membrane can be subdivided into two major types, namely, NMDA and AMPA [99].

At an AMPA receptor the postsynaptic channels open very rapidly. The resulting increase in conductance peaks within a few hundred microseconds, with an exponential decay of around 1 ms. The time course of the synaptic conductance change can be modeled in terms of an nth state Markov process [154] (see Sect. 1.3.3). Usually a simplified representation of $g_{\mathrm{syn}}(t)$ is used that is given by the difference of exponentials

$$g_{\mathrm{syn}}(t) = \bar{g}\left(\frac{1}{\tau_2} - \frac{1}{\tau_1}\right)(\mathrm{e}^{-t/\tau_1} - \mathrm{e}^{-t/\tau_2})H(t), \tag{1.40}$$

with $H(t)$ the Heaviside function. In many cases, the rise time is much shorter than the fall time ($\tau_1 \ll \tau_2$) so that we have an exponential synapse with $g_{\mathrm{syn}}(t) = \bar{g}\mathrm{e}^{-t/\tau_d}$. In the limit $\tau_2 \to \tau_1 = \alpha^{-1}$, (1.40) reduces to the well-known α function

$$g_{\mathrm{syn}}(t) = \bar{g}\alpha^2 t\mathrm{e}^{-\alpha t}H(t). \tag{1.41}$$

These expressions for the conductance are also used for GABA inhibitory synapses (see below). In contrast to an AMPA receptor, the NMDA receptor operates about ten times slower and the amplitude of the conductance change depends on the postsynaptic membrane potential. If the postsynaptic potential is at rest and glutamate is bound to the NMDA receptor, then the channel opens, but it is physically obstructed by Mg^{2+} ions. As the membrane is depolarized, the Mg^{2+} ions move out and the channel becomes permeable to Na^+ and Ca^{2+} ions. The NMDA conductance can be derived from a model in which the binding rate constant of Mg^{2+} varies as an exponential function of the local voltage v [304]:

$$g_{\mathrm{syn}}(t) = g\frac{\mathrm{e}^{-t/\tau_1} - \mathrm{e}^{-t/\tau_2}}{1 + \eta[Mg^{2+}]\mathrm{e}^{-\gamma v(t)}}, \quad t > 0, \tag{1.42}$$

where $[X]$ denotes concentration of X and η is a rate constant. The rapid influx of calcium ions due to the opening of NMDA channels is thought to be the critical trigger for the onset of *long-term potentiation* or LTP, a major component of synaptic plasticity (see also Sect. 4.1).

The most common inhibitory neurotransmitter in the central nervous system of both vertebrates and invertebrates appears to be GABA. There are two major forms of postsynaptic receptors termed A and B. The GABA_A receptors open channels selective to chloride ions, whose reversal potential $V_{\mathrm{syn}} = -70\,\mathrm{mV}$ is close to that of the resting potential. The postsynaptic conductance change is quite fast, rising within 1 ms and decaying within 10–20 ms. GABA_B receptors are at least 10 times slower and open ion channels selective for K^+ ions. Thus they tend to be considerably more hyperpolarizing with $V_{\mathrm{syn}} \approx -100\,\mathrm{mV}$. The two receptor classes tend to be segregated with GABA_A occurring at or close to the soma and GABA_B further out on the dendrites. Another way to distinguish between GABA_A and GABA_B receptors is that the former are *ionotropic* (as are NMDA and AMPA) while the latter are *metabotropic* [99]. Neurotransmitter binding to an ionotropic receptor directly opens an ion channel through a series of conformational changes of the receptor. On the other hand, neurotransmitter binding to a metabotropic receptor indirectly opens an ion channel elsewhere in the membrane through a sequence of biochemical steps mediated by G proteins.

1.3.2 Synaptic Depression

A single synaptic event due to the arrival of an action potential at time T induces a synaptic current of the form (1.39). As a crude approximation we might try summing individual responses to model the synaptic current arising from a train of action potentials arriving at times T^m, integer m:

$$I_{\rm syn}(t) = \sum_m g_{\rm syn}(t - T^m)(V_{\rm syn} - v(t)). \tag{1.43}$$

Note that this sum only includes spikes for which $T^m < t$ since $g_{\rm syn}(t) = 0$ for $t < 0$ (causality condition). For many synapses such a simple ansatz does not hold, since some form of short-term synaptic depression causes the amplitude of the response to depend on the previous history of presynaptic firing [4, 405]. One way to incorporate this history-dependent effect is to take [1]

$$I_{\rm syn}(t) = \left[\sum_m q(T^m) g_{\rm syn}(t - T^m)\right](V_{\rm syn} - v(t)), \tag{1.44}$$

where the factor $q(T^m)$ reduces the response evoked by an action potential by an amount that depends upon the details of the previous spike train data. One interpretation of the factor q is that it represents a short-term (reversible) reduction in the release probability for synaptic transmission due to a depletion in the number of vesicles that can readily fuse with the cell membrane [700]. In certain cases, it is also possible for a synapse to undergo a temporary facilitation in response to activation, which may be due to the presence of residual calcium in the axonal terminal [700].

A common phenomenological model of synaptic depression is to assume that between spikes $q(t)$ relaxes at a rate τ_q to its steady-state value of one, but that directly after the arrival of a spike it changes discontinuously, that is, $q \to \gamma q$ with $\gamma < 1$. The depression time constant τ_q can vary between around 100 ms and a few seconds [4]. The model for synaptic depression may be written succinctly as

$$\frac{dq}{dt} = \frac{(1-q)}{\tau_q} - (1-\gamma)\sum_n q(T^n)\delta(t - T^n), \quad q(0) = 1, \tag{1.45}$$

which has the solution of the form

$$q(T^m) = 1 - (1-\gamma)\sum_{n<m} \gamma^{[m-n-1]\beta} \mathrm{e}^{-(T^m - T^n)/\tau_q}.$$

Assuming a regular sequence of incoming spikes $T^n - T^{n-1} = \Delta$ for all n we find that the asymptotic amplitude $q_\infty(\Delta) \equiv \lim_{m\to\infty} q(T^m)$ is given by

$$q_\infty(\Delta) = \frac{1 - e^{-\Delta/\tau_q}}{1 - \gamma e^{-\Delta/\tau_q}}. \tag{1.46}$$

One possible computational role for synaptic depression is as a mechanism for cortical gain control [4]. The basic idea can be understood from the dependence of the asymptotic amplitude $q_\infty(\Delta)$ on the stimulus frequency $f = \Delta^{-1}$. Assuming that $\tau_q \gg \Delta$, we can Taylor expand q_∞ in (1.46) to find that $q_\infty(f) \approx \Gamma/f$, where $\Gamma = \tau_q/(1-\gamma)$. The main point to note is that the postsynaptic response per unit time is approximately independent of f (assuming that each spike elicits the same response in the steady state). This means that the synapse is very sensitive to changes in the stimulus frequency. The instantaneous response to a rapid increase Δf in the stimulus rate is given by $\Gamma \Delta f/f$. In other words, the synapse responds to relative rather than absolute changes in the rate of input stimulation.

1.3.3 Kinetic Model of a Synapse

Let $g_{\text{syn}}(t) \sim s(t)$ where $s(t)$ is the fraction of synaptic receptor channels that are in an open conducting state. The probability of being in an open state depends on the presence and concentration $\mathscr{T}$ of neurotransmitter released by the presynaptic neuron. Assuming a first-order kinetic scheme, in which a closed receptor in the presence of a concentration of neurotransmitter $\mathscr{T}$ equilibrates with the open receptor state, we have

$$C \underset{r_2(v)}{\overset{r_1(v,\mathscr{T})}{\rightleftharpoons}} O,$$

where C and O represent the closed and open states of the channel and $r_1(v,\mathscr{T})$ and $r_2(v)$ are the associated rate constants. However, in many cases synaptic channels are found to have time-dependent properties that are more accurately modeled with a second-order kinetic scheme. In fact the presence of one or more receptor sites on a channel allows the possibility of transitions to *desensitized states*. Such states are equivalent to the inactivated states of voltage-dependent ion channels. The addition of such a desensitized state to the first-order process generates a second-order scheme:

$$\begin{aligned}\frac{ds}{dt} &= r_1(v,\mathscr{T})(1-s-z) - [r_2(v)+r_3(v)]s + r_4(v)z,\\ \frac{dz}{dt} &= r_6(v,\mathscr{T})(1-s-z) - [r_4(v)+r_5(v)]z + r_3(v)s,\end{aligned} \tag{1.47}$$

where z is the fraction of channels in the desensitized state. All neurotransmitter-dependent rate constants have the form $r_i(v,\mathscr{T}) = r_i(v)\mathscr{T}$. It is common for detailed Markov models of voltage-gated channels to assume that the voltage dependence of all rates takes a simple exponential form. However, it has been shown that the number of states needed by a model to more accurately reproduce the behavior of a channel may be reduced by adopting sigmoidal functions for the voltage-dependent transition rates (see Destexhe et al. [154] for a discussion), so that we write

$$r_i(v) = \frac{a_i}{1+\exp[-(v-c_i)/b_i]}. \tag{1.48}$$

The a_i set the maximum transition rate, b_i the steepness of the voltage dependence, and c_i the voltage at which the half-maximal rate is reached. Furthermore, the concentration of neurotransmitter can often be successfully approximated by a sigmoidal function of the presynaptic potential v_{pre}:

$$\mathscr{T}(v_{\text{pre}}) = \frac{\mathscr{T}_{\max}}{1+\exp[-(v_{\text{pre}}-v_\Delta)/\Delta]}. \tag{1.49}$$

Here, $\mathscr{T}_{\max}$ is the maximal concentration of transmitter in the synaptic cleft, v_{pre} is the presynaptic voltage, Δ gives the steepness, and v_Δ sets the value at which the function is half activated. It is common to take $\Delta = 5\,\text{mV}$ and $v_\Delta = 2\,\text{mV}$. One of the main advantages of using an expression such as (1.49) is that it provides a smooth transformation between presynaptic voltage and transmitter concentration from which postsynaptic currents can easily be calculated from (1.39), (1.47), (1.48), and (1.49).

Now consider the following second-order gating scheme

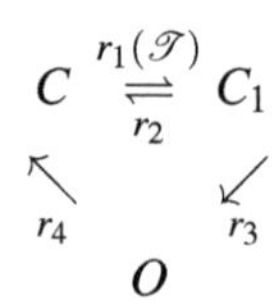

where C and C_1 are the closed forms of the receptor, O is the open (conducting) form, and the r_i are voltage-independent transition rates. Under certain assumptions it may be shown that this particular second-order scheme describes the so-called alpha function response commonly used in synaptic modeling. The following approximations are required: (i) The transmitter concentration $\mathscr{T}$ occurs as a pulse $\delta(t-t_0)$ for a release event occurring at time $t = t_0$, that is, $r_1(\mathscr{T}) = r_1\delta(t-t_0)$; (ii) The fraction of channels in C is considered constant and ~ 1. The kinetic equation (1.47) then reduce to

$$\frac{d\mathbf{c}(t)}{dt} = \mathbf{Q}\mathbf{c}(t) + \mathbf{I}(t),$$

(assuming $\mathbf{c}(0) = \mathbf{0}$), where

$$\mathbf{Q} = \begin{pmatrix} -\frac{1}{\tau_1} & 0 \\ r_3 & -\frac{1}{\tau_2} \end{pmatrix}, \quad \mathbf{I}(t) = \begin{pmatrix} r_1\delta(t-t_0) \\ 0 \end{pmatrix}, \quad \mathbf{c} = \begin{pmatrix} z \\ s \end{pmatrix},$$

and $\tau_1 = 1/(r_2 + r_3)$, $\tau_2 = 1/r_4$. Here z and s represent the fraction of receptors in the forms C_1 and O, respectively. This Markov chain system has a solution of the form

$$\mathbf{c}(t) = \int_0^t \mathbf{G}(t-s)\mathbf{I}(s)\mathrm{d}s, \quad \mathbf{G}(t) = \mathrm{e}^{t\mathbf{Q}}.$$

The eigenvectors of $\mathbf{Q}$ are $(1, r_3/(\tau_2^{-1} - \tau_1^{-1}))$ and $(0,1)$ with associated eigenvalues $-1/\tau_1$ and $-1/\tau_2$, respectively. Hence, one finds that

$$s(t) = r_1 r_3 \left(\frac{1}{\tau_2} - \frac{1}{\tau_1}\right)^{-1} (\mathrm{e}^{-(t-t_0)/\tau_1} - \mathrm{e}^{-(t-t_0)/\tau_2}), \quad t > t_0.$$

In the limit $\tau_2 \to \tau_1 \to \tau_s$ this reduces to an alpha function

$$s(t) = r_1 r_3 (t - t_0)\mathrm{e}^{-(t-t_0)/\tau_s}, \quad t > t_0.$$

This kinetic derivation of the alpha function only holds for $s \ll 1$ in order to remain consistent with condition (ii).

The time course of some ion-channel open and closed states seems to follow a power law rather than multiexponential law at large times [429]. In order to understand such power-law behavior, consider an ion channel with N closed states such that the transition to an open state can only take place from state 1 at one end of a chain

$$0 \overset{\alpha}{\leftarrow} 1 \underset{\beta_1}{\overset{\gamma_1}{\rightleftharpoons}} 2 \ldots \underset{\beta_{N-1}}{\overset{\gamma_{N-1}}{\rightleftharpoons}} N.$$

The corresponding kinetic equations are

$$\begin{aligned}
\frac{dc_1}{dt} &= \beta_1 c_2 - (\gamma_1 + \alpha)c_1, \\
\frac{dc_n}{dt} &= \gamma_{n-1}c_{n-1} + \beta_n c_{n+1} - (\gamma_n + \beta_{n-1})c_n, \quad 1 < n < N, \\
\frac{dc_N}{dt} &= \gamma_{N-1}c_{N-1} - \beta_{N-1}c_N.
\end{aligned}$$

In the following we take $\gamma_n = \beta_n = 1$ for all n and $\alpha = 1$, so that the system of equations describes a discrete diffusion process along a chain with a reflecting boundary at $n = N$ and an absorbing boundary at $n = 0$. In the large N limit, it can be shown that given the initial condition $p_n(0) = \delta_{n,1}$, the exact solution is

$$c_n(t) = \mathrm{e}^{-2t}[I_{n-1}(t) - I_{n+1}(t)], \tag{1.50}$$

where $I_n(t)$ is the modified Bessel function of integer order:

$$I_n(t) = \int_{-\pi}^{\pi} \mathrm{e}^{ink}\mathrm{e}^{2t\cos(k)}\frac{dk}{2\pi}.$$

By carrying out an asymptotic expansion for large t, it can be shown that

$$c_n(t) \approx \frac{n}{2\pi^{1/2}t^{3/2}}.$$

Define $F(t)$ to be the total probability of finding the system in a closed state:

$$F(t) = \sum_{n=1}^{N} c_n(t).$$

It follows that $dF/dt = -\alpha c_1$ and, hence, $F(t) \approx (\pi t)^{-1/2}$ for large N, t and $\alpha = 1$. More recently, it has been suggested that synapses with multiple states, which exhibit dynamics over a wide range of time scales and show power-law-like behavior, could have some interesting computational properties [208, 219]. For example, it has been suggested that such synapses could provide a way of combining high levels of memory storage with long retention times [208].

1.4 Dendritic Processing

Typically, a single neuron in cerebral cortex has up to 10,000 synapses, which are spatially distributed along the dendritic tree (and perhaps on the cell body and proximal part of the axon). In order to find the total synaptic current $u(t)$ entering the cell body, it is necessary to determine how the various local currents flow along the dendritic tree and combine at the soma. We will show that if the dendrites are modeled as passive electrical cables, then the dendritic tree acts as a linear spatiotemporal filter of synaptic currents.

1.4.1 The Cable Equation

Neurons display a wide range of dendritic morphologies, ranging from compact arborizations to elaborate branching patterns. At the simplest level, the dendritic tree can be treated as a passive electrical medium that filters incoming synaptic stimuli in a diffusive manner. The current flow and potential changes along a branch of the tree may be described with a second-order, linear partial differential equation commonly known as the *cable equation*. (The application of cable theory to the study of passive, spatially extended dendrites was pioneered by Wilfrid Rall in the 1960s and 1970s. For more recent accounts of this work see [515] and the annotated collection of papers edited by Segev, Rinzel, and Shepherd [299].) The cable equation is based on a number of approximations: (1) magnetic fields due to the movement of electric charge can be neglected, (2) changes in ionic concentrations are sufficiently small so that Ohm's law holds, (3) radial and angular components of voltage can be ignored so that the cable can be treated as one-dimensional medium, and (4) dendritic membrane properties are voltage-independent, that is, there are no active elements.

A nerve cable consists of a long thin, electrically conducting core surrounded by a thin membrane whose resistance to transmembrane current flow is much greater than that of either the internal core or the surrounding medium. Injected current can travel long distances along the dendritic core before a significant fraction leaks out across the highly resistive cell membrane. Linear cable theory expresses conservation of electric current in infinitesimal cylindrical elements of nerve fiber modeled using the equivalent circuit shown in Fig. 1.11. Define $v(x,t)$ as the membrane potential at position x along a cable at time t (measured relative to the resting potential of the membrane). Let C_m be the capacitance per unit area of the cell membrane, R the resistivity of the intracellular fluid (in units of resistance $\times$ length), R_m the cell membrane resistance (in units of resistance $\times$ area), and a the cable radius. Note that C_m, R_m, R are independent of cable radius—the corresponding quantities expressed per unit length of cable are

$$r = \frac{R}{\pi a^2}, \quad \frac{1}{r_m} = \frac{2\pi a}{R_m}, \quad c_m = 2C_m \pi a. \tag{1.51}$$

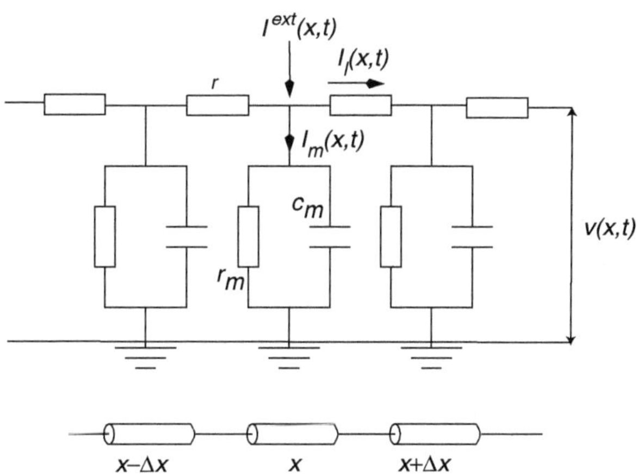

Fig. 1.11 Equivalent circuit for successive cylindrical segments of passive dendritic membrane

Current conservation implies that (see Fig. 1.11)

$$I^{ext}(x,t) - I_m(x,t) = \frac{I_l(x,t) - I_l(x-\Delta x,t)}{\Delta x} \approx \frac{\partial I_l(x,t)}{\partial x}, \tag{1.52}$$

where $I^{ext}(x,t)$ is an external injected current density. From Ohm's law we also have

$$I_l(x,t)r = \frac{v(x,t) - v(x+\Delta x,t)}{\Delta x} \approx -\frac{\partial v(x,t)}{\partial x}, \tag{1.53}$$

and

$$I_m(x,t) = c_m \frac{\partial v(x,t)}{\partial t} + \frac{v(x,t)}{r_m}. \tag{1.54}$$

Combining these three equations yields the uniform cable equation

$$\tau_m \frac{\partial v(x,t)}{\partial t} = -v(x,t) + \lambda_m^2 \frac{\partial^2 v(x,t)}{\partial x^2} + r_m I_{ext}(x,t), \qquad t \geq 0, \tag{1.55}$$

where $\tau_m = R_m C_m$ is the membrane time constant and $\lambda_m = (R_m a/2R)^{1/2}$ is the membrane space constant. (It follows that the coupling constant appearing in the Hodgkin–Huxley equation (1.8) is $K = \lambda_m^2/R_m$.)

Infinite Cable. In the case of an infinite uniform cable, $x \in \mathbb{R}$, we can solve (1.55) by Fourier transforming with respect to x. That is, define the Fourier transform of v (and other quantities) as

$$\tilde{v}(k,t) = \int_{-\infty}^{\infty} \mathrm{e}^{-ikx} v(x,t) dx,$$

with inverse transform

$$v(x,t) = \int_{-\infty}^{\infty} \mathrm{e}^{ikx} \tilde{v}(k,t) \frac{dk}{2\pi}.$$

Then

$$\tau_m \frac{\partial \tilde{v}(k,t)}{\partial t} = -\tilde{v}(k,t) - \lambda_m^2 k^2 \tilde{v}(k,t) + r_m \tilde{I}_{ext}(k,t), \quad x \in \mathbb{R}, \quad t \geq 0. \tag{1.56}$$

This first-order ODE can now be solved straightforwardly as

$$\tilde{v}(k,t) = \frac{r_m}{\tau_m} \int_{-\infty}^{t} \tilde{G}_0(k,t-t') \tilde{I}^{ext}(k,t') dt', \tag{1.57}$$

where $\tilde{G}_0(k,t) = \mathrm{e}^{-(1+\lambda_m^2 k^2)t/\tau_m}$. Taking the inverse Fourier transform and using the convolution theorem shows that

$$v(x,t) = r_m \int_{-\infty}^{t} \left[\int_{-\infty}^{\infty} G_0(x-x',t-t') I^{ext}(x',t') dx' \right] dt', \tag{1.58}$$

where we have absorbed τ_m into r_m and

$$\begin{aligned} G_0(x,t) &= \int_{-\infty}^{\infty} \frac{\mathrm{d}k}{2\pi} \mathrm{e}^{ikx} \mathrm{e}^{-(1+\lambda_m^2 k^2)t/\tau_m} \\ &= \frac{1}{2\lambda_m \sqrt{\pi t/\tau_m}} \mathrm{e}^{-t/\tau_m} \mathrm{e}^{-\tau_m x^2/4\lambda_m^2 t}. \end{aligned} \tag{1.59}$$

The function $G_0(x,t)$ is the fundamental solution or Green's function for the cable equation with unbounded domain. It is positive and symmetric and satisfies the homogeneous cable equation

$$\left(\tau_m \frac{\partial}{\partial t} + 1 - \lambda_m^2 \frac{\partial^2}{\partial x^2}\right) G_0(x,t) = 0, \tag{1.60}$$

with initial condition

$$G_0(x,0) = \delta(x). \tag{1.61}$$

Moreover, for any $0 < s < t$, it satisfies the Markovian property

$$G_0(x-y,t) = \int_{-\infty}^{\infty} G_0(x-z,s) G_0(z-y,t-s) \mathrm{d}z. \tag{1.62}$$

The Green's function is plotted as a function of time in Fig. 1.12a for a range of separations x.

Semi-infinite Cable. Using Fourier cosine or sine transforms with respect to x it is straightforward to determine the Green's function for a semi-infinite cable $x \in [0,\infty)$ with either an open-end boundary condition

$$v(0,t) = 0 \tag{1.63}$$

or a closed-end boundary condition (zero current flow)

$$\left.\frac{\partial v(x,t)}{\partial x}\right|_{x=0} = 0 \tag{1.64}$$

One finds that

$$v(x,t) = r_m \int_{-\infty}^{t} \left[\int_0^{\infty} G_{\pm}(x,x',t-t') I^{ext}(x',t') dx' \right] dt' \tag{1.65}$$

where

$$G_{\pm}(x,y,t) = G_0(x-y,t) \pm G_0(x+y,t) \tag{1.66}$$

for the open $(-)$ and closed $(+)$ cases. For a discussion of finite-length cables see [342].

Single Branching Node. Let us now consider a single branching node and label each semi-infinite segment by the index $i = 1,\ldots,N$. (Typically $N = 3$.) We shall assume that the cables only differ in their radius a_i. In order to simplify the analysis we will measure the distance along the ith cable from the branch point at $x = 0$ in units of $\lambda_{m,i} = \sqrt{R_m a_i / 2R}$ such that the cable equation on each branch can be written as

$$\tau_m \frac{\partial v_i(X,t)}{\partial t} = -v_i(x,t) + \frac{\partial^2 v_i(x,t)}{\partial x^2} + I_i(x,t). \tag{1.67}$$

The boundary conditions are continuity of the potential at the node

$$v_i(0,t) = v_j(0,t), \tag{1.68}$$

and conservation of current

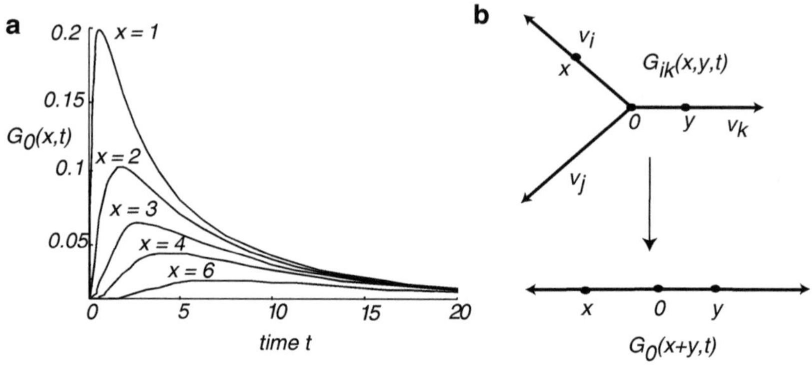

Fig. 1.12 (**a**) Green's function of an infinite cable as a function of time t (in units of τ_m) for various values of distance x (in units of λ). (**b**) Branching dendritic tree of an idealized single neuron and an equivalent cylinder representation

$$\sum_{j=1}^{N} a_j^{3/2} \left.\frac{\partial v_j(x,t)}{\partial x}\right|_{x=0} = 0. \tag{1.69}$$

The factor $a_i^{3/2}$ arises from the fact that we have rescaled length and that the longitudinal resistance varies as the cross-sectional area of the cable. The solution of these equations is

$$v_i(x,t) = \sum_{j=1}^{N} \int_{-\infty}^{t} \left[\int_0^{\infty} G_{ij}(x,x',t-t') I_j(x',t') dx' \right] dt', \tag{1.70}$$

where

$$G_{ij}(x,y,t) = \delta_{i,j} G_0(x-y,t) + (2p_j - \delta_{i,j}) G_o(x+y,t), \tag{1.71}$$

and

$$p_k = \frac{a_k^{3/2}}{\sum_m a_m^{3/2}}. \tag{1.72}$$

It is interesting to note that if $p_k = 1/2$ for a certain segment, then the response to current injected into that segment can be represented by a single equivalent cylinder; see Fig. 1.12b. That is, if $i \neq j$, then $G_{ij}(x,y,t) = G_0(x+y)$, where $x+y$ is the distance of the direct path between x and y on the tree. Hence, the node in some sense acts no differently than an ordinary point on a single cable. This is the basis of Rall's equivalent cylinder concept [515].

Dendritic Tree. An arbitrary dendritic tree Γ may be considered as a set of branching nodes linked by finite pieces of uniform cable. Given an external current density $I^{ext}(x,t)$ the voltage response has the formal solution (absorbing r_m into the definition of G)

$$v(x,t) = \int_{-\infty}^{t} \left[\int_{\Gamma} G(x,y,t-s) I_{\text{ext}}(y,s) dy \right] ds. \tag{1.73}$$

The associated Green's function $G(x,y,t-s)$ satisfies the homogeneous cable equation on each segment together with boundary conditions at the branching and terminal nodes of the tree. Rules for constructing such a Green's function have been developed by Butz and Cowan [95] using a graphical calculus and by Abbott et al. [3] using path-summing methods. The latter approach can be understood in terms of a compartmental model obtained by spatially discretizing the cable equation. The construction of the discretized Green's function involves summing over paths of a random walk on the tree with the corresponding Green's function for the cable equation recovered in the continuum limit [77].

1.4.2 Dendritic Filtering of Synaptic Inputs

So far we have considered the linear response of a dendritic cable to external current injection as determined by the Green's function or transfer function. Suppose that we replace the external current by a synaptic current of the form discussed in Sect. 1.3. That is, $I_{\text{ext}}(x,t) \to I(x,t)$, where $I(x,t)$ is the synaptic current density at location x at time t:

$$I(x,t) = \rho(x) \sum_{m} g_{\text{syn}}(t - T^m(x))[V_{\text{syn}} - v(x,t)] \equiv g(x,t)[V_{\text{syn}} - v(x,t)], \tag{1.74}$$

where $g(x,t) = \rho(x) \sum_m g_{\text{syn}}(t - T^m(x))$. Here $\rho(x)$ is the density of synapses (assuming that they have identical properties) and $\{T^m(x)\}$ is the sequence of spikes arriving into the synapses located at x. In the case of a discrete set of identical synapses at dendritic locations $\{x_j, j = 1, \ldots, M\}$, we have $\rho(x) = \sum_j \delta(x - x_j)$ and $T^m(x_j) = T_j^m$. The formal solution for the membrane potential is now [see (1.73)]

$$v(x,t) = \int_{-\infty}^{t} \left[\int_{\Gamma} G(x,x',t-t') g(x',t') [V_{\text{syn}} - v(x',t')]] dx' \right] dt' \tag{1.75}$$

which is a Volterra integral equation of the second kind. In order to solve this integral equation, we introduce the convolution operator $*$,

$$[G * f](x,t) := \int_{-\infty}^{t} \left[\int_{\Gamma} G(x,x',t-t') f(x',t') dx' \right] dt' \tag{1.76}$$

for any function $f(x,t)$. We can then iterate (1.75) to obtain a series solution for v:

$$\begin{aligned} v &= V_{\text{syn}} G * g - G * (gv) \\ &= V_{\text{syn}} G * g - V_{\text{syn}} G * (g[V_{\text{syn}} G * g - G * (gv)]) \\ &= V_{\text{syn}} G * g - V_{\text{syn}}^2 G * [gG * g] + G * [gG * (gv)] \end{aligned}$$

$$= V_{\text{syn}} G * g - V_{\text{syn}}^2 G * (gG * g) + V_{\text{syn}}^3 G * (gG * [gG * g]) - \dots$$
$$= V_{\text{syn}} (G - V_{\text{syn}} G * gG + V_{\text{syn}}^2 G * gG * gG - \dots) * g$$
$$= V_{\text{syn}} \hat{G} * g, \tag{1.77}$$

where

$$\hat{G} := G - V_{\text{syn}} G * gG + V_{\text{syn}}^2 G * gG * gG - \dots \tag{1.78}$$

is a Neumann series expansion for the effective Green's function $\hat{G}$, which is convergent for a passive cable [344]. More explicitly, we can write the solution as

$$v(x,t) = V_{\text{syn}} \int_{-\infty}^{t} \left[\int_{\Gamma} \hat{G}(x,t;x',t') g(x',t') dx' \right] dt', \tag{1.79}$$

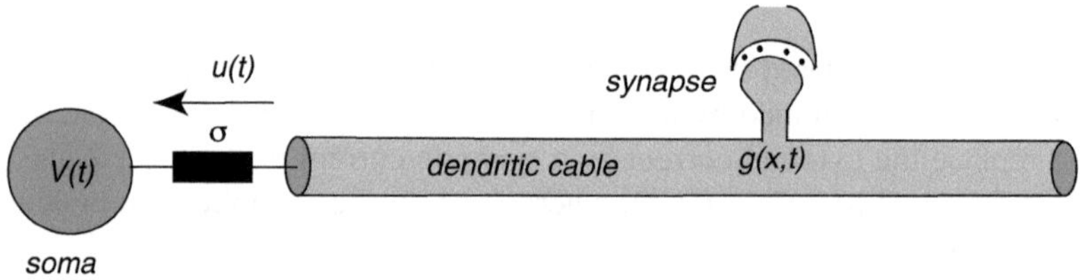

Fig. 1.13 Schematic diagram of a neuron consisting of a soma resistively coupled to one end of a dendritic cable. A synaptic conductance change $g(x,t)$ at position x on the cable induces a synaptic current into the soma at $x = 0$

with $\hat{G}$ satisfying the Volterra integral equation

$$\hat{G}(x,t;x',t') = G(x,x',t-t') \tag{1.80}$$
$$-V_{\text{syn}} \int_{t'}^{t} \left[\int_{0}^{\infty} G(x,x'',t-t'') g(x'',t'') \hat{G}(x'',t'';x',t') dx'' \right] dt''.$$

One can check that iteration of this equation recovers the series expansion for $\hat{G}$. The major point to note is that the resulting series involves interactions between synaptic conductances at different points on the dendritic cable. For simplicity, however, we shall assume that $V_{\text{syn}} \gg v(x,t)$ so that $\hat{G} \to G$ and synaptic interactions become negligible.

Given a distribution of synaptic inputs innervating the dendritic tree, what is the net synaptic current $u(t)$ entering the soma? In order to address this problem, consider a semi-infinite uniform dendritic cable, $0 \le x < \infty$, with the soma located at the end $x = 0$. The soma is modeled as a conductance-based point process that is passively coupled to the dendritic cable as illustrated in Fig. 1.13:

$$C \frac{dv}{dt} = -I_{\text{con}} + \sigma[v(0,t) - v(t)], \tag{1.81}$$

and

$$\tau_m \frac{\partial v(x,t)}{\partial t} = -v(x,t) + \lambda_m^2 \frac{\partial^2 v(x,t)}{\partial x^2} + r_m V_{\text{syn}} g(x,t). \tag{1.82}$$

Here $u(t) = \sigma[v(0,t) - v(t)]$ is the net current density flowing into the soma from the dendritic cable at $x = 0$. (We are neglecting direct synaptic inputs into the soma.) Current conservation implies the boundary condition

$$-\frac{1}{r}\frac{\partial v}{\partial x}(0,t) = \sigma[v(0,t) - v(t)]. \tag{1.83}$$

Since we can eliminate the term $-\sigma v(t)$ in (1.81) by shifting the linear term in v, it follows that the total synaptic current into the soma is $u(t) = \sigma v(0,t)$.

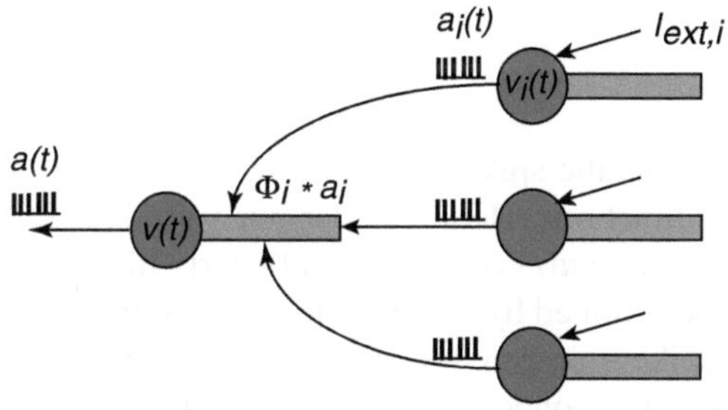

Fig. 1.14 Schematic diagram of a feedforward network showing a set of afferent neurons labeled by i synaptically driving another neuron

The inhomogeneous boundary problem for $v(0,t)$ can be solved formally using the Green's function G_+ for the semi-infinite cable with a closed boundary:

$$\begin{aligned} v(0,t) &= r_m V_{\text{syn}} \int_{-\infty}^{t} \left[\int_0^{\infty} G_+(0,x',t-t') g(x',t') dx' \right] dt' \\ &\quad -\sigma r \int_{-\infty}^{t} G_+(0,0,t-t')[v(0,t') - v(t')] dt' \end{aligned} \tag{1.84}$$

This shows that the effective synaptic current $u(t)$ flowing into the soma will itself be affected by the cell firing an action potential, due to the dependence of $v(0,t)$ on the somatic potential $v(t)$. However, suppose that the second term on the r.h.s. is negligible compared to the first term arising from synaptic inputs. This approximation corresponds to imposing the homogeneous boundary condition $\partial v/\partial x(0,t) = 0$. It then follows that the total synaptic input into the soma is

$$u(t) = \sigma r_m V_{\text{syn}} \int_{-\infty}^{t} \left[\int_0^{\infty} G_+(0,x',t-t') g(x',t') dx' \right] dt' \tag{1.85}$$

A similar analysis can also be carried out for more general dendritic topologies with the soma coupled to one of the terminals of the tree. We conclude that under the given approximations, the passive dendritic tree acts like a spatiotemporal linear filter on incoming spike trains, whose properties are determined by the underlying Green's function on the tree.

Recall from (1.74) that $g(x,t) = \rho(x)\sum_m g_{\text{syn}}(t - T^m(x))$. Suppose that there exists a discrete set of synapses along the dendrite so that $\rho(x) = \sum_j \delta(x - x_j)$ and $T^m(x_j) = T_j^m$. Substituting into (1.85) then gives

$$u(t) = \sum_j \sum_m \Phi_j(t - T_j^m) = \sum_j \int_{-\infty}^{t} \Phi_j(t - t')a_j(t')dt', \tag{1.86}$$

where

$$\Phi_j(t) = \sigma r_m V_{\text{syn}} \int_0^t G_+(0, x_j, t - \tau) g_{\text{syn}}(\tau) d\tau, \tag{1.87}$$

and

$$a_j(t) = \sum_m \delta(t - T_j^m). \tag{1.88}$$

Note that $a_j(t)$ represents the spike train arriving into the jth synapse in terms of a sum of Dirac delta functions. Hence, under our various approximations, we can view the total synaptic input $u(t)$ as the sum of linearly filtered spike trains, with the kernel of each filter determined by synaptic and dendritic processing. Now suppose that each incoming spike train is associated with a distinct afferent neuron, so that there is a one-to-one correspondence between synaptic and afferent neuron labels; see Fig. 1.14. Then $T_j^m = \hat{T}_j^m + \Delta\tau_j$, where $\hat{T}_j^m$ is the mth firing time of the jth afferent neuron and $\Delta\tau_j$ is an axonal propagation time delay. Unless stated otherwise, we will ignore axonal propagation delays and set $T_j^m = \hat{T}_j^m$. Let us also assume that each afferent neuron is described by a conductance-based model of the form

$$C_i \frac{dv_i}{dt} = -I_{\text{con},i} + I_{\text{ext},i} \tag{1.89}$$

where v_i is the somatic membrane potential of the ith afferent neuron, each of which is driven by an external input $I_{\text{ext},i}$. If we associate with each neuron a firing threshold κ, then the spike times T_i^m are determined according to

$$T_i^m = \inf\{t, t > T_i^{m-1} | v_i(t) = \kappa, \dot{v}_i(t) > 0\}. \tag{1.90}$$

Finally, given $u(t)$, the spike train $a(t)$ of the output neuron is determined by (1.1) and (1.9). In summary, the feedforward network involves a mapping $\{I_{\text{ext},i}(t)\} \to \{T_i^m\} \to \{T^m\}$.

1.4.3 Active Dendrites

It has been known for more than twenty years that the dendrites of cortical neurons do not simply act as passive electrical cables but also exhibit a variety of active physiological processes [608]. For example, thick apical dendrites of pyramidal neurons express voltage-gated Na^+, K^+, and Ca^{2+} channels, which support the back propagation of action potentials (APs) from the soma into the dendritic tree [397, 606]; back-propagating APs are thought to play an important role in spike-timing-dependent synaptic plasticity (STDP) [585]. In addition, sufficient local stimulation of active apical dendrites can initiate regenerative membrane depolarizations known as *dendritic spikes* [333, 555]. Some dendritic spikes are restricted to the local initiation zone rather than invading the cell body and are thus well placed to mediate the long-term potentiation of synaptic inputs in the absence of output spiking of the neuron [226]. On the other hand, Ca^{2+} action potentials initiated in apical dendrites can propagate towards the soma, which provides a mechanism for actively amplifying the effects of distal synapses on AP generation in the cell body [361]. Following advances in imaging techniques and methods of dendritic stimulation, Schiller et al. [556] established in vitro that active processes can also occur in thin basal and apical dendritic branches of pyramidal neurons, where the majority of synapses occur; see Fig. 1.15a. In particular, they found stimulus-evoked dendritic spikes whose major ionic component involved ligand-gated and voltage-gated N-methyl-D-aspartate receptor (NMDAR) channels; see also [362, 400, 520] and the review [12]. When glutamate binds to an NMDAR, it modifies the voltage sensitivity of the corresponding ion-channel current, which develops a negative slope conductance due to removal of a magnesium block [414, 466]. This means that in the presence of high levels of glutamate, the current–voltage (I–V) characteristics of an NMDAR channel are very similar to the voltage-gated Na channel. Hence, during

Fig. 1.15 (**a**) Schematic illustration of a pyramidal neuron showing the thick apical dendrite and various thin dendrites. The latter support the initiation of dendritic NMDA spikes. (**b**) Typical waveform of a dendritic NMDA spike. Weak glutamatergic inputs generate EPSP-like (subthreshold) depolarizations. A stronger input can trigger a dendritic plateau potential, consisting of a rapid onset that is often associated with a Na spikelet, a long-lasting plateau phase that can have a duration of several hundred ms, and a sudden collapse at the end of the plateau phase. The plateau potential consists of several dendritic conductances, the most predominant being due to NDMAR channels. Pharmacologically blocking Na and Ca^{2+} channels reveals the pure dendritic NMDA spike [556]

strong stimulation of a thin dendrite due to the local uncaging of glutamate or high frequency stimulation of a cluster of synapses, the NMDARs can fire a regenerative dendritic spike, just as Na channels support the initiation of an action potential following membrane depolarization. However, the duration of the dendritic spike is of the order 100 ms rather than 1 ms; see Fig. 1.15b. Finally, active processes can also be found in dendritic spines, which can support the propagation of saltatory waves; see Sect. 3.1

For relatively small deviations of the membrane potential from some constant value, a linearization of the channel kinetics can be performed. The resulting system has a membrane impedance that displays resonant-like behavior due to the additional presence of effective inductances [61, 141, 341, 342]. We sketch how inductance-like behavior can arise from active neural membrane by considering a generic ion current of the form $I(v,x_1,\ldots,x_M)$, where v is membrane voltage and x_k are gating variables that satisfy

$$\tau_k(v)\frac{dx_k}{dt} = x_{k,\infty}(v) - x_k, \quad k = 1,\ldots,M. \tag{1.91}$$

It is convenient to set $\tau_k(v) = (\alpha_k(v)+\beta_k(v))^{-1}$ and $x_{k,\infty}(v) = \alpha_k(v)\tau_k(v)$. Linearizing around a fixed point $\mathbf{z} = \mathbf{z}^*$ with vectors defined by $\mathbf{z} = (v,x_1,\ldots,x_M)^T$ and $\mathbf{z}^* = (v^*,x_{1,\infty}(v^*),\ldots,x_{M,\infty}(v^*))$, we have

$$\delta I = \frac{\delta v}{R} + \sum_{k=1}^{M} \left.\frac{\partial I}{\partial x_k}\right|_{\mathbf{z}=\mathbf{z}^*} \delta x_k, \tag{1.92}$$

where R is an effective resistance such that $R^{-1} = \partial I/\partial V|_{\mathbf{z}=\mathbf{z}^*}$. From (1.91) it follows that

$$\left(\frac{d}{dt} + \alpha_k + \beta_k\right)\delta x_k = \left(\frac{d\alpha_k}{dV} - x_{k,\infty}\frac{d[\alpha_k+\beta_k]}{dV}\right)\delta V_k. \tag{1.93}$$

Combining (1.92) and (1.93) we arrive at the following equation for the first-order variation of the current:

$$\delta I = \frac{\delta V}{R} + \sum_{k=1}^{M} \delta I_k, \tag{1.94}$$

where

$$\left(r_k + L_k\frac{d}{dt}\right)\delta I_k = \delta V, \tag{1.95}$$

and

$$r_k^{-1} = \tau_k \frac{\partial I}{\partial x_k}\left(\frac{d\alpha_k}{dV} - x_{k,\infty}\frac{d[\alpha_k+\beta_k]}{dV}\right)\bigg|_{\mathbf{z}=\mathbf{z}^*} \tag{1.96}$$

$$L_k = \tau_k r_k \tag{1.97}$$

Hence, for a small perturbation around the steady state, the current I responds as though the resistance R is in parallel with M impedance lines, each of which is a resistance r_k that is itself in series with an inductance L_k (see Fig. 1.16). Such inductive terms account for the oscillatory *overshoot* commonly seen in response to depolarising current steps or even after the firing of an action potential. This form of equivalent linear membrane circuit is typically called *quasi-active* in order to distinguish it from a truly *active* (i.e. nonlinear) membrane [341].

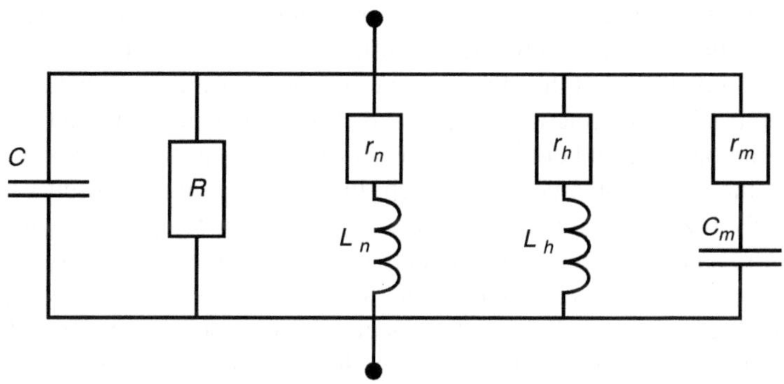

Fig. 1.16 Electrical LRC circuit representing the linearized response of the Hodgkin–Huxley equations

Quasilinear membrane can be incorporated into the cable equation (1.55) by introducing the space-dependent voltage $v(x,t)$ and currents $I_k(x,t)$, $x \in \mathbb{R}$, such that

$$\frac{\partial V}{\partial t} = -\frac{V}{\tau_m} + D_m \frac{\partial^2 V}{\partial x^2} - \frac{1}{C_m}\left[\sum_k I_k - I_{\text{ext}}\right] \tag{1.98a}$$

$$L_k \frac{dI_k}{dt} = -r_k I_k + V. \tag{1.98b}$$

Here $D_m = \lambda_m^2/\tau_m$. Laplace transforming (1.98) with $v(x,0) = 0, I_k(x,0) = 0$ yields the ODE

$$-\frac{d\tilde{V}^2}{dx^2} + \gamma^2(\omega)\tilde{V} = \tilde{I}, \tag{1.99}$$

with $\tilde{V} = \tilde{V}(x,\omega)$, $\tilde{I} = \tilde{I}(x,\omega) = \tilde{I}_{\text{ext}}(x,\omega)/C_m$, and

$$\gamma^2(\omega) = \frac{1}{D_m}\left[\frac{1}{\tau_m} + \omega + \frac{1}{C_m}\sum_k \frac{1}{r_k + \omega L_k}\right]. \tag{1.100}$$

It follows that in Laplace space,

$$\tilde{V}(x,\omega) = \int_0^\infty \tilde{G}(x-y,\omega)\tilde{I}(y,\omega)dy, \quad \tilde{G}(x,\omega) = \frac{\mathrm{e}^{-\gamma(\omega)|x|}}{2D_m\gamma(\omega)}, \tag{1.101}$$

where $\tilde{G}$ is the Laplace transform of the Green's function on an infinite quasi-active cable. Having obtained G, it is then possible to extend Green's function methods outlined for passive branching cables to the quasi-active case [141].

1.5 Stochastic Ion Channels

In the standard conductance-based model of a neuron (1.6), it is assumed that the number of voltage-gated ion channels is sufficiently large so that one can represent the opening and closing of the channels in terms of deterministic kinetic equations. These keep track of the fraction of open and closed channels as a function of time. However, the opening and closing of a single channel is a stochastic process. This then raises the important issue of how such stochasticity manifests itself when there are relatively few ion channels. (A similar issue applies to synapses with a small number of receptor-mediated ion channels.) In this section, we develop the theory of stochastic ion channels in some detail, since it provides an excellent platform for introducing various methods and ideas in stochastic processes that will be used throughout the book. For reviews on noise in ion channels see [227, 588, 670].

1.5.1 Ensemble of Two-State Ion Channels

First, consider a single ion channel that can exist either in a closed state (C) or an open state (O). Transitions between the two states are governed by a continuous-time jump Markov process

$$C(\text{closed}) \underset{\beta(v)}{\overset{\alpha(v)}{\rightleftharpoons}} O(\text{open}). \tag{1.102}$$

with voltage-dependent transition rates $\alpha(v), \beta(v)$. For the moment, we assume that v is fixed. In order to understand what such a process means, let $Z(t)$ be a discrete random variable taking values $Z \in \{C, O\}$ and set $P_z(t) = \text{Prob}[Z(t) = z]$. From conservation of probability,

$$P_C(t) + P_O(t) = 1.$$

The transition rates then determine the probability of jumping from one state to the other in a small interval Δt:

$$\alpha \Delta t = \text{Prob}[Z(t+\Delta t) = O | Z(t) = C], \quad \beta \Delta t = \text{Prob}[Z(t+\Delta t) = C | Z(t) = O].$$

It follows that there are two possible ways for the ion channel to enter or leave the closed state:

$$P_C(t+\Delta t) = P_C(t) - \alpha P_C(t)\Delta t + \beta P_O(t)\Delta t.$$

Writing down a similar equation for the open state, dividing by Δt, and taking the limit $\Delta t \to 0$ leads to the pair of equations

$$\frac{dP_C}{dt} = -\alpha P_C + \beta P_O \tag{1.103a}$$

$$\frac{dP_O}{dt} = \alpha P_C - \beta P_O. \tag{1.103b}$$

Now suppose that there are N identical, independent two-state ion channels. In the limit $N \to \infty$ we can reinterpret P_C and P_O as the mean fraction of closed and open ion channels within the population, and fluctuations can be neglected. After setting $P_O = X$ and $P_C = 1 - X$, we recover the kinetic equation (1.4). (An identical argument can be applied to the kinetic model of a synapse considered in Sect. 1.3.)

In order to take into account fluctuations in the case of finite N, it is necessary to keep track of the probability $P(n,t)$ that there are n open channels at time t, $0 \leq n \leq N$. (If there are n open channels, then it immediately follows that there are $N - n$ closed channels, so we do not need to keep track of the latter as well.) Consider a time interval $[t, t+\Delta t]$ with Δt sufficiently small so that only one channel has a significant probability of making a $C \to O$ or $O \to C$ transition. There are four possible events that can influence $P(n,t)$ during this interval, two of which involve transitions into the state of n open ion channels, and two of which involve transitions out of the state. Collecting these terms and taking the limit $\Delta t \to 0$ leads to the *master equation*

$$\frac{d}{dt}P(n,t) = \alpha(N-n+1)P(n-1,t) + \beta(n+1)P(n+1,t) - [\alpha(N-n) + \beta n]P(n,t). \tag{1.104}$$

The first term on the right-hand side represents the probability flux that one of $N - (n-1)$ closed channels undergoes the transition $C \to O$, whereas the second term represents the probability flux that one of $n+1$ open channels undergoes the transition $O \to C$. The last two terms represent transitions $n \to n \pm 1$. Define the mean number of open channels at time t by

$$\overline{n}(t) = \sum_{n=0}^{N} nP(n,t).$$

By differentiating both sides of this equation with respect to t and using the master equation (1.104), it can be shown that in the limit of large N (where the upper limit in the sum can be taken to be ∞) we recover the kinetic equation (1.4) with $X = \overline{n}/N$.

The steady-state solution $P_s(n)$ of the master equation (1.104) satisfies $J(n) = J(n+1)$ with

$$J(n) = \omega_-(n)P_s(n) - \omega_+(n-1)P_s(n-1),$$

and

$$\omega_+(n) = (N-n)\alpha, \quad \omega_-(n) = n\beta.$$

Using the fact that n is a nonnegative integer, that is, $P_s(n) = 0$ for $n < 0$, it follows that $J(n) = 0$ for all n. Hence, by iteration,

$$P_s(n) = P_s(0) \prod_{m=1}^{n} \frac{\omega_+(m-1)}{\omega_-(m)} = P_s(0) \left[\frac{\alpha}{\beta}\right]^n \frac{N!}{n!(N-n)!}. \tag{1.105}$$

Taking logs of both sides of this equation and using Stirling's formula $\log(n!) \approx n\log n - n$ it can be shown that for large n, N, $P_s(n) \approx p_s(x)$ where $x = n/N$,

$$p_s(x) = \mathcal{N} \mathrm{e}^{-N\Phi(x)} \tag{1.106}$$

with normalization factor $\mathcal{N}$ and $\Phi(x)$ is the effective potential

$$\Phi(x) = -x\log(\alpha/\beta) + x\log(x) + (1-x)\log(1-x). \tag{1.107}$$

Let x^* be the unique critical point of the effective potential $\Phi(x)$, that is, $\Phi'(x^*) = 0$. Note that x^* coincides with the fixed point of the corresponding deterministic kinetic equations,

$$x^* = \frac{\alpha}{\alpha+\beta}. \tag{1.108}$$

Since N is large, we can make the Gaussian approximation

$$p_s(x) \approx p(0)\exp\left[-N\Phi(x^*) - N\Phi''(x^*)(x-x^*)^2/2\right]. \tag{1.109}$$

Under this approximation, the mean and variance of the fraction of open channels are given by

$$\frac{\bar{n}}{N} = x^* = \frac{\alpha}{\alpha+\beta}, \quad \frac{\langle (n-\bar{n})^2 \rangle}{N^2} = \frac{x^*(1-x^*)}{N}. \tag{1.110}$$

It is clear that fluctuations become negligible in the large-N limit.

1.5.2 Diffusion Approximation

A useful diffusion approximation of the master equation (1.104) for large but finite N can be obtained by carrying out a Kramers–Moyal or system-size expansion to second order in N^{-1} [210, 309], which was originally applied to ion-channel models by Fox and Lu [200]. This yields a *Fokker–Planck* (FP) equation describing the evolution of the probability density of a corresponding continuous stochastic process that is the solution to a *stochastic differential equation* (SDE), which in the physics literature is often called a *Langevin equation*; we will take these terms to be interchangeable. Further details concerning SDEs can be found in appendix section 1.7 and in various references [210, 476]. Moreover, a rigorous analysis of the diffusion approximation and its relationship to the system-size expansion has been carried out by Kurtz [352].

First, introduce the rescaled variable $x = n/N$ and transition rates $N\Omega_\pm(x) = \omega_\pm(Nx)$. Equation (1.104) can then be rewritten in the form

$$\frac{dp(x,t)}{dt} = N[\Omega_+(x-1/N)p(x-1/N,t) + \Omega_-(x+1/N)p(x+1/N,t) \\ -(\Omega_+(x)+\Omega_-(x))p(x,t)].$$

Treating $x, 0 \leq x \leq 1$, as a continuous variable and Taylor expanding terms on the right-hand side to second order in N^{-1} leads to the FP equation

$$\frac{\partial p(x,t)}{\partial t} = -\frac{\partial}{\partial x}[A(x)p(x,t)] + \frac{1}{2N}\frac{\partial^2}{\partial x^2}[B(x)p(x,t)] \tag{1.111}$$

with

$$A(x) = \Omega_+(x) - \Omega_-(x) \equiv \alpha - (\alpha+\beta)x, \tag{1.112a}$$
$$B(x) = \Omega_+(x) + \Omega_-(x) \equiv \alpha + (\beta-\alpha)x. \tag{1.112b}$$

The FP equation takes the form of a conservation equation

$$\frac{\partial p}{\partial t} = -\frac{\partial J}{\partial x}, \tag{1.113}$$

where $J(x,t)$ is the probability flux,

$$J(x,t) = -\frac{1}{2N}\frac{\partial}{\partial x}[B(x)p(x,t)] + A(x)p(x,t). \tag{1.114}$$

The FP equation is supplemented by the no-flux or reflecting boundary conditions at the ends $x = 0, 1$ and a normalization condition,

$$J(0,t) = J(1,t) = 0, \quad \int_0^1 p(x,t)dx = 1. \tag{1.115}$$

The FP equation has a unique steady-state solution obtained by setting $J(x,t) = 0$ for all $0 \leq x \leq 1$. The resulting first-order ODE can be solved to give a steady-state probability density of the form (1.107), with corresponding potential

$$\Phi_{\mathrm{FP}}(x) = -2\int^x \frac{A(x')}{B(x')}dx' = -2\int^x \frac{\Omega_+(x') - \Omega_-(x')}{\Omega_+(x') + \Omega_-(x')}dx'. \tag{1.116}$$

The mean and variance of the fraction of open channels close to the fixed point x^* can again be determined by carrying out a Gaussian approximation, and the results agree with those obtained from the steady-state solution of the master equation. An alternative way of calculating the mean and variance is to note that the solution to the FP equation (1.111) determines the probability density function for a corresponding stochastic process $X(t)$, which evolves according to the SDE or Langevin equation [210]

$$dX = A(X)dt + \frac{1}{\sqrt{N}} b(X)dW(t). \tag{1.117}$$

with $b(x)^2 = B(x)$. Here $W(t)$ denotes a Wiener process with $dW(t)$ distributed according to a Gaussian process with mean and covariance

$$\langle dW(t) \rangle = 0, \quad \langle dW(t)dW(s) \rangle = \delta(t-s)dtds. \tag{1.118}$$

Note that the noise term in (1.117) is multiplicative, since it depends on the current state $X(t)$. It is well known that there is an ambiguity in how one integrates multiplicative noise terms, which relates to the issue of Ito versus Stratonovich versions of stochastic calculus [210]; see appendix section1.7. However, for this particular example, based on the reduction of a master equation, the explicit form of the corresponding FP equation (1.111) ensures that the noise should be interpreted in the sense of Ito.

Thus, one can view the SDE as describing a stochastic path in phase space that involves a deterministic trajectory converging to the unique stable fixed point x^* that is perturbed by Gaussian-like fluctuations of order $1/\sqrt{N}$. Substituting $X - x^* = Y/\sqrt{N}$ into the SDE equation (1.117) and formally Taylor expanding to lowest order in $1/\sqrt{N}$ yields the so-called linear noise approximation

$$dY = -kYdt + b(x^*)dW(t), \tag{1.119}$$

with

$$k \equiv -A'(x^*) = \alpha + \beta, \quad b(x^*) = \sqrt{B(x^*)} = \sqrt{\frac{2\alpha\beta}{\alpha+\beta}}.$$

This takes the form of an Ornstein–Uhlenbeck equation [210], which can be solved as

$$Y(t) = \mathrm{e}^{-kt} Y_0 + b(x^*) \int_0^t \mathrm{e}^{-k(t-t')} dW(t'), \tag{1.120}$$

given the initial condition $Y(0) = Y_0$. It follows that the mean and covariance of $Y(t)$ are

$$\langle Y(t) \rangle = \mathrm{e}^{-kt} Y_0, \tag{1.121}$$

and

$$\begin{aligned} \mathrm{cov}(Y(t), Y(s)) &\equiv \langle [Y(t) - \langle Y(t) \rangle][Y(s) - \langle Y(s) \rangle] \rangle \\ &= \left\langle b(x^*)^2 \left[\int_0^t \mathrm{e}^{-k(t-t')} dW(t') \right] \left[\int_0^t \mathrm{e}^{-k(s-t'')} dW(t'') \right] \right\rangle \\ &= b(x^*)^2 \mathrm{e}^{-k(t+s)} \int_0^s \mathrm{e}^{2kt'} dt' = \frac{b(x^*)^2}{2k} \mathrm{e}^{-k(t-s)} \left[1 - \mathrm{e}^{-2ks} \right]. \end{aligned} \tag{1.122}$$

Here

$$\frac{b(x^*)^2}{2k} = \frac{\alpha\beta}{(\alpha+\beta)^2} = x^*(1-x^*),$$

and, without loss of generality, we have assumed that $t \geq s$. Thus, in the stationary limit $t \to \infty$,

$$\langle Y(t) \rangle \to 0, \quad \mathrm{cov}(Y(t), Y(s)) \to \frac{b(x^*)^2}{2k} \mathrm{e}^{-k|t-s|}.$$

Since $Y(t)/\sqrt{N} = X(t) - x^*$, we recover the results of (1.110).

Note that (1.116) differs from the effective potential (1.107) obtained directly from the master equation for large N, given that the latter can be rewritten in the form

$$\Phi(x) = \int^x \ln \frac{\Omega_-(x')}{\Omega_+(x')} dx'. \tag{1.123}$$

Although, this discrepancy is not much of an issue when the underlying kinetic equations have a unique fixed point, it can lead to exponentially large errors when there are multiple stable fixed points. The Langevin or diffusion approximation still accounts for the effects of fluctuations well within the basin of attraction of a locally stable fixed point. However, there is now a small probability that there is a noise-induced transition to the basin of attraction of another fixed point. Since the probability of such a transition is usually of order $\mathrm{e}^{-\tau N}$ with $\tau = \mathscr{O}(1)$, except close to the boundary of the basin of attraction, such a contribution cannot be analyzed accurately using standard Fokker–Planck methods [309]. These exponentially small transitions play a crucial role in allowing the network to approach the unique stationary state (if it exists) in the asymptotic limit $t \to \infty$. In other words, for multistable systems, the limits $t \to \infty$ and $N \to \infty$ do not commute [24, 256, 641]. One example where multistability is important is when considering the effects of stochastic ion channels on membrane voltage fluctuations (see below).

1.5.3 Membrane Voltage Fluctuations

Let us now return to a conductance-based model of a neuron, in which the stochastic opening of ion channels generates a stochastic ionic current that drives the membrane voltage. We are particularly interested in how fluctuations affect the initiation of an action potential due to the opening of a finite number of Na^+ channels. Therefore, we imagine freezing the slow K^+ channels, so that they effectively act as a leak current and simplify the sodium channels by treating each as a single activating subunit. The stochastic membrane voltage then evolves according to the piecewise deterministic equation

$$\frac{dV}{dt} = F(V,n) \equiv \frac{1}{N} f(V) n(t) - g(V), \tag{1.124}$$

where $f(v) = g_{\mathrm{Na}}(V_{\mathrm{Na}} - v)$ represents the gated sodium currents, $g(v) = -g_{\mathrm{eff}}[V_{\mathrm{eff}} - v] - I$ represents the sum of effective leakage currents and external inputs I, and $n(t)$ is the number of open sodium channels. Note that (1.124) only holds between jumps in the number of open ion channels, with the latter described by the master equation (1.104). The stochastic process defined by (1.104) and (1.124) is an example of a so-called stochastic hybrid system with piecewise deterministic dynamics. There has been a lot of recent interest in such systems, particularly within the context of conductance-based models [88, 321, 484, 654]. The associated probability density $p(v,n,t)$, which is defined according to

$$p(v,n,t)dv = \text{Prob}[n(t) = n, v \leq V(t) \leq v + dv],$$

given an initial condition $V(0) = V_0, n(0) = n_0$, satisfies the differential Chapman–Kolmogorov (CK) equation

$$\begin{aligned}\frac{\partial p}{\partial t} &= -\frac{\partial}{\partial v}\left[\left(\frac{n}{N}f(v) - g(v)\right)p\right] + \omega_+(v,n-1)p(v,n-1,t) \\ &+\omega_-(v,n+1)p(v,n+1,t) - [\omega_+(v,n) + \omega_-(v,n)]p(v,n,t),\end{aligned} \tag{1.125}$$

with

$$\omega_+(v,n) = \alpha(v)(N-n), \quad \omega_-(v,n) = \beta(v)n. \tag{1.126}$$

Note that the right-hand side of (1.124) is negative for large V and positive for small V, which implies that the voltage V is confined to some bounded domain $[V_1, V_2]$.

In order to investigate action potential initiation, we will assume that N is sufficiently large so that we can approximate the jump Markov process for the ion channels by a continuous Markov process using a diffusion approximation, and (ii) we assume that the transitions between different discrete states is much faster than the voltage dynamics so we can assume that, for fixed v, the number of open ion channels is close to the quasi-equilibrium $x^*(v) = \alpha(v)/(\alpha(v)+\beta(v))$. This limiting case was originally considered by Chow and White [116]. Under these approximations, the voltage dynamics is described by an SDE of the form [see (1.119)]

$$dV = \left[f(V)(x^* + Y(t)/\sqrt{N}) - g(V)\right]dt, \quad dY = -kY dt + b(x^*)dW(t). \tag{1.127}$$

Thus the stochastic voltage is coupled to a fast Ornstein–Uhlenbeck or colored noise process $Y(t)$. If we make the further assumption that the latter is in quasi-equilibrium for a given V (fast ion channels), $Y(t)dt \approx k^{-1}b(x^*)dW(t)$, then we obtain a scalar SDE for the voltage:

$$dV = [f(V)x^*(V) - g(V)]dt + \frac{1}{\sqrt{N}}\sigma(V)f(V)dW(t), \tag{1.128}$$

where

$$\sigma(V) = \frac{b(x^*(V))}{k(V)} = \frac{1}{\alpha(V)+\beta(V)}\sqrt{\frac{2\alpha(V)\beta(V)}{\alpha(V)+\beta(V)}}. \tag{1.129}$$

Taking $\alpha, \beta = \mathcal{O}(1/\varepsilon)$ for some dimensionless parameter $0 < \varepsilon \ll 1$, we see that $\sigma(V) = \mathcal{O}(\varepsilon^{1/2})$. In deriving (1.128), we have effectively taken a zero correlation limit of a colored noise process. It can be shown that the multiplicative noise term should be interpreted in the sense of Stratonovich, and the associated FP equation is given by [210, 321]

$$\frac{\partial p(v,t)}{\partial t} = -\frac{\partial}{\partial v}[\mathscr{A}(v)p(v,t)] + \frac{1}{N}\frac{\partial}{\partial v}\left[\mathscr{B}(v)\frac{\partial}{\partial v}p(v,t)\right], \tag{1.130}$$

with

$$\mathscr{A}(v) = f(v)x^*(v) - g(v), \quad \mathscr{B}(v) = [\sigma(v)f(v)]^2/2.$$

We have ignored an $\mathscr{O}(\varepsilon)$ contribution to the drift term of the form $\mathscr{B}'(v)/N$. The FP equation is supplemented by reflecting boundary conditions at $v = V_1, V_2$:

$$J(V_1,t) = J(V_2,t) = 0, \tag{1.131}$$

with

$$J(v,t) = \mathscr{A}(v)p(v,t) - \frac{\mathscr{B}(v)}{N}\frac{\partial}{\partial v}p(v,t). \tag{1.132}$$

1.5.4 First Passage Time Problem

A key property that one would like to calculate is the mean time to fire an action potential (MFPT) as a function of the stimulus current I. In the absence of noise, the system evolves according to the deterministic equation

$$\frac{dv}{dt} = \mathscr{A}(v) = \frac{\alpha(v)}{\alpha(v)+\beta(v)}f(v) - g(v) \equiv -\frac{d\Phi_0(v)}{dv}, \tag{1.133}$$

where $\Phi_0(v)$ is a deterministic potential. In Fig. 1.17, we plot $\Phi_0(v)$ as a function of v for various values of the external input current and the particular transition rates

$$\alpha(v) = \beta \exp\left(\frac{2(v-v_1)}{v_2}\right), \quad \beta = \text{constant}.$$

The minima and maxima of the potential correspond to stable and unstable fixed points of the deterministic dynamics, respectively. It can be seen that below a threshold current I_*, $I < I_*$, there exist two stable fixed points $v_\pm$ (minima) separated by an unstable fixed point at v_0 (maximum), that is, the system exhibits bistability. The left-hand fixed point represents the resting state, whereas the right-hand fixed point corresponds to an excited state. Thus, in the bistable regime the deterministic system requires an external perturbation in order to generate an action potential starting from the resting state. On the other hand, for the stochastic system it is possible that fluctuations in the opening and closing of Na^+ ion channels induce a transition from the resting state to the excited state by crossing over the potential hill at v_0. This is directly analogous to a diffusing particle escaping from the left to the right well in a double well potential, which is a classical example of a first passage time (FPT) problem in physics [210, 257]. (Of course, once such an event occurs, one has to take into account the K^+ dynamics in order to incorporate the effects of repolarization that return the system to the resting state. If one includes the slow opening and closing of these channels, then the underlying deterministic system becomes excitable rather than bistable; see Sect. 2.1. For the moment, we will assume that this does not significantly affect the noise-induced initiation of an action potential. It turns out that such an assumption breaks down if fluctuations in the opening and closing of K^+ channels become significant [456].)

We now outline the basic calculation of the mean time to escape from the resting state using the diffusion approximation. Since the voltage will rapidly approach the excited state v_+ once it has passed the maximum at v_0, the major contribution to the escape time will be due to the fluctuation-driven transition from v_- to v_0. We can model this process by replacing the reflecting boundary condition at $v = V_2$ with an absorbing boundary condition at $v = v_0 < V_2$:

$$p(v_0,t) = 0.$$

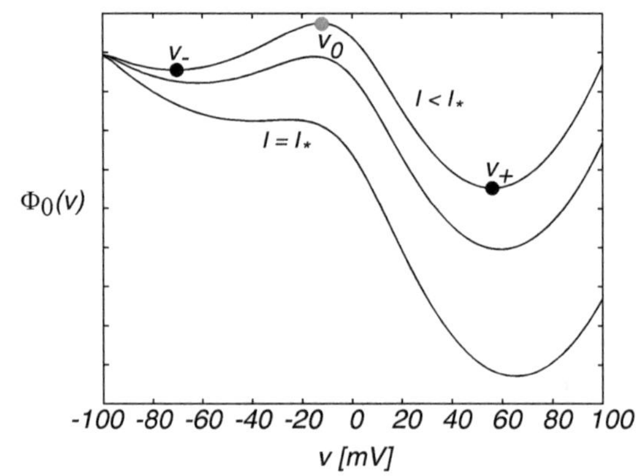

Fig. 1.17 Plot of deterministic potential $\Phi_0(v)$ as a function of voltage v for different values of the external stimulus current I. Parameter values are $N = 10$, $v_{\rm Na} = 120\,$mV, $v_{\rm eff} = -62.3\,$mV, $g_{\rm Na} = 4.4\,$mS/cm^2, $g_{\rm eff} = 2.2\,$mS/cm^2, $\beta = 0.8\,$s^{-1}, and $\alpha(v) = \beta \exp[(v+1.2)/9]$

We also shift the voltage v so that the left-hand boundary is at $V_1 = 0$. Let $T(v)$ denote the stochastic time for the particle to exit the right-hand boundary at v_0, given that it starts at location $v \in [0, v_0]$ at time $t = 0$. As a first step, we introduce the survival probability $\mathbb{P}(v,t)$ that the particle has not yet exited at time t:

$$\mathbb{P}(v,t) = \int_0^{v_0} p(x,t|v,0)dx. \tag{1.134}$$

It follows that $\text{Prob}[T(v) \leq t] = 1 - \mathbb{P}(v,t)$ and we can define the FPT density according to

$$f(v,t) = -\frac{\partial \mathbb{P}(v,t)}{\partial t}. \tag{1.135}$$

It can be shown from (1.130) and the Markovian nature of the stochastic process that the FPT density satisfies a backward FP equation of the form [210]

$$\frac{\partial \mathbb{P}(v,t)}{\partial t} = \mathscr{A}(v)\frac{\partial \mathbb{P}(v,t)}{\partial v} + \frac{\partial}{\partial v}\left(\mathscr{B}(v)\frac{\partial}{\partial v}\mathbb{P}(v,t)\right), \tag{1.136}$$

where we have absorbed the factor of $1/N$ into $\mathscr{B}$.

A quantity of particular interest is the mean first passage time (MFPT) $\tau(v)$ defined according to

$$\begin{aligned}\tau(v) = \langle T(v)\rangle &\equiv \int_0^\infty f(v,t)t dt \\ &= -\int_0^\infty t\frac{\partial \mathbb{P}(v,t)}{\partial t}dt = \int_0^\infty \mathbb{P}(v,t)dt,\end{aligned} \tag{1.137}$$

after integration by parts. Hence, integrating both sides of (1.136) shows that the MFPT satisfies the ODE

$$[\mathscr{A}(v) + \mathscr{B}'(v)]\frac{d\tau(v)}{dv} + \left(\mathscr{B}(v)\frac{d^2}{dv^2}\tau(v)\right) = -1. \tag{1.138}$$

Equation (1.138) is supplemented by reflecting and absorbing boundary conditions for the backward FP equation:

$$\tau'(0) = 0, \quad \tau(v_0) = 0. \tag{1.139}$$

It is straightforward to solve (1.138) by direct integration [210]. First, introducing an integration factor and integrating once gives

$$e^{\Psi(v)}\tau'(v) = -\int_0^v \frac{e^{\Psi(v')}}{\mathscr{B}(v')}dv',$$

where

$$\Psi(v) = \int_0^v \frac{\mathscr{A}(v') + \mathscr{B}'(v')}{\mathscr{B}(v')}dv'. \tag{1.140}$$

and we have used the boundary condition $\tau'(0) = 0$. Integrating once more with respect to v and using $\tau(v_0) = 0$ then gives

$$\tau(v) = \int_v^{v_0} e^{-\Psi(v')}dv' \int_0^{v'} \frac{e^{\Psi(v'')}}{\mathscr{B}(v'')}dv''. \tag{1.141}$$

There is now a standard procedure for approximating this double integral based on Kramers reaction rate theory [210, 257]. We simply quote the result here: $\tau(v_-) = 1/\lambda$ where

$$\lambda \approx \frac{\mathscr{B}(v_0)}{\pi}\sqrt{\left|\frac{\mathscr{A}'(v_-)}{\mathscr{B}(v_-)}\right|\frac{\mathscr{A}'(v_0)}{\mathscr{B}(v_0)}}\exp\left[\int_{v_-}^{v_0}\frac{\mathscr{A}(v)}{\mathscr{B}(v)}dv.\right]. \tag{1.142}$$

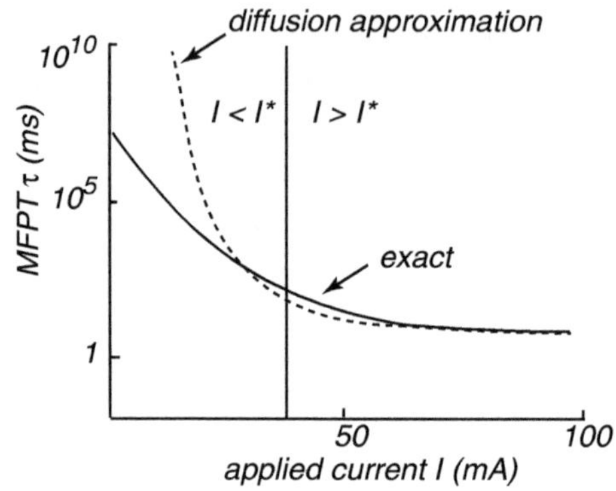

Fig. 1.18 Schematic diagram comparing MFPT calculated using the diffusion approximation with the MFPT of the full system. The scales of the axes are based on numerical results carried out in [321] for $N = 10$. Other parameter values as in Fig. 1.17

Keener and Newby [321] calculated the MFPT using the diffusion approximation and compared it with Monte Carlo simulations of the full stochastic model whose probability density evolves according to the CK equation (1.126). A summary of their findings is shown schematically in Fig. 1.18. The main observation is that although the Gaussian-like diffusion approximation does well in the superthreshold regime ($I > I_*$), it deviates significantly from the full model results in the subthreshold regime ($I < I_*$), where it overestimates the mean time to spike. This is related to the fact that the effective potential of the steady-state density under the diffusion approximation generates exponentially large errors in the MFPT, as discussed at the end of Sect. 1.5.2. Finally, note that the model of stochastic ion channels and membrane voltage fluctuations presented above is based on a number of simplifications. First, only the initiation of an action potential was considered, which involved the opening of sodium channels, while the termination of the action potential due to Na^+ inactivation and K^+ activation was ignored. Indeed, in stochastic versions of the Hodgkin–Huxley equations spontaneous activity can be observed due to fluctuations in K^+ ion channels [116, 456]. That is, if some K^+ channels spontaneously close, then a regenerative Na^+ current can induce an action potential. The second simplification was to treat each ion channel as a single unit rather than as a cluster of subunits. In other words, the Markov chain of events associated with opening and closing of an ion channel involves transitions between more than two internal states.

1.6 Stochastic Hybrid System with Fast Kinetics

As shown by Keener and Newby [321], it is possible to obtain a much more accurate analytical expression for the MFPT in the subthreshold regime by analyzing the CK equation (1.126) in the limit of fast sodium channels. This analysis applies equally well to small and large N. First, it is convenient to rewrite the CK equation in a more compact and general form:

$$\frac{\partial p}{\partial t} = -\frac{\partial[F(v,n)p(v,n,t)]}{\partial v} + \frac{1}{\varepsilon}\sum_{m=0}^{N} A(n,m;v)p(v,m,t) \tag{1.143}$$

with F given by (1.124) and A a tridiagonal matrix (for fixed v): $A(n,n-1;v) = \omega_+(v,n-1)$, $A(n,n;v) = -\omega_+(v,n) - \omega_-(n)$, $A(n,n+1;v) = \omega_-(n+1)$. We are now making explicit the condition that the open and closing of Na^+ channels occurs on a much faster time scale than the voltage dynamics by scaling the transition rates according to the small parameter $\varepsilon \ll 1$. In the limit $\varepsilon \to 0$, (1.124) reduces to the deterministic or mean field (1.133) with $\mathscr{A}(v) = \sum_n F(v,n)\rho(v,n)$, where $\rho(v,n)$ is the unique steady-state density satisfying $\sum_m A(n,m;v)\rho(v,m) = 0$ (see (1.105)):

$$\rho(v,n) = \frac{N!}{(N-n)!n!}a(v)^n b(v)^{N-n}, \quad a(v) = \frac{\alpha(v)}{\alpha(v)+\beta}, b(v) = \frac{\beta}{\alpha(v)+\beta}. \tag{1.144}$$

The mean number of open channels is $\langle n \rangle = \sum_{n=1}^{N} n\rho(v,n) = Na(v)$. In the regime $0 < \varepsilon \ll 1$, for which there are typically a large number of transitions between different channel states n while the voltage v hardly changes at all. This suggests that the system rapidly converges to the (quasi-)steady-state $\rho(v,n)$, which will then be perturbed as v slowly evolves. The resulting perturbations can be analyzed using a quasi-steady-state (QSS) diffusion or adiabatic approximation [210, 454, 487], in which the CK equation (1.143) is approximated by a Fokker–Planck (FP) equation; see also Sect. 6.4

The QSS approximation was first developed from a probabilistic perspective by Papanicolaou [487]; see also [210]. It has subsequently been applied to a wide range of problems in biology, including cell movement [274, 481], wavelike behavior in models of slow axonal transport [206, 207, 518], and molecular motor-based models of random intermittent search [453, 454]. The basic idea of the QSS reduction is to decompose the probability density as

$$p(v,n,t) = C(v,t)\rho(v,n) + \varepsilon w(v,n,t), \tag{1.145}$$

where $\sum_n p(v,n,t) = C(v,t)$ and $\sum_n w(v,n,t) = 0$. Carrying out an asymptotic expansion in ε, it can be shown that C evolves according to the FP equation [81, 453, 454]

$$\frac{\partial C}{\partial t} = -\frac{\partial}{\partial v}(\mathscr{A}C) + \varepsilon\frac{\partial}{\partial v}\left(\mathscr{D}\frac{\partial C}{\partial v}\right), \tag{1.146}$$

with the drift term given by (1.133), and diffusion coefficient

$$\mathscr{D}(v) = \sum_{n=0}^{N} Z(v,n)F(v,n), \tag{1.147}$$

where $Z(v,n)$ is the unique solution to

$$\sum_m A(n,m;v)Z(v,m) = [\mathscr{A}(v) - F(v,n)]\rho(v,n) \tag{1.148}$$

with $\sum_m Z(v,m) = 0$. As in the case of the diffusion approximation for large N, the FP equation (1.146) captures the Gaussian-like fluctuations within the basin of attraction of a fixed point of the mean field (1.133), but breaks down when considering rare event transitions between metastable states since it yields exponentially large errors in the escape rates. Therefore, one has to deal with the full CK equation (1.143).

1.6.1 First Passage Time Problem and the Projection Method

In order to revisit the FPT problem considered in Sect. 1.5.4, we supplement (1.143) with the following absorbing boundary conditions at v_0:

$$p(v_0,n,t) = 0, \text{ for all } n \in \Sigma = \{0,\ldots,k-1\}. \tag{1.149}$$

Here, Σ denotes the set of integers for which $F(v_0,n) < 0$. The initial condition is taken to be

$$p(v,n,0) = \delta(v - v_-)\delta_{n,n_0}. \tag{1.150}$$

Let T denote the (stochastic) FPT for which the system first reaches v_0, given that it started at v_-. The distribution of FPTs is related to the survival probability that the system has not yet reached v_0:

$$S(t) \equiv \int_0^{v_0} \sum_{n=0}^{N} p(v,n,t)dv. \tag{1.151}$$

That is, Prob$\{t > T\} = S(t)$ and the FPT density is

$$f(t) = -\frac{dS}{dt} = -\int_0^{v_0} \sum_{n=0}^{N} \frac{\partial p}{\partial t}(v,n,t)dv. \tag{1.152}$$

Substituting for $\partial p/\partial t$ using the CK equation (1.143) shows that

$$f(t) = \int_0^{v_0} \left[\sum_{n=0}^{N} \frac{\partial[F(v,n)p(v,n,t)]}{\partial v}\right] dv = \sum_{n=0}^{N} p(v_0,n,t)F(v_0,n). \tag{1.153}$$

We have used $\sum_n A(n,m;v) = 0$ and $\lim_{v\to-\infty} F(v,n)p(v,n,t) = 0$. The FPT density can thus be interpreted as the probability flux $J(v,t)$ at the absorbing boundary, since we have the conservation law

$$\sum_{n=0}^{N} \frac{\partial p(v,n,t)}{\partial t} = -\frac{\partial J(v,t)}{\partial v}, \quad J(v,t) = \sum_{n=0}^{N} F(v,n)p(v,n,t). \tag{1.154}$$

We wish to solve the FPT problem in the weak noise limit $\varepsilon \ll 1$. One of the characteristic features of the weak noise limit is that the flux through the absorbing boundary and the inverse of the MFPT $\langle T\rangle$ are exponentially small, that is, $\langle T\rangle \sim \mathrm{e}^{-C/\varepsilon}$ for some constant C. This means that standard singular perturbation theory cannot be used to solve the resulting boundary value problem, in which one matches inner and outer solutions of a boundary layer around the point $v = v_0$. Instead, one proceeds by finding a quasistationary solution using a Wentzel–Kramers–Brillouin (WKB) approximation. Recently, this approach has been extended by Keener and Newby [321] to a CK equation of the form (1.143), using a so-called projection method [660].

In order to apply the projection method, it is necessary to assume certain properties of the non-self-adjoint linear operator $-\hat{L}$ on the right-hand side of (1.143) with respect to the Hilbert space of functions $h : [0,v_0] \times \{0,\ldots,N\} \to \mathbb{R}$ with inner product defined according to

$$\langle h,g\rangle = \int_0^{v_0} \sum_{n=0}^{N} h(v,n)g(v,n)dv. \tag{1.155}$$

1. $\hat{L}$ has a complete set of eigenfunctions ϕ_r with

$$\hat{L}\phi_r(v,n) \equiv \frac{d}{dv}(F(v,n)\phi_r(v,n)) - \frac{1}{\varepsilon}\sum_{m=0}^{N} A(n,m;v)\phi_r(v,m) = \lambda_r\phi_r(v,n), \tag{1.156}$$

together with the boundary conditions

$$\phi_r(v_0,n) = 0, \text{for} n = 0,\ldots,k-1. \tag{1.157}$$

2. The real part of each eigenvalue λ_r is positive definite and the smallest eigenvalue λ_0 is real and simple. Thus we can introduce the ordering $0 < \lambda_0 < \mathrm{Re}[\lambda_1] \leq \mathrm{Re}[\lambda_2] \leq \ldots$.
3. λ_0 is exponentially small, $\lambda_0 \sim \mathrm{e}^{-C/\varepsilon}$, whereas $\mathrm{Re}[\lambda_r] = \mathscr{O}(1)$ for $r \geq 1$. In particular, $\lim_{\varepsilon\to 0}\lambda_0 = 0$ and $\lim_{\varepsilon\to 0}\phi_0(v,n) = \rho(v,n)$.

Under the above assumptions, we can introduce the eigenfunction expansion

$$p(v,n,t) = \sum_{r=0}^{N} C_r \mathrm{e}^{-\lambda_r t}\phi_r(v,n),. \tag{1.158}$$

with $\lambda_0 \ll \mathrm{Re}[\lambda_r]$ for all $r \geq 1$. Thus, at large times we have the quasistationary approximation

$$p(v,n,t) \sim C_0 \mathrm{e}^{-\lambda_0 t}\phi_0(v,n). \tag{1.159}$$

Substituting such an approximation into (1.153) implies that

$$f(t) \sim \mathrm{e}^{-\lambda_0 t}\sum_{n=0}^{N}\phi_0(v_0,n)F(v_0,n), \quad \lambda_1 t \gg 1. \tag{1.160}$$

Equation (1.156) implies that

$$\sum_{n=0}^{N}\int_0^{v_0}\widehat{L}\phi_0(v,n)dv \equiv \sum_{n=0}^{N}F(v_0,n)\phi_0(v_0,n)$$
$$=\lambda_0\sum_{n=0}^{N}\int_0^{v_0}\phi_0(v,n)dv.$$

In other words,

$$\lambda_0 = \frac{\sum_{n=0}^{N}F(v_0,n)\phi_0(v_0,n)}{\langle 1,\phi_0\rangle}. \tag{1.161}$$

Combining (1.161) and the quasistationary approximation (1.160) shows that the (normalized) FPT density reduces to

$$f(t) \sim \lambda_0 \mathrm{e}^{-\lambda_0 t} \tag{1.162}$$

and, hence, $\langle T\rangle = \int_0^{\infty} tf(t)dt \sim 1/\lambda_0$.

It remains to obtain an approximation ϕ_ε of the principal eigenfunction ϕ_0, which can be achieved using the WKB method as described in Sect. 1.6.2. This yields a quasistationary density that approximates ϕ_0 up to exponentially small terms at the boundary, that is,

$$\hat{L}\phi_\varepsilon = 0, \quad \phi_\varepsilon(u_*,n) = \mathscr{O}(\mathrm{e}^{-C/\varepsilon}). \tag{1.163}$$

In order to express λ_0 in terms of the quasistationary density ϕ_ε, we consider the eigenfunctions of the adjoint operator, which satisfy the equation

$$\hat{L}^{\dagger}\xi_r(v,n) \equiv -F(v,n)\frac{d\xi_r(v,n)}{dv} - \frac{1}{\varepsilon}\sum_m A(m,n;v)\xi_r(v,m) = \lambda_r\xi_r(v,n), \tag{1.164}$$

and the boundary conditions

$$\xi_r(v_0,n) = 0, \quad n \geq k. \tag{1.165}$$

The two sets of eigenfunctions $\{\phi_r\}$ and $\{\xi_r\}$ form a biorthogonal set with respect to the underlying inner product,

$$\langle \phi_r, \xi_s\rangle = \delta_{r,s}. \tag{1.166}$$

Now consider the identity

$$\langle \phi_\varepsilon, \hat{L}^{\dagger}\xi_0\rangle = \lambda_0\langle \phi_\varepsilon, \xi_0\rangle. \tag{1.167}$$

Integrating by parts the left-hand side of (1.164) picks up a boundary term so that

$$\lambda_0 = -\frac{\sum_{n=0}^{N}\phi_\varepsilon(v_0,n)\xi_0(v_0,n)F(v_0,n)}{\langle \phi_\varepsilon, \xi_0\rangle}. \tag{1.168}$$

The calculation of the principal eigenvalue λ_0 thus reduces to the problem of determining the quasistationary density ϕ_ε and the adjoint eigenfunction ξ_0.

1.6.2 The WKB Method and the Quasistationary Density

We now show how the WKB method [160, 256, 398, 445, 559] can be used to compute the quasistationary density ϕ_ε. We seek a solution of the form

$$\phi_\varepsilon(v,n) \sim R(v,n)\exp\left(-\frac{\Phi(v)}{\varepsilon}\right), \tag{1.169}$$

where $\Phi(v)$ is a scalar potential. Substituting into $\hat{L}\phi_\varepsilon = 0$ gives

$$\sum_{m=0}^{N}\left(A(n,m;v) + \Phi'(v)\delta_{n,m}F(v,m)\right)R(v,m) = \varepsilon\frac{dF(v,n)R(v,n)}{dx}, \tag{1.170}$$

where $\Phi' = d\Phi/dx$. Introducing the asymptotic expansions $R \sim R^{(0)} + \varepsilon R^{(1)}$ and $\Phi \sim \Phi_0 + \varepsilon\Phi_1$, the leading-order equation is

$$\sum_{m=0}^{N} A(n,m;v)R^{(0)}(v,m) = -\Phi_0'(v)F(v,n)R^{(0)}(v,n). \tag{1.171}$$

(Note that since $F(v,n)$ is nonzero almost everywhere for $v < v_0$, we can identify $-\Phi_0'$ and $R^{(0)}$ as an eigenpair of the matrix operator $\hat{A}(n,m;v) = A(n,m;v)/F(v,n)$ for fixed v.) Positivity of the probability density ϕ_ε requires positivity of the corresponding solution $R^{(0)}$. One positive solution is $R^{(0)} = \rho$, for which $\Phi_0' = 0$. However, such a solution is not admissible since $\Phi_0 =$ constant. It can be proven using linear algebra that if $F(v,n)$ for fixed $v < v_0$ changes sign as n increases from zero, then there exists one other positive solution, which also has the appropriate functional form to describe the potential well. That is, $\Phi_0'(x)$ has the correct sign and vanishes at the fixed points. Hence, it can be identified as the appropriate WKB solution.

Proceeding to the next order in the asymptotic expansion of (1.170), we have

$$\begin{aligned}\sum_{m=0}^{N}&\left(A(n,m;v) + \Phi_0'(v)\delta_{n,m}F(v,m)\right)R^{(1)}(v,m)\\ &= \frac{dF(v,n)R^{(0)}(v,n)}{dx} - \Phi_1'(v)F(v,n)R^{(0)}(v,n).\end{aligned} \tag{1.172}$$

For fixed v and WKB potential Φ_0, the matrix operator $\bar{A}(n,m;v) = A(n,m;v) + \Phi_0'(v)\delta_{n,m}F(v,m)$ on the left-hand side of this equation has a one-dimensional null space spanned by the positive WKB solution $R^{(0)}$. The Fredholm alternative theorem[1] then implies that the right-hand side of (1.172) is orthogonal to the left null vector S of $\bar{A}$. That is, we have the solvability condition

$$\sum_{n=0}^{N} S(v,n)\left[\frac{dF(v,n)R^{(0)}(v,n)}{dv} - \Phi_1'(v)F(v,n)R^{(0)}(v,n)\right] = 0, \tag{1.173}$$

with S satisfying

$$\sum_{n=0}^{N} S(v,n)\left(A(n,m;v) + \Phi_0'(v)\delta_{n,m}F(v,m)\right) = 0. \tag{1.174}$$

Given $R^{(0)}, S$ and Φ_0, the solvability condition yields the following equation for Φ_1:

$$\Phi_1'(x) = \frac{\sum_{n=0}^{N} S(v,n)[F(v,n)R^{(0)}(v,n)]'}{\sum_{n=0}^{N} S(v,n)F(v,n)R^{(0)}(v,n)}. \tag{1.175}$$

Combining the various results, and defining

$$k(v) = \exp\left(-\int_{v_-}^{v} \Phi_1'(y)dy\right), \tag{1.176}$$

gives to leading order in ε,

$$\phi_\varepsilon(v,n) \sim \mathcal{N} k(v) \exp\left(-\frac{\Phi_0(v)}{\varepsilon}\right) R^{(0)}(v,n), \tag{1.177}$$

where we choose $\sum_n R^{(0)}(v,n) = 1$ for all v and $\mathcal{N}$ is the normalization factor,

$$\mathcal{N} = \left[\int_0^{v_0} k(v) \exp\left(-\frac{\Phi_0(v)}{\varepsilon}\right)\right]^{-1}. \tag{1.178}$$

The latter can be approximated using Laplace's method to give

$$\mathcal{N} \sim \frac{1}{k(v_-)}\sqrt{\frac{|\Phi_0''(v_-)|}{2\pi\varepsilon}} \exp\left(\frac{\Phi_0(v_-)}{\varepsilon}\right). \tag{1.179}$$

[1] Consider an M-dimensional linear inhomogeneous system $\mathbf{A}\mathbf{x} = \mathbf{b}$ with $\mathbf{x}, \mathbf{b} \in \mathbb{R}^M$. Suppose that the $M \times M$ matrix $\mathbf{A}$ has a nontrivial null space and let $\mathbf{v}$ be a null vector of the adjoint matrix $\mathbf{A}^\dagger$, that is, $\mathbf{A}^\dagger \mathbf{v} = 0$. The Fredholm alternative theorem states that the inhomogeneous equation has a (nonunique) solution if and only if $\mathbf{v} \cdot \mathbf{b} = 0$ for all null vectors $\mathbf{v}$.

1.6.3 Calculation of the Principal Eigenvalue

In order to evaluate the principal eigenvalue λ_0 satisfying (1.168), it is necessary to determine the adjoint eigenfunction ξ_0. Following [321, 455], this can be achieved using singular perturbation methods. Since λ_0 is exponentially small in ε, (1.164) yields the leading-order equation

$$\varepsilon F(v,n)\frac{d\xi_0(v,n)}{dx} + \sum_{m=0}^{N} A(m,n;v)\xi_0(v,m) = 0, \tag{1.180}$$

supplemented by the absorbing boundary condition

$$\xi_0(v_0,n) = 0, \quad n \geq k. \tag{1.181}$$

A first attempt at obtaining an approximate solution that also satisfies the boundary conditions is to construct a boundary layer in a neighborhood of the unstable fixed point v_0 by performing the change of variables $v = v_0 - \varepsilon z$ and setting $Q(z,n) = \xi_0(v_0 - \varepsilon z)$. Equation (1.180) then becomes

$$F(v_0,n)\frac{dQ(z,n)}{dz} + \sum_{m=0}^{N} A(m,n;v_0)Q(z,m) = 0. \tag{1.182}$$

This inner solution has to be matched with the outer solution $\boldsymbol{\xi}_0 = \mathbf{1}$, which means that

$$\lim_{z\to\infty} Q(z,n) = 1 \tag{1.183}$$

for all n. Consider the eigenvalue equation

$$\sum_{n=0}^{N} \left(A(n,m;v) - \mu_r(v)\delta_{n,m}F(v,m)\right) S_r(v,n) = 0, \tag{1.184}$$

with $r = 0,\ldots,N$. We take $S_0(v,n) = 1$ so that $\mu_0 = 0$ and set $S_1(v,n) = S(v,n), \mu_1(v) = -\Phi_0'(v)$, where S satisfies (1.174). We then introduce the eigenfunction expansion

$$Q(z,n) = c_0 + \sum_{r=1}^{N} c_r S_r(v_0,n)\mathrm{e}^{-\mu_r(v_0)z}. \tag{1.185}$$

In order that the solution remains bounded as $z \to \infty$ we require that $c_r = 0$ if $\mathrm{Re}[\mu_r(v_0)] < 0$. The boundary conditions (1.181) generate a system of linear equations for the coefficients c_r with codimension k. One of the unknowns is determined by matching the outer solution, which suggests that there are $k-1$ eigenvalues with negative real part. The eigenvalues are ordered so that $\mathrm{Re}[\mu_r(v_0)] < 0$ for $r > N+1-k$.

There is, however, one problem with the above eigenfunction expansion, namely, that $\mu_1(v_0) \equiv -\Phi_0'(v_0) = 0$ so that the zero eigenvalue is degenerated at $v = v_0$. Hence, the solution needs to include a secular term involving the generalized eigenvector $\hat{S}$,

$$\sum_{n=0}^{N} A(n,m;v_0)\hat{S}(v_0,n) = -F(v_0,m). \tag{1.186}$$

The Fredholm alternative theorem ensures that $\hat{S}$ exists and is unique, since the stationary density $\rho(v_0,m)$ is the right null vector of $A(n,m;v_0)$ and $\sum_n \rho(v_0,n)F(v_0,n) \equiv \mathscr{A}(v_0) = 0$; see (1.133). The solution for $\mathbf{Q}(z)$ is now

$$Q(z,n) = c_0 + c_1(\hat{S}(v_0,n) - z) + \sum_{r=2}^{N+1-k} c_r S_r(v_0,n)\mathrm{e}^{-\mu_r(v_0)z}. \tag{1.187}$$

The presence of the secular term means that the solution is unbounded in the limit $z \to \infty$, which means that the inner solution cannot be matched with the outer solution. One way to remedy this situation is to introduce an alternative scaling in the boundary layer of the form $v = v_0 + \varepsilon^{1/2}z$, as detailed in [455]. One can then eliminate the secular term $-c_1 z$ and show that

$$c_1 \sim \sqrt{\frac{2|\Phi_0''(v_0)|}{\pi}} + \mathscr{O}(\varepsilon^{1/2}), \quad c_r = \mathscr{O}(\varepsilon^{1/2})\,\mathrm{for}\, r \geq 2 \tag{1.188}$$

It turns out that we only require the first coefficient c_1 in order to evaluate the principal eigenvalue λ_0 using (1.168). This follows from (1.171) and (1.184) and the observation that the left and right eigenvectors of the matrix $\hat{A}(n,m;v) = A(n,m;v)/F(v,n)$ are biorthogonal. In particular, since the quasistationary approximation ϕ_ε is proportional to $R^{(0)}$ (see (1.177)), it follows that ϕ_ε is orthogonal to all eigenvectors S_r, $r \neq 1$. Simplifying the denominator of (1.168) by using the outer solution $\xi_0 \sim 1$, we obtain

$$\begin{aligned}
\lambda_0 &\sim -\frac{\sum_n \xi_0(v_0,n)F(v_0,n)\phi_\varepsilon(v_0,n)}{\langle \phi_\varepsilon, \mathbf{1}\rangle} \\
&\sim c_1 \frac{k(v_0)B(v_0)}{k(v_-)}\sqrt{\frac{|\Phi''(v_-)|}{2\pi}} \exp\left(-\frac{\Phi_0(v_0) - \Phi_0(v_-)}{\varepsilon}\right),
\end{aligned} \tag{1.189}$$

with

$$B(v_0) = -\sum_{n=0}^{\infty} \hat{S}(v_0,n)F(v_0,n)\rho(v_0,n). \tag{1.190}$$

Substituting for c_1

$$\lambda_0 \sim \frac{1}{\pi}\frac{k(v_0)B(v_0)}{k(v_-)}\sqrt{\Phi_0''(v_-)|\Phi_0''(v_0)|}\exp\left(-\frac{\Phi_0(v_0) - \Phi_0(v_-)}{\varepsilon}\right). \tag{1.191}$$

Finally, comparison of (1.186) and (1.190) with (1.148) and (1.147) establishes that $B(v_0) \equiv \mathscr{D}(v_0)$.

The above analysis holds for any CK equation of the form (1.143). There are essentially three basic steps needed in order to evaluate the escape rate formula (1.191), which we now apply to the specific model of membrane voltage fluctuations.

1. Find the unique nontrivial positive eigenfunction $\psi_n(v) = R^{(0)}(v,n)$ and associated eigenvalue $\mu(v) = -\Phi_0'(v)$. In the case of the stochastic ion-channel model, (1.171) takes the explicit form

$$(N-n+1)\alpha\psi_{n-1} - [n\beta + (N-n)\alpha]\psi_n + (n+1)\beta\psi_{n+1}$$
$$= \mu\left(\frac{n}{N}f(v) - g(v)\right)\psi_n \quad (1.192)$$

Motivated by the form of the stationary density $\rho(v,n)$, we try the solution

$$\psi_n(v) = \frac{\Lambda(v)^n}{(N-n)!n!}, \quad (1.193)$$

which yields the following equation relating Λ and μ:

$$\frac{n\alpha}{\Lambda} + \Lambda\beta(N-n) - n\beta - (N-n)\alpha = \mu\left(\frac{n}{N}f(v) - g(v)\right).$$

We now fix μ in terms of Λ so that the terms linear in n vanish:

$$\mu = \frac{N}{f}\left[\alpha\left(\frac{1}{\Lambda}+1\right) - \beta(\Lambda+1)\right].$$

Eliminating μ then shows that

$$\Lambda(v) = \frac{g(v)}{f(v) - g(v)}.$$

We deduce that

$$\mu(v) = N\frac{\alpha(v)f(v) - (\alpha(v)+\beta)g(v)}{g(v)(f(v)-g(v))}, \quad (1.194)$$

and the normalized eigenvector is

$$\psi_n(v) = \frac{N!}{(N-n)!n!}\frac{(f(v)-g(v))^{N-n}g(v)^n}{f(v)^N}. \quad (1.195)$$

Note that $\mu(v)$ vanishes at the fixed points v_-, v_0 of the mean field (1.133) with $\mu(v) > 0$ for $0 < v < v_-$ and $\mu(v) > 0$ for $v_- < v < v_0$. Moreover, $\psi_n(v) = \rho(v,n)$ at the fixed points $v_0, v_\pm$. In conclusion $R^{(0)}(v,n) = \psi_n(v)$ and the effective potential Φ_0 is given by

$$\Phi_0(v) = -\int_{v_-}^{v} \mu(y)dy. \quad (1.196)$$

The effective potential is defined up to an arbitrary constant, which has been fixed by setting $\Phi_0(v_-)=0$.

2. Determine the null eigenfunction $\eta_n(v)=S(v,n)$ of (1.174), which becomes

$$\begin{aligned}(N-m)\alpha\eta_{m+1} &- [(N-m)\alpha+m\beta]\eta_m+m\beta\eta_{m-1}\\ &=\mu\left(\frac{m}{N}f(v)-g(v)\right)\eta_m.\end{aligned} \tag{1.197}$$

Trying a solution of the form $\eta_m(v)=\Gamma(v)^m$ yields

$$(N-m)\alpha\Gamma-((N-m)\alpha+m\beta)+m\beta\Gamma^{-1}=\mu\left(\frac{m}{N}f(v)-g(v)\right). \tag{1.198}$$

Γ is then determined by canceling terms independent of m, which gives

$$\eta_n(v)=\left(\frac{b(v)g(v)}{a(v)(f(v)-g(v)))}\right)^n. \tag{1.199}$$

The prefactor $k(v)$ may now be determined using (1.175) and (1.176).

3. Calculate the generalized eigenvector $\zeta_n=\hat{S}(v_0,n)$ of (1.186), which reduces to

$$(N-n)\alpha(v_0)\zeta_{n+1}+n\beta\zeta_{n-1}-((N-n)\alpha(v_0)+n\beta)\zeta_n=g(v_0)-\frac{n}{N}f(v_0). \tag{1.200}$$

It is straightforward to show that this has the solution

$$\zeta_n=\frac{f(v_0)}{N(\alpha(v_0)+\beta)}n. \tag{1.201}$$

It follows that the factor $B(v_0)$ defined by (1.190) is

$$\begin{aligned}B(v_0) &= -\frac{f(v_0)}{N(\alpha(v_0)+\beta)}\sum_{n=0}^{N}\rho(v_0,n)\left[-g(v_0)n+\frac{f(v_0)}{N}n^2\right]\\ &= -\frac{f(v_0)}{N(\alpha(v_0)+\beta)}\left[-g(v_0)\langle n\rangle+\frac{f(v_0)}{N}\langle n^2\rangle\right]\\ &= \frac{f(v_0)^2\alpha(v_0)\beta}{N(\alpha(v_0)+\beta)^3},\end{aligned} \tag{1.202}$$

where have used the fixed point condition $g(v_0)=f(v_0)a(v_0)$.

Keener and Newby [321] showed that the WKB and asymptotic methods outlined above yield a MFPT that was in excellent agreement with numerical simulations in both the superthreshold and subthreshold regimes.

1.7 Appendix: Stochastic Calculus

In this appendix we give an informal introduction to stochastic calculus, following along the lines of Jacobs [302]. A more detailed treatment can be found in Gardiner [210], and a rigorous mathematical account can be found in [476]. The basic approach is to treat a continuous-time stochastic process as the limit of a discrete time process. That is, an SDE prescribes how a stochastic variable $X(t)$ changes in each infinitesimal time step dt. Determining changes over finite times then requires evaluating an associated stochastic integral. In order to make sense of this, we discretize time into small, but finite, intervals of duration Δt and consider a corresponding stochastic difference equation for $X_n = X(n\Delta t)$.

1.7.1 Ito Calculus

Suppose that we divide the time interval $[0,T]$ into N increments of size $\Delta t = T/N$ and set $t_n = n\Delta t$. Consider the stochastic difference equation

$$\Delta X(t_n) \equiv X(t_{n+1}) - X(t_n) = \Delta W_n,$$

where each ΔW_n, $n = 0,\ldots,N-1$, is an independent, identically distributed (i.i.d) Gaussian variable with zero mean and variance $\sigma^2 = \Delta t$:

$$P(\Delta W) = \frac{1}{\sqrt{2\pi\Delta t}} \mathrm{e}^{-(\Delta W)^2/2\Delta t}. \tag{1.203}$$

Iterating the difference equation starting from $X(0) = 0$ yields

$$X_n \equiv X(n\Delta t) = \sum_{j=0}^{n-1} \Delta W_j.$$

Using the fact that the sum of Gaussian random variables is also a Gaussian, it follows that the probability density for X_n is a Gaussian. Thus, we only need to determine its mean and variance. Since the ΔW_j are all independent, we have

$$\langle X_n \rangle = \sum_{j=0}^{n-1} \langle \Delta W_j \rangle = 0, \quad \mathrm{Var}(X_n) = \sum_{j=0}^{n-1} \mathrm{Var}(\Delta W_j) = N\Delta t,$$

and

$$P(X_n) = \frac{1}{\sqrt{2\pi n\Delta t}} \mathrm{e}^{-X_n^2/(2n\Delta t)}.$$

We can now construct a corresponding continuous-time process by taking the limit $N \to \infty$ such that $\Delta t \to 0$ with $N\Delta T = T$ fixed. In particular,

$$X(T) = \lim_{N\to\infty} \sum_{j=0}^{N-1} \Delta W_j \equiv \int_0^T dW(t) \equiv W(T),$$

where $W(T)$ is identified as a Wiener process. (A rigorous treatment would be more precise with regard to what is meant by the convergence of random variables.) It is still a Gaussian, whose mean and variance are obtained by taking the limit $N \to \infty$ of the results for X_n. We deduce that $W(t)$ has the Gaussian probability density

$$P(w(t)) = \frac{1}{\sqrt{2\pi t}} \mathrm{e}^{-w(t)^2/2t}.$$

Now consider the modified stochastic difference equation

$$X_{n+1} - X_n = f(t_n)\Delta W_n,$$

where $f(t)$ is a deterministic function of time. Once again X_n is a Gaussian random variable, with

$$\langle X_n \rangle = \sum_{j=0}^{n-1} \langle f(t_j)\Delta W_j \rangle = 0, \quad \mathrm{Var}(X_n) = \sum_{j=0}^{n-1} \mathrm{Var}(f(t_j)\Delta W_j) = \sum_{j=0}^{n-1} f(t_j)^2 \Delta t.$$

Taking the continuum limit along identical lines to the previous case yields the continuous-time Gaussian variable

$$X(T) = \lim_{N\to\infty} \sum_{j=0}^{N-1} f(t_j)\Delta W_j \equiv \int_0^T f(t)dW(t), \tag{1.204}$$

with zero mean and variance

$$\mathrm{Var}(X(T)) = \int_0^T f(s)^2 ds. \tag{1.205}$$

Substituting for $X(T)$ into this equation gives

$$\left\langle \int_0^T f(t)dW(t) \int_0^T f(s)dW(s) \right\rangle = \int_0^T f(s)^2 ds,$$

which can be captured by the rule

$$\langle dW(t)dW(s) \rangle = \delta(t-s)dt\,ds. \tag{1.206}$$

However, care must be taken with this rule when $\delta(t-s)$ appears inside an integral having t or s as one of its limits. For example, consider the double stochastic integral

$$\int_0^T \left[\int_0^t f(s)dW(s) \right] g(t)dW(t) \equiv \lim_{N\to\infty} \sum_{n=0}^{N-1} \left[\sum_{m=0}^{n-1} f(t_m)dW_m \right] g(t_n)dW_n.$$

We see that there are no terms in the double sum on the right-hand side that have a product of Wiener increments in the same time interval. Thus, taking the expectation of both sides,

$$\left\langle \int_0^T \left[\int_0^t f(s)dW(s) \right] g(t)dW(t) \right\rangle = 0.$$

Hence, we require

$$\int_0^t f(s)\delta(t-s)ds = 0, \quad \int_0^t f(s)\delta(s)ds = f(0). \tag{1.207}$$

Following the previous examples, let us turn to a discretized version of the general SDE for $X(t)$,

$$dX = a(X,t)dt + b(X,t)dW(t), \tag{1.208}$$

which takes the form

$$X_{n+1} - X_n = a(X_n,t_n)\Delta t + b(X_n,t_n)\Delta W_n. \tag{1.209}$$

Iterating this equation starting from a fixed $X(0) = x_0$ yields

$$X_N = x_0 + \sum_{n=0}^{N-1} a(X_n,t_n)\Delta t + \sum_{n=0}^{N-1} b(X_n,t_n)\Delta W_n.$$

The continuum limit then gives the stochastic integral equation

$$X(T) = x_0 + \int_0^T a(X(t),t)dt + \int_0^T b(X(t),t)dW(t), \tag{1.210}$$

with the final term defined as the Ito stochastic integral

$$\int_0^T b(X(t),t)dW(t) = \lim_{N\to\infty} \sum_{n=0}^{N-1} b(X_n,t_n)\Delta W_n. \tag{1.211}$$

The integral equation is not very useful for generating an explicit solution for $X(t)$. However, from the definition of the Ito stochastic integral, it immediately follows that

$$\left\langle \int_0^T b(X(t),t)dW(t) \right\rangle = 0, \tag{1.212}$$

since X_n is a function of previous Wiener increments $\Delta W_{n-1}, \ldots, \Delta W_0$ so it is uncorrelated with ΔW_n. The stochastic difference equation (1.209) is the starting point for developing numerical schemes for solving an SDE. However, if one is interested in carrying out explicit calculations, it is usually more useful to go to the associated Fokker–Planck equation for the probability density. In order to derive the FP equation from the corresponding SDE, we first need to consider the object $(dW)^2$.

In terms of Wiener increments,

$$\int_0^T (dW(t))^2 = \lim_{N\to\infty} \sum_{n=0}^{N-1} (\Delta W_n)^2.$$

Taking the expectation of both sides and using the fact that each ΔW_n is an i.i.d., gives

$$\left\langle \int_0^T (dW(t))^2 \right\rangle = \int_0^T \langle (dW(t))^2 \rangle = \int_0^T dt = T. \tag{1.213}$$

What about the variance? Using the Gaussian probability density (1.203), it is simple to show that

$$\mathrm{Var}[(\Delta W)^2] = 2(\Delta t)^2 = 2T^2/N^2.$$

Hence,

$$\mathrm{Var}\left[\int_0^T (dW(t))^2\right] = \lim_{N\to\infty} \mathrm{Var}\left[\sum_{n=0}^{N-1} (\Delta W_n)^2\right] = \lim_{N\to\infty} \sum_{n=0}^{N-1} \mathrm{Var}\left[(\Delta W_n)^2\right]$$
$$= \lim_{N\to\infty} \frac{2T^2}{N} = 0.$$

We thus obtain the surprising result that the integral of $(dW)^2$ is deterministic and thus equal to its mean:

$$\int_0^T (dW(t))^2 = T = \int_0^T dt. \tag{1.214}$$

In other words, we can set $(dW)^2 = dt$, a result known as Ito's rule. Using similar arguments, it can also be shown that $dW^m = 0$ for $m > 2$.

1.7.2 Ito's Formula and the Fokker–Planck Equation

The result $dW(t)^2 = dt$ has important implications for how one carries out a change of variables in stochastic calculus. This is most directly established by considering the SDE for an arbitrary function $f(X(t))$ with $X(t)$ evolving according to (1.208):

$$\begin{aligned} df(X(t)) &= f(X(t) + dX(t)) - f(X(t)) \\ &= f'(X(t))dX(t) + \frac{1}{2}f''(X(t))dX(t)^2 + \ldots \\ &= f'(X(t))[a(X,t)dt + b(X,t)dW(t)] + \frac{1}{2}f''(X(t))b(X,t)^2 dW(t)^2, \end{aligned}$$

where all terms of higher order than dt have been dropped. Now using $dW(t)^2 = dt$, we obtain the following SDE for f, which is known as Ito's formula:

$$df(X(t)) = \left[a(X(t),t)f'(X(t)) + \frac{1}{2}b(X,t)^2 f''(X(t))\right] dt + b(X,t)f'(X(t))dW(t). \tag{1.215}$$

Hence, changing variables in Ito calculus is not given by ordinary calculus unless f is a constant or a linear function.

We can now use Ito's formula to derive the FP equation for an Ito SDE. First,

$$\begin{aligned}
\frac{\langle df(X(t))\rangle}{dt} &= \left\langle a(X(t),t)f'(X(t)) + \frac{1}{2}b(X(t),t)^2 f''(X(t))\right\rangle \\
&= \int \left[a(x,t)f'(x) + \frac{1}{2}b(x,t)^2 f''(x)\right] p(x,t)dx, \\
&= \int f(x)\left[-\frac{\partial}{\partial x}(a(x,t)p(x,t)) + \frac{1}{2}\frac{\partial^2}{\partial x^2}(b(x,t)^2 p(x,t))\right] dx. \quad (1.216)
\end{aligned}$$

after integration by parts, where $p(x,t)$ is the probability density of the stochastic process $X(t)$ under the initial condition $X(t_0) = x_0$. However, we also have

$$\begin{aligned}
\frac{\langle df(X(t))\rangle}{dt} &= \left\langle \frac{df(X(t))}{dt} \right\rangle \\
&= \frac{d}{dt}\langle f(X(t),t)\rangle \\
&= \int f(x)\frac{\partial}{\partial t}p(x,t)dx. \quad (1.217)
\end{aligned}$$

Comparing (1.216) and (1.217) and using the fact that $f(x)$ is arbitrary, we obtain the Ito version of the FP equation

$$\frac{\partial}{\partial t}p(x,t) = -\frac{\partial}{\partial x}(a(x,t)p(x,t)) + \frac{1}{2}\frac{\partial^2}{\partial x^2}(b(x,t)^2 p(x,t)). \quad (1.218)$$

1.7.3 Multiplicative Noise and Stratonovich Integrals

It turns out that there is more than one way to define a stochastic difference equation driven by an incremental Wiener process and thus more than one way to obtain an SDE in the continuum limit. This issue only arises in the case of multiplicative noise, that is, when the term multiplying $dW(t)$ depends on the state variable $X(t)$. Recall that in the Ito integral (1.211), it is the value of $b(x,t)$ at the start of the nth time step that multiplies ΔW_n, so that there are no contributions of the form $(\Delta W_n)^2$. An alternative definition of a stochastic integral is the Stratonovich integral

$$\oint_0^T b(X(t),t)dW(t) = \lim_{N\to\infty} \sum_{n=0}^{N-1} b\left(\frac{X_{n+1}+X_n}{2}, t_n\right)\Delta W_n, \quad (1.219)$$

where we have used $\oint$ to distinguish it from the Ito integral. Now b depends on the value X_{n+1} at the end of the nth time step, which means there will be an extra term involving $(\Delta W_n)^2$. In order to compare the Ito and Stratonovich integrals, suppose that X_n evolves according to the stochastic difference equation (1.209). Thus, in the continuum limit $X(t)$ is the solution to an Ito SDE. Suppose that we Taylor expand the nth term in the sum defining the Stratonovich integral about the point X_n and set $b_n = b(X_n, t_n)$:

$$b\left(\frac{X_{n+1}+X_n}{2}, t_n\right) = b_n + \frac{\Delta X_n}{2}\frac{\partial b_n}{\partial x} + \frac{1}{2}\left(\frac{\Delta X_n}{2}\right)^2 \frac{\partial^2 b_n}{\partial x^2} + \ldots$$

Substituting for ΔX_n using (1.209) and dropping terms that are higher order than Δt shows that

$$b\left(\frac{X_{n+1}+X_n}{2}, t_n\right) = b_n + \left(\frac{a_n}{2}\frac{\partial b_n}{\partial x} + \frac{b_n^2}{8}\frac{\partial^2 b_n}{\partial x^2}\right)\Delta t + \left(\frac{b_n}{2}\frac{\partial b_n}{\partial x}\right)\Delta W_n.$$

Applying this result to the sum appearing in the definition of the Stratonovich integral, (1.219), and again dropping higher-order terms in Δt, yields the result

$$\sum_{n=0}^{N-1} b\left(\frac{X_{n+1}+X_n}{2}, t_n\right)\Delta W_n = \sum_{n=0}^{N-1} b_n \Delta W_n + \sum_{n=0}^{N-1} \frac{b_n}{2}\frac{\partial b_n}{\partial x}(\Delta W_n)^2.$$

Finally, taking the continuum limit with $dW(t)^2 = dt$, we have

$$\oint_0^T b(X(t),t)dW(t) = \int_0^T b(X(t),t)dW(t) + \frac{1}{2}\int_0^T \frac{\partial b(X(t),t)}{\partial x} b(X(t),t)dt. \tag{1.220}$$

Now suppose that $Y(t)$ is a stochastic process evolving according to the Stratonovich SDE

$$dY = a(Y,t) + b(Y,t)dW(t). \tag{1.221}$$

This means that the integral equation satisfied by $Y(t)$ is based on the Stratonovich integral, that is,

$$Y(t) = y_0 + \int_0^t a(Y(s),s)ds + \oint_0^t b(Y(t),t)dW(t). \tag{1.222}$$

Using (1.220), we can rewrite the solution in terms of an Ito integral according to

$$Y(t) = y_0 + \int_0^t \left[a(Y(s),s) + \frac{1}{2}\frac{\partial b(Y(s),s)}{\partial y} g(Y(s),s)\right] ds + \int_0^t b(Y(s),s)dW(s). \tag{1.223}$$

The latter is the solution to an equivalent Ito SDE of the form

$$dY = \left[a(Y(t),t) + \frac{b(Y(t),t)}{2}\frac{\partial b(Y(t),t)}{\partial y}\right] dt + b(Y(t),t)dW(t). \tag{1.224}$$

Finally, given that we know the FP equation corresponding to an Ito SDE, we can immediately write down the FP equation corresponding to the Stratonovich SDE equation (1.221):

$$\frac{\partial}{\partial t}p(y,t) = -\frac{\partial}{\partial y}(a(y,t)p(y,t)) + \frac{1}{2}\frac{\partial}{\partial y}\left(b(y,t)\frac{\partial}{\partial y}[b(y,t)p(y,t)]\right). \tag{1.225}$$

Chapter 2
Traveling Waves in One-Dimensional Excitable Media

We now consider our first example of wave propagation in neural media, namely, the propagation of an action potential along the axon of a neuron. Such a phenomenon can be formulated mathematically in terms of finding a traveling pulse solution of the spatially extended Hodgkin–Huxley equations (1.8) and (1.4). Formally speaking, a traveling wave is a solution of a PDE on an infinite domain that travels at constant velocity and fixed shape. For one-dimensional systems, one can distinguish two types of solitary traveling wave: a traveling front linking a stable resting state to a stable excited state and a traveling pulse that begins and ends at the resting state; see Fig. 2.1. For the Hodgkin–Huxley model, a traveling front would occur if the recovery variable n associated with K^+ channels were frozen; otherwise repolarization ensures that the trailing edge of the wave returns to the resting state. In order to develop the basic theory of wave propagation in one-dimensional excitable media, we will consider a simplified version of the Hodgkin–Huxley equations given by the FitzHugh–Nagumo (FN) equations [192, 446]:

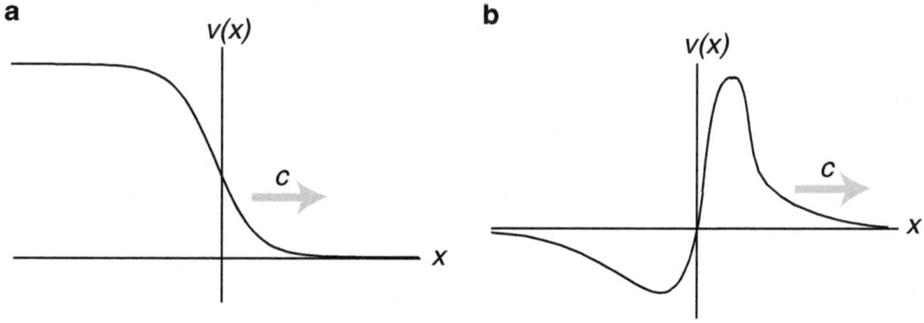

Fig. 2.1 Schematic illustration of (**a**) a traveling front, and (**b**) a traveling pulse

P.C. Bressloff, *Waves in Neural Media: From Single Neurons to Neural Fields*, Lecture Notes on Mathematical Modelling in the Life Sciences,
DOI 10.1007/978-1-4614-8866-8_2,

$$\frac{\partial v}{\partial t} = \frac{\partial^2 v}{\partial x^2} + f(v) - w \equiv f(v,w), \tag{2.1a}$$

$$\frac{\partial w}{\partial t} = \varepsilon(v - w) \equiv \varepsilon g(v,w). \tag{2.1b}$$

with $0 < \varepsilon \ll 1$ and

$$f(v) = v(v-a)(1-v). \tag{2.2}$$

Here v represents a fast voltage variable and w is a slow recovery variable. The time and space scales have been non-dimensionalized so that the effective diffusivity of the cable is unity. A number of excellent reviews of waves in excitable systems can be found elsewhere [242, 316, 322, 444].

2.1 Excitable Systems

Before consider traveling wave solutions, it is instructive to consider the excitable behavior of the space-clamped (x-independent) FN model. The space-clamped FN model takes the form of a planar dynamical system

$$\frac{dv}{dt} = f(v,w), \tag{2.3a}$$

$$\frac{dw}{dt} = \varepsilon g(v,w). \tag{2.3b}$$

The fast variable has a cubic nullcline (along which $\dot{v} = 0$) and the slow variable has a monotonically increasing nullcline (along which $\dot{w} = 0$). It is assumed that the nullclines have a single intersection point at (v^*, w^*). This corresponds to a fixed point of the system, which we identify with the resting state. A schematic diagram of the phase plane is shown in Fig. 2.2. For a finite range of values of w, there exist three solutions $v = v(w)$ of the equation $f(v,w) = 0$, which we denote by $V_-(w), V_0(w)$, and $V_+(w)$. Whenever these solutions coexist, we have the ordering $V_-(w) \leq V_0(w) \leq V_+(w)$. Let W_* denote the minimal value of w for which $V_-(w)$ exists, and let W^* denote the maximal value of w for which $V_+(w)$ exists.

First, suppose that the fixed point is located on the left-hand branch close to the minimum of the cubic. It is straightforward to show that the fixed point is linearly stable by evaluating the eigenvalues of the corresponding Jacobian. Moreover, the system is excitable in the sense that sufficiently large perturbations of the resting state result in a time-dependent trajectory taking a prolonged excursion through state space before returning to the resting state; see Fig. 2.3. Such a trajectory rapidly transitions to the right branch V_+, after which it slowly moves upward in a neighborhood of the branch before reaching the maximum. It then rapidly transitions back to the left branch V_- followed by a slow returns to the resting state along this branch. The time-dependent plot of the variable v can be interpreted as an action potential. Since the resting state is linearly stable, small perturbations simply result in small

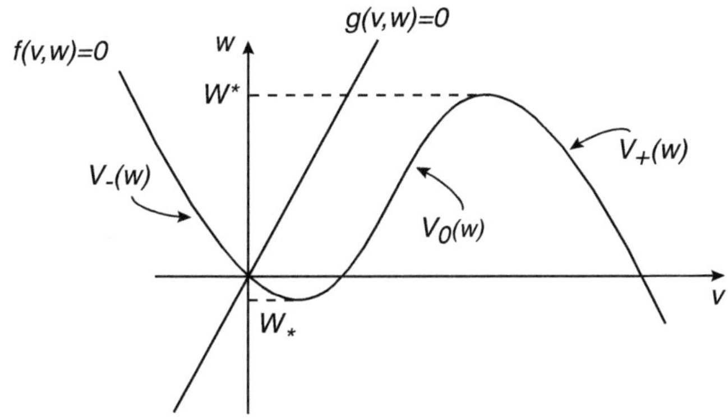

Fig. 2.2 Schematic diagram of the phase plane for the FitzHugh–Nagumo equations

excursions that decay exponentially in time. Hence, there is effectively a threshold phenomenon in which subthreshold perturbations result in a simple return to the resting state, whereas superthreshold perturbations generate an action potential.

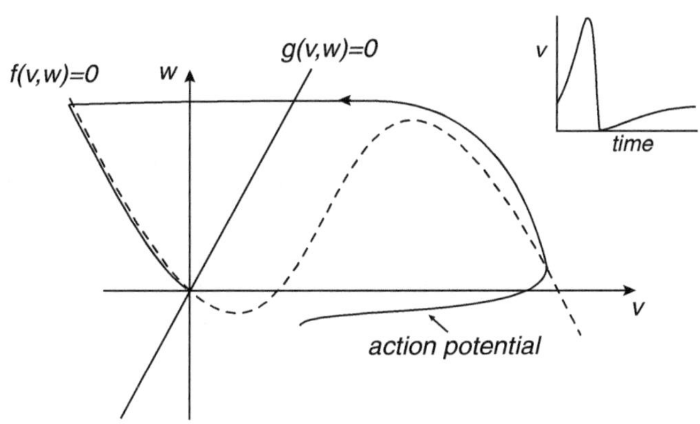

Fig. 2.3 Schematic diagram illustrating the trajectory of a single action potential in the phase plane for the FitzHugh–Nagumo equations. The unique rest point is stable. Inset shows the action potential as a function of time

A more mathematical description of the above events can be developed in terms of singular perturbation theory [242, 322]. Due to the separation of time scales with $\varepsilon \ll 1$, the fast variable v rapidly adjusts whenever it can to maintain the quasi-equilibrium $f(v,w) = 0$. This can be captured by introducing the slow time scale $\tau = \varepsilon t$ such that (2.3) become

$$\varepsilon \frac{dv}{d\tau} = f(v,w), \quad \frac{dw}{d\tau} = g(v,w). \tag{2.4}$$

Now setting $\varepsilon = 0$ and assuming that v is moving along the stable branches $V_{\pm}(w)$ of $f(v,w) = 0$, the dynamics of the recovery variable reduces to

$$\frac{dw}{d\tau} = g(V_{\pm}(w),w) \equiv G_{\pm}(w). \tag{2.5}$$

In the case of rapid transitions between the left and right branches, the dynamics with respect to the fast time scale can be approximated by setting $\varepsilon = 0$ in (2.3),

$$\frac{dv}{dt} = f(v,w), \quad \frac{dw}{dt} = 0. \tag{2.6}$$

Thus, on this time scale, w is constant and v converges to a stable solution of $f(v,w) = 0$. Suppose that the system starts from a superthreshold initial condition (v_0, w_0) such that $v_0 > V_0(w_0)$. After rapidly reaching the right branch, it takes a finite time to reach the upper "knee" of the nullcline $f(v,w)$ and is obtained by integrating (2.5):

$$T_e = \int_{w_0}^{W^*} \frac{dw}{G_+(w)}. \tag{2.7}$$

On the other hand, the time taken to return to the resting state along the left branch is infinite, since $G_-(w)$ vanishes at the fixed point.

It is possible to convert the FN equations from an excitable to an oscillatory regime by adding a constant external current I_{ext} to the right-hand side of the voltage equation in (2.3). For an intermediate range of values of I_{ext} one finds that the fixed point shifts to the middle branch $V_0(w)$ where it is unstable. The fixed point now coexists with a limit cycle, along which the trajectory alternates periodically between the left and right branches, while w varies between W_* and W^*; see Fig. 2.4. The resulting limit cycle behavior with fast jumps alternating with slow dynamics is known as a *relaxation oscillator*. For small ε, the period T of the oscillator is dominated by the times to move along the left and right branches. Hence

$$T = \int_{W_*}^{W^*} \left(\frac{1}{G_+(w)} - \frac{1}{G_-(w)} \right) dw, \tag{2.8}$$

with $G_+ > 0$ and $G_- < 0$.

Another well-known planar model of an excitable neuron is the Morris–Lecar (ML) model [440] (see Eq. (1.28)) which we write in the form

$$\frac{dv}{dt} = a(v) f_{\text{Na}}(v) + w f_{\text{K}}(v) - g(v) \tag{2.9a}$$

$$\frac{dw}{dt} = \frac{w_\infty(v) - w}{\tau_w(v)}, \tag{2.9b}$$

where $f_i(v) = g_i(v_i - v)$ and w represents the fraction of open K^+ channels. The fraction of Na^+ channels (or Ca^{2+} channels in the original formulation of the model) is assumed to be in quasi steady state. Again we can analyze the generation of action

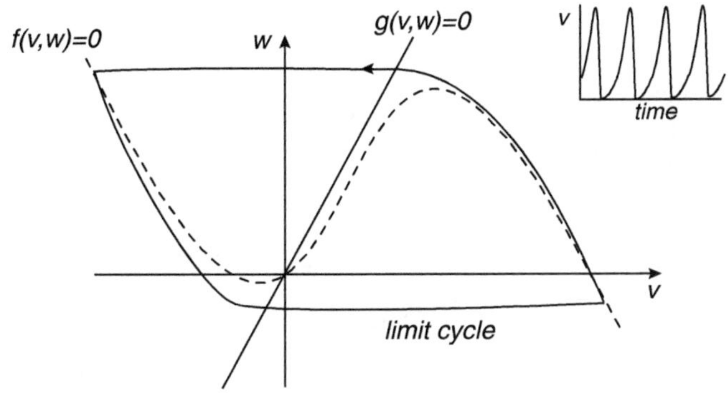

Fig. 2.4 Schematic diagram illustrating the trajectory of a globally stable periodic orbit in the phase plane for the FitzHugh–Nagumo equations. The unique rest point is unstable. Inset shows the periodic orbit as a function of time

potentials using a slow/fast analysis of the deterministic system. However, it turns out that this adiabatic approximation breaks down when stochastic fluctuations in the opening and closing of K^+ channels are taken into account. This can be established by considering a stochastic version of the ML model [456] consisting of N sodium and M potassium channels (see also Sect. 1.5):

$$\frac{dv}{dt} = F(v,m,n) \equiv \frac{n}{N} f_{\text{Na}}(v) + \frac{m}{M} f_{\text{K}}(v) - g(v). \tag{2.10}$$

We assume that each channel can either be open or closed and can switch between each state according to the kinetic scheme

$$C \underset{\beta_i(v)}{\overset{\alpha_i(v)}{\rightleftarrows}} O, \quad i = \text{Na, K}, \tag{2.11}$$

The Na^+ channels open and close rapidly relative to the voltage and K^+ dynamics. The probability density function $p(v,m,n,t)$ of the resulting stochastic hybrid system (see Sect. 1.6) evolves according to the differential Chapman–Kolmogorov (CK) equation,

$$\frac{\partial p}{\partial t} = -\frac{\partial (Fp)}{\partial v} + \mathbb{L}_{\text{K}} p + \mathbb{L}_{\text{Na}} p. \tag{2.12}$$

The jump operators $\mathbb{L}_j$, $j = \text{Na}, \text{K}$, are defined according to

$$\mathbb{L}_j = (\mathbb{E}_n^- - 1)\omega_j^+(n,v) + (\mathbb{E}_n^+ - 1)\omega_j^-(n,v), \tag{2.13}$$

with $\mathbb{E}_n^{\pm} f(n) = f(n \pm 1)$, $\omega_j^-(n,v) = n\beta_j(v)$ and $\omega_j^+(n,v) = (N-n)\alpha_j(v)$.

Introducing the small parameter ε with $\alpha_{\text{Na}}, \beta_{\text{Na}}, M = \mathcal{O}(1/\varepsilon)$, one can extend the WKB approximation method of Sect. 1.6 to analyze noise-induced transitions in the phase plane [456]. The WKB potential Φ can be interpreted as the action of an effective Hamiltonian dynamical system whose solutions determine charac-

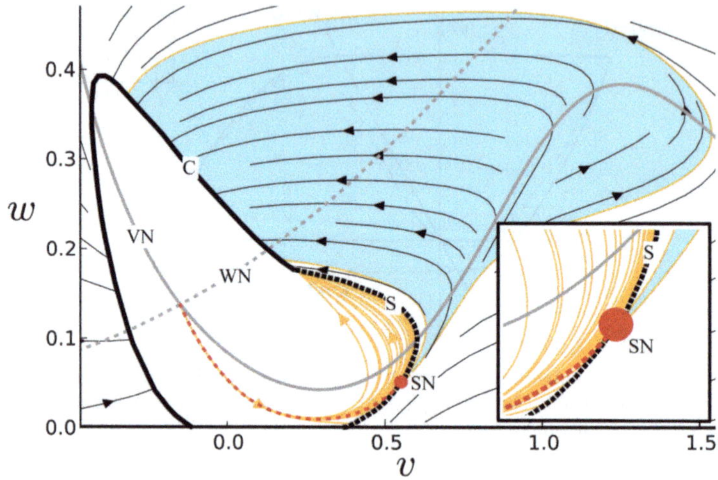

Fig. 2.5 Illustration of spontaneous action potentials (SAPs) for the stochastic ML model (2.10) with a finite number of sodium and potassium ion channels [456]. Orange curves are SAP trajectories, shown until they reach the effective metastable separatrix (S). The dashed red curve is a SAP that reaches S near the metastable saddle (SN). All of the SAP trajectories in the shaded region—containing the most probable, observable SAP trajectories—are visually indistinguishable from the dashed red line before crossing S. Deterministic trajectories are shown as black streamlines. Also shown are a caustic (C), caustic formation point (CP), v nullcline (VN), and w nullcline (WN). Parameter values are $N = M = 10$ and $\varepsilon = 0.1$

teristic paths in the phase plane (see also Sect. 4.4). The latter correspond to the paths a stochastic trajectory is most likely to follow during a metastable transition (i.e., a path of maximum likelihood [160]). Based on the fast/slow analysis of the deterministic system (2.9), one might expect w to be approximately constant along a stochastic trajectory that jumps between the left and right branches of the voltage nullcline, since the K^+ channels open and close slowly. In fact this does not hold for spontaneous action potentials arising from K^+ channel fluctuations [456] (see Fig. 2.5), which is in contrast to the effects of noise in the voltage or fast sodium channels. In general, it is difficult to solve FPT problems in more than one dimension. In the case of a metastable state with a well-defined basin of attraction, one has to calculate the MFPT to cross the separatrices forming the boundary of the basin of attraction. There is an additional level of complexity for an excitable system, due to the fact that there is no well-defined deterministic separatrix. Interestingly, as illustrated in Fig. 2.5, the stochastic ML model has an effective separatrix that any stochastic trajectory has to cross in order to generate a stochastic action potential [456]; see also [327]. Another commonly observed feature of the WKB approximation in two or more dimensions is the formation of caustics, where characteristic projections of the Hamiltonian intersect. There is now quite an extensive literature on the effects of noise in excitable systems, as reviewed in [383]. Most of these studies consider extrinsic Gaussian noise in the voltage dynamics and phenomena such as stochastic and coherence resonance. In Sect. 2.6 we will consider the effects of

Gaussian noise on wave propagation in a spatially extended excitable system. One can motivate this form of noise by considering diffusion approximations of models of stochastic ion channels. However, as the above example shows, certain caution must be exercised when considering such approximations.

Finally, we note that fast/slow decomposition has been applied extensively in recent years to the study of rhythmic activity patterns in single neurons and in synaptically coupled relaxation oscillators (see also Sect. 5). In the latter case, if synapses turn on and off on the fast time scale, then geometric singular perturbation theory can be used to investigate how synaptic inputs modify geometric structures such as null surfaces in the phase spaces of individual neurons, as reviewed by Rubin and Terman [537]; see also Chap. 9 of [173]. Moreover, such methods have been used to study a three-dimensional version of the Hodgkin–Huxley model, in which there is one fast variable and two slow variables [539, 540]. In addition to the transition from excitable to regular oscillatory behavior, as observed in classical relaxation oscillators, the model neuron also exhibits more complex dynamics such as mixed-mode oscillations that are associated with slow action potentials.

2.2 Traveling Fronts in a Scalar Bistable Equation

In the absence of a recovery variable, the FN equations reduce to the so-called scalar bistable equation

$$\frac{\partial v}{\partial t} = \frac{\partial^2 v}{\partial x^2} + f(v), \quad -\infty < x < \infty \tag{2.14}$$

with $f(v)$ given by the cubic (2.2). For such a choice of nonlinearity, the corresponding ODE, $dv/dt = f(v)$, has stable equilibria at $v = 0, 1$ separated by an unstable equilibrium at $x = a$. We define a traveling front solution according to

$$v(x,t) = v(x-ct) = V(\xi), \quad \xi = x - ct \tag{2.15}$$

for some yet to be determined wave speed c, supplemented by asymptotic boundary conditions ensuring that the front links the two stable fixed points of the x-independent system. For concreteness, we take

$$V(\xi) \to 1 \text{ as } \xi \to -\infty, \quad V(\xi) \to 0 \text{ as } \xi \to \infty. \tag{2.16}$$

Substituting the traveling front solution into the bistable Eq. (2.14) yields the ODE

$$V_{\xi\xi} + cV_{\xi} + f(V) = 0, \tag{2.17}$$

where $V_{\xi} = dV/d\xi$.

Classical phase-plane analysis can be used to find a traveling front solution by rewriting the second-order equation in the form

$$V_{\xi} = Z, \quad Z_{\xi} = -cZ - f(V). \tag{2.18}$$

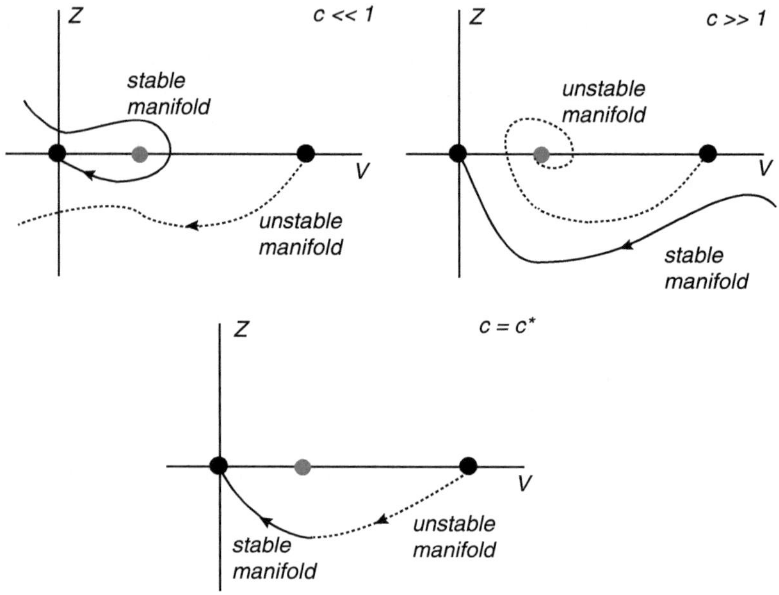

Fig. 2.6 Shooting method for constructing a front solution in the (V,Z) phase plane with $Z = V_\xi$. See text for details

One now has to look for a heteroclinic connection that links the excited state $(V,Z) = (1,0)$ at $\xi \to -\infty$ to the resting state at $\xi \to \infty$. This can be achieved using a geometric argument based on a shooting method, as illustrated in Fig. 2.6. For the sake of illustration, suppose that $0 < a < 1/2$ so that $c > 0$ (see below). First note that irrespective of the speed c, the fixed points $(1,0)$ and $(0,0)$ are saddles, each with one-dimensional stable and unstable manifolds. By looking at trajectories in the phase plane, it is straightforward to see that when $c \ll 1$, the unstable manifold of $(1,0)$ lies below the stable manifold of $(0,0)$ when $0 < V < 1$, whereas the opposite holds when c is very large. Since these manifolds depend continuously on c, it follows that there must exist at least one value of c for which the manifolds cross, and this corresponds to the heteroclinic connection that represents the traveling front solution. It can also be established that this front is unique. A useful formula for determining the sign of the wave speed can be obtained by multiplying both sides of (2.17) by V_ξ and integrating with respect to ξ:

$$\begin{aligned} c\int_{-\infty}^{\infty}(V_\xi)^2 d\xi &= -\int_{-\infty}^{\infty} V_\xi f(V(\xi))d\xi - \int_{-\infty}^{\infty} V_\xi V_{\xi\xi} d\xi, \\ &= \int_0^1 f(V)dV, \end{aligned} \tag{2.19}$$

since $V(\xi)$ is monotone, and $\int_{-\infty}^{\infty} V_\xi V_{\xi\xi} d\xi = \int_{-\infty}^{\infty} \frac{d[V_\xi^2/2]}{d\xi} d\xi = 0$. As the integral on the left-hand side is positive, it follows that the sign of c is determined by the sign of the area of f between the two stable equilibria. If $0 < a < 1/2$, then the latter is positive and the wave moves to the right, converting the medium from the resting state to the excited state. On the other hand, a left-moving front occurs when $1/2 < a < 1$, converting the medium from the excited state to the resting state. If the negative and positive areas exactly cancel, then the front is stationary.

The numerical construction of the traveling front using a shooting method does not depend on the precise form of f. However, if f is given by the cubic (2.2), then it is possible to construct the front explicitly. That is, we substitute the ansatz $Z = -AV(1-V)$ into (2.17) to obtain the condition $A^2(2V-1)+cA-(V-a)=0$. Collecting terms linear in V and terms independent of V requires $A = 1/\sqrt{2}$ and $c = (1-2a)/\sqrt{2}$. This immediately establishes that c switches sign at $a = 1/2$. Since $W = V_\xi$, it follows that the corresponding wave profile is

$$V(\xi) = \frac{1}{2}\left[1-\tanh(\xi/2\sqrt{2})\right].$$

Finally, recall that we have non-dimensionalized the units of space and time in the bistable equation by setting the membrane time and space constants of the cable to unity ($\tau_m = 1, \lambda_m = 1$); see (1.55). Hence, in physical units, the speed of the wave is

$$\hat{c} = \frac{c\lambda_m}{\tau_m} = \frac{c}{2C_m}\sqrt{\frac{d}{R_m R}}, \tag{2.20}$$

where d is the cable diameter. Based on empirical estimates, one finds that $\hat{c} \sim \sqrt{d}\,\mathrm{mm/sec}$. In the case of a squid axon of diameter $d = 500\,\mu\mathrm{m}$, the estimated propagation speed is of around $20\,\mathrm{mm/ms}$.

Another choice of nonlinearity for which an explicit front can be calculated is the piecewise linear function

$$f(v) = -v + H(v-a). \tag{2.21}$$

Substituting into Eq. (2.17) gives

$$V_{\xi\xi} + cV_\xi - V + H(V-a) = 0. \tag{2.22}$$

Translation symmetry of the system means that we are free to choose V to cross the threshold a at $\xi = 0$ so that $V(\xi) > a$ for $\xi < 0$ and $V(\xi) < a$ for $\xi > 0$. Solving the resulting linear equation on either side of the threshold point $\xi = 0$ and imposing the threshold condition $V(0) = a$ yields the explicit solution

$$V(\xi) = \begin{cases} ae^{\lambda_-\xi}, & \xi > 0 \\ 1+(a-1)e^{\lambda_+\xi}, & \xi < 0, \end{cases} \tag{2.23}$$

where $\lambda_\pm$ are the roots of the characteristic equation $\lambda^2 + c\lambda - 1 = 0$. The wave speed is then obtained by imposing continuity of V_ξ at $\xi = 0$, $(a-1)\lambda_+ = a\lambda_-$, which after rearranging gives

$$c = \frac{1-2a}{\sqrt{a-a^2}}. \tag{2.24}$$

2.3 Traveling Pulses in the FitzHugh–Nagumo Equations

The bistable equation cannot support a traveling pulse solution because there is no recovery variable, that is, it does describe an excitable system. In order to obtain traveling pulse solutions, it is necessary to consider the full FitzHugh–Nagumo equations (2.1). Suppose that the unique fixed point (v^*, w^*) lies on the left-hand branch as in Fig. 2.2. Assume a traveling wave solution of the form $(v(x,t), w(x,t)) = (V(\xi), W(\xi))$ with $\xi = x - ct$ so that (2.1) reduce to the system of ODEs

$$V_{\xi\xi} + cV_\xi + f(V,W) = 0, \tag{2.25a}$$

$$cW_\xi + \varepsilon g(V,W) = 0. \tag{2.25b}$$

These are supplemented by the asymptotic boundary conditions

$$\lim_{\xi\to\pm\infty} (V(\xi), W(\xi)) = (v^*, w^*). \tag{2.26}$$

Mathematically speaking, one needs to find a trajectory in the phase space (V,Z,W) with $Z = V_\xi$ that is homoclinic to the resting state $(v^*, 0, w^*)$. (Although the resting state is stable in the space-clamped system, it is a saddle in the phase plane (V,Z,W) of the spatially extended system.) The existence of such an orbit can be demonstrated using geometric singular perturbation theory [104, 267]. The basic idea is to formally set $\varepsilon = 0$ and construct a singular homoclinic orbit. This will consist of four parts: the jump-up from the resting state to the right branch $V_+(w)$, an active phase along $V_+(w)$, the jump-down to the left branch $V_-(w)$, and a quiescent phase as it returns to the resting state along $V_-(w)$; see Fig. 2.7. Given the existence of a singular homoclinic orbit, one can then prove that such an orbit persists for $\varepsilon > 0$ provided that ε is sufficiently small. In this section we will focus on the construction of the singular solution, exploiting the fact that the jumps take place on a fast spatial scale ξ, whereas the active and quiescent phases occur on a slow spatial scale $z = \varepsilon\xi$.x

In order to analyze the jump-up from the resting state to the active phase, we set $\varepsilon = 0$ in (2.25) to obtain the reduced system

$$V_{\xi\xi} + cV_\xi + f(V,W) = 0, \tag{2.27a}$$

$$W_\xi = 0. \tag{2.27b}$$

Thus the recovery variable is a constant w and V evolves according to the bistable equation

$$V_{\xi\xi} + cV_\xi + f(V,w) = 0, \tag{2.28}$$

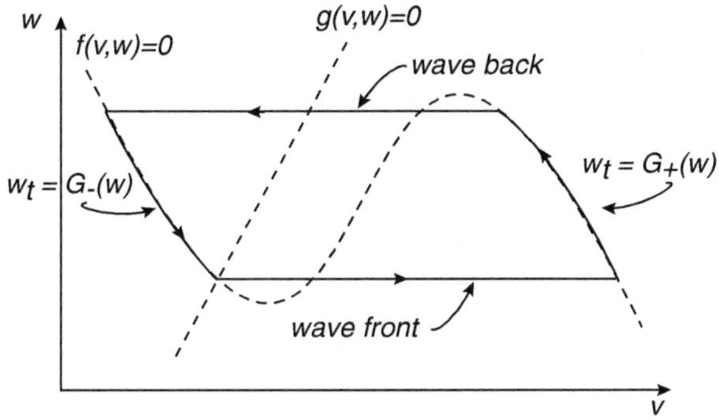

Fig. 2.7 Phase portrait of the fast traveling solitary pulse for the FN equations in the singular limit $\varepsilon \to 0$. The arrow directions are for increasing time (decreasing ξ)

for $f(V,w) = v(v-a)(1-v) - w$. Following Sect. 2.2, we know that for fixed w there exists a unique traveling front solution of (2.28) with associated wave speed $c = c(w)$ that is a heteroclinic connection with $V \to V_L(w)$ as $\xi \to \infty$ and $v \to V_R(w)$ as $\xi \to -\infty$. Moreover,

$$c(w) = \frac{\int_{V_-(w)}^{V_+(w)} f(V,w)dV}{\int_{-\infty}^{\infty} V_\xi^2 d\xi}. \tag{2.29}$$

Setting $w = w^*$ with $V_L(w^*) = v^*$ we obtain the wave speed $c^* = c(w^*)$. Next consider the slow active phase with $c = c^*$. Introducing the slow time scale $z = \varepsilon\xi$ we have

$$\varepsilon^2 V_{zz} + c\varepsilon V_z + f(V,W) = 0, \tag{2.30a}$$
$$cW_z + g(V,W) = 0. \tag{2.30b}$$

Setting $\varepsilon = 0$ then leads to the reduced system

$$f(V,W) = 0, \tag{2.31a}$$
$$W_z = \frac{1}{c^*} g(V,W). \tag{2.31b}$$

Taking the solution $V = V_R(W)$ of $f(V,W) = 0$, it follows that the trajectory moves along the right branch at a rate determined by $G_+(W) = g(V_R(W),W)$.

Now suppose that the trajectory leaves the branch at some value $W = W_d < W^*$ (where W^* is at the maximum of the cubic $f(V,W) = 0$) and jumps back down to the left branch. Similar to the jump-up phase, the recovery variable is constant, and V evolves according to the bistable Eq. (2.28) with $w = W_d$. Again using Sect. 2.2, we can construct a unique traveling wave solution with associated wave speed $c(W_d) < 0$ that is a heteroclinic connection from $V_R(W_d)$ to $V_L(W_d)$. The wave speed is negative, since the jump-down starts from an active state rather than a quiescent state; it is thus referred to as a wave back. The requirement that the solution be a steadily propagating pulse means that the speeds of the jump-up and jump-down solutions must be the same, which implies that

$$c(w^*) = -c(W_d). \tag{2.32}$$

This condition uniquely determines the transition point W_d. Finally, the trajectory returns to the resting state by moving along the left branch at a rate determined by $G_-(W)/c^*$.

A number of comments are in order. First, from the perspective of matched asymptotics in the slow variable z, the active and quiescent phases correspond to outer solutions that are slowly varying, whereas the jump-up and jump-down phases correspond to transition layers or inner solutions. Denote the wave-front and wave-back solutions by $V_\pm(\xi)$. Matching the inner and outer solutions then leads to the following asymptotic conditions:

$$\begin{aligned}
&\lim_{\xi\to-\infty} V_+(\xi) = V_R(W(0)), \quad W(0) = w^* \\
&\lim_{\xi\to\infty} V_+(\xi) = \lim_{z\to\infty} V_L(W(z)) = v^* \\
&\lim_{\xi\to-\infty} V_-(\xi) = V_L(W(z_T)), \quad W(z_T) = W_d \\
&\lim_{\xi\to\infty} V_-(\xi) = V_R(W(z_T))
\end{aligned}$$

The location z_T where the jump-down occurs is determined from the slow dynamics according to

$$z_T = c^* \int_{w^*}^{W_d} \frac{dW}{G_+(W)} dW. \tag{2.33}$$

One can interpret z_T as the width of the pulse. Second, it may be the case that there is no solution of $c(w^*) = -c(W_d)$ such that $W_d < W^*$. The jump-down transition then occurs at the upper knee, and the solution is referred to as a phase wave, that is, the jump occurs at a time or phase determined solely by the outer dynamics. The wave behavior is then qualitatively different, since the wave can travel at any speed above some minimum, analogous to the well-known Fisher–KPP equation of population genetics [191, 345]; see Sect. 3.3.

2.3.1 Periodic Wave Trains

One of the characteristic features of excitable systems is that they exhibit refractoriness. That is, once the system has responded to a superthreshold stimulus by generating an action potential, there is a refractory period during which no subsequent stimuli can be evoked. From the singular construction of a traveling pulse, the refractory period can be estimated as follows: after the jump-down there is a range of values of the recovery variable, $W_0 \leq w \leq W_d$ for which the front solution to the bistable Eq. (2.2) has negative wave speed, $c(w) \leq 0$, with $c(W_0) = 0$. The time taken to reach W_0 from W_d along the left branch is (for fixed x and rescaled time)

$$T_{\text{ref}} = \int_{W_0}^{W_d} \frac{dW}{|G_-(W)|}. \tag{2.34}$$

We can identify T_{ref} as the refractory period. Once excitability has been restored, it is possible to evoke another wave of excitation. However, the speed of the subsequent pulse is expected to be slower due to the residual refractoriness of the system. Now suppose that we periodically initiate action potentials at one end of a semi-infinite cable. This will asymptotically produce a periodic wave train with the time between successive pulses equal to the forcing period T. Assuming that the wave train travels at a speed c, the spatial separation or wavelength of the pulses will be $\lambda = cT$. The speed of the wave train is expected to be smaller than the speed of an isolated pulse due to refractoriness, although this effect should decrease as T increases. In other words, there exists a dispersion curve $c = c(T)$ with $c(T)$, a monotonically increasing function of T.

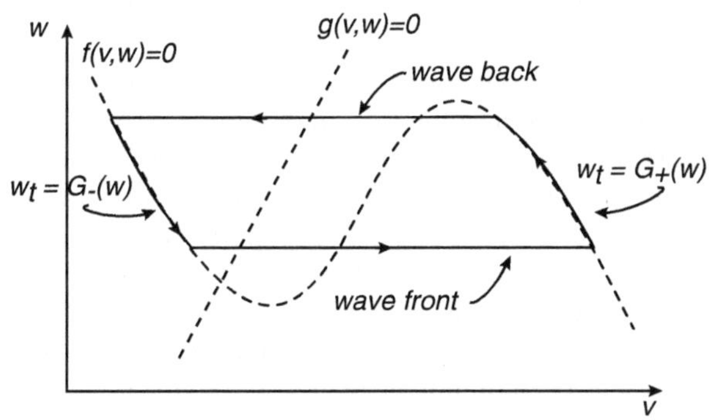

Fig. 2.8 Phase portrait for the fast periodic wave train for the FN equations in the singular limit $\varepsilon \to 0$

It is possible to estimate the dispersion curve for the FitzHugh–Nagumo equations using the previous singular construction. A periodic wave train consists of an alternating sequence of jump-ups and jump-downs, separated by regions of slow dynamics. A phase portrait for such a solution is shown in Fig. 2.8. The major difference from an isolated pulse (see Fig. 2.7) is that the jump-up occurs before reaching the resting state, with $W = W_P > w_*$. Denoting the corresponding value at the jump-down by W_Q, we require that the speeds of the corresponding wave front and wave back are the same, that is, $c(W_P) = -c(W_Q)$. Since the time taken for the jumps is negligible, the major contributions to the period T come from the time spent traversing the right and left branches:

$$T = \int_{W_P}^{W_Q} \frac{dW}{G_+(W)} + \int_{W_Q}^{W_P} \frac{dW}{G_-(W)}. \tag{2.35}$$

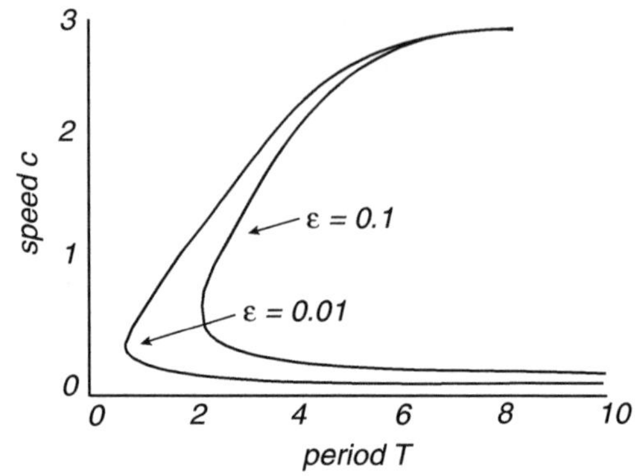

Fig. 2.9 Illustration of typical dispersion curves $c = c(T)$ for T-periodic wave-train solutions of the FN equations. A stable fast branch coexists with an unstable slow branch. They annihilate in a saddle–node bifurcation

Solving for W_Q in terms of W_P using the matching of speeds and inverting the relation $c = c(W_P)$ then generate the dispersion curve $c(T)$. It turns out that the dispersion curve breaks down when the wavelength $\lambda \equiv cT = \mathcal{O}(\varepsilon)$, because the transitional fronts and backs become arbitrarily close together so that it is no longer possible to carry out matched asymptotics. In order to construct the true dispersion curve, it is necessary to carry out a regular perturbation analysis in ε with $c = \mathcal{O}(\varepsilon)$ [156]. One then finds that the dispersion curve consists of two branches that annihilate in a saddle–node bifurcation at $T = T_c$; see Fig 2.9. Hence there are no traveling pulses for $T < T_c$. It can be shown that the upper branch of fast traveling pulses is stable, whereas the lower branch of slow pulses is unstable. (The issue of wave stability will be addressed in Sect. 2.4.) Note that as $T \to \infty$, $c \to c_\infty$ on the upper branch, where c_∞ is the speed of an isolated pulse. Interestingly, c also approaches a finite limit as $T \to \infty$ on the lower branch, suggesting that there also exists a slow unstable isolated pulse; this is indeed found to be the case.

2.3.2 *Kinematics*

It is also possible to have wave trains consisting of action potentials that are irregularly spaced and move at different speeds. Rinzel and Maginu [523] developed a kinematic theory of wave propagation that uses the dispersion relation to determine the instantaneous speed of a pulse. That is, suppose an initial pulse is generated at $x = 0$ at time t_1. The time at which the pulse reaches a given point x will be $T_1(x) = t_1 + x/c_\infty$, where c_∞ is the speed of a solitary pulse. Suppose that a second spike is initiated at $x = 0$ at time t_2. The instantaneous speed of the new pulse at x will depend on the time difference $T_2(x) - T_1(x)$ due to refractoriness. The Rinzel and Maginu approximation is to take the instantaneous speed to be $c(T_2(x) - T_1(x))$, where $c(T)$ is the dispersion curve for a periodic wave train. It then follows that

$$\frac{dT_2(x)}{dx} = \frac{1}{c(T_2(x) - T_1(x))}, \tag{2.36}$$

and the time interval $\phi = T_2(x) - T_1(x)$ between two action potentials initiated at $x = 0$ will evolve in space according to

$$\frac{d\phi}{dx} = \frac{1}{c(\phi)} - \frac{1}{c_\infty} \equiv \Gamma(\phi). \tag{2.37}$$

If the function $\Gamma(\phi)$ has one or more zeroes $\bar{\phi}$, then the phase difference between the two action potentials will lock at $\bar{\phi}$; the phase-locked state will be stable if $\Gamma'(\phi) < 0$. Note the kinematic approximation can be extended to multiple action potentials under the assumption that the instantaneous speed only depends on the relative phase of the preceding action potential. If $T_{n+1}(x)$ is the arrival time of the $n+1$th action potential, then

$$\frac{dT_{n+1}(x)}{dx} = \frac{1}{c(T_{n+1}(x) - T_n(x))}. \tag{2.38}$$

Note that an explicit version of kinematic theory can be derived using the singular solution of the FN equations [322], assuming that recovery always occurs via a phase wave at the value W^*. Suppose that $w_n(x)$ is the value of the recovery variable along the wave front of the nth pulse when it is located at x. The instantaneous speed of the front is thus $c(w_n)$. The time between fronts of successive action potentials is then given by

$$T_{n+1}(x) - T_n(x) = \int_{w_n}^{w^*} \frac{dw}{G_+(w)} + \int_{w_*}^{w_{n+1}} \frac{dw}{G_-(w)}. \tag{2.39}$$

Differentiating both sides with respect to x and using $dT_n/dx = 1/c(w_n)$ yields

$$\frac{1}{G_-(w_{n+1})} \frac{dw_{n+1}}{dx} = \frac{1}{G_-(w_n)} \frac{dw_n}{dx} + \frac{1}{c(w_{n+1})} - \frac{1}{c(w_n)}. \tag{2.40}$$

This generates an iterative equation for $w_n(x)$, which can be solved to determine the speed and arrival time of each successive action potential.

2.4 Wave Stability and Evans Functions

This section requires some basic definitions and results in functional analysis, in particular, with regard to Banach spaces and the spectrum of linear differential operators; see appendix section 2.7.

2.4.1 Stability of Fronts in the Bistable Equation

In order to introduce some of the basic principles of wave stability, let us return to the simple case of the scalar bistable Eq. (2.14). Let $V(\xi)$, $\xi = x - ct$, denote the unique traveling front solution with speed c such that $V(\xi) \to 1$ as $\xi \to -\infty$ and $V(\xi) \to 0$ as $\xi \to \infty$. In order to investigate the linear stability of such a solution, we set

$$v(x,t) = V(\xi) + \phi(\xi,t), \tag{2.41}$$

where ϕ is some small perturbation belonging to an appropriately defined Banach space $\mathscr{B}$ (complete, normed vector space). It is convenient to use the moving coordinate ξ so that we may see how the perturbation evolves in the moving frame of the front. Substituting for v in (2.14) and keeping only terms linear in ϕ gives

$$\frac{\partial \phi}{\partial t} = \mathbb{L}\phi \equiv \frac{\partial^2 \phi}{\partial \xi^2} + c\frac{\partial \phi}{\partial \xi} + f'(V)\phi, \quad \xi \in \mathbb{R}, \quad t > 0. \tag{2.42}$$

Equation (2.42) takes the form of a linear equation with associated linear differential operator $\mathbb{L} : \mathscr{D}(\mathbb{L}) \to \mathscr{B}$ with domain $\mathscr{D}(\mathbb{L}) \subseteq \mathscr{B}$. Determining the linear stability of the front thus reduces to the problem of calculating the spectrum $\sigma(\mathbb{L})$ of $\mathbb{L}$. That is, the front will be asymptotically stable if $\|\phi\| \to 0$ as $t \to \infty$ for all $\phi \in \mathscr{B}$, with $\|\cdot\|$ the norm on $\mathscr{B}$. This is guaranteed if $\sigma(\mathbb{L})$ lies strictly in the left-hand side of the complex plane, that is, there exists $\beta > 0$ such that $Re(\lambda) \leq -\beta$ for all $\lambda \in \sigma(\mathbb{L})$. The longtime asymptotics is then $\|\phi\| \sim e^{-\beta t}$. However, differentiating both sides of (2.17) with respect to ξ gives

$$V_{\xi\xi\xi} + cV_{\xi\xi} + f'(V)V_{\xi} \equiv \mathbb{L}V_{\xi} = 0,$$

which implies that zero is an eigenvalue of $\mathbb{L}$ with associated eigenfunction V_{ξ}. This is not a major issue, once one notices that V_{ξ} is the generator of infinitesimal translations of the front solution:

$$V(\xi + h) = V(\xi) + hV_{\xi}(\xi) + \mathscr{O}(h^2).$$

Hence, such perturbations only cause a phase shift of the original front and can thus be discounted. This motivates defining stability of the solution V in terms of the stability of the family of waves obtained by rigid translations of V. In other words, V is said to be stable if and only if $v(x,t) = V(\xi) + \phi(\xi,t)$ converges to $V(\xi + h)$ for some constant, finite h as $t \to \infty$. This will hold provided that zero is a simple eigenvalue of $\mathbb{L}$ and the remainder of the spectrum lies in a half-space $\{\lambda, Re(\lambda) \leq -\beta\}$ for some real $\beta > 0$. It is important to note that the spectrum of $\mathbb{L}$ will depend on the choice of Banach space $\mathscr{B}$. Restricting the class of admissible functions can push the spectrum to the left-half complex plane. However, this may exclude classes of perturbations that are physically relevant. A common choice is thus $L^2(\mathbb{R})$, which includes all normalizable, continuous functions on $\mathbb{R}$ with respect to the L_2 norm:

$$\|\phi\| = \int_{-\infty}^{\infty} |\phi(\xi)|^2 d\xi < \infty.$$

We now turn to the spectrum of the differential operator appearing in (2.42). As explained in appendix section 2.7, $\sigma(\mathbb{L})$ consists of isolated eigenvalues of finite multiplicity combined with the essential spectrum, which includes any continuous and residual spectrum. $\mathbb{L}$ is of the general form of the second-order operator considered in appendix section 2.7 (see (2.106)) with constant coefficient $p = c$ and x-dependent coefficient $q(x) = f'(V(x))$. Moreover, $q(x) \to q_\pm$ as $x \to \pm\infty$ with $q_+ = f'(0) = -a$ and $q_- = f'(1) = -1 + a$. Since the essential spectrum is located to the left of the parabolas defined by (2.108), and $q_\pm < 0$, we deduce that the essential spectrum is bounded to the left of $Re(\lambda) = \min\{-a, a-1\}$ and thus does not contribute to any instabilities. The stability of the front thus depends on the eigenvalues λ of $\mathbb{L}$, where

$$\mathbb{L}\phi \equiv \phi_{\xi\xi} + c\phi_\xi + f'(V)\phi = \lambda\phi, \tag{2.43}$$

with $\phi \in L^2(\mathbb{R})$. Suppose that $Re(\lambda) \geq 0$ so $\phi(\xi) \sim \mathrm{e}^{-\beta\xi}$ as $\xi \to \infty$ with $\beta \geq c$. (This follows from noting $f'(V) \to -a$ as $\xi \to \infty$ and analyzing the resulting constant-coefficient characteristic equation.) Performing the change of variables $\psi(\xi) = \phi(\xi)\mathrm{e}^{c\xi/2}$ yields the modified eigenvalue problem

$$\mathbb{L}_1\psi \equiv \psi_{\xi\xi} + \left(f'(V) - \frac{c^2}{4}\right)\psi = \lambda\psi, \tag{2.44}$$

with $\psi \in L^2(\mathbb{R})$, since it also decays exponentially as $|\xi| \to \infty$. The useful feature of the modified operator is that it is self-adjoint, implying that any eigenvalues in the right-half complex plane are real. Multiplying both sides of the self-adjoint eigenvalue Eq. (2.44) by ψ and integrating over $\mathbb{R}$, we have

$$\lambda \int_{-\infty}^{\infty} \psi^2 d\xi = -\int_{-\infty}^{\infty} \left[\psi_\xi^2 - \left(f'(V) - \frac{c^2}{4}\right)\psi^2\right] d\xi. \tag{2.45}$$

Recall that V_ξ is an eigenfunction of $\mathbb{L}$ with $\lambda = 0$, so that if $\Phi(\xi) = V_\xi(\xi)e^{c\xi/2}$, then $\Phi_{\xi\xi} + (f'(\psi) - c^2/4)\Phi = 0$. Hence, (2.45) can be rewritten as

$$\begin{aligned}
\lambda \int_{-\infty}^{\infty} \psi^2 d\xi &= -\int_{-\infty}^{\infty} \left[\psi_\xi^2 + \frac{\Phi_{\xi\xi}\psi^2}{\Phi}\right] d\xi \\
&= -\int_{-\infty}^{\infty} \left[\psi_\xi^2 - \frac{2\psi\psi_\xi\Phi_\xi}{\Phi} + \frac{\Phi_\xi^2\psi^2}{\Phi^2}\right] d\xi \\
&= -\int_{-\infty}^{\infty} \Phi^2 \left(\frac{d}{d\xi}(\psi/\Phi)\right)^2 d\xi.
\end{aligned}$$

This last result implies that $\lambda \leq 0$, and if $\lambda = 0$, then $\psi \sim \Phi = V_\xi$. We conclude that there are no eigenvalues in the right-half complex plane and $\lambda = 0$ is a simple eigenvalue. Thus the traveling front of the scalar bistable equation is stable.

2.4.2 The Evans Function

Determining the stability of traveling pulse solutions of the FN equations (2.1) or the Hodgkin–Huxley equations (1.8) is much more complicated. One general result, however, is that the discrete spectrum of the differential operator obtained by linearizing about a traveling wave solution may be associated with the zeros of a complex analytic function known as the Evans function. Indeed, Evans [177] originally developed the formalism within the context of the stability of solitary pulses in Hodgkin–Huxley-type equations for action potential propagation. Since then, the Evans function construction has been extended to a wide range of PDEs; see the review [551]. The basic construction of the Evans function can be illustrated relatively easily by considering a higher-dimensional version of the bistable equation [173].

Consider a general class of reaction–diffusion equations of the form

$$\frac{\partial u}{\partial t} = D\frac{\partial^2 u}{\partial x^2} + F(u), \tag{2.46}$$

where $u(x,t) \in \mathbb{R}^N$ and $F : \mathbb{R}^N \to \mathbb{R}^N$. Moreover, D is assumed to be a diagonal matrix with positive definite entries corresponding to the diffusion coefficients of the various component fields. Suppose that the system exhibits bistability, that is, there are two stable fixed points $u = u_j$, $j = 1,2$, with $F(u_j) = 0$. We will assume that there exists a traveling front solution $U(\xi)$ with speed c that connects u_1 and u_2. Linearizing about the wave solution along identical lines to the scalar case by setting $u(x,t) = U(\xi) + p(\xi)\mathrm{e}^{\lambda t}$ leads to the eigenvalue problem

$$\mathbb{L}p \equiv Dp_{\xi\xi} + cp_\xi + \partial F(U)p = \lambda p, \tag{2.47}$$

where $\partial F(U)$ denotes the matrix with components $\partial F_i/\partial U_j$. It is convenient to rewrite this as a system of $2N$ first-order equations

$$\begin{pmatrix} p_\xi \\ q_\xi \end{pmatrix} = \begin{pmatrix} 0 & I_n \\ D^{-1}(\lambda - \partial F(U)) & -cD^{-1} \end{pmatrix} \begin{pmatrix} p \\ q \end{pmatrix}. \tag{2.48}$$

Any eigensolution of this equation must satisfy the asymptotic conditions

$$\lim_{\xi \to \pm\infty} (p(\xi), q(\xi)) = (0,0).$$

Setting $z = (p,q)^T \in \mathbb{R}^n$, $n = 2N$, the associated ODE takes the general form

$$\mathscr{T}(\lambda)z(\xi) \equiv \frac{dz}{d\xi} - A(\xi;\lambda)z = 0, \tag{2.49}$$

with $A(\xi;\lambda) = A(\xi) + \lambda B(\xi)$. Thus we have a family of linear operators $\mathscr{T}(\lambda) : \mathscr{D} \to \mathscr{B}$ parameterized by λ. We take $\mathscr{B} = L^2(\mathbb{R}, \mathbb{R}^n)$ and $\mathscr{D}$ to be the space of admissible functions such that $z \in \mathscr{B}, \mathscr{T}(\lambda)z \in \mathscr{B}$. The basic form of the linear Eq. (2.49) holds for a wide range of PDEs supporting solitary traveling waves [551]. The discrete spectrum of the operator $\mathbb{L}$ thus corresponds to the values of λ for which $\mathscr{T}(\lambda)$ is not invertible.

An important concept for analyzing (2.49) is that of *exponential dichotomies*. First, consider the linear constant-coefficient equation

$$\frac{dz}{d\xi} = A(\lambda)z, \tag{2.50}$$

for which $A(\lambda)$ is independent of ξ. Suppose that the matrix $A(\lambda)$ is hyperbolic, that is, all its eigenvalues have nonzero real part. We can then decompose $\mathbb{R}^n$ (or its complexification) as

$$\mathbb{R}^n = E_s(\lambda) \oplus E_u(\lambda),$$

where $E_{s,u}(\lambda)$ are the generalized stable and unstable eigenspaces of the matrix $A(\lambda)$. Thus E_s is spanned by $n_+(\lambda)$ eigenfunctions that decay exponentially as $\xi \to \infty$ and E_u is spanned by $n_-(\lambda)$ eigenfunctions that decay exponentially as $\xi \to -\infty$ with $n_+(\lambda) + n_-(\lambda) = n$. The notion of exponential dichotomies can now be extended to (2.49) by noting that

$$\lim_{\xi \to \pm\infty} A(\xi;\lambda) \to A_\pm(\lambda) \tag{2.51}$$

where $A_\pm$ correspond to the matrix appearing in (2.48) in the limits $U(\xi) \to u_1$ and $U(\xi) \to u_2$, respectively. Moreover, the spectral properties of $\mathscr{T}(\lambda)$ can be expressed in terms of these exponential dichotomies. We summarize the main results below:

- Equation (2.49) is said to have an exponential dichotomy on $\mathbb{R}^+$ if and only if the matrix $A_+(\lambda)$ is hyperbolic. Let $\mathscr{V}_+(\lambda)$ denote thc linear subspace spanned by solutions of (2.49) that decay as $\xi \to \infty$. The codimension of $\mathscr{V}_+(\lambda)$ is defined to be the Morse index $i_+(\lambda)$ of the exponential dichotomy on $\mathbb{R}^+$, and $i_+(\lambda) = \dim E_u^+(\lambda)$.
- Equation (2.49) is said to have an exponential dichotomy on $\mathbb{R}^-$ if and only if the matrix $A_-(\lambda)$ is hyperbolic. Let $\mathscr{V}_-(\lambda)$ denote the linear subspace spanned by solutions of (2.49) that decay as $\xi \to -\infty$. The dimension of $\mathscr{V}_-(\lambda)$ is defined to be the Morse index $i_-(\lambda)$ of the exponential dichotomy on $\mathbb{R}^-$, and $i_-(\lambda) = \dim E_u^-(\lambda)$.
- λ is in the discrete spectrum if and only if $A_\pm(\lambda)$ are both hyperbolic with the same Morse index $i_+(\lambda) = i_-(\lambda)$ such that $\mathscr{V}_+(\lambda) \cap \mathscr{V}_-(\lambda) \neq \{0\}$.
- λ is in the essential spectrum if either at least one of the two asymptotic matrices $A_\pm(\lambda)$ is not hyperbolic or else if both are hyperbolic but their Morse indices differ.

In most applications the essential spectrum Σ_{ess} lies in the left-half complex plane and thus does not contribute to instabilities of a wave solution. Therefore, suppose that $\lambda \notin \Sigma_{\mathrm{ess}}$. It then follows that $i_+(\lambda) = i_-(\lambda) = k$, say. (For the multicomponent bistable equation $k = N = n/2$.) In order to construct the Evans function, introduce a basis for the subspaces $\mathscr{V}_\pm(\lambda)$ according to

$$\mathscr{V}_-(\lambda) = \mathrm{span}\,\{Q_1^-, Q_2^-, \ldots, Q_k^-\}, \quad \mathscr{V}_+(\lambda) = \mathrm{span}\,\{Q_1^+, Q_2^+, \ldots, Q_{n-k}^+\}$$

where each $Q_j^{\mp}(\xi)$ is an n-dimensional basis vector. Now form the $n \times n$ matrix $\mathscr{M}(\xi)$, in which the first k columns are given by the vectors $Q_j^-(\xi)$, $j = 1, \ldots, k$, and the next $n-k$ columns are given by Q_j^+, $j = 1, \ldots, n-k$. The Evans function is then defined according to

$$\mathscr{E}(\lambda) = \det \mathscr{M}(\xi_0), \tag{2.52}$$

for an arbitrary point ξ_0 which can be taken to be zero. The Evans function has a number of important properties. First, λ is an eigenvalue if and only if $\mathscr{E}(\lambda) = 0$. Second, if λ is a zero of $\mathscr{E}(\lambda)$, then the order of this zero is equal to the algebraic multiplicity of λ viewed as an eigenvalue. Third, the Evans function is analytic. The first property is simple to establish. For $\mathscr{E}(\lambda) = 0$ if and only if $\det \mathscr{M}(\xi_0) = 0$, and the latter holds if and only if there exist constant coefficients c_i^+, c_j^- such that

$$\sum_{j=1}^{k} c_j^- Q_j^-(\xi_0) + \sum_{i=1}^{n-k} c_i^+ Q_i^+(\xi_0) = 0,$$

that is,

$$\sum_{j=1}^{k} c_j^- Q_j^-(\xi_0) = -\sum_{i=1}^{n-k} c_i^+ Q_i^+(\xi_0).$$

Hence, λ is a zero of $\mathscr{E}(\lambda)$ if and only if $\mathscr{V}_+(\lambda)$ and $\mathscr{V}_-(\lambda)$ have a nonzero intersection, which means that λ is an eigenvalue, since the corresponding eigensolution decays at both $\xi = \infty$ and $\xi = -\infty$. Finally, note that one of the powerful features of the Evans function construction is that it can be applied to a wide variety of wave phenomena beyond fronts, including pulses, periodic wave trains, and multi-bump pulses [551]. In a sense, a single pulse is a special case of a front, since the exponential dichotomies on $\mathbb{R}^+$ and R^- are the same, that is,

$$\lim_{\xi \to \pm\infty} A(\xi;\lambda) \to A_0(\lambda),$$

with $A_0(\lambda)$ evaluated at the same resting state.

Finally, a word of caution: linear stability does not necessarily imply nonlinear stability. When considering perturbations about a traveling wave solution of a nonlinear PDE, $p(x,t) = u(x,t) - U(\xi)$, one can decompose the PDE as

$$\frac{\partial p}{\partial t} = \mathbb{L}p + \mathscr{N}(p), \tag{2.53}$$

where $\mathcal{N}(p) = \mathcal{O}(|p|^2)$. Determining linear stability of the traveling wave in terms of the spectrum $\sigma(\mathbb{L})$ assumes that the perturbation p remains small with respect to the given norm. A challenging mathematical problem is determining whether or not this is true. In the simpler case of ODEs one can use the stable manifold theorem to show that linear (in)stability implies nonlinear (in)stability in the case of hyperbolic fixed points. In the case of non-hyperbolic fixed points one has to use the center manifold theorem and bifurcation theory. A number of techniques have been used to study nonlinear stability of traveling waves, including center manifold reductions, Lyapunov functionals, and energy methods [551]. For example, the nonlinear stability of traveling wave solutions of the reaction–diffusion system (2.46) can be proven using a center manifold reduction. Nonlinear stability theory is also important if one wishes to determine what happens when a traveling wave becomes unstable, since perturbations grow and the linear approximation breaks down.

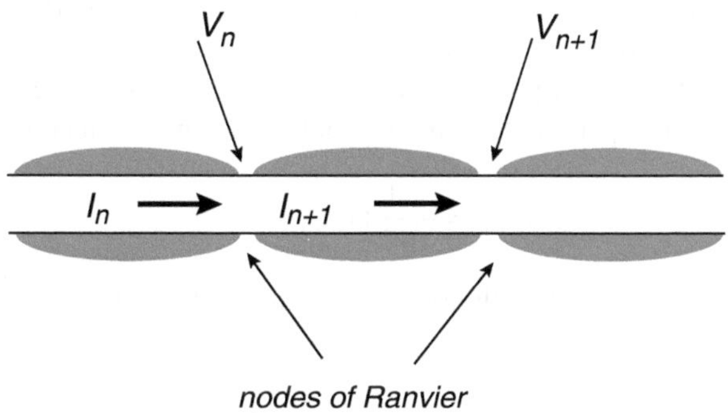

Fig. 2.10 Schematic diagram of a myelinated axon. Currents in myelinated region are confined to the axial direction. Potentials at the nodes are governed by active currents

2.5 Myelinated Axons and Discrete Diffusion

Many vertebrate axons are coated with a lipid material called *myelin*, which is made up of the membranes of *glial cells* (see Sect. 4.5) that wrap around the axon many times to form a thick insulating layer. This wrapping increases the effective membrane resistance and decreases the membrane capacitance by a factor of around 100. At regularly spaced intervals the axon is exposed to the extracellular medium at the so-called nodes of Ranvier, where there is a high density of sodium channels; see Fig. 2.10. The length L of myelin sheath between successive nodes is typically 1–2 mm, and the width l of a single node of Ranvier is around 1 μm. Propagation of an action potential along a myelinated axon is considerably faster than along a nonmyelinated axon. In terms of the cable equation, this can be understood as consequence of the fact that the transmembrane currents in the myelinated sections are negligible so that the myelinated sheath acts like a simple resistor. In effect, the action potential does not propagate continuously along the axon but rather jumps from node to node as a saltatory (leaping) wave.

Wave propagation along a myelinated axon can be modeled by spatially discretizing the diffusion term in the spatially extended Hodgkin–Huxley model (1.8). Suppose that the membrane voltage does not vary within each node of Ranvier, that is, the nodes are isopotential and denote the voltage of the nth node by V_n. Treating each myelin sheath as a pure Ohmic resistor with resistance rL, where r is the intracellular resistance per unit length, the axial current between nodes n and $n+1$ is

$$I_{n+1} = -\frac{1}{rL}(V_{n+1} - V_n). \tag{2.54}$$

Conservation of current at the nth node of Ranvier then implies that the total transmembrane current into the node is

$$2\pi al\left(C_m \frac{\partial V_n}{\partial t} + I_{\text{ion}}\right) = I_n - I_{n+1} = \frac{1}{rL}(V_{n+1} - 2V_n + V_{n-1}), \tag{2.55}$$

where a is the radius of the cable. It follows that

$$\frac{\partial V_n}{\partial t} = -\hat{I}_{\text{ion}} + D(V_{n+1} - 2V_n + V_{n-1}), \tag{2.56}$$

with coupling coefficient

$$D = \frac{R_m}{(2\pi ar)lL\tau_m} = \frac{\lambda_m^2}{lL\tau_m}.$$

We have used $R = \pi a^2 r$, $\tau_m = R_m C_m$ and $\lambda_m = (R_m a/2R)^{1/2}$; see Sect. 1.4.1.

2.5.1 The Discrete Bistable Equation

In order to investigate the effects of myelination on propagation speed, let us consider the simpler case of the spatially discrete bistable equation

$$\frac{\partial V_n}{\partial t} = f(V_n) + D(V_{n+1} - 2V_n + V_{n-1}), \tag{2.57}$$

where $f(V) = V(V-a)(1-V)$. Proving the existence of traveling wave solutions of (2.57) is nontrivial [699]. However, assuming that such a solution does exist, one can use perturbation methods to estimate the wave speed [317, 321]. A traveling wave solution is defined according to the iterative relationship $V_{n+1}(t) = V_n(t+\tau_d)$ where τ_d represents the time delay for the wave to jump between successive nodes. The corresponding invariant wave profile $\Phi(t)$ satisfies a delay differential equation that is obtained by substituting $V_n(t) = \Phi(t)$ into (2.57):

$$\frac{d\Phi}{dt} = D(\Phi(t-\tau_d) - 2\Phi(t) + \Phi(t+\tau_d)) + f(\Phi(t)). \tag{2.58}$$

Suppose that we fix the time units by setting $\tau_m = 1$. If the function $\Phi(t)$ is sufficiently smooth and τ_d is sufficiently small, then we can expand $\Phi(t \pm \tau_d)$ as a Taylor series in τ_d. Keeping only terms up to τ_d^2 yields the leading-order equation

$$D\tau_d^2 \Phi_{tt} - \Phi_t + f(\Phi) = 0. \tag{2.59}$$

This is identical in form to the bistable Eq. (2.17) for a traveling front of (dimensionless) wave speed c, provided that we set $\xi = -ct$ and $D\tau_d^2 = c^{-2}$. It follows that $\Phi(-\xi/c)$ is the traveling front solution of the continuous bistable equation and $c = (1-2a)/\sqrt{2}$. The wave speed of the saltatory wave (in physical time units) is then

$$\hat{c} = \frac{L+l}{\tau_d} = (L+l)c\sqrt{\frac{D}{\tau_m}} = \frac{L+l}{\sqrt{lL}}\left(\frac{c\lambda_m}{\tau_m}\right). \tag{2.60}$$

Comparison with (2.20) shows that myelination increases wave speed by a factor $(L+l)/\sqrt{lL} \approx 10$ assuming that $L \approx 100l$.

Higher-order corrections to the wave speed can be determined using regular perturbation theory [317, 321]. Thus, in terms of the small parameter $\varepsilon = D^{-1}$, we introduce the series expansions (with $\tau_m = 1$)

$$\Phi(t) = \Phi_0(t) + \varepsilon\Phi_1(t) + \ldots, \quad \tau_d{}^2 = \frac{\varepsilon}{c^2} + \varepsilon^2\tau_1 + \ldots \tag{2.61}$$

Substituting into the discrete bistable Eq. (2.58) and collecting terms of equal powers in ε yields a hierarchy of equations for Φ_n, with Φ_0 satisfying (2.59) for $D\tau_d^2 = c^{-2}$, and

$$\mathbb{L}\Phi_1 \equiv \frac{1}{c^2}\Phi_1'' - \Phi_1' + f'(\Phi_0)\Phi_1 = -\frac{\Phi_0^{(4)}}{12c^4} - \tau_1\Phi_0''. \tag{2.62}$$

Here $\mathbb{L}$ denotes a linear differential operator acting on the space $L^2(\mathbb{R})$ of square-integrable function on $\mathbb{R}$. The operator $\mathbb{L}$ is not invertible, since $\mathbb{L}\Phi_0' = 0$, which follows from differentiating (2.58) with respect to t. (A similar operator arose in our discussion of wave stability in Sect. 2.4.) It follows from the Fredholm alternative (see appendix section 2.7) that a solution for Φ_1 exists if and only if the right-hand side of (2.62) is orthogonal to the null-space of the adjoint operator $\mathbb{L}^\dagger$. The latter is

$$\mathbb{L}^\dagger\mathscr{V} = \frac{1}{c^2}\mathscr{V}'' + \mathscr{V}' + f'(\Phi_0)\mathscr{V}, \tag{2.63}$$

which has a one-dimensional null-space spanned by $\mathscr{V}(t) = \mathrm{e}^{-c^2 t}\Phi_0'(t)$. We thus have a solvability condition for the leading-order correction τ_1 to the delay:

$$\int_{-\infty}^{\infty} \mathrm{e}^{-c^2 t}\Phi_0'(t)\left[\frac{\Phi_0^{(4)}}{12c^4} + \tau_1\Phi_0''\right] dt = 0.$$

Once τ_1 has been determined, the propagation speed is (in physical time units)

$$\hat{c} = c(L+l)\sqrt{\frac{D}{\tau_m}}\left(1 - \frac{\tau_1 c^2}{2D\tau_m} + \mathscr{O}([c^2/D\tau_m]^2)\right). \tag{2.64}$$

2.5.2 Propagation Failure

The above perturbation analysis establishes that for sufficiently large coupling D, there are traveling front solutions that approach the speed of the appropriately scaled continuous model. Another important property of the discrete bistable equation is that if D is sufficiently small, then wave propagation failure occurs, reflecting the fact that there are nontrivial standing front solutions even when $\int_0^1 f(v)dv > 0$ [165, 183, 317, 318]. Here we describe one method for estimating the critical value of coupling D_c below which propagation fail, which has been developed by Keener [318] using averaging theory. The first step is to rewrite the discrete bistable equation in the form

$$(1+g'(x))[v_t - f(v)] = Dv_{xx}, \tag{2.65}$$

where $g(x)$ is the periodic sawtooth function

$$g(x) = \frac{L}{2} - x, \quad 0 < x < L, \quad g(x+nL) = g(x), \tag{2.66}$$

and

$$1 + g'(x) = L\sum_n \delta(x - nL). \tag{2.67}$$

Equation (2.65) implies that $v_{xx} = 0$ between nodes, that is, $v(x,t) = A_n x + B_n$ for $nL < x < (n+1)L$. Matching $v(x,t)$ with $V_n(t)$ and $V_{n+1}(t)$ at both ends shows that $A_n = L^{-1}[V_{n+1} - V_n]$ and hence $v_x((n+1)L,t) = L^{-1}(V_{n+1} - V_n)$. If we now integrate (2.65) over the interval $[nL,(n+1)L)$, then we recover the discrete bistable equation

$$\partial_t V_n - f(V_n) = \frac{D}{L}[v_x((n+1)L,t) - v_x(nL,t)] = D[V_{n+1} - 2V_n + V_{n-1}]. \tag{2.68}$$

In order to proceed, we will smooth out the function $g(x)$ by taking

$$1 + g'(x) = \frac{L}{\sqrt{2\pi\sigma^2}}\sum_n \exp\left(-\frac{(x-nL)^2}{2\sigma^2}\right), \tag{2.69}$$

so that $1 + g'(x) > 0$ for all x. At the end of the calculation we will take the limit $\sigma \to 0$ to recover the sawtooth function. We then have the scalar bistable equation with an inhomogeneous diffusion coefficient,

$$v_t = f(v) + \frac{D}{1+g'(x)}v_{xx}.$$

Following Keener [318], it is convenient to carry out the coordinate transformation $y = y(x) \equiv x + g(x) - L/2$ so that

$$v_t = f(v) + \partial_y \left([1 + g'(x)]\partial_y v\right).$$

If $g(x)$ is a sawtooth-like function, then y is a steplike function with $y = nL$ for $nL < x < (n+1)L$ in the limit $\sigma \to 0$. Hence $1 + g'(y(x)) = \sum_n \delta(y(x) - nL)$ blows up for all $nL < x < (n+1)L$, whereas $1 + g'(y(x)) = 0$ when $x = nL$. That is, $1 + g'(x) = 1/(1 + g'(y(x)))$, so that (after rewriting y as x),

$$v_t = f(v) + \partial_x \left(\frac{D}{1 + g'(x)} v_x\right).$$

Fixing the spatial units by setting $L = 1$ (having already non-dimensionalized time), and using the fact that waves exist for sufficiently large D, we introduce the small dimensionless parameter $\varepsilon = 1/\sqrt{D}$. Rescaling space according to $x \to x/\sqrt{D}$, we finally obtain the modified bistable equation

$$v_t = f(v) + \partial_x \left(\frac{1}{1 + g'(x/\varepsilon)} v_x\right). \tag{2.70}$$

Thus the problem of wave propagation failure for the discrete bistable equation has been reduced to the problem of calculating the mean wave speed $\bar{c}(\varepsilon)$ of a wavelike solution of an inhomogeneous continuous bistable equation and determining how $\bar{c}(\varepsilon)$ vanishes as ε increases (the coupling D decreases).

Equation (2.70) is now in a suitable form to apply the averaging method of Keener [318]; see below. The basic result is that the wavelike solution takes the form

$$v(x,t) = V(x - \phi(t)) + \mathcal{O}(\varepsilon), \tag{2.71}$$

where $V(x)$ is the wave solution in the homogeneous case ($g = 0$). The phase $\phi(t)$ evolves according to

$$\frac{d\phi}{dt} = c - \Phi(\phi/\varepsilon) \tag{2.72}$$

with c the speed of the unmodulated wave,

$$\Phi(\phi/\varepsilon) = \frac{1}{\Lambda} \int_{-\infty}^{\infty} g'([\xi + \phi]/\varepsilon) V''(\xi) V'(\xi) \mathrm{e}^{c\xi} d\xi, \tag{2.73}$$

and

$$\Lambda = \int_{-\infty}^{\infty} \mathrm{e}^{c\xi} V'(\xi)^2 d\xi. \tag{2.74}$$

Equation (2.72) implies that the solution is not translationally invariant; rather, it moves with a time-dependent velocity ϕ'. If $c - \Phi(\phi/\varepsilon)$ is strictly positive, then $\phi'(t)$ is a positive, periodic function of t with period

$$T = \int_0^{\varepsilon} \frac{d\phi}{c - \Phi(\phi/\varepsilon)}. \tag{2.75}$$

The mean speed of the wave is $\bar{c} = \varepsilon/T$. On the other hand, if $c - \Phi(\phi/\varepsilon)$ vanishes for some ϕ, then propagation failure is expected to occur:

Averaging method for discrete bistable equation. Standard applications of the averaging theorem in dynamical systems theory [248] only apply to structurally stable solutions, whereas traveling wave solutions are not structurally stable. Therefore, it is necessary to consider a modified averaging procedure as outlined by Keener [318, 319]. The first step is to rewrite (2.70) as the second-order system

$$v_x = (1 + g'(x/\varepsilon))u, \quad u_x = v_t - f(v). \tag{2.76}$$

Introducing the exact change of variables $v = w + \varepsilon u g(x/\varepsilon)$ yields the new system

$$w_x = u - \varepsilon u_x g(x/\varepsilon), \quad u_x = w_t + \varepsilon u_t g(x/\varepsilon) - f(w + \varepsilon u g(x/\varepsilon)). \tag{2.77}$$

It can be seen that if we ignore terms of order ε, then the system of equations is independent of x/ε. This lowest-order averaged system reduces to the standard bistable equation, which we know supports a traveling front solution $V(x - ct)$. Including the inhomogeneous factors $g(x/\varepsilon)$ means that the system is no longer translationally invariant. However, we can look for solutions that are in some sense close to a traveling front by going to a traveling coordinate system by setting $\xi = x - \phi(t)$, with

$$w_\xi = u - \varepsilon u_\xi g([\xi + \phi]/\varepsilon) \tag{2.78a}$$

$$u_\xi = -\phi' w_\xi - \varepsilon \phi' u_\xi g([\xi + \phi]/\varepsilon) - f(w + \varepsilon u g([\xi + \phi]/\varepsilon). \tag{2.78b}$$

We now seek a perturbative solution of the form

$$w = w_0(\xi) + \varepsilon w_1(\xi) + \ldots, \quad v(\xi) = v_0(\xi) + \varepsilon v_1(\xi) + \ldots, \quad \phi'(t) = c + \varepsilon \phi_1'(t) + \ldots$$

Substituting into (2.78) and collecting terms of equal powers in ε yields a hierarchy of equations, the first two of which are

$$\partial_\xi w_0 = u_0, \quad \partial_\xi u_0 = -c u_0 - f(w_0), \tag{2.79}$$

which recovers the homogeneous bistable equation for w_0, and

$$\partial_\xi w_1 - u_1 = -(\partial_\xi u_0) g([\xi + \phi]/\varepsilon), \tag{2.80a}$$

$$\partial_\xi u_1 + f'(w_0) w_1 + c u_1 = -\phi_1' u_0 - f'(w_0) u_0 g([\xi + \phi]/\varepsilon). \tag{2.80b}$$

Let us take the solution of the lowest-order equation to be the traveling front solution $V(\xi)$ constructed in Sect. 2.2: $w_0 = V, u_0 = V'$. The next-order system of Eq. (2.80) can be rewritten in the vector form

$$\mathbb{L} \begin{pmatrix} w_1 \\ u_1 \end{pmatrix} = \begin{pmatrix} h_w \\ -\phi_1' V' + h_u \end{pmatrix}, \quad \mathbb{L} = \begin{pmatrix} \partial_\xi & -1 \\ f'(V) & \partial_\xi + c \end{pmatrix}, \tag{2.81}$$

with h_w and h_u determined by inhomogeneous terms on the right-hand side of (2.80a) and (2.80b), respectively. Following our analysis of the linear operator (2.62), we know that the matrix operator appearing in (2.81) has a null-space spanned by (V', V''). Similarly, the adjoint operator

$$\mathbb{L}^\dagger = \begin{pmatrix} -\partial_\xi & f'(w_0) \\ -1 & -\partial_\xi + c \end{pmatrix}$$

has the null vector $e^{c\xi}(-V'', V')$. Hence, applying the Fredholm alternative, we see that $\phi'(t)$ must satisfy

$$\begin{aligned}
\phi_1'(t)\int_{-\infty}^{\infty} e^{c\xi}V'(\xi)^2 d\xi &= \int_{-\infty}^{\infty} e^{c\xi}[-V''(\xi)h_w(\xi)+V'(\xi)h_u(\xi)d\xi \\
&= \int_{-\infty}^{\infty} e^{c\xi}[V''(\xi)^2 - f'(V)V'(\xi)^2]g([\xi+\phi]/\varepsilon)d\xi \\
&= \int_{-\infty}^{\infty} g([\xi+\phi]/\varepsilon)\frac{d}{d\xi}\left[V''(\xi)V'(\xi)e^{c\xi}\right]d\xi, \\
&= -\frac{1}{\varepsilon}\int_{-\infty}^{\infty} g'([\xi+\phi]/\varepsilon)V''(\xi)V'(\xi)e^{c\xi}d\xi,
\end{aligned}$$

after using $V''' + cV'' + f'(V)V' = 0$ and performing integration by parts. We thus obtain the phase Eq. (2.72) with $\phi(t) = c + \varepsilon\phi_1(t)$.

It turns out that solving the phase equation in the case of a cubic nonlinearity is rather involved [318]. Therefore, for the sake of illustration, we will consider the simpler case of the piecewise linear function (2.21). There then exists a unique traveling front solution of the homogeneous bistable equation given by (2.23) with corresponding wave speed (2.24). Substituting (2.23) into (2.74) gives

$$\begin{aligned}
\Lambda &= (a\lambda_-)^2\int_0^{\infty} e^{c\xi}e^{2\lambda_-\xi}d\xi + ([a-1]\lambda_+)^2\int_{-\infty}^{0} e^{c\xi}e^{2\lambda_+\xi}d\xi \\
&= (a\lambda_-)^2\left[\frac{1}{c+2\lambda_+} - \frac{1}{c+2\lambda_-}\right] \\
&= 2\sqrt{a-a^2}(a-a^2),
\end{aligned}$$

where we have used the results $a\lambda_- = (a-1)\lambda_+$,

$$c+2\lambda_{\pm} = \pm\sqrt{c^2+4} = \frac{\pm 1}{\sqrt{a-a^2}}, \quad \lambda_- = -\frac{1}{2}\left[c+\sqrt{c^2+4}\right] = \frac{a-1}{\sqrt{a-a^2}}.$$

Similarly, substituting (2.23) into (2.73) gives, to leading order in ε,

$$\begin{aligned}
\Phi(\phi/\varepsilon) &= \frac{1}{\Lambda}a^2\lambda_-^3\int_0^{\infty} g'([\xi+\phi]/\varepsilon)e^{c\xi}e^{2\lambda_-\xi}d\xi \\
&\quad + \frac{1}{\Lambda}[a-1]^2\lambda_+^3\int_{-\infty}^{0} g'([\xi+\phi]/\varepsilon)e^{c\xi}e^{2\lambda_+\xi}d\xi \\
&\approx \frac{1}{\Lambda}\left[a^2\lambda_-^3\int_0^{\infty} g'([\xi+\phi]/\varepsilon)d\xi + [a-1]^2\lambda_+^3\int_{-\infty}^{0} g'([\xi+\phi]/\varepsilon)d\xi\right] \\
&= \frac{\varepsilon}{\Lambda}(a\lambda_-)^2[\lambda_+ - \lambda_-]g(\phi/\varepsilon) + \mathcal{O}(\varepsilon^2) \\
&= \frac{\varepsilon}{2[a-a^2]}g(\phi/\varepsilon) + \mathcal{O}(\varepsilon^2).
\end{aligned}$$

Applying this to the sawtooth function (2.66), the phase Eq. (2.72) reduces to

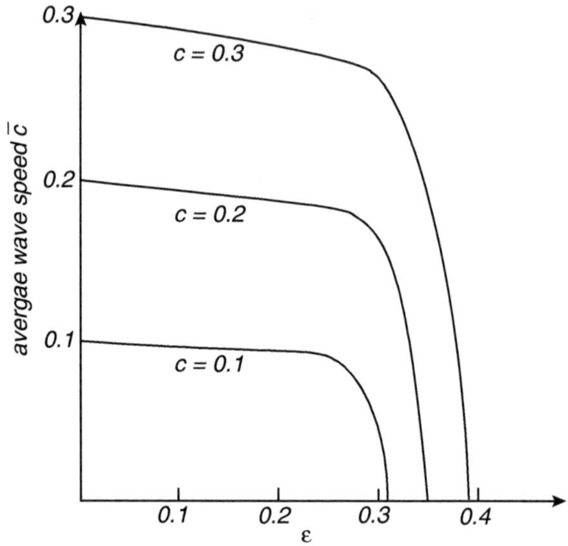

Fig. 2.11 Illustrative sketch of mean wave speed as a function of ε for the discrete bistable Eq. (2.65) with $L = 2\pi$

$$\frac{d\phi}{dt} = c - \frac{\varepsilon}{2[a-a^2]}\left[\frac{1}{2} - \frac{\phi}{\varepsilon}\right],$$

and

$$T = 2\ln\left(\frac{c+\varepsilon/\chi}{c-\varepsilon/\chi}\right), \quad \chi = 4[a-a^2].$$

We deduce that the mean wave speed is

$$\bar{c} = \frac{\varepsilon}{2}\frac{1}{\ln\left(\frac{c+\varepsilon/\chi}{c-\varepsilon/\chi}\right)}. \tag{2.82}$$

This predicts propagation failure when $\varepsilon \geq \chi c$. In particular, propagation failure is more likely to occur as ε increases, which is equivalent to reducing the coupling strength D. Finally, note that the cubic nonlinearity yields different behavior, both qualitatively and quantitatively. The main reason for this is that the above averaging procedure results in an expression for the mean wave speed that involves exponentially small terms of the form $e^{-\pi/\varepsilon}$ [318]. This has two major implications. First, it is necessary to include higher-order terms in the perturbation analysis in order to obtain sufficient accuracy. Second, the rapid increase in $e^{-\pi/\varepsilon}$ as ε increases can result in a sharp transition to propagation failure at relatively small values of ε, as illustrated in Fig. 2.11.

2.6 Stochastic Traveling Waves

In Sect. 1.5 we showed how the stochastic opening and closing of a finite number of ion channels can generate a multiplicative noise term in a space-clamped, conductance-based model of a neuron. The membrane voltage then evolves according to a stochastic differential equation (SDE) such as (1.128). This suggests that ion channel fluctuations could also affect the propagation of action potentials in a spatially extended model. Such an observation motivates the more general issue of how to analyze traveling wave solutions of stochastic partial differential equations (SPDEs). In this section we review a formal perturbation method for analyzing traveling front solutions in the case of weak noise [14, 494, 546, 557], which we apply to a stochastic version of the bistable equation. (The rigorous treatment of SPDEs is much more complicated than SDEs because one has to keep track of the regularity of solutions with respect to both time and space. Nevertheless, there have been some recent mathematical studies of a stochastic version of spatially extended excitable membranes based on the Hodgkin–Huxley equations [17, 88].)

Consider a scalar SPDE of the form

$$\frac{\partial}{\partial t}V(x,t) = \frac{\partial^2}{\partial x^2}V(x,t) + f(V(x,t)) + \sqrt{\varepsilon}g(V(x,t))\eta(x,t), \tag{2.83}$$

Here $\eta(x,t)$ is a Gaussian random function with zero mean and correlation

$$\langle \eta(x,t)\eta(x',t')\rangle = 2C([x-x']/\lambda)\delta(t-t') \tag{2.84}$$

The parameter λ is the spatial correlation length of the noise such that $C(x/\lambda) \to \delta(x)$ in the limit $\lambda \to 0$, and ε determines the strength of the noise, which is assumed to be weak. Note that we can formally set $\eta(x,t)dt = dW(x,t)$, where $W(x,t)$ is a space-dependent Wiener process with zero mean and $\langle dW(x,t)dW(x',t)\rangle = 2C([x-x']/\lambda)$. We have also included a multiplicative noise factor $g(V)$, and based on our analysis of channel fluctuations, we treat the multiplicative noise in the sense of Stratonovich (see Sect. 1.7). The starting point of the perturbation method is the observation that multiplicative noise in the Stratonovich sense leads to a systematic shift in the speed of the front (assuming a front of speed c exists when $\varepsilon = 0$) [14]. This is a consequence of the fact that $\langle g(V)\eta\rangle \neq 0$ even though $\langle \eta \rangle = 0$. The former average can be calculated using Novikov's theorem [465, 486]:

$$\varepsilon^{1/2}\langle g(U)\eta\rangle = \varepsilon C(0)\langle g'(U)g(U)\rangle dt, \tag{2.85}$$

Note that in the limit $\lambda \to 0$, $C(0) \to 1/\Delta x$ where Δx is a lattice cutoff, which can be identified with the step size of the spatial discretization scheme used in numerical simulations.

Novikov's theorem. Suppose that $X(s)$ is a Gaussian random function with zero mean and correlation

$$\langle X(s)X(s')\rangle = C(s,s'),$$

then for any functional $\mathscr{L}[h]$,

$$\langle X(s)\mathscr{L}[X]\rangle = \int_{-\infty}^{\infty} C(s,s')\left\langle \frac{\delta \mathscr{L}[X]}{\delta X(s')}\right\rangle ds'.$$

Applying this theorem to $\langle g(U)\eta\rangle$, we have

$$\langle g(V)\eta\rangle = 2\int_{-\infty}^{\infty} C([x-x']/\lambda)\left\langle g'(V(x,t))\frac{\delta V(x,t)}{\delta\eta(x',t)}\right\rangle dx'.$$

The stochastic voltage is a functional of the noise term $\eta(x,t)$, as can be seen by formally integrating Eq. (2.83) with respect to time:

$$V(x,t) = \int_{-\infty}^{\infty}\left[\frac{\partial^2}{\partial x^2}V(x,t'') + f(V(x,t'')) + \sqrt{\varepsilon}g(V(x,t''))\eta(x,t'')\right]H(t-t'')dt''.$$

It follows that

$$\frac{\delta V(x,t)}{\delta\eta(x',t')} = \sqrt{\varepsilon}g(V(x,t'))H(t-t')\delta(x-x'),$$

which yields (2.85) after using $H(0)=1/2$.

An alternative derivation of (2.85) is based on Fourier transforming (2.83) [546]. It is convenient to restrict x to a bounded domain, $-L/2 \leq x \leq L/2$, and to impose periodic boundary conditions. We can then introduce the discrete Fourier series

$$V(x,t) = \frac{1}{L}\sum_n \mathrm{e}^{ik_n x}V_n(t), \quad W(x,t) = \frac{1}{L}\sum_n \mathrm{e}^{ik_n x}W_n(t) \tag{2.86}$$

with $k_n = 2\pi n/L$ and $W_n(t)$, an independent Wiener process, such that

$$\langle dW_n(t)\rangle = 0, \quad \langle dW_n(t)dW_m(t)\rangle = 2L\delta_{m+n,0}dt. \tag{2.87}$$

Fourier transforming (2.83) gives in differential form

$$dU_n(t) = [-k_n^2 V_n(t) + F_n(t)]dt + \frac{\varepsilon^{1/2}}{L}\sum_m g_{n-m}(t)dW_m(t), \tag{2.88}$$

where F_n, g_n are the Fourier coefficients of the time-dependent functions $F\circ U(t)$ and $g\circ U(t)$, respectively. The associated Stratonovich Fokker–Planck equation takes the form [209] (see also Sect. 1.7)

$$\frac{\partial P}{\partial t} = -\sum_l \frac{\partial}{\partial u_l}[(-k_l^2 V_n(t) + F_l(t))P] + \frac{\varepsilon}{L}\sum_{l,m,q}\frac{\partial}{\partial V_l}g_{l-q}\frac{\partial}{\partial V_m}g_{m+q}P. \tag{2.89}$$

Multiplying both sides of this equation by V_n and integrating with respect to V_m, integer m, leads to the following evolution equation for the mean:

$$\frac{d\langle V_n\rangle}{dt} = -k_n^2\langle V_n\rangle + \langle F_n\rangle + \frac{\varepsilon}{L}\sum_{m,q}\left\langle \frac{\partial g_{n-q}}{\partial V_m}g_{m+q}\right\rangle. \tag{2.90}$$

Finally, taking the inverse transform of (2.90) gives

$$\frac{d\langle U(x,t)\rangle}{dt} = \frac{\partial^2}{\partial x^2}\langle V(x,t)\rangle + \langle F(V(x,t))\rangle + \frac{\varepsilon}{\Delta x}\langle g(U(x,t))g'(U(x,t))\rangle, \tag{2.91}$$

where we have used the result $\partial g_n/\partial U_m = [g'(U)]_{n-m}$. Note that it is necessary to introduce a cutoff in the frequencies, which is equivalent to introducing a fundamental lattice spacing of Δx. Alternatively, the multiplicative noise can be taken to have a small but finite correlation length in space so that $C(0) = 1/\Delta x$. Comparison of (2.90) with the mean of (2.83) yields the desired result.

Following [14], it is convenient to rewrite (2.83) so that the fluctuating term has zero mean:

$$\frac{\partial}{\partial t}V(x,t) = \frac{\partial^2}{\partial x^2}V(x,t) + h(V(x,t)) + \sqrt{\varepsilon}R(V,x,t), \tag{2.92}$$

where

$$h(V) = f(V) + \varepsilon C(0)g'(V)g(V) \tag{2.93}$$

and

$$R(V,x,t) = g(V)\eta(x,t) - \varepsilon^{1/2}C(0)g'(U)g(U). \tag{2.94}$$

The stochastic process R has zero mean (so does not contribute to the effective wave speed) and correlation:

$$\langle R(V,x,t)R(V,x',t')\rangle = \langle g(V(x,t))\eta(x,t)g(V(x',t')\eta(x',t')\rangle + \mathscr{O}(\varepsilon^{1/2}). \tag{2.95}$$

The next step in the analysis is to assume that the fluctuating term in (2.92) generates two distinct phenomena that occur on different time scales: a diffusive-like displacement of the front from its uniformly translating position at long time scales and fluctuations in the front profile around its instantaneous position at short time scales [14, 494, 546, 557]. In particular, following [14], we express the solution V of (2.92) as a combination of a fixed wave profile V_0 that is displaced by an amount $\Delta(t)$ from its uniformly translating position $\xi = x - \bar{c}t$ and a time-dependent fluctuation Φ in the front shape about the instantaneous position of the front:

$$V(x,t) = V_0(\xi - \Delta(t)) + \varepsilon^{1/2}\Phi(\xi - \Delta(t),t). \tag{2.96}$$

The wave profile V_0 and associated wave speed $\bar{c}$ are obtained by solving the modified deterministic equation

$$\bar{c}\frac{dV_0}{d\xi} + \frac{d^2V_0^2}{d\xi^2} + h(V_0(\xi)) = 0. \tag{2.97}$$

Both $\bar{c}$ and V_0 depend nontrivially on the noise strength ε due to the ε dependence of the function h; see (2.93). As an example, suppose that $f(V) = V(V-a)(1-V)$ and $g(V) = V(1-V)$ [14]. The form of multiplicative noise is chosen so that it preserves the stationary states $V = 0, 1$. Hence, the noise is most important in regions close to the front but vanishes asymptotically at $\xi \pm \infty$. The effective nonlinearity h is also a cubic with

$$h(V) = V(1-V)(a' - k'V), \quad a' = a - \varepsilon C(0), \quad k' = 1 - 2\varepsilon C(0).$$

Thus, from the analysis of the bistable equation in Sect. 2.2, we find that $\bar{c} = (k' - 2a')/\sqrt{2k'}$.

It turns out that if V_0 is chosen to satisfy (2.97) then to leading order, the stochastic variable $\Delta(t)$ undergoes unbiased Brownian motion (a Wiener process):

$$\langle \Delta(t) \rangle = 0, \quad \langle \Delta(t)^2 \rangle = 2D(\varepsilon)t \tag{2.98}$$

with a diffusion coefficient $D(\varepsilon) = \mathcal{O}(\varepsilon)$ (see below). Thus $\Delta(t)$ represents the effects of slow fluctuations, whereas $\boldsymbol{\Phi}$ represents the effects of fast fluctuations. Note that since $\Delta(t) = \mathcal{O}(\varepsilon^{1/2})$, (2.96) implies that $V(x,t) = V_0(x - \bar{c}t) + \mathcal{O}(\varepsilon^{1/2})$. Hence, averaging with respect to the noise shows that $\langle V(x,t) \rangle = V_0(x - \bar{c}t) + \mathcal{O}(\varepsilon^{1/2})$. Thus, in the case of weak noise, averaging over many realizations of the stochastic wave front generates a mean front whose speed is approximately equal to $\bar{c}$.

Calculation of diffusion coefficient. Substitute the decomposition (2.96) into (2.92) and expand to first order in $\mathcal{O}(\varepsilon^{1/2})$ (exploiting the fact that the usual rules of calculus apply in the case of Stratonovich noise, see Sect. 1.7):

$$\begin{aligned} &-[\bar{c} + \dot{\Delta}(t)]V_0'(\xi_t) + \varepsilon^{1/2}\left[\partial_t \Phi(\xi_t,t) - [\bar{c} + \dot{\Delta}(t)]\partial_\xi \Phi(\xi_t,t)\right] \\ &\quad = h(V_0(\xi_t) + \varepsilon^{1/2} h'(V_0(\xi_t))\Phi(\xi_t,t) \\ &\quad + \partial_\xi^2 \left[V_0(\xi_t) + \varepsilon^{1/2}\Phi(\xi_t,t)\right] + \varepsilon^{1/2} R(V_0(\xi_t),x,t) + \mathcal{O}(\varepsilon), \end{aligned}$$

where $\xi_t \equiv \xi - \Delta(t)$. Imposing (2.97) and dividing through by $\varepsilon^{1/2}$ then gives

$$\frac{\partial \Phi(\xi,t)}{\partial t} + \mathbb{L}_\xi \Phi(\xi,t) = \varepsilon^{-1/2} V_0'(\xi)\dot{\Delta}(t) + R(V_0(\xi),\xi,t) + \mathcal{O}(\varepsilon^{1/2}), \tag{2.99}$$

where $\mathbb{L}_\xi$ is the non-self-adjoint linear operator

$$\mathbb{L}_\xi A(\xi) = A''(\xi) + \bar{c}A'(\xi) + h'(V_0(\xi))A(\xi) \tag{2.100}$$

for any function $A(\xi) \in L^2(\mathbb{R})$. We have also made the approximation $\xi_t \approx \xi$, since $\Delta(t) = \mathcal{O}(\varepsilon^{1/2})$. The linear differential operator $\mathbb{L}_\xi$ has a zero eigenvalue with associated eigenfunction $V_0'(\xi)$, which can be seen by differentiating (2.97) with respect to ξ, and reflects the fact that the underlying system is equivariant with respect to uniform translations. We then have the solvability condition for the existence of a bounded solution of (2.99), namely, that the inhomogeneous part on the right-hand side is orthogonal to all elements of the null-space of the adjoint operator $\mathbb{L}_\xi^\dagger$. The latter is defined with respect to the inner product

$$\int_{-\infty}^{\infty} B(\xi)\mathbb{L}_\xi A(\xi) d\xi = \int_{-\infty}^{\infty} \left[\mathbb{L}_\xi^\dagger B(\xi)\right] A(\xi) d\xi \tag{2.101}$$

where $A(\xi)$ and $B(\xi)$ are arbitrary integrable functions. Hence,

$$\mathbb{L}_\xi^\dagger B(\xi) = B''(\xi') - \bar{c}B'(\xi) + h'(V_0(\xi))B(\xi). \tag{2.102}$$

The linear operator $\mathbb{L}^\dagger$ also has a zero eigenvalue, with corresponding eigenfunction $\mathcal{V}(\xi) = e^{\bar{c}\xi} V_0'(\xi)$. Thus taking the inner product of both sides of (2.99) with respect to $\mathcal{V}(\xi)$ leads to the solvability condition

$$\int_{-\infty}^{\infty} \mathcal{V}(\xi) \left[V_0'(\xi)\dot{\Delta}(t) + \varepsilon^{1/2} R(V_0,\xi,t)\right] d\xi = 0, \tag{2.103}$$

which implies that $\Delta(t)$ satisfies the stochastic differential equation (SDE)

$$d\Delta(t) = -\varepsilon^{1/2} \frac{\int_{-\infty}^{\infty} \mathscr{V}(\xi) dR(V_0,\xi,t) d\xi}{\int_{-\infty}^{\infty} \mathscr{V}(\xi) V_0'(\xi) d\xi}. \tag{2.104}$$

Using the lowest-order approximation $dR(V_0,\xi,t) = g(V_0(\xi))dW(\xi,t)$, we deduce that (for$\Delta(0) = 0$) $\Delta(t)$ is a Wiener process with diffusion coefficient

$$D(\varepsilon) = \varepsilon \frac{\int_{-\infty}^{\infty}\int_{-\infty}^{\infty} \mathscr{V}(\xi)\mathscr{V}(\xi')g(V_0(\xi))g(V_0(\xi'))C([\xi-\xi']/\lambda)d\xi d\xi'}{\left[\int_{-\infty}^{\infty} \mathscr{V}(\xi)V_0'(\xi)d\xi\right]^2}. \tag{2.105}$$

Although the above analysis is based on a formal perturbation calculation, rather than rigorous analysis, it does appear to capture well the effects of weak external noise on front propagation in a variety of reaction–diffusion models [486, 546]. In Sect. 7.4, we will show how the method can be extended to study stochastic traveling waves in nonlocal neural field equations, which represent large-scale continuum models of spatially structured neural networks. Note, however, that one class of front solution where the method breaks down is a so-called pulled front, which propagates into an unstable rather than a metastable state and whose dynamics is dominated by the linear spreading of small perturbations within the leading edge of the front [544]. One well-known reaction–diffusion model that supports pulled fronts is the Fisher–KPP equation [191, 345]. As we will describe later, pulled fronts also arise in a PDE model of CaMKII translocation waves along spiny dendrites [72, 161] (Sect. 3.2), in certain neural field models (Sect. 7.4), and in a model of protein aggregation (Sect. 9.6).

2.7 Appendix: Linear Differential Operators

Throughout this book, we will encounter linear operators acting on some function space. As already demonstrated in this chapter, linear differential operators arise when analyzing the stability of a traveling wave solution of some PDE, or when carrying out a formal perturbation expansion. In this appendix, we summarize some of the basic results regarding linear differential operators acting on a function space, viewed from the perspective of a linear map acting on an infinite-dimensional vector space. For simplicity, we will restrict ourselves to real-valued functions $f : \mathbb{R} \to \mathbb{R}$, although it is straightforward to generalize the results to complex-valued functions.

2.7.1 Function Spaces

Consider the set of all real functions $f(x)$ on the interval $[a,b]$. This is a vector space over the set of real numbers: given two functions $f_1(x), f_2(x)$ and two real numbers a_1, a_2, we can form the sum $f(x) = a_1 f_1(x) + a_2 f_2(x)$ such that $f(x)$ is also a function on $[a,b]$. Either on physical grounds or for mathematical convenience, we usually restrict ourselves to a subspace of functions that are differentiable to some given order. For example, the space of functions on $[a,b]$ with n continuous derivatives is denoted by $C^n[a,b]$, and the space of analytic functions (those whose Taylor expansion converges to the given function) is denoted by $C^\omega[a,b]$.

In order to describe the convergence of a sequence of functions $f_n, n = 1,2,\ldots$ to a limit function f, we need to introduce the concept of a *norm*, which is a generalization of the usual measure of the length of a finite-dimensional vector. The norm $\|f\|$ of a function f is a real number with the following properties:

(i) Positivity: $\|f\| \geq 0$, and $\|f\| = 0$ if and only if $f = 0$
(ii) The triangle inequality: $\|f+g\| \leq \|f\| + \|g\|$
(iii) Linearity: $\|\lambda f\| = |\lambda| \|f\|$ for $\lambda \in \mathbb{R}$

Common examples of norms are the "sup" norm

$$\|f\|_\infty = \sup_{x\in[a,b]} |f(x)|,$$

and the L^p norm

$$\|f\|_p = \left(\int_a^b |f(x)|^p dx\right)^{1/p}.$$

Given the L^p norm, we can introduce another important function space $L^p[a,b]$, which is the space of real-valued functions on $[a,b]$ for which $\|f\|_p < \infty$. However, there is one subtlety here, namely, that it is possible for $\|f\| = 0$ without f being identically zero. For example, f may vanish at all but a finite set of points (set of measure zero). This violates the positivity property of a norm. Therefore, one should really treat elements of $L^p[a,b]$ as equivalence classes of functions, where functions differing on a set of measure zero are identified.

Given a normed function space, convergence of a sequence $f_n \to f$ can be expressed as

$$\lim_{n\to\infty} \|f_n - f\| = 0.$$

In the case of the "sup" norm, f_n is said to converge *uniformly* to f, whereas for the L^1 norm, it is said to converge *in the mean*. An important property of a function space is that of being *complete*. First, consider the following definition of a *Cauchy sequence*: A sequence f_n in a normed vector space is Cauchy if for any $\varepsilon > 0$, we can find an integer N such that $n,m > N$ implies that $\|f_m - f_n\| < \varepsilon$. In other words, elements of the sequence become arbitrarily close together as $n \to \infty$. A normed vector space is then *complete* with respect to its norm if every Cauchy sequence converges to some element in the space. A complete normed vector space is called a Banach space $\mathscr{B}$. In many applications, the norm of the function space is taken to be the so-called natural norm obtained from an underlying *inner product*. For example, if we define an inner product for $L^2[a,b]$ according to

$$\langle f, g \rangle = \int_a^b f(x)g(x)dx,$$

then the $L^2[a,b]$ norm can be written as

$$\|f\|_2 = \sqrt{\langle f, f \rangle}.$$

A Banach space with an inner product is called a *Hilbert space* $\mathscr{H}$.

2.7.2 Fredholm Alternative Theorem

In the case of one-dimensional traveling wave solutions, it is more natural to consider functions on the real line $\mathbb{R}$ rather than a finite interval. Suppose that $\mathbb{L}$ is a linear differential operator acting on a subspace of $L^2(\mathbb{R})$, which we denote by the domain $\mathscr{D}(\mathbb{L})$. Linearity of the operator means that for $f_1, f_2 \in \mathscr{D}(\mathbb{L})$ and $a_1, a_2 \in \mathbb{R}$,

$$\mathbb{L}(a_1 f_1 + a_2 f_2) = a_1 \mathbb{L} f_1 + a_2 \mathbb{L} f_2.$$

Given the standard inner product on $L^2(\mathbb{R})$, we define the adjoint linear operator $\mathbb{L}^\dagger$ according to

$$\langle f, \mathbb{L}g \rangle = \langle \mathbb{L}^\dagger f, g \rangle, \quad f, g \in \mathscr{D}(L).$$

The operator is said to be self-adjoint if $\mathbb{L}^\dagger = \mathbb{L}$. Note that, in practice, one determines $\mathbb{L}^\dagger$ using integration by parts. For functions defined on finite intervals, this generates boundary terms that only vanish if appropriate boundary conditions are imposed. In general, this can result in different domains for $\mathbb{L}$ and $\mathbb{L}^\dagger$. Therefore, the condition for self-adjointness becomes $\mathbb{L} = \mathbb{L}^\dagger$ and $\mathscr{D}(\mathbb{L}) = \mathscr{D}(\mathbb{L}^\dagger)$. Given a differential operator $\mathbb{L}$ on $L^2(\mathbb{R})$, we can now state the *Fredholm alternative theorem*: The inhomogeneous equation

$$\mathbb{L} f = h$$

has a solution if and only if

$$\langle h, v \rangle = 0 \quad \text{for all v satisfying} \quad \mathbb{L}^\dagger v = 0.$$

2.7.3 Spectrum of a Linear Differential Operator

Let $\mathscr{B}$ be a Banach space and $\mathbb{L} : \mathscr{D}(\mathbb{L}) \to \mathscr{B}$ be a linear operator with domain $\mathscr{D}(\mathbb{L}) \subseteq \mathscr{B}$. For any complex number λ, introduce the new operator

$$\mathbb{L}_\lambda = \mathbb{L} - \lambda \mathbb{I},$$

where $\mathbb{I}$ is the identity operator on $\mathscr{B}$. If $\mathbb{L}_\lambda$ has an inverse, then $R_\lambda(\mathbb{L}) = \mathbb{L}_\lambda^{-1}$ is called the resolvent of $\mathbb{L}$. Given these definitions, λ is said to be a regular point for $\mathbb{L}$ if the following hold:

(i) R_λ exists.
(ii) R_λ is bounded.
(iii) R_λ is defined on a dense subset of $\mathscr{B}$.

The spectrum $\sigma(\mathbb{L})$ is then the set of points that are not regular, which generally consists of three disjoint parts:

(a) The *point spectrum* of eigenvalues is the set of values of λ for which R_λ does not exist.
(b) The *continuous spectrum* is the set of values of λ for which R_λ exists but is unbounded.
(c) The *residual spectrum* is the set of values of λ for which R_λ exists, is bounded, but is not defined on a dense subset of $\mathscr{B}$.

The continuous spectrum and residual spectrum are contained in the essential spectrum, which is any point in $\sigma(\mathbb{L})$ that is not an isolated eigenvalue of finite multiplicity.

We will illustrate how to calculate the essential spectrum of a simple linear operator acting on $\mathscr{B} = L^2(\mathbb{R})$ [242]:

$$\mathbb{L}u = u_{xx} + pu_x - qu \tag{2.106}$$

for constant positive coefficients p,q and $\mathscr{D}(\mathbb{L}) = \{u : u \in L^2(\mathbb{R}), \mathbb{L}u \in L^2(\mathbb{R})\}$. Firstly, suppose that $\mathbb{L}_\lambda$ is not invertible for some λ. This means that there exists $\phi \in \mathscr{B}$ such that $\mathbb{L}_\lambda \phi = 0$. The latter equation is a linear second-order ODE with constant coefficients and thus has solutions of the form $e^{\nu_\pm x}$ with $\nu_\pm$, the roots of the characteristic polynomial $\nu^2 + p\nu - (q+\lambda) = 0$. Such a solution cannot decay at both $x = \pm\infty$ and so does not belong to $\mathscr{B}$. It follows that $\mathbb{L}$ has no eigenvalues and the resolvent R_λ exists. We can then represent R_λ in terms of the Green's function G defined according to $\mathbb{L}_\lambda^\dagger G(x-x') = \delta(x-x')$, where $\mathbb{L}^\dagger$ is the adjoint of $\mathbb{L}$ with respect to the standard inner product on $L^2(\mathbb{R})$:

$$\mathbb{L}_\lambda^\dagger u = u_{xx} - pu_x - (q+\lambda)u.$$

For any $h \in \mathscr{D}(R_\lambda) \subseteq \mathscr{B}$ we can express the solution $u = R_\lambda h$ to the inhomogeneous equation $\mathbb{L}_\lambda u = h$ as

$$u(x) = \int_{-\infty}^{\infty} h(y)G(y-x)dy.$$

For constant coefficients, the Green's function can be solved explicitly according to

$$G(y) = \begin{cases} \alpha\, e^{\mu_+ y} & y \leq 0 \\ \alpha\, e^{\mu_- y} & y \geq 0, \end{cases}$$

where $\mu_{\pm}$ are the roots of the characteristic polynomial

$$P(\mu) = \mu^2 - p\mu - (\lambda + q),$$

and α is chosen such that $-1 = \alpha(\mu_+ - \mu_-)$.

If $P(\mu)$ has one root μ_+ with positive real part and one root μ_- with negative real part, then clearly $G \in L_1(\mathbb{R})$ so that R_λ is bounded with dense domain equal to $\mathscr{B}$. This situation holds, for example, when λ is real and $\lambda > -p$. The roots of $P(\mu)$ vary continuously with λ in the complex plane. Hence, the boundedness of R_λ will break down when one of the roots crosses the imaginary axis at ik, say, with $\lambda = -q - k^2 - ika$. This is a parabola in the complex λ plane (λ_r, λ_i) given by $\lambda_r = -q - \lambda_i^2/p^2$. If λ_r is to the right of this parabola,

$$\lambda_r > -q - \frac{\lambda_i^2}{p^2},$$

then $P(\mu)$ has a root on either side of the imaginary axis and R_λ is bounded. We conclude that the essential spectrum lies to the left of the parabola,

$$\sigma(\mathbb{L}) \subseteq \{\lambda : \mathrm{Re}(\lambda) \leq -q - \mathrm{Im}(\lambda)^2/p^2\}. \tag{2.107}$$

It can be shown that the essential spectrum includes the parabola itself. It immediately follows that the essential spectrum lies in the left-half complex plane if $q > 0$.

In Sect. 2.4, we considered the linear stability of a traveling front, which required finding the spectrum of a second-order linear operator with nonconstant coefficients; see (2.42). It turns out that one can generalize the above analysis to an operator with x-dependent coefficients $p(x), q(x)$. Suppose that $p(x), q(x) \to p_\pm, q_\pm$ as $x \to \pm\infty$. Introduce the parabolas

$$S_\pm = \{\lambda : \lambda = -q_\pm - k^2 - ikp_\pm\}. \tag{2.108}$$

Let A denote the union of the regions to the left of the curves $S_\pm$ that includes the curves themselves. Then the essential spectrum of $\mathbb{L}$ lies in A and includes $S_\pm$.

Chapter 3
Wave Propagation Along Spiny Dendrites

The majority of excitatory synapses that occur in the cerebral cortex are located on tiny specialized protoplasmic protuberances called dendritic *spines* [607]; see Fig. 3.1. They typically occupy 20–70 % of the total dendritic membrane. Since the input impedance of a spine head is typically large, a small excitatory synaptic current can produce a large local depolarization. Moreover, the thin stem neck connecting the spine to the main body of the dendrite provides an axial resistance that partially decouples the spine-head dynamics from the dendritic tree. Hence, it has long been theorized that the dendritic spine is a favorable site for the initiation of an action potential [427, 572] and is thus a possible substrate for local computations [571]. Modeling studies also suggest that if the heads of dendritic spines have excitable membrane properties, then the spread of current from one spine along the dendrites could bring adjacent spines to their thresholds for impulse generation. The result would be a sequence of spine-head action potentials, representing a saltatory propagating wave in the distal dendritic branches [19, 129]. Calcium-imaging experiments provide strong evidence that the spine heads are endowed with voltage-dependent Na^+ and Ca^{2+} channels that can indeed support an all-or-nothing response to an excitatory synaptic input [694].

Early theoretical studies of spines also considered their potential role in synaptic plasticity and Hebbian learning. This was motivated by the idea that small changes in spine morphology, such as changes in the width of the spine neck, could lead to large changes in the amplitude of response to excitatory synaptic inputs on to the spine. It is now known that spines are rich in actin filaments, which have the capacity to drive such changes in spine shape [413]. Moreover, there is increasing experimental evidence that the growth and removal of spines provides an important substrate for structural changes during brain development [410, 480, 693]. It is less clear whether changes in spine morphology play a significant role in adult plasticity. Nevertheless, the basic geometry of a spine does provide an isolated biochemical microenvironment for Ca^{2+} to accumulate, and Ca^{2+} is thought to be a major chemical signal for the induction of synaptic plasticity [343, 462]. The dynamics of calcium diffusion in dendritic spines has been explored in a number of computational models [202, 283].

P.C. Bressloff, *Waves in Neural Media: From Single Neurons to Neural Fields*, Lecture Notes on Mathematical Modelling in the Life Sciences,
DOI 10.1007/978-1-4614-8866-8_3,

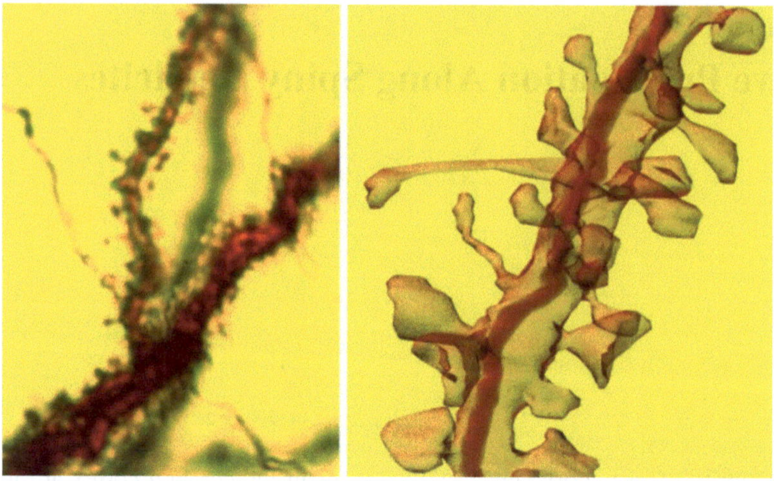

Fig. 3.1 An example of a piece of spine-studded dendritic tissue (from rat hippocampal region CA1 stratum radiatum). Magnified view on right-hand side shows a dendrite $\sim 5\,\mu m$ in length. Taken with permission from SynapseWeb, Kristen M. Harris, PI, http://synapses.clm.utexas.edu/

Another important signaling molecule involved in the induction of synaptic plasticity is Ca^{2+}–calmodulin-dependent protein kinase II (CaMKII) [295, 386]. CaMKII is also found to be abundant within dendritic spines, where it can detect changes in the local levels of Ca^{2+} entering the synapse following plasticity-inducing stimuli, via binding of CaMKII to Ca^{2+}/CaM. Confinement of CaMKII within spines arises from the geometry of the spine and through interactions with protein receptors and cytoskeletal elements within the postsynaptic density (PSD), which is the protein-rich region at the tip of the spine head. Activated CaMKII phosphorylates substrates responsible for the expression of synaptic plasticity, namely, the number and the conductivity of synaptic AMPA receptors [151, 370]. Moreover, once activated, CaMKII can transition into a Ca^{2+}/CaM-independent, hyper-activated state via the autophosphorylation of neighboring enzymatic subunits and thus continue to phosphorylate its substrates even after the plasticity-inducing Ca^{2+} signal has ended [261, 427, 547, 689]. Experimentally, translocation of CaMKII into spines can be induced in a local region of dendrite by exposing it to a puff of glutamate, and this can initiate a wave of CaMKII translocation that spreads towards the distal end of the dendrite with an average speed of $\sim 1\,\mu m/s$ [532]. It is found that the wave is preceded by a much faster Ca^{2+}-mediated spike that propagates along the dendrite (see above), which could provide a mechanism for priming CaMKII outside the stimulus region for diffusion-based activation. Moreover, the CaMKII translocation wave is associated with an increase in AMPA receptor numbers at both stimulated and nonstimulated synapses [532]. This suggests that it could provide a possible molecular substrate for heterosynaptic plasticity.

In this chapter we consider two mathematical models of wave propagation along spiny dendrites: (I) a spike–diffuse–spike model of spine-mediated spikes [129, 139] and (II) a reaction–diffusion model of CaMKII translocation waves [72, 161]. The former model introduces methods for analyzing solitary waves propagating in spiking networks; see also Sect. 5.4. The latter model turns out to be identical in form to the diffusive SI model introduced by Noble [464] to explain the spread of bubonic plague through Europe in the fourteenth century. This, in turn, is a generalization of the scalar Fisher–KPP equation [191, 345] which was originally introduced to model the invasion of a gene into a population. One characteristic feature of such equations is that they support traveling fronts propagating into an unstable steady state, in which the wave speed and longtime asymptotics are determined by the dynamics in the leading edge of the wave—so-called pulled fronts [544]. In particular, a sufficiently localized initial perturbation will asymptotically approach the traveling front solution that has the minimum possible wave speed. Hence, pulled fronts have very different properties from those found for the bistable equation in Sect. 2.2. Another important topic raised by these models is how to use homogenization methods to approximate the discrete distribution of spines by a smooth distribution.

3.1 Solitary Waves in a Spike–Diffuse–Spike Model of Active Spines

The first theoretical study of active wave propagation along a dendritic cable that is mediated by dendritic spines was carried out by Baer and Rinzel [19]. They considered a continuum model of a dendritic tree coupled to a distribution of excitable dendritic spines. The active spine-head dynamics is modeled with Hodgkin–Huxley kinetics, while the (distal) dendritic tissue is modeled with the cable equation. The spine head is coupled to the cable via a spine-stem resistance that delivers a current proportional to the number of spines at the contact point. There is no direct coupling between neighboring spines; voltage spread along the cable is the only way for spines to interact. Numerical studies of the Baer–Rinzel model [19] show both smooth and saltatory traveling wave solutions, the former arising in the case of uniform spine distributions and the latter when spines are clustered in groups. The saltatory nature of a propagating wave may be directly attributed to the fact that active spine clusters are physically separated. In this section we describe an alternative, analytically tractable treatment of saltatory waves based on the so-called spike–diffuse–spike (SDS) model of active dendritic spines [129, 138, 624, 625], which reduces the spine-head dynamics to an all-or-nothing action potential response.

In order to formulate the model, we first consider a continuum of spines with $\rho(x)$ representing the spine density per unit length along a uniform, passive dendritic cable. Denoting the voltage at position x on the cable at time t by $V = V(x,t)$, the associated cable equation is given by (see Sect. 1.4)

$$\frac{\partial V}{\partial t} = -\frac{V}{\tau_m} + D_m \frac{\partial^2 V}{\partial x^2} + \rho(x) \frac{a - V}{C_m r_s}, \tag{3.1}$$

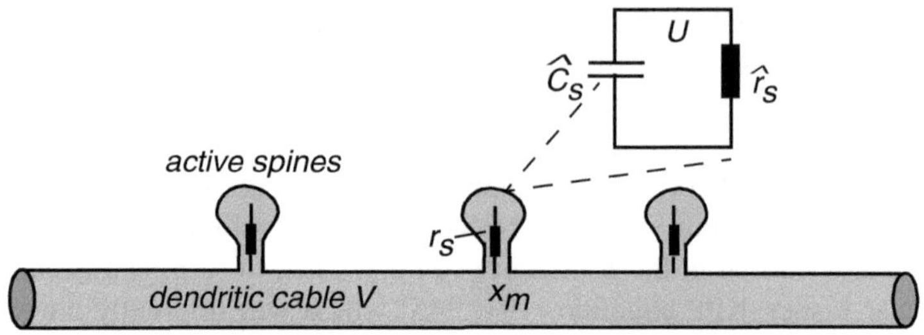

Fig. 3.2 Schematic diagram of a dendritic cable with active spines

where $D_m = \lambda_m^2/\tau_m$ with τ_m and λ_m the membrane time and space constants of the cable, C_m is the membrane capacitance per unit length, and r_s is the spine-stem resistance of an individual spine. In the SDS model, the function $a(x,t)$ represents the sequence of action potentials generated in the spine head at x whenever the associated subthreshold spine-head potential $U(x,t)$, driven by current from the shaft, crosses some threshold h. Given the high resistance of the spine stem, we neglect subthreshold currents into the cable. The voltage U evolves according to the integrate-and-fire (IF) equation (see Sect. 5.3)

$$\hat{C}_s \frac{\partial U}{\partial t} = -\frac{U}{\hat{r}_s} + \frac{V-U}{r_s}, \tag{3.2}$$

such that whenever U crosses the threshold h it is immediately reset to zero. Here $\hat{C}_s$ and $\hat{r}_s$ are the membrane capacitance and resistance of the spine head; see Fig. 3.2. Let $T^j(x)$ denote the jth firing time of the spine head at position x such that $U(x,T^j(x)) = h$. Then

$$a(x,t) = \sum_j \eta(t - T^j(x)), \tag{3.3}$$

with $\eta(t) = 0$ for all $t < 0$. The shape of the action potential is specified by the function $\eta(t)$, which can be fitted to the universal shape of an action potential.

3.1.1 Existence of a Solitary Wave for a Uniform Density

Let us begin by considering the simplest case of a uniform spine density $\rho(x) = \rho_0$ and a rectangular pulse shape of height η_0 and duration τ_d. We define a solitary wave as one that causes the spine head at x to reach threshold only once at the time $t = T(x) \equiv x/c$. We recognize c as the speed of the wave so that $a(x,t) = \eta(t - x/c)$, which suggests adopting a moving coordinate frame $\xi = ct - x$. Equation (3.1) becomes

$$D_m V_{\xi\xi}(\xi) - cV_\xi(\xi) - (\tau_m^{-1} + \beta)V(\xi) = -\beta a(\xi/c), \tag{3.4}$$

where $\beta = \rho_0/(C_m r_s)$, and

$$a(\xi) = \begin{cases} 0 & -\infty < \xi < 0 \\ \eta_0 & 0 < \xi < c\tau_d \\ 0 & \xi > c\tau_d. \end{cases} \tag{3.5}$$

If one is looking for traveling pulses which satisfy $\lim_{\xi \to \pm\infty} V(\xi) = 0$, then the solution to (3.4) takes the form

$$V(\xi) = \begin{cases} \alpha_1 \exp(m_+ \xi), & -\infty < \xi < 0 \\ \alpha_2 \exp(m_+ \xi) + \alpha_3 \exp(m_- \xi) + \frac{\beta \eta_0}{\tau_m^{-1} + \beta}, & 0 < \xi < c\tau_d \\ \alpha_4 \exp(m_- \xi), & \xi > c\tau_d \end{cases} \tag{3.6}$$

with

$$m_\pm = \frac{c \pm \sqrt{c^2 + 4D_m(\tau_m^{-1} + \beta)}}{2D_m}. \tag{3.7}$$

By ensuring the continuity of the solution and its first derivative at $\xi = 0$ and $\xi = c\tau_d$ one may solve for the unknowns $\alpha_1 \ldots \alpha_4$ as

$$\alpha_1 = \alpha_3 \frac{m_-}{m_+}[1 - \exp(-m_+ c\tau_d)], \tag{3.8}$$

$$\alpha_2 = -\alpha_3 \frac{m_-}{m_+} \exp(-m_+ c\tau_d), \tag{3.9}$$

$$\alpha_3 = \frac{\beta \eta_0}{\tau_m^{-1} + \beta} \frac{m_+}{(m_- - m_+)}, \tag{3.10}$$

$$\alpha_4 = \alpha_3 [1 - \exp(-m_- c\tau_d)]. \tag{3.11}$$

As yet the speed of the pulse is undetermined. However, by demanding that the IF process in the spine head reaches threshold at $\xi = 0$ and that $\lim_{\xi \to \pm\infty} U(\xi) = 0$, one can determine a self-consistent value for the speed of the traveling wave along similar lines to the analysis of solitary waves in one-dimensional networks of IF neurons [62, 63, 167]; see Sect. 5.4. In the traveling coordinate frame, (3.2) becomes

$$c\hat{C}_s U_\xi = -g_s U + \frac{V}{r_s} \tag{3.12}$$

with $U(0) = h$ and $g_s = r_s^{-1} + \hat{r}_s^{-1}$. This first-order system may be solved as

$$U(\xi) = \exp(-\xi/[c\hat{\tau}]) \left[h - \frac{1}{c\hat{C}_s r_s} \int_\xi^0 V(\xi') \exp(\xi'/[c\hat{\tau}]) d\xi' \right], \tag{3.13}$$

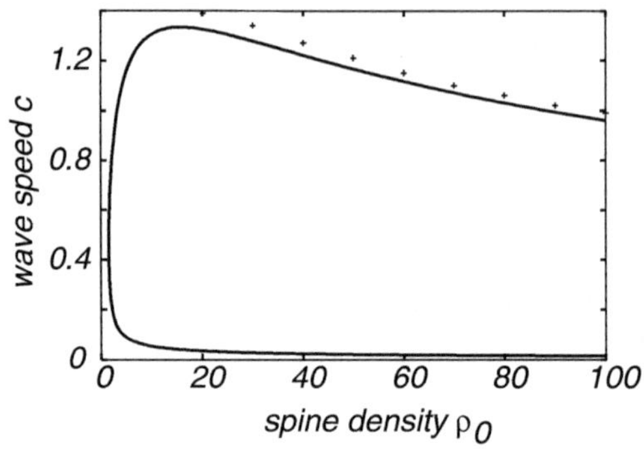

Fig. 3.3 Speed of a traveling pulse as a function of uniform spine density ρ_0 for $\tau_m = D_m = 1$, $h = 2.5$, $r_s = 2$, $\widehat{r_s} = 0.8$, $\hat{C}_s = 1$, $C_m = 1$, $\tau_d = 2$, $\eta_0 = 100$. The crosses show the results of a direct numerical simulation of the SDS model with $N = 200$ discrete spines uniformly distributed along a cable of length $L = 10$

where $\hat{\tau} = \hat{C}_s/g_s$. In order for this to be bounded as $\xi \to -\infty$, the term inside the large parentheses must vanish as $\xi \to -\infty$. This yields the dispersion relationship for the speed of the pulse as a function of system parameters:

$$h = \frac{1}{\hat{C}_s r_s} \frac{\alpha_1}{\hat{\tau}^{-1} + cm_+} \tag{3.14}$$

In Fig. 3.3 we plot dispersion curves for the speed c of a traveling pulse as a function of the spine density ρ_0, which are obtained by numerically solving (3.14). Note that there are two solution branches for a given density ρ_0. Direct simulations suggest that it is the upper (faster) branch that is stable, which can be confirmed analytically using linear stability analysis (see below). Figure 3.3 also shows that for a wide range of ρ_0 the speed of the stable wave is approximately λ_m/τ_m in physical units, in agreement with the original observations of Baer and Rinzel [19].

One useful feature of the SDS model is that it yields an exact expression for the speed of the wave that can easily be solved to obtain the dependence in terms of other system parameters such as the spine-stem resistance. Hence, one can find the minimum spine density capable of supporting a traveling pulse as well as extract information about how the speed decays as a function of spine density. In Fig. 3.4(a) we plot the speed of a traveling pulse as a function of the spine-stem resistance r_s. It is clear that for realistic choices of the biophysical parameters in the model, that propagation failure can occur for too large a choice of the spine-stem resistance. Moreover, for small r_s, the speed of a stable pulse is very sensitive to r_s, demonstrating that a modifiable value of the spine-stem resistance could have important ramifications for neural processing. Finally in Fig. 3.4(b) we show the dependence of the wave speed on the width, τ_d, of a rectangular pulse generated in the spine head. Interestingly, for a fixed pulse height, there is a minimum duration time below which propagation cannot occur. This highlights the fact that it is crucial to model the shape of an action potential in the reduced IF model with biologically realistic choices for the amplitude and duration of the spine-head pulse. For large values of τ_d the speed of the wave approaches a constant value (i.e., the speed of the wave becomes insensitive to the precise choice of τ_d).

3.1.2 Linear Stability Analysis

For the purposes of linear stability analysis it is more convenient to work in terms of the original variables (x,t) rather than in the moving frame. Since the shape of the traveling pulse is fixed by the function $\eta(t)$, it is natural to consider local perturbations of the firing times given by $T(x) = x/c + \Delta(x)$. A similar approach is used to analyze the stability of traveling waves in IF networks with synaptic and passive dendritic interactions [62, 63]; see Sect. 5.4. The membrane potential $V(x,t)$ satisfying (3.1) with $a(t) = \eta(t - T(x))$ can be evaluated in terms of the Green's function G_0 for the infinite cable equation. That is, setting $C_m = 1$ and taking $\tau_m \gg \rho_0/r_s$,

$$V(x,t) = \frac{\rho_0}{r_s}\int_{-\infty}^{t}\left[\int_{-\infty}^{\infty} G_0(x-y,t-s)\eta(s-T(y))dy\right]ds \tag{3.15}$$

with G_0 given by (1.59). If we now demand that the IF process (3.2) driven by the potential $V(x,t)$ reaches threshold at time $T(x)$, then we obtain the following self-consistency condition for a traveling pulse (with $\hat{C}_s = 1$):

$$h = U(x,T(x)) = \frac{1}{r_s}\int_{-\infty}^{0} e^{t/\hat{\tau}}V(x,t+T(x))dt. \tag{3.16}$$

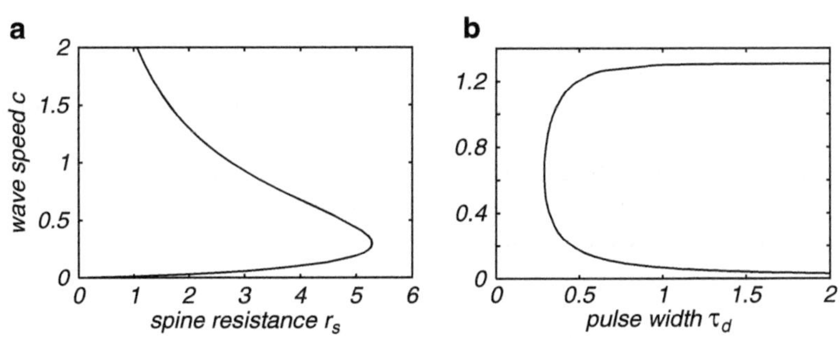

Fig. 3.4 (**a**) Speed of a traveling pulse as a function of spine-stem resistance r_s with $\rho_0 = 25$ and $\tau_d = 2$. (**b**) Speed of a traveling pulse as a function of pulse width τ_d with $\rho_0 = 25$ and $r_s = 2$. All other parameters are as Fig. 3.3. Note that for sufficiently large r_s or small τ_d solitary waves cannot propagate

We now expand (3.15) and (3.16) in powers of the perturbation $\Delta(x)$. The zeroth-order term generates the self-consistency condition for the speed c of the unperturbed traveling pulse:

$$h = \frac{1}{r_s} \int_{-\infty}^{0} e^{t/\hat{\tau}} \overline{V}(x, t + x/c) dt, \tag{3.17}$$

where

$$\overline{V}(x,t) = \frac{\rho_0}{r_s} \int_{-\infty}^{t} \left[\int_{-\infty}^{\infty} G_0(x-y, t-s) \eta(s - y/c) dy \right] ds. \tag{3.18}$$

We can evaluate (3.18) using Fourier transforms without needing to restrict η to be a rectangular pulse. That is, expand $\eta(s)$ as

$$\eta(t) = \int_{-\infty}^{\infty} e^{i\omega t} \tilde{\eta}(\omega) \frac{d\omega}{2\pi}, \tag{3.19}$$

and then perform the integrations over y and s with

$$\int_{-\infty}^{\infty} e^{-ikx} G_0(x,t) dx = e^{-\varepsilon(k)t}, \quad \varepsilon(k) = \tau_m^{-1} + D_m k^2, \tag{3.20}$$

to obtain

$$\overline{V}(x,t) = \frac{\rho_0}{r_s} \int_{-\infty}^{\infty} e^{i\omega(t - x/c)} \frac{\tilde{\eta}(\omega)}{\varepsilon(\omega/c) + i\omega} \frac{d\omega}{2\pi}. \tag{3.21}$$

Substitution of (3.21) into (3.17) finally gives

$$h = \frac{\rho_0}{r_s^2} \int_{-\infty}^{\infty} \frac{\tilde{\eta}(\omega)}{[\varepsilon(\omega/c) + i\omega][\hat{\tau}^{-1} + i\omega]} \frac{d\omega}{2\pi}. \tag{3.22}$$

When one considers a rectangular pulse shape for the action potential waveform of height η_0 and duration τ_d such that

$$\tilde{\eta}(\omega) = \eta_0 \frac{1 - e^{-i\omega\tau_d}}{i\omega}, \tag{3.23}$$

then it is a simple matter to check that the dispersion relationship (3.22) reduces to (3.14).

The first-order term in the expansion of (3.16) yields a linear equation for the perturbations $\Delta(x)$ from which the linear stability of the traveling pulse can be deduced. This linear equation takes the form

$$0 = \frac{1}{r_s} \int_{-\infty}^{0} e^{t/\hat{\tau}} \delta_\Delta V(x, t + x/c) dt, \tag{3.24}$$

where (after integration by parts)

$$\delta_\Delta V(x,t) = \frac{\rho_0}{r_s} \int_{-\infty}^{t} \left[\int_{-\infty}^{\infty} G_0(x-y, t-s) \eta'(s - y/c) [\Delta(x) - \Delta(y)] dy \right] ds. \tag{3.25}$$

Equations (3.24) and (3.25) have solutions of the form $\Delta(x) = e^{\lambda x}$ with λ satisfying the characteristic equation

$$I(\lambda) \equiv \int_{-\infty}^{\infty} \frac{ik\tilde{\eta}(\omega)}{[\varepsilon(\omega/c+i\lambda)+i\omega][\hat{\tau}^{-1}+i\omega]} \frac{d\omega}{2\pi} - I(0) = 0. \tag{3.26}$$

Asymptotic stability holds if all nonzero solutions of the characteristic equation have negative real part. (The existence of a solution $\lambda = 0$ reflects the translation invariance of the underlying system; see Sect. 2.4.) Equation (3.26) can be evaluated by closing the contour in the lower-half complex ω-plane. Since $\eta(s) = 0$ for $s < 0$ it follows that any poles of $\tilde{\eta}(\omega)$ lie in the upper-half complex plane so that we only have to consider poles arising from the zeros of the function $\varepsilon(\omega/c+i\lambda)+i\omega$. The latter are given explicitly by $\omega = i\omega_{\pm}(\lambda)$ where

$$\frac{\omega_{\pm}(\lambda)}{c} = -\left(\lambda + \frac{c}{2D_m}\right) \pm \sqrt{R(\lambda)}, \tag{3.27}$$

where

$$R(\lambda) = \frac{c^2}{4D_m^2} + c\lambda/D_m + 1/(D_m\tau_m). \tag{3.28}$$

Let us decompose λ into real and imaginary parts according to $\lambda = \alpha + i\beta$. Then

$$\omega_{\pm}(\lambda) = -u_{\pm}(\alpha,\beta) - iv_{\pm}(\alpha,\beta), \tag{3.29}$$

with

$$\frac{u_{\pm}(\alpha,\beta)}{c} = \alpha + \frac{c}{2D_m} \mp A(\alpha,\beta), \quad \frac{v_{\pm}(\alpha,\beta)}{c} = \beta \mp B(\alpha,\beta), \tag{3.30}$$

and (for $\beta > 0$)

$$A(\alpha,\beta) = \sqrt{\frac{1}{2}\left[R(\alpha) + \sqrt{R(\alpha)^2 + c^2\beta^2/D_m^2}\right]},$$

$$B(\alpha,\beta) = \sqrt{\frac{1}{2}\left[-R(\alpha) + \sqrt{R(\alpha)^2 + c^2\beta^2/D_m^2}\right]}. \tag{3.31}$$

One may now determine the linear stability of a solitary pulse by simultaneously solving $\mathrm{Re}[I(\lambda)] = \mathrm{Re}[I(0)] = 0$ and $\mathrm{Im}[I(\lambda)] = \mathrm{Im}[I(0)] = 0$ for α and β, with c determined by (3.22) (or equivalently (3.14) for the specific case of a rectangular pulse shape). This is illustrated in Fig. 3.5 for a rectangular pulse, which shows solutions with $\beta = 0$ and $\alpha < 0$ along the upper branch of Fig. 3.3. Moreover, α changes sign as it passes through the point where $d\rho_0/dc = 0$ in the (c, ρ_0) plane while moving from the upper solution branch to the lower. Hence, of the two possible traveling wave solutions, the faster one is stable. Other solutions with $\alpha < 0$ and $\beta > 0$ are also found for both the fast and slow branches but do not affect the above stability argument.

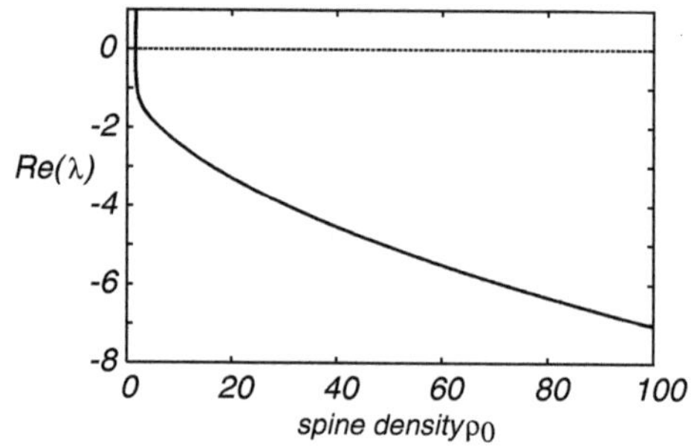

Fig. 3.5 A plot of the eigenvalues arising in the linearization of the SDS model shows that solutions with $\alpha < 0$ and $\beta = 0$ can be found for the branch of solutions with greatest speed. The graph above shows the behavior of $\alpha = \mathrm{Re}(\lambda)$ with $\beta = 0$ for the upper solution branch of Fig. 3.3. Note that $\alpha = 0$ at the point in the (c, ρ) plane at which the speed of the two possible traveling waves becomes equal. For the slower branch one finds that there exists solutions in the (α, β) plane with $\alpha > 0$ indicating that the slower branch is unstable

3.1.3 Discrete Distribution of Spines

So far we have assumed that the spine density is uniform. Although wave propagation failure is known to occur if the spine density is below some critical level, the numerical studies of Baer and Rinzel [19] suggest that propagation may be recovered by redistributing the spines into equally spaced dense clusters. Since interspine distances are of the order of μm and electronic length λ_m is typically measured in mm, we shall consider spine-head voltage at a cluster site to be the local spatial average of membrane potential in adjacent spines. Hence, we consider a discrete distribution of spines for which

$$\rho(x) = \bar{n} \sum_m \delta(x - x_m), \tag{3.32}$$

where x_m is the location of the mth spine cluster and $\bar{n}$ is the number of spines in a cluster. Such a distribution breaks continuous translation symmetry so that saltatory or lurching waves are expected rather than traveling waves of constant profile. (Saltatory waves are also found in myelinated axons; see Sect. 2.5.) We define a saltatory wave as an ordered sequence of firing times $\ldots T_{m-1} < T_m < T_{m+1}$ in which each spine cluster only fires once. Substituting the discrete density (3.32) into (3.1) with

$$\rho(x)a(x,t) = \bar{n} \sum_m \eta(t - T_m)\delta(x - x_m),$$

and making the approximation $\overline{n} \ll r_s$ so that the term $\rho(x)V/r_s$ can be neglected, we obtain the equation (using physical units)

$$\frac{\partial V}{\partial t} = -\frac{V}{\tau_m} + D_m \frac{\partial^2 V}{\partial x^2} + \frac{\bar{n}}{r_s} \sum_n \delta(x - x_n)\eta(t - T_n), \tag{3.33}$$

where we have absorbed a factor of r_m/τ_m into η and $D_m = \lambda_m^2/\tau_m$. This can be solved using the Green's functions G_0 of the cable equation:

$$V(x,t) = \frac{\overline{n}}{r_s} \sum_{m=0}^{\infty} H(x - x_m, t - T_m), \tag{3.34}$$

where we assumed the initial condition $V(x,0) = 0$ for all x.

$$H(x,t) = \int_0^t G_0(x,t-s)\eta(s)ds, \quad G_0(x,t) = \frac{1}{\sqrt{4\pi D_m t}} \mathrm{e}^{-t/\tau_m - x^2/4D_m t}. \tag{3.35}$$

Suppose that the spine clusters are uniformly distributed along the cable such that $x_m = md$, where d is the spacing between clusters. We will derive a condition for the existence of a saltatory wave solution given by $T_m = m\Delta$ for large m. The parameter Δ measures the time between successive threshold crossings at adjacent spine-head clusters such that the speed $c = d/\Delta$. First, using the causality condition $H(x,t) = 0$ for $t < 0$, it follows from (3.34) that

$$V(Nd, N\Delta) = \frac{\overline{n}}{r_s} \sum_{n=1}^{N} H(nd, n\Delta), \tag{3.36}$$

The wave speed of a saltatory wave (if it exists) is then determined self-consistently from the threshold condition

$$h = \lim_{N \to \infty} V(Nd, N\Delta) = \frac{\overline{n}}{r_s} \sum_{n=1}^{\infty} H(nd, n\Delta). \tag{3.37}$$

In order to calculate the wave speed, it is useful to rewrite (3.35) in the form

$$H(x,t) = \int_{-\infty}^{\infty} \frac{dk}{2\pi} \mathrm{e}^{ikx} \mathrm{e}^{-\varepsilon(k)t} \eta(k,t), \tag{3.38}$$

where $\eta(k,t) = \int_0^t \eta(s) \mathrm{e}^{\varepsilon(k)s} ds$. One may then exploit the convolution structure of (3.38) to evaluate it in closed form for a given $\eta(t)$. For the sake of illustration, consider again the rectangular pulse shape η, for which

$$\eta(k,t) = \eta_0 (\mathrm{e}^{\varepsilon(k)\min(t,\tau_d)} - 1)/\varepsilon(k),$$

so that $H(x,t) = \eta_0[A(x, t - \min(t, \tau_d)) - A(x,t)]$, with

$$A(x,t) = \int_{-\infty}^{\infty} \frac{dk}{2\pi} \frac{\mathrm{e}^{ikx - \varepsilon(k)t}}{\varepsilon(k)}. \tag{3.39}$$

This is a standard integral given explicitly by

$$A(x,t) = \frac{1}{4}\sqrt{\frac{\tau_m}{D_m}}\left\{ e^{-|x|/\sqrt{\tau_m D_m}} \mathrm{erfc}\left(-\frac{|x|}{\sqrt{4D_m t}} + \sqrt{\frac{t}{\tau_m}} \right) \right.$$
$$\left. + e^{|x|/\sqrt{\tau_m D_m}} \mathrm{erfc}\left(\frac{|x|}{\sqrt{4D_m t}} + \sqrt{\frac{t}{\tau_m}} \right) \right\}. \tag{3.40}$$

Finally, we may write the threshold condition in the form

$$h = \frac{\overline{n}}{r_s^2} \sum_{m=1}^{\infty} \hat{H}(md, m\Delta), \tag{3.41}$$

where $\hat{H}(x,t) = \eta_0[\hat{A}(x,t-\tau_d) - \hat{A}(x,t)]$, and

$$\hat{A}(x,t) = \int_{-\infty}^{0} e^{s/\hat{\tau}} A(x,t-s)ds. \tag{3.42}$$

The sum in (3.41) can then be performed numerically to obtain the speed of a saltatory wave $c = c(d,h)$.

In Fig. 3.6(a) we plot the speed c as a function of cluster spacing d for fixed threshold h, which shows that if the spine clusters are separated beyond some critical spacing, on the order of the electronic length $\lambda_m = 1$, a saltatory pulse will fail to propagate. Again linear stability analysis establishes that, as in the continuum model, it is the faster of the two branches that is stable. It is also instructive to consider the region in the (d,h) parameter plane where saltatory pulses exist. This may be obtained by continuing the limit point defining propagation failure of a saltatory pulse in the (d,c) plane as a function of h. The resulting phase diagram is shown in Fig. 3.6(b) and establishes that with increasing d the critical threshold for propagation failure decreases. Interestingly, the minimum wave speed of a stable saltatory pulse is found to be relatively insensitive to variation in cluster spacing d and threshold h. Now that the speed of a saltatory pulse has been determined as a function of system parameters it is possible to close the expression for the shape of a solitary pulse given by $V(x,t) = (\overline{n}/r_s)\sum_m H(x - md, t - md/v)$. A plot of this analytical expression is shown in Fig. 3.7, which clearly demonstrates that the saltatory pulse has a nonconstant profile.

Finally, note that there have been a number of recent extensions of the spike–diffuse–spike model. These include taking into account the branching structure of the dendritic tree [625], in which the Green's function G_0 is replaced by the Green's function of the tree, and incorporating active processes within the dendritic cable. In the latter case, each infinitesimal cable compartment is modeled as a quasi-linear LRC circuit, where the membrane resistance r_m is in series with an inductance L [624]; see Fig. 1.11. It should also be pointed out that from a mathematical perspective, the spike–diffuse–spike model is very similar in structure to the fire–diffuse–fire model of calcium puffs; see Sect. 4.3.

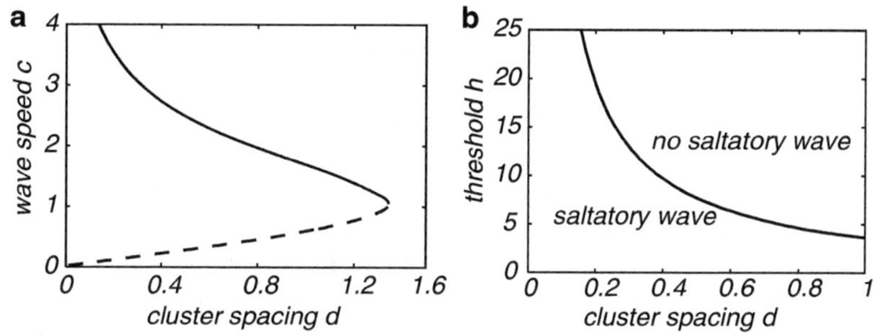

Fig. 3.6 (**a**) Plot of wave speed c for a saltatory pulse as a function of cluster spacing d. Here, $\hat{\tau} = \tau_d = 1$, $h = 2.5$, $r_s = 2$, $\eta_0 = 100$, and $\bar{n} = 2$. The upper (lower) branch is stable (unstable). (**b**) Continuation of the limit point in (**a**) showing the region in the (d,h)-plane where stable saltatory traveling waves exist

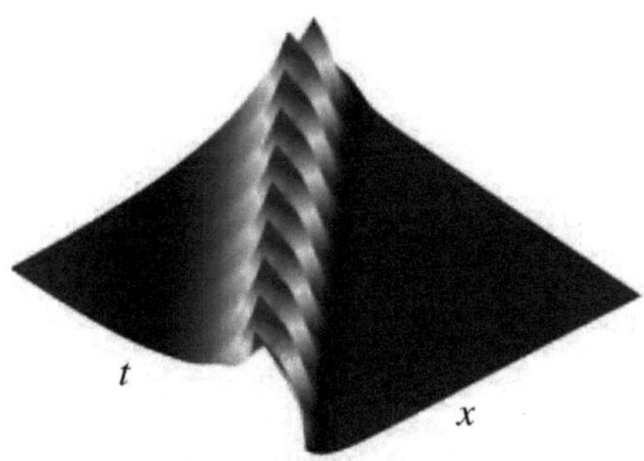

Fig. 3.7 Plot of the analytically obtained saltatory solution $V(x,t)$ in the dendritic cable with parameters as in Fig. 3.6 and $d = 1$. The x axis covers 10 lattice sites and the t axis $10d/c$

3.2 Reaction–Diffusion Model of CaMKII Translocation Waves

As we indicated at the beginning of this chapter, CaMKII (Ca^{2+}–calmodulin-dependent protein kinase II) is a key regulator of glutamatergic synapses and plays an essential role in many forms of synaptic plasticity. It has recently been observed experimentally that chemically stimulating a local region of dendrite not only induces the local translocation of CaMKII from the dendritic shaft to synaptic targets within spines, but also initiates a wave of CaMKII translocation that spreads distally through the dendrite with an average speed of order 1 μm/s [532]. In Fig. 3.8, we provide a cartoon of the mechanism for translocation waves hypothesized by Rose et al. [532]. Before local stimulation using a glutamate/glycine puff, the majority of CaMKII is in an inactive state and distributed uniformly throughout the dendrite. Upon stimulation, all CaMKII in the region of the puff (∼30 μm of dendrite) is converted to an active state, probably the autonomous state of CaMKII (see Fig. 3.8a), and begins translocating into spines. Simultaneously, a Ca^{2+} spike is initiated and rapidly travels the length of the dendrite (as modeled in Sect. 3.1), causing CaMKII to bind Ca^{2+}/CaM along the way. In this primed or partially phosphorylated state, CaMKII does not yet translocate into spines. In the meantime, a portion of the activated CaMKII from the stimulated region diffuses into the region of primed CaMKII and the two types interact, with the result that primed CaMKII is activated. Some of these newly activated holoenzymes translocate into spines while others diffuse into more distal regions of the dendrite containing primed CaMKII, and the wave proceeds in this fashion. In certain cases one also finds a second wave propagating proximally from the stimulated site to the soma [532]. A schematic diagram illustrating the progression of a translocation wave along a dendrite following the rapid priming phase is shown in Fig. 3.8b.

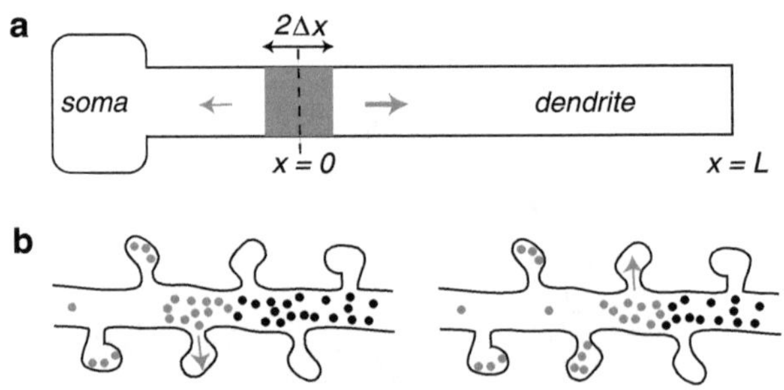

Fig. 3.8 Proposed mechanism of CaMKII translocation waves. (**a**) A glutamate/glycine puff activates CaMKII locally and initiates a fast Ca^{2+} spike that propagates distally (*indicated by larger horizontal arrow*) and primes CaMKII in the remainder of the dendrite. In certain cases one also finds a second wave propagating proximally from the stimulated site to the soma (*indicated by smaller horizontal arrow*). (**b**) Activated CaMKII (*gray dots*) both translocates into spines and diffuses into distal regions of the dendrite where it activates primed CaMKII (*black dots*). The net effect is a wave of translocated CaMKII propagating along the dendrite

A simple mathematical model of the above mechanism can be constructed using a system of reaction–diffusion equations for the concentrations of activated and primed CaMKII in the dendrite and spines [72, 161]. These equations incorporate three major components of the dynamics: *diffusion* of CaMKII along the dendrite, *activation* of primed CaMKII, and *translocation* of activated CaMKII from the dendrite to spines. For simplicity, consider a uniform one-dimensional, nonbranching dendritic cable as shown in Fig. 3.8a. Suppose that a region of width 30 μm is stimulated with a glutamate/glycine puff at time $t = 0$. The center of the stimulated region is taken to be at $x = 0$ and the distal end of the dendrite is at $x = L = 150$ μm. The diffusion, activation, and translocation of CaMKII along the dendrite following stimulation are modeled according to the following system of equations:

$$\frac{\partial P}{\partial t} = D\frac{\partial^2 P}{\partial x^2} - k_0 AP \tag{3.43a}$$

$$\frac{\partial A}{\partial t} = D\frac{\partial^2 A}{\partial x^2} + k_0 AP - hA \tag{3.43b}$$

$$\frac{\partial S}{\partial t} = hA, \tag{3.43c}$$

where D is the diffusivity of CaMKII within the cytosol. Here $P(x,t)$ and $A(x,t)$ denote the concentration of primed and activated CaMKII at time $t > 0$ and location x along the dendrite. $S(x,t)$ denotes the corresponding concentration of CaMKII in the population of spines at the same time and distance. For simplicity, all parameters are constant in space and time. The reaction term kAP represents the conversion of CaMKII from its primed to active state based on the irreversible first-order reaction scheme

$$A + P \to 2A$$

with mass action kinetics, where k_0 is the rate at which primed CaMKII is activated per unit concentration of activated CaMKII. The decay term hA represents the loss of activated CaMKII from the dendrite due to translocation into a uniform distribution of spines at a rate h. The model assumes that translocation is irreversible over the time scale of simulations, which is reasonable given that activated CaMKII accumulation at synapses can persist for several minutes [569].

As a further simplification we will only consider the distal transport of CaMKII from the stimulated region by taking $0 \leq x \leq L$ and imposing closed or reflecting boundary conditions at the ends $x = 0, L$. Hence, no CaMKII can escape from the ends. In reality activated CaMKII could also diffuse in the proximal direction and trigger a second proximal translocation wave. However, the choice of boundary condition has little effect on the properties of the wave. Taking the distal half of the stimulated region to be $0 \leq x \leq 15\,\mu$m, consider the following initial conditions: $P(x,0) = 0$ and $A(x,0) = P_0$ for $0 \leq x \leq 15\,\mu$m, whereas $P(x,0) = P_0$ and $A(x,0) = 0$ for $x \geq 15\,\mu$m, where P_0 is the uniform resting concentration of CaMKII in the dendrite. Typical values of C range from 0.1 to 30μM [605], covering two orders of magnitude. We also set $S(x,0) = 0$ everywhere. In other words, we assume that all the CaMKII is activated within the stimulated region at $t = 0$, but none has yet translocated into spines nor diffused into the nonstimulated region. We also neglect any delays associated with priming CaMKII along the dendrite. This is a reasonable approximation, since the Ca^{2+} spike travels much faster than the CaMKII translocation wave [532]; see Sect. 3.1. Thus by the time a significant amount of activated CaMKII has diffused into nonstimulated regions of the dendrite, any CaMKII encountered there will already be primed. The benefit of this assumption is that it eliminates the need to model the Ca^{2+} spike. However, a more detailed model that takes into account the initial transient associated with the priming phase could be constructed by coupling the reaction–diffusion equations with the spike–diffuse–spike model of Sect. 3.1.

Note that the system of equations (3.43) is identical in form to the diffusive SI model introduced by Noble [464] to explain the spread of bubonic plague through Europe in the fourteenth century. In the latter model, $P(x,t)$ and $A(x,t)$ would represent the densities of susceptible and infective people at spatial location x at time t, respectively; k_0 would be the transmission rate and h the death rate. In the absence of translocation into spines ($h = 0$), the total amount of CaMKII is conserved so that $A(x,t) + P(x,t) = P_0$ for all x and $t \geq 0$. Equations (3.43) then reduce to the scalar Fisher–KPP equation

$$\frac{\partial A}{\partial t} = D\frac{\partial^2 A}{\partial x^2} + k_0 A(P_0 - A), \tag{3.44}$$

which was originally introduced to model the invasion of a gene into a population. The Fisher–KPP equation and its generalizations have been widely used to describe the spatial spread of invading species including plants, insects, genes, and diseases; see, for example, [285, 444, 575, 648] and references therein. One characteristic feature of such equations is that they support traveling fronts propagating into an unstable steady state, in which the wave speed and longtime asymptotics are determined by the dynamics in the leading edge of the wave—so-called pulled fronts [544]. In particular, a sufficiently localized initial perturbation (such as the stimulus used to generate CaMKII waves) will asymptotically approach the traveling front solution that has the minimum possible wave speed. (If we perform the change of variables $Q = P_0 - P$ in the CaMKII model, then the traveling wave solution constructed below propagates into the unstable state $A = 0, Q = 0$.) An overview of the theory of pulled fronts is presented in appendix section 3.3.

3.2.1 Translocation Waves for a Uniform Distribution of Spines

A traveling wave solution of (3.43a) and (3.43b) is $P(x,t) = P(\xi)$ and $A(x,t) = A(\xi)$, $\xi = x - ct$, where $c, c > 0$, is the wave speed, such that

$$P(\xi) \to P_0, \quad A(\xi) \to 0 \quad \text{as}\, \xi \to \infty$$

and

$$P(\xi) \to P_1 < P_0, \quad A(\xi) \to 0 \quad \text{as}\, \xi \to -\infty.$$

Here P_1 is the residual concentration of primed CaMKII following translocation of activated CaMKII into spines. The minimum wave speed can be calculated by substituting the traveling wave solution into (3.43a) and (3.43b) and linearizing near the leading edge of the wave where $P \to P_0$ and $A \to 0$. In the traveling wave coordinate frame (3.43a) and (3.43b) are transformed to

$$-c\frac{dP}{d\xi} = D\frac{d^2P}{d\xi^2} - k_0 AP \tag{3.45a}$$

$$-c\frac{dA}{d\xi} = D\frac{d^2A}{d\xi^2} + k_0 AP - hA \tag{3.45b}$$

This is a system of two second-order ordinary differential equations in the variable ξ. A global view of the nature of traveling wave solutions can be obtained by identifying (3.45a) and (3.45b) with the equation of motion of a classical particle in two spatial dimensions undergoing damping due to "friction" and subject to an "external force." Thus we identify A and P with the "spatial" coordinates of the particle, ξ with the corresponding "time" coordinate, and the speed c as a "friction coefficient." If we ignore boundary effects by taking $-\infty < \xi < \infty$, then we can view a traveling wave solution as a particle trajectory that connects the point $(P,A) = (0,0)$ at $\xi = -\infty$ to the point $(P,A) = (P_0,0)$ at $\xi = \infty$. A restriction on the allowed values of c can now be obtained by investigating how the point $(1,0)$ is approached in the large-ξ limit.

Linearizing Eqs. (3.45a) and (3.45b) about the point $(P,A) = (P_0,0)$ we obtain a pair of second-order linear equations, which have solutions of the form $(P-P_0,A) = \mathbf{V}e^{-\lambda\xi}$ where λ and $\mathbf{V}$ satisfy the matrix equation

$$c\lambda\mathbf{V} = \begin{pmatrix} D\lambda^2 & -k \\ 0 & D\lambda^2+k-h \end{pmatrix}\mathbf{V}, \tag{3.46}$$

where $k = k_0P_0$. Solving for the eigenvalue λ leads to the four solutions

$$\lambda = 0, \quad \frac{c}{D}, \quad \frac{c \pm \sqrt{c^2-4D(k-h)}}{2D} \tag{3.47}$$

and these, along with their corresponding eigenvectors $\mathbf{V}$, determine the shape of the wave as it approaches the point $(1,0)$. Note that the last two eigenvalues have a nonzero imaginary part when $c^2 < 4D(k-h)$, implying that as ξ becomes large the wave oscillates about the point $(1,0)$. This cannot be allowed since it would imply that the activated CaMKII concentration A takes on negative values (inspection of the corresponding eigenvectors shows that their components in the A-direction are nonzero and so A would indeed oscillate). Therefore, we must have

$$c \geq c_{\min} = 2\sqrt{D(k-h)}, \tag{3.48}$$

which implies that $k > h$. Note that the minimum wave speed can be identified with the linear spreading velocity of a pulled front; see appendix section 3.3. This then yields a more direct method for obtaining the minimum wave speed. That is, the characteristic equation obtained from (3.46) yields the dispersion relation

$$c(\lambda) = D\lambda + \frac{k-h}{\lambda}. \tag{3.49}$$

The theory of pulled fronts shows that the minimum wave speed is obtained by minimizing $c(\lambda)$. The equation $c'(\lambda) = 0$ gives $D\lambda = (k-h)/\lambda$, which has the solution $\lambda_0 = \sqrt{(k-h)/D}$, so that $c_{\min} = c(\lambda_0) = 2D\lambda_0 = 2\sqrt{D(k-h)}$.

An example of a numerically determined traveling wave solution with minimal speed $c_{\min}$ is shown in Fig. 3.9 for parameter values consistent with experimental studies of CaMKIIα, which is one of the two main isoforms of CaMKII. In its inactive state CaMKIIα tends to be located in the cytosol, whereas the other isoform, CaMKIIβ, is weakly actin bound [570]. One finds empirically that $D \sim 1\,\mu\text{m}^2/\text{s}$, $h \sim 0.05\,\text{s}^{-1}$ and $c \sim 1\,\mu\text{m/s}$ [532, 569, 570] for CaMKIIα. (CaMKIIβ has a diffusivity and translocation rate an order of magnitude smaller but exhibits compa-

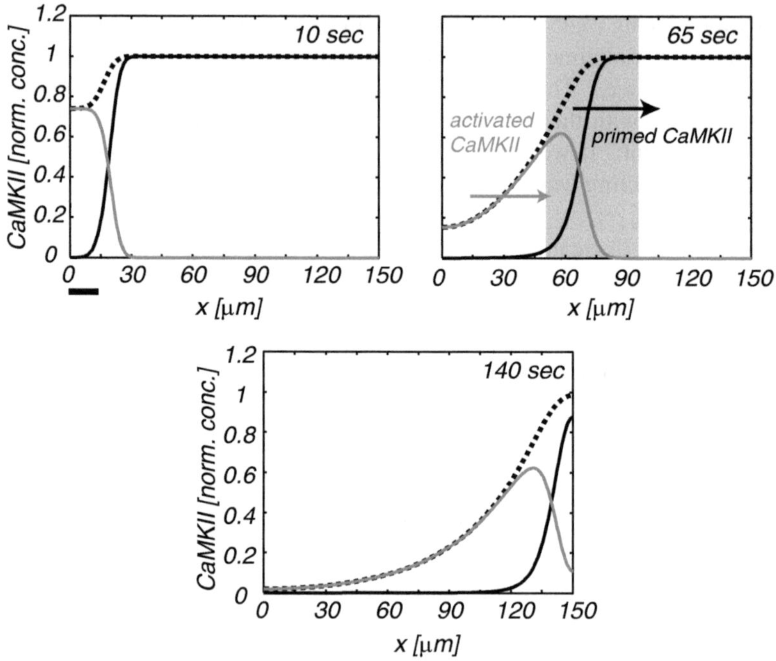

Fig. 3.9 Three successive snapshots of a numerically simulated translocation wave propagating along a homogeneous dendrite. Solutions of (3.43a) and (3.43b) are plotted for parameter values consistent with experimental data on CaMKII [532, 569, 570]. The translocation rate $h = 0.05$ s, diffusivity $D = 1\,\mu\text{m}^2$/s and the activation rate $k_0P_0 = 0.21$ s. At time $t = 0$ all of the CaMKII within the stimulated region (*indicated by thick bar*) is in the activated state, whereas all of the CaMKII within the nonstimulated region is in the primed state. Concentrations are normalized with respect to the initial concentration of primed CaMKII. Composite wave consists of a pulse of activated CaMKII (*gray curve*) moving at the same speed as a front of primed CaMKII (*black curve*). Also shown is the total CaMKII concentration along the dendrite (*dashed black curve*), which decreases with time due to translocation into spines. As indicated in the center plot, the front forms an interface between a quiescent region containing a uniform concentration of primed CaMKII and a region dominated by translocation of activated CaMKII into spines. The dynamics in the interfacial (*shaded*) region is dominated by diffusion–activation of primed CaMKII

rable wave speeds.) The formula for the wave speed then gives an estimate for the unknown activation rate, $k \sim 0.2\,\text{s}^{-1}$. It can be seen in Fig. 3.9 that the wave profile of primed CaMKII is in the form of a front, whereas the co-moving wave profile of activated CaMKII is a localized pulse.

The above analysis predicts wave propagation failure when the translocation rate h is greater than the effective activation rate k. Experimentally, h is determined by globally activating CaMKII along a dendrite and determining the rate at which the level of CaMKII decays [569, 570]. The detailed microscopic mechanism whereby CaMKII is translocated into spines is currently not known, so it is difficult to relate h to individual spine properties. A simple hypothesis is that the translocation rate depends on the spine density according to $h = \rho_0 v_0$, where v_0 is an effective "velocity" associated with translocation into an individual spine. Since the activation rate $k = k_0 P_0$, where P_0 is the initial concentration of primed CaMKII in the nonstimulated region of the dendrite, the model predicts that CaMKII translocation waves will fail to propagate when

$$\rho_0 v_0 > k_0 P_0. \tag{3.50}$$

For example, this inequality predicts that dendrites with a high density of spines are less likely to exhibit translocation waves than those with a low spine density. It also predicts that dendrites with a larger initial concentration of primed CaMKII in the shaft are more likely to exhibit translocation waves than those with a smaller initial concentration. Since the initial concentration P_0 of primed CaMKII depends on the effectiveness of the Ca^{2+} spike in both propagating along the dendrite and priming the inactive state, the model agrees with the experimental finding that translocation waves fail to propagate when L-type Ca^{2+} channels are blocked [532]. One also finds that Ca^{2+} spikes are less likely to propagate towards the soma, which could explain why translocation waves are more often observed propagating towards the distal end of a dendrite.

3.2.2 *Pulsating Waves in the Presence of Discretely Distributed Spines*

One of the major simplifications of the above model is that the discrete nature of dendritic spines is ignored by treating the spine density ρ and, hence, the translocation rate h as uniform. Given the fact that the radius of the spine neck is typically at the submicron level, which is much smaller than any other length scale of the system, one can take into account the discreteness of spines by representing the spine density as the sum of Dirac delta functions (3.32), which represents clusters of $\bar{n}$ spines at discrete locations x_n; see Sect. 3.1. It immediately follows that the translocation rate h is itself space-dependent and (3.43a) and (3.43b) become heterogeneous. For the sake of illustration, consider the case of a set of spine clusters that are uniformly spaced with $x_n = n\Delta$ where Δ is the spine cluster spacing. In order to separate the effects of discreteness from the effects of spine density, we will assume that the size of a cluster scales with Δ so that $\overline{n} = \rho_0 \Delta$ with ρ_0 fixed. Thus, setting $\overline{h} = \rho_0 v_0$, we have the space-dependent translocation rate

$$h(x) = \overline{h}\Delta \sum_{n\in\mathbb{Z}} \delta(x - n\Delta), \tag{3.51}$$

such that $L^{-1}\int_0^L h(x)dx = \overline{h}$ for $L \gg \Delta$.

In recent years, there has been an increasing interest in studying biological invasion in heterogeneous environments using reaction–diffusion equations [40, 102, 212, 335, 574, 575, 587, 664, 687]. Heterogeneity is often incorporated by assuming that the diffusion coefficient and the growth rate of a population are periodically varying functions of space. One of the simplest examples of a single population model in a periodic environment was proposed by Shigesada et al.

[574, 575], in which two different homogeneous patches are arranged alternately in one-dimensional space so that the diffusion coefficient and the growth rate are given by periodic step functions. The authors showed numerically that an invading population starting from a localized perturbation evolves to a traveling periodic wave in the form of a pulsating front. The population density $u(x,t)$ of such a wave is defined by the condition $u(x,t) = u(x+\sigma, t+T)$ such that $\lim_{x\to\infty} u(x,t) = 0$ and $\lim_{x\to-\infty} u(x,t) = p(x)$, where $p(x)$ is a spatially periodic stationary solution of the corresponding reaction–diffusion equation. This form of solution repeats itself in a time interval T if it is observed at two successive points separated by a distance σ. The speed of the wave is then taken to be $c = \sigma/T$. Shigesada et al. [574] also used linearized information within the leading edge of the pulsating front to derive wave speed estimates, generalizing the analysis of pulled fronts in homogeneous media [544]. An interesting recent extension of this approach has been used to study pulsating fronts in periodically modulated nonlocal neural field equations [132]. The theory of pulsating fronts has also been developed in a more general and rigorous setting [40, 212, 587, 664, 687].

The analysis of CaMKII translocation waves developed in [72] follows the basic formulation of Shigesada et. al. [574] by linearizing (3.43b) at the leading edge of the wave where $A(x,t) \to 0$ and $\mathscr{P}(x,t) \to P_0$:

$$\frac{\partial A}{\partial t} = D\frac{\partial^2 A}{\partial x^2} + kA - h(x)A, \tag{3.52}$$

with $h(x)$ given by the Δ-periodic function (3.51). Assume a solution of the form $A(x,t) = a(\xi)\mathscr{P}(x)$, $\xi = x - ct$, and set

$$\frac{\partial}{\partial t} \to -c\frac{\partial}{\partial \xi}, \quad \frac{\partial}{\partial x} \to \frac{\partial}{\partial x} + \frac{\partial}{\partial \xi}.$$

Substitution into (3.52) then gives

$$-c\mathscr{P}(x)a'(\xi) = D\left[a''(\xi)\mathscr{P}(x) + 2a'(\xi)\mathscr{P}'(x) + a(\xi)\mathscr{P}''(x)\right] + [k - h(x)]a(\xi)\mathscr{P}(x).$$

Dividing through by $a(\xi)\mathscr{P}(x)$ and rearranging yields

$$D\frac{a''(\xi)}{a(\xi)} + \left[2D\frac{\mathscr{P}'(x)}{\mathscr{P}(x)} + c\right]\frac{a'(\xi)}{a(\xi)} = -D\frac{\mathscr{P}''(x)}{\mathscr{P}(x)} - k + h(x). \tag{3.53}$$

Applying the operator $\partial_x \partial_\xi$ to both sides of (3.53) implies that either $\mathscr{P}'(x)/\mathscr{P}(x)$ is a constant or $a'(\xi)/a(\xi)$ is a constant. Only the latter condition is consistent with $P(x)$ being a periodic function. Thus, $a(\xi) = A_0 \mathrm{e}^{-\lambda\xi}$ with λ determined by solutions to the damped Hill equation

$$\mathscr{P}''(x) - 2\lambda\mathscr{P}'(x) + \left[\lambda^2 + \frac{k - h(x) - c\lambda}{D}\right]\mathscr{P}(x) = 0. \tag{3.54}$$

Note that if $\mathscr{P}(x) = \mathrm{e}^{\lambda x}U(x)$ then $U(x)$ satisfies the undamped Hill equation

$$DU''(x) + [k - h(x) - c\lambda]U(x) = 0. \tag{3.55}$$

In order to determine the minimal wave speed $c_{\min}$, it is necessary to find a bounded periodic solution $P(x)$ of (3.54), which yields a corresponding dispersion relation $c = c(\lambda)$, whose minimum with respect to λ can then be determined (assuming it exists). Unfortunately, for general periodic functions $h(x)$, it is not possible to solve (3.54) explicitly, and some form of approximation scheme is required as described in the next section.

3.2.3 Homogenization of Pulsating Waves for a Fast Periodic Modulation of Spine Density

Since the spine cluster spacing Δ is at least an order of magnitude smaller than the width of the traveling wave of the homogeneous system, one can use homogenization theory to approximate the discrete effects of spines by a corresponding continuum model [72]. Such a method has also been applied to studying variations in electrical voltage/conductance [424] and the distribution of protein receptors [68] along spiny dendrites. Interestingly, Smaily et al. [587] independently applied the same homogenization procedure to analyze wave speed in the population model of Shigesada et. al. [574]. A more general discussion of homogenization techniques applied to traveling fronts can be found in the review by Xin [687].

As a first step, we introduce a macroscopic length scale σ and set $\Delta = \varepsilon\sigma$ with $\varepsilon \ll 1$. We identify σ with the effective width of the primed CaMKII front, which turns out to be around 20–30 μm in the given parameter regimes. Equation (3.55) can then be rewritten in the form

$$\frac{d^2U}{dx^2} + \left[\overline{\Gamma} - \Delta\Gamma\left(\frac{x}{\varepsilon}\right)\right]U(x) = 0, \tag{3.56}$$

where

$$\overline{\Gamma} = [k - c\lambda - \overline{h}]/D \tag{3.57}$$

and

$$\Delta\Gamma(y) = \frac{\overline{h}}{D}\left(\sigma\sum_{n\in\mathbb{Z}}\delta(y - n\sigma) - 1\right) \tag{3.58}$$

such that $\Delta\Gamma(y)$ is a σ-periodic function of y. Applying the homogenization procedure outlined below leads to the result

$$U(x) = \langle U(x)\rangle\left[1 + \varepsilon^2\Delta U(x/\varepsilon)\right] + \mathcal{O}(\varepsilon^3), \tag{3.59}$$

where $\langle U \rangle$ satisfies the averaged equation

$$\frac{d^2\langle U \rangle}{dx^2} + \Gamma_\varepsilon \langle U \rangle = 0, \tag{3.60}$$

and

$$\Gamma_\varepsilon = \overline{\Gamma} - \varepsilon^2 \Gamma_2 + \mathscr{O}(\varepsilon^3), \quad \Gamma_2 = \frac{1}{12}\left(\frac{\overline{h}\sigma}{D}\right)^2, \tag{3.61}$$

$$\Delta U(y) = \frac{\overline{h}\sigma^2}{D}\left[\frac{y}{2\sigma} - \frac{y^2}{2\sigma^2} - \frac{1}{12}\right]. \tag{3.62}$$

Homogenization method for discrete spines. The basic idea of multi-scale homogenization is to expand the solution of Eq. (3.56) as a power series in ε, with each term in the expansion depending explicitly on the "slow" (macroscopic) variable x and the "fast" (microscopic) variable $y = x/\varepsilon$ [68, 424, 587]:

$$U(x,y) = U_0(x) + \varepsilon U_1(x,y) + \varepsilon^2 U_2(x,y) + \ldots, \tag{3.63}$$

where $U_j(x,y), j = 1, \ldots$ are σ-periodic in y. The perturbation series expansion is then substituted into Eq. (3.56) with x, y treated as independent variables so that derivatives with respect to x are modified according to $\partial_x \to \partial_x + \varepsilon^{-1}\partial_y$. This generates a hierarchy of equations corresponding to successive powers of ε:

$$\frac{\partial^2 U_1}{\partial y^2} = 0 \tag{3.64}$$

$$\frac{d^2 U_0}{dx^2} + 2\frac{\partial^2 U_1}{\partial x \partial y} + \frac{\partial^2 U_2}{\partial y^2} + \left[\overline{\Gamma} - \Delta\Gamma(y)\right] U_0 = 0 \tag{3.65}$$

at powers $\varepsilon^{-1}, 1$ and

$$\frac{\partial^2 U_n}{\partial x^2} + 2\frac{\partial^2 U_{n+1}}{\partial x \partial y} + \frac{\partial^2 U_{n+2}}{\partial y^2} + \left[\overline{\Gamma} - \Delta\Gamma(y)\right] U_n = 0 \tag{3.66}$$

at $\mathscr{O}(\varepsilon^n), n \geq 1$.

Equation (3.64) and boundedness of U_1 imply that U_1 is independent of y and can thus be absorbed into $U_0(x)$. Thus the leading-order corrections arising from small-scale fluctuations in the spine density occur at $\mathscr{O}(\varepsilon^2)$. Define the spatial average of a periodic function $F(y)$, denoted by $\langle F \rangle$, according to

$$\langle F \rangle = \frac{1}{\sigma}\int_0^\sigma F(y)dy. \tag{3.67}$$

Taking the spatial average of (3.65) with $U_0 = \langle U_0 \rangle$ then gives

$$\frac{d^2 U_0}{dx^2} + \overline{\Gamma} U_0 = 0. \tag{3.68}$$

We have exploited the fact that U_2 is periodic in y so $\langle \partial^2 U_2/\partial y^2 \rangle = 0$. In order to calculate U_2, we first subtract the averaged Eq. (3.68) from (3.65) to obtain

$$\frac{\partial^2 U_2}{\partial y^2} = \Delta\Gamma(y) U_0(x). \tag{3.69}$$

It follows that $U_2(x,y) = U_0(x)\chi(y)$ with $\chi''(y) = \Delta\Gamma(y)$ and χ a σ-periodic function of y. Integrating once with respect to y gives $\chi'(y) = \chi'(0) + \int_0^y \Delta\Gamma(z)dz$. We can eliminate the unknown $\chi'(0)$ by spatially averaging with respect to y and using $\langle\chi'\rangle = 0$. This gives $\chi'(y) = \oint_0^y \Delta\Gamma(z)dz$ with

$$\oint_0^y f(z)dz \equiv \int_0^y f(z)dz - \left\langle \int_0^y f(z)dz \right\rangle \tag{3.70}$$

for any integrable function f. Another integration with respect to y shows that

$$\chi(y) = \chi(0) + \int_0^y \oint_0^{y'} \Delta\Gamma(z)dzdy'.$$

Spatially averaging this equation with respect to y in order to express $\chi(0)$ in terms of $\langle\chi\rangle$ and multiplying through by $U_0(x)$ finally gives

$$\begin{aligned} \Delta U_2(x,y) &\equiv U_2(x,y) - \langle U_2\rangle(x) \\ &= U_0(x)\oint_0^y \oint_0^{y'} \Delta\Gamma(z)dzdy'. \end{aligned} \tag{3.71}$$

It remains to determine the equation satisfied by $\langle U_2\rangle$. Spatially averaging Eq. 3.66) for $n = 2$ gives

$$\frac{d^2\langle U_2\rangle}{dx^2} + \overline{\Gamma}\langle U_2\rangle = \langle \Delta\Gamma(y)U_2(x,y)\rangle. \tag{3.72}$$

Substituting (3.71) into (3.72) and reordering the resulting multiple integral yields the result

$$\frac{d^2\langle U_2\rangle}{dx^2} + \overline{\Gamma}\langle U_2\rangle = -\left\langle \left(\oint_0^y \Delta\Gamma(z)dz\right)^2 \right\rangle U_0(x). \tag{3.73}$$

Finally, writing $\langle U\rangle = U_0 + \varepsilon^2\langle U_2\rangle + \ldots$ we obtain the homogenized version of the Hill equation (3.55):

$$\frac{d^2\langle U\rangle}{dx^2} + \Gamma_\varepsilon\langle U\rangle = 0, \tag{3.74}$$

where

$$\Gamma_\varepsilon = \overline{\Gamma} + \varepsilon^2\Gamma_2 + \mathcal{O}(\varepsilon^3), \quad \Gamma_2 = \left\langle \left(\oint_0^y \Delta\Gamma(z)dz\right)^2 \right\rangle. \tag{3.75}$$

It is straightforward to calculate the integrals in (3.71) and (3.75) for a periodic spine density with $\Delta\Gamma(y)$ given by (3.58) [424]:

$$\Gamma_2 = \left\langle \left(\oint_0^y \Delta\Gamma(z)dz\right)^2 \right\rangle = \frac{1}{12}\left(\frac{\overline{h}\sigma}{D}\right)^2, \tag{3.76}$$

$$\oint_0^y \oint_0^{y'} \Delta\Gamma(z)dzdy' = \frac{\overline{h}\sigma^2}{D}\left[\frac{y}{2\sigma} - \frac{y^2}{2\sigma^2} - \frac{1}{12}\right]. \tag{3.77}$$

We thus obtain (3.59) and (3.60).

Note that it is possible to extend the above homogenization scheme to the case of randomly rather than periodically distributed spines, provided that the resulting heterogeneous medium is *ergodic*. That is, the result of averaging over all realizations of the ensemble of spine distributions is equivalent to averaging over the length L of the dendrite in the infinite-

L limit. If such an ergodic hypothesis holds and L is sufficiently large so that boundary terms can be neglected, then the above analysis carries over with $\langle \cdot \rangle$ now denoting ensemble averaging [424].

Recall from our discussion of (3.54) that the solution $\mathscr{P}(x) = e^{\lambda x} U(x)$ has to be a bounded Δ-periodic function of x. It follows from (3.59) and (3.60) that $e^{\lambda x}\langle U(x)\rangle$ should be a finite constant. Writing the solution of (3.60) as $\langle U(x)\rangle \sim e^{-\sqrt{-\Gamma_\varepsilon} x}$, yields the characteristic equation

$$\lambda = \sqrt{\frac{c\lambda - k + \overline{h}}{D} - \varepsilon^2 \Gamma_2}, \tag{3.78}$$

where we have substituted for $\overline{\Gamma}$ using (3.57). Squaring both sides and rearranging thus leads to the following dispersion relation for the wave speed c:

$$c = c(\lambda) \equiv D\lambda + \frac{k - \overline{h} + \varepsilon^2 D\Gamma_2}{\lambda}. \tag{3.79}$$

Minimizing with respect λ then shows that

$$c_{\min} = 2\sqrt{D(k - \overline{h}) + \varepsilon^2 D^2 \Gamma_2}, \tag{3.80}$$

For sufficiently small ε, we can Taylor expand (3.80) to obtain

$$c_{\min} \approx \overline{c} + \frac{2D^2\Gamma_2}{\overline{c}}\varepsilon^2, \tag{3.81}$$

with $\overline{c} = 2\sqrt{D(k - \overline{h})}$ the wave speed of the corresponding homogeneous distribution of spines. Hence, a periodic variation in the spine density due to clustering leads to an $\mathscr{O}(\varepsilon^2)$ increase in the wave speed. An analogous result was obtained by Smaily et. al. [587] for the Shigesada et. al. model [574]. Note that (3.59) also implies that there are small-scale fluctuations of the wave profile in the leading edge of the wave

$$\frac{\Delta P(x/\varepsilon)}{\langle P \rangle} = \varepsilon^2 \Delta U(x/\varepsilon) + \mathscr{O}(\varepsilon^3). \tag{3.82}$$

Since $\varepsilon = \Delta/\sigma$, it follows from (3.62) that fluctuations in the wave profile vary between $-\overline{h}\Delta^2/(12D)$ at spine clusters and $\overline{h}\Delta^2/(24D)$ between spine clusters. In terms of physical parameters, the minimum wave speed is

$$c_{\min} = 2\sqrt{D(k - \overline{h}) + \Delta^2 \overline{h}^2/12}. \tag{3.83}$$

It immediately follows that for fixed $\overline{h}, k, D$ (and hence $\overline{c}$), spine clustering increases the speed of a translocation wave. This is illustrated in Fig. 3.10, where we plot the minimal wave speed $c_{\min}$ given by (3.83) as a function of the activation rate k for various values of the cluster spacing Δ. An additional important consequence of clustering is that it reduces the threshold for the existence of a translocation wave.

That is, there exists a critical value of the activation rate, $k = \kappa(\Delta)$, below which translocation waves do not exist and $\kappa(\Delta)$ is a decreasing function of Δ. In the homogenization limit $\Delta \to 0$, we recover the result $\kappa = \overline{h}$.

The existence of a pulsating wave due to spine clustering and the associated increase (decrease) in the minimal speed (threshold) of the wave can also be confirmed numerically. For the sake of illustration, consider a dendrite of length $L = 300\,\mu\text{m}$ with reflecting boundary conditions at both ends $x = 0, L$. The initial conditions are taken to be

$$\begin{aligned} P(x,0) &= P_0, A(x,0) = 0 \text{ for all } x \notin [0, \delta L] \\ P(x,0) &= 0, A(x,0) = P_0 \text{ for all } x \in [0, \delta L], \end{aligned}$$

with $\delta L = 15\,\mu\text{m}$. We discretize space by setting $x = m\delta x$, where δx is the step length and $m = 0, 1, \ldots, M$ with $M = L/\delta x$. In discrete spatial units the spine cluster spacing is taken to be $\Delta = P\delta x$. The spine cluster distribution is then represented numerically by the discrete sum

$$\rho(m\delta x) = \frac{1}{\delta x} \sum_{j=0}^{M/P} \delta_{m,jP},$$

where $\delta_{m,j}$ is the Kronecker delta and δx is chosen so that M, P and M/P are integers. An example of a pulsating wave is shown in Fig. 3.11. Comparison between waves for a spatially uniform distribution of spines and a spatially discrete distribution of spine clusters shows that the wave is periodically modulated and faster in the latter case. This is a consequence of the fact that translocation is less effective

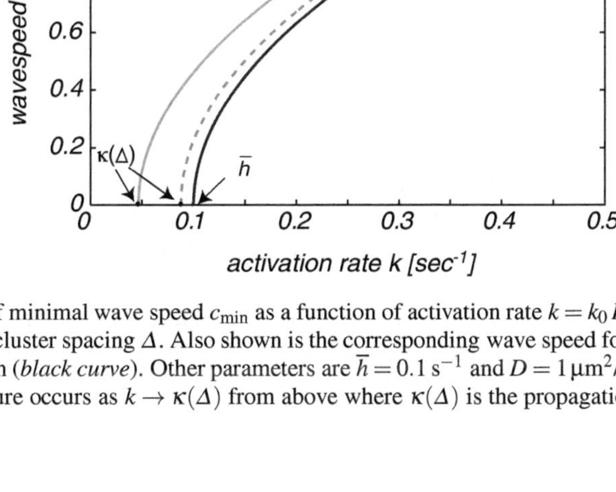

Fig. 3.10 Plot of minimal wave speed $c_{\min}$ as a function of activation rate $k = k_0 P_0$ for various values of the spine cluster spacing Δ. Also shown is the corresponding wave speed for a homogeneous spine distribution (*black curve*). Other parameters are $\overline{h} = 0.1\,\text{s}^{-1}$ and $D = 1\,\mu\text{m}^2/\text{s}$. Note that wave propagation failure occurs as $k \to \kappa(\Delta)$ from above where $\kappa(\Delta)$ is the propagation threshold

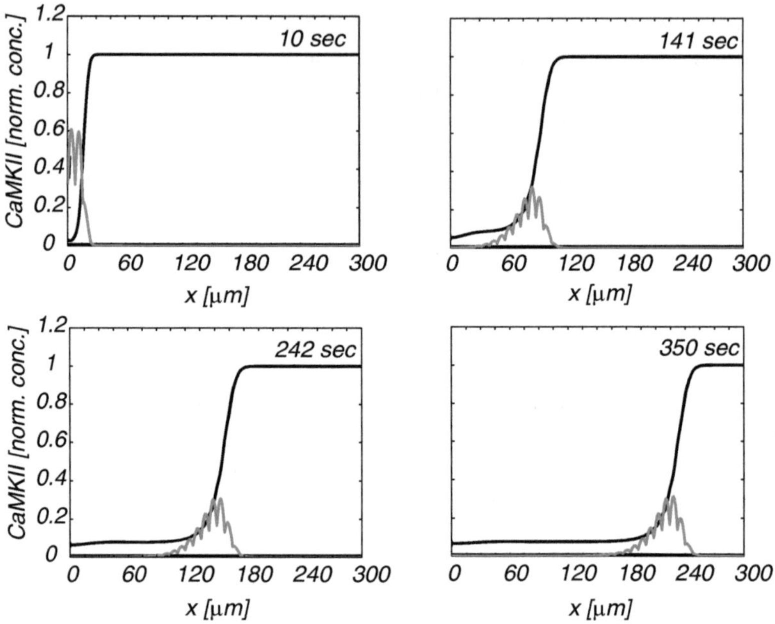

Fig. 3.11 Numerical traveling wave solution of (3.43a) and (3.43b) for an inhomogeneous distribution of spine clusters with $\Delta = 8\,\mu\text{m}$. The translocation rate $\overline{h} = 0.1$ s, diffusivity $D = 1\,\mu\text{m}^2/\text{s}$ and the activation rate $k = 0.19$ s. At time $t = 0$ all of the CaMKII within the stimulated region *(indicated by thick bar)* is in the activated state, whereas all of the CaMKII within the nonstimulated region is in the primed state. Concentrations are normalized with respect to the initial concentration of primed CaMKII. The resulting wave profiles for activated *(gray curve)* and primed *(black curve)* CaMKII along the dendrite are shown at four successive snapshots in time. The numerically estimated wave speed $c_{\min} \approx 0.66\,\mu\text{m/s}$, which is faster than the wave speed $\overline{c} = 0.6\,\mu\text{m/s}$ of the corresponding uniform spine distribution

in the presence of spine clusters. Although doubling the degree of clustering only leads to a change in wave speed of order $0.05\,\mu\text{m/s}$ (consistent with the analytical calculations), it leads to a significant difference in propagation times along a $300\,\mu\text{m}$ dendrite.

3.2.4 Wave Speed for a Slowly Modulated Spine Density

So far we have considered the effects of heterogeneity at a microscopic length scale comparable to the spacing of individual spines. In particular, we took the homogenized translocation rate $\overline{h}$ to be constant over the length of a dendrite. However, it is found experimentally that there is a slow proximal to distal variation in the density of spines [22, 346]. An illustration of a typical spine density found in pyramidal neurons of mouse cortex [22] is shown in Fig. 3.12. Such a variation in spine density can be incorporated into (3.43a) and (3.43b) by setting $h = \overline{h} + \Delta h(\varepsilon x)$, where $\overline{h}$ denotes the translocation rate at the initiation point x_0 of the wave and $\Delta h(\varepsilon x)$ represents the slow modulation of the (homogenized) translocation rate over the length of a dendrite with $\varepsilon \ll 1$. The general problem of biological invasion in slowly modulated heterogeneous environments can be analyzed using a Hamilton–Jacobi method for front velocity selection [421, 687]; see Sect. 3.3. This method was originally applied to homogeneous media by Freidlin using large deviation theory [203, 204, 212] and was subsequently formulated in terms of PDEs by Evans and Sougandis [178]. We will illustrate the method by applying it to the reaction–diffusion model of CaMKII translocation waves with slow periodic modulation; see also [72]

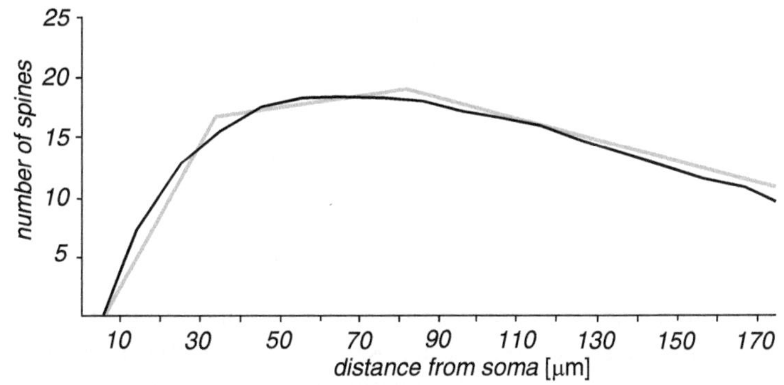

Fig. 3.12 Illustrative example of the spine density variation along the basal dendrite of a pyramidal cell in mouse cortex (*black curve*). Density is calculated as the number of spines per 10 μm segment of the dendrite from the soma to the tip of the dendrite. Abstracted from experimental data in [22]. Also shown is a simplified piecewise linear approximation of the spine density variation (*gray curve*)

The first step in the analysis is to rescale space and time in (3.43a) and (3.43b) according to $t \to t/\varepsilon$ and $x \to x/\varepsilon$; see [178, 204, 421]:

$$\varepsilon \frac{\partial P}{\partial t} = D\varepsilon^2 \frac{\partial^2 P}{\partial x^2} - k_0 AP \tag{3.84a}$$

$$\varepsilon \frac{\partial A}{\partial t} = D\varepsilon^2 \frac{\partial^2 A}{\partial x^2} + k_0 AP - [\overline{h} + \Delta h(x)]A. \tag{3.84b}$$

Under the spatial rescaling the front region where A (P) rapidly increases (decreases) as x decreases from infinity becomes a step as $\varepsilon \to 0$. This motivates the introduction of solutions of the form

$$P(x,t) \sim P_0 \left[1 - e^{-G_\varepsilon(x,t)/\varepsilon}\right], \quad A(x,t) \sim A_0(x) e^{-G_\varepsilon(x,t)/\varepsilon} \tag{3.85}$$

with $G_\varepsilon(x,t) > 0$ for all $x > x(t)$ and $G_\varepsilon(x(t),t) = 0$. The point $x(t)$ determines the location of the front and $c = \dot{x}$. Substituting (3.85) into (3.84a) and (3.84b) gives

$$-\frac{\partial G_\varepsilon}{\partial t} = D\left[\frac{\partial G_\varepsilon}{\partial x}\right]^2 - D\varepsilon\frac{\partial^2 G_\varepsilon}{\partial x^2} - k_0 A_0(x)\left[1-\mathrm{e}^{-G_\varepsilon(x,t)/\varepsilon}\right]$$

$$-A_0(x)\frac{\partial G_\varepsilon}{\partial t} = A_0(x)\left[D\left[\frac{\partial G_\varepsilon}{\partial x}\right]^2 - D\varepsilon\frac{\partial^2 G_\varepsilon}{\partial x^2} + k_0 P_0\left[1-\mathrm{e}^{-G_\varepsilon(x,t)/\varepsilon}\right] - [\bar{h}+\Delta h(x)]\right]$$
$$+\varepsilon^2 A_0''(x) G_\varepsilon - 2\varepsilon A_0'(x)\frac{\partial G_\varepsilon}{\partial x}.$$

Since $\mathrm{e}^{-G_\varepsilon(x,t)/\varepsilon} \to 0$ as $\varepsilon \to 0$ for $G_\varepsilon > 0$, it follows that the limiting function $G(x,t) = \lim_{\varepsilon\to 0} G_\varepsilon(x,t)$ satisfies

$$-\frac{\partial G}{\partial t} = D\left[\frac{\partial G}{\partial x}\right]^2 - k_0 A_0(x) \tag{3.86a}$$

$$-\frac{\partial G}{\partial t} = D\left[\frac{\partial G}{\partial x}\right]^2 + k - [\bar{h}+\Delta h(x)], \tag{3.86b}$$

where $k = k_o P_0$ as before. It immediately follows that

$$A_0(x) = \left[\frac{k-\bar{h}-\Delta h(x)}{k}\right] P_0. \tag{3.87}$$

The remaining equation (3.86b) can be analyzed along identical lines to a previous study of the heterogeneous Fisher–KPP equation [421]. Formally comparing (3.86b) with the Hamilton–Jacobi equation $\partial_t G + H(\partial_x G, x) = 0$, we define the Hamiltonian

$$H = Dp^2 + k - [\bar{h}+\Delta h(x)], \tag{3.88}$$

where $p = \partial_x G$ is interpreted as the conjugate momentum of x. It now follows that (3.86b) can be solved in terms of the Hamilton equations

$$\frac{dx}{ds} = 2Dp, \quad \frac{dp}{ds} = \frac{d\Delta h}{dx}. \tag{3.89}$$

Combining these equations yields the second-order ODE

$$\ddot{x} - 2D\Delta h(x)' = 0. \tag{3.90}$$

This takes the form of a Newtonian particle moving in a "potential" $V(x) = -2D\Delta h(x)$. Given the solution $x(s) = \phi(s;x,t)$ with $\phi(0;x,t) = x_0$ and $\phi(t;x,t) = x$, we can then determine $G(x,t)$ according to

$$G(x,t) = -E(x,t)t + \frac{1}{2D}\int_0^t \dot{\phi}(s;x,t)^2 ds. \tag{3.91}$$

Here

$$E(x,t) = H(\dot{\phi}(s;x,t)/2D, \phi(s;x,t)), \tag{3.92}$$

which is independent of s due to conservation of energy.

For certain choices of the modulation function $\Delta h(x)$, (3.90) can be solved explicitly [421]. In particular, suppose that the spine density curve in Fig. 3.3 is approximated by a piecewise linear function, in which the density increases linearly with distance from the soma to some intermediate location κ along the dendrite and then decreases linearly towards the distal end. Assuming that the right-moving wave is initiated beyond the point κ, $x_0 > \kappa$, then we can simply take $\Delta h(x) = -\beta(x - x_0)$ for $\beta > 0$. Substituting into (3.90) and integrating twice with respect to s using the Cauchy conditions gives

$$\phi(s;x,t) = x_0 + (x - x_0)s/t + D\beta ts - D\beta s^2. \tag{3.93}$$

The corresponding "energy" function is then

$$E(x,t) = \frac{(x-x_0)^2}{4Dt^2} + k - \overline{h} + \frac{\beta}{2}(x - x_0) + \frac{\beta^2}{4}Dt^2 \tag{3.94}$$

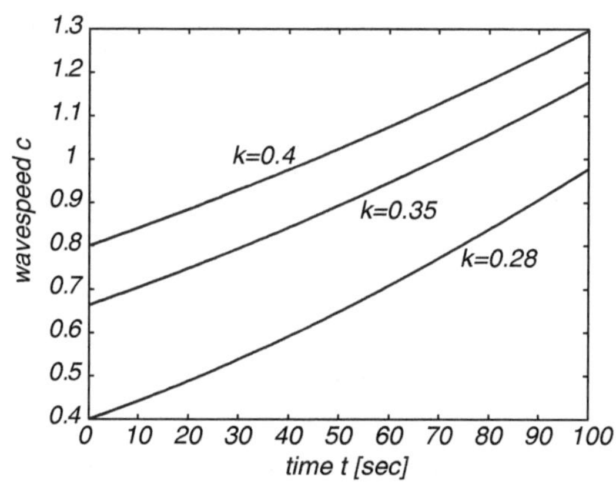

Fig. 3.13 Plot of time-dependent variation in wave speed c given by (3.96) for various values of the activation rate k. Other parameters are $\overline{h} = 0.24\ \mathrm{s}^{-1}$ and $D = 1\ \mu\mathrm{m}^2/\mathrm{s}$. At $t = 0$, $c(0) = 2\sqrt{D(k - \overline{h})}$

and (3.91) shows that

$$G(x,t) = \frac{(x-x_0)^2}{4Dt} - [k - \overline{h}]t - \frac{\beta}{2}(x - x_0)t - \frac{\beta^2}{12}Dt^3. \tag{3.95}$$

We can now determine the wave speed c by imposing the condition $G(x(t),t) = 0$. This leads to a quadratic equation with positive solution

$$x(t) = x_0 + D\beta t^2 + 2Dt\sqrt{\frac{k-\overline{h}}{D} + \frac{\beta^2}{3}t^2}$$
$$= x_0 + \overline{c}t\sqrt{1 + \frac{4\beta^2 D^2 t^2}{3\overline{c}^2}} + D\beta t^2$$

with $\overline{c} = 2\sqrt{D(k-\overline{h})}$. Finally, differentiating both sides with respect to t yields

$$c \equiv \dot{x}(t) = \overline{c}\sqrt{1+\Gamma_0\beta^2t^2} + \frac{\overline{c}\Gamma_0\beta^2t^2}{\sqrt{1+\Gamma_0\beta^2t^2}} + 2D\beta t, \tag{3.96}$$

where $\Gamma_0 = 4D^2/(3\overline{c}^2)$. For sufficiently small times such that $D\beta t \ll 1$, we have the approximation

$$c \approx \overline{c} + 2D\beta t + \frac{2(D\beta t)^2}{\overline{c}}. \tag{3.97}$$

Figure 3.13 shows example plots of the time-dependent wave speed for various choices of the activation rate k. It can be seen that there are significant changes in speed over a time course of 100 s, which is comparable to the time a wave would travel along a dendrite of a few hundred microns. In principle, it should be possible to test experimentally the predictions of the above analysis by initiating a translocation wave at different points along a dendrite and determining the corresponding wave speed.

3.3 Appendix: Pulled and Pushed Fronts

In this appendix, we review some of the basic properties of fronts propagating into unstable states. For a much more detailed account, see the review by van Saarloos [544] and Chap. 4 of [422]. For concreteness, we focus on a slight generalization of the Fisher–KPP equation

$$\frac{\partial u}{\partial t} = \frac{\partial^2 u}{\partial x^2} + f(u), \quad f \in C^1[0,1], \quad f(0) = f(1) = 0, \tag{3.98}$$

for which the homogeneous fixed point $u = 0$ is unstable $(f'(0) > 0)$ and $u = 1$ is stable $(f'(1) < 0)$. We also assume that $f(u) > 0$ for all $u \in (0,1)$. We are interested in determining the longtime asymptotics of a front propagating to the right into the unstable state $u = 0$, given initial conditions for which $u(x,0) = 0$ for sufficiently large x. It is not possible to carry out an asymptotic analysis by simply moving to a traveling coordinate frame, since there is a continuous family of front solutions. However, considerable insight into the evolution of a localized initial condition can be obtained by linearizing about the unstable state.

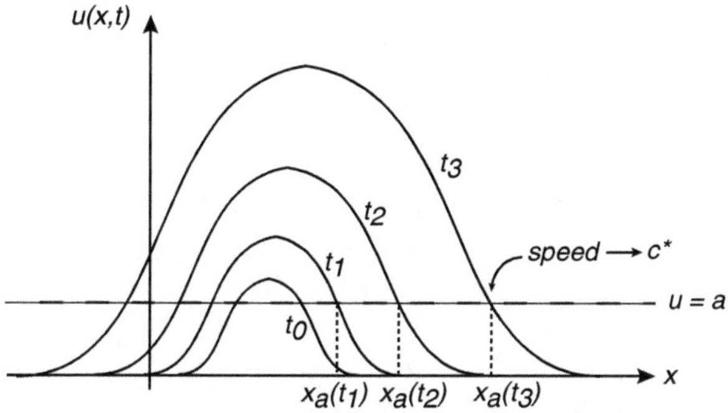

Fig. 3.14 Illustrative sketch of the growth and spreading of a solution $u(x,t)$ of the Fisher equation linearized about the unstable state $u = 0$, given a localized initial condition $u(x,t_0)$

3.3.1 The Linear Spreading Velocity

Linearizing (3.98) about $u = 0$,

$$\frac{\partial u}{\partial t} = \frac{\partial^2 u}{\partial x^2} + f'(0)u. \tag{3.99}$$

Substitution of the Fourier mode $e^{-i\omega t + ikx}$ gives the dispersion relation

$$\omega(k) = i(f'(0) - k^2). \tag{3.100}$$

The state $u = 0$ is then said to be linearly unstable if $\mathrm{Im}[\omega(k)] > 0$ for some range of k-values. In the case of the Fisher–KPP equation, after writing $k = k_r + ik_i$, this will occur when $f'(0) + k_i^2 > k_r^2$. Consider some generic initial condition $u(x,0)$ that is sufficiently localized in space (to be made precise later). Since there exists a range of unstable linear eigenmodes, we expect the localized initial condition to grow and spread out within the linear regime, as illustrated in Fig. 3.14. Tracking the evolution of a level curve $x_a(t)$ with $u(x_a(t),t) = a$, the *linear spreading velocity* c^* is defined to be the asymptotic speed of the point $x_a(t)$ in the rightward moving edge (assuming it exists):

$$c^* = \lim_{t\to\infty} \frac{dx_a(t)}{dt}. \tag{3.101}$$

The linearity of the underlying evolution equation (3.99) means that c^* is independent of the value a. (Note that for an isotropic medium, the leftward moving edge moves with the same asymptotic speed but in the opposite direction.) Suppose that c^* is finite. If we were to move in the traveling coordinate frame $\xi = x - c^*t$, then the leading rightward edge would neither grow nor decay exponentially. Imposing this condition on the Fourier expansion of the solution $u(x,t)$ then determines c^* in terms of the dispersion curve $\omega(k)$. More specifically, denoting the Fourier transform of the initial condition $u_0(x) = u(x,0)$ by $\tilde{u}_0(k)$, we have

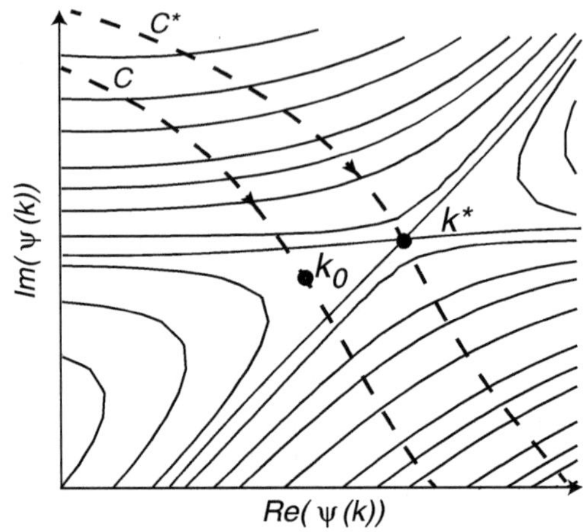

Fig. 3.15 Method of steepest descents. Sketch of Re(ψ) contours in the complex k-plane for an analytic function in the region of a saddle point at k^*. The integration contour C is deformed so that it passes through the saddle point

$$\begin{aligned} u(x,t) &= \int_{-\infty}^{\infty} \tilde{u}_0(k) e^{i[kx-\omega(k)t])} \frac{dk}{2\pi} \\ &= \int_{-\infty}^{\infty} \tilde{u}_0(k) e^{ik\xi} e^{-i[\omega(k)-c^*k]t)} \frac{dk}{2\pi} \\ &= \int_{-\infty}^{\infty} \tilde{u}_0(k) e^{ik\xi} e^{\psi(k)t} \frac{dk}{2\pi}, \end{aligned} \tag{3.102}$$

where

$$\psi(k) = -i[\omega(k) - c^*k] = \omega_i(k) - c^*k_i - i[\omega_r(k) - c^*k_r] \tag{3.103}$$

In the limit $t \to \infty$ with ξ finite, we can approximate this integral using steepest descents. For the moment, we assume that $\tilde{u}(k)$ is an entire function (analytic in every finite region of the complex plane) so that we can deform the contour in the complex k-plane, that is, $(-\infty,\infty) \to C$, with C linking points at infinity in the complex plane where $\text{Re}(\psi) < 0$.

Method of steepest descents. We briefly describe the method of steepest descents for a general analytic function $\psi(k)$; see also Fig. 3.15 and [275]. First, one would like to choose C so that the maximum of $\psi_r \equiv \text{Re}(\psi)$ along the contour at k_0, say, is as large as possible, since this point will dominate the integral. Recall, however, that one of the Cauchy–Riemann conditions on an analytic function is that $\nabla^2(\text{Re}(\psi)) = 0$, which means that $\text{Re}(\psi)$ cannot have any maxima or minima (except at singularities or branch points where ψ would be nonanalytic). Therefore $\nabla(\text{Re}(\psi)) = 0$ only at saddle points. Second, for a general integration contour, evaluating the integral in a neighborhood of the point k_0 will overestimate the value of the integral, since it does not take into account cancellations due to the rapidly oscillating function $e^{i\text{Im}(\psi(k))}$. The latter issue can be eliminated by choosing the contour that is the path of steepest ascent to a saddle point and the steepest descent away from the saddle. (If there exists more than one saddle, then one chooses the "highest" one). By construction, the path is parallel to $\text{Re}(\psi)$. Hence, from the Cauchy–Riemann conditions, $\nabla(\text{Im}(\psi)) = 0$

so that $\psi_i \equiv \text{Im}(\psi)$ is constant along the contour. In other words, there are no fast oscillations along the path of steepest ascent and descent and one can obtain a good estimate of the integral by Taylor expanding about the saddle point. Thus, taking $\psi_i(k) = \psi_i(k^*)$ along C^*, we have

$$\begin{aligned} I &\equiv \int_{-\infty}^{\infty} \tilde{u}_0(k)\mathrm{e}^{ik\xi}\mathrm{e}^{\psi(k)t}\frac{dk}{2\pi} \\ &= \int_{C^*} \tilde{u}_0(k)\mathrm{e}^{ik\xi}\mathrm{e}^{\psi(k)t}\frac{dk}{2\pi} \\ &\approx \tilde{u}_0(k^*)\mathrm{e}^{i\psi_i(k^*)t}\int_{C^*} \mathrm{e}^{ik\xi}\mathrm{e}^{\psi_r(k)t}\frac{dk}{2\pi}. \end{aligned}$$

Finally, we Taylor expand $\psi_r(k)$ to second order in $\Delta k = k - k^*$, noting that $\psi'(k^*) = 0$ and $\psi_r''(k^*) < 0$ at the saddle,

$$\psi_r(k) = \psi_r(k^*) + \frac{(\Delta k)^2}{2}\psi_r''(k^*),$$

and approximate the remaining contour integral by a Gaussian. This then gives

$$I \approx \frac{1}{\sqrt{4\pi Dt}}\tilde{u}_0(k^*)\mathrm{e}^{i[k^*\xi + \psi_i(k^*)t]}\mathrm{e}^{-\xi^2/4Dt}\mathrm{e}^{\psi_r(k^*)t}, \tag{3.104}$$

where $D = -\psi_r''(k^*)/2$.

Let us now apply steepest descent to the integral (3.102) for $\psi(k)$ given by (3.103) such that $\psi_r(k) = \omega_i(k) - c^* k_i$ and $\psi_i(k) = -[\omega_r(k) - c^* k_r]$. At the (unique) saddle point k^* at which $\psi'(k^*) = 0$, we have

$$c^* = \left.\frac{d\omega(k)}{dk}\right|_{k=k^*}. \tag{3.105}$$

Moreover, equation (3.104) becomes

$$I \approx \frac{1}{\sqrt{4\pi Dt}}\tilde{u}_0(k^*)\mathrm{e}^{i[k^*\xi - [\omega_r(k^*) - c^* k_r^*]t]}\mathrm{e}^{-\xi^2/4Dt}, \tag{3.106}$$

where

$$D = -\frac{1}{2}\omega_i''(k^*). \tag{3.107}$$

Finally, we can determine the linear spreading velocity c^* by requiring that the asymptotic solution neither grows nor decays with time, $\psi_r(k^*) = 0$, which implies

$$c^* = \frac{\omega_i(k^*)}{k_i^*}. \tag{3.108}$$

Note that equating real and imaginary parts in (3.105) and combining with (3.108) means that we have three equations in the three unknowns c^*, k_r^*, k_i^*. In the particular case of the Fisher–KPP equation (3.98),

$$k_r^* = 0, \quad k_i^* = \sqrt{f'(0)}, \quad c^* = 2\sqrt{f'(0)}, \quad D = 1. \tag{3.109}$$

Since $k_r^* = 0$, we can combine (3.105) and (3.108) into the single condition $c^* = c(\lambda^*)$, $\lambda^* = k_i^*$, where

$$c(\lambda) = \frac{\omega_i(i\lambda)}{\lambda}, \quad \left.\frac{dc(\lambda)}{d\lambda}\right|_{\lambda=\lambda^*} = 0. \tag{3.110}$$

It can also be seen that the modulus of u falls off as

$$|u(x,t)| \sim \frac{1}{\sqrt{t}} \mathrm{e}^{-\lambda^* \xi} \mathrm{e}^{-\xi^2/4t}. \tag{3.111}$$

A more direct way to derive the function $c(\lambda)$ is to consider the linearized PDE in the moving frame $\xi = x - ct$,

$$-c\frac{dU}{d\xi} = \frac{d^2U}{d\xi^2} + f'(0)U,$$

and to assume the leading-edge solution $U(\xi) \sim \mathrm{e}^{-\lambda\xi}$.

In the above analysis, it was assumed that the Fourier transform of the initial condition $u_0(x)$ was an entire function. This would apply to cases for which $u_0(x)$ is a Dirac delta function, has compact support, or decays faster than any exponential for large enough x (e.g., a Gaussian). Now suppose that $u_0(x)$ falls off exponentially for large x, $u_0(x) \sim \mathrm{e}^{-\lambda x}$ for some λ. Then $\tilde{u}_0(k)$ has a pole in the upper-half complex plane at $k = k'$ with $\mathrm{Im}(k') = \lambda$. It follows that when deforming the contour C in the complex k-plane in order to perform steepest descents, we pick up a contribution from the pole. Taking this into account, it can be shown that, within the linear regime, initial conditions whose exponential decay rate $\lambda > \lambda^*$ lead to profiles that asymptotically spread with the linear spreading velocity v^*. On the other hand, if $\lambda < \lambda^*$ then the profile advances at a speed faster than c^* [544].

So far we have investigated the evolution of a localized initial condition in the linear regime. It still remains to determine whether or not there are classes of initial conditions under which the full nonlinear system converges to a unique asymptotic front solution and how the speed of the front c is related to the linear spreading velocity c^*. It turns out that for front propagation into a linearly unstable state, there are only two possibilities when starting from sufficiently steep initial conditions, that is, initial conditions that fall off faster than $\mathrm{e}^{-\lambda^* x}$ [544]:

Pulled front: $c = c^*$ so that the front dynamics is determined by the behavior in the leading edge of the front where $u(x,t) \approx 0$, that is, the front is pulled along by the linear spreading of small perturbations into the linearly unstable state.

Pushed front: $c > c^*$ so that nonlinearities play an important role in determining the velocity of the front, pushing it into the unstable state.

In the special case of initial conditions with compact support, it can be proven that the solution evolves into a front that propagates at the minimal possible wave speed $c_{\min}$, which is bounded above and below [15]:

$$c^* = 2\sqrt{f'(0)} \leq c_{\min} < 2\sqrt{\sup_u \left[\frac{f(u)}{u}\right]}. \tag{3.112}$$

For any concave function, $f(u) \leq uf'(0)$, the lower and upper bounds coincide and we have a pulled front; this applies to the standard Fisher–KPP equation where $f(u) = u(1-u)$. On the other, the upper and lower bounds do not coincide for concave $f(u)$. The minimal front velocity can then be larger than the linear velocity indicative of a pushed front. An example of the latter is the Ginzburg–Landau term $f(u) = u(1-u)(1+\alpha u)$ with $\alpha > 0$. One finds that for compact initial conditions, a pulled front is selected when $\alpha \leq 2$, whereas a pushed front is selected when $\alpha > 2$ [33].

3.3.2 Variational Method for Front Velocity Selection

Over recent years a number of methods have been developed to tackle the general problem of front velocity selection, that is, which of the infinity of possible wave speeds is selected by a given initial condition. One method analyzes the dynamics of the front position using Hamilton–Jacobi theory. Although it is only applicable to pulled fronts, it has the advantage of being able to tackle the effects of heterogeneities, as illustrated in Sect. 3.2.4 for CaMKII translocation waves. Here we briefly describe another method due to Benguria and Depassier [36], which is based on a variational principle. The latter can be applied to both pulled and pushed fronts, although in many cases it only provides lower and upper bounds for the front velocity. Consider a front solution of the reaction–diffusion Eq. (3.98) in traveling wave coordinates,

$$u_{\xi\xi} + cu_{\xi} + f(u) = 0, \tag{3.113}$$

with $\lim_{\xi\to-\infty} u(\xi) = 1$ and $\lim_{\xi\to\infty} u(\xi) = 0$, $\xi = x - ct$. Set $q(u) = -u_{\xi} > 0$ for $u \in (0,1)$, so that (3.113) becomes

$$q(u)\frac{dq}{du} - cq(u) + f(u) = 0, \quad q(0) = 0, \quad q(1) = 0. \tag{3.114}$$

Let $g \in C^1[0,1]$ be a positive, monotonically decreasing function with $\int_0^1 g(u)du < \infty$, and set $h = -g' > 0$. We denote the space of admissible functions g by the domain $\mathscr{D}$. Multiplying (3.114) by g/q and integrating with respect to u gives (after integration by parts)

$$\int_0^1 \left(h(u)q(u) + \frac{f(u)}{u}g(u)\right) du = c\int_0^1 g(u)du. \tag{3.115}$$

For fixed u, the functional

$$\Phi[q] = qh + \frac{f}{q}g$$

has a minimum at $q_0 = \sqrt{fg/h}$ with $\Phi[q_0] = 2\sqrt{fgh}$. It follows that

$$c\int_0^1 g(u)du = \int_0^1 \Phi[q(u)]du \geq 2\int_0^1 \sqrt{f(u)g(u)h(u)}du,$$

that is,

$$c \geq 2\frac{\int_0^1 \sqrt{f(u)g(u)h(u)}du}{\int_0^1 g(u)du} \equiv I[g]. \tag{3.116}$$

As we show below, there exists a function $\hat{g} \in \mathscr{D}$ that maximizes the lower bound for c in such a way that the equality holds. Hence, we have the variational principle

$$c = \max_{g\in\mathscr{D}} \left(2\frac{\int_0^1 \sqrt{f(u)g(u)h(u)}du}{\int_0^1 g(u)du} \right), \tag{3.117}$$

which can be used to estimate c using a parameterized set of trial functions. This result can also be used to derive the upper bound of (3.112) [36].

In order to establish the existence of the function $\hat{g}$, we first require $\Phi[q] = \Phi[q_0]$ for all u, that is, $hq = f\hat{g}/q$. Combined with (3.114) this gives

$$-\frac{\hat{g}'}{\hat{g}} = \frac{c}{q} - \frac{q'}{q},$$

which can be integrated to yield

$$\hat{g}(u) = q(u)\exp\left(\int_u^{u_0} \frac{c}{q(u')}du'\right) \tag{3.118}$$

for some u_0, $0 < u_0 < 1$. Since $\hat{g} > 0$ on $(0,1)$ and $h = \hat{g}f/q^2 > 0$ we deduce that $\hat{g} \in C^1[0,1]$ and is a positive, decreasing function. It remains to check that $\int_0^1 \hat{g}(u)du < \infty$, which requires determining the behavior near $u = 0$. Linearizing (3.113) around $u = 0$ shows that if $u \sim e^{-\lambda\xi}$ for $\xi \to \infty$ with $\lambda = (c + \sqrt{c^2 - 4f'(0)})/2$, then $q \sim \lambda u$ for $u \sim 0$. The integral solution for $\hat{g}$ then implies that

$$\hat{g} \sim \lambda \frac{1}{u^{c/\lambda - 1}}, \quad u \sim 0.$$

Therefore, if $c < 2\lambda$, that is, $c > 2\sqrt{f'(0)}$, then $\int_0^1 \hat{g}(u)du < \infty$ and $\hat{g} \in \mathscr{D}$. Finally, in the special case $c = 2\sqrt{f'(0)}$, one can take the set of trial functions $g_\alpha(u) = \alpha u^{\alpha-1}$ with $g_\alpha \in \mathscr{D}$ for $0 < \alpha < 1$ and show that $I[g_\alpha] \to 2\sqrt{f'(0)}$ as $\alpha \to 0$. This means $c = 2\sqrt{f'(0)} \geq \max_g I[g] \geq 2\sqrt{f'(0)}$, that is, $c = \max_g I[g]$.

Chapter 4
Calcium Waves and Sparks

Calcium (Ca^{2+}) is one of the most important and well-studied cellular signaling molecules. From a modeling perspective, it attracts a great deal of interest due to the fact that calcium signaling often involves complex spatiotemporal dynamics, including oscillations and waves. There are a number of excellent general reviews on the modeling of calcium dynamics within cells, including Chap. 7 of Keener and Sneyd [322] and Falcke [181]. In this chapter, we focus on the mathematical modeling and analysis of calcium waves and their potential significance in neuronal calcium signaling; see also the reviews by Berridge [43] and Ross [534]. Although there are a number of very detailed whole-cell models of calcium signaling, we will focus on simplified models that are analytically tractable. In particular, we will highlight some of the mathematical methods used to characterize calcium waves.

4.1 Calcium Signaling in Neurons

In vertebrates, most of the Ca^{2+} is stored in bones, from where it can be released by hormonal stimulation to maintain a high extracellular Ca^{2+} concentration (around 1 mM). On the other hand, active ion pumps and exchangers maintain the cytoplasmic Ca^{2+} concentration at relatively low levels (around 10–100 nM). The resulting steep concentration gradient across the plasma membrane means that cells are able to increase their cytoplasmic Ca^{2+} concentration rapidly by opening either voltage-gated or ligand-gated Ca^{2+} ion channels. Within the context of neuronal processing, the activation of voltage-gated calcium channels (VGCCs) at axon terminals triggers the release of synaptic neurotransmitter (see Sect. 1.3); VGCCs also contribute to the generation of dendritic action potentials (see Sects. 1.4.3 and 3.1). A classical example of ligand gating in neurons is the entry of Ca^{2+} through NMDA receptors on postsynaptic dendritic spines. The resulting transient Ca^{2+} signal is thought to play a major role in the induction of changes in synaptic strength; see below.

Another major mechanism whereby cells, including neurons, regulate their cytoplasmic Ca^{2+} concentration is via the intracellular supply of Ca^{2+} from internal

P.C. Bressloff, *Waves in Neural Media: From Single Neurons to Neural Fields*, Lecture Notes on Mathematical Modelling in the Life Sciences,
DOI 10.1007/978-1-4614-8866-8_4,

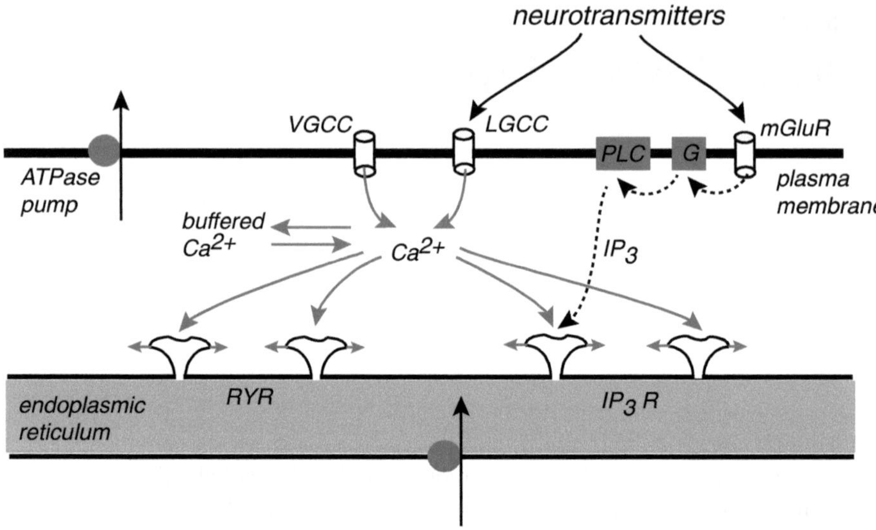

Fig. 4.1 Neural calcium signaling. The entry of Ca^{2+} from outside the cell is mediated by voltage-gated (VGCC) and ligand-gated (LGCC) calcium channels. Stimulation of metabotropic glutamate receptors (mGluRs) produces IP_3 second messengers that bind to IP_3 receptors (IP_3Rs), which subsequently release Ca^{2+} from the endoplasmic reticulum. Both IP_3Rs and Ryanodine receptors (RyRs) are sensitive to Ca^{2+}, resulting in Ca^{2+}-induced Ca^{2+}-release (CICR). The latter can sometimes result in the propagation of a Ca^{2+} wave along the dendrites

stores such as the *endoplasmic reticulum* (ER) and *mitochondria. Inositol (1,4,5)-trisphosphate* (IP_3) receptors and *Ryanodine* (Ry) receptors distributed throughout the ER, for example, mediate the release of Ca^{2+} into the cytoplasm, whereas Ca^{2+} ion pumps maintain the relatively high Ca^{2+} concentration within the ER. The Ry receptor plays a critical role in excitation–contraction coupling in skeletal and cardiac muscle cells, but is also found in nonmuscle cells such as neurons. One important feature of Ry receptors is that they can undergo Ca^{2+}-induced Ca^{2+}-release (CICR), in which elevated cytoplasmic Ca^{2+} activates Ry receptors that release further Ca^{2+}, which then activates other Ry receptors, resulting in a nonlinear regenerative feedback mechanism. The IP_3 receptor is similar in structure to the Ry receptor, but is found predominantly in nonmuscle cells and is sensitive to the second messenger IP_3. The binding of an extracellular ligand such as a hormone or a neurotransmitter to a metabotropic receptor results in the activation of a G-protein and the subsequent activation of phospholipase C (PLC). This then cleaves phosphatidylinositol bisphosphate (PIP_2) into diacylglycerol (DAG) and IP_3. The water soluble IP_3 is free to diffuse throughout the cell cytoplasm and bind to IP_3 receptors located on the ER membrane, which then open and release Ca^{2+} from the ER. The opening and closing of an IP_3 receptor is also modulated by the concentration of cytoplasmic Ca^{2+}, so it too can undergo CICR. Yet another mechanism for controlling cytoplasmic Ca^{2+} is through buffering (binding) to large proteins. Indeed, it is estimated that

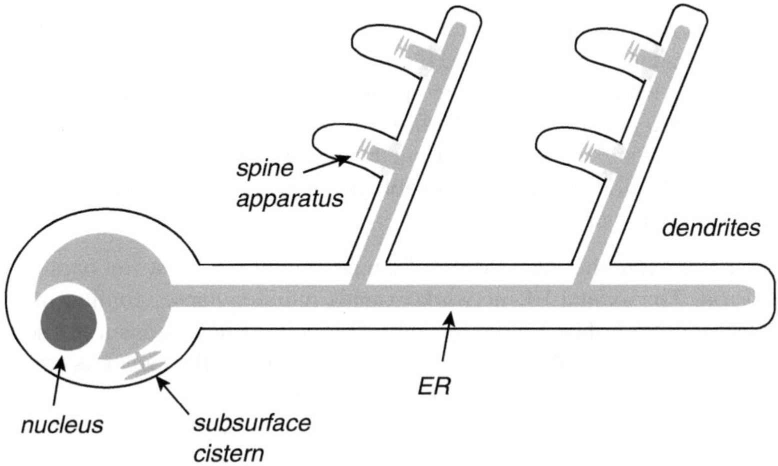

Fig. 4.2 The endoplasmic reticulum (ER) is a single and continuous membrane-bound organelle that is distributed throughout the dendrites and axon (not shown) of a neuron [516]. It is responsible for the synthesis and posttranslational modification of most secretory and membrane proteins, as well as the regulation of Ca^{2+} levels. The shape of the ER is heterogeneous, but can be divided into three domains: the nuclear envelope, the ribosome-bound rough ER (RER), and the ribosome-free smooth ER (SER). The RER is constituted primarily by sheets or cisternae, whereas the SER is predominantly composed of 60–100 nm diameter tubules that form irregular polygons with a common luminal space connected via three-way junctions. The ER present in the soma and proximal dendritic compartment is rich in ribosomes, corresponding to the RER, whereas the ER distributed in distal dendrites and axons corresponds mostly to SER and contains only a few sparse ribosomes. The ER comes into close proximity with the plasma membrane at various locations, forming subsurface cisternae within the soma and axon initial segment, and the spine apparatus within dendritic spines

at least 99 % of the total cytoplasmic Ca^{2+} is bound to buffers. A summary of the basic extracellular and intracellular mechanisms for controlling the concentration of cytoplasmic Ca^{2+} is shown in Fig. 4.1.

The signaling mechanisms and dynamics of Ca^{2+} release from the ER in neurons have been much less well studied than extracellular sources of Ca^{2+}. However, the ER is a continuous network that is distributed throughout a neuron, extending all the way into the dendrites and axon and coming into close contact with the plasma membrane and synapses at many locations; see Fig. 4.2. Hence, as suggested by Berridge [43], it is likely that the integrative and regenerative properties of the ER and plasma membrane provide a binary membrane system that regulates a variety of neuronal process via Ca^{2+} signaling, including excitability, neurotransmitter release, synaptic plasticity, and gene transcription.

(a) Within the soma and initial axonal segment, the ER forms flattened sheets known as *subsurface cisternae* that come into close contact with the plasma membrane. These internal structures are thought to play an important role in regulating neuronal excitability. For example, many neurons exhibit significant after-hyperpolarizations (AHPs) following either a single action potential or a burst of action potentials, which suppress subsequent firing of the cell (spike frequency adaptation). Often AHPs have both a fast component and a slow component, which arise from the opening of two separate Ca^{2+}-activated K^+ channels (gK_{Ca1} and gK_{Ca2}). The fast activation of gK_{Ca1} is due to the entry of extracellular Ca^{2+} via VCCNs during the course of an action potential, whereas the slow activation of gK_{Ca2} is probably due to CICR from Ry receptors located in the subsurface cisternae.

(b) Within the axon, the ER consists of connecting tubules that run parallel along the axon. The axonal ER network extends into a synapse, coming into close contact with the plasma membrane and associated neurotransmitter secretory vesicles. It is thought that calcium release from Ry and IP_3 receptors contributes to the formation of localized high-concentration bursts of Ca^{2+} necessary for the exocytosis (transport and fusion) of vesicles to the plasma membrane.

(c) The ER network is also distributed throughout the dendritic tree and into spines, where it terminates at the so-called spine apparatus. The latter consists of closely packed plates of ER that are located within the middle of the spine and is thus well placed to contribute to Ca^{2+} signaling during synaptic plasticity. The latter refers to the rapid and long-lasting changes in synaptic strength that are thought to provide the subcellular substrate for learning and memory. Much of the information regarding synaptic plasticity in mammalian neurons has been obtained from studies of hippocampal and cerebellar neurons [120, 123, 385, 402]. In particular, it has been found that the same synapses can be switched rapidly and reversibly from an enhanced state of *long-term potentiation* (LTP) to a reduced state of *long-term depression* (LTD). Interestingly, both LTP and LTD are induced by a transient increase in the local Ca^{2+} concentration within a spine. It was originally thought that the bidirectional switch only depended on the amplitude of the Ca^{2+} signal, with a stronger stimulus inducing LTP and a weaker one LTD. Now, however, it appears likely that the detailed spatial and temporal structure of the calcium signal may be important. Although the influx of extracellular Ca^{2+} through NMDA receptors or voltage-gated channels is still the major contributor to the induction of LTP and LTD, there is increasing evidence that there is also a component arising from intracellular stores [428, 447, 534]. Activation of the ER could be mediated either by Ca^{2+} itself via CICR or by stimulation of metabotropic glutamate (mGlu) receptors in the plasma membrane, resulting in the production of IP_3. The fact that CICR requires both cytoplasmic IP_3 and Ca^{2+} suggests that IP_3 receptors could act as *coincidence detectors* for mGluR activation and postsynaptic Ca^{2+} signaling arising from back-propagating action potentials or dendritic Ca^{2+} spikes [658].

(d) One of the most dramatic consequences of CICR is the propagation of intracellular Ca^{2+} waves mediated primarily by the opening of IP_3 receptors. These waves were first observed in nonneuronal cells such as *Xenopus laevis* oocytes [367, 490], where the resulting changes in Ca^{2+} concentration across the whole cell provided a developmental signal. The discovery of Ca^{2+} waves in neurons is more recent [303, 428, 447, 534, 658]. Since these cells have extensive dendritic and axonal arborizations, the distribution of the components of the Ca^{2+} signaling apparatus within the cell plays a crucial role in determining whether or not local Ca^{2+} release transforms into a propagating wave, and how far it travels. For example, in pyramidal neurons of the hippocampus and cortex, Ca^{2+} waves are usually observed in the primary apical dendrite and perhaps the soma, rarely reaching beyond the point where thick dendrites begin to branch. Since the ER network is distributed throughout the cell, this suggests that there is a heterogeneous distribution of mGlu receptors and IP_3 receptors along the dendrite. Ca^{2+} waves rarely propagate from the dendrite to the soma, unless there is a sufficiently strong and sustained stimulation of a neuron that results in high levels of cytoplasmic IP_3 or higher concentrations of internally stored Ca^{2+}. If a Ca^{2+} wave did reach the soma, then it would generate a strong Ca^{2+} signal in the soma and nucleus due to the high concentration of IP_3 receptors in these regions. It has been suggested that a Ca^{2+} wave could allow a strongly stimulated synapse in the dendrites to signal to the nucleus, where the large increase in Ca^{2+} concentration could activate gene transcription, which is a necessary step in more persistent forms of synaptic plasticity such as late long-term potentiation (L-LTP). However, the role of Ca^{2+} waves in initiating gene transcription is still controversial. On the other hand, the likely contribution of IP_3 receptor-mediated Ca^{2+} release during normal LTP and LTD indicates that Ca^{2+} waves could be important in determining the spatiotemporal pattern of synaptic changes.

(e) Many nonneuronal cell types exhibit spontaneous localized Ca^{2+} release events known as *sparks* [110]. Calcium sparks, which are thought to be the building blocks of the large regenerative Ca^{2+} signal that controls contraction in cardiac and skeletal muscle cells, arise from the opening of clusters of RyRs by local CICR. The frequency of calcium spark events is sensitive to changes in membrane potential, although they rarely induce calcium waves [111]. Similar events known as calcium *puffs* have been found in *Xenopus laevis* oocytes [489, 490, 690]. They are also spatially and temporally localized, are fast, and occur stochastically. However, in contrast to sparks, they are mediated by IP_3Rs and coalesce more easily to form calcium waves. Localized events in neurons with spark-like and puff-like properties have recently been observed in slice preparations, and their frequency can be modulated by synaptic activity and changes in membrane potential [389, 403]. Although it is likely that they also occur in vivo, it is not yet clear whether or not they have a specific function.

4.2 Reaction–Diffusion Models of Ca^{2+} Dynamics

One of the major challenges in modeling calcium waves is that the ER and cytoplasm comprise two highly interconnected three-dimensional domains (ignoring, for simplicity, other intracellular compartments such as the mitochondria). Assuming that Ca^{2+} undergoes normal diffusion within each compartment, together with buffering, we can write down the following pair of diffusion equations for the Ca^{2+} concentrations c, c_e in the cytoplasm and ER [322]:

$$\frac{\partial c}{\partial t} = \nabla \cdot (D_c \nabla c) - J_{\text{on}} + J_{\text{off}}, \quad \mathbf{r} \in \Omega_c \tag{4.1}$$

and

$$\frac{\partial c_e}{\partial t} = \nabla \cdot (D_e \nabla c_e) - \hat{J}_{\text{on}} + \hat{J}_{\text{off}}, \quad \mathbf{r} \in \Omega_e, \tag{4.2}$$

where Ω_c and Ω_e denote the cytoplasmic and ER domains, respectively, D_c and D_e are the corresponding diffusivities in the two domains, and the on/off fluxes take into account the binding/unbinding of Ca^{2+} to buffering proteins. These equations are supplemented by various boundary conditions that take into account the exchange of Ca^{2+} between the cytoplasm and the plasma membrane or ER. Let J_{in} denote the total inward flux of Ca^{2+} from the plasma membrane to the cytoplasm via voltage-gated and ligand-gated ion channels, and let J_{p1} be the outward flux due to ATPase ion pumps in the plasma membrane. Then

$$D_c \nabla c \cdot \mathbf{n} = J_{\text{in}} - J_{\text{p1}}, \quad \mathbf{r} \in \partial \Omega_{c,m}, \tag{4.3}$$

where $\partial \Omega_{c,m}$ is the surface of the plasma membrane with unit normal $\mathbf{n}$. Similarly, let J_{IP} and J_{Ry} denote the inward fluxes from the ER to the cytoplasm via IP_3Rs and RyRs, respectively, and let J_{p2} denote the outward flux due to ion pumps in the ER. This yields

$$D_c \nabla c \cdot \hat{\mathbf{n}} = -D_e \nabla c_e \cdot \hat{\mathbf{n}} = J_{\text{IP}} + J_{\text{Ry}} - J_{\text{p2}}, \quad \mathbf{r} \in \partial \Omega_e, \tag{4.4}$$

where $\partial \Omega_e$ is the boundary of the ER with unit normal $\hat{\mathbf{n}}$. A schematic diagram of all the fluxes is shown in Fig. 4.3.

Let us first consider a space-clamped version of this model, in which one treats the cytoplasm and ER as two well-mixed homogeneous compartments. In that case, the various surface and volume fluxes are combined as follows:

$$\frac{dc}{dt} = -J_{\text{on}} + J_{\text{off}} + \frac{|\partial \Omega_{c,m}|}{|\Omega_c|} [J_{\text{in}} - J_{\text{p1}}] + \frac{|\partial \Omega_e|}{|\Omega_c|} [J_{\text{IP}} + J_{\text{Ry}} - J_{\text{p2}}] \tag{4.5}$$

and

$$\frac{dc_e}{dt} = -\hat{J}_{\text{on}} + \hat{J}_{\text{off}} - \frac{|\partial \Omega_e|}{|\Omega_e|} [J_{\text{IP}} + J_{\text{Ry}} - J_{\text{p2}}]. \tag{4.6}$$

Each of the fluxes in these equations corresponds to a component of the so-called Ca^{2+}-signaling toolkit [44]. Thus, the equations for Ca^{2+} dynamics have to be coupled to additional biochemical equations describing the dynamics of ion pumps, the Ca^{2+}-dependent gating of the Ry and IP_3 receptors, and how the latter are regulated by the second messenger IP_3. Often the external flux J_{in} is simply taken to be a linear increasing function of the IP_3 concentration. There have been many studies of space-clamped Ca^{2+} models with particular focus on calcium oscillations

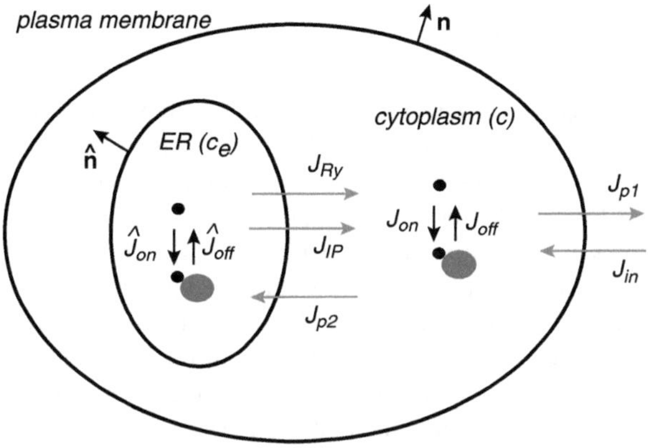

Fig. 4.3 Schematic diagram showing various fluxes in reaction–diffusion model of Ca^{2+} dynamics

[181, 322]. Although the models differ in their degree of complexity with regards the component fluxes, they can often be reduced to simpler models that have similar structure to the FitzHugh–Nagumo (FN) model of excitability. For example, suppose that we ignore fluxes through the plasma membrane so that the total intracellular Ca^{2+} is conserved, that is, $|\Omega_c|c + |\Omega_e|c_e$ is a constant. Furthermore, suppose that there are only two intracellular fluxes, J_{IP} and J_{p2}, and the IP_3 concentration is fixed. Ignoring the effects of buffering, one can write down an analogous equation to (2.3) of the form

$$\frac{dc}{dt} = f(c,h), \quad \tau(c)\frac{dh}{dt} = h_\infty(c) - h, \tag{4.7}$$

where c is the concentration of free cytosolic Ca^{2+}, $f(c,h)$ represents the net flux into the cytoplasm from the ER, and h is an inactivation variable that plays a similar role to the recovery variable w of the original FN equations. One interpretation of h is the fraction of IP_3 receptors that are not inactivated by bound Ca^{2+}. Li and Rinzel [381] derived such a two-variable model by reducing the classical kinetic model of Ca^{2+}-gated IP_3 receptors due to De Young and Keizer [692]; see Sect. 4.4.1.

Unfortunately, the complex geometry and heterogeneity of the ER means that the spatially distributed model given by (4.1)–(4.4) is far too complicated to analyze, even when simplified models of the various fluxes are considered. A common strategy is to ignore the detailed structure of the ER by assuming that the concentrations c and c_e coexist at every point in space. This is motivated by the idea that since diffusion is fast over short distances, local variations due to heterogeneities are smoothed out. Phenomenologically speaking, this leads to the *calcium bidomain equations*

$$\frac{\partial c}{\partial t} = \nabla \cdot (\overline{D}_c \nabla c) - J_{\mathrm{on}} + J_{\mathrm{off}} + \chi_c[J_{\mathrm{IP}} + J_{\mathrm{Ry}} - J_{\mathrm{p2}}] \tag{4.8}$$

and

$$\frac{\partial c_e}{\partial t} = \nabla \cdot (\overline{D}_e \nabla c_e) - \hat{J}_{\text{on}} + \hat{J}_{\text{off}} - \chi_e [J_{\text{IP}} + J_{\text{Ry}} - J_{\text{p2}}]. \tag{4.9}$$

Here $\chi_c = |\partial\Omega_e|/|\Omega_c|, \chi_e = |\partial\Omega_e|/|\Omega_e|$ are surface to volume ratios, and $\overline{D}_c, \overline{D}_e$ are effective diffusion coefficients. Note, however, that in order to determine the effective diffusion coefficients, which will depend on the geometry of the ER, and justify the precise form of the factors χ_c, χ_e, it is necessary to derive these equations from first principles using homogenization theory [224, 322]. Elements of the theory are described in appendix section 4.6. Now suppose that only the fluxes J_{IP} and J_{p2} are included along the lines of the space-clamped model. However, in contrast to the latter, it is now necessary to keep track of both the cytoplasmic and ER Ca^{2+} concentrations, since they are spatially varying. A one-dimensional bidomain model with constant diffusion coefficients then takes the form

$$\frac{\partial c}{\partial t} = D_c \frac{\partial^2 c}{\partial x^2} + f(c, c_e, h) \tag{4.10a}$$

$$\frac{\partial c_e}{\partial t} = D_e \frac{\partial^2 c}{\partial x^2} - \chi_e f(c, c_e, h) \tag{4.10b}$$

$$\tau(c) \frac{\partial h}{\partial t} = h_\infty(c) - h, \tag{4.10c}$$

where $f(c, c_e, h)$ represents the net Ca^{2+} flux from the ER to the cytoplasm the spatial dimensions have been fixed such that $\chi_c = 1$. Analysis of traveling wave solutions of (4.10) proceeds along very similar lines to that of the Hodgkin–Huxley equations for membrane voltage excitability; see Chap. 2. Depending on parameter values, the model system can support solitary waves when the physiological state of the cytoplasm is excitable, periodic waves when it is oscillatory, and traveling fronts when it is bistable [322].

However, one important feature that needs to be taken into account in the case of calcium excitability is buffering. Insights into the effects of buffering can be obtained by looking at reduced models of excitability such as the bistable equation or the FN equations [592, 635, 636, 653]. Following [592, 653], consider a simple extension of the bistable equation (Sect. 2.2). First, note that the basic chemical reaction for Ca^{2+} buffering takes the form

$$P + \text{Ca}^{2+} \underset{k_-}{\overset{k_+}{\rightleftharpoons}} B,$$

where P is the buffering protein and B is the protein/Ca^{2+} complex. Let c and b denote, respectively, the concentration of free cytosolic Ca^{2+} and buffered Ca^{2+}. The extended bistable equation is then (in one spatial dimension)

$$\frac{\partial c}{\partial t} = D_c \frac{\partial^2 c}{\partial x^2} + f(c) + k_- b - k_+ c(b_t - b), \tag{4.11a}$$

$$\frac{\partial b}{\partial t} = D_b \frac{\partial^2 b}{\partial x^2} - k_- b + k_+ c(b_t - b). \tag{4.11b}$$

Here b_t is the total buffer concentration, $f(c)$ takes into account all Ca^{2+} fluxes, and the diffusion coefficients of free and buffered Ca^{2+} are assumed to be constants. Suppose that the buffer reaction rates $k_\pm$ are faster than the time constants of Ca^{2+} dynamics. The buffer concentration b can be taken to be in the quasi steady state $k_- b - k_+ c(b_t - b) = 0$, that is,

$$b = \frac{b_t c}{K+c}, \quad K = \frac{k_-}{k_+}. \tag{4.12}$$

Adding (4.11a) and (4.11b) and eliminating b using (4.12) gives [653]

$$\begin{aligned}\frac{\partial c}{\partial t} &= \frac{1}{1+\theta(c)}\left(\frac{\partial^2 w(c)}{\partial x^2} + f(c)\right) \\ &= \frac{D_c + D_b\theta(c)}{1+\theta(c)}\frac{\partial^2 c}{\partial x^2} - \frac{2D_b\theta(c)}{(K+c)(1+\theta(c))}\left(\frac{\partial c}{\partial x}\right)^2 + \frac{f(c)}{1+\theta(c)},\end{aligned}$$

where

$$w(c) = D_c c + D_b b_t \frac{c}{K+c}, \quad \theta(c) = \frac{b_t K}{(K+c)^2}. \tag{4.13}$$

It is clear that a mobile buffer ($D_b > 0$) generates a nonlinear advection–diffusion equation for c; the advection term vanishes if the buffer is immobile ($D_b = 0$).

The form of the effective reaction–diffusion equation for c suggests making then change of variables $w = w(c)$, with w, a monotone increasing function of c since $w'(c) = D_c + D_b\theta(c) > 0$. Denoting the unique inverse of the function $w(c)$ by $c = \phi(w)$, we can write

$$\frac{\partial w}{\partial t} = \frac{D_c + D_b\Theta(w)}{1+\Theta(w)}\left(\frac{\partial^2 w}{\partial x^2} + f(\phi(w))\right), \tag{4.14}$$

where $\Theta(w) = b_t K/(K+\phi(w))^2$. Now suppose that $f(c)$ has two stable zeros $c_\pm$ separated by an unstable zero c_0 with $c_- < c_0 < c_+$. It immediately follows that $f(\Theta(w))$ has three corresponding zeros $w_\pm, w_0$ with $w_- < w_0 < w_+$ and $w_\pm$ stable. A traveling front solution of wave speed v, $w(x,t) = W(x - vt)$, satisfies the ODE

$$V_{\xi\xi} + \frac{v}{D_{eff}(W)}W_\xi + f(\phi(W)). \tag{4.15}$$

Proceeding as in Sect. 2.2, multiply both sides of this equation by W_ξ and integrate to show that

$$v\int_{-\infty}^{\infty}\frac{W_\xi^2}{D_{\text{eff}}(W(\xi))}d\xi = \int_{w_-}^{w_+} f(\phi(w))dw.$$

Since the integral on the left-hand side is positive definite, the sign of v is determined by the sign of the integral on right-hand side. In particular, a right-moving wave ($v > 0$) exists provided that

$$\int_{c_-}^{c_+}(D_c + D_b\theta(c))f(c)dc > 0, \tag{4.16}$$

where we have converted back to the original variable c. It can be seen that immobile buffers ($D_b = 0$) have no effect on the existence of a traveling wave; this result also holds true for wave stability and uniqueness. On the other hand, mobile buffers can eliminate a traveling wave when they are of high affinity (small K) or have a large diffusion coefficient (large D_b).

4.3 Discrete Calcium Sources and the Fire–Diffuse–Fire Model

4.3.1 Discrete Calcium Sources and Averaging

The above reaction–diffusion models assumed that Ca^{2+} release from the ER is homogeneous in space. This neglects the fact that IP_3 receptors (as well as Ry receptors) tend to be arranged in clusters. For example, in *Xenopus* oocytes, the density of clusters is around 1 per $30\,\mu m^2$, with each cluster containing about 25 IP_3 receptors [322]. Consequently, the propagation of Ca^{2+} waves tends to be saltatory, jumping from one cluster to the next. (Clustering of IP_3 receptors is also thought to occur along dendrites [534]). We have previously encountered saltatory waves due to discreteness effects, namely, action potential propagation along myelinated axons (Sect. 2.5) and solitary Ca^{2+} spikes propagating along spiny dendrites (Sect. 3.1). The latter waves depend on the action of VGCCs and other voltage-gated ion channels and are thus distinct from the Ca^{2+} waves considered in this chapter, which depend on Ca^{2+} release from internal pools. Nevertheless, the analytical techniques used to study these other examples can be adapted to the present case. First, we will discuss how the averaging method of Sect. 2.5 can be used to study wave propagation failure in the presence of discrete release sites, following [318]. We will then describe an alternative, *fire–diffuse–fire* (FDF) model of saltatory Ca^{2+} waves, which is identical in structure to the spike–diffuse–spike (SDS) model of Sect. 3.1. (In contrast to other chapters, we will use c to denote cytosolic Ca^{2+} concentration and v to denote wave speed.)

Consider a one-dimensional model of Ca^{2+} diffusion and release of the form [318]

$$\frac{\partial c}{\partial t} = -k_c c + D_c \frac{\partial^2 c}{\partial x^2} + L \sum_n \delta(x - nL) f(c), \tag{4.17}$$

where $f(c)$ represents the Ca^{2+}-dependent rate of release from each cluster (taken to be identical), and the decay term represents the effects of ion pumps. The clusters are assumed to be evenly spaced with spatial separation L. For simplicity, the ER concentration c_e and IP_3 concentration p are taken to be fixed. It is tempting to apply homogenization theory to this problem (see appendix section 4.6), assuming diffusion is fast on the length scale L, that is, $L^2 k_c / D_c \ll 1$. Intuitively, one would expect the L-periodic function $g(x) = L \sum_n \delta(x - nL)$ to be replaced by its spatial average $\int_0^L g(x) dx / L = 1$. However, as highlighted by Keener [318, 319] and discussed in Sect. 2.5, the homogenized system cannot account for the fact that discreteness effects can lead to wave propagation failure in regimes that the continuous model would predict traveling waves exist. This reflects the fact that wave solutions are not structurally stable. The occurrence of propagation failure can be established by constructing standing wave-front solutions [318, 322]. The latter are stationary solutions of (4.17). On the intervals $nL < x < (n+1)L$, this gives

$$0 = -k_c c + D_c \frac{\partial^2 c}{\partial x^2}, \tag{4.18}$$

which is supplemented by jump conditions at $x = nL$,

$$D_c \frac{\partial c}{\partial x}\bigg|_{nL_-}^{nL_+} + Lf(c_n) = 0. \tag{4.19}$$

The latter is obtained by integrating the stationary equation on either side of $x = nL$. The general solution of (4.18) is

$$c(x) = c_n \cosh(\beta(x - nL)/L) + \hat{c}_n \sinh(\beta(x - nL)/L), \quad nL < x < (n+1)L,$$

where $\beta^2 = k_c L^2/D_c$, $c_n = c(nL)$ and $\beta \hat{c}_n/L = c'(nL_+)$. Continuity of the solution at $x = (n+1)L$ shows that

$$c_n \cosh(\beta) + \hat{c}_n \sinh(\beta) = c_{n+1},$$

that is,

$$\hat{c}_n = \frac{c_{n+1} - c_n \cosh(\beta)}{\sinh(\beta)}.$$

It follows that

$$c'(nL_+) = (c_{n+1} - c_n \cosh(\beta)) \frac{\beta}{L \sinh(\beta)}$$

and

$$c'(nL_-) = -(c_{n-1} - c_n \cosh(\beta)) \frac{\beta}{L \sinh(\beta)}.$$

Substituting these results into the jump condition yields the difference equation

$$\frac{k_s}{\beta \sinh(\beta)} (c_{n+1} - 2c_n \cosh(\beta) + c_{n-1}) + f(c_n) = 0. \tag{4.20}$$

It is usually not possible to obtain an explicit solution of a nonlinear difference equation. One exception is if $f(c)$ is piecewise linear,

$$f(c) = f_0 H(c - c^*), \tag{4.21}$$

where c^* is a release threshold. Without loss of generality, suppose that the standing wave solution crosses threshold from below in the interval $0 < x < L$. Try a solution of the form $c_n = A_- \mu^{-n}$ for $n \leq 0$ and $c_n = C - A_+ \mu^n$ for $n \geq 0$ with $0 < \mu < 1$. The constant C is determined by taking the limit $n \to \infty$ in (4.20) with $c_n \to C > c^*$:

$$\frac{k_s}{\beta \sinh(\beta)}(C - 2C\cosh(\beta) + C) + f_0 = 0,$$

so that

$$C = \frac{\beta \sinh(\beta)}{2\cosh(\beta) - 2}\frac{f_0}{k_c} = \frac{\beta \cosh(\beta/2)}{2\sinh(\beta/2)}\frac{f_0}{k_c}.$$

The constant μ then satisfies the quadratic equation $\mu - 2\lambda + \mu^{-1} = 0$ with $\lambda = \cosh(\beta)$. Choosing the root that is less than one, we have $\mu = \lambda - \sqrt{\lambda^2 - 1} = e^{-\beta}$. Finally, the factors $A_\pm$ can be determined by considering the difference equation for $n = 0, 1$, which gives [322]

$$A_-(\beta) = \frac{\mu(\beta)C(\beta)}{1 + \mu(\beta)}, \quad A_+(\beta) = e^{\beta} A_-(\beta).$$

Given that C,$A_\pm$ and $\mu_\pm$ are all functions of β (assuming that f_0/k_c is fixed), the conditions for the existence of a standing wave-front solution are

$$A_-(\beta) < c^*, \quad C(\beta) - A_+(\beta)\mu(\beta) = e^{\beta} A_-(\beta) \geq c^*. \tag{4.22}$$

For a given β, a traveling wave exists if the threshold c^* lies beneath the curve $A_-(\beta)$, whereas a standing wave exists if c^* lies between $A_-(\beta)$ and $e^{\beta}A_-(\beta)$; in the latter case propagation failure occurs. Note, in particular, that increasing the threshold c^* or increasing β makes propagation failure more likely. Moreover, increasing β corresponds to increasing the cluster spacing L or decreasing the Ca^{2+} diffusion coefficient D_c.

Having established that wave propagation failure can occur, we now describe how to calculate the wave speed as a function of "discreteness" using the averaging method of Keener [318, 319]. The analysis proceeds along identical lines to the study of inhomogeneous diffusivity in the discrete bistable equation; see Sect. 2.5. First, rewrite (4.17) in the dimensionless form

$$\frac{\partial c}{\partial t} = -c + \frac{\partial^2 c}{\partial x^2} + (1 + g'(x/\varepsilon))f(c), \tag{4.23}$$

where $\varepsilon = \sqrt{L^2 k_c/D} \ll 1$, a factor of k_c^{-1} has been absorbed into f, and $g(x)$ is the periodic sawtooth function, $g(x) = 1/2 - x$ for $0 < x < 1$ and $g(x+n) = g(x)$. (More precisely, $g(x)$ is assumed to be the sum of Gaussians (2.69) of width σ, with the limit $\sigma \to 0$ taken at the end of the calculation.) Suppose that the nonlinear function $f(c)$ is one for which there exists a traveling front solution $c(x,t) = C(x - vt)$ of the homogeneous equation ($g \equiv 0$). The main result of the averaging method is that for ε sufficiently small, there exists a wavelike solution or pulsating wave of the form [318]

$$c(x,t) = C(x - \phi(t)) + \mathcal{O}(\varepsilon), \tag{4.24}$$

where the phase $\phi(t)$ evolves according to [see also (2.72)]

$$\frac{d\phi}{dt} = v - \Phi(\phi/\varepsilon) \tag{4.25}$$

with

$$\Phi(\phi/\varepsilon) = \frac{1}{\Lambda}\int_{-\infty}^{\infty} g'([\xi+\phi]/\varepsilon)f(C(\xi))C'(\xi)e^{v\xi}d\xi, \tag{4.26}$$

and

$$\Lambda = \int_{-\infty}^{\infty} e^{v\xi}C'(\xi)^2 d\xi. \tag{4.27}$$

Equation (4.25) implies that the solution is not translationally invariant, rather it moves with a time-dependent velocity ϕ'. If $v-\Phi(\phi/\varepsilon)$ is strictly positive then $\phi'(t)$ is a positive, periodic function of t with period

$$T = \int_0^{\varepsilon} \frac{d\phi}{v-\Phi(\phi/\varepsilon)}. \tag{4.28}$$

The mean speed of the wave is $\bar{v} = \varepsilon/T$. On the other hand. if $v-\Phi(\phi/\varepsilon)$ vanishes for some ϕ, then propagation failure is expected to occur.

Averaging method for discrete Ca^{2+} release sites. The first step is to rewrite (4.23) as the first-order system

$$c_x = u, \quad u_x = c_t - (1+g'(x/\varepsilon))f(c) + c. \tag{4.29}$$

Introducing the exact change of variables $u = w - \varepsilon g(x/\varepsilon)f(u)$ yields the new system

$$c_x = w - \varepsilon g(x/\varepsilon)f(c), \quad w_x = c_t - f(c) + c + \varepsilon g(x/\varepsilon)f'(c)c_x. \tag{4.30}$$

It can be seen that if the $\mathcal{O}(1)$ system of equations is independent of x/ε, then the lowest-order averaged system reduces to the homogeneous equation, which we are assuming supports a traveling front solution $C(x-vt)$. Including the inhomogeneous factors $g(x/\varepsilon)$ means that the system is no longer translationally invariant. However, we can look for solutions that are close to a traveling front by going to a traveling coordinate system with $\xi = x - \phi(t)$ to give [318]

$$c_\xi - w = -\varepsilon g([\xi+\phi]/\varepsilon)f(c) \tag{4.31a}$$

$$-c_t + w_\xi + \phi' c_\xi + f(c) - c = \varepsilon g([\xi+\phi]/\varepsilon)f'(c)c_\xi. \tag{4.31b}$$

We now seek a perturbative solution of the form

$$c = C(\xi) + \varepsilon c_1(\xi,t) + \ldots, \quad w(\xi) = C'(\xi) + \varepsilon w_1(\xi,t) + \ldots, \quad \phi'(t) = v + \varepsilon\phi_1'(t) + \ldots$$

Substituting into equations (4.31) and collecting terms of equal powers in ε yields a hierarchy of equations, the first of which is

$$\partial_\xi c_1 - w_1 = -g([\xi+\phi]/\varepsilon)f(C) \tag{4.32a}$$

$$-\partial_t c_1 + \partial_\xi w_1 + v w_1 + f'(C)c_1 - c_1 = vg([\xi+\phi]/\varepsilon)f(C) - \phi_1'(t)C' + g([\xi+\phi]/\varepsilon)f'(C)C'. \tag{4.32b}$$

Equations (4.32) can be rewritten as

$$-\begin{pmatrix} 0 \\ \partial_t c_1 \end{pmatrix} + \mathbb{L}\begin{pmatrix} c_1 \\ w_1 \end{pmatrix} = \begin{pmatrix} h_c \\ -\phi_1' C' + h_w \end{pmatrix}, \quad \mathbb{L} = \begin{pmatrix} \partial_\xi & -1 \\ f'(C) - c & \partial_\xi + c \end{pmatrix}, \tag{4.33}$$

with h_c and h_w determined by inhomogeneous terms on the right-hand side of equations (4.32a) and (4.32b), respectively. Following our analysis of the linear operator (2.62), we know that the matrix operator appearing in equation (4.33) has a null space spanned by (C', C''). Similarly, the adjoint operator

$$\mathbb{L}^\dagger = \begin{pmatrix} -\partial_\xi & f'(C) - C \\ -1 & -\partial_\xi + c \end{pmatrix}$$

has the null vector $e^{c\xi}(-C'', C')$. Hence, applying the Fredholm alternative, it follows that $\phi'(t)$ must satisfy the phase equation (4.25) with $\phi(t) = v + \varepsilon\phi_1(t)$.

For the particular choice $f(c) = H(c - c^*)$ used in the analysis of standing waves with $f_0/k_c = 1$, the homogeneous equation is identical to the one analyzed at the end of Sect. 2.2. Thus, there exists a unique traveling front solution of the form (2.23) with corresponding wave speed (2.24). The mean wave speed can then be calculated explicitly along similar lines to Sect. 2.5, so that

$$\bar{v} = \frac{\varepsilon}{2} \frac{1}{\ln\left(\frac{v+\varepsilon/\chi}{v-\varepsilon/\chi}\right)}. \tag{4.34}$$

with $v = (1 - 2c^*)/\sqrt{c^* - c^{*2}}$ and $\chi = 4[c^* - c^{*2}]$. This would predict propagation failure when $\varepsilon \geq \chi v$, that is,

$$\varepsilon \geq 4(1 - 2c^*)(\sqrt{c^* - c^{*2}}). \tag{4.35}$$

Since the analysis is based on the assumption that ε is small, we expect this condition to be reasonable when $c^* \approx 0.5$. Taylor expanding about this point the condition for propagation failure is approximately $\varepsilon > 2(1 - 2c^*)$. Similarly, the condition (4.22) based on the standing wave analysis predicts propagation failure when $\varepsilon/(2(e^\varepsilon - 1)) < c^*$, which reduces to $\varepsilon > 2(1 - 2c^*)$ when $c^* \approx 1/2$. Thus the two approaches agree for small ε and yield qualitatively similar results for large ε. Finally, as in the analysis of myelinated axons, a cubic nonlinearity for the release function $f(c)$ yields an expression for the mean wave speed that involves exponentially small terms so it is necessary to include higher-order terms in the perturbation analysis in order to obtain sufficient accuracy [318].

4.3.2 *The Fire–Diffuse–Fire Model of* Ca^{2+} *Release*

An alternative approach to studying the effects of discrete Ca^{2+} release sites is to consider the so-called *FDF* model [127, 137, 140, 325, 497, 511]. The basic idea of the model is that once the Ca^{2+} concentration reaches a threshold value c^* at a given release site, that site fires and instantaneously releases a fixed amount σ of Ca^{2+} into the cytoplasm [325, 497, 511]. It is also straightforward to take into account the effects of ion pumps by including a decay term and replacing instantaneous release by a fixed pulse of finite width [127]. The FDF model is then almost identical to the SDS model of Ca^{2+} spikes propagating along spiny dendrites [129, 130]. Since the latter model was analyzed in detail in Sect. 3.1, we can carry over the previous results after minor modifications. Therefore, consider the following version of the FDF model [127]:

$$\frac{\partial c}{\partial t} = -\frac{c}{\tau_c} + D_c\frac{\partial^2 c}{\partial x^2} + \sum_n \delta(x-x_n)\eta(t-T_n), \tag{4.36}$$

where T_n is the firing time of the site x_n, which only fires once:

$$T_n = \inf\{t\,|\,c(x_n,t) \geq c^*\}. \tag{4.37}$$

For concreteness, let

$$\eta(t) = (\sigma/\tau_d)H(t)H(\tau_d - t),$$

so that each Ca^{2+} puff or spark is of size σ and duration τ_d. Suppose that the Ca^{2+} release sites are regularly spaced, $x_m = mL$, and consider a saltatory wave solution $T_m = m\Delta$ with speed $v = L/\Delta$. The existence of such a wave can be determined straightforwardly by noting that (4.36) is identical to (3.33) under the mapping

$$V \to c, \quad \frac{\bar{n}}{r_s}\eta_0 \to \frac{\sigma}{\tau_d}.$$

The one major difference between the two models is that thresholding in the SDS model is mediated by an IF process (3.2), which introduces an additional time constant $\hat{\tau}$. Therefore, in order to apply the analysis of saltatory waves along a spiny dendrite to the FDF model (see in Sect. 3.1.3), we need to take the limit $\hat{\tau} \to 0$. This then gives the following threshold condition for the existence of a saltatory wave:

$$c^* = \sum_{n=1}^{\infty} H(nL, n\Delta), \tag{4.38}$$

with

$$H(x,t) = \frac{\sigma}{\tau_d}\int_0^{\tau_d} G_0(x,t-s)ds, \quad G_0(x,t) = \frac{1}{\sqrt{4\pi D_c t}}e^{-t/\tau_c - x^2/4D_c t}. \tag{4.39}$$

In the limit $\tau_d \to 0$, $H(x,t) \to \sigma G_0(x,t)$ and the threshold condition becomes

$$\begin{aligned} c^* &= \sigma \sum_{n=1}^{\infty} G_0(nL, n\Delta) \\ &= \sigma \sum_{n=1}^{\infty} \frac{1}{\sqrt{4\pi D_c n\Delta}} \exp\left(-n\left[\frac{L^2 t}{4D_c\Delta} + \frac{\Delta}{\tau_c}\right]\right) \equiv g(\Delta). \end{aligned} \tag{4.40}$$

In the additional limit $\tau_c \to \infty$ (i.e., neglecting the flux due to ion pumps), the results of [497] are recovered. In particular, $g(\Delta)$ becomes a monotone function of Δ, and the speed of the traveling waves scales linearly as D_c/L. The latter follows after introducing the dimensionless threshold $\tilde{c}^* = c^*L/\sigma$ and noting that Lg/σ is then only a function of the dimensionless quantity $\Delta D_c/L^2$. However, this result is inconsistent with the analysis of standing waves, which showed that propagation failure can occur by decreasing the diffusion coefficient. The discrepancy is resolved by taking into account the effects of ion pumps, that is, taking τ_c to be finite. Moreover, $g(\Delta)$

is now non-monotone and consequently one finds two solution branches with the faster one corresponding to stable waves [127]. This result also holds in the case of finite pulse width τ_d, for which

$$H(x,t) = \frac{\sigma}{\tau_d}\left[A(x,t-\min(t,\tau_d)) - A(x,t)\right], \tag{4.41}$$

with $A(x,t)$ given by (3.40). In summary, the main qualitative results of the SDS model carry over to the FDF model, for example, one finds that propagation failure occurs if the spacing between release sites becomes too large.

One of the limitations of the above FDF model is that it neglects variations in the Ca^{2+} concentration within the ER. However, in many cells, Ca^{2+} release can result in a significant local depletion of ER Ca^{2+}. In order to take this into account, the FDF model has been extended in the form of a bidomain threshold-release model [621, 622]:

$$\frac{\partial c}{\partial t} = D_c\frac{\partial^2 c}{\partial x^2} + J_{\mathrm{rel}}(c,c_e) - J_2(c,c_e) \tag{4.42a}$$

$$\frac{\partial c_e}{\partial t} = D_e\frac{\partial^2 c}{\partial x^2} - \chi_e\left[J_{\mathrm{IP}}(c,c_e) - J_2(c,c_e)\right], \tag{4.42b}$$

where the release and uptake fluxes, $J_{rel}(c,c_e) - J_2(c,c_e)$, are functions of the cytosolic and ER Ca^{2+} concentrations. For simplicity, assume that there is a continuous and uniform distribution of Ca^{2+} release sites and that the release flux depends on the difference in the cytosolic and ER Ca^{2+} concentrations:

$$J_{\mathrm{rel}}(x,t) \equiv J_{\mathrm{rel}}(c(x,t),c_e(x,t)) = (c_e(x,t) - c(x,t))\eta(t - T(x)), \tag{4.43}$$

with

$$T(x) = \inf\{t\,|\,c(x,t) \geq c^*\}. \tag{4.44}$$

The release sites could be either RyRs or IP_3Rs; in the latter case the IP_3 concentration p is held fixed. Finally, the reuptake flux due to the action of ion pumps is taken to have the linear form

$$J_2(c,c_e) = \frac{c}{\tau_c} - \frac{c_e}{\tau_e}. \tag{4.45}$$

The existence and stability of a solitary pulse solution of (4.42) can be investigated along similar lines to Sect. 3.1.1.

4.4 Stochastic Models of Ca^{2+} Release

The fluorescent imaging of localized Ca^{2+} puffs and sparks has established that Ca^{2+} release is a stochastic process that occurs at spatially discrete sites consisting of clusters of IP_3Rs and RyRs, respectively. In *Xenopus* oocytes, Ca^{2+} puffs have an amplitude ranging from around 50 to 600 nM, a spatial spread of approximately

6 μm and a typical duration of 1 s [489, 490, 690]. For sufficiently high levels of IP_3 concentration, the amplification of Ca^{2+} puffs by CICR can lead to the formation of Ca^{2+} waves. Ca^{2+} sparks in heart and skeletal muscle tend to be of shorter duration and less spatial spread and are less likely to result in wave propagation [110, 111]. As we have already mentioned Ca^{2+} puffs and sparks have also been observed in neurons [534], suggesting that one needs to take into account stochastic release events when modeling Ca^{2+} waves. Typically, one combines a stochastic model of localized release through RyRs or IP_3Rs with a deterministic reaction–diffusion model of Ca^{2+} waves such as the FDF model [137, 320]. The stochastic modeling of the Ca^{2+}-dependent opening and closing of these receptors proceeds along analogous lines to the modeling of voltage-gated ion channels considered in Sect. 1.5.

4.4.1 Stochastic Model of Ca^{2+} Puffs in a Cluster of IP_3Rs

Stochastic models of Ca^{2+} puffs typically treat a cluster of IP_3Rs as a set of N channels that open and close independently, but are indirectly coupled by the common cytoplasmic Ca^{2+} concentration [180, 182, 579, 612]. Models differ in the level of detail regarding individual receptors. The first deterministic kinetic model of Ca^{2+}-gated IP_3Rs was proposed by De Young and Keizer, in their study of agonist-induced Ca^{2+} oscillations. This model assumes that the IP_3 receptor consists of three equivalent receptor subunits, all of which have to be in a conducting state in order to generate a Ca^{2+} flux. Each subunit is taken to have an IP_3-binding site, an activating Ca^{2+}-binding site, and an inactivating Ca^{2+}-binding site; the conducting state corresponds to the state in which all subunits have the first two binding sites occupied but the third unoccupied. Although the De Young–Keizer model is simple to describe, it involves a relatively large number of variables that have to be coupled to the Ca^{2+} and IP_3 concentrations. A simplified version of the model was subsequently developed by Li and Rinzel [381]. They exploited the fact that the binding of IP_3 and activating Ca^{2+} are fast relative to inactivating Ca^{2+} and used a quasi-steady-state argument to reduce the eight-state subunit model to a model that simply keeps track of whether or not the inactivating Ca^{2+}-binding site of a subunit is occupied. More specifically, the Li–Rinzel model is a two-variable model given by

$$\frac{dc}{dt} = J_{\rm IP} + J_{\rm leak} - J_{\rm p2} \tag{4.46a}$$

$$\frac{dh}{dt} = \alpha_h(1-h) - \beta_h h, \tag{4.46b}$$

where c is the cytoplasmic Ca^{2+} concentration, h is the fraction of receptors in a cluster not inactivated by Ca^{2+}, and p is the IP_3 concentration, which is assumed fixed. The three Ca^{2+} fluxes included in the model are the channel flux $J_{\rm IP}$ and leakage flux $J_{\rm leak}$ from the ER to the cytoplasm, and the flux $J_{\rm p2}$ pumped back into the ER; see also Fig. 4.2. The expressions for the various fluxes are

$$J_{\rm IP} = f(c,p)h^3[c_e - c], \quad J_{\rm leak} = v_0[c_e - c], \quad J_{\rm P2} = \frac{v_1 c^2}{k_3^2 + c^2}, \tag{4.47}$$

with

$$f(c,p) = \left(\frac{p}{p+k_4}\right)^3 \cdot \left(\frac{c}{c+k_5}\right)^3, \quad \alpha_h = v_2\frac{p+k_4}{p+k_6}, \quad \beta_h = v_3 c. \tag{4.48}$$

The various cubic terms reflect the existence of three subunits. Parameter values of the model can be found in [381]. We see that the simplified model resembles the Hodgkin–Huxley model (see Sect. 1.1), after replacing Ca^{2+} concentration c by membrane voltage v and c_e by a reversal potential.

We now describe a stochastic version of the Li–Rinzel model for a cluster of IP_3Rs due to Shuai and Jung [579]. For stochastic versions of the full De Young–Keizer model, see, for example, [180, 182, 244, 612]. The deterministic equations (4.46) describe the mean behavior of a large cluster of Ca^{2+} channels, just as the Hodgkin–Huxley equations for membrane voltage apply to a large number of voltage-gated ion channels. As we discussed in Sect. 1.5, if the number of channels is relatively small, then it is necessary to take into account thermally driven fluctuations in the opening and closing of individual channels. In the case of the Li–Rinzel model, one only needs to consider the opening and closing process for the gating variable h of each subunit. The latter is modeled as the two-state Markov process

$$C(\text{closed}) \underset{\beta_h(c)}{\overset{\alpha_h}{\rightleftharpoons}} O(\text{open}). \tag{4.49}$$

Suppose that there are N independent IP_3Rs, each with three independent subunits labeled $i = 1,2,3$ that are described by the above two-state Markov process. Let $N_i(t)$ $(i = 1,2,3)$ denote the number of receptors at time t that have the ith subunit open. Under the adiabatic assumption that the Ca^{2+} concentration c evolves much more slowly than the state transitions of the channels, we can write down a master equation for the probability $P(n_i,t) = \text{Prob}[N_i(t) = n_i|N_i(0) = n_0]$ according to

$$\begin{aligned}\frac{dP(n_i,t)}{dt} &= (N - n_i + 1)\alpha_h P(n_i - 1,t) + (n_i + 1)\beta_h P(n_i + 1,t) \\ &\quad - (n_i\beta_h + (N - n_i)\alpha_h)P(n_i,t), \quad i = 1,2,3.\end{aligned} \tag{4.50}$$

As with voltage-gated ion channels, we have a stochastic hybrid system, since the transition rate β_h depends on the Ca^{2+} concentration $c(t)$, which evolves according to a piecewise deterministic equation of the form (4.46a). The latter, in turn, couples to the discrete stochastic variables $N_i(t)$ through the flux

$$J_{\rm IP} = f(c(t),p)[c_e - c(t)]\prod_{i=1}^{3}\frac{N_i(t)}{N}. \tag{4.51}$$

[Note that one should really write down a differential Chapman–Kolmogorov equation for the joint probability density $p(n_1,n_2,n_3,c,t)$ along the lines of (1.126)]. Finally, for large N, one can obtain a further simplification by carrying out a Kramers–Moyal expansion of the master equation (4.50) along the lines of Sect. 1.5.2. This yields the following SDE for $H_i(t) = N_i(t)/N$ with H_i treated as a continuous stochastic variable:

$$dH_i = \alpha_h(1-H_i) - \beta_h H_i + \frac{1}{\sqrt{N}} b(H_i) dW_i, \tag{4.52}$$

where

$$b(H_i) = \sqrt{\alpha_h(1-H_i) + \beta_h H_i},$$

and $W_i(t)$ is an independent Wiener process with

$$\langle dW_i(t)\rangle = 0, \quad \langle dW_i(t)dW_j(t')\rangle = \delta(t-t')dt\,dt'\delta_{i,j}.$$

Shuai and Jung [579] simulated the stochastic Li–Rinzel model in order to investigate the effects of noise on Ca^{2+} oscillations in a space-clamped model. They assumed that the deterministic system (4.46) was monostable at low and high IP_3 concentrations and exhibited limit cycle oscillations (occurring via a Hopf bifurcation) at intermediate concentrations. They showed that noise can enlarge the range of IP_3 concentrations over which oscillations occur—an effect known as *coherence resonance*. They also found a broad distribution of puff amplitudes, lifetimes, and interpuff intervals. In particular, at low IP_3 concentrations, the amplitude distribution is a monotonically decaying function, whereas at higher concentrations, it is unimodal. This suggests that Ca^{2+} puffs become more significant as IP_3 concentration is increased and hence could impact the spontaneous generation of Ca^{2+} waves. This issue was investigated numerically by Falcke [180] using a stochastic version of the De Young–Keizer model that was incorporated into a reaction–diffusion model of spatially distributed channel clusters. He showed that there is indeed a transition from Ca^{2+} puffs to waves as the IP_3 concentration is increased. At low concentrations, only puffs occur, since there is not enough Ca^{2+} released to stimulate neighboring clusters, which means that the response is purely local. However, as IP_3 concentration increases, global Ca^{2+} waves can emerge from local nucleation sites of high Ca^{2+} concentration. At intermediate levels of IP_3, global events are rare and waves only progress a short distance before dying out. On the other hand, for higher IP_3 concentrations, global waves occur regularly with a well-defined period. Again this oscillatory-like behavior can occur in parameter regimes for which the deterministic model is non-oscillatory.

4.4.2 Stochastic Model of Ca^{2+} Sparks in a Cluster of RyRs

We now turn to a stochastic model of Ca^{2+} sparks due to Hinch [276]; see also [241]. This model was originally developed for *cardiac myocytes* (heart muscle cells) and includes details of the geometry of Ca^{2+} release units, in particular, the narrow junctional gap known as the diadic space that separates the sarcoplasmic reticulum (SR) from the plasma membrane; see Fig. 4.4. (In smooth muscle cells the smooth ER is referred to as the sarcoplasmic reticulum.) However, it is possible that a similar mechanism occurs in neurons at the specialized subsurface cisternae of the ER, which also form narrow junctions with the plasma membrane; see Fig. 4.2. In a typical myocyte, there could be up to 10,000 Ca^{2+} release units, each one containing a cluster of around $N = 50$ RyRs on the surface of the SR. The cluster of RyRs is apposed to L-type Ca^{2+} channels located on so-called t-tubules, which are invaginations of the plasma membrane into the myocyte. (The Ca^{2+} channels are not involved in the spontaneous generation of Ca^{2+} sparks so are ignored in the model.) The diadic space separating the SR from the t-tubules is a region of the mytoplasm (intracellular fluid of myocytes), which is approximately cylindrical in shape with width 10 nm and radius 100 nm. Since the diadic space is a small enclosed volume, it supports an elevation in Ca^{2+} concentration relative to the bulk mytoplasm following the release of Ca^{2+} from an RyR. Such a local elevation plays a crucial role in the Ca^{2+}-induced Ca^{2+}-release (CICR) that results in a Ca^{2+} spark. The SR in a neighborhood of the RyRs is known as the junctional SR (JSR), which may have a different Ca^{2+} concentration from the bulk or network SR (NSR).

We now briefly introduce the model of Hinch in nondimensional form; details of model approximations and estimates of experimentally based model parameters can be found in [276]. First, the diadic space is modeled as a single compartment with Ca^{2+} concentration c satisfying the current conservation equation

$$\tau_D \frac{dc}{dt} = J_{RyR} - J_{\mathrm{D}}. \tag{4.53}$$

Here τ_D is a time constant, J_{RyR} is the total Ca^{2+} current through the RyRs, and J_{D} is the diffusive current from the diadic space to the bulk mytoplasm. The latter is modeled as the Fickian current

$$J_{\mathrm{D}} = c - c_m, \tag{4.54}$$

where c_m is the bulk mytoplasm Ca^{2+} concentration. The total current through the RyRs is taken to be proportional to the number n of open RyRs times the Ca^{2+} concentration c_{sr} in the JSR:

$$J_{\mathrm{RyR}} = c_{\mathrm{sr}} x, \quad x = \frac{n}{N}, \tag{4.55}$$

with N the total number of RyRs in the cluster. Each RyR has Ca^{2+}-binding sites, which can be activating or deactivating. When an RyR is in an activated state it is promoted to a mode where it continuously opens and closes according to a Markov process (see Sect. 1.5), with a mean open time of 1 ms [695]. The opening of an RyR channel results in an extra Ca^{2+} current flowing into the diadic space, which increases the rate at which Ca^{2+} binds to the other RyRs via CICR, thus creating a positive feedback loop. This feedback loop provides a mechanism for bistability.

Note that the RyRs also contain inactivating Ca^{2+}-binding sites, but these do not play a role in initiating a Ca^{2+} spark so are not included in the model. For simplicity, the RyRs are modeled using a two-state Markov process involving a single closed state and a single open state (see also [324]):

$$C(\text{closed}) \underset{k_-}{\overset{k_+(c)}{\rightleftharpoons}} O(\text{open}), \tag{4.56}$$

with transition rates

$$k_+(c) = \frac{1}{k\tau_o}\frac{c^\alpha}{c^\alpha+1}, \quad k_- = \frac{1}{\tau_o}. \tag{4.57}$$

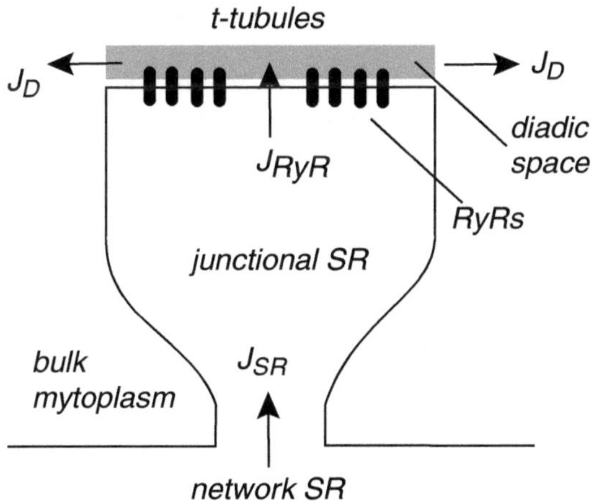

Fig. 4.4 Schematic diagram of a Ca^{2+} release unit in a cardiac myocyte. It is divided into four compartments: the network SR, the junctional SR, the diadic space, and the bulk mytoplasm. See text for details

Here τ_o is the mean open time of an RyR, α is the number of Ca^{2+} ions that are needed to open an RyR, and k determines the proportion of time the RyRs are open. It is assumed that the RyRs are gated independently of each other. They are, however, indirectly coupled via the Ca^{2+} concentration in the diadic space. The time constant τ_D of diffusive flux from the diadic space is several orders of magnitude smaller than the mean open time τ_o of an RyR, that is, $\tau_D \sim 3\,\mu\text{s}$ whereas $\tau_o \sim 1$ ms. Therefore, the Ca^{2+} concentration in the diadic space can be taken to be in quasi-equilibrium, $\tau_D \to 0$, so that

$$c = c_m + c_{sr}x. \tag{4.58}$$

It follows that the transition rate can be reexpressed as a function of the fraction of open channels and the Ca^{2+} concentration in the SR, $k_+ = k_+(c_m + c_{sr}x)$.

Now consider N independent RyRs within a Ca^{2+} release unit, each described by the above two-state Markov process. Let $N(t)$ be the number of open channels at time t and set $p_n(t) = \text{Prob}[N(t) = n|N(0) = n_0]$. Following along similar lines to Sect. 1.5.1, it can be shown that $p_n(t)$ evolves according to a master equation with transitions rates $\omega_\pm(n/N) : n \to n \pm 1$:

$$\omega_+(x) = N(1-x)\frac{(c_m + c_{\text{sr}}x)^\alpha}{k((c_m + c_{\text{sr}}x)^\alpha + 1)}, \quad \omega_-(x) = Nx. \tag{4.59}$$

For the moment, it is assumed that c_m and c_{sr} are fixed so that ω_+ can be treated as a function of x alone. (In Sect. 4.4.4, the dynamics of c_{sr} following initiation of a Ca^{2+} spark will also be taken into account.) The units of time are fixed by setting $\tau_o = 1$. The master equation for the RyR cluster is then

$$\frac{dp_n}{dt} = \omega_+((n-1)/N)p_{n-1} + \omega_-((n+1)/N)p_{n+1} - (\omega_+(n/N) + \omega_-(n/N))p_n, \tag{4.60}$$

with boundary conditions $p_{N+1} = p_{-1} \equiv 0$. Multiplying both sides of the master equation (4.60) by n/N and summing over n gives

$$\frac{d\langle n/N\rangle}{dt} = \langle \Omega_+(n/N)\rangle - \langle \Omega_-(n/N)\rangle, \tag{4.61}$$

where $\omega_\pm = N\Omega_\pm$, and the brackets $\langle \ldots \rangle$ denote a time-dependent ensemble averaging over realizations of the stochastic dynamics, that is, $\langle A(n/N)\rangle = \sum_n p_n A(n/N)$ for any function of state $A(n/N)$. In the thermodynamic limit $N \to \infty$ (infinite cluster size), statistical correlations can be ignored so that one can take the mean-field limit $\langle \Omega_\pm(n/N)\rangle \to \Omega_\pm(\langle n/N\rangle)$. This then yields a deterministic equation for the fraction x of open RyRs:

$$\frac{dx}{dt} = \Omega_+(x) - \Omega_-(x) = (1-x)\frac{(c_m + c_{\text{sr}}x)^\alpha}{k((c_m + c_{\text{sr}}x)^\alpha + 1)} - x. \tag{4.62}$$

It can be shown that, for physiologically reasonable parameter values, this equation exhibits bistability [276], that is, there exists a pair of stable fixed points $x_\pm$ separated by an unstable fixed point x_0. The fixed point $x_- \approx 0$ represents a quiescent state, whereas the other fixed point x_+ represents a Ca^{2+} spark in which a significant fraction of RyRs are in the active mode and can be interpreted as a burst phase. Hence, this model operates in a completely different regime to the model considered by Shuai and Jung [579], which was in a monostable state (a fixed point or limit cycle) in the deterministic limit. Bistability of the Hinch model means that a Ca^{2+} spark is a distinct event rather than a broad distribution of events.

Noise-induced transitions from x_- to x_+ determine the distribution of inter-spark intervals, just as noise-induced transitions from x_+ to x_- determine the distribution of spark lifetimes. Hence, estimating the mean time for the occurrence of a spark event reduces to the problem of calculating the MFPT to reach x_+, starting from a neighborhood of x_-, by crossing x_0. (The dominant contribution to this MFPT is the time to reach x_0, since the system then quickly relaxes to x_+.) We encountered an analogous problem in Sect. 1.5, where we considered the mean time for spontaneous action potential initiation due to ion-channel fluctuations. We first estimated the MFPT by carrying out a diffusion approximation of the underlying master equation (1.104), and calculating the MFPT of the corresponding Fokker–Planck equation. However, the diffusion approximation resulted in exponentially large errors. Therefore, we considered an alternative approach in Sect. 1.6, which was based on WKB methods and asymptotic analysis. Here we consider a corresponding WKB approximation of the master equation (4.60). Such an approach has been used increasingly to analyze escape problems in chemical and biological systems [160, 164, 175, 176, 256, 339, 340], including the Ca^{2+} release model of Hinch [277]. Note, however, that in contrast to the analysis of membrane voltage fluctuations in Sect. 1.6, the analysis of calcium sparks is performed with respect to the small parameter $1/N$ rather than ε. That is, we carry out a system-size expansion rather than assuming fast kinetics. For the moment, we simply state the results and leave the details to Sect. 4.4.3.

If x_- is not on the boundary $x = 0$ then a standard analysis shows that the mean time τ_i to initiate a Ca^{2+} spark starting from the quiescent state x_- is

$$\tau_i = \frac{2\pi}{\Omega_+(x_-)\sqrt{|\gamma(x_0)|\gamma(x_-)}} e^{N[\Phi(x_0)-\Phi(x_-)]}. \tag{4.63}$$

with

$$\gamma(x) = \frac{\Omega_-'(x)}{\Omega_-(x)} - \frac{\Omega_+'(x)}{\Omega_+(x)}, \quad \Phi(x) = \int^x \ln\left(\frac{\Omega_-(y)}{\Omega_+(y)}\right) dy.$$

Similarly, the mean duration τ_f of a spark, which corresponds to the mean time to transition back from x_+ to x_-, is given by

$$\tau_f = \frac{2\pi}{\Omega_+(x_+)\sqrt{|\gamma(x_0)|\gamma(x_+)}} e^{N[\Phi(x_0)-\Phi(x_+)]}. \tag{4.64}$$

Note that at a fixed point x_j, $\Omega_+(x_j) = \Omega_-(x_j)$ so that $\text{sign}(\gamma(x_j)) = \text{sign}(\Omega_-'(x_j) - \Omega_+'(x_j))$. Thus, $\gamma(x_0) < 0$ (unstable) and $\gamma(x_\pm) > 0$ (stable). One also finds that $\Phi(x_0) > \Phi(x_\pm)$ so that τ_i, τ_f are exponentially large. It turns out that in the case of Ca^{2+} release, the quiescent state x_- is in an $\mathcal{O}(1/N)$ neighborhood of the boundary $x = 0$, so that the prefactor of the MFPT has to be modified accordingly; see [276, 277] for details. Nevertheless, the leading order exponential is unchanged. From the perspective of modeling stochastic Ca^{2+} waves using the FDF model of Sect. 4.3.2, the main result that emerges from the analysis of Ca^{2+} sparks is that the deterministic threshold mechanism needs to be replaced by a probabilistic mechanism. In the deterministic case, a cluster of receptors fires whenever the intracellular Ca^{2+} crosses a fixed threshold. On the other hand, in a stochastic FDF model, Ca^{2+} release is a probabilistic process. Given the MFPT τ_i, the probability of a spark event in a time interval t is given by

$$P_i(t) = 1 - e^{-t/\tau_i}. \tag{4.65}$$

Recall that the transition rates $\Omega_{\pm}$, and thus τ_i, are functions of the Ca^{2+} concentration c_m of the mytoplasm, which has so far been assumed fixed. Indeed, one finds that the probability of release in a given time interval t is a sigmoidal function of c_m [276]. This stochastic release process can now be incorporated into a stochastic version of the FDF model by taking the sequence of release times T_n in (4.36) to be generated according to a sigmoidal probability distribution that is parameterized by the Ca^{2+} concentration in the mytoplasm, which itself evolves according to (4.36); see [137, 320] for details. The resulting model can be used to simulate the spontaneous occurrence of Ca^{2+} waves.

4.4.3 WKB Method and Quasistationary Approximation

We now present the details of how to calculate the rate of escape from a metastable state. Although, we focus here on the particular problem of Ca^{2+} release, the basic approach applies to the master equation of any bistable system in the large N (weak noise) limit. Throughout the analysis we will switch between n/N and x, with x treated as a continuous variable; this is a reasonable approximation when N is large. First, note that the master equation (4.60) with the given reflecting boundary conditions at $x = 0, 1$ has a unique stationary state given by (see also Sect. 1.5.1)

$$p_n^* = p_k^* \prod_{m=k+1}^{n} \frac{\Omega_+((m-1)/N)}{\Omega_-(m/N)} = p_k^* \exp\left(\sum_{m=1}^{n} \ln\left(\frac{\Omega_+((m-1)/N)}{\Omega_-(m/N)} \right) \right). \tag{4.66}$$

The functions $\Omega_{\pm}(x)$ are smooth and the second derivatives of $\ln \Omega_{\pm}(x)$ are bounded (except in the limits $x \to 0$ and $x \to 1$). Therefore, away from the boundaries, the sums can be approximated by the trapezium rule [276]

$$\sum_{m=k}^{n} \ln(f(m/N)) = \frac{\ln(f(k/N) + \ln f(n/N)}{2} + N \int_{k/N}^{n/N} \ln(f(y))dy + \mathcal{O}(1/N),$$

so that

$$p_n^* = \frac{\mathcal{N}}{\sqrt{\Omega_+(x)\Omega_-(x)}} e^{-N\Phi(x)}, \quad x = n/N, \tag{4.67}$$

where $\Phi(x)$ is the effective potential

$$\Phi(x) = \int^x \ln\left(\frac{\Omega_-(y)}{\Omega_+(y)} \right) dy, \tag{4.68}$$

and $\mathcal{N}$ is a normalization factor such that $\sum_{n=0}^{N} p_n^* = 1$.

Now suppose that we place an absorbing boundary at the unstable fixed point $x_0 = n_0/N$. Although there no longer exists a stationary solution, the flux through the absorbing boundary is exponentially small for large N, so that we can use a spectral projection method analogous to Sect. 1.6. First, rewrite the master equation (4.60) for $n = 0, \ldots, n_0$ as the linear system

$$\frac{d\mathbf{p}}{dt} = \mathbf{Q}\mathbf{p}, \tag{4.69}$$

where $\mathbf{p} = (p_0(t), p_1(t), \ldots, p_{n_0}(t))^T$, $\mathbf{Q}$ is the matrix of transition rates, and $p_{n_0}(t) = 0$ (absorbing boundary condition). Suppose that the eigenvalues of $\mathbf{Q}$ are ordered according to $0 > \lambda_0 \geq \mathrm{Re}[\lambda_1] \geq \mathrm{Re}[\lambda_2] \geq \ldots$ with corresponding eigenfunctions $\phi_n^{(j)}$, and consider the eigenfunction expansion

$$p_n(t) = \sum_{r=0}^{n_0} C_r \mathrm{e}^{-\lambda_r t} \phi_n^{(r)}. \tag{4.70}$$

We assume that $|\lambda_0| \sim \mathrm{e}^{-\eta N}$ for $\eta = \mathscr{O}(1)$, whereas λ_r for $r > 0$ are only weakly dependent on N. It follows that all other eigenmodes decay to zero much faster than the perturbed stationary density. Thus at large times, we have the quasistationary approximation

$$p_n(t) \sim C_0 \mathrm{e}^{-\lambda_0 t} \phi_n^{(0)}. \tag{4.71}$$

One can now use a WKB ansatz (see below) to generate a quasistationary solution ϕ_n^ε for which $\mathbf{Q}\phi^\varepsilon = 0$ and $\phi_{n_0}^\varepsilon \sim \mathscr{O}(\mathrm{e}^{-\eta N})$. Since the WKB solution does not satisfy the absorbing boundary condition, it is necessary to perform an asymptotic expansion in order to match the quasistationary solution with the solution in a neighborhood of x_0. In the process this determines λ_0, whose inverse can be identified as the MFPT to escape from the basin of attraction of the metastable state x_-.

Dropping exponentially small terms and writing $\phi_n^\varepsilon = \phi^\varepsilon(x)$ with x treated as a continuous variable, we have

$$\begin{aligned} 0 &= \Omega_+(x - 1/N)\phi^\varepsilon(x - 1/N) \\ &\quad + \Omega_-(x + 1/N)\phi^\varepsilon(x + 1/N) - (\Omega_+(x) + \Omega_-(x))\phi^\varepsilon(x). \end{aligned} \tag{4.72}$$

We seek a WKB solution of the form

$$\phi^\varepsilon(x) \sim K(x;\varepsilon)\mathrm{e}^{-\Phi(x)/\varepsilon}, \tag{4.73}$$

with $K(x;\varepsilon) \sim \sum_{m=0}^\infty \varepsilon^m K_m(x)$. Substituting (4.73) into (4.72), Taylor expanding with respect to ε, and collecting the $\mathscr{O}(1)$ terms gives

$$\Omega_+(x)(\mathrm{e}^{\Phi'(x)} - 1) + \Omega_-(x)(\mathrm{e}^{-\Phi'(x)} - 1) = 0, \tag{4.74}$$

where $\Phi' = d\Phi/dx$. Solving this quadratic equation in $\mathrm{e}^{\Phi'}$ shows that

$$\Phi = \int^x \ln \frac{\Omega_-(y)}{\Omega_+(y)} dy \quad \text{or} \quad \Phi = \text{constant}. \tag{4.75}$$

Proceeding to the next level, equating terms at $\mathcal{O}(\varepsilon)$ gives

$$\Omega_+ \mathrm{e}^{\Phi'} \left(-\frac{K_0'}{K_0} + \frac{\phi''}{2} \right) + \Omega_- \mathrm{e}^{-\Phi'} \left(\frac{K_0'}{K_0} + \frac{\phi''}{2} \right) - \Omega_+' \mathrm{e}^{\Phi'} + \Omega_- \mathrm{e}^{-\Phi'} = 0.$$

Substituting for Φ using (4.74) and solving for K_0 yields the following leading order forms for ϕ^ε:

$$\phi^\varepsilon(x) = \frac{A}{\sqrt{\Omega_+(x)\Omega_-(x)}} \mathrm{e}^{-N\Phi(x)}. \tag{4.76}$$

with Φ given by (4.75), which is sometimes called the activation solution, and

$$\phi^\varepsilon(x) = \frac{B}{\Omega_+(x) - \Omega_-(x)}, \tag{4.77}$$

which is sometimes called the relaxation solution. The constants A, B are determined by matching solutions around x_0. Clearly, (4.77) is singular at any fixed point x_j, where $\Omega_+(x_j) = \Omega_-(x_j)$, so is not a valid solution for the required quasistationary density. On the other hand, it does have an interpretation in terms of a corresponding Hamiltonian–Jacobi formulation of the WKB ansatz.

The WKB ansatz was also used in the analysis of metastability in stochastic ion-channel models in Sect. 1.6 and in the analysis of sharp interfaces for slowly modulated pulled fronts; see Sect. 3.2.4. Following along similar lines to the latter example, we can formally interpret (4.74) as a stationary Hamilton–Jacobi equation $H(x, \Phi'(x)) = 0$ for Φ, with Hamiltonian

$$H(x,p) = \sum_{r=\pm} \Omega_r(x) \left[\mathrm{e}^{rp} - 1 \right]. \tag{4.78}$$

This suggests a corresponding classical mechanical interpretation, in which H determines the motion of a particle with position x and conjugate momentum p. A trajectory of the particle is given by the solution of Hamilton's equations

$$\dot{x} = \frac{\partial H}{\partial p} = \sum_{r=\pm 1} r\Omega_r(x)\mathrm{e}^{rp} \tag{4.79}$$

$$\dot{p} = -\frac{\partial H}{\partial x} = \sum_{r=\pm 1} \frac{\partial \Omega_r}{\partial x}(x) \left[\mathrm{e}^{rp} - 1 \right]. \tag{4.80}$$

Here the time t should be viewed as a parameterization of paths rather than as a real time variable. Introducing the Lagrangian

$$L(x, \dot{x}) = p \cdot \dot{x} - H(x,p), \tag{4.81}$$

it follows that $\Phi(x)$ with $\Phi(\bar{x}) = 0$ corresponds to the classical action evaluated along the least-action trajectory from $\bar{x}$ to x:

$$\Phi(x) = \inf_{x(t_0)=\bar{x}, x(T)=x} \int_0^T L(x, \dot{x}) dt. \tag{4.82}$$

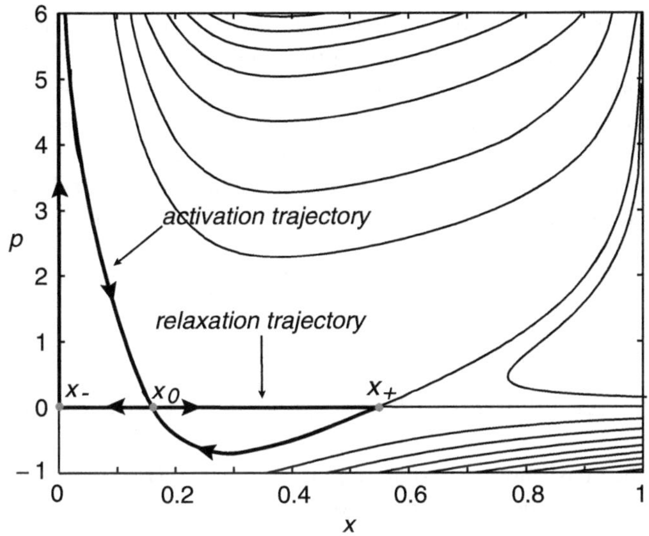

Fig. 4.5 Phase portrait of Hamiltonian equations of motion for $\Omega_{\pm} = \omega_{\pm}/N$ given by equation (4.59) with $c_m = 0.01, c_e = 4, \alpha = 4$ and $k = 0.8$. The zero-energy solutions are shown as *thicker curves*

In terms of the underlying stochastic process $X(t)$, the least-action path can be interpreted as the most probable fluctuational path from $\bar{x}$ to x (in the large-N limit) [205, 398]. Since $p = S'$ everywhere along this path, we have

$$\Phi(x) = \int_{\bar{x}}^{x} p(x')dx', \tag{4.83}$$

with the integral taken along the trajectory. It follows that the leading order term in the WKB approximation is determined by finding zero-energy solutions $p = p(x)$ such that $H(x, p(x)) = 0$. One solution is $p = 0$ or $\Phi =$ constant, which represents the classical action along a deterministic (or relaxation) trajectory. For example, once the system escapes from the metastable state x_-, it tends to rapidly converge to the other metastable state x_+ along such a deterministic path. (The contribution of relaxation trajectory to the mean escape time is usually neglected.) Another solution for Φ is (4.75), which can be interpreted as the action along a non-deterministic path that represents the most probable path of escape from x_- to x_0 [160, 164, 175, 176]. In Fig. 4.5 we illustrate the Hamiltonian phase space for the model showing the constant energy solutions of the Hamiltonian given by (4.78); the zero-energy activation and relaxation trajectories through the fixed points of the deterministic system are highlighted as thicker curves

Given the quasistationary approximation, the rate of escape from the metastable state centered about $x = x_-$ can be calculated by matching it with an appropriate inner solution in a neighborhood of the point $x = x_0$ [160, 164, 175, 256, 277]. This is necessary since the quasistationary solution (4.76) does not satisfy the absorbing boundary condition at the point x_0 separating the two metastable states. There are a number of different ways of carrying out the matched asymptotics; see, for example, [277]. Here we will follow an approach based on fixing the probability flux J_0 through x_0 and then matching the activation solution for $x < x_0$ with the relaxation solution for $x > x_0$ using a diffusion approximation of the full master equation (4.60) in the vicinity of x_0 [164, 175, 256]. The latter yields the Fokker–Planck equation (1.111), which can be rewritten in the form of a conservation equation

$$\frac{\partial}{\partial t}P(x,t) = -\frac{\partial}{\partial x}J(x,t) \tag{4.84}$$

with

$$J(x,t) = (\Omega_+(x) - \Omega_-(x))P(x,t) - \frac{1}{2N}\frac{\partial}{\partial x}[(\Omega_+(x) + \Omega_-(x))P(x,t)].$$

Substituting the quasistationary solution $p(x,t) = C_0 e^{-\lambda_0 t}\Pi(x)$ into (4.84) and using the fact that λ_0 is exponentially small give

$$J_0 = (\Omega_+(x) - \Omega_-(x))\Pi(x) - \frac{1}{2N}\frac{\partial}{\partial x}[(\Omega_+(x) + \Omega_-(x))\Pi(x)],$$

where J_0 is the constant flux through x_0. In a neighborhood of x_0, this equation can be Taylor expanded to leading order in $x - x_0$ and integrated to obtain the solution

$$\Pi(x) = \frac{J_0 N}{\Omega_+(x_0)} e^{(x-x_0)^2/\sigma^2} \int_x^\infty e^{-(y-x_0)^2/\sigma^2} dy, \tag{4.85}$$

where

$$\sigma = \sqrt{\frac{2\Omega_+(x_0)}{N[\Omega_+'(x_0) - \Omega_-'(x_0)]}} \tag{4.86}$$

determines the size of the boundary layer around x_0.

In order to match the activation and relaxation solutions, the following asymptotic behavior of the inner solution (4.85) is used:

$$\Pi(x) = \begin{cases} \dfrac{NJ_0\sigma^2}{(x-x_0)\Omega_+(x_0)}, & x - x_0 \gg \sigma \\[2ex] \dfrac{NJ_0\sigma\sqrt{\pi}}{\Omega_+(x_0)} e^{(x-x_0)^2/\sigma^2}, & x_0 - x \gg \sigma. \end{cases} \tag{4.87}$$

The solution to the right of the saddle matches the relaxation solution (4.77) since $\Omega_+(x) - \Omega_-(x) \approx (x - x_0)[\Omega_+'(x_0) - \Omega_-'(x_0)]$ for $x \approx x_0$ such that $B = J_0$. In order to match the solution on the left-hand side of x_0 with the activation solution (4.76), Taylor expand $\Phi(x)$ about x_0 using $\Phi'(x_0) = 0$ and $\Phi''(x_0) = 2/N\sigma^2$. It follows that

$$J_0 = \frac{A\Omega_+(x_0)}{\sqrt{\Omega_+(x_0)\Omega_-(x_0)}}\sqrt{\frac{|\Phi''(x_0)|}{2\pi N}} e^{-N\Phi(x_0)}. \tag{4.88}$$

The final step in the analysis is to link the flux J_0 with the escape rate λ_0. This is achieved by substituting the quasistationary solution into the continuity equation (4.84) and integrating over the interval $x \in [0, x_0]$ with a reflecting boundary condition at $x = 0$:

$$\frac{1}{\lambda_0} = \frac{1}{J_0} \int_0^{x_0} \phi^{\varepsilon}(y)dy. \tag{4.89}$$

Since the activation solution is strongly peaked around the fixed point x_-, a Gaussian approximation of $\phi^{\varepsilon}(x)$ around x_- yields the final result

$$\lambda_0 = \frac{\Omega_+(x_-)}{2\pi}\sqrt{|\Phi''(x_0)|\Phi''(x_-)}\mathrm{e}^{-N[\Phi(x_0)-\Phi(x_-)]}. \tag{4.90}$$

Hence, we obtain (4.63) with $\tau_i = \lambda_0^{-1}$ and

$$\Phi''(x) = \frac{d}{dx}\ln\left(\frac{\Omega_-(x)}{\Omega_+(x)}\right) = \frac{\Omega_-'(x)}{\Omega_-(x)} - \frac{\Omega_+'(x)}{\Omega_+(x)} = \gamma(x).$$

Similarly, we can obtain (4.64) for the mean time τ_f to terminate a spark.

4.4.4 Stochastic Phase-Plane Analysis

In the above analysis of Ca^{2+} sparks, the concentration c_{sr} in the JSR was held fixed. This is a reasonable approximation when considering the initiation of a Ca^{2+} spark. However, following Ca^{2+} release from the RyRs, the Ca^{2+} concentration c_{sr} slowly changes according to

$$\tau_{sr}\frac{dc_{sr}}{dt} = -c_{sr}x + k_{sr}[c_0 - c_{sr}]. \tag{4.91}$$

where $\tau_{sr} \gg \tau_o \gg \tau_D$. The first term on the right-hand side is the loss of Ca^{2+} through the RyRs, whereas the second terms are the influx J_{SR} of Ca^{2+} from the NSR with fixed Ca^{2+} concentration c_0; see Fig. 4.4. The variation of c_{sr} means that one has to modify the analysis of the time to terminate the Ca^{2+} spark. Following Hinch [276], this can be achieved by combining the theory of stochastic transitions outlined in Sect. 4.4.2 with the classical phase-plane analysis of slow–fast excitable systems such as the FitzHugh–Nagumo equations (see Sect. 2.1). That is, (4.62) and (4.91) form an excitable system with the fraction x of open RyRs acting as the fast variable and c_{sr} acting as the slow variable. In Fig. 4.6 we sketch the nullclines of the deterministic system in a parameter regime where there is a single, stable fixed point (x^*, c_{sr}^*). In the full stochastic model, the initiation of a Ca^{2+} spark induces a transition to the right-hand x-nullcline according to $x_-(c_{sr}^*) \to x_+(c_{sr}^*)$ as outlined in Sect. 4.4.2. The slow variable then moves down the right-hand nullcline $x_+(c_{sr})$ according to the equation

$$\tau_{sr}\frac{dc_{sr}}{dt} = -c_{sr}x_+(c_{sr}) + k_{sr}[c_0 - c_{sr}]. \tag{4.92}$$

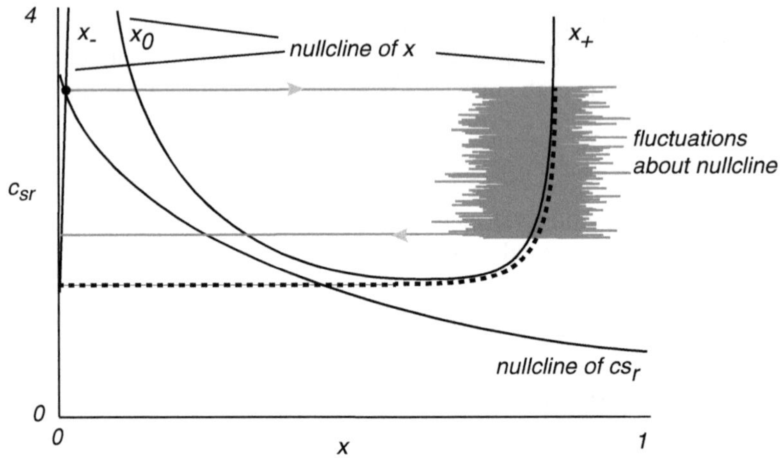

Fig. 4.6 Sketch of nullclines in the deterministic planar Ca^{2+} spark model with x denoting the fraction of open RyRs and c_{SR} the Ca^{2+} concentration in the junctional SR. The c_{sr} nullcline is a monotonically decreasing function $x(c_{sr})$, whereas the x nullcline is cubic-like with three branches $x_{\pm}(c_{sr})$ and $x_0(c_{sr})$. (Note that the branch $x_-(c_{sr}) \approx 0$; we have moved it away from the vertical axis for the sake of illustration.) In the given diagram there is a single, stable fixed point on the left-hand branch. In the stochastic version of the model a Ca^{2+} spark initiates a jump to the right-hand branch $x_+(c_{sr})$ along the lines outlined in Sect. 4.4.2. This is followed by a stochastic trajectory in which the slow variable $c_{sr}(t)$ moves down the nullcline until it undergoes a noise-induced transition back to the left-hand branch before the knee at $x = x_c$. In the deterministic case, the return transition occurs at the knee (*dashed curve*)

That is, although x is a stochastic variable, it fluctuates much faster than the dynamics of c_{sr} so one can substitute a time-averaged value of x in (4.91).

Suppose that $c_{sr}(t)$ is the solution of (4.92) with $c_{sr}(0) = c_{sr}^*$, that is, the Ca^{2+} spark occurs at $t = 0$. In principle, the spark can terminate at any time $t > 0$ due to fluctuations in the number of open RyRs. Again using a separation of time scales, we can estimate the rate of transition back to the left-hand branch at time t using

$$\lambda(t) = \frac{\Omega_+(x_+(t))\sqrt{|\gamma(x_0(t))|\gamma(x_+(t))}}{2\pi}e^{N[\Phi(x_+(t))-\Phi(x_0(t))]}, \tag{4.93}$$

where $x_+(t) = x_+(c_{sr}(t))$, etc. One can now calculate the distribution of spark durations T. Let $P(\tau) = \mathbb{P}(T > \tau)$ and introduce the spark duration probability density

$$p(\tau) = -\frac{dP}{d\tau}.$$

The probability that a spark terminates in an infinitesimal time interval $\delta\tau$ is $\lambda(\tau)\delta\tau$, so that

$$P(\tau + \delta\tau) = P(\tau)(1 - \lambda(\tau)\delta\tau).$$

Taking the limit $\delta\tau \to 0$ and integrating gives $P(\tau) = \exp\left(-\int_0^\tau \lambda(t)dt\right)$, and hence

$$p(\tau) = \lambda(\tau)\exp\left(-\int_0^\tau \lambda(t)dt\right). \tag{4.94}$$

4.4.5 Whole-Cell Model of Ca^{2+} Sparks

One of the major simplifications of the Hinch model [276] is that the background Ca^{2+} concentrations in the mytoplasm (c_m) and the NSR (c_{nsr}) are held fixed. It thus fails to capture the collective behavior of a large population of Ca^{2+} release units (CaRUs) that are coupled via global changes in these background concentrations (assuming diffusion can be neglected on the relevant time scales). This has motivated the development of a whole-cell model of calcium-induced calcium release in cardiac myocytes, based on a system of $\mathcal{N}$ globally coupled CaRUs [672, 673]. We will describe a simplified version of the model in which Ca^{2+} pumps, leakage currents, and voltage-gated Ca^{2+} channels are not modeled explicitly. Let c_j and $\hat{c}_j$ denote the Ca^{2+} concentration in the dyadic space and JSR of the jth CaRU, $j = 1,\ldots,\mathcal{N}$. Then

$$\tau_D\frac{dc_j}{dt} = J^j_{\text{R}yR} - J^j_{\text{D}}, \qquad \tau_{\text{s}r}\frac{d\hat{c}_j}{dt} = J^j_{\text{s}r} - J^j_{\text{R}yR}, \tag{4.95}$$

where J^j_{RyR} is the total Ca^{2+} current through the RyRs, J^j_{D} is the diffusive current from the diadic space of the jth CaRU to the bulk mytoplasm, and J^j_{sr} is the current from the NSR to the JSR of the jth CaRU. The various fluxes are given by

$$J^j_{\text{D}} = c_j - c_m, \quad J^j_{\text{R}yR} = x_j\hat{c}_j, \quad J^j_{\text{s}r} = k_{\text{s}r}[c_{\text{nsr}} - \hat{c}_j], \tag{4.96}$$

where x_j is the fraction of open RyRs in the jth CaRU. Finally, from conservation of Ca^{2+} ions, the bulk concentrations c_m and c_{nsr} evolve according to

$$\tau_{\text{m}}\frac{dc_m}{dt} = -\sum_{j=1}^{\mathcal{N}} J^j_D + J_m, \qquad \tau_{\text{s}r}\frac{dc_{\text{nsr}}}{dt} = J_{\text{nsr}} - \sum_{j=1}^{\mathcal{N}} J^j_{\text{s}r}. \tag{4.97}$$

Here J_m and J_{nsr} are the total external currents into the mytoplasm and NSR, respectively, arising from Ca^{2+} pumps and other factors.

When the number $\mathcal{N}$ of CaRUs is large, one can develop a probability density version of the above model, which takes the form of a differential Chapman–Kolmogorov (CK) equation describing the evolution of a single equivalent stochastic CaRU [672]. The basic idea is to introduce a probability density $p(c,\hat{c},n,t)$, with $p(c,\hat{c},n,t)dcd\hat{c}$ interpreted as the fraction of CaRUs at time t that have Ca^{2+} concentrations in the range $c \le c(t) \le c + dc, \hat{c} \le \hat{c}(t) \le \hat{c} + d\hat{c}$ and are in the internal state n, with the latter determined by the number of open RyRs, for example. Introduce the single CaRU equations

$$
\begin{aligned}
\tau_D \frac{dc}{dt} &\equiv F(c,\hat{c},n) = \frac{n}{N}\hat{c} + c_m - c \\
\tau_{sr} \frac{d\hat{c}}{dt} &\equiv \hat{F}(\hat{c},n) = k_{sr}[c_{\rm nsr} - \hat{c}] - \frac{n}{N}\hat{c}.
\end{aligned} \tag{4.98}
$$

The probability density p then evolves as

$$
\frac{\partial p}{\partial t} = -\frac{\partial (Fp)}{\partial c} - \frac{\partial (\hat{F}p)}{\partial \hat{c}} + \sum_{n'} A(n,n';c,\hat{c}) p(c,\hat{c},n',t), \tag{4.99}
$$

with $A(n,n',c,\hat{c})$ constructed from the transition rates of the master equation describing the opening and closing of RyRs within each CaRU. Note that the CK equation (4.99) is formally similar to the CK equation (1.143) for membrane voltage fluctuations considered; see Sects. 1.5 and 1.6. However, the solution of (1.126) represents the distribution of state trajectories with respect to repeated trails of a single neuron, whereas the solution of (4.99) represents the distribution of state trajectories with respect to a large ensemble of identical CaRUs. Moreover, the latter are globally coupled due to the dependence of $F, \hat{F}$ and A on the Ca^{2+} concentrations $c_m, c_{\rm nsr}$. Within the probabilistic framework these evolve according to the equations

$$
\tau_{\rm m} \frac{dc_m}{dt} = -\int_0^\infty \int_0^\infty \rho(c,\hat{c},t)[c_m - c] dc d\hat{c} + J_m \tag{4.100}
$$

$$
\tau_{sr} \frac{dc_{\rm nsr}}{dt} = J_{\rm nsr} - k_{sr} \int_0^\infty \int_0^\infty \rho(c,\hat{c},t)[c_{\rm nsr} - \hat{c}] dc d\hat{c}, \tag{4.101}
$$

with $\rho(c,\hat{c},t) = \sum_n p(c,\hat{c},n,t)$. The probability density formulation of CaRUs can increase computational efficiency compared to Monte Carlo simulations of the full ODE model [672, 673].

4.5 Intercellular Calcium Waves in Astrocytes

So far we have considered intracellular Ca^{2+} waves that propagate within a single cell. However, one also finds intercellular Ca^{2+} waves (ICWs) consisting of increases in cytoplasmic Ca^{2+} concentration that are communicated between cells and appear as waves that spread out from an initiating or trigger cell. The speed and size of ICWs depend on the initiating stimulus and the particular propagation mechanism. An ICW often propagates at a speed of 10–20 μm/s and lasts for periods of up to tens of seconds, indicating that it can involve the recruitment of hundreds of contiguous cells. ICWs were first observed in cultures of astrocytes (discussed below) in response to extracellular glutamate [143], and in airway epithelial cells following mechanical stimulation [550]. They have subsequently been found in a wide range of cell types and under a variety of stimulus conditions, as reviewed in [380, 554].

4.5.1 Astrocytes and the Tripartite Synapse

Within the context of the central nervous system (CNS), ICWs are emerging as a major long-range signaling mechanism of a particular type of glial (nonneuronal) cell known as an *astrocyte*; see Fig. 4.7. Other types of glial cells include *oligodendrocytes* and *radial glial cells*. The former are responsible for myelination in the CNS just as Schwann cells are responsible for myelination in the peripheral nervous system; see also Sect. 2.5. On the other hand, radial glial cells play a pivotal role during development, with regard to both neuronal migration and neurogenesis [101]; see Sect. 9.1. Radial glial cells usually disappear post-development, but some cells, such as Bergmann glia in the cerebellum and Muller glia in the retina, maintain their radial glia-like properties into adulthood.

Traditionally, astrocytes were thought to be physiologically passive cells that only play a supporting role in the CNS by regulating and optimizing the environment within which neurons operate. Their housekeeping duties include maintaining local ion and pH homeostasis, delivering glucose and providing metabolic support via contacts with the vasculature, and clearing neuronal waste such as metabolic products and excess neurotransmitters in the synaptic cleft. However, over the past 20 years there has been an increasing amount of empirical data indicating that astrocytes play an active role in modulating synaptic transmission [253, 268, 449]. That is, astrocytes respond to synaptic neurotransmitters such as glutamate by elevating their intracellular Ca^{2+} levels. This then results in the release of glutamate and other signaling molecules from the astrocytes (gliotransmission), which in turn influences synaptic activity. Thus, astrocytes could potentially contribute to higher brain function and, consequently, behavior. This is consistent with the fact that the number of astrocytes relative to the number of neurons increases dramatically with

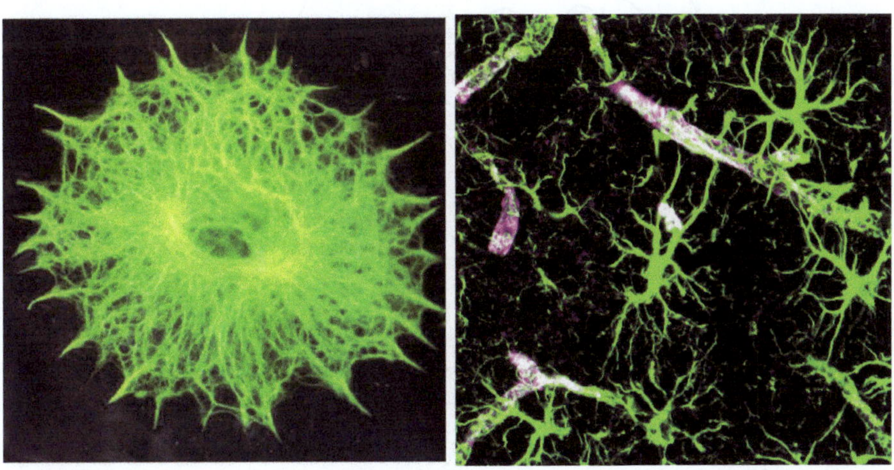

Fig. 4.7 *Left*: Astrocyte in vitro stained with GFAP to show filaments. *Right*: Astrocytes stained for GFAP, with end-feet ensheathing blood vessels [Public domain figure from Wikipedia Commons]

brain complexity, ranging from 1:25 in the leech, 1:6 in *C elegans*, 1:3 in lower mammals, to 1.4:1 in the human cortex. Conversely, it is known that following injury to the nervous system or under pathological conditions such as Alzheimer's disease, epilepsy, and schizophrenia, the structure and protein expression of astrocytes are altered [564]. One interpretation of these structural changes is that they are indicative of the supportive action of astrocytes following injury. On the other hand, given that many CNS disorders are correlated with alterations in synaptic transmission and gliotransmission regulates synaptic processing, it is possible that changes in gliotransmission contribute to these disorders; see Chap. 9.

In many higher organisms, astrocytes establish nonoverlapping anatomical and functional domains of size around 100 μm, and their distribution is highly organized.

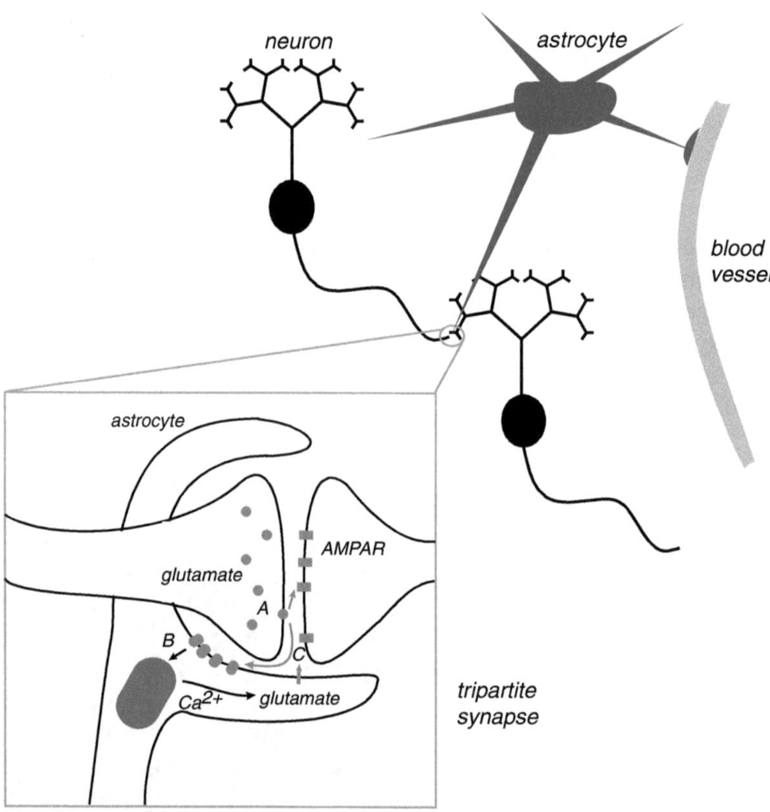

Fig. 4.8 Astrocytes modulate synaptic transmission and blood flow via Ca^{2+} signaling. *Inset*: the tripartite synapse. Synaptic transmission of glutamate (A) activates not only postsynaptic receptors such as AMPA but also astrocytic receptors that elevate astrocytic Ca^{2+} levels (B). This then induces the local release of glutamate from the astrocyte (C), which can modulate the synapse. Astrocytes also form a link between a neuron and the local vasculature. Increases in astrocytic Ca^{2+} concentration propagate via astrocyte end feet to small blood vessels, where they control vessel diameter and blood flow

The resulting astrocytic network parcellates the whole neutrophil such that a single astrocyte can make contact with thousands of synapses. The processes of a single astrocyte make contact with a synapse by wrapping around both the presynaptic and postsynaptic domains to form what is known as a *tripartite synapse* [253, 268, 449]; see Fig. 4.8. Hence, from a geometrical perspective, it is well placed to modulate synaptic transmission. The first component of glial–neuronal interactions at a synapse, namely, the metabotropic glutamate receptor 5 (mGluR5)-dependent elevation in astrocytic Ca^{2+} concentration in response to the release of glutamate neurotransmitters, was discovered in the early 1990s using Ca^{2+} imaging to study cultured glial cells [106, 143]. Both intracellular Ca^{2+} oscillations and ICWs propagating between neighboring astrocytes were observed. Subsequent studies have established that neuroglial signaling also occurs in astrocytes of isolated retina and brain slices [265, 460]. (Note, however, that a recent study of adult rodents has called into question the link between neuroglial Ca^{2+} signaling at the tripartite synapse [610]. It would appear that the astrocytic expression of mGluR5 receptors is developmentally regulated and, in particular, is undetectable in mature rodent brains; almost all previous studies of the tripartite synapse have been in immature animals.) Following the observation that neuronal activity increases Ca^{2+} levels in astrocytes, it was subsequently discovered that the resulting Ca^{2+} elevation induces gliotransmission of glutamate and a variety of other signaling molecules including ATP [249, 491]. Gliotransmission has been demonstrated to modulate synaptic transmission presynaptically and postsynaptically. For example, glutamate can enhance neurotransmitter release presynaptically by binding to metabotropic (mGlu) receptors or kainate receptors [189, 388]. Another example of enhanced synaptic transmission is the binding of ATP to postsynaptic receptors, which induces elevations in Ca^{2+}. This, in turn, can drive the insertion of additional AMPA receptors, thus temporarily increasing synaptic efficacy [236]. (The binding of ATP to postsynaptic receptors also provides a mechanism for the propagation of ICWs via paracrine signaling; see below.) On the other hand, hydrolysis of ATP results in the accumulation of adenosine, causing a reduction of synaptic transmission [493].

In addition to directly modulating synaptic transmission, astrocytes appear to regulate NMDA-receptor function. Recall from Sect. 1.3 that NMDA receptors are crucial for the induction of long-term synaptic plasticity, requiring both glutamate and postsynaptic depolarization in order to activate. The opening of an NMDA channel leads to a temporary increase of intracellular Ca^{2+} within the postsynaptic domain, whose amplitude and temporal profile determine whether the synapse is potentiated or depressed. It turns out that there is an additional glycine-binding site that regulates NMDA-receptor activity, whose co-agonist D-serine is released by astrocytes during gliotransmission. Increases in the level of D-serine appear to shift the balance point of long-term plasticity from depression to potentiation [485], a process known as metaplasticity. Finally, not only does astrocytic Ca^{2+} signaling modulate synaptic transmission and plasticity, it also controls blood flow, since intracellular waves propagate to small blood vessels via astrocyte end feet, where they control vessel diameter; see Fig. 4.8.

Following the original studies of ICWs in astrocyte cultures [106, 143], intracellular Ca^{2+} wave propagation between astrocytes has been observed in vitro in brain slices and whole retina preparations [265, 460, 492]. These empirical findings, combined with the structure of the tripartite synapse, have led to the suggestion that networks of astrocytes could form an extraneuronal pathway for rapid long-distance signaling within the CNS, which interacts reciprocally with neuronal networks. In particular, ICWs could provide a potential mechanism for coordinating and synchronizing the activity of a large group of neuronal and nonneuronal cells. It should be pointed out, however, that the function of ICWs in astrocytes remains controversial, since there have been very few in vivo observations of ICWs in the CNS, excluding the case of spontaneous ICWs during early development where astrocytes play a role in the generation, differentiation, and migration of neurons [666]; see Sect. 9.1. One recent exception is a study demonstrating in vivo spontaneous ICWs (termed glissandi) in mouse hippocampal astrocytes. (ICWs have also been observed in vivo in other glial networks, e.g., Bergmann glia in the cerebellum of rodents [287].) Another concern regarding the functional role of ICWs, either in vitro or in vivo, is whether they require strong inducing stimuli that do not have a normal physiological counterpart. On the other hand, certain pathological conditions such as brain trauma, brain ischemia (oxygen deprivation), and seizures are often associated with strong stimuli. They are all coupled with the release of ATP and glutamate and reductions in the level of extracellular Ca^{2+} that collectively could promote ICW generation. Such ICWs could exacerbate an injury by propagating signals that initiate

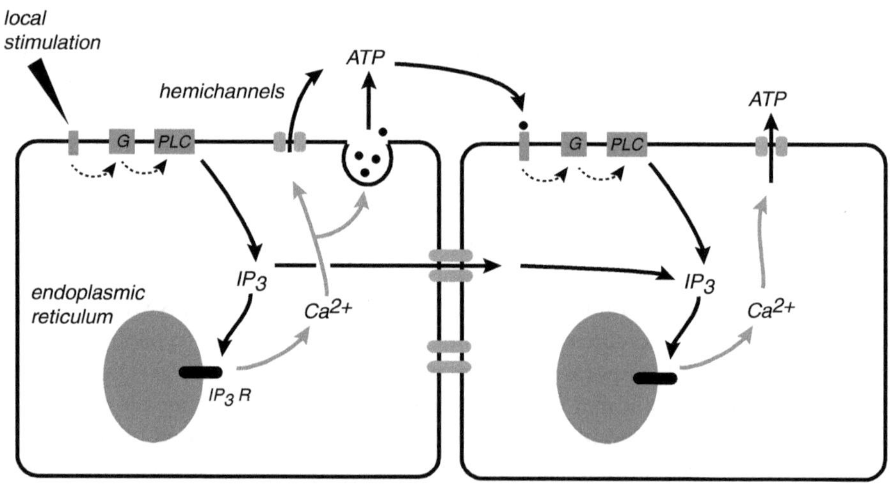

Fig. 4.9 Two mechanisms for the propagation of ICWs. Local stimulation of a single cell induces an elevation in intracellular IP_3 that generates an intracellular Ca^{2+} wave. Diffusion of IP_3 through gap-junction coupling with an adjacent cell initiates a second intracellular Ca^{2+} wave. In addition, or alternatively, stimulation of the first cell releases ATP via hemichannels in the plasma membrane or vesicular release; this could also be dependent on intracellular Ca^{2+}. The extracellular diffusion of ATP to neighboring cells can activate metabotropic receptors that then stimulate the production of IP_3 and the production of downstream intracellular Ca^{2+} waves

cell apoptosis (cell death) in surrounding cells [681]. There is also growing evidence that ICWs provide a source of excitation during epileptic episodes [564, 647, 668]; see Sect. 9.4. For example, studies of a brain slice that exhibits epileptiform activity, showed that there was also a corresponding increase in the frequency of Ca^{2+} oscillations in astrocytes [623]. Moreover, the release of glutamate from astrocytes has been observed to cause synchronous neuronal depolarizations [186].

4.5.2 Mechanisms and Models of Intercellular Calcium Waves

There are thought to be two basic mechanisms for the propagation of ICWs between adjacent cells that occur in astrocytes and other cell types: (a) diffusion of the second messenger IP_3 and, to a lesser extent, Ca^{2+} via gap junctions; (b) paracrine signaling via a diffusing extracellular messenger such as ATP. These mechanisms are illustrated in Fig. 4.9.

Gap-Junction Signaling

Gap junctions are arrays of transmembrane channels that connect the cytoplasm of two neighboring cells and thus provide a direct diffusion pathway between the cells. Cells sharing a gap-junction channel each provides a hemichannel (also known as a connexon) that connect head-to-head [545]. Each hemichannel is composed of proteins called connexins that exist as various isoforms named Cx23 through Cx62, with Cx43 being the most common. The physiological properties of a gap junction, including its permeability and gating characteristics, are determined by the particular connexins forming the channel. Although gap junctions are readily permeable to Ca^{2+} and IP_3, the diffusion of Ca^{2+} through a channel is greatly restricted due to buffering, so that the propagation of ICWs is primarily mediated by the diffusion of IP_3 when gap junctions are the main mechanism. The possible role of gap junctions in the propagation of ICWs was initially suggested by the observation in epithelial cells that waves only spread in adjacent cells from localized membrane sites in direct contact with the initiating cell. Moreover, the ICWs were not influenced by fluid flow over the apical surface of cells, that would have interfered with extracellular signals [259]. Further evidence for the role of gap junctions came from the study of glioma cells that lacked gap junctions [107]. These cells only supported ICWs when they were transfected with connexins; however, the presence of connexins could also enhance ATP and thus contribute to paracrine signaling (see below). Subsequent experimental and modeling studies have established that the diffusion of IP_3 through gap junctions can support the propagation of ICWs [379, 591]. In particular, ICWs could be initiated by an elevation of IP_3 in a single cell followed by diffusion to other cells without any regeneration of IP3. On the other hand, the range of the waves tends to be limited to only a few cells, suggesting that some other mechanism may be needed in order to generate longer-range waves.

The first model of ICWs mediated by gap junctions was based on studies of mechanical stimulation of cultured epithelial cells [590, 591]. It has subsequently been extended to models of ICWs in networks of astrocytes of varying complexity [163, 281, 310]. The models all have the same basic structure, but differ in terms of whether or not regenerative mechanisms or extracellular signaling mechanisms are included. A simple 1D version of an ICW model is shown in Fig. 4.10. Each cell is treated as a square of size L with local coordinates (x,y), $0 \leq x,y \leq L$. There are N cells labeled $n = 1,\ldots,N$ arranged on a regular lattice with nearest neighbor coupling. The intracellular dynamics within each cell is described by reaction–diffusion equations of the form given in Sect. 4.2. For example, in the original model of Sneyd et al. [591], this included equations for cytoplasmic Ca^{2+} concentration (c_n), IP_3 concentration p_n, and a slow variable h_n representing the fraction of IP3 receptors that have not been inactivated by Ca^{2+}. An example of an intracellular model for the nth cell is (see, e.g., [280])

$$\frac{\partial p_n}{\partial t} = D_p \nabla^2 p_n - k_p p_n, \tag{4.102a}$$

$$\frac{\partial c_n}{\partial t} = D_c \nabla^2 p_n - k_c c_n + f(c_n, h_n, p_n), \tag{4.102b}$$

$$\tau_h \frac{dh_n}{dt} = h_\infty(c_n) - h_n. \tag{4.102c}$$

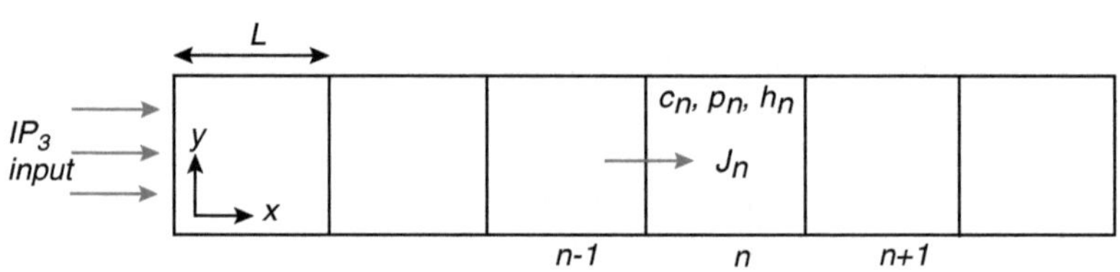

Fig. 4.10 Schematic diagram of a 1D model of intercellular Ca^{2+} waves mediated by gap junctions

Here the nonlinear function f represents the net flux due to Ca^{2+} release from IP_3 receptors. It is usually assumed that $D_p \gg D_c$ due to Ca^{2+} buffering. Coupling between adjacent cells in incorporated into the model using the boundary conditions for IP_3. Thus along the common vertical edge between cells $n-1$ and n, $1 < n < N$, the IP_3 flux J_n is taken to be of the form

$$-D_p \left.\frac{\partial p_{n-1}(x,y)}{\partial x}\right|_{x=L} = D_p \left.\frac{\partial p_n(x,y)}{\partial x}\right|_{x=0} \equiv \mathscr{K}[p_{n-1}(L,y) - p_n(0,y)]. \tag{4.103}$$

These are supplemented by no-flux boundary conditions for IP_3 across all other edges; all cell borders are assumed to be impermeable to Ca^{2+}. In a 2D version of the model arranged on a square grid, IP_3 can also flow across common horizontal edges of neighboring cells, while no-flux boundary conditions are imposed along the exterior boundary of the grid. A typical simulation of such a model is to inject IP_3 in the cell at one end of the lattice, which is then allowed to diffuse from cell to cell, resulting in the propagation of an intracellular calcium wave. It is important to note that the ICW is not a traveling wave in the mathematical sense used throughout this book. Rather, the spread of Ca^{2+} is mediated by passive diffusion of intracellular Ca^{2+} and intra-/intercellular IP_3. Thus the rate of spread decreases with distance from the stimulated cell, and the arrival time of the wave increases exponentially with distance. In order to create more persistent ICWs, it is necessary to incorporate some regenerative mechanism. One example of such a mechanism is to assume that the rate of production of IP_3 can be amplified by intracellular Ca^{2+} so that (4.102a) becomes [281]

$$\frac{\partial p_n}{\partial t} = D_p \nabla^2 p_n + \frac{\nu c_n^2}{K + c_n^2} - k_p p_n.$$

Hofer et al. [281] showed that for an intermediate range of values of ν, partial regeneration of IP_3 supports ICWs that propagate further than those relying solely on passive diffusion of IP_3. Finally, an alternative mechanism for increasing the range of ICWs, which does not require some form of regeneration or active process, is to combine gap-junction coupling with an extracellular pathway involving ATP [163, 310]; see below.

Paracrine Signaling

Early evidence for the involvement of an extracellular component to ICWs arose from the observation that ICWs can propagate between disconnected groups of astrocytes, provided that the degree of separation does not exceed around 100 μm [266]. More recently, advanced imaging techniques have revealed that ICWs are associated with a cloud of extracellular ATP [249, 457]. There are a variety of pathways that could mediate ATP release, including hemichannels in the plasma membrane and vesicular discharge. Modeling studies suggest that a combination of gap-junction and paracrine signaling can significantly increase the range of ICWs [298]. It is also possible that there is a regenerative mechanism that boosts the propagation of ICWs. One candidate for active rather than passive (diffusive) wave propagation is Ca^{2+}-dependent ATP release via connexin hemichannels [652]. In the case of astrocytes, the dominant propagation mechanism depends on where they are located within the nervous system. For example, ICWs in the neocortex tend to be mediated by gap junctions, whereas in the hippocampus, both gap-junction and paracrine signaling play a role.

Here we will describe a mathematical model of paracrine signaling in astrocyte networks due to Bennett et al. [37]. Suppose that $a(t)$ denotes the extracellular concentration of ATP in the vicinity of a single cell. ATP binds to metabotropic P2Y receptors in the cell membrane according to the first-order kinetic scheme

$$ATP + R \underset{k_-}{\overset{k_+}{\rightleftharpoons}} R.$$

In the case of fast binding kinetics, the fraction of P2Y receptors that are bound by ATP is given by the quasi-equilibrium solution

$$r = \frac{a}{K_R + a}, \quad K_R = \frac{k_-}{k_+}. \tag{4.104}$$

This then drives activation of G-proteins according to the kinetic scheme

$$G_0 \underset{k_d}{\overset{k_a(r+r_0)}{\rightleftharpoons}} G_1,$$

where $G_{0,1}$ represent the inactive and active states of the G-protein. Denoting the amount of activated protein by g and the total G-protein by g_T, we have

$$\frac{dg}{dt} = k_a(r + r_0)(g_T - g) - k_d g. \tag{4.105}$$

Again assuming fast kinetics, the quasi-equilibrium solution is

$$g = g_T \frac{r + r_0}{K_G + r + r_0}, \quad K_G = \frac{k_d}{k_a}. \tag{4.106}$$

A major simplification of the model is to assume that the release of IP_3 within the cell is mainly driven by activation of G-protein so that Ca^{2+} dependence is ignored. If p is the intracellular IP_3 concentration then

$$\frac{dp}{dt} = \kappa g - k_{\text{deg}} p, \tag{4.107}$$

for constants κ, k_{deg}. The next stage of paracrine signaling, namely, the release of ATP by astrocytes is not yet well understood. However, IP_3 as a triggering mechanism, possibly via Ca^{2+} release from intracellular stores, is probably involved. In the model of Bennett et al. [37], IP_3 is assumed to directly cause ATP release from the cell surface into the extracellular space at the rate

$$J_{\text{ATP}}(p,t) = \kappa_{\text{ATP}} \chi(t) \left[\frac{p - p_{\min}}{K_{\text{rel}} + p} \right]_0. \tag{4.108}$$

Here κ_{ATP} and K_{rel} are constants, $[z]_0 = z$ for $z > 0$ and $[z]_0 = 0$ for $z < 0$, and $\chi(t)$ takes into account depletion of ATP stores within the cell. That is $\chi(0) = 1$ and

$$\frac{d\chi}{dt} = -\gamma J_{\text{ATP}}(t). \tag{4.109}$$

The rate of ATP release is thresholded in order to insure that very small levels of ATP are not amplified, and the depletion factor $\chi(t)$ is included in order to terminate

ATP release; the latter is motivated by the observation that ATP release is suppressed by repeated stimulation of a cell.

Finally, coupling between cells is modeled in terms of a diffusion equation for the extracellular ATP concentration that takes into account the IP_3-mediated release of ATP from the astrocytes. One way to proceed would be to use a continuum bidomain model, in which astrocytes and extracellular ATP coexist at every point in space. Taking $a = a(\mathbf{x},t), p = p(\mathbf{x},t)$ and denoting the density of astrocytes by $\rho(\mathbf{x})$, we would then have

$$\frac{\partial a}{\partial t} = D_{\mathrm{ATP}}\nabla^2 a + \rho(\mathbf{x})J_{\mathrm{ATP}}(p,t) - \frac{a}{K_{\mathrm{deg}} + a}. \tag{4.110}$$

Alternatively, one could explicitly distinguish between extracellular space and the intracellular space of astrocytes, in which case the fluxes J_{ATP} only exist at the boundaries of the cells. The latter approach is used in the computer simulations of Bennett et al. [37], who take a simplified cubic geometry for the astrocytes and also allow for intracellular diffusion of IP_3. Note that under a quasi-steady-state approximation, (4.107) and (4.110) form a closed system of differential equations, since (4.104) and (4.107) can be used to express g in terms of a. Hence, one can solve for the ATP and IP_3 concentrations independently of Ca^{2+} and then incorporate ICWs by using the IP_3 to drive a standard model of Ca^{2+} release from intracellular stores such as Li–Rinzel. However, if there is regenerative Ca^{2+}-dependent production of IP_3 or a Ca^{2+}-dependent mechanism for gliotransmission of ATP, then it is necessary to couple the Ca^{2+} dynamics to that of IP_3 and ATP.

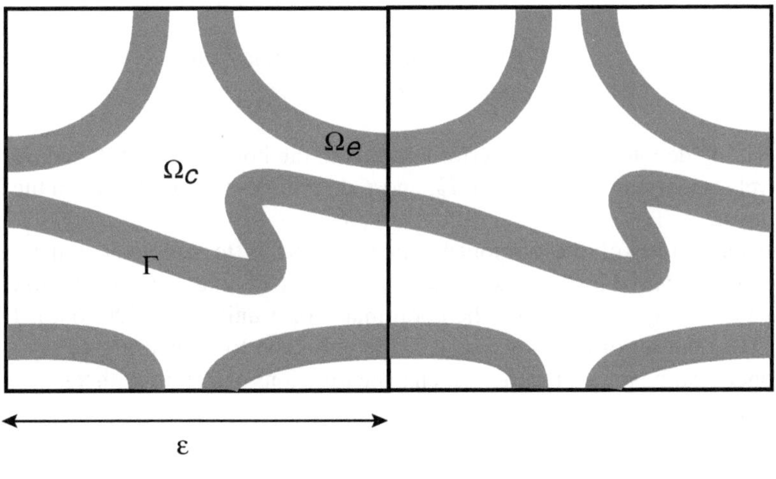

Fig. 4.11 Illustration of a 2D periodic heterogeneous medium consisting of a cytosolic domain Ω_c and an ER domain Ω_e separated by a boundary Γ

4.6 Appendix: Homogenization

As we noted in Sect. 4.2, it is not practical to try modeling the detailed microstructure of the cytoplasmic and ER boundaries, which are highly interleaved. This motivates consideration of a mean-field description of the Ca^{2+} concentrations, in which the concentrations c, c_e are taken to coexist at every point in space with effective diffusion coefficients. As shown by Goel et al. [224], the resulting bidomain model can be derived from first principles using homogenization theory. In this appendix we sketch the basic steps in their derivation. Note that we also encountered a 1D example of homogenization theory in Sect. 3.2.3, where we spatially averaged over a discrete distribution of dendritic spines.

Consider a domain $\Omega \subset \mathbb{R}^3$ in which the ER is treated as a periodic network that occupies a fraction of Ω, as illustrated in Fig. 4.11. For simplicity, assume that the unit cell is a cube of length ε. The ER Ca^{2+} occupies a connected domain Ω_e^ε, and the complementary connected domain $\Omega_c^\varepsilon = \Omega/\Omega_e^\varepsilon$ is occupied by cytosolic Ca^{2+}. The concentrations c^ε and c_e^ε satisfy the 3D diffusion equation in their respective domains:

$$\frac{\partial c}{\partial t} = \nabla \cdot (A_\varepsilon(\mathbf{x}/\varepsilon)\nabla c(\mathbf{x})), \quad \mathbf{x} \in \Omega_c^\varepsilon \tag{4.111a}$$

$$\frac{\partial c_e}{\partial t} = \nabla \cdot (B_\varepsilon(\mathbf{x}/\varepsilon)\nabla c_e(\mathbf{x})), \quad \mathbf{x} \in \Omega_e^\varepsilon, \tag{4.111b}$$

where A_ε and B_ε are the diffusion tensors for Ca^{2+} in the cytosol and ER, respectively. The boundary conditions on the membrane Γ_ε separating the cytosol from the ER take the form

$$A_\varepsilon(\mathbf{x}/\varepsilon)\nabla c(\mathbf{x}) \cdot \mathbf{n}_c^\varepsilon = \varepsilon f(c^\varepsilon, c_e^\varepsilon), \quad \text{on } \Gamma_\varepsilon \tag{4.112a}$$

$$-B_\varepsilon(\mathbf{x}/\varepsilon)\nabla c_e(\mathbf{x}) \cdot \mathbf{n}_c^\varepsilon = \varepsilon f(c^\varepsilon, c_e^\varepsilon), \quad \text{on } \Gamma_\varepsilon, \tag{4.112b}$$

where $n_c^\varepsilon, n_e^\varepsilon$ denote the unit exterior normals to the boundaries $\partial\Omega_c^\varepsilon$ and $\partial\Omega_e^\varepsilon$, respectively, satisfying $n_c^\varepsilon = -n_e^\varepsilon$ on Γ_ε, and f determines the net flux from the ER to the cytosol.

The basic idea of the homogenization method is to supplement the macroscopic spatial variables $\mathbf{x} = (x_1, x_2, x_3)$ with a set of microscopic variables $\mathbf{y} = (y_1, y_2, y_3), 0 \leq y_i \leq 1$, that are the coordinates of a unit cube [496, 630]. Denote by Ω_c (Ω_e) the set of points $y = x/\varepsilon$ in the unit cube for which $\mathbf{x} \in \Omega_c^\varepsilon$ ($\mathbf{x} \in \Omega_e^\varepsilon$). The concentrations $c^\varepsilon, c_e^\varepsilon$ are then taken to be functions of x and y with $\mathbf{x} \in \Omega, \mathbf{y} \in \Omega_c$ for c^ε, and $\mathbf{x} \in \Omega, \mathbf{y} \in \Omega_e$ for c_e^ε:

$$c^\varepsilon = c(\mathbf{x}, \mathbf{y}, t), \quad c_e^\varepsilon = c_e(\mathbf{x}, \mathbf{y}, t) \tag{4.113}$$

with $y = x/\varepsilon$. The next step is to introduce formal asymptotic expansion for c^ε and c_e^ε of the form

$$c^\varepsilon = c^0(\mathbf{x}, \mathbf{y}, t) + \varepsilon c^1(x, y, t) + \varepsilon^2 c^2(\mathbf{x}, \mathbf{y}, t) + \ldots \tag{4.114a}$$

$$c_e^\varepsilon = c_e^0(\mathbf{x}, \mathbf{y}, t) + \varepsilon c_e^1(\mathbf{x}, \mathbf{y}, t) + \varepsilon^2 c_e^2(\mathbf{x}, \mathbf{y}, t) + \ldots, \tag{4.114b}$$

where c^k and c_e^k are 1-periodic functions of $\mathbf{y}$. Setting

$$\nabla u = \nabla_{\mathbf{x}} u(\mathbf{x},\mathbf{y}) + \varepsilon^{-1}\nabla_{\mathbf{y}} u(\mathbf{x},\mathbf{y}),$$

it follows that (4.111) becomes

$$\frac{\partial c^\varepsilon}{\partial t} = \left(\varepsilon^{-2}\nabla_{\mathbf{y}}\cdot(A_\varepsilon(\mathbf{y})\nabla_{\mathbf{y}}) + \varepsilon^{-1}[\nabla_{\mathbf{x}}\cdot(A_\varepsilon(\mathbf{y})\nabla_{\mathbf{y}}) + \nabla_{\mathbf{y}}\cdot(A_\varepsilon(\mathbf{y})\nabla_{\mathbf{x}})] + A_\varepsilon(\mathbf{y})\nabla_{\mathbf{x}}^2\right) c^\varepsilon$$
$$\text{for } \mathbf{x}\in\Omega, \mathbf{y}\in\Omega_c \tag{4.115a}$$
$$\frac{\partial c_e^\varepsilon}{\partial t} = \left(\varepsilon^{-2}\nabla_{\mathbf{y}}\cdot(B_\varepsilon(\mathbf{y})\nabla_{\mathbf{y}}) + \varepsilon^{-1}[\nabla_{\mathbf{x}}\cdot(B_\varepsilon(\mathbf{y})\nabla_{\mathbf{y}}) + \nabla_{\mathbf{y}}\cdot(B_\varepsilon(\mathbf{y})\nabla_{\mathbf{x}})] + A_\varepsilon(\mathbf{y})\nabla_{\mathbf{x}}^2\right) c_e^\varepsilon$$
$$\text{for } \mathbf{x}\in\Omega, \mathbf{y}\in\Omega_e, \tag{4.115b}$$

and the boundary conditions (4.112) become

$$A_\varepsilon(\mathbf{y})\left[\nabla_{\mathbf{x}} + \varepsilon^{-1}\nabla_{\mathbf{y}}\right] c\cdot\mathbf{n}_c^\varepsilon = \varepsilon f(c^\varepsilon, c_e^\varepsilon), \quad \text{for } \mathbf{x}\in\Omega, \mathbf{y}\in\Gamma_\varepsilon \tag{4.116a}$$
$$-B_\varepsilon(\mathbf{y})\left[\nabla_{\mathbf{x}} + \varepsilon^{-1}\nabla_{\mathbf{y}}\right] c_e\cdot\mathbf{n}_c^\varepsilon = \varepsilon f(c^\varepsilon, c_e^\varepsilon), \quad \text{for } \mathbf{x}\in\Omega, \mathbf{y}\in\Gamma_\varepsilon. \tag{4.116b}$$

Substituting the asymptotic expansions for c^ε and c_e^ε and collecting terms of the same order in ε then yields a hierarchy of equations for c^k, c_e^k. At $\mathcal{O}(\varepsilon^{-2})$, we have

$$\nabla_{\mathbf{y}}\cdot(A_\varepsilon(\mathbf{y})\nabla_{\mathbf{y}})c^0 = 0, \quad \mathbf{y}\in\Omega_c \tag{4.117a}$$
$$\nabla_{\mathbf{y}}\cdot(B_\varepsilon(\mathbf{y})\nabla_{\mathbf{y}})c_e^0 = 0, \quad \mathbf{y}\in\Omega_e. \tag{4.117b}$$

together with the boundary conditions

$$A_\varepsilon(\mathbf{y})\nabla_{\mathbf{y}} c\cdot\mathbf{n} = 0 = B_\varepsilon(\mathbf{y})\nabla_{\mathbf{y}} c_e^0\cdot\mathbf{n}, \quad \mathbf{y}\in\Gamma, \tag{4.118}$$

where $\mathbf{n}$ is a unit normal on Γ. For fixed x, the only periodic solution to these equations is $c^0 = \text{constant}, c_e^0 = \text{constant}$, that is,

$$c^0 = c^0(x,t), \quad c_e^0 = c_e^0(x,t). \tag{4.119}$$

The $\mathcal{O}(\varepsilon^{-1})$ equations are

$$\nabla_{\mathbf{y}}\cdot\left(A_\varepsilon(\mathbf{y})\left[\nabla_{\mathbf{x}} c^0 + \nabla_{\mathbf{y}} c^1\right]\right) = 0, \quad \mathbf{y}\in\Omega_c \tag{4.120a}$$
$$\nabla_{\mathbf{y}}\cdot\left(B_\varepsilon(\mathbf{y})\left[\nabla_{\mathbf{x}} c_e^0 + \nabla_{\mathbf{y}} c_e^1\right]\right) = 0, \quad \mathbf{y}\in\Omega_e, \tag{4.120b}$$

together with the boundary conditions

$$A_\varepsilon(\mathbf{y})[\nabla_{\mathbf{x}} c^0 + \nabla_{\mathbf{y}} c^1]\cdot\mathbf{n} = 0 = B_\varepsilon(\mathbf{y})[\nabla_{\mathbf{x}} c_e^0 + \nabla_{\mathbf{y}} c_e^1]\cdot\mathbf{n}, \quad \mathbf{y}\in\Gamma. \tag{4.121}$$

The solutions for c^1 and c_e^1 can be written in the form

$$c^1(x,y,t) = \sum_{i=1}^{3} V_i^c(y)\frac{\partial c^0(x,t)}{\partial x_i} + \bar{c}^1(x,t), \tag{4.122a}$$

$$c_e^1(x,y,t) = \sum_{i=1}^{3} V_i^e(y) \frac{\partial c_e^0(x,t)}{\partial x_i} + \bar{c}_e^1(x,t), \tag{4.122b}$$

with the 1-periodic vectors $\mathbf{V}^c$ and $\mathbf{V}^e$ satisfying the "cell equations"

$$\nabla_{\mathbf{y}} \cdot (A_\varepsilon(\mathbf{y}) [\nabla_{\mathbf{y}} V_k^c(\mathbf{y}) + \mathbf{e}_k]) = 0, \quad \mathbf{y} \in \Omega_c \tag{4.123a}$$

$$\nabla_{\mathbf{y}} \cdot (B_\varepsilon(\mathbf{y}) [\nabla_{\mathbf{y}} V_k^e(\mathbf{y}) + \mathbf{e}_k]) = 0, \quad \mathbf{y} \in \Omega_e, \tag{4.123b}$$

and

$$A_\varepsilon(\mathbf{y})[\nabla_{\mathbf{y}} V_k^c(\mathbf{y}) + \mathbf{e}_k] \cdot \mathbf{n} = 0 = B_\varepsilon(\mathbf{y})[\nabla_{\mathbf{y}} V_k^e(\mathbf{y}) + \mathbf{e}_k] \cdot \mathbf{n}, \quad \mathbf{y} \in \Gamma, \tag{4.124}$$

for $k = 1,2,3$, where $\mathbf{e}_k$ is the vector with components $e_{jk} = \delta_{jk}$.

In order to derive mean-field equations for the homogenized concentrations c^0, c_e^0, it is necessary to proceed to $\mathscr{O}(1)$, for which

$$\nabla_{\mathbf{x}} \cdot \left[A_\varepsilon(\mathbf{y})(\nabla_{\mathbf{x}} c^0 + \nabla_{\mathbf{y}} c^1)\right] + \nabla_{\mathbf{y}} \cdot \left[A_\varepsilon(\mathbf{y})(\nabla_{\mathbf{x}} c^1 + \nabla_{\mathbf{y}} c^2)\right] = \frac{\partial c^0}{\partial t}, \quad \mathbf{y} \in \Omega_c \tag{4.125a}$$

$$\nabla_{\mathbf{x}} \cdot \left[B_\varepsilon(\mathbf{y})(\nabla_{\mathbf{x}} c_e^0 + \nabla_{\mathbf{y}} c_e^1)\right] + \nabla_{\mathbf{y}} \cdot \left[B_\varepsilon(\mathbf{y})(\nabla_{\mathbf{x}} c_e^1 + \nabla_{\mathbf{y}} c_e^2)\right] = \frac{\partial c_e^0}{\partial t}, \quad \mathbf{y} \in \Omega_e \tag{4.125b}$$

with boundary conditions

$$A_\varepsilon(\mathbf{y})[\nabla_{\mathbf{x}} c^1 + \nabla_{\mathbf{y}} c^2] \cdot \mathbf{n} = f(c^0, c_e^0), \quad \mathbf{y} \in \Gamma \tag{4.126a}$$

$$B_\varepsilon(\mathbf{y})[\nabla_{\mathbf{x}} c^1 + \nabla_{\mathbf{y}} c^2] \cdot \mathbf{n} = -f(c^0, c_e^0), \quad \mathbf{y} \in \Gamma. \tag{4.126b}$$

Integrating (4.125a) over Ω_c gives

$$\int_{\Omega_c} \left(\nabla_{\mathbf{x}} \cdot \left[A_\varepsilon(\mathbf{y})(\nabla_{\mathbf{x}} c^0 + \nabla_{\mathbf{y}} c^1)\right] + \nabla_{\mathbf{y}} \cdot \left[A_\varepsilon(\mathbf{y})(\nabla_{\mathbf{x}} c^1 + \nabla_{\mathbf{y}} c^2)\right]\right) d\mathbf{y} = |\Omega_c| \frac{\partial c^0}{\partial t}, \tag{4.127}$$

where $|\Omega_c|$ is the volume fraction of the unit cell occupied by the cytosol. Using (4.122a), the first term on the left-hand side becomes

$$\begin{aligned} &\int_{\Omega_c} \nabla_{\mathbf{x}} \cdot \left[A_\varepsilon(\mathbf{y})(\nabla_{\mathbf{x}} c^0 + \nabla_{\mathbf{y}} c^1)\right] d\mathbf{y} \\ &= \int_{\Omega_c} \nabla_{\mathbf{x}} \cdot \left[A_\varepsilon(\mathbf{y})(\nabla_{\mathbf{x}} c^0 + \nabla_{\mathbf{y}} \left(\sum_{i=1}^{3} V_i^c(y) \frac{\partial c^0}{\partial x_i}\right)\right] d\mathbf{y} \\ &= \nabla_{\mathbf{x}} \cdot \left(\tilde{A}_\varepsilon \nabla_{\mathbf{x}} c^0\right), \end{aligned}$$

where

$$[\tilde{A}_\varepsilon]_{ik} = \int_{\Omega_c} \sum_j [A_\varepsilon(\mathbf{y})]_{ij} \left[\frac{\partial V_k^c(\mathbf{y})}{\partial y_j} + \delta_{j,k}\right] d\mathbf{y}. \tag{4.128}$$

Finally, applying the divergence theorem to the second term on the left-hand side of (4.127),

$$\begin{aligned}\int_{\Omega_c} \nabla_{\mathbf{y}} \cdot \left[A_\varepsilon(\mathbf{y})(\nabla_{\mathbf{x}} c^1 + \nabla_{\mathbf{y}} c^2)\right] d\mathbf{y} &= \int_\Gamma A_\varepsilon(\mathbf{y})(\nabla_{\mathbf{x}} c^1 + \nabla_{\mathbf{y}} c^2) \cdot \mathbf{n} d\Gamma(\mathbf{y}) \\ &= \int_\Gamma f(c^0, c_e^0) d\Gamma(y) \\ &= |\Gamma| f(c^0, c_e^0).\end{aligned}$$

Combining all of the results, we obtain a macroscopic equation for c^0 given by

$$|\Omega_c| \frac{\partial c^0}{\partial t} = \nabla_{\mathbf{x}} \cdot \left(\tilde{A}_\varepsilon \nabla_{\mathbf{x}} c^0\right) + |\Gamma| f(c^0, c_e^0). \tag{4.129}$$

Proceeding in an identical fashion, we also obtain a macroscopic equation for c_e^0:

$$|\Omega_e| \frac{\partial c_e^0}{\partial t} = \nabla_{\mathbf{x}} \cdot \left(\tilde{B}_\varepsilon \nabla_{\mathbf{x}} c_e^0\right) - |\Gamma| f(c^0, c_e^0), \tag{4.130}$$

where $\tilde{B}_\varepsilon$ is defined according to (4.128) with A_ε replaced by B_ε on the right-hand side. Equations (4.129) and (4.130) constitute a bidomain model with homogenized diffusion tensors $\tilde{A}_\varepsilon$ and $\tilde{B}_\varepsilon$. The latter can be determined by solving the cell equations (4.123) and (4.124) for $\mathbf{V}^c$ and $\mathbf{V}^e$. Goel et al. [224] calculated effective diffusion tensors for several different ER microstructures. They found that diffusion in the cytoplasm could be halved as the ER volume fraction increased from 0 to 0.9.

Part II
Networks

Chapter 5
Waves in Synaptically Coupled Spiking Networks

In the second part of this book, we consider a variety of neural network models that have been used to investigate the mechanisms and properties of wave propagation in the cortex and other parts of the nervous system. There has been a rapid increase in the number of computational studies of network spatiotemporal dynamics, including oscillations and waves, which are based on biophysically detailed conductance-based models of synaptically (and possibly electrically) coupled neurons; see, for example, [153]. These models provide considerable insights into the role of ionic currents, synaptic processing, and network structure on spatiotemporal dynamics, but they tend to be analytically intractable. This has motivated an alternative approach to network dynamics, involving simplified neuron models that hopefully capture important aspects of wave phenomena, while allowing a more concise mathematical treatment. In the case of oscillatory networks, such a simplification can be achieved by reducing a conductance-based neuron model to a phase model along the lines of Sect. 1.2. Alternatively, one can use a simplified spiking neuron model such as integrate-and-fire (Sect. 3.1) in order to investigate waves in excitable and oscillatory neural media. Both of these approaches are considered in this chapter. (Note that we do not discuss another important method for analyzing the dynamics of synaptically coupled spiking networks, namely, fast/slow decomposition and geometric singular perturbation theory, which is reviewed extensively elsewhere [173, 537, 538].) In Chaps. 6 and 7 we turn to population-based approaches, for which the fundamental active element is taken to be a local population of cells rather than a single neuron, and the cortex is treated as a continuum neural field. The mathematical advantage of neural fields is that many of the PDE methods from Part I can be adapted to study traveling wave solutions. An illustrative application of neural field theory to binocular rivalry waves and models of visual cortex is presented in Chap. 8.

P.C. Bressloff, *Waves in Neural Media: From Single Neurons to Neural Fields*, Lecture Notes on Mathematical Modelling in the Life Sciences,
DOI 10.1007/978-1-4614-8866-8_5,

5.1 Waves in Cortical and Subcortical Systems

We begin by summarizing the basic structure of the brain and then describe various experimental studies of waves in different regions of the brain, focusing on cortical and subcortical structures. The brain can be divided into three main parts: the forebrain, midbrain, and hindbrain; see Fig. 5.1. The forebrain consists of the cerebral cortex, thalamus, and limbic system, the midbrain consists of the tectum and tegmentum, and the hindbrain is made of the cerebellum, pons, and medulla. Often the midbrain, pons, and medulla are referred to collectively as the *brainstem*, which is responsible for vital life functions such as breathing, blood pressure, and heart pumping. The *cerebral cortex* is the largest part of the human brain and is responsible for higher brain function. The cortex is subdivided into four main sections or "lobes": (i) the frontal lobe, associated with planning, rational and emotional thinking, and problem solving; (ii) the parietal lobe, associated with motor control and somatosensory processing; (iii) the occipital lobe, associated with visual processing; (iv) the temporal lobe, associated with auditory processing, language, and memory. Cortical folding increases the effective size of the cortex, which when unfolded forms a thin sheet of neural tissue only a few millimeters thick. Neurons are arranged into six distinct layers through the depth of the cortex, each with its own characteristic set of interneurons, input sources (afferents), and output targets (efferents). For example, layer IV is the primary input layer of the sensory cortex, receiving sensory information via the thalamus, superficial layers I–III mediate intracortical connections, while deep layers V and VI send feedback to subcortical regions including the thalamus; see Fig. 5.2. For a more detailed description of the

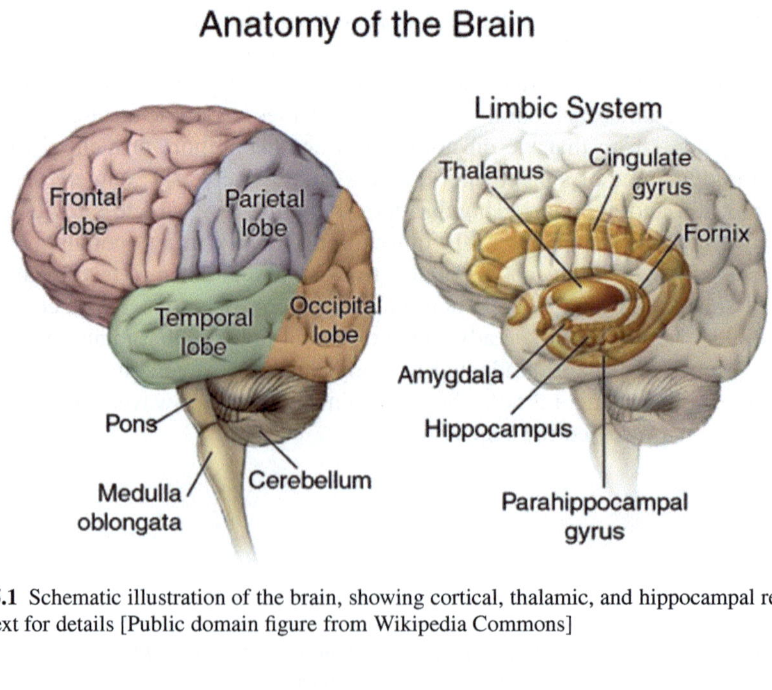

Fig. 5.1 Schematic illustration of the brain, showing cortical, thalamic, and hippocampal regions. See text for details [Public domain figure from Wikipedia Commons]

functional structure of primary visual cortex, see Sect. 8.1. *The limbic system* is a complex set of brain structures, including the hippocampus and amygdala, which is located on either side of the thalamus and adjacent to the inner border of the cortex. The limbic system appears to be primarily involved in memory and emotions. For example, the *hippocampus* is required for converting short-term memories to long-term memories and is also implicated in maintaining cognitive maps for spatial navigation. The *amygdala*, on the other hand, is involved in processing emotionally significant stimuli such as those related to fear and reward. Finally, moving on from the limbic system, the *cerebellum* of the hindbrain is associated with the regulation and coordination of fine movements, as well as posture and balance.

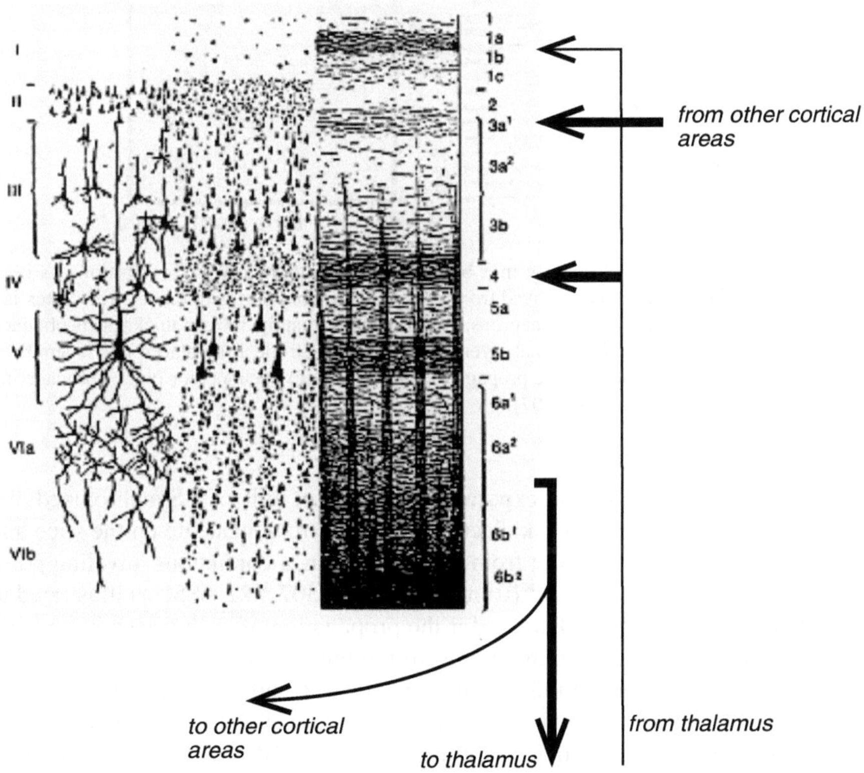

Fig. 5.2 Schematic illustration of the layered structure of primate sensory cortex. See text for details [Public domain figure from Wikipedia Commons]

Cortical Waves

A common in vitro method for studying the propagation of electrical activity in the cortex (or hippocampus) is to remove a vertical brain slice and bathe it in a pharmacological medium that blocks the effects of inhibition. This type of slice preparation

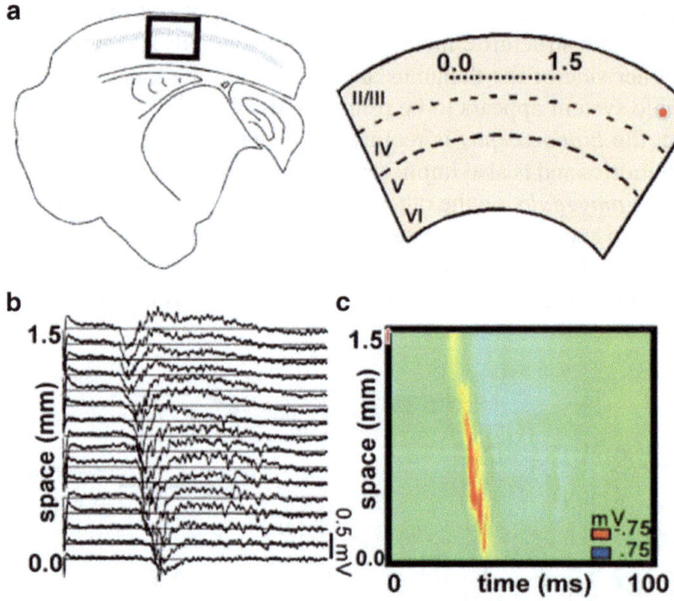

Fig. 5.3 Propagating wave of activity in a brain slice preparation in which inhibition has been blocked. (**a**) Schematic of a slice removed from the somatosensory cortex of the rat. The cortex is a thin (submillimeter) layered sheet of neurons. A slice approximately 500 μm thickness is obtained by cutting vertically through the cortical layers. (**b**) A multielectrode array placed into layers II/III of the slice records extracellularly the propagation of a wave. (**c**) Pseudocolor plot of propagating activity (Adapted from Pinto et al. [507])

was originally developed as an experimental model of epilepsy. Synchronized discharges can be evoked by a weak electrical stimulus to a local site on the slice and each discharge propagates away from the stimulus as a continuous traveling pulse at a characteristic speed of about 10 cm/s [112, 228, 507, 521, 685], as illustrated in Fig. 5.3. The underlying mechanism for the propagation of such waves is synaptic in origin rather than diffusive, in contrast to the much faster propagation of action potentials along the axons of individual neurons; see Chap. 2. The pharmacological dependence of propagating activity in vitro has been well characterized and implicates polysynaptic fast glutamatergic transmission (AMPA) as the major factor in sustaining propagating activity. On the other hand, if a cortical slice is bathed in normal cerebral fluid so that inhibition is present, then the speed of propagating activity is slowed by a factor of 10, and the propagating pulse of ensemble activity has significantly more fluctuations consistent with the complex oscillation profile seen in the corresponding local field potential [124, 549, 685]. This type of oscillatory wave consists of several oscillation cycles embedded in a single propagating wave envelope. The resulting activity is non-epileptic, with individual neurons firing at a relatively low rate of around 10 Hz, and is mediated by slower NMDA-based excitation. It has been suggested that these waves are the in vitro analog of low-frequency oscillations observed during slow-wave sleep [602]; see below. An alternative type of oscillatory wave, which is often referred to as a "one-cycle-one-wave," has been observed in disinhibited cortical slices or in slices for which NMDA receptor-mediated excitation has been enhanced. In this case, the cortical slice acts like an oscillatory medium with a frequency around 10 Hz, and the traveling pulse takes the form of a phase wave [684]; see Sect. 5.2.2. In two-dimensional (tangential) slices, these oscillations can develop into two-dimensional waves, including spiral waves [292].

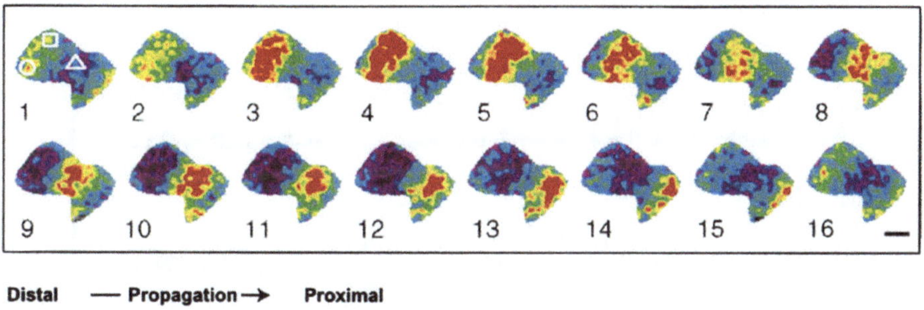

Fig. 5.4 Propagating wave of activity in the central olfactory lobe of the mollusk *Limax*. Successive images of the membrane potential in the olfactory lobe shown for one cycle. A band of depolarization can be seen propagating from the distal to the proximal end of the lobe. The time scale is 112 ms/frame and the length-scale bar is 100 μm (Adapted from Kleinfeld et al. [338])

A variety of sensory stimuli have been linked to propagating waves in vivo [442, 553, 683]. For example, a number of studies of vertebrate and invertebrate olfactory bulbs have shown that they support propagating phase waves [149, 338, 357]; see Fig. 5.4. These waves occur spontaneously, but transiently switch to standing oscillations during the presentation of an odorant stimulus, suggesting that the spatiotemporal patterns may play a role in olfactory processing. Propagating waves consisting of low-frequency oscillations superimposed on a traveling pulse have been observed in the visual cortex of anesthetized rats [255, 688], cats [39], and primates [243]. One feature of these visually evoked waves is that they exhibit anisotropy. For example, Xu et al. [688] found that the stimulus-evoked wave propagated fast in V1 but then slowed down as it approached the V1/V2 border, causing wave compression. This was followed by the initiation of a reflected wave that propagated backward into V1 and a transmitted wave propagating across V2. In addition to these single-cycle events, it is also possible to observe traveling phase waves in the visual cortex [512]. The precise functional role of spontaneous and stimulus-evoked waves in the cortex is not currently known. However, one possible contribution has been proposed in the case of sensory brain regions [170, 683]. More specifically, synchrony could represent a state of perceptual recognition, while waves could correspond to a state of perceptual scanning. In the latter case, a phase wave would provide a background depolarization in successive cortical regions, in which neurons have an increased probability of firing and a higher synaptic transmission efficacy (enhanced sensitivity). Thus, rotating waves such as spiral waves could act as an organizing center that periodically sensitizes an area without the need for any pacemaker. Interestingly, it has recently been hypothesized that in the motor cortex, synchrony and phase waves switch roles, with synchrony encoding the absence of motor action while waves with different frequencies and directions of propagation encode different motor actions [269].

Sleep–Wake Cycle

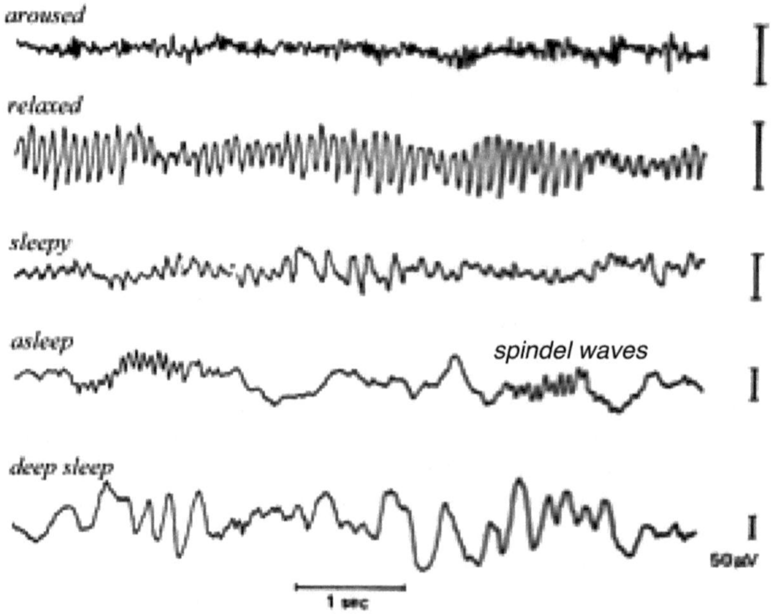

Fig. 5.5 EEG recordings from cortex at different stages of the sleep–wake cycle [Adapted from wiki.brown.edu]

Cortical oscillations and waves as measured using *electroencephalogram* (EEG) and microelectrode recordings have different signatures depending on the general state of the brain [153, 272, 442, 602]; see Fig. 5.5. Such rhythms either are generated directly in cortex or emerge from interactions with the thalamus and with a sheet of inhibitory neurons between the thalamus and cortex known as the thalamic reticular nucleus (TRN); see Fig. 5.6. The waking state and the rapid eye movement (REM) stage of sleep are both characterized by low-amplitude fast oscillations that tend to have a low level of spatiotemporal coherence. On the other hand, during non-REM sleep and anesthesia, the brain exhibits significant oscillatory activity at a variety of frequencies and with striking long-range synchrony. Of

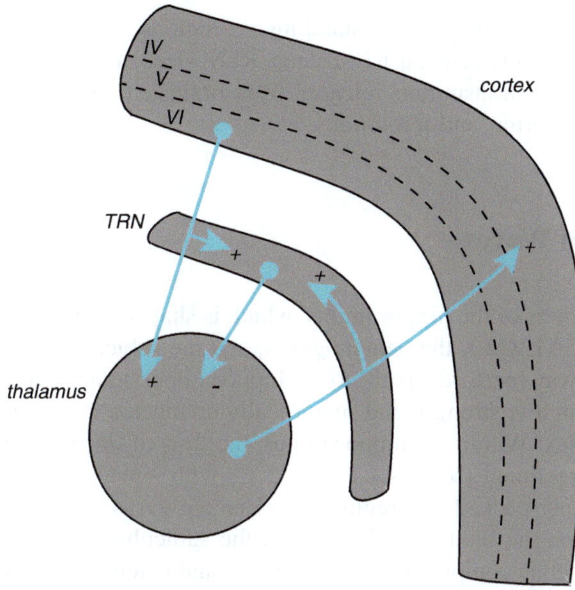

Fig. 5.6 Three principal regions thought to be responsible for the various rhythms and waves observed during the sleep–wake cycle. Surrounding the thalamus and interposed between the thalamus and cortex is a thin layer of GABAergic neurons known as the thalamic reticular nucleus (TRN). These neurons are reciprocally coupled to excitatory neurons in the thalamus, forming the reticulothalamic loop. Thalamic neurons innervate pyramidal cells in input layer IV of cortex, while pyramidal cells in cortical layer VI send feedback connections to thalamus, resulting in the thalamocortical loop. There are also axon collaterals from layer VI neurons to neurons in the TRN

particular significance is a slow oscillation (< 1 Hz) of the membrane potential of cortical neurons, comprising a hyperpolarization phase or *down state*, during which the majority of neurons are silent for several seconds, followed by a depolarization phase or *up state* of smaller duration in which neurons fire at a rate of around 10 Hz. Cortical slice studies [549] combined with computational modeling [27, 122] suggest that the transition to the up state is due to the spontaneous activation of a small subset of kindling neurons in deep cortical layers, followed by propagation of activity vertically through the layers and horizontally across the cortical network. The transition to the down state involves some combination of synaptic depression and the opening of a slow Na^+-dependent K^+ conductance. Superimposed on the slow oscillations is another rhythm known as *spindle waves*, which appear in the EEG during non-REM sleep as 7–14 Hz oscillations that wax and wane over a period of a few seconds. Unlike the slow oscillations, which are cortical in origin, the spindle waves are generated by de-inactivation of a low-threshold Ca^{2+} current in thalamic cells following inhibition from TRN cells, which is sufficient to generate a rebound burst of action potentials that then excites the TRN cells; see Sect. 5.4.3. Propagating spindle waves with speeds of 1 cm/s can be observed in thalamic slices following local current injection [334]. (Note that various EEG rhythms are referred to as “waves” even though this does not necessarily imply that they are actually propagating as waves—they can be standing oscillations, e.g., spindle waves, sharp waves.) Transitions between non-REM sleep, REM sleep, and arousal are controlled by a variety of neurotransmitters released from brainstem nuclei including acetylcholine, norepinephrine, and serotonin.

Hippocampal Waves

The primary regions of the hippocampus, which is shaped like a ram's horn (*cornu ammonis*), are CA1, CA3, the dentate gyrus, and the subiculum; see Fig. 5.7. The main input pathway (perforant path) to the hippocampus comes from the entorhinal cortex (EC), which is strongly and reciprocally connected to many other parts of the cerebral cortex. Within the hippocampus, the flow of information from the EC is largely unidirectional (see inset of Fig. 5.7), with signals propagating through a series of densely packed cell regions: the dentate gyrus $\rightarrow$ CA3 via the mossy fibers $\rightarrow$ CA1 via the Shaffer collaterals $\rightarrow$ the subiculum $\rightarrow$ EC. Each of these regions also contains complex intrinsic circuitry and extensive lateral connections. In particular, both the dentate gyrus and CA3 have extensive recurrent connections that are thought to be important in the storage of memories within the hippocampus [10]. The hippocampus exhibits two major modes of physiological activity, known as the *theta rhythm* and *large irregular activity* (LIA), each of which corresponds to distinct patterns of population activity and EEG rhythms [98]. The properties of these activity states have been determined primarily from studies of the rat, although analogous states have been found in primates.

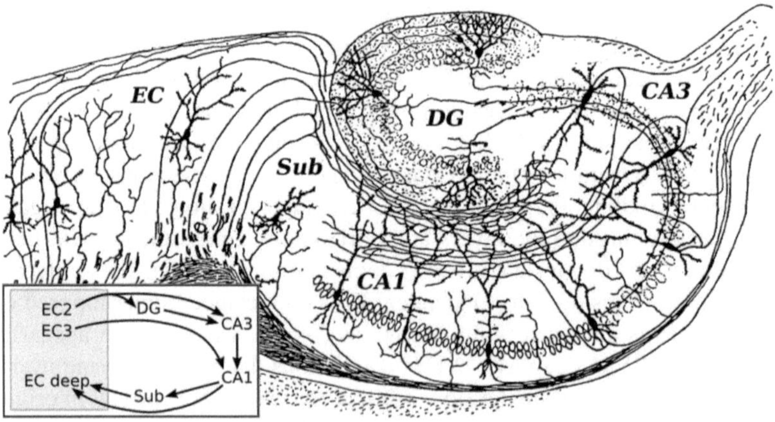

Fig. 5.7 Modified drawing of the neural circuitry of the rodent hippocampus by Santiago Ramn y Cajal. Histologie du Systeme Nerveux de l'Homme et des Vertebretes, Vols. 1 and 2. A. Maloine. Paris. 1911. DG: dentate gyrus. Sub: subiculum. EC: entorhinal cortex [Public domain figure from Wikipedia Commons]

During the theta phase, the EEG is dominated by large regular oscillations with a frequency range of 6–9 Hz [97]. It is associated with sparse population activity, in which the vast majority of neurons are silent within any short time window, with the small remaining fraction of cells firing at relatively high rates (up to 50 Hz) for around a second. Theta rhythms occur during REM sleep and during periods of active, alert behavior (particularly locomotion), where cell activity is primarily determined by the spatial location of the animal, so that the set of active neurons is constantly changing. This is exemplified by the selective activation of CA1 pyramidal cells at particular locations in space called *place fields* [474]. A given configuration of place cells provides an internal representation of the environment that affords an animal knowledge of its position relative to important locations. A key observation is that the spike activity of a hippocampal CA1 cell advances to earlier phases of the theta cycle as the animal passes through the cell's place field [475]. One possible interpretation of this phase advance is that it reflects a prediction of the sequence of upcoming locations based on the rat's current position, that is, it is part of a path integration system [384]. It was originally thought that theta oscillations were synchronized across the hippocampus, but it has subsequently been shown that theta oscillations in freely behaving rats propagate as traveling waves [393].

The LIA mode occurs during non-REM sleep and during periods of waking immobility such as resting or eating. The pattern of irregular activity is occasionally interrupted by large amplitude surges called *sharp waves* [96], which coincide with 50–100 ms bursts of spiking activity in pyramidal cells of CA3 and CA1. Superimposed on the sharp waves are short-lasting high-frequency oscillations called *ripples*, with frequencies in the range 150–200 Hz in rats. An in vitro study indicates that the sharp wave–ripple complexes (SPW-R) propagate from CA3 to CA1 and then to the subiculum with a speed around 5 cm s^{-1} [399]. One of the interesting features of sharp waves is that they have been implicated in memory consolidation during sleep [611]. For example, place cells exhibit experience-dependent reactivation during sleep that is representative of previous behavior [677]. More specifically, neurons with overlapping place fields during spatial exploration within an environment show increased coactivity during subsequent sleep, firing in the same sequence as they fired when the rodent was exploring its surroundings in the awake state.

Pathological States

Propagating waves of activity are also associated with a variety of pathological brain states, including cortical spreading depression and epileptic seizures. Cortical spreading depression is a complex pulse-like wave of transient depolarization of neurons and glial cells that propagates slowly across cortical and subcortical gray matter at speeds of 1–5 mm/min [596]. Spreading depression, which is thought to play a major role in migraine with aura [358, 639], appears to involve the diffusion and active transport of extracellular ions such as K^+ and is thus modeled in terms of reaction–diffusion equations rather than spiking networks or neural field equations;

see Sect. 9.3. On the other hand, the latter models have been used to study the initiation, propagation, and termination of epileptiform activity [158, 507]. Epilepsy is characterized by spontaneously occurring seizures that may involve a significant portion of the brain at onset (primary generalized seizures) or originate from a circumscribed brain region and may or may not spread to other regions (focal seizures) [125, 348, 415, 430]; see Sect. 9.4. In the case of focal seizures, a localized region of pacemaker neurons within deep layers of cortex can drive hypersynchronous activity that rapidly spreads vertically through the layers and then spreads horizontally as a traveling pulse. There is also a relationship between epilepsy and sleep, which is not surprising given that sleep also involves changes in levels of synchrony. For example, the increased synchronization during non-REM sleep might facilitate the abnormal synchronization of the epileptic brain, while the relatively desynchronized state during REM sleep or waking might suppress it. Unfortunately, the relationship between sleep and epilepsy is far more complex and heterogeneous [584].

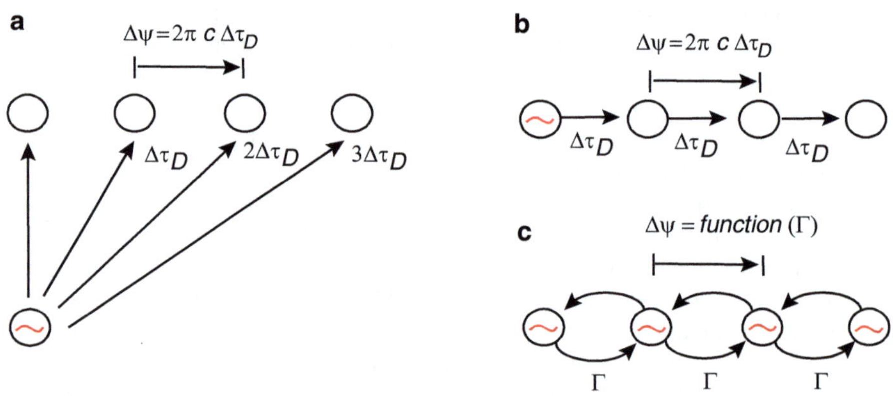

Fig. 5.8 Three mechanisms for generating and sustaining propagating waves. (**a**) Fictive waves due to a local oscillator driving adjacent cortical regions with increasing time delays $\Delta\tau_D$. (**b**) A local oscillator drives a chain of excitable cells. (**c**) Traveling phase wave in a chain of neuronal oscillators with local coupling parameter Γ

Basic Wave Propagation Mechanisms

Ermentrout and Kleinfeld [170] have highlighted three basic mechanisms for wave propagation in the cortex and other neural circuits; see Fig. 5.8:

1. A single oscillator directly excites neighboring areas with increasing time delays that generates a fictive propagating wave. This has been suggested as one mechanism for traveling waves in anesthetized visual cortex, with the spread of the wave determined by axonal propagation delays in long-range horizontal connections [39, 243]. The predominant wave mechanism is monosynaptic.

2. A single pacemaker oscillator (or local group of oscillators) drives an excitable neural medium. Each time the pacemaker fires, it generates a solitary propagating wave of activity. If the pacemaker ceases to be active, then waves no longer occur. This polysynaptic mechanism underlies the propagation of solitary pulses in disinhibited cortical slices and is one framework for understanding the spread of synchronous activity in focal epilepsy or during slow-wave sleep. We will consider traveling waves in excitable spiking networks in Sect. 5.4.
3. The whole medium is in an oscillatory state so that wave propagation represents a spatial gradient in the phase of the oscillators, as has been observed in various sensory cortices as well as central pattern generators (CPGs) for locomotion. In these cases one can interpret the resulting wave as a phase-locked solution of a coupled phase oscillator model; see Sect. 5.2.

One important issue is determining what the fundamental oscillatory unit is in each of these mechanisms—a single neuron as considered here or a local population of neurons (see Chap. 6). A related issue is the time scale over which phase-locking occurs—does it occur at the level of individual spikes, or bursts of action potentials, or population firing rates?

5.2 Phase Reduction of Synaptically Coupled Neural Oscillators

In Sect. 1.4 we considered a feedforward neuronal network in which a single output neuron was synaptically driven by a set of afferent neurons that were themselves driven by external inputs; see Fig. 1.14. However, in order to model the dynamics of neuronal networks in the cerebral cortex, for example, it is necessary to take into account the fact that around 80 % of inputs into a cortical neuron are from other cortical neurons, indicative of a highly recurrent network architecture. This means that the output spike train of one neuron feeds back as input to a subset of other neurons that themselves innervate the given neuron; see Fig. 5.9. Therefore, consider a network of N synaptically coupled cortical neurons labeled $i = 1,\ldots,N$. Let $v_i(t)$ denote the somatic membrane potential of the ith neuron, evolving according to a conductance-based model of the form

$$C\frac{\mathrm{d}v_i}{\mathrm{d}t} = -I_{\mathrm{con,i}}(v_i,\ldots) + u_i(t), \tag{5.1}$$

supplemented by additional equations for the ionic gating variables. (For the moment we are ignoring any external currents.) The total synaptic current $u_i(t)$ entering the soma is generated by the incoming spike trains from other neurons in the network that are filtered by synaptic and dendritic processing. Denote the sequence of firing times of the jth neuron by $\{T_j^m, m \in \mathbf{Z}\}$, with

$$T_j^m = \inf\{t, t > T_j^{m-1} | v_j(t) = \kappa, \dot{v}_j(t) > 0\}, \tag{5.2}$$

where κ is the firing threshold. Following Sect. 1.4, the net synaptic current into postsynaptic neuron i due to innervation by the spike train from presynaptic neuron j is taken to have the general form $\sum_m \Phi_{ij}(t - T_j^m)$, where $\Phi_{ij}(t)$ represents the temporal filtering effects of synaptic and dendritic processing on the spike train from neuron j innervating neuron i. (For the moment we ignore short-term synaptic depression and the effects of axonal propagation delays.) Assuming that all synaptic inputs sum linearly, we find that the total synaptic input to the soma of the ith neuron is

$$u_i(t) = \sum_{j=1}^N \sum_m \Phi_{ij}(t - T_j^m) = \sum_{j=1}^N \int_{-\infty}^t \Phi_{ij}(t-t')a_j(t')dt', \tag{5.3}$$

where we have set

$$a_j(t) = \sum_{m\in\mathbf{Z}} \delta(t - T_j^m). \tag{5.4}$$

That is, $a_j(t)$ represents the output spike train of the jth neuron in terms of a sum of Dirac delta functions. In general, (5.1)–(5.3) are difficult to analyze, particularly when N is large, so that some form of approximation scheme is needed. One approach is to use the phase reduction procedure of Sect. 1.2.

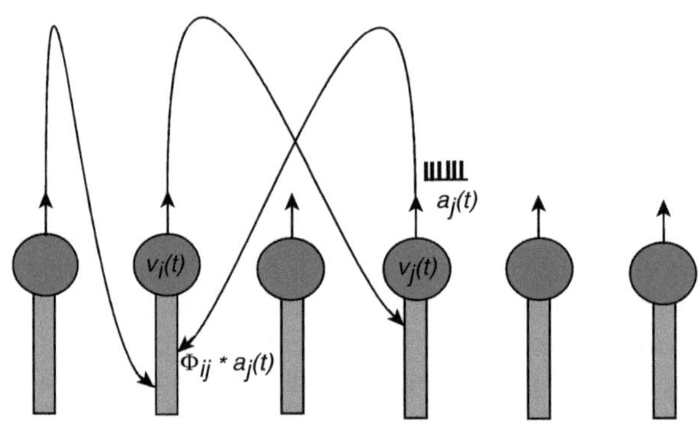

Fig. 5.9 Schematic diagram of a recurrent neuronal network. Not all connections are shown

Consider a network of N identical conductance-based neurons, each of which is acting as a limit cycle oscillator. (It is straightforward to extend the analysis to nonidentical oscillators with different intrinsic frequencies, e.g.) If the synaptic interactions between the neurons are sufficiently weak, then the phase reduction method outlined in Sect. 1.2 can be applied at the network level. This leads to a system of coupled equations for the phases of the neural oscillators in the network, in which the effective phase interaction function H depends on both the nature of the synaptic interactions and the intrinsic properties of each oscillator as expressed through its phase-resetting curve (see Fig. 5.10). Denoting the state of the ith neuron by $\mathbf{x}_i \in \mathbb{R}^k$, with $x_{1,i} = v_i$, and taking $\Phi_{ij}(t) = w_{ij}\Phi(t)$, the network dynamics given by (5.1) and (5.3) can be written as the coupled system of ODEs

$$\frac{dx_{\mu,i}}{dt} = f_\mu(\mathbf{x}_i) + \varepsilon \delta_{\mu,1} \sum_{j=1}^{N} w_{ij} \sum_m \Phi(t - T_j^m) \tag{5.5}$$

for $\mu = 1, \ldots, k$. Suppose that in the absence of interactions each neuron has a stable limit cycle $\mathbf{X}$ with natural frequency ω_0, and introduce the phase variable θ_i such that

$$\frac{d\theta_i}{dt} = \omega_0. \tag{5.6}$$

There is a one-to-one mapping $\theta_i \to \mathbf{X}(\theta_i)$ on the limit cycle. Treating $p_i(t) = \varepsilon \sum_{j=1}^{N} w_{ij} \sum_m \Phi(t - T_j^m)$ as a weak time-dependent perturbation of the voltage of the ith oscillator, we can use the method of isochrones in Sect. 1.2, to obtain the phase equation

$$\frac{d\theta_i}{dt} = \omega_0 + \varepsilon \sum_{j=1}^{N} w_{ij} Z(\theta_i(t)) \sum_m \Phi(t - T_j^m), \tag{5.7}$$

where $Z(\theta_i)$ is the phase-resetting curve (PRC)—more precisely, the component of the PRC with respect to voltage perturbations.

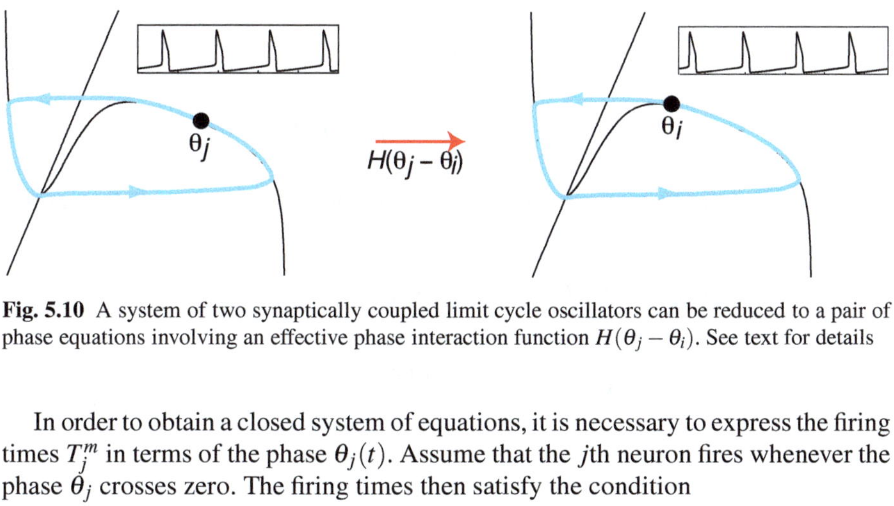

Fig. 5.10 A system of two synaptically coupled limit cycle oscillators can be reduced to a pair of phase equations involving an effective phase interaction function $H(\theta_j - \theta_i)$. See text for details

In order to obtain a closed system of equations, it is necessary to express the firing times T_j^m in terms of the phase $\theta_j(t)$. Assume that the jth neuron fires whenever the phase θ_j crosses zero. The firing times then satisfy the condition

$$\theta_j(T_j^m) = 2\pi m, \quad \dot{\theta}_j(T_j^m) > 0. \tag{5.8}$$

If we now set $\psi_j(t) = \theta_j(t) - \omega_0 t$ then ψ_i is a slowly varying function of time (for small ε) so that to a first approximation

$$T_j^m = (2\pi m - \psi_j(t))/\omega_0 = t + (2\pi m - \theta_j(t))/\omega_0, \tag{5.9}$$

that is, the current phase determines the most recent sequence of firing times. Hence, (5.7) reduces to the closed system of equations

$$\frac{d\theta_i}{dt} = \omega_0 + \varepsilon \sum_{j=1}^{N} w_{ij} Z(\theta_i(t)) P(\theta_j(t)), \tag{5.10}$$

where

$$P(\theta_j(t)) = \sum_m \Phi[(\theta_j(t) - 2\pi m)/\omega_0].$$

The final step of the analysis is to use the method of averaging to obtain effective interactions that depend only on phase differences. Substituting $\psi_i = \theta_i - \omega_0 t$ into the phase equation (5.10) gives

$$\frac{d\psi_i}{dt} = \varepsilon \sum_{j=1}^{N} w_{ij} Z(\psi_i(t) + \omega_0 t) P(\psi_j(t) + \omega_0 t). \tag{5.11}$$

The corresponding averaged equation is then

$$\frac{d\psi_i}{dt} = \varepsilon \sum_{j=1}^{N} w_{ij} H(\psi_j - \psi_i), \tag{5.12}$$

where

$$\begin{aligned} H(\psi_j - \psi_i) &= \frac{1}{\Delta_0} \int_0^{\Delta_0} Z(\psi_i + \omega_0 t) P(\psi_j + \omega_0 t]) dt \\ &= \frac{1}{2\pi} \int_0^{2\pi} Z(\theta + \psi_i - \psi_j) P(\theta) d\theta, \end{aligned} \tag{5.13}$$

with $\Delta_0 = 2\pi/\omega_0$. We have exploited the periodicity of the functions R, P. A useful alternative expression for H is obtained by substituting for P:

$$\begin{aligned} H(\psi) &= \frac{1}{2\pi} \int_0^{2\pi} Z(\theta - \psi) \sum_m \Phi([\theta - 2\pi m]/\omega_0) d\theta \\ &= \frac{1}{2\pi} \int_0^{\infty} Z(\theta - \psi) \Phi(\theta/\omega_0) d\theta. \end{aligned} \tag{5.14}$$

5.2.1 *Phase-Locking*

We define a 1:1 phase-locked solution of the phase equation (5.12) to be of the form

$$\psi_i(t) = (\omega - \omega_0)t + \overline{\psi}_i, \tag{5.15}$$

where ω is the *collective frequency* of the coupled oscillators and $\Psi = (\bar{\psi}_1, \ldots, \bar{\psi}_n)$ is a set of constant phases. Substitution into (5.12) leads to the fixed-point equations

$$\omega = \omega_0 + \varepsilon \sum_j w_{ij} H(\overline{\psi}_j - \overline{\psi}_i), \quad i = 1, \ldots, N. \tag{5.16}$$

After choosing some reference oscillator, the N phase equations determine the collective period ω and $N-1$ relative phases with the latter independent of ε. The local stability of a phase-locked solution can be analyzed by setting $\psi_i(t) = \overline{\psi}_i + [\omega - \omega_0]t + \delta\psi_i(t)$ and expanding to first order in $\delta\psi_i(t)$:

$$\frac{\mathrm{d}\delta\psi_i}{\mathrm{d}t} = \varepsilon \sum_{j=1}^{N} \mathscr{H}_{ij}(\Phi)\delta\psi_j, \tag{5.17}$$

where

$$\mathscr{H}_{ij}(\Phi) = w_{ij}H'(\overline{\psi}_j - \bar{\psi}_i) - \delta_{i,j}\sum_k w_{ik}H'(\overline{\psi}_k - \overline{\psi}_i), \tag{5.18}$$

and $H'(\psi) = dH(\psi)/\mathrm{d}\psi$. One of the eigenvalues of the Jacobian $\mathscr{H}$ is always zero, and the corresponding eigenvector points in the direction of the flow, that is $(1,1,\ldots,1)$. The phase-locked solution will be stable provided that all other eigenvalues have a negative real part [166].

Consider, for example, two identical oscillators ($N = 2$) that are symmetrically coupled, with $w_{12} = w_{21} = 1$ and $w_{11} = w_{22} = 0$. The phase difference $\psi = \psi_2 - \psi_1$ satisfies the simple equation

$$\frac{d\psi}{dt} = \varepsilon H^-(\psi), \tag{5.19}$$

where $H^{\pm}(\psi) = H(-\psi) \pm H(\psi)$. Phase-locked states are then given by the zeros of the odd function H^-:

$$H^-(\overline{\psi}) = 0. \tag{5.20}$$

A given phase-locked state is stable provided that

$$\varepsilon \left.\frac{\mathrm{d}H^-(\psi)}{\mathrm{d}\psi}\right|_{\overline{\psi}} < 0. \tag{5.21}$$

Note that by symmetry both the inphase ($\overline{\psi} = 0$) and antiphase ($\overline{\psi} = \pi$) states are guaranteed to exist. However, there may be additional fixed points depending on the parameters. The collective frequency of oscillation is determined by the even part of H:

$$\omega = \omega_0 + \frac{\varepsilon}{2} H^+(\overline{\psi}). \tag{5.22}$$

It turns out that the stability of the inphase state for a pair of neurons depends on a number of factors [258, 651]: whether the synapses are excitatory or inhibitory, whether the kinetics are fast or slow, whether the phase-resetting curve is type I or type II, and the size of any axonal delays. We shall illustrate this by considering a simple example.

Example: Pair of Hodgkin–Huxley neurons. In Fig. 5.11 we show a typical phase-resetting curve $R(\phi)$ for the Hodgkin–Huxley model (1.6) together with the resulting phase interaction function $H(\psi)$ for $\Phi(t)$ given by an α function:

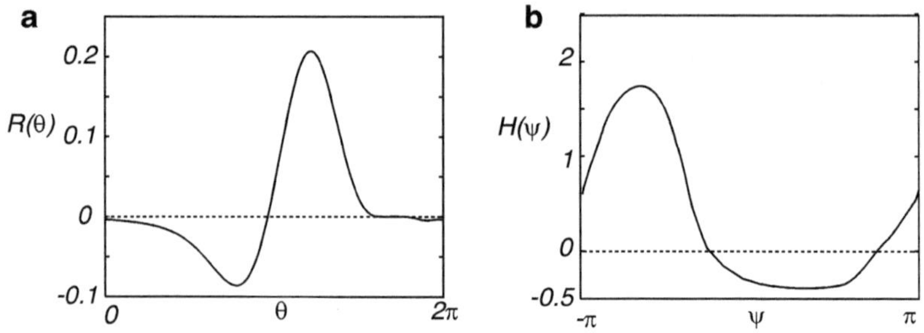

Fig. 5.11 Plot of type II phase-resetting curve $R(\theta)$ for the Hodgkin–Huxley model together with effective phase interaction function $H(\psi)$

$$\Phi(t) = \alpha^2 t e^{-\alpha t}. \tag{5.23}$$

If we approximate the PRC by $Z(\phi) = -\sin\phi$ then

$$\begin{aligned} H(\psi) &= \frac{1}{2\pi}\int_0^{2\pi} Z(\theta-\psi)P(\theta)d\theta \\ &= \frac{\alpha^2}{2\pi}\frac{d}{d\alpha}\int_0^{\infty} \sin(\theta-\psi)e^{-\alpha\theta/\omega_0}d\theta \\ &= \frac{\alpha^2}{2\pi}\mathrm{Im}\frac{d}{d\alpha}\int_0^{\infty} e^{i(\theta-\psi)}e^{-\alpha\theta/\omega_0}d\theta. \end{aligned}$$

Now

$$\frac{d}{d\alpha}\int_0^{\infty} e^{i\theta}e^{-\alpha\theta/\omega_0}d\theta = -\frac{\omega_0}{(\alpha-i\omega_0)^2}$$

and so

$$H(\psi) = A\sin\psi - B\cos\psi, \tag{5.24}$$

where

$$A = [\alpha^2-\omega_0^2]K, \quad B = 2\alpha\omega_0 K, \quad K = \frac{\alpha^2\omega_0}{2\pi[\alpha^2+\omega_0^2]^2}.$$

The odd part is then $H^-(\psi) = -A\sin\psi$ with $A \neq 0$ provided that $\alpha \neq \omega_0$, and then there are phase-locked states at $\psi = 0, \pi$ as expected. The synchronous solution is stable if $\varepsilon H^{-'}(0) < 0$. This implies that synchronization will occur for excitatory coupling if $\alpha > \omega_0$ and for inhibitory coupling if $\alpha < \omega_0$. Thus, *inhibition* rather than excitation synchronizes the neurons when the synapses have sufficiently slow kinetics. The reverse is true for the antiphase solution $\psi = \pi$. If higher-order terms in the Fourier series expansion of H are included, $H(\psi) = \sum_n h_n e^{in\psi}$, then the transition from synchronous to antisynchronous behavior is smoother, that is, the system no longer makes a sudden jump from one to the other at a critical value of α.

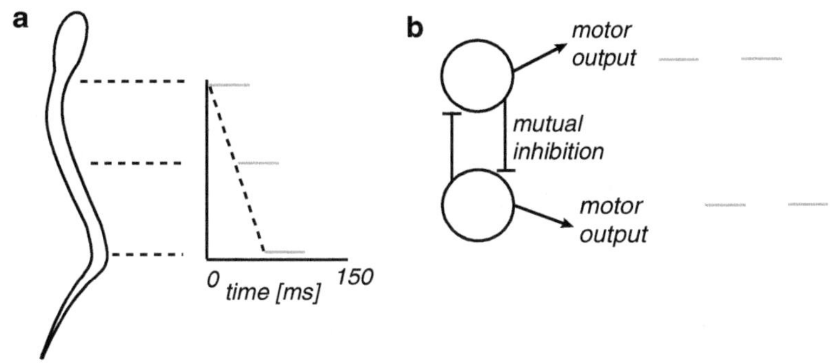

Fig. 5.12 (**a**) Swimming locomotion is associated with a phase gradient along the spinal cord of a lamprey. (Gray bars indicate active phase of a given neuron at different spatial locations or segments along the spinal cord.) (**b**) Half-center oscillator consisting of a pair of neurons that have reciprocal inhibition and fire in antiphase

5.2.2 Traveling Phase Waves on a Chain

The idea that a network of synaptically coupled spiking neurons can be reduced to a corresponding network of coupled phase oscillators is the starting point for considering a different type of wave than previous chapters, which can arise when the network is endowed with spatial structure. That is, suppose that the oscillators are arranged on a regular 1D lattice with sites labeled $n = 1, \ldots N$, and assume that the synaptic weight w_{nm} of the connection from lattice site m to lattice site n depends on the spatial separation of the two sites, $w_{nm} = w(n-m)$. It is then possible for the lattice of oscillators to support a phase-locked solution in which there is an approximately constant phase gradient Γ along the bulk of the lattice. In other words $\theta_n(t) \approx \omega t + n\Delta\theta$. Such a solution is known as a *phase wave*. Considerable insight into the conditions under which spatial phase gradients occur have been obtained in studies of central pattern generators (CPGs) for swimming locomotion in invertebrates such as the leech and primitive vertebrates such as the lamprey [121, 273, 404, 443]; see Fig. 5.12. At the simplest level, this may be modeled in terms of a chain of N phase oscillators θ_n with nearest-neighbor synaptic coupling as shown in Fig. 5.13. Each oscillator represents the CPG of an individual segment of the spinal cord. In the case of the lamprey, each oscillatory unit is a so-called half-center oscillator; see Fig. 5.12b. A half-center oscillator consists of two neurons that are reciprocally coupled by (intrasegmental) inhibition and consequently oscillate in antiphase. The swimming movement is thus produced by alternating neural activity between the left and right side of the body that propagates along the body as a phase wave to produce a side-to-side undulating wave; see Fig. 5.12a. Note that the individual CPG neurons of a half-center oscillator need not be intrinsic oscillators, in which case reciprocal inhibition is required in order to generate the rhythm, for example, via post-inhibitory rebound. In the case of the lamprey, however, it now appears that excitation from an external drive combined with membrane properties is sufficient for rhythm generation in an individual CPG neuron. In particular, the primary mechanism for terminating a burst is the activation of fast and slow Ca^{2+}-dependent K^+ channels [273, 443]. In the case of the leech, intrasegmental neurons on either side of the body axis fire synchronously, which is consistent with the fact that the undulatory motion during swimming is dorsoventral rather than side-to-side.

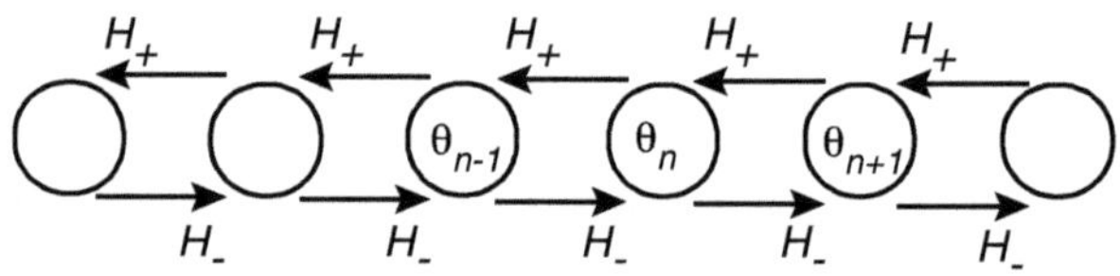

Fig. 5.13 A chain of *N* phase oscillators

During swimming, approximately one full body-length wave is maintained by period-independent intersegmental phase lags; see Fig. 5.12a. That is, there is a phase wave that propagates along the spinal cord such that the phase lag $\Delta\theta = \theta_{n+1} - \theta_n$ between adjacent segments is a fixed percentage of the cycle period. For example, a lamprey typically has 100 segments so that the phase lag is around 1 %. At least two symmetry-breaking mechanisms have been identified for maintaining a phase gradient along a chain of oscillators [171, 347, 443]:

(a) Asymmetry in the form of a gradient in intrinsic frequencies of the individual neurons, that is, $\omega_n = \omega_0 + n\omega_1$ where $\omega_{0,1}$ are constants, will induce a gradient in the relative phases of the coupled system. The resulting phase wave propagates opposite to the direction of the gradient, that is, from high to low frequencies. Such a frequency gradient has been found in the lamprey CPG for swimming, with frequency decreasing from the head (rostral end) to the tail (caudal end). On the other hand, the distribution of intrinsic frequencies appears to be "U-shaped" in the leech, with the highest frequencies occurring mid-body [443]. Interestingly, a spatial gradient of frequencies has also been observed experimentally in slices of molluscan olfactory lobe, with the wave propagating from high to low frequencies as predicted by theory [167]; see Sect. 7.5.
(b) Asymmetry between the ascending and descending synaptic interactions so that $H_+ \neq H_-$. In this case, a constant phase lag can be obtained in the bulk of the chain provided that $H_+(\theta)$ or $H_-(\theta)$ vanishes at a nonzero value of θ, which is known as the edge condition [171, 347]; see below. Asymmetric coupling has been found in both the leech and the lamprey. In the latter case, the descending interactions have stronger functional coupling than ascending connections, whereas in the leech, the coupling strength is symmetric while the specific pattern of connections is asymmetric. More specifically, leech CPG neurons within a hemi-segment (segment on one side of the body) are active at three different phases separated by 120°, and there are asymmetries between ascending and descending fibers associated with a given phase group.

We will illustrate the above two mechanisms using a simple model of a chain with $H_{\pm}(\theta) = W_{\pm}H(\theta)$. Such a network is described by the following system of coupled phase equations:

$$\begin{aligned}\dot{\theta}_1 &= \omega_1 + W_+ H(\theta_2 - \theta_1)\\ \dot{\theta}_n &= \omega_n + W_+ H(\theta_{n+1} - \theta_n) + W_- H(\theta_{n-1} - \theta_n), \quad n = 2, \ldots, N-1\\ \dot{\theta}_N &= \omega_N + W_- H(\theta_{N-1} - \theta_N).\end{aligned} \tag{5.25}$$

Introducing the intersegmental phase differences $\psi_n = \theta_{n+1} - \theta_n$ leads to the $N-1$-dimensional system of equations

$$\dot{\psi}_n = \Delta\omega_n + W_+[H(\psi_{n+1}) - H(\psi_n)] + W_-[H(-\psi_n) - H(-\psi_{n-1})], \quad n = 1, \ldots, N-1 \tag{5.26}$$

with boundary conditions

$$H(-\psi_0) = 0 = H(\psi_N) \tag{5.27}$$

and $\Delta\omega_n = \omega_{n+1} - \omega_n$. Suppose that there is a gradient of frequencies along the chain, that is, $\Delta\omega_n$ has the same sign for all n. Also take an isotropic, odd interaction function, $W_{\pm} = 1$ and $H(\psi) = -H(-\psi)$. Then

$$\dot{\psi}_n = \Delta\omega_n + H(\psi_{m+1}) + H(\psi_{n-1}) - 2H(\psi_n), \quad n = 1, \ldots, N-1. \tag{5.28}$$

The phase-locked solutions $\Psi = (\psi_1, \ldots, \psi_{N-1})$ then satisfy the matrix equation

$$\mathbf{H}(\Psi) = -\mathbf{A}^{-1}\mathbf{D}, \tag{5.29}$$

where $\mathbf{H}(\Psi) = (H(\psi_1), \ldots, H(\psi_{N-1}))^T$, $\mathbf{D} = (\Delta\omega_1, \ldots \Delta\omega_{N-1})$ and $\mathbf{A}$ is a tridiagonal matrix with elements $A_{nn} = -2, A_{n,n+1} = 1 = A_{n+1,n}$. Suppose, for concreteness, that $H(\psi) = \sin\psi$. Then a solution Θ will exist only if every component of $\mathbf{A}^{-1}\mathbf{D}$ lies between ± 1. Let $a_0 = \max\{|\mathbf{A}^{-1}\mathbf{D}_i|\}$. If $a_0 < 1$, then for each $n = 1, \ldots, N-1$ there are two distinct solutions $\psi_n^{\pm}$ in the interval $[0, 2\pi)$ with $H'(\psi_n^-) > 0$ and $H'(\psi_n^+) < 0$. In other words, there are 2^N phase-locked solutions. Linearizing about each phase-locked solution and exploiting the structure of the matrix $\mathbf{A}$, it can be proven that only the solution $\Psi^- = (\psi_1^-, \ldots \psi_{N-1}^-)$ is stable [171]. Assuming that the frequency gradient is monotonic, this solution corresponds to a stable traveling wave. When the frequency gradient becomes too steep to allow phase-locking, that is, $a_0 > 1$, two or more pools of oscillators (frequency plateaus) tend to form and oscillate at different frequencies. Waves produced by a frequency gradient do not have a constant speed or, equivalently, constant phase lags along the chain.

Constant speed waves can be generated from (5.26) by considering phase-locked solutions defined by $\psi_n = \psi$ for all n with a collective period of oscillation determined using $\dot{\psi}_n = \Omega$ such that

$$\Omega = \omega_1 + W_+ H(\psi). \tag{5.30}$$

The steady-state solutions are then

$$
\begin{aligned}
\Delta\omega_1 + W_- H(-\psi) &= 0 \\
\Delta\omega_n &= 0, \quad n = 2, \ldots, N-2 \\
\Delta\omega_{N-1} - W_+ H(\psi) &= 0.
\end{aligned}
\tag{5.31}
$$

This implies that all internal oscillators ($i = 2,\ldots,N-2$) have the same intrinsic frequency. If we further assume that $\Delta w_1 = 0 = \Delta\omega_{N-1}$, then a constant speed wave will exist provided that, (i) the interactions are unidirectional with $W_- = 0$ and (ii) H satisfies the edge Property $H(\psi) = 0$ for some $\psi \neq 0$. For example, if $H(\psi) = \sin(\psi + \sigma)$ for a fixed phase shift σ, then the steady-state phase solution is $\psi = -\sigma$. It also follows that although the speed of the wave can change by changing the intrinsic frequency ω_1, the phase lag ψ remains the same. This is consistent with the swimming motion of invertebrates. Finally, note that if coupling is in both directions, as in the lamprey CPG, one finds that for sufficiently long chains and $H^+ \neq H^-$, the system typically behaves like the unidirectional case except that there is a small *boundary layer* close to one end [347].

Although the mathematical analysis of phase waves along a chain of oscillators with nearest-neighbor coupling has provided important insights into CPGs and swimming locomotion, there are still a number of outstanding issues. One of the most important is understanding how the phase lags maintain a one body-length wave irrespective of the cycle period. Is this somehow built into the phase interaction function or the distribution of intrinsic frequencies, or does sensory feedback play an important role? For example, it is known that leeches and lampreys can sense their own body movements via proprioceptors and can adjust their cycle period and phase lags accordingly. Moreover, their swimming behavior is strongly affected by changes in their fluid environment, consistent with the idea that fluid resistance forces provided by water are also sensed by proprioceptors [443]. Yet another factor in determining phase lags is the range of synaptic interactions. The chain model assumed nearest-neighbor coupling, which is consistent with experimental observations that short-range coupling plays an important role in determining the phase. However, longer-range projections do also exist, which in lamprey can extend up to 20 segments.

5.2.3 *Ring of Phase Oscillators and the Sync Basin*

In the above example, a stable phase wave solution was generated through a combination of network heterogeneities (in natural frequencies or synaptic interactions) and boundary effects. An alternative network architecture that supports traveling phase waves is a ring of identical oscillators [166, 671]. However, in this case, one or more stable phase waves can coexist with a stable synchronous state. This then raises the interesting question as to the size of the basin of attraction of the synchronous state (the sync basin) or a particular phase wave as network parameters change. In other words, what is the set of initial conditions that induces a particular phase-locked state? From a dynamical systems perspective, finding such basins of attraction is nontrivial. However, Wiley et al. [671] have made some progress by analyzing a ring of N identical phase oscillators, each of which is coupled to its k nearest neighbors on either side using a sinusoidal interaction function:

$$\dot{\theta}_i = \omega + \sum_{j=i-k}^{i+k} \sin(\theta_j - \theta_i), \quad i = 1, \ldots, N, \tag{5.32}$$

where $N \gg 1$ and the index i is taken to be periodic mod N. In the case $k = N/2$ (or its integer part), (5.32) reduces to the Kuramoto model of globally coupled oscillators [350]. It is straightforward to show that the system (5.32) always supports a stable synchronous state $\theta_i(t) = \omega t$ for all $i = 1, \ldots, N$. Moreover, numerically it is found that the synchronous state is a global attractor for sufficiently long-range coupling, that is, for $k > k_c \approx 0.34N$ [671]. However, as k decreases below k_c, other competing attractors emerge, which take the form of traveling phase waves (also known as splay states or uniformly twisted waves):

$$\theta_j = \omega t + \frac{2\pi q j}{N} + C, \quad j = 1, \ldots, N. \tag{5.33}$$

Here C is a constant and the integer q is a winding number that specifies the number of complete rotations or twists in phase that occur as the ring is traversed once. We summarize the analysis of (5.32) carried out by Wiley et al. [671].

Moving to a rotating frame by setting $\theta_i = \phi_i - \omega t$ and allowing for a nonuniform coupling $W(i-j)$ between oscillators i and j that depends on the distance $|i-j|$ gives

$$\dot{\phi}_i = \sum_{j=i-k}^{i+k} W(i-j) \sin(\phi_j - \phi_i), \quad i = 1, \ldots, N. \tag{5.34}$$

It is assumed that $W(s)$ is a positive, even, monotonically decreasing function of $|s|$. Equation (5.34) is an example of a gradient system, since it can be rewritten in the form

$$\dot{\phi}_i = -\frac{\partial V}{\partial \phi_i}, \quad V = -\frac{1}{2} \sum_{i,j=1}^{N} W(i-j) \cos(\phi_j - \phi_i), \tag{5.35}$$

where V is a potential function. It follows that all trajectories of (5.34) flow monotonically downhill on the resulting potential surface $V(\phi_1, \ldots, \phi_N)$, and asymptotically approaches fixed points. In other words the only attractors of the system are phase-locked states. Since N is assumed to be large, it is convenient to consider a spatially continuous version of (5.34), which takes the form (see also Sect. 5.4.1)

$$\frac{\partial \phi}{\partial t} = \int_{-\pi}^{\pi} w(x-y) \sin[\phi(y,t) - \phi(x,t)] dy. \tag{5.36}$$

This can be derived by setting $\Delta = \pi/N$, $x = (i - N/2)\Delta$, $\phi_i = \phi(x)$, $W(i-j) = w(x-y)\Delta$ and taking the limit $N \to \infty$. For convenience, w is normalized to unity. It is simple to show that $\phi(x,t) = qx$, integer q, is a phase-locked solution of (5.36) by direct substitution:

$$0 = \int_{-\pi}^{\pi} w(x-y)\sin[q(x-y)]dy,$$

which holds since w is an even function while sine is an odd function. The ring topology requires that q is an integer, with $q=0$ corresponding to a synchronous state and $q \neq 0$ to a phase wave with winding number q.

The linear stability of a phase-locked state is determined by setting $\phi(x,t) = qx + \eta(x,t)$ and keeping only linear terms in η:

$$\frac{\partial \eta}{\partial t} = \int_{-\pi}^{\pi} w(x-y)\cos[q(x-y)][\eta(y,t)-\eta(x,t)]dy. \tag{5.37}$$

The corresponding eigenvalue equation is obtained by setting $\eta(x,t) = \mathrm{e}^{\lambda t}\mathrm{e}^{imx}$ for integer m:

$$\lambda = \lambda(m,q) = \frac{\hat{w}(q+m) + \hat{w}(q-m) - 2\hat{w}(q)}{2}, \tag{5.38}$$

where $\hat{w}$ is the Fourier transform of the weight distribution:

$$\hat{w}(q) = \int_{-\pi}^{\pi} w(y)\mathrm{e}^{iqy}dy. \tag{5.39}$$

From the properties of w, all eigenvalues are real and $\lambda(-m,q) = \lambda(m,q)$ so that the condition for stability of the q-phase wave is

$$\hat{w}(q+m) + \hat{w}(q-m) - 2\hat{w}(q) < 0 \tag{5.40}$$

for all $m = 1,2,\ldots$. Note that the existence of a zero eigenvalue ($\lambda(0,q) = 0$) reflects the fact the system is time-translation invariant, that is, it corresponds to a perturbation consisting of a uniform shift of all phases. In general, it is difficult to establish that (5.40) holds for all positive integers m even when q is fixed. However, for a particular choice of w, which yields the continuum version of (5.32), the countably infinite set of stability conditions reduces to a single condition [671].That is, consider the weight distribution

$$w(x) = \begin{cases} (2\pi r)^{-1}, & |x| \leq \pi r \\ 0, & |x| > \pi r \end{cases} \tag{5.41}$$

for $0 < r < 1$, which means that each oscillator is coupled to all oscillators within a fraction r of the ring with coupling strength $1/(2\pi r)$. The Fourier transform of w is

$$\tilde{w}(q) = f(\pi q r) \equiv \frac{\sin(\pi q r)}{\pi q r}, \tag{5.42}$$

so that (5.40) becomes

$$S_Q(M) \equiv \frac{1}{2}[f(Q+M)+f(Q-M)] - f(Q) < 0, \quad M = r, 2r, \ldots, \tag{5.43}$$

where $Q = qr, M = mr$. Treating M as a continuous, real variable and plotting the function $S_Q(M)$ for various Q, one finds that $S_Q(M)$ is negative for all $M > 0$ when Q is sufficiently small, which ensures that the corresponding phase waves are stable. At a critical value $Q = \mu$, $S_Q''(0)$ changes sign from negative to positive; $S_Q''(0) = f''(Q)$ so μ is the smallest positive root of $f''(Q)$ and can be obtained by solving the transcendental equation

$$\tan(\pi\mu) = \frac{2\pi\mu}{2-(\pi\mu)^2},$$

which yields $\mu \approx 0.66$. Since $S_Q(0) = 0$, it follows that the continuous function $S_Q(M)$ becomes positive in a neighborhood of $M = 0$ when $Q > \mu$, and one can no longer guarantee stability.

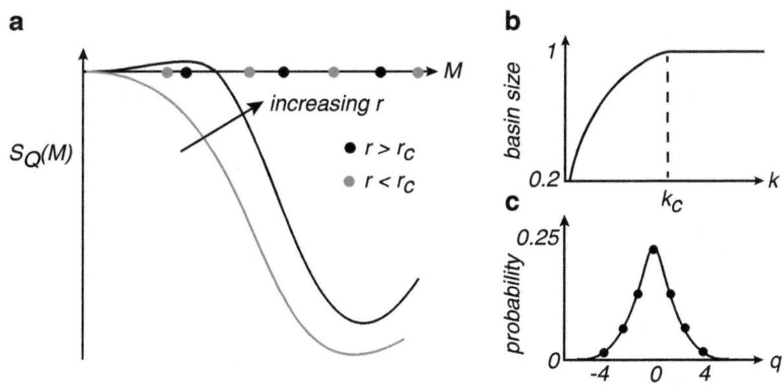

Fig. 5.14 Phase-locked states on a ring of N identical phase oscillators. (**a**) Graphical construction to determine the necessary condition for the stability of a q-phase wave with q fixed. A sketch of the function $S_Q(M)$, $Q = qr$, is shown for $r < r_c$ or $Q < \mu$ (*gray curve*) and $r > r_c$ or $Q > Q_c$ (*black curve*). The points $M = nr$, integer n are also shown for $r < r_c$ (*gray dots*) and $r > r_c$ (*black dots*). (**b**) Variation of the size of the sync basin as a function of k for large N. One way to numerically estimate the basin size is to consider the fraction Γ of uniformly random initial conditions that converge to the synchronous state. For $k > k_c$, the synchronous state becomes a global attractor with $\Gamma = 1$. (**c**) For $k = 1$, the probability that a uniformly random condition converges to a q-phase wave can be fitted by a Gaussian function of q

The above analysis provides a sufficient condition for stability of a q-phase wave assuming a given range of coupling r [671]: the q-phase wave is stable if $|q|r < \mu$. The stability condition is not necessary, however, because one has to take into account the fact that M is actually discrete. That is, it is possible that the discrete points $M = r, 2r, \ldots,$ lie outside the interval where $S_Q(M)$ is positive when $Q > \mu$. A simple graphical argument then establishes that for fixed q, increasing r from near zero means that Q crosses μ from below, resulting in an increasing interval over which $S_Q(M)$ is positive. Even though the points $M = r, 2r, \ldots,$ are also moving away from the origin as r increases, the expanding interval eventually encloses the point $M = r$ at a critical value r_c with $Q_c = qr_c(q) > \mu$; see Fig. 5.14a. This argument also establishes that the first eigenmode to become unstable is $M = r$. In summary, given $r > 0$, the q-phase wave is stable if and only if [671]

$$\frac{1}{2}[f(qr+r)+f(qr-r)] < f(qr). \tag{5.44}$$

For a given choice of q, the critical range of coupling $r_c(q)$ where the q-phase wave becomes unstable is the solution to the equation

$$\frac{1}{2}[f(qr_c+r_c)+f(qr_c-r_c)] = f(qr_c). \tag{5.45}$$

Numerically evaluating $r_c(q)$, one finds that $r_c(q)$ is a decreasing function of q. In particular, $r_c(1) \approx 0.68$, which means that all phase waves are unstable for $r > r_c(1)$. In terms of the original discrete network (5.32) with $r = 2k/N$, this means that the synchronous solution is globally stable when $k > k_c(N) \approx 0.34N$; see Fig. 5.14b. (To be more careful, it has not been established that there are no other phase-locked states other than synchrony and phase waves. However, numerical simulations suggest that any additional states are unstable [671].) Finally, one can estimate the relative size of the basin of attraction of a phase-locked state by determining the fraction of uniformly random initial conditions that converge to that particular state; see Fig. 5.14b, c.

Clearly the idealized ring model considered by Wiley et al. [671] is not biologically realistic. However, it provides a useful framework for exploring an issue that has been neglected in almost all studies of oscillations and waves in neural networks. That is, in addition to determining existence and stability, it is also important to consider the types of initial conditions that generate such solutions, particularly when trying to relate models to experiments in which oscillations and waves are initiated by specific environmental conditions or external stimuli such as locally injected currents.

5.2.4 Two-Dimensional Network of Phase Oscillators

Traveling phase waves have recently garnered attention beyond the context of central pattern generators. In particular, Ermentrout and Kleinfeld [170] have argued that the theory of phase waves in spatially structured networks provides a basis for understanding traveling electrical waves that have been observed across a variety of sensory cortices [149, 292, 357, 683, 684]; see Fig. 5.4. The waves are often expressed during periods of rest, whereas synchronous activity tends to dominate during the presentation of a strong stimulus. On the other hand, it has been suggested that in motor cortex it is phase waves that encode motor actions, whereas synchrony represents an idling state [269]. In contrast to CPGs, the cortex is effectively a two-dimensional oscillatory medium (assuming activity through the cortical layers can be lumped together). There have been relatively few studies of coupled phase oscillators in discrete or continuum two-dimensional arrays [173, 495]. Here we follow Heitmann et al. [269] and consider a two-dimensional sheet of identical, locally coupled Kuramoto oscillators:

$$\frac{\partial \theta}{\partial t} = \Omega + \int_{\mathbb{R}^2} w(|\mathbf{r}-\mathbf{r}'|)\sin[\theta(\mathbf{r}',t)-\theta(\mathbf{r},t)]d\mathbf{r}'. \tag{5.46}$$

For simplicity, the weight kernel is assumed to be isotropic and the domain is taken to be unbounded. Following the analysis of the ring network in Sect. 5.2.3, one expects planar phase wave solutions of the form $\theta(\mathbf{r},t) = \Omega t + \mathbf{k}\cdot\mathbf{r}$. This can be verified by substituting such a solution into (5.46) and using polar coordinates $\mathbf{r} = r(\cos\phi, \sin\phi), \mathbf{k} = k(\cos\phi_k, \sin\phi_k)$ together with the Jacobi–Anger expansion of Bessel functions

$$\mathrm{e}^{iz\cos\phi} = \sum_{n=-\infty}^{\infty} i^n J_n(z)\mathrm{e}^{in\phi}. \tag{5.47}$$

Thus

$$\begin{aligned} 0 &= \int_{\mathbb{R}^2} w(|\mathbf{r}-\mathbf{r}'|)\sin[\mathbf{k}\cdot(\mathbf{r}'-\mathbf{r})]d\mathbf{r}' \\ &= \mathrm{Im}\int_0^\infty rw(r)\left[\int_0^{2\pi} \mathrm{e}^{ikr\cos(\phi-\phi_k)}d\phi\right]dr \\ &= \mathrm{Im}\int_0^\infty rw(r)J_0(kr), \end{aligned}$$

which holds since the right-hand side integral is real.

Since the domain is unbounded we expect the spectrum of the linear operator obtained by linearizing about the phase wave solution to be continuous. That is, writing $\theta(\mathbf{r},t) = \Omega t + \mathbf{k}\cdot\mathbf{r} + \psi(\mathbf{r},t)$ and Taylor expanding (5.46) to first in order in ψ yields the linear equation

$$\frac{\partial \psi}{\partial t} = \int_{\mathbb{R}^2} w(|\mathbf{r}-\mathbf{r}'|)\cos[\mathbf{k}\cdot(\mathbf{r}'-\mathbf{r})](\psi(\mathbf{r}',t)-\psi(\mathbf{r},t))d\mathbf{r}'. \tag{5.48}$$

This has solutions of the form

$$\psi(\mathbf{r},t) = \mathrm{e}^{\lambda t}\mathrm{e}^{i\mathbf{q}\cdot\mathbf{r}}, \tag{5.49}$$

with $\mathbf{q} = q(\cos\phi_q, \sin\phi_q)$ and λ satisfying

$$\lambda = \int_{\mathbb{R}^2} w(|\mathbf{r}-\mathbf{r}'|)\cos[\mathbf{k}\cdot(\mathbf{r}'-\mathbf{r})]\left(\mathrm{e}^{i\mathbf{q}\cdot(\mathbf{r}'-\mathbf{r})}-1\right)d\mathbf{r}'. \tag{5.50}$$

Again using polar coordinates and the Jacobi–Anger expansion, we a obtain a dispersion relation for the continuous spectrum

$$\begin{aligned}\lambda(\mathbf{q}) &= \frac{1}{2}\int_0^\infty rw(r)\int_0^{2\pi}\left[\sum_{n=-\infty}^{\infty}\left(i^n J_n(kr)\mathrm{e}^{in(\phi-\phi_k)}+\text{c.c.}\right)\right] \\ &\quad\times\left[\sum_{m=-\infty}^{\infty} i^m J_m(qr)\mathrm{e}^{im(\phi-\phi_q)}-1\right]d\phi dr \\ &= \frac{1}{2}\int_0^\infty rw(r)\sum_{n=-\infty}^{\infty}\left[J_n(kr)J_{-n}(qr)\mathrm{e}^{in(\phi_q-\phi_k)}+J_n(kr)J_n(qr)\mathrm{e}^{-in(\phi_q-\phi_k)}\right]dr \\ &\quad-\int_0^\infty rw(r)J_0(kr).\end{aligned}$$

Using the identity $J_{-n}(z)=(-1)^nJ_n(z)$ then gives

$$\lambda(\mathbf{q})=\int_0^\infty rw(r)\left[\sum_{n=1}^{\infty}(1+(-1)^n)J_n(kr)J_n(qr)\cos(n\Delta\phi)+J_0(kr)(J_0(qr)-1)\right]dr, \tag{5.51}$$

where $\Delta\phi=\phi_q-\phi_k$. It immediately follows that $\lambda(\mathbf{q})$ is real for all $\mathbf{q}$, the spectrum is independent of the direction of $\mathbf{k}$ (reflecting isotropy of the system), and $\lambda(0)=0$ (reflecting time-translation symmetry). The phase wave of wave number k is then stable (modulo uniform phase shifts) if $\lambda(\mathbf{q})<0$ for all $q>0$ and $\Delta\phi\in[0,2\pi]$. Note that it is not sufficient to analyze stability by considering a reduced 1D ring network as was assumed by Heitmann et al. [269], even if one were to consider a bounded 2D domain.

In the case of the synchronous solution $k=0$, (5.51) reduces to

$$\lambda(q)=\int_0^\infty rw(r)(J_0(qr)-1)dr, \tag{5.52}$$

since $J_n(0)=\delta_{n,0}$. Given that $J_0(z)\le 1$ for all z, it follows that $\lambda(q)<0$ for all $q>0$ when the coupling is purely excitatory, that is, $w(r)\ge 0$ for all r. Hence the synchronous state of a 2D network of Kuramoto phase oscillators with excitatory coupling is stable. Heitmann et al. used numerical simulations to explore pattern forming instabilities in the case of center-surround coupling for which (see Fig. 5.15a)

$$w(r)=\mathrm{e}^{-br^2}+4h\mathrm{e}^{-br^2}\left(\frac{1}{3}b^2r^4-br^2\right). \tag{5.53}$$

They found that when $h=0$ (no surround inhibition), the synchronous state is stable, but increasing h destabilizes the synchronous state resulting in regular or irregular phase waves. Performing a numerical continuation of stable solutions as h varied, Heitmann et al. [269] also observed hysteresis in the level of phase coherence R as a function of h, with the latter order parameter defined according to [350]

$$R(t)e^{i\psi(t)} = \int_{\mathbb{R}^2} e^{i\theta(\mathbf{r},t)} d\mathbf{r}. \tag{5.54}$$

A sketch of a numerically determined hysteresis curve is shown in Fig. 5.15b. Synchronous states are identified by high phase coherence ($R \approx 1$), while phase waves are identified by low phase coherence ($R \approx 0$). We summarize the main findings of Heitmann et al. [269]. Starting at low levels of inhibition (small h), the system is in a synchronous state (upper branch of Fig. 5.15b), which then develops small amplitude spatial modulations or ripple patterns as h increases, with intermediate phase coherence, until the ripples destabilize and collapse to a full wave solution (lower branch of Fig. 5.15b). The transition from the lower to upper branch occurs at a smaller value of h then the opposite transition, that is, there is hysteresis. It immediately follows that there exists a range of h values over which there is bistability between the synchronous state and phase waves. Moreover, in the bistable regime, it is found numerically that phase waves tend to develop spatial heterogeneities, which can destabilize the phase waves at lower values of h than regular planar waves.

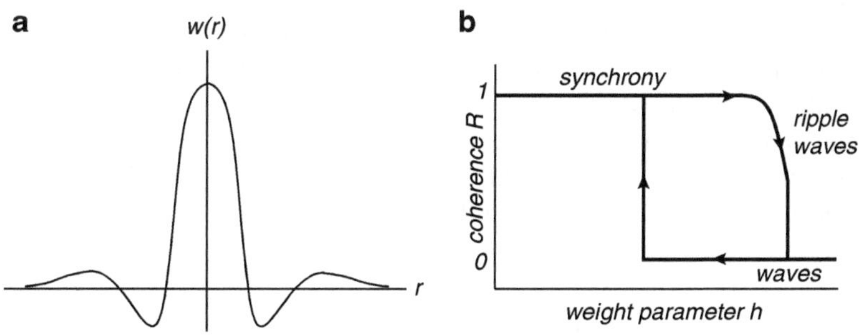

Fig. 5.15 (**a**) Radially symmetric center-surround weight kernel $w(r)$. (**b**) Sketch of hysteresis curve for phase coherence R as a function of weight parameter h, indicating transitions between the synchronous state ($R \approx 1$) and phase waves ($R \approx 0$)

Under the hypothesis that synchrony represents an idling state, whereas a phase wave somehow encodes a voluntary movement, there has to be some mechanism for switching between the two modes. Heitmann et al. [269] exploit the bistability of the model system by proposing that such a switch could be achieved in terms of a state-dependent perturbation that repels each oscillator from the global mean phase ψ (see (5.54)) according to

$$\Delta\theta(x) = k\sin(\theta(x) - \psi). \tag{5.55}$$

One possible physiological implementation of such a perturbation could be a temporary activation of a dense network of inhibitory neurons within the motor cortex by excitatory thalamocortical projections. Interestingly, it is known that voluntary movement is initiated in the cortex when these projections are disinhibited by the basal ganglia [6]. A partial validation of the model was achieved by comparing the output of the network under perturbation-driven transitions with movement-related modulations of beta oscillations (20–25 Hz), which were obtained from magnetoencephalogram (MEG) recordings of human motor cortex during a repetitive finger-tapping task [54]. The network output was generated by band-pass filtering the real part of the order parameter (5.54). It would be interesting to understand some of these numerical results analytical, using the dispersion relation (5.51), for example, as well as investigating the effects of boundaries (where the spectrum becomes discrete), network heterogeneities such as spatial variations in the natural frequencies of the oscillators, and anistropic weight distributions. As might be expected, the latter tends to select planar waves in particular directions [269]. It would also be interesting to further explore the similarities and differences between phase waves in sensory and motor cortices.

5.3 Synaptically Coupled Integrate-and-Fire (IF) Oscillators

The *integrate-and-fire* (IF) model of a spiking neuron provides a caricature of the capacitative nature of the cell membrane at the expense of a detailed model of the spike generation process [323]. The basic circuit of the IF model consists of a capacitor C in parallel with a resistor R driven by a current $I_i(t)$, where i is a neuron label. The latter is decomposed as $I_i(t) = I_{\text{ext},i}(t) + u_i(t)$ where $I_{\text{ext},i}(t)$ is an external current and $u_i(t)$ is the synaptic current (5.3) generated by the arrival of spikes from other neurons. The voltage $v_i(t)$ of an IF neuron evolves according to the first-order equation

$$C\frac{dv_i}{dt} = -\frac{v_i(t)}{R} + I_i(t). \tag{5.56}$$

The form of the action potential is not described explicitly. Spikes are formal events characterized by the ordered sequence of firing times $\{T_i^m, m \in \mathbf{Z}\}$ determined by the threshold crossing conditions (5.2). However, immediately after firing, the potential is *reset* to a value $v_r < \kappa$,

$$\lim_{t \to [T_i^m]^+} v_i(t) = v_r. \tag{5.57}$$

For simplicity, we set $\kappa = 1$, $v_r = 0$ and $C = R = 1$. We also take $I_{\text{ext},i}(t) = I_0$, where I_0 is a constant background input. A network of synaptically coupled IF neurons is then given by (5.56), (5.2), and (5.3) together with the reset condition (5.57). Taking the synaptic kernel $\Phi_{ij}(t) = w_{ij}\Phi(t)$, $u_i(t)$ satisfies the equation

$$\frac{du_i}{dt} = -u_i + I_0 + \varepsilon \sum_{j=1}^{M} w_{ij} \sum_{m \in \mathbf{Z}} \Phi(t - T_j^m), \tag{5.58}$$

for $T_i^n < t < T_i^{n+1}$. Two network operating regimes can be distinguished according to the value of the background input relative to the threshold $\kappa = 1$. If $I_0 > 1$, then the network is in an oscillatory regime, since each IF neuron acts as an intrinsic oscillator when $\varepsilon = 0$. On the other hand, if $I_0 < 1$, then the network is in an excitable regime, since synaptic inputs are necessary in order to push a neuron over the threshold. We will consider excitable IF networks in Sect. 5.4

Suppose that $I_0 > 1$. For sufficiently weak coupling ε, the oscillatory IF network can be reduced to a coupled phase model following Sect. 5.2. All that is required is to determine the PRC of a single IF neuron. Therefore, set $I_i(t) = I_0 + \varepsilon I(t)$ in (5.56) and perform the change of variables

$$\theta_i(t) \quad \text{mod } 1 = \frac{1}{\Delta_0} \int_0^{v(t)} \frac{dV}{I_0 - V} = \frac{1}{\Delta_0} \log \frac{I_0}{I_0 - v(t)}, \tag{5.59}$$

where $\Delta_0 = \log(I_0/[I_0 - 1])$ is the natural period of oscillation. [It is convenient to take $\theta \in [0,1)$ rather than $[0,2\pi)$.] Differentiating both sides with respect to t shows that

$$\frac{d\theta}{dt} = \omega_0 + \varepsilon \frac{I(t)}{I_0 - v(t)} = \omega_0 + \varepsilon I(t) \frac{e^{\theta(t)\Delta_0}}{I_0 \Delta_0}, \tag{5.60}$$

where $\omega_0 = 1/\Delta_0$. Thus the phase-resetting curve is the periodic function

$$Z(\theta) = \frac{e^{\theta \Delta_0}}{I_0 \Delta_0}, \quad 0 \leq \theta < 1. \tag{5.61}$$

The phase interaction function may now be calculated using (5.13), rewritten in the form

$$H(\psi) = \int_0^1 Z(\theta) P(\theta + \psi) d\theta. \tag{5.62}$$

Here $P(\theta) = \sum_m \Phi([\theta - m]\Delta_0)$ is a periodic function of θ that in the case of the alpha function (5.23) can be summed as a geometric series to give

$$P(\theta) = \frac{\alpha^2 \Delta_0 e^{-\alpha\theta\Delta_0}}{1 - e^{-\alpha\Delta_0}} \left[\theta\Delta_0 + \Delta_0 \frac{e^{-\alpha\Delta_0}}{(1 - e^{-\alpha\Delta_0})} \right], \quad 0 < \theta < 1. \tag{5.63}$$

Using the above phase interaction function, the α-dependence of phase-locked solutions in the case of a symmetric pair of excitatory or inhibitory IF neurons can be determined [651]. The results are summarized in Fig. 5.16a. For excitatory coupling ($\varepsilon > 0$) the synchronous state is unstable for all $0 < \alpha < \infty$. On the other hand, the antisynchronous solution is stable for $\alpha < \alpha_c$ but loses stability when $\alpha > \alpha_c$ with the creation of two stable, partially synchronized or asynchronous states. The emergence of these two additional states can be understood in terms of the behavior of the odd part of the interaction function $H^-(\psi)$ as shown in Fig. 5.16b. In the limit $\alpha \to \infty$ the two asynchronous states asymptotically approach the synchronous state so that the neurons are almost perfectly synchronized. This is consistent with the analysis of Mirollo and Strogatz [431] who proved rigorously that globally coupled IF oscillators almost always synchronize in the presence of instantaneous excita-

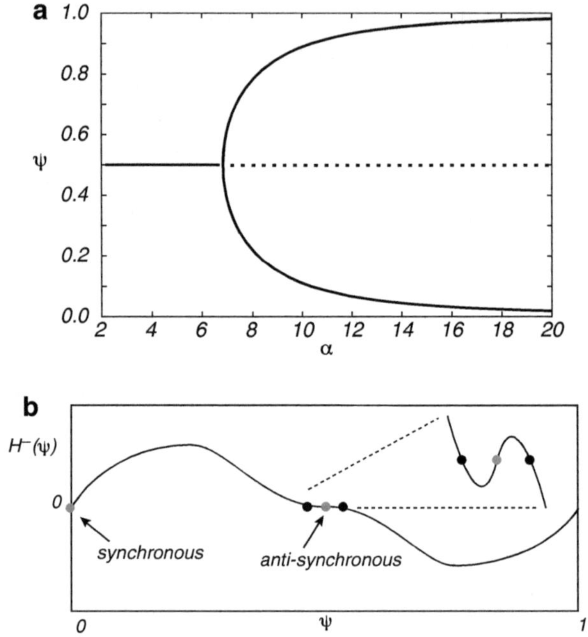

Fig. 5.16 (**a**) Relative phase $\psi = \psi_2 - \psi_1$ for a pair of IF oscillators with symmetric excitatory coupling as a function of α with $I_0 = 2$. In each case the antiphase state undergoes a bifurcation at a critical value of $\alpha = \alpha_c$ where it becomes unstable and two additional stable solutions ψ, $1-\psi$ are created. The synchronous state remains unstable for all α. (**b**) Odd part of the phase interaction function for α just above the critical value α_c showing two new phase-locked solutions

tory interactions. The stability properties of all solutions are reversed in the case of inhibitory coupling ($\varepsilon < 0$) so that, in particular, the synchronous state is now stable for all α. Introduction of a discrete axonal delay τ_a produces a checkerboard pattern of alternating stable/unstable solution branches that can overlap to produce multistable solutions [133].

One of the useful features of the IF model (5.58) is that it is possible to derive phase-locking equations without the assumption of weak coupling [62, 73, 651]. This can be achieved by solving equation (5.58) directly under the ansatz that the firing times are of the form $T_j^n = (n - \psi_j)\Delta$ for some self-consistent period Δ and constant phases ψ_j. Integrating over the interval $t \in (-\Delta\psi_i, \Delta - \Delta\psi_i)$ and incorporating the reset condition leads to the result

$$1 = (1 - \mathrm{e}^{-\Delta})I_0 + \varepsilon \sum_{j=1}^{N} w_{ij} H(\psi_j - \psi_i), \tag{5.64}$$

with

$$H(\psi) = \mathrm{e}^{-\Delta} \int_0^{\Delta} dt\, \mathrm{e}^t \sum_{k\in\mathbf{Z}} \Phi(t + (k+\psi)\Delta). \tag{5.65}$$

Note that up to a positive constant factor, $H(\psi)$ reduces to the weak coupling phase interaction function of equation (5.62) for $\Delta \to \Delta_0$. It can be seen that (5.64) has an identical structure to that of (5.16). However, the former is exact whereas the latter is only strictly valid to $\mathscr{O}(\varepsilon)$, since it is derived under the assumption of weak coupling. Moreover, the collective period of oscillations Δ must be determined self-consistently in (5.64), since H depends on Δ rather than the natural period Δ_0. It is also possible to study the stability of phase-locked solutions directly in terms of the firing times [73, 215, 649]. In the case of slow synapses, one can carry out a perturbation analysis in terms of the small rate parameter α. One then finds that there are at least two distinct types of local instabilities: (i) instabilities that are independent of the strength of coupling $|\varepsilon|$, which also occur in the corresponding phase reduced mode; (ii) strong coupling instabilities that occur at some critical value of the coupling strength $|\varepsilon|$ [73]. (Note that there could be additional global instabilities arising from discontinuous bifurcations [138].)

Linear stability analysis of phase-locked states. For the sake of illustration, consider the linear stability of the synchronous solution $\psi_i = 0$ for all i. In order to ensure that such a solution exists, we impose the condition $\sum_j w_{ij} = \Gamma$. Defining perturbed firing times according to $T_i^n = n\Delta + \delta_i^n$, we integrate (5.58) from T_i^n to T_i^{n+1} using the reset condition. This leads to a mapping of the firing times that can be expanded to first order in the perturbations δ_i^n [73, 215, 649]:

$$\{I_0 - 1 + \varepsilon\Gamma P(0)\}\left[\delta_i^{n+1} - \delta_i^n\right] = \varepsilon \sum_{j=1}^{N} w_{ij} \sum_{m=-\infty}^{\infty} G_m(0)\left[\delta_j^{n-m} - \delta_i^n\right], \tag{5.66}$$

where $P(\phi) = \sum_m \Phi([\phi - m]\Delta)$ and

$$G_m(\phi) = \mathrm{e}^{-\Delta}\int_0^{\Delta} \mathrm{e}^t \Phi'(t + (m+\phi)\Delta)dt. \tag{5.67}$$

Note that in contrast to the weak coupling regime, P now depends on the collective period Δ rather than the natural period Δ_0 The resulting linear delay-difference equation (5.66) has solutions of the form $\delta_j^n = \mathrm{e}^{n\lambda}\delta_j$ with $0 \leq \mathrm{Im}(\lambda) < 2\pi$. Taking $(\delta_1, \ldots \delta_N)$ to be an eigenvector of the weight matrix $\mathbf{w}$ with associated eigenvalue μ_p yields the characteristic equation

$$[I_0 - 1 + \varepsilon\Gamma P(0)]\,(\mathrm{e}^{\lambda} - 1) = \varepsilon[\tilde{G}(\lambda)\mu_p - \Gamma\tilde{G}(0)], \tag{5.68}$$

with

$$\tilde{G}(\lambda) = \sum_{m=-\infty}^{\infty} \mathrm{e}^{-m\lambda} G_m(0, \Delta). \tag{5.69}$$

One solution to (5.68) is $\lambda = 0$ with $\delta_i = \delta$ for all $i = 1, \ldots, N$. This reflects the invariance of the dynamics with respect to uniform phase shifts in the firing times, $T_i^n \to T_i^n + \delta$. Thus the condition for linear stability of a phase-locked state is that all remaining solutions λ of (5.68) have negative real part. This ensures that $\delta_j^n \to 0$ as $n \to \infty$ and, hence, that the synchronous solution is asymptotically stable. [Note that the above analysis does not take into account the possibility of instabilities arising from discontinuous bifurcations. These additional instabilities shift the stability boundaries calculated using linear theory [138].]

It is not possible to obtain exact solutions of (5.68). However, considerable insight into the nature of phase-locking instabilities can be obtained by taking $\Phi(t)$ to be the alpha function (5.23) and performing a perturbation expansion of equation (5.68) with respect to the inverse rise-time α, that is, by working in the regime of slow synapses. First, using the Fourier transform of $\Phi(t)$ gives

$$P(\theta) = \frac{\alpha^2}{\Delta} \sum_{m\in\mathbf{Z}} \frac{e^{i\omega_m \theta \Delta}}{(\alpha + i\omega_m)^2} = \frac{1}{\Delta} + \mathscr{O}(\alpha^2), \tag{5.70}$$

and

$$\tilde{G}(\lambda) = \frac{\alpha^2}{\Delta} e^{-\Delta} \sum_{m\in\mathbf{Z}} \frac{i\omega_m + \lambda/T}{(\alpha + i\omega_m + \lambda/\Delta)^2} \frac{1}{(1 + i\omega_m + \lambda/\Delta)} + \mathscr{O}(\alpha^3), \tag{5.71}$$

with $\omega_m = 2\pi m/\Delta$. Substitute these into the eigenvalue equation (5.68) and expand λ as a power series in α. Interestingly, one finds that there exist two distinct types of solution for λ, one of which is $\mathscr{O}(\alpha)$ and the other is $\mathscr{O}(\alpha^2)$. This is illustrated schematically in Fig. 5.17.

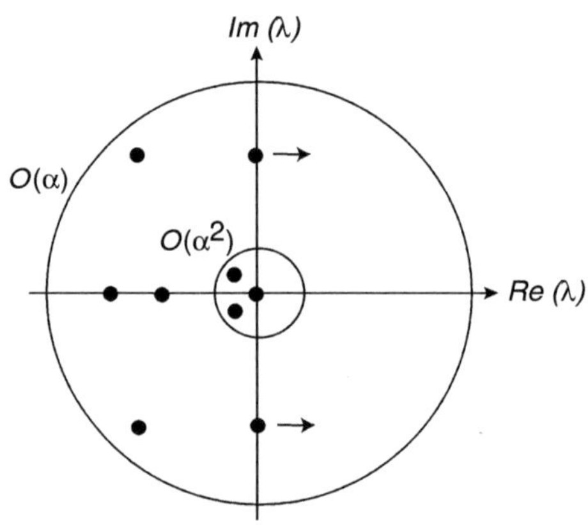

Fig. 5.17 Schematic diagram illustrating the spectrum of the linearized firing time map in the limit of slow synapses. There are N eigenvalues λ^{ω} of $\mathscr{O}(\alpha^2)$, one of which is a zero eigenvalue associated with phase-shift invariance, and $2N$ eigenvalues λ^s of $\mathscr{O}(\alpha)$. (The case $N = 3$ is shown.) Destabilization of the synchronous state can occur in the strong coupling regime due to a complex conjugate pair of $\mathscr{O}(\alpha)$ eigenvalues crossing the imaginary axis signaling the onset of a discrete Hopf bifurcation in the firing times

$\lambda = \mathscr{O}(\alpha^2)$ *and weak coupling instabilities.* Performing a perturbation expansion of equation (5.68) under the assumption $\lambda = \mathscr{O}(\alpha^2)$ yields the N eigenvalues

$$\lambda^w(\mu_p) \sim \varepsilon[\mu_p - \Gamma]\tilde{G}(0), \tag{5.72}$$

$p = 1,\ldots,N$. Hence, to the lowest order in α, a necessary condition for stability of the synchronous state at a given coupling strength ε is Re $\lambda^w(\mu_p) < 0$ for all $p = 1,\ldots,N-1$ (with $\mu_N = \Gamma$). In the weak coupling regime this stability condition is also sufficient, as can be shown by performing a perturbation expansion of equation (5.68) with respect to the coupling strength ε rather than α. Comparison of (5.67) and (5.69) with (5.65) shows that $\tilde{G}(0) \sim H'(0)$). It follows that the $\mathscr{O}(\alpha^2)$ stability condition is equivalent to the one based on the Jacobian of the phase-averaged model, (5.17). That is, $\varepsilon H'(0)[\text{Re}\,\mu_p - \Gamma] < 0,\ p = 1,\ldots,N-1$ (with $\mu_N = \Gamma$). Such a stability condition only depends on the sign of ε. We wish to determine whether or not this stability condition breaks down as $|\varepsilon|$ is increased (with the sign of ε fixed).

$\lambda = \mathscr{O}(\alpha)$ *and strong coupling instabilities.* Performing a perturbation expansion of equation (5.68) with $\lambda = \mathscr{O}(\alpha)$ gives the $2N$ eigenvalues

$$\lambda_{\pm}^{s}(\mu_p) = \left[-1 \pm \sqrt{\frac{\varepsilon\mu_p[1-\mathrm{e}^{-\Delta}]}{(I_0 - 1 + \varepsilon\Gamma/\Delta)\Delta^2}}\,\right]\alpha\Delta, \tag{5.73}$$

$p = 1,\ldots,N$. In contrast to the previous case, these solutions are guaranteed to lie in the left-half complex plane when ε is sufficiently small so they do not play a role in weak coupling instabilities. However, as ε is increased from zero one or more eigenvalues may cross the imaginary axis from left to right in the complex λ-plane signaling the onset of a discrete Hopf bifurcation in the firing times and the breakdown of 1:1 frequency-locking. The corresponding Hopf frequency is of the form $\omega_H = \omega_1\alpha + \omega_2\alpha^2 + \mathscr{O}(\alpha^3)$ where $w_1 = \text{Im}\lambda_c^s$ at the bifurcation point and ω_2 is determined by higher-order terms in the perturbation expansion. If $\omega_1 \neq 0$, then the instability induces a time periodic variation in the instantaneous firing rates of the neurons, whereas if $\omega_1 = 0$, then the firing rates remain constant but may differ between neurons.

It turns out that the strong coupling instabilities for slow synapses can be understood in terms of a corresponding rate model [73]. First, rewrite (5.58) as

$$\frac{\mathrm{d}v_i}{\mathrm{d}t} = -v_i + I_0 + u_i(t)$$

with

$$u_i(t) = \varepsilon \sum_j w_{ij} \int_0^\infty \Phi(\tau) a_j(t-\tau)d\tau, \quad a_j(t) = \sum_{m\in\mathbf{Z}} \delta(t - T_j^m).$$

If the synapses are sufficiently slow (small α) then $\Phi(t)$ is effectively performing a short-term temporal average over the incoming spike trains, and the synaptic current varies slowly in time. Thus, over time intervals of $\mathscr{O}(\alpha^{-1})$, we can treat the total input to each IF neuron as approximately constant, so that the neuron acts like a regular oscillator with instantaneous firing rate

$$F(u_i) = [\log(I_0 + u_i)/(I_0 + u_i - 1))]^{-1}. \tag{5.74}$$

Hence, under the approximation

$$\int_0^\infty \Phi(\tau) a_j(t-\tau)d\tau \approx \int_0^\infty \Phi(\tau) F(u_j(t-\tau))d\tau,$$

we obtain the closed set of rate equations

$$u_i(t) = \varepsilon \sum_j w_{ij} \int_0^\infty \Phi(\tau) F(u_j(t-\tau)d\tau. \tag{5.75}$$

Finally, introducing the differential operator $\mathbb{L}$ for which $\mathbb{L}\Phi_i(t) = \delta(t)$, we obtain the second-order equation

$$\frac{1}{\alpha^2}\frac{d^2u_i(t)}{dt^2} + \frac{1}{\alpha}\frac{du_i(t)}{dt} + u_i(t) = \varepsilon \sum_j w_{ij} F(u_j(t)). \tag{5.76}$$

(A related derivation of a rate model based on population averaging will be considered in Sect. 6.1.) Under the condition $\sum_j w_{ij} = \Gamma$ for all i, there exists a homogeneous fixed point $u_i(t) = \bar{u}$ such that $\bar{u} = \varepsilon F(I_0 + \bar{u})\Gamma$. Linearizing (5.76) about the fixed point and substituting a solution of the form $u_i(t) = \mathrm{e}^{\lambda t}\delta_i$, where δ is an eigenvector of the weight matrix $\mathbf{w}$, we obtain the set of eigenvalues

$$\frac{\lambda_\pm(\mu_p)}{\alpha} = -1 \pm \sqrt{\varepsilon F'(\bar{u})\mu_p}, \qquad p = 1,\ldots,N. \tag{5.77}$$

The fixed point will be asymptotically stable if and only if $\mathrm{Re}\,\lambda_\pm(\mu_p) < 0$ for all p. It is now straightforward to establish that the $\mathcal{O}(\alpha)$ eigenvalues associated with the linear stability of the synchronous state (see (5.73)) are equivalent to those obtained from the linear stability analysis of the homogeneous state in the firing rate model (5.77). For a direct comparison, take the firing rate to be the same in the two models by setting $F(I_0 + \bar{u}) = 1/\Delta$ so that $\varepsilon\Gamma/\Delta = \bar{u}$ and $1 - \mathrm{e}^{-\Delta} = 1/(\bar{u} + I_0)$. Differentiating $F(I_0 + \bar{u})$ with respect to $\bar{u}$ then shows that $F(I_0 + \bar{u})^2/F'(I_0 + \bar{u}) = (\bar{u} + I_0)(\bar{u} + I_0 - 1)$ and the result follows.

The above analysis establishes that, in the limit of slow synapses, a strong coupling instability in the synchronous state of the spiking model is reflected by a corresponding instability in the homogeneous fixed point of the rate model. As $|\varepsilon|$ is increased from zero, a codimension one bifurcation of the rate model occurs in two distinct ways. If a single real eigenvalue crosses the origin in the complex λ-plane ($\omega_1 = 0$) then a static bifurcation can occur, leading to the emergence of additional fixed-point solutions with inhomogeneous firing rates. On the other hand, if a pair of complex conjugate eigenvalues crosses the imaginary axis from left to right in the complex plane ($\omega_1 \neq 0$), then a continuous Hopf bifurcation can occur, leading to the formation of periodic solutions, that is, time-dependent firing rates. This suggests that the spiking model undergoes a corresponding transition to a non-phase-locked state, consisting of spatiotemporal modulations of the interspike intervals that are well described by the firing-rate model, at least in the small α regime. Moreover, the fine temporal structure of these modulations, which are averaged out in the firing-rate model, should become significant with increasing α in the spiking model. This is indeed found to be the case [73].

In Fig. 5.18 we illustrate a strong coupling instability in an excitatory–inhibitory pair of IF neurons. It is straightforward to show that the rate-based version of this network exhibits periodic variations in the mean firing rates in the strong coupling regime. It can be established from (5.72) and (5.73) that the synchronous state of the IF network is stable for sufficiently weak coupling but destabilizes due to a Hopf bifurcation in the firing time map at a critical coupling $\varepsilon_c(\alpha)$. A direct numerical simulation of the system shows that beyond the bifurcation point, the neurons jumps to a state in which they exhibit periodic bursting patterns. This can be understood in terms of mode-locking associated with periodic variations of the interspike intervals (ISIs) $\Delta_k^n = T_k^{n+1} - T_k^n$, $k = E, I$, on closed attracting orbits. Suppose that the kth oscillator has a periodic solution of length M_k so that $\Delta_k^{n+pM_k} = \Delta_k^n$ for all integers p. If $\Delta_k^1 \gg \Delta_k^n$ for all $n = 2, \ldots, M_k$, say, then the resulting spike train exhibits bursting with the interburst interval equal to Δ_k^1 and the number of spikes per burst equal to M_k. Although both oscillators have different interburst intervals ($\Delta_1^1 \neq \Delta_2^1$) and numbers of spikes per burst ($M_1 \neq M_2$), their spike trains have the same total period, that is, $\sum_{n=1}^{M_1} \Delta_1^n = \sum_{n=1}^{M_2} \Delta_2^n$. Due to the periodicity of the activity, the ISIs only fall on a number of discrete points on the orbit. The variation of the ISIs Δ_i^n with n is compared directly with the corresponding variation of the firing rates of the reduced model in Fig. 5.18b, which shows that the instantaneous firing rate captures the modulation in the ISIs.

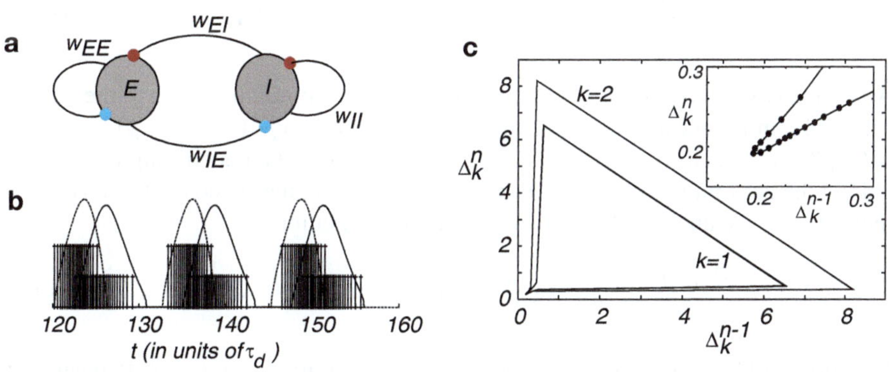

Fig. 5.18 Strong coupling instability in an excitatory–inhibitory pair of IF neurons. (**a**) Basic network. Synaptic weights are $w_{EE} = w_{II} = 0$, $w_{IE} = 1$ and $w_{EI} = -2$. The synaptic rate constant is $\alpha = 1$ and the coupling strength is $\varepsilon = 2$. (**b**) Spike train dynamics of the two neurons. For the sake of illustration the firing times of the two oscillators are represented with lines of different heights (marked with a +). *Smooth curves* represent variations in the firing rates of the corresponding rate model. (**c**) A plot of the interspike intervals (ISIs) $(\Delta_k^{n-1}, \Delta_k^n)$, $k = E, I$, of the spike trains shown in (**b**) showing how they lie on closed periodic orbits. *Points* on an orbit are connected by lines. Each triangular region is associated with only one of the neurons, highlighting the difference in interburst intervals. The inset is a blowup of orbit points for one of the neurons within a burst

5.4 Spiking Network Models of Cortical and Subcortical Waves

5.4.1 Traveling Pulses in an Excitatory Spiking Network

In the previous section we showed how oscillations and phase-locking in a synaptically coupled IF network can be analyzed in terms of an underlying firing time map. It turns out that the firing times can also be used to investigate solitary wave propagation in an excitable IF network, for which $I_0 < 1$ [62, 63, 167]. Consider a network of IF neurons distributed along an infinite 1D lattice:

$$\frac{dv_j}{dt} = -v_j + I_0 + g\sum_{k\in\mathbf{Z}} W(j-k)\Phi(t-T_k), \tag{5.78}$$

with the threshold condition $v_j(T_j) = 1$ and the reset condition $\lim_{\varepsilon\to 0^+} v_j(T_j+\varepsilon) = 0$. For simplicity, we are assuming that each neuron fires only once during passage of a traveling pulse. (Numerical simulations indicate that wave propagation is not dramatically altered if each neuron fires multiple spikes [167].) We now define a traveling pulse solution as one for which there is a constant time lag between the firing of adjacent neurons, that is, $T_k = k\Delta$ (up to an arbitrary constant). (The same definition was used for intracellular Ca^{2+} waves in the fire–diffuse–fire model of Sect. 4.3.2.) Substituting into (5.78), gives

$$\frac{dv_j}{dt} = -v_j + I_0 + g\sum_{k\in\mathbf{Z}} W(j-k)\Phi(t-k\Delta). \tag{5.79}$$

Analytically speaking, it is convenient to consider a continuum version of the model. Therefore, introduce a lattice spacing d and set $v(x,t) = v_{x/d}(t)$, $w(x-y) = W([x-y]/d)$, $T(y) = T_{y/d}$. Moreover, rescale the coupling constant according to $g \to gd$. Taking the continuum limit $d \to 0$ and $\Delta \to 0$ such that the speed $d/\Delta = c$ is fixed, then $T(y) = y/c$ and we obtain the continuum IF equation

$$\frac{\partial v(x,t)}{\partial t} = I_0 - v(x,t) + g\int_{-\infty}^{\infty} w(x-y)\Phi(t-y/c)dy. \tag{5.80}$$

Equation (5.80) is supplemented by the reset condition $v(x,t^+) = 0$ whenever $v(x,t) = 1$. In the following, we set $I_0 = 0$, assume that g is a positive coupling parameter, and $W(x)$ is a positive, symmetric, monotonically decreasing function of $|x|$ with $\int_0^\infty W(x)dx < \infty$. Typical choices for such a weight function are

$$w(x) = \frac{1}{\sqrt{2\pi\sigma^2}}\mathrm{e}^{-x^2/2\sigma^2} \quad \text{(Gaussian)}, \quad w(x) = \frac{1}{2\sigma}\mathrm{e}^{-|x|/\sigma} \quad \text{(exponential)}. \tag{5.81}$$

For concreteness, we choose the exponential function, although the basic results will not depend on the precise form of $w(x)$.

In order to determine the velocity c of a solitary pulse, multiply both sides of (5.80) by e^t and integrate over the interval $-\infty < t \leq T(x) = x/c$ using the threshold condition $v(x,T(x)) = 1$. This generates a self-consistency condition for c of the form [62, 167]

$$\begin{aligned} 1 &= g\int_{-\infty}^{x/c} \mathrm{e}^t \left[\int_{-\infty}^{\infty} w(x-y)\Phi(t-y/c)dy\right] dt \\ &= g\int_0^{\infty} w(x)\mathrm{e}^{-x/c}\int_0^{x/c} \mathrm{e}^t\Phi(t)dtdx. \end{aligned} \tag{5.82}$$

The second line is obtained by noting that $\Phi(t-y/c)=0$ for $y>ct$, reversing the order of integration,

$$\int_{-\infty}^{x/c} dt \int_{-\infty}^{ct} dy \to \int_{-\infty}^{x} dy \int_{y/c}^{x/c} dy,$$

and performing a change of variables. In the case of an exponential weight function, the integration over x and t can be performed explicitly by Fourier expanding $\Phi(t)$ and using the analyticity properties of $\tilde{\Phi}(\omega)=\int_{-\infty}^{\infty} e^{-i\omega t}\Phi(t)dt$, namely, that $\tilde{\Phi}(\omega)$ is analytic in the lower-half complex ω-plane since $\Phi(t)=0$ for $t \leq 0$. After performing a contour integral, we obtain the result

$$1=\frac{gc}{2(\sigma+c)}\tilde{\Phi}(-ic/\sigma). \tag{5.83}$$

Suppose that $\Phi(t)$ is given by the α function (5.23) together with a finite axonal propagation delay τ_a:

$$\Phi(t)=\alpha^2(t-\tau_a)e^{-\alpha(t-\tau_a)}H(t-\tau_a), \tag{5.84}$$

where H is the Heaviside function. Substitution into (5.83) leads to the following implicit equation for the propagation velocity of a solitary pulse:

$$1=\frac{g\alpha^2 c}{2(\sigma+c)}\frac{e^{-c\tau_a/\sigma}}{(\alpha+c/\sigma)^2}. \tag{5.85}$$

In the absence of axonal delays ($\tau_a=0$) and large c, the velocity scales according to a power law $c \sim \alpha\sigma\sqrt{g/2}$ [167]. This form of power law is consistent with the behavior observed in more detailed computational models [228]. The velocity c is plotted as a function of the coupling g in Fig. 5.19. It can be seen that for fixed α and τ_a there exists a critical coupling $g_s(\alpha,\tau_a)$ such that there are no traveling pulse solutions for $g<g_s$ and two solutions for $g>g_s$. Using linear stability analysis (see below), one finds that for the given parameter values, the lower (upper) solution branch is unstable (stable).

The (local) asymptotic stability of a solitary wave can be determined by considering perturbations of the firing times. For simplicity, we will only consider linear stability analysis based upon perturbations traveling at the same speed as the wave itself, that is, perturbations of the single neuron firing times $T(x)$. It is possible that there are further instabilities associated with global bifurcations where, for example, certain neurons fire more than once, fail to fire, or fire out of sequence. Suppose that $T(x)=x/c+\phi(x)$ with $\phi(x)=\phi_0(x)$ for $x\leq 0$ and $\phi_0(x)$ a prescribed, bounded function on $(-\infty,0]$. Asymptotic stability then corresponds to the condition $\phi(x)\to 0$ as $x\to\infty$ for arbitrary nonuniform initial data $\phi_0(x)$. Integrating (5.80) over the interval $(-\infty,T(x)]$ gives

$$e^{x/c+\phi(x)}=g\int_{-\infty}^{x/c+\phi(x)} e^t \int_{-\infty}^{\infty} w(x-x')\Phi(t-x'/c-\phi(x'))dx'dt.$$

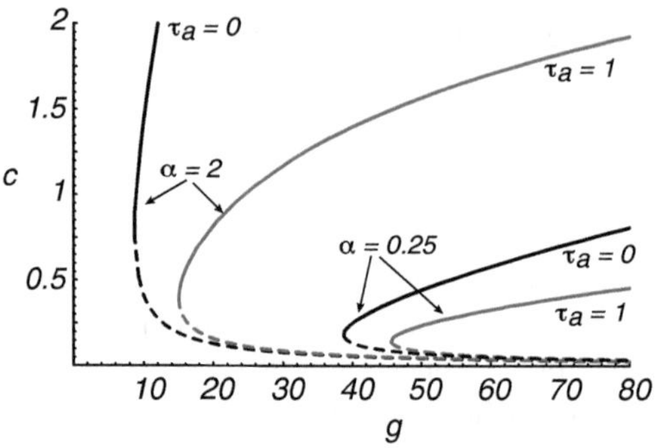

Fig. 5.19 Variation of the speed c of a traveling pulse as a function of the synaptic coupling g for different values of the inverse rise-time α and axonal delay τ_a. *Solid (dashed) lines* correspond to stable (unstable) branches. The external input is taken to be $I_0 = 0$ and the range of the exponential weight function is $\sigma = 1$

Expanding to first order in $\phi(x)$ and imposing (5.82) leads to the linear integral equation

$$\int_{-\infty}^{0} \mathrm{e}^t \left[\int_{-\infty}^{\infty} w(x-x')\Phi'(t+[x-x']/c)\left[\phi(x)-\phi(x')\right]dx' \right] dt = 0. \tag{5.86}$$

Substitution of a solution of the form $\phi(x) = \mathrm{e}^{\lambda x}$ then yields the characteristic equation (after reversing the order of integration and performing a change of variables)

$$\mathscr{H}(\lambda) \equiv \int_0^{\infty} \left[\mathrm{e}^{-\lambda x} - 1\right] w(x) f(x/c) dx = 0, \tag{5.87}$$

with

$$f(\tau) = \int_0^{\tau} \mathrm{e}^{-(\tau - t)} \Phi'(t) dt. \tag{5.88}$$

Asymptotic stability then holds if all nonzero solutions λ have negative real part. Such solutions depend indirectly on the coupling g via the speed c, which satisfies (5.82). Equation (5.87) can be evaluated using Fourier analysis and, up to an irrelevant constant factor,

$$\mathscr{H}(\lambda) = P(c[\sigma^{-1} + \lambda]) - P(c/\sigma), \tag{5.89}$$

$$P(z) = \frac{z}{2(1+z)} \tilde{\Phi}(-iz). \tag{5.90}$$

Note that $P(z)$ is real when z is real, and $P(z) \to 0$ as $|z| \to \infty$. The stability of the fast and slow solution branches displayed in Fig 5.19. can be determined from the following theorem [62]:

Theorem 5.1. *Let $\mathscr{C}_+$ and $\mathscr{C}_-$ denote, respectively, the fast and slow solution branches $c = c_\pm(g)$, $g \geq g_s$, of the self-consistency condition (5.83) for the velocity of a solitary pulse, assuming a unimodal response function $\Phi(t)$. Here g_s is the critical coupling for the existence of a solitary wave. Denote the time to peak of $\Phi(t)$ by τ_{max}. The following stability results then hold: (i) The branch $\mathscr{C}_-$ is unstable for all τ_{max} and τ_a. (ii) The branch $\mathscr{C}_+$ is stable for all τ_{max} in the case of zero axonal delays, $\tau_a = 0$. (iii) In the case of large axonal delays and fast synapses (small τ_{max}), there exists a Hopf bifurcation point g_h such that $c_+(g)$ is stable (unstable) for $g > g_h$ ($g_s \leq g < g_h$).*

We sketch a proof. First, note that (5.83) can be written in the more compact form

$$1 = g(c)P(c/\sigma), \tag{5.91}$$

where the dependence on the coupling as a function of c has been made explicit. Now decompose $\lambda = a + ib$ and differentiate (5.89) with respect to a for $b = 0$:

$$\partial\mathscr{H}(a,0)/\partial a|_{a=0} = \sigma^{-1}P'(c/\sigma).$$

Similarly, differentiating (5.91) with respect to c shows that

$$0 = g'(c)P(c/\sigma) + g(c)P'(c/\sigma).$$

It follows that

$$\text{sign}\,[\partial\mathscr{H}(a,0)/\partial a|_{a=0}] = -\text{sign}[g'(c)].$$

Hence, $\partial H(a,0)/\partial a|_{a=0} > 0$ when $dg/dc < 0$, that is, when $c \in \mathscr{C}_-$. Since $\lim_{a\to\infty}\mathscr{H}(a,0) = -P(c) < 0$, it follows that $H(a,0)$ must cross the positive a-axis at least once when $c \in \mathscr{C}_-$. This implies that $H(a,0) = 0$ has at least one solution for $a > 0$ and, hence, that the branch $\mathscr{C}_-$ is unstable. Second, assuming that the delay kernel $\Phi(t)$ is unimodal with a maximum at τ_{max}, it follows from (5.88) that $f(\tau) > 0$ for $0 < \tau < \tau_{max}$. The real part of (5.89) can then be decomposed as

$$\text{Re}[\mathscr{H}(\lambda)] = \int_0^{c\tau_{\max}} \left[e^{-ax} - 1\right] w(x)f(x/c)dx + \int_{c\tau_{\max}}^{\infty} \left[e^{-\lambda x} - 1\right] w(x)f(x/c)dx.$$

with the first term on the right-hand side positive definite. Since $w(x)$ is a monotonically decreasing function of x, it follows that the second term becomes negligible when $c\tau_{max}$ is sufficiently large. In other words, $\mathscr{H}(\lambda) \neq 0$ and hence the upper branch $\mathscr{C}_+$ is stable in the large-c limit. It can also be established that $\mathscr{H}(a,0) = 0$ does not have any positive definite solutions when $c \in \mathscr{C}_+$ so that the upper branch is stable with respect to static bifurcations. The proof is completed by establishing conditions for the occurrence of a Hopf instability on the upper branch. This requires searching for solutions $a = 0, b \neq 0$ of (5.89).

It turns out that for the parameter values chosen in Fig. 5.19, the upper branch is stable. However, one finds that for larger values of the axonal delay τ_a, a continuous solitary pulse can become unstable as the axonal delay is increased. This is illustrated in Fig. 5.20, where the fast and slow solution branches are plotted as a function of τ_a. Interestingly, Golomb and Ermentrout [230] have demonstrated that in the regime where the continuous wave is unstable, lurching (pulsating) waves propagate with discontinuous, periodic spatiotemporal characteristics; see Fig 5.21. Space is now spontaneously divided into discrete spatial intervals of length Δx, and

$$T(x + \Delta x) = T(x) + \Delta t, \tag{5.92}$$

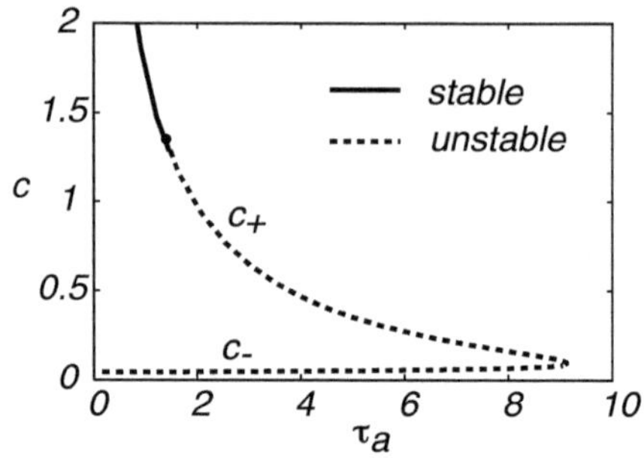

Fig. 5.20 Fast and slow velocity branches as a function of the axonal delay τ_a for fixed α and g. The stable slow branch undergoes a Hopf instability at a critical value of axonal delay

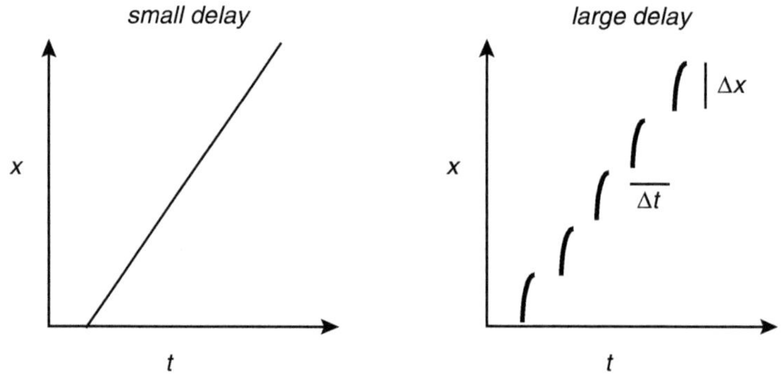

Fig. 5.21 Schematic illustration of the difference between continuous and lurching pulses in an excitatory IF network with delays. Each IF neuron only fires a single spike at a time as indicated by the *solid curves*

where Δt is the period of a lurching cycle. Hence, the average speed of the pulse is $c = \Delta x/\Delta t$. Suppose that the lurching wave starts at $x = 0$ and $T(0) = 0$. The firing time of the neuron at x is then

$$T(x) = n\Delta t + f(\hat{x}), \tag{5.93}$$

where n is the integer part of $T(x)/\Delta t$ and $\hat{x} = x - n\Delta x$. The function f specifies how the firing times vary within a single lurch, and is defined on the interval $[0.L)$ with $f(0) = 0$. One also finds that lurching waves can persist in regions where the continuous wave does not exist and, for certain choices of weight kernel $w(x)$ such as a step function, the two types of wave may coexist (bistability) [230]. It follows that $T(x) - x/c = f(\hat{x}) - \hat{x}/c$ is a Δx-periodic function of x. Interestingly, the possible correlation between the nature of wave propagation and the effective size of discrete delays is suggestive of the observed differences between the spread of activity in cortical and thalamic slices; see Sect. 5.4.3.

There have also been a number of studies of 2D spiking networks. For example, Kistler et al. analyzed solitary waves and periodic wave trains in an excitable network based on the so-called spike response model [337], which is essentially an integrated version of the IF model. However, there were a few important differences between their treatment and the analysis of IF networks reviewed above. First, they effectively took the activity of a solitary pulse propagating in the x-direction, say, to be $c^{-1}\Phi(t-x/c)$ rather than $\Phi(ct-x)$. The additional factor of c^{-1} results in only one solution branch for $c=c(g)$ rather than two as found in [62, 167]. It is not clear that there is any biophysical justification for this additional factor, given the interpretation of $\Phi(t)$ as a synaptic impulse response function; see Sect. 1.4. Second, their stability analysis was mainly based on studying changes in local field potentials, which is not sufficient to guarantee asymptotic stability with respect to more general perturbations of the firing times as considered above. In particular, it does not take into account the full spectrum of the linearized firing time map, which is crucial in establishing the occurrence of Hopf instabilities. The numerical results presented in [337] did, however, reveal a rich repertoire of geometrical waves such as spirals and target patterns. Spontaneous spiral waves have also been observed in a stochastic IF network [117]. Finally, the existence of stable traveling pulses in a synaptically coupled IF network has been linked to the empirical observation of traveling pulses in disinhibited brain slices and with epileptiform activity following brain trauma. However, as in the case of oscillatory IF networks, it is not clear what the appropriate level of modeling should be. That is, whether the fundamental physiological element of a synaptically coupled network, discrete or continuous, is a single neuron or a local population of neurons. We will consider the latter case in Chaps. 6 and 7.

5.4.2 Traveling Pulses in Networks of Excitatory and Inhibitory Spiking Neurons

It is also possible to extend the above analysis to a 1D network of excitatory (E) and inhibitory IF neurons, in which there are fast (AMPA) and slow (NMDA) excitatory synapses [229]; see Fig. 5.22. The generalization of the continuum model (5.80) takes the form

$$\begin{aligned}\frac{\partial v_\alpha(x,t)}{\partial t} &= I_\alpha - v_\alpha(x,t) + \sum_{\gamma=f,s} g^\gamma_{\alpha E}\int_{-\infty}^{\infty} w_{\alpha E}(x-y)\Phi^\gamma_{\alpha E}(t-T_E(y))dy \\ &\quad -g_{\alpha I}\int_{-\infty}^{\infty} w_{\alpha I}(x-y)\Phi_{\alpha I}(t-T_I(y))dy \qquad (5.94)\end{aligned}$$

for $\alpha = E, I$. Here $v_\alpha(x,t)$ describes the time-dependent membrane potential of the α-type neuron at x, $w_{\alpha\beta}(x)$ represents the synaptic weight density from neurons of type β to neurons of type α, and the synaptic filters are taken to be exponentials

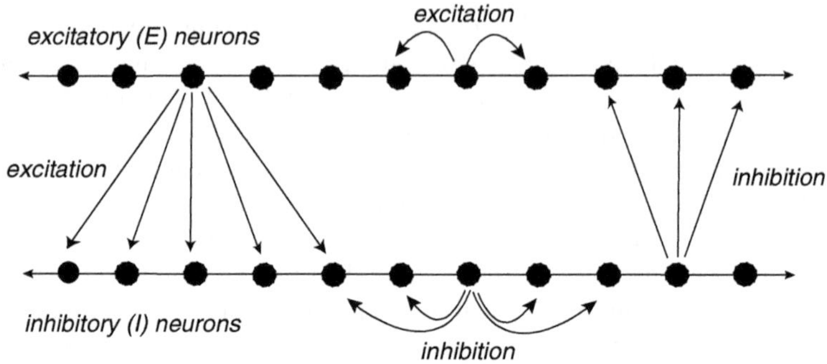

Fig. 5.22 Cortical network model consisting of a 1D layer of excitatory neurons reciprocally coupled to a 1D layer of inhibitory layers

$$\Phi_{\alpha\beta} = \frac{1}{\tau_{\alpha\beta}} \exp(-t/\tau_{\alpha\beta})H(t). \tag{5.95}$$

The additional superscript γ for excitatory synapses specifies whether the excitatory synapse is fast ($\gamma = f$) or slow ($\gamma = s$). For concreteness, the synaptic weights are also taken to be exponentials,

$$w_{\alpha\beta} = \frac{1}{\sigma_{\alpha\beta}} \exp(-|x|/\sigma_{\alpha\beta}). \tag{5.96}$$

Again it is assumed that each neuron at x only fires once and the corresponding firing times are denoted by $T_\alpha(x)$ for $\alpha = E, I$. The spatial and temporal scales are fixed by taking $\sigma_{EE} = 1$ and the membrane time constants to be unity. Numerical simulations establish that there are two types of stable traveling pulses. One is a *fast pulse* that is similar to the continuous pulse found in the excitatory network—its speed increases as inhibition decreases. The other is a continuous *slow pulse* that is characterized by a much smaller velocity and by the fact that the inhibitory cells fire before their neighboring excitatory cells [229]. Moreover, in a narrow range of parameters, the two solutions coexist (bistability).

The analysis of existence and stability of traveling pulses proceeds along similar lines to the excitatory case [229], so we will only summarize the results. Integrating (5.94) with respect to t and imposing the threshold conditions $v_\alpha(T_\alpha) = \kappa_\alpha$ gives the pair of equations

$$\begin{aligned}\kappa_\alpha = &\sum_{\gamma=f,s} g_{\alpha E}^{\gamma} \int_{-\infty}^{\infty} w_{\alpha E}(x') G_{\alpha E}^{\gamma}[T_\alpha(x) - T_E(x-x')]dx' \\ &- g_{\alpha I} \int_{-\infty}^{\infty} w_{\alpha I}(x') G_{\alpha I}[T_\alpha(x) - T_I(x-x')]dx',\end{aligned} \tag{5.97}$$

where

$$G_{\alpha\beta}(t) = \int_0^t \Phi_{\alpha\beta}(s)ds. \tag{5.98}$$

It is also required that the voltage be below threshold prior to firing, $V_\alpha(x,t) < \kappa_\alpha$ for $t < T_\alpha(x)$. A traveling pulse solution with velocity c is defined according to

$$T_E(x) = \frac{x}{c}, \quad T_I(x) = \frac{x}{c} + \zeta, \tag{5.99}$$

where ζ is the shift in the firing times of the inhibitory neurons relative to the excitatory neurons, so that $\zeta < 0$ means that the I cells fire before the corresponding E cells. Substituting such a solution into (5.97) and using the explicit expressions for w and G yield two algebraic equations for the two unknowns c, ζ. Existence of a traveling pulse then reduces to the problem of solving for c, ζ. Similarly, stability of a solution is determined by substituting

$$T_E(x) = \frac{x}{c} + \phi_E(x), \quad T_I(x) = \frac{x}{c} + \zeta + \phi_I(x), \tag{5.100}$$

into (5.97), expanding to first order in $\phi_{E,I}$ and analyzing the spectrum of the resulting linear operator. The quantitative results that emerge from such an analysis are summarized in Fig. 5.23. A number of additional points are as follows [229]:

1. The fast pulses are the continuation of the waves found in excitatory networks. The E neuron typically fires before or just after neighboring I cells. They are robust solutions that can be generated by a strong localized current injection.
2. Slow waves are characterized by I cells firing significantly before neighboring E cells. Although slow waves can exist in the absence of slow excitation, the presence of the latter greatly expands the basin of attraction and the possibility of inducing a slow wave with a localized perturbation.
3. There exists a bistable regime and a regime in which there is a continuous transition from fast to slow waves.
4. It is also possible to find parameter regimes in which there are lurching waves and other more irregular solutions.

The possibility for a cortical network to sustain two propagation velocities has also been observed in a much more detailed conductance-based model [122], with slow waves observed in the presence of inhibition and fast waves in the absence of inhibition. The more detailed model also showed that inhibitory neurons always fired before excitatory neurons during the propagation of slow waves. One limitation of the IF network study was that, in order to make the mathematical analysis tractable, it was necessary to impose the constraint that each neuron fired only one spike. On the other hand, the biophysical model can simulate the propagation of a pulse that represents the onset of a long-lasting active spiking (up) state and the transition to an inactive (down) state. Neurons within the pulse fire coherently at a rate of around 10 Hz. Repetitive stimulus-invoked or spontaneous activation of a local population resulted in the rhythmic slow oscillations (< 1 Hz) between up and down states observed in EEG recordings during slow-wave sleep [602] and in vitro slice studies [549]. There is still a debate regarding the precise mechanisms for the spontaneous transitions between the up and down states, although they are likely to involve some combination of intrinsic membrane properties and slow/fast synaptic interactions [27, 122, 419, 488, 549].

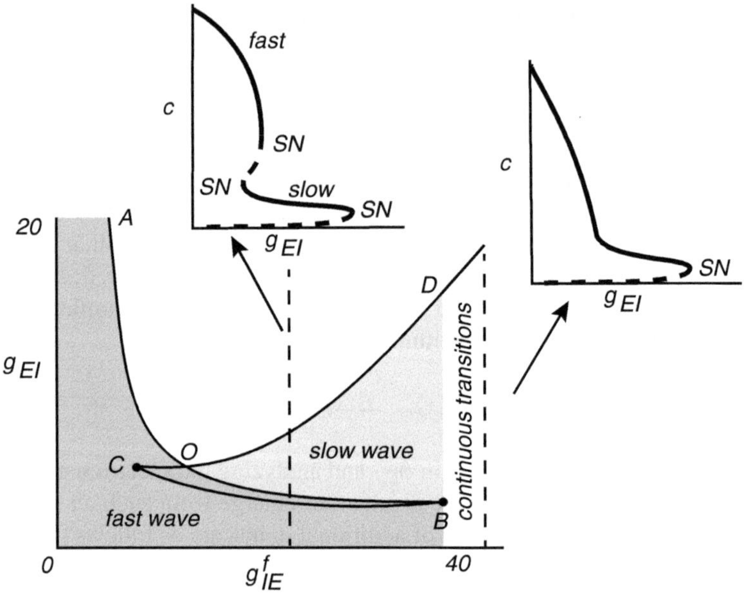

Fig. 5.23 Sketch of phase diagram showing the existence of stable fast and slow waves in the (g^f_{IE}, g_{EI}) plane for $g^s_{EE} = 10$, $g^f_{EE} = 12$, and $g^s_{IE} = 0$. The ranges of the synaptic weights are taken to be $\sigma_{EE} = 1 = \sigma_{IE}, \sigma_{II} = 0.5 = \sigma_{IE}$, and the thresholds are $\kappa_E = \kappa_I = 1$. Assuming the membrane time constants are unity, $\tau^s_{EE} = \tau^s_{IE} = 5/3$, $\tau^f_{EE} = \tau^f_{IE} = 1/12$, and $\tau_{II} = \tau_{EI} = 4/15$. The *solid lines* show saddle–node (SN) bifurcation curves: along AB a stable/unstable pair of fast waves coalesce, along CD a stable/unstable pair of slow waves coalesce, and along OB the unstable fast wave coalesces with the stable slow wave. A stable fast wave exists in the *dark-shaded domain*, a stable slow wave exists in the *light-shaded domain*, and the waves coexist in the region OCB (bistability). The points B, C are codimension two-cusp bifurcation points. The insets show the bifurcation diagram with respect to g_{EI} for two different values of g^f_{IE} (not shown to scale). In particular, the right-hand inset illustrates the case of a continuous transition from a fast to a slow wave with the latter disappearing in an SN. The parameter regime over which slow waves exists shrinks significantly in the absence of slow excitation ($g^s_{IE} \to 0$), partly due to the fact that the slow wave can destabilize via a Hopf bifurcation

5.4.3 Spiking Model of Thalamic Spindle Waves

As we have already highlighted, computational and experimental studies of disinhibited neocortical slices indicate that neuronal discharges propagate continuously at a velocity $c \sim 10\,\text{cm/s}$ [228]. Axonal delays are relatively small. On the other hand, in models of thalamic slices, composed of excitatory thalamocortical (TC) neurons and inhibitory reticular (RE) thalamic neurons, waves propagate in a lurching manner at a velocity $c \sim 1\,\text{cm/s}$ [232, 334, 349]. This is thought to form the basic mechanism for the generation of 7–14 Hz spindle oscillations during the onset of sleep. Each recruitment cycle of the lurching waves has two stages (see Fig. 5.24): (i) a new group of inhibitory RE cells is excited by synapses from TC cells, and this

RE group then inhibits a new group of TC cells; (ii) the new recruited TC cells rebound from hyperpolarization and fire a burst of spikes, which further recruit more RE cells during the next cycle. One can reduce the two-layer thalamic model to a single-layer excitatory network with a large effective delay ($\tau_a \approx 100\,\text{ms}$) caused by the time needed for a TC cell to rebound from inhibition [230, 525]. If such a network is modeled in terms of an IF network with delays, then one can obtain lurching waves, as discussed at the end of Sect. 5.3. A more detailed analysis of the post-inhibitory rebound mechanism underlying the generation of continuous and lurching waves in thalamic networks has been carried out using conductance-based [153, 232, 620] and integrate-and-fire-or-burst models [128]. We briefly describe the latter model.

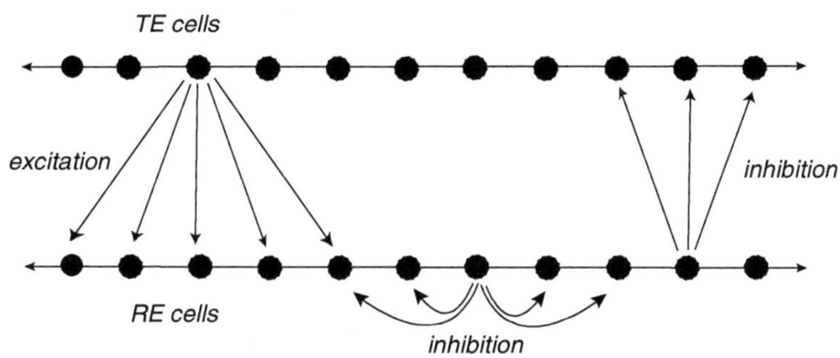

Fig. 5.24 Two-layer thalamic model. Inhibitory reticular (RE) cells inhibit excitatory thalamocortical (TC) cells, and TC cells excite RE cells. RE cells also mutually inhibit each other but this has a relatively small effect on discharge propagation

Several models of bursting neurons include a low-threshold calcium T-current [657]. Such a current is slowly activated when the neuron is sufficiently hyperpolarized so that it can cause the neuron to fire a burst of action potentials when released from sustained inhibition, so-called *post-inhibitory* rebound. Recently the integrate-and-fire model has been extended to incorporate this slow acting T-current leading to the aptly named *integrate-and-fire-or-burst* (IFB) model [128, 138, 589]. The current balance equation for the model is

$$C\frac{dv}{dt} = -I - I_L - I_T, \tag{5.101}$$

where C is a membrane capacitance, I represents external or synaptic current, and $g_L(v - v_L)$ is a leakage current with constant conductance g_L and reversal potential v_L. The low-threshold Ca^{2+} current is given by

$$I_T = g_T h(t)(v - v_T)H(v - v_h), \tag{5.102}$$

where H is a Heaviside function and the slow variable h has the dynamics

$$\frac{dh}{dt} = \begin{cases} -h/\tau_-, & v \geq v_h \\ (1-h)/\tau_+, & v < v_h. \end{cases} \tag{5.103}$$

The variable h, which represents slow de-inactivation of the low-threshold Ca^{2+} conductance, relaxes to zero with time constant τ_- when $v \geq v_h$ and relaxes to unity with time constant τ_+ when $v < v_h$. Since $v_h < v_r < \kappa$, where κ is the firing threshold and v_r is the rest level, it follows that sustained hyperpolarization is required in order that the potential crosses the v_h boundary from above, resulting in an increase in h. If the cell is then released from inhibition so that v recrosses the threshold v_h from below, a nonzero I_T current is generated that can depolarize the cell and produce a burst of spikes. In the absence of such hyperpolarization, the system acts as a regular integrate-and-fire neuron. For an appropriate choice of parameters, it has been shown that the IFB model under periodic stimulation exhibits excellent agreement with experimental results obtained from intracellular recordings of thalamic relay cells [589]. In the case of a constant external input, it is possible to obtain an exact solution of the IFB model in terms of characteristic curves [105]. Taking the ratio of (5.101) and (5.103) in the Ca^{2+} de-inactivated region ($v < v_h$), we have (for $C = 1$)

$$\frac{dv}{dh} = \tau_+ \frac{I_0 + g_L(v_L - v)}{1 - h}, \tag{5.104}$$

whereas, in the Ca^{2+} inactivated region ($v < v_h$),

$$\frac{dv}{dh} = -\tau_- \frac{I_0 + g_L(v_L - v) + g_T h(v_T - v)}{h}. \tag{5.105}$$

A sketch of a typical solution for $v = v(h)$ in both regimes is shown in Fig. 5.25; see [105] for a more detailed analysis.

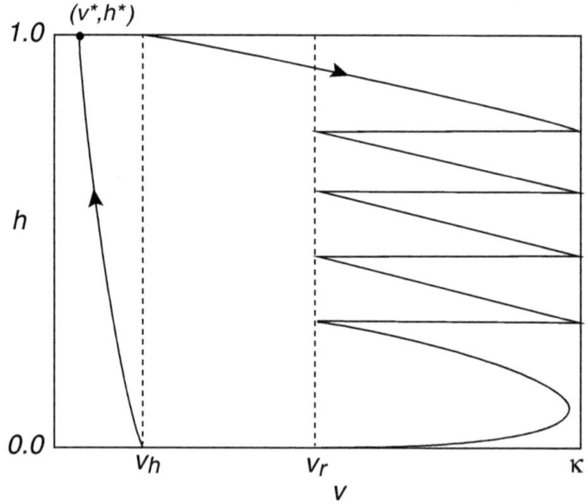

Fig. 5.25 Illustrative sketch of typical characteristic curves for an IFB neuron driven by a subthreshold constant input. A burst of four spikes arises when the neuron crosses the threshold v_h from below. Bursting ceases when h has decreased sufficiently so that the voltage v can no longer cross the firing threshold κ. Once v decreases below v_h, h increases to its equilibrium value $h^* = 1$ where $v = v^* < v_h$

A network version of the IFB model can be constructed along similar lines to a standard IF model [128]. Let (v_i, h_i) denote the voltage and slow inactivation variable of neuron i. Equation (5.101) becomes (for $C = 1$)

$$\frac{dv_i}{dt} = u_i(t) - g_L(v_i - v_L) - g_T h_i(t)(v_i - v_T)H(v_i - v_h), \tag{5.106}$$

where h_i evolves according to (5.103) and $u_i(t)$ is the total synaptic current into neuron i:

$$u_i(t) = \sum_j w_{ij} \sum_m \Phi(t - T_j^m), \tag{5.107}$$

with the firing times T_j^m determined by (5.2). Equation (5.106) is supplemented by the reset condition (5.57). In principle, one could now investigate the existence and stability of continuous and lurching waves in a one-dimensional network of IFB neurons by taking a continuum limit along the lines of Sect. 5.4.1. However, the analysis is still nontrivial unless some further approximation is made such as taking the synapses to be slow so that the spiking model can be reduced to a rate-based model [128]. Nevertheless, it can be shown that a single-layer inhibitory network of IFB neurons with v_h below the resting potential can support lurching waves, whereas a corresponding excitatory network with v_h above the resting potential supports continuous waves. In terms of the two-layer network shown in Fig. 5.24, the former can be interpreted as a thalamic network with TRN-mediated inhibition while the former corresponds to a TRN network with thalamus-mediated excitation.

Chapter 6
Population Models and Neural Fields

In this chapter we show how to construct population-based models of synaptically coupled neural networks, distinguishing between voltage-based and activity-based versions [71, 168]. We also consider the important issue of how noise at the single-cell level (Poisson-like spiking statistics) relates to noise at the population level. In particular, we describe a neural master equation formulation of stochastic population dynamics. One biological motivation for population-based models is that neurons in cortex tend to be organized into tightly coupled groups known as cortical columns, in which the cells share similar functional properties (Sects. 5.1 and 8.1). Another important feature of cortex is that it is spatially organized at both the anatomical and functional levels, which can be modeled in terms of spatially structured networks of interacting populations. In the continuum limit, such models can be described in terms of neural field equations. The advantage of a continuum rather than a discrete representation of spatially structured networks is that various techniques from the analysis of PDEs presented in Part I can be adapted to the study of neural field models of cortex (Chaps. 7 and 8).

6.1 Population Averaging and Rate Models

Suppose that a network of synaptically coupled spiking neurons is partitioned into a set of P homogeneous populations labeled $\alpha = 1,\ldots,P$, with $N_\alpha = N$ neurons in each population. (It is straightforward to relax this assumption by taking $N_\alpha = \mathcal{O}(N)$.) Let χ denote the population function that maps the single neuron index $i = 1,\ldots,N$ to the population index α to which neuron i belongs: $\chi(i) = \alpha$. Furthermore, suppose the synaptic interactions between populations are the same for all neuron pairs. (Relaxing this assumption can lead to additional sources of stochasticity as explored in [184, 632].) Denote the sequence of firing times of the jth neuron by $\{T_j^m, m \in \mathbf{Z}\}$. The net synaptic current into postsynaptic neuron i due to innervation by the spike train from presynaptic neuron j, with $\chi(i) = \alpha, \chi(j) = \beta$, is taken to have the general form $N^{-1}\sum_m \Phi_{\alpha\beta}(t - T_j^m)$, where $N^{-1}\Phi_{\alpha\beta}(t)$ represents

P.C. Bressloff, *Waves in Neural Media: From Single Neurons to Neural Fields*, Lecture Notes on Mathematical Modelling in the Life Sciences,
DOI 10.1007/978-1-4614-8866-8_6,

the temporal filtering effects of synaptic and dendritic processing of inputs from any neuron of population β to any neuron of population α; see Sect. 1.4.2. Assuming that all synaptic inputs sum linearly, the total synaptic input to the soma of the ith neuron is

$$u_i(t) = \sum_{\beta} \frac{1}{N} \sum_{j;\chi(j)=\beta} \Phi_{\alpha\beta}(t - T_j^m) = \int_{-\infty}^{t} \sum_{\beta} \Phi_{\alpha\beta}(t-t') \frac{1}{N} \sum_{j;\chi(j)=\beta} a_j(t')dt' \tag{6.1}$$

for all $\chi(i) = \alpha$, where $a_j(t) = \sum_{m\in\mathbf{Z}} \delta(t - T_j^m)$. That is, $a_j(t)$ represents the output spike train of the jth neuron in terms of a sum of Dirac delta functions. In order to obtain a closed set of equations, we have to determine the firing times T_i^m given by Eq. (5.2), where $v_i(t)$ evolves according to conductance-based model (5.1). It follows that, since the right-hand side of (6.1) is independent of i, we can set $u_i(t) = u_\alpha(t)$ for all $p(i) = \alpha$ with

$$u_\alpha(t) = \sum_{\beta=1}^{P} \int_{-\infty}^{t} \Phi_{\alpha\beta}(t-t')a_\beta(t')dt', \tag{6.2}$$

and $a_\alpha(t)$ is the output activity of the αth population:

$$a_\alpha(t) = \frac{1}{N} \sum_{j;\chi(j)=\alpha} a_j(t). \tag{6.3}$$

We now assume that each homogeneous population is close to a so-called asynchronous state, in which the spike trains of different neurons within a homogeneous population are uncorrelated (see Sect. 6.3.1). The population activity a_α is then approximately constant, which means that the synaptic currents are also slowly varying functions of time. It follows that a_α can track the input current according to $a_\alpha(t) = F(u_\alpha(t))$, where F is a population gain function [2, 86, 87, 213]. Substituting this approximation into (6.2) leads to the closed system of equations

$$u_\alpha(t) = \sum_{\beta=1}^{P} \int_{-\infty}^{t} \Phi_{\alpha\beta}(t-t')F[u_\beta(t')]dt', \tag{6.4}$$

(Note, however, the asynchronous state only exists in the thermodynamic limit so that for finite N we expect statistical fluctuations around the asynchronous state; see Sect. 6.4.) A rate equation identical in form to (6.4) was derived in Sect. 5.3 for an IF network with slow synapses, except that it involved single neuron labels rather than the population labels α,β. As highlighted elsewhere [71, 168], Eq. (6.4) can be reduced to a system of ordinary differential equations provided that we place rather mild restrictions on the time dependence of $\Phi_{\alpha\beta}(t)$. First, suppose that $\Phi_{\alpha\beta}(t) = w_{\alpha\beta}\Phi_\alpha(t)$ where $w_{\alpha\beta}$ denotes the mean synaptic strength of connections from population β to neuron α and $\Phi_\alpha(t)$ determines the time course of the input, which is assumed to depend only on properties of the postsynaptic population

α. Furthermore, suppose that there exists a differential operator $\mathbb{L}_\alpha$ such that (see also (5.76))

$$\mathbb{L}_\alpha \Phi_\alpha(t) = \delta(t). \tag{6.5}$$

Applying the operator $\mathbb{L}_\alpha$ to both sides of equation (6.4) then leads to a system of differential equations for the population currents $u_\alpha(t)$:

$$\mathbb{L}_\alpha u_\alpha(t) = \sum_{\beta=1}^{P} w_{\alpha\beta} F_\beta(u_\beta(t)). \tag{6.6}$$

Note that we could easily convert the synaptic current $u_\alpha(t)$ into an input voltage $v_\alpha(t) = u_\alpha(t)/\sigma$ using an effective input conductance σ, for example. Thus Eq. (6.6) is often referred to as a voltage equation and forms the basis of most classical neural networks such as the Hopfield model [288]. On the other hand, if the time course of the inputs depends only on presynaptic parameters, $\Phi_{\alpha\beta}(t) = w_{\alpha\beta}\Phi_\beta(t)$, with Φ_β having inverse differential operator $\mathbb{L}_\beta$, then we obtain a system of differential equations for the so-called synaptic drives:

$$z_\alpha(t) = \int_{-\infty}^{t} \Phi_\alpha(t-t') F_\alpha(u_\alpha(t'))dt'. \tag{6.7}$$

That is, applying the differential operator $\mathbb{L}_\alpha$ to Eq. (6.7) and using $u_\alpha(t) = \sum_{\beta=1}^{P} w_{\alpha\beta} z_\beta(t)$ leads to the activity-based model

$$\mathbb{L}_\alpha z_\alpha(t) = F_\alpha\left(\sum_{\beta=1}^{P} w_{\alpha\beta} z_\beta(t)\right). \tag{6.8}$$

The differential operator $\mathbb{L}_\alpha$ appearing in Eqs. (6.6) and (6.8) is often taken to be first order in time:

$$\mathbb{L}_\alpha = \frac{\partial}{\partial t} + \frac{1}{\tau_\alpha}, \tag{6.9}$$

with inverse kernel $\Phi_\alpha(t) = H(t)\mathrm{e}^{-t/\tau_\alpha}$. In order to relate the effective time constant τ_α to membrane and synaptic time constants, let us assume for simplicity that all synapses are sufficiently close to the soma so that the dendrite simply acts as a first-order low-pass filter and set (with $V_{\text{rest}} = 0$)

$$\Phi_{\alpha\beta}(t) = \sigma_\alpha r_{m,\alpha} V_{\text{syn},\beta} \bar{g}_{\alpha\beta} H(t) \int_0^t \mathrm{e}^{-(t-s)/\tau_{m,\alpha}} h_\beta(s) ds,$$

with

$$h_\beta(s) = \frac{\tau_{d,\beta}}{\tau_{d,\beta} - \tau_{r,\beta}} (\mathrm{e}^{-s/\tau_{d,\beta}} - \mathrm{e}^{-s/\tau_{r,\beta}}).$$

We have made explicit that the reversal potential V_{syn} and synaptic rise/fall times $\tau_{r,d}$ only depend on the particular class of synapses innervated by the presynaptic population β, whereas the membrane time constant τ_m, resistance r_m, and conductance σ are solely properties of the postsynaptic population α. Only the maximum conductance $\bar{g}$ is specific to the particular synapses $\beta \to \alpha$. The various constant factors can be combined to define the synaptic weight $w_{\alpha\beta}$. In particular, $w_{\alpha\beta} \sim V_{\text{syn},\beta}$ so that the sign of $V_{\text{syn},\beta}$ (relative to the resting potential) determines whether the synapse is excitatory or inhibitory. If $\tau_m \gg \tau_r, \tau_d$, then the time course is effectively independent of the presynaptic label β, and we have the voltage-based model (6.6) with first-order operator $\mathbb{L}_\alpha$ and $\tau_\alpha = \tau_m$. On the other hand, if $\tau_d \gg \tau_m, \tau_r$, then we obtain the activity-based model with $\tau_\alpha = \tau_d$. Although the reduction to a rate-based model is a major simplification of the full conductance-based model, it is still possible to incorporate various additional physiological features.

1. *Synaptic depression.* In order to incorporate synaptic depression we need to return to the single neuron level. Equation (6.1) is modified according to

$$u_i(t) = \sum_\beta \frac{1}{N} \sum_{j;\chi(j)=\beta} \Phi_{\alpha\beta}(t - T_j^m) q_{ij}(t - T_j^m)$$

$$= \int_{-\infty}^t \sum_\beta \Phi_{\alpha\beta}(t-t') \left[\frac{1}{N} \sum_{j;\chi(j)=\beta} a_j(t') q_{ij}(t') \right] dt'$$

with q_{ij} evolving according to an equation of the form (1.45), which we write as

$$\frac{dq_{ij}}{dt} = \frac{1-q_{ij}}{\tau_q} - (1-\gamma) q_{ij}(t) a_j(t).$$

Averaging the latter equation with respect to $j, \chi(j) = \beta$, and introducing the compact notation

$$\langle f(t) \rangle_\beta = \frac{1}{N} \sum_{j;\chi(j)=\beta} f_j(t),$$

we have the pair of equations

$$u_i(t) = \int_{-\infty}^t \sum_\beta \Phi_{\alpha\beta}(t-t') \langle a(t') q_i(t') \rangle_\beta dt' \tag{6.10}$$

and

$$\frac{d\langle q_i \rangle_\beta}{dt} = \frac{1 - \langle q_i \rangle_\beta}{\tau_q} - (1-\gamma) \langle a(t) q_i(t) \rangle_\beta.$$

We now make the mean-field approximation

$$\langle a(t) q_i(t) \rangle_\beta = a_\beta(t) \langle q_i(t) \rangle_\beta \tag{6.11}$$

Since all averaged depression variables $\langle q_i(t) \rangle_\beta, i = 1, \ldots, N\}$ for fixed β now have a common input drive $a_\beta(t)$, it follows that

$$\tau_q \frac{d(\langle q_i(t) \rangle_\beta - \langle q_{i'}(t) \rangle_\beta)}{dt} = -[\langle q_i(t) \rangle_\beta - \langle q_{i'}(t) \rangle_\beta],$$

and thus $\langle q_i(t)\rangle_\beta \to \langle q_{i'}(t)\rangle_\beta = q_\beta(t)$ for all $i,i' = 1,\ldots,N$. In other words, after an initial transient of duration τ_q, we can identify all depression variables associated with a given presynaptic population β. If we now assume that $\Phi_{\alpha\beta}(t) = w_{\alpha\beta}\Phi_\beta(t)$, we can introduce the synaptic drives (6.7) and derive the modified activity-based model [25, 617, 637, 644]:

$$\mathbb{L}_\alpha z_\alpha(t) = F_\alpha\left(\sum_{\beta=1}^{P} w_{\alpha\beta} q_\beta(t) z_\beta(t)\right), \tag{6.12}$$

with

$$\frac{dq_\alpha}{dt} = \frac{1-q_\alpha(t)}{\tau_q} - (1-\gamma)q_\alpha(t)F_\alpha\left(\sum_{\beta=1}^{P} w_{\alpha\beta} q_\beta(t) z_\beta(t)\right). \tag{6.13}$$

The corresponding voltage-based model is

$$\mathbb{L}_\alpha u_\alpha(t) = \sum_{\beta=1}^{P} w_{\alpha\beta} q_\beta(t) F_\beta(u_\beta(t)). \tag{6.14}$$

with

$$\frac{dq_\alpha}{dt} = \frac{1-q_\alpha(t)}{\tau_q} - (1-\gamma)q_\alpha(t)F_\alpha(u_\alpha(t)). \tag{6.15}$$

2. *Axonal propagation delays.* In the above derivation of rate-based models, we have assumed that the spiking of a presynaptic neuron has an instantaneous effect on downstream postsynaptic neurons. This neglects the fact that action potentials take time to propagate along an axon to innervate a synaptic terminal. Let us denote the corresponding axonal delay of a synapse $\alpha \to \beta$ by $\tau_{\alpha\beta}$. The integral equation (6.2) is modified according to

$$u_\alpha(t) = \sum_{\beta=1}^{P} \int_{-\infty}^{t} \Phi_{\alpha\beta}(t-t') a_\beta(t'-\tau_{\alpha\beta}) dt'. \tag{6.16}$$

The corresponding voltage-based model then takes the form of a system of delay differential equations,

$$\mathbb{L}_\alpha u_\alpha(t) = \sum_{\beta=1}^{P} w_{\alpha\beta} F_\beta(u_\beta(t-\tau_{\alpha\beta})), \tag{6.17}$$

and similarly for the activity-based model.

3. *Adaptive threshold dynamics.* Another biophysical process that can be incorporated into rate-based models is spike frequency adaptation. Spike frequency adaptation causes a neuron's firing rate to decay to a submaximal level and occurs when a potassium current, presumably activated by elevated intracellular

calcium, hyperpolarizes the membrane voltage [35, 396, 604]. This afterhyperpolarization current has a time constant of around 40–120 ms. Spike frequency adaptation can be introduced as a negative current $-c_i$ on the right-hand side of the conductance-based model equation (5.1). Assuming that $c_i(t)$ varies slowly relative to the voltage $v_i(t)$, it can be shown that c_i effectively acts as an adaptive threshold that varies linearly with the firing rate [35]. In the case of a network of homogeneous populations, each neuron within a given population fires at the same mean rate so that we can identify $c_i = c_\alpha$ for all i such that $\chi(i) = \alpha$. Hence, the voltage-based model becomes

$$\mathbb{L}_\alpha u_\alpha(t) = \sum_{\beta=1}^{P} w_{\alpha\beta} F_\beta(u_\beta(t) - c_\beta(t)). \tag{6.18}$$

with

$$\frac{dc_\alpha}{dt} = -\frac{c_\alpha(t)}{\tau_c} + \gamma_c F_\alpha(u_\alpha(t) - c_\alpha(t)). \tag{6.19}$$

6.2 E–I Oscillator Networks

One of the simplest population-based networks, which is often taken to be a fundamental module in large-scale models of cortex, is a pair of mutually coupled local populations of excitatory and inhibitory neurons known as an E–I network; see Fig. 6.1. An E–I network has the important property of exhibiting limit cycle oscillations and can thus act as a basic oscillatory element in network models of cortical phase waves, as an alternative to single spiking neurons (Sect. 5.2). An activity-based version of an E–I network takes the form (for first-order synapses)

$$\begin{aligned}\frac{da_E}{dt} &= -a_E + F\left(w_{EE}a_E - w_{EI}a_I + h_E\right)\\ \frac{da_I}{dt} &= -a_I + F\left(w_{IE}a_E - w_{II}a_I + h_I\right),\end{aligned} \tag{6.20}$$

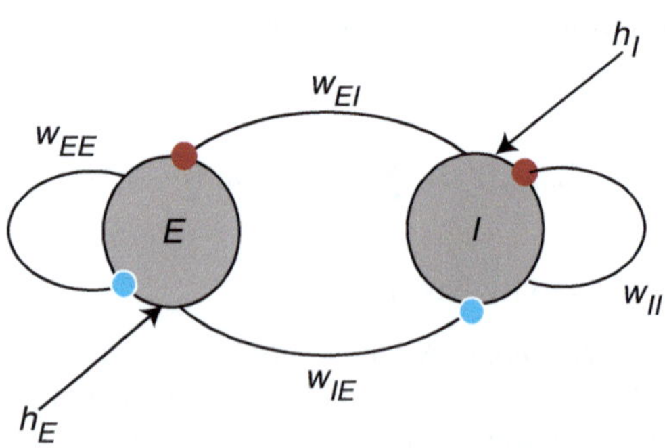

Fig. 6.1 Two-population E–I network

where h_E, h_I represent constant background inputs. For simplicity, we set $\tau_E = \tau_I = 1$. The bifurcation structure of the two-population model given by Eq. (6.20) has been analyzed in detail elsewhere [55]. An equilibrium (a_E^*, a_I^*) is obtained as a solution of the pair of equations

$$\begin{aligned} a_E^* &= F\left(w_{EE}a_E^* - w_{EI}a_I^* + h_E\right) \\ a_I^* &= F\left(w_{IE}a_E^* - w_{II}a_I^* + h_I\right). \end{aligned} \tag{6.21}$$

These can be inverted to yield

$$\begin{aligned} h_E &= F^{-1}(a_E^*) - w_{EE}a_E^* + w_{EI}a_I^* \\ h_I &= F^{-1}(a_I^*) - w_{IE}a_E^* + w_{II}a_I^*. \end{aligned} \tag{6.22}$$

As a further simplification, take the gain function F to be the simple sigmoid $F(u) = (1 + \mathrm{e}^{-u})^{-1}$. Using the fact that the sigmoid function then satisfies $F' = F(1 - F)$ and applying the fixed-point equations allows one to represent the associated Jacobian in the form

$$\Delta = \begin{pmatrix} -1 + w_{EE}a_E^*(1-a_E^*) & -w_{EI}a_E^*(1-a_E^*) \\ w_{IE}a_I^*(1-a_I^*) & -1 - w_{II}a_I^*(1-a_I^*) \end{pmatrix}.$$

An equilibrium will be stable provided that the eigenvalues $\lambda_\pm$ of Δ have negative real parts, where

$$\lambda_\pm = \frac{1}{2}\left(\mathrm{Tr}\,\Delta \pm \sqrt{[\mathrm{Tr}\,\Delta]^2 - 4\mathrm{Det}\,\Delta}\right). \tag{6.23}$$

This leads to the stability conditions $\mathrm{Tr}\,\Delta < 0$ and $\mathrm{Det}\,\Delta > 0$. In order to construct a phase diagram in the (h_E, h_I)-plane for a fixed weight matrix $\mathbf{w}$, we express a_I^* as a function of a_E^* by imposing a constraint on the eigenvalues $\lambda_\pm$ and then substitute the resulting function into Eq. (6.22). This yields bifurcation curves in the (h_E, h_I)-plane that are parameterized by a_E^*, $0 < a_E^* < 1$; see Fig. 6.2. For example, the constraint

$$\mathrm{Tr}\,\Delta \equiv -2 + w_{EE}a_E^*(1-a_E^*) - w_{II}a_I^*(1-a_I^*) = 0 \tag{6.24}$$

with $\mathrm{Det}\,\Delta > 0$ determines Hopf bifurcation curves where a pair of complex conjugate eigenvalues cross the imaginary axis. Since the trace is a quadratic function of a_E^*, a_I^*, we obtain two Hopf branches. Similarly, the constraint $\mathrm{Det}\,\Delta = 0$ with $\mathrm{Tr}\,\Delta < 0$ determines saddle–node or fold bifurcation curves where a single real eigenvalue crosses zero. The saddle–node curves have to be determined numerically, since the determinant is a quartic function of a_E^*, a_I^*.

Now consider a network of synaptically coupled E–I modules or subnetworks, with each module labeled by the discrete index n:

$$\frac{da_E^n}{dt} = -\frac{a_E^n}{\tau_E} + F\left(\sum_m [w_{EE}(n,m)a_E^m - w_{EI}(n,m)a_I^m + h_E]\right) \tag{6.25a}$$

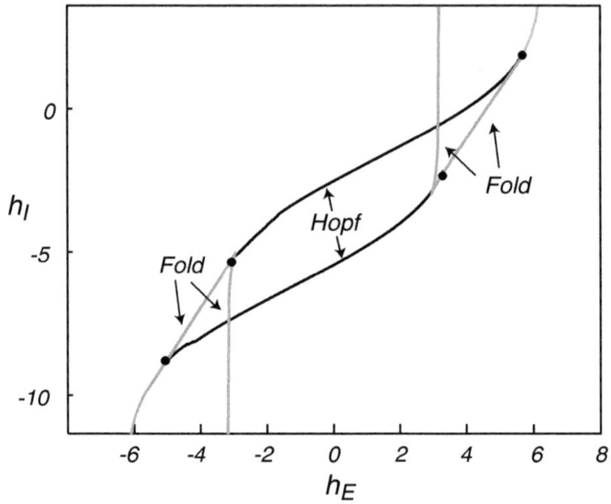

Fig. 6.2 Phase diagram of two-population Wilson–Cowan model (6.20) for fixed set of weights $w_{EE} = 11.5, w_{IE} = w_{EI} = 10, , w_{II} = -2$. The *dots* correspond to Takens–Bogdanov bifurcation points

$$\frac{da_I^n}{dt} = -\frac{a_I^n}{\tau_I} + F\left(\sum_m [w_{IE}(n,m)a_E^m - w_{II}(n,m)a_I^m + h_I]\right), \tag{6.25b}$$

Suppose that synaptic interactions within a local E–I network are stronger than those between E–I networks. That is, we write

$$w_{ab}(n,m) = w_{ab}\delta_{nm} + \varepsilon J_{ab}(n,m)(1-\delta_{nm})$$

with $\varepsilon \ll 1$. Substituting into (6.25) and Taylor expanding to $\mathscr{O}(\varepsilon)$ then gives

$$\frac{da_E^n}{dt} = -a_E^n + F\left([w_{EE}a_E^n - w_{EI}a_I^n + h_E]\right) \tag{6.26a}$$
$$+\varepsilon F'\left([w_{EE}a_E^n - w_{EI}a_I^n + h_E]\right)\sum_{m\neq n}[J_{EE}(n,m)a_E^m - J_{EI}(n,m)a_I^m]$$

$$\frac{da_I^n}{dt} = -a_I^n + F\left([w_{IE}a_E^n - w_{II}a_I^n + h_I]\right) \tag{6.26b}$$
$$+\varepsilon F'\left([w_{IE}a_E^n - w_{II}a_I^n + h_E]\right)\sum_{m\neq n}[J_{IE}(n,m)a_E^m - J_{II}(n,m)a_I^m]$$

One can view the synaptic interaction between E–I modules as weak perturbations of the underlying limit cycle oscillators, which suggests carrying out a phase reduction of (6.26) along the lines of Sects. 1.2 and 5.2. In order to proceed, it is first necessary to determine the phase-resetting curve (PRC) of an individual E–I network. Suppose that each uncoupled E–I network operates in a parameter regime where the mean-field equations (6.20) support a stable limit cycle. For concreteness, take a point in parameter space between the two Hopf curves in Fig. 6.2, namely,

$(h_E, h_I) = (0, -4)$. A plot of the oscillation in phase space is shown in Fig. 6.3(a) and the components Z_E, Z_I of the corresponding PRC are shown in Fig. 6.3(b). Note that both components are approximately sinusoidal so that the E–I network acts as a type II oscillator.

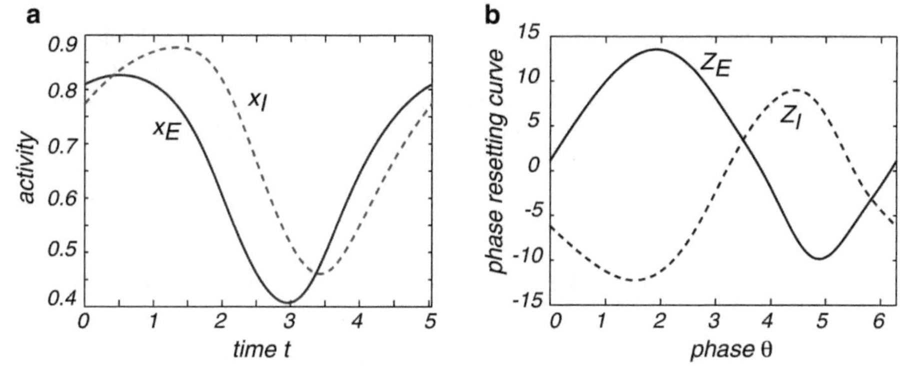

Fig. 6.3 (**a**) Trajectories along the limit cycle of an E–I network: $x_E(t)$ (*solid curve*) and $x_I(t)$ (*dashed curve*). Parameters are $w_{EE} = 11.5, w_{IE} = -w_{EI} = 10$, $w_{II} = -2$, $h_E = 0$, and $h_I = -4$. Also $F(u) = 1/(1+\mathrm{e}^{-u})$. (**b**) Corresponding components Z_E and Z_I of the phase-resetting curve

The next step is to rewrite (6.26) in the more compact form

$$\frac{d\mathbf{a}^n}{dt} = -\mathbf{a}^n + \mathbf{f}(\mathbf{a}^n) + \varepsilon \mathbf{g}(\mathbf{a}^n)^T \sum_{m \neq n} \mathbf{J}(n,m)\mathbf{a}^m, \tag{6.27}$$

where $\mathbf{a}^n = (a_E^n, a_I^n)^T$, $\mathbf{f} = (f_E, f_I)$, $\mathbf{g} = (g_E, g_I)$,

$$f_p(\mathbf{a}) = F(w_{pE}a_E - w_{pI}a_I + h_p), \quad g_p(\mathbf{a}) = F'(w_{pE}a_E - w_{pI}a_I + h_p), \quad p = E, I.$$

Phase reduction now proceeds using the method of isochrones as described in Sect. 1.2. Let $\Theta(\mathbf{a})$ represent the isochronal mapping in a neighborhood of an uncoupled limit cycle oscillator and denote the natural frequency by ω_0. Then

$$Z_E(\theta) = \frac{\partial \Theta(\mathbf{a}^*(\theta))}{\partial a_E}, \quad Z_I(\theta) = \frac{\partial \theta(\mathbf{a}^*(\theta))}{\partial a_I}, \tag{6.28}$$

where $\mathbf{a}^*$ is a point on the limit cycle. Applying the phase reduction procedure to (6.27) with $\theta_n = \Theta(\mathbf{a}^n)$, we have

$$\frac{d\theta_n}{dt} = \omega_0 + \varepsilon \sum_{p,q=E,I} Z_p(\theta_n) g_p(\theta_n) \sum_m J_{pq}(n,m) a_q^*(\theta_m). \tag{6.29}$$

Here, all quantities are evaluated on the limit cycles so that $g_p(\theta_n) = g_p(a^*(\theta_n))$ etc. Finally, averaging over one period $\Delta_0 = 2\pi/\omega_0$ gives the phase equations

$$\frac{d\theta_n}{dt} = \omega_0 + \varepsilon \sum_{p,q=E,I} \left[\sum_m J_{pq}(n,m) H_{pq}(\theta_m - \theta_n) \right], \tag{6.30}$$

with phase interaction functions

$$H_{pq}(\phi) = \frac{1}{2\pi}\int_0^{2\pi} Z_p(\theta-\phi)g_p(\theta-\phi)a_q^*(\theta)d\theta. \tag{6.31}$$

Given the phase equations, one can investigate the existence and stability of phase-locked states, including phase waves, along identical lines to the analysis of waves on a chain in Sect. 5.2.2. Note that phase-coupled E–I networks have been used to study the phenomenon of stimulus-induced oscillations and synchrony in primary visual cortex model, where each E–I network is interpreted as a cortical column consisting of reciprocally coupled populations of excitatory and inhibitory neurons [239, 560].

6.3 Firing Rates, Asynchronous States, and Spiking Statistics

6.3.1 The Asynchronous State in a Homogeneous Spiking Network

One of the major assumptions in the derivation of rate-based population models in Sect. 6.1 was that each homogeneous population is in an asynchronous state. Here we consider the existence and stability of an asynchronous state in a large, globally coupled network of IF neurons [2, 213]. Consider the following synaptically coupled network of nonlinear IF neurons (see also Sect. 5.3):

$$\frac{dv_i}{dt} = G(v_i) + \frac{\varepsilon}{N}\sum_{j=1}^{N}\int_{-\infty}^{\infty}\Phi(t'-\tau_d)\sum_m \delta(t-t'-T_j^m)dt', \tag{6.32}$$

with Φ given by the alpha function (5.23), τ_d a discrete axonal delay and ε determines the strength of coupling. We take a threshold $v_\kappa = 1$ and a reset $v_r = 0$. In the case of global coupling, we can express the sum over delta functions in terms of the population activity variable

$$a(t) = \sum_{j=1}^{N}\sum_m \delta(t-T_j^m) \tag{6.33}$$

so that

$$\frac{dv_i}{dt} = G(v_i) + \varepsilon u(t), \tag{6.34}$$

where $u(t) = \int_{-\infty}^{\infty}\Phi(t'-\tau_d)a(t-t')dt'$. Now suppose that there exists an asynchronous state $a(t) = a_0$. (Strictly speaking, such a state only exists in the thermodynamic limit $N \to \infty$.) Since Φ is normalized to unity it follows that $u(t) = a_0$ as well (ignoring transients). An implicit equation for a_0 is then obtained by integrating equation (6.34) between successive firing times:

$$\frac{1}{a_0} = \int_0^1 \frac{du}{G(u) + \varepsilon a_0}. \tag{6.35}$$

We will assume that there exists a unique solution to this equation for given G and ε. If G depends on a uniform external input I, then this yields the population gain function F with $a_0 = F(I)$.

In order to study the stability of the asynchronous state, it is convenient to carry out the change of variables

$$y_i = a_0 \int_0^{v_i} \frac{du}{G(u) + \varepsilon a_0}, \tag{6.36}$$

with $0 < y_i < 1$ such that Eq. (6.34) becomes

$$\frac{dy_i}{dt} = a_0 + \Gamma(y_i)[u(t) - a_0] \tag{6.37}$$

and

$$\Gamma(y) = \frac{a_0 \varepsilon}{G(y) + a_0 \varepsilon}. \tag{6.38}$$

We also incorporate the effects of synaptic noise by including an additive white noise term $\xi_i(t)$,

$$\frac{dy_i}{dt} = a_0 + \Gamma(y_i)[u(t) - a_0] + \xi_i(t), \tag{6.39}$$

with

$$\langle \xi_i(t) \rangle = 0, \quad \langle \xi_i(t)\xi_j(t') \rangle = \sigma^2 \delta_{ij} \delta(t - t'). \tag{6.40}$$

(Note that diffusive fluctuations of the membrane potential due to stochastic background activity would lead to an additive white noise term in Eq.(6.34) rather than in Eq. (6.40). The corresponding stochastic equation for y_i would then involve multiplicative noise, which is much harder to analyze.) In the presence of noise the variable y_i can become negative so $-\infty < y_i < 1$. The Langevin equation (6.40) has an associated Fokker–Planck equation

$$\frac{\partial}{\partial t} p(y,t) = -\frac{\partial}{\partial y} J(y,t), \tag{6.41}$$

where $J(y,t)$ is the probability flux

$$J(y,t) = [a_0 + \Gamma(y)[u(t) - a_0]]\, p(y,t) - \frac{\sigma^2}{2} \frac{\partial}{\partial y} p(y,t). \tag{6.42}$$

This is supplemented by the boundary conditions arising from reset:

$$p(1,t) = 0, \quad J(1,t) = a(t), \tag{6.43}$$

$$p(0^+,t) = p(0^-,t), \quad J(0^+,t) - J(0^-,t) = a(t). \tag{6.44}$$

We also require $p(-\infty,t) = 0$ and $J(-\infty,t) = 0$. The steady-state solution of the Fokker–Planck equation is $J(y,t) = a_0$ and $p(y,t) = p_0(y)$ with

$$p_0(y) = \begin{cases} e^{2a_0 y/\sigma^2} - e^{2a_0(y-1)/\sigma^2}, & y < 0 \\ 1 - e^{2a_0(y-1)/\sigma^2}, & 0 < y < 1 \end{cases}. \tag{6.45}$$

The stability of the steady state can be determined by setting

$$p(y,t) = p_0(y) + \rho(y)e^{\lambda t}, \quad a(t) = a_0 + a_1 e^{\lambda t} \tag{6.46}$$

and expanding to first order in ρ, a_1. This gives the eigenvalue equation

$$\lambda \rho(y) = \frac{\sigma^2}{2}\frac{\partial^2}{\partial y^2}\rho(y) - a_0 \frac{\partial}{\partial y}\rho(y) - a_1 \tilde{\Phi}(\lambda)\frac{\partial}{\partial y}[\Gamma(y)p_0(y)], \tag{6.47}$$

where $\tilde{\Phi}(\lambda)$ is the Laplace transform

$$\tilde{\Phi}(\lambda) = \int_0^\infty \Phi(t - \tau_d)e^{-\lambda t} = \frac{\alpha^2}{(\lambda + \alpha)^2}e^{\tau_d \lambda}. \tag{6.48}$$

Defining the function

$$h(y) = a_1 \tilde{\Phi}(\lambda)\frac{\partial}{\partial y}[\Gamma(y)p_0(y)], \tag{6.49}$$

we can write Eq. (6.47) as the inhomogeneous equation

$$[\mathscr{L}\rho(y) \equiv \left[\frac{\sigma^2}{2}\frac{\partial^2}{\partial y^2} - a_0\frac{\partial}{\partial y}\right]\rho(y) = \lambda \mathbf{1}]\rho(y) + h(y;\lambda). \tag{6.50}$$

This inhomogeneous equation can be solved in terms of the associated one-dimensional Green's function satisfying $[\mathscr{L} - \lambda\mathbf{1}]\mathscr{G}(y,y';\lambda) = \delta(y - y')$ and $\mathscr{G}(1,y',\lambda) = 0$:

$$\rho(y) = \int_{-\infty}^1 \mathscr{G}(y,y';\lambda)h(y';\lambda)dy' - a_1\mathscr{G}(y,0;\lambda) \tag{6.51}$$

with

$$\mathscr{G}(y,y';\lambda) = \begin{cases} A\left(e^{\mu_+(\lambda)[y-1]} - e^{-\mu_-(\lambda)[y-1]}\right)e^{\mu_-(\lambda)[y'-1]}, & y' < y < 1 \\ A\left(e^{\mu_-(\lambda)[y'-1]} - e^{-\mu_+(\lambda)[y'-1]}\right)e^{\mu_+(\lambda)[y-1]}, & y < y', \end{cases} \tag{6.52}$$

where

$$A = \frac{2}{\sigma^2}\frac{1}{\mu_+ + \mu_-}, \tag{6.53}$$

$$\mu_\pm(\lambda) = \frac{1}{\sigma^2}\left[\sqrt{a_0^2 + 2\lambda\sigma^2} \pm a_0\right]. \tag{6.54}$$

Note that the term $a_1\mathscr{G}(y,0;\lambda)$ ensures that the flux discontinuity at $y=0$ is satisfied. Finally, an implicit equation for the eigenvalues λ can be obtained by substituting Eq. (6.49) into (6.51) and imposing the boundary condition $J(1,t)=a(t)$, which corresponds to the following first-order condition:

$$-\frac{\sigma^2}{2}\frac{\partial}{\partial y}\rho(y,t)|_{y=1}=a_1. \tag{6.55}$$

The resulting characteristic equation is [2]

$$\left(e^{\mu_-(\lambda)}-1\right)=\mu_-(\lambda)\tilde{\Phi}(\lambda)\int_{-\infty}^{1}p_0(y)\Gamma(y)e^{\mu_-(\lambda)y}dy. \tag{6.56}$$

In the zero noise limit $\sigma\to 0$, we have $\mu_-(\lambda)\to\lambda/a_0$ and $p_0(y)\to 1$ for $0<y<1$ and is zero otherwise. Thus, Eq. (6.56) becomes

$$\left(e^{\lambda/a_0}-1\right)=\frac{\lambda}{a_0}\tilde{\Phi}(\lambda)\int_0^1\Gamma(y)e^{\lambda y/a_0}dy. \tag{6.57}$$

In the weak coupling regime, solutions of Eq. (6.57) are of the form $\lambda=2\pi i n a_0+\Lambda_n$ for integer n with $\Lambda_n=\mathscr{O}(\varepsilon)$. The term Λ_n can be calculated by performing a perturbation expansion in the coupling ε. The lowest-order contribution is simply determined by setting $\lambda=2\pi i n a_0$ on the right-hand side of Eq. (6.57). In the case of a linear IF model with $G(u)=I-u$, we have $\Gamma(y)=e^{y/a_0}$ so that

$$\Lambda_n=\varepsilon\left(\frac{2\pi i n a_0}{1+2\pi i n a_0}\right)\tilde{\Phi}(2\pi i n a_0)+\mathscr{O}(\varepsilon^2). \tag{6.58}$$

We then have the following stability results in the absence of noise [2, 213]:

(i) For zero axonal delays ($\tau_d=0$) and excitatory coupling ($\varepsilon>0$), the asynchronous state is stable with respect to excitation of the nth mode if and only if $\alpha<\alpha_n$ where

$$\alpha_n=-1+\sqrt{1+4n^2\pi^2a_0^2} \tag{6.59}$$

Hence, it is stable for sufficiently slow synapses, that is, $\alpha<\alpha_1$. The asynchronous state is always unstable in the case of inhibitory coupling since the condition for stability with respect to the nth harmonic is now $\alpha>\alpha_n$, which cannot be satisfied for all n.

(ii) The asynchronous state is almost always unstable for nonzero delays (in the noise-free case).

(iii) For large n, $|\Lambda_n|\sim 1/n^2$ so that higher harmonics grow or decay slowly.

Note that although the zero delay case is a singular limit in the absence of noise, it becomes non-singular for arbitrarily small amounts of noise, where instabilities with respect to higher harmonics are suppressed [2, 213]. One finds that for sufficiently high noise levels the asynchronous state is always stable. Reducing the noise for fixed delay induces an instability due to excitation of one of the harmonic modes with frequency $\omega \approx \omega_n = 2\pi n a_0$. A bifurcation at $\omega \approx \omega_1$ implies that the period of the resulting collective oscillation is identical to the period of the individual oscillators. Higher harmonics correspond to instabilities of the asynchronous state that lead to the formation of cluster states [213, 231]: each neuron fires with mean rate a_0, but the population of neurons splits up into several groups that fire in sequence so that the overall activity exhibits faster oscillations. Finally, note that fast oscillations are also found in sparsely connected random networks [87].

6.3.2 Balanced Networks and Neural Variability

It is well known that the spike trains of individual cortical neurons in vivo tend to be very noisy, having interspike interval (ISI) distributions that are close to Poisson [179, 593]. Indeed, one observes trial-to-trial variability in spike trains, even across trials in which external stimuli are identical. This raises a number of important issues. First, neurons are continuously bombarded by thousands of synaptic inputs, many of which are uncorrelated, so that an application of the law of large numbers would suggest that total input fluctuations are small. This would make it difficult to account for the Poisson-like behavior of individual neurons, even when stochastic ion channel fluctuations (Sect. 1.5) or random synaptic background activity is taken into account. For example, in the homogeneous spiking network analyzed above, the spike trains of individual neurons can be quite regular even when the population activity is asynchronous. Conversely, irregularity in spiking at the single-cell level can coexist with regular firing rates at the population level. However, there is growing evidence that noise can play a significant role at the network level. For example, noise appears to be present during perceptual decision making [656] and bistable perception, the latter being exemplified by perceptual switching during binocular rivalry [436, 578, 663]; see Sect. 8. Noise also contributes to the generation of spontaneous activity during resting states [148, 152]. At the level of large-scale neural systems, as measured with functional MRI (fMRI) imaging, this ongoing spontaneous activity reflects the organization of a series of highly coherent functional networks that may play an important role in cognition. Another issue is the possible computational role of pairwise and higher-order statistical correlations between the spike trains of distinct neurons, that is, whether or not it is necessary to go beyond firing rates.

One paradigm for exploring these various issues is the so-called balanced network [567, 645, 650]. In such networks, each neuron is driven by a combination of strong excitation and strong inhibition, which mainly cancel each other out, so that the remaining fluctuations occasionally and irregularly push the neuron over the firing threshold. Even in the absence of any external sources of noise, the resulting deterministic dynamics is chaotic and neural outputs are Poisson-like. Interestingly, there is some experimental evidence that cortical networks can operate in a balanced regime [391]. Another emergent feature of balanced networks is that they can support an asynchronous state characterized by large variability in single neuron spiking and yet arbitrarily small pairwise correlations even in the presence of substantial amounts of shared inputs [519]. Thus there is a growing consensus that the trial-to-trial irregularity in the spiking of individual neurons is often unimportant, and that information is typically encoded in firing rates. (Such rates could be a measure of population activity as in Sect. 6.1 or correspond to the rates of individual neurons modeled as inhomogeneous Poisson processes; see below.) There is then another level of neural variability, namely, trail-to-trial variations in the firing rates themselves. Recent physiological data shows that the onset of a stimulus reduces firing rate fluctuations in cortical neurons while having little or no effect on spiking variability [119]; see Fig. 6.4. Litwin-Kumar and Doiron have recently shown how these two levels of stochastic variability can emerge in a chaotic balanced network of randomly connected spiking neurons, in which a small amount of clustered connections induces firing rate fluctuations superimposed on spontaneous spike fluctuations [387].

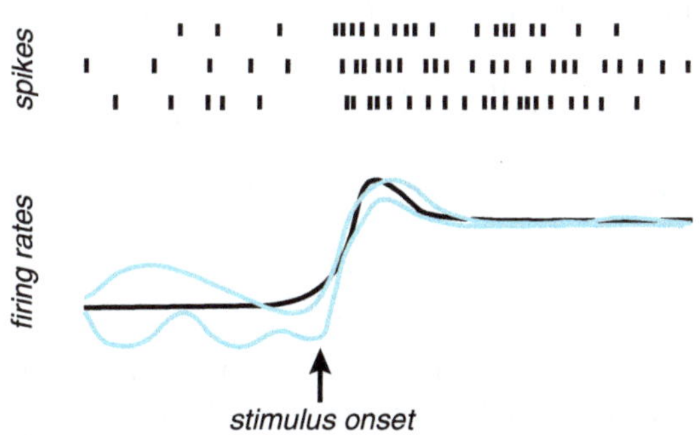

Fig. 6.4 Neurons can exhibit double stochasticity. Trial-to-trial variability in spike trains (shown as raster plots) can coexist with trial-to-trial variability in the firing rate (shown as continuous *gray curves*). Trial-averaged firing rate is the *black curve*. Following stimulus onset, the variability in the firing rate is reduced

6.3.3 Spike Statistics and the Poisson Process

Given the ubiquity of Poisson processes in spiking neuron models and its relevance to neural master equations (see Sect. 6.4), it is useful to consider these processes in a little more detail. Suppose that a neuron emits a sequence of spikes at times $t_1, t_2, \ldots, t_n$. One way to characterize this sequence statistically is in terms of the probability density $\rho(t_1, \ldots t_n)$ of finding such a sequence over many experimental trials. In other words, the probability of having a sequence of n spikes in the interval $[0,T]$ with the ith spike falling between the times t_i and $t_i + \Delta t$

is $P[t_1,\ldots t_n] = \rho(t_1,\ldots t_n)\Delta t^n$. In principle, the probability of an event occurring, namely, the firing of the next spike, could depend on the entire history of past spikes. If this dependence only extends to the previous spike so that the interspike intervals are statistically independent, then the stochastic process is said to be a *renewal process*. If there is no dependence at all on preceding spikes so that the firing times are themselves independent, then we have a *Poisson process*.

Consider a sequence of spikes generated by a homogeneous Poisson process, that is, one with a time-independent mean firing rate r. Divide a given time interval T into M bins of size $\Delta t = T/M$ and assume that ΔT is small enough so that the probability of finding two spikes within any one bin can be neglected. Then the probability $P_T[n]$ of finding n spikes over the interval T is given by

$$P_T[n] = \lim_{\Delta t \to 0} \frac{M!}{(M-n)!n!}(r\Delta t)^n(1-r\Delta t)^{M-n}$$

This consists of the probability $(r\Delta t)^n$ of finding n spikes in n specific bins multiplied by the probability $(1-r\Delta t)^{M-n}$ of not finding spikes in the remaining bins. The binomial factor is the number of ways of choosing n out of M bins with spikes. Using the approximation $M-n \approx M = T/\Delta t$ and defining $\varepsilon = -r\Delta t$, we have that

$$\lim_{\Delta t \to 0}(1-r\Delta t)^{M-n} = \lim_{\varepsilon \to 0}\left((1+\varepsilon)^{1/\varepsilon}\right)^{-rT} = \mathrm{e}^{-rT}.$$

For large M, $M!/(M-n)! \approx M^n = (T/\Delta t)^n$, so that we obtain the Poisson distribution

$$P_T[n] = \frac{(rT)^n}{n!}\mathrm{e}^{-rT}$$

Given that there are n independent spikes over the interval $[0,T]$, the probability that these spikes lie within specified bins of size Δt is $n!(\Delta t/T)^n$. Hence

$$\rho(t_1,\ldots t_n) = n!\left(\frac{1}{T}\right)^n P_T[n] = r^n\mathrm{e}^{-rT} \tag{6.60}$$

Also note that the fastest way to generate a sequence of Poisson spikes for constant r is to iterate the firing times $t_{n+1} = t_n - \log(x_{rand})/r$ with x_{rand} uniformly distributed over $[0,1]$.

A simple method for calculating the moments of the Poisson distribution is to introduce the moment-generating function

$$g(s) = \sum_{n=0}^{\infty} P_T[n]\mathrm{e}^{sn}$$

Differentiating with respect to s shows that

$$\left.\frac{d^k g(s)}{ds^k}\right|_{s=0} = \langle n^k \rangle$$

The generating function for the Poisson process can be evaluated explicitly as

$$g(s) = \exp(-rT)\exp(rTe^s)$$

from which we deduce that

$$\langle n \rangle = rT, \quad \sigma_n^2 = rT$$

Another useful quantity is the interspike interval (ISI) distribution. Suppose that a spike was last fired at time T^n. The probability of a homogeneous Poisson process generating the next spike in the interval $T^n + \tau \leq T^{n+1} \leq T^n + \tau + \Delta\tau$ is equal to the probability that no spike is fired for a time τ, which is $P_\tau[0] = e^{-r\tau}$ multiplied by the probability $r\Delta\tau$ of generating a spike within the following interval $\Delta\tau$:

$$\mathbf{Pr}[\tau \leq T^{n+1} - T^n \leq \tau + \Delta\tau] = r\Delta\tau e^{-r\tau}$$

The ISI probability density is thus an exponential, $\rho(\tau) = re^{-r\tau}$. It follows that the mean interspike interval is

$$\langle \tau \rangle = \int_0^\infty re^{-r\tau}\tau d\tau = \frac{1}{r}$$

and the variance is

$$\sigma_\tau^2 = \int_0^\infty re^{-r\tau}\tau^2 d\tau - \langle \tau \rangle^2 = \frac{1}{r^2}$$

The ratio of the standard deviation to the mean is called the *coefficient of variation*

$$C_V = \frac{\sigma_\tau}{\langle \tau \rangle}$$

It follows that for a homogeneous Poisson process $C_V = 1$.

How well do Poisson statistics describe experimentally measured spike trains? One often finds that for ISIs longer than about 10 msec, the ISI distribution is indeed approximately exponential. However, for shorter intervals, there is a rapid decrease in the distribution reflecting the fact that neurons are refractory immediately after

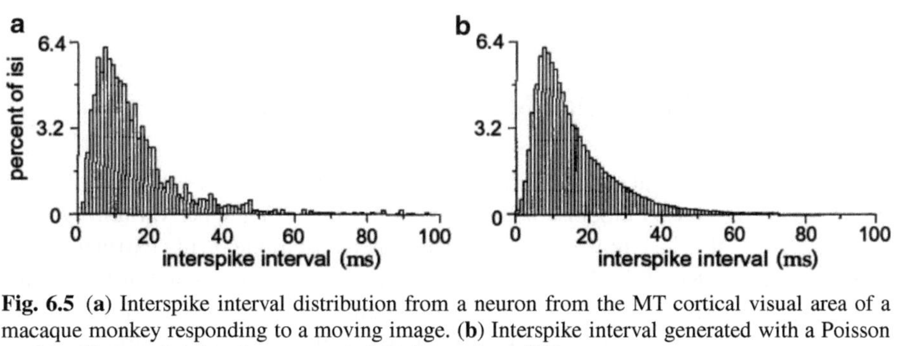

Fig. 6.5 (**a**) Interspike interval distribution from a neuron from the MT cortical visual area of a macaque monkey responding to a moving image. (**b**) Interspike interval generated with a Poisson model with a stochastic refractory period. Redrawn from [20]

firing. This is illustrated in Fig. 6.5. The data can be fitted more accurately by a gamma distribution

$$p[\tau] = \frac{r(r\tau)^k e^{-t\tau}}{k!} \tag{6.61}$$

Alternatively, one can introduce a refractory period into a standard Poisson model. Note that C_V values extracted from the spike trains of cortical neurons also take values around unity, provided that the mean interspike interval is not too small [593].

It is possible to generalize the above Poisson model to the case of a time-dependent rate $r(t)$. The simplest way to analyze this inhomogeneous Poisson process is to consider the probability distribution $\rho(t_1,\ldots t_n)$. This is given by the product of the probabilities $r(t_i)\Delta t$ that the neuron fires within the time intervals $t_i \leq t \leq t_i + \Delta t$ and the probabilities of not firing during the interspike intervals. The latter is given by

$$\mathbf{Pr}[\text{no spikes in } (t_i,t_{i+1})] = \prod_{m=1}^{M} (1 - r(t_i + m\Delta t)\Delta t)$$

where we have partitioned the interval (t_i,t_{i+1}) into M bins of size Δt. Taking the logarithm,

$$\log \mathbf{Pr}[\text{no spikes in } (t_i,t_{i+1})] = \sum_{m=1}^{M} \log(1 - r(t_i + m\Delta t)\Delta t) \approx -\sum_{m=1}^{M} r(t_i + m\Delta t)\Delta t$$

Taking the limit $\Delta t \to 0$ and exponentiating again shows that

$$\mathbf{Pr}[\text{no spikes in } (t_i,t_{i+1})] = \exp\left(-\int_{t_i}^{t_{i+1}} r(t)dt\right)$$

Hence

$$\rho(t_1,\ldots t_n) = \prod_{i=1}^{n} r(t_i) \exp\left(-\int_0^T r(t)dt\right) \tag{6.62}$$

In the case of a time-dependent rate, one generates x_{rand} at each time step and a spike is fired if $r(t)\Delta t > x_{rand}$.

6.4 Stochastic Population Models

The discussion of neural variability in Sect. 6.3 motivates the incorporation of noise directly into rate-based models, without explicitly modeling spike variability. One approach is to introduce noise into a rate-based network model using extrinsic noise sources [21, 84, 184, 297]. An alternative approach is to assume that noise arises intrinsically as a collective population effect. A number of methods involve carrying out some form of dimension reduction of a network of synaptically coupled spiking neurons. These include population density methods [395, 468, 477], mean-field theories [21, 86, 87], and Boltzmann-like kinetic theories [90, 100, 517]. However, such methods tend to consider either fully connected or sparsely connected networks and simplified models of spiking neurons such as the integrate-and-fire (IF) model. Nevertheless, as discussed in Sect. 6.3, one important feature of spiking networks is that they can support an asynchronous state in which individual neurons exhibit Poisson-like statistics, whereas the total population activity can be expressed in terms of a characteristic activation or gain function [2, 86, 87, 213, 519]. Formally speaking, the asynchronous state only exists in the thermodynamic limit $N \to \infty$, where N determines the size of the population. This then suggests a possible source of intrinsic noise at the network level arises from fluctuations about the asynchronous state due to finite size effects [57, 220, 412, 425, 597]; this is distinct from intrinsic noise at the single-cell level due to channel fluctuations and it is assumed that the latter is negligible at the population level. The presence of finite size effects has motivated the development of a closer analogy between intrinsic noise in biochemical networks (including ion channel kinetics) and neural networks [68, 69, 72], based on extensions of the neural master equation introduced by Buice et al. [91, 92]; see also [473]. In this section, we describe the master equation framework for stochastic neural population dynamics.

6.4.1 Neural Master Equations

One way to incorporate intrinsic noise at the population level is to treat the output activity of a local homogeneous population as a discrete stochastic variable $A_\alpha(t)$ rather than the instantaneous firing rate $a_\alpha = F(u_\alpha)$ [68, 91, 92]:

$$A_\alpha(t) = \frac{N_\alpha(t)}{N\Delta t}, \tag{6.63}$$

where $N_\alpha(t)$ is the number of neurons in the αth population that fired in the time interval $[t-\Delta t, t]$ and Δt is the width of a sliding window that counts spikes. The discrete stochastic variables $N_\alpha(t)$ are taken to evolve according to a one-step jump Markov process:

$$N_\alpha(t) \to N_\alpha(t) \pm 1: \quad \text{transition rate } \omega_\pm(U_\alpha(t), N_\alpha(t)), \tag{6.64}$$

with the synaptic current $U_\alpha(t)$ given by (for exponential synapses)

$$\tau dU_\alpha(t) = \left[-U_\alpha(t) + \sum_{\beta=1}^{M} w_{\alpha\beta} A_\beta(t) \right] dt, \tag{6.65}$$

where, for convenience, we have rescaled the weights according to $w_{\alpha\beta} \to w_{\alpha\beta}/\tau$. The transition rates are taken to be (cf. [68])

$$\omega_+(u_\alpha, n_\alpha) \to \omega_+(u_\alpha) = \frac{N\Delta t}{\tau_a} F(u_\alpha), \quad \omega_-(u_\alpha, n_\alpha) \to \omega_-(n_\alpha) = \frac{n_\alpha}{\tau_a}. \tag{6.66}$$

We see that the transition rate ω_+ depends on U_α, with the latter itself coupled to the associated jump Markov according to Eq. (6.65), which is only defined between jumps, during which $U_\alpha(t)$ evolves deterministically. Hence, the resulting stochastic process defined by Eqs. (6.63), (6.64), (6.65) and (6.66) provides an example of a stochastic hybrid system at the population neuron level, just as stochastic ion channel gating provides an example at the molecular level (Sect. 1.5). It is important to note that the time constant τ_a cannot be identified directly with membrane or synaptic time constants. Instead, it determines the relaxation rate of a local population to the instantaneous firing rate.

A. Case $\tau \to 0$ (fast synapses). In the limit $\tau \to 0$, Eq. (6.65) implies that the continuous variables $U_\alpha(t)$ can be eliminated by setting $U_\alpha(t) = \sum_\beta w_{\alpha\beta} A_\alpha(t)$. This then leads to a pure birth–death process for the discrete variables $N_\alpha(t)$. That is, let $P(\mathbf{n},t) = \text{Prob}[\mathbf{N}(t) = \mathbf{n}]$ denote the probability that the network of interacting populations has configuration $\mathbf{n} = (n_1, n_2, \ldots, n_M)$ at time $t, t > 0$, given some initial distribution $P(\mathbf{n}, 0)$ with $0 \le n_\alpha \le N$. The probability distribution then evolves according to the birth–death master equation [68, 91, 92]

$$\frac{dP(\mathbf{n},t)}{dt} = \sum_\alpha \left[(\mathbb{T}_\alpha - 1)\left(\omega_\alpha^-(\mathbf{n}) P(\mathbf{n},t)\right) + (\mathbb{T}_\alpha^{-1} - 1)\left(\omega_\alpha^+(\mathbf{n}) P(\mathbf{n},t)\right)\right], \tag{6.67}$$

where

$$\omega_\alpha^+(\mathbf{n}) = \frac{N\Delta t}{\tau_a} F\left(\sum_\beta w_{\alpha\beta} n_\alpha / N\Delta t\right), \quad \omega_\alpha^-(\mathbf{n}) = \frac{n_\alpha}{\tau_a}, \tag{6.68}$$

and $\mathbb{T}_\alpha$ is a translation operator: $\mathbb{T}_\alpha^{\pm 1} F(\mathbf{n}) = F(\mathbf{n}_{\alpha\pm})$ for any function F with $\mathbf{n}_{\alpha\pm}$ denoting the configuration with n_α replaced by $n_\alpha \pm 1$. Equation (6.67) is supplemented by the boundary conditions $P(\mathbf{n},t) \equiv 0$ if $n_\alpha = N+1$ or $n_\alpha = -1$ for some α. The neural master equation (6.67) has two versions depending on the choice of Δt. First, fixing $\Delta t = 1$ leads to the Bressloff version of the master equation. For large but finite N, the master equation (6.67) can be approximated by a Fokker–Planck equation using a Kramers–Moyal or system-size expansion, so that the population activity A_α evolves according to a Langevin equation [68]. Introduce the rescaled variables $a_\alpha = n_\alpha / N$ and corresponding transition rates $\Omega_\alpha^-(\mathbf{a}) = a_\alpha / \tau_a$ and $\Omega_\alpha^+(\mathbf{a}) = \tau_\alpha^{-1} F\left(\sum_\beta w_{\alpha\beta} a_\beta\right)$. Carrying out a Kramers–Moyal expansion to second order then leads to the multivariate FP equation

$$\frac{\partial P(\mathbf{a},t)}{\partial t} = -\sum_{\alpha=1}^{M} \frac{\partial}{\partial a_\alpha}\left[\mathscr{A}_\alpha(\mathbf{a}) P(\mathbf{a},t)\right] + \frac{1}{2N}\sum_{\alpha=1}^{M} \frac{\partial^2}{\partial a_\alpha^2}\left[\mathscr{B}_\alpha(\mathbf{a}) P(\mathbf{a},t)\right] \tag{6.69}$$

with

$$\mathscr{A}_\alpha(\mathbf{a}) = \Omega_\alpha^+(\mathbf{a}) - \Omega_\alpha^-(\mathbf{a}), \quad \mathscr{B}_\alpha(\mathbf{a}) = \Omega_\alpha^+(\mathbf{a}) + \Omega_\alpha^-(\mathbf{a}). \tag{6.70}$$

The solution to the Fokker–Planck equation (6.69) determines the probability density function for a corresponding stochastic process $\mathbf{A}(t)=(A_1(t),\ldots,A_M(t))$, which evolves according to a neural Langevin equation or SDE of the form

$$dA_\alpha = \mathscr{A}_\alpha(\mathbf{A})dt + \frac{1}{\sqrt{N}} b_\alpha(\mathbf{A})dW_\alpha(t). \tag{6.71}$$

with $b_\alpha(\mathbf{x})^2 = \mathscr{B}_\alpha(\mathbf{x})$. Here $W_\alpha(t)$ denotes an independent Wiener process such that

$$\langle dW_\alpha(t)\rangle = 0, \quad \langle dW_\alpha(t)dW_\beta(s)\rangle = \delta_{\alpha,\beta}\delta(t-s)dtds. \tag{6.72}$$

In the thermodynamic limit $N\to\infty$, we recover the activity-based mean-field equation

$$\tau_\alpha \frac{da_\alpha}{dt} = \mathscr{A}_\alpha(\mathbf{a}) = -a_\alpha(t) + F(\sum_\beta w_{\alpha\beta}a_\alpha(t)). \tag{6.73}$$

Note that the multiplicative noise in Eq. (6.71) is interpreted in the sense of Ito, which follows from the form of the FP equation (6.69); see Sect. 1.7.

A rigorous probabilistic treatment of the thermodynamic limit of the neural master equation has also been developed [89], extending previous work on chemical master equations [351]. However, as we noted within the context of stochastic ion channels in Sect. 1.5, the diffusion approximation breaks down when considering noise-induced transitions between multistage states. It is then necessary to use a WKB approximation of the master equation together with matched asymptotics [69]. The second version of the neural master equation, which was introduced by Buice et al. [91, 92], is obtained by taking the limit $N\to\infty, \Delta t\to 0$ such that $N\Delta t = 1$. In this case there is no small parameter that allows one to construct a Langevin approximation to the master equation. Nevertheless, it is possible to determine the moment hierarchy of the master equation using path integral methods or factorial moments, based on the observation that the network operates in a Poisson-like regime. The role of the sliding window size Δt is crucial in understanding the difference between the two versions of the master equation. First, it should be emphasized that the stochastic models are keeping track of *changes* in population spiking activity. If the network is operating close to an asynchronous state for large N, then one-step changes in population activity could occur relatively slowly, so there is no need to take the limit $\Delta t\to 0$. On the other hand, if population activity is characterized by a Poisson process, then it is necessary to take the limit $\Delta t\to 0$ in order to maintain a one-step process. However, given the existence of an arbitrarily small time scale Δt, it is no longer clear that one is justified in ignoring synaptic dynamics by taking the limit $\tau\to 0$ in Eq. (6.65).

B. Case $\tau \gg \tau_a > 0$ (stochastic hybrid system). Now suppose that $\tau > 0$ in the full stochastic model given by (6.63)–(6.66), with $N\to\infty$, $\Delta t\to 0$ such that $N\Delta t = 1$. Denote the random state of the full model at time t by $\{(U_\alpha(t),N_\alpha(t));\alpha = 1,\ldots,M\}$. Introduce the corresponding probability density

$$\text{Prob}\{U_\alpha(t) \in (u_\alpha, u_\alpha + du), N_\alpha(t) = n_\alpha; \alpha = 1, \ldots, M\} = p(\mathbf{u}, \mathbf{n}, t)d\mathbf{u}, \tag{6.74}$$

with $\mathbf{n} = (n_1, \ldots, n_M)$ and $\mathbf{u} = (u_1, \ldots, u_M)$. It follows from Eqs. (6.63)–(6.66) that the probability density evolves according to the Chapman–Kolmogorov (CK) equation [82]

$$\begin{aligned} &\frac{\partial p}{\partial t} + \frac{1}{\tau} \sum_\alpha \frac{\partial [v_\alpha(\mathbf{u}, \mathbf{n}) p(\mathbf{u}, \mathbf{n}, \mathbf{t})]}{\partial u_\alpha} \\ &\quad = \frac{1}{\tau_a} \sum_\alpha \left[(\mathbb{T}_\alpha - 1)(\omega_-(n_\alpha) p(\mathbf{u}, \mathbf{n}, t)) + (\mathbb{T}_\alpha^{-1} - 1)(\omega_+(u_\alpha) p(\mathbf{u}, \mathbf{n}, t)) \right], \end{aligned} \tag{6.75}$$

with

$$\omega_+(u_\alpha) = F(u_\alpha), \quad \omega_-(n_\alpha) = n_\alpha, \quad v_\alpha(\mathbf{u}, \mathbf{n}) = -u_\alpha + \sum_\beta w_{\alpha\beta} n_\beta. \tag{6.76}$$

Equation (6.75) can be rewritten in the more compact form (see also (1.143))

$$\frac{\partial p}{\partial t} = -\frac{1}{\tau} \sum_{\alpha=1}^M \frac{\partial}{\partial u_\alpha} (v_\alpha(\mathbf{u}, \mathbf{n}) p(\mathbf{u}, \mathbf{n}, t)) + \frac{1}{\tau_a} \sum_{\mathbf{m}} A(\mathbf{n}, \mathbf{m}; \mathbf{u}) p(\mathbf{u}, \mathbf{m}, t). \tag{6.77}$$

The drift "velocities" $v_\alpha(\mathbf{u}, \mathbf{n})$ for fixed $\mathbf{n}$ represent the piecewise deterministic synaptic dynamics according to

$$\tau \frac{du_\alpha}{dt} = v_\alpha(\mathbf{u}, \mathbf{n}), \quad \alpha = 1, \ldots, M, \tag{6.78}$$

and A represents the $\mathbf{u}$-dependent transition matrix for the jump Markov process.

For fixed $\mathbf{u}$, the matrix $A(\mathbf{n}, \mathbf{m}; \mathbf{u})$ is irreducible (which means that there is a nonzero probability of transitioning, possibly in more than one step, from any state to any other state in the jump Markov process) and has a simple zero eigenvalue. In particular, $\sum_{\mathbf{n}} A(\mathbf{n}, \mathbf{m}; \mathbf{u}) = 0$ for all $\mathbf{m}$, that is, $\mathbf{n} = (1, 1, \ldots, 1)^T$ is the left null vector of A. The Perron–Frobenius theorem (for large but finite N) ensures that all other eigenvalues of A are negative and the continuous-time Markov process for fixed $\mathbf{u}$,

$$\frac{dp(\mathbf{u}, \mathbf{n}, t)}{dt} = \frac{1}{\tau_a} \sum_{\mathbf{m} \in I} A(\mathbf{n}, \mathbf{m}; \mathbf{u}) p(\mathbf{u}, \mathbf{m}, t),$$

has a globally attracting steady-state $\rho(\mathbf{u}, \mathbf{n})$ such that $p(\mathbf{u}, \mathbf{n}, t) \to \rho(\mathbf{u}, \mathbf{n})$ as $t \to \infty$. The steady-state equation is

$$\begin{aligned} 0 &= \sum_{\mathbf{m}} A(\mathbf{n}, \mathbf{m}; \mathbf{u}) \rho(\mathbf{u}, \mathbf{m}) \\ &= \sum_{\alpha=1}^M \left[(n_\alpha + 1) \rho(\mathbf{u}, \mathbf{n} + \mathbf{e}_\alpha) - n_\alpha \rho(\mathbf{u}, \mathbf{n}) + F(u_\alpha)(\rho(\mathbf{u}, \mathbf{n} - \mathbf{e}_\alpha) - \rho(\mathbf{u}, \mathbf{n})) \right], \end{aligned}$$

where $[\mathbf{e}_\alpha]_\beta = \delta_{\alpha,\beta}$. The solution can be factorized as $\rho(\mathbf{u},\mathbf{n}) = \prod_{\beta=1}^M \rho_1(u_\beta,n_\beta)$ with

$$0 = \sum_{\alpha=1}^{M} \left[\prod_{\beta\neq\alpha} \rho_1(u_\beta,n_\beta)\right] [J(u_\alpha,n_\alpha+1) - J(u_\alpha,n_\alpha)],$$

where

$$J(u,n) = n\rho_1(u,n) - F(u)\rho_1(u,n-1).$$

Since $\rho_1(u,-1) \equiv 0$, it follows that $J(u,n) = 0$ for all n. Hence,

$$\rho_1(u,n) = \rho(u,0)\prod_{m=1}^{n} \frac{F(u)}{m} = \rho(u,0)\frac{F(u)^n}{n!}, \tag{6.79}$$

and the corresponding normalized density is a Poisson process with rate $F(u)$

$$\rho_1(u,n) = \mathrm{e}^{-F(u)}\frac{F(u)^n}{n!}. \tag{6.80}$$

There are two time scales in the CK equation (6.75), the synaptic time constant τ and the time constant τ_a, which characterizes the relaxation rate of population activity. In the limit $\tau \to 0$ for fixed τ_a, Eq. (6.75) reduces to the neural master equation (6.67) with $\mathbf{u} = \mathbf{u}(\mathbf{n})$ such that $v_\alpha(\mathbf{u}(\mathbf{n}),\mathbf{n}) = 0$. On the other hand, if $\tau_a \to 0$ for fixed τ, then we obtain deterministic voltage- or current-based mean-field equations

$$\begin{aligned}\tau\frac{du_\alpha}{dt} &= \langle v_\alpha\rangle(\mathbf{u}(t)) \equiv \sum_{\mathbf{n}} v_\alpha(\mathbf{u}(t),\mathbf{n})\rho(\mathbf{u}(t),\mathbf{n}) \\ &= -u_\alpha(t) + \sum_{\beta=1}^{M} w_{\alpha\beta}\sum_{\mathbf{n}} n_\beta\rho(\mathbf{u}(t),\mathbf{n}).\end{aligned} \tag{6.81}$$

Since $\rho(\mathbf{u},\mathbf{n})$ is given by product of independent Poisson processes with rates $F(u_\alpha)$, consistent with the operating regime of the Buice *et. al.* master equation [91, 92], it follows that

$$\langle n_\beta\rangle = F(u_\beta), \tag{6.82}$$

and (6.81) reduces to the standard voltage- or current-based activity equation. Now suppose that the network operates in the regime $0 < \tau_a/\tau \equiv \varepsilon \ll 1$, for which there are typically a large number of transitions between different firing states $\mathbf{n}$ while the synaptic currents $\mathbf{u}$ hardly change at all. This suggests that the system rapidly converges to the (quasi) steady state $\rho(\mathbf{u},\mathbf{n})$, which will then be perturbed as $\mathbf{u}$ slowly evolves. The resulting perturbations can be analyzed using a quasi-steady-state (QSS) diffusion or adiabatic approximation, in which the CK equation (6.75) is approximated by a Fokker–Planck equation [82]; see Sect. 1.6. However, when considering escape from a metastable state, it is necessary to use the WKB method outlined in Sect. 1.6, as will be illustrated below.

6.4.2 Metastability in a One-Population Model

One of the useful features of the master equation formulation of stochastic population dynamics is that one can apply methods previously used to analyze jump Markov processes at the cellular level. We will illustrate this by considering a first passage time (FPT) problem for a one-population version of the CK equation (6.75), which exhibits bistability in the deterministic limit [82]. This is analogous to the problem of spontaneous action potential generation presented in Sect. 1.6. In the case of a single homogeneous population of excitatory neurons, (6.75) becomes

$$\frac{\partial p}{\partial t}+\frac{\partial[v(u,n)p(u,n,t)]}{\partial u}=\frac{1}{\varepsilon}\sum_{m}A(n,m;u)p(u,m,t) \tag{6.83}$$

with drift term

$$v(u,n)=-u+n, \tag{6.84}$$

and tridiagonal transition matrix

$$A(n,n-1;u)=F(u), \quad A(n,n;u)=-F(u)-n, \quad A(n,n+1;u)=n+1. \tag{6.85}$$

As shown in Sect. 6.4.1, the steady-state density for A is given by a Poisson process,

$$\rho(u,n)=\frac{[F(u)]^n\mathrm{e}^{-F(u)}}{n!}, \tag{6.86}$$

and the mean-field equation obtained in the $\varepsilon\to 0$ limit is

$$\frac{du}{dt}=\sum_{n=0}^{\infty}v(u,n)\rho(u,n)=-u+F(u). \tag{6.87}$$

If F is given by the sigmoid (1.14), it is straightforward to show that (6.87) exhibits bistability for a range of thresholds and gains; see Fig. 6.6.

The CK equation (6.83) is identical in form to (1.143) under the change of variables $u\to v$, where v is voltage, and $v(u,n)\to F(v,n)$. It follows that the general analysis developed in Sect. 1.6 can be applied directly to the stochastic neural population model. In particular, the escape rate λ_0 from the low activity state u_- is given by (1.191). As in the case of stochastic ion channels, there are three basic steps needed in order to evaluate λ_0 using the particular form of the drift v and transition matrix A [72]:

(1) Find the unique nontrivial positive eigenfunction $\psi_n(u)=R^{(0)}(u,n)$ and associated eigenvalue $\mu(u)=-\Phi_0'(u)$. In the case of the neural population model, Eq. (1.171) takes the explicit form

$$F(u)\psi_{n-1}(u)-(F(u)+n)\psi_n(u)+(n+1)\psi_{n+1}(u)=\mu(-u+n)\psi_n(u). \tag{6.88}$$

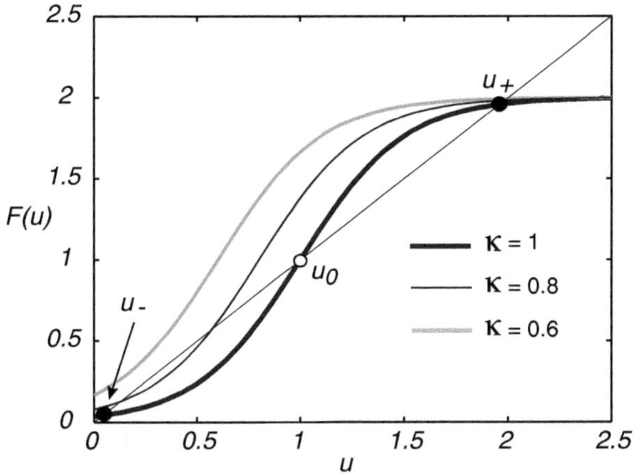

Fig. 6.6 Bistability in the deterministic network satisfying $\dot{u} = -u + F(u)$ with F given by the sigmoid (1.14) for gain $\eta = 4$ and threshold $\kappa = 1.0$, $F_0 = 2$. There exist two stable fixed points $u_\pm$ separated by an unstable fixed point u_0. As the threshold κ is reduced the network switches to a monostable regime

Trying a solution for ψ of the form

$$\psi_n(u) = \frac{\Lambda(u)^n}{n!} \tag{6.89}$$

yields the following equation relating Λ and the corresponding eigenvalue μ:

$$\left[\frac{F(u)}{\Lambda} - 1\right] n + \Lambda - F(u) = \mu(u)(-u+n).$$

We now collect terms independent of n and linear in n, respectively, to obtain the pair of equations

$$\mu = \left[\frac{F(u)}{\Lambda} - 1\right], \quad \Lambda = F(u) - \mu u.$$

We deduce that

$$\Lambda = u, \quad \mu = \left[\frac{F(u)}{u} - 1\right], \tag{6.90}$$

and the normalized eigenfunction is

$$\psi_n(u) = \frac{u^n}{n!} e^{-u}. \tag{6.91}$$

Note that $\mu(u)$ vanishes at the fixed points u_-, u_* of the mean-field equation (6.87) with $\mu(u) > 0$ for $0 < u < u_-$ and $\mu(u) < 0$ for $u_- < u < u_*$. Moreover, comparing Eq. (6.86) with (6.91) establishes that $\psi_n(u) = \rho(u,n)$ at the fixed points $u_*, u_\pm$. In conclusion $R^{(0)}(u,n) = \psi_n(u)$ and the effective potential Φ_0 is given by

$$\Phi_0(u) = -\int_{u_-}^{u^*} \mu(y)dy. \tag{6.92}$$

The effective potential is defined up to an arbitrary constant, which has been fixed by setting $\Phi_0(u_-) = 0$.

(2) Determine the null eigenfunction $\eta_n(u) = S(u,n)$ of Eq. (1.174), which becomes

$$F(u)\eta_{m+1} - (F(u)+m)\eta_m + m\eta_{m-1} = \mu(u)[-u+m]\eta_m. \tag{6.93}$$

Trying a solution of the form $\eta_m = \Gamma^m$ yields

$$(F(u))\Gamma - (F(u)+m) + m\Gamma^{-1} = \mu(u)[-u+m]. \tag{6.94}$$

Γ is then determined by canceling terms linear in m, which finally gives

$$\eta_n(u) = \left(\frac{u}{F(u)}\right)^n. \tag{6.95}$$

(3) Calculate the generalized eigenvector $\zeta_n = \hat{S}(u_*,n)$ of Eq. (1.186), which reduces to

$$F(u_*)\zeta_{n+1} + n\zeta_{n-1} - (F(u_*)+n)\zeta_n = u_* - n. \tag{6.96}$$

It is straightforward to show that this has the solution $\zeta_n = n$ (up to an arbitrary constant that does not contribute to the principal eigenvalue). It follows from Eq. (6.86) that the factor $B(u_*)$ defined by (1.190) is

$$\begin{aligned} B(u_*) &= \sum_{n=0}^{\infty} \rho(u_*,n)\left[-u_*n+n^2\right] \\ &= \left[-u_*\langle n\rangle + \langle n^2\rangle\right]. \end{aligned} \tag{6.97}$$

Recall that $\rho(u,n)$ is given by a Poisson density with rate $F(u)$, which implies that $\langle n^2\rangle = \langle n\rangle + \langle n\rangle^2$ with $\langle n\rangle = F(u)$. Therefore,

$$B(u_*) = F(u_*)\left[2F(u_*) - u_*\right], \tag{6.98}$$

which reduces to $B(u_*) = F(u_*)^2$ since $u^* = F(u^*)$ at a fixed point.

It is instructive to compare the effective potential Φ_0 obtained using the WKB approximation with the potential obtained using the FP equation (1.146) based on the QSS approximation. First, substitute (6.86), (6.84), and (6.85) into Eqs. (1.147) and (1.148) under the change of variables $u \to v$ and $v(u,n) \to F(v,n)$. We find that $Z(u,n) = n\rho(u,n)$ so that

$$\mathscr{D}(u) = [-u\langle n\rangle + w\langle n^2\rangle] = B(u). \tag{6.99}$$

The steady-state solution of the FP equation (1.146) takes the form $C(u) \sim \exp^{-\hat{\Phi}_0(u)/\varepsilon}$ with stochastic potential

$$\hat{\Phi}_0(u) = -\int^u \frac{\mathscr{F}(y)}{\mathscr{D}(y)}dy = -\int^u \frac{-y+F(y)}{wF(y)[2F(y)-y]}dy. \qquad (6.100)$$

Note that $\hat{\Phi}_0$ differs from the potential Φ_0, Eq. (6.92), obtained using the more accurate WKB method. Equations (6.90) and (6.92) show that the latter has the integral form

$$\Phi_0(u) = -\int^u \frac{1}{w}\left[\frac{wF(y)}{y} - 1\right]dy. \qquad (6.101)$$

Thus, there will be exponentially large differences between the steady states for small ε.

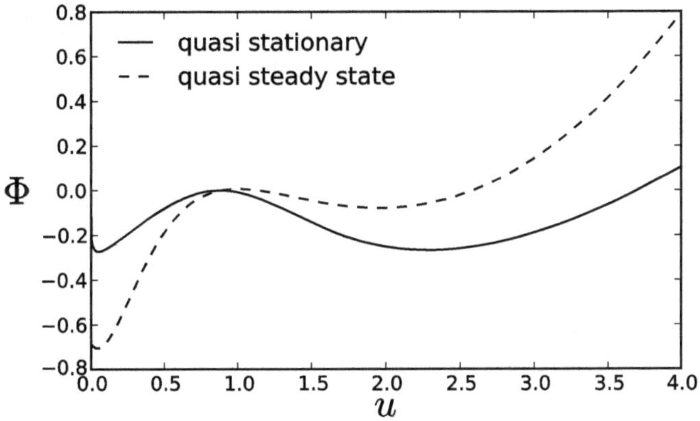

Fig. 6.7 Comparison of the double-well potentials $\Phi_0(u)$ and $\hat{\Phi}_0(u)$ obtained using the quasistationary approximation and the QSS diffusion approximation, respectively [82]. Parameter values are chosen so that deterministic network is bistable: $F_0 = 2.3$, $\gamma = 4$, and $\kappa = 1$

In Fig. 6.7, we plot the potential function Φ_0 of Eq. (6.101), which is obtained using the quasistationary approximation in a parameter regime for which the underlying deterministic network is bistable. We also plot the corresponding potential function $\hat{\Phi}_0$ of Eq. (6.100), under the QSS diffusion approximation. The differences between the two lead to exponentially large differences in estimates for the mean exit times when ε is small. The mean exit time from the left and right well is shown in Fig. 6.8. Solid curves show the analytical approximation $T \sim 1/\lambda_0$, where λ_0 is given by (1.191), as a function of $1/\varepsilon$. For comparison, the mean exit time computed from averaged Monte Carlo simulations of the full stochastic system is shown as symbols. As expected, the log of the mean exit time is an asymptotically linear function of $1/\varepsilon$, and this is confirmed by Monte Carlo simulations. The slope is determined by the depth of the potential well, and the vertical shift is determined by the prefactor. Also shown is the corresponding MFPT calculated using the QSS diffusion approximation (dashed curves), which is typically several orders of magnitude out and validates the relative accuracy of the quasistationary approximation.

One of the applications of noise-induced transitions between metastable states in a stochastic population model is to the study of switching between up and down states during slow-wave sleep; see Sect. 5.1. For example, Holcman and Tsodyks

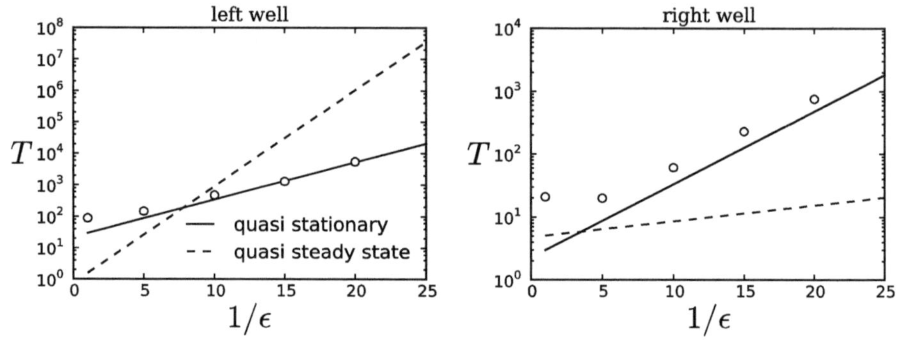

Fig. 6.8 Mean exit time from the left and right well calculated using the quasistationary approximation (*solid line*) and the QSS diffusion approximation (*dashed line*) [82]. The *open circles* represent data points obtained by numerically solving the corresponding jump velocity Markov process using the Gillespie algorithm. Parameter values are the same as in Fig. 6.7

[282] consider an extension of the deterministic one-population model (6.87) by including the effects of synaptic depression and extrinsic noise. The voltage-based model equations take the form

$$\tau \frac{du}{dt} = -u + qw\chi F(u) + \sqrt{\tau}\sigma\xi(t) + I(t) \tag{6.102a}$$

$$\frac{dq}{dt} = \frac{1-q}{\tau_q} - \chi q F(V), \tag{6.102b}$$

where $q(t)$ is the depression variable, $\xi(t)$ is a white noise term, $I(t)$ is an external input, and the constant χ is known as a utilization parameter. The firing rate function is taken to be a linear threshold function $F(V) = \eta[u-\kappa]_+$. In the absence of noise ($\sigma = 0$) and external inputs ($I = 0$), the deterministic system undergoes a series of bifurcations as the weight w is increased. For sufficiently small w, there exists a globally attracting stable fixed point which is a low activity or down state. At a critical value of w, a saddle and unstable node appear, with the latter undergoing a subcritical Hopf bifurcation at a second critical value of w—this generates a stable up state inside an unstable limit cycle. The system then exhibits bistability such that noise-induced transitions from the down to the up state have to cross both the separatrix of the saddle and the unstable limit cycle. The resulting fluctuations in activity are suggestive of slow-wave oscillations observed in cortical slices [549].

6.4.3 Metastability in a Two-Population Model

It turns out that the analysis of metastability in a one-population model can be extended to higher-dimensional networks [72]. Consider, for example, the E–I network of Sect. 6.2. The corresponding CK equation (6.75) becomes

$$\frac{\partial p}{\partial t} = -\frac{\partial}{\partial x}(vp) - \frac{\partial}{\partial y}(\tilde{v}p) + \frac{1}{\varepsilon}\sum_{\mathbf{m}} A(\mathbf{n},\mathbf{m};\mathbf{x})p(\mathbf{x},\mathbf{n}) \tag{6.103}$$

where $\mathbf{x} = (x,y)$, $\mathbf{n} = (n_x, n_y)$, the drift terms are

$$v(x,n_x,n_y) = -x + [w_{EE}n_x - w_{EI}n_y], \tag{6.104}$$

$$\tilde{v}(y,n_x,n_y) = -y + [w_{IE}n_x - w_{II}n_y], \tag{6.105}$$

and A has the nonzero components

$$A(n_x,n_y,n_x-1,n_y;\mathbf{x}) = F(x), \quad A(n_x,n_y,n_x,n_y-1;\mathbf{x}) = F(y), \tag{6.106}$$

$$A(n_x,n_y,n_x+1,n_y;\mathbf{x}) = n_x+1, \quad A(n_x,n_y,n_x,n_y+1;\mathbf{x}) = n_y+1, \tag{6.107}$$

and

$$A(n_x,n_y,n_x,n_y;\mathbf{x}) = -[F(x)+F(y)+n_x+n_y]. \tag{6.108}$$

Here x and y denote the excitatory and inhibitory population variables a_E, a_I, respectively. In the limit $\varepsilon \to 0$, we recover the deterministic equations (6.20). However, in contrast to Sect. 6.2, it is now assumed that the E–I network operates in a bistable regime as illustrated in Fig. 6.9.

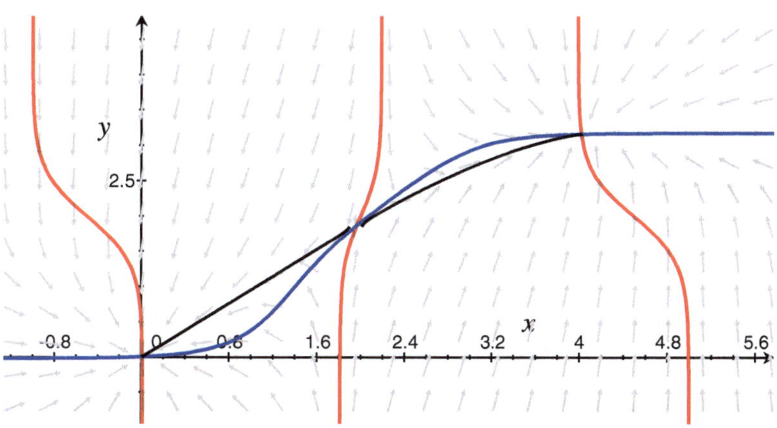

Fig. 6.9 Bistability in an E–I network. The x-nullcline through the saddle is its stable manifold and acts as the separatrix Σ between the two stable fixed points. Two deterministic trajectories are shown (directed *black curves*), starting from either side of the unstable saddle and ending at a stable fixed point. Parameter values are $F_0 = 1$, $\gamma = 3$, $\kappa = 2$, $w_{EE} = 5$, $w_{EI} = 1$, $w_{IE} = 9$, and $w_{II} = 6$

Again the MFPT can be identified as the inverse of the principal eigenvalue λ_0 of the associated linear operator on the right-hand side of (6.103). However, now the analysis is more complicated, since one has to consider stochastic trajectories crossing different points along the separatrix Σ between the two metastable states.

Nevertheless, the various steps in the calculation of λ_0 proceed along similar lines to the one-population model. That is, a spectral projection method can be used to express λ_0 in terms of the inner product of a quasistationary density with a corresponding adjoint eigenfunction. The former can be calculated using a WKB approximation, except that now one has to determine the stochastic potential Φ_0 by solving an equation of the form [82]

$$\mathscr{H} \equiv -x\mathscr{P}_x - y\mathscr{P}_y - F(x) - F(y) + \Lambda_x(x, \mathscr{P}_x, \mathscr{P}_y) + \Lambda_y(y, \mathscr{P}_x, \mathscr{P}_y) = 0, \quad (6.109)$$

where

$$\mathscr{P}_x = \frac{\partial \Phi_0}{\partial x} \quad \mathscr{P}_y = \frac{\partial \Phi_0}{\partial y}, \quad (6.110)$$

and

$$\Lambda_x = \frac{F(x)}{1 - w_{EE}\mathscr{P}_x - w_{IE}\mathscr{P}_y}, \quad \Lambda_y = \frac{F(y)}{1 + w_{EI}\mathscr{P}_x + w_{II}\mathscr{P}_y} \quad (6.111)$$

Mathematically speaking, Eq. (6.109) is identical to a stationary Hamilton–Jacobi equation for a classical particle with $\mathscr{H}$ identified as the Hamiltonian (see also Sect. 4.4). A trajectory of the particle is given by the solution of Hamilton's equations

$$\frac{dx}{dt} = \frac{\partial \mathscr{H}}{\partial \mathscr{P}_x}, \quad \frac{dy}{dt} = \frac{\partial \mathscr{H}}{\partial \mathscr{P}_y},$$
$$\frac{d\mathscr{P}_x}{dt} = -\frac{\partial \mathscr{H}}{\partial x}, \quad \frac{d\mathscr{P}_y}{dt} = -\frac{\partial \mathscr{H}}{\partial y} \quad (6.112)$$

Here t is treated as a parameterization of trajectories rather than as a real-time variable. Given a solution curve $(x(t), y(t))$, known as a ray, the potential Φ_0 can be determined along the ray by solving the equation

$$\frac{d\Phi_0}{dt} \equiv \frac{\partial \Phi_0}{\partial x}\frac{dx}{dt} + \frac{\partial \Phi_0}{\partial y}\frac{dy}{dt} = \mathscr{P}_x\frac{dx}{dt} + \mathscr{P}_y\frac{dy}{dt}. \quad (6.113)$$

Thus, Φ_0 can be identified as the action along a zero-energy trajectory. One numerically solves for Φ_0 by considering Cauchy data in a neighborhood of the stable fixed point (x_-, y_-) [82].

The rays $(x(t), y(t))$ (i.e., solutions to Hamilton's equations (6.112) in the (x, y) plane) have an important physical meaning. The trajectory of the ray is the most likely trajectory or path leading away from a point in the neighborhood of a stable fixed point [160]. The rays shown in Fig. 6.10 are obtained by integrating Hamilton's equations (6.112) . These trajectories are only valid in one direction: away from the stable fixed points. For parameter values considered in Fig. 6.10, rays originating from the neighborhood of each stable fixed point cover separate regions, so that the most likely paths between points in each region are connected by deterministic trajectories starting at the boundary between the two regions. Note that this boundary is not the separatrix (gray curve). For example, a trajectory initially at the left fixed point which crosses the separatrix at the saddle would most likely follow

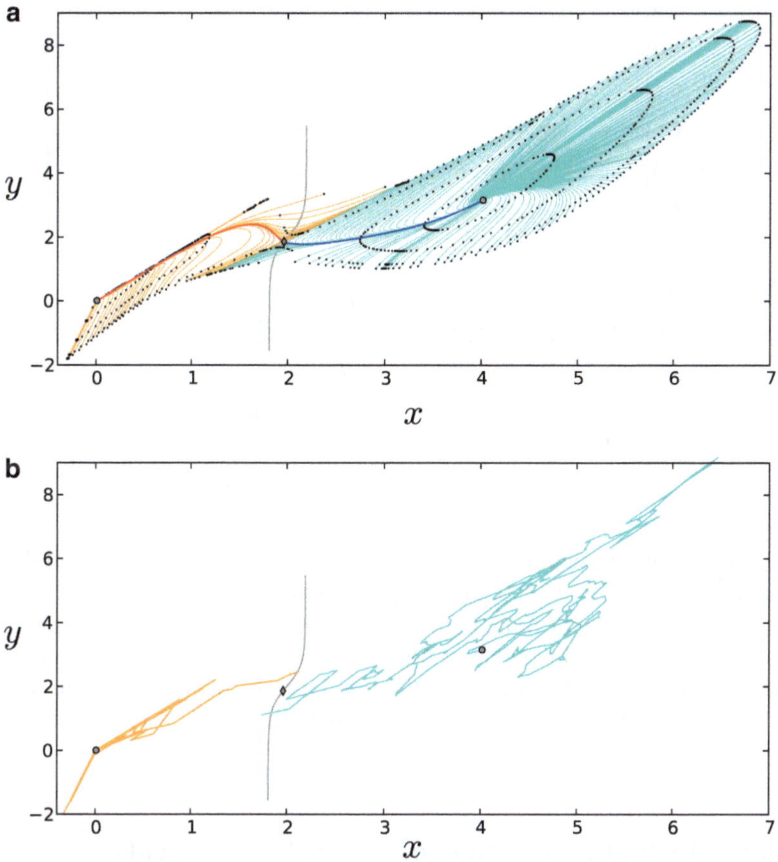

Fig. 6.10 (**a**) Characteristic paths of maximum likelihood for the 2D model. Rays originating from the left (right) stable fixed point are shown in *orange* (*cyan*), with the ray connecting to the saddle shown in *red* (*blue*). The *gray curve* is the separatrix Σ. Level curves of constant $\Phi_0(x,y)$ are shown as *black dots*. Each ray has four dots for different values of $\Phi_0(x,y)$. Rays originating from the left fixed point have dots at $\Phi = 0.1, 0.2, \Phi_* + 0.01, \Phi_* + 0.02$, and rays originating from the right fixed point have dots at $\Phi_0 = 0.19, 0.23, 0.28, 0.30$, where $\Phi_* = \Phi(x_*, y_*) = 0.28$. All rays terminate at $\Phi_0 = \Phi_* + 0.02$. (**b**) Sample trajectories of the two-population velocity jump Markov process, whose associated probability density evolves according to (6.83), are computed using the Gillespie algorithm with $\varepsilon = 0.05$ and $N\Delta t = 1$. (The maximum likelihood paths are independent of ε.) Other parameter values are the same as in Fig. 6.9 [72]

a ray towards the saddle and then follow a deterministic trajectory to the right fixed point. If a trajectory crosses the separatrix away from the saddle, it is most likely to cross the separatrix above the saddle when starting from the left fixed point and below the saddle when starting from the right fixed point. In Fig. 6.11, the probability density function for the y coordinate of the point on the separatrix reached by an exit trajectory is shown for each well (square symbols show the histogram for exit from the left well and likewise, 'o' symbols for the right well). Each density function is

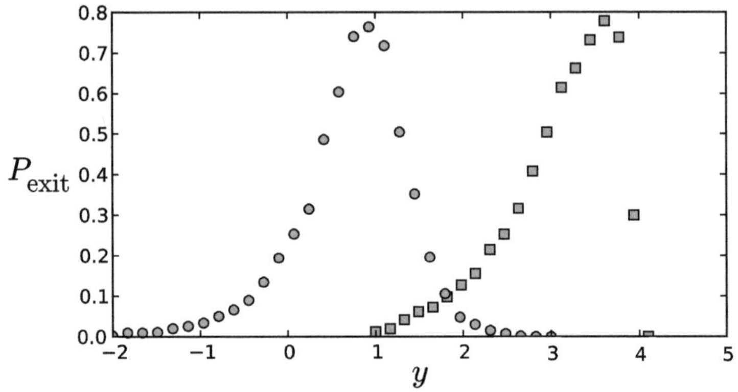

Fig. 6.11 The probability density for the exit point (y coordinate) where the separatrix is crossed by an exiting trajectory. Results are obtained by 10^2 Monte Carlo simulation with the same parameters as used in Fig. 6.9, with $\varepsilon = 0.08$. The *square* symbols show trajectories from the left well, and 'o' symbols show trajectories from the right well

peaked away from the saddle point, showing a phenomena known as saddle point avoidance [398, 559]. As $\varepsilon \to 0$, the two peaks merge at the saddle point. Although one might expect the saddle point to be the most likely exit point, since it is the point on the separatrix where the potential Φ_0 takes its minimum value, Fig. 6.11 shows that this is not necessarily true. Even though the most likely exit point is shifted from the saddle, the value of the potential around the saddle point still dominates the mean first exit time.

6.5 Spatially Structured Networks and Neural Fields

So far we have not made any assumptions about the topology of the underlying neural network, that is, the structure of the weight matrix $\mathbf{W}$ with components w_{ij}. If one looks at a region of cortex such as primary visual cortex (V1), one finds that it has a characteristic spatial structure, in which a high density of neurons (10^5 per mm^3 in primates) are distributed according to an approximately two-dimensional (2D) architecture. That is, the physical location of a vertical column of neurons within the two-dimensional cortical sheet often reflects the specific information processing role of that population of neurons. For example, in V1 there is an orderly retinotopic mapping of the visual field onto the cortical surface, with left and right halves of the visual field mapped onto right and left visual cortices, respectively. Superimposed upon this are additional two-dimensional maps reflecting the fact that neurons respond preferentially to stimuli with particular features such as local orientation [613]. (A more detailed description of the functional architecture of V1 is given in Sect. 8.1.) This suggests labeling neurons according to their spatial location in cortex. We now give a heuristic argument for how such labeling leads to a continuum neural field model of cortex, following along similar lines to Gerstner and Kistler [214].

For simplicity, consider a population of neurons distributed along a one-dimensional axis. (Extensions to higher dimensions proceed in a similar fashion.) Suppose that we partition space into segments of length d such that the number of neurons in segment $[nd,(n+1)d]$ is $N=\rho d$ where ρ is the cell density. We treat neurons in that interval as a homogeneous population of cells (cortical column) labeled by the integer n and assume that synaptic interactions between the nth and mth populations only depend on the discrete locations of the populations on the line. Writing $\Phi_{nm}(t)=\rho d\Phi(nd,md,t)$ and $u_n(t)=u(nd,t)$, Eq. (6.4) becomes

$$u(nd,t)=\rho d\sum_m\int_{-\infty}^t \Phi(nd,md,t-t')F(u(md,t'))dt'.$$

Taking the limit $d\to 0$, the summation on the right-hand side can be replaced by an integral to give

$$u(x,t)=\int_{-\infty}^{\infty}\int_{-\infty}^{t}\Phi(x,y,t-t')F(u(y,t'))dt'dy, \tag{6.114}$$

where we have absorbed the factor ρ into Φ. Following our derivation of the discrete voltage-based model (6.6), suppose that we can decompose the integral kernel as

$$\Phi(x,y,t)=w(x,y)\Phi(t),\quad \Phi(t)=\mathrm{e}^{-t/\tau}H(t).$$

That is, we assume that there is only one type of neuron so that the temporal kernel $\Phi(t)$ is independent of the presynaptic label y and the postsynaptic label x. Applying the differential operator $\mathbb{L}_t=\partial_t+\tau^{-1}$ to the integral equation for u then leads to the scalar neural field equation

$$\frac{\partial}{\partial t}u(x,t)=-\frac{u(x,t)}{\tau}+\int_{-\infty}^{\infty}w(x,y)F(u(y,t))dy. \tag{6.115}$$

Alternatively, we could have applied the differential operator $\mathbb{L}_t$ to the corresponding synaptic drive $z(x,t)=\int_{-\infty}^t\Phi(t-t')F(u(x,t'))dt'$ to obtain the activity-based neural field model

$$\frac{\partial}{\partial t}z(x,t)=-\frac{z(x,t)}{\tau}+F\left(\int_{-\infty}^{\infty}w(x,y)z(y,t)dy\right). \tag{6.116}$$

Following the same basic procedure, it is straightforward to incorporate into the neural field equation (6.115) or (6.116) additional features such as synaptic depression [329–331], adaptive thresholds [135, 329], and axonal propagation delays [16, 139, 296, 307, 382, 528, 603]. For example, a voltage-based neural field equation with synaptic depression takes the form

$$\begin{aligned}\frac{\partial}{\partial t}u(x,t)&=-\frac{u(x,t)}{\tau}+\int_{-\infty}^{\infty}w(x,y)q(y,t)F(u(y,t))dy,\\ \frac{\partial}{dt}q(x,t)&=\frac{1-q(x,t)}{\tau_q}-\beta q(x,t)F(u(x,t)),\end{aligned} \tag{6.117}$$

with $\beta = 1-\gamma$. In the case of axonal delays, $\tau_{ij} \to \tau(x,y)$ in the continuum limit. Assuming that an action potential propagates with constant speed v along the axon, then $\tau(x,y) = |x-y|/v$ so that the voltage-based equation (6.115) becomes

$$\frac{\partial}{\partial t}u(x,t) = -\frac{u(x,t)}{\tau} + \int_{-\infty}^{\infty} w(x,y)F(u(y,t-|x-y|/v))dy. \tag{6.118}$$

Two-dimensional versions of these various models are obtained by taking $x \to \mathbf{x} = (x_1,x_2)$ and $y \to \mathbf{y} = (y_1,y_2)$ with $d\mathbf{y} = dy_1dy_2$.

Now suppose that there are M classes of neuron distributed along the line labeled by the population index $a = 1,\dots M$. Equation (6.114) then generalizes to the multi-population integral equation

$$u_a(x,t) = \int_{-\infty}^{\infty}\int_{-\infty}^{t}\sum_{b=1}^{M}\Phi_{ab}(x,y,t-t')F_b(u_b(y,t'-|x-y|/v_{ab}))dt'dy. \tag{6.119}$$

We have included axonal delays with v_{ab} the conduction velocity along axonal projections from neurons of type b to neurons of type a. Assuming that $\Phi_{ab}(x,y,t) = w_{ab}(x,y)\Phi(t)$ with $\Phi(t) = \mathrm{e}^{-t/\tau}H(t)$, we obtain multi-population neural field equations:

$$\frac{\partial u_a}{\partial t} = -\frac{u_a(x,t)}{\tau} + \sum_{b=1}^{M}\int_{-\infty}^{\infty} w_{ab}(x,y)F_b(u_b(y,t-|x-y|/v_{ab}))dy, \tag{6.120}$$

and

$$\frac{\partial z_a}{\partial t} = -\frac{z_a(x,t)}{\tau} + F_b\left(\sum_{b=1}^{M}\int_{-\infty}^{\infty} w_{ab}(x,y)z_b(y,t-|x-y|/v_{ab})dy\right) \tag{6.121}$$

for $a = 1,\dots,M$. The latter is a version of the *Wilson–Cowan equations* for cortical dynamics [675, 676]. Note that all synapses innervated by a particular type of neuron have the same sign. That is, if type b neurons are excitatory (inhibitory), then $w_{ab}(x,y) \geq 0$ ($w_{ab}(x,y) \leq 0$) for all $a = 1,\dots,M$ and (x,y). Thus, one of the major reasons for considering more than one class of neuron is to incorporate both excitatory and inhibitory synapses. It can be argued that since excitatory and inhibitory synapses tend to have different time courses in response to action potentials, one should take $\Phi_{ab}(x,y,t) = w_{ab}(x,y)\Phi_b(t)$, suggesting that the activity-based model (6.121) with $\tau \to \tau_a$ is more biologically realistic than the voltage-based model, at least in the case of excitatory–inhibitory networks [173]. However, in practice, both versions of the neural field equations are used to model cortical dynamics. Since both versions exhibit very similar types of solution, and since most analytical results have been obtained for voltage-based neural fields, we will mainly focus on the latter.

Under certain additional simplifying assumptions, it is possible to incorporate inhibition into the scalar neural field equations (6.115) or (6.116) [505]. For example, consider a two-population model ($M = 2$) of excitatory ($a = E$) and inhibitory ($a = I$) neurons evolving according to the pair of continuum voltage-based equations

$$\frac{\partial u_E}{\partial t} = -\frac{u_E(x,t)}{\tau_E} + \int_{-\infty}^{\infty} w_{EE}(x,y) F_E(u_E(y,t)) dy + \int_{-\infty}^{\infty} w_{EI}(x,y) F_I(u_I(y,t)) dy \tag{6.122a}$$

$$\frac{\partial u_I}{\partial t} = -\frac{u_I(x,t)}{\tau_I} + \int_{-\infty}^{\infty} w_{IE}(x,y) F_E(u_E(y,t)) dy + \int_{-\infty}^{\infty} w_{II}(x,y) F_I(u_I(y,t)) dy, \tag{6.122b}$$

with $w_{EE}, w_{IE} \geq 0$ and $w_{EI}, w_{II} \leq 0$. Now suppose that $w_{II} \equiv 0$, $F_I(u_I) = u_I/\tau_I$ and $\tau_I \ll \tau_E$. It follows that we can eliminate u_I by setting

$$u_I(x) \sim \tau_I \int_{-\infty}^{\infty} w_{IE}(x,y) F_E(u_E(y,t)) dy,$$

which leads to a scalar equation for u_E of the form (6.115) with effective weight distribution

$$w(x,y) = w_{EE}(x,y) + \int_{-\infty}^{\infty} w_{EI}(x,y') w_{IE}(y',y) dy'. \tag{6.123}$$

It is then possible for $w(x,y)$ to change sign as a function of x, y. (Often w is modeled as a difference of Gaussians or exponentials—the so-called Mexican hat weight distribution.) The reduced model can be used to investigate the effects of inhibition on stationary solutions and propagating waves. However, in contrast to the full two-population model, it does not support oscillatory solutions (in the absence of axonal delays, higher-order synapses, or some form of adaptation such as synaptic depression).

It is important to emphasize that there does not currently exist a multi-scale analysis of conductance-based neural networks that provides a rigorous derivation of neural field equations, although some progress has been made in this direction [100, 147, 184, 307, 675, 676]. One crucial step in the derivation of neural field equations presented here was the assumption of slowly varying synaptic currents, which is related to the assumption that there is not significant coherent activity at the level of individual spikes. This allowed us to treat the output of a neuron (or population of neurons) as an instantaneous firing rate. A more rigorous derivation would need to incorporate the mean-field analysis of local populations of stochastic spiking neurons into a larger-scale cortical model and to carry out a systematic form of coarse graining or homogenization in order to generate a continuum neural field model. Nevertheless, the heuristic approach does provide a framework for relating parameters of neural field equations to biophysical parameters such as membrane/synaptic time constants and axonal delays and also prescribes how to incorporate additional physiological processes such as synaptic depression and spike frequency adaptation. Moreover, neural field models make it possible to explore the dependence of cortical dynamics on the detailed anatomy of local and long-range synaptic connections. It is often assumed that w depends on the Euclidean distance between interacting cells within the 2D cortical sheet so that

$w(\mathbf{x},\mathbf{y}) = w(|\mathbf{x}-\mathbf{y}|)$. However, this is an oversimplification of the detailed architecture of cortex [64–66, 308, 526]; see Sect. 8.1. A related simplification is to take axonal delays to depend on Euclidean distance according to $|\mathbf{x}-\mathbf{y}|/v$, where v is the speed of propagation.

It is also possible to construct continuum neural field equations for stochastic population models. For the sake of illustration, consider the Langevin equation (6.71), except we simplify the multiplicative noise by additive noise σdW_α for constant noise strength σ. The continuum limit of Eq. (6.71) proceeds as follows. First, set $A_\alpha(t) = A(\alpha\Delta d, t)$ and $w_{\alpha\beta} = \rho\Delta d w(\alpha\Delta d, \beta\Delta d)$ where ρ is a synaptic density and Δd is an infinitesimal length scale. We also assume that the noise strength $\sigma_\alpha = \sigma/\sqrt{\Delta d}$ and define $W_\alpha(t)/\sqrt{\Delta d} = W(\alpha\Delta d, t)$. Taking the limit $\Delta d \to 0$ with $\tau_\alpha = \hat{\tau}$ for all α gives

$$\hat{\tau} dA(x,t) = \left[-A(x,t) + F\left(\int w(x-y)A(y,t)dy\right)\right] dt + \sigma dW(x,t) \tag{6.124}$$

with

$$\langle dW(x,t)\rangle = 0, \quad \langle dW(x,t)dW(y,t)\rangle = \delta(x-y)dt. \tag{6.125}$$

Similarly, we can write down a stochastic version of a voltage-based neural field equation, namely,

$$\tau dU(x,t) = \left[-U(x,t) + \int w(x-y)F(U(y,t))dy\right]dt + \sigma dW(x,t) \tag{6.126}$$

From a numerical perspective, any computer simulation would involve rediscretizing space and then solving a time-discretized version of the resulting stochastic differential equation. On the other hand, in order to investigate analytically the effects of noise on spatiotemporal dynamics, it is more useful to work directly with stochastic neural fields. One can then adapt various PDE methods for studying noise in spatially extended systems [546], as illustrated in Sect. 7.4. Recently, Buice and Cowan [91] have used path integral methods and renormalization group theory to establish that a stochastic neural field model based on a continuum version of a birth–death master equation belongs to the universality class of directed percolation and consequently exhibits power law behavior, suggestive of many measurements of spontaneous cortical activity in vitro and in vivo [30, 508]; see Sect. 9.5. Although the existence of power law behavior is still controversial [28], the application of path integral methods provides yet another example of how analytical techniques familiar in the study of PDEs are being adapted to studies of continuum neural fields. (For reviews on path integral methods for stochastic differential equations see [115, 618, 698].)

Finally, note that several groups have constructed equivalent PDE models for neural fields with axonal propagation delays, which take the form of damped inhomogeneous wave equations [307, 382, 467, 527, 528, 603]. The basic idea is to assume a particular form for the synaptic weight distribution and to use Fourier transforms. Consider, for example, a 2D version of the multi-population integral equation (6.119). Suppose that $\Phi_{ab}(\mathbf{x},\mathbf{y},t) = w_{ab}(|\mathbf{x}-\mathbf{y}|)\Phi(t)$ and introduce the auxiliary field

$$\Psi_{ab}(\mathbf{x},t) = \int_{\mathbb{R}^2} w_{ab}(|\mathbf{x}-\mathbf{y}|)F_b(\mathbf{y},t-|\mathbf{x}-\mathbf{y}|/v_{ab})d\mathbf{y},$$

where we have set $F_b(\mathbf{y},t) = F_b(u_b(\mathbf{y},t))$. Fourier transforming this equation with

$$\hat{\Psi}_{ab}(k,\omega) = \int_{\mathbb{R}^2}\int_{-\infty}^{\infty} \mathrm{e}^{-i(\mathbf{k}\cdot\mathbf{r}+\omega t)}\Psi_{ab}(\mathbf{x},t)dt\,d\mathbf{x},$$

and

$$w_{ab}(r) = \frac{w_{ab}^0}{2\pi}\mathrm{e}^{-r/\sigma_{ab}},$$

we find that

$$\hat{\Psi}_{ab}(k,\omega) = w_{ab}^0\frac{\mathscr{A}_{ab}(\omega)}{(\mathscr{A}_{ab}(\omega)^2+k^2)^{3/2}}\hat{F}_b(k,\omega)$$

with $\mathscr{A}_{ab}(\omega) = 1/\sigma_{ab} + i\omega/v_{ab}$. If one now makes a long-wavelength approximation by Taylor expanding the denominator of the above equation about $k=0$ and rearranging to give $(\mathscr{A}_{ab}(\omega)^2 + 3k^2/2)\hat{\Psi}_{ab}(k,\omega) = \hat{F}_b(k,\omega)$, one can then apply the inverse Fourier transform to derive the damped wave equation

$$\left[\left(\frac{1}{\sigma_{ab}} + \frac{1}{v_{ab}}\partial_t\right)^2 - \frac{3}{2}\nabla^2\right]\Psi_{ab}(\mathbf{x},t) = w_{ab}^0 F_b(u_b(\mathbf{x},t)).$$

The current $u_a(\mathbf{x},t)$ is then related to the field $\Psi_{ab}(\mathbf{x},t)$ according to

$$u_a(\mathbf{x},t) = \int_{-\infty}^{t}\Phi(t-t')\sum_b \Psi_{ab}(\mathbf{x},t')dt',$$

which can itself be converted to a PDE by applying the inverse operator $\mathbb{L}_t$. There have also been various extensions of the PDE theory including improvements upon the long-wavelength approximation [142] and incorporation of network inhomogeneities [526]. The damped wave equation (6.127) and its generalizations have been used extensively to study large-scale phenomena such as EEG rhythms, where axonal propagation delays are important [60, 467]; see also Sect. 9.4. PDE models have also been used to study single and multi-bump stationary solutions of scalar neural field equations, in which the time-independent equations reduce to fourth-order differential equations with a Hamiltonian structure [355, 356].

Chapter 7
Waves in Excitable Neural Fields

In Sect. 5, we showed how both solitary pulses and oscillatory phase waves could occur in a synaptically coupled network of spiking neurons, where the fundamental element of the network was a single neuron. Hence, whether the network acted as an excitable or an oscillatory medium depended primarily on the intrinsic properties of the individual neurons. In this chapter, we focus on waves in excitable neural fields, where the fundamental network element is a local population of cells (see Sect. 6), and show how many of the PDE methods and results for the analysis of waves in reaction–diffusion equations (see part I) can be extended to the nonlocal integrodifferential equations of neural field theory. We begin by analyzing the existence and stability of solitary traveling fronts in a 1D scalar neural field. (Since there is strong vertical coupling between layers of a cortical column, it is possible to treat a thin vertical cortical slice as an effective 1D medium.) In order to relate the models to experiments on disinhibited cortical slices (Sect. 5.1), we assume that the weight distribution is purely excitatory. This is also motivated by the observation that epileptic seizures are often associated with greatly enhanced levels of recurrent excitation [430] (Sect. 9.4). We then extend the analysis to the case of traveling pulses, which requires the inclusion of some form of local negative feedback mechanism such as synaptic depression or spike frequency adaptation. Next we describe two approaches to analyzing wave propagation failure in inhomogeneous neural media: one based on homogenization theory [79, 332] and the other on interfacial dynamics [132]. This is followed by a discussion of wave propagation in stochastic neural fields.

It is useful to emphasize here that there are two main approaches to analyzing the spatiotemporal dynamics of neural field equations. The first method is based on the original work of Amari [8], in which one establishes the existence of nonlinear traveling wave solutions by explicit construction. This is possible if one takes the firing rate function F to be the Heaviside (1.15). It is also possible to study the linear stability of such solutions by constructing an associated Evans function, whose zeros determine the spectrum of the resulting linear operator [134, 552, 696]. The constructive approach of Amari [8] has been particularly useful in providing explicit insights into how spatiotemporal network dynamics depends on the structure

P.C. Bressloff, *Waves in Neural Media: From Single Neurons to Neural Fields*, Lecture Notes on Mathematical Modelling in the Life Sciences,
DOI 10.1007/978-1-4614-8866-8_7,

of the synaptic weight kernel as well as various physiological parameters. Moreover, in certain cases, it is possible to use singular perturbation methods [504, 505] or fixed-point theorems [172, 336] to extend results for neural fields with Heaviside nonlinearities to those with more realistic sigmoidal nonlinearities; see also [136]. The second method is based on bifurcation theory, following the original work of Ermentrout and Cowan [169], in which one investigates the emergence of spatially periodic stationary and oscillatory patterns through a combination of linear stability analysis, weakly nonlinear analysis, symmetric bifurcation theory, and numerical simulations, as reviewed in [67, 71, 167]. Rigorous functional analytical techniques combined with numerical bifurcation schemes have also been used to study the existence and (absolute) stability of stationary solutions for a general class of neural field models with smooth firing rate functions [185, 642]. As far as we are aware, these methods have not yet been applied to traveling wave solutions of neural field equations.

7.1 Traveling Fronts in a Scalar Neural Field

7.1.1 Propagating Fronts in a Bistable Neural Field

We begin by using Amari's constructive method [8] to analyze the existence of traveling front solutions in a scalar neural field equation. Similar analyses are found in Refs. [76, 134, 504]. We assume a Heaviside rate function (1.15) and an excitatory weight distribution of the form $w(x,y) = w(x-y)$ with $w(x) \geq 0$ and $w(-x) = w(x)$. We also assume that $w(x)$ is a monotonically decreasing function of x for $x \geq 0$. A common choice is the exponential weight distribution

$$w(x) = \frac{1}{2\sigma}\mathrm{e}^{-|x|/\sigma}, \tag{7.1}$$

where σ determines the range of synaptic connections. The latter tends to range from $100\,\mu$m to 1 mm. The resulting neural field equation is

$$\frac{\partial u(x,t)}{\partial t} = -u(x,t) + \int_{-\infty}^{\infty} w(x-x')F(u(x',t))dx', \tag{7.2}$$

with $F(u) = H(u-\kappa)$. We have fixed the units of time by setting $\tau = 1$. If τ is interpreted as a membrane time constant, then $\tau \sim 10$ msec. In order to construct a traveling front solution of (7.2), we introduce the traveling wave coordinate $\xi = x - ct$, where c denotes the wave speed, and set $u(x,t) = U(\xi)$ with $\lim_{\xi\to-\infty} U(\xi) = U_+ > 0$ and $\lim_{\xi\to\infty} U(\xi) = 0$ such that $U(\xi)$ only crosses the threshold κ once. Here $U_+ = \int_{-\infty}^{\infty} w(y)dy$ is a spatially uniform fixed-point solution of (7.2). Since Eq. (7.2) is equivariant with respect to uniform translations, we are free to take the threshold crossing point to be at the origin, $U(0) = \kappa$, so that $U(\xi) < \kappa$ for $\xi > 0$ and $U(\xi) > \kappa$ for $\xi < 0$. Substituting this traveling front solution into Eq. (7.2) then gives

$$-cU'(\xi)+U(\xi)=\int_{-\infty}^{0}w(\xi-\xi')d\xi'=\int_{\xi}^{\infty}w(x)dx\equiv W(\xi), \tag{7.3}$$

where $U'(\xi)=dU/d\xi$. Multiplying both sides of the above equation by $e^{-\xi/c}$ and integrating with respect to ξ leads to the solution

$$U(\xi)=e^{\xi/c}\left[\kappa-\frac{1}{c}\int_{0}^{\xi}e^{-y/c}W(y)dy\right]. \tag{7.4}$$

Finally, requiring the solution to remain bounded as $\xi\to\infty$ ($\xi\to-\infty$) for $c>0$ (for $c<0$) implies that κ must satisfy the condition

$$\kappa=\frac{1}{|c|}\int_{0}^{\infty}e^{-y/|c|}W(\text{sign}(c)y)dy. \tag{7.5}$$

Thus, one of the useful aspects of the constructive method is that it allows us to derive an explicit expression for the wave speed as a function of physiological parameters such as firing threshold and range of synaptic connections. In the case of the exponential weight distribution (7.1), the relationship between wave speed c and threshold κ is

$$c=\frac{\sigma}{2\kappa}[1-2\kappa]\ (\text{for } \kappa<0.5),\quad c=\frac{\sigma}{2}\frac{1-2\kappa}{1-\kappa}\ (\text{for } 0.5<\kappa<1). \tag{7.6}$$

This establishes the existence of a unique front solution for fixed κ, which travels to the right ($c>0$) when $\kappa<0.5$ and travels to the left ($c<0$) when $\kappa>0.5$. As we will show below, the traveling front is stable.

Given the existence of a traveling front solution for a Heaviside rate function, it is possible to extend the analysis to a smooth sigmoid nonlinearity using a continuation method [172]. We briefly summarize the main result. Consider the scalar neural field equation (7.2) with F given by the sigmoid function (1.14) and $w(x)$ nonnegative and symmetric with normalization $\int_{-\infty}^{\infty}w(x)dx=1$. Suppose that the function $\tilde{F}(u)=-u+F(u)$ has precisely three zeros at $u=U_{\pm},U_0$ with $U_-<U_0<U_+$ and $\tilde{F}'(U_{\pm})<0$. It can then be shown that (modulo uniform translations) there exists a unique traveling front solution $u(x,t)=U(\xi)$, $\xi=x-ct$, with

$$-cU'(\xi)+U(\xi)=\int_{-\infty}^{\infty}w(\xi-\xi')F(U(\xi'))d\xi', \tag{7.7}$$

and $U(\xi)\to U_{\pm}$ as $\xi\to\mp\infty$ [172]. Moreover, the speed of the wave satisfies

$$c=\frac{\Gamma}{\int_{-\infty}^{\infty}U'(\xi)^2F'(U(\xi))d\xi}, \tag{7.8}$$

where $F'(U)=dF/dU$ and

$$\Gamma=\int_{U_-}^{U_+}\tilde{F}(U)dU. \tag{7.9}$$

Since the denominator of Eq. (7.8) is positive definite, the sign of c is determined by the sign of the coefficient Γ. In particular, if the threshold $\kappa = 0.5$ and the gain of the sigmoid $\eta > 4$ (see Eq. (1.14), then there exists a pair of stable homogeneous fixed points with $U_- = -U_+$, which in turn implies that $\Gamma = 0$ and the front solution is stationary. Note that this analysis has been extended to a more general form of nonlocal equations by Chen [108].

7.1.2 Wave Stability and Evans Functions

Suppose that the scalar neural field equation (7.2) has a traveling wave solution $u(x,t) = U(\xi), \xi = x - ct$ with $c > 0$. Following Coombes and Owen [134], it is convenient to rewrite the neural field equation in the integral form

$$u(x,t) = \int_{-\infty}^{\infty}\int_{0}^{\infty} w(y)\Phi(s)F(u(x-y,t-s))dsdy, \tag{7.10}$$

with $\Phi(t) = \mathrm{e}^{-t}H(t)$. For this representation, the front solution satisfies

$$U(\xi) = \int_{-\infty}^{\infty}\int_{0}^{\infty} w(y)\Phi(s)F(U(\xi - y + cs)dsdy. \tag{7.11}$$

In order to determine the stability of the front solutions, we transform to traveling wave coordinates by setting $u(x,t) = U(\xi,t) = U(\xi) + \varphi(\xi,t)$, and Taylor expand to first order in φ. This leads to the linear integral equation

$$\varphi(\xi,t) = \int_{-\infty}^{\infty}\int_{0}^{\infty} w(y)\Phi(s)F'(U(\xi - y + cs))\varphi(\xi - y + cs, t - s)dsdy. \tag{7.12}$$

We now seek solutions of Eq. (7.12) of the form $\varphi(\xi,t) = \varphi(\xi)\mathrm{e}^{\lambda t}$, $\lambda \in \mathbb{C}$, which leads to the eigenvalue equation $\varphi = \mathbb{L}(\lambda)\varphi$. That is,

$$\varphi(\xi) = \int_{-\infty}^{\infty}\int_{\xi-y}^{\infty} w(y)\Phi((s+y-\xi)/c)\mathrm{e}^{-\lambda(s+y-\xi)/c}F'(U(s))\varphi(s)\frac{ds}{c}dy, \tag{7.13}$$

where we have performed the change of variables $cs + \xi - y \to s$. The linear stability of the traveling front can then be determined in terms of the spectrum $\sigma(\mathbb{L}(\lambda))$.

Following appendix section 2.7, we assume that the eigenfunctions $\varphi \in L^2(\mathbb{R})$ and introduce the resolvent operator $\mathscr{R}(\lambda) = [\mathbb{L}(\lambda) - I]^{-1}$, where I denotes the identity operator. We can then decompose the spectrum $\sigma(\mathscr{L})$ into the disjoint sum of the discrete spectrum and the essential spectrum. Given the spectrum of the linear operator defined by Eq. (7.13), the traveling wave is said to be linearly stable if [551]

$$\max\{\mathrm{Re}(\lambda) : \lambda \in \sigma(\mathscr{L}), \lambda \neq 0\} \leq -K \tag{7.14}$$

for some $K > 0$, and $\lambda = 0$ is a simple eigenvalue of λ. The existence of at least one zero eigenvalue is a consequence of translation invariance. Indeed, differentiating equation (7.11) with respect to ξ shows that $\varphi(\xi) = U'(\xi)$ is an eigenfunction solution of Eq. (7.13) with $\lambda = 0$. As in the case of PDEs (see Sect. 2.4), the discrete spectrum may be associated with the zeros of an Evans function. A number of authors have applied the Evans function construction to neural field equations [134, 198, 506, 536, 552, 696], as well as more general nonlocal problems [314]. Moreover, for neural fields with Heaviside firing rate functions, the Evans function can be calculated explicitly. This was first carried out by Zhang [696], who applied the method of variation of parameters to the linearized version of the integrodifferential Eq. (7.2), and was subsequently extended using a more direct integral formulation by Coombes and Owen [134].

Construction of Evans function. Setting $F(U) = H(U - \kappa)$ in Eq. (7.12) and using the identity

$$H'(U(\xi) - \kappa) = \delta(U(\xi) - \kappa) = \frac{\delta(\xi)}{|U'(0)|} \tag{7.15}$$

gives

$$\varphi(\xi) = \frac{\varphi(0)}{c|U'(0)|} \int_{-\infty}^{\infty} w(y)\Phi((y-\xi)/c)e^{-\lambda(y-\xi)/c} dy. \tag{7.16}$$

In order to obtain a self-consistent solution at $\xi = 0$, we require that

$$\varphi(0) = \frac{\varphi(0)}{c|U'(0)|} \int_0^{\infty} w(y)\Phi(y/c)e^{-\lambda y/c} dy, \tag{7.17}$$

We have used the fact that $\Phi(y) = 0$ for $y < 0$, which is a consequence of causality. Hence, a nontrivial solution exists provided that $\mathscr{E}(\lambda) = 0$, where

$$\mathscr{E}(\lambda) = 1 - \frac{1}{c|U'(0)|} \int_0^{\infty} w(y)\Phi(y/c)e^{-\lambda y/c} dy. \tag{7.18}$$

Equation (7.18) can be identified with the Evans function for the traveling front solution of the scalar neural field equation (7.10). It is real valued if λ is real. Furthermore, (i) the complex number λ is an eigenvalue of the operator $\mathscr{L}$ if and only if $\mathscr{E}(\lambda) = 0$, and (ii) the algebraic multiplicity of an eigenvalue is equal to the order of the zero of the Evans function [134, 552, 696]. We briefly indicate the proof of (i) for $\Phi(t) = e^{-t}H(t)$. Equation (7.16) becomes

$$\begin{aligned}
\varphi(\xi) &= \frac{\varphi(0)}{c|U'(0)|} e^{(\lambda+1)\xi/c} \int_{\xi}^{\infty} w(y)e^{-(\lambda+1)y/c} dy, \\
&= \varphi(0)\left[1 - \frac{1}{c|U'(0)|} \int_0^{\xi} w(y)e^{-(\lambda+1)y/c} dy\right] e^{(\lambda+1)\xi/c},
\end{aligned}$$

which in the limit $\xi \to \infty$ gives

$$\lim_{\xi\to\infty} \varphi(\xi) = \varphi(0)\mathscr{E}(\lambda) \lim_{\xi\to\infty} e^{(\lambda+1)\xi/c}.$$

Assuming that $\mathrm{Re}\lambda > -1$ (which turns out to be to the right of the essential spectrum), then $\varphi(\xi)$ will be unbounded as $\xi \to \infty$ unless $\mathscr{E}(\lambda) = 0$. That is, if $\mathscr{E}(\lambda) = 0$, then $\varphi(\xi)$ is normalizable, the resolvent operator is not invertible and λ is an eigenvalue.

It is also straightforward to show that $\mathcal{E}(0)=0$, which we expect from translation invariance. First, setting $F(U)=H(U-\kappa)$ in Eq. (7.11) and differentiating with respect to ξ show that

$$U'(\xi)=-\frac{1}{c}\int_{-\infty}^{\infty}w(y)\Phi((y-\xi)/c)dy. \tag{7.19}$$

Thus, defining

$$\mathcal{H}(\lambda)=\int_{0}^{\infty}w(y)\Phi(y/c)\mathrm{e}^{-\lambda y/c}dy, \tag{7.20}$$

we see that $c|U'(0)|=\mathcal{H}(0)$ and, hence,

$$\mathcal{E}(\lambda)=1-\frac{\mathcal{H}(\lambda)}{\mathcal{H}(0)}. \tag{7.21}$$

It immediately follows that $\mathcal{E}(0)=0$.

In order to determine the essential spectrum, consider the inhomogeneous equation

$$\varphi(\xi)-\frac{\varphi(0)}{c|U'(0)|}\int_{-\infty}^{\infty}w(y)\Phi((y-\xi)/c)\mathrm{e}^{-\lambda(y-\xi)/c}dy=h(\xi) \tag{7.22}$$

for some normalizable smooth function h on $\mathbb{R}$. Assuming that λ does not belong to the discrete spectrum, $\mathcal{E}(\lambda)\neq 0$, we can expresses the constant $\varphi(0)$ in terms of $h(0)$ by setting $\xi=0$ in Eq. (7.22): $\varphi(0)=h(0)/\mathcal{E}(\lambda)$. Thus,

$$\varphi(\xi)=h(\xi)+\frac{1}{\mathcal{E}(\lambda)}\frac{h(0)}{c|U'(0)|}\int_{-\infty}^{\infty}w(y)\Phi((y-\xi)/c)\mathrm{e}^{-\lambda(y-\xi)/c}dy. \tag{7.23}$$

Fourier transforming this equation using the convolution theorem gives

$$\hat{\varphi}(k)=\hat{h}(k)+\frac{1}{\mathcal{E}(\lambda)}\frac{h(0)}{c|U'(0)|}\hat{\omega}(k)\hat{\Phi}(kc+i\lambda), \tag{7.24}$$

where

$$\hat{\varphi}(k)=\int_{-\infty}^{\infty}\varphi(y)\mathrm{e}^{iky}dy \tag{7.25}$$

etc. Now suppose that for a given value of k, there exists $\lambda=\lambda(k)$ for which $[\hat{\Phi}(kc+i\lambda(k))]^{-1}=0$. It follows that the right-hand side of Eq. (7.24) blows up if $\lambda=\lambda(k)$, that is, the dispersion curve belongs to the essential spectrum.

For the sake of illustration, let us calculate the zeros of the Evans function in the case of the exponential weight function (7.1). Substituting $\Phi(t)=\mathrm{e}^{-t}$ and $w(y)=\mathrm{e}^{-|y|/\sigma}/2\sigma$ in Eq. (7.20) gives

$$\mathcal{H}(\lambda)=\frac{1}{2\sigma}\frac{1}{\sigma^{-1}+\lambda/c+1/c}$$

so that [134]

$$\mathcal{E}(\lambda)=\frac{\lambda}{c/\sigma+1+\lambda}. \tag{7.26}$$

It follows that $\lambda=0$ is the only zero of the Evans function and it is a simple root (since $\mathcal{E}'(0)>0$). Furthermore, in the particular case $\Phi(t)=\mathrm{e}^{-t}$, we have $[\tilde{\Phi}(kc+i\lambda)]^{-1}=1-ikc+\lambda$ so that the essential spectrum is $\lambda(k)=-1+ikc$, that is, a vertical line in the complex plane at $\mathrm{Re}\lambda=-1$. It follows that the corresponding traveling front (it it exists) is stable. This example illustrates one of the powerful features of the constructive method based on Heavisides. Not only is it possible to construct exact traveling wave solutions and derive formulae for the speed of the wave, but one can also explicitly construct the Evans function that determines wave stability. The method extends to multi-population neural field models, neural fields with axonal propagation delays, and adaptive neural fields [134]. (Although taking the high-gain limit of a smooth firing rate function is not very realistic from a biological perspective, one finds that many of the basic features of traveling waves persist for finite gain.) In the particular case of axonal delays, it can be shown that delays reduce the speed of a wave but do not affect its stability properties. For example, given a right-moving traveling front solution of the scalar neural field equation (6.118) with $\tau = 1$ and exponential weights, one finds that the speed of the wave is [134, 139]

$$c = \sigma \frac{1-2\kappa}{2\kappa + \sigma(1-2\kappa)/v},$$

where v is the propagation speed along an axon, and the Evans function is

$$\mathscr{E}(\lambda) = \frac{\lambda}{c/\sigma + (1-c/v) + \lambda}.$$

7.1.3 Pulled Fronts

So far we have assumed that the scalar neural field operates in a bistable regime analogous to the FitzHugh–Nagumo equations; see Sect. 2. However, as we explored within the context of CaMKII translocation waves (Sect. 3.2), Fisher-like reaction–diffusion equations support traveling waves propagating into unstable states, resulting in pulled fronts (Sect. 3.3). It turns out that it is also possible to observe pulled fronts in an activity-based version of a scalar neural field equation [70, 132]:

$$\tau \frac{\partial a(x,t)}{\partial t} = -a(x,t) + F\left(\int_{-\infty}^{\infty} w(x-x')a(x',t)dx'\right). \tag{7.27}$$

with $a(x,t) \geq 0$ for all (x,t). Note that the restriction to positive values of a is a feature shared with population models in ecology or evolutionary biology, for example, where the corresponding dependent variables represent number densities. Indeed, Eq. (7.27) has certain similarities with a nonlocal version of the Fisher–KPP equation, which takes the form [238]

$$\tau \frac{\partial p(x,t)}{\partial t} = D\frac{\partial^2 p(x,t)}{\partial x^2} + \mu p(x,t)\left(1 - \int_{-\infty}^{\infty} K(x-x')p(x',t)dx'\right). \tag{7.28}$$

One major difference from a mathematical perspective is that Eq. (7.28) supports traveling fronts even when the range of the interaction kernel K goes to zero, that is, $K(x) \to \delta(x)$, since we recover the standard local Fisher–KPP equation (3.44) [191, 345]. In particular, as the nonlocal interactions appear nonlinearly in Eq. (7.28), they do not contribute to the linear spreading velocity in the leading edge of the front. On the other hand, nonlocal interactions play a necessary role in the generation of fronts in the neural field equation (7.27).

Suppose that $F(a)$ in Eq. (7.27) is a positive, bounded, monotonically increasing function of a with $F(0) = 0$, $\lim_{a\to 0^+} F'(a) = 1$ and $\lim_{a\to\infty} F(a) = \kappa$ for some positive constant κ. For concreteness, we take

$$F(a) = \begin{cases} 0, & a \le 0 \\ a, & 0 < a \le \kappa \\ \kappa, & a > \kappa. \end{cases} \tag{7.29}$$

A homogeneous fixed-point solution a^* of Eq. (7.27) satisfies

$$a^* = F(W_0 a^*), \quad W_0 = \int_{-\infty}^{\infty} w(y)dy. \tag{7.30}$$

In the case of the given piecewise linear firing rate function, we find that if $W_0 > 1$, then there exists an unstable fixed point at $a^* = 0$ (absorbing state) and a stable fixed point at $a^* = \kappa$; see Fig. 7.1(a). The construction of a front solution linking the stable and unstable fixed points differs considerably from that considered in neural fields with sigmoidal or Heaviside nonlinearities [8, 167], where the front propagates into a metastable state; see Fig 7.1(b). Following the PDE theory of fronts propagating into unstable states [544] (see Sect. 3.3), we expect there to be a continuum of front velocities and associated traveling wave solutions.

Recall that a conceptual framework for studying such solutions is the linear spreading velocity c^*, which is the asymptotic rate with which an initial localized perturbation spreads into an unstable state based on the linear equations obtained by linearizing the full nonlinear equations about the unstable state. Therefore, linearizing equation (7.27) about $a = 0$ gives

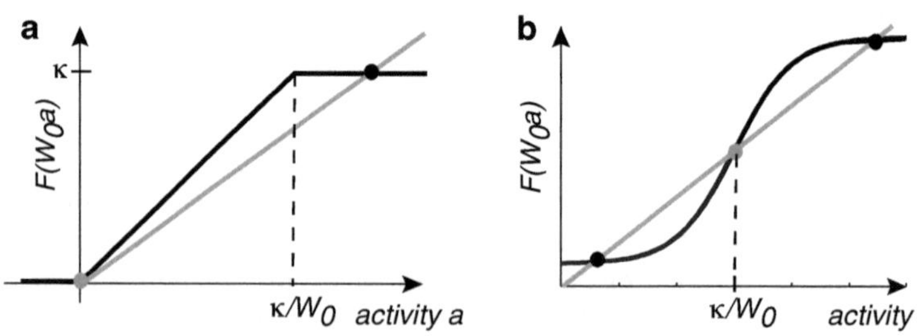

Fig. 7.1 Plots of firing rate function. Intercepts of $y = F(W_0 a)$ with the straight line $y = a$ determine homogeneous fixed points. (**a**) Piecewise linear rate function (7.29) showing the existence of an unstable fixed point at $a = 0$ and a stable fixed point at $a = \kappa$. (**b**) Sigmoidal rate function $F(a) = 2/(1 + e^{-2[a-\kappa]})$ showing the existence of two stable fixed points separated by an unstable fixed point

$$\frac{\partial a(x,t)}{\partial t} = -a(x,t) + \int_{-\infty}^{\infty} w(x-x')a(x',t)dx'. \tag{7.31}$$

Note that if $a(x,0) \geq 0$ for all x, then Eq. (7.31) ensures that $a(x,t) \geq 0$ for all x and $t > 0$. One way to see this is to note from Eq. (7.31) that $a(x,t+\Delta t) = (1-\Delta t)a(x,t) + \Delta t \int_{-\infty}^{\infty} w(x-x')a(x',t)dx'$. Assuming positivity of the solution at time t and using the fact that the neural field is purely excitatory ($w(x) \geq 0$ for all x), it follows that $a(x,t+\Delta t)$ is also positive. An arbitrary initial condition $a(x,0)$ will evolve under Eq. (7.31) as

$$a(x,t) = \int_{-\infty}^{\infty} G(x-y,t)a(y,0)dy, \tag{7.32}$$

where $G(x,t)$ is Green's function

$$G(x,t) = \int_{-\infty}^{\infty} \mathrm{e}^{ikx-i\omega(k)t} \frac{dk}{2\pi}, \quad \omega(k) = i[\tilde{w}(k) - 1], \tag{7.33}$$

and $\tilde{w}(k)$ is the Fourier transform of the weight distribution $w(x)$. Hence, the solution can be written in the form of Eq. (3.102):

$$a(x,t) = \int_{-\infty}^{\infty} \tilde{a}_0(k)\mathrm{e}^{i[kx-\omega(k)t])} \frac{dk}{2\pi}. \tag{7.34}$$

with $\tilde{a}_0$, the Fourier transform of the initial condition $a(x,0)$.

Given a sufficiently steep initial condition, for which the Fourier transform $\tilde{a}(k)$ is analytic, the asymptotic behavior of $a(x,t)$ can be obtained from the large-time asymptotics of $G(x,t)$ based on steepest descents. It immediately follows from the analysis of the Fisher equation in Sect. 3.3 that the linear spreading velocity c^* is given by $c^* = c(\lambda^*)$ where

$$c(\lambda) = \frac{\mathrm{Im}(\omega(i\lambda))}{\lambda}, \quad \left.\frac{dc(\lambda)}{d\lambda}\right|_{\lambda=\lambda^*} = 0. \tag{7.35}$$

Using the fact that the Fourier transform of the weight distribution is real valued, we find that

$$c(\lambda) = \frac{1}{\lambda}\left[\mathscr{W}(\lambda) - 1\right], \tag{7.36}$$

where $\mathscr{W}(\lambda) = \hat{W}(\lambda) + \hat{W}(-\lambda)$ and $\hat{W}(\lambda)$ is the Laplace transform of $w(x)$:

$$\hat{W}(\lambda) = \int_0^{\infty} w(y)\mathrm{e}^{-\lambda y}dy. \tag{7.37}$$

We are assuming that $w(y)$ decays sufficiently fast as $|y| \to \infty$ so that the Laplace transform $\hat{W}(\lambda)$ exists for bounded, negative values of λ. This holds in the case of a Gaussian weight distribution

$$w(x) = \frac{W_0}{\sqrt{2\pi\sigma^2}} e^{-x^2/2\sigma^2}, \tag{7.38}$$

since

$$\mathscr{W}(\lambda) = \int_{-\infty}^{\infty} w(y) e^{-\lambda y} dy = \frac{W_0}{\sqrt{2\pi\sigma^2}} \int_{-\infty}^{\infty} e^{-y^2/2\sigma^2} e^{-\lambda y} dy = W_0 e^{\lambda^2\sigma^2/2}.$$

Hence,

$$c(\lambda) = \frac{W_0 e^{\lambda^2\sigma^2/2} - 1}{\lambda}. \tag{7.39}$$

If $W_0 > 1$ (necessary for the zero activity state to be unstable), then $c(\lambda)$ is a positive unimodal function with $c(\lambda) \to \infty$ as $\lambda \to 0$ or $\lambda \to \infty$ and a unique minimum at $\lambda = \lambda_0$ with λ_0 the solution to the implicit equation

$$\lambda_0{}^2 = \frac{W_0 - e^{-\lambda_0{}^2\sigma^2/2}}{\sigma^2 W_0}. \tag{7.40}$$

Example dispersion curves are shown in Fig. 7.2(a) for various values of the Gaussian weight amplitude W_0. Combining Eqs. (7.39) and (7.40) shows that

$$\frac{c_0}{\lambda_0} = \sigma^2 W_0 e^{\lambda_0{}^2\sigma^2/2} = \sigma^2(\lambda_0 c_0 + 1), \tag{7.41}$$

so that

$$\lambda_0 = \frac{1}{2}\left[-\frac{1}{c_0} + \sqrt{\frac{1}{c_0{}^2} + \frac{4}{\sigma^2}}\right]. \tag{7.42}$$

Assuming that the full nonlinear system supports a pulled front (see Sect. 3.3), then a sufficiently localized initial perturbation (one that decays faster than $e^{-\lambda_0 x}$) will asymptotically approach the traveling front solution with the minimum wave speed $c_0 = c(\lambda_0)$. Note that $c_0 \sim \sigma$ and $\lambda_0 \sim \sigma^{-1}$. In Fig. 7.2(b), we show an asymptotic front profile obtained by numerically solving the neural field equation (7.27) when $W_0 = 1.2$. The corresponding displacement of the front is a linear function of time with a slope consistent with the minimal wave speed $c_0 \approx 0.7$ of the corresponding dispersion curve shown in Fig. 7.2(a). This wave speed is independent of κ.

The asymptotic analysis of the linear equation (7.31) also shows that, given a sufficiently localized initial condition, $|a(x,t)| \sim e^{-\lambda^*\xi}\psi(\xi,t)$ as $t \to \infty$, where $\xi = x - c^*t$ and the leading-edge variable $\psi(\xi,t)$ is given by

$$\psi(\xi,t) \approx \frac{e^{-\xi^2/(4\mathscr{D}t)}}{\sqrt{4\pi\mathscr{D}t}} \tag{7.43}$$

with

$$\mathscr{D} = -\frac{\omega_i''(i\lambda^*)}{2} = \frac{\lambda^*}{2}\left.\frac{d^2c(\lambda)}{d\lambda^2}\right|_{\lambda^*}. \tag{7.44}$$

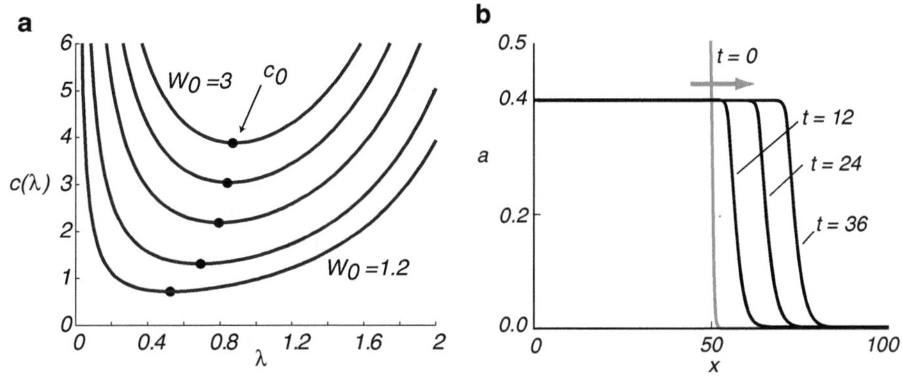

Fig. 7.2 (**a**) Velocity dispersion curves $c = c(\lambda)$ for a pulled front solution of the neural field equation (7.27) with piecewise linear firing rate function (7.29) and a Gaussian weight distribution with amplitude W_0 and width σ. Here $\sigma = 1.0$, $\kappa = 0.4$ and $W_0 = 1, 2, 1.5, 2.0, 2.5, 3.0$. *Black dots* indicate minimum wave speed c_0 for each value of W_0. (**b**) Snapshots of the front profile evolving from an initial condition consisting of a steep sigmoid function of unit amplitude (*gray curve*). Here $W_0 = 1.2$

Positivity of $\mathscr{D}$ follows from the fact that λ^* is a minimum of $c(\lambda)$. However, as shown by Ebert and van Saarloos [162], although the spreading of the leading edge under linearization gives the right qualitative behavior, it fails to match correctly the traveling front solution of the full nonlinear system. In particular, the asymptotic front profile takes the form $\mathscr{A}(\xi) \sim \xi e^{-\lambda^* \xi}$ for $\xi \gg 1$. The factor of ξ reflects the fact that at the saddle point the two branches of the velocity dispersion curve $c(\lambda)$ meet, indicating a degeneracy. In order to match the $\xi e^{-\lambda^* \xi}$ asymptotics of the front solution with the leading-edge solution, it is necessary to take the leading-edge function $\psi(x,t)$ to be the so-called dipole solution of the diffusion equation [162]:

$$\psi(x,t) = -\partial_\xi \frac{e^{-\xi^2/(4\mathscr{D}t)}}{\sqrt{4\pi\mathscr{D}t}} = \xi \frac{e^{-\xi^2/(4\mathscr{D}t)}}{\sqrt{2\pi}(2\mathscr{D}t)^{3/2}}. \tag{7.45}$$

Putting all of this together, if the neural field equation supports a pulled front, then the leading edge should relax asymptotically as

$$|a| \sim \xi e^{-\lambda^* \xi} e^{-\xi^2/(4\mathscr{D}t)} t^{-3/2} \tag{7.46}$$

with $\xi = x - c^* t$. Finally, writing

$$e^{-\lambda^* \xi} t^{-3/2} = e^{-\lambda^* [x - v^* t - X(t)]}, \quad X(t) = -\frac{3}{2\lambda^*} \ln t \tag{7.47}$$

suggests that to leading order, the velocity relaxes to the pulled velocity c^* according to (see also [162])

$$v(t) = c^* + \dot{X}(t) = c^* - \frac{3}{2\lambda^* t} + h.o.t. \tag{7.48}$$

7.2 Traveling Pulses in Adaptive Neural Fields

Traveling fronts are not particularly realistic, since populations of cells do not stay in the excited state forever. Hence, rather than a traveling front, propagating activity in cortex is usually better described as a traveling pulse. (One example where fronts rather than pulses occur is wave propagation during binocular rivalry [83, 312, 369, 678]; see Sect. 8.) One way to generate a traveling pulse is to include some form of synaptic inhibition, provided that it is not too strong [8]. However, even in the absence of synaptic inhibition, most neurons possess intrinsic negative feedback mechanisms that slowly bring the cell back to resting voltages after periods of high activity. Possible nonlinear mechanisms include synaptic depression or spike frequency adaptation as discussed in Sect. 6.1. However, most analytical studies of traveling pulses in neural field models have been based on a simpler linear form of adaptation introduced by Pinto and Ermentrout [504]. (For an analysis of waves in neural fields with nonlinear adaptation, see, e.g., [135, 329].) The linear adaptation model is given by

$$\frac{\partial u(x,t)}{\partial t} = -u(x,t) + \int_{-\infty}^{\infty} w(x-x')F(u(x',t))dx' - \beta q(x,t) \tag{7.49a}$$

$$\frac{1}{\varepsilon}\frac{\partial q(x,t)}{\partial t} = -q(x,t) + u(x,t), \tag{7.49b}$$

where ε and β determine the rate and amplitude of linear adaptation. We first show how to construct a traveling pulse solution of Eq. (7.49) in the case of a Heaviside rate function $F(u) = H(u-\kappa)$, following the particular formulation of [198, 696]. We then indicate how singular perturbation methods can be used to construct a traveling pulse for smooth F, as carried out by Pinto and Ermentrout [504]. The introduction of adaptation means that the neural field can support fronts or pulses, depending on whether there exist one or two stable homogeneous fixed points; see Fig. 7.3. We will focus on the latter here. Note, however, that linear (or nonlinear) adaptation can have a nontrivial effect on the propagation of traveling fronts [76, 80]. This is due to the occurrence of a symmetry breaking front bifurcation analogous to that found in reaction–diffusion systems [251, 252, 524, 561]. That is, a stationary front can undergo a supercritical pitchfork bifurcation at a critical rate of adaptation, leading to bidirectional front propagation. As in the case of reaction–diffusion systems, the front bifurcation acts as an organizing center for a variety of nontrivial dynamics including the formation of oscillatory fronts or breathers. The latter can occur, for example, through a Hopf bifurcation from a stationary front in the presence of a weak stationary input inhomogeneity [76].

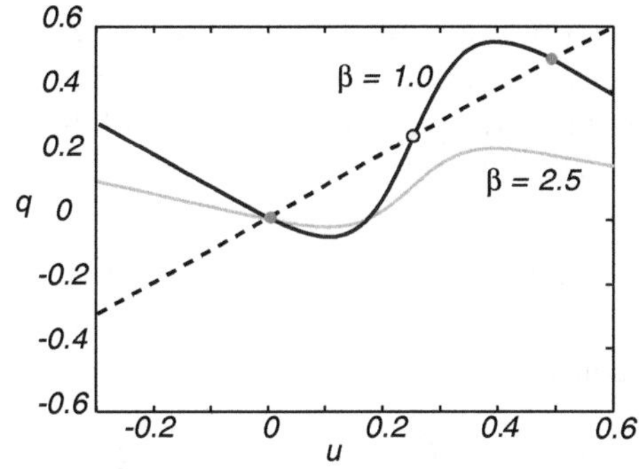

Fig. 7.3 Plot of nullclines for space-clamped planar system $\dot{u} = -u + F(u) - \beta u$, $\varepsilon^{-1}\dot{q} = -q + u$ with $F(u) = 1/(1 + \mathrm{e}^{-\eta(u-\kappa)})$. Nullcline $q = -u + F(u)]/\beta$ for $\beta = 1.0$ ($\beta = 2.5$) intercepts straight nullcline $q = u$ at three fixed points (one fixed point) and the corresponding spatially extended network acts as a bistable (excitable) medium. Other parameters are $\eta = 20, \kappa = 0.25$

7.2.1 Exact Traveling Pulse Solution

Without loss of generality, let us consider a right-moving traveling pulse solution of the form $(u(x,t), q(x,t)) = (U(x - ct), Q(x - ct))$ with $U(\pm\infty), Q(\pm\infty) = 0$ and $U(-\Delta) = U(0) = \kappa$; see Fig. 2.1(b). Here c, Δ denote the speed and width of the wave, respectively. We also assume that $U(\xi) > \kappa$ for $\xi \in (-\Delta, 0)$ and $U(\xi) < \kappa$ for $\xi < -\Delta$ and $\xi > 0$. Substituting this solution into Eq. (7.49) with $\xi = x - ct$ then gives

$$
\begin{aligned}
-cU'(\xi) + U(\xi) + \beta Q(\xi) &= \int_{-\Delta}^{0} w(\xi - \xi')d\xi' \\
-cQ'(\xi) + \varepsilon[Q(\xi) - U(\xi)] &= 0.
\end{aligned} \tag{7.50}
$$

It is useful to rewrite Eq. (7.50) in the matrix form

$$
\begin{pmatrix} 1 & \beta \\ -\varepsilon & \varepsilon \end{pmatrix} \begin{pmatrix} U \\ Q \end{pmatrix} - c\partial_\xi \begin{pmatrix} U \\ Q \end{pmatrix} = [W(\xi) - W(\xi + \Delta)] \begin{pmatrix} 1 \\ 0 \end{pmatrix} \tag{7.51}
$$

with $W(\xi) = \int_\xi^\infty w(x)dx$. We proceed by diagonalizing the left-hand side of Eq. (7.51) using the right eigenvectors $\mathbf{v}$ of the matrix

$$
\mathbf{M} = \begin{pmatrix} 1 & \beta \\ -\varepsilon & \varepsilon \end{pmatrix}. \tag{7.52}
$$

These are given by $\mathbf{v}_\pm = (\varepsilon - \lambda_\pm, \varepsilon)^T$ with corresponding eigenvalues

$$
\lambda_\pm = \frac{1}{2}\left[1 + \varepsilon \pm \sqrt{(1+\varepsilon)^2 - 4\varepsilon(1+\beta)}\right]. \tag{7.53}
$$

We will assume that ε is sufficiently small so that $\beta < (1-\varepsilon)^2/4\varepsilon$ and consequently $\lambda_\pm$ are real. (For a discussion of the effects of complex eigenvalues $\lambda_\pm$ see [580].) Note that $\mathbf{v}_\pm e^{\lambda_\pm \xi/c}$ are the corresponding null vectors of the linear operator on the left-hand side of Eq. (7.51). Performing the transformation

$$\begin{pmatrix} \tilde{U} \\ \tilde{Q} \end{pmatrix} = \mathbf{T}^{-1} \begin{pmatrix} U \\ Q \end{pmatrix}, \quad \mathbf{T} = \begin{pmatrix} \mathbf{v}_+ & \mathbf{v}_- \end{pmatrix}, \tag{7.54}$$

then gives the pair of equations

$$-c\partial_\xi \tilde{U} + \lambda_+ \tilde{U} = \eta_+[W(\xi) - W(\xi+\Delta)]$$
$$-c\partial_\xi \tilde{Q} + \lambda_- \tilde{Q} = \eta_-[W(\xi) - W(\xi+\Delta)]$$

with $\eta_\pm = \mp 1/(\lambda_+ - \lambda_-)$. Integrating the equation for $\tilde{U}$ from $-\Delta$ to ∞, we have

$$\tilde{U}(\xi) = e^{\lambda_+ \xi/c}\left[\tilde{U}(-\Delta)e^{\Delta\lambda_+/c} - \frac{\eta_+}{c}\int_{-\Delta}^{\xi} e^{-\lambda_+ \xi'/c}[W(\xi') - W(\xi'+\Delta)]d\xi'\right].$$

Finiteness of $\tilde{U}$ in the limit $\xi \to \infty$ requires the term in square brackets to cancel. Hence, we can eliminate $\tilde{U}(-\Delta)$ to obtain the result

$$\tilde{U}(\xi) = \frac{\eta_+}{c}\int_0^\infty e^{-\lambda_+ \xi'/c}[W(\xi'+\xi) - W(\xi'+\xi+\Delta)]d\xi'. \tag{7.55}$$

Similarly,

$$\tilde{Q}(\xi) = \frac{\eta_-}{c}\int_0^\infty e^{-\lambda_- \xi'/c}[W(\xi'+\xi) - W(\xi'+\xi+\Delta)]d\xi'. \tag{7.56}$$

Performing the inverse transformation $U = (\varepsilon - \lambda_+)\tilde{U} + (\varepsilon - \lambda_-)\tilde{Q}$ we have

$$U(\xi) = \frac{1}{c}\int_0^\infty \left[\chi_+ e^{-\lambda_+ \xi'/c} + \chi_- e^{-\lambda_- \xi'/c}\right][W(\xi'+\xi) - W(\xi'+\xi+\Delta)]d\xi', \tag{7.57}$$

with $\chi_\pm = (\varepsilon - \lambda_\pm)\eta_\pm$. The threshold conditions $U(-\Delta) = \kappa$ and $U(0) = \kappa$ then yield a pair of equations whose solutions determine existence curves relating the speed c and width Δ of a pulse to the threshold κ [134, 198, 504].

For the sake of illustration, let w be given by the exponential function (7.1). In the domain $\xi > 0$, there is a common factor of $e^{-\xi/\sigma}$ in the integrand of Eq. (7.57) so that $U(\xi) = \kappa e^{-\xi/\sigma}$ for $\xi > 0$ provided that

$$\kappa = \frac{1}{2}\frac{\sigma(c+\varepsilon\sigma)(1-e^{-\Delta/\sigma})}{c^2 + c\sigma(1+\varepsilon) + \sigma^2\varepsilon(1+\beta)}. \tag{7.58}$$

(Note that for zero negative feedback ($\beta = 0$), Eq. (7.58) reduces to the formula for wave speed of a front in the limit $\Delta \to \infty$.) On the other hand, when $\xi < 0$, one has to partition the integral of Eq. (7.57) into the separate domains $\xi' > |\xi|$,

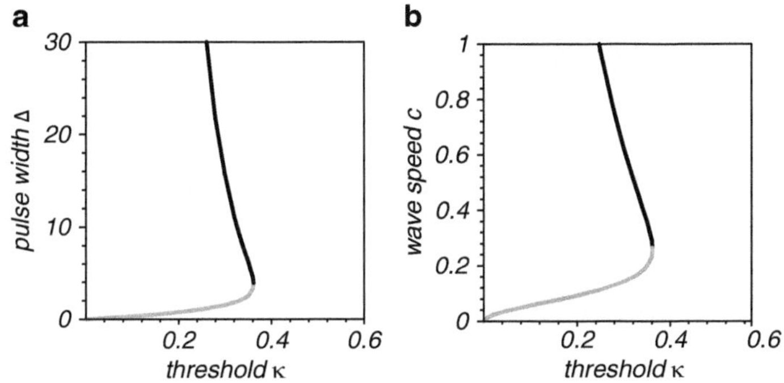

Fig. 7.4 Existence of right-moving traveling pulses in the case of the excitatory network (7.49) with linear adaptation for an exponential weight distribution (7.1). Here $\sigma = 1, \varepsilon = 0.01$ and $\beta = 2.5$. (**a**) Plot of pulse width Δ against threshold κ. (**b**) Plot of wave speed c against threshold κ. Stable (unstable) branches indicated by *black* (*gray*) curves

$|\xi| - \Delta < \xi' < |\xi|$ and $\xi' < |\xi| - \Delta$. This then determines the second threshold condition as well as the asymptotic behavior of $U(\xi)$ in the limit $\xi \to -\infty$:

$$U(\xi) = A_+ e^{\lambda_+ \xi/c} + A_- e^{\lambda_- \xi/c} + A_0 e^{\sigma\xi}. \tag{7.59}$$

where the amplitudes $A_\pm$ and A_0 can be determined from matching conditions at the threshold crossing points [198, 504]. Note that the leading edge of the pulse is positive, whereas the trailing edge is negative due to the effects of adaptation. One finds that for sufficiently slow negative feedback (small ε) and large β there exist two pulse solutions: one narrow and slow and the other wide and fast. This is illustrated in Fig. 7.4. Note that a numerical value of $c \sim 1$ in dimensionless units ($\sigma = \tau = 1$) translates into a physical speed of 60–90 mm/s if the membrane time constant $\tau = 10$ msec and the range of synaptic connections is $\sigma = 600$–$900\,\mu$m. Numerically, the fast solution is found to be stable [504], and this can be confirmed analytically using an Evans function construction [134, 198, 507]; see below. Finally, note that one of the predictions of the neural field model is that the speed of wave propagation should increase as the threshold decreases [504]. Interestingly, this has been confirmed experimentally by applying electric fields to a disinhibited rat cortical slice [521]. The experimental setup is shown in Fig. 7.5. A positive (negative) electric field increases (decreases) the speed of wave propagation by altering the effective excitability of layer V pyramidal neurons. Such neurons have long apical dendrites and are easily polarizable by the electric field.

Construction of Evans function. Rewrite the neural field equation (7.49) in the integral form

$$u(x,t) = \int_{-\infty}^{\infty} \int_0^{\infty} w(y)\Phi(s)F(u(x-y,t-s))dsdy - \beta \int_0^{\infty} \Psi(s)u(x,t-s)ds, \tag{7.60}$$

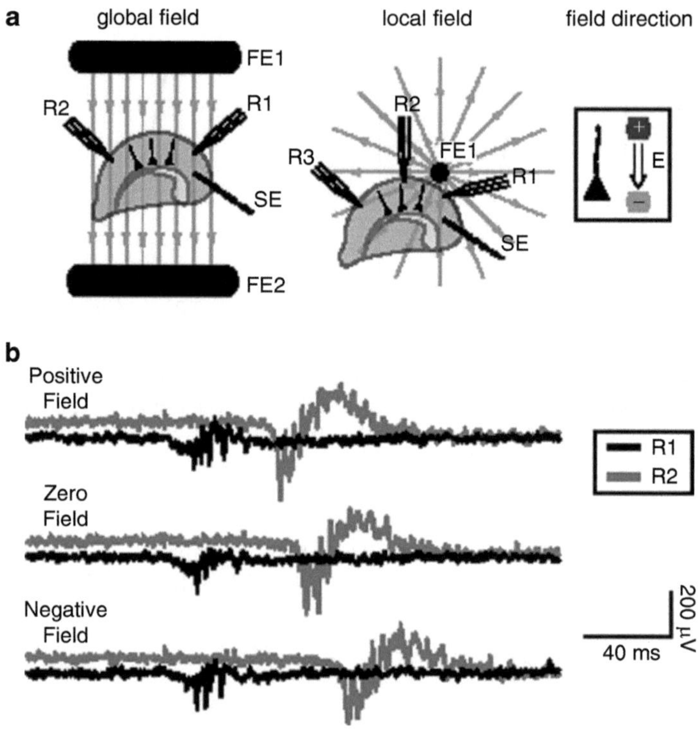

Fig. 7.5 (**a**) Rat cortical slices are bathed in picrotoxin (a GABA$_A$ blocker) and a stimulation electrode (SE) is placed in layers 5–6 to initiate epileptiform bursts. An electric field is applied globally or locally across the slice using Ag/AgCl electrodes (FE1,FE2). Layer five neurons have long apical dendrites and are easily polarizable by an electric field, which controls the effective firing threshold of the neuron. (**b**) The time for an activity pulse to travel between two recording electrodes R1 and R2 depends on the applied electric field, reflecting the dependence of wave speed on the effective firing threshold. [Adapted from Richardson, Schiff and Gluckman [521]]

with $\Phi(t) = \mathrm{e}^{-t}H(t)$ and $\Psi(t) = \int_0^t \Phi(s)\mathrm{e}^{-\varepsilon(t-s)}ds$. Linearizing about the pulse solution by setting $u(x,t) = U(\xi) + \varphi(\xi)\mathrm{e}^{\lambda t}$ gives

$$\varphi(\xi) = \int_{-\infty}^{\infty}\int_{\xi-y}^{\infty} w(y)\Phi((s+y-\xi)/c)\mathrm{e}^{-\lambda(s+y-\xi)/c}F'(U(s))\varphi(s)\frac{ds}{c}dy$$
$$-\beta\int_{\xi}^{\infty}\Psi((s-\xi)/c)\mathrm{e}^{-\lambda(s-\xi)/c}\varphi(s)\frac{ds}{c}. \tag{7.61}$$

Proceeding along similar lines to the analysis of front stability in Sect. 7.1, we set $F(U) = H(U-\kappa)$ and use the identity

$$H'(U(\xi)-\kappa) = \delta(U(\xi)-\kappa) = \frac{\delta(\xi)}{|U'(0)|} + \frac{\delta(\xi+\Delta)}{|U'(-\Delta)|}. \tag{7.62}$$

This gives

$$\varphi(\xi)+\beta\int_{\xi}^{\infty}\Psi((s-\xi)/c)\mathrm{e}^{-\lambda(s-\xi)/c}\varphi(s)\frac{ds}{c} \tag{7.63}$$
$$=\frac{\varphi(0)}{c|U'(0)|}\mathscr{H}(\lambda,\xi)+\frac{\varphi(-\Delta)}{c|U'(-\Delta)|}\mathscr{H}(\lambda,\xi+\Delta)$$

where

$$\mathscr{H}(\lambda,\xi)=\int_{\xi}^{\infty}w(y)\Phi((y-\xi)/c)\mathrm{e}^{-\lambda(y-\xi)/c}dy. \tag{7.64}$$

Let $\hat{\mathscr{H}}(\lambda,k)$ denote the Fourier transform of $\mathscr{H}(\lambda,\xi)$ and $\hat{\mathscr{G}}(\lambda,k)$ denote the Fourier transform of $\Psi(\xi/c)\mathrm{e}^{-\xi/c}$. Using Fourier transforms and the convolution theorem, Eq. (7.63) can then be rewritten as

$$\varphi(\xi)=\frac{\varphi(0)}{c|U'(0)|}\mathscr{B}(\lambda,\xi)+\frac{\varphi(-\Delta)}{c|U'(-\Delta)|}\mathscr{B}(\lambda,\xi+\Delta), \tag{7.65}$$

with $\mathscr{B}(\lambda,\xi)$ the inverse transform of

$$\hat{\mathscr{B}}(\lambda,k)=\frac{\hat{\mathscr{H}}(\lambda,k)}{[1+\beta\hat{\mathscr{G}}(\lambda,-k)/c]}. \tag{7.66}$$

Finally, the eigenvalues λ are determined by setting $\xi=0,-\Delta$ and solving the resulting matrix equation $\mathbf{f}=\mathscr{M}(\lambda)\mathbf{f}$ with $\mathbf{f}=(\varphi(0),\varphi(-\Delta))$ and

$$\mathscr{M}(\lambda)=\frac{1}{c}\begin{pmatrix}\frac{\mathscr{B}(\lambda,0)}{|U'(\xi_1)|} & \frac{\mathscr{B}(\lambda,\Delta)}{|U'(-\Delta)|}\\ \frac{\mathscr{B}(\lambda,-\Delta)}{|U'(0)|} & \frac{\mathscr{B}(\lambda,0)}{|U'(-\Delta)|}\end{pmatrix}. \tag{7.67}$$

It follows that the eigenvalues λ are zeros of the Evans function

$$\mathscr{E}(\lambda)=\mathrm{Det}[\mathbf{1}-\mathscr{M}(\lambda)], \tag{7.68}$$

where $\mathbf{1}$ denotes the identity matrix.

7.2.2 Singularly Perturbed Pulse Solution

In the case of slow adaptation ($\varepsilon \ll 1$), Pinto and Ermentrout [504] showed how to construct a traveling pulse solution of Eq. (7.49) for a smooth firing rate function F by exploiting the existence of traveling front solutions of the corresponding scalar equation (7.2). The method is analogous to the construction of traveling pulses in the FitzHugh–Nagumo equation [316]; see Sect.2.3. The basic idea is to analyze separately the fast and slow time behavior of solutions to Eq. (7.49) expressed in traveling wave coordinates:

$$-c\frac{dU(\xi)}{d\xi}=-U(\xi)-\beta Q(\xi)+\int_{-\infty}^{\infty}w(\xi-\xi')F(U(\xi'))d\xi', \tag{7.69}$$

$$-c\frac{dQ(\xi)}{d\xi}=\varepsilon[-Q(\xi)+U(\xi)]. \tag{7.70}$$

We will assume the normalization $\int_{-\infty}^{\infty} w(y)dy = 1$. In the case of fast time, the slow adaptation is taken to be constant by setting $\varepsilon = 0$ so that we have the inner layer equations

$$-c\frac{dU(\xi)}{d\xi} = -U - \beta Q_0 + \int_{-\infty}^{\infty} w(\xi - \xi')F(U(\xi'))d\xi', \tag{7.71}$$

$$-c\frac{dQ(\xi)}{d\xi} = 0. \tag{7.72}$$

Since $Q(\xi) = Q_0$ is a constant, the term βQ_0 can be absorbed into the threshold of the firing rate function F by making the shift $U(\xi) \to U(\xi) + \beta Q_0$. Hence Eq. (3.55) is equivalent to the scalar equation (7.7), which supports the propagation of traveling fronts. In the case of slow time, we introduce the compressed variable $\zeta = \varepsilon\xi$ so that

$$-c\varepsilon\frac{dU(\zeta)}{d\zeta} = -U(\zeta) - \beta Q(\zeta) + \frac{1}{\varepsilon}\int_{-\infty}^{\infty} w([\zeta - \zeta']/\varepsilon)F(U(\zeta'))d\zeta', \tag{7.73}$$

$$-c\frac{dQ(\zeta)}{d\zeta} = -Q(\zeta) + U(\zeta). \tag{7.74}$$

In the limit $\varepsilon \to 0$, we have

$$\frac{1}{\varepsilon}w([\zeta - \zeta']/\varepsilon) \to \delta(\zeta - \zeta')$$

so that first equation becomes

$$\beta Q(\zeta) = -U(\zeta) + F(U(\zeta)) \tag{7.75}$$

Inverting this equation yields two branches $U = g_\pm(Q)$. Hence we obtain a slow time or outer layer equation on each branch (see Fig. 7.6):

$$\frac{dQ}{d\zeta} = \frac{1}{c}[Q - g_\pm(Q)] \tag{7.76}$$

The construction of the traveling pulse now proceeds by matching inner and outer solutions [504]. This can be visualized by considering the nullclines of the space-clamped version of Eq. (7.49); see Fig. 7.6. We assume that the gain of F and the strength β of adaptation are such that there is only a single fixed point of the space-clamped system:

I Starting at the unique fixed point, use the fast inner equations and the existence results of [172] to construct a leading front solution at $Q = Q_0$ with speed c_0 and matching conditions $\lim_{\xi \pm\infty} U(\xi) = g_\pm(Q_0)$.

II Use the slow outer equations to determine dynamics of Q along upper branch $U = g_+(Q)$

III The solution leaves upper branch at some point Q_1. Once again use the fast inner equations and [172] to construct a trailing front solution with speed c_1 and matching conditions

$$\lim_{\xi \pm\infty} U(\xi) = g_\mp(Q_1)$$

IV Finally, use the slow outer equations to determine the return to the fixed point along the lower branch.

In order to establish the existence of a traveling pulse solution, it remains to find a value Q_1 for which $c_1 = -c_0$ so that the leading and trailing edges of the pulse move at the same speed and thus the pulse maintains its shape as it propagates. (Since Q_0 is known, so is c_0.) Adapting the formula for the wave speed obtained in [172], we have

$$c_1 = -\frac{\Gamma}{\int_{-\infty}^{\infty} U'^2(\xi)F'(U(\xi))d\xi}, \qquad \Gamma = \int_{g_-(Q_1)}^{g_+(Q_1)} [-U - Q_1 + F(U)]dU. \quad (7.77)$$

Unfortunately, it is not possible to derive a closed form expression for the wave speed. However, the existence of a matching speed can be established provided that certain additional assumptions are made regarding the shape of the firing rate function; see [504] for more details.

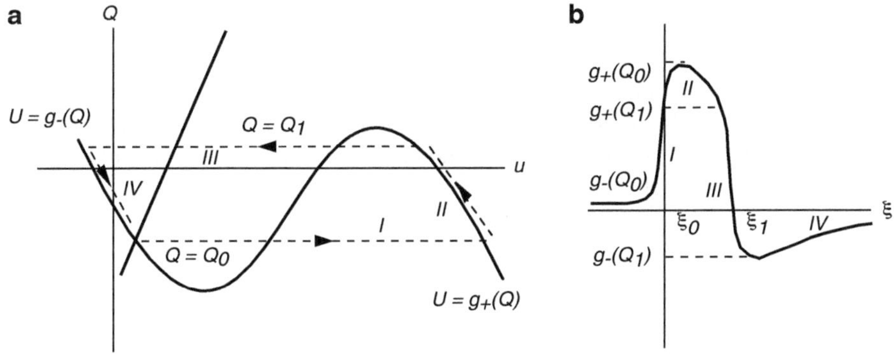

Fig. 7.6 Singular perturbation construction of a traveling pulse in (**a**) the phase plane and (**b**) traveling wave coordinates. See text for details

7.3 Wave Propagation in Heterogeneous Neural Fields

Most studies of neural field theory assume that the synaptic weight distribution only depends upon the distance between interacting populations, that is, $w(x,y) = w(|x-y|)$. This implies translation symmetry of the underlying integrodifferential equations (in an unbounded or periodic domain) and an excitatory network can support the propagation of solitary traveling waves. However, if one looks more closely at the anatomy of cortex, it is clear that its detailed microstructure is far from homogeneous. For example, to a first approximation, primary visual cortex (V1) has a periodic-like microstructure on the millimeter length scale, reflecting the existence of various stimulus feature maps; see Sect. 8.1. This has motivated a number of studies concerned with the effects of a periodically modulated weight distribution on wave propagation in neural fields [64, 132, 332].

We first consider the voltage-based neural field equation (6.115) with periodically modulated weight distribution

$$w(x,y) = w(x-y)[1+\rho K(y/\varepsilon)], \tag{7.78}$$

where ρ is the amplitude of the periodic modulation and ε is the period with $K(x) = K(x+1)$ for all x. It will also be assumed that if $\rho = 0$ (no periodic modulation), then the resulting homogeneous network supports a traveling front solution of speed c_0 as analyzed in Sect. 7.1.1. We will describe two alternative methods for analyzing the effects of periodic wave modulation: one based on averaging theory for small ε [79], which adapts the method use to study the propagation failure in myelinated axons (Sect. 2.5), and the other based on analyzing interfacial dynamics [132]. Both approaches make use of the observation that for sufficiently small ρ, numerical simulations of the inhomogeneous network show a front-like wave separating high and low activity states. However, the wave does not propagate with constant speed, but oscillates periodically in an appropriately chosen moving frame. This pulsating front solution satisfies the periodicity condition $u(x,t) = u(x+\varepsilon, t+T)$ so that we can define the mean speed of the wave to be $c = \varepsilon/T$. We will then consider the effects of periodically modulated weights on the propagation of pulled fronts in an activity-based neural field equation, extending the Hamilton–Jacobi method used in the analysis of CaMKII waves in Sect. 3.2.

7.3.1 Averaging Theory

Suppose that the period ε of weight modulations is much smaller than the range of synaptic interactions $\varepsilon \ll \sigma$. (We fix the length scales by setting $\sigma = 1$.) Following the analysis of saltatory waves along myelinated axons (Sect. 2.5) and [318, 319], we want any inhomogeneous terms to be $\mathcal{O}(\varepsilon)$. Therefore, after substituting Eq. (7.78) into (6.115), we integrate by parts to obtain the equation

$$\begin{aligned}\frac{\partial u(x,t)}{\partial t} &= -u(x,t) + \int_{-\infty}^{\infty} w(x-x')F(u(x',t))dx' \\ &\quad +\varepsilon \int_{-\infty}^{\infty} \mathcal{K}(x'/\varepsilon)\left[w'(x-x')F(u(x',t)) - w(x-x')\frac{\partial F(u(x',t))}{\partial x'}\right]dx'.\end{aligned} \tag{7.79}$$

Here $\mathcal{K}'(x) = \rho K(x)$ with $\mathcal{K}$ only having to be defined up to an arbitrary constant. Motivated by the existence of pulsating front solutions, we perform the change of variables $\xi = x - \phi(t)$ and $\tau = t$. Equation (7.79) becomes

$$\frac{\partial u}{\partial \tau} = -u(\xi,\tau)+\int_{-\infty}^{\infty} w(\xi-\xi')F(u(\xi',\tau))d\xi'+\phi'\frac{\partial u(\xi,\tau)}{\partial \xi} \tag{7.80}$$
$$+\varepsilon\int_{-\infty}^{\infty}\mathscr{K}\left(\frac{\xi'+\phi}{\varepsilon}\right)\left[w'(\xi-\xi')F(u(\xi',\tau))-w(\xi-\xi')\frac{\partial F(u(\xi',\tau))}{\partial \xi'}\right]d\xi'.$$

Next perform the perturbation expansions

$$u(\xi,\tau) = U(\xi)+\varepsilon u_1(\xi,\tau)+\varepsilon^2 u_2(\xi,\tau)+\ldots, \tag{7.81}$$

$$\phi'(\tau) = c_0+\varepsilon\phi_1'(\tau) \tag{7.82}$$

where $U(\xi)$ is the unique traveling wave solution of the corresponding homogeneous equation (7.7) with unperturbed wave speed $c = c_0$. The first-order term u_1 satisfies the inhomogeneous linear equation

$$-\frac{\partial u_1(\xi,\tau)}{\partial \tau}+\mathbb{L}u_1(\xi,\tau) = -\phi_1'(\tau)U'(\xi)+h_1(\xi,\phi/\varepsilon) \tag{7.83}$$

where

$$\mathbb{L}u(\xi) = c_0\frac{du(\xi)}{d\xi}-u(\xi)+\int_{-\infty}^{\infty} w(\xi-\xi')F'(U(\xi'))u(\xi')d\xi' \tag{7.84}$$

and

$$h_1 = \int_{-\infty}^{\infty}\mathscr{K}\left(\frac{\xi'+\phi}{\varepsilon}\right)\left[-w'(\xi-\xi')F(U(\xi'))+w(\xi-\xi')\frac{dF(U(\xi'))}{d\xi'}\right]d\xi'. \tag{7.85}$$

The linear operator $\mathbb{L}$ has a one-dimensional null-space spanned by U'. The existence of U' as a null vector follows immediately from differentiating both sides of Eq. (7.7) with respect to ξ, whereas its uniqueness can be shown using properties of positive linear operators [172]. Therefore, a bounded solution of Eq. (7.83) with respect to ξ and τ will only exist if the right-hand side of Eq. (7.83) is orthogonal to all elements of the null-space of the adjoint operator $\mathbb{L}^{\dagger}$. The latter is defined with respect to the inner product

$$\int_{-\infty}^{\infty} u(\xi)\mathbb{L}v(\xi)d\xi = \int_{-\infty}^{\infty}\left[\mathbb{L}^{\dagger}u(\xi)\right]v(\xi)d\xi \tag{7.86}$$

where $u(\xi)$ and $v(\xi)$ are arbitrary integrable functions. Hence,

$$\mathbb{L}^{\dagger}u(\xi) = -c\frac{du(\xi)}{d\xi}-u(\xi)+F'(U(\xi))\int_{-\infty}^{\infty} w(\xi-\xi')u(\xi')d\xi'. \tag{7.87}$$

It can be proven that $\mathbb{L}^\dagger$ also has a one-dimensional null-space [172], that is, it is spanned by some function $V(\xi)$. Equation (7.83) thus has a bounded solution if and only if

$$B_0\phi_1'(\tau) = \int_{-\infty}^{\infty} V(\xi)h_1(\xi,\phi/\varepsilon)d\xi \tag{7.88}$$

where

$$B_0 = \int_{-\infty}^{\infty} V(\xi)U'(\xi)d\xi. \tag{7.89}$$

Note that B_0 is strictly positive since V and U' can be chosen to have the same sign [172]. Substituting for h_1 using Eqs. (7.85) and (7.82) and performing an integration by parts leads to a differential equation for the phase ϕ:

$$\frac{d\phi}{d\tau} = c + \varepsilon\Phi_1\left(\frac{\phi}{\varepsilon}\right), \tag{7.90}$$

where

$$\begin{aligned}\Phi_1\left(\frac{\phi}{\varepsilon}\right) = \frac{1}{B_0}\int_{-\infty}^{\infty}\int_{-\infty}^{\infty} w(\xi-\xi')\mathscr{K}\left(\frac{\xi'+\phi}{\varepsilon}\right) \\ \times\left[V'(\xi)F(U(\xi'))+V(\xi)\frac{dF(U(\xi'))}{d\xi'}\right]d\xi' d\xi.\end{aligned} \tag{7.91}$$

The phase equation (7.90) is identical in form to the one derived in Sect. 2.5 for wave propagation along myelinated axons; see Eq. (2.72). It implies that there are two distinct types of behavior. If the right-hand side of Eq. (7.90) is strictly positive, then there exists a pulsating front of the approximate form $U(x-\phi(t))$ and the average speed of propagation is $c = \varepsilon/T$ with

$$T = \int_0^{\varepsilon} \frac{d\phi}{c+\varepsilon\Phi_1\left(\frac{\phi}{\varepsilon}\right)}. \tag{7.92}$$

On the other hand, if the right-hand side of Eq. (7.90) vanishes for some ϕ, then there is wave propagation failure.

In the case of a Heaviside firing rate function $F(u) = H(u-\kappa)$, it is possible to derive an explicit expression for the wave speed c [64]. The solution for the unperturbed wave front $U(\xi)$ was derived in Sect. 7.1, so it is only necessary to determine the solution $V(\xi)$ of the adjoint equation (7.87), which becomes

$$cV'(\xi)+V(\xi) = -\frac{\delta(\xi)}{U'(0)}\int_{-\infty}^{\infty} w(\xi')V(\xi')d\xi'. \tag{7.93}$$

This can be integrated to give

$$V(\xi) = -H(\xi)\mathrm{e}^{-\xi/c}. \tag{7.94}$$

Given the solutions for $U(\xi)$ and $V(\xi)$, it can then be shown that (7.91) reduces to the form

$$B_0\Phi_1\left(\frac{\phi}{\varepsilon}\right) = W(0)\mathscr{K}\left(\frac{\phi}{\varepsilon}\right) + \int_0^\infty \mathscr{K}\left(\frac{\phi-\xi}{\varepsilon}\right)\left[\frac{W(\xi)}{c} - w(\xi)\right]d\xi, \quad (7.95)$$

where

$$W(\xi) = \int_0^\infty \mathrm{e}^{-y/c_0}w(y+\xi)dy \equiv -cU'(\xi), \quad (7.96)$$

and

$$B_0 = \frac{1}{c_0}\int_0^\infty \mathrm{e}^{-\xi/c_0}W(\xi)d\xi. \quad (7.97)$$

Keeping only the lowest-order contribution to Φ_1, Eq. (7.92) reduces to

$$T = \int_0^\varepsilon \frac{d\phi}{c_0 + \varepsilon\Gamma(c_0)A\left(\frac{\phi}{\varepsilon}\right)} \quad (7.98)$$

with $\Gamma(c_0) = W(0)/B_0$. For the sake of illustration, suppose that the periodic modulation functions K and A are pure sinusoids. Setting $A(x) = \rho\sin(2\pi x)/(2\pi)$ in Eq. (7.98) we find that

$$T = \frac{\varepsilon}{\sqrt{c_0^2 - \varepsilon^2\rho^2\Gamma(c_0)^2}} \quad (7.99)$$

and, hence,

$$c = \sqrt{c_0^2 - \varepsilon^2\rho^2\Gamma(c_0)^2/(2\pi)^2}. \quad (7.100)$$

This establishes that a sinusoidally varying heterogeneous neural medium only supports a propagating wave if the velocity c_0 of the (unique) solution of the corresponding homogeneous medium satisfies the inequality

$$c_0 \geq \varepsilon\rho\Gamma(c_0). \quad (7.101)$$

For the particular example of an exponential distribution (7.1) with $\sigma = 1$, we have $c_0 = (1-2\kappa)/(2\kappa)$ and $\Gamma(c_0) = 1 + c_0$ so that

$$c = c_0\sqrt{1-\gamma_0\rho^2\varepsilon^2}, \quad \gamma_0 = \frac{1}{2\pi(2\kappa-1)}. \quad (7.102)$$

The above averaging method can also be extended to the case of periodically modulated traveling pulses (pulsating pulses) (see [332]), in which there are two

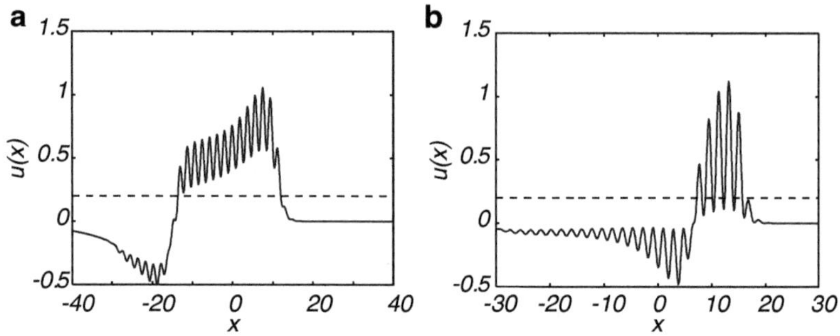

Fig. 7.7 Pulsating pulse solutions in a 1D excitatory neural field with linear adaptation and Heaviside firing rate function; see Eq. (7.49). The threshold is $\kappa = 0.2$, strength of adaptation is $\beta = 2.0$, and adaptation rate constant is $\varepsilon = 0.04$. The weight distribution is given by $w(x,y) = \rho w(x-y)\sin(2\pi x/\varepsilon)$ with $2\pi\varepsilon = 0.3$ and $w(x)$ an exponential weight function. (**a**) Single-bump solution for $\rho = 0.3$. The interior of the pulse consists of non-propagating, transient ripples. (**b**) Multi-bump solution for $\rho = 0.8$. The solitary pulse corresponds to the envelope of a multiple bump solution, in which individual bumps are non-propagating and transient. The disappearance of bumps at one end and the emergence of new bumps at the other end generate the propagation of activity [332]

threshold crossing points. One simplifying assumption of the analysis is that, in the presence of periodically modulated weights, additional threshold crossing points do not occur. However, numerical solutions of a neural field equation with linear adaptation have shown that in the case of large amplitude modulations, a pulsating pulse can develop multiple threshold crossing points [332]. That is, the traveling wave represents the envelope of a multi-bump solution, in which individual bumps are non-propagating and transient; see Fig. 7.7. The appearance (disappearance) of bumps at the leading (trailing) edge of the pulse generates the coherent propagation of the pulse. Wave propagation failure occurs when activity is insufficient to maintain bumps at the leading edge.

7.3.2 Interfacial Dynamics

The averaging method provides a reasonable estimate for the mean wave speed and the critical amplitude ρ for wave propagation failure, provided that the spatial period $\varepsilon \ll 1$. As shown by Coombes and Laing [132] in the case of a Heaviside firing rate function, a more accurate estimate for the wave speed for larger values of ε can be obtained by analyzing the dynamics of the interface between high and low activity states, provided that the amplitude of periodic modulations is not too large [132]. The basic idea is to change to a co-moving frame of the unperturbed system, $u = u(\xi,t)$ with $\xi = x - c_0 t$ such that Eq. (6.115) becomes

$$-c_0 u_\xi + u_t = -u + \int_{-\infty}^{\infty} w(\xi + c_0 t, y) F(u(y - c_0 t, t) dy, \tag{7.103}$$

with w given by Eq. (7.78) and $F(u) = H(u-\kappa)$. The moving interface (level set) is then defined according to the threshold condition

$$u(\xi_0(t),t) = \kappa. \tag{7.104}$$

Differentiating with respect to t then determines the velocity of the interface in the co-moving frame according to

$$\frac{d\xi_0}{dt} = -\frac{u_t(\xi_0(t),t)}{u_\xi(\xi_0(t),t)}. \tag{7.105}$$

As in the previous averaging method, suppose that for $\rho = 0$, there exists a traveling front solution $U(\xi)$ of the homogeneous equation (7.7) with speed c_0. Now make the approximation $u_\xi(\xi_0(t),t) = U'(0)$, which is based on the assumption that for small amplitudes ρ, the slope of the traveling front varies sufficiently slowly. Setting $\xi = \xi_0(t)$ in Eq. (7.103) and using Eq. (7.3), it is then straightforward to show that [132]

$$\frac{d\xi_0}{dt} = \rho c_0 \frac{\int_0^\infty w(y)K(\xi_0 + c_0 t - y)}{\kappa - \int_0^\infty w(y)dy}. \tag{7.106}$$

In order to match up with the previous method, let $K(x) = \sin(2\pi x/\varepsilon)$ and $w(x) = \mathrm{e}^{-|x|}/2$. Then $c_0 = (1-2\kappa)/(2\kappa)$ and [132]

$$\frac{d\xi_0}{dt} = c_0 \rho \gamma(\varepsilon) \sin\left[\frac{2\pi}{\varepsilon}(\xi_0(t) + c_0 t) + \phi_0(\varepsilon)\right], \tag{7.107}$$

with

$$\gamma(\varepsilon) = \frac{1}{2\kappa - 1}\frac{1}{\sqrt{1+(2\pi/\varepsilon)^2}}, \quad \tan\phi_0(\varepsilon) = \frac{2\pi}{\varepsilon}. \tag{7.108}$$

The final step is to look for a T-periodic solution of Eq. (7.107) such that $\xi_0(t) = \xi_0(t+T)$. Setting $x_0 = \xi_0 + c_0 t$ with $x_0 \in [0,\varepsilon]$ and integrating gives

$$\int_0^{x_0} \frac{dx}{1+\rho\gamma\sin(2\pi x/\sigma + \phi)} = c_0 t. \tag{7.109}$$

This may be evaluated using a half-angle substitution,

$$c_0 t = \frac{\varepsilon}{\pi}\frac{1}{\sqrt{1-\rho^2\gamma^2}} \tan^{-1} \left.\frac{z}{\sqrt{1-\rho^2\gamma^2}}\right|_{z_0(0)+\rho\gamma}^{z_0(t)+\rho\gamma}, \tag{7.110}$$

where $z_0(t) = \tan[(2\pi x_0(t)/\varepsilon + \phi)/2]$ and $x_0(0) = 0$. A self-consistent pulsating front solution is then obtained by imposing the condition $\varepsilon = x_0(T)$, which then determines the effective speed $c = \varepsilon/T$ to be

$$c = c_0\sqrt{1-\rho^2\gamma(\varepsilon)^2}. \tag{7.111}$$

Note that on Taylor expanding $\gamma(\varepsilon)$ to first order in ε, Eq. (7.111) recovers the corresponding result (7.102) obtained using averaging theory. However, the expression derived using interfacial dynamics is more accurate when the period ε increases, provided that the amplitude ρ does not become too large.

7.3.3 Hamilton–Jacobi Dynamics and Slow Spatial Heterogeneities

We now turn to the effects of periodically modulated weights on the propagation of pulled front solutions of the activity-based neural field equation (7.27). In the case of high-frequency modulation, Coombes and Laing [132] adapted previous work by Shigesada et al. on pulsating fronts in reaction–diffusion models of the spatial spread of invading species into heterogeneous environments [574, 575]. (In Sect. 3.2.2 we applied the theory of pulsating fronts to CaMKII translocation waves along spiny dendrites.) We briefly sketch the basic steps in the analysis. First, substitute the periodically modulated weight distribution (7.78) into Eq. (7.27) and linearize about the leading edge of the wave where $a(x,t) \sim 0$:

$$\frac{\partial a(x,t)}{\partial t} = -a(x,t) + \int_{-\infty}^{\infty} w(x-y)[1+K(y/\varepsilon)]a(y,t)dy. \tag{7.112}$$

Now assume a solution of the form $a(x,t) = A(\xi)P(x)$, $\xi = x - ct$ with $A(\xi) \to 0$ as $\xi \to \infty$ and $P(x+2\pi\varepsilon) = P(x)$. Substitution into Eq. (7.112) then gives

$$-cP(x)A'(\xi) = -P(x)A(\xi) + \int_{-\infty}^{\infty} w(x-y)[1+K(y/\varepsilon)]P(y)A(\xi-[x-y])dy. \tag{7.113}$$

Taking $A(\xi) \sim e^{-\lambda\xi}$ and substituting into the above equation yields a nonlocal version of the Hill equation:

$$(1+c\lambda)P(x) = \int_{-\infty}^{\infty} e^{\lambda[x-y]} w(x-y)[1+K(y/\varepsilon)]P(y)dy. \tag{7.114}$$

In order to determine the minimal wave speed, it is necessary to find a bounded periodic solution $P(x)$ of Eq. (7.114), which yields a corresponding dispersion relation $c = c(\lambda)$, whose minimum with respect to λ can then be determined (assuming it exists). One way to obtain an approximate solution to Eq. (7.114) is to use Fourier methods to derive an infinite matrix equation for the Fourier coefficients of

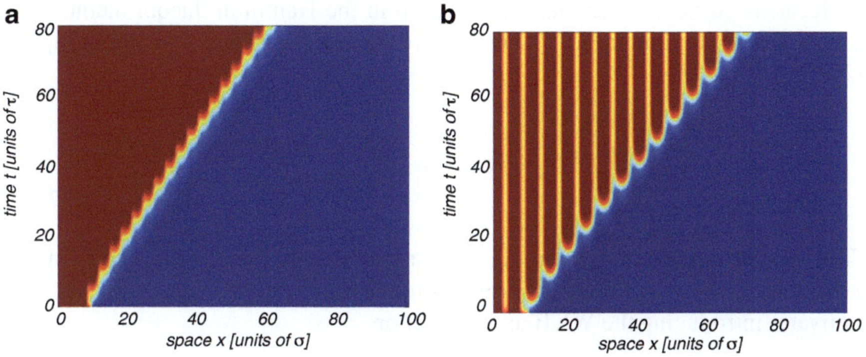

Fig. 7.8 Space–time contour plots of a pulsating front solution of the neural field equation (7.112) with piecewise linear firing rate function (7.29), Gaussian weight distribution (7.38), and a $2\pi\varepsilon$-periodic modulation of the synaptic weights, $K(x) = \cos(x/\varepsilon)$. (**a**) $\varepsilon = 0.5$ and (**b**) $\varepsilon = 0.8$. Other parameters are $W_0 = 1.2$, $\sigma = 1.0$, and $\kappa = 0.4$

the periodic function $P(x)$ and then to numerically solve a finite truncated version of the matrix equation. This is the approach followed in [132]. The matrix equation takes the form

$$(1 + c\lambda)P_m = \mathscr{W}(\lambda - im/\varepsilon)P_m + \mathscr{W}(\lambda - im/\varepsilon)\sum_l K_l P_{m-l}, \tag{7.115}$$

where $K(x/\varepsilon) = \sum_n K_n e^{imx/\varepsilon}$, $P(x) = \sum_n P_n e^{imx/\varepsilon}$, and $\mathscr{W}(p) = \hat{W}(p) + \hat{W}(-p)$ with $\hat{W}(p)$, the Laplace transform of $w(x)$. One finds that the mean velocity of a pulsating front increases with the period $2\pi\varepsilon$ of the synaptic modulations [132]. This is illustrated in Fig. 7.8, which shows space–time plots of a pulsating front for $\varepsilon = 0.5$ and $\varepsilon = 0.8$.

Now suppose that there is a slowly varying spatial modulation of the synaptic weight distribution (relative to the range of synaptic interactions). (Although we do not have a specific example of long-wavelength modulations in mind, we conjecture that these might be associated with inter-area cortical connections. For example, it has been shown that heterogeneities arise as one approaches the V1/V2 border in visual cortex, which has a number of effects including the generation of reflected waves [688].) In the case of slow modulations, one can extend the Hamilton–Jacobi theory of sharp interfaces developed originally for PDEs (see [178, 203, 204, 212, 421] and Sect. 3.3.2) to the case of neural fields [70]. In order to illustrate this, consider a heterogeneous version of the activity-based neural field equation (7.27) of the form

$$\frac{\partial a(x,t)}{\partial t} = -a(x,t) + F\left(\int_{-\infty}^{\infty} w(x-x')J(\varepsilon x')a(x',t)dx'\right), \tag{7.116}$$

in which there is a slow (nonperiodic) spatial modulation $J(\varepsilon x)$ of the synaptic weight distribution with $\varepsilon \ll 1$.

Recall from Sect. 3.3.2 that the first step in the Hamilton–Jacobi method is to rescale space and time in Eq. (7.116) according to $t \to t/\varepsilon$ and $x \to x/\varepsilon$ [178, 204, 421]:

$$\varepsilon \frac{\partial a(x,t)}{\partial t} = -a(x,t) + F\left(\frac{1}{\varepsilon}\int_{-\infty}^{\infty} w([x-x']/\varepsilon)J(x')a(x',t)dx'\right). \tag{7.117}$$

Under this hyperbolic rescaling, the front region where the activity $a(x,t)$ rapidly increases as x decreases from infinity becomes a step as $\varepsilon \to 0$; see Fig. 7.2(b). This motivates introducing the WKB approximation

$$a(x,t) \sim \mathrm{e}^{-G(x,t)/\varepsilon} \tag{7.118}$$

with $G(x,t) > 0$ for all $x > x(t)$ and $G(x(t),t) = 0$. The point $x(t)$ determines the location of the front and $c = \dot{x}$. Substituting (3.85) into Eq. (7.117) gives

$$-\partial_t G(x,t) = -1 + \frac{1}{\varepsilon}\int_{-\infty}^{\infty} w([x-x']/\varepsilon)J(x')\mathrm{e}^{-[G(x',t)-G(x,t)]/\varepsilon}dx'. \tag{7.119}$$

We have used the fact that for $x > x(t)$ and $\varepsilon \ll 1$, the solution is in the leading edge of the front so that F can be linearized. Equation (7.119) can be simplified using the method of steepest descents [70]; see below. This yields the equation

$$-\partial_t G(x,t) = -1 + \tilde{w}(i\partial_x G(x,t))J(x), \tag{7.120}$$

where $\tilde{w}(k)$ is the Fourier transform of $w(x)$:

$$w(x) = \int_{-\infty}^{\infty} \tilde{w}(k)\mathrm{e}^{ikx}\frac{dk}{2\pi}. \tag{7.121}$$

Equation (7.120) is formally equivalent to the Hamilton–Jacobi equation

$$\partial_t G + H(\partial_x G, x) = 0 \tag{7.122}$$

with corresponding Hamiltonian

$$H(p,x) = -1 + \tilde{w}(ip)J(x) \tag{7.123}$$

where $p = \partial_x G$ is interpreted as the conjugate momentum of x, and $\tilde{w}(ip) = \mathscr{W}(p)$. It follows that the Hamilton–Jacobi equation (7.122) can be solved in terms of the Hamilton equations

$$\frac{dx}{ds} = \frac{\partial H}{\partial p} = J(x)\mathscr{W}'(p) = J(x)[\hat{W}'(p) - \hat{W}'(-p)] \tag{7.124}$$

$$\frac{dp}{ds} = -\frac{\partial H}{\partial x} = -J'(x)\mathscr{W}(p). \tag{7.125}$$

Let $X(s;x,t), P(s;x,t)$ denote the solution with $x(0)=0$ and $x(t)=x$. We can then determine $G(x,t)$ according to

$$G(x,t) = -E(x,t)t + \int_0^t P(s;x,t)\dot{X}(s;x,t)ds. \tag{7.126}$$

Here

$$E(x,t) = H(P(s;x,t), X(s;x,t)), \tag{7.127}$$

which is independent of s due to conservation of "energy," that is, the Hamiltonian is not an explicit function of time.

Steepest descent calculation of G. The derivation of Eq. (7.120) using steepest descents proceeds as follows. First, substituting the Fourier transfer (7.121) into (7.119) and reversing the order of integration gives

$$-\partial_t G(x,t) = -1 + \frac{1}{\varepsilon}\int_{-\infty}^{\infty}\int_{-\infty}^{\infty} \tilde{w}(k)J(x')\mathrm{e}^{-S(k,x';x,t)/\varepsilon}dx'\frac{dk}{2\pi}, \tag{7.128}$$

where

$$S(k,x';x,t) = ik(x'-x) + G(x',t) - G(x,t). \tag{7.129}$$

Exploiting the fact that ε is small, we perform steepest descents with respect to the x' variable with (k,x,t) fixed. Let $x' = z(k,t)$ denote the stationary point for which $\partial S/\partial x' = 0$, which is given by the solution to the implicit equation

$$ik + \partial_x G(x',t) = 0. \tag{7.130}$$

Taylor expanding S about this point (assuming it is unique) gives to second order

$$\begin{aligned} S(k,x';x,t) &\approx S(k,z(k,t);x,t) + \frac{1}{2}\frac{\partial^2 S}{\partial x'^2}\bigg|_{x'=z(k,t)}(x'-z(k,t))^2 \\ &= ik[z(k,t)-x] + G(z(k,t),t) - G(x,t) \\ &\quad - \frac{1}{2}\partial_{xx}G(z(k,t),t)(x'-z(k,t))^2. \end{aligned} \tag{7.131}$$

Substituting into Eq. (7.128) and performing the resulting Gaussian integral with respect to x' yields the result

$$\begin{aligned} -\partial_t G(x,t) = -1 + \frac{1}{\varepsilon}\int_{-\infty}^{\infty} \sqrt{\frac{2\pi\varepsilon}{\partial_{xx}G(z(k,t),t)}}\tilde{w}(k)J(z(k,t)) \\ \times \mathrm{e}^{-\left(ik[z(k,t)-x] + G(z(k,t),t) - G(x,t)\right)/\varepsilon}\frac{dk}{2\pi}. \end{aligned} \tag{7.132}$$

This can be rewritten in the form

$$-\partial_t G(x,t) = -1 + \frac{1}{\sqrt{2\pi\varepsilon}}\int_{-\infty}^{\infty}\tilde{w}(k)J(z(k,t))\mathrm{e}^{-\hat{S}(k;x,t)/\varepsilon}dk, \tag{7.133}$$

where

$$\hat{S}(k;x,t) = ik[z(k,t)-x] + G(z(k,t),t) - G(x,t) + \frac{\varepsilon}{2}\ln\partial_{xx}G(z(k,t),t). \tag{7.134}$$

The integral over k can also be evaluated using steepest descents. Thus, Taylor expand $\hat{S}$ to second order about the stationary point $k = k(x,t)$, which is the solution to the equation

$$0 = \frac{\partial \hat{S}}{\partial k} = i[z(k,t) - x] + \frac{\partial z(k,t)}{\partial k}\left[ik + \partial_x G(z(k,t),t) + \frac{\varepsilon}{2}\frac{\partial_{xxx} G(z(k,t),t)}{\partial_{xx} G(z(k,t),t)}\right]. \tag{7.135}$$

It follows from Eqs. (7.130) and (7.135) that $z(k(x,t),t) = x + \mathscr{O}(\varepsilon)$ and so

$$k(x,t) = i\partial_x G(x,t) + \mathscr{O}(\varepsilon). \tag{7.136}$$

Moreover,

$$\hat{S}(k;x,t) \approx \frac{1}{2}\left.\frac{\partial^2 \hat{S}}{\partial k^2}\right|_{k=k(x,t)} (k - k(x,t))^2. \tag{7.137}$$

Substituting into Eq. (7.133) and performing the Gaussian integral with respect to k gives to leading order in ε

$$-\partial_t G(x,t) = -1 + \frac{1}{\sqrt{i\partial_{xx} G(x,t)\partial_k z(k(x,t),t)}}\tilde{w}(k(x,t))J(x). \tag{7.138}$$

Finally, setting $x' = z(k,t)$ in Eq. (7.130) and differentiating with respect to k show that $\partial_{xx} G(z(k,t),t)\partial_k z(k,t) = -i$ and we obtain Eq. (7.120).

Given $G(x,t)$, the location $x(t)$ of the front at time t is determined by the equation $G(x(t),t) = 0$. Differentiating with respect to t shows that $\dot{x}\partial_x G + \partial_t G = 0$. Let us begin by rederiving the wave speed for a homogeneous neural field by setting $J(x) \equiv 1$. In this case, $dp/ds = 0$ so that $p = \lambda_0$ independently of s. Hence, $x(s) = xs/t$, which implies that

$$\dot{x} = \frac{dx}{ds} = \mathscr{W}'(\lambda_0). \tag{7.139}$$

On the other hand,

$$\dot{x} = -\frac{\partial_t G}{\partial_x G} = \frac{-1 + \mathscr{W}(\lambda_0)}{\lambda_0}. \tag{7.140}$$

Combining these two results means that λ_0 is given by the minimum of the function

$$c(\lambda) = \frac{-1 + \mathscr{W}(\lambda)}{\lambda} \tag{7.141}$$

and $c_0 = c(\lambda_0)$. This recovers the result of Sect. 7.1.3. Thus, in the case of a Gaussian weight distribution, λ_0 is related to c_0 according to Eq. (7.42). Now suppose that there exists a small-amplitude, slow modulation of the synaptic weights $J(x) = 1 + \beta f(x)$ with $\beta \ll 1$. We can then obtain an approximate solution of Hamilton's Eqs. (7.124) and (7.125) and the corresponding wave speed using regular perturbation theory along analogous lines to a previous study of the F–KPP equation [421]. We find (see below) that

$$x(t) = c_0 t + \frac{\beta \mathscr{W}(\lambda_0)}{c_0 \lambda_0} \int_0^{c_0 t} f(y) dy + \mathscr{O}(\beta^2). \tag{7.142}$$

Here c_0 is the wave speed of the homogeneous neural field ($\beta = 0$), which is given by $c_0 = c(\lambda_0)$ with λ_0 obtained by minimizing the function $c(\lambda)$ defined by Eq. (7.141); see Eq. (7.42). Finally, differentiating both sides with respect to t and inverting the hyperbolic scaling yields

$$c \equiv \dot{x}(t) = c_0 + \frac{\beta \mathscr{W}(\lambda_0)}{\lambda_0} f(\varepsilon c_0 t) + \mathscr{O}(\beta^2). \tag{7.143}$$

The analytical results agree reasonably well with numerical simulations, provided that ε is sufficiently small [70]. In Fig. 7.9(a) we show snapshots of a pulled front in the case of a homogeneous network with Gaussian weights (7.38) and piecewise linear firing rate function (7.29). Space and time units are fixed by setting the range of synaptic weights $\sigma = 1$ and the time constant $\tau = 1$. A corresponding space–time plot is given in Fig. 7.9(b), which illustrates that the speed of the front asymptotically approaches the calculated minimal wave speed c_0. (Note that pulled fronts take an extremely long time to approach the minimal wave speed at high levels of numerical accuracy, since the asymptotics are algebraic rather than exponential in time [162].) In Figs. 7.9(c,d) we plot the corresponding results in the case of an inhomogeneous network. For the sake of illustration, the synaptic heterogeneity is taken to be a linear function of displacement, that is, $J(x) = 1 + \varepsilon(x - l)$, and $\beta = \varepsilon$. Equation (7.142) implies that

$$\begin{aligned} x(t) &= l + c_0 t + \frac{\varepsilon^2 \mathscr{W}(\lambda_0)}{2 c_0 \lambda_0} [(c_0 t)^2 - 2 c_0 l t] \\ &= l + \left[c_0 - \frac{\varepsilon^2 l (c_0 \lambda_0 + 1)}{\lambda_0} \right] t + \frac{\varepsilon^2 c_0 (c_0 \lambda_0 + 1)}{2 \lambda_0} t^2, \end{aligned} \tag{7.144}$$

where we have used Eq. (7.141) and assumed that the initial position of the front is $x(0) = l$. Hence, the perturbation theory predicts that a linearly increasing modulation in synaptic weights results in the leading edge of the front tracing out a downward parabola in a space–time plot for times $t \ll \mathscr{O}(1/\varepsilon^2)$. This is consistent with numerical simulations for $\varepsilon^2 = 0.005$, as can be seen in the space–time plot of Fig. 7.9(d).

Perturbation calculation of wave speed. Introduce the perturbation expansions

$$x(s) = x_0(s) + \beta x_1(s) + \mathscr{O}(\beta^2), \quad p(s) = p_0(s) + \beta p_1(s) + \mathscr{O}(\beta^2) \tag{7.145}$$

and substitute into Eqs. (7.124) and (7.125). Taylor expanding the nonlinear function $f(x)$ about x_0 and $\mathscr{W}(p) = \hat{W}(p) + \hat{W}(-p)$ about p_0 then leads to a hierarchy of equations, the first two of which are

$$\dot{p}_0(s) = 0, \quad \dot{x}_0(s) = \mathscr{W}'(p_0), \tag{7.146}$$

and

$$\dot{p}_1(s) = -f'(x_0)\mathscr{W}(p_0), \quad \dot{x}_1(s) = \mathscr{W}''(p_0) p_1(s) + f(x_0)\mathscr{W}'(p_0), \tag{7.147}$$

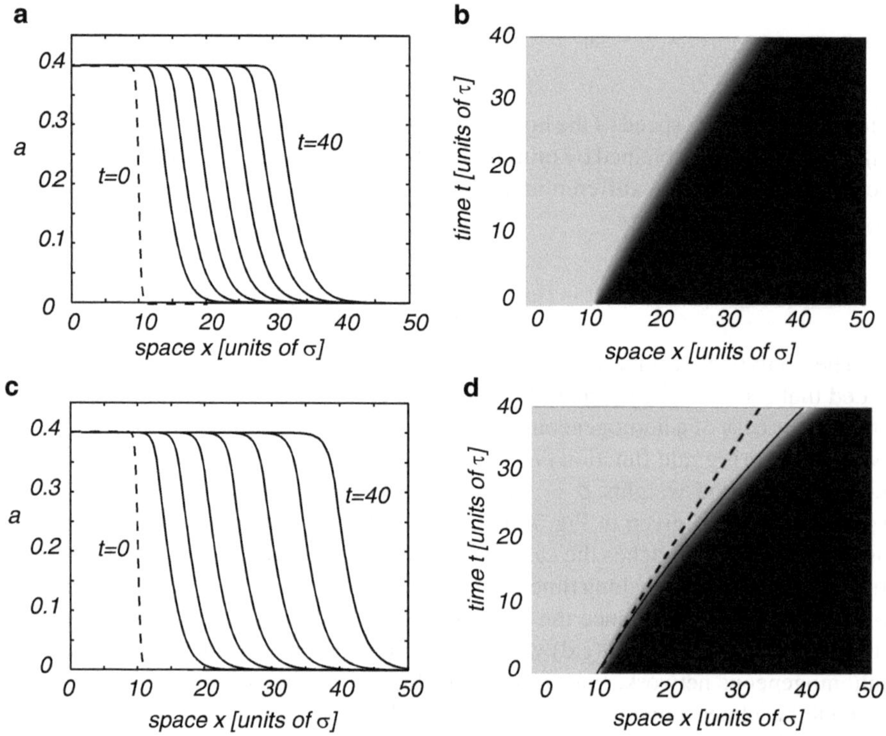

Fig. 7.9 (**a**) Propagating front in a homogeneous network with Gaussian weights (7.38) and piecewise linear rate function (7.29). Parameter values are $W_0 = 1.2, \sigma = 1, \kappa = 0.4$. The initial condition is taken to be a steep sigmoid $a(x,0) = 0.5/(1+\exp(-\eta(x-l)))$ with $\eta = 5$ and $l = 10$. (**a**) Snapshots of wave profile at time intervals of width $\Delta t = 5$ from $t = 10$ to $t = 40$. (**b**) Space–time contour plot. Wave speed asymptotically approaches the minimum c_0 of the velocity dispersion curve given by Eq. (7.39). (**c,d**) Propagating front in a network with a linear heterogeneity in the synaptic weights, $J(x) = 1+\varepsilon(x-l)$, $l = 10$, and $\varepsilon^2 = 0.005$. Other parameters as in (**a,b**). (**c**) Snapshots of wave profile at time intervals of width $\Delta t = 5$ from $t = 10$ to $t = 40$. (**d**) Space–time contour plot. Wave speed increases approximately linearly with time, so the position $x(t)$ of front evolves according to a downward parabola. Theoretical curve based on the perturbation calculation is shown by the *solid curve*. The trajectory of the front in the corresponding homogeneous case is indicated by the *dashed curve*

These are supplemented by the Cauchy conditions $x_0(0) = 0, x_0(t) = x$ and $x_n(0) = x_n(t) = 0$ for all integers $n \geq 1$. Equations (7.146) have solutions of the form

$$p_0(s) = \lambda, \quad x_0(s) = \mathscr{W}'(\lambda)s + B_0 \tag{7.148}$$

with λ, B_0 independent of s. Imposing the Cauchy data then implies that $B_0 = 0$ and λ satisfies the equation

$$\mathscr{W}'(\lambda) = x/t. \tag{7.149}$$

At the next order,

$$p_1(s) = -\mathscr{W}(\lambda)\frac{t}{x}f(xs/t) + A_1, \tag{7.150}$$

$$x_1(s) = -\mathscr{W}''(\lambda)\mathscr{W}(\lambda)\frac{t^2}{x^2}\int_0^{xs/t} f(y)dy + \int_0^{xs/t} f(y)dy + \mathscr{W}''(\lambda)A_1 s + B_1, \tag{7.151}$$

with A_1, B_1 independent of s. Imposing the Cauchy data then implies that $B_1 = 0$ and

$$A_1 = A_1(x,t) = \mathscr{W}(\lambda)\frac{t}{x^2}\int_0^x f(y)dy - \frac{1}{t\mathscr{W}''(\lambda)}\int_0^x f(y)dy. \tag{7.152}$$

Given these solutions, the energy function $E(x,t)$ is

$$\begin{aligned} E(x,t) &= -1 + [1 + \beta f(x_0 + \beta x_1 + \ldots)]\mathscr{W}(\lambda + \beta p_1 + \ldots) \\ &= -1 + \mathscr{W}(\lambda) + \beta[\mathscr{W}'(\lambda)p_1(s) + f(x_0(s))\mathscr{W}(\lambda)] + \mathscr{O}(\beta^2). \end{aligned} \tag{7.153}$$

Substituting for $x_0(s)$ and $p_1(s)$ and using the condition $\mathscr{W}'(\lambda) = x/t$, we find that

$$E(x,t) = -1 + \mathscr{W}(\lambda) + \beta\frac{x}{t}A_1(x,t) + \mathscr{O}(\beta^2), \tag{7.154}$$

which is independent of s as expected. Similarly,

$$\begin{aligned} \int_0^t p(s)\dot{x}(s)ds &= \lambda x + \beta\mathscr{W}'(\lambda)\int_0^t p_1(s)ds + \mathscr{O}(\beta^2) \\ &= \lambda x + \beta\frac{\mathscr{W}'(\lambda)}{\mathscr{W}''(\lambda)}\int_0^t \left[\dot{x}_1(s) - \mathscr{W}'(\lambda)f(\mathscr{W}'(\lambda)s)\right] ds + \mathscr{O}(\beta^2) \\ &= \lambda x - \beta\frac{\mathscr{W}'(\lambda)}{\mathscr{W}''(\lambda)}\int_0^x f(y)dy + \mathscr{O}(\beta^2). \end{aligned} \tag{7.155}$$

Hence, to first order in β,

$$G(x,t) = t - \mathscr{W}(\lambda)t + \lambda x - \beta\mathscr{W}(\lambda)\frac{t}{x}\int_0^x f(y)dy. \tag{7.156}$$

We can now determine the wave speed c by imposing the condition $G(x(t),t) = 0$ and performing the perturbation expansions $x(t) = x_0(t) + \beta x_1(t) + \mathscr{O}(\beta^2)$ and $\lambda = \lambda_0 + \beta\lambda_1 + \mathscr{O}(\beta^2)$. Substituting into Eq. (7.156) and collecting terms at $\mathscr{O}(1)$ and $\mathscr{O}(\beta)$ leads to Eq. (7.142).

7.4 Wave Propagation in Stochastic Neural Fields

In Sect. 6.4 we constructed stochastic neuronal population models based on a master equation formulation. However, continuum versions of these models are difficult to analyze even under a diffusion approximation, due to the nonlocal nature of the multiplicative noise terms; see Eq. (6.71). Therefore, in this section, we analyze the effects of noise on wave propagation in stochastic neural fields with local multiplicative noise, extending the PDE methods outlined in Sect. 2.6. This form of noise can also be interpreted in terms of parametric fluctuations in the firing threshold [58].

7.4.1 Spontaneous Front Propagation

Consider the following stochastic neural field equation: $U(x,t)$

$$dU(x,t) = \left[-U(x,t) + \int_{-\infty}^{\infty} w(x-y)F(U(y,t))dy\right] dt + \varepsilon^{1/2} g(U(x,t))dW(x,t). \tag{7.157}$$

We assume that $dW(x,t)$ represents an independent Wiener process such that

$$\langle dW(x,t)\rangle = 0, \quad \langle dW(x,t)dW(x',t')\rangle = 2C([x-x']/\lambda)\delta(t-t')dtdt', \tag{7.158}$$

where $\langle\cdot\rangle$ denotes averaging with respect to the Wiener process. Here λ is the spatial correlation length of the noise such that $C(x/\lambda) \to \delta(x)$ in the limit $\lambda \to 0$, and ε determines the strength of the noise, which is assumed to be weak. Moreover, the multiplicative noise term is taken to be of Stratonovich form. The analysis of Eq. (7.157) proceeds along similar lines to the analysis of the stochastic bistable equation in Sect. 2.6; see also [70]. First, using Novikov's theorem, we rewrite Eq. (7.157) so that the fluctuating term has zero mean:

$$dU(x,t) = \left[h(U(x,t)) + \int_{-\infty}^{\infty} w(x-y)F(U(y,t))dy\right] dt + \varepsilon^{1/2} dR(U,x,t), \tag{7.159}$$

where

$$h(U) = -U + \varepsilon C(0) g'(U) g(U) \tag{7.160}$$

and

$$dR(U,x,t) = g(U)dW(x,t) - \varepsilon^{1/2} C(0) g'(U) g(U) dt. \tag{7.161}$$

The stochastic process R has the variance

$$\langle dR(U,x,t)dR(U,x',t)\rangle = \langle g(U(x,t))dW(x,t)g(U(x',t)dW(x',t)\rangle + \mathcal{O}(\varepsilon^{1/2}). \tag{7.162}$$

The next step in the analysis is to express the solution U of Eq. (7.159) as a combination of a fixed wave profile U_0 that is displaced by an amount $\Delta(t)$ from its uniformly translating position $\xi = x - c_\varepsilon t$ and a time-dependent fluctuation Φ in the front shape about the instantaneous position of the front:

$$U(x,t) = U_0(\xi - \Delta(t)) + \varepsilon^{1/2} \Phi(\xi - \Delta(t), t). \tag{7.163}$$

The wave profile U_0 and associated wave speed c_ε are obtained by solving the modified deterministic equation

$$-c_\varepsilon \frac{dU_0}{d\xi} - h(U_0(\xi)) = \int_{-\infty}^{\infty} w(\xi - \xi')F(U_0(\xi'))d\xi', \tag{7.164}$$

As in Sect. 2.6, Eq. (7.164) is chosen so that that to leading order, the stochastic variable $\Delta(t)$ undergoes unbiased Brownian motion with

$$\langle \Delta(t) \rangle = 0, \quad \langle \Delta(t)^2 \rangle = 2D(\varepsilon)t \tag{7.165}$$

where the diffusion coefficient $D(\varepsilon) = \mathcal{O}(\varepsilon)$ can be calculated using perturbation analysis (see below).

Perturbation calculation of diffusivity $D(\varepsilon)$. Substitute the decomposition (7.163) into Eq. (7.159) and expand to first order in $\mathcal{O}(\varepsilon^{1/2})$:

$$\begin{aligned}
&-c_\varepsilon U_0'(\xi-\Delta(t))dt - U_0'(\xi-\Delta(t))d\Delta(t) + \varepsilon^{1/2}\left[d\Phi(\xi-\Delta(t),t) - c_\varepsilon \Phi'(\xi-\Delta(t),t)dt\right] \\
&\quad -\varepsilon^{1/2}\Phi'(\xi-\Delta(t),t)d\Delta(t) \\
&\quad = h(U_0(\xi-\Delta(t)))dt + h'(U_0(\xi-\Delta(t)))\varepsilon^{1/2}\Phi(\xi-\Delta(t),t)dt \\
&\quad + \int_{-\infty}^{\infty} w(\xi-\xi')\left(F(U_0(\xi'-\Delta(t))) + F'(U_0(\xi'-\Delta(t)))\varepsilon^{1/2}\Phi(\xi'-\Delta(t),t)\right)d\xi' dt \\
&\quad + \varepsilon^{1/2} dR(U_0(\xi-\Delta(t)),\xi,t) + \mathcal{O}(\varepsilon).
\end{aligned}$$

Imposing Eq. (7.164), after shifting $\xi \to \xi - \Delta(t)$, and dividing through by $\varepsilon^{1/2}$ then gives

$$\begin{aligned}
d\Phi(\xi-\Delta(t),t) &= \hat{L}\circ\Phi(\xi-\Delta(t),t)dt + \varepsilon^{-1/2}U_0'(\xi-\Delta(t))d\Delta(t) \\
&\quad + dR(U_0(\xi-\Delta(t),\xi,t) + \mathcal{O}(\varepsilon^{1/2}),
\end{aligned} \tag{7.166}$$

where $\mathbb{L}$ is the non-self-adjoint linear operator

$$\mathbb{L}\circ A(\xi) = c_\varepsilon \frac{dA(\xi)}{d\xi} + h'(U_0(\xi))A(\xi) + \int_{-\infty}^{\infty} w(\xi-\xi')F'(U_0(\xi'))A(\xi')d\xi' \tag{7.167}$$

for any function $A(\xi) \in L^2(\mathbb{R})$. Finally, for all terms in Eq. (7.166) to be of the same order, we require that $\Delta(t) = \mathcal{O}(\varepsilon^{1/2})$. It then follows that $U_0(\xi - \Delta(t)) = U_0(\xi) + \mathcal{O}(\varepsilon^{1/2})$ and Eq. (7.166) reduces to

$$d\Phi(\xi,t) = \mathbb{L}\circ\Phi(\xi,t)dt + \varepsilon^{-1/2}U_0'(\xi)d\Delta(t) + dR(U_0(\xi),\xi,t) + \mathcal{O}(\varepsilon^{1/2}). \tag{7.168}$$

It can be shown that for a sigmoid firing rate function and exponential weight distribution, the operator $\mathbb{L}$ has a 1D null-space spanned by $U_0'(\xi)$ [172]. (The fact that $U_0'(\xi)$ belongs to the null-space follows immediately from differentiating equation (7.164) with respect to ξ.) We then have the solvability condition for the existence of a nontrivial solution of Eq. (7.168), namely, that the inhomogeneous part is orthogonal to all elements of the null-space of the adjoint operator $\mathbb{L}^\dagger$. The latter is almost identical in form to Eq. (7.87):

$$\mathbb{L}^\dagger B(\xi) = -c_\varepsilon \frac{dB(\xi)}{d\xi} + h'(U_0(\xi))B(\xi) + F'(U_0(\xi))\int_{-\infty}^{\infty} w(\xi-\xi')B(\xi')d\xi'. \tag{7.169}$$

Hence, $\mathbb{L}^\dagger$ has a one-dimensional null-space that is spanned by some function $\mathcal{V}(\xi)$. Taking the inner product of both sides of Eq. (7.168) with respect to $\mathcal{V}(\xi)$ then leads to the solvability condition

$$\int_{-\infty}^{\infty} \mathcal{V}(\xi)\left[U_0'(\xi)d\Delta(t) + \varepsilon^{1/2}dR(U_0,\xi,t)\right]d\xi = 0. \tag{7.170}$$

Thus $\Delta(t)$ satisfies the stochastic differential equation (SDE)

$$d\Delta(t) = -\varepsilon^{1/2} \frac{\int_{-\infty}^{\infty} \mathscr{V}(\xi) dR(U_0,\xi,t) d\xi}{\int_{-\infty}^{\infty} \mathscr{V}(\xi) U_0'(\xi) d\xi}. \tag{7.171}$$

Using the lowest-order approximation $dR(U_0,\xi,t) = g(U_0(\xi))dW(\xi,t)$, we deduce that $\Delta(t)$ is a Wiener process with effective diffusion coefficient

$$D(\varepsilon) = \varepsilon \frac{\int_{-\infty}^{\infty}\int_{-\infty}^{\infty} \mathscr{V}(\xi)\mathscr{V}(\xi')g(U_0(\xi))g(U_0(\xi'))\langle dW(\xi,t)dW(\xi',t)\rangle d\xi d\xi'}{\left[\int_{-\infty}^{\infty} \mathscr{V}(\xi) U_0'(\xi) d\xi\right]^2} \tag{7.172}$$

In the case of a Heaviside rate function $F(U) = H(U-\kappa)$ and multiplicative noise $g(U) = g_0 U$, the effective speed c_ε and diffusion coefficient $D(\varepsilon)$ can be calculated explicitly [70]. (The constant g_0 has units of $\sqrt{\text{length/time.}}$) The deterministic equation (7.164) for the fixed profile U_0 then reduces to

$$-c_\varepsilon \frac{dU_0}{d\xi} + U_0(\xi)\gamma(\varepsilon) = \int_{-\infty}^{\infty} w(\xi-\xi')H(U_0(\xi')-\kappa)d\xi', \tag{7.173}$$

with

$$\gamma(\varepsilon) = (1-\varepsilon g_0^2 C(0)), \tag{7.174}$$

This is identical in structure to Eq. (7.3) for the deterministic neural field modulo the rescaling of the decay term. The analysis of the wave speeds proceeds along similar lines to Sect. 7.1. Thus, multiplying both sides of Eq. (7.173) by $\mathrm{e}^{-\xi\gamma(\varepsilon)/c_\varepsilon}$ and integrating with respect to ξ gives

$$U_0(\xi) = \mathrm{e}^{\xi\gamma(\varepsilon)/c_\varepsilon}\left[\kappa - \frac{1}{c_\varepsilon}\int_0^{\xi} \mathrm{e}^{-y\gamma(\varepsilon)/c_\varepsilon} W(y)dy\right]. \tag{7.175}$$

Finally, requiring the solution to remain bounded as $\xi \to \infty$ ($\xi \to -\infty$) for $c_\varepsilon > 0$ (for $c_\varepsilon < 0$) implies that κ must satisfy the condition

$$\kappa = \frac{1}{|c_\varepsilon|}\int_0^{\infty} \mathrm{e}^{-y\gamma(\varepsilon)/|c_\varepsilon|} W(\mathrm{sign}(c_\varepsilon)y)dy. \tag{7.176}$$

Hence, in the case of the exponential weight distribution (7.1), we have

$$c_\varepsilon = \frac{\sigma}{2\kappa}[1-2\kappa\gamma(\varepsilon)] \tag{7.177}$$

for $c_\varepsilon > 0$, and

$$c_\varepsilon = \frac{\sigma\gamma(\varepsilon)}{2}\frac{1-2\kappa\gamma(\varepsilon)}{1-\kappa\gamma(\varepsilon)}, \tag{7.178}$$

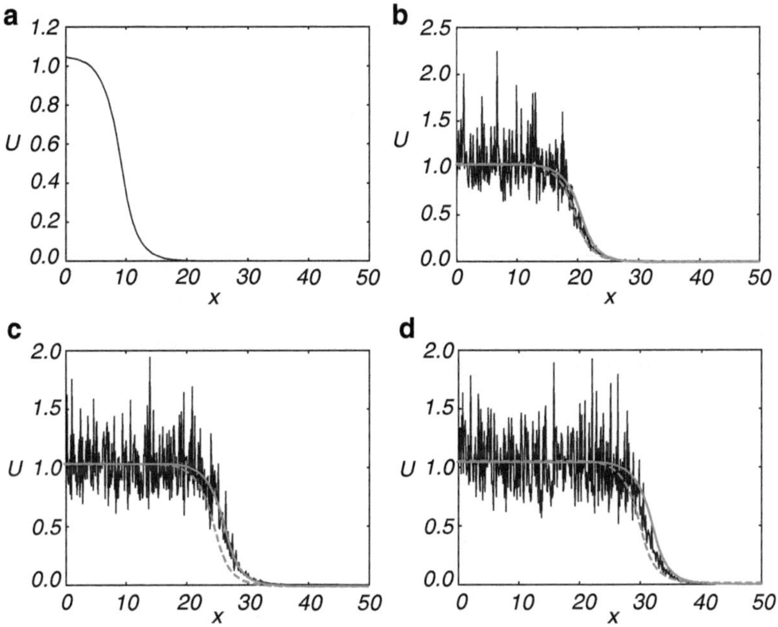

Fig. 7.10 Numerical simulation showing the propagation of a front solution of the stochastic neural field equation (7.157) for Heaviside weight function $F(U) = H(U - \kappa)$ with $\kappa = 0.35$, exponential weight function (7.1) with $\sigma = 2$, and multiplicative noise $g(U) = U$. Noise strength $\varepsilon = 0.005$ and $C(0) = 10$. The wave profile is shown at successive times (**a**) $t = 0$ (**b**) $t = 12$ (**c**) $t = 18$ and (**d**) $t = 24$, with the initial profile at $t = 0$ given by the solution to Eq. (7.164). In numerical simulations we take the discrete space and time steps $\Delta x = 0.1, \Delta t = 0.01$. The deterministic part U_0 of the stochastic wave is shown by the *solid gray curves* and the corresponding solution in the absence of noise ($\varepsilon = 0$) is shown by the *dashed gray curves*

for $c_\varepsilon < 0$. Assuming that $0 \leq \gamma(\varepsilon) \leq 1$, we see that multiplicative noise shifts the effective velocity of front propagation in the positive ξ direction.

In order to calculate the diffusion coefficient, it is first necessary to determine the null vector $\mathscr{V}(\xi)$ of the adjoint linear operator $\mathbb{L}^\dagger$ defined by Eq. (7.169). Setting $F(U) = H(U - \kappa)$ and $g(U) = g_0 U$, we obtain an adjoint equation almost identical to (7.93):

$$c_\varepsilon \mathscr{V}'(\xi) + \gamma(\varepsilon)\mathscr{V}(\xi) = -\frac{\delta(\xi)}{U_0'(0)} \int_{-\infty}^{\infty} w(\xi')\mathscr{V}(\xi')d\xi'. \tag{7.179}$$

Hence, this has the solution

$$\mathscr{V}(\xi) = -H(\xi)\exp(-\Gamma(\varepsilon)\xi), \quad \Gamma(\varepsilon) = \frac{\gamma(\varepsilon)}{c_\varepsilon}, \tag{7.180}$$

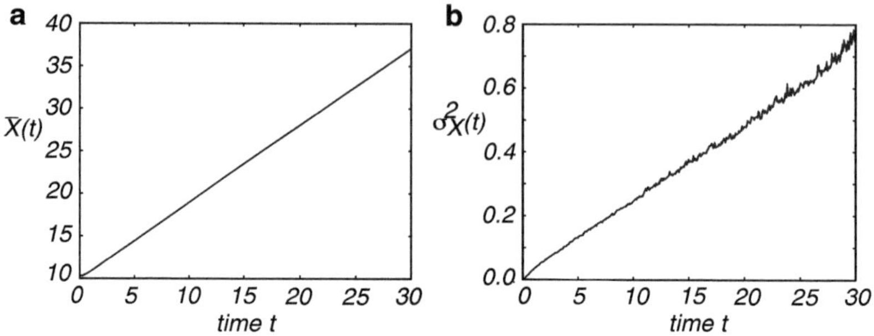

Fig. 7.11 Plot of (**a**) mean $\overline{X}(t)$ and (**b**) variance $\sigma_X^2(t)$ of front position as a function of time, averaged over $N = 4096$ trials. Same parameter values as Fig. 7.10

and Eq. (7.172) reduces to the form

$$D(\varepsilon) = \varepsilon \frac{\displaystyle\int_0^\infty \mathrm{e}^{-2\Gamma(\varepsilon)\xi} U_0(\xi)^2 d\xi}{\left[\displaystyle\int_0^\infty \mathrm{e}^{-\Gamma(\varepsilon)\xi} U_0'(\xi) d\xi\right]^2}. \tag{7.181}$$

In the case of an exponential weight distribution, $U_0(\xi)$ has the explicit form

$$U_0(\xi) = \begin{cases} \dfrac{1}{2c_\varepsilon} \dfrac{\sigma e^{-\xi/\sigma}}{1+\sigma\Gamma(\varepsilon)} & \xi \geq 0 \\[2ex] \dfrac{1}{2c_\varepsilon}\left[\dfrac{2e^{\xi\Gamma(\varepsilon)}}{\Gamma(\varepsilon)\left(-1+\sigma^2\Gamma(\varepsilon)^2\right)} + \dfrac{2}{\Gamma(\varepsilon)} + \dfrac{\sigma e^{\xi/\sigma}}{1-\sigma\Gamma(\varepsilon)}\right] & \xi < 0, \end{cases} \tag{7.182}$$

and the integrals in (7.181) can be evaluated explicitly to give

$$D(\varepsilon) = \frac{1}{2}\varepsilon\sigma g_0^2(1+\sigma\Gamma(\varepsilon)). \tag{7.183}$$

In Fig. 7.10 we show the temporal evolution of a single stochastic wave front, which is obtained by numerically solving the stochastic neural field equation (7.157) for $F(U) = H(U-\kappa)$, $g(U) = U$ and an exponential weight distribution w. In order to numerically calculate the mean location of the front as a function of time, we carry out a large number of level set position measurements. That is, we determine the positions $X_a(t)$ such that $U(X_a(t),t) = a$, for various level set values $a \in (0.5\kappa, 1.3\kappa)$ and then define the mean location to be $\overline{X}(t) = \mathbb{E}[X_a(t)]$, where the expectation is first taken with respect to the sampled values a and then averaged over N trials. The corresponding variance is given by $\sigma_X^2(t) = \mathbb{E}[(X_a(t) - \bar{X}(t))^2]$. In Fig. 7.11 we plot $\overline{X}(t)$ and $\sigma_X^2(t)$ as a function of t. It can be seen that both vary linearly with t, consistent with the assumption that there is a diffusive-like displacement of the front from its uniformly translating position at long time scales. The slopes of these curves then determine the effective wave speed and diffusion coefficient according

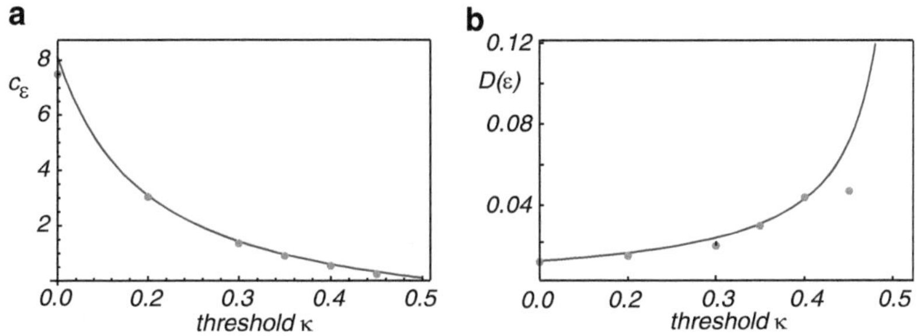

Fig. 7.12 Plot of (**a**) wave speed c_ε and (**b**) diffusion coefficient $D(\varepsilon)$ as a function of threshold κ. Numerical results (*solid dots*) are obtained by averaging over $N = 4,096$ trials starting from the initial condition given by Eq. (7.182). Corresponding theoretical predictions (*solid curves*) for c_ε and $D(\varepsilon)$ are based on Eqs. (7.177) and (7.181), respectively. Other parameters as in Fig. 7.10

to $\overline{X}(t) \sim c_\varepsilon t$ and $\sigma_X^2(t) \sim 2D(\varepsilon)t$. In Fig. 7.12 we plot the numerically estimated speed and diffusion coefficient for various values of the threshold κ and compare these to the corresponding theoretical curves obtained using the above analysis. It can be seen that there is excellent agreement with our theoretical predictions provided that κ is not too large. As $\kappa \to 0.5$, the wave speed decreases towards zero so that the assumption of relatively slow diffusion breaks down.

7.4.2 Stimulus-Locked Fronts

So far we have assumed that the underlying deterministic neural field equation is homogeneous in space so that there exists a family of traveling front solutions related by a uniform shift. Now suppose that there exists an external front-like input that propagates at a uniform speed v, so that the deterministic equation (7.2) becomes

$$\frac{\partial u(x,t)}{\partial t} = -u(x,t) + \int_{-\infty}^{\infty} w(x-x')F(u(x',t))dx' + I(x-vt), \tag{7.184}$$

where the input is taken to be a positive, bounded, monotonically decreasing function of amplitude $I_0 = I(-\infty) - I(\infty)$. The resulting inhomogeneous neural field equation can support a traveling front that locks to the stimulus, provided that the amplitude of the stimulus is sufficiently large [198]. Consider, in particular, the case of a Heaviside firing rate function $F(u) = H(u-\kappa)$. (See [174] for an extension to the case of a smooth sigmoid function F.) We seek a traveling wave solution $u(x,t) = \mathscr{U}(\xi)$ where $\xi = x - vt$ and $\mathscr{U}(\xi_0) = \kappa$ at a single threshold crossing point $\xi_0 \in \mathbb{R}$. The front is assumed to travel at the same speed as the input (stimulus-locked front). If $I_0 = 0$, then we recover the homogeneous equation (7.2) and ξ_0 becomes a free parameter, whereas the wave propagates at the natural speed $c(\kappa)$ given by Eq. (7.6). Substituting the front solution into Eq. (7.184) yields

$$-v\frac{d\mathscr{U}(\xi)}{d\xi} = -\mathscr{U}(\xi) + \int_{-\infty}^{\xi_0} w(\xi-\xi')d\xi' + I(\xi). \qquad (7.185)$$

This can be solved for $v > 0$ by multiplying both sides by the integrating factor $v^{-1}e^{-v\xi}$ and integrating over the interval $[\xi,\infty)$ with $U(\xi) \to 0$ as $\xi \to \infty$ to give

$$\mathscr{U}(\xi) = \frac{1}{v}\int_{\xi}^{\infty} e^{(\xi-\xi')/v}[W(\xi'-\xi_0) + I(\xi')]d\xi',$$

with $W(\xi)$ defined according to Eq. (7.3). Similarly, for $v < 0$, we multiply by the same integrating factor and then integrate over $(-\infty,\xi]$ with $U(\xi) \to W_0$ as $\xi \to -\infty$ to find

$$\mathscr{U}(\xi) = -\frac{1}{v}\int_{-\infty}^{\xi} e^{(\xi-\xi')/v}[W(\xi'-\xi_0) + I(\xi')]d\xi'.$$

The threshold crossing condition $\mathscr{U}(\xi_0) = \kappa$ then determines the position ξ_0 of the front relative to the input as a function of speed v, input amplitude I_0, and threshold κ.

One of the interesting features of stimulus-locked fronts is that they are much more robust to noise [70]. In order to show this, consider the following stochastic version of Eq. (7.184):

$$dU(x,t) = \left[-U(x,t) + \int_{-\infty}^{\infty} w(x-y)F(U(y,t))dy + I(x-vt)\right]dt$$
$$+\varepsilon^{1/2}g(U(x,t))dW(x,t). \qquad (7.186)$$

Proceeding along identical lines to the case of freely propagating fronts, Eq. (7.186) is first rewritten so that the fluctuating term has zero mean:

$$dU(x,t) = \left[h(U(x,t)) + \int_{-\infty}^{\infty} w(x-y)F(U(y,t))dy + I(x-vt)\right]dt + \varepsilon^{1/2}dR(U,x,t), \qquad (7.187)$$

and h and R are given by Eqs. (7.160) and (7.161), respectively. The stochastic field $U(x,t)$ is then decomposed according to Eq. (7.163) with U_0 a front solution of

$$-v\frac{dU_0}{d\xi} - h(U_0(\xi)) - I(\xi) = \int_{-\infty}^{\infty} w(\xi-\xi')F(U_0(\xi'))d\xi'. \qquad (7.188)$$

It is assumed that the fixed profile U_0 is locked to the stimulus (has speed v). However, multiplicative noise still has an effect on U_0 by generating an ε-dependent threshold crossing point ξ_ε such that $U_0(\xi_\varepsilon) = \kappa$.

Proceeding to the next order and imposing equation (7.188), we find that $\Delta(t) = \mathscr{O}(\varepsilon^{1/2})$ and

$$d\Phi(\xi,t) = \mathbb{L} \circ \Phi(\xi,t)dt + \varepsilon^{-1/2}U_0'(\xi)d\Delta(t) + dR(U_0,\xi,t) + \varepsilon^{-1/2}I'(\xi)\Delta(t)dt \tag{7.189}$$

where $\mathbb{L}$ is the non-self-adjoint linear operator (7.167) with $c_\varepsilon \to v$. The last term on the right-hand side of Eq. (7.189) arises from the fact that in Eq. (7.163), U_0 and Φ are expressed as functions of $\xi - \Delta(t)$ and $I(\xi) = I(\xi - \Delta(t) + \Delta(t)) \approx I(\xi - \Delta(t)) + I'(\xi - \Delta(t))\Delta(t)$. A nontrivial solution of Eq. (7.189) exists if and only if the inhomogeneous part is orthogonal to the null vector $\mathscr{V}(\xi)$ of the adjoint operator $\mathbb{L}^\dagger$ defined by Eq. (7.169) with $c_\varepsilon \to v$. Taking the inner product of both sides of Eq. (7.189) with respect to $\mathscr{V}(\xi)$ thus leads to the solvability condition

$$\int_{-\infty}^{\infty} \mathscr{V}(\xi)\left[U_0'(\xi)d\Delta(t) + I'(\xi)\Delta(t)dt + \varepsilon^{1/2}dR(U_0,\xi,t)\right]d\xi = 0. \tag{7.190}$$

It follows that, to leading order, $\Delta(t)$ satisfies the Ornstein–Uhlenbeck equation

$$d\Delta(t) + A\Delta(t)dt = d\hat{W}(t), \tag{7.191}$$

where

$$A = \frac{\int_{-\infty}^{\infty} \mathscr{V}(\xi)I'(\xi)d\xi}{\int_{-\infty}^{\infty} \mathscr{V}(\xi)U_0'(\xi)d\xi}, \tag{7.192}$$

and

$$\hat{W}(t) = -\varepsilon^{1/2}\frac{\int_{-\infty}^{\infty} \mathscr{V}(\xi)g(U_0(\xi))W(\xi,t)d\xi}{\int_{-\infty}^{\infty} \mathscr{V}(\xi)U_0'(\xi)d\xi}. \tag{7.193}$$

Note that $A > 0$ for $I_0 > 0$, since both $U_0(\xi)$ and $I(\xi)$ are monotonically decreasing functions of ξ. Moreover

$$\langle d\hat{W}(t)\rangle = 0, \quad \langle d\hat{W}(t)d\hat{W}(t)\rangle = 2D(\varepsilon)dt \tag{7.194}$$

with $D(\varepsilon)$ given by Eq. (7.172). Using standard properties of an Ornstein–Uhlenbeck process [210], we conclude that

$$\langle \Delta(t)\rangle = \Delta(0)\mathrm{e}^{-At}, \quad \langle \Delta(t)^2\rangle - \langle \Delta(t)\rangle^2 = \frac{D(\varepsilon)}{A}\left[1 - \mathrm{e}^{-2At}\right]. \tag{7.195}$$

In particular, the variance approaches a constant $D(\varepsilon)/A$ in the large t limit, rather than increasing linearly with time as found for freely propagating fronts.

In order to illustrate the above analysis, take $g(U) = g_0U$ for the multiplicative noise term and set $F(U) = H(U - \kappa)$. The deterministic Eq. (7.188) for the profile U_0 then reduces to

$$-v\frac{dU_0}{d\xi} + U_0(\xi)[1 - \varepsilon g_0^2 C(0)] + I(\xi) = \int_{-\infty}^{\infty} w(\xi - \xi')H(U_0(\xi') - \kappa)d\xi'. \tag{7.196}$$

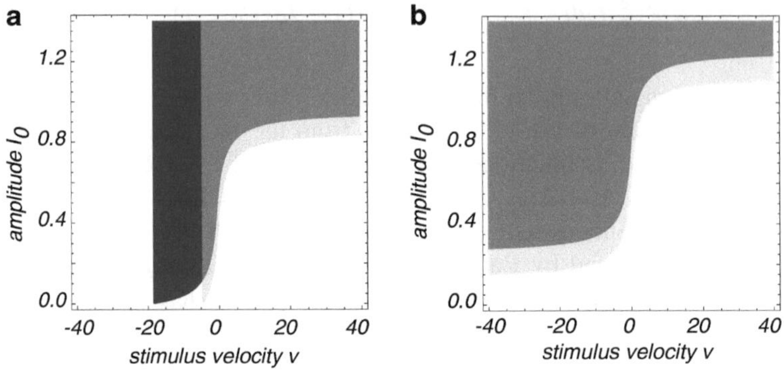

Fig. 7.13 Plot of existence regions of a stimulus-locked front without noise ($\gamma = 1$, *dark gray*) and in the presence of noise ($\gamma = 0.9$, *light gray*) with overlapping regions indicated by *medium gray*. Stimulus taken to be of the form $I(x,t) = I_0 H(-\xi), \xi = x - vt$ with amplitude I_0 and speed v. Other parameter values as in Fig. 7.10. (**a**) $\kappa = 0.95$: spontaneous fronts exist in the absence of a stimulus ($I_0 = 0$). (**b**) $\kappa = 1.25$: there are no spontaneous fronts

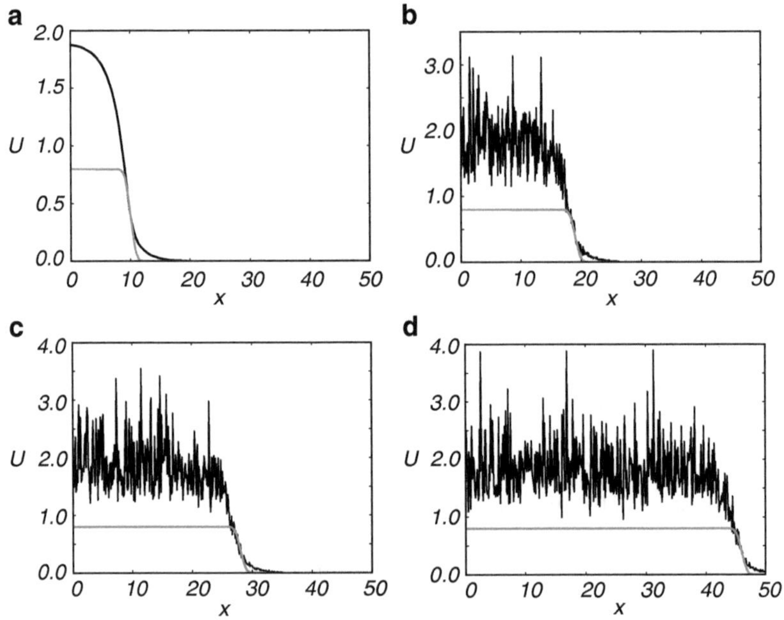

Fig. 7.14 Numerical simulation showing the propagation of a stimulus-locked wave-front solution (*black curves*) of the stochastic neural field equation (7.186) for Heaviside weight function $F(U) = H(U - \kappa)$ with $\kappa = 0.35$, exponential weight function (7.1) with $\sigma = 2$, and multiplicative noise $g(U) = U$. The external input (*gray curves*) is taken to be of the form $I(x,t) = I_0 \mathrm{Erfc}[x - vt]$ with amplitude $I_0 = 0.4$ and speed $v = 1.5$. Noise strength $\varepsilon = 0.005$ and $C(0) = 10$. The wave profile is shown at successive times (**a**) $t = 0$ (**b**) $t = 6$ (**c**) $t = 12$ and (**d**) $t = 24$, with the initial profile at $t = 0$ given by the solution U_0 of Eq. (7.188). In numerical simulations we take the discrete space and time steps $\Delta x = 0.1, \Delta t = 0.01$

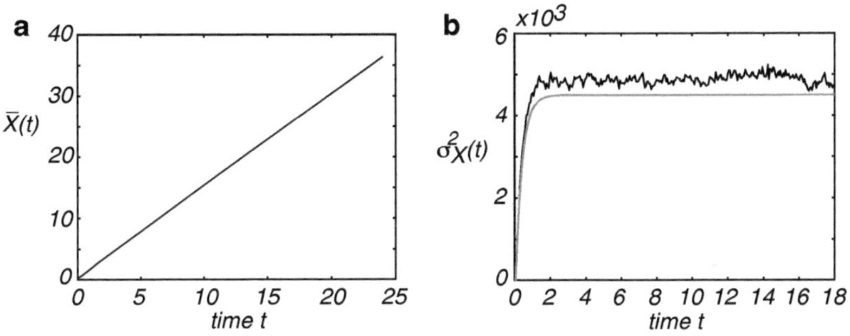

Fig. 7.15 Plot of (**a**) mean $\overline{X}(t)$ and (**b**) variance $\sigma_{\overline{X}}^2(t)$ of the position of a stimulus-locked front as a function of time, averaged over $N = 4096$ trials. Smooth gray curve in (b) indicates theoretical prediction of variance. Stimulus taken to be of the form $I(x,t) = I(x - ct) = I_0\text{Erfc}[x - vt]$ with amplitude $I\_0 = 0.4$ and speed $v = 1.5$. Other parameter values as in Fig. 7.10

Existence of front solution proceeds along identical lines to Sect. 7.1, except now the speed v is fixed, whereas the threshold crossing point ξ_0, say, is no longer arbitrary due to the breaking of translation symmetry. The point ξ_0 is determined by the threshold condition $U_0(\xi_0) = \kappa$ and will depend on the noise strength ε. In Fig. 7.13 we show existence regions in the (v, I_0)-plane for stimulus-locked fronts when $I(\xi) = I_0 H(-\xi)$, that is, for a step function input of speed v and amplitude I_0. This illustrates the fact that multiplicative noise leads to an ε-dependent shift in the existence regions. In Fig. 7.14 we show the temporal evolution of a single stimulus-locked front, which is obtained by numerically solving the Langevin equation (7.186) for $F(U) = H(U - \kappa)$, $g(U) = U$ and an exponential weight distribution w. Numerically speaking, it is convenient to avoid discontinuities in the input by taking $I(x,t) = I_0\text{Erfc}[x - vt]$ rather than a Heaviside. The corresponding mean $\overline{X}(t)$ and variance $\sigma_{\overline{X}}^2(t)$ of the position of the front, which are obtained by averaging over level sets as outlined in Sect. 7.4.1, are shown in Fig. 7.15. It can be seen that, as predicted by the analysis, $\overline{X}(t)$ varies linearly with t with a slope equal to the stimulus speed $v = 1.5$. Moreover, the variance $\sigma_{\overline{X}}^2(t)$ approaches a constant value as $t \to \infty$, which is comparable to the theoretical value $D(\varepsilon)/A$ evaluated for the given input. Thus, we find that stimulus-locked fronts are much more robust to noise than freely propagating fronts, since the variance of the mean position saturates as $t \to \infty$. Consequently, stimulus locking persists in the presence of noise over most of the parameter range for which stimulus locking is predicted to occur.

7.4.3 Stochastic Pulled Fronts

In the case of the F–KPP equation with multiplicative noise, one finds that the stochastic wandering of a pulled front about its mean position is subdiffusive with $\text{var}\Delta(t) \sim t^{1/2}$, in contrast to the diffusive wandering of a front propagating into a

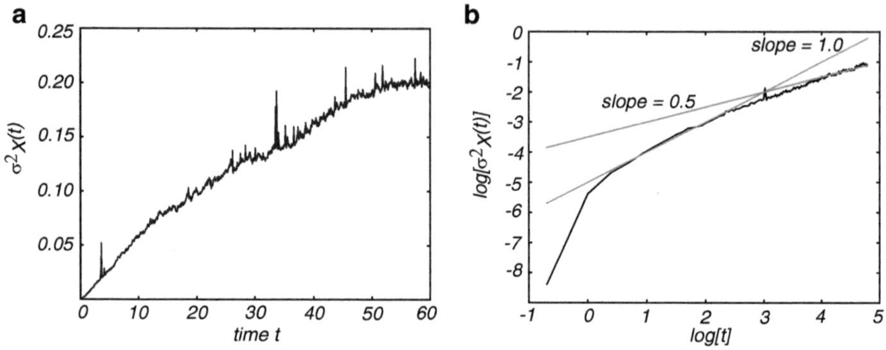

Fig. 7.16 (**a**) Plot of variance $\sigma_X^2(t)$ of the position of a stochastic pulled front as a function of time. (**b**) Log-log plot of variance $\sigma_X^2(t)$ as a function of time t. Noise amplitude $\varepsilon = 0.005$ and $\kappa = 0.8$. Other parameter values as in Fig. 7.2

metastable state for which $\mathrm{var}\Delta(t) \sim t$ [530]. Such scaling is a consequence of the asymptotic relaxation of the leading edge of the deterministic pulled front. Since pulled front solutions of the neural field equation (7.27) exhibit similar asymptotic dynamics (see Eq. (7.46)), it suggests that there will also be subdiffusive wandering of these fronts in the presence of multiplicative noise. In order to illustrate this, consider the stochastic neural field equation

$$dA(x,t) = \left[-A(x,t)+F\left(\int_{-\infty}^{\infty} w(x-y)A(y,t)dy\right)\right] dt+\varepsilon^{1/2}g_0A(x,t)dW(x,t) \quad (7.197)$$

with $W(x,t)$ a Wiener process satisfying Eq. (2.84). Note that the noise term has to vanish when $A(x,t)=0$, since the firing rate A is restricted to be positive. Hence, the noise has to be multiplicative. Formally speaking, one can carry over the analysis of the Langevin equation (7.157). First, decompose the solution along similar lines to Eq. (2.96):

$$A(x,t) = A_0(\xi - \Delta(t)) + \varepsilon^{1/2}\Phi(\xi - \Delta(t),t) \quad (7.198)$$

with $\xi = x - c_\varepsilon t$, with the fixed front profile A_0 satisfying the deterministic equation

$$-c_\varepsilon \frac{dA_0}{d\xi} + A_0(\xi)[1-\varepsilon g_0^2 C(0)] = F\left(\int_{-\infty}^{\infty} w(\xi-\xi')A_0(\xi')d\xi'\right). \quad (7.199)$$

The effective velocity c_ε of the front is given by the minimum of the dispersion curve

$$c_\varepsilon(\lambda) = \frac{1}{\lambda}\left[\hat{W}(\lambda)+\hat{W}(-\lambda)-[1-\varepsilon g_0^2 C(0)]\right]. \quad (7.200)$$

Fluctuations thus shift the dispersion curve to higher velocities. However, it is no longer possible to derive an expression for the diffusion coefficient $D(\varepsilon)$ along the lines of Eq. (7.172), since both numerator and denominator would diverge for a pulled front. This reflects the asymptotic behavior of the leading edge of the front.

It is also a consequence of the fact that there is no characteristic time scale for the convergence of the front velocity to its asymptotic value, which means that it is not possible to separate the fluctuations into a slow wandering of front position and fast fluctuations of the front shape [162, 486]. Nevertheless, numerical simulations of Eq. (7.197) with F given by the piecewise linear firing rate (7.29) are consistent with subdiffusive wandering of the front. In Fig. 7.16(a), we plot the variance $\sigma_X^2(t)$ of the position of a pulled front solution of Eq. (7.197), which are obtained by averaging over level sets along identical lines to Sect. 7.4.1. It can be seen that the variance appears to exhibit subdiffusive behavior over longer time scales. This is further illustrated by plotting a log-log plot of $\sigma_X^2(t)$ against time t; see Fig. 7.16(b). At intermediate time scales, the slope of the curve is approximately equal to one, consistent with normal diffusion, but at later times the slope decreases, indicating subdiffusive behavior.

7.5 Traveling Waves in 2D Oscillatory Neural Fields

Troy and Shusterman [580, 634] have shown how a neural field model with strong linear adaptation (see Eq. (7.49)) can act as an oscillatory network that supports 2D target patterns and spiral waves consistent with experimental studies of tangential cortical slices [292]. (For the analysis of spiral waves in the corresponding excitable regime, see [354].) However, since the linear form of adaptation used in these studies is not directly related to physiological models of adaptation, it is difficult to ascertain whether or not the strength of adaptation required is biologically reasonable. This motivated a more recent study of spiral waves in a 2D neural

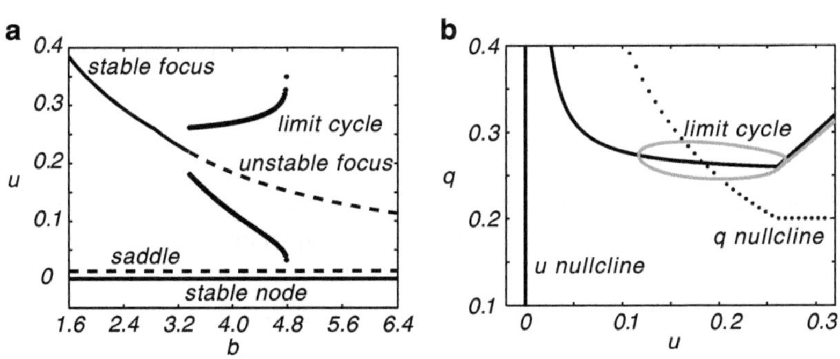

Fig. 7.17 Limit cycle oscillations in the space-clamped system (7.202) for a piecewise linear firing rate function (1.16) with threshold $\kappa = 0.01$ and gain $\eta = 4$. (**a**) Bifurcation diagram showing fixed points u of the system as a function of β for $\tau_q = 80$. (**b**) Corresponding phase-plane plot of q versus u (*gray curve*) for $\beta = 4$, showing that the system supports a stable limit cycle [329]

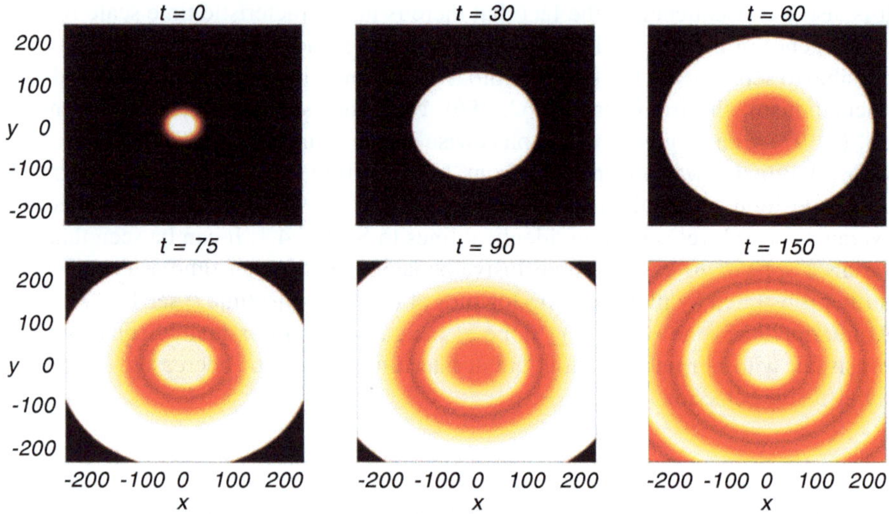

Fig. 7.18 Target patterns in a 2D neural field with synaptic depression induced by an initial conditional stimulus specified by Eq. (7.203) at $t = 0$, where $\chi = 1$ and $\zeta = 25$. Initially, an activated state spreads radially outward, across the entire medium as a traveling front. Then, the localized oscillating core of activity emits a target wave with each oscillation cycle. Eventually, these target waves fill the domain. Each target wave can be considered as a phase shift in space of the oscillation throughout the medium; they travel with the same speed as the initial front. Parameters are $\tau_q = 80$, $\beta = 4$, $\eta = 4$, and $\kappa = 0.01$ [330]

medium involving a nonlinear, physiologically based form of adaptation, namely, synaptic depression [330]. The latter model is given by

$$\begin{aligned}\frac{\partial u(\mathbf{r},t)}{\partial t} &= -u(\mathbf{r},t) + \int w(|\mathbf{r}-\mathbf{r}'|)q(\mathbf{r}',t)F(u(\mathbf{r}',t))d\mathbf{r}' \\ \frac{\partial q(\mathbf{r},t)}{\partial t} &= \frac{1-q(\mathbf{r},t)}{\tau_q} - \beta q(\mathbf{r},t)F(u(\mathbf{r},t)).\end{aligned} \tag{7.201}$$

The radially symmetric excitatory weight distribution is taken to be an exponential, $w(r) = \mathrm{e}^{-r}/2\pi$. It can be shown that the space-clamped model

$$\dot{u}(t) = -u(t) + q(t)F(u(t)), \quad \dot{q}(t) = \frac{1-q(t)}{\tau_q} - \beta q(t)F(u(t)), \tag{7.202}$$

supports limit cycle oscillations provided that the firing rate function has finite gain. For example, in the case of the piecewise linear firing rate function (1.16), oscillations arise via a subcritical Hopf bifurcation of a high activity fixed point; see Fig. 7.17. One then finds that the full network model (7.201) supports a spatially localized oscillating core that periodically emits traveling pulses [330]. Such dynamics can be induced by taking an initial condition of the form

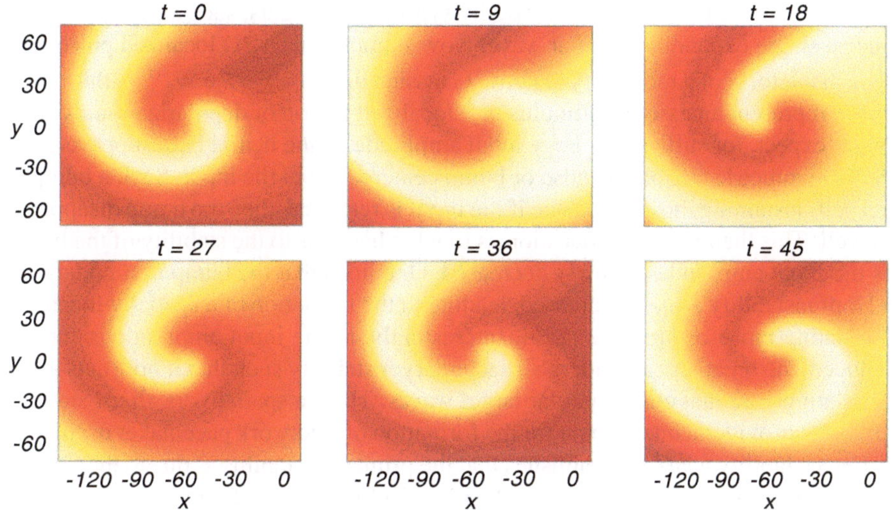

Fig. 7.19 Spiral wave generated by shifting the phase of the top and bottom halves of the target pattern shown in Fig. 7.18. The period of the spiral wave oscillation is roughly the same as the period of the oscillation in the space-clamped system. All patches of neurons are oscillating at the same frequency, but phase-shifted as coordinates are rotated about the central phase singularity [330]

$$(u(\mathbf{r},0),q(\mathbf{r},0)) = (\chi \mathrm{e}^{-(x^2+y^2)/\zeta^2},1), \tag{7.203}$$

where χ and ζ parameterize the amplitude and spatial constant of the initial state. An example of a pulse-emitting core is shown in Fig. 7.18, which oscillates at a frequency of roughly 3 Hz. Pulses are emitted each cycle and travel at a speed of roughly 30 cm/s, which is determined by the period of the oscillations; the latter is set by the time constant of synaptic depression. The initial emission of spreading activity appears as a traveling front which propagates from the region activated by the input current into the surrounding region of zero activity; it travels at the same speed as the subsequent target waves. The front converts each region of the network into an oscillatory state that is phase-shifted relative to the core, resulting in the appearance of a radially symmetric target pattern. Spiral waves can also be induced by breaking the rotational symmetry of pulse emitter solutions [330]. More specifically, if the target pattern produced by the emitter has the top and bottom halves of its domain phase-shifted, then the dynamics evolves into two counterrotating spirals on the left and right halves of the domain. Closer inspection of one of these spirals reveals that it has a fixed center about which activity rotates indefinitely as shown in Fig. 7.19.

A very different mechanism for generating periodic waves in a 1D or 2D neural field model is through the combination of adaptation and a spatially localized input [196, 197]. Recall from Sect. 7.2 that a 1D excitatory neural field with adaptation supports the propagation of solitary traveling pulse, which can be induced by perturbing the system with a transient localized pulse. (In contrast to the previous example, we are assuming that the neural field operates in an excitable regime.)

In the case of a 2D network with radially symmetric weights, such a pulse will produce a single expanding circular wave. Now suppose that a 2D localized pulse persists in the form of a radially symmetric Gaussian input $I(\mathbf{r}) = I_0 e^{-r^2/\sigma^2}$—this could either represent an external stimulus or a localized region of depolarization. As one might expect, for sufficiently large input amplitude I_0, the neural field supports a radially symmetric stationary pulse or bump centered about the input. Such a bump is not self-sustaining, however, since if the input is removed, then the bump disappears as well. This then raises the question as to what happens to the stability of the bump as the input amplitude is slowly decreased. One finds that the bump first undergoes a Hopf instability as I_0 is decreased, leading to the formation of a spatially localized oscillating pulse or *breather* [196]. Interestingly, as the input amplitude is further reduced, the breather can undergo a secondary instability such that it now acts as an oscillating core that emits circular target waves. Thus, a spatially localized stationary input provides a mechanism for the formation of a network pacemaker oscillator. A linear stability analysis establishes that the primary instability is due to the growth of radially symmetric eigenmodes. A similar bifurcation scenario also occurs in a neural field with lateral inhibition, except that now the Hopf bifurcation typically involves the growth of nonradially symmetric eigenmodes, resulting in asymmetric breathers and rotating waves [197].

Chapter 8
Neural Field Model of Binocular Rivalry Waves

Binocular rivalry is the phenomenon whereby perception switches back and forth between different images presented to the two eyes. The resulting fluctuations in perceptual dominance and suppression provide a basis for noninvasive studies of the human visual system and the identification of possible neural mechanisms underlying conscious visual awareness [46, 47]. Various psychophysical experiments have demonstrated that the switch between a dominant and suppressed visual percept propagates as a traveling front for each eye [312, 369, 678]. In this chapter we show how the neural field theory presented in Chap. 7 can be used to model the phenomenon of binocular rivalry waves [83, 663].

8.1 Binocular Rivalry Waves and the Structure of Primary Visual Cortex

We begin by summarizing several important properties of neurons and their functional organization within primary visual cortex (V1), and then review some of the experimental findings regarding binocular rivalry.

8.1.1 Functional Architecture of V1

V1 is the first cortical area to receive visual information from the retina (see Fig. 8.1). The output from the retina is conveyed by ganglion cells whose axons form the optic nerve. The optic nerve conducts the output spike trains of the retinal ganglion cells to the lateral geniculate nucleus (LGN) of the thalamus, which acts as a relay station between retina and primary visual cortex (V1). Prior to arriving at the LGN, some ganglion cell axons cross the midline at the optic chiasm. This allows the left and right sides of the visual fields from both eyes to be represented on the right and left sides of the brain, respectively. Note that signals from the left and

P.C. Bressloff, *Waves in Neural Media: From Single Neurons to Neural Fields*, Lecture Notes on Mathematical Modelling in the Life Sciences,
DOI 10.1007/978-1-4614-8866-8_8, © Springer Science+Business Media New York 2014

right eyes are segregated in the LGN and in input layers of V1. This means that the corresponding LGN and cortical neurons are monocular, in the sense that they only respond to stimuli presented to one of the eyes but not the other (ocular dominance).

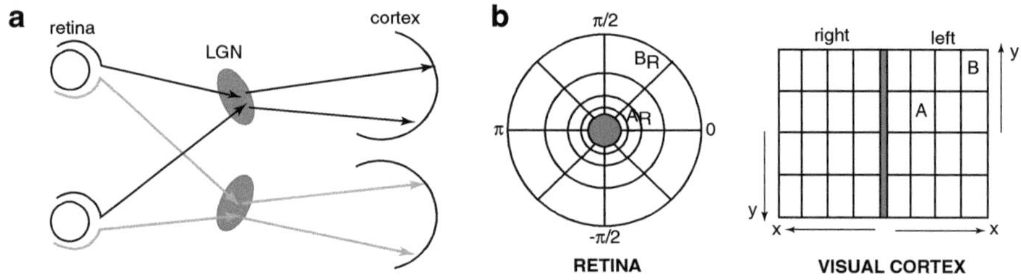

Fig. 8.1 (**a**) Visual pathways from the retina through the lateral geniculate nucleus (LGN) of the thalamus to the primary visual cortex (V1). (**b**) Schematic illustration of the complex logarithmic mapping from retina to V1. Foveal region in retina is indicated by *gray disk*. Regions A_R and B_R in the visual field are mapped to regions A and B in cortex

One of the striking features of the early visual system is that the visual world is mapped onto the cortical surface in a topographic manner. This means that neighboring points in a visual image evoke activity in neighboring regions of visual cortex. Moreover, one finds that the central region of the visual field has a larger representation in V1 than the periphery, partly due to a nonuniform distribution of retinal ganglion cells. The retinotopic map is defined as the coordinate transformation from points in the visual world to locations on the cortical surface and can be approximated by a complex logarithm [562]. Superimposed upon the retinotopic map are additional maps reflecting the fact that neurons respond preferentially to stimuli with particular features [613]. Neurons in the retina, LGN, and primary visual cortex respond to light stimuli in restricted regions of the visual field called their classical receptive fields (RFs). Patterns of illumination outside the RF of a given neuron cannot generate a response directly, although they can significantly modulate responses to stimuli within the RF via long-range cortical interactions (see below). The RF is divided into distinct ON and OFF regions. In an ON (OFF) region illumination that is higher (lower) than the background, light intensity enhances firing. The spatial arrangement of these regions determines the selectivity of the neuron to different stimuli. For example, one finds that the RFs of most V1 cells are elongated so that the cells respond preferentially to stimuli with certain preferred orientations (see Fig. 8.2). Similarly, the width of the ON and OFF regions within the RF determines the optimal spacing of alternating light and dark bars to elicit a response, that is, the cell's spatial frequency preference.

In recent years much information has accumulated about the spatial distribution of orientation selective cells in V1 [216]. One finds that orientation preferences rotate smoothly over the surface of V1, so that approximately every 300 μm the same preference reappears, i.e., the distribution is π-periodic in the orientation preference angle. One also finds that cells with similar feature preferences tend to arrange themselves in vertical columns so that to a first approximation the layered structure of cortex can be ignored. A more complete picture of the two-dimensional distribution of both orientation preference and ocular dominance in layers 2/3 has been obtained using optical imaging techniques [50, 51, 53]. The basic experimental procedure involves shining light directly on to the surface of the cortex. The degree of light absorption within each patch of cortex depends on the local level of activity. Thus, when an oriented image is presented across a large part of the visual field, the regions of cortex that are particularly sensitive to that stimulus will be differentiated. The topography revealed by these methods has a number of characteristic features [469]; see Fig. 8.2b: (i) Orientation preference changes continuously as a function of cortical location, except at singularities or *pinwheels*. (ii) There exist *linear zones*, approximately $750 \times 750\,\mu\text{m}^2$ in area (in primates), bounded by pinwheels, within which iso-orientation regions form parallel slabs. (iii) Linear zones tend to cross the borders of ocular dominance stripes at right angles; pinwheels tend to align with the centers of ocular dominance stripes. These experimental findings suggest that there is an underlying periodicity in the microstructure of V1 with a period of approximately 1 mm (in cats and primates). The fundamental domain of this approximate periodic (or quasiperiodic) tiling of the cortical plane is the hypercolumn [293, 294, 377], which contains two sets of orientation preferences $\theta \in [0, \pi)$ per eye, organized around a pair of singularities; see Fig. 8.2b.

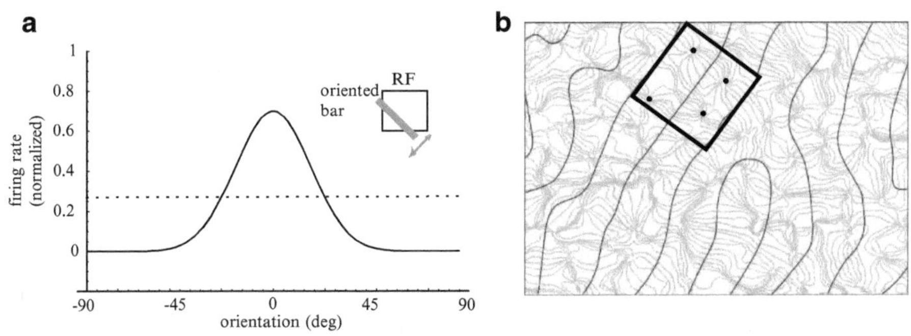

Fig. 8.2 (**a**) Schematic illustration of an orientation tuning curve of a V1 neuron. Average firing rate is plotted as a function of the orientation of a bar stimulus that is moved back and forth within the receptive field (RF) of the neuron. The peak of the orientation tuning curve corresponds to the orientation preference of the cell. (**b**) Schematic illustration of iso-orientation (*light*) and ocular dominance (*dark*) contours in a region of primate V1. A cortical hypercolumn consists of two orientation singularities or pinwheels per ocular dominance column

Within each hypercolumn, neurons with sufficiently similar orientations tend to excite each other, whereas those with sufficiently different orientations inhibit each other, and this serves to sharpen a particular neuron's orientation preference [34, 188]. There are also longer-range horizontal or lateral connections that link cells in different hypercolumns [218, 531]. Optical imaging combined with labeling techniques has generated considerable information concerning the pattern of

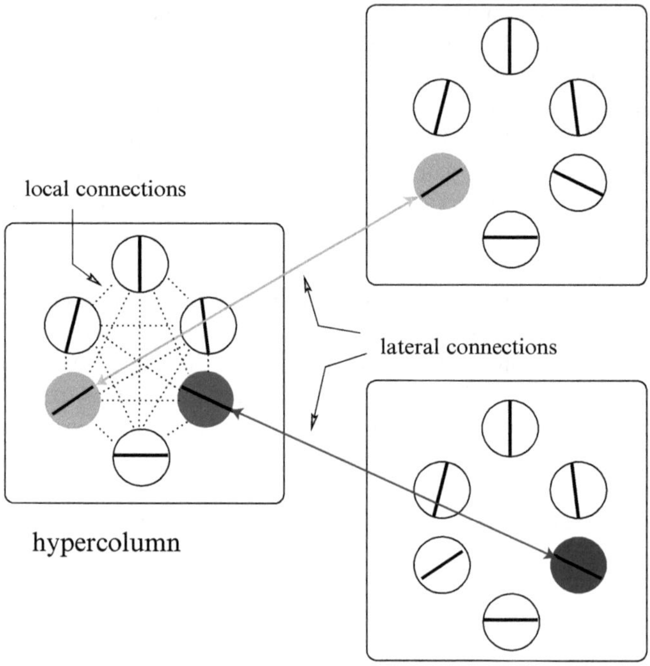

Fig. 8.3 Schematic illustration of anisotropic horizontal connections. Orientation selective cells within a hypercolumn tend to connect to all neighbors in a roughly isotropic fashion. On the other hand, longer-range horizontal connections link cells between hypercolumns with similar orientation preferences along a particular visuotopic axis

these connections in superficial layers of V1 [56, 401, 691]. In particular, one finds that the horizontal connections are patchy and tend to link cells with similar feature preferences. Moreover, in certain animals such as tree shrew and cat, there is a pronounced anisotropy in the distribution of patchy connections, with differing iso-orientation patches preferentially connecting to neighboring patches in such a way as to form continuous contours following the topography of the retinotopic map [56]. This is illustrated schematically Fig. 8.3. That is, the major axis of the horizontal connections tends to run parallel to the visuotopic axis of the connected cells' common orientation preference. There is also a clear anisotropy in the patchy connections of primates [11, 583]. However, in these cases most of the anisotropy can be accounted for by the fact that V1 is expanded in the direction orthogonal to ocular dominance columns [11]. Nevertheless, it is possible that when this expansion is factored out, there remains a weak anisotropy correlated with orientation selectivity. Moreover, patchy feedback connections from higher-order visual areas in primates are strongly anisotropic [11]. Stimulation of a hypercolumn via lateral connections modulates rather than initiates spiking activity [278], suggesting that the long-range interactions provide local cortical processes with contextual information about the global nature of stimuli. As a consequence horizontal and feedback connections have been invoked to explain a wide variety of context-dependent visual processing phenomena [11, 217]. In summary, the functional relationship between stimulus feature preferences and synaptic connections within V1 suggests that V1 is a likely substrate of many simple examples of binocular rivalry, such as those involving sinusoidal grating stimuli.

8.1.2 Binocular Rivalry

The perception of ambiguous figures is a well-known phenomenon where, in response to a single fixed image, a subject switches back and forth between two rival percepts; see Fig. 8.4. In the case of binocular rivalry, fixed but different images are presented to the two eyes. Rather than perceiving a superposition of the two images, a subject typically sees the left-eye image, say, which then switches to the right-eye image in an alternating fashion. Although binocular rivalry has been studied for hundreds of years, only recently have experimentalists clarified some of its specific statistical properties using a combination of psychophysics and brain imaging [47, 48]:

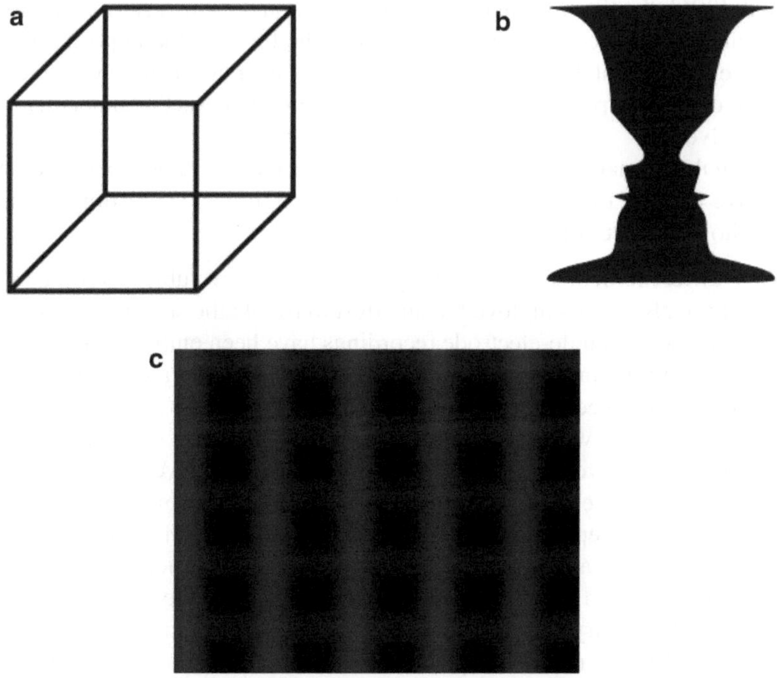

Fig. 8.4 Ambiguous figures. (**a**) Necker cube. (**b**) Faces/vase. (**c**) Monocular rivalry between a horizontal orange grating and a vertical green grating. [Reproduced from Wikimedia Commons (public domain)]

1. Increasing the "strength" of one of the rival figures (brighter, moving rather than stationary, more densely contoured) increases the percentage of time that it is dominant. However, this tends to occur due to a decrease in the periods of suppression of the enhanced figure rather than an increase in its dominance periods [199, 390]. The stimulus dependence of rivalry was summarized in a series of postulates by Levelt [378].
2. The switching from one percept to the next during binocular rivalry occurs stochastically, with the switching times distributed according to a gamma distribution [390]. A number of studies suggest that there are little or no correlations between one dominance time and the next [59, 372].
3. Attention to a particular stimulus may prolong the dominance time of the attended stimulus [270]. Experimental evidence suggests that this correlation may hold true when the subject practices repeatedly [114]. However, the non-attended stimulus will eventually appear in the observer's perception, no matter how strongly attention is focused on the other. The fact that attention can bias the outcomes of binocular rivalry suggests that higher level visual areas may play a modulatory role in the phenomenon [674]. Attention may increase the apparent contrast of the attended stimulus or simply boost the activity of one of two competing neural populations [114].
4. The "tilt aftereffect" is still observed when vertical and tilted lines are rivalrous stimuli [306]; the tilt aftereffect is the phenomenon by which staring at a slightly tilted line will make a vertical line appear to be tilted in the opposite direction if viewed immediately after. Since the neural substrate of this phenomenon is known to reside in primary visual cortex (V1), this suggests that binocular rivalry can involve V1.
5. Perceptual dominance transitions are not instantaneous; rather they emerge in a wavelike fashion, originating at one region of a figure and spreading from there throughout the rest of the figure [312, 369, 678].

Several different methods of recording neural activity in subjects during binocular rivalry have also been employed in an effort to isolate the specific sites of its encoding. In monkeys, single-electrode recordings have been employed to track electrical activity during binocular rivalry tasks [47]. Evidence of an elevation in some cells' firing rates has been found in the V1, V2, and V4 regions of visual cortex, and these correspond well with the monkey's reported perception of a stimulus [376]. Thus, it appears that several areas of visual cortex may be involved. However, single-unit recordings have yet to reveal changes in the firing rate of LGN neurons that correspond to the perceptual changes of binocular rivalry [373]. In humans, less invasive techniques such as scalp recordings and functional magnetic resonance imaging (fMRI) have helped to localize brain regions whose activity reflects the experience of binocular rivalry. Visually evoked potentials measured on the scalp during a binocular rivalry task reveal that potential waveforms associated with each stimulus possess a time course very closely linked to the subject's perceptual observation [85]. A number of fMRI studies have verified there are changes in the blood-oxygen-level-dependent (BOLD) signals in V1 that correspond to the perceived switching in stimulus dominance [369, 510, 626]. It is likely that the site of binocular rivalry depends on the nature of the rivalrous images. Indeed, some theories propose that there may be a hierarchy of visual areas involved so that there is a competition between either eye's input as well as between differing patterns [627, 674].

One way to observe and measure the speed of perceptual waves in psychophysical experiments of binocular rivalry [369, 678] is to take the rival images to be a low-contrast radial grating presented to one eye and a high-contrast spiral grating presented to the other eye. Each image is restricted to an annular region of the visual field centered on the fixation point of the observer, thus effectively restricting wave propagation to the one dimension around the annulus. Switches in perceptual dominance can then be triggered using a brief rapid increase in stimulus contrast within a small region of the suppressed low-contrast grating. This induces a perceptual traveling wave in which the observer perceives the local dominance of the low-contrast image spreading around the annulus. The observer presses a key when the perceptual wave reaches a target area at a fixed distance from the trigger zone, and this determines the wave speed [369, 678]. Since the rival images consist of oriented gratings, one might expect that primary visual cortex (V1) plays some role in the generation of binocular rivalry waves. Indeed, it has been shown using fMRI that there is a systematic correspondence between the spatiotemporal dynamics of activity in V1 and the time course of perceptual waves [369]. However, it has not been established whether the waves originate in V1 or are evoked by feedback from extrastriate cortex. Recently Kang et al. [312, 313] have developed a new psychophysical method for studying binocular rivalry waves that involves periodic perturbations of the rival images consisting of linear gratings. An observer tracks rivalry within a small, central region of spatially extended rectangular grating patterns, while alternating contrast triggers are presented repetitively in the periphery of the rival patterns. The basic experimental setup is illustrated in Fig. 8.5. A number of interesting results have been obtained from these studies. First, over a range of trigger frequencies, the switching between rival percepts within the central regions is entrained to the triggering events. Moreover, the optimal triggering frequency depends on the natural frequency of spontaneous switching (in the absence of triggers). Second, the latency between triggering event and perceptual switching increases approximately linearly with the distance between the triggering site and the central region being tracked by the observer, consistent with the propagation of a traveling front. Third, the speed of the traveling wave across observers covaries with the spontaneous switching rate.

8.2 Competitive Network Models of Binocular Rivalry

The above psychophysical experiments suggest that binocular rivalry waves consist of two basic components: the switching between rivalrous left-/right-eye states and the propagation of the switched state across a region of cortex. Let us first focus on the switching mechanism by neglecting spatial effects. Suppose, for the sake of illustration, that a horizontally oriented grating is presented to the left eye and a vertically oriented grating is presented to the right eye. This triggers rivalry due to the combination of orientation and eye-specific cross-inhibition in V1. During left-eye

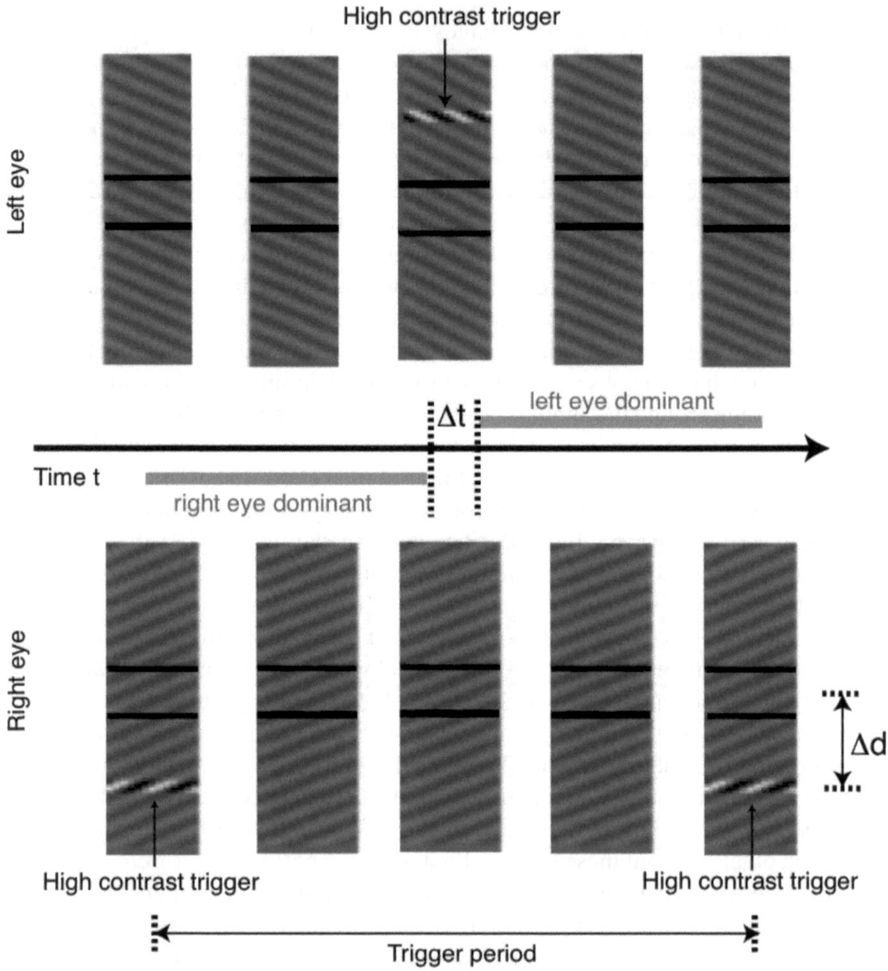

Fig. 8.5 Schematic diagram illustrating experimental protocol used to study binocular rivalry waves [312]. High-contrast triggers are presented periodically in anti-phase within the upper extended region of one grating pattern and within the lower region of the rival pattern. Subject simply reports perceptual alternations in rival dominance within the central monitoring region indicated by the *horizontal black lines* on each pattern. The monitoring region is a distance Δd from the trigger region, which can be adjusted. If Δt is the latency between the triggering event and the subsequent observation of a perceptual switch, then the speed c of the wave is given by the slope of the plot $\Delta d = c\Delta t$

stimulus dominance, it is assumed that a group of the left-eye neurons that respond to horizontal orientations are firing persistently, while right-eye neurons are suppressed by cross-inhibitory connections. Of course, there may still be some low-rate firing of the right-eye neurons, but it will be less than the firing rate of the left-eye horizontally tuned neurons [47]. Following this, some slow adaptive process causes a switch so that right-eye vertical orientation neurons fire persistently, suppressing the left-eye neurons resulting in a repetitive cycle of perceptual dominance between the left- and right-eye stimuli. The competitive network architecture of reciprocal inhibition paired with slow adaptation (Fig. 8.6) has been used extensively to model oscillations in binocular rivalry [199, 328, 355, 577, 578, 586, 619, 678]. (In some versions of the model, recurrent excitation is omitted.) In most cases a firing rate model appears sufficient to capture the elevation in neuronal spiking associated with the dominant stimulus.

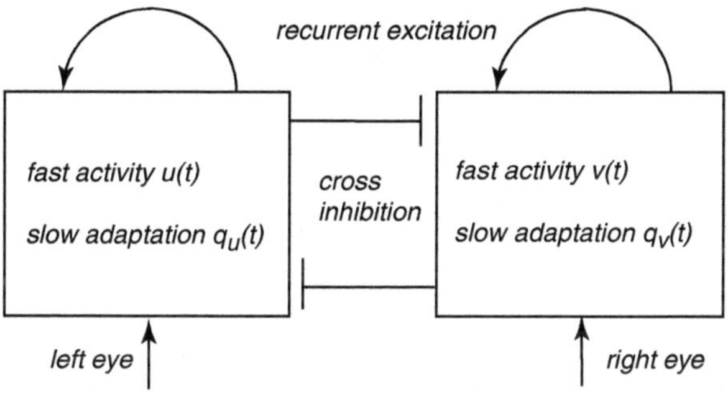

Fig. 8.6 Schematic diagram of a competitive network architecture for rivalry oscillations [355, 577, 678], consisting of two homogeneous populations of cells: one driven by left-eye stimuli and the other by right-eye stimuli. Recurrent connections within each population are assumed to be excitatory, whereas connections between the two populations are inhibitory (cross-inhibition). Each network is described by two variables: a fast activity variable and a slow adaptation variable

It remains an open question as to which slow adaptive process is most responsible for the eventual switching of one stimulus dominance to the other [577]. The mechanism of spike frequency adaptation has been suggested, since it can curtail excitatory activity in a single neuron [355, 678]. Spike frequency adaptation is the process by which a hyperpolarizing current is switched on due to a buildup of a certain ion, like calcium, within the cell due to repetitive firing [604]; see Sect. 4.1. The maximal firing rate of a neuron is lowered as a result. In the case of binocular rivalry, this may cause the dominant population to eventually drop its firing rate so that cross-inhibition suppressing the other population is then low enough for the suppressed populations to rekindle its firing rate into dominance. Since the recently released population is not adapted, it can then remain in dominance and suppress the other population for a period of time roughly equal to the time constant of spike frequency adaptation [355, 392, 678]. Another proposed switching mechanism is that the inhibitory synapses from one eye's neurons to the other's undergo synaptic depression; see Sect. 1.3. (More precisely, synaptic depression tends to be associated only with excitatory synapses, so that inclusion of depressing inhibitory connections would have to be mediated by excitatory connections innervating local interneurons, for example.) If inhibitory synapses remain repeatedly active, due to one eye's neurons suppressing the others, eventually most of those synapses' resources will be used up, the effect of inhibition will be weakened and the suppressed population will escape [328, 355, 577]. Synaptic depression could also occur in the excitatory recurrent connections. For concreteness, we will take slow adaptation to arise from synaptic depression in all synaptic weights; however, the specific choice of adaptation does not affect the main results.

Let $u(t)$ and $v(t)$ denote the activity variables of the left- and right-eye populations at time t. The rate-based equations for a competitive network model with synaptic depression can be constructed as follows [328]:

$$\tau\frac{du(t)}{dt} = -u(t)+I_u(t)+\overline{w}_e q_u(t)F(u(t))-\overline{w}_i q_v(t)F(v(t)), \tag{8.1}$$

$$\tau_s\frac{dq_u(t)}{dt} = 1-q_u(t)-\beta q_u(t)F(u(t)), \tag{8.2}$$

and

$$\tau\frac{dv(t)}{dt} = -v(t)+I_v(t)+\overline{w}_e q_v(t)F(v(t))-\overline{w}_i q_u(t)F(u(t)), \tag{8.3}$$

$$\tau_s\frac{dq_v(t)}{dt} = 1-q_v(t)-\beta q_v(t)F(v(t)), \tag{8.4}$$

where the positive constants $\overline{w}_e$ and $\overline{w}_i$ denote the strengths of recurrent excitatory and cross-inhibitory connections, I_u, I_v denote the input stimuli from the left and right eyes, and F is a firing rate function that is taken to be the sigmoid (1.14). The depression variables q_u, q_v satisfy a rate-based version of (1.45); see also (6.15). In anticipation of the analysis of traveling front solutions of a spatially extended version of the model, we will take the high gain limit of F to obtain the Heaviside function $F(u) = H(u-\kappa)$. In this limit, the existence and stability of fixed-point solutions of (8.1) and (8.4) can be analyzed explicitly [328].

Suppose that $I_u = I_v = I$ with I constant. It turns out that there are four possible homogeneous fixed points and all are stable. First, there is the *off state* $U^* = V^* = I$, which occurs when $I < \kappa$, that is, the input is not strong enough to activate either population. Second there is the *on state* or *fusion state*, where both populations are simultaneously active:

$$(U^*,V^*) = \left(\frac{\overline{w}_e-\overline{w}_i}{1+\beta}+I, \frac{\overline{w}_e-\overline{w}_i}{1+\beta}+I\right), \quad (Q_u^*,Q_v^*) = \left(\frac{1}{1+\beta}, \frac{1}{1+\beta}\right),$$

and occurs when $I > \kappa-(\overline{w}_e-\overline{w}_i)/(1+\beta)$. This case is more likely for very strong depression (β large), since cross-inhibition will be weak, or when the local connections are strong and excitation-dominated. Finally there are two *winner-takes-all* (WTA) states in which one population dominates the other: the left-eye dominant state

$$(U^*,V^*) = \left(\frac{\overline{w}_e}{1+\beta}+I, I-\frac{\overline{w}_i}{1+\beta}\right), \quad (Q_u^*,Q_v^*) = \left(\frac{1}{1+\beta}, 1\right),$$

and the right-eye dominant state

$$(U^*,V^*) = \left(I-\frac{\overline{w}_i}{1+\beta}, \frac{\overline{w}_e}{1+\beta}+I\right), \quad (Q_u^*,Q_v^*) = \left(1, \frac{1}{1+\beta}\right).$$

The WTA states exist provided that

$$I > \kappa - \frac{\overline{w}_e}{1+\beta}, \quad I < \kappa + \frac{\overline{w}_i}{1+\beta}.$$

This will occur in the presence of weak depression (β small) and strong cross-inhibition such that depression cannot exhaust the dominant hold one population has on the other. It can also be shown that (8.1)–(8.4) also support homogeneous limit cycle oscillations in which there is periodic switching between left- and right-eye dominance consistent with binocular rivalry [328]. Since all the fixed points are stable, it follows that such oscillations cannot arise via a standard Hopf bifurcation. Indeed, one finds bistable regimes where a rivalry state coexists with a fusion state as illustrated in Fig. 8.7. (Such behavior persists in the case of smooth sigmoid firing rate functions, at least for sufficiently high gain [328].) For a detailed discussion of the various bifurcation scenarios in competitive neural networks see [435, 563, 577].

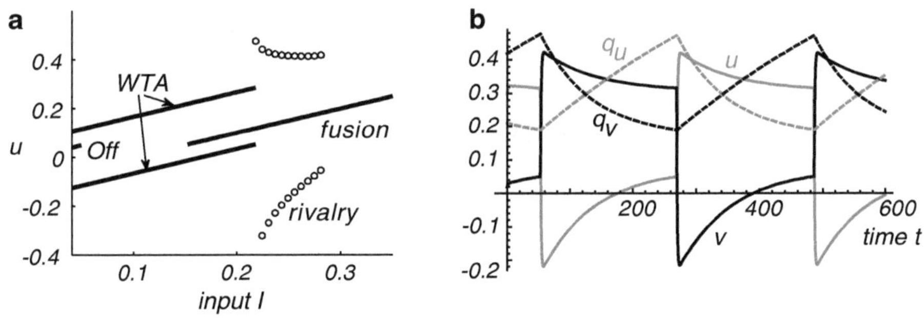

Fig. 8.7 (**a**) Bifurcation diagram showing homogeneous solutions for the left population activity u as a function of the input amplitude I. *Solid lines* represent stable states, whereas *circles* represent the maximum and minimum of rivalry oscillations. It can be seen that there are regions of off/WTA bistability, WTA/fusion bistability, and fusion/rivalry bistability. Parameters are $\tau_s = 500$, $\kappa = 0.05$, $\beta = 5$, $\overline{w}_e = 0.4$, and $\overline{w}_i = 1$. (**b**) Homogeneous oscillatory solution in which there is spontaneous periodic switching between left- and right-eye dominance. Plot against time of the left-eye neural field u (*solid gray*) and the right-eye neural activity v (*solid black*) together with the corresponding depression variables q_u (*dashed gray*) and q_v (*dashed black*)

In order to take into account the propagation of activity seen in binocular rivalry waves, it is necessary to introduce a spatially extended network model. Therefore, suppose that the neuronal populations responding to stimuli from the left eye, say, are distributed on a one-dimensional (1D) lattice and are labeled according to the integer n; a second 1D network responds to stimuli from the right eye; see Fig. 8.8. In terms of the functional architecture of V1, one can interpret the nth left-/right-eye populations as consisting of neurons in a given hypercolumn that respond maximally to the distinct stimuli presented to the two eyes at a particular coarse-grained location in space. In the case of the orientated grating stimuli used by Kang et al. [312, 313] (see Fig. 8.5), this would mean neurons whose orientation

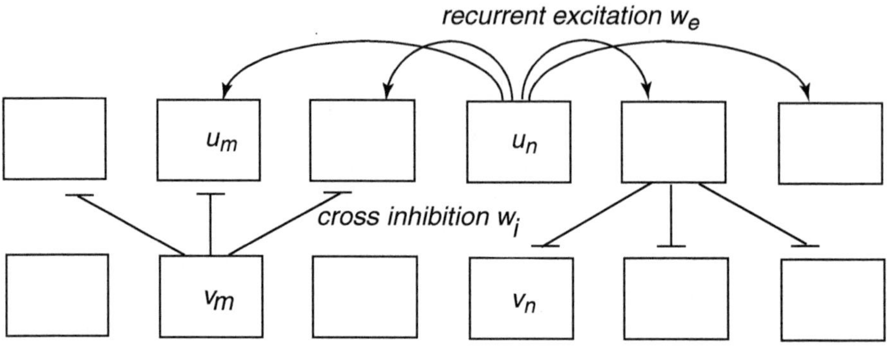

Fig. 8.8 Schematic diagram of a competitive neural network consisting of two sets of populations distributed along a 1D lattice. Recurrent connections within each 1D network are assumed to be excitatory, whereas connections between the two networks are inhibitory (cross-inhibition). Slow adaptation is incorporated into the model by taking the network connections to exhibit synaptic depression

preference coincides with the stimulus orientation presented to a given eye. Since the orientation does not change along the length of the grating, all neurons receive the same external drive. Letting $u_n, v_n, q_{u,n}, q_{v,n}$ denote the activity and depression variables of the nth left- and right-eye networks, we have

$$\tau \frac{du_m}{dt} = -u_m + I_u + \sum_n [[w_e]_{mn} q_{u,n} F(u_n) - [w_i]_{mn} q_{v,n} F(v_n)] \tag{8.5}$$

$$\tau_s \frac{dq_{u,m}}{dt} = 1 - q_{u,m} - \beta q_{u,m} F(u_m) \tag{8.6}$$

and

$$\tau \frac{dv_m}{dt} = -v_m + I_v + \sum_n [[w_e]_{mn} q_{v,n} F(v_n) - [w_i]_{mn} q_{u,n} F(u_n)] \tag{8.7}$$

$$\tau_s \frac{dq_{v,m}}{dt} = 1 - q_{v,m} - \beta q_{v,m} F(v_m). \tag{8.8}$$

Here $[w_e]_{mn}$ is the strength of excitation from the nth to the mth population with the same eye preference, and $[w_i]_{mn}$ is the strength of cross-inhibition between populations with opposite eye preferences. The weights are typically assumed to decrease with distance of separation $|m-n|$ according to an exponential or Gaussian distribution. Note that a similar network architecture was previously considered in a computational model of binocular rivalry waves [313, 678], in which cross-inhibition was mediated explicitly by interneurons and, rather than including depressing synapses, the excitatory neurons were taken to exhibit spike frequency adaptation. Numerical simulations of the model showed that the network supported traveling waves consistent with those observed numerically under physiologically reasonable parameter regimes.

8.3 Neural Field Model

Although the discrete lattice model presented in Sect. 8.2 is directly amenable to numerical simulations, it is difficult to derive any analytical results for the model that generate, for example, explicit expressions for how the wave speed depends on network parameters. Therefore, in this section we turn to a neural field model that can be constructed by taking an appropriate continuum limit of (8.5) and (8.8) [83, 663]. Introduce a lattice spacing a, and write $[w_e]_{nm} = aw_e(na,ma)$, $u_n(t) = u(na,t)$, etc. Substituting into (8.5)–(8.8) and taking the continuum limit $a \to 0$ such that $na \to x$ and $ma \to y$, the discrete sums can be replaced by integrals to give the neural field model

$$\tau \frac{du(x,t)}{dt} = -u(x,t) + I_u + \int_{-\infty}^{\infty} w_e(x-x')q_u(x',t)F(u(x',t))dx' - \int_{-\infty}^{\infty} w_i(x-x')q_v(x',t)F(v(x',t))dx' \tag{8.9a}$$

$$\tau_s \frac{dq_u(x,t)}{dt} = 1 - q_u(x,t) - \beta q_u(x,t)F(u(x,t)) \tag{8.9b}$$

and

$$\tau \frac{dv(x,t)}{dt} = -v(x,t) + I_v + \int_{-\infty}^{\infty} w_e(x-x')q_v(x',t)F(v(x',t))dx' - \int_{-\infty}^{\infty} w_i(x-x')q_u(x',t)F(u(x',t))dx' \tag{8.10a}$$

$$\tau_s \frac{dq_v(x,t)}{dt} = 1 - q_v(x,t) - \beta q_v(x,t)F(v(x,t)), \tag{8.10b}$$

assuming that the weights only depend on the distance between interacting populations. For concreteness, the distributions w_e and w_i are both taken to be Gaussians:

$$w_e(r) = \frac{\overline{w}_e}{\sqrt{2\pi\sigma_e^2}} e^{-\frac{r^2}{2\sigma_e^2}}, \quad w_i(r) = \frac{\overline{w}_i}{\sqrt{2\pi\sigma_i^2}} e^{-\frac{r^2}{2\sigma_i^2}}. \tag{8.11}$$

Following [83], assume that $\sigma_e > \sigma_i$ (longer-range excitation) and fix the length scale by setting $\sigma_e = 2$, $\sigma_i = 1$. Assuming that excitation spans a single hypercolumn, σ_e should be the same approximate size as a hypercolumn, that is, of the order 200 µm. We also take $\tau = 1$ in units of the membrane time constant, which is typically of the order 10 ms.

8.3.1 Rivalry Waves in a Deterministic Neural Field Model

The next step is to interpret the binocular rivalry wave seen in the experiments of Kang et al. [312, 313] as a traveling wave-front solution of the neural field equations (8.9) and (8.10), in which a high activity state invades the suppressed left-eye network, say, while retreating from the dominant right-eye network; see Fig. 8.9. Such a wave is defined as

$$u(x,t) = U(x-ct), \quad v(x,t) = V(x-ct), \tag{8.12}$$

where c is the wave speed and $\xi = x - ct$ is a traveling wave coordinate together with the asymptotic conditions

$$(U(\xi),V(\xi)) \to \mathbf{X}_L \text{ as } \xi \to -\infty, \quad (U(\xi),V(\xi)) \to \mathbf{X}_R \text{ as } \xi \to \infty \tag{8.13}$$

with $U(\xi)$ a monotonically decreasing function of ξ and $V(\xi)$ a monotonically increasing function of ξ. Here $\mathbf{X}_L$ ($\mathbf{X}_R$) represents a homogeneous left-eye (right-eye) dominant state. Given the asymptotic behavior of the solution and the requirements of monotonicity, we see that $U(\xi)$ and $V(\xi)$ each cross threshold at a single location, which may be different for the two eyes. Exploiting translation invariance, we take $U(0) = \kappa$ and $V(d) = \kappa$.

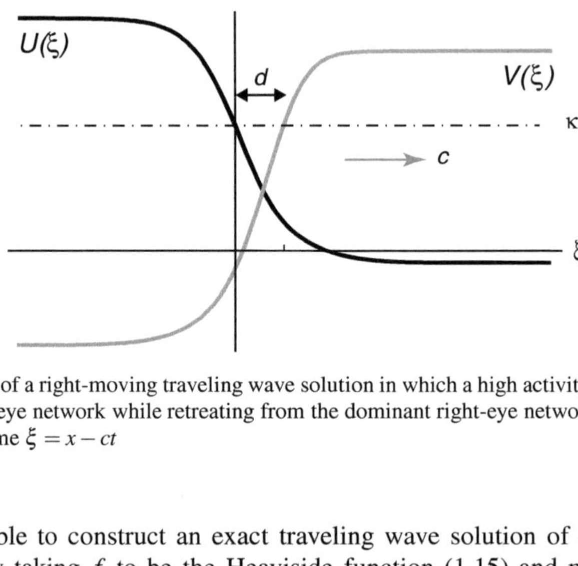

Fig. 8.9 Sketch of a right-moving traveling wave solution in which a high activity state invades the suppressed left-eye network while retreating from the dominant right-eye network. *Wave* is shown in a moving frame $\xi = x - ct$

It is possible to construct an exact traveling wave solution of equations (8.9) and (8.10) by taking f to be the Heaviside function (1.15) and making the adiabatic approximation that $q_u(x,t) = Q_u, q_v(x,t) = Q_v$ with Q_u, Q_v constants [83]. The latter is based on the assumption that adaptation is sufficiently slow so that the wave traverses cortex in a time $T_l \ll \tau_s$. Substituting the traveling wave solution (8.12) into (8.9a) and (8.10a) with fixed Q_u, Q_v and $F(u) = H(u - \kappa)$ leads to the equations

$$-c\frac{dU}{d\xi} + U = Q_u \int_{-\infty}^{0} w_e(\xi - \xi')d\xi' - Q_v \int_d^{\infty} w_i(\xi - \xi')d\xi' + I \tag{8.14a}$$

$$-c\frac{dV}{d\xi} + V = Q_v \int_d^{\infty} w_e(\xi - \xi')d\xi' - Q_u \int_{-\infty}^{0} w_i(\xi - \xi')d\xi' + I. \tag{8.14b}$$

where, for simplicity, we have set $I_u = I_v = I$. Multiplying both sides of (8.14a) and (8.14b) by $\mathrm{e}^{-\xi/c}$, integrating with respect to ξ and imposing the threshold conditions gives

$$U(\xi) = \mathrm{e}^{\xi/c}\left[\kappa - \frac{1}{c}\int_0^{\xi} \mathrm{e}^{-z/c}\Psi_d(z)dz - I(1-\mathrm{e}^{-\xi/c})\right], \tag{8.15a}$$

$$V(\xi) = \mathrm{e}^{(\xi-d)/c}\left[\kappa - \frac{1}{c}\int_0^{\xi-d} \mathrm{e}^{-z/c}\Phi_d(-z)dz - I(1-\mathrm{e}^{-(\xi-d)/c})\right] \tag{8.15b}$$

for $\xi > 0$, with Ψ and Φ defined by

$$\Psi_d(z) = Q_u \int_z^{\infty} w_e(y)dy - Q_v \int_{-\infty}^{z-d} w_i(y)dy. \tag{8.16}$$

$$\Phi_d(z) = Q_v \int_z^{\infty} w_e(y)dy - Q_u \int_{-\infty}^{z-d} w_i(y)dy. \tag{8.17}$$

Finally, requiring that the wave solution remain bounded as $\xi \to \infty$ (assuming $c > 0$) yields the pair of threshold conditions

$$\kappa = \int_0^{\infty} \mathrm{e}^{-s}\Psi_d(cs)ds + I, \quad \kappa = \int_0^{\infty} \mathrm{e}^{-s}\Phi_d(-cs)ds + I. \tag{8.18}$$

In the particular case $Q_u = Q_v = 1$ (no synaptic depression), the above equations have no solution for $c \neq 0$, that is, there does not exist a traveling wave solution [83]. This follows from setting $Q_u = Q_v = 1$ in (8.18) and subtracting the pair of equations to obtain

$$\int_0^{\infty} \mathrm{e}^{-s}\left[\Psi_d(cs) - \Psi_d(-cs)\right] ds = 0.$$

The latter equation has no solution for $c \neq 0$, since

$$\Psi_d(cs) - \Psi_d(-cs) = -\int_{-cs}^{cs} w_e(y)dy - \int_{-cs-d}^{cs-d} w_i(y)dy < 0$$

for all $s \in [0,\infty)$. On the other hand, slow synaptic depression with $Q_u \neq Q_v$ breaks the symmetry of the threshold crossing conditions, leading to a unique solution for c, d as a function of the network parameters. Moreover, one can establish that the traveling front is linearly stable by constructing the corresponding Evans function along similar lines to Sect. 7.1.2 [83]. The threshold conditions can also be used to simplify the expressions for the wave profile, namely,

$$U(\xi) = \frac{1}{c}\int_0^{\infty} \mathrm{e}^{-z/c}\Psi_d(z+\xi)dz + I \tag{8.19a}$$

$$V(\xi) = \frac{1}{c}\int_0^{\infty} \mathrm{e}^{-z/c}\Phi_d(-z-\xi+d)dz + I. \tag{8.19b}$$

Example plots of the wave speed are shown in Fig. 8.10. Baseline parameter values are chosen so that spontaneous oscillations and traveling fronts coexist as found experimentally [312, 313]. The model wave speed is of the order $c = 1$ in dimensionless units, that is, $c = \sigma_e/2\tau$ where σ_e is the range of excitation and τ is the membrane time constant. Taking σ_e to be of the order 200 μm and τ to be of the order 10 ms yields a wave speed of around 10 mm/s, which is consistent with the speeds observed experimentally. (In the psychophysical experiments of Kang et al., binocular rivalry waves took approximately 0.8 s to traverse 2 degrees of the visual field. The magnification factor in humans throughout the foveal region is approximately 0.4 cm/deg, which corresponds to 0.8 cm of cortex.) Figure 8.10 shows that the speed of the wave is a decreasing function of the threshold κ and an increasing function of the input amplitude I; the latter is consistent with what is found experimentally when the stimulus contrast is increased [312, 313]. Yet another experimental result that emerges from the model is that the wave speed covaries with the frequency of spontaneous rivalry oscillations [83].

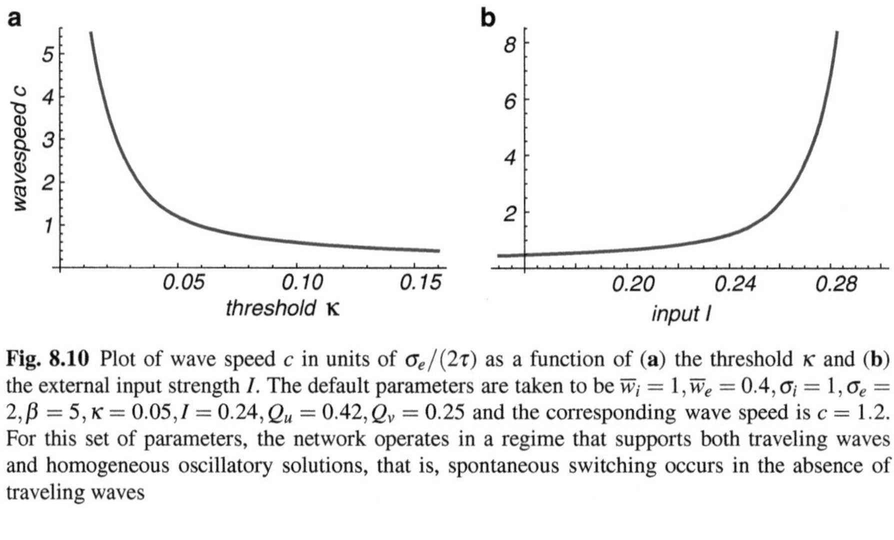

Fig. 8.10 Plot of wave speed c in units of $\sigma_e/(2\tau)$ as a function of (**a**) the threshold κ and (**b**) the external input strength I. The default parameters are taken to be $\overline{w}_i = 1, \overline{w}_e = 0.4, \sigma_i = 1, \sigma_e = 2, \beta = 5, \kappa = 0.05, I = 0.24, Q_u = 0.42, Q_v = 0.25$ and the corresponding wave speed is $c = 1.2$. For this set of parameters, the network operates in a regime that supports both traveling waves and homogeneous oscillatory solutions, that is, spontaneous switching occurs in the absence of traveling waves

8.3.2 Effects of Extrinsic Noise in the Fast Activity Variables

As highlighted in Sect. 8.1, the switching between rivalrous images has a stochastic component. This motivates studying a stochastic version of the neural field model (8.9), which, in the adiabatic limit of fixed depression variables, takes the form of an SPDE for the stochastic activity variables $U(x,t)$ and $V(x,t)$ [663]:

$$dU = \left[-U + Q_u \int_{-\infty}^{\infty} w_e(x-y)F[U(y,t)]dy - Q_v \int_{-\infty}^{\infty} w_i(x-y)F[V(y,t)]dy + I_u\right] dt + \varepsilon^{\frac{1}{2}} g(U) \circ dW_u \tag{8.20a}$$

$$dV = \left[-V + q_v \int_{-\infty}^{\infty} w_e(x-y) F[V(y,t)] dy - Q_u \int_{-\infty}^{\infty} w_i(x-y) F[U(y,t)] dy + I_v \right] dt$$
$$+\varepsilon^{\frac{1}{2}} g(V) \circ dW_v. \tag{8.20b}$$

with Q_u, Q_v fixed. Here W_u, W_v represent independent Wiener processes such that [see (7.158)]

$$\langle dW_{\{u,v\}}(x,t) \rangle = 0, \quad \langle dW_i(x,t) dW_j(x',t') \rangle = 2\delta_{i,j} C([x-x']/\lambda) \delta(t-t') dt dt', \tag{8.21}$$

where $i, j = u, v$ and $\langle \cdot \rangle$ denotes averaging with respect to the Wiener processes. The effects of extrinsic noise on binocular rivalry waves can be analyzed using the multiple time-scale method described in Sect. 2.6 for PDEs and Sect. 7.4 for neural fields [663]. First, (8.20) and (8.20b) are rewritten so that the fluctuating terms have zero mean according to Novikov's theorem. Then (U,V) is expressed as a fixed wave profile (U_0, V_0) that is displaced by an amount $\Delta(t)$ from its uniformly translating position $\xi = x - c_\varepsilon t$ and a time-dependent fluctuation (U_1, V_1) in the wave shape about its instantaneous position:

$$U(x,t) = U_0(\xi - \Delta(t)) + \varepsilon^{1/2} U_1(\xi - \Delta(t), t), \tag{8.22a}$$
$$V(x,t) = V_0(\xi - \Delta(t)) + \varepsilon^{1/2} V_1(\xi - \Delta(t), t). \tag{8.22b}$$

The wave profile (U_0, V_0) and associated wave speed c_ε are obtained by solving the deterministic equations (8.14) with the decay terms U and V replaced by $h(U)$ and $h(V)$, where $h(U) = -U + \varepsilon C(0) g'(U) g(U)$. By carrying out a perturbation expansion in ε and imposing a solvability condition, one finds that $\Delta(t)$ is given by a Wiener process with

$$\langle \Delta(t) \rangle = 0, \quad \langle \Delta(t)^2 \rangle = 2D(\varepsilon) t, \tag{8.23}$$

The diffusion coefficient $D(\varepsilon)$ is

$$D(\varepsilon) = \varepsilon \frac{\int_{-\infty}^{\infty} \left(\mathscr{V}_1(\xi)^2 g(U_0(\xi))^2 + \mathscr{V}_2(\xi)^2 g(V_0(\xi))^2 \right) d\xi}{\left[\int_{-\infty}^{\infty} \left(\mathscr{V}_1(\xi) U_0'(\xi) + \mathscr{V}_2(\xi) V_0'(\xi) \right) d\xi \right]^2}, \tag{8.24}$$

where $\mathscr{V} = (\mathscr{V}_1, \mathscr{V}_2)^T$ is the null vector of the adjoint linear operator $\mathbb{L}^\dagger = (\mathbb{L}_u^\dagger, \mathbb{L}_v^\dagger)^T$

$$\mathbb{L}_u^\dagger(B_1, B_2) = -c_\varepsilon \frac{dB_1}{d\xi} + h'(U_0) B_1 + F'(U_0) Q_u \int_{-\infty}^{\infty} w_e(\xi - \xi') B_1(\xi') d\xi'$$
$$-F'(V_0) Q_v \int_{-\infty}^{\infty} w_i(\xi - \xi') B_2(\xi') d\xi' \tag{8.25}$$

and

$$\mathbb{L}_v^\dagger(B_1,B_2) = -c_\varepsilon \frac{dB_2}{d\xi} + h'(V_0)B_2 + F'(V_0)Q_v \int_{-\infty}^{\infty} w_e(\xi-\xi')B_2(\xi')d\xi' \\ -F'(U_0)Q_u \int_{-\infty}^{\infty} w_i(\xi-\xi')B_1(\xi')d\xi'. \tag{8.26}$$

It is important to point out that, in contrast to traveling front solutions of scalar neural field equations (see Sect. 7.4), we are now considering composite wave solutions consisting of an invading front in the left-eye network, say, co-moving with a retreating front in the right-eye network. Thus in addition to the center-of-mass of the composite wave, which moves with speed c in the absence of noise, there is an additional degree of freedom corresponding to the "width" of the composite wave. [In the case of a Heaviside rate function, the width is determined by the threshold crossing point d; see (8.18).] For simplicity, we assume that the width of the composite wave is only weakly affected by the noise; this is consistent with what is found numerically [663].

8.3.3 Explicit Results for a Heaviside Rate Function

In order to illustrate the above analysis, we consider a particular example where the mean speed c_ε and diffusion coefficient $D(\varepsilon)$ can be calculated explicitly. That is, set $g(U) = g_0 U$ for the multiplicative noise term and take $F(U) = H(u-\kappa)$. The deterministic equations for U_0, V_0 can be solved along identical lines to (8.14a) and (8.14b) and lead to the modified threshold conditions

$$\kappa = \frac{1}{\gamma(\varepsilon)} \int_0^\infty \mathrm{e}^{-s} \Psi_d(cs/\gamma(\varepsilon))ds + I, \quad \kappa = \frac{1}{\gamma(\varepsilon)} \int_0^\infty \mathrm{e}^{-s} \Phi_d(-cs/\gamma(\varepsilon))ds + I, \tag{8.27}$$

where $\gamma(\varepsilon) = 1 - \varepsilon C(0) g_0^2$. The corresponding wave profiles are

$$U(\xi) = \frac{1}{c} \int_0^\infty \mathrm{e}^{-z\gamma/c} \Psi_d(z+\xi)dz + I \tag{8.28a}$$

$$V(\xi) = \frac{1}{c} \int_0^\infty \mathrm{e}^{-z\gamma/c} \Phi_d(-z-\xi+d)dz + I. \tag{8.28b}$$

It immediately follows that both the speed c and displacement d depend on the noise strength ε. In order to calculate the diffusion coefficient, it is first necessary to determine the null vector $\mathscr{V}(\varepsilon)$ of the adjoint linear operator $\mathbf{L}^\dagger$. Substituting $F(U) = H(U-\kappa)$ and $g(U) = g_0 U$ in (8.25) and (8.26) shows that the components of $\mathscr{V}$ satisfy the simultaneous equations

$$c\frac{d\mathscr{V}_1}{d\xi} + \gamma(\varepsilon)\mathscr{V}_1 = \frac{\delta(\xi)}{|U_0'(0)|} Q_u \int_{-\infty}^{\infty} w_e(z)\mathscr{V}_1(z)dz \\ -\frac{\delta(\xi-d)}{|V_0'(d)|} Q_v \int_{-\infty}^{\infty} w_i(z-d)\mathscr{V}_2(z)dz \tag{8.29}$$

$$c\frac{d\mathscr{V}_2}{d\xi}+\gamma(\varepsilon)\mathscr{V}_2=-\frac{\delta(\xi)}{|U_0'(0)|}Q_u\int_{-\infty}^{\infty}w_i(z)\mathscr{V}_1(z)dz$$
$$+\frac{\delta(\xi-d)}{|V_0'(d)|}Q_v\int_{-\infty}^{\infty}w_e(z-d)\mathscr{V}_2(z)dz. \quad (8.30)$$

Proceeding along similar lines to Sect. 7.4, we make the ansatz that

$$\mathscr{V}_1(\xi) = \mathscr{A}_1\mathrm{e}^{-\gamma\xi/c}H(\xi)-\mathscr{B}_1\mathrm{e}^{-\gamma[\xi-d]/c}H(\xi-d),$$
$$\mathscr{V}_2(\xi) = -\mathscr{A}_2\mathrm{e}^{-\gamma\xi/c}H(\xi)+\mathscr{B}_2\mathrm{e}^{-\gamma[\xi-d]/c}H(\xi-d).$$

Substituting back into the adjoint equations yields algebraic conditions for the constant coefficients $\mathscr{A}_j,\mathscr{B}_j$:

$$\mathscr{A}_1 = \frac{Q_u}{|U_0'(0)|}\left(\mathscr{A}_1\Omega_e[0]-\mathscr{B}_1\Omega_e[d]\right) \quad (8.31)$$

$$\mathscr{B}_1 = \frac{Q_v}{|V_0'(d)|}\left(\mathscr{B}_2\Omega_i[0]-\mathscr{A}_2\Omega_i[-d]\right) \quad (8.32)$$

$$\mathscr{A}_2 = \frac{Q_u}{|U_0'(0)|}\left(\mathscr{A}_1\Omega_i[0]-\mathscr{B}_1\Omega_i[d]\right) \quad (8.33)$$

$$\mathscr{B}_2 = \frac{Q_v}{|V_0'(d)|}\left(\mathscr{B}_2\Omega_e[0]-\mathscr{A}_2\Omega_e[-d]\right), \quad (8.34)$$

where

$$\Omega_j[x]=\int_0^{\infty}\mathrm{e}^{-z\gamma/c}w_j(x+z)dz, \quad j=e,i. \quad (8.35)$$

Differentiating (8.28a) and (8.28b) with respect to ξ and using (8.16) and (8.17) show that

$$U_0'(0)=-Q_u\Omega_e[0]-Q_v\Omega_i[-d]<0, \quad (8.36)$$

and

$$V_0'(d)=Q_v\Omega_e[0]+Q_u\Omega_i[d]>0. \quad (8.37)$$

We have also used the fact that the weight distributions $w_e(x),w_i(x)$ are even functions of x. Substituting these derivatives into (8.31) and (8.34) gives

$$\mathscr{B}_2=-\frac{Q_v\Omega_e[-d]}{Q_u\Omega_i[d]}\mathscr{A}_2, \quad \mathscr{B}_1=-\frac{Q_v\Omega_i[-d]}{Q_u\Omega_e[d]}\mathscr{A}_1. \quad (8.38)$$

It follows from (8.33) and (8.38) that $\mathscr{A}_2,\mathscr{B}_1,\mathscr{B}_2$ can all be expressed as constant multiples of $\mathscr{A}_1$, with the latter determined by normalizing the null vector.

It can be verified that the various assumptions used in the above analysis are reasonable by carrying out numerical simulations of the full system given by (8.20a), (8.20b), (8.9a) and (8.9b) with $Q_{u,v}\to q_{u,v}$ [663]. In particular it can be checked that fluctuations in the separation between the left- and right-eye fronts are negligible in the long time limit. An illustration of the progression of a stochastic composite front is shown in Fig. 8.11. The corresponding mean $\overline{X}(t)$ and variance $\sigma_X^2(t)$ of wave position as a function of time are plotted in Fig. 8.12 using level sets as

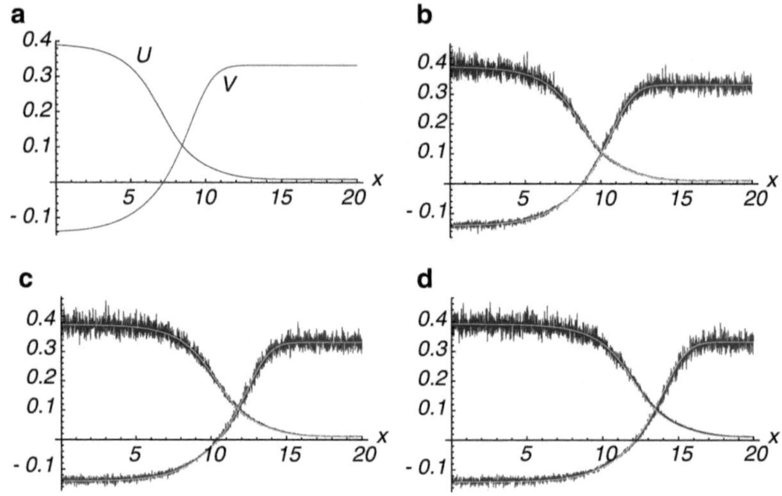

Fig. 8.11 Effects of multiplicative noise in the activity variables. Simulation of the activity variables $u(x,t)$ and $v(x,t)$ for (**a**) t=0, (**b**) t=1, (**c**) t=2, (**d**) t=3. The continuous lines are the functions $U_0(x-ct)$ and $V_0(x-ct)$ which were found by solving (2.97). The functions U_0 and V_0 were also used for the initial condition. Parameter values were $a_i = 1, a_e = 0.4, \sigma_i = 1, \sigma_e = 2, \beta = 5, \kappa = 0.05, I = 0.24, \varepsilon = 0.006$ with spatial and temporal grid sizes both being 0.01

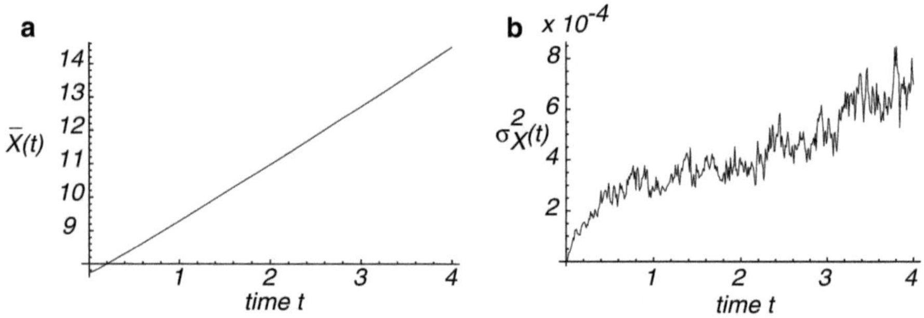

Fig. 8.12 (**a**) Mean and (**b**) variance of the wave position for an ensemble of 128 stochastic simulations. Parameters were the same as Fig. 8.11

explained in Sect. 7.4.1. It can be seen that $\overline{X}(t)$ varies linearly with t, consistent with the assumption that there is constant speed wave, $\overline{X}(t) \sim c_\varepsilon t$. The variance initially increases rapidly with respect to t, but then follows linear behavior consistent with a diffusive-like displacement of the wave from its uniformly translating position at long time scales, $\sigma_X^2(t) \sim 2D(\varepsilon)t$. The initial sharp increase in the variance results from the fact that the left and right fronts do move slightly with respect to each other, resulting in higher-order behavior. However, one finds numerically that this component of the variance is small and bounded so that it becomes negligible as time increases.

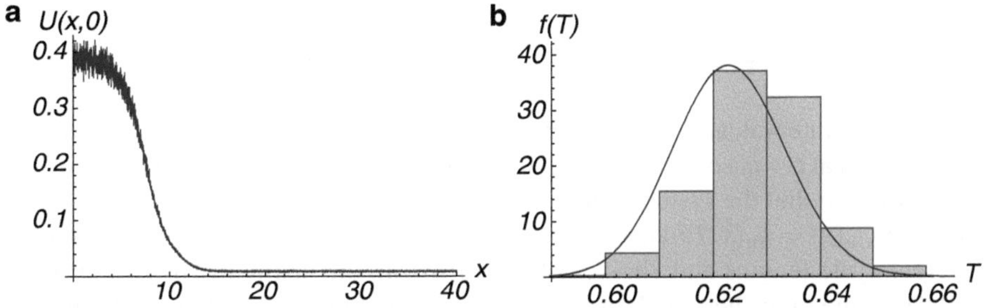

Fig. 8.13 (**a**) Snapshot of a stochastic traveling wave (only left-eye activity is shown). (**b**) First passage time distribution for the wave to travel a distance $L = 1$, starting from the wave profile in (**a**). Parameter values were $a_i = 1, a_e = 0.4, \sigma_i = 1, \sigma_e = 2, \beta = 5, \kappa = 0.05, I = 0.24, \varepsilon = 0.006$ with spatial and temporal grid sizes both being 0.01.The best fit inverse Gaussian $\mathscr{F}(T;\mu,\lambda)$ for the histogram gives the parameters $\mu = 0.62, \lambda = 2200$ which is in very good agreement with the theoretical predictions of $\mu = L/c = 0.6, \lambda = L^2/D = 2100$

The above analysis predicts that the wave position will follow a Brownian motion, which implies that the time taken for a wave to travel a distance $L > 0$ has a distribution given by the standard first passage time formula for Brownian motion with drift c. That is, let T_L denote the first passage time for the wave to travel a distance L: $cT_L + \Delta(T_L) = L$ given $\Delta(0) = 0$. Then the first passage time density is given by an inverse Gaussian or Wald distribution:

$$f(T_L) = \mathscr{F}(T_L; \frac{L}{c}, \frac{L^2}{D}), \tag{8.39}$$

where

$$\mathscr{F}(T;\mu,\lambda) = \left[\frac{\lambda}{2\pi T^3}\right]^{1/2} \exp\left(-\frac{\lambda(T-\mu)^2}{2\mu^2 T}\right). \tag{8.40}$$

Figure 8.13 shows the first passage time distribution generated over a large number of simulations for an initial condition given by a fully developed stochastic traveling wave. Consistent with the analysis, the distribution can be fitted by an inverse Gaussian. However, it has been assumed that the predominant source of noise is in the activity variables. Now suppose that the main source of noise is in the depression variables. Once again assume that a traveling wave crosses cortex significantly faster than the relaxation time of synaptic depression, so that q_u and q_v can be taken to be constant with respect to time during passage of the wave. However, they are no longer constant with respect to space nor with respect to trials, that is, $q_u = Q_u(x)$ and $q_v = Q_v(x)$ with Q_u, Q_v random variables of x. So although the wave itself will travel deterministically in a given trial, the functions $Q_u(x)$ and $Q_v(x)$ will be different across trials. In several experimental studies [312, 313] of binocular rivalry waves, a dominance switch was induced about halfway through a normal dominance cycle by locally changing the contrast in the depressed stimulus. The latter can be represented by increasing the input strength I over a small region of the model. This suggests taking q_u and q_v to evolve according to

$$\tau_s dq_u(x,t) = [1 - q_u(x,t)]\, dt + \eta dW_{q_u}(x,t), \tag{8.41a}$$

$$\tau_s dq_v(x,t) = [1 - q_v - \beta q_v]\, dt + \eta dW_{q_v}(x,t) \tag{8.41b}$$

over the time interval $[t_0, T]$. It is assumed that a switch from left- to right-eye dominance occurs at $t = t_0$ so that $u(x,t) < \kappa$ and $v(x,t) > \kappa$ for $t \in (t_0, T)$. The time T is then chosen so that the system is about $2/3$ of the way through a rivalry oscillation, such that $Q_u(x) = q_u(x,T)$ and $Q_v(x) = q_v(x,T)$. (Similar results were obtained for different cycle lengths and choices of T away from the beginning or end of a cycle.) Thus, the quenched random variables $Q_u(x)$ and $Q_v(x)$ are obtained by taking a snapshot of two lines of independent Ornstein–Uhlenbeck processes. Averaging over many trials, it is found numerically that the first passage time distribution is still well approximated by an inverse Gaussian distribution; see Fig. 8.14 and [663].

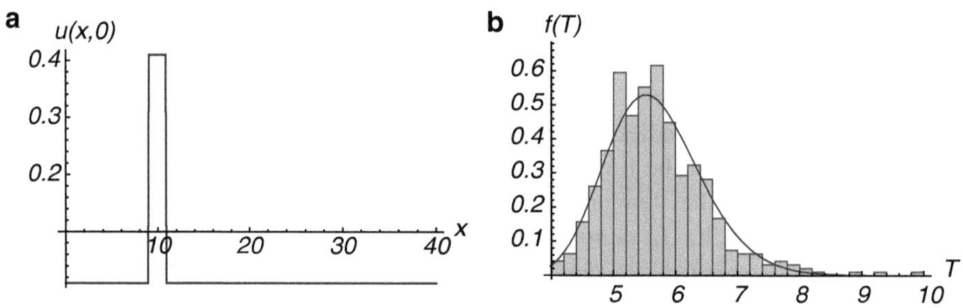

Fig. 8.14 Effects of quenched noise in the depression variables. (**a**) Initial condition given by a right-eye dominant homogeneous steady state perturbed by a narrow square pulse of left-eye activity. (**b**) First passage time distribution for a wave to travel a distance $L = 6$ starting from the initial condition shown in (**a**). Parameter values were $a_i = 1, a_e = 0.4, \sigma_i = 1, \sigma_e = 2, \beta = 5, \kappa = 0.05, I = 0.24, \eta = 0.037\langle Q\rangle$, where $\langle Q\rangle$ is the spatial average of the quenched depression variables (to make the results comparable with multiplicative noise). The spatial grid size $\Delta x = 0.1$ and the temporal grid step is $\Delta t = 0.01$. The *solid curve* in (**b**) is the best fit inverse Gaussian

In summary, we have illustrated one application of neural field theory, namely, to the phenomenon of binocular rivalry waves. Formulating the problem in terms of continuum neural field equations allows one to study the short time behavior associated with the propagation of eye dominance from an analytical perspective, both for deterministic and stochastic waves. In the deterministic case, one finds that some form of slow adaptation such as synaptic depression is needed in order to provide a symmetry breaking mechanism that allows propagation of a binocular rivalry wave [83]. That is, the equations for the left- and right-eye networks have to be different on the time scales during which traveling waves propagate. One implication of this is that purely noise-driven switching between rivalrous states in the absence of adaptation could not by itself generate rivalry waves, since it would not produce bias in the appropriate direction. However, noise does have a significant effect on binocular rivalry waves in the presence of slow adaptation. In particular, the analysis predicts that motion of the wave in the presence of noise in the activity variables has a diffusive component that leads to a distribution of first passage times given by an inverse Gaussian. A similar distribution of first passage times occurs for quenched noise in the depression variables [83].

8.4 Modeling Primary Visual Cortex Using 2D Neural Fields

The neural field model of binocular rivalry waves introduced in Sect. 8.3 is based on a very simple model of V1. In particular, it is a 1D model that ignores any correlations between long-range horizontal connections and stimulus feature maps. Such simplifications are reasonable as a first step, particularly when modeling the psychophysical experiments of Kang et al. [312, 313]. However, in order to account for the dependence of wave velocity on the orientation patterns of annular stimuli in the experiments of Wilson et al. [678], for example, it is necessary to consider a more detailed, 2D model of V1 that takes into account such correlations. In the final section of this chapter, we briefly describe one possible approach, which is based on a neural field model of coupled hypercolumns [71, 79].

8.4.1 2D Planar Model

For the sake of illustration, consider a 2D scalar neural field (without any slow adaptation) of the form

$$\frac{\partial u(\mathbf{r},t)}{\partial t} = -u(\mathbf{r},t) + \int w(\mathbf{r},\mathbf{r}')F(u(\mathbf{r}',t))d\mathbf{r}'. \tag{8.42}$$

Almost all 2D neural field models take $w(\mathbf{r},\mathbf{r}') = w(|\mathbf{r}-\mathbf{r}'|)$ so that w is invariant with respect to the Euclidean group $\mathbf{E}(2)$ of rigid body transformations in the plane. That is,

$$\gamma \cdot w(\mathbf{r},\mathbf{r}') = w(\gamma^{-1}\cdot\mathbf{r},\gamma^{-1}\cdot\mathbf{r}') = w(\mathbf{r},\mathbf{r}')$$

for all $\gamma \in \mathbf{E}(2)$. The Euclidean group is composed of the (semi-direct) product of $\mathbf{O}(2)$, the group of planar rotations $\mathbf{r} \to R_\varphi \mathbf{r}$ and reflections $(x,y) \to (x,-y)$, with $\mathbb{R}^2$, the group of planar translations $\mathbf{r} \to \mathbf{r}+\mathbf{s}$. Here

$$R_\varphi = \begin{pmatrix} \cos\varphi & -\sin\varphi \\ \sin\varphi & \cos\varphi \end{pmatrix}, \quad \varphi \in [0,2\pi). \tag{8.43}$$

However, one of the immediate implications of the existence of regularly repeating feature maps and patchy horizontal (or feedback) connections (Sect. 8.1.1) is that one can no longer treat the weight distribution w in the neural field equation (8.42) as Euclidean invariant. That is, one has to consider a more general weight distribution of the form

$$w(\mathbf{r},\mathbf{r}') = w(|\mathbf{r}-\mathbf{r}'|) + \rho w^{\Delta}(\mathscr{F}(\mathbf{r}),\mathscr{F}(\mathbf{r}')), \tag{8.44}$$

where $\mathscr{F}(\mathbf{r})$ denotes a cortical feature map, w^{Δ} represents the dependence of excitatory horizontal connections on the feature preferences of the presynaptic and postsynaptic neuron populations, and ρ is a positive coupling parameter. The local connections span a single hypercolumn, whereas the patchy horizontal connections link cells with similar feature preferences in distinct hypercolumns.

In the absence of long-range connections ($\rho = 0$), the resulting weight distribution is invariant under the action of the Euclidean group. However, the long-range connections break Euclidean symmetry due to correlations with the feature map $\mathscr{F}(\mathbf{r})$. A certain degree of symmetry still remains under the approximation that the feature map is periodic. For example, Fig. 8.2b suggests partitioning V1 into a set of hypercolumns organized around a lattice of orientation pinwheels. Therefore, suppose we treat the distribution of pinwheels as a regular planar lattice L. The resulting weight distribution for $\rho \neq 0$ is then doubly periodic with respect to L:

$$w(\mathbf{r}+\ell, \mathbf{r}'+\ell) = w(\mathbf{r}, \mathbf{r}') \tag{8.45}$$

for all $\ell \in L$. Additional symmetries may also exist depending on the particular form of w^{Δ}. There are number of distinct ways in which w^{Δ} may depend on the underlying feature map $\mathscr{F}$. The first reflects the "patchiness" of the horizontal connections that link cells with similar feature preferences. In the case of a periodic feature map, this may be implemented by taking [65, 66, 526]

$$w^{\Delta}(\mathscr{F}(\mathbf{r}), \mathscr{F}(\mathbf{r}')) = \sum_{\ell \in \mathscr{L}} J_{\ell} \Delta(\mathbf{r}-\mathbf{r}'-\ell), \tag{8.46}$$

where $\Delta(\mathbf{r})$ is some localized unimodal function that is maximal when $\mathbf{r} = 0$, thus ensuring that presynaptic and postsynaptic cells with similar feature preferences are connected. The width of Δ determines the size of the patches and $J_{\ell}, \ell \neq 0$ is a monotonically decreasing function of ℓ. In this particular example, the patchy horizontal connections break continuous rotation symmetry down to the discrete rotation symmetry of the lattice. On the other hand, continuous translation symmetry (homogeneity) still holds, since w^{Δ} only depends on the relative separation $\mathbf{r}-\mathbf{r}'$ in cortex.

However, if the anisotropy of horizontal connections is taken into account (Fig. 8.3), then continuous translation symmetry is broken as well. That is, the direction of anisotropy is correlated with the orientation preference map and thus rotates periodically across cortex [56]. Anisotropy can be incorporated into the model by modifying the weight distribution w^{Δ} along the following lines [75, 79]:

$$w^{\Delta}(\mathscr{F}(\mathbf{r}), \mathscr{F}(\mathbf{r}')) = \sum_{\ell \in \mathscr{L}} J_{\ell} \mathscr{A}_{\mathscr{F}(\mathbf{r}')}(\ell) \Delta(\mathbf{r}-\mathbf{r}'-\ell), \tag{8.47}$$

with

$$\mathscr{A}_{\mathscr{F}(\mathbf{r})}(\ell) = \frac{1}{4\eta(\mathbf{r})}\left(H[\eta(\mathbf{r}) - |\arg \ell - \theta(\mathbf{r})|] + H[\eta(\mathbf{r}) - |\arg \ell - \theta(\mathbf{r}) - \pi|]\right), \tag{8.48}$$

where $\theta(\mathbf{r})$ denotes the orientation preference map. The second term takes account of the fact that $\theta \in [0, \pi)$ whereas $\arg \ell \in [0, 2\pi)$. The parameter $\eta(\mathbf{r})$ determines the degree of anisotropy, that is, the angular spread of the horizontal connections around the axis joining cells with similar orientation preferences. The degree of anisotropy is also likely to depend on position $\mathbf{r}$ relative to pinwheels, since populations of cells around pinwheels have zero average orientation preference so that one expects the corresponding distribution of weights to be isotropic, in contrast to cells in the linear zones of the orientation preference map; see Fig. 8.2. In conclusion, at the submillimeter length scale, there is an approximately periodic modulation of the synaptic connections that can lead to a periodically varying heterogeneous medium. We have already explored one implication of this in a simpler 1D model of traveling waves in a heterogeneous neural field (see Sect. 7.3), namely, that it can cause wave propagation failure [79, 132, 332]. Another consequence of such an inhomogeneity is that it can lead to the pinning of a spatially periodic pattern to the underlying lattice of pinwheels [65, 66].

8.4.2 Coupled Hypercolumn Model

Treating the distribution of pinwheels as a regular lattice does not take into account the considerable degree of disorder in the distribution of feature preferences across cortex. One way to avoid such complexity is to collapse each hypercolumn into a single point (through some form of spatial coarse graining) and to treat V1 as a continuum of hypercolumns [78, 79]. Thus cortical position $\mathbf{r}$ is replaced by the pair $\{\mathbf{r}, \mathscr{F}\}$ with $\mathbf{r} \in \mathbb{R}^2$ now labeling the hypercolumn at (coarse-grained) position $\mathbf{r}$ and $\mathscr{F}$ labeling the feature preferences of neurons within the hypercolumn. Let $u(\mathbf{r}, \mathscr{F}, t)$ denote the activity of a neuronal population at $(\mathbf{r}, \mathscr{F})$, and suppose that u evolves according to the neural field equation

$$\frac{\partial u(\mathbf{r}, \mathscr{F}, t)}{\partial t} = -u(\mathbf{r}, \mathscr{F}, t) + \int_{\mathbb{R}^2} \int w(\mathbf{r}, \mathscr{F} | \mathbf{r}', \mathscr{F}') F(u(\mathbf{r}', \mathscr{F}', t)) D\mathscr{F}' d\mathbf{r}', \tag{8.49}$$

with $D\mathscr{F}'$ an appropriately defined measure on feature space. We decompose w into local and long-range parts by assuming that the local connections mediate interactions within a hypercolumn, whereas the patchy horizontal connections mediate interactions between hypercolumns:

$$w(\mathbf{r}, \mathscr{F} | \mathbf{r}', \mathscr{F}') = \delta(\mathbf{r} - \mathbf{r}') w(\mathscr{F}, \mathscr{F}') + \rho J(|\mathbf{r} - \mathbf{r}'|) \mathscr{A}_{\mathscr{F}'}(\mathbf{r} - \mathbf{r}') w^{\Delta}(\mathscr{F}, \mathscr{F}'), \tag{8.50}$$

where $w(\mathscr{F}, \mathscr{F}')$ and $w^{\Delta}(\mathscr{F}, \mathscr{F}')$ represent the dependence of the local and long-range interactions on the feature preferences of the pre- and postsynaptic cell populations, and $J(\mathbf{r})$ with $J(0) = 0$ is a positive function that determines the variation in the strength of the long-range interactions with cortical distance. We have also included the anisotropy factor $\mathscr{A}_{\mathscr{F}}$ of (8.48). The advantage of collapsing each hypercolumn to a single point in the cortical plane is that a simpler representation of the internal structure of a hypercolumn can be developed that captures the essential tuning properties of the cells as well as incorporating the modulatory effects of long-range connections.

For the sake of illustration, suppose that we identify $\mathscr{F}$ in (8.49) and (8.50) with the orientation preference $\theta \in [0,\pi)$ of cells within a hypercolumn. The weight distribution (8.50) is taken to have the form [79]

$$w(\mathbf{r},\theta|\mathbf{r}',\theta') = \delta(\mathbf{r}-\mathbf{r}')w(\theta-\theta') + \rho J(|\mathbf{r}-\mathbf{r}'|)\mathscr{P}(\arg(\mathbf{r}-\mathbf{r}')-\theta)w^{\Delta}(\theta-\theta') \tag{8.51}$$

with

$$\mathscr{P}(\psi) = \frac{1}{4\eta}[H(\eta-|\psi|)+H(\eta-|\psi-\pi)|]. \tag{8.52}$$

(Note that the direction $\arg(\mathbf{r}-\mathbf{r}')$ can be taken to be correlated with either θ or θ', since w^{Δ} is a sharply peaked function.) The functions $w(\theta)$ and $w^{\Delta}(\theta)$ are assumed to be even, π-periodic functions of θ, with corresponding Fourier expansions

$$\begin{aligned} w(\theta) &= w_0 + 2\sum_{n\geq 1} w_n \cos 2n\theta \\ w^{\Delta}(\theta) &= w_0^{\Delta} + 2\sum_{n\geq 1} w_n^{\Delta} \cos 2n\theta. \end{aligned} \tag{8.53}$$

The distribution $w^{\Delta}(\theta)$ is taken to be a positive, narrowly tuned distribution with $w^{\Delta}(\theta) = 0$ for all $|\theta| > \theta_c$ and $\theta_c \ll \pi/2$; the long-range connections thus link cells with similar orientation preferences. Equation (8.49) then describes a continuum of coupled ring networks, each of which corresponds to a version of the so-called ring model of orientation tuning [34, 74, 595]. If there is no orientation-dependent anisotropy, then the weight distribution (8.51) is invariant with respect to the symmetry group $\mathbf{E}(2)\times\mathbf{O}(2)$ where $\mathbf{O}(2)$ is the group of rotations and reflections on the ring S^1 and $\mathbf{E}(2)$ is the Euclidean group acting on $\mathbb{R}^2$. The associated group action is

$$\begin{aligned} \zeta\cdot(\mathbf{r},\theta) &= (\zeta\mathbf{r},\theta), \qquad \zeta\in\mathbf{E}(2) \\ \xi\cdot(\mathbf{r},\theta) &= (\mathbf{r},\theta+\xi) \\ \kappa\cdot(\mathbf{r},\theta) &= (\mathbf{r},-\theta). \end{aligned} \tag{8.54}$$

Invariance of the weight distribution can be expressed as

$$\gamma\cdot w(\mathbf{r},\theta|\mathbf{r}',\theta) = w(\gamma^{-1}\cdot(\mathbf{r},\theta)|\gamma^{-1}\cdot(\mathbf{r}',\theta')) = w(\mathbf{r},\theta|\mathbf{r}',\theta')$$

for all $\gamma\in\Gamma$ where $\Gamma = \mathbf{E}(2)\times\mathbf{O}(2)$. Anisotropy reduces the symmetry group Γ to $\mathbf{E}(2)$ with the following *shift–twist* action on $\mathbb{R}^2\times S^1$ [64, 78]:

$$\begin{aligned} \mathbf{s}\cdot(\mathbf{r},\theta) &= (\mathbf{r}+\mathbf{s},\theta) \\ \xi\cdot(\mathbf{r},\theta) &= (R_\xi\mathbf{r},\theta+\xi) \\ \kappa\cdot(\mathbf{r},\theta) &= (R_\kappa\mathbf{r},-\theta), \end{aligned} \tag{8.55}$$

where R_ξ denotes the planar rotation through an angle ξ and R_κ denotes the reflection $(x_1,x_2)\mapsto(x_1,-x_2)$. It can be seen that the discrete rotation operation comprises a translation or *shift* of the orientation preference label θ to $\theta+\xi$, together with a rotation or *twist* of the position vector $\mathbf{r}$ by the angle ξ. The fact that the weight distribution is invariant with respect to this shift–twist action has important consequences for the global dynamics of V1 in the presence of anisotropic horizontal connections [79].

Finally, one can write down a 2D neural field model for binocular rivalry waves based on a continuum of coupled hypercolumns for left- and right-eye preferences. Ignoring the effects of anisotropy and taking an adiabatic approximation for the depression variables, such a model takes the form

$$\begin{aligned}\frac{\partial u}{\partial t} &= -u(\mathbf{r},\theta,t) + \int_0^{2\pi} w(\theta-\theta')F(u(\mathbf{r},\theta',t)d\theta' \\ &\quad + Q_u\int_{-\infty}^{\infty} J_e(|\mathbf{r}-\mathbf{r}'|)F(u(\mathbf{r}',\theta,t)d\mathbf{r}' - Q_v\int_{-\infty}^{\infty} J_i(|\mathbf{r}-\mathbf{r}'|)F(v(\mathbf{r}',\theta,t)d\mathbf{r}'\end{aligned}$$

$$\begin{aligned}\frac{\partial v}{\partial t} &= -v(\mathbf{r},\theta,t) + \int_0^{2\pi} w(\theta-\theta')F(v(\mathbf{r},\theta',t)d\theta' \\ &\quad + Q_v\int_{-\infty}^{\infty} J_e(|\mathbf{r}-\mathbf{r}'|)F(v(\mathbf{r}',\theta,t)d\mathbf{r}' - Q_u\int_{-\infty}^{\infty} J_i(|\mathbf{r}-\mathbf{r}'|)F(u(\mathbf{r}',\theta,t)d\mathbf{r}'.\end{aligned}$$

An interesting issue is then how to define a traveling wave solution for such a model that combines orientation tuning within a hypercolumn due to the effects of the local weight distribution w with a variation of activity across hypercolumns in the form of a traveling front.

Part III
Development and Disease

Chapter 9
Waves in the Developing and the Diseased Brain

In this final chapter we discuss various wave phenomena in the developing and diseased brain. We begin by describing two examples of intercellular developmental waves: Ca^{2+} waves in radial glial cells, which coordinate cell division and cell migration during early development, and spontaneous retinal waves, which are thought to contribute to the activity-dependent development of circuits in visual cortex. We then consider the possible role of traveling waves in the maturation of neurons from sphere-like cells to polarized cells consisting of a single axon and several shorter dendrites. In particular, we consider some new wave phenomena including wave-pinning and cytoskeletal actin waves. Switching to the case of pathological waves in the diseased brain, we focus on two examples: cortical spreading depression, which is often associated with migraine auras, and epilepsy. In the former case, we review reaction–diffusion models of spreading depression that track changes in the concentration of extracellular ions such as K^+, and show how kinematic wave theory can be used to model the progression of a migraine aura across the visual field. It is also necessary to determine the effective diffusion of ions in the extracellular space, which can be carried out using the theory of porous media, for example. Epilepsy is a vast field in its own right, so we will focus on how changes in network structure can result in abnormal cortical states such as seizure-like activity or population bursting. We also consider recent work on characterizing normal and abnormal brain states according to how close the cortex is to a critical state. This leads to another approach to studying the propagation of activity in cortex, based on the theory of branching processes. We finish by discussing the spread of protein aggregates in neurodegenerative diseases such as Alzheimer's disease, Parkinson's disease, and Huntington's disease. There is growing experimental evidence that such spread has parallels with the neuropathogenesis of prion diseases, in which the infected agent is a misfolded protein aggregate.

P.C. Bressloff, *Waves in Neural Media: From Single Neurons to Neural Fields*, Lecture Notes on Mathematical Modelling in the Life Sciences,
DOI 10.1007/978-1-4614-8866-8_9, © Springer Science+Business Media New York 2014

9.1 Intercellular Waves in Neural Development

9.1.1 Calcium Waves in Early Development

There is growing experimental evidence that intercellular Ca^{2+} waves play an important role in early neural development [146, 368, 498, 666, 697]. In order to describe such findings, it is first necessary to give a short overview of the early stages of development. During embryogenesis a strip of specialized cells called the *notchcord* induces epithelial cells of the *ectoderm* directly above it to transform into the primitive nervous system known as the *neuroepithelium*; see Fig. 9.1. The latter then wrinkles and folds over to form a hollow tube (the *neural tube*), which is the precursor of the brain and spinal cord. Meanwhile, the ectoderm and endoderm continue to curve around and fuse beneath the embryo to create the body cavity, completing the transformation of the embryo from a flattened disk to a three-dimensional body. Cells originating from the fused tips of the neuroectoderm (*neural crest cells*) migrate to various locations throughout the embryo, where they initiate the development of the peripheral nervous system including the retina. As neurogenesis

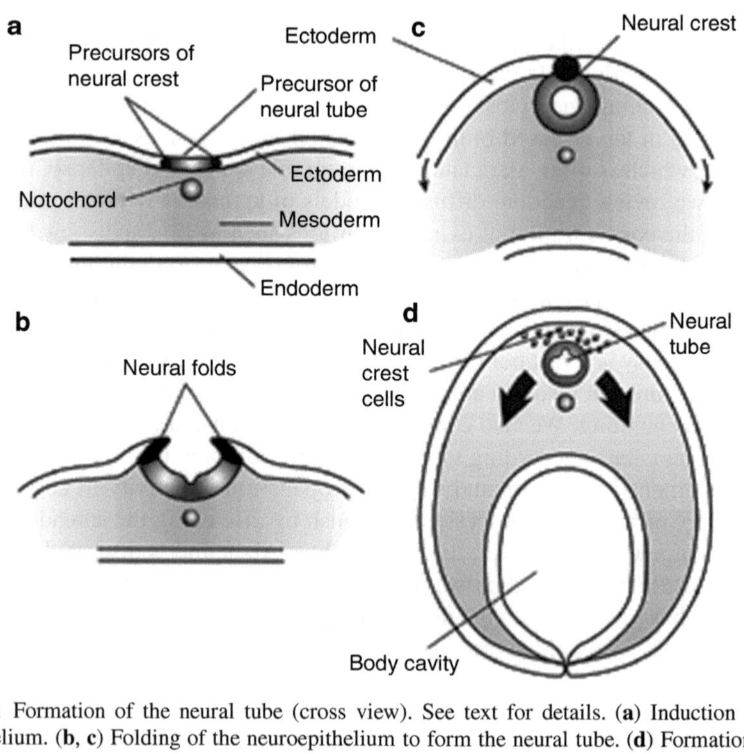

Fig. 9.1 Formation of the neural tube (cross view). See text for details. (**a**) Induction of neuroepithelium. (**b, c**) Folding of the neuroepithelium to form the neural tube. (**d**) Formation of 3D body structure and migration of neural crest cells [Reproduced from Wikimedia Commons (public domain)]

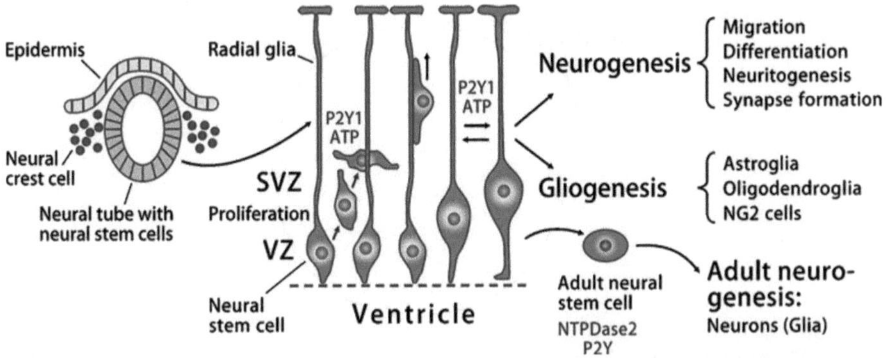

Fig. 9.2 Transformation of neuroepithelial cells in the neural tube into radial glial cells, which are the progenitors of neurons and glial cells. Their radial projections also guide cell migration from the ventricular zone (*VZ*) or inner surface of the neural tube to the outer surface. Cortical cells are also produced in the subventricular zone (*SVZ*). During early neurogenesis, intercellular ATP signaling is essential for the migration of SVZ progenitors from the VZ to the SVZ. During late neurogenesis, ATP released from cortical radial glia mediates Ca^{2+} waves that are essential for VZ proliferation (Adapted from [697])

proceeds, neuroepithelial cells of the neural tube transform into *radial glial cells*, also known as progenitor cells, which ultimately give rise to neurons and glial cells; see Fig. 9.2 and [514]. The cell body of radial glial cells is found in the *ventricular zone* (VZ), or inner surface of the neural tube, and has a radial extension that projects near the border of the marginal zone, or outer surface. After asymmetrical division, their daughter cells migrate along the radial projections and differentiate into neurons. At the onset of cortical neurogenesis, a second proliferating population of progenitor cells appears in a new layer known as the subventricular zone (SVZ), which generates both neurons and glial cells. Finally, towards the end of cortical neurogenesis, radial glial cells retract their extensions to become astrocytes. In summary, radial glial cells function as stem cells for neurogenesis and as scaffolding that guides the migration of neurons out of the ventricular zone. Moreover, the process of radial migration occurs in an "inside-out manner." In other words, cells that arrive the earliest (the oldest ones) form the deepest layer of the cortex, whereas the late-arriving (the youngest) neurons form the outermost layer.

An experimental study of embryonic rat cortical slices provides strong evidence that ATP-mediated calcium waves, which propagate between radial glial cells in the ventricular zone, are involved in the modulation of cell proliferation [666]. The predominant mechanism for the intracellular Ca^{2+} waves appears to be paracrine signaling, in which gliotransmission leads to the release of ATP via gap hemichannels, which diffuses and binds to metabotropic P2Y receptors. This activates a G-protein-mediated IP_3 signaling pathway, which leads to an increase in intracellular Ca^{2+}; see Sect. 4.5. As suggested in [666], one possible function of ATP-mediated Ca^{2+} waves during neurogenesis is that it could coordinate the cell cycle. Recall that the cell cycle consists of four distinct phases or stages G1–S–G2–M; see Fig. 9.3. Here G1 and G2 are *gap* phases, while S is the *synthesis* stage where DNA replication

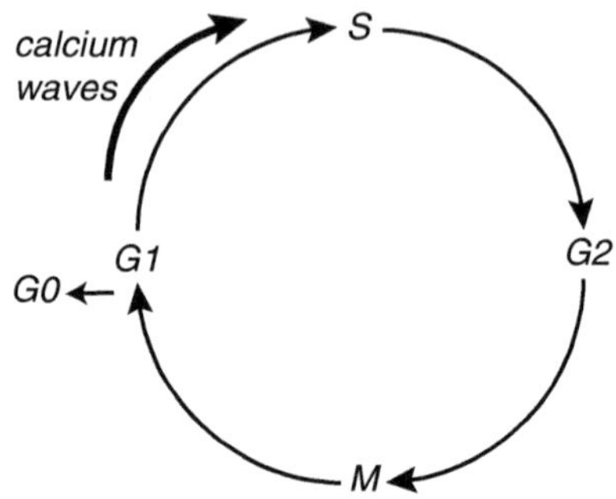

Fig. 9.3 Phases of the cell cycle: mitosis (*M*), Gap1 (*G1*), Gap2 (*G2*), and synthesis (*S*). *G0* represents the quiescent phase where cells exit the cell cycle either temporarily or permanently. Calcium waves can coordinate cells to enter the S phase

occurs. The M phase consists of two tightly coupled processes: *mitosis*, in which the cell's chromosomes are divided into two, and *cytokinesis*, in which the cell's cytoplasm divides in half forming distinct cells. Activation of each phase is dependent on the proper progression and completion of the previous one. Cells that have temporarily or reversibly stopped dividing are said to have entered a state of quiescence called the G0 phase. It is known that increases in intracellular Ca^{2+} are required for a cell to enter the S and M phases [42]. For example, during the G1 phase Ca^{2+} may drive a cell to reenter the S phase rather than exit the cell cycle. Experimentally blocking intercellular Ca^{2+} waves in the ventricular zone using ATP receptor antagonists appears to decrease cell proliferation without blocking Ca^{2+} elevations in individual cells [666]. Hence, one function of the intercellular waves could be to generate Ca^{2+} increases that help a nearby cluster of cells to enter the S phase in a coordinated manner. This hypothesis is also supported by the experimental observation that ventricular zone waves appear to be triggered by cells in the S phase [666]. In addition to S phase coordination, it is possible that intercellular Ca^{2+} waves also regulate differential gene expression [246], as suggested by the observation that the frequency of Ca^{2+} waves increases as neurogenesis proceeds.

Intercellular Ca^{2+} waves are also found in early developmental stages of structures in the peripheral nervous system such as the retina [146, 498, 499]. The developing retina is a layered structure whose outermost layer, the *retinal pigment epithelium* (RPE), is essential for the development of the underlying neural layers. The layer closest to the RPE is called the ventricular zone where, similar to the developing cortex, progenitor cells divide and differentiate to form the neurons and glia of the fully developed retina. ATP-mediated Ca^{2+} waves propagate in the RPE, possibly triggered by a single epithelial cell, and arise via the same paracrine signaling mechanism as Ca^{2+} waves in developing cortex; see Fig. 9.4. Diffusing ATP molecules released from the RPE also bind to P2Y receptors on the progenitor cells within the ventricular zone, which enhances cell proliferation and speeds up mitosis. It also appears that ATP from the RPE, possibly in combination with the action of neurotransmitters from differentiated neurons, can induce

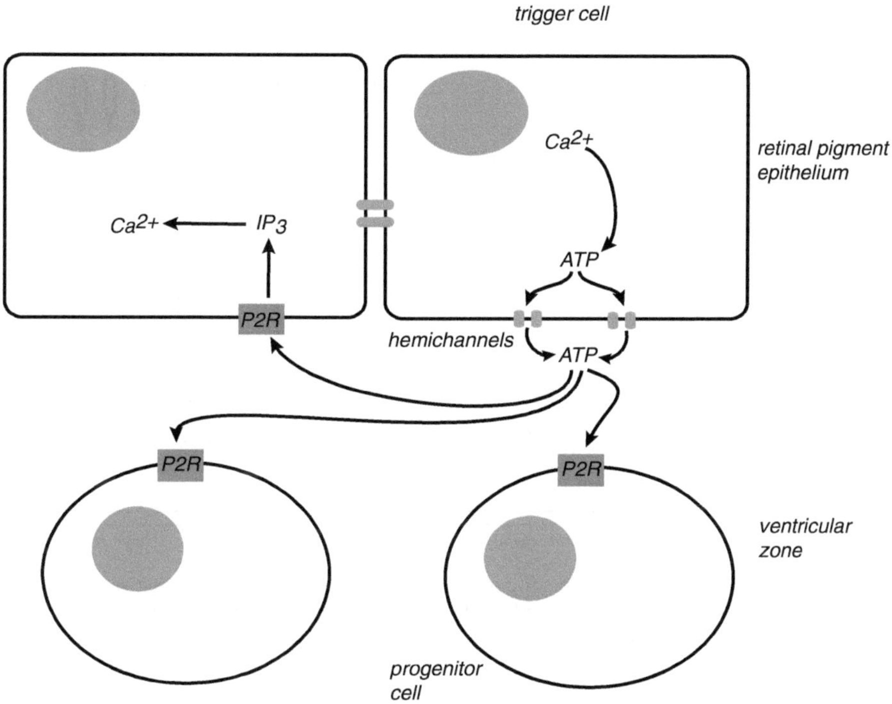

Fig. 9.4 ATP-mediated Ca^{2+} wave in radial pigment epithelium. ATP released from hemichannels diffuses and binds to metabotropic receptors (P2Rs) in neighboring epithelial cells leading to the release of intracellular Ca^{2+} from internal stores. The ATP also binds to P2Rs of progenitor cells in the ventricular zone, thus promoting cell proliferation

Ca^{2+} waves in the ventricular zone [498, 614]. The propagation of ventricular Ca^{2+} waves, which is probably mediated by gap-junction coupling between progenitors rather than paracrine signaling, can also modulate cell proliferation within a cluster of progenitor cells.

9.1.2 Retinal Waves in Later Development

The fully developed retina located at the back of the eye converts patterns of light activity from a visual image into neural spiking activity. The first stage is the conversion of light into graded neural activity by photoreceptors. Such activity is then modulated by both vertical and horizontal neuronal pathways via synaptic and gap-junction coupling, before reaching the *retinal ganglion cells* (RGCs). These form the output layer of the retina, which transmits the resulting pattern of spiking activity along the optic nerve to the brain; see also Sect. 8.1.1. Following the very early stages of development discussed at the end of the previous section, but before photoreceptors are visually responsive, the retina has already formed a layered

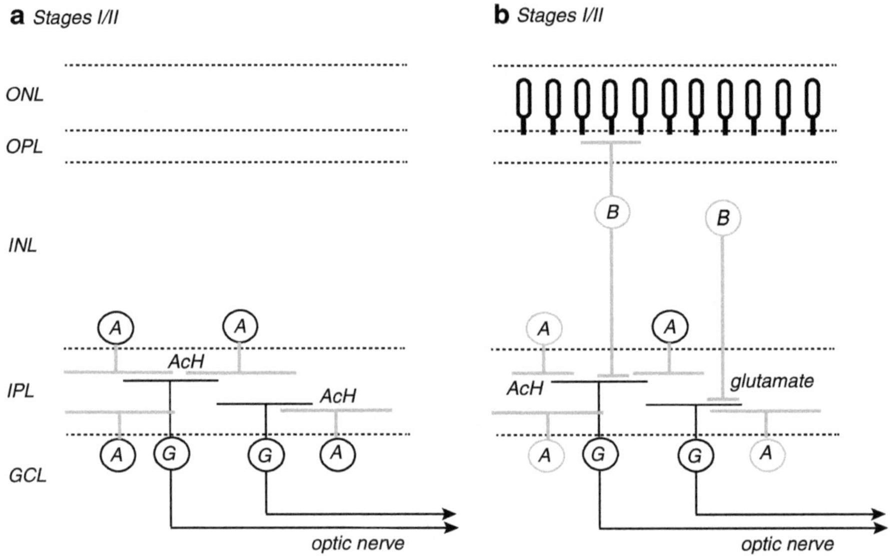

Fig. 9.5 Simplified schematic of neural circuits that mediate retinal waves during (**a**) stages I/II and (**b**) stage III. The retina is partitioned into three main layers: outer nuclear layer (*ONL*), inner nuclear layer (*INL*), and ganglion cell layer (*GCL*). There are also two layers of interconnecting processes: inner plexiform layer (*IPL*) and outer plexiform layer (*OPL*). During stages I/II, spontaneous retinal waves between retinal ganglion cells (*G*) are driven mainly by the interactions between ganglion cells and cholinergic amacrine cells (*A*). In stage III, waves are predominantly driven by glutamatergic bipolar cells (*B*), which are connected to photoreceptors in the ONL

structure with distinct classes of differentiated neurons, including RGCs. Although the RGCs are not yet driven by visual stimuli, they are far from silent, firing spontaneous bursts of action potentials that travel across the retinal surface as retinal waves; see the reviews [190, 222, 682]. Interestingly, such spontaneous rhythmic activity occurs in many other developing neural circuits [49, 471], including the cochlea [633], spinal cord [260, 472], cerebellum [661], hippocampus [32], and cortex [326]. There is growing evidence that the resulting spontaneous depolarizations, and corresponding increases in intracellular Ca^{2+} (via voltage-gated ion channels rather than release from intracellular stores), play an important role in driving many aspects of development, including axon guidance, cell differentiation, local circuit formation, and the emergence of sensory maps. These phenomena have been extensively studied in the retina, where upregulating or blocking retinal waves has a major affect on the development of visual circuits [190, 315, 416, 682] and has led to the suggestion that retinal waves are crucial for the normal development of the retinotopic map and the segregation of eye dominance in the cortex. However, their precise role still remains controversial. We briefly summarize some of the main properties of retinal waves at various stages of development and describe a number of theoretical models; see also [222].

Retinal waves can initiate at any retinal location due to spontaneous activation of a local cluster of cells and form non-repeating propagation boundaries that are determined, in part, by refractory effects. The waves can be categorized into three distinct developmental stages [615]. During stages I and II, the retina has only a few functional cell types and the waves arise via a coupled network of RGCs and *amacrine cells* forming the inner retina; see Fig. 9.5a. Stage I waves are mediated by the neuromodulator adenosine and by gap junctions, since very few synapses have formed. Stage II waves first appear at the onset of synaptogenesis and are primarily mediated by the neurotransmitter *acetylcholine*; they have a characteristic refractory period so that following passage of a retinal wave over a section of retina, another wave can only propagate across the same section after tens of seconds have elapsed. Stage I waves tend to be faster and occur more frequently than stage II waves. The transition to stage III occurs when vertical *bipolar cells* form synaptic connections with RGCs and the neurotransmitter glutamate becomes involved in mediating retinal waves; see Fig. 9.5b. Although acetylcholine continues to be involved as well, the nicotinic receptors found in stage II are replaced by muscarinic receptors. One finds that stage III waves are comparable in speed to stage II waves, but are smaller and less frequent. In certain species (e.g., the mouse), the retina first becomes sensitive to light near the onset of stage III, whereas in other species (e.g., the rabbit) light sensitivity occurs later. Spontaneous retinal waves generally disappear after eye opening, although in the turtle they can persist in diminished form for several weeks. Finally, retinal waves tend to have speeds in the range 100–300 μm/s, which are much slower than waves found in disinhibited cortical slices, for example; see Chap. 7. On the other hand, the speeds are much faster than processes based on the extracellular diffusion of signaling molecules.

Most computational studies of retinal waves have focused on developmental stage II and are based on networks of cholinergically coupled amacrine cells. The first such model also included a layer of RGCs that acted as a low-pass filter of the spontaneous activity generated in the amacrine layer [94]. The basic mechanism of the model was that amacrine cells would spontaneously depolarize and excite neighboring cells, after which they would enter a refractory state of fixed duration. A propagating wave was initiated whenever a sufficiently large local cluster of amacrine cells spontaneously became active and the surrounding medium contained a sufficient number of non-refractory cells. In spite of the model's simplicity, containing only two free parameters (an activation threshold and the probability of spontaneous activation), it could generate patterns of non-repeating waves that were statistically similar to experimental data. However, the model was not robust to changes in parameters, that is, it had to be finely tuned. An alternative model of retinal waves was proposed by Godfrey and Swindale [223], who took the refractory period of amacrine cells to be activity dependent. This provided a more robust model and did not require an RGC layer to generate biologically plausible waves. A more biophysically detailed model of the roles of cholinergic coupling, noise, and refractory period in the generation of retinal waves in amacrine circuits has subsequently been developed by Hennig et al. [271]. An individual amacrine cell is now described in terms of a conductance-based model, which consists of a voltage-gated Ca^{2+} current that supports spontaneous bursting, and a slow after-hyperpolarization

(AHP) current that prevents further bursting. Moreover, the duration of the AHP current is taken to increase with the level of depolarization. Thus, during passage of a wave, strong synaptic coupling between adjacent amacrine cells generates a stronger depolarization than intrinsic bursts in an isolated cell, which in turn causes a longer-lasting refractory period due to AHP currents. Thus, the effective network connectedness is dynamically regulated in such a way that the network can operate close to a percolation threshold, which means that the distribution of propagation events (frequency and duration) exhibits power-law behavior consistent with some experimental observations [271]; see also Sect. 9.5.

9.2 Traveling Waves, the Cytoskeleton, and Cell Polarization

9.2.1 Neuronal Polarization and the Cytoskeleton

Cell polarization refers to the breaking of cell symmetry and is an essential process in various cell types, including migration of fibroblasts during wound healing, asymmetric cell growth or division, and the functional maturation of neurons. Neurons are among the most polarized cells due to their compartmentalization into a thin, long axon and several shorter, tapered dendrites (Sect. 1.1). Experimental studies of neuronal polarization have mainly been performed on dissociated, embryonic cortical and hippocampal neurons or on postnatal cerebellar granule neurons. Such studies have identified three basic stages of polarization [13, 452, 509, 581]; see Fig. 9.6. Cultured neurons initially attach to their substrate as round spheres

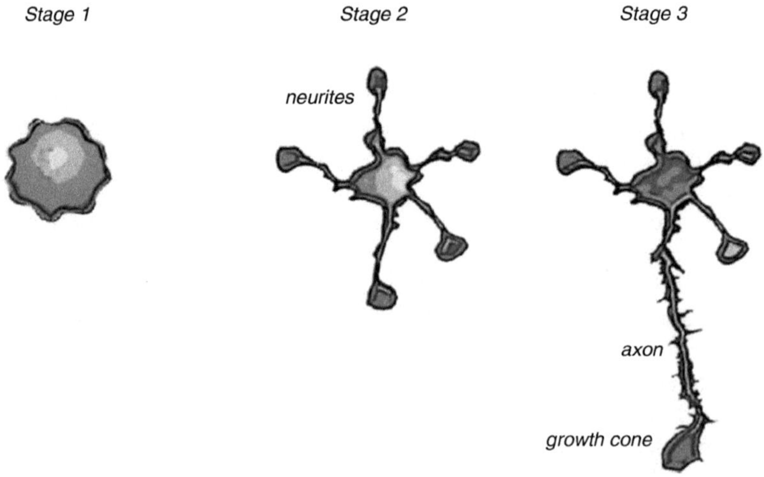

Fig. 9.6 Stages of neuronal polarization. A neuron attaches to the substrate as a round sphere surrounded by actin-rich lamellipodia and filopodia (stage 1). Growth cones formation by the consolidation of lamellipodia leads to the establishment of several neurites (stage 2). One neurite starts to elongate rapidly and forms the axon (stage 3)

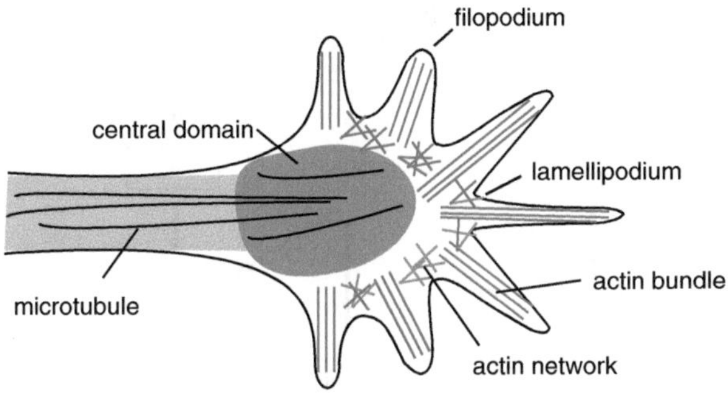

Fig. 9.7 Schematic diagram of growth cone showing cytoskeletal structures

surrounded by actin-rich structures such as *lamellipodia* (mesh-like actin sheets) and *filopodia* (packed actin bundles) (stage 1). Lamellipodia then coalesce to form *growth cones*, followed by the establishment of several short processes, called *neurites* (stage 2). Eventually one of the neurites starts to grow more rapidly to become the axon (stage 3), while the other neurites remain short and develop into dendrites at later stages of maturation. During the development of a neuron, the cytoskeleton plays a fundamental role (as does intracellular transport [289]). Since neurite formation is a key event in neuronal polarization, it is instructive to describe the role of the cytoskeleton during the elongation of neurites.

The growth cone at the mobile tip of an elongating neurite or axon contains microtubules within a central domain (C-domain) and actin filaments within the peripheral domain (P-domain); see Fig. 9.7. The microtubules provide the structural backbone of the shaft and a substrate for intracellular transport to the growth cone. They polymerize with their growing ends pointed towards the leading edge of the growth cone. Actin filaments within the P-domain form filopodia and lamellipodia that shape and direct the motility of the growth cone. In both structures, the actin filaments face with their barbed (growing) ends towards the plasma membrane. Polymerization of actin filaments towards the leading edge causes the extension and protrusion of the growth cone. This creates a force that pushes the actin network and the tightly linked plasma membrane backwards (retrograde flow) and hinders the invasion of the microtubules into the P-domain. The retrograde flow is also enhanced by the action of myosin molecular motors, which drag the actin cytoskeleton back towards the C-domain where actin filaments depolymerize at their pointed ends. If there is a balance between actin polymerization in the P-domain and retrograde flow, then there is no elongation. However, signals from surface adhesion receptors bound to a substrate can suppress the retrograde flow of actin filaments, shifting the balance towards polymerization-driven forward motion that involves both actin filaments and microtubules. (This signaling mechanism can in turn be modulated by external chemoattractant cues during axon guidance [394, 470].) The growth of a neurite proceeds in three stages [225, 432]. First, enhanced polymerization in the P-domain pushes the growth cone forward (protrusion phase). Second, an actin-free

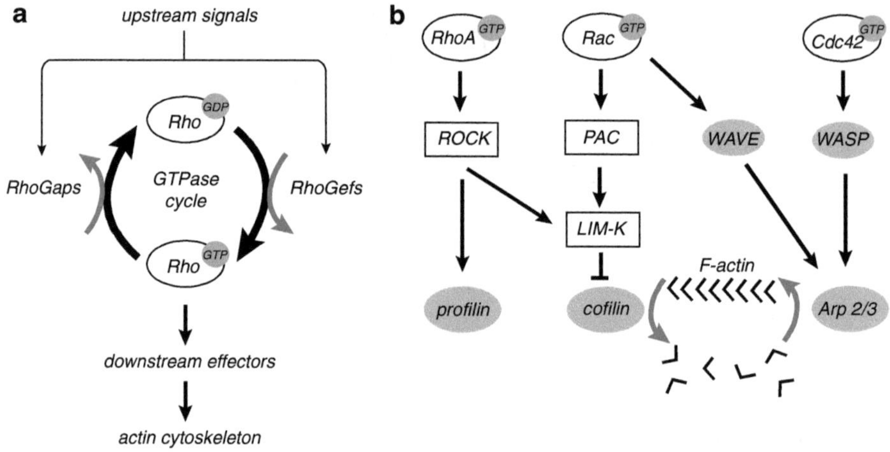

Fig. 9.8 (**a**) The Rho GTPase cycle. (**b**) Simplified signaling pathways from GTPases RhoA, Rac, and Cdc42 to actin accessory proteins cofilin and Arp 2/3 that regulate actin polymerization in the growth cone (Redrawn from Luo [394])

zone is created by a combination of cytoskeletal restructuring and actin disassembly so that the retrograde flow no longer hinders the advancement of microtubules into this zone (engorgement phase). Third, vesicles and organelles enter the newly created zone via intracellular transport, resulting in an elongation of the neurite shaft (consolidation phase).

One class of signaling molecule that appears to play an important role in the polarization and migration of many different cell types is *Rho GTPases* [235, 254]. These act as intracellular molecular switches that cycle between an active GTP-bound form and an inactive GDP-bound form; see Fig. 9.8a. Guanine nucleotide exchange factors (RhoGEFs) facilitate the conversion from GDP-bound to GTP-bound form, whereas GTPase-activating proteins (RhoGAPs) enhance GTP hydrolysis and are thus negative regulators. RhoGEFs and RhoGAPs are both regulated by upstream signals. A major downstream target of the Rho GTPase signaling pathways is the actin cytoskeleton within the growth cone [254, 394, 452]. (There is growing evidence that the GTPases also regulate microtubules [13], although we will not discuss this here.) A variety of actin accessory proteins mediate the different components of actin dynamics within the growth cone and can be activated by the various signaling pathways associated with neuronal polarization. We describe a few of the major players; see [452, 483] for more details.

First, the *actin-related proteins 2 and 3* (Arp2/3) *complex* stimulates actin polymerization by creating new nucleation cores. The Arp2/3 complex is activated by members of the *Wiskott–Aldrich syndrome protein* (WASP) family protein (WAVE) complex, which localizes to lamellipodia where it facilitates actin polymerization. The WAVE complex is activated by the small GTPase *Rac1*, which modulates the actin cytoskeleton dynamics by controlling the formation of lamellipodia. Filopodia formation, on the other hand, is regulated by another member of the small GTPases, called *Cdc42*. A second important accessory protein is *cofilin*, a member of the actin-depolymerizing factor (ADF)/cofilin family, which modifies actin dynamics by increased severing and depolymerization of actin filaments via its binding to the non-barbed (pointed) ends. Cofilin is inhibited when phosphorylated by LIM-kinase, which is itself activated by Pak-kinase. Since the latter is a downstream target of Rac1 and Cdc42, it follows that Rac1 and Cdc42 inhibit actin depolymerization by downregulating cofilin. At first sight, one would expect the reduction of cofilin to counter the affects of Arp2/3. However, the extent to which cofilin enhances or reduces cell protrusion depends on the spatial and temporal scale over which it operates [150]. It turns out that the overall effect of Rac1 and Cdc42 is to increase actin dynamics thus promoting neurite growth. This is opposed by the action of a third type of GTPase known as *RhoA*, which tends to stabilize the actin network by activating ROCK-kinase, which promotes *profilin* (an actin-binding protein that catalyzes polymerization) and suppresses cofilin. A summary of the basic signaling pathways is given in Fig. 9.8b.

9.2.2 Wave-Pinning and Cell Polarity in a Bistable Reaction–Diffusion Model

We now turn to a reaction–diffusion model of cell polarity developed and analyzed by Keshet et al. [305, 438, 439]; see also [478, 482]. Although the model was originally constructed in order to understand the mechanisms of cell polarization in motile cells such as fibroblasts, it is possible that similar mechanisms occur prior to the sprouting of a new neurite during stage 2 (neuritogenesis). The basic idea is to use reaction–diffusion equations to describe the spatiotemporal dynamics of Rho GTPases such as Rac1, Cdc42, and RhoA, which are known to cycle between active membrane-bound forms and inactive cytosolic forms. One finds that a transient and localized stimulus can be spatially amplified, leading to a robust partitioning of the cell into a front and back, where the concentration of the active forms is high and low, respectively. Mathematically speaking, the emergence of cell polarization can be understood in terms of front propagation in a bistable reaction–diffusion system (Sect. 2.2) with the following additional features [438, 439]: (1) the inactive and active states have unequal rates of diffusion; (2) the total amount of each GTPase is conserved. Consequently, a local stimulus induces a propagating front that decelerates as it propagates across the cell so that it becomes stationary, a process known as *wave-pinning*; the stationary front persists in the absence of the stimulus and represents a polarized cell. Note that the wave-pinning mechanism is distinct from the diffusion-driven Turing mechanism found in other reaction–diffusion models of cell polarization [417, 482, 609]. One of the basic differences is that in the latter class of models, a homogeneous state becomes unstable to arbitrarily small fluctuations and the nonlinear reaction terms support the growth of a spatially varying pattern via a Turing stability. For a recent application of the Turing mechanism to neurite formation, see [420].

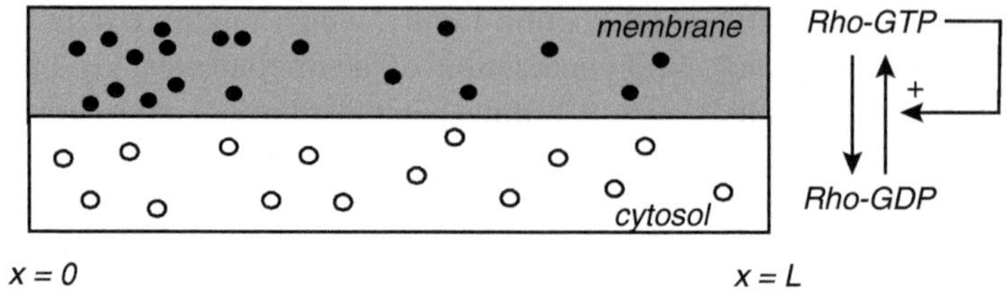

Fig. 9.9 Schematic diagram of wave-pinning model showing spatial distribution of membrane-bound active and cytosolic inactive forms of Rho GTPase

The simplest version of the model considers a single Rho GTPase that can transition between inactive and active forms diffusing in a bounded 1D domain of length L [438]; see Fig. 9.9. Let $a(x,t)$ and $b(x,t)$ be the concentrations of the active/inactive states. Then

$$\frac{\partial a}{\partial t} = D_a \frac{\partial^2 a}{\partial x^2} + f(a,b), \tag{9.1a}$$

$$\frac{\partial b}{\partial t} = D_b \frac{\partial^2 b}{\partial x^2} - f(a,b), \tag{9.1b}$$

Since the rate of diffusion of the membrane-bound (active) state is significantly slower than that of the cytosolic (inactive) state, $D_a \ll D_b$. The nonlinear function $f(a,b)$ represents the difference between the rates of activation and inactivation of the Rho GTPase. Assuming there is cooperative positive feedback in the activation of the protein, which is modeled as a Hill function of index 2, then

$$f(a,b) = b\left(k_0 + \frac{\gamma a^2}{K^2 + a^2}\right) - k_{-}a. \tag{9.2}$$

It can be checked that for a range of uniform concentrations of the inactive state, $b_{\min} < b < b_{\max}$, the space-clamped version of the model exhibits bistability with two stable fixed points $a_{\pm}(b)$ separated by an unstable fixed point $a_0(b)$. Equation (9.1) is supplemented by no-flux boundary conditions at the ends of the domain:

$$\left.\frac{\partial a}{\partial x}\right|_{x=0,L} = 0, \quad \left.\frac{\partial b}{\partial x}\right|_{x=0,L} = 0. \tag{9.3}$$

It follows that there is mass conservation of the total amount of Rho GTPase, that is,

$$\int_0^L (a+b)dx = C. \tag{9.4}$$

The mathematical explanation of wave-pinning proceeds as follows [438]. First, since $D_b \gg D_a$ and there are no-flux boundary conditions, one can assume that b rapidly diffuses to establish a uniform concentration within the bounded domain $[0,L]$; b then changes on a slower time scale as the a dynamics evolves (quasi-steady-state approximation). Thus, on short time scales b can be treated as a fixed global parameter of a scalar equation for $a(x,t)$ given by (9.1a). Suppose that initially $b_{\min} < b < b_{\max}$, so (9.1a) is bistable. On an infinite domain, the bistable equation supports the propagation of a traveling front linking the stable fixed point $a_+(b), a_-(b)$ (see Sect. 2.2). That is, for $-\infty < x < \infty$ there exists a monotonically decreasing solution $a(x,t) = A(\xi)$, $\xi = x - ct$ with $\lim_{\xi \to -\infty} A(\xi) = a_+(b)$ and $\lim_{\xi \to \infty} A(\xi) = a_-(b)$. Moreover the wave speed satisfies $c = c(b)$ with [see also (2.19)]

$$c(b) = \frac{\displaystyle\int_{a_-}^{a_+} f(a,b)da}{\displaystyle\int_{-\infty}^{\infty} (\partial A/\partial \xi)^2 d\xi}. \tag{9.5}$$

Note that the wave speed depends on the global parameter b. Since the denominator of (9.5) is always positive, the sign of $c(b)$ will depend on the sign of $I(b) \equiv \int_{a_-}^{a_+} f(a,b)da$, which has a geometrical interpretation in terms of the difference between the area of the curve $y = f(a,b)$ above the straight line $y = k_- a$ and the area below; see Fig. 9.10. In the case of a sufficiently sharp front that is away from the boundaries, these results carry over to the bounded domain $[0,L]$.

Now suppose that a transient stimulus near the edge of the cell at $x = 0$ triggers at time $t = 0$ a traveling front as described above. This implies that $b_{\min} < b(0) < b_{max}$ and $I(b(0)) > 0$. As the front starts to propagate into the interior of the cell, it converts a greater fraction of the domain from $a \approx a_-(b)$ to $a \approx a_+(b)$. From the conservation condition (9.4), it follows that the approximately uniform concentration $b(t)$ of the inactive state decreases, eventually reaching a critical value b_c, $b_{\min} < b_c < b_{\max}$, for which

$$I(b_c) \equiv \int_{a_-}^{a_+} f(a,b_c)da = 0, \tag{9.6}$$

and wave-pinning occurs. The basic steps are illustrated in Fig. 9.10.

Asymptotic Analysis of Wave-Pinning. As shown by Mori et al. [439], it is possible to analyze wave-pinning in more detail using a multi-time-scale analysis. The first step is to nondimensionalize (9.1):

$$\varepsilon \frac{\partial a}{\partial t} = \varepsilon^2 \frac{\partial^2 a}{\partial x^2} + f(a,b), \tag{9.7a}$$

$$\varepsilon \frac{\partial b}{\partial t} = D \frac{\partial^2 b}{\partial x^2} - f(a,b), \tag{9.7b}$$

with $x \in [0,1]$ and $0 < \varepsilon \ll 1$. In order to look at the dynamics on short time scales, set $\tau = t/\varepsilon$ and introduce the asymptotic expansions $a \sim a_0 + \varepsilon a_1 + \ldots$ and $b \sim b_0 + \varepsilon b_1 + \ldots$. This yields the following pair of equations for a_0 and b_0:

$$\frac{\partial a_0}{\partial \tau} = f(a_0, b_0), \tag{9.8a}$$

$$\frac{\partial b_0}{\partial \tau} = D \frac{\partial^2 b_0}{\partial x^2} - f(a_0, b_0). \tag{9.8b}$$

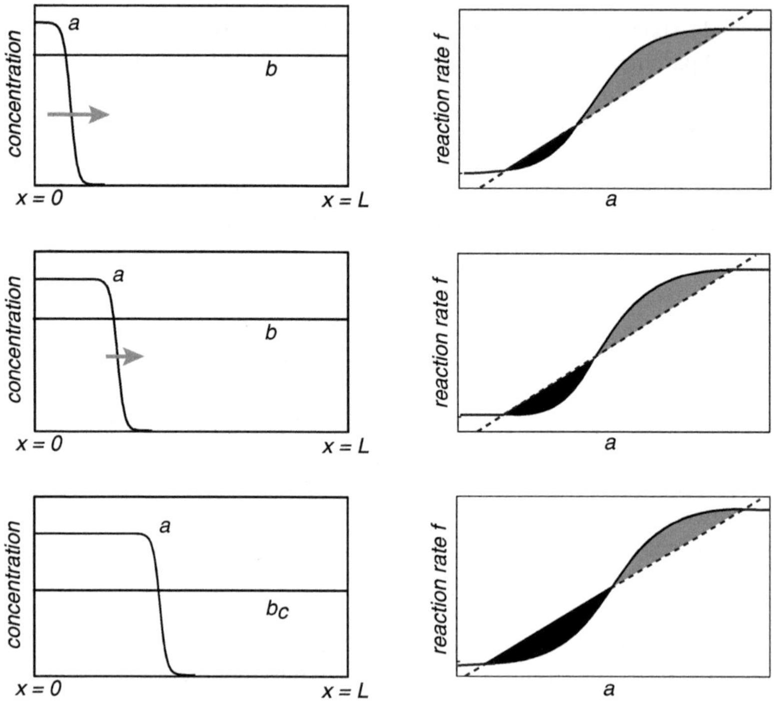

Fig. 9.10 Schematic diagram explaining the mechanism of wave-pinning developed in [438]. A sequence of snapshots of the traveling front (*left column*) showing that as the front advances into the domain, the background concentration b of the inactive state decreases so that the front decelerates until it becomes stationary. The corresponding geometric construction of $I(b)$ (*right column*), which is given by the difference of the *shaded areas*, shows that $I(b)$ is initially positive but vanishes at the critical value b_c

Suppose that $b_{\min} < b_0 < b_{\max}$ so that $f(a_0, b_0)$ is bistable with respect to a_0. Equation (9.8a) implies that $a_0 \to a_+(b_0)$ or $a_0 \to a_-(b_0)$ at each point $x \in [0,1]$. Assume initial conditions in which there exists a single sharp transition layer linking a_+ to a_- (a traveling front).

Let $\phi(t)$ be the position of the transition layer with respect to the slower time scale t. Divide the domain $[0,1]$ into inner and outer regions, with the latter given by $[0, \phi(t) - \mathscr{O}(\varepsilon)) \cup (\phi(t) + \mathscr{O}(\varepsilon), 1]$. Taking the limit $\varepsilon \to 0$ in (9.1), with $a \sim a_0 + \varepsilon a_1 + \ldots$, $b \sim b_0 + \varepsilon b_1 + \ldots$, and restricting x to the outer layer, we have

$$0 = f(a_0, b_0), \tag{9.9a}$$

$$0 = D\frac{\partial^2 b_0}{\partial x^2} - f(a_0, b_0). \tag{9.9b}$$

Adding these two equations immediately implies that $\partial^2 b_0/\partial x^2 = 0$. Combining this with the no-flux boundary conditions, it follows that

$$b_0(x,t) = \begin{cases} b_L(t) & 0 \leq x < \phi(t) - \mathscr{O}(\varepsilon), \\ b_R(t) & \phi(t) + \mathscr{O}(\varepsilon) < x \leq 1. \end{cases} \tag{9.10}$$

with $b_{L,R}(t)$ independent of x. The corresponding solution for a_0 is

$$a_0(x,t) = \begin{cases} a_+(b_L) & 0 \leq x < \phi(t) - \mathcal{O}(\varepsilon), \\ a_-(b_R) & \phi(t) + \mathcal{O}(\varepsilon) < x \leq 1. \end{cases} \tag{9.11}$$

In order to determine the solution in the inner layer near the front, introduce the stretched coordinate $\xi = (x - \phi(t))/\varepsilon$, and set

$$A(\xi,t) = a([x - \phi(t)]/\varepsilon, t), \quad B(\xi,t) = b([x - \phi(t)]/\varepsilon, t) \tag{9.12}$$

Introduce the asymptotic expansion $A \sim A_0 + \varepsilon A_1 + \ldots$ and similarly for B and ϕ. Substitution into (9.7) and expansion to lowest order in ε then gives

$$\frac{\partial^2 A_0}{\partial \xi^2} + \frac{d\phi_0}{dt}\frac{\partial A_0}{\partial \xi} + f(A_0, B_0) = 0, \tag{9.13a}$$

$$\frac{\partial^2 B_0}{\partial \xi^2} = 0. \tag{9.13b}$$

The solution of (9.13b) is of the form

$$B_0 = \alpha_1(t)\xi + \alpha_2(t), \tag{9.14}$$

where $\alpha_{1,2}(t)$ are determined by matching the inner (B_0) and outer (b_0) solutions according to

$$\lim_{\xi \to -\infty} B_0(\xi) = b_L, \quad \lim_{\xi \to \infty} B_0(\xi) = b_R. \tag{9.15}$$

The matching conditions can only be satisfied if B_0 is constant in the inner layer, which means that the inactive state is uniform throughout the whole domain, $B_0 = b_0$. The next step is to determine the inner solution for A_0, given that B_0 is ξ-independent. That is, we have to solve the boundary value problem (9.13a) with matching conditions:

$$\lim_{\xi \to -\infty} A_0(\xi) = a_+(b_0), \quad \lim_{\xi \to \infty} A_0(\xi) = a_-(b_0). \tag{9.16}$$

The inner solution thus corresponds to the standard front solution of a bistable equation with

$$\frac{d\phi_0}{dt} \equiv c(b_0) = \frac{\displaystyle\int_{a_-(b_0)}^{a_+(b_0)} f(a, b_0)\,da}{\displaystyle\int_{-\infty}^{\infty} (\partial A_0/\partial \xi)^2 d\xi}, \tag{9.17}$$

see (9.5).

The final step is to incorporate the conservation condition (9.4), which to lowest order in ε becomes

$$\int_0^1 a_0 dx + b_0 = C. \tag{9.18}$$

The integral term can be approximated by substituting for a_0 using the outer solution:

$$\begin{aligned} \int_0^1 a_0 dx &= \int_0^{\phi(t)-\mathcal{O}(\varepsilon)} a_0 dx + \int_{\phi(t)+\mathcal{O}(\varepsilon)}^1 a_0 dx + \mathcal{O}(\varepsilon) \\ &= a_+(b_0)\phi_0(t) + a_-(b_0)(1 - \phi_0(t)) + \mathcal{O}(\varepsilon). \end{aligned}$$

Combining the various results, the analysis of a traveling front solution of (9.7) reduces to the problem of solving the ODE system [439]

$$\frac{d\phi_0}{dt} = c(b_0), \quad b_0 = C - a_+(b_0)\phi_0(t) - a_-(b_0)(1-\phi_0(t)). \tag{9.19}$$

These equations can be used to establish that the front slows down and eventually stops to form a stable stationary front. In particular, differentiating the second relation in (9.19) with respect to t shows that

$$\left(1 + \frac{da_+(b_0)}{db}\phi_0 + \frac{da_-(b_0)}{db}(1-\phi_0)\right)\frac{db_0}{dt} = -[a_+(b_0) - a_-(b_0)]\frac{d\phi_0}{dt}. \tag{9.20}$$

Differentiating the condition $f(a_\pm(b), b) = 0$ with respect to b and imposing the condition that $f(a,b)$ is bistable, it can be established that $1 + da_\pm/db > 0$. Since $0 < \phi_0 < 1$ and $a_+(b_0) > a_-(b_0)$, it follows from (9.20) that db_0/dt and $d\phi_0/dt$ have opposite signs. Hence, as the front advances, b_0 decreases until the front stalls at a critical value b_c for which $c(b_c) = 0$. The corresponding front position is ϕ_c with

$$b_c = C - a_+(b_c)\phi_c - a_-(b_c)(1-\phi_c). \tag{9.21}$$

Explicit conditions for wave-pinning can be obtained if the reaction term f is taken to be a cubic

$$f(a,b) = a(1-a)(a-1-b), \tag{9.22}$$

rather than the Hill function (9.2). In this case, the speed of the front solution in the inner layer can be calculated explicitly (see Sect. 2.2), so that (9.19) becomes

$$\frac{d\phi_0}{dt} = \frac{b_0 - 1}{\sqrt{2}}, \quad b_0 = C - (1+b_0)\phi_0. \tag{9.23}$$

It follows that the wave stops when $b_0 = 1 \equiv b_c$ and the stall position is $\phi_c = (C-1)/2$. Finally, note that Mori et al. [439] also extend the asymptotic analysis to the case of multiple transition layers and carry out a bifurcation analysis to determine parameter regimes for which wave-pinning occurs.

One interesting issue is to what extent the wave-pinning mechanism differs from the phenomenon of wave propagation failure due to spatial discretization, as occurs in myelinated axons, for example (Sect. 2.5). This is particularly important given that any numerical simulation of the wave-pinning model involves the introduction of a spatial grid, and the wave becomes more sensitive to discretization effects as it slows down. A careful numerical study has shown that wave-pinning and propagation failure are distinct effects [655]. In the same study, a stochastic version of the wave-pinning model was also considered, which takes into account fluctuations in the number of active and inactive molecules at low concentrations. It was found that when the total number of molecules is lowered, wave-pinning behavior is lost due to a broadening of the transition layer as well as increasing fluctuations in the pinning position.

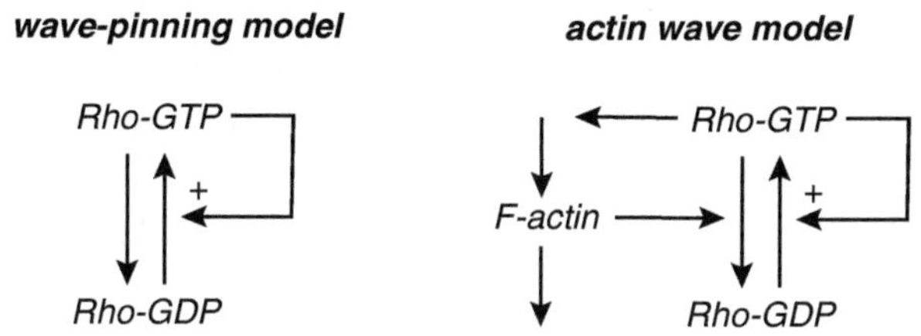

Fig. 9.11 Schematic diagram showing the basic pathways involved in the wave-pinning model [438] and actin wave model [286]

9.2.3 Actin Waves

There are a number of intracellular wavelike phenomena associated with the maturation of single neurons. First, localized Ca^{2+} transients occur in the growth cone of developing neurons [233, 234, 245]. Although they spread passively by diffusion rather than actively, they are often referred to as growth cone Ca^{2+} waves. These waves regulate the rate of neurite extension, which is inversely proportional to their frequency, since elevations in intracellular Ca^{2+} concentration above baseline slow any growth. Second, a neuron-wide Ca^{2+} wave propagating from the growth cone of an axon to the soma mediates reversal of neuronal migration in response to a repulsive extracellular cue (Slit-2) [247]. It is hypothesized that Slit-2 receptor activation at the leading edge of the growth cone triggers the Ca^{2+} wave, which propagates to the soma, resulting in a front-to-rear gradient of Ca^{2+} concentration. This gradient in turn causes (by some yet-to-be-identified mechanism) a redistribution of the GTPase RhoA towards the rear of soma, a process necessary for establishing the reversed direction of migration. (One possibility is that the calcium wave acts as a trigger for a polarization wave similar to the one discussed in Sect. 9.2.2.) We will focus on another type of wave, namely, an actin wave propagating along the axon shaft of neurites during cell maturation. In developing hippocampal neurons, neurites undergo periodic extensions and retractions, which correlate with the propagation of growth cone-like structures that originate in the soma and propagate down the neurites [194, 542, 543]. The waves travel at a speed of around 3 μm/min, transporting actin and associated proteins such as cofilin down the neurite shaft to the growth cone. Waves occur in all neurites during maturation stage 2, but become more frequent in the emerging axon during the transition from stage 2 to stage 3 [194]. This coincides with an engorgement and increased dynamics of the axon's growth cone. Interestingly, growth cone-like actin waves have also been observed during the regeneration of injured axons in cultured hippocampal neurons [155].

Actin waves are found in a number of other cell types including *Dictyostelium* [643], neutrophils [665], and fibroblasts (see the review [7]). Indeed, most current models of actin waves are based on studies of nonneuronal cells. They also tend to be rather complicated, containing details regarding the distribution of actin lengths or orientations, for example [103, 157, 643]. Here we will consider a simpler model of actin waves that extends the wave-pinning model of Mori et al. by coupling the Rho

GTPase signaling module to a module representing the assembly and disassembly of actin polymers (F-actin) [286]; see Fig. 9.11. More specifically, active Rho GTPase promotes the polymerization of F-actin which feeds back to enhance the inactivation of Rho GTPase (negative feedback). In order to incorporate the effects of negative feedback, the kinetic rate function (9.2) is modified according to

$$f(a,b,u) = b\left(k_0 + \frac{\gamma a^3}{K^3 + a^3}\right) - k_-\left(s_1 + s_2\frac{u}{u_0 + u}\right)a, \tag{9.24}$$

where u is the concentration of F-actin and s_1, s_2, u_0 are additional parameters. If $s_2 = 0$ and $s_1 = 1$, then we recover the previous wave-pinning model with f having a hill coefficient of $n = 3$ rather than $n = 2$. (Although $n = 2$ also works, the choice $n = 3$ gives more robust waves). It remains to specify the F-actin dynamics. It is assumed that F-actin is nucleated from an infinite reservoir of actin monomers with concentration M, at a rate that is proportional to the concentration of active Rho GTPase. Thus, neglecting the diffusion of F-actin,

$$\frac{\partial u}{\partial t} = \varepsilon[k_n M - k_s u], \tag{9.25}$$

where εk_n is the nucleation rate and εk_s is the disassembly rate.

Extensive numerical simulations show that the actin wave model supports a wide range of spatiotemporal dynamics not found in the wave-pinning model, including solitary traveling pulses, periodic wave trains, persistent reflecting waves, and more exotic localized patterns [286]. In order to characterize the parameter dependence of these various patterns, Holmes et al. [286] considered the nonlinear stability of homogeneous steady states with respect to non-infinitesimal localized perturbations. This analysis identified two distinct stability regimes: an *excitable* regime where a pattern can only be triggered by a perturbation whose amplitude exceeds some threshold and an *unstable* regime where arbitrarily small noise can induce a pattern. Traveling pulses occur in the excitable regime, where the negative feedback due to F-actin plays a role analogous to the refractory variable in the FitzHugh–Nagumo equations (see Chap. 2). On the other hand, wave trains occur in the unstable regime, and reflecting waves occur in both regimes. As with other models of actin waves, the motivating examples are from nonneuronal cells. However, it is possible that the solitary traveling pulse could represent the growth cone-like wave seen in developing neurites [194, 542, 543].

9.2.4 Diffusion-Based Model of Neurite Symmetry Breaking

A major unresolved issue is how a single axon is specified among equally potential neurites, that is, what is the underlying symmetry breaking mechanism for neurite differentiation? During stage 3, the cytoskeleton of the emerging axonal growth cone becomes more dynamic and less dense (due to the action of Rac1 and Cdc42, say), whereas the actin cytoskeleton of the other growth cones remains more stable and maintains a denser network (due to the action of RhoA, say). Consequently, microtubules are hindered from entering the growth cone of the inhibited neurites, whereas they can advance into the P-domain of the axon thus stimulating elongation. An emerging hypothesis [9, 13, 193, 629] is that neurite extension and retraction are controlled by positive and negative signaling molecules (morphogens) respectively, which diffuse within neurites to regulate the behavior of actin filaments and microtubules or to influence cargo transport. In particular, it is hypothesized that prior to neurite differentiation, positive and negative feedback mechanisms are balanced; this balance can be broken by an external cue or by spontaneous intrinsic fluctuations, resulting in the enhancement of positive feedback within a single neurite that elongates to form an axon; simultaneously, the activated neurite generates a strong negative feedback signal that prevents other neurites from forming a second axon. A discussion of various candidate signaling cascades that could support autocatalytic (positive feedback) processes in growth cones can be found elsewhere [9, 13]. Here we will describe a particular model of neurite differentiation due to Toriyama et al. [629], which is based on the transport of a protein called *Shootin1*. Shootin1 is thought to be an upstream regulator of various signaling molecules responsible for neuronal polarization [576]. The model is motivated by the experimental observation that Shootin1 is actively transported from the soma to growth cones as discrete boluses within actin waves [628]. The active anterograde transport combines with diffusive retrograde transport in such a way that Shootin1 accumulates in the growth cones in a neurite length-dependent manner [629]. Moreover, during the stage 2/3 transition, Shootin1 predominantly accumulates in the nascent axon. The observation that Shootin1 can induce neurite outgrowth by regulating actin retrograde flow [576] suggests that the asymmetric accumulation of Shootin1 is one of the critical events in neurite differentiation.

Consider a developing neuron that consists of N neurites labeled $i = 1,\ldots,N$, each having length L_i; see Fig. 9.12a. Let C_0 denote the concentration of Shootin1 in the soma and C_i the concentration in the ith growth cone. The diffusion coefficient of Shootin is around $350\,\mu\text{m}^2$/min so that the diffusion time constant $\tau = L^2/D$ is around 1 min for a typical neurite length of $L = 20\,\mu\text{m}$. Given that the volumes of the cell body (V_S) and growth cone (V) are large compared to the volume of a thin neurite shaft, it follows that changes in Shootin1 concentration within the growth cone and soma occur on a slower time scale than the formation of a concentration gradient along a neurite. Hence, one can assume as a first approximation that the diffusive retrograde flow is in quasi-equilibrium. This means that the diffusive flux J_i along the jth neurite is approximately uniform so that the net flux from the growth cone is

$$J_i(t) = \frac{D}{L_i}(C_i(t) - C_0(t)). \tag{9.26}$$

For simplicity, the propagation of actin waves is not modeled explicitly; rather, the waves are treated as sequences of discrete stochastic events corresponding to

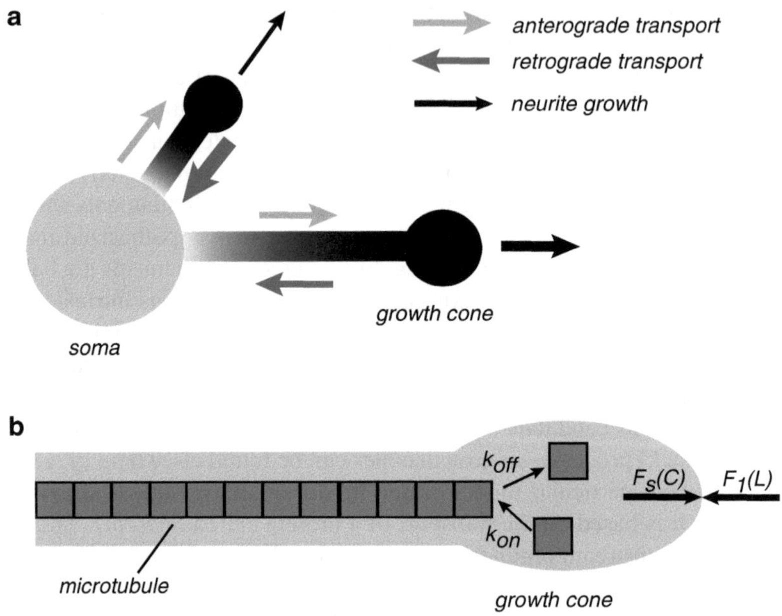

Fig. 9.12 Model of diffusion-based neurite growth due to Toriyama et al. [629]. (**a**) A concentration gradient of Shootin1 is set up along each dendrite due to the combination of active anterograde flow and length-dependent retrograde diffusion. Accumulation of Shootin1 in a growth cone sets up a positive feedback loop for neurite growth. Neurite competition for resources leads to symmetry breaking where one neurite grows at the expense of all the others. (**b**) Neurite growth is regulated by the balance between a Shootin1-induced traction force $F_S(C)$ and a length-dependent neurite tensional force $F_1(L)$

Gaussian-like packets of Shootin1 delivered to a neurite's growth cone at random times T_j^n and amplitudes a_j^n. Thus, the active transport in neurite j is represented as a sum of discrete events

$$\Gamma_j(t) = \frac{V_S}{V}\sum_n a_j^n g(t - T_j^n). \tag{9.27}$$

with $g(t)$ a normalized Gaussian of width σ. The amplitudes and delivery times are modeled by gamma distributions fitted to the experimental data [629]. Combining (9.26) and (9.27), the Shootin1 concentration in the jth growth cone evolves as

$$\frac{dC_i}{dt} = -\frac{AD}{VL_i}(C_i - C_0) + \Gamma_j(t). \tag{9.28}$$

(For simplicity, the growth cone volume V and cross-sectional shaft area A is taken to be the same for each neurite.)

The next component of the model specifies how the accumulation of Shootin1 in the growth cone regulates neurite growth; see Fig. 9.12b. It is assumed that growth is determined primarily by the balance between two mechanical forces: a Shootin1-dependent traction force $F_S(C)$ and a neurite length-dependent tensional force $F_1(L)$. The former is taken to be a Hill function

$$F_S(C) \sim \frac{C^h}{K_s^h + C^h}, \tag{9.29}$$

for constants K_s, h, while the latter is approximated as a logarithmic Hill-like function,

$$F_1(L) \sim \frac{\ln(L/L_0)}{\ln(L/L_0) + \ln(K_1/L_0)} \tag{9.30}$$

for constants K_1, L_0. The growth velocity of the ith neurite is determined as

$$\frac{dL_i}{dt} = \delta(k_i^{\text{on}} M - k_i^{\text{off}}), \tag{9.31}$$

where δ is the size of a tubulin monomer (the basic building block of microtubules), M is the intracellular concentration of tubulin, and k_i^{on} and k_i^{off} are the rates of polymerization and depolymerization, respectively. From the thermodynamics analysis of a simple ratchet model of polymerization [434], it can be shown that the ratio of the on and off rates depends on the net force $F = F_1(L) - F_s(C)$ according to

$$\frac{k^{\text{off}}(F)}{k^{\text{on}}(F)} = \frac{k^{\text{off}}(0)}{k^{\text{on}}(0)} e^{F\delta/k_B T}, \tag{9.32}$$

where k_B is the Boltzmann constant and T is absolute temperature. Fitting to experimental data suggests that the on rate is approximately independent of F. Hence, setting $F_i = F_1(L_i) - F_s(C_i)$, $k_i^{\text{on}} = k_{\text{on}}$ and $k_i^{\text{off}} = k^{\text{off}}(F_i) = k_{\text{off}} e^{F_i\delta/k_B T}$, we have

$$\frac{dL_i}{dt} = \delta \left[k_{\text{on}} M - k_{\text{off}} e^{-[F_s(C_i) - F_1(L_i)]\delta/k_B T} \right]. \tag{9.33}$$

As it stands, if C_0 and M were fixed (reflecting an unlimited supply of tubulin and Shootin1), then each neurite would independently undergo a positive feedback loop in which an increase in neurite length L_i would slow the rate of retrograde diffusion, the growth cone concentration C_i of Shootin1 would increase, and this would in turn speed up the rate of growth. However, the neurites actually compete for a finite amount of resources. If S is the total amount of Shootin1 in the neuron and V_0 is the volume of the soma, then

$$S = V_0 C_0 + V \sum_{j=1}^{N} C_j. \tag{9.34}$$

Hence, as a neurite grows, it depletes the level of Shootin1, which inhibits the growth of the other neurites. There is also competition for tubulin monomers, which can be expressed as

$$T = V_{\text{tot}} M + \rho \sum_{j=1}^{N} L_j, \tag{9.35}$$

where $V_{\text{tot}} = V_0 + NV$, T is the total amount of cellular tubulin and ρ is a parameter that determines how much tubulin is needed to build a neurite of a given length. Finally, experimental data indicates that both T and S are upregulated, that is, they are slowly increasing functions of time that can be fitted by a time-dependent function of the form $S(t) \sim \alpha_s + \gamma_s/(1+\beta_s t^{-2.5})$ for constants $\alpha_s, \beta_s, \gamma_s$, and similarly for $T(t)$. Numerical simulations of the full model, with parameter values based on studies of cultured hippocampal neurons, robustly exhibits spontaneous symmetry breaking and the formation of an axon [629]. The model also reproduces various atypical behaviors observed in cultured neurons. For example, applying mechanical tension to a single neurite during stage 2 results in its axonal specification, over-expression of Shootin1 induces the formation of multiple axons, and repression of Shootin1 inhibits polarization.

Finally, note that the Toriyama et al. model can be viewed as one of a class of models for neurite differentiation that involve competition among neurites for a pool of proteins involved in neuronal polarization. The first models of this type [479, 548, 640] simply took the rate of elongation to be proportional to the intracellular concentration of the protein (possibly tubulin) at the neurite tip. One prediction of these models is that the rate at which a neuron polarizes to form an axon should increase with the number of neurites. However, a recent experiment establishes that the time to polarize is independent of the number of neurites [680], which is consistent with another recent model of neurite differentiation based on the transport of the protein HRas rather than Shootin1 [193]. It turns out that all of the models exhibit independence of the number of neurites N, provided that the expression of the relevant protein increases with N [680], which was one of the assumptions in [193].

9.3 Spreading Depression

Cortical spreading depression (SD) is a complex wave of transient depolarization of neurons and glial cells that propagates slowly across cortical and subcortical gray matter at speeds of 1–5 mm/min [93, 365, 407, 461, 596]. It occurs as a response to various forms of brain damage or pathology [237], including focal ischemia, traumatic brain injury, and seizure activity. There is also strong evidence that SD is a correlate or cause of *migraine with aura*, where a light scintillation in the visual field propagates before disappearing to be replaced by a sustained headache [358, 639]. Although SD does not usually damage brain tissue, if it arises repeatedly during stroke or head trauma, then it may promote neuronal damage. One of the characteristic features of SD is a large increase in extracellular K^+ and a dramatic decrease in extracellular Ca^{2+} and other ions. Simultaneously, the DC surface potential of electroencephalogram (EEG) undergoes a negative shift of around 19 mV that lasts 30 s, followed by a smaller but longer positive voltage shift. At the cellular level, the slow shift in potential coincides with a period of membrane depolarization followed by a longer period of hyperpolarization. A comprehensive picture of the precise physiological and chemical processes underlying the generation of SD is still lacking. However, a number of mechanisms are thought to contribute, including diffusion and membrane transport of ions, chemical synapses and neurotransmitters, spatial buffering, gap junctions, and osmotic effects [433].

Although SD waves are associated with populations of cells, they have several properties in common with propagating action potentials in the axons of single neurons (Chap. 2): (1) they are essentially all-or-none phenomena; (2) they travel with approximately constant speed and wave form; (3) they involve membrane depolarization; (4) they exhibit refractoriness; (5) multiple waves can be generated from a sustained stimulus and colliding waves can annihilate. Phenomenologically speaking, the general characteristics of a solitary SD wave can be reproduced by appropriately parameterized Fitzhugh–Nagumo equations, for example, with the principal dependent variable being extracellular K^+ concentration rather than membrane potential. However, such a model fails to give insights into the underlying physiological mechanisms and is thus not useful for developing clinical treatments for associated pathologies such as migraine. Indeed, there are a number of significant differences between SD waves and action potentials: (1) space and time scales are much longer for SD waves; (2) SD wave propagation mainly involves ionic fluxes through synaptic rather than non-synaptic membrane; (3) there are substantial changes in extracellular ion concentrations during passage of an SD wave; (4) the recovery phase of SD involves active transport mechanisms such as ion pumps.

9.3.1 Tuckwell–Miura Model

The first physiologically based mathematical model of SD waves was introduced by Tuckwell and Miura [638]. It centers on the idea that an initial large increase in extracellular K^+ would depolarize presynaptic terminals, resulting in the release of neurotransmitter. This would then open K^+ channels in the postsynaptic membrane of a neuron, leading to the flux of additional K^+ into the extracellular space. The resulting regenerative process would support the propagation of an SD wave. The model of Tuckwell and Miura is a bidomain continuum model (see Sect. 4.2), in which the concentration of ions in the extracellular and intracellular domains are taken to coexist at every point in space. This is a reasonable approximation given that the spatial scale of an SD wave is much larger than the size of individual neurons. Ions are free to diffuse over significant distances within the extracellular space, whereas intracellular ions can only diffuse within a cellular domain that is negligible compared to the scale of SD waves. Thus the effective diffusivity of intracellular ions is taken to be zero. Another assumption of the model is that the equilibrium membrane potentials for neighboring presynaptic and postsynaptic membranes are approximately the same and are thus taken to have the common value $V(x,t)$. This can be justified by first noting that during passage of an SD wave, the changes in intracellular ion concentrations are negligible, since the ratio of the intracellular to

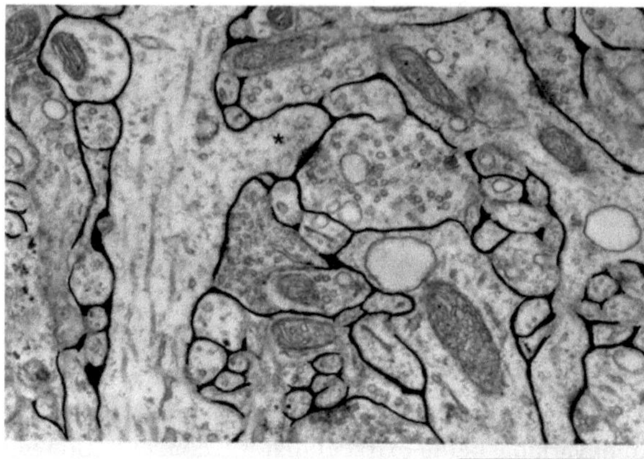

Fig. 9.13 Electron micrograph of a small region of the cerebral cortex of a rat illustrating the small volume fraction of the extracellular space, which is indicated by the black areas between cells. (Note, however, that the extracellular space may have been reduced in size as a consequence of the histological processing.) The *asterisk* indicates a dendritic spine, which is joined by a synaptic cleft to a presynaptic terminal containing several small round vesicles filled with neurotransmitter. Some of the round profiles adjoining the dendrite are likely to be unmyelinated axons cut in cross section. The *scale bar* represents a distance of approximately 1 μm (Adapted from [461])

extracellular volumes is large; see Fig. 9.13. Second, the local equilibrium membrane potential of a neuron is given by the Hodgkin–Katz formula

$$V = \frac{k_B T}{q} \ln \left[\frac{K^{\text{out}} + p_{\text{Na}} Na^{\text{out}} + p_{\text{Cl}} Cl^{\text{in}}}{K^{\text{in}} + p_{\text{Na}} Na^{\text{in}} + p_{\text{Cl}} Cl^{\text{out}}} \right]. \tag{9.36}$$

Here (K^{out}, Na^{out}, Cl^{out}) and (K^{in}, Na^{in}, Cl^{in}) denote the extracellular and intracellular concentrations of potassium, sodium, and chloride, respectively, q is the charge of a proton, and p_{Na}, p_{Cl} are the ratios of the sodium and chloride membrane permeabilities to the potassium permeability. It follows that the membrane potential $V(x,t)$ simply tracks the extracellular ion concentrations, which are common to presynaptic and postsynaptic neurons.

Let c_j^{out} and c_j^{in} denote the extracellular and intracellular concentration of ion species j, with $j = 1,2,3,4$ corresponding to sodium, chloride, potassium, and calcium, respectively. The basic model equations are then of the general form (in one space dimension) [638]

$$\frac{\partial c_j^{out}}{\partial t} = D_j \frac{\partial^2 c_j^{out}}{\partial x^2} + g_j(V - V_j) + P_j, \tag{9.37a}$$

$$\frac{\partial c_j^{in}}{\partial t} = -\frac{\alpha}{1-\alpha} \left[g_j(V - V_j) + P_j \right]. \tag{9.37b}$$

Here D_j is the extracellular diffusion coefficient of the jth ion species, V_j is the corresponding Nernst potential for an ion with charge modulus q_j,

$$V_j = \frac{k_B T}{q_j} \ln \left(\frac{c_j^{out}}{c_j^{in}} \right), \tag{9.38}$$

g_j is an effective conductance, and P_j represents the contribution of active transport mechanisms (pumps). Conservation of ions implies that the non-diffusive rate of change of each extracellular concentration is the negative of the rate of change of the corresponding intracellular concentration, scaled by a factor $\alpha/(1-\alpha)$, where α is the "volume" fraction of the extracellular space. In principle, the contribution to the conductance g_j could come from channels at both synaptic and non-synaptic sites. However, it is assumed that exchange of ions between the extracellular and intracellular domains during passage of an SD wave occurs predominantly at synaptic sites. First, the Ca^{2+} conductance is taken to be presynaptic and a sigmoidal function of voltage

$$g_4 = g_4(V) \equiv g_0 \left(1 + \tanh[\eta(V+\kappa)]\right), \tag{9.39}$$

for constants g_0, κ, η. A large increase in the extracellular K^+ concentration generates a positive shift in the equilibrium potential V given by (9.36). The resulting depolarization leads to an influx of presynaptic Ca^{2+}, which causes the release of neurotransmitter. (For simplicity, only one type of neurotransmitter is considered here.) The rate of release is taken to be proportional to the calcium current. However, since such an event lasts a small time interval of size τ_r, the effective concentration of extracellular neurotransmitter is approximated by

$$T(x,t) = k_4 \tau_r g_4(V(x,t))(V(x,t) - V_4), \tag{9.40}$$

where k_4 is a constant. On the other hand, the conductance associated with the other ions is assumed to arise from the binding of neurotransmitter to postsynaptic receptors according to

$$g_j(x,t) = k_j T(x,t), \quad j = 1,2,3, \tag{9.41}$$

with constants k_j. Finally, the pump rates are given by exponential functions of the ion concentrations:

$$P_j = f_j(1 - \mathrm{e}^{-r_j(c_j^{\text{out}} - \bar{c}_j)}), \tag{9.42}$$

for constants $f_j, r_j, \bar{c}_j$ with $\bar{c}_j$, the resting extracellular concentration. Note that $f_1 < 0$ for potassium and $f_j > 0$ for all the other ions, since the potassium pump maintains a higher intracellular concentration. Equations (9.36)–(9.42) determine the full version of the Tuckwell–Miura model [638]. It is also possible to consider a simplified version of the model, in which the sodium and chloride concentrations are kept fixed.

Using physiologically based parameter values, the Tuckwell–Miura model produces a wave speed of around 1 mm/min and a wave form consistent with experimental findings. However, it is necessary to take into account the effects of tortuosity, that is, the effective increase in path length for diffusion of ions around cells. This introduces a scaling factor of around 1.5 for the diffusion coefficient of ions in aqueous solution. One way to estimate the tortuosity of brain tissue as well as the volume fraction of the extracellular space is to use the theory of porous media [461, 616]; see Sect. 9.3.2. The heterogeneous and anisotropic nature of brain tissue means that extending the Tuckwell–Miura model to higher dimensions is nontrivial, since one would need to determine an effective diffusion tensor. Nevertheless, the model has the advantage of being relatively simple and amenable to mathematical analysis. For example, one of the interesting mathematical features of the model is that there exists a family of rest states that depend continuously on a set of parameters. This degeneracy implies that traveling waves have a number of properties that are not present in the case of isolated rest points [686]. For example, a localized disturbance can lead to a pair of outward propagating waves that slow down and reverse direction before ultimately colliding and annihilating each other. Since the introduction of the Tuckwell–Miura model, which focused on extracellular diffusion and membrane ion currents, a number of other mechanisms have been identified as possible contributors to the initiation and propagation of SD waves:

1. *Gap junctions.* Experiments have shown that pharmacological blockade of gap junctions slows or even prevents the propagation of retinal SD waves [359, 406, 448]. As we described in Sect. 4.5, there is extensive gap-junction coupling between astrocytes. However, it would appear that SD waves still occur in the absence of astrocytes [360]. More specifically, destroying many astrocytes using the toxin fluorocitrate only reduces the effectiveness of SD by 50%. This motivated Shapiro [568] to develop a detailed biophysical model of SD waves, in which gap-junction coupling between neurons played a major role in the transport of K^+. Unfortunately, there is no evidence for such widespread gap-junction coupling between neurons. In addition, it is possible that after application of toxins, there are still a sufficient number of intact astrocytes to support some level of SD.

2. *Neuron–astrocyte networks.* There is growing experimental evidence that SD waves involve both neurons and astrocytes acting together via the release of ATP and glutamate. First, intercellular Ca^{2+} waves precede SD but blocking these waves by gap-junction inhibitors does not prevent SD [501]. On the other hand, SD is blocked by antagonists to NMDA receptors suggesting a role for glutamate signaling. Second, a large amount of ATP is released from astrocytes during SD [118, 558]. This is consistent with the observation that the ATP antagonist heptanol completely prevents SD in hippocampus [360], presumably by blocking paracrine signaling. It is also known that glutamate can stimulate the release of ATP from astrocytes [667] and, conversely, activation of purinergic receptors by ATP can induce glutamate (and possibly ATP) release from astrocytes [187]. These experimental results have led to a model of SD waves [38], in which a wave of ATP propagates with the astrocyte Ca^{2+} wave, resulting in an accompanying wave of glutamate release from the astrocytes. The glutamate acts on NMDA receptors in the postsynaptic membrane of neurons, which triggers the large depolarization characteristic of SD together with the presynaptic release of additional glutamate from neurons. The latter acts on neighboring astrocytes to initiate the release of additional ATP. Hence, one has a regenerative mechanism for the propagation of ATP and glutamate, resulting in an SD wave. Note that one

major assumption of the model is that glutamate and ATP can be released from astrocytes via Ca^{2+}-independent pathways [250, 659].

3. *Spatial buffering.* Another well-known role of astrocytes is in the regulation of extracellular K^+ by several mechanisms including the action of Na^+/K^+–ATPase pumps and K^+ ion channels [109, 211, 601]. The spatial buffering of K^+ by astrocytes has only been studied within the context of small perturbations away from the resting state rather than the large excursions that occur during the passage of an SD wave. Nevertheless, it is possible that spatial buffering can account for the range of wave speeds observed during SD. The central mechanism for spatial K^+ buffering via ion channels is that a local increase in extracellular K^+ causes a local depolarization of glial cell membranes, which then spreads electrotonically to other glial cells via gap junctions. There is thus an influx of extracellular K^+ into glia at the original site of enhanced extracellular K^+ and an efflux of K^+ into the extracellular space at distal sites where extracellular K^+ is close to the resting level. In other words, there is a rapid net transport of K^+ from sites of high extracellular concentration to sites of low extracellular concentration, which is mediated by ion channels and gap-junction coupling of astrocytes.
4. *Cell swelling.* A number of models of SD waves have taken into account the effects of cell swelling that results from ions crossing neuronal membranes leading to osmotic stresses [568, 596]. This reduces the volume fraction of the extracellular space.

9.3.2 Brain Tissue as a Porous Medium

Diffusion in the extracellular space occurs in the narrow gaps between cells (neurons and glia) and is analogous to diffusion in a porous medium consisting of two phases such as water in the sandy bed of a river [503]. The connection with porous media can be exploited to estimate the effective diffusivity of ions moving in the extracellular space; see the reviews [461, 616]. One immediate observation is that if one were to draw a line through a section of brain tissue (see Fig. 9.13), one would find that the extracellular concentration is a discontinuous function of position along the line. In order to construct continuum reaction–diffusion models such as Tuckwell–Miura, it is necessary to carry out some form of spatial averaging using either homogenization theory (see Sect. 4.6) or volume averaging. Although the former approach is more rigorous, we will describe the latter approach here following [461], since it is more intuitive. The first step is to partition brain tissue into a set of representative elementary volumes (REVs) that are sufficiently large to include a sufficient range of cellular elements for averaging, yet sufficiently small to allow a local description of averaged quantities. Consider an arbitrary REV with center $\mathbf{x}$ relative to a fixed origin and local coordinates $\boldsymbol{\xi}$ about the center; see Fig. 9.14. (One can view ξ and $\mathbf{x}$ as microscopic and macroscopic coordinates, respectively.) Suppose that each REV is composed of two phases σ, ω corresponding, respectively, to the extracellular and intracellular domains and occupying the volumes V_σ, V_ω. Let $V = V_\sigma + V_\omega$ be the total volume of the REV and let $\alpha = V_\sigma/V$ be the volume fraction of the extracellular space. Derivation of an effective diffusion equation in the extracellular space now proceeds by applying a spatial averaging theorem to Fick's law and the particle conservation equation.

Spatial averaging theorem. This relates the average of the divergence of a microscopic quantity to the divergence of a macroscopic (averaged) quantity [240, 503]. Let **F** denote a microscopic state vector that is a function of time t and position $\mathbf{x}+\xi$ within the REV centered at **x**. Applying the divergence theorem to the volume occupied by extracellular space V_σ, we have

$$\int_{V_\sigma} \nabla\cdot\mathbf{F}dV_\xi = \int_{S_\sigma} \mathbf{F}\cdot\mathbf{n}_\sigma dS_\xi + \int_{S_{\sigma\omega}} \mathbf{F}\cdot\mathbf{n}_\sigma dS_\xi, \tag{9.43}$$

where $\mathbf{n}_\sigma$ is the unit normal vector on the boundary of the σ phase pointing outwards. Introduce the index function χ such that $\chi(\mathbf{x}+\xi) = 1$ if $\mathbf{x}+\xi$ lies in the σ phase and is zero otherwise. We can then write the first term on the right-hand side as $\int_{\partial V}(\mathbf{F}\chi)\cdot\mathbf{n}_\sigma dS_\xi$ where ∂V is the exterior boundary of the REV. Applying the divergence theorem to the whole volume V then gives

$$\int_{V_\sigma} \nabla\cdot\mathbf{F}dV_\xi = \int_{V} \nabla_\xi\cdot(\mathbf{F}\chi)dV_\xi + \int_{S_{\sigma\omega}} \mathbf{F}\cdot\mathbf{n}_\sigma dS_\xi. \tag{9.44}$$

Since $\mathbf{F}\chi$ is a function of $\mathbf{x}+\xi$, we can replace ∇_ξ by $\nabla_\mathbf{x}$ and take the divergence operator outside the volume integral with respect to ξ:

$$\int_{V_\sigma} \nabla\cdot\mathbf{F}dV_\xi = \nabla_\mathbf{x}\cdot\int_{V} (\mathbf{F}\chi)dV_\xi + \int_{S_{\sigma\omega}} \mathbf{F}\cdot\mathbf{n}_\sigma dS_\xi. \tag{9.45}$$

Finally, noting that the index function χ restricts the first integral on the right-hand side to the domain V_σ, we have (after dividing through by V), the spatial averaging theorem [240]

$$\langle\nabla\cdot\mathbf{F}\rangle = \nabla\cdot\langle\mathbf{F}\rangle + \frac{1}{V}\int_{S_{\sigma\omega}} \mathbf{F}\cdot\mathbf{n}_\sigma dS_\xi, \tag{9.46}$$

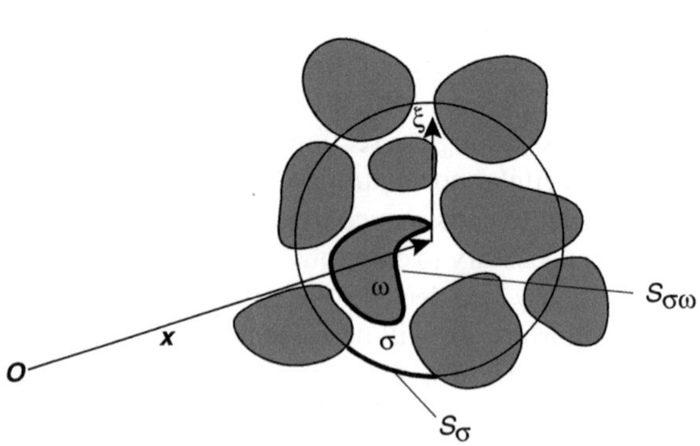

Fig. 9.14 A representative elementary volume (*REV*) composed of σ and ω phases and centered at **x**. The interface between the phases is denoted by $S_{\sigma\omega}$ and the surfaces that make up the boundary of the REV are denoted by S_σ and S_ω, respectively

where

$$\langle f \rangle = \frac{1}{V} \int_{V_\sigma} f dV \tag{9.47}$$

for any function f. The spatial averaging theorem for a scalar quantity ψ may then be obtained by setting $\mathbf{F}(\mathbf{x}+\xi) = \psi(\mathbf{x}+\xi)\mathbf{a}$ with $\mathbf{a}$ an arbitrary constant vector:

$$\langle \nabla \psi \rangle = \nabla \langle \psi \rangle + \frac{1}{V} \int_{S_{\sigma\omega}} \psi \mathbf{n}_\sigma dS_\xi. \tag{9.48}$$

Fick's law for the diffusive flux $\mathbf{J}$ in a homogeneous medium takes the form

$$\mathbf{J} = -D\nabla c, \tag{9.49}$$

where D is the diffusion coefficient and $c(\mathbf{x},t)$ is the concentration of the diffusing particles at position $\mathbf{x}$ and time t. Combining this with particle conservation,

$$\frac{\partial c}{\partial t} + \nabla \cdot \mathbf{J} = 0, \tag{9.50}$$

then recovers the standard diffusion equation. Now suppose that Fick's law only applies to particles located within the extracellular space, which is a multiply connected 3D domain. One would then need to solve the associated diffusion equation within each simply connected component of extracellular space supplemented by surface boundary conditions. An alternative approach is to apply the spatial averaging theorem to Fick's law and the conservation equation to obtain a macroscopic version of the diffusion equation. First, we have

$$\langle \mathbf{J} \rangle = -D\langle \nabla c \rangle = -D\nabla \langle c \rangle + \frac{1}{V} \int_A c\mathbf{n} dS, \tag{9.51}$$

where the area integral is over the interfaces A between the extracellular and intracellular spaces. The nontrivial part of the analysis is determining the surface integrals, which corresponds to the problem of determining the effective diffusion tensor by solving the cell equations in homogenization theory; see Sect. 4.6. Nevertheless, the surface integral reduces to the general form [374]

$$\frac{1}{V} \int_A c\mathbf{n} dS = (\mathbf{K} - \mathbf{I})\nabla \langle c \rangle, \tag{9.52}$$

where $\mathbf{K}$ is a second-order tensor and $\mathbf{I}$ is the identity. It follows that Fick's law for a porous medium takes the form

$$\mathbf{J} = -\mathbf{D}^* \nabla \langle c \rangle, \tag{9.53}$$

where $\mathbf{D}^* = D\mathbf{K}$ is an effective diffusion tensor. In the special case of an isotropic medium, this reduces to scalar diffusion coefficient $D^* = D/\lambda^2$, where λ is the tortuosity. Similarly, volume averaging the particle conservation equation gives

$$\left\langle \frac{\partial c}{\partial t} \right\rangle + \langle \nabla \cdot \mathbf{J} \rangle = 0. \tag{9.54}$$

Applying the spatial averaging theorem to the divergence of the flux implies that

$$\langle \nabla \cdot \mathbf{J} \rangle = \nabla \cdot \langle \mathbf{J} \rangle + \frac{1}{V} \int_A \mathbf{J} \cdot \mathbf{n} dS.$$

In the case of impermeable cell membranes, $\mathbf{J} \cdot \mathbf{n} = 0$, whereas for permeable membrane

$$\mathbf{J} \cdot \mathbf{n} = g(c - c_{in}) \tag{9.55}$$

where g is an effective permeability and $c_{\rm in}$ is the local mean intracellular concentration. For simplicity, we focus on the impermeable case. Assuming that the interfaces A are fixed, that is, ignoring osmotic effects and cell swelling, then

$$\left\langle \frac{\partial c}{\partial t} \right\rangle = \frac{\partial \langle c \rangle}{\partial t}.$$

Collecting the various results finally gives in the isotropic case

$$\frac{\partial \langle c \rangle}{\partial t} = D^* \nabla^2 \langle c \rangle - \phi k(\langle c \rangle - c_{\rm in}), \tag{9.56}$$

where $\phi = \int_A dS/V$ is a surface-to-volume ratio.

9.3.3 Migraine Auras

In this section we describe the relationship between SD and migraine aura. See also the review by Dahlem and Chronicle [144]. A characteristic feature of migraine is an episodic headache of moderate to severe intensity, which is usually unilateral (one side of the brain) and pulsating in nature. In the West, the lifetime prevalence of migraine is around 33 % in women and 13 % in men. In the case of migraine with aura, the migraine attack is preceded by a sensory hallucination called aura that usually lasts for less than half an hour. Consider, in particular, the classical visual aura that occurs in approximately 25–30 % of patients. Such an aura typically starts with a *scotoma* forming near the central region of the visual field; a scotoma (Greek for darkness) is an area of visual field in which there is partially or severely reduced visual acuity that is surrounded by a field of normal vision. The scotoma, which initially has indistinct boundaries, expands into a sickle or C-shaped region whose advancing edge contains a zigzag pattern of bright lines and shapes oriented at 60° to each other. The patterns are often called fortification patterns, since they are reminiscent of baroque city ramparts. An illustration of a fortification pattern is shown in Fig. 9.15, which has been superimposed upon a photograph of moose taken in the author's backyard. The fortification pattern becomes more distinct as it expands into the visual periphery, before disappearing at the edge of the visual field. The aura rarely returns during a single migraine event and leaves no noticeable permanent disruption of vision. Note that the fortification pattern at the leading edge of

Fig. 9.15 Illustration of a fortification pattern within the scotoma of a classical migraine aura

a scotoma reflects a short first phase of intense neuronal excitation, which precedes the prolonged phase of depressed neural activity that is associated with the scotoma. The existence of two phases is consistent with early studies of SD waves [93].

One of the important features of the migraine aura is that it clearly corresponds to a physiological wave propagating across the visual cortex. Moreover, the detailed structure of the fortification pattern appears to reflect the functional organization of the cortex, including its topographic organization and finer-scale feature maps; see Sect. 8.1.1. Indeed, it is now generally accepted that the fortification pattern provides a direct view of the layout of orientation columns in human visual cortex. Interestingly, many of these observations were anticipated by the psychologist Lashley in 1941 [363]. The first link between migraine aura and SD waves was made in 1945 [366], and there is now considerable experimental evidence supporting such a link [237]. There is also growing evidence that patients susceptible to SD waves have hyperexcitable visual cortices [441].

The progression of the scotoma across the visual cortex can be modeled using a kinematic theory of wave propagation in two-dimensional excitable media [145]. Such a model describes the propagation of a wave front in terms of the motion of curves with free ends, without detailed knowledge of the underlying reaction–diffusion equations [426]. Wave propagation is analyzed by assuming that the normal velocity c of the wave depends solely on the front curvature and by taking into account an additional tangential velocity γ that describes the growth rate of the front at its open ends. To motivate the curvature dependence of the normal velocity, consider a 2D version of the FitzHugh–Nagumo equations (Sect. 2.3)

$$\frac{\partial v}{\partial t} = f(v,w) + D\nabla^2 v, \tag{9.57a}$$

$$\varepsilon^{-1}\frac{\partial w}{\partial t} = -w + v. \tag{9.57b}$$

In particular, consider a circular target pattern in which the leading edge of a traveling pulse forms a circle of radius R at time t, with R much larger than the width of the pulse. $K = 1/R$ is then the curvature of the circular front. Introducing polar coordinates (r, ϕ) with the origin at the center of the target pattern, we have

$$\frac{\partial v}{\partial t} = f(v,w) + \frac{D}{r}\frac{\partial v}{\partial r} + D\frac{\partial^2 v}{\partial r^2}, \tag{9.58a}$$

$$\varepsilon^{-1}\frac{\partial w}{\partial t} = -w + v. \tag{9.58b}$$

Since $\partial v/\partial r$ is negligible outside the transition layers of the pulse, we can replace D/r by D/R on the right-hand side of (9.58a). Writing the circular wave solution as $v = V(\xi)$, $w = W(\xi)$ with $\xi = r - ct$, yields the pair of equations

$$-(c+DK)V' = f(V,W) + DV'', \tag{9.59a}$$

$$-\frac{c}{\varepsilon}W' = -W + V, \tag{9.59b}$$

with $V' = dV/d\xi$, etc. These equations are equivalent to the 1D FN equations in the moving frame of a pulse traveling with speed $c_0 = c + DK$, under the rescaling $\varepsilon \to (c+DK)\varepsilon/c$. Approximating c_0 by the singular solution of the 1D FN equations (see Sect. 2.3), we deduce that the speed of the circular front is a linear function of the curvature [426]

$$c(K) = c_0 - DK. \tag{9.60}$$

This result extends locally to a more general curved front, by partitioning the curve into small segments approximated by circular arcs. Assuming that there are no self-intersections, the curve can be represented in polar coordinates as $r = r(\phi)$ and there is now a phase-dependent curvature $K(\phi)$. Unfortunately, there is no general theory for the form of the tangential velocity γ. Following [145], we will take γ to be a linearly decreasing function of the curvature at the end point (Fig. 9.16):

$$\gamma(K) = \gamma_0(1 - K/K_c), \tag{9.61}$$

up to some critical curvature K_c.

If both $c(K)$ and $\gamma(K)$ are known, then one can determine the evolution of an arbitrary initial excitation pattern, at least numerically. An analytical solution is only possible when c and γ are linear function of the curvature and the initial front is circular symmetric. The basic kinematic theory proceeds as follows [145, 426]. Suppose that the curvature of a point a on the front at time t is K_a, and introduce a local polar coordinate system (r, ϕ) with the origin at the center of curvature from point a. Note that $r' = r'' = 0$ at a where $r' = \partial r/\partial\phi$. The curve in a neighborhood of a is given by a function $r(\phi,t)$ with $r(0,t) = r_a = 1/K_a$. Over a short time interval dt, the neighborhood of point a moves to a neighborhood of point b with the new curve segment

$$r(\phi,t+dt) = r(\phi,t) + c(\phi,t)dt, \quad c(\phi,t) = c(K(\phi,t)). \tag{9.62}$$

Given a local polar representation of a curve, its local curvature is given by the formula

$$K = \frac{r^2 + r'^2 - rr''}{[r^2 + r'^2]^{3/2}}. \tag{9.63}$$

The curvature K_b can then be determined by substituting for $r(\phi, t+dt)$ using (9.62), expanding to first order in dt, and setting $\phi = 0$:

$$K_b = K_a - \left(c(K_a)K_a^2 + K_a^2 \left. \frac{\partial^2 c}{\partial \phi^2} \right|_{\phi=0} \right) dt. \tag{9.64}$$

In order to integrate this equation, it is necessary to replace ϕ by the intrinsic arc-length coordinate l using $dl = r_a d\phi$. Assuming that a is an intrinsic distance l from one free end and taking $K_a = K(l,t)$, $K_b - K_a = dK(l,t)$ gives

$$dK = -\left[K^2 c(K) + \frac{\partial^2 c}{\partial l^2} \right] dt. \tag{9.65}$$

Note that the points a, b have different values for their intrinsic coordinate, since a circular wave propagating outwards increases its arc length and the arc length is measured with respect to an end point of a curve that may be growing or shrinking. It follows that the difference in arc length between the points a and b is

$$dl = \left[\int_0^l K(s,t)c(K(s,t))ds \right] dt + \gamma(0,t)dt. \tag{9.66}$$

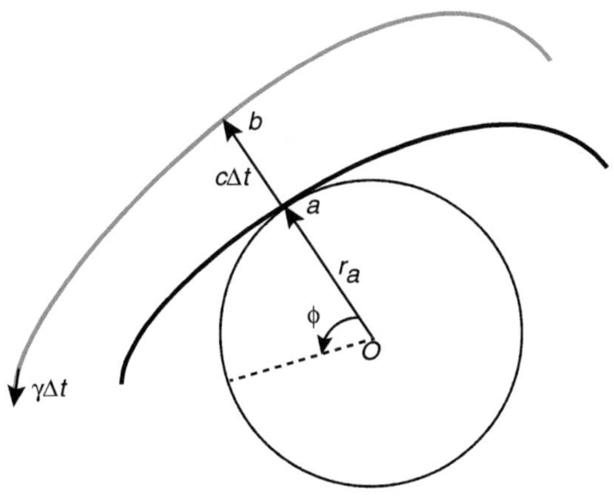

Fig. 9.16 Front position at two successive times $t, t+\Delta t$. Normal velocity at a given point on the front at time t is c and the tangential velocity at an end point is γ

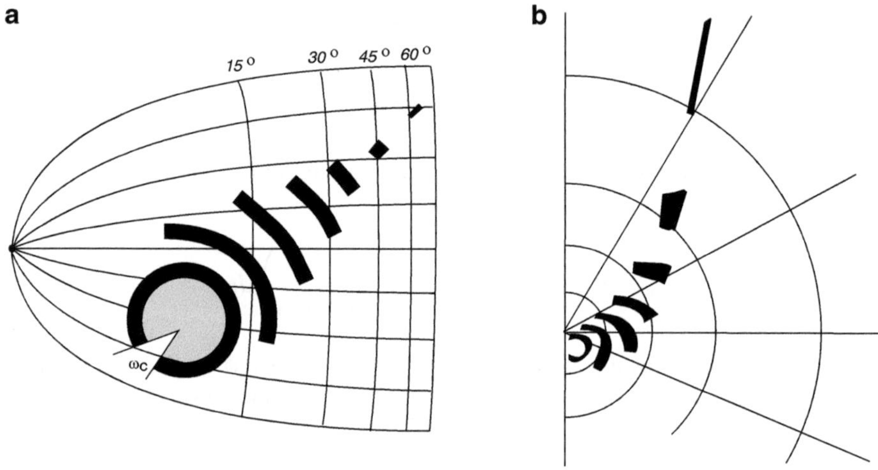

Fig. 9.17 (**a**) Propagation of a front across one half of primary visual cortex. Away from the focus, horizontal and vertical lines in cortex transform under the inverse retinotopic map to radii and concentric circles in the corresponding visual hemifield. (**b**) Progression of scotoma across the visual hemifield, which is obtained by applying the inverse retinotopic map to the propagating cortical front in (a)

Since $K = K(l,t)$, we have the alternative expression for dK,

$$dK = \frac{\partial K}{\partial l} dl + \frac{\partial K}{\partial t} dt. \tag{9.67}$$

with dl given by (9.66). Comparing (9.65) with (9.66) and (9.67) finally yields the integrodifferential equation [145, 426]

$$\frac{\partial K}{\partial t} + \left[\int_0^l K(s,t)c(K(s,t))ds + \gamma(0,t) \right] \frac{\partial K}{\partial l} = -c(K)K^2 - \frac{\partial^2 c}{\partial l^2}. \tag{9.68}$$

Dahlem and Muller applied the above kinematic theory to a circular arc propagating in the visual cortex under the assumptions that $c \approx c_0$ and γ are given by (9.61). The initial front is taken to be an arc of small radius r_0, which means that $K > K_c$ and $\gamma < 0$, that is, the end points, shrink as a function of time. Since the normal velocity is approximately independent of curvature, the front at time t is still a circular arc, but its arc length $l(t)$ has shrunk. Setting $K(t) = 1/r(t)$ with $r(t) = r_0 + c_0 t$ and $c(K) = c_0$, (9.66) implies that

$$\frac{dl}{dt} = \frac{V_0 l}{r} + G_0 \left(1 - \frac{1}{rk_c}\right).$$

Let $\omega = l/r$ be the angle subtended by the arc. Then

$$\frac{d\omega}{dr} = G_0 \left(\frac{1}{r} - \frac{1}{r^2 k_c}\right),$$

so that integrating with respect to r gives

$$\omega(r) = \omega(1/k_c) + G_0 \left(\ln(rk_c) + \frac{1}{rk_c} - 1 \right). \tag{9.69}$$

Using the analytical expression for the propagation of a circular arc across visual cortex, it is possible to determine the progression of a corresponding scotoma across the visual hemifield by applying the inverse retino-cortical map; see Sect. 8.1.1. This construction was carried out by Dahlem and Muller [145], and an illustrative sketch of their results is shown in Fig. 9.17. Finally, note that it is also possible to reproduce fortification patterns by introducing a region of enhanced neural excitability in the leading edge of the propagating front and superimposing this onto an orientation preference map. The resulting activity pattern of orientation-selective neurons was used to generate a set of population vectors that determined the distribution of oriented edges in the visual field [144].

9.4 Epilepsy

Epilepsy is the world's most common serious brain disorder, affecting around 50 million people [348, 430, 594]. Although there have been major advances in the medical and surgical treatment of epileptic seizures, almost a third of patients are untreatable and, even in successful cases, there can be serious side effects. Traditionally, epilepsy has been divided into two main categories: *primary generalized seizures* that arise simultaneously across large parts of the brain and *focal* or *partial* seizures that initially occur within a localized area of the brain but may subsequently propagate to other brain regions. If the latter does not cause a disruption of consciousness or cognitive abilities, then it is referred to as a simple seizure, otherwise it is called complex. In reality, there is probably a spectrum of different types of seizure with overlapping or related mechanisms for the initiation, propagation, and termination of the seizure [507]. Most studies of epilepsy have focused on the molecular, anatomical, and cellular processes underlying the development of epilepsy (epileptogenesis) and the initiation of seizures [125, 415]. Much of this work has been based on in vitro and in vivo animal models.

In the cortex and hippocampus, a stereotypical signature of focal epileptiform activity is the *interictal spike* that occurs between seizures (*ictal* activity). Interictal spikes are brief bursts of synchronized activity that last around 100 ms and which show up as sharp spikes in EEG recordings and as large amplitude depolarizations at the intracellular level. The spike appears to be initiated in deep layer V of the cortex, which contains a subset of pyramidal neurons that can intrinsically generate high-frequency bursts of up to five action potentials at 200–350 Hz and resemble in some ways the up states of the slow (< 1Hz) oscillations seen in cortical slices [549]; see Sect. 5.1. Recurrent excitation mediated by glutamatergic synapses then leads to a rapid spread of activity both vertically and horizontally through a large local network of neurons. Activity is finally terminated by a combination of local recurrent GABA inhibition, Ca^{2+}-dependent AHP currents, and possibly synaptic depression [415]. In the hippocampus, pyramidal neurons in the CA3 region appear to be the substrate for intrinsic bursters. There are two competing theories for the primary cause of the large depolarization that initiates an interictal spike. The first assumes that there is an excessive synchronous synaptic activation of essentially normal pyramidal neurons due to rewiring of excitatory connections or removal of inhibition, for example, whereas the second proposes that the excitability properties of the neurons exhibiting strong depolarizations are altered such that normal synaptic drive results in an abnormal response. Until recently, the first theory was predominant, but now there is growing experimental evidence that several forms of epilepsy involve genetically determined ion channel defects that alter intrinsic excitability [41]. A third possible cause of epileptiform activity has also recently emerged, namely, neuron–astrocyte interactions [311, 564, 647, 668]. As discussed in Sect. 4.5, astrocytes are in a position to induce local synchronization of neurons via gliotransmission of glutamate and gap-junction coupling, which could be a precursor of larger-scale epileptiform activity. Alternatively, excessive swelling of astrocytes through the intake of water could reduce the volume of the extracellular space, thus increasing the concentration of extracellular K^+ and resulting in a depolarization of surrounding neurons, see also our discussion of spreading depression in Sect. 9.3.

Another ongoing debate concerns the precise role of interictal spikes in epileptogenesis? At the cellular level, the transition to seizure-like activity is typically associated with a prolonged depolarization and high-frequency repetitive firing around 10–20 Hz [415]. This *tonic phase* of a seizure is often followed by a period of irregular periodic bursts known as the *clonic phase*, in which there is very little action potential activity between bursts. Finally, the clonic phase is succeeded by a relatively quite *postictal depression phase*, during which the membrane potential is hyperpolarized. One hypothesis is that interictal spikes promote epileptogenesis, which is motivated by the observation that in certain animal models interictal spikes arise in a latent period prior to spontaneous seizures [599]. The main idea is that persistent abnormal interictal activity over a period of time leads to the formation of new excitatory connections and a strengthening of existing excitatory circuits via Hebbian plasticity, both of which ultimately reinforce the epileptic state. A completely opposite proposal is that although interictal spikes resemble patterns of spiking activity during seizures, they are independent events such that, in certain cases, the occurrence of interictal activity can actually reduce the likelihood of seizures; the latter effect has been observed in some brain slice experiments [18]. Yet another hypothesis is that deafferentation following brain trauma, for example, engages homeostatic synaptic plasticity mechanisms, which strengthen connections between intact neurons ultimately leading to seizures; interictal spikes could then be an intermediate stage in the progression to full seizure-like activity [290, 463]. In order to help resolve these issues, a number of computational models have explored how epileptogenesis depends on changes in network structure [159, 437, 451, 646].

Recent advances in human brain imaging such as functional magnetic resonance imaging (fMRI) and diffusion-tensor imaging, together with more traditional electrophysiological methods such as *electroencephalogram* recordings and chronic multielectrode implants, have led to a rapid increase in large-scale brain models of epilepsy [522, 594]. Broadly speaking, such models focus either on the global temporal dynamics of primary generalized seizures [60, 529, 582] or on characterizing the types of large-scale brain networks that support generalized seizures and the propagation of focal seizures [348, 375]. Two major types of generalized seizure are *tonic–clonic* (grand mal) seizures (see above) and *absence* (petit mal) seizures. The latter are generated in the thalamocortical loop, where reticular thalamic neurons are recruited by the neocortex and then inhibit thalamocortical neurons; see Sect. 5.4.3. The subsequent activation of various ionic currents generates post-inhibitory rebound bursting in the thalamocortical neurons, which project onto and excite neocortical neurons, resulting in a repeating cycle of spiking activity at around 3 Hz [415]. Since primary generalized seizures occur across large parts of the brain, it is reasonable as a first approximation to consider a nonspatial (mean-field) model of interacting neural populations (Chap. 6). The resulting model can be described by a system of ODEs, and bifurcation theory can be used to analyze the onset of seizures and to identify electrophysiological differences between tonic–clonic and absence seizures [60, 529, 582]. At least two different bifurcation scenarios have been identified [522]: (1) a change in some control parameter transitions the dynamical system from a normal state to a seizure state or (2) a change in a control parameter moves the brain from a normal state to a bistable state, in which both normal and seizure states coexist. In the latter case, the switch to the seizure state would require changes in some intrinsic physiological process such as the sleep–wake cycle or an external drive such as a stroboscope. One potential clinical application of these ODE models is that they generate time series that can be directly compared to EEG recordings and, in principle, could be used to predict the onset of seizures.

In this section, we focus on network aspects of epilepsy. (For a comprehensive review of computational models of epilepsy, see the book by Soltesz and Staley [594].) We begin by summarizing some basic network theory that is particularly relevant to brain networks. We then review recent work on identifying structural and functional brain networks that support epileptiform bursts and seizures. Ultimately, one would like to incorporate such findings in neural field models of wave propagation (Chap. 7), in order to better understand the spread of epileptiform activity in the human brain.

9.4.1 Basics of Network Theory

Much of the modern theory of networks was stimulated by the discovery of small-world networks in 1998 [662] and scale-free networks in 1999 [23], both of which will be defined below. There are a number of excellent reviews of modern network theory [52, 459], including several that are tailored towards brain networks [541,

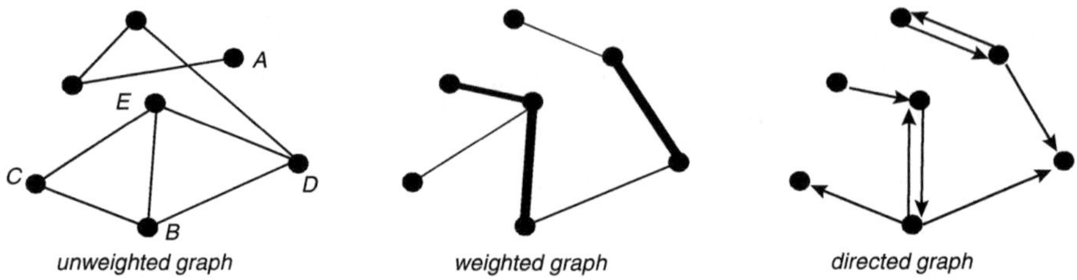

Fig. 9.18 Graph representation of a network in terms of vertices and edges. The vertex B in the unweighted graph has degree $k = 3$ with neighbors C, D, E. The clustering coefficient of B is 2/3, since there is a maximum of three edges between the neighbors but only CE and ED exist, whereas CD is missing. The shortest path length from A to C is $d = 5$

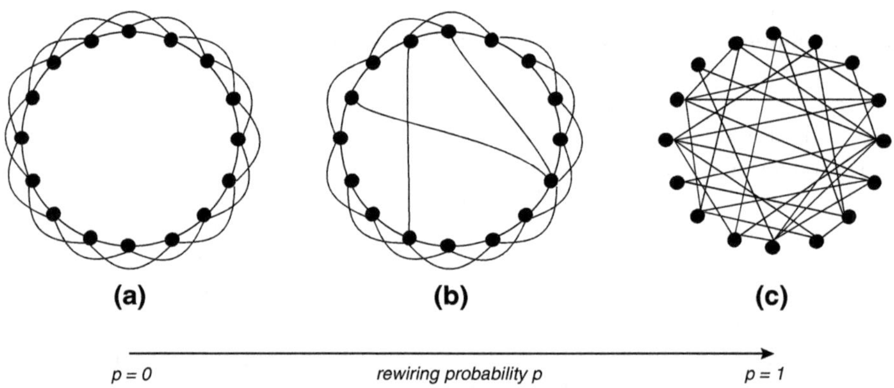

Fig. 9.19 Three basic network types illustrated by the ring network of Watts and Strogatz [662]. (**a**) The ordered network has $N = 16$ vertices and each vertex is connected to $k = 4$ neighbors. (**b**) and (**c**) The small-world and random networks are constructed by randomly reconnecting edges with rewiring probability p. For small values of p one obtains a small-world network, whereas a completely random network is created when $p = 1$

600]; we will follow [600]. An abstract representation of a network, whether it is a neural network, a social network, or a communications network, is a graph. The latter consists of a set or *vertices* (or nodes) and set of *edges* or connections. The connectivity structure of the graph is specified by an *adjacency matrix* **A** with $A_{ij} = 1$ if the vertices i and j are linked by an edge; otherwise $A_{ij} = 0$. The number of edges ending at given vertex is called the *degree k* of the vertex. The probability $P(k)$ that a randomly chosen vertex has degree k is called the degree distribution and is an important factor in determining network properties. If the direction of information flow along each edge is unimportant, then the graph is said to be *undirected* and **A** is a symmetric matrix; otherwise it is *directed*; in the latter case the degree distribution of incoming and outgoing edges may be distinct. Graphs where all edges are treated equally are said to be *unweighted*, whereas if each edge is assigned a weight to signify its strength or efficacy, then the graph is *weighted*; see Fig. 9.18.

There are two measures that are frequently used to characterize the structure of an unweighted graph, namely, the *clustering coefficient C* and the *characteristic path*

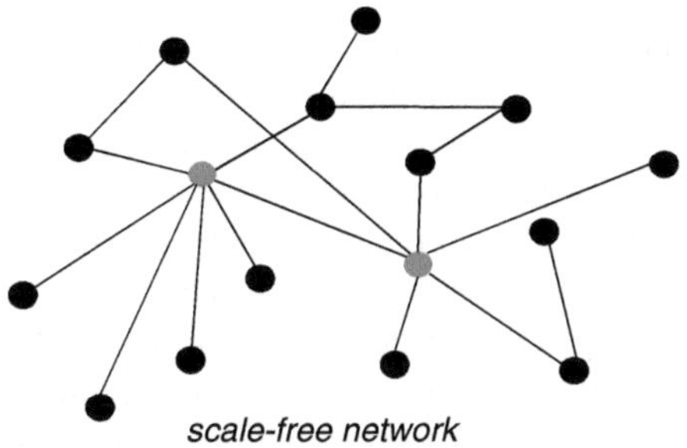

Fig. 9.20 Scale-free network with two hubs *shaded gray*

length L. The clustering coefficient C_i of a vertex i with degree k_i is typically defined as the ratio of the number of existing edges between neighbors of i (those directly connected to i) and the maximum possible number of such edges. Thus,

$$C_i = \frac{\sum_{j,k \in \mathcal{N}} A_{ij} A_{jk} A_{ki}}{k_i(k_i - 1)}. \tag{9.70}$$

with $0 \leq C_i \leq 1$. An equivalent definition of C_i is that it determines the number of completed triangles with i as one of the vertices. The clustering coefficient of the graph is then obtained by averaging C_i over all vertices:

$$C = \frac{1}{N} \sum_{i=1}^{N} C_i. \tag{9.71}$$

The clustering coefficient is a local measure of network structure that characterizes the degree of resilience to random errors, since if $C \approx 1$, then removal of a single vertex has a minimal affect on the connectivity pattern of the other vertices. The second important measure of an unweighted graph is the characteristic path length. The path length or distance d_{ij} between two vertices i and j is the minimal number of edges that have to be traversed in order to go from i to j. The characteristic path length of a graph is then defined to be the mean of the path lengths between all vertex pairs:

$$L = \frac{1}{N(N-1)} \sum_{i,j,i \neq j} d_{ij}. \tag{9.72}$$

Using the degree distribution, clustering coefficient, and path length, one can distinguish between four different types of graph: ordered, small world, random, and scale-free. The first three cases can be illustrated using a 1D network on a ring, which was introduced by Watts and Strogatz [662], see Fig. 9.19. Initially each vertex is connected to k nearest neighbors ($k/2$ on either side) to form an *ordered network*. (Higher-dimensional versions of an ordered network could be generated using a regular lattice.) Ordered networks have a high C and a high L; in the ring network of Fig. 9.19a with $k = 4$ and $N = 16$, we have $C = 3/4$ and $L = N/2k$. Now suppose that with a probability p, edges are chosen at random, detached at one end, and then reconnected to another, randomly chosen vertex. For small p, one thus obtains a *small-world network*, which has a C close to an ordered network but a very small path length close to that of a random network. As p is increased, more and more edges are randomly reconnected until finally all connections are random when $p = 1$. A random network has small C and small L. The fourth type of network is a *scale-free network*, which has a power-law degree distribution $P(k)$. This was originally formulated in terms of model for network growth by Barabasi and Albert [23]. They assumed that the probability that a newly added edge will connect to a vertex depends upon the current degree of the edge. In other words, vertices with a large number of edges (large degree) are more likely to be assigned even more edges. A network with this form of growth are characterized by a degree distribution of the form $P(k)) \sim k^{-1/\alpha}$; the original scale-free network of Barabasi and Albert has $\alpha = 3$. It turns out that many real-world networks appear to be scale-free, including the World Wide Web and collaboration networks of researchers. Scale-free networks not only can have very small path lengths [$L \sim \ln(\ln(N))$], but may also have a smaller clustering coefficient than small-world networks. Another interesting feature of scale-free networks is that they typically have a few vertices with extremely high degrees, which act as network *hubs*; see Fig. 9.20.

9.4.2 Epilepsy and Network Structure

One major research topic in modern network theory is the relationship between topological network characteristics and network dynamics, particularly synchronization [52]. A number of studies have investigated how the synchronization of coupled conductance-based or integrate-and-fire (IF) neurons depends on the rewiring probability of a small-world network [353, 409, 535]. Here the neural oscillators are placed at the vertices of the graph and the edges correspond to synaptic connections. Roughly speaking, ordered networks exhibit coherent oscillations but slow system response, whereas random networks have fast response properties but cannot support coherent oscillations. On the other hand, small-world networks support coherent oscillations and a fast system response. The dependence of synchronization on network topology has also been investigated within the context of epilepsy [159, 437, 451, 646]. In particular, Netoff et al. [451] based their study on results from hippocampal slice experiments indicating that short synchronous bursts tend to be seen in the CA3 region, where there is a relatively high degree of recurrent excitation, whereas seizure-like activity is found in the more sparsely connected CA1 region [450]. They modeled a small-world ring network (with $N = 3000$; $k = 30$ for CA1 and $k = 90$ for CA3) consisting of various types of stochastic model neurons such as IF and Hodgkin–Huxley. The neurons were assumed to be initially in a quiescent state, so that network activity was kindled by spontaneous action potentials that propagated as waves around the ring; the pattern of activity that emerged then depended on the number of long-range connections, as well as the interplay between noise and refractoriness. It was found that as the rewiring probability was increased, the model network displayed normal spiking behavior (low, sustained population firing rate), then seizure-like activity (significantly higher, sustained firing rates with a relatively low level of coherence), and finally bursting (waxing and waning of synchronous activity). Moreover, the transition from seizures

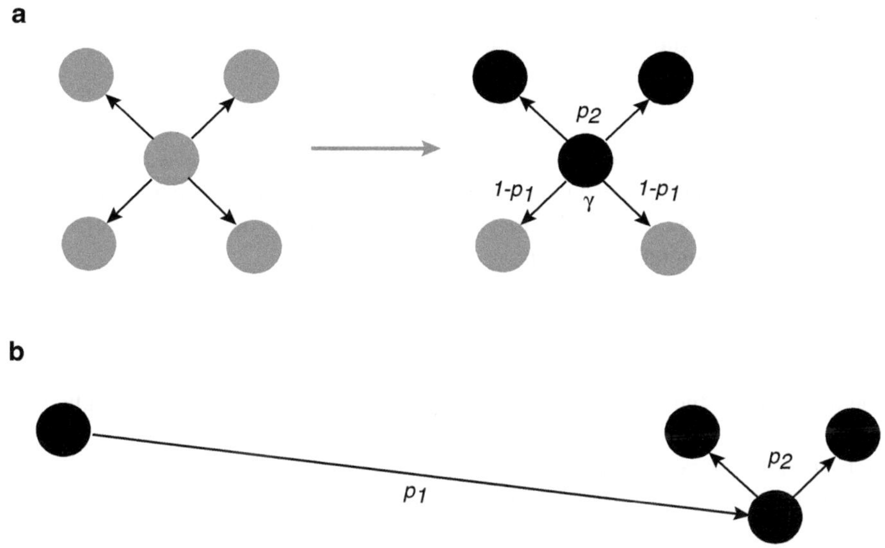

Fig. 9.21 Two mechanisms for spontaneous wave generation in the model of Netoff et al. [451]. Excitable (active) neurons are indicated by *gray* (*black*) disks. (**a**) An excitable neuron fires spontaneously at a rate γ and triggers at least two of its neighbors to fire with probability p_2: the probability that a neighbor does not fire is $1-p_1$. (**b**) An active neuron with a random long-range connection induces an excitable neuron to fire with probability p_1, which in turn triggers at least two of its neighbors to fire with probability p_2

to synchronized bursting occurred at a lower value of p for higher degree k, that is, for the CA3 model.

Netoff et al. [451] also introduced a reduced model of propagating activity, under the assumption that the number of neurons N is much larger than the number of refractory or active neurons. This model takes the form of a discrete birth–death process for the number of spontaneous waves X_n at the nth time step:

$$X_{n+1} = f(X_n) = X_n + \omega_+(X_n) - \omega_-(X_n), \tag{9.73}$$

where $\omega_\pm$ are the birth and death rates. Suppose that each neuron in the network has k neighbors and if a given presynaptic neuron fires then there is a probability p_1 that this induces a given postsynaptic neuron to fire. A local spontaneous wave is generated if at least two neighboring postsynaptic neurons fire (see Fig. 9.21), with associated probability

$$p_2 = 1-(1-p_1)^k - kp_1(1-p_1)^{k-1}.$$

It follows that the rate at which new waves are generated in one time step by local interactions is

$$\Gamma_n = \gamma E_n p_2,$$

where E_n is the number of excitable neurons and γ is the spontaneous rate at which each neuron fires. The latter is taken to be

$$E_n = N - \alpha X_n(1+R). \tag{9.74}$$

Here $\alpha = k/2 - 1$ is the mean number of active neurons in a wave front and R is the number of time steps each neuron remains refractory in the wake of a wave. A second contribution to the birth rate comes from the long-range connections in the ring network, which takes the form

$$\hat{\Gamma}_n = 2\alpha X_n kp \left(\frac{p_1 p_2 E_n}{N} \right).$$

The first factor is the number of long-range connections with active presynaptic neurons for a given rewiring probability p and the second factor is the probability that such a connection results in the formation of a wave emerging from the postsynaptic neuron. Putting these results together, we have

$$\omega_+(X_n) = \Gamma_n + \hat{\Gamma}_n = E_n \left[\frac{2\alpha X_n kpp_1 p_2}{N} + \gamma p_2 \right]. \tag{9.75}$$

Since E_n is an affine function of X_n, the total birth rate is a quadratic function of X_n. Finally, the death rate is assumed to be

$$\omega_-(X_n) = \frac{2\alpha X_n}{E_n}, \tag{9.76}$$

and is based on an estimate of how often waves collide and annihilate.

Sketches of the function f for various rewiring probabilities are shown in Fig. 9.22. For sufficiently small rewiring probability p there exists a unique stable

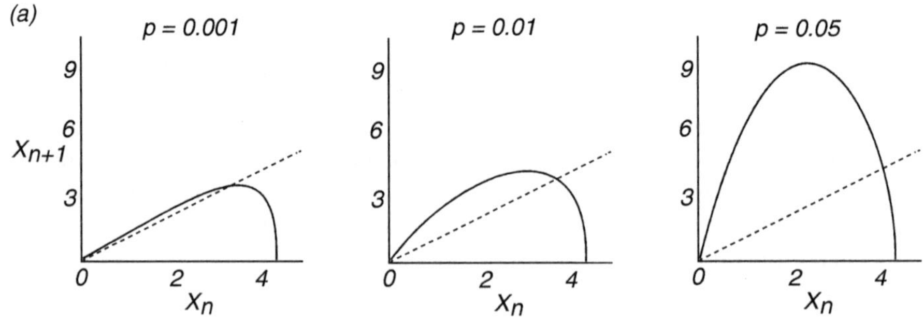

Fig. 9.22 Birth–death process for activity in a ring network [451]. The map $f : X_n \to X_{n+1}$ describing the evolution of the number X_n of spontaneous waves is sketched for several values of the rewiring probability p. Other parameter values are $\gamma = 0.0315, p_1 = 0.025, k = 90, N = 3000$. There is a unique fixed point, which is weakly attracting for small p, strongly attracting for intermediate p, and unstable for large p

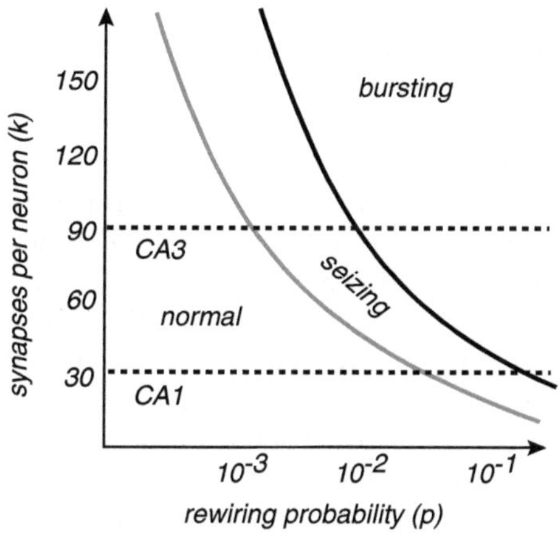

Fig. 9.23 Sketch of phase diagram for the reduced model of Netoff et al. [451]. The phase boundaries separating normal, seizing, and bursting states are plotted in the (p,k)-plane. The *black* (*gray*) boundary curve is obtained from the 1D map [$(R+1)$-dimensional map]. For a range of rewiring probabilities the CA3 network with $k=90$ is in a bursting state, whereas the CA1 network with $k=30$ is in a seizing state

fixed point $X^* = f(X^*)$, which is only weakly attracting. At each time step the number of waves remains approximately constant, that is, waves are rarely born or destroyed. This implies that there is a low rate of spontaneous wave generation consistent with normal brain activity. For larger p, the fixed point becomes strongly attracting and the number of waves increases, which is interpreted as a seizure state. Finally, for even larger values of p, the fixed point becomes unstable via a flip bifurcation, and new waves are generated rapidly, suggestive of a bursting state. (True population bursting cannot be realized in the reduced model under the restriction that the number of refractory neurons is much smaller than N, since termination of a population burst requires most of the network to enter a refractory state.)The one-dimensional map captures quite well the transition boundary between the seizing and bursting states. However, in order to capture the transition between the normal and oscillatory seizing state, it is necessary to consider a more accurate model of refractory dynamics by replacing (9.74) with

$$E_n = N - \alpha X_n - \alpha \sum_{m=1}^{R} X_{n-m}. \tag{9.77}$$

This leads to an $(R+1)$th-order map that can undergo a Hopf bifurcation resulting in stable oscillations. A typical phase diagram for the model is shown in Fig. 9.23.

Changes in network structure are not necessary, however, for abnormal rhythms to occur. Motivated by the observation that epileptiform bursting activity often arises following severe head trauma, Volman et al. [646] studied a computational model of traumatized cortex, in which a significant fraction of neurons had their afferent inputs reduced (deafferented neurons). Hence, the model could be viewed as a population of intact neurons embedded in a sea of traumatized neurons. Homeostatic synaptic plasticity was incorporated into the model so that following deafferentation, the remaining intact neurons strengthened their connections within the intact subnetwork, as well as their efferents to traumatized neurons. The authors found that this resulted in the emergence of interictal bursts propagating across the cortical network, provided that the density of intact neurons was relatively high so that they could also trigger activity in deafferented neurons. Thus the transition to bursting involved a strengthening of the intact subnetwork rather than a topological restructuring. Indeed, the level of bursting was found to be weakly dependent on properties of the subnetwork such as its clustering coefficient and path length [646]; a much more critical factor was the density of intact neurons.

The above examples correspond to structural or anatomical network models, in which the vertices of a given network are taken to be individual neurons and the edges correspond to physical synaptic connections. At this microscopic scale, the only complete structural network that has been mapped is for the 302 neurons of the nematode worm *C. elegans* [669]. Such a feat remains intractable for the human brain, which consists of more than 10^{10} neurons and 10^{12} synapses, although non-invasive neuroimaging techniques such as *magnetic resonance imaging* (MRI) can be used to construct coarser-grained macroscopic structural networks in humans. An alternative representation of brain networks has been gaining increasing attention, namely, one involving functional networks [5, 598, 600]. A functional network represents the coupling between dynamical activity recorded from separate brain areas. It is based on the assumption that statistical correlations between the time series of electrophysiological or metabolic activity reflect functional interactions between neurons or populations of neurons in different brain regions. Clearly functional connectivity depends on the underlying structural connectivity, but they need not be identical. A variety of neuroimaging methods have been used to construct functional networks at different spatial and temporal scales, ranging from single unit recordings to fMRI. One of the ongoing challenges in establishing a functional brain network is choosing the most appropriate coupling measure (two-point correlations, Granger causality) and determining the threshold value above which the level of coupling constitutes an edge [500]. Roughly speaking, experimental observations of focal seizures indicate that functional networks acquire larger path lengths and clustering coefficients at seizure onset and become more small world during seizure propagation and more random at seizure termination [348]. However, contradictory results have also been obtained, probably reflecting the fact that epilepsy is a complex, heterogeneous phenomenon.

9.5 Neuronal Avalanches, Criticality, and Branching Processes

Another characterization of large-scale brain activity that has gained prominence in recent years is the extent to which the cortex operates close to criticality [29, 31, 113, 508, 573]. Roughly speaking, criticality at the network level means that the network is in a balanced state with regard to the propagation of activity induced by the spontaneous spiking of individual neurons. As a simple *gedanken* experiment [573], imagine a fully connected network of N neurons and let p denote the probability that each neuron in the network fires in response to a single spike. The parameter p represents a variety of neurobiological factors that could contribute to the efficacy of a single action potential, including the relative number of excitatory and inhibitory neurons, the strength of excitatory versus inhibitory synapses, the intrinsic excitability of neurons, and the strength of neuromodulators. Clearly, if $p \ll 1/N$, then activity in the network quickly dies out, and it is said to be in the *subcritical regime*. On the other hand, if $p \gg 1/N$, then a single spike almost surely ignites a runaway chain of spiking events that never terminates, and we have a *supercritical regime*. Finally, if $p = 1/N$, then one spike is likely to cause one subsequent spike, so that there is reverberating activity in the network that does not overwhelm the network with a hypersynchronous state—the *critical* regime. Within the context of epilepsy, the basic idea is that the normal brain operates close to a critical state, whereas the epileptic brain deviates significantly from criticality during seizures, due to a failure in the underlying homeostatic mechanisms that maintain the critical state [291, 418].

The concept of criticality has its origins in the statistical physics of equilibrium phase transitions. A classical example is the phase transition from paramagnetism to ferromagnetism in an iron bar magnet as the temperature T crosses a critical threshold from above, known as the Curie temperature T_c. The *Ising model* treats the bar magnet as a lattice of electrons, each of which can be in one of two "spin" states represented by an arrow pointing up (north) or pointing down (south)—the arrows can be viewed as microscopic bar magnets. Nearest-neighbor interactions between the spins favor spin alignment (order), whereas thermal fluctuations favor disalignment (disorder). Both processes occur in order to minimize the free energy of the system, the first by reducing the interaction energy and the second by increasing the entropy. At low temperatures ($T < T_c$), the ordering effects of coupling dominate and the iron bar exhibits macroscopic magnetization due to most spins pointing in the same direction; see Fig. 9.24. On the other hand, at high temperatures ($T > T_c$) the disordering effects of thermal fluctuations win out and the spins point in random directions so there is no net magnetization. At the critical point $T = T_c$ there is a

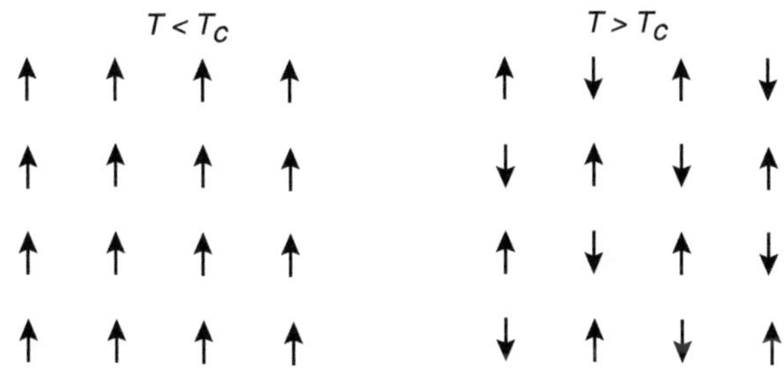

Fig. 9.24 Diagram illustrating an ordered ($T < T_c$) and a disordered ($T > T_c$) alignment of spins in the 2D Ising model, where T_c is the Curie temperature at criticality

balance between the two effects. (Technically speaking, this so-called second-order phase transition occurs on a 2D or 3D lattice but not a 1D lattice.) Although there is now no global order, there will be local domains where all the spins point up and other domains where all the spins point down. At the critical temperature there is a broad distribution of domain sizes, and the domains are themselves dynamically shifting and changing size. An important signature of criticality in equilibrium phase transitions is that various physical quantities exhibit power-law behavior. In the case of the Ising model, this includes the distribution of domain sizes and the two-point correlation matrix, which is defined according to

$$C_{ij} = \langle (s_i - \langle s_i \rangle)(s_j - \langle s_j \rangle) \rangle. \tag{9.78}$$

Here $s_i = \pm 1$ denotes the up/down state of the spin at lattice site i, and the angled brackets indicate time averaging. Thus $(s_i - \langle s_i \rangle)$ represents the size of fluctuations of the ith spin about its average at a given time. It follows that C_{ij} will only be large if s_i and s_j both fluctuate (some disorder) and these fluctuations are coordinated in the same direction (some order). As one might expect, C_{ij} only becomes large in a domain where these two effects are balanced, that is, close to criticality. One finds that $C_{ij} \sim \mathrm{e}^{-|i-j|/\Gamma}$, where Γ is known as the *correlation length* and

$$\Gamma \sim \frac{1}{|T - T_c|^{\alpha}}. \tag{9.79}$$

That is, the correlation length diverges as the system approaches the critical temperature T_c according to a power law with critical exponent α. (In fact, the critical exponent α differs on either side of the critical point.) An important feature of power laws is that they show no characteristic scale—they are scale-free. Moreover, they are a signature of universality, in the sense that a large class of systems that differ at the microscopic level exhibit the same macroscopic behavior close to criticality—they belong to the same *universality class*.

Extending the notion of criticality to biological networks such as the cortex is nontrivial, not only given the fact that these networks are complex and heterogeneous but also because they operate far from thermodynamic equilibrium. Nevertheless, criticality can be understood in a dynamical sense, as suggested by the previous

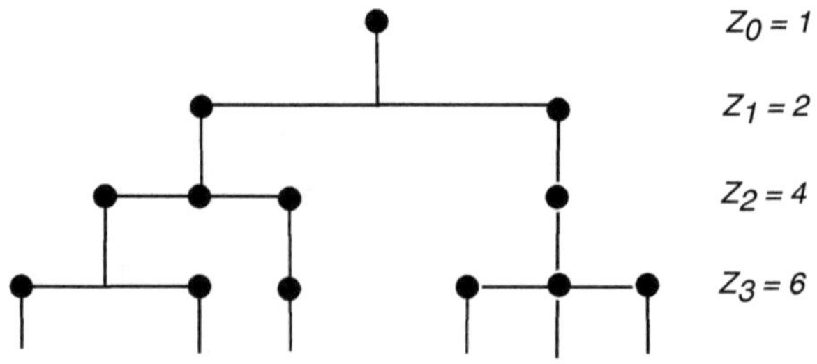

Fig. 9.25 Illustration of a branching process with three generations

gedanken experiment concerning spike avalanches, and evidence (but not proof) of criticality can be obtained by observing power laws in experimental data. The first substantive study of power-law distributions at the network level arose from multielectrode array recordings in cell cultures and acute brain slices [29]. Bursts of spontaneous activity were observed such that if the number of electrodes activated in each distinct burst were counted, then the burst sizes were distributed according to a power law. That is, denoting the burst size by l,

$$P(l) \sim l^{-\alpha} \tag{9.80}$$

with $\alpha \approx 1.5$. The index value of 1.5 can be understood using the theory of branching processes [264, 364]; see below. Since these initial findings, power-law distributions of these so-called neuronal avalanches have been observed in a variety of animal models, both in vitro and in vivo, using both multielectrode recordings and imaging methods such as fMRI; see the reviews [31, 573]. In spite of these results, there is still some controversy regarding their interpretation, both in terms of whether or not the statistical data is truly given by power laws and in terms of the causes of power-law-like behavior [28, 631]. Nevertheless, neuronal avalanches and the theory of branching processes provide another interesting framework for understanding the propagation of activity in networks. In the following we sketch how to obtain the observed power law from a classical branching process, following [364].

A simple branching process is a discrete Markov model for the evolution of a population, which in the neural context would correspond to the progression of spiking events in a neuronal avalanche. Let Z_n denote the number of members (events) of the nth generation (time step). Each member of the nth generation gives birth to a family, which could be empty, of members of the $(n+1)$th generation with the following assumption: the family sizes of the individuals of the branching process are independent identically distributed random variables [264]. An example realization of a branching process is shown in Fig. 9.25. A useful method for analyzing a branching process is to use generating functions. Let $G_n(s) = \mathbb{E}(s^{Z_n})$ be the generating function of the random variable Z_n with probability distribution $\mathbb{P}(Z_n = m)$. A number of results can now be established.

1. *Recursive structure of generating functions.* Each member of the $(n+1)$th generation has a unique ancestor in the nth generation such that

$$Z_{n+1} = X_1 + X_2 + \ldots + X_{Z_n},$$

where X_i is the size of the family produced by the ith member of the mth generation. It follows that

$$\begin{aligned} G_{n+1}(s) &= \mathbb{E}(s^{Z_{n+1}}) = \mathbb{E}(s^{X_1+\ldots+X_{Z_n}}) \\ &= \sum_{n=0}^{\infty} \mathbb{E}(s^{X_1+\ldots+X_m}|Z_n = m)\mathbb{P}(Z_n = m) \end{aligned}$$

$$= \sum_{n=0}^{\infty} \mathbb{E}(s^{X_1} s^{X_2} \ldots^{X_m} | Z_n = m) \mathbb{P}(Z_n = m)$$

$$= \sum_{n=0}^{\infty} \left[\prod_{j=1}^{m} \mathbb{E}(s^{X_j}) \right] \mathbb{P}(Z_n = m)$$

$$= \sum_{n=0}^{\infty} [G_1(s)]^m \mathbb{P}(Z_n = m) = G_n(G_1(s)).$$

Iterating this resulting and dropping the subscript on G_1, we have the recurrence relation

$$G_n(s) = G_{n-1}(G(s)) = G(G(\ldots(G(s))\ldots)). \tag{9.81}$$

2. *Mean and variance.* Let $\mu = \mathbb{E}(Z_1)$ and $\sigma^2 = \mathrm{var}(Z_1)$. In order to determine the mean and variance of Z_n, we use the recursive structure of the generating functions. First,

$$\mathbb{E}(Z_n) = G_n'(1) = \frac{d}{ds} G(G_{n-1}(s)) \Big|_{s=1} = G'(1) G_{n-1}'(s) = \mu G_{n-1}'(s).$$

Iterating this result shows that

$$\mathbb{E}(Z_n) = \mu^n. \tag{9.82}$$

Similarly,

$$\mathbb{E}(Z_n(Z_n - 1)) = G_n''(1) = G''(1) G_{n-1}'(1)^2 + G'(1) G_{n-1}''(1).$$

This gives the iterative result

$$\mathrm{var}(Z_n) = \sigma^2 \mu^{2n-2} + \mu \, \mathrm{var}(Z_{n-1}),$$

from which one finds that

$$\mathrm{var}(Z_n) = \begin{cases} n\sigma^2 & \text{if } \mu = 1 \\ \dfrac{\sigma^2(\mu^n - 1)\mu^{n-1}}{\mu - 1} & \text{if } \mu \neq 1. \end{cases} \tag{9.83}$$

3. *Total number of individuals.* Let T_n be the total number of individuals up to and including the nth generation. Then

$$\begin{aligned} \mathbb{E}(T_n) &= \mathbb{E}(Z_0 + Z_1 + Z_2 + \ldots + Z_n) \\ &= 1 + \mathbb{E}(Z_1) + \mathbb{E}(Z_2) + \ldots + \mathbb{E}(Z_n) \\ &= 1 + \mu + \mu^2 + \ldots + \mu^n \\ &= \begin{cases} \frac{\mu^{n+1} - 1}{\mu - 1}, & \mu \neq 1, \\ n + 1, & \mu = 1. \end{cases} \end{aligned}$$

It follows that

$$\lim_{n\to\infty} \mathbb{E}(T_n) = \begin{cases} \infty, & \mu \geq 1, \\ \frac{1}{1-\mu}, & \mu < 1. \end{cases}$$

Let $H_n(s) = \mathbb{E}(s^{T_n})$ be the generating function for the random variable T_n. The generating functions satisfy the recurrence relation

$$H_{n+1}(s) = sG(H_n(s)). \tag{9.84}$$

4. *Probability of extinction.* One important property of a branching process is whether or not it eventually becomes extinct, that is, $Z_n = 0$ for some finite n. We can define the probability of extinction η according to

$$\eta = \lim_{n\to\infty} \mathbb{P}(Z_n = 0). \tag{9.85}$$

It can be shown that η is the smallest nonnegative root of the equation $s = G(s)$. Moreover, $\eta = 1$ if $\mu < 1$ (extinction occurs almost surely), whereas $\eta < 1$ if $\mu > 1$ (there is a nonzero probability that there is persistent growth). Moreover, if $\mu = 1$, then $\eta = 1$ provided that the variance of the family size distribution has a strictly positive variance. In light of our discussion of criticality, $\mu = 1$ is a critical point separating a subcritical regime ($\mu < 1$) from a supercritical regime ($\mu > 1$).

For the sake of illustration, consider a geometric branching process where the distribution of family sizes is given by $\mathbb{P}(Z_1 = k) \equiv f(k) = qp^k$ with $q = 1 - p$. In this case, one can calculate the generating function and other quantities explicitly:

$$G(s) = q(1-ps)^{-1}, \quad \mu = \frac{p}{q}, \quad \sigma^2 = \frac{p^2}{q^2} + \frac{p}{q}. \tag{9.86}$$

Moreover, it can be shown by induction that

$$G_n(s) = \begin{cases} \dfrac{n-(n-1)s}{n+1-ns} & \text{if } p = q = \frac{1}{2}, \\[2ex] \dfrac{q[p^n - q^n - ps(p^{n-1} - q^{n-1})]}{p^{n+1} - q^{n+1} - ps(p^n - q^n)} & \text{if } p \neq q. \end{cases} \tag{9.87}$$

It follows that

$$\mathbb{P}(Z_n = 0) = G_n(0) = \begin{cases} \dfrac{n}{n+1} & \text{if } p = q, \\[2ex] \dfrac{q(p^n - q^n)}{p^{n+1} - q^{n+1}} & \text{if } p \neq q, \end{cases} \tag{9.88}$$

and, hence, $\eta = 1$ if $p \leq q$ and $\eta = q/p$ if $p > q$. We conclude that for a geometric branching process, extinction occurs almost surely if $\mathbb{E}(Z_1) = \mu = p/q \leq 1$,

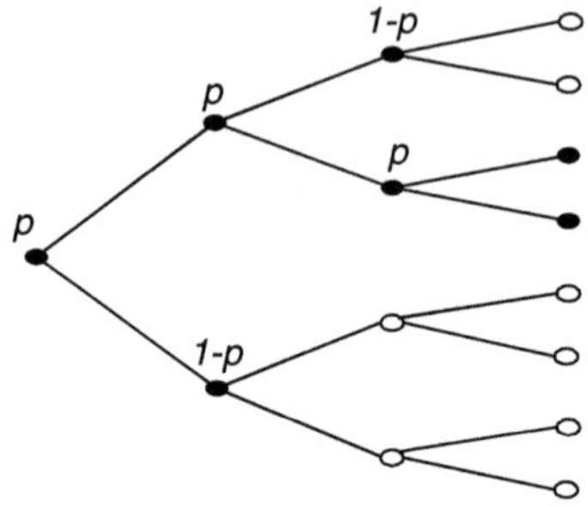

Fig. 9.26 Branching process for the propagation of activity during an avalanche. Activity propagates to a pair of descendants with probability p, whereas it fails to propagate with probability $1-p$. *Black* nodes are active and *white* nodes are inactive

otherwise there is a finite probability of persistent growth. In light of our discussions of criticality and avalanches, we can identify the regime $p < q$ as subcritical, the regime $p > q$ as supercritical, and the point $p = q = 1/2$ as critical.

In the case of neuronal avalanches, a more appropriate branching process is one where each node has two descendants and activity can propagate to the descendants with probability p [29, 364]; see Fig. 9.26. Thus Z_n denotes the number of active members of the nth generation (burst size) with $0 \leq Z_n \leq 2^n$. For this process $f(k) = (1-p)\delta_{k,0} + p\delta_{k,2}$. It follows that

$$G(s) = 1 - p + ps^2, \quad \mu = 2p, \quad \sigma^2 = 4p(1-p), \tag{9.89}$$

and the critical point is at $p = 1/2$. The total number of events T_n is the burst size, whose distribution can be determined from the generating function $H_n(s)$. For the given branching process, $H_n(s)$ satisfies the recurrence relation (see (9.84))

$$H_{n+1}(s) = s[(1-p) + pH_n(s)^2]. \tag{9.90}$$

The solution of this recurrence relation for $n \geq 1$ is approximately [264, 364]

$$H_n(s) = \frac{1 - \sqrt{1 - 4s^2 p(1-p)}}{2sp}. \tag{9.91}$$

Expanding as a power series in s and comparing with the definition of $H_n(s)$, one finds that for $1 \ll l < n$, the distribution of avalanche sizes l is given by

$$P(l) \equiv \mathbb{P}(T_n = l) = \frac{\sqrt{2(1-p)/p}}{\sqrt{\pi} l^{3/2}} \mathrm{e}^{-l/l_c(p)}, \tag{9.92}$$

with

$$l_c(p) = -\frac{2}{\ln[4p(1-p)]}. \tag{9.93}$$

For $l > n$ one finds that $P(l)$ decays exponentially. This then establish the 3/2 power-law behavior at the critical point $p = 1/2$, since $1/l_c(1/2) = 0$ and

$$P(l) \sim \sqrt{\frac{2}{\pi}} l^{-3/2} \tag{9.94}$$

for $1 \ll l < n$ and n is large. Also note that at criticality $\mu = 1$, which is consistent with the experimental observation that the ratio of the number of active descendant electrodes to the number of active ancestor electrodes is around one [29].

9.6 Protein Aggregate Spreading in Neurodegenerative Diseases

There is growing experimental evidence suggesting that there are parallels between prion disease pathogenesis and other, more common neurodegenerative diseases that are typically not thought to be infectious, including Alzheimer's disease, Parkinson's disease, frontal-lobe dementia, and Huntington's disease [284, 300, 371]. Prion diseases are fatal neurodegenerative diseases that rose to prominence during the BSE epidemic in England in the 1980s, where there was thought to be a possible transmission pathway from infected animals to humans. Unlike classical infective agents such as bacteria or viruses, prion diseases involve an agent consisting solely of a misfolded protein or *prion* [513]. Pathologically folded prion protein (PrP*) corrupts normally folded PrP via a fixed conformational change, which usually leads to the formation of protein aggregates, and this process propagates across brain cells at a speed of around 0.1–1mm per day. The initial cause could be an external infective agent or, more commonly, the spontaneous formation of PrP* aggregates. Recent molecular, cellular, and animal studies suggest that the intercellular propagation of protein misfolding also occurs for a variety of aggregate-prone proteins that are linked to noninfective neurodegenerative diseases [262, 533, 565, 566]. These include amyloid-β and tau (Alzheimer's disease), α-synuclein (Parkinson's diseases), and huntingtin (Huntington's disease).

The conformational change of a normally folded protein tends to occur via direct contact with a misfolded protein aggregate. This process is commonly called *aggregate seeding* or *seeded polymerization* [300]. If the resulting protein aggregate is small, then it remains soluble, whereas larger aggregates precipitate out of solution under physiological conditions. Two distinct morphological types of aggregate can be identified: *amorphous aggregates* and *amyloid fibrils*. The former have an irregular granular appearance when viewed under an electron microscope, whereas the latter consist of highly ordered and repetitive structures in which all the polypeptides adopt a common fold. Amyloid fibrillogenesis consists of multiple stages, involving nucleation, polymerization, elongation, and aggregate seeding. It used to be thought that mature amyloid fibrils were ultimately responsible for the toxic effects of protein aggregates. However, there is a widening consensus that it is the non-fibrillar assemblies in early stages of amyloid formation that are particularly toxic. The monomers of proteins such as tau and α-synuclein are found within the cytosol of cells. This means that the spread of such protein aggregates requires the internalization of the corresponding fibrils in order to seed the formation of new aggregates. The newly formed aggregates must then be released back into the extracellular space. The precise mechanisms of internalization and externalization are currently unknown. Such processes are not required for prions and amyloid-β within cells, since they are both exposed to the extracellular space. Hence, the latter proteins can diffuse intracellularly and then infect a neighboring cell directly.

So far there has been very little modeling work on the spread of protein aggregates. Here we will review a reaction–diffusion model for the propagation of prion disease in the brain [411]. The simplest version of the model uses a heterodimer model for prion–prion interactions, in which one molecule of normal PrP (denoted by A) combines with one molecule of PrP* (denoted by B) to form two molecules of B:

$$A + B \to AB \to BB \to B + B. \tag{9.95}$$

For simplicity, this is represented as a single-step reaction with rate k_{AB}:

$$A + B \underset{k_{AB}}{\to} 2B. \tag{9.96}$$

The molecules also degrade at rates k_A and k_B, respectively, with $k_A > k_B$ since PrP* is more resistant to the action of proteases. If the reaction kinetics are combined with classical diffusion on a 1D domain $x \in [0,L]$, then we obtain the reaction–diffusion system

$$\frac{\partial a}{\partial t} = \gamma_0 - k_A a - k_{AB} ab + D_A \frac{\partial^2 a}{\partial x^2} \tag{9.97a}$$

$$\frac{\partial b}{\partial t} = k_{AB} ab - k_B b + D_B \frac{\partial^2 b}{\partial x^2}, \tag{9.97b}$$

where $a(x,t)$ and $b(x,t)$ denote the concentration of normal and misfolded PrP at time $t > 0$ and location x. The corresponding diffusivities are D_A, D_B, and γ_0 represents a constant rate of production of PrP. Equations (9.97) are supplemented by no-flux boundary conditions at the ends of the domain. The resulting model belongs to the same class of PDE as the Fisher–KPP equation (3.44) and the CaMKII model (3.43) of Sect. 3.2. In particular, it supports a spreading wave in the form of a pulled front. Suppose that the system is initially in a stable, homogeneous steady-state $a(x,0) = a^* = \gamma_0/k_A, b(x,0) = 0$. A local transient increase in b at one end of the domain will induce a propagating front whose speed can be determined by linearizing about the steady state in the leading edge of the wave where $a \approx a^*$ (Sects. 3.2 and 3.3). The resulting linear equation for b is

$$\frac{\partial b}{\partial t} = Kb + D_B \frac{\partial^2 b}{\partial x^2} \tag{9.98}$$

with $K = k_{AB}a^* - k_B$. Taking $b(x,t) = B(x - ct) = B(\xi)$ gives the ODE $-cB' = Kb + Db''$. The latter has a solution of the form (neglecting boundary effects) $b(\xi) =$

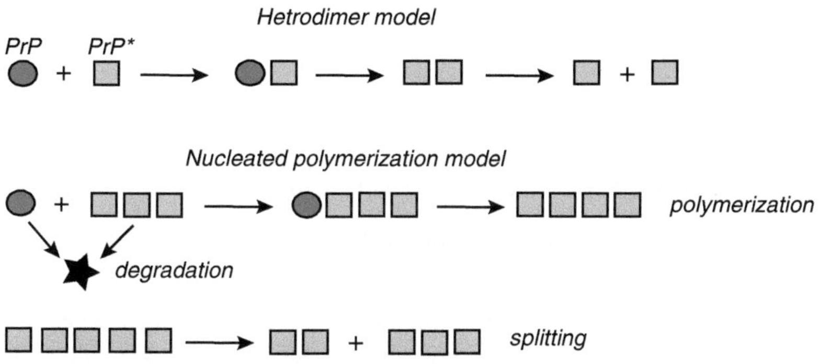

Fig. 9.27 Schematic illustration of the heterodimer and nucleated polymerization models of prion reaction kinetics

$Ce^{-\lambda\xi}$ with

$$c = c(\lambda) \equiv \frac{K}{\lambda} + D_B\lambda. \tag{9.99}$$

Finally, minimizing $c(\lambda)$ with respect to λ yields the minimum speed

$$c_{\min} = \sqrt{D_B K} = \sqrt{D_B(k_{AB}\gamma_0/k_A - k_B)}. \tag{9.100}$$

This is the asymptotic speed of the front induced by a localized perturbation.

The above model suffers from at least two major simplifications, as highlighted by Matthaus [411]. First, the heterodimer model is not a good description of aggregate seeding, which involves the interaction between PrP* polymers of various lengths and normal PrP monomers. Second, as in the case of cortical spreading depression (Sect. 9.3), the brain should really be treated as a highly heterogeneous medium. Focusing on the first issue, one can replace the heterodimer reaction scheme by a nucleated polymerization model [263, 408], in which the infective agent is not a single molecule of PrP* but a PrP* oligomer (short polymer of misfolded PrP). Consider, in particular, the model of Masel et al. [408], which is illustrated in Fig. 9.27. There are a number of steps in the formulation of the kinetic model:

1. Denote a PrP* polymer of size m (m monomer subunits) by B_m. Polymers below a critical size n (the nucleation length) are unstable and are assumed to disintegrate instantaneously into m monomers ($m < n$). The nucleation of new stable PrP* polymers is neglected.
2. When a PrP molecule A comes in contact with a PrP* polymer of size m, $m \geq n$, it forms a polymer of size $m+1$:

$$B_m + A \underset{k_{AB}}{\rightarrow} B_{m+1}$$

with the rate k_{AB} independent of m.

3. A polymer of size $m, m \geq n$, can split into two pieces of sizes $m-k$ and k:

$$B_m \underset{\beta}{\rightarrow} B_{m-k} + B_k$$

with the rate β independent of m and k. If the resulting pieces are smaller than the nucleation length, then they immediately disintegrate.

4. PrP is produced at a rate γ_0 and degrades at a rate k_A. All PrP* polymers degrade at a rate k_B.

Let a be the concentration of PrP monomers and b_m denote the concentration of PrP* polymers of size m. Using the law of mass action, the above kinetic scheme can be expressed in terms of the following system of ODES [408]:

$$\frac{da}{dt} = \gamma_0 - k_A a - k_{AB} a \sum_{m\geq n} b_m + 2\beta \sum_{k=1}^{n-1} k \sum_{m\geq k+1} b_m, \tag{9.101a}$$

$$\frac{db_m}{dt} = k_{AB} a(b_{m-1} - b_m) - k_B b_m - \beta(m-1)b_m + 2\beta \sum_{l\geq m+1} b_l \tag{9.101b}$$

for $m \geq n$, and $b_m \equiv 0$ for $m < n$. It turns out that this infinite hierarchy of equations can be reduced to a closed system of three ODES by summation. Defining

$$b = \sum_{m\geq n} b_m, \quad z = \sum_{m\geq n} m b_m, \tag{9.102}$$

one finds that

$$\frac{da}{dt} = \gamma_0 - k_A a - k_{AB} ab + n(n-1)\beta b, \tag{9.103a}$$

$$\frac{db}{dt} = -k_B b - \beta z - (2n-1)\beta b, \tag{9.103b}$$

$$\frac{dz}{dt} = k_{AB} ab - k_B z - n(n-1)\beta b. \tag{9.103c}$$

The reaction kinetics given by the system of (9.101) can be incorporated into a reaction–diffusion model for the spread of prion disease by adding the terms $D_A \partial^2 a/\partial x^2$ and $D_m \partial^2 b_m/\partial x^2$ to the right-hand sides of (9.101a) and (9.101b), respectively [411]. Numerical simulations confirm that the extended model also supports propagating fronts. (Note that since the cytosolic diffusion coefficient D_m will depend on polymer size m, it is not possible to use summation to reduce (9.103) to a small closed system of coupled RD equations.)

Finally, we briefly consider the issue of diffusion. The 1D model can capture the intracellular diffusion of prions along a single nerve. However, applying a homogeneous and isotropic 3D diffusion equation to the spread on larger spatial scales fails to take into account the fact that neurons form complex networks, which include both local connections to neighbors and long-range connections. That is, one has to consider epidemic diseases on complex networks such as the scale-free and small-world networks defined in Sect. 9.4.1. Some insights into the spread of diseases on networks have been obtained by considering standard models of epidemics such as the susceptible–infected (SI) model [26, 411, 458]. In the latter model, each individual is represented by a vertex of the network, and the edges represent the connections between individuals along which the infection may spread. (In a prion model each vertex would be a neuron and each edge a nerve fiber, say.) The total population N (number of vertices) is kept fixed so that $N = S(t) + I(t)$, where $S(t)$ and $I(t)$ are the number of susceptible and infected individuals, respectively.

A first approach to analyzing the SI model is to use a mean-field description [26]. That is, all the vertices are treated equivalently, which means that the average density $i(t) = I(t)/N$ of infected individuals satisfies

$$\frac{di(t)}{dt} = \lambda \langle k \rangle i(t)[1 - i(t)]. \tag{9.104}$$

The above equation can be interpreted as follows: the growth rate of infected individuals is proportional to a constant spreading rate, λ, the density of susceptible vertices, $s(t) = 1 - i(t)$, and the number of infected individuals in contact with any susceptible vertex. The mean-field approximation takes the latter to be the product of the mean number of neighbors $\langle k \rangle$ and the average density $i(t)$. Note that if the initial density i_0 of infected individuals is small, then one can drop the quadratic term in (9.104) and deduce that the leading behavior at small times is $i(t) \sim i_0 \mathrm{e}^{t/\tau_H}$ with $\tau_H = 1/(\lambda \langle k \rangle)$. The time constant τ_H determines the initial rate at which the disease spreads. The mean-field approximation does not take into account fluctuations arising from fluctuations in the degree k of vertices. In order to proceed further, it is necessary to consider the evolution equation for the average density of infected vertices of degree k, $i_k(t) = I_k(t)/N_k$, where N_k and I_k are the total number and infected number of k-vertices. Thus

$$\frac{di_k(t)}{dt} = \lambda k [1 - i_k(t)] \Gamma_k(t), \tag{9.105}$$

where $\Gamma_k(t)$ is the density of infected individuals that are neighbors of a k-vertex. As a further simplification, suppose that $\Gamma_k = \Gamma$ for all k (a so-called uncorrelated network). Let $P(k)$ be the degree distribution of the network (Sect. 9.4.1). The probability that a susceptible vertex is connected to an infected vertex of degree k' is proportional to the number of edges emanating from the latter. Since at least one edge is likely to be connected to another infected vertex, it follows that this number is $k' - 1$. One thus obtains the result [26]

$$\Gamma(t) = \frac{1}{\langle k \rangle} \sum_{k'} (k' - 1) P(k') i_{k'}(t). \tag{9.106}$$

Differentiate this equation with respect to t and use (9.105). Dropping quadratic terms in the early stages of the epidemic yields the pair of equations

$$\frac{di_k}{dt} = \lambda k \Gamma(t), \quad \frac{d\Gamma}{dt} = \lambda \left(\frac{\langle k^2 \rangle}{\langle k \rangle} - 1 \right) \Gamma(t). \tag{9.107}$$

Finally, assume uniform initial conditions $i_k(0) = i_0$ for all k. The pair of (9.107) can then be solved in terms of $i(t) = \sum_k P(k) i_k(t)$ according to

$$i(t) = i_0 \left[1 + \frac{\langle k \rangle^2 - \langle k \rangle}{\langle k^2 \rangle - \langle k \rangle} \left(e^{t/\tau} - 1 \right) \right], \tag{9.108}$$

with new time constant

$$\tau = \frac{\langle k \rangle}{\lambda(\langle k^2 \rangle - \langle k \rangle)}. \tag{9.109}$$

In the case of a Poisson degree distribution for which $\mathrm{var}[k] = \langle k \rangle$, we recover the mean-field result $\tau = 1/(\lambda \langle k \rangle)$. On the other hand, for a scale-free network with $P(k) \sim k^{-\gamma}$, $2 < \gamma < 3$, we find that $\langle k^2 \rangle \to \infty$ as $N \to \infty$. In other words, for large scale-free networks, there is approximately an instantaneous rise in the number of infected.

References

1. Abbott, L.F., Marder, E.: Modelling small networks. In: Koch, C., Segev, I. (eds.) Methods in Neuronal Modelling, pp. 361–410, 2nd edn. MIT Press, Cambridge (1998)
2. Abbott, L.F., van Vresswijk, C.: Asynchronous states in networks of pulse–coupled oscillators. Phys. Rev. E **48**(2), 1483–1490 (1993)
3. Abbott, L.F., Fahri, E., Gutmann, S.: The path integral for dendritic trees. Biol. Cybern. **66**, 49–60 (1991)
4. Abbott, L.F., Varela, J.A., Sen, K., Nelson, S.B.: Synaptic depression and cortical gain control. Science **275**, 220–224 (1997)
5. Aertsen, A.M.H.J., Gerstein, G.L., Habib, M.K., Palm, G.: Dynamics of neuronal firing correlation: modulation of 'effective connectivity'. J. Neurophysiol. **61**, 900–917 (1989)
6. Alexander, G., Crutcher, M.: Functional architecture of basal ganglia circuits: neural substrates of parallel processing. Trends. Neurosci. **13**, 266–271 (1990)
7. Allard, J., Moginler, A.: Traveling waves in actin dynamics and cell motility. Curr. Opin. Cell Biol. **25**, 107–115 (2013)
8. Amari, S.: Dynamics of pattern formation in lateral inhibition type neural fields. Biol. Cybern. **27**, 77–87 (1977)
9. Andersen, S.S., Bi, G.Q.: Axon formation: a molecular model for the generation of neuronal polarity. Bioessays **22**, 172–179 (2000)
10. Andersen, P., Morris, R., Amaral, D., Bliss, T., O'Keefe, J.: The Hippocampus Book. Oxford University Press, Oxford (2008)
11. Angelucci, A., Levitt, J.B., Walton, E.J.S., Hupe, J.M., Bullier, J., Lund, J.S.: Circuits for local and global signal integration in primary visual cortex. J. Neurosci. **22**, 8633–8646 (2002)
12. Antic, S.D., Zhou, W.L., Moore, A.R., Short, S.M., Ikonomu, K.D.: The decade of the dendritic NMDA spike. J. Neurosci. Res. **88**, 2991–3001 (2010)
13. Arimura, N., Kaibuchi, K.: Neuronal polarity: from extracellular signals to intracellular mechanisms. Nat. Rev. Neurosci **8**, 194–205 (2007)
14. Armero, J., Casademunt, J., Ramirez-Piscina, L., Sancho, J.M.: Ballistic and diffusive corrections to front propagation in the presence of multiplicative noise. Phys. Rev. E **58**, 5494–5500 (1998)
15. Aronson, D.G., Weinberger, H.F.: Multidimensional nonlinear diffusion arising in population genetics. Adv. Math. **30**, 33–76 (1978)
16. Atay, F.M., Hutt, A.: Stability and bifurcations in neural fields with finite propagation speed and general connectivity. SIAM J. Appl. Math. **65**, 644–646 (2005)
17. Austin, T.D.: The emergence of the deterministic Hodgkin-Huxley equations as a limit from the underlying stochastic ion-channel mechanism. Ann. Appl. Prob. **18**, 1279–1325 (2006)
18. Avoli, M., de Curtis, G.B.M.: Do interictal spikes sustain seizures and epileptogenesis. Epilepsy Currents **6**, 203–207 (2006)

P.C. Bressloff, *Waves in Neural Media: From Single Neurons to Neural Fields*, Lecture Notes on Mathematical Modelling in the Life Sciences,
DOI 10.1007/978-1-4614-8866-8,

19. Baer, S.M., Rinzel, J.: Propagation of dendritic spikes mediated by excitable spines: a continuum theory. J. Neurophysiol. **65**, 874–890 (1991)
20. Bair, W., Koch, C., Newsome, W.T., Britten, K.H.: Power spectrum analysis of bursting cells in area MT in the behaving monkey. J. Neurosci. **14**, 2870–2893 (1994)
21. Baladron, J., Fasoli, D., Faugeras, O., Touboul, J.: Mean field description of and propagation of chaos in recurrent multipopulation networks of Hodgkin-Huxley and FitzHugh-Nagumo neurons. J. Math. Neuro. **2**, 10 (2012)
22. Ballesteros-Yanez, I., Benavides-Piccione, R., Elston, G.N., Yuste, R., DeFelipe, J.: Density and morphology of dendritic spines in mouse neocortex. Neurosci. **138**, 403–409 (2006)
23. Barabasi, A.L., Albert, R.: Emergence of scaling in random networks. Science **286**, 509–512 (1999)
24. Baras, F., Mansour, M., Malek, M., Pearson, J.E.: Microscopic simulation of chemical bistability in homogeneous systems. J. Chem. Phys. **105**, 8257–8261 (1996)
25. Bart, E., Bao, S., Holcman, D.: Modeling the spontaneous activity of the auditory cortex. J. Comput. Neurosci. **19**, 357–378 (2005)
26. Barthelemy, M., Barrat, A., Pastor-Satorras, R., Vespignani2, A.: Velocity and hierarchical spread of epidemic outbreaks in scale-free networks. Phys. Rev. Lett. **92**, 178701 (2004)
27. Bazhenov, M., Timofeev, I., Steriade, M., Sejnowski, T.J.: Model of thalamocortical slow-wave sleep oscillations and transitions to activated states. J. Neurosci. **22**, 8691–8704 (2002)
28. Bedard, C., Destexhe, A.: Macroscopic models of local field potentials the apparent 1/f noise in brain activity. Biophys. J. **96**, 2589–2603 (2009)
29. Beggs, J.M., Plenz, D.: Neuronal avalanches in neocortical circuits. J. Neurosci. **23**, 11167–11177 (2003)
30. Beggs, J.M., Plenz, D.: Neuronal avalanches are diverse and precise activity patterns that are stable for many hours in cortical slice cultures. J. Neurosci. **24**, 5216–5229 (2004)
31. Beggs, J.M., Timme, N.: Being critical of criticality in the brain. Front. Physiol. **3 (163)**, 1–14 (2012)
32. Ben-Ari, Y.: Developing networks play a similar melody. Trend. Neurosci. **24**, 353–360 (2001)
33. Ben-Jacob, E., Brand, H., Dee, G., Kramer, L., Langer, J.S.: Pattern propagation in nonlinear dissipative systems. Physica D **14**, 348–364 (1985)
34. Ben-Yishai, R., Bar-Or, R.L., Sompolinsky, H.: Theory of orientation tuning in visual cortex. Proc. Nat. Acad. Sci. **92**, 3844–3848 (1995)
35. Benda, J., Herz, A.V.M.: A universal model for spike-frequency adaptation. Neural Comput. **15**, 2523–2564 (2003)
36. Benguria, R., Depassier, M.: Variational characterization of the speed of propagation of fronts for the nonlinear diffusion equation. Comm. Math. Phys. **175**, 221–227 (1996)
37. Bennett, M.R., Farnell, L., Gibson, W.G.: A quantitative model of purinergic junctional transmission of calcium waves in astrocyte networks. Biophys. J. **89**, 2235–2250 (2005)
38. Bennett, M.R., Farnell, L., Gibson, W.G.: A quantitative model of cortical spreading depression due to puringenic and gap-junction transmission in astrocyte networks. Biophys. J. **95**, 5648–5660 (2008)
39. Benucci, A., Frazor, R., Carandini, M.: Standing waves and traveling waves distinguish two circuits in visual cortex. Neuron **55**, 103–17 (2007)
40. Berestycki, H., Hamel, F., Roques, L.: Analysis of the periodically fragmented environment model: II-biological invasions and pulsating travelling fronts. J. Math. Biol. **51**, 75–113 (2005)
41. Bernard, C., Anderson, A., Becker, A., Poolos, N.P., Beck, H., Johnston, D.: Acquired dendritic channelopathy in temporal lobe epilepsy. Science **305**, 532–535 (2004)
42. Berridge, M.J.: Calcium signalling and cell proliferation. Bioessays **17**, 491–500 (1995)
43. Berridge, M.J.: Neuronal calcium signaling. Neuron **21**, 13–26 (1998)
44. Berridge, M.J., Lipp, P., Bootman, M.D.: The versatility and universality of calcium signalling. Nat. Rev. Mol. Cell Biol. **1**, 11–21 (2000)

45. Bertram, R., Smith, G.D., Sherman, A.: Modeling study of the effects of overlapping calcium microdomains on neurotransmitter release. Biophys. J. **76**, 735–750 (1999)
46. Blake, R.: A primer on binocular rivalry, including current controversies. Brain Mind **2**, 5–38 (2001)
47. Blake, R., Logothetis, N.: Visual competition. Nat. Rev. Neurosci. **3**, 1–11 (2002)
48. Blake, R., Wilson, H.R.: Binocular vision. Vis. Res. **51**, 754–770 (2011)
49. Blakenship, A.G., Feller, M.B.: Mechanisms underlying spontaneous patterned activity in developing neural circuits. Nat. Rev. Neurosci. **11**, 18–29 (2010)
50. Blasdel, G.G.: Orientation selectivity, preference, and continuity in monkey striate cortex. J. Neurosci. **12**, 3139–3161 (1992)
51. Blasdel, G.G., Salama, G.: Voltage-sensitive dyes reveal a modular organization in monkey striate cortex. Nature **321**, 579–585 (1986)
52. Boccaletti, S., Latora, V., Moreno, Y., Chavez, M., Hwang, D.U.: Complex networks: Structure and dynamics. Phys. Rep. **424**, 175–308 (2006)
53. Bonhoeffer, T., Grinvald, A.: Orientation columns in cat are organized in pinwheel like patterns. Nature **364**, 166–146 (1991)
54. Boonstra, T., Daffertshofer, A., Breakspear, M., Beek, P.: Multivariate time-frequency analysis of electromagnetic brain activity during bimanual motor learning. Neuroimage **36**, 370–377 (2007)
55. Borisyuk, R., Kirillov, A.B.: Bifurcation analysis of a neural network model. Biol. Cybern. **66**, 319–325 (1992)
56. Bosking, W.H., Zhang, Y., Schofield, B., Fitzpatrick, D.: Orientation selectivity and the arrangement of horizontal connections in tree shrew striate cortex. J. Neurosci. **17**, 2112–2127 (1997)
57. Boustani, S.E., Destexhe, A.: A master equation formalism for macroscopic modeling of asynchronous irregular activity states. Neural Comput. **21**, 46–100 (2009)
58. Brackley, C.A., Turner, M.S.: Random fluctuations of the firing rate function in a continuum neural field model. Phys. Rev. E **75**, 041913 (2007)
59. Brascamp, J.W., van Ee, R., Noest, A.J., Jacobs, R.H., van den Berg, A.V.: The time course of binocular rivalry reveals a fundamental role of noise. J. Vision **6**, 1244–1256 (2006)
60. Breakspear, M., Roberts, J.A., Terry, J.R., Rodrigues, S., Mahant, N., Robinson, P.A.: A unifying explanation of primary generalized seizures through nonlinear brain modeling and bifurcation analysis. Cereb. Cortex **16**, 1296–1313 (2006)
61. Bressloff, P.C.: Resonantlike synchronization and bursting in a model of pulse-coupled neurons with active dendrites. J. Comput. Neurosci. **6**, 237–249 (1999)
62. Bressloff, P.C.: Synaptically generated wave propagation in excitable neural media. Phys. Rev. Lett. **82**, 2979–2982 (1999)
63. Bressloff, P.C.: Traveling waves and pulses in a one-dimensional network of integrate-and-fire neurons. J. Math. Biol. **40**, 169–183 (2000)
64. Bressloff, P.C.: Traveling fronts and wave propagation failure in an inhomogeneous neural network. Physica D **155**, 83–100 (2001)
65. Bressloff, P.C.: Bloch waves, periodic feature maps and cortical pattern formation. Phys. Rev. Lett. **89**, 088101 (2002)
66. Bressloff, P.C.: Spatially periodic modulation of cortical patterns by long-range horizontal connections. Physica D **185**, 131–157 (2003)
67. Bressloff, P.C.: Pattern formation in visual cortex. In: Chow, C.C., Gutkin, B., Hansel, D., Meunier, C., Dalibard, J. (eds.) Les Houches 2003: Methods and Models in Neurophysics, pp. 477–574. Elsevier, Amsterdam (2005)
68. Bressloff, P.C.: Stochastic neural field theory and the system-size expansion. SIAM J. Appl. Math. **70**, 1488–1521 (2009)
69. Bressloff, P.C.: Metastable states and quasicycles in a stochastic Wilson-Cowan model of neuronal population dynamics. Phys. Rev. E **85**, 051903 (2010)
70. Bressloff, P.C.: From invasion to extinction in heterogeneous neural fields. J. Math. Neurosci. **2**(6), 1–27 (2012)
71. Bressloff, P.C.: Spatiotemporal dynamics of continuum neural fields. J. Phys. A **45**, 033001 (109 pp.) (2012)

72. Bressloff, P.C.: Propagation of CaMKII translocation waves in heterogeneous spiny dendrites. J. Math. Biol. **66**, 1499–1525 (2013)
73. Bressloff, P.C., Coombes, S.: Dynamics of strongly coupled spiking neurons. Neural Comput. **12**, 91–129 (2000)
74. Bressloff, P.C., Cowan, J.D.: Amplitude equation approach to contextual effects in visual cortex. Neural Comput. **14**, 493–525 (2002)
75. Bressloff, P.C., Cowan, J.D.: The visual cortex as a crystal. Physica D **173**, 226–258 (2002)
76. Bressloff, P.C., Folias, S.E.: Front bifurcations in an excitatory neural network. SIAM J. Appl. Math. **65**, 131–151 (2005)
77. Bressloff, P.C., Dwyer, V.M., Kearney, M.J.: Sum-over-paths approach to diffusion on trees. J. Phys. A **29**, 1881–1896 (1996)
78. Bressloff, P.C., Cowan, J.D., Golubitsky, M., Thomas, P.J.: Scalar and pseudoscalar bifurcations: pattern formation on the visual cortex. Nonlinearity **14**, 739–775 (2001)
79. Bressloff, P.C., Cowan, J.D., Golubitsky, M., Thomas, P.J., Wiener, M.: Geometric Visual Hallucinations, Euclidean Symmetry and the Functional Architecture of Striate Cortex. Phil. Trans. Roy. Soc. Lond. B **356**, 299–330 (2001)
80. Bressloff, P.C., Folias, S.E., Prat, A., Li, Y.X.: Oscillatory waves in inhomogeneous neural media. Phys. Rev. Lett. **91**, 178101 (2003)
81. Bressloff, P.C., Newby, J.: Quasi-steady state analysis of motor-driven transport on a two-dimensional microtubular network. Phys. Rev. E **83**, 061139 (2011)
82. Bressloff, P.C., Newby, J.M.: Metastability in a stochastic neural network modeled as a velocity jump Markov process. SIAM J. Appl. Dyn. Sys. **12**, 1394–1435 (2013)
83. Bressloff, P.C., Webber, M.: Neural field model of binocular rivalry waves. J. Comput. Neurosci. **32**, 233–252 (2012)
84. Bressloff, P.C., Webber, M.A.: Front propagation in stochastic neural fields. SIAM J. Appl. Dyn. Syst. **11**, 708–740 (2012)
85. Brown, R.J., Norcia, A.M.: A method for investigating binocular rivalry in real-time with the steady state VEP. Vision Res. **37**, 2401–2408 (1997)
86. Brunel, N.: Dynamics of sparsely connected networks of excitatory and inhibitory spiking neurons. J. Comput. Neurosci **8**, 183–208 (2000)
87. Brunel, N., Hakim, V.: Fast global oscillations in networks of integrate–and–fire neurons with low firing rates. Neural Comput. **11**, 1621–1671 (1999)
88. Buckwar, E., Riedler, M.G.: An exact stochastic hybrid model of excitable membranes including spatio-temporal evolution. J. Math. Biol. **63**, 1051–1093 (2011)
89. Buckwar, E., Riedler, M.G.: Laws of large numbers and langevin approximations for stochastic neural field equations. J. Math. Neurosci. **3**(1), 1–54 (2013)
90. Buice, M.A., Chow, C.C.: Effective stochastic behavior in dynamical systems with incomplete information. Phys. Rev. E **84**, 051120 (2011)
91. Buice, M., Cowan, J.D.: Field-theoretic approach to fluctuation effects in neural networks. Phys. Rev. E **75**, 051919 (2007)
92. Buice, M., Cowan, J.D., Chow, C.C.: Systematic fluctuation expansion for neural network activity equations. Neural Comp. **22**, 377–426 (2010)
93. Bures, J., Buresova, O., Krivanek, J.: The Mechanisms and Applications of Leao's Spreading Depression of Electrical Activity. Academia, Prague (1974)
94. Butts, D.A., Feller, M.B., Shatz, C.J., Rokshar, D.S.: Retinal waves are governed by collective network properties. J. Neurosci. **19**, 3580–3593 (1999)
95. Butz, E.G., Cowan, J.D.: Transient potentials in dendritic systems of arbitrary geometry. Biophys. J. **14**, 661–689 (1974)
96. Buzsaki, G.: Hippocampal sharp waves: their origin and significance. Brain Res. **398**, 242–252 (1986)
97. Buzsaki, G.: Theta oscillations in the hippocampus. Neuron **33**, 325–40 (2002)
98. Buzsaki, G.: Rhythms of the Brain. Oxford University Press, Oxford (2006)
99. Byrne, J.H., Roberts, J.L.: From Molecules to Networks: An Introduction to Cellular and Molecular Neuroscience. Elsevier, Amsterdam (2004)

100. Cai, D., Tao, L., Shelley, M., McLaughlin, D.W.: An effective kinetic representation of fluctuation–driven neuronal networks with application to simple and complex cells in visual cortex. Proc. Natl. Acad. Sci. USA **101**, 7757–7562 (2004)
101. Campbell, K., Gotz., M.: Radial glia: multi-purpose cells for vertebrate brain development. Trend. Neurosci. **25**, 235–238 (2002)
102. Cantrell, R.S., Cosner, C.: Spatial Ecology Via Reaction-Diffusion Equations. Wiley, Chichester (2003)
103. Carlsson, A.E.: Dendritic actin filament nucleation causes traveling waves and patches. Phys. Rev. Lett. **104**, 228102 (2010)
104. Carpenter, G.: A geometric approach to singular perturbation problems with applications to nerve impulse equations. J. Diff. Eqns. **23**, 335–367 (1977)
105. Casti, A.R.R., Ormutag, A., Sornborger, A., Kaplan, E., Knight, B., Victor, J., Sirovich, L.: A population study of integrate–and–fire–or burst neurons. Neural Comput. **14**, 957–986 (2002)
106. Charles, A.C., Merrill, J.E., Dickson, E.R., Sanderson, M.J.: Intercellular signaling in glial cells: calcium waves and oscillations in response to mechanical stimulation and glutamate. Neuron **6**, 983–992 (1991)
107. Charles, A.C., Naus, C.C., Zhu, D., Kidder, G.M., Dirksen, E.R., Sanderson, M.J.: Intercellular calcium signaling via gap junctions in glioma cells. J. Cell. Biol. **118**, 195–201 (1992)
108. Chen, X.: Existence, uniqueness, and asymptotic stability of traveling waves in nonlocal evolution equations. Adv. Diff. Eqns. **2**, 125 (1997)
109. Chen, K.C., Nicholson, C.: Spatial buffering of potassium ions in brain extracellular space. Biophys. J. **78**, 2776–2797 (2000)
110. Cheng, H., Lederer, W.J.: Calcium sparks. Physiol. Rev. **88**, 1491–1545 (2008)
111. Cheng, H., Lederer, M.R., Lederer, W.J., Cannell, M.B.: Calcium sparks and waves in cardiac myocytes. Am. J. Physiol. **270**, C148–C159 (1996)
112. Chervin, R.D., Pierce, P.A., Connors, B.W.: Periodicity and directionality in the propagation of epileptiform discharges across neocortex. J. Neurophysiol. **60**, 1695–1713 (1988)
113. Chialvo, D.: Emergent complex neural dynamics. Nat. Phys. **6**, 744–750 (2010)
114. Chong, S.C., Tadin, D., Blake, R.: Endogenous attention prolongs dominance durations in binocular rivalry. J. Vision **5**, 1044–1012 (2005)
115. Chow, C.C., Buice, M.: Path integral methods for stochastic differential equations. arXiv nlin/105966v1 (2011)
116. Chow, C.C., White, J.A.: Spontaneous action potentials due to channel fluctuations. Biophys. J. **71**, 3013–3021 (1996)
117. Chu, P.H., Milton, J.G., Cowan, J.D.: Connectivity and the dynamics of integrate-and-fire neural networks. Int. J. Bif. Chaos **4**, 237–243 (1994)
118. Chuquet, J., Hollander, L., Nimchinsky., E.A.: High-resolution in vivo imaging of the neurovascular unit during spreading depression. J. Neurosci. **27**, 4036–4044 (2007)
119. Churchland, M.M., et al.: Stimulus onset quenches neural variability: a widespread cortical phenomenon. Nat. Neurosci. **13**, 369–378 (2010)
120. Citri, A., Malenka, R.C.: Synaptic plasticity: multiple forms, functions, and mechanisms. Neuropsychopharmacol. **33**, 18–41 (2008)
121. Cohen, A.H., Ermentrout, G.B., Kiermel, T., Kopell, N., Sigvardt, K.A., Williams, T.L.: Modeling of intersegmental coordination in the lamprey central pattern generator for motion. Trend. Neurosci. **15**, 434–438 (1992)
122. Compte, A., Sanchez-Vives, M.V., McCormick, D.A., Wang, X.J.: Cellular and network mechanisms of slow oscillatory activity (<1 Hz) and wave propagations in a cortical network model. J. Neurophysiol. **89**, 2707–2725 (2003)
123. Collingridge, G.L., Peineau, S., Howland, J.G., Wang, Y.T.: Long-term depression in the CNS. Nat. Rev. Neurosci. **11**, 459–473 (2010)
124. Compte, A., Reig, R., Descalzo, V.F., Harvey, M.A., Puccini, G.D., Sanchez-Vives, M.V.: Spontaneous high-frequency (10–80 Hz) oscillations during up states in the cerebral cortex in vitro. J. Neurosci. **28**, 13828–13844 (2008)

125. Connors, B.W., Amitai, Y.: Epilespy: Models, mechanisms and concepts. In: Schwartkroin, P.A. (ed.) Generation of Epileptiform Discharge by Local Circuits of Neocortex, pp. 388–423. Cambridge University Press, Cambridge (1993)
126. Connors, B.W., Long, M.A.: Electrical synapses in the mammalian brain. Annu. Rev. Neurosci. **27**, 393–418 (2004)
127. Coombes, S.: The effects of ion pumps on the speed of traveling waves in the fire-diffuse-fire model of Ca^{2+} release. Bull. Math. Biol. **63**, 1–20 (2001)
128. Coombes, S.: Dynamics of synaptically coupled integrate-and-fire-or-burst neuron. Phys. Rev. E **67**, 041910 (2003)
129. Coombes, S., Bressloff, P.C.: Solitary waves in a model of dendritic cable with active spines. SIAM J. Appl. Math. **61**, 432–453 (2000)
130. Coombes, S., Bressloff, P.C.: Saltatory waves in the spike-diffuse-spike model of active dendritic spines. Phys. Rev. Lett. **91**, 028102 (2003)
131. Coombes, S., Bressloff, P.C. (eds.): Bursting: The Genesis of Rhythm in the Nervous System. World Scientific Press, London (2005)
132. Coombes, S., Laing, C.R.: Pulsating fronts in periodically modulated neural field models. Phys. Rev. E **83**, 011912 (2011)
133. Coombes, S., Lord, G.J.: Intrinsic modulation of pulse-coupled integrate-and-fire neurons. Phys. Rev. E **56**(5), 5809–5818 (1997)
134. Coombes, S., Owen, M.R.: Evans functions for integral neural field equations with Heaviside firing rate function. SIAM J. Appl. Dyn. Syst. **4**, 574–600 (2004)
135. Coombes, S., Owen, M.R.: Bumps, breathers, and waves in a neural network with spike frequency adaptation. Phys. Rev. Lett. **94**, 148102 (2005)
136. Coombes, S., Schmidt, H.: Neural fields with sigmoidal firing rates: approximate solutions. Discret. Contin. Dyn. Syst. Ser. A **28**, 1369–1379 (2010)
137. Coombes, S., Timofeeva, Y.: Sparks and waves in a stochastic fire-diffue-fire model of Ca^{2+} release. Phy. Rev. E **68**, 021915 (2003)
138. Coombes, S., Owen, M.R., Smith, G.D.: Mode-locking in a periodically forced integrate-and-fire-or-burst neuron model. Phys. Rev. E **64**, 041914 (2001)
139. Coombes, S., Lord, G.J., Owen, M.R.: Waves and bumps in neuronal networks with axo-dendritic synaptic interactions. SIAM J. Appl. Dyn. Sys. **5**, 552–574 (2003)
140. Coombes, S., Hinch, R., Timofeeva, Y.: Receptors, sparks and waves in a fire-diffuse-fire framework for Ca^{2+} release. Prog. Biophys. Mol. Biol. **85**, 197–216 (2004)
141. Coombes, S., Timofeeva, Y., Svensson, C.M., Lord, G.J., Josic, K., Cox, S.J., Colbert, C.M.: Branching dendrites with resonant membrane: a "sum-over-trips" approach. Biol. Cybern. **97**, 137–149 (2007)
142. Coombes, S., Venkov, N.A., Bojak, I., Liley, D.T.J., Laing, C.R.: Modeling electrocortical activity through improved local approximations of integral neural field equations. Phys. Rev. E **76**, 051901 (2007)
143. Cornell-Bell, A.H., Finkbeiner, S.M., Cooper, M.S., Smith, S.J.: Glutamate induces calcium waves in cultured astrocytes: long-range glial signaling. Science **247**, 470–473 (1990)
144. Dahlem, M.A., Chronicle, E.P.: A computational perspective on migraine aura. Prog. Neurobiol. **74**, 351–361 (2004)
145. Dahlem, M.A., Muller, S.C.: Migraine aura dynamics after reverse retinotopic mapping of weak excitation waves in the primary visual cortex. Biol. Cybern. **88**, 419–424 (2003)
146. Dale, N.: Dynamic ATP signalling and neural development. J. Physiol. **586**, 2429–2436 (2008)
147. Deco, G., Jirsa, V., Robinson, P.A., Breakspear, M., Friston, K.: The dynamic brain: from spiking neurons to neural masses and cortical fields. PLoS Comput. Biol. **4**, e1000092 (2008)
148. Deco, G., Jirsa, V.K., McIntosh., A.R.: Emerging concepts for the dynamical organization of resting-state activity in the brain. Nat. Rev. Neurosci. **12**, 43–56 (2011)
149. Delaney, K.R., Galperin, A., Fee, M.S., Flores, J.A., Gervais, R., Tank, D.W., Kleinfeld, D.: Waves and stimulus-modulated dynamics in an oscillating olfactory network. Proc. Natl. Acad. Sci. USA **91**, 669–673 (1994)

150. Delorme, V., Machacek, M., DerMardirossian, C., Anderson, K.L., Wittmann, T., Hanein, D., Waterman-Storer, C., Danuser, G., Bokoch, G.M.: Cofilin activity downstream of Pak1 regulates cell protrusion efficiency by organizing lamellipodium and lamella actin networks. Dev. Cell. **13**, 646–662 (2007)
151. Derkach, V., Barria, A., Soderling, T.R.: Ca^{2+}/calmodulin-kinase II enhances channel conductance of α-amino-3-hydroxyl-5-methyl-4-isoxazoleproprionate type glutamate receptors. Proc. Nat. Acad. Sci. (USA) **96**, 3269–3274 (1999)
152. Destexhe, A., Contreras., D.: Neuronal computations with stochastic network states. Science **314**, 85–90 (2006)
153. Destexhe, A., Sejnowski, T.J.: Thalamocortical Assemblies: How Ion Channels, Single Neurons and Large-Scale Networks Organize Sleep Oscillations. Oxford University Press, Oxford (2001)
154. Destexhe, A., Mainen, Z.F., Sejnowski, T.J.: Synthesis of models for excitable membranes synaptic transmission and neuromodulation using a common kinetic formalism. J. Comput. Neurosci. **1**, 195–231 (1994)
155. Difato, F., Tsushima, H., Pesce, M., Benfenati, F., Blau, A., Chieregatti, E.: The formation of actin waves during regeneration after axonal lesion is enhanced by BDNF. Sci. Rep. **1 (183)**, 1–18 (2011)
156. Dockery, J.D., Keener, J.P.: Diffusive effects on dispersion in excitable media. SIAM J. Appl. Dyn. **49**, 539–566 (1989)
157. Doubrovinski, K., Kruse, K.: Cytoskeletal waves in the absence of molecular motors. Europhys. Lett. **83**, 18003 (2008)
158. van Drongelen, W., Lee, H.C., Stevens, R.L., Hereld, M.: Propagation of seizure-like activity in a model of neocortex. J. Clin. Neurophysiol. **24**, 182–188 (2007)
159. Dyhrfjeld-Johnsen, J., Santhakumar, V., Morgan, R.J., Huerta, R., Tsimring, L., Soltesz, I.: Topological determinants of epileptogenesis in large-scale structural and functional models of the dentate gyrus derived from experimental data. J. Neurophysiol. **97**, 1566–1587 (2007)
160. Dykman, M.I., Mori, E., Ross, J., Hunt, P.M.: Large fluctuations and optimal paths in chemical kinetics. J. Chem. Phys. A **100**, 5735–5750 (1994)
161. Earnshaw, B.A., Bressloff, P.C.: Diffusion-activation model of CaMKII translocation waves in dendrites. J. Comput. Neurosci. **28**, 77–89 (2010)
162. Ebert, U., van Saarloos, W.: Front propagation into unstable states: universal algebraic convergence towards uniformly translating pulled fronts. Physica D **146**, 1–99 (2000)
163. Edwards, J.R., Gibson, W.G.: A model for Ca^{2+} waves in networks of glial cells incorporating both intracellular and extracellular communication pathways. J. Theor. Biol. **263**, 45–58 (2010)
164. Elgart, V., Kamenev, A.: Rare event statistics in reaction–diffusion systems. Phys. Rev. E **70**, 041106 (2004)
165. Elmer, C.E., Vleck, E.S.V.: Analysis and computation of traveling waves solutions of bistable differential-difference equations. Nonlinearity **12**, 771–798 (1999)
166. Ermentrout, G.B.: The behaviour of rings of coupled oscillators. Journal of Mathematical Biology **23**, 55–74 (1985)
167. Ermentrout, G.B.: The analysis of synaptically generated travelling waves. J. Comput. Neurosci. **5**, 191–208 (1998)
168. Ermentrout, G.B.: Neural networks as spatio-temporal pattern-forming systems. Rep. Prog. Phy. **61**, 353–430 (1998)
169. Ermentrout, G.B., Cowan, J.: A mathematical theory of visual hallucination patterns. Bio. Cybern. **34**, 137–150 (1979)
170. Ermentrout, G.B., Kleinfeld, D.: Traveling electrical waves in cortex: insights from phase dynamics and speculation on a computational role. Neuron **29**, 33–44 (2001)

171. Ermentrout, G.B., Kopell, N.: Frequency plateaus in a chain of weakly coupled oscillators. SIAM J. Appl. Math. **15**, 215–237 (1984)
172. Ermentrout, G.B., McLeod, J.B.: Existence and uniqueness of travelling waves for a neural network. Proc. Roy. Soc. Edin. A **123**, 461–478 (1993)
173. Ermentrout, G.B., Terman, D.: Mathematical Foundations of Neuroscience. Springer, Berlin (2010)
174. Ermentrout, G.B., Jalics, J.Z., Rubin, J.E.: Stimulus-driven traveling solutions in continuum neuronal models with a general smooth firing rate function. SIAM J. Appl. Math. **70**, 3039–3064 (2010)
175. Escudero, C., Kamanev, A.: Switching rates of multistep reactions. Phys. Rev. E **79**, 041149 (2009)
176. Escudero, C., Rodriguez, J.A.: Persistence of instanton connections in chemical reactions with time–dependent rates. Phys. Rev. E **77**, 011130 (2008)
177. Evans, J.: Nerve axon equations IV: The stable and unstable impulse. Ind. Univ. Math. J. **24**, 1169–1190 (1975)
178. Evans, L.C., Sougandis, P.E.: A PDE approach to geometric optics for certain semilinear parabolic equations. Ind. Univ. Math. J. **38**, 141–172 (1989)
179. Faisal, A.A., Selen, L.P.J., Wolpert, D.M.: Noise in the nervous system. Nat. Rev. Neurosci. **9**, 292 (2008)
180. Falcke, M.: On the role of stochastic channel behavior in intracellular Ca^{2+} dynamics. Biophys. J. **84**, 42–56 (2003)
181. Falcke, M.: Reading the patterns in living cells - the physics of Ca^{2+} signaling. Adv. Phys. **53**, 255–440 (2004)
182. Falcke, M., Tsimiring, L., Levine, H.: Stochastic spreading of intracellular Ca^{2+} release. Phys. Rev. E **62**, 2636–2643 (2000)
183. Fath, G.: Propagation failure of traveling waves in a discrete bistable medium. Physica D **116**, 176–190 (1998)
184. Faugeras, O., Touboul, J., Cessac, B.: A constructive mean–field analysis of multi–population neural networks with random synaptic weights and stochastic inputs. Frontiers in Comp. Neurosci. **3**, 1–28 (2009)
185. Faugeras, O., Veltz, R., Grimbert, F.: Persistent neural states: Stationary localized activity patterns in nonlinear continuous n-population, q-dimensional neural networks. Neural Comput. **21**, 147–187 (2009)
186. Fellin, T., Pascual, O., Gobbo, S., Pozzan, T., Haydon, P.G., Carmignoto, G.: Neuronal synchrony mediated by astrocytic glutamate through activation of extrasynaptic NMDA receptors. Neuron **43**, 729–743 (2004)
187. Fellin, T., Pozzan, T., Carmignoto., G.: Purinergic receptors mediate two distinct glutamate release pathways in hippocampal astrocytes. J. Biol. Chem. **281**, 4274–4284 (2006)
188. Ferster, D., Miller, K.: Neural mechanisms of orientation selectivity in the visual cortex. Annual review of neuroscience **23**(1), 441–471 (2000)
189. Fianco, T.A., McCarthy, K.D.: Intracellular astrocyte calcium waves in situ increase the frequency of spontaneous AMPA receptor currents in CA1 pyramidal cells. J. Neurosci. **24**, 722–732 (2004)
190. Firth, S.I., Wang, C.T., Feller, M.B.: Retinal waves: mechanisms and function in visual system development. Cell Calcium **37**, 425–432 (2005)
191. Fisher, R.A.: The wave of advance of advantageous genes. Ann. Eugenics **7**, 353–369 (1937)
192. FitzHugh, R.: Impulses and physiological states in theoretical models of nerve membrane. Biophys. J. **1**, 445–466 (1960)
193. Fivaz, M., Bandara, S., Inoue, T., Meyer, T.: Robust neuronal symmetry-breaking by Ras-triggered local positive feedback. Curr. Biol. **18**, 44–50 (2008)
194. Flynn, K.C., Pak, C.W., Shaw, A.E., Bradke, F., Bamburg, J.R.: Growth cone-like waves transport actin and promote axonogenesis and neurite branching. Dev. Neurobiol. **69**, 761–779 (2009)

195. Fogelson, A., Zucker, R.S.: Presynaptic calcium diffusion from various arrays of single channels. implications for transmitter release and synaptic facilitation. Biophys. J. **48**, 1003–1017 (1985)
196. Folias, S.E., Bressloff, P.C.: Breathing pulses in an excitatory neural network. SIAM J. Appl. Dyn. Syst. **3**, 378–407 (2004)
197. Folias, S.E., Bressloff, P.C.: Breathers in two–dimensional neural media. Phys. Rev. Lett. **95**, 208107 (2005)
198. Folias, S.E., Bressloff, P.C.: Stimulus–locked traveling pulses and breathers in an excitatory neural network. SIAM J. Appl. Math. **65**, 2067–2092 (2005)
199. Fox, R., Rasche, F.: Binocular rivalry and reciprocal inhibition. Attention Percept. Psychophys. **5**(4), 215–217 (1969)
200. Fox, R.F., Lu, Y.N.: Emergent collective behavior in large numbers of globally coupled independent stochastic ion channels. Phys. Rev. E **49**, 3421–3431 (1994)
201. Franci, A., Drion, G., Sepulchre, R.: An organizing center in a planar model of neuronal excitability. SIAM J. Appl. Dyn. Sys. **11**, 1698–1722 (2013)
202. Franks, K.M., Sejnowski, T.J.: Complexity of calcium signaling in synaptic spines. Bioessays **24**, 1130–1144 (2002)
203. Freidlin, M.I.: Limit theorems for large deviations and reaction-diffusion equations. Ann. Prob. **13**, 639–675 (1985)
204. Freidlin, M.I.: Geometric optics approach to reaction–diffusion equations. SIAM J. Appl. Math. **46**, 222–232 (1986)
205. Freidlin, M.I., Wentzell, A.D.: Random Perturbations of Dynamical Systems. Springer, New York (1984)
206. Friedman, A., Craciun, G.: A model of intracellular transport of particles in an axon. J. Math. Biol. **51**, 217–246 (2005)
207. Friedman, A., Hu, B.: Uniform convergence for approximate traveling waves in linear reaction-hyperbolic systems. Ind. Univ. Math. J. **56**, 2133–2158 (2007)
208. Fusi, S., Drew, P.J., Abbott, L.F.: Cascade models of synaptically stored memories. Neuron **45**, 599–611 (2005)
209. Garcia-Ojalvo, J., Sancho, J.M.: External fluctuations in a pattern-forming instability. Phys. Rev. E **53**, 5680–5689 (1996)
210. Gardiner, C.W.: Handbook of Stochastic Methods, 4th edn. Springer, Berlin (2009)
211. Gardner-Medwin, A.R.: Analysis of potassium dynamics in mammalian brain tissue. J. Physiol. (Lond) **335**, 393–462 (1983)
212. Gartner, J., Freidlin, M.I.: On the propagation of concentration waves in periodic and random media. Soviet Math. Dokl. **20**, 1282–1286 (1979)
213. Gerstner, W., van Hemmen, J.L.: Coherence and incoherence in a globally coupled ensemble of pulse–emitting units. Phys. Rev. Lett. **71**(3), 312–315 (1993)
214. Gerstner, W., Kistler, W.: Spiking Neuron Models. Cambridge University Press, Cambridge (2002)
215. Gerstner, W., van Hemmen, J.L., Cowan, J.D.: What matters in neuronal locking? Neural Comput. **8**, 1689–1712 (1996)
216. Gilbert, C.D.: Horizontal integration and cortical dynamics. Neuron **9**, 1–13 (1992)
217. Gilbert, C.D., Das, A., Ito, M., Kapadia, M., Westheimer, G.: Spatial integration and cortical dynamics. Proc. Nat. Acad. Sci. USA **93**, 615–622 (1996)
218. Gilbert, C.D., Wiesel, T.N.: Clustered intrinsic connections in cat visual cortex. J. Neurosci. **3**, 1116–1133 (1983)
219. Gilboa, G., Chen, R., Brenner, N.: History–dependent multiple-time-scale dynamics in a single-neuron model. J. Neurosci. **25**, 6479–6489 (2005)
220. Ginzburg, I., Sompolinsky, H.: Theory of correlations in stochastic neural networks. Phys. Rev. E **50**, 3171–3191 (1994)
221. Glass, L., Mackey, M.C.: From Clocks to Chaos. Princeton University Press, Princeton (1988)
222. Godfrey, K.B., Eglen, S.J.: Theoretical models of spontaneous activity generation and propagation in the developing retina. Mol. Biosyst. **5**, 1527–1535 (2009)

223. Godfrey, K.B., Swindale, N.V.: Retinal wave behavior through activity-dependent refractory periods. PLoS Comp. Biol. **3 (11)**, 2408–2420 (2007)
224. Goel, P., Sneyd, J., Friedman, A.: Homogenization of the cell cytoplasm: the calcium bidomain equations. SIAM Nultiscal. Model Simul. **5**, 1045–1062 (2006)
225. Goldberg, J.L.: How does an axon grow? Genes Dev. **17**, 941–958 (2003)
226. Golding, N.L., Staff, N.P., Spruston, N.: Dendritic spikes as a mechanism for cooperative long-term potentiation. Nature **418**, 326–331 (2002)
227. Goldwyn, J.H., Shea-Brown, E.: The what and where of adding channel noise to the Hodgkin-Huxley equations. PLoS Comp. Biol. **7 (11)**, e1002247 (2011)
228. Golomb, D., Amitai, Y.: Propagating neuronal discharges in neocortical slices: Computational and experimental study. J. Neurophysiol. **78**, 1199–1211 (1997)
229. Golomb, D., Ermentrout, G.: Bistability in pulse propagation in networks of excitatory and inhibitory populations. Phys. Rev. Lett. **86**, 4179–4182 (2001)
230. Golomb, D., Ermentrout, G.B.: Continuous and lurching traveling waves in neuronal networks with spatially-decaying connectivity and delay. Proc. Natl. Acad Sci. USA **96**, 13480–13485 (1999)
231. Golomb, D., Rinzel, J.: Clustering in globally coupled inhibitory neurons. Physica D **72**, 259–282 (1994)
232. Golomb, D., Wang, X.J., Rinzel, J.: Propagation of spindle waves in a thalamic slice model. J. Neurophysiol. **75**, 750–769 (1996)
233. Gomez, T.M., Spitzer, N.C.: Regulation of growth cone behavior by calcium: New dynamics to earlier perspectives. J. Neurobiol. **44**, 174–183 (1999)
234. Gomez, T.M., Zheng, J.Q.: The molecular basis for calcium-dependent axon pathfinding. Nat. Rev. Neurosci. **7**, 115–125 (2006)
235. Gonzalez-Billault, C., Munoz-Llancao, P., Henriquez, D.R., Wojnacki, J., Conde, C., Caceres, A.: The role of small GTPases in neuronal morphogenesis and polarity. Cytoskeleton **69**, 464–485 (2012)
236. Gordon, G.R., Baimoukhametova, D.V., Hewitt, S.A., Kosala, W.R.A., Rajapaksha, J.S., Fisher, T.E., Bains, J.S.: Noropinephrine triggers release of glial ATP to increase postsynaptic efficacy. Nat. Neurosci. **8**, 1078–1086 (2005)
237. Gorji, A.: Spreading depression: A review of the clinical relevance. Brain Res. Revs. **38**, 33–60 (2001)
238. Gourley, S.A.: Travelling front solutions of a nonlocal fisher equation. J. Math Biol. **41**, 272–284 (2000)
239. Grannan, E.R., Kleinfeld, D., Sompolinsky, H.: Stimulus dependent synchronization of neuronal assemblies. Neural Comput. **5**, 550–569 (1993)
240. Gray, W.G., Lee, P.C.Y.: On the theorems for local volume averaging of multiphase systems. Int. J. Multiphase Flow **3**, 333–340 (1977)
241. Greenstein, J.L., Hinch, R., Winslow, R.L.: Mechanisms of excitation-contraction coupling in an integrative model of the cardiac ventricular myocyte. Biophys. J. **90**, 77–91 (2006)
242. Grindrod, P.: Patterns and Waves: The Theory and Application of Reaction-Diffusion Equations. Clarendon Press, Oxford (1991)
243. Grinvald, A., Lieke, E.E., Frostig, R.D., Hildesheim, R.: Cortical point-spread function and long-range lateral interactions revealed by real-time optical imaging of macaque monkey primary visual cortex. J. Neurosci. **14**, 2545–2568 (1994)
244. Groff, J.R., DeRemigio, H., Smith, G.D.: Stochastic methods in neuroscience, chap. 2. In: Markov Chain Models of Ion Channels and Calcium Release Sites, pp. 29–64. Oxford University Press, Oxford (2009)
245. Gu, X., Olson, E.C., Spitzer, N.: Spontaneous neuronal calcium spikes and waves during early differentiation. . J. Neurosci. **14**, 6325–6335 (1994)
246. Gu, X., Spitzer, N.C.: Breaking the code: regulation of neuronal differentiation by spontaneous calcium transients. Dev. Neurosci. **19**, 33–41 (1997)
247. Guan, C.B., Xu, H.T., Jin, M., Yuan, X.B., Poo, M.M.: Long-range Ca^{2+} signaling from growth cone to soma mediates reversal of neuronal migration induced by Slit-2. Cell **129**, 385–395 (2007)

248. Guckenheimer, J., Holmes, P.J.: Nonlinear Oscillations, Dynamical Systems and Bifurcations of Vector Fields. Springer, New York (1983)
249. Guthrie, P.B., Knappenberger, J., Segal, M., Bennett, M.V., , Charles, A.C., Kater, S.B.: ATP released from astrocytes mediates glial calcium waves. J. Neurosci. **19**, 520–528 (1999)
250. Haas, B., Schipke, C.G., Peters, O., Sohl, G., Willecke, K., Kettenmann, H.: Activity-dependent ATP-waves in the mouse neocortex are independent from astrocytic calcium waves. Cereb. Cortex **16**, 237–246 (2006)
251. Hagberg, A., Meron, E.: Pattern formation in non–gradient reaction–diffusion systems: the effects of front bifurcations. Nonlinearity **7**, 805–835 (1994)
252. Hagberg, A., Meron, E., Rubinstein, I., Zaltzman., B.: Controlling domain patterns far from equilibrium. Phys. Rev. Lett. **76**, 427–430 (1996)
253. Halassa, M.M., Fellin, T., Haydon, P.G.: The tripartite synapse: roles for gliotransmission in health and disease. Trend. Mol. Med. **13**, 54–63 (2007)
254. Hall, A.: Rho GTPases and the actin cytoskeleton. Science **279**, 509–514 (1998)
255. Han, F., Caporale, N., Dan, Y.: Reverberation of recent visual experience in spontaneous cortical waves. Neuron **60**, 321–327 (2008)
256. Hanggi, P., Grabert, H., Talkner, P., Thomas, H.: Bistable systems: master equation versus Fokker–Planck modeling. Z. Physik B **28**, 135 (1984)
257. Hanggi, P., Talkner, P., Borkovec, M.: Reaction-rate theory: fifty years after Kramers. Rev. Mod. Phys. **62**, 251–341 (1990)
258. Hansel, D., Mato, G., Meunier, C.: Synchrony in excitatory neural networks. Neural Comput. **7**, 2307–2337 (1995)
259. Hansen, M., Boltano, S., Dirksen, E.R., Sanderson, M.J.: Intercellular calcium signaling induced by extracellular adenosine 5-triphosphate and mechanical stimulation in airway epithelial cells. J. Cell. Sci. **106**, 995–1004 (1993)
260. Hanson, M.G., Landmesser, L.T.: Chracaterization of the circuits that generate spontaneous episodes of activity in the early embryonic mouse spinal cord. J. Neurosci. **23**, 587–600 (2003)
261. Hanson, P.I., Meyer, T., Stryer, L., Schulman, H.: Dual role of calmodulin in autophosphorylation of multifunctional cam kinase may underlie decoding of calcium signal. Neuron **12**, 943–956 (1994)
262. Hardy, J.: A hundred years of Alzheimer's disease research. Neuron **52**, 3–13 (2006)
263. Harper, J.D., Lansbury, P.T. Jr.: Models of amyloid seeding in Alzheimer's disease and scrapie: mechanistic truths and physiological consequences of the time-dependent solubility of amyloid proteins. Annu. Rev. Biochem. **66**, 385–407 (1997)
264. Harris, T.E.: The theory of branching processes. Dover, New York (1989)
265. Harris-White, M.E., Zanotti, S.A., Frautschy, S.A., Charles, A.C.: Spiral intercellular calcium waves in hippocampal slice cultures. J. Neurophysiol. **79**, 1045–1052 (1998)
266. Hassinger, T.D., Guthrie, P.B., Atkinson, P.B., Bennett, M.V., Kater, S.B.: An extracellular signaling component in propagation of astrocytic calcium waves. Proc. Natl. Acad. Sci. USA **93**, 13268–13273 (1996)
267. Hastings, S.P.: The existence of progressive wave solutions to the Hodgkin-Huxley equations. Arc. Rat. Mech. Anal. **60**, 229–257 (1975)
268. Haydon, P.G.: Glia: listening and talking to the synapse. Nat. Rev. Neurosci. **2**, 185–193 (2000)
269. Heitmann, S., Gong, P., Breakspear, M.: A computational role for bistability and traveling waves in motor cortex. Front. Comp. Neurosci. **6 (67)**, 1–15 (2012)
270. von Helmholtz, H.: Treatise on physiological optics. Dover, New York (1866)
271. Hennig, M.H., Adams, C., Willshaw, D., Sernagor, E.: Early-stage waves in the retinal network emerge close to a critical state transition between local and global functional connectivity. J. Neurosci. **29**, 1077–1086 (2009)
272. Hill, S., Tononoi, G.: Modeling sleep and wakefulness in the thalamocortical system. J. Neurophysiol. **93**, 1671–1698 (2005)

273. Hill, A.A.V., Masino, M.A., Calabrese, R.L.: Intersegmental coordination of rhythmic motor patterns. J. Neurophysiol. **90**, 531–538 (2003)
274. Hillen, T., Othmer, H.: The diffusion limit of transport equations derived from velocity-jump processes. SIAM J. Appl. Math. **61**, 751–775 (2000)
275. Hinch, E.J.: Perturbation Methods. Cambridge University Press, Cambridge (1991)
276. Hinch, R.: A mathematical analysis of the generation and termination of calcium sparks. Biophys. J. **86**, 1293–1307 (2004)
277. Hinch, R., Chapman, S.J.: Exponentially slow transitions on a Markov chain: the frequency of calcium sparks. Eur. J. Appl. Math. **16**, 427–446 (2005)
278. Hirsch, J.D., Gilbert, C.D.: Synaptic physiology of horizontal connections in the cat's visual cortex. J. Physiol. Lond. **160**, 106–154 (1991)
279. Hodgkin, A.L., Huxley, A.F.: A quantitative description of membrane and its application to conduction and excitation in nerve. J. Physiol. **117**, 500–544 (1952)
280. Hofer, T., Politi, A., Heinrich, R.: Intercellular Ca^{2+} wave propagation through gap-junctional Ca^{2+} diffusion: a theoretical study. Biophys. J. **80**, 75–87 (2001)
281. Hofer, T., Venance, L., Giaume, C.: Control and plasticity of intercellular calcium waves in astrocytes: a modeling approach. J. Neurosci. **22**, 4850–4859 (2002)
282. Holcman, D., Tsodyks, M.: The emergence of up and down states in cortical networks. PLoS Comp. Biol. **2(3)**, 0174–0181 (2006)
283. Holcman, D., Schuss, Z., Korkotian, E.: Calcium dynamics in dendritic spines and spine motility. Biophys. J. **87**, 81–91 (2004)
284. Holmes, B.B., Diamond, M.I.: Cellular mechanisms of protein aggregate propagation. Curr. Opin. Neuro. **25**, 721–726 (2012)
285. Holmes, E.E., Lewis, M.A., Banks, J.E., Veit, R.R.: Partial differential equations in ecology: spatial interactions and population dynamics. Ecology **75**, 17–29 (1994)
286. Holmes, W.R., Carlsson, A.E., Edelstein-Keshet, L.: Regimes of wave type patterning driven by refractory actin feedback: transition from static polarization to dynamic wave behaviour. Phys. Biol. **9**, 046005 (2012)
287. Hoogland, T.M., Kuhn, B., Gobel, W., Huang, W., Nakai, J., Helmchen, F., Flint, J., Wang, S.S.: Radially expanding transglial calcium waves in the intact cerebellum. Proc. Natl. Acad. Sci. USA **106**, 3496–3501 (2009)
288. Hopfield, J.J.: Neurons with graded response have collective computational properties like those of two–state neurons. Proc. Natl. Acad. Sci. USA **81**, 3088–3092 (1984)
289. Horton, A.C., Ehlers, M.D.: Neuronal polarity and trafficking. Neuron **40**, 277–295 (2003)
290. Houweling, A.R., Bazhenov, M., Timofeev, I., Steriade, M., Sejnowski, T.J.: Homeostatic synaptic plasticity can explain post-traumatic epileptogenesis in chronically isolated neocortex. Cereb. Cortex **15**, 834–845 (2005)
291. Hsu, D., Tang, A., Hsu, M., Beggs, J.M.: Simple spontaneously active Hebbian learning model: homeostatsis of activity and connectivity, and consequences for learining and epileptogenesis. Phys. Rev. E **76**, 041909 (2007)
292. Huang, X., Troy, W.C., Yang, Q., Ma, H., Laing, C.R., Schiff, S.J., Wu, J.: Spiral waves in disinhibited mammalian neocortex. J. Neurosci. **24**, 9897–9902 (2004)
293. Hubel, D.H., Wiesel, T.N.: Sequence regularity and geometry of orientation columns in the monkey striate cortex. J. Comp. Neurol. **158**, 267–294 (1974)
294. Hubel, D.H., Wiesel, T.N.: Uniformity of monkey striate cortex: A parallel relationship between field size, scatter, and magnification factor. J. Comp. Neurol. **158**, 295–306 (1974)
295. Hudson, A., Schulman, H.: Neuronal Ca^{2+}/calmodulin-dependent protein kinase II: the role of structure and autoregulation in cellular function. Annu. Rev. Biochem. **71**, 473–510 (2002)
296. Hutt, A., Bestehorn, M., Wennekers, T.: Pattern formation in intracortical neuronal fields. Network **14**, 351–368 (2003)
297. Hutt, A., Longtin, A., Schimansky-Geier, L.: Additive noise-induces Turing transitions in spatial systems with application to neural fields and the Swift-Hohenberg equation. Physica D **237**, 755–773 (2008)
298. Iacobas, D.A., Suadicani, S.O., Spray, D.C., Scemes, E.: A stochastic two-dimensional model of intercellular Ca^{2+} wave spread in glia. Biophys. J. **90**, 24–41 (2006)

299. Idan, S., Rinzel, J., Gordon, M.S. (eds.): The Theoretical Foundations of Dendritic Function: Selected Papers of Wilfrid Rall with Commentaries. MIT Press, Cambridge (1995)
300. Invernizzi, G., Papaleo, E., Sabate, R., Ventura, S.: Protein aggregation: Mechanisms and functional consequences. Int. J. Biochem. Cell Biol. **44**, 1541–1554 (2012)
301. Izhikevich, E.: Dynamical Systems in Neuroscience: The Geometry of Excitability and Bursting. MIT Press, Cambridge (2006)
302. Jacobs, K.: Stochastic Processes for Physicists. Cambridge University Press, Cambridge (2010)
303. Jaffe, D.B., Brown, T.H.: Metabotropic glutamate receptor activation induces calcium waves within hippocampal dendrites. J. Neurophysiol. **72**, 471–474 (1994)
304. Jahr, C., Stevens, C.: A quantitative description of NMDA receptor-channel kinetic behavior. J. Neurosci. **10**, 1830–1837 (1990)
305. Jilkine, A., Maree, A.F.M., Edelstein-Keshet, L.: Mathematical model for spatial segregation of the Rho-family GTPases based on inhibitory crosstalk. Bull. Math. Biol. **68**, 1169–1211 (2007)
306. Jin, D.Z., Dragoi, V., Sur, M., Seung, H.S.: Tilt aftereffect and adaptation-induced changes in orientation tuning and visual cortex. J. Neurosci. **94**, 4038–4050 (2005)
307. Jirsa, V.K., Haken, H.: A derivation of a macroscopic field theory of the brain from the quasi-microscopic neural dynamics. Physica D **99**, 503–526 (1997)
308. Jirsa, V.K., Kelso, J.A.S.: Spatiotemporal pattern formation in neural systems with heterogeneous topologies. Phys. Rev. E **62**, 8462–8465 (2000)
309. van Kampen, N.G.: Stochastic Processes in Physics and Chemistry. North-Holland, Amsterdam (1992)
310. Kang, M., Othmer, H.G.: Spatiotemporal characteristics of calcium dynamics in astrocytes. Chaos **19**, 037116 (2009)
311. Kang, N., Xu, J., Xu, Q., Nedergaard, M., Kang, J.S.: Astrocytic glutamate release-induced transient depolarization and epileptiform discharges in hippocampal CA1 pyramidal neuron. J. Neurophysiol. **94**, 4121–4130 (2005)
312. Kang, M., Heeger, D.J., Blake, R.: Periodic perturbations producing phase-locked fluctuations in visual perception. J. Vision **9(2):8**, 1–12 (2009)
313. Kang, M., Lee, S.H., Kim, J., Heeger, D.J., Blake, R.: Modulation of spatiotemporal dynamics of binocular rivalry by collinear facilitation and pattern-dependent adaptation. J. Vision **10(11):3**, 1–15 (2010)
314. Kapitula, T., Kutz, N., Sandstede, B.: The Evans function for nonlocal equations. Ind. Univ. Math. J. **53**, 1095–1126 (2004)
315. Katz, L.C., Shatz, C.J.: Synaptic activity and the construction of cortical circuits. Science **274**, 1133–1138 (1996)
316. Keener, J.P.: Waves in excitable media. SIAM J. Appl. Math. **39**, 528–548 (1981)
317. Keener, J.P.: Propagation and its failure in coupled systems of discrete excitable cells. SIAM J. Appl. Math **31**, 269–276 (1987)
318. Keener, J.P.: Homogenization and propagation in the bistable equation. Physica D **136**, 1–17 (2000)
319. Keener, J.P.: Propagation of waves in an excitable medium with discrete release sites. SIAM J. Appl. Math. **61**, 317–314 (2000)
320. Keener, J.P.: Stochastic calcium oscillations. Math. Med. Biol. **23**, 1–25 (2006)
321. Keener, J.P., Newby, J.M.: Perturbation analysis of spontaneous action potential initiation by stochastic ion channels. Phy. Rev. E **84**, 011918 (2011)
322. Keener, J., Sneyd, J.: Mathematical Physiology I: Cellular Physiology, 2nd edn. Springer, New York (2009)
323. Keener, J.P., Hoppensteadt, F.C., Rinzel, J.: Integrate–and–fire models of nerve membrane response to oscillatory input. SIAM J. Appl. Math. **41**(3), 503–517 (1981)
324. Keizer, J., Smith, G.D.: Spark-to-wave transition: saltatory transmission of calcium waves in cardiac myocytes. Biophys. Chem. **72**, 87–100 (1998)
325. Keizer, J., Smith, G.D., Ponce-Dawson, S., Pearson, J.E.: Saltatory propagation of Ca^{2+} waves by Ca^{2+} sparks. Biophys. J. **75**, 595–600 (1998)

326. Khazipov, R., Luhmann, H.J.: Early patterns of electrical activity in the developing cerebral cortex of humans and rodents. Trend. Neurosci. **29**, 414–419 (2006)
327. Khovanov, I.A., Polovinkin, A.V., Luchinsky, D.G., McClintock, P.V.E.: Noise-induced escape in an excitable system. Phys. Rev. E **87**, 032116 (2013)
328. Kilpatrick, Z.P., Bressloff, P.C.: Binocular rivalry in a competitive neural network with synaptic depression. SIAM J. Appl. Dyn. Syst. **9**, 1303–1347 (2010)
329. Kilpatrick, Z.P., Bressloff, P.C.: Effects of synaptic depression and adaptation on spatiotemporal dynamics of an excitatory neuronal network. Physica D **239**, 547–560 (2010)
330. Kilpatrick, Z.P., Bressloff, P.C.: Spatially structured oscillations in a two-dimensional neuronal network with synaptic depression. J. Comp. Neurosci. **28**, 193–209 (2010)
331. Kilpatrick, Z.P., Bressloff, P.C.: Stability of bumps in piecewise smooth neural fields with nonlinear adaptation. Physica D **239**, 1048–1060 (2010)
332. Kilpatrick, Z.P., Folias, S.E., Bressloff, P.C.: Traveling pulses and wave propagation failure in inhomogeneous neural media. SIAM J. Appl. Dyn. Syst. **7**, 161–185 (2008)
333. Kim, H.G., Connors, B.W.: Apical dendrites of the neocortex: correlation between sodium and calcium-dependent spiking and pyramidal cell morphology. J. Neurosci. **13**, 5301–5311 (1993)
334. Kim, U., Bal, T., McCormick, D.A.: Spindle waves are propagating synchronized oscillations in the ferret LGN in vitro. J. Neurophysiol. **74**, 1301–1323 (1995)
335. Kinezaki, N., Kawasaki, K., Takasu, F., Shigesada, N.: Modeling biological invasions into periodically fragmented environments. Theor. Popul. Biol. **64**, 291–302 (2003)
336. Kishimoto, K., Amari, S.: Existence and stability of local excitations in homogeneous neural fields. J. Math. Biol. **7**, 303–318 (1979)
337. Kistler, W.M., Seitz, R., van Hemmen, J.L.: Modeling collective excitations in cortical tissue. Physica D **114**, 273–295 (1998)
338. Kleinfeld, D., Delaney, K.R., Fee, M.S., Flores, J.A., Tank, D.W., Gelperin, A.: Dynamics of propagating waves in the olfactory network of a terrestrial mollusc: an electrical and optical study. J. Neurophysiol. **72**, 1402–1419 (1994)
339. Knessl, C., Mangel, M., Matkowsky, B.J., Schuss, Z., Tier, C.: An asymptotic solution of the Kramers-Moyal equation and first passage time problems for Markov jump processes. Phys. Rev. E **29**, 3359–3369 (1984)
340. Knessl, C., Matkowsky, B.J., Schuss, Z., Tier, C.: An asymptotic theory of large deviations for Markov jump processes. SIAM J. Appl. Math. **46**, 1006–1028 (1985)
341. Koch, C.: Cable theory in neurons with active, linearized membranes. Biol. Cybern. **50**, 15–33 (1984)
342. Koch, C.: Biophysics of Computation. Oxford University Press, New York (1999)
343. Koch, C., Zador, A.M.: The function of dendritic dendritic spines. devices subserving biochemical rather than electrical compartmentalization. J. Neurosci. **13**, 413–422 (1993)
344. Koch, C., Poggio, T., Torre, V.: Nonlinear interactions in a dendritic tree: localization, timing and role in information processing. Proc. Nat. Acad. Sci. (USA) **80**, 2799–2802 (1983)
345. Kolmogorff, A., Petrovsky, I., Piscounoff, N.: Étude de l'équation de la diffusion avec croissance de la quantité de matière et son application à un problème biologique. Moscow Univ. Bull. Math. **1**, 1–25 (1937)
346. Konur, S., Rabinowitz, D., Fenstermaker, V.L., Yuste, R.: Systematic regulation of spine sizes and densities in pyramidal neurons. J. Neurobiol. **56**, 95–112 (2003)
347. Kopell, N., Ermentrout, G.B.: Symmetry and phase-locking in chains of weakly coupled oscillators. Comm. Pure Appl. Math. **39**, 623–660 (1986)
348. Kramer, M.A., Cash, S.S.: Epilepsy as a disorder of cortical network organization. Neuroscientist **18**, 360–372 (2012)
349. von Krosigk, M., Bal, T., McCormick, D.: Cellular mechanisms of a synchronized oscillation in the thalamus. Science **261**, 361–364 (1993)
350. Kuramoto, Y.: Chemical Oscillations, Waves and Turbulence. Springer, New-York (1984)
351. Kurtz, T.G.: Limit theorems for a sequence of jump Markov processes approximating ordinary differential equations. J. Appl. Prob. **8**, 344–356 (1971)

352. Kurtz, T.G.: Limit theorems and diffusion approximations for density dependent Markov chains. Math. Prog. Stud. **5**, 67 (1976)
353. Lago-Fernandez, L.F., Huerta, R., Corbacho, F., Siguenza, J.A.: Fast response and temporal coherent oscillations in small-world networks. Phys. Rev. Lett. **84**, 2758–2761 (2000)
354. Laing, C.R.: Spiral waves in nonlocal equations. SIAM J. Appl. Dyn. **4**, 588–606 (2005)
355. Laing, C.R., Chow, C.C.: A spiking neuron model for binocular rivalry. J. Comput. Neurosci. **12**, 39–53 (2002)
356. Laing, C.R., Troy, W.C.: PDE methods for nonlocal models. SIAM J. Appl. Dyn. Syst. **2**, 487–516 (2003)
357. Lam, Y.W., Cohen, L.B., Wachowiak, M., Zochowski, M.R.: Odors elicit three different oscillations in the turtle olfactory bulb. J. Neurosci. **20**, 749–762 (2000)
358. Lance, J.W.: Current concepts of migraine pathogenesis. Neurology **43**, 11–15 (1993)
359. Largo, C., Cuevas, P., Somjen, G., Rio, R.M., Herreras, O.: The effect of depressing glial function in rat brain in situ on ion homeostasis, synaptic transmission, and neuronal survival. J. Neurosci. **16**, 1219–1229 (1996)
360. Largo, C., Ibarz, J.M., Herreras, O.: Effects of the gliotoxin fluorocitrate on spreading depression and glial membrane potential in rat brain in situ. J. Neurophysiol. **78**, 295–307 (1997)
361. Larkum, M.E., Zhu, J.J., Sakmann, B.: A new cellular mechanism for coupling inputs arriving at different cortical layers. Nature **398**, 338–341 (1999)
362. Larkum, M.E., Nevian, T., Sandler, M., Polsky, A., Schiller, J.: Synaptic integration in tuft dendrites of layer 5 pyramidal neurons: a new unifying principle. Science **325**, 756–760 (2009)
363. Lashley, K.: Patterns of cerebral integration indicated by scotomas of migraine. Arch. Neurol. Pyschiatry **46**, 331–339 (1941)
364. Lauritsen, K.B., Zapperi, S., Stanley, H.E.: Self-organized branching processes: avalanche models with dissipation. Phys. Rev. E **54**, 2483–2488 (1996)
365. Leao, A.A.P.: Spreading depression of activity in the cerebral cortex. J. Neurophysiol. **10**, 409–414 (1944)
366. Leao, A.A.P., Morrisson, R.S.: Propagation of spreading cortical depression. J. Neurophysiol. **8**, 33–45 (1945)
367. Lechleiter, J., Girard, S., Peralta, E., Clapham, D.: Spiral calcium wave propagation and annihilation in *xenopus laevis* oocytes. Science **252**, 123–126 (1991)
368. Leclerc, C., Neant, I., Moreau, M.: The calcium: an early signal that initiates the formation of the nervous system during early embryogenesis. Front. Mol. Neurosci. **5 (64)**, 1–12 (2012)
369. Lee, S.H., Blake, R., Heeger, D.J.: Traveling waves of activity in primary visual cortex during binocular rivalry. Nat. Neurosci. **8**, 22–23 (2005)
370. Lee, S.J., Escobedo-Lozoya, Y., Szatmari, E.M., Yasuda, R.: Activation of CaMKII in single dendritic spines during long-term potentiation. Nature **458**, 299–304 (2009)
371. Lee, S.J., Lim, H.S., Masliah, E., Lee, H.J.: Protein aggregate spreading in neurodegerative diseases: problems and perspectives. Neuro. Res. **70**, 339–348 (2011)
372. Lehky, S.R.: Binocular rivalry is not chaotic. Proc. Roy. Soc. Lond. Biol Sci. **259**, 71–76 (1995)
373. Lehky, S.R.: No binocular rivalry in the LGN of alert macaque monkeys. Vis. Res. **36**, 1225–1234 (1996)
374. Lehner, F.K.: On the validity of Fick's law for transient diffusion through a porous medium. Chem. Eng. Sci. **34**, 821–825 (1977)
375. Lemieux, L., Daunizeau, J., Walker, M.C.: Concepts of connectivity and human epileptic activity. Front. Syst. Neurosci. **5 (12)**, 1–13 (2011)
376. Leopold, D.A., Logothetis, N.K.: Activity changes in early visual cortex reflect monkeys' percepts during binocular rivalry. Nature **379**, 549–553 (1996)
377. LeVay, S., Nelson, S.B.: Columnar organization of the visual cortex. In: Leventhal, A.G. (ed.) The Neural Basis of Visual Function, pp. 266–315. CRC Press, Boca Raton (1991)
378. Levelt, W.J.M.: On Binocular Rivalry. Institute for Perception RVO–TNO, Soesterberg, The Netherlands (1965)
379. Leybaert, L., Paemeleire, K., Strahonja, A., Sanderson, M.J.: Inositol-trisphosphate-dependent intercellular calcium signaling in and between astrocytes and endothelial cells. Glia **24**, 398–407 (1998)

380. Leybaert, L., Sanderson, M.J.: Intercellular Ca^{2+} waves: mechanisms and function. Physiol. Rev. **92**, 1359–1392 (2012)
381. Li, Y., Rinzel, J.: Equations for $InsP_3$ receptor-mediated calcium oscillations derived from a detailed kinetic model: a Hodgkin-Huxley like formalism. J. Theor. Biol. **166**, 461–473 (1994)
382. Liley, D.J.T., Cadusch, P.J., Dafilis, M.P.: A spatially continuous mean field theory of electrocortical activity. Network **13**, 67–113 (2002)
383. Linder, B., Garcia-Ojalvo, J., Neiman, A., Schimansky-Geier, L.: Effects of noise in excitable systems. Phys. Rep. **392**, 321–424 (2004)
384. Lisman, J.E.: Relating hippocampal circuitry to function: recall of memory sequences by reciprocal dentate-CA3 interactions. Neuron **22**, 233–242 (1999)
385. Lisman, J.: Long-term potentiation: outstanding questions and attempted synthesis. Phil. Trans. R. Soc. Lond. B **358**, 829–842 (2003)
386. Lisman, J.E., Schulman, H., Cline, H.: The molecular basis of CaMKII function in synaptic and behavioral memory. Nat. Rev. Neurosci. **3**, 175–190 (2002)
387. Litwin-Kumar, A., Doiron, B.: Slow dynamics and high variability in balanced cortical networks with clustered connections. Nat. Neurosci. **15**, 1498–1505 (2012)
388. Liu, Q.S., Xu, Q., Kang, G.A.J., Nedergaard, M.: Astrocyte-mediated activation of neuronal kainate receptors. Proc. Natl. Acad Sci. USA **101**, 3172–3177 (2004)
389. Llano, I., González, J., Caputo, C., Lai, F.A., Blayney, L.M., Tan, Y.P., Marty, A.: Presynaptic calcium stores underlie large-amplitude miniature IPSCs and spontaneous calcium transients. Nat. Neurosci. **3**, 1256–1265 (2000)
390. Logothetis, N.K., Leopold, D.A., Sheinberg, D.L.: What is rivalling during binocular rivalry? Nature **380**, 621–624 (1996)
391. London, M., Roth, A., Beeren, L., Hausser, M., Latham, P.E.: Sensitivity to perturbations in vivo implies high noise and suggests rate coding in cortex. Nature **466**, 123–127 (2010)
392. Loxley, P., Robinson, P.: Soliton model of competitive neural dynamics during binocular rivalry. Phys. Rev. Lett. **102**(25), 258701 (2009)
393. Lubenov, E.V., Siapas, A.G.: Hippocampal theta oscillations are travelling waves. Nature **459**, 534–539 (2009)
394. Luo, L.: Actin cytoskeleton regulation in neuronalmorphogenesis and structural plasticity. Ann. Rev. Cell Dev. Biol. **18**, 601–635 (2002)
395. Ly, C., Tranchina, D.: Critical analysis of a dimension reduction by a moment closure method in a population density approach to neural network modeling. Neural Comput. **19**, 2032–2092 (2007)
396. Madison, D.V., Nicoll, R.A.: Control of the repetitive discharge of rat CA1 pyramidal neurones in vitro. J. Physiol. **354**, 319–331 (1984)
397. Magee, J.C., Johnston, D.: Characterization of single voltage-gated Na^{+} and Ca^{2+} channels in apical dendrites of rat CA1 pyramidal neurons. J. Physiol. **487**, 67–90 (1995)
398. Maier, R.S., Stein, D.L.: Limiting exit location distribution in the stochastic exit problem. SIAM J. Appl. Math **57**, 752–790 (1997)
399. Maier, N., Nimmrich, V., Draguhn, A.: Cellular and network mechanisms underlying spontaneous sharp wave–ripple complexes in mouse hippocampal slices. J. Physiol. **550**, 873–887 (2003)
400. Major, G., Polsky, A., Denk, W., Schiller, J., Tank, D.W.: Spatiotemporally graded NMDA spike/plateau potentials in basal dendrites of neocortical pyramidal neurons. J. Neurophysiol. **99**, 2584–2601 (2008)
401. Malach, R., Harel, Y.A.M., Grinvald, A.: Relationship between intrinsic connections and functional architecture revealed by optical imaging and in vivo targeted biocytin injections in primate striate cortex. Proc. Natl. Acad. Sci. USA **90**, 0469–10473 (1993)
402. Malenka, R.C., Bear, M.F.: LTP and LTD: an embarrassment of riches. Neuron **44**, 5–21 (2004)
403. Manita, S., Ross, W.N.: Synaptic activation and membrane potential changes modulate the frequency of spontaneous elementary Ca^{2+} releases events in the dendrites of pyramidal neurons. J. Neurosci. **29**, 7833–7845 (2009)

404. Marder, E., Calabrese, R.L.: Principles of rhythmic motor pattern generation. Physiol. Rev. **76**, 687–717 (1996)
405. Markram, H., Tsodyks, M.: Redistribution of synaptic efficacy between neocortical pyramidal neurons. Nature **382**, 807–810 (1996)
406. Martins-Ferreira, H., Ribeiro, I.J.: Biphasic effects of gap junctional uncoupling agents on the propagation of retinal spreading depression. Braz. J. Med. Biol. Res. **28**, 991–994 (1995)
407. Martins-Ferreira, H., Nedergaard, M., Nicholson, C.: Perspectives on spreading depression. Brain Res. Rev. **32**, 215–234 (2000)
408. Masel, J., Jansen, V.A.A., Nowak, M.A.: Quantifying the kinetic parameters of prion replication. Biophys. Chem. **77**, 139–15 (1999)
409. Masuda, N., Aihara, K.: Global and local synchrony of coupled neurons in small-world networks. Biol. Cybern. **90**, 302–309 (2004)
410. Mataga, N., Mizuguchi, Y., Hensch, T.K.: Experience-dependent pruning of dendritic spines in visual cortex by tissue plasminogen activator. Neuron **44**, 1031–1041 (2004)
411. Matthaus, F.: Diffusion versus network models as descriptions for the spread of prion diseasess in the brain. J. Theor. Biol. **240**, 104–113 (2006)
412. Mattia, M., Guidice, P.D.: Population dynamics of interacting spiking neurons. Phys. Rev. E **66**, 051917 (2002)
413. Matus, A.: Actin plasticity in dendritic spines. Science **290**, 754–758 (2000)
414. Mayer, M.L., Westbrook, G.L., Guthrie, P.B.: Voltage-dependent block by Mg2+ of NMDA responses in spinal cord neurone. Nature **309**, 261–263 (1984)
415. McCormick, D.A., Contreras, D.: On the cellular and network bases of epileptic seizures. Annu. Rev. Physiol. **63**, 815–846 (2001)
416. McLaughlin, T., Torborg, C.L., Feller, M.B., O'Leary, D.D.M.: Retinotopic map refinement requires spontaneous retinal waves during a brief critical period of development. Neuron **40**, 1147–1160 (2003)
417. Meinhardt, H.: Orientation of chemotactic cells and growth cones: models and mechanisms. J. Cell. Sci. **112**, 2867–2874 (1999)
418. Meisel, C., Storch, A., Hallmeyer-Elgner, S., Bullmore, E., Gross, T.: Failure of adaptive self-organized criticality during epileptic seizure attacks. PLoS Comp. Biol. **8**, e1002312 (2012)
419. Melamed, O., Barak, O., Silberberg, G., Markram, H., Tsodyks, M.: Slow oscillations in neural networks with facilitating synapses. J. Comp. Neurosci. **25**, 308–316 (2008)
420. Menchon, S.A., Gartner, A., Roman, P., Dotti, C.G.: Neuronal (bi)polarity as a self-organized process enhanced by growing membrane. PLoS One **6**(9), e24190 (2011)
421. Mendez, V., Fort, J., Rotstein, H.G., Fedotov, S.: Speed of reaction-diffusion fronts in spatially heterogeneous media. Phys. Rev. E **68**, 041105 (2003)
422. Mendez, V., Fedotov, S., Horsthemke, W.: Reaction-Transport Systems. Springer, Berlin (2010)
423. Meng, X., Huguet, G., Rinzel, J.: Type III excitability, slope sensitivity and coincidence detection. Disc. Cont. Dyn. Systs, **32**, 2729–2757 (2012)
424. Meunier, C., d'Incamps, B.L.: Extending cable theory to heterogeneous dendrites. Neural Comput. **20**, 1732–1775 (2008)
425. Meyer, C., van Vreeswijk, C.: Temporal correlations in stochastic networks of spiking neurons. Neural Comput. **14**, 369–404 (2002)
426. Mikhailov, A.S., Davydov, V.A., Zykov, V.S.: Complex dynamics of spiral waves and motion of curves. Physica D **70**, 1–39 (1994)
427. Miller, J.P., Rall, W., Rinzel, J.: Synaptic amplification by active membrane in dendritic spines. Brain Res. **325**, 325–330 (1985)
428. Miller, L.D., Golarai, J.J.P., Connor, J.: Ca^{2+} release from intracellular stores induced by afferent stimulation of CA3 pyramidal neurons in hippocampal slices. J. Neurophysiol. **76**, 554–562 (1996)
429. Millhauser, G.L., Salpeter, E.E., Oswald, R.E.: Diffusion model of ion–channel gating and the origin of power–law distributions from single–channel recordings. Proc. Natl. Acad. Sci. USA **85**, 1503–1507 (1988)

430. Milton, J., Jung, P.: Epilepsy as a Dynamic Disease. Springer, Berlin (2003)
431. Mirollo, R.E., Strogatz, S.H.: Synchronisation of pulse–coupled biological oscillators. SIAM J. Appl. Math. **50**(6), 1645–1662 (1990)
432. Mitchison, T., Kirschner., M.: Cytoskeletal dynamics and nerve growth. Neuron **1**, 761–772 (1988)
433. Miura, R., Huang, H., Wylie, J.: Cortical spreading depression: An enigma. Eur. Phys. J. Special Topics **147**, 287–302 (2007)
434. Mogilner, A., Oster, G.: Cell motility driven by actin polymerization. Biophys. J. **71**, 3030–3045 (1996)
435. Moldakarimov, S., Rollenhagen, J.E., Olson, C.R., Chow, C.C.: Competitive dynamics in cortical responses to visual stimuli. J. Neurophysiol. **94**, 3388–3396 (2005)
436. Moreno-Bote, R., Rinzel, J., Rubin, N.: Noise-induced alternations in an attractor network model of perceptual bistability. J. Neurophysiol. **98**, 1125–1139 (2007)
437. Morgan, R.J., Soltesz, I.: Nonrandom connectivity of the epileptic dentate gyrus predicts a major role for neuronal hubs in seizures. Proc. Natl. Acad. Sci. USA **105**, 6179–6184 (2008)
438. Mori, Y., Jilkine, A., Edelstein-Keshet, L.: Wave-pinning and cell polarity from a bistable reaction-diffusion system. Biophys. J. **94**, 3684–3697 (2008)
439. Mori, Y., Jilkine, A., Edelstein-Keshet, L.: Asymptotic and bifurcation analysis of wave-pinning in a reaction-diffusion model for cell polarization. SIAM J. Appl. Math. **71**, 1401–1427 (2011)
440. Morris, C., Lecar, H.: Voltage oscillations in the barnacle giant muscle fiber. J. Biophys. **35**, 193–213 (1981)
441. Mulleners, W.M., Chronicle, E.P., Palmer, J.E., Koehler, P.J., Vredeveld, J.W.: Visual cortex excitability in migraine with and without aura. Headache **41**, 565–572 (2001)
442. Muller, L., Destexhe, A.: Propagating waves in thalamus, cortex and the thalamocortical system: experiments and models. J. Physiol. (Paris) **106**, 222–238 (2012)
443. Mullins, O.J., Hackett, J.T., Buchanan, J.T., Friesen, W.O.: Neuronal control of swimming behavior: Comparison of vertebrate and invertebrate model systems. Prog. Neurobiol. **93**, 244–269 (2011)
444. Murray, J.D.: Mathematical Biology, vol. I, II. Springer, Berlin (2002)
445. Naeh, T., Klosek, M.M., Matkowsky, B.J., Schuss, Z.: A direct approach to the exit problem. SIAM J. Appl. Math. **50**, 595–627 (1990)
446. Nagumo, J., Armoto, S., Yoshizawa, S.: An active pulse transmission line simulating nerve axon. Proc. Inst. Radio Eng. **50**, 2061–2070 (1964)
447. Nakamura, T., Barbara, J.G., Nakamura, K., Ross, W.N.: Synergistic release of Ca^{2+} from IP_3 sensitive stores evoked by synaptic activation of mGluRs paired with backpropagating action potentials. Neuron **24**, 727–737 (1999)
448. Nedergaard, M., Cooper, A.J., Goldman, S.A.: Gap junctions are required for the propagation of spreading depression. J. Neurobiol. **28**, 433–444 (1995)
449. Nedergaard, M., Ransom, B., Goldman, S.A.: New roles for astrocytes: redefining the functional architecture of the brain. Trend. Neurosci. **26**, 523–530 (2003)
450. Netoff, T.I., Schiff, S.J.: Decreased neuronal synchronization during experimental seizures. J. Neurosci. **22**, 7297–7307 (2002)
451. Netoff, T.I., Clewely, R., Arno, S., Keck, T., White, J.A.: Epilepsy in small-world networks. J. Neurosci. **24**, 8075–8083 (2004)
452. Neukirchen, D., Bradke, F.: Neuronal polarization and the cytoskeleton. Sem. Cell. Dev. Biol. **22**, 825–833 (2011)
453. Newby, J.M., Bressloff, P.C.: Local synaptic signalling enhances the stochastic transport of motor-driven cargo in neurons. Phys. Biol. **7**, 036004 (2010)
454. Newby, J.M., Bressloff, P.C.: Quasi-steady state reduction of molecular-based models of directed intermittent search. Bull. Math. Biol. **72**, 1840–1866 (2010)
455. Newby, J., Chapman, J.: Metastable behavior in Markov processes with internal states: breakdown of model reduction techniques. J. Math. Biol. published online, DOI: 10.1007/s00285-013-0723-1 (2013)

456. Newby, J.M., Bressloff, P.C., Keeener, J.P.: The effect of potassium channels on spontaneous action potential initiation by stochastic ion channels. Phys. Rev. Lett. **111**, 128–101 (2013)
457. Newman, E.A.: Propagation of intercellular calcium waves in retinal astrocytes and Muller cells. J. Neurosci. **21**, 2215–2223 (2001)
458. Newman, M.E.J.: Spread of epidemic disease on networks. Phys. Rev. E **66**, 016128 (2002)
459. Newman, M.E.J.: The structure and function of complex networks. SIAM Rev. **45**, 167–256 (2003)
460. Newman, E.A., Zahs, K.R.: Calcium waves in retinal glial cells. Science **275**, 844–847 (1997)
461. Nicholson, C.: Diffusion and related transport mechanisms in brain tissue. Rep. Prog. Phys. **64**, 815–884 (2001)
462. Nimchinsky, E.A., Sabatini, B.L., Svoboda, K.: Structure and function of dendritic spines. Ann. Rev. Physiol. **64**, 313–353 (2002)
463. Nita, D.A., Cisse', Y., Timofeev, I., Steriade, M.: Increased propensity to seizures after chronic cortical deafferentation in vivo. J. Neurophysiol. **95**, 902–913 (2006)
464. Noble, J.V.: Geographic and temporal development of plagues. Nature **250**, 726–729 (1974)
465. Novikov, E.A.: Functionals and the random-force method in turbulence theory. Sov. Phys. JETP **20**, 1290 (1965)
466. Nowak, L., Bregestovski, P., Ascher, P., Herbet, A., Prochiantz, A.: Magnesium gates glutamate-activated channels in mouse central neurones. Nature **307**, 462–465 (1984)
467. Nunez, P.I.: Neocortical dynamics and human EEG rhythms. Oxford University Press, New York (1995)
468. Nykamp, D., Tranchina, D.: A population density method that facilitates large–scale modeling of neural networks: analysis and application to orientation tuning. J. Comp. Neurosci. **8**, 19–50 (2000)
469. Obermayer, K., Blasdel, G.G.: Geometry of orientation and ocular dominance columns in monkey striate cortex. J. Neurosci. **13**, 4114–4129 (1993)
470. O'Donnell, M., Chance, R.K., Bashaw, G.J.: Axon growth and guidance: Receptor regulation and signal transduction. Ann. Rev. Neurosci. **32**, 383–412 (2009)
471. O'Donovan, M.J.: The origin of spontaneous activity in developing networks of the vertebrate nervous system. Curr. Opin. Neuro. **9**, 94–104 (1999)
472. O'Donovan, M.J., Chub, N., Wenner, P.: Mechanisms of spontaneous activity in the developing retina. J. Neurobiol. **37**, 131–145 (1998)
473. Ohira, T., Cowan, J.D.: Stochastic neurodynamics and the system size expansion. In: Ellacott, S., Anderson, I.J. (eds.) Proceedings of the First International Conference on Mathematics of Neural Networks, pp. 290–294. Academic, New York (1997)
474. O'Keefe, J., Dostrovsky, J.: The hippocampus as a spatial map. preliminary evidence from unit activity in the freely moving rat. Brain Res. **34**, 171–174 (1971)
475. O'Keefe, J., Recce, M.L.: Phase relationship between hippocampal place units and the EEG theta rhythm. Hippocampus **3**, 317–330 (1993)
476. Oksendal, B.: Stochastic Differential Equations: An Introduction with Applications. Springer, New York (2007)
477. Omurtag, A., Knight, B.W., Sirovich, L.: On the simulation of large populations of neurons. J. Comput. Neurosci. **8**, 51–63 (2000)
478. Onsum, M., Rao, C.V.: A mathematical model for neutrophil gradient sensing and polarization. PLoS Comput. Biol. **3**, e36 (2007)
479. Ooyen, A.V., Graham, B.P., Ramakers, G.J.A.: Competition for tubulin between growing neurites during development. Neurocomputing **38**, 73–78 (2001)
480. Oray, S., Majewska, A., Sur, M.: Dendritic spine dynamics are regulated by monocular deprivation and extracellular matrix degredation. Neuron **44**, 1021–1030 (2004)
481. Othmer, H., Dunbar, S., Alt, W.: Models of dispersal in biological systems. J. Math. Biol. **26**, 263–298 (1988)
482. Otsuji, M., Ishihara, S., Co, C., Kaibuchi, K., Mochizuki, A., Kuroda, S.: A mass conserved reaction-diffusion system captures properties of cell polarity. PLoS Comput. Biol. **3**, e108 (2007)

483. Pak, C.W., Flynn, K.C., Bamburg, J.R.: Actin-binding proteins take the reins in growth cones. Nat. Rev. Neurosci. **9**, 136–147 (2008)
484. Pakdaman, K., Thieullen, M., Wainrib, G.: Fluid limit theorems for stochastic hybrid systems with application to neuron models. J. Appl. Prob. **24**, 1 (2010)
485. Panatier, A., Theodosis, D.T., Mothet, J.P., Touquet, B., Pollegioni, L., Poulain, D.A., Oliet, S.H.: Glia-derived D-serine controls NMDA receptor activity and synaptic memory. Cell **125**, 775–784 (2006)
486. Panja, D.: Effects of fluctuations on propagating fronts. Phys. Rep. **393**, 87–174 (2004)
487. Papanicolaou, G.C.: Asymptotic analysis of transport processes. Bull. Amer. Math. Soc. **81**, 330–392 (1975)
488. Parga, N., Abbott, L.F.: Network model of spontaneous activity exhibiting synchronous transitions between up and down states. Front. Neurosci. **1**, 57–66 (2007)
489. Parker, I., Ivorra, I.: Localized all-or-none calcium liberation by inositol triphosphate. Science **250**, 977–979 (1990)
490. Parker, I., Yao, Y.: Regenerative release of calcium from functionally discrete subcellular stores by inositol triphosphate. Proc. Roy. Soc. Lond. B **246**, 269–274 (1991)
491. Parpura, V., Basarsky, T.A., Liu, F., Jeftinija, K., Jeftinija, S., Haydon, P.G.: Glutamate-mediated astrocyte-neuron signalling. Nature **369**, 744–747 (1994)
492. Parri, H.R., Gould, T.M., Crunelli, V.: Spontaneous astrocytic Ca^{2+} oscillations in situ drive NMDAR-mediated neuronal excitation. Nat. Neurosci. **4**, 803–81 (2001)
493. Pascual, O., Casper, K.B., Kubera, C., Zhang, J., Revilla-Sanchez, R., Sul, J.Y., Takano, H., Moss, S.J., McCarthy, K., Haydon, P.G.: Astrocytic purinergic signaling coordinates synaptic networks. Science **310**, 113–116 (2005)
494. de Pasquale, F., Gorecki, J., Poielawski., J.: On the stochastic correlations in a randomly perturbed chemical front. J. Phys. A **25**, 433 (1992)
495. Paullet, J.E., Ermentrout, G.B.: Stable rotating waves in two-dimensional discrete active media. SIAM J. Appl. Math. **54**, 1720–1744 (1994)
496. Pavliotis, G.A., Stuart, A.M.: Multiscale Methods: Averaging and Homogenization. Springer, New York (2008)
497. Pearson, J.E., Ponce-Dawson, S.: Crisis on skid row. Physica A **257**, 141–148 (1998)
498. Pearson, R., Catsicas, M., Becker, D., Mobbs, P.: Purinergic and muscarinic modulation of the cell cycle and calcium signaling in the chick retinal ventricular zone. J. Neurosci. **22**, 7569–7579 (2002)
499. Pearson, R.A., Dale, N., Llaudet, E., Mobbs, P.: ATP released via gap junction hemichannels from the pigment epithelium regulates neural retinal progenitor proliferation. Neuron **46**, 731–744 (2005)
500. Pereda, E., Quiroga, R., Bhattacharya, J.: Nonlinear multivariate analysis of neurophysiological signals. Prog. Neurobiol. **77**, 1–37 (2005)
501. Peters, O., Schipke, C.G., Hashimoto, Y., Kettenmann, H.: Different mechanisms promote astrocyte Ca^{2+} waves and spreading depression in the mouse neocortex. J. Neurosci. **23**, 9888–9896 (2003)
502. Pikovsky, A., Rosenblum, M., Kurths, J.: Synchronization: A Universal Concept in Nonlinear Sciences. Cambridge University Press, Cambridge (2003)
503. Pinder, G.F., Gray, W.G.: Essentials of Multiphase Flow and Transport in Porous Media. Wiley, New York (2008)
504. Pinto, D., Ermentrout, G.B.: Spatially structured activity in synaptically coupled neuronal networks I: traveling fronts and pulses. SIAM J. Appl. Math **62**, 206–225 (2001)
505. Pinto, D., Ermentrout, G.B.: Spatially structured activity in synaptically coupled neuronal networks II: lateral inhibition and standing pulses. SIAM J. Appl. Math. **62**, 226–243 (2001)
506. Pinto, D., Jackson, R.K., Wayne, C.E.: Existence and stability of traveling pulses in a continuous neuronal network. SIAM J. Appl. Dyn. Syst. **4**, 954–984 (2005)
507. Pinto, D., Patrick, S.L., Huang, W.C., Connors, B.W.: Initiation, propagation, and termination of epileptiform activity in rodent neocortex in vitro involve distinct mechanisms. J. Neurosci. **25**, 8131–8140 (2005)
508. Plenz, D., Thiagarajan, T.C.: The organizing principles of neuronal avalanches: cell assemblies in the cortex? Trend. Neurosci. **30**, 101–110 (2007)

509. Polleux, F., Snider, W.: Initiating and growing an axon. Cold Spring Harb Perspect Biol **2**, a001925 (2010)
510. Polonsky, A., Blake, R., Braun, J., Heeger, D.J.: Neuronal activity in human primary visual cortex correlates with perception during binocular rivalry. Nat. Neurosci. **3**, 1153–1159 (2000)
511. Ponce, S.J., Keizer, J., Pearson, J.E.: Fire-diffuse-fire model of dynamics of intracellular calcium waves. Proc. Nat. Acad. Sci. USA **96**, 6060–6063 (1999)
512. Prechtl, J.C., Cohen, L.B., Pesaran, B., Mitra, P.P., Kleinfeld, D.: Visual stimuli induce waves of electrical activity in turtle cortex. Proc. Natl. Acad. Sci. USA **94**, 7621–7626 (1997)
513. Prusiner, S.B.: Prions. Proc. Nat. Acad. Sci. (USA) **95**, 13363–13383 (1998)
514. Rakic, P.: Elusive radial glial cells: historical and evolutionary perspective. Glia **43**, 19–32 (2003)
515. Rall, W., Snir, H.A.: Cable theory for dendritic neurons. In: Koch, C., Segev, I. (eds.) Methods in Neuronal Modelling, 2nd edn. MIT Press, Cambridge (1998)
516. Ramirez, O.A., Couve, A.: The endoplasmic reticulum and protein trafficking in dendrites and axons. Trend. Cell. Biol. **21**, 219–227 (2011)
517. Rangan, A.V., Kovacic, G., Cai, D.: Kinetic theory for neuronal networks with fast and slow excitatory conductances driven by the same spike train. Phys. Rev. E **77**, 041915 (2008)
518. Reed, M.C., Venakides, S., Blum, J.J.: Approximate traveling waves in linear reaction-hyperbolic equations. SIAM J. Appl. Math. **50**(1), 167–180 (1990)
519. Renart, A., de la Rocha, J., Bartho, P., Hollender, L., Parga, N., Reyes, A., Harris, K.D.: The asynchronous state in cortical circuits. Science **327**, 587–590 (2010)
520. Rhodes, P.: The properties and implications of NMDA spikes in neocortical pyramidal cells. J. Neurosci. **26**, 6704–6715 (2006)
521. Richardson, K.A., Schiff, S.J., Gluckman, B.J.: Control of traveling waves in the mammalian cortex. Phys. Rev. Lett. **94**, 028103 (2005)
522. Richardson, M.P.: Large scale brain models of epilepsy: dynamics meets connectomics. Epilepsy (2012)
523. Rinzel, J., Maginu, K.: Kinematic analysis of wave pattern formation in excitable media. In: Pacault, A., Vidal, C. (eds.) Non-Equilibrium Dynamics in Chemical Systems. Springer, Berlin (1984)
524. Rinzel, J., Terman, D.: Propagation phenomena in a bistable reaction–diffusion system. SIAM J. Appl. Math. **42**, 1111–1137 (1982)
525. Rinzel, J., Terman, D., Wang, X., Ermentrout, B.: Propagating activity patterns in large-scale inhibitory neuronal networks. Science **279**, 1351–1355 (1998)
526. Robinson, P.A.: Patchy propagator, brain dynamics, and the generation of spatially structured gamma oscillations. Phys. Rev. E **73**, 041904 (2006)
527. Robinson, P.A., Rennie, C.J., Wright, J.J.: Propagation and stability of waves of electrical activity in the cerebral cortex. Phys. Rev. E **56**, 826–840 (1997)
528. Robinson, P.A., Rennie, C.J., Wright, J.J., Bahramali, H., Gordon, E., Rowe, D.I.: Prediction of electroencephalographic spectra from neurophysiology. Phys. Rev. E **63**, 021903 (2001)
529. Robinson, P.A., Rennie, C.J., Rowe, D.L.: Dynamics of large-scale brain activity in normal arousal states and epileptic seizures. Phys. Rev. E **65**, 041924 (2002)
530. Rocco, A., Ebert, U., van Saarloos, W.: Subdiffusive fluctuations of "pulled" fronts with multiplicative noise. Phys. Rev. E. **65**, R13–R16 (2000)
531. Rockland, K.S., Lund, J.: Intrinsic laminar lattice connections in primate visual cortex. J. Comp. Neurol. **216**, 303–318 (1983)
532. Rose, J., Jin, S.X., Craig, A.M.: Heterosynaptic molecular dynamics: locally induced propagating synaptic accumulation of CaM Kinase II. Neuron **61**, 351–358 (2009)
533. Ross, C.A., Poirier, M.A.: Protein aggregation and neurodegenerative disease. Nat. Med. **10**(Suppl.), S10–S17 (2004)
534. Ross, W.N.: Understanding calcium waves and sparks in central neurons. Nat. Rev. Neurosci. **13**, 157–168 (2012)
535. Roxin, A., Riecke, H., Solla, S.A.: Self-sustained activity in a small-world network of excitable neurons. Phys. Rev. Lett. **92**, 198101 (2004)

536. Rubin, J.E.: A nonlocal eigenvalue problem for the stability of a traveling wave in a neuronal medium. Discret. Contin. Dyn. Syst. **10**, 925–940 (2004)
537. Rubin, J., Terman, D.: Geometric singular perturbation analysis of neuronal dynamics. In: Fiedler, B. (ed.) Handbook of Dynamical Systems: Towards Applications, vol. 2. Elsevier, Amsterdam (2002)
538. Rubin, J.E., Terman, D.: Explicit maps to predict activation order in multiphase rhythms of a coupled cell network. J. Math. Neurosci. **2**, 4 (2012)
539. Rubin, J., Wechselberger, M.: Giant squid-hidden canard: the 3d geometry of the Hodgkin–Huxley model. Biol. Cybern. **97**, 5–32 (2007)
540. Rubin, J., Wechselberger, W.: The selection of mixed-mode oscillations in a Hodgkin-Huxley model with multiple timescales. Chaos **18**, 015105 (2008)
541. Rubinov, M., Sporns, O.: Complex network measures of brain connectivity: Uses and interpretations. Neuroimage **52**, 1059–1069 (2010)
542. Ruthel, G., Banker, G.: Actin-dependent anterograde movement of growth-cone-like structures along growing hippocampal axons: A novel form of axonal transport? Cell Motil. Cytoskel. **40**, 160–173 (1998)
543. Ruthel, G., Banker, G.: Role of moving growth cone-like wave structures in the outgrowth of cultured hippocampal axons and dendrites. J. Neurobiol. **39**, 97–106 (1999)
544. van Saarloos, W.: Front propagation into unstable states. Phys. Rep. **386**, 29–222 (2003)
545. Saez, J.C., Berthoud, V.M., Branes, M.C., Martinez, A.D., Beyer., E.C.: Plasma membrane channels formed by connexins: their regulation and functions. Physiol. Rev. **83**, 1359–1400 (2003)
546. Sagues, F., Sancho, J.M., Garcia-Ojalvo, J.: Spatiotemporal order out of noise. Rev. Mod. Phys. **79**, 829–882 (2007)
547. Saitoh, T., Schwartz, J.H.: Phosphorylation-dependent subcellular translocation of a Ca^{2+}/calmodulin-dependent protein kinase produces an autonomous enzyme in aplysia neurons. J. Cell. Biol. **100**, 835–842 (1985)
548. Samuels, D.C., Hentschel, H.G.E., Fine, A.: The origin of neuronal polarization: a model of axon formation. Phil. Trans. R. Soc. Lond. B **351**, 1147–1156 (1996)
549. Sanchez-Vives, M.V., McCormick, D.A.: Cellular and network mechanisms of rhythmic recurrent activity in neocortex. Nat. Neurosci. **3**, 1027–1034 (2000)
550. Sanderson, M.J., Charles, A.C., Dirksen, E.R.: Mechanical stimulation and intercellular communication increases intracellular Ca^{2+} in epithelial cells. Cell. Regul. **1**, 585–596 (1990)
551. Sandstede, B.: Stability of travelling waves. In: Handbook of Dynamical Systems, vol. 2, pp. 983–1055. North-Holland, Amsterdam (2002)
552. Sandstede, B.: Evans functions and nonlinear stability of traveling waves in neuronal network models. Int. J. Bifur. Chaos Appl. Sci. Eng. **17**, 2693–2704 (2007)
553. Sato, T.K., Nauhaus, I., Carandini, M.: Traveling waves in visual cortex. Neuron **75**, 218–229 (2012)
554. Scemes, E., Giaume, C.: Astrocyte calcium waves: what they are and what they do. Glia **54**, 716–725 (2006)
555. Schiller, J., Schiller, Y., Stuart, G., Sakmann, B.: Calcium action potentials restricted to distal apical dendrites of rat neocortical pyramidal neurons. J. Physiol. **505**, 605–616 (1997)
556. Schiller, J., Major, G., Koester, H.J., Schiller, Y.: NMDA spikes in basal dendrites of cortical pyramidal neurons. Nature **404**, 285–289 (2000)
557. Schimansky-Geier, L., Mikhailov, A.S., Ebeling., W.: Effects of fluctuations on plane front propagation in bistable nonequilibrium systems. Ann. Phys. **40**, 277 (1983)
558. Schock, S.C., Munyao, N., Yakubchyk, Y., Sabourin, L.A., Hakim, A.M., Ventureyra, E.C.G., Thompson, C.S.: Cortical spreading depression releases ATP into the extracellular space and purinergic receptor activation contributes to the induction of ischemic tolerance. Brain Res. **1168**, 129–138 (2007)
559. Schuss, Z.: Theory and applications of stochastic processes: an analytical approach. In: Applied Mathematical Sciences, vol. 170. Springer, New York (2010)
560. Schuster, H.G., Wagner, P.: A model for neuronal oscillations in visual cortex. Biol. Cybern. **64**, 77–82 (1990)

561. Schutz, P., Bode, M., Purwins, H.G.: Bifurcations of front dynamics in a reaction-diffusion system with spatial inhomogeneities. Physica D **82**, 382–397 (1995)
562. Schwartz, E.: Spatial mapping in the primate sensory projection: analytic structure and relevance to projection. Biol. Cybern. **25**, 181–194 (1977)
563. Seely, J., Chow, C.C.: Role of mutual inhibition in binocular rivalry. J. Neurophysiol. **106**, 2136–2150 (2011)
564. Seifert, G.: Astrocyte dysfunction in neurological disorders; a molecular perspective. Nat. Rev. Neurosci. **7**, 194–206 (2006)
565. Selkoe, D.J.: Folding proteins in fatal way. Nature **426**, 900–904 (2003)
566. Selkoe, D.J.: Cell biology of protein misfolding: the examples of Alzheimer's and Parkinson's diseases. Nat. Cell Biol. **6**, 1054–1061 (2004)
567. Shadlen, M.N., Newsome, W.T.: Noise, neural codes and cortical organization. Curr. Opin. Neurobiol. **4**, 569–579 (1994)
568. Shapiro, B.E.: Osmotic forces and gap junctions in spreading depression: A computational model. J. Comp. Neurosci. **10**, 99–120 (2001)
569. Shen, K., Meyer, T.: Dynamic control of CaMKII translocation and localization in hippocampal neurons by NMDA receptor simulation. Science **284**, 162–166 (1999)
570. Shen, K., Tereul, M.N., Subramanian, K., Meyer, T.: CaMKIIβ functions as an F-actin targeting module that localizes CaMKIIα/β heterooligomers to dendritic spines. Neuron **21**, 593–606 (1998)
571. Shepherd, G.M.: The dendritic spine: a multifunctional unit. J. Neurophysiol. **75**, 2197–2210 (1996)
572. Shepherd, G.M., Brayton, R.K., Miller, J.P., Segev, I., Rinzel, J., Rall, W.: Signal enhancement in distal cortical dendrites by means of interactions between active dendritic spines. Proc. Natl. Acad. Sci. USA **82**, 2192–2195 (1985)
573. Shew, W.L., Plenz, D.: The functional benefits of criticality in the cortex. Neuroscientist **19**, 88–100 (2012)
574. Shigesada, N., Kawasaki, K., Teramoto, E.: Traveling periodic waves in heterogeneous environments. Theor. Popul. Biol. **30**, 143–160 (1986)
575. Shigesada, N., Kawasaki, K.: Biological invasions: theory and practice. Oxford University Press, Oxford (1997)
576. Shimada, T., Toriyama, M., Uemura, K., Kamiguchi, H., Sugiura, T., Watanabe, N., Inagaki, N.: Shootin1 interacts with actine retrograde flow and L1-CAM to promote axon outgrowth. J. Cell. Biol. **181**, 817–829 (2008)
577. Shpiro, A., Moreno-Bote, R., Rubin, N., Rinzel, J.: Dynamical characteristics common to neuronal competition models. J. Neurophysiol. **97**, 37–54 (2007)
578. Shpiro, A., Curtu, R., Rinzel, J., Rubin, N.: Balance between noise and adaptation in competition models of perceptual bistability. J Comp. Neurosci. **27**, 462–473 (2009)
579. Shuai, J.W., Jung, P.: Stochastic properties of Ca^{2+} release of Inositol 1,4,5-Triphosphate receptor clusters. Biophys. J. **83**, 87–97 (2002)
580. Shusterman, V., Troy, W.C.: From baseline to epileptiform activity: a path to synchronized rhythmicity in large-scale neural networks. Phys. Rev. E **77**, 061911 (2008)
581. Silva, J.S.D., Dotti, C.G.: Breaking the neuronal sphere: regulation of the actin cytoskeleton in neuritogenesis. Nat. Rev. Neurosci. **3**, 694–704 (2002)
582. Silva, F.H.L.D., Blanes, W., Kalitzin, S.N., Parra, J., Suffczynski, P., Velis, D.N.: Dynamical diseases of brain systems: different routes to epileptic seizures. IEEE Trans. Biomed. Eng. **50**, 540–548 (2003)
583. Sincich, L.C., Blasdel, G.G.: Oriented axon projections in primary visual cortex of the monkey. J. Neurosci. **21**, 4416–4426 (2001)
584. Sinha, S.R.: Basic mechanisms of sleep and epilespy. J. Clin. Neurophysiol. **28**, 103–110 (2011)
585. Sjostrom, P.J., Rancz, E.A., Roth, A., Hausser, M.: Dendritic excitability and synaptic plasticity. Physiol. Rev. **88**, 769–840 (2008)
586. Skinner, F., Kopell, N., Marder, E.: Mechanisms for oscillation and frequency control in reciprocally inhibitory model neural networks. J. Comput. Neurosci. **1**, 69–87 (1994)

587. Smaily, M.E., Hamel, F., Roques, L.: Homogenization and influence of fragmentation in a biological invasion model. Discret. Contin. Dyn. Syst. Ser. A **25**, 321–342 (2009)
588. Smith, G.D.: Modeling the stochastic gating of ion channels. In: Fall, C., Marland, E.S., wagner, J.M., Tyson, J.J. (eds.) Computational Cell Biology, chap. 11. Springer, New York (2002)
589. Smith, G.D., Cox, C.L., Sherman, S.M., Rinzel, J.: Fourier analysis of sinusoidally driven thalamocortical relay neurons and a minimal integrate-and-burst model. J. Neurophysiol. **83**, 588–610 (2000)
590. Sneyd, J., Charles, A.C., Sanderson, M.J.: A model for the propagation of intercellular calcium waves. Amer. J. Physiol. cell physiol. **266**, C293–C302 (1994)
591. Sneyd, J., Wetton, B.T., Charles, A.C., Sanderson, M.J.: Intercellular calcium waves mediated by diffusion of inositol trisphosphate: a two-dimensional model. Am. J. Physiol. Cell Physiol. **268**, C1537–C1545 (1995)
592. Sneyd, J., Dale, P.D., Duffy, A.: Traveling waves in buffered systems: applications to calcium waves. SIAM J. Appl. Math. **58**, 1178–1192 (1998)
593. Softky, W.R., Koch, C.: Cortical cell should spike regularly but do not. Neural Comput. **4**, 643–646 (1992)
594. Soltesz, I., Staley, K.: Computational Neuroscience in Epilepsy. Academic, San Diego (2008)
595. Somers, D.C., Nelson, S., Sur, M.: An emergent model of orientation selectivity in cat visual cortical simple cells. J. Neurosci. **15**, 5448–5465 (1995)
596. Somjen, G.G.: Mechanisms of spreading depression and hypoxic spreading depression-like depolarization. Physiol. Rev. **81**, 1065–1096 (2001)
597. Soula, H., Chow, C.C.: Stochastic dynamics of a finite-size spiking neural network. Neural Comput. **19**, 3262–3292 (2007)
598. Sporns, O.: Networks of the Brain. MIT Press, Cambridge (2010)
599. Staley, K.J., Dudek, F.E.: Interictal spikes and epileptogenesis. Epilepsy Currents **6**, 199–202 (2006)
600. Stam, C.J., Reijeneveld, J.C.: Graph theoretical analysis of complex networks in the brain. Nonlin. Biomed. Phys. **1 (3)**, 1–19 (2007)
601. Steinberg, B., Wang, Y.Q., Huang, H., Miura, R.: Spatial buffering mechanism: Mathematical model and computer simulations. Math. Biosci. Eng. **2**, 675–702 (2005)
602. Steriade, M., McCormick, D.A., Sejnowski, T.J.: Thalamocortical oscillations in the sleeping and aroused brain. Science **262**, 679–685 (1993)
603. Steyn-Ross, M.L., Sleigh, D.A.S.R.J.W., Whiting, D.R.: Theoretical predictions for spatial covariance of the electroencephalographic signal during the anesthetic-induced phase transition: Increased correlation length and emergence of spatial self-organization. Phys. Rev. E **68**, 021902 (2003)
604. Stocker, M., Krause, M., Pedarzani, P.: An apamin-sensitive Ca^{2+}-activated K^{+} current in hippocampal pyramidal neurons. Proc. Natl. Acad. Sci. USA **96**, 4662–4667 (1999)
605. Strack, S., Choi, S., Lovinger, D.M., Colbran, R.J.: Translocation of autophosphorylated calcium/calmodulin-dependent protein kinase II to the postsynaptic density. J. Biol. Chem. **272**, 13467–13470 (1997)
606. Stuart, G.J., Sakmann, B.: Active propagation of somatic action potentials into neocortical pyramidal cell dendrites. Nature **367**, 69–72 (1994)
607. Stuart, G., Spruston, N., Hausser, M.: Dendrites. Oxford University Press, New York (1999)
608. Stuart, G., Spruston, N., Hausser, M. (eds.): Dendrites. Oxford University Press, Oxford (2007)
609. Subramanian, K., Narang, A.: A mechanistic model for eukaryotic gradient sensing: spontaneous and induced phosphoinositide polarization. J. Theor. Biol. **231**, 49–67 (2004)
610. Sun, W., McConnell, E., Pare, J.F., Xu, Q., Chen, M., Peng, W., Lovatt, D., Han, X., Y.Smith, Nedergaard, M.: Glutamate-dependent neuroglial calcium signaling differs between young and adult brain. Science **339**, 197–200 (2013)
611. Sutherland, G.R., McNaughton, B.: Memory trace reactivation in hippocampal and neocortical neuronal ensembles. Curr. Opin. Neurobiol. **10**, 180–186 (2000)

612. Swillens, S., Dupont, G., Combettes, L., Champeil, P.: From calcium blips to calcium puffs: theoretical analysis of the requirement for interchannel communication. Proc. Nat. Acad. Sci. (USA) **96**, 13750–13755 (1999)
613. Swindale, N.V.: The development of topography in the visual–cortex: A review of models. Network **7**, 161–274 (1996)
614. Syed, M.M., Lee, S., He, S., Zhou, Z.J.: Spontaneous waves in the ventricular zone of developing mammalian retina. J. Neurophysiol. **91**, 1999–2009 (2004)
615. Syed, M.M., Lee, S., Zheng, J., Zhou, Z.J.: Stage-dependent dynamics and modulation of spontaneous waves in the developing rabbit retina. J. Physiol. **560**, 533–549 (2004)
616. Sykova, E., Nicholson, C.: Diffusion in brain extracellular space. Physiol. Rev. **88**, 1277–1340 (2008)
617. Tabak, J., Senn, W., O'Donovan, M.J., Rinzel, J.: Modeling of spontaneous activity in developing spinal cord using activity-dependent depression in an excitatory network. J. Neurosci. **20**, 3041–3056 (2000)
618. Tauber, U.C.: Field-theory approaches to nonequilibrium dynamics. Lect. Notes Phys. **716**, 295–348 (2007)
619. Taylor, A., Cottrell, G., Kristan Jr, W.: Analysis of oscillations in a reciprocally inhibitory network with synaptic depression. Neural comput. **14**(3), 561–581 (2002)
620. Terman, D.H., Ermentrout, G.B., Yew, A.C.: Propagating activity patterns in thalamic neuronal networks. SIAM J. Appl. Math **61**, 1578–1604 (2003)
621. Thul, R., Coombes, S., Smith, G.D.: Sensitisation waves in a bidomain fire-diffuse-fire model of intracellular Ca^{2+} dynamics. Physica D **238**, 2142–2152 (2009)
622. Thul, R., Smith, G.D., Coombes, S.: A bidomain threshold model of propagating calcium waves. J. Math. Biol. **56**, 435–463 (2008)
623. Tian, G.F., , Hooman, A., Takano, T., Xu, Q., Peng, W., Lin, J., Oberheim, N., Lou, N., Zielke, R., Kang, J., Nedergaard, M.: An astrocytic basis of epilepsy. Nat. Med. **11**, 973–981 (2005)
624. Timofeeva, Y.: Traveling waves in a model of quasi-active dendrites with active spines. Physica D **239**, 494–503 (2010)
625. Timofeeva, Y., Lord, G.J., Coombes, S.: Spatio-temporal filtering properies of a dendritic cable with active spines: a modeling study in the spike-diffuse-spike framework. J. Comput. Neurosci. **21**, 293–306 (2006)
626. Tong, F., Engel, S.A.: Interocular rivalry revealed in the human cortical blind-spot representation. Nature **411**, 195–199 (2001)
627. Tong, F., Meng, M., Blake, R.: Neural bases of binocular rivalry. Trend. Cog. Sci. **10**, 502–511 (2006)
628. Toriyama, M., Shimada, T., Kim, K.B., Mitsuba, M., Nomura, E., Katsuta, K., Sakumura, Y., Roepstorff, P., Inagaki, N.: Shootin1: a protein involved in the organization of an asymmetric signal for neuronal polarization. J. Cell Biol. **175**, 147–157 (2006)
629. Toriyama, M., Sakumura, Y., Shimada, T., Ishii, S., Inagaki, N.: A diffusion-based neurite length-sensing mechanism involved in neuronal symmetry breaking. Mol. Syst. Biol. **6**, 1–16 (2010)
630. Torquato, S.: Random Heterogeneous Materials. Springer, New York (2002)
631. Touboul, J., Destexhe, A.: Can power-law scaling and neuronal avalanches arise from stochastic dynamics? PLoS One **5**, 14 (2010)
632. Touboul, J., Hermann, G., Faugeras, O.: Noise–induced behaviors in neural mean field dynamics. SIAM J. Appl. Dyn. Sys. **11**, 49–81 (2012)
633. Tritsch, N.X., Yi, E., Gale, J.E., Glowatzki, E., Bergles, D.E.: The origin of spontaneous activity in the developing auditory system. Nature **450**, 50–55 (2007)
634. Troy, W.C., Shusterman, V.: Patterns and features of families of traveling waves in large-scale neuronal networks. SIAM J. Appl. Dyn. Syst. **6**, 263–292 (2007)
635. Tsai, J.C., Sneyd, J.: Are buffers boring? uniqueness and asymptotical sttability of traveling wave fronts in the buffered bistable equation. J. Math. Biol. **54**, 513–553 (2007)
636. Tsai, J.C., Sneyd, J.: Traveling waves in the buffered FitzHugh-Nagumo model. SIAM J. Appl. Math. **71**, 1606–1636 (2011)

637. Tsodyks, M., Pawelzik, K., Markram, H.: Neural networks with dynamic synapses. Neural Comput. **10**, 821–835 (1998)
638. Tuckwell, H.C., Miura, R.M.: A mathematical model for spreading cortical depression. Biophys. J. **23**, 257–276 (1978)
639. Vecchia, D., Pietrobon, D.: Migrain: a disorder of brain excitatory-inhibitory balance? Trend. Neurosci. **35**, 507–520 (2012)
640. Veen, M.V., Pelt, J.V.: Neuritic growth rate described by modelling microtubule dynamics. Bull. Math. Biol. **56**, 249–273 (1994)
641. Vellela, M., Qian, H.: Stochastic dynamics and non-equilibrium thermodynamics of a bistable chemical system: the Schlögl model revisited. J. R. Soc. Interface **6**, 925–940 (2009)
642. Veltz, R., Faugeras, O.: Local/global analysis of the stationary solutions of some neural field equations. SIAM J. Appl. Dyn. Syst. **9**, 954–998 (2010)
643. Vicker, M.G.: Eukaryotic cell locomotion depends on the propagation of self-organized reaction-diffusion waves and oscillations of actin filament assembly. J. Expt. Cell Res. **275**, 54–66 (2002)
644. Vladimirski, B.B., Tabak, J., ODonovan, M.J., Rinzel, J.: Episodic activity in a heterogeneous excitatory network, from spiking neurons to mean field. J. Comp. Neurosci. **25**, 39–63 (2008)
645. Vogels, T.P., Abbott, L.F.: Signal propagation and logic gating in networks of integrate-and-fire neurons. J. Neurosci. **25**, 10786–10795 (2005)
646. Volman, V., Sejnowski, T.J., Bazhenov, M.: Topological basis of epileptogenesis in a model of severe cortical trauma. J. Neurophysiol. **106**, 1933–1942 (2011)
647. Volman, V., Bazhenov, M., Sejnowski, T.J.: Computational models of neuron-astrocyte interaction in epilepsy. Front. Comp. Neurosci. **6**(58), 1–10 (2012)
648. Volpert, V., Petrovskii, S.: Reaction-diffusion waves in biology. Phys. Life Rev. **6**, 267–310 (2009)
649. van Vreeswijk, C.: Partial synchronization in populations of pulse-coupled oscillators. Phys. Rev. E. **54**(5), 5522–5537 (1996)
650. van Vreeswijk, C., Sompolinsky, H.: Chaotic balanced state in a model of cortical circuits. . Neural Comput. **10**, 1321–1371 (1998)
651. van Vreeswijk, C., Ermentrout, G.B., Abbott, L.F.: When inhibition not excitation synchronizes neural firing. J. Comput. Neurosci. **1**, 313–321 (1994)
652. Vuyst, E.D., Wang, N., Decrock, E., Bock, M.D., Vinken, M., Moorhem, M.V., Lai, C., Culot, M., Rogiers, V., Cecchelli, R., Naus, C.C., Evans, W.H., Leybaert, L.: Ca^{2+} regulation of connexin 43 hemichannels in C6 glioma and glial cells. Cell Calcium **46**, 176–187 (2009)
653. Wagner, J., Sneyd, J.: Effects of rapid buffers on Ca^{2+} diffusion and Ca^{2+} oscillations. Biophys. J. **67**, 447–456 (1994)
654. Wainrib, G., Thieullen, M., Pakdaman, K.: Reduction of stochastic conductance-based neuron models with time-scales separation. J. Comput. Neurosci. **32**, 327–46 (2012)
655. Walther, G.R., Maree, A.F., Edelstein-Keshet, L., Grieneisen, V.A.: Deterministic versus stochastic cell polarisation through wave-pinning. Bull. Math. Biol. **74**, 2570–2599 (2012)
656. Wang, X.J.: Decision making in recurrent neuronal networks. Neuron **60**, 215–234 (2008)
657. Wang, X.J., Rinzel, J.: Alternating and synchronous rhythms in reciprocally inhibitory model neurons. Neural Comput. **4**, 84–97 (1992)
658. Wang, S.S., Denk, W., Hausser, M.: Coincidence detection in single dendritic spines mediated by calcium release. Nat. Neurosci. **3**, 1266–1273 (2000)
659. Wang, Z., Haydon, P.G., Yeung, E.S.: Direct observation of calcium-independent intercellular ATP signaling in astrocytes. . Anal. Chem. **72**, 2001–2007 (2000)
660. Ward, M.J., Lee, J.: On the asymptotic and numerical-analyses of exponentially ill-conditioned singularly perturbed boundary value problems. Stud. Appl. Math. **94**, 271–326 (1995)
661. Watt, A.J.: Traveling waves in developing cerebellar cortex mediated by asymmetrical purkinje cell connectivity. Nat. Neurosci. **12**, 463–473 (2009)

662. Watts, D.J., Strogatz, S.H.: Collective dynamics of "small-world" networks. Nature **393**, 440–442 (1998)
663. Webber, M., Bressloff, P.C.: The effects of noise on binocular rivalry waves: a stochastic neural field model. J. Stat. Mech **3**, P03001 (2013)
664. Weinberger, H.F.: On spreading speeds and traveling waves for growth and migration in a periodic habitat. J. Math. Biol. **45**, 511–548 (2002)
665. Weiner, O.D., Marganski, W.A., Wu, L.F., Altschuler, S.J., Kirschner, M.W.: An actin-based wave generator organizes cell motility. PLoS Biol. **5**, 2053–2063 (2007)
666. Weissman, T.A., Riquelme, P.A., Ivic, L., Flint, A.C., Kriegstein, A.R.: Calcium waves propagate through radial glial cells and modulate proliferation in the developing neocortex. Neuron **43**, 647–661 (2004)
667. Werry, E.L., Liu, G.J., Bennett., M.R.: Glutamate-stimulated ATP release from spinal cord astrocytes is potentiated by substance P. J. Neurochem. **99**, 924–936 (2006)
668. Wetherington, J., Serranoa, G., Dingledine, R.: Astrocytes in the epileptic brain. Neuron **58**, 168–178 (2008)
669. White, J., Southgate, E., Thomson, J., Brenner, S.: The structure of the nervous system of the nematode Caenorhabditis elegans. Phil. Trans. R. Soc. Lond. B Biol. Sci. **314**, 1–340 (1986)
670. White, J.A., Rubinstein, J.T., Kay, A.R.: Channel noise in neurons. Trends. Neurosci. **23**, 131–137 (2000)
671. Wiley, D.A., Strogatz, S.H., Girvan, M.: The size of the sync basin. Chaos **16**, 015103 (2006)
672. Williams, G.S.B., Huertas, M.A., Sobie, E.A., Jafri, M.S., Smith, G.D.: A probability density approach to modeling local control of calcium-induced calcium release in cardiac myocytes. Biophys. J. **92**, 2311–2328 (2007)
673. Williams, G.S.B., Huertas, M.A., Sobie, E.A., Jafri, M.S., Smith, G.D.: Moment closure for local control models of calcium-induced calcium release in cardiac myocytes. Biophys. J. **95**, 1689–1703 (2008)
674. Wilson, H.R.: Computational evidence for a rivalry hierarchy in vision. Proc. Natl. Acad Sci. USA **100**, 14499–144503 (2003)
675. Wilson, H.R., Cowan, J.D.: Excitatory and inhibitory interactions in localized populations of model neurons. Biophys. J. **12**, 1–23 (1972)
676. Wilson, H.R., Cowan, J.D.: A mathematical theory of the functional dynamics of cortical and thalamic nervous tissue. Kybernetik **13**, 55–80 (1973)
677. Wilson, M.A., McNaughton, B.L.: Reactivation of hippocampal ensemble memories during sleep. Science **265**, 676–679 (1994)
678. Wilson, H.R., Blake, R., Lee, S.H.: Dynamics of traveling waves in visual perception. Nature **412**, 907–910 (2001)
679. Winfree, A.: The Geometry of Biological Time. Springer, New York (1980)
680. Wissner-Gross, Z.D., Scott, M.A., Steinmeyer, J.D., Yanik, M.F.: Synchronous symmetry breaking in neurons with different neurite counts. PLoS One **8 (2)**, e54905 (2013)
681. Wolszon, L.R., Rehder, V., Kater, S.B., Macagno, E.R.: Calcium wave fronts that cross gap junctions may signal neuronal death during development. J. Neurosci. **14**, 3437–3448 (1994)
682. Wong, R.O.L.: Retinal waves and visual development. Ann. Rev. Neurosci. **22**, 29–47 (1999)
683. Wu, J.: Propagating waves of activity in the neocortex: what they are, what they do. Neuroscientist **14**, 487–502 (2008)
684. Wu, J.Y., Guan, L., Tsau, Y.: Propagating activation during oscillations and evoked responses in neocortical slices. J. Neurosci. **19**, 5005–5015 (1999)
685. Wu, J., Guan, L., Bai, L., Yang, Q.: Spatiotemporal properties of an evoked population activity in rat sensory cortical slices. J. Neurophysiol. **86**, 2461–74 (2001)
686. Wylie, J.J., Miura, R.M.: Traveling waves in coupled reaction-diffusion models with degenerate sources. Phys. Rev. E **74**, 021909 (2006)
687. Xin, J.: Front propagation in heterogeneous media. SIAM Rev. **42**, 161–230 (2000)
688. Xu, W., Huang, X., Takagaki, K., Wu, J.Y.: Compression and reflection of visually evoked cortical waves. Neuron **55**, 119–129 (2007)
689. Yang, E., Schulman, H.: Structural examination of autoregulation of multifunctional calcium/calmodulin-dependent protein kinase II. J. Biol. Chem. **274**, 26199–26208 (1999)

690. Yao, Y., Choi, J., Parker, I.: Quantal puff of intracellular Ca^{2+} evoked by inositol triphosphate in Xenopus oocytes. J. Physiol. **482**, 533–553 (1995)
691. Yoshioka, T., Blasdel, G.G., Levitt, J.B., Lund, J.S.: Relation between patterns of intrinsic lateral connectivity, ocular dominance, and cytochrome oxidase–reactive regions in macaque monkey striate cortex. Cereb. Cortex **6**, 297–310 (1996)
692. Young, G.W.D., Keizer, J.: A single pool IP_3-receptor model for agonist stimulated Ca^{2+} oscillations. Proc. Nat. Acad. Sci. (USA) **89**, 9895–9899 (1992)
693. Yuste, R., Bonhoeffer, T.: Morphological changes in dendritic spines associated with LTP. Ann. Rev. Neurosci. **24**, 1071–1089 (2001)
694. Yuste, R., Denk, W.: Dendritic spines as a basic functional units of neuronal integration. Nature **375**, 682–684 (1995)
695. Zahradnikova, A., Zahradnik, I.: A minimal gating model for the cardiac Ca^{2+} release channel. Biophys. J. **71**, 2996–3012 (1996)
696. Zhang, L.: On the stability of traveling wave solutions in synaptically coupled neuronal networks. Diff. Int. Eq. **16**, 513–536 (2003)
697. Zimmermann, H.: Purinergic signaling in neural development. Sem. Cell Dev. Biol. **22**, 194–204 (2011)
698. Zinn-Justin, J.: Quantum Field Theory and Critical Phenomena, 4th edn. Oxford University Press, Oxford (2002)
699. Zinner, B.: Existence of traveling wavefront solutions for the discrete Nagumo equation. J. Diff. Eqns. **96**, 1–27 (1992)
700. Zucker, R.S.: Short term synaptic plasticity. Ann. Rev. Neurosci. **12**, 13–31 (1989)

Index

P.C. Bressloff, *Waves in Neural Media: From Single Neurons to Neural Fields*,
Lecture Notes on Mathematical Modelling in the Life Sciences,
DOI 10.1007/978-1-4614-8866-8,

MIX
Papier aus verantwortungsvollen Quellen
Paper from responsible sources
FSC® C105338

If you have any concerns about our products,
you can contact us on
ProductSafety@springernature.com

In case Publisher is established outside the EU,
the EU authorized representative is:
Springer Nature Customer Service Center GmbH
Europaplatz 3, 69115 Heidelberg, Germany

Printed by Libri Plureos GmbH
in Hamburg, Germany

Lessons from Animal Diabetes VI

75th Anniversary of the Insulin Discovery

Related Birkhäuser Titles

From the *Hormones in Health and Disease* Series

Steroid Hormone Receptors:
Basic and Clinical Aspects
V.K. Moudgil, Editor
ISBN 0-8176-3694-3
© 1994, 536 pp.

Hormones and Cancer
W.V. Vedeckis, Editor
ISBN 0-8176-3797-4
© 1996, 632 pp.

Estrogens, Progestins, and Their Antagonists
Volume 1: Health Issues
E.J. Pavlik, Editor
ISBN 0-8176-3854-7
© 1996, approx 344 pp.

Estrogens, Progestins, and Their Antagonists
Volume 2: Functions and Mechanisms of Action
E.J. Pavlik, Editor
ISBN 0-8176-3947-0
© 1996, approx 256 pp.

Lessons from Animal Diabetes VI

75th Anniversary of the Insulin Discovery

Eleazar Shafrir
Editor

Birkhäuser
Boston • Basel • Berlin

Israel M. Gelfand
James Lepowsky
Department of Mathematics
Rutgers University
New Brunswick, NJ 08903

Mikhail M. Smirnov
Department of Mathematics
Columbia University
New York, NY 10027

Library of Congress Cataloging-in-Publication Data

The Gelfand Mathematical Seminars, 1993–1995

ISSN 1068-7122

Printed on acid-free paper

Birkhäuser
Softcover reprint of the hardcover 1st edition

ISBN-13: 978-1-4612-8658-5 e-ISBN-13: 978-1-4612-4112-6
DOI: 10.1007/978-1-4612-4112-6

Reformatted by Texniques, Inc., Brighton, MA

Contents

II TRANSGENIC ANIMALS IN DIABETES RESEARCH

III NON-INSULIN-DEPENDENT DIABETES: NEW MODELS AND UPDATE ON EXISTING MODELS. DIABETES COMPLICATIONS AND THERAPEUTIC MODALITIES

Editor's Preface: Introduction to Lessons from Animal Diabetes VI

We are pleased to announce that this special anniversary volume of *Lessons from Animal Diabetes* is published by Birkhäuser. The year 1996 is the 75th Anniversary of the Discovery of Insulin with many symposia and special meetings devoted to this momentous event. This is an opportunity to remember and underscore that the major breakthroughs in the progress toward the cure of diabetes have been made by investigation of animal diabetes.

There is no need to emphasize to our readers the enormous contribution of research in animal diabetes as a channel of information for the understanding of the pathogenesis of human diabetes. The first documented use of animals in research of diabetes and demonstration that the liver glycogen is the source of the circulating glucose can be attributed to Claude Bernard around 1850. The very discovery of the link between pancreas and diabetes became evident by the dog experiments of Oscar Minkowski in 1899. The isolation of insulin by Banting and Best in 1921 and the first preparations of insulin for treatment of patients were again from dogs. In the last two decades the progress in the understanding of the autoimmune B-cell failure in IDDM has been decisively advanced by a multitude of studies in BB rats and NOD mice. Similarly, important landmarks in the study of NIDDM have been made in many other species such as corpulent SHR rats, Zucker rats, *Psammomys obesus* rhesus monkeys and particularly *db/db* and *ob/ob* mice, to mention just a few. The most recent isolation of the obesity gene and synthesis of its product leptin at the Rockefeller University from *ob/ob* mice by one of the past Editors of LAD, Doug Coleman from Bar Harbor, ME represents another breakthrough towards eradicating diabesity.

New developments covered in part in this Volume pose a challenge and give hope for the prevention of IDDM in the near future. They also signify an intensified effort to contain and prevent the increasing spread of NIDDM which at present occurs at a ratio 9 to 1 compared with IDDM and is particularly rampant in populations emerging into nutritional affluence from scarcity.

Volume VI of LAD comprises invited and peer refereed reviews on subjects of contemporary interest in experimental diabetes rather than proceedings of continuing Workshops on Lessons from Animal Diabetes. The main topics are divided into sections, the first one comprising a group of articles related to B-cell function and IDDM. This section includes an article on the expression of glutamic acid decarboxylase (GAD) and its role as an antigen in IDDM, the role of macrophages, cytokines and retroviruses in the pathogenesis of IDDM, molecular genetics of IDDM as well as new perceptions on the unified mechanism leading to the final B-cell death as a result of both cytotoxic and immune agents.

There is an expanded group of chapters dealing with new animal models rising in importance in the research of diabetes and its related phenomena, the transgenic constructs. Five articles are devoted to this topic ranging from the B-cell function in transgenic animals to the various effects on diabetes complications.

The section on NIDDM, comprising of 10 articles, deals both with new and existing models, their particular widely varying pathogenesis, genetic characteristics and complications. The animals reviewed include: spontaneously diabetic OLETF rats, Chinese hamsters, Goto-Kakizaki rats, *db/db* mice, rhesus monkeys, dogs and an article demonstrating the genetic link between the Zucker *fa/fa* and corpulent *cp/cp* obese interstrains.

We wish to welcome the new members to our Editorial Board, Dr. Hubert Kolb from Düsseldorf, Dr. Alex Rabinovitch from Edmonton, Dr. Takayoshi Toyota from Sendai and Dr. Soroku Yagihashi from Hirosaki. At the same time we would like to thank Dr. Douglas L. Coleman from Bar Harbor and Dr. George Eisenbarth from Denver for their editorial contribution to the previous LAD Volumes who have retired from the Editorial Board.

With great sadness we have to mention Dr. Otho Michaelis IV from Beltsville, MD who suddenly passed away this year. Otho, or Mike as he was called by friends, made a significant contribution in developing and metabolically defining several strains of corpulent *cp* rats and was extremely helpful in the editorial work of the LAD series. His contribution will remain in a lasting memory of all researchers of animal diabetes.

Eleazar Shafrir
Jerusalem, January 1996

Abbreviations used in this volume

Ab	Antibody
ACTH	Adrenocorticotropic hormone
ADH	Antidiuretic hormone
ANOVA	Analysis of Variance
ANP	Atrial Natriuretic Peptide
APC	Antigen Presenting Cells
ARI	Aldose Reductase Inhibitors
BAT	Brown Adipose Tissue
BB	Bio Breeding (rat)
BMC	Bone Marrow Cells
bp	base pairs
CBA	Competitive Binding Assay
cDNA	Complementary DNA
CHO	Carbohydrate
CFU	Colony Forming Units
CMV	CytoMegaloVirus
CNS	Central Nervous System
Con A	Concavalin A
CRF	Corticotropin Releasing Factor
CSF	CerebroSpinal Fluid
CTL	Cytotoxic Thymus Lymphocytes
CTS	Cataract Shionogi Mouse
CY	Cyclophosphamide
DAG	DiAcyl Glycerophosphate
DAP	Diabetes Associated Peptide
2-DG	2-DeoxyGlucose
DHEA	DeHydroEpiAndrosterone
DMSO	DiMethyl SulphOxide
DP	Diabetes Prone
DR	Diabetes Resistant
ELISA	Enzyme Linked Immunosorbent Assay
EM	Electron Microscopy
EMC	EncephaloMyoCarditis (virus)
FACS	Fluorescence Activated Cell Sorter
FAD	Flavin Adenine Dinucleotide
FBS	Fetal Bovine Serum
FFA	Free Fatty Acids
FGF	Fibroblast Growth Factor
FFM	Fat Free Mass
FMF	Flow MicroFluorimetry
FITC	Fluorescein IsoThioCyanate (label)
FSH	Follicle Stimulating Hormone
GAF	Gomori Aldehyde Fuchsin (stain)
GBM	Glomerular Basement Membrane
GFR	Glomerular Filtration Rate
GHb	Glycated Hemoglobin (glycohemoglobin, HBA_{1C})
GHRH	Growth Hormone Releasing Hormone
GIP	Gastric Inhibitory Peptide
GT	Glucose Tolerance
HDL	High Density Lipoprotein
HLA	Histocompatibility Locus Antigen
HPLC	High Power Liquid Chromatography
HSV	Herpes Simplex Virus

IAPP	Islet Amyloid PolyPeptide
ICA	Islet Cell Antibody
ICAM	Intercellular Adhesion Molecule
ICSA	Islet Cell Surface Antibody
IDL	Intermediate Density Lipoprotein
IDDM	Insulin Dependent Diabetes Mellitus
IFN	Interferon
IGF	Insulin-like Growth Factor
IL	Interleukin
IR	Insulin Receptor
IRK	Insulin Receptor Kinase
kb	Kilobase
kDa	kilodalton (molecular mass unit)
LDL	Low Density Lipoprotein
LFA	Lymphocyte Function Associated Antigen
LH	Luteinizing Hormone
LHRH	Luteinizing Hormone Releasing Hormone
LPL	Lipoprotein Lipase
LPS	Lipopolysaccharide
MBP	Myelin Basic Protein
ME	Median Eminence
MHC	Major Histocompatibility Complex
MLDS	Multiple Low Dose Streptozotocin
MNCV	Motor Nerve Conduction Velocity
MoAb	Monoclonal Autobody
MW	Molecular Weight
NIDDM	Non–Insulin-Dependent Diabetes
NMR	Nuclear Magnetic Resonance
NOD	Non-Obese Diabetic (mice)
NON	Non-Obese Nondiabetic (mice)
NCV	Nerve Conduction Velocity
PAGE	PolyAcrylamide Gel Electrophoresis
PAS	Periodic Acid Schiff
PBC	Peripheral Blood Cells
PBL	Peripheral Blood Leukocytes
PCR	Polymerase Chain Reaction
PP	Pancreatic Polypeptide
PDGF	Platelet Derived Growth Factor
PFU	Plaque Forming Units
PHA	Phytohemagglutinin
PHSC	Pluripotent Hemopoietic Stem Cells
PKC	Protein Kinase C
PL	Phospholipids
PMA	Phorbol 12-Myristate-B Acetate (PKC activator)
Poly-I:C	PolyInosinic-Cytidylic acid
RBC	Red Blood Cells
RFLP	Restriction Fragment Length Polymorphism
RIA	RadioImmunoAssay
RIN	Rat Insulinoma (cell line)
SD	Standard Deviation
SDS	Sodium Dodecyl Sulphate
SE	Standard Error
SNS	Sympathetic Nervous System
SPF	Specific Pathogen Free
STZ	STreptoZotocin
T_3	Triiodothyronine
T_4	Thyroxine
TCR	T-Cell Receptor

TG	TriGlycerides
TK	Tyrosine Kinase
TNF	Tumor Necrosis Factor
TRH	Thyrotropin Releasing Hormone
VAF	Viral Antibody Free
VIP	Vasoactive Intestinal Peptide
VLDL	Very Low Density Lipoproteins
VMN	Ventro-Medial Nucleus
WBC	White Blood Cell
wk	week
wt	weight
x^2	chi square

1

Expression of glutamic acid decarboxylase (GAD) and its role as an autoantigen in insulin-dependent diabetes

BEVERLY E. FAULKNER-JONES*, MICHELLE B. FRENCH, ROBERT S. SCHMIDLI, SUSAN WILLIAMSON*, DAVID S. CRAM, MARGO C. HONEYMAN, AND LEONARD C. HARRISON

Burnet Clinical Research Unit, The Walter and Eliza Hall Institute of Medical Research, Parkville, Melbourne, Victoria 3050, Australia.

Key words: glutamic acid decarboxylase, tissue expression, rodent, human, insulin-dependent diabetes, autoimmunity.

Insulin-dependent diabetes mellitus (IDDM) results from autoimmune, T-lymphocyte–dependent destruction of pancreatic islet B-cells in the two rodent models of human IDDM, the non-obese diabetic (NOD) mouse and the Bio-Breeding (BB) rat.[1,2] Similar pathogenic mechanisms are thought to result in human IDDM.[3] In all three species, autoimmune B-cell destruction appears to be a highly selective process, with relative sparing of the other islet cell types. The primary target antigen and/or the presentation of primary antigenic peptide is thus likely to be either unique to the B-cell, or at least B-cell-restricted within the islets.[3]

The enzyme glutamic acid decarboxylase (GAD) has recently been recognized as an autoantigen in IDDM.[4–6] GAD is now thought to be synonymous with a diabetes-associated 64 kDa antigen discovered in the early 1980s,[7] and as such has strong credentials for being pathogenic.[3,8–10] GAD was originally identified in the central nervous system (CNS) and has been intensively investigated in this context.[11,12] Much less is known about the biology of GAD in peripheral tissues such as the pancreas. We first review GAD and its expression patterns in the CNS and in peripheral tissues, and then autoimmunity to GAD in both rodents and humans, and discuss the role of GAD as an autoantigen in IDDM.

Correspondence: Leonard C. Harrison, Burnet Clinical Research Unit, The Walter and Eliza Hall Institute of Medical Research, Parkville, Melbourne, Victoria 3050, Australia.

* Present address: Department of Anatomy and Cell Biology, University of Melbourne, Parkville 3052, Australia.

Lessons from Animal Diabetes VI. Ed. E. Shafrir.

GAD enzymes

GAD belongs to an ancient family of highly conserved decarboxylases that all use pyridoxal-5′-phosphate (PLP) as a cofactor. These enzymes appear to have a common evolutionary origin and include histidine decarboxylase and dopa-decarboxylase.[13] In addition to mammals, GAD has been identified in bacteria (*Escherichia coli*),[14,15] yeast (*Neurospora crassa*),[16] plants (petunia),[17] and *Drosophila*.[18,19]

GAD (EC. 4.1.1.15) catalyzes the α-decarboxylation of L-glutamate to form carbon dioxide and gamma-amino butyric acid (GABA).[11] GABA is the major inhibitory neurotransmitter in the mammalian brain. Since its discovery in 1950, active, native GAD has been purified from the brains of several mammals, and multiple forms have been found to coexist.[6,20] The molecular mass of GADs reported by different groups has varied between 59 and 67 kDa, but there was agreement that there are two major forms, which differ in size by 2–4 kDa.[6]

In vivo, GAD exists as both the apoenzyme and holoenzyme. The apoenzyme is unbound to pyridoxal-5-phosphate (PLP) and consequently inactive. At least 50% of the GAD in the rat brain exists as the apoenzyme, and most of the apoenzyme is the smaller of the two GAD forms (see also later).[12,20,21] The regulated binding of PLP to apo-GAD is thought to be one of the mechanisms controlling the rate of GABA synthesis in neurons.[20,21]

GAD genes

Cloning of GAD from the CNS and peripheral tissues. Much of the controversy surrounding the identity of the multiple forms of GAD present in the mammalian brain was resolved by cloning and sequencing two distinct GAD cDNAs. These are derived from separate genes and have been isolated from the brain, pancreas, and/or testis of humans, pigs, cats, rats, and mice.

The first GAD to be cloned from a mammal was isolated from the feline occipital cortex.[22,23] This cDNA encodes a protein with a predicted molecular mass of 67 kDa. It has been designated GAD 67. GAD 67 cDNAs were subsequently cloned from human,[24–26] pig,[27] rat,[28–30] and mouse brains[31,32]; from human[24,33,34–36] and rat[33,37] islets; and a partial cDNA from human testis.[38] Comparison of the nucleotide sequences obtained from the same organ by different workers suggests that minor allelic variation occurs in both human and rat GAD 67 genes. However, most nucleotide substitutions are conservative or silent. GAD 67 cDNAs obtained from different organs in the same species have the same nucleotide sequence, indicating that they are derived from the same gene.[33,35–37]

The other major form of GAD has a predicted molecular mass of 65 kDa and is now designated GAD 65. Since 1991, GAD 65 cDNAs have been cloned from human,[26] pig,[27] rat,[30] and mouse[31,39] brains, and from human islets.[34,40] As for GAD 67, CNS and peripheral tissue GAD 65 cDNAs also appear to be derived from the same gene (D. Cram, unpublished data). The database code and accession numbers for the GADs cloned from different species and organs are given in Table 1. A cDNA that encodes a protein with GAD-like enzymatic activity in

Table 1. Cloning of mammalian glutamate decarboxylase genes: Database codes and accession numbers

Species	Organ	Ref.	Date	Author	Database	Code/Accession
GAD 67						
Cat	Brain	23	1987	Kobayashi Y	Genbank	CATGAD/M18629
Human	Brain	24	1991	Cram DS	Genbank	HUMGADBR/M70434
		25	1992	Kelly CD	EMBL	HSGAD67/Z22750
		26	1992	Bu DF	Genbank	HUMGAD67A/M81883
	Islets	24	1991	Cram DS	Genbank	HUMGADPAN/M70435
		35	1993	Kawasaki E	EMBL	S61898/S61898
		36	1993	Yamashita K	EMBL	S61897/S61897
	Testis	38	1990	Persson H	Genbank	HUMGAD/M55574
Pig	Brain	27	1995	Suzuki R	Genbank	PIGGAD67B/D31849
Rat	Brain	28	1990	Wyborski RJ	Genbank	RATGADA/M34445
		29	1990	Julien JF	EMBL	RNGADBRA/X57572
	Islets	33	1991	Michelsen BK	Genbank	RATGAD67/M76177
Mouse	Brain	31	1993	Faulkner-Jones BE	Genbank	S67453/S67453
		32	1990	Katarova Z	Genbank	MMGAD67C/Z49976
GAD65						
Human	Brain	26	1992	Bu DF	Genbank	HUMGAD65A/M81882
	Islets	40	1991	Karlsen AE	Genbank	HUMGAD2A/M74826
Pig	Brain	27	1995	Suzuki R	Genbank	PIGGAD65A/D31848
Rat	Brain	30	1991	Erlander MG	Genbank	RATGAD65/M72422
Mouse	Brain	31	1993	Faulkner-Jones BE	Genbank	S67454/S67454
		39	1993	Lee DS	Genbank	MUSGAD65A/L16980

vitro has been cloned from mouse brain.[41] Its nucleotide sequence is distinct from that of GAD 65 and GAD 67, and the protein has a predicted molecular mass of 83 kDa. The in vivo significance of "GAD 83" remains unclear.

To date, no islet-specific form of GAD has been identified. We previously reported[24] the isolation of a partial cDNA from human islets that is homologous to the equivalent region of human brain GAD 67 cDNA, but differed by 8% at the nucleotide level. We have failed to isolate the full-length version of this cDNA to confirm its authenticity. This cDNA may represent an extreme example of allelic polymorphism in the human GAD 67 gene; however, the possibility of a sequencing error must also be considered.

The sequences of both GAD forms are highly conserved between species. Cat and mouse GAD 67 are 95% identical at the amino acid level.[32] The values are 97% and 98% for rat vs. human and human vs. cat, respectively.[26] Human and rat GAD 65 are 97% identical. The primary amino acid sequences of GAD 67 and GAD 65 protein are also highly related. Human brain GAD 67 and GAD 65 are 65% identical and 81% similar, with the majority of differences being found in the first 100 amino acid residues (Fig. 1).[26,40] The same is true for rat brain GAD 67 and GAD 65.[30] Interestingly, *Drosophila* GAD[19] has significant homology, both with human GAD 67 (53%) and GAD 65 (51%), but at different

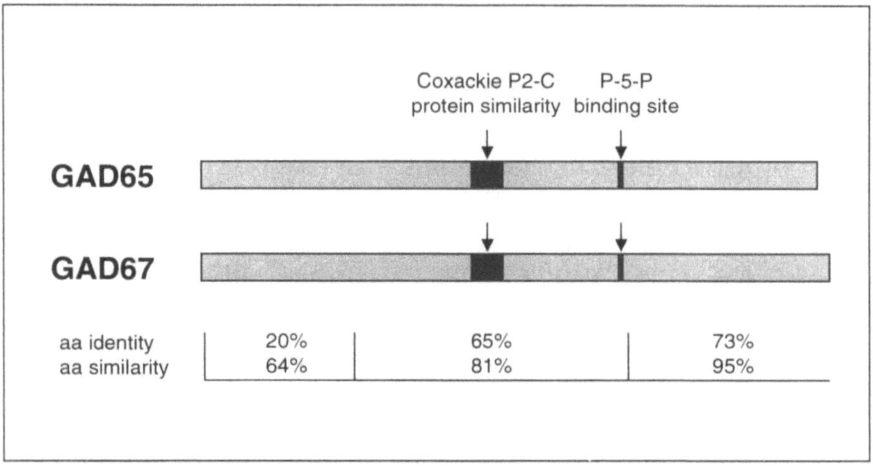

Figure 1. Amino acid (aa) sequence comparison, Coxsackie P2-C protein similarity, and pyridoxal-5′-phosphate P-5-P binding site of GAD 65 and GAD 67.

amino acid positions. It is therefore likely that mammalian GAD 65 and GAD 67 genes were derived from a common ancestral gene, possibly by gene duplication, and then diverged as a consequence of different selection pressures. The chromosomal location of both GAD genes has been determined in human, rat, and mouse. In mouse and human, GAD 67 maps to chromosome 2 (2q31 in the human) while GAD 65 maps to chromosome 10 (10p11.23 in the human).[26,40,42] In the rat, GAD 67 maps to chromosome 3 and GAD 65 to chromosome 17.[43]

GAD messenger RNAs. The major mRNA transcript encoding GAD 67 is 3.7 kb in adult human,[26] rat,[30,44] and mouse brains,[31,32] and in fetal human brain.[26] Transcripts of 7–9 kb and 2 kb have also been found in the developing rat[44] and mouse.[45] The predominant GAD 67 mRNA in human testis is 2.5 kb, although a less abundant 3.7 kb transcript is detectable.[38] Six species of GAD 67 mRNA, ranging from 1.1 to 3.4 kb, were detected in rat testis, and two, of 3.4 and 1.7 kb, were found in rat oviduct.[46] There is some evidence for developmental regulation of the different mRNA transcripts expressed in testis.[46] The major mRNA transcript encoding GAD 65 is 5.6–5.7 kb in adult human,[26] rat,[30] and mouse brains,[31] in fetal human brain,[26] in human islets,[40] and in rat oviduct.[46] Less abundant 2.5 kb and 1.8 kb GAD 65 transcripts have also been reported in human islets[40] and in rat testis.[46]

As discussed in the previous section, the same GAD genes are expressed in the CNS and in peripheral tissues.[31] Thus, the presence of multiple mRNA transcripts of different sizes for each GAD gene suggests that either alternative splicing and/or alternative polyadenylation occurs for both forms. In developing rat brain, the alternative splicing of an exon containing a stop codon produces a 25 kDa GAD, which is truncated upstream of the PLP binding site and is consequently inactive.[47] Truncated, non–GABA-synthesizing 25 kDa GADs have also been observed in developing rat spinal cord.[48,49] Similarly, truncated, alternatively spliced GAD 67 mRNA transcripts have been demonstrated in the developing mouse.[45] Alternative splicing, confirmed by sequencing, has been reported for GAD 67 in a human insulinoma.[33] However, alternative splicing of either of the GAD forms has yet to be demonstrated under physiological conditions in the islet. The basis of GAD 67 mRNA heterogeneity is still unknown but may be clarified when the complete, genomic DNA sequences for GAD 67 become available.

Summary. GAD heterogeneity in mammals is partly explained by the existence of two distinct GAD genes that are located on separate chromosomes. Each of the GAD forms encoded by these two genes is highly conserved between different species. In addition, the two GADs are closely related and are likely to have been derived from a common ancestral gene. Allelic variation of GAD 67 apparently occurs in humans. The same GAD gene is expressed in the CNS and in peripheral tissues; in particular, no islet-specific GAD has yet been identified. Further GAD heterogeneity may be generated by the alternative splicing and/or polyadenylation of GAD mRNAs. Both developmentally regulated and tissue-specific alternative splicing may occur. No islet-restricted splice variant of either GAD form has been reported to date.

GAD proteins

The sequences for human GAD 67 and GAD 65 are 65% identical and 81% similar at the amino acid level.[26,40] Consequently, their hydropathy plots are also very similar. Neither GAD form has a conventional leader sequence or a specific transmembrane domain[40]; thus a cytosolic distribution would be predicted for both proteins. However, although over half of the total GAD enzymatic activity in brain is recovered from the cytosol,[6] a significant proportion is membrane associated.

Most of the membrane-associated GAD appears to be GAD 65. It is predominantly localized to synaptic vesicles in neurons and to synaptic-like microvesicles (SLMVs) in islet beta cells.[50,51] An additional pool of GAD 65 is found in the Golgi complex region of both cell types.[52,53] GAD 65 is synthesized as a hydrophilic, soluble protein.[51] Subsequent modification of its amino-terminal domain results in a firm, but reversible, membrane anchoring. Membrane anchoring was thought to involve hydroxylamine-sensitive palmitoylation[51] of the first two cysteines in the amino-terminal domain.[54] However, mutation of these cysteine residues results in the loss of palmitoylation but does not prevent membrane-association in vivo.[54] Deletion of a separate group of eight amino acids in the amino terminus results in a hydrophilic GAD 65 with a cytosolic distribution.[54] The signal for targeting to the Golgi complex resides in an overlapping region.[53]

Unlike GAD 65, GAD 67 has a predominantly cytosolic distribution. Transfection of GAD 67 into COS cells generates a hydrophilic, cytosolic protein.[33] However, a small proportion of GAD 67 is found associated with nerve terminals in vivo.[12] The mechanism of membrane association of GAD 67 is different from that of GAD 65. Cotransfection studies have shown that GAD 67 first associates with the amino terminus of GAD 65 and that this interaction is required for subsequent membrane targeting of GAD 67.[55] The GAD 67–GAD 65 interaction does not appear to involve palmitoylation.[55] There is no direct evidence for cell surface expression of either GAD form.

Differential detection of the two GAD forms

It is already clear that GAD 67 and GAD 65 are expressed both in the CNS and in peripheral tissues. From the perspective of IDDM, it is likely that only GAD 65 is synonymous with the original 64 kDa islet antigen.[4,7] Thus, it became necessary to map the expression patterns of GAD 67 and GAD 65 independently.

GAD activity assays. The traditional method of detecting GAD activity in a tissue relies on the release of ^{14}C-labeled carbon dioxide from L-glutamate. Although relatively specific for glutamic acid decarboxylases, the assay does not distinguish between the two GAD forms. Several workers have determined the relative contribution of the two GADs to the total GAD activity by using antisera that selectively deplete one or other forms.[12,46] This approach, however, gives no information on cellular localization.

Immunoreactivity. The high degree of sequence conservation between the two GADs has made it relatively difficult to generate antibodies that recognize either one alone. The early GAD antisera, for example, Oertel's S3,[56] did not differentiate between the two GAD forms. Discriminatory antibodies have been reported[12,57,58] (R. Schmidli, unpublished data), and the two most widely used GAD-form–specific antibodies are K-2[12] and GAD-6.[58] K-2 is a rabbit antiserum that recognizes GAD 67, and GAD-6 is a mouse monoclonal antibody that recognizes GAD 65 (Fig. 2).

Both K-2 and GAD-6 are believed to be GAD-form specific when used to detect denatured GADs by Western blotting. However, when used to detect native GAD proteins by immunoprecipitation, crossreactivity can occur.[46,58–60] This is relatively easy to detect after the immunoprecipitated proteins have been size-fractionated by polyacrylamide gel electrophoresis. Both K-2[12] and GAD-6[4] antibodies can be used to detect GAD immunoreactivity in fixed tissue sections, but under these conditions it is difficult to be certain that the antibodies are genuinely GAD-form specific.

Nucleic acid hybridization. An alternative approach to the immunological differentiation of GAD 67 from GAD 65 is nucleic acid hybridization. Both cDNA and cRNA probes (riboprobes) have been shown to be form specific when used to analyze GAD mRNAs by Northern blot hybridization,[40,46] RNase protection assay,[31,34] and in situ hybridization.[31,34] Localization of mRNA by in situ hybridization has the further advantage over immunohistochemistry of localizing GAD to its site of synthesis.

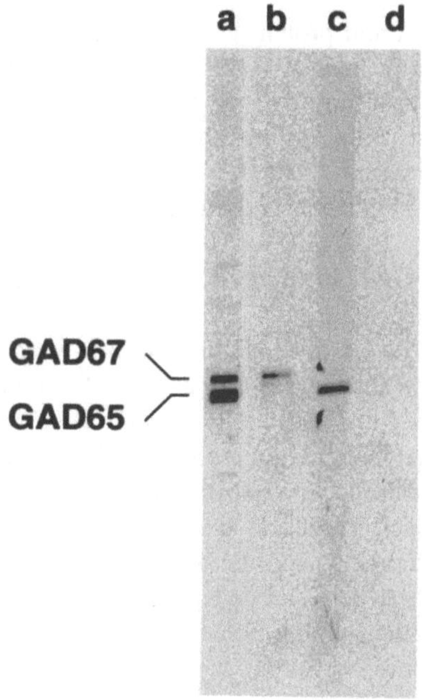

Figure 2. Immunoblot of GAD 65 and GAD 67 in isolated rat islets. Islets lysates were fractionated by SDS-PAGE gel, electrophoretically transferred to nitrocellulose membrane, and incubated with primary Ab. Horseradish peroxidase-labeled secondary Ab and ECL reagent (Amerhsam) were used to detect bound primary Ab. Primary Ab: (a) Rabbit antiserum 1877 raised against CDIDFLIEEIER-LGQ, the shared C-terminal sequence of rat GAD (1:100). (b) GAD 67-specific rabbit polyclonal Ab K2 (1:1000). (c) GAD 65-specific mouse monoclonal Ab GAD-6 (1:200) ascites. (d) Nonimmune rabbit serum 1881 (1:100).

Relationship between GAD activity, GAD protein, and GAD mRNA levels. Hybridization to GAD mRNAs can be both a specific and sensitive method for discriminating between the two GAD forms. However, GAD protein, and not GAD mRNA, is the functional gene product and thus the potential autoantigen. To date, there are no studies that directly correlate GAD enzymatic activity, GAD protein, and GAD mRNA levels. In the CNS, two studies on rat brain provide indirect evidence that GAD protein and mRNA levels are expressed in parallel.[30,61] In the periphery, GAD enzymatic activity[62] and GAD mRNA levels[31] in the rat pancreas are ~ 2.6 times greater than those in mouse pancreas. Immunohistochemical analyses further demonstrate that there is more immunoreactive GAD in rat than mouse islets.[63] On the basis of this limited evidence, GAD enzymatic activity, GAD protein, and GAD mRNA levels within the CNS and the pancreas appear to be parallel.

Summary. The necessity of differentiating the expression of GAD 67 from that of GAD 65 has led to the development of GAD-form–specific antibodies and nucleic acid probes. Discriminatory antibodies have the advantage over DNA or RNA probes of detecting the GAD autoantigen itself, and not a "precursor." They are specific for denatured, but not native GAD proteins, which under most circumstances is only problematic for immunohistochemistry. Discriminatory RNA probes are specific for denatured and "native" GAD 67 and GAD 65 mRNAs, and can also be used to localize the site of mRNA synthesis. In addition, as GAD mRNA and protein levels appear to be roughly parallel, the relative differences found in mRNA levels are likely to be reflected in the tissue content of GAD protein. Many data on the expression of GAD 67 and GAD 65 have been obtained using various combinations of the GAD detection methods. These have generated a detailed, although still incomplete, picture of GAD expression patterns in peripheral tissues.

GAD in the pancreas: Physiological role and expression patterns

High levels of GAD and GABA were demonstrated in the pancreatic islets of several species as early as 1976.[64] Superficially, the discovery that a highly specialized and terminally differentiated cell, such as the B-cell, expressed GAD, an enzyme strongly associated with the CNS, was surprising. However, islet cells and neurones have several features in common. They express a number of markers, such as neuron-specific enolase, PGP 9.5, and tyrosine hydroxylase during development, and the temporal profile of expression suggests a developmental link between islet cells and neurones.[65] B-cells generate spontaneous, calcium-dependent action potentials and thus share some physiological properties with neurones.[66,67]

Physiological role. The islet is a highly organized functional unit and because the constituent cells produce hormones with opposing effects, it is advantageous for them to communicate to coordinate their responses. This requirement is met in part by GABA, synthesized by GAD and acting in a paracrine fashion via $GABA_A$ receptors.[66,68,69] Glucose inhibits glucagon release from A-cells, and it has been proposed that this effect is mediated by the glucose-induced release of GABA from B-cells. Two studies have demonstrated inhibition of glucagon release from guinea pig[66] and mouse and rat[68] A-cells, by GABA. However, it is first worth noting that glucose itself has a far greater inhibitory effect than GABA, and, second, that there is still no conclusive evidence that the release of GABA from B-cells is induced by insulin.[70]

For GABA to act as an inhibitory signaling molecule within the islet, it makes most physiological sense if GABA is synthesized by and released from B-cells, and acts via $GABA_A$ receptors on A- and D-cells. In the rodent pancreas, particularly the rat, this orderly arrangement is found for both GAD and GABA.[66,70] $GABA_A$ receptors, however, can occasionally be found on B-cells.[70]

Expression in rat pancreas. The early data on the differential expression of the two GAD forms were derived from immunoblot or immunoprecipitation analyses of isolated rat islets.[33,37,71,72] These data indicated that GAD 65 is usually the more abundant form. However, it is worth noting that the isolation and subsequent metabolic labeling of islets subjects them to hours or days of manipulation ex vivo. At the end the procedure, their GAD content may not necessarily reflect that found in vivo. Indeed, the conditions used for islet culture have been shown to alter the expression of GAD and other proteins in primate[60,73] and rodent islets.[62,74]

In the *unmanipulated* adult rat pancreas, GAD 67 mRNA is expressed at higher levels than that of GAD 65, but the expression of both mRNAs is restricted to the islet.[31] Within the islet, GAD mRNA expression is confined to B-cells (Fig. 3). Using the S3 GAD antiserum,[56] GAD immunoreactivity has been found by some workers to be islet restricted and B-cell specific.[63,68,75] In addition, GAD immunoreactivity in the islet core could be abolished by pretreatment with the selective B-cell toxins, alloxan and streptozotocin.[75] Interestingly, other workers using the same S3 antiserum reported immunoreactive GAD within the zymogen granules of the rat acinar pancreas[76] and suggested that the rat islet is surrounded by GABAergic nerve cell bodies that extend processes into the islet mantle.[77] Whilst this may be the case in the rat, the autonomic innervation of the islets has a highly species-specific and variable pattern.[67] The consensus of the available data is that GAD expression is islet restricted and B-cell specific under physiological conditions in the rat pancreas.

Expression in mouse pancreas. The results of immunoblotting and immunoprecipitation analyses of isolated mouse islet were reported in two studies.[62,72] Unlike the rat, GAD 67 protein was shown to be the predominant form ex vivo. In the first report,[62] only GAD 67 could be detected by immunoprecipitation, Western blotting, or immunohistochemistry, even after culture of islets in high glucose, a maneuver known to upregulate GAD 65 expression.[60,62,73,74] In addition, mouse islets contained only 40% of the GAD activity of rat islets. In the second study,[72] low levels of GAD 65 protein expression in mouse islets were detected by Western blotting.

In the *unmanipulated* adult mouse pancreas, GAD 67 mRNA is the physiologically predominant form.[31] The expression level in mouse pancreas was approximately half that found in the rat. Very low levels of GAD 65 mRNA could be detected in only one of six pancreas RNA samples by the sensitive RNase protection assay.[78] GAD 67 mRNA was found to be islet restricted within the pancreas and was confined to B-cells within the islet (Fig. 4). GAD 65 mRNA could not be localized by in situ hybridization. GAD immunoreactivity has also been localized to the B-cells of normal mouse pancreas by immunohistochemistry (B.F. Jones, unpublished data), although the level of immunostaining found was less than in the rat.[63]

The two transformed mouse B-cell lines, β-TC3[79] and NIT-1,[80] have attracted attention from those interested in studying the biology of GAD. They represent relatively homogenous populations of islet B-cells that are easy to obtain. We have reported that both cell lines differ in their GAD mRNA expression patterns from each other, as well as from freshly isolated pancreas.[31] GAD 65 was relatively more abundant than GAD 67 mRNA in β-TC3 cells, but only GAD 67 mRNA could be detected in NIT-1 cells. Culture conditions were also found to influence the level of GAD mRNA expression.

Both cell lines resulted from the transformation of B-cells using the SV-40 large T-antigen. A variable population of cells within each culture will revert to a less mature (insulin and glucagon-producing) phenotype.[65] The departure from physiological GAD expression patterns may be a consequence of either trans-

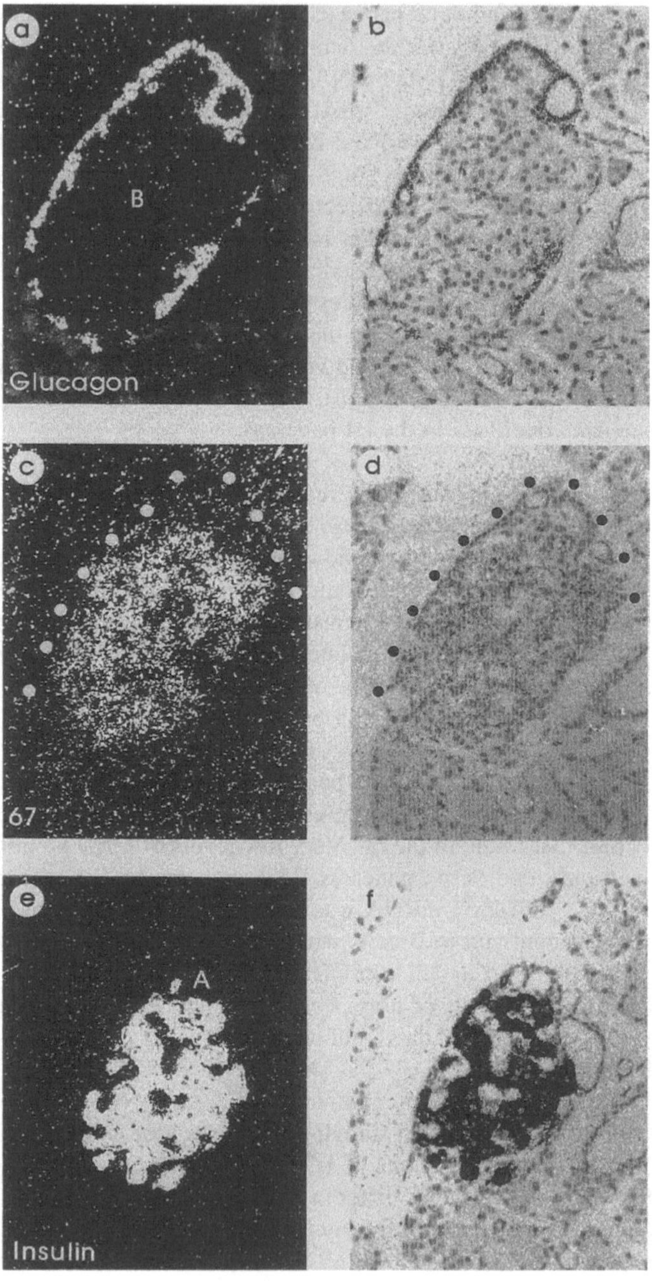

Figure 3. GAD 67 mRNA is expressed in rat islet B-cells and not A-cells. 3-1: Dark-field views of adjacent 5 μm sections of the same islet in perfusion-fixed rat pancreas hybridized with ^{3}H-labeled glucagon (a), ^{33}P-labeled mouse-brain GAD 67 cRNA (c), and ^{3}H-labeled insulin antisense cRNA (e). The corresponding lightfield views are shown in (b), (d), and (f), respectively. The border of the islet, and hence the position of A-cells, is marked in (c). The positions of A-cells (A) and B-cells (B) are marked in (e) and (a), respectively. The magnification for all panels is the same.

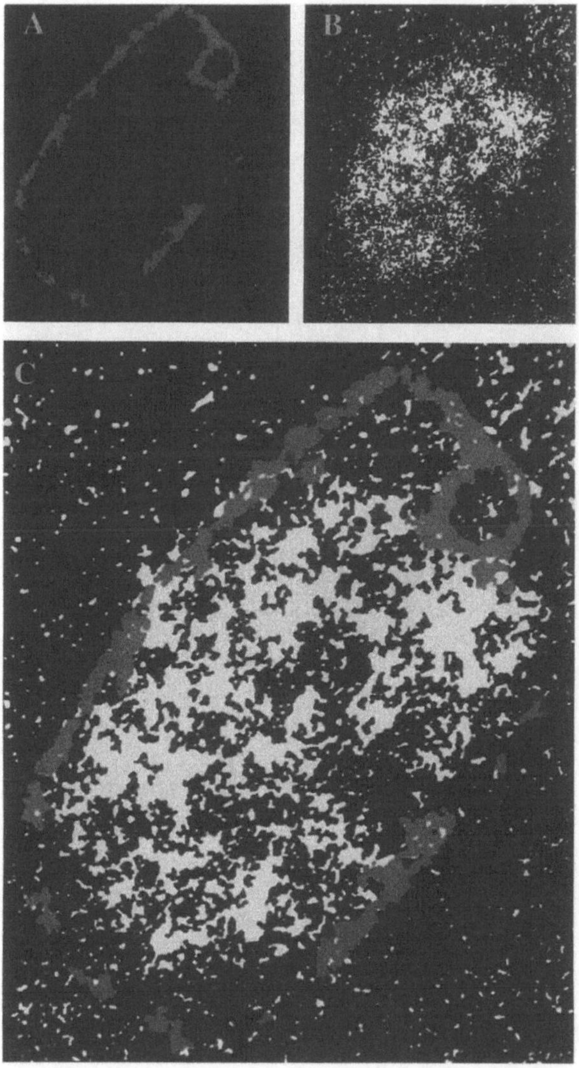

Figure 3 (continued). 3-2: The darkfield views of glucagon [panel (a) in 3-1] and GAD 67 [panel (c) in 3-2] mRNA expression have been pseudo-colored in purple (glucagon-A) and yellow (GAD-B). Superimposition of the two pseudo-colored images confirms the mutually exclusive expression of glucagon and GAD 67 mRNA by islet cells.

formation or phenotypic reversion, or both. Alterations in GAD protein expression patterns in cultured islets and in transformed B-cells have been reported by others.[71] As with the data derived from isolated islets, care must be taken when extrapolating from the findings in cultured B-cell lines to physiological or even pathological conditions in vivo.

Expression in human pancreas. The cloning of a GAD 65 cDNA from human islets[40] was coincident with the partial cloning of a GAD 67 cDNA from human

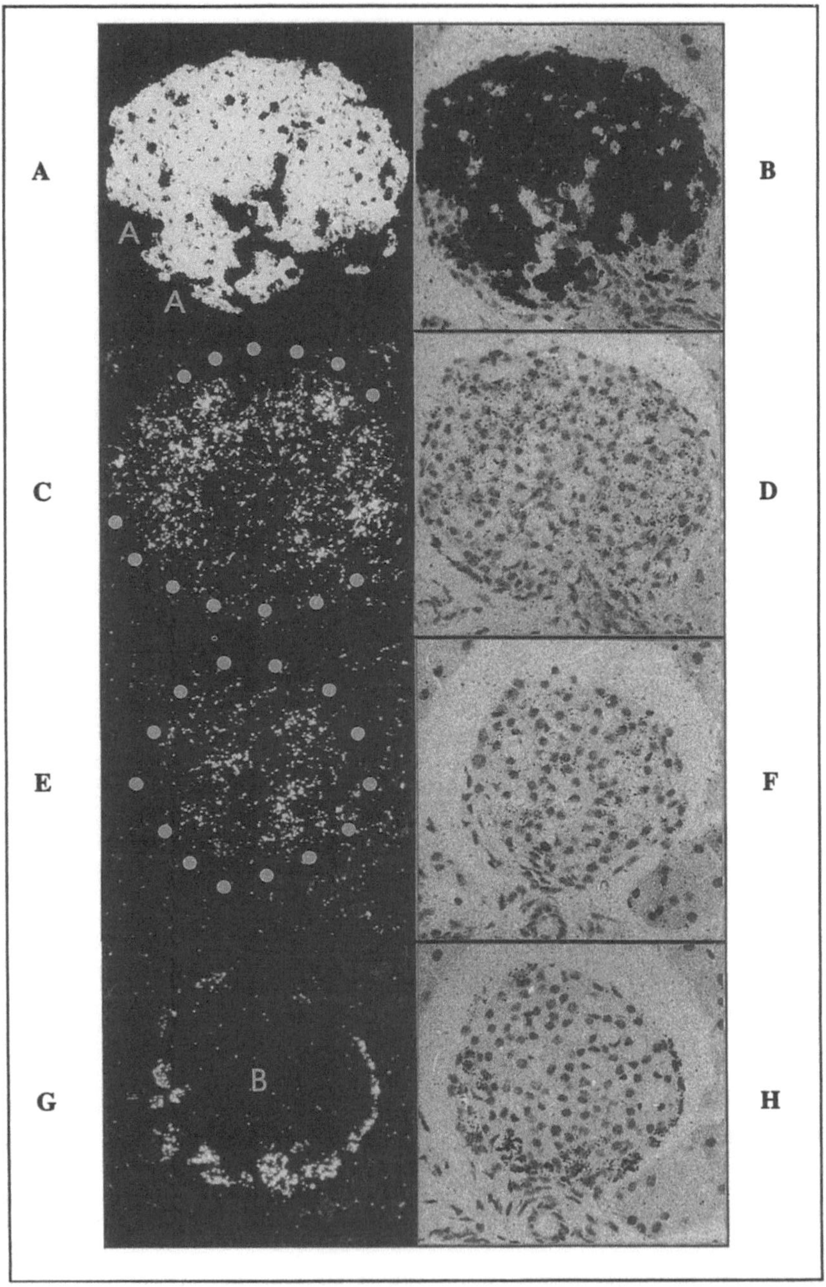

Figure 4. GAD 67 mRNA is expressed in mouse islet B-cells and not A-cells. 4-1: Darkfield and corresponding lightfield views of adjacent 5 μm sections of the same islet in perfusion-fixed mouse pancreas hybridized with ^{3}H-labeled insulin (A, B), ^{33}P-labeled mB-GAD 67 cRNA (C–F), and ^{3}H-labeled glucagon antisense cRNAs (G, H). The two sections in A and D, and in E–H, are serial sections. The border of the islet, and hence the position of A-cells, is marked in C and F. The positions of A-cells (A) and B-cells (B) are marked in panels A and G, respectively. The magnification of all eight panels is the same.

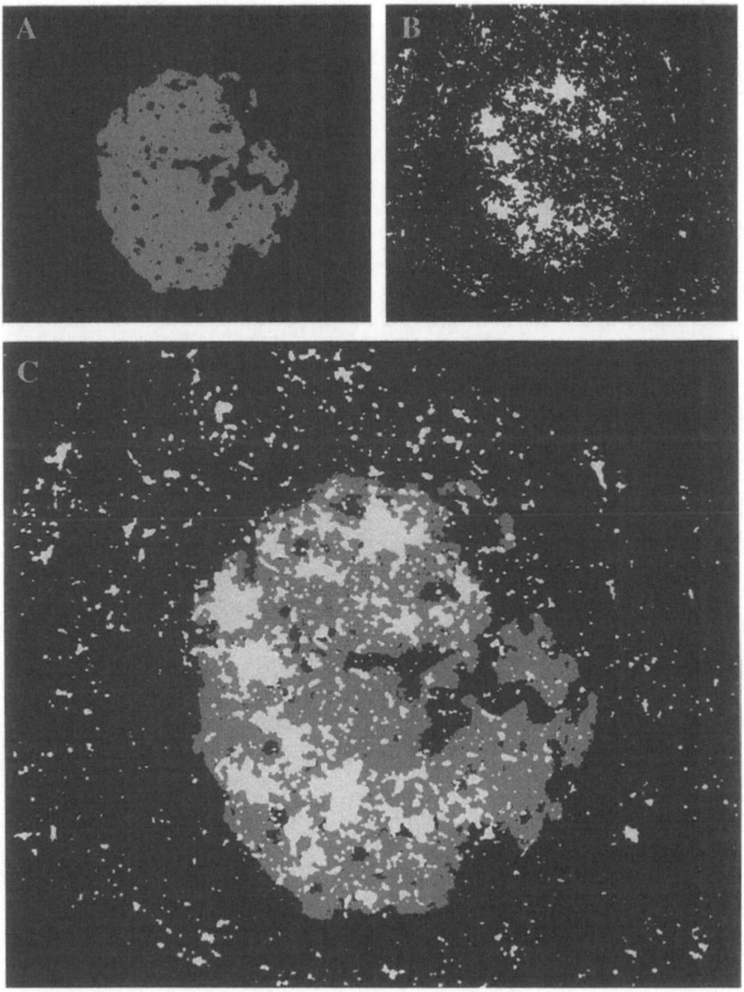

Figure 4 (continued). 4.2: The darkfield views of insulin (A in 4-1) and GAD 67 (C in 4-1) mRNA expression have been pseudo-colored in purple (insulin, A) and yellow (GAD 67, B). Superimposition of the two pseudo-color images (C) confirms the colocalization of insulin and GAD 67 mRNA expression in islet cells.

insulinoma cells.[33] However, GAD 67 mRNA was undetectable in normal human islets by Northern blot analysis.[40] This observation led Karlsen et al.[40] to conclude that GAD 65 was the only form expressed. This conclusion was reinforced by further studies on GAD protein and mRNA expression.[59,60,71–73,81]

The cloning of a GAD 67 cDNA from human islets using reverse transcriptase PCR[24,36] demonstrated that GAD 67 mRNA is expressed, although at low levels. RNase protection assays, which are $\sim$ 50-fold more sensitive than Northern blot hybridizations,[78] have confirmed this finding. GAD 67 mRNA is ex-

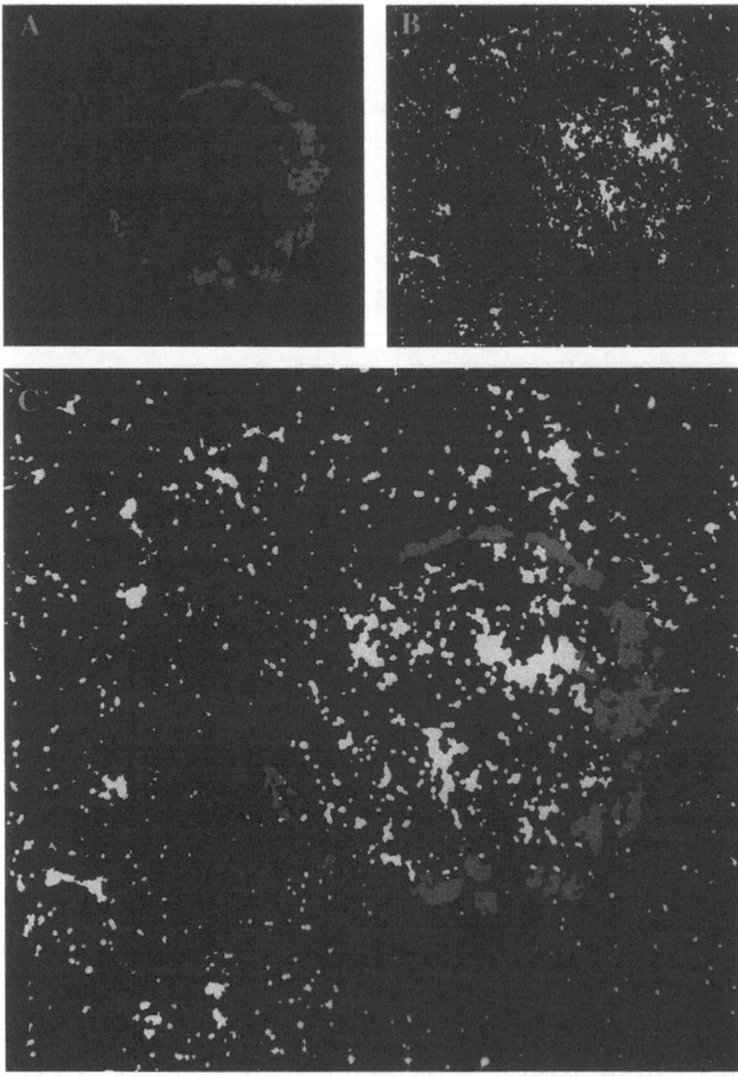

Figure 4 (continued). 4-3: The darkfield views of glucagon (G in 4-1) and GAD 67 (E in 4-1) mRNA expression have been pseudo-colored in purple (glucagon, A) and yellow (GAD 67, B). Superimposition of the two pseudo-color images (C) confirms the mutually exclusive expression of glucagon and GAD 67 mRNA by islet cells.

pressed at low levels in human pancreas, although GAD 65 mRNA is 200 times more abundant[34] (Fig. 5).

Only GAD 65 mRNA could be localized by in situ hybridization, and its expression was restricted to islets (Fig. 6). The site of GAD 67 mRNA expression could not be localized. However, GAD 67 mRNA was found in isolated islets and in whole pancreas. The same relative levels of the two GAD mRNAs were present in both, which implies that GAD 67 expression is also limited to islets.

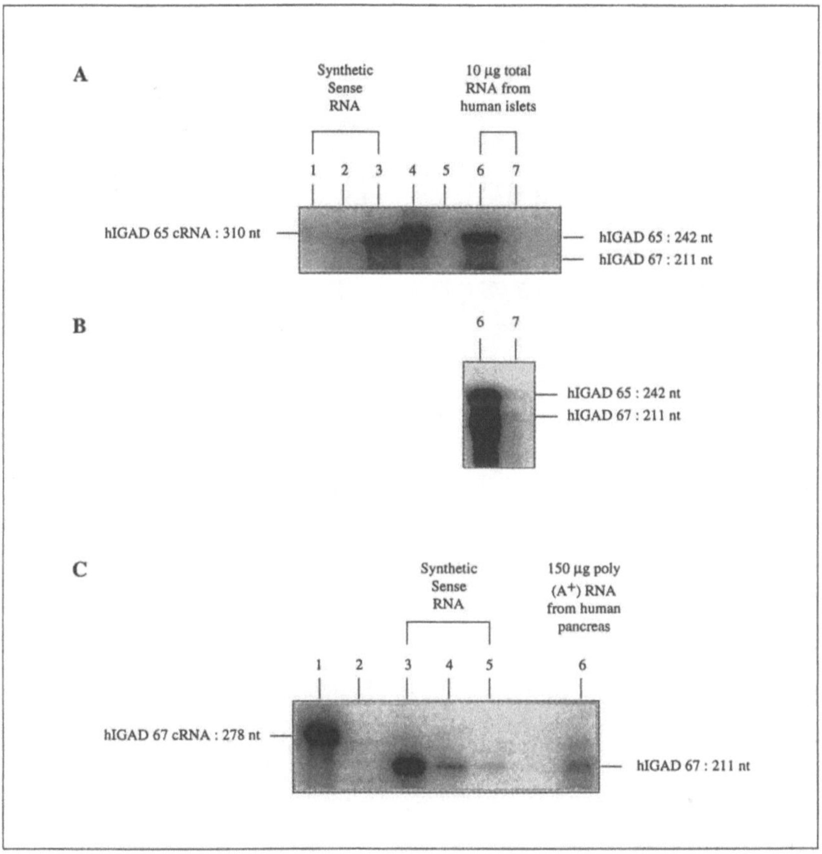

Figure 5. Analysis by RNase protection assay of GAD 67 and GAD 65 mRNA expression in human pancreas. (A,B,C) Phosphorimages of digestion products separated on denaturing polyacrylamide gels from RNase protection assay analysis of GAD 67 and GAD 65 mRNA. In A, the signals from a graded series of synthetic GAD 65 sense cRNAs hybridized to a GAD 65 cRNA probe are shown in lane 1 (0.1 pg), lane 2 (1 pg), and lane 3 (10 pg). Undigested GAD 65 cRNA probe was run in lane 4. The signal from 10 μg of total RNA from human islets hybridized with GAD 65 or GAD 67 CRNA probes is shown in lanes 6 and 7, respectively. No signal is seen after hybridization of the GAD 65 cRNA probe with tRNA (lane 5). Panel B is a longer exposure of lanes 6 and 7 from panel A. The sizes of the two protected fragments are indicated on the right. A faint signal is visible in lane 7, indicating that GAD 67 is expressed in human islets, although at a level 200-fold less than that of GAD 65. In panel C, the signals from a graded series of synthetic GAD 67 sense cRNAs hybridized to a GAD 65 cRNA probe are shown in lane 3 (10 pg), lane 4 (1 pg), and lane 5 (0.1 pg). Undigested GAD 65 CRNA probe was run in lane 1. The signal from 150 μg of poly(A+) RNA from whole human pancreas hybridized with the GAD 67 cRNA probe is shown in lane 6. No signal is seen after hybridization of the GAD probe with tRNA (lane 2).

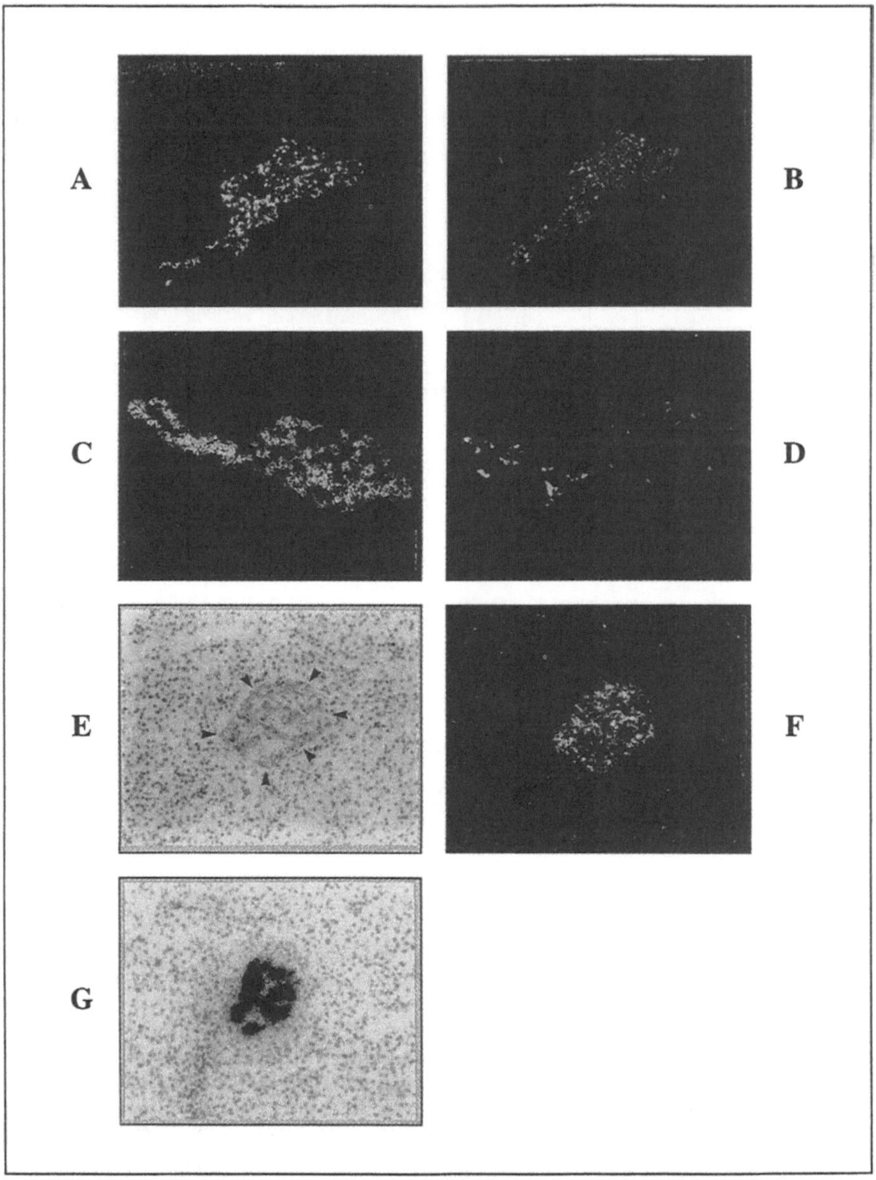

Figure 6. GAD 65 mRNA expression in human pancreas is restricted to islets. In situ hybridization for protein hormone and GAD 65 mRNAs in human pancreas. Five to 6 μm paraformaldehyde-fixed frozen sections of human pancreas were hybridized with ^{3}H-labeled insulin (A, C), glucagon (B), and somatostatin (D), and ^{33}P-labeled insulin (G) and hIGAD 65 antisense cRNA probes (E, F). Darkfield views are shown in A–D and F, and lightfield views in E and G. The magnifications and exposures in A–D and E–G are identical.

GAD 65 but not GAD 67 protein could be detected immunohistochemically[34] (Fig. 7). In small islets,[82,83] all cells contained immunoreactive GAD 65. The staining intensity varied between cells, with the most intense staining in peripherally located cells with the morphology of D-cells. In large islets, most cells

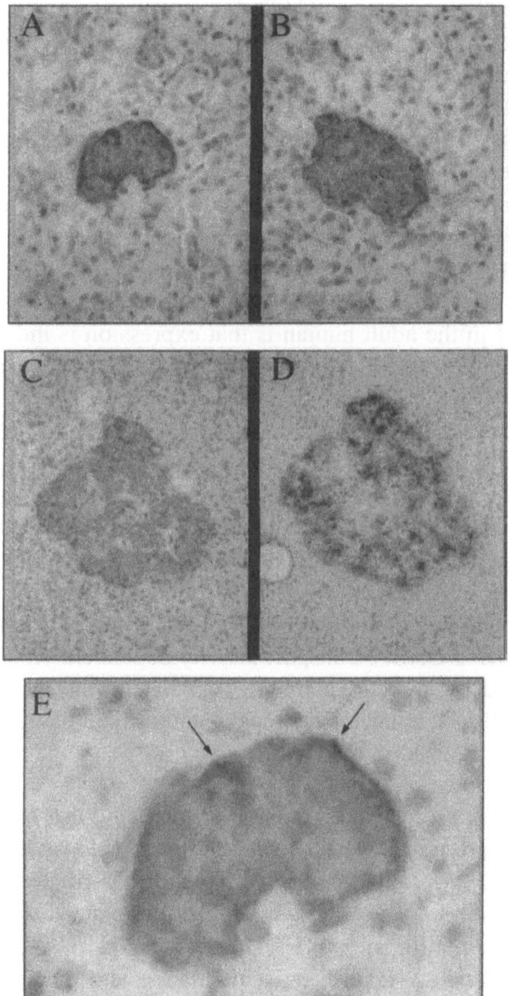

Figure 7. GAD 65 protein in the human pancreas is expressed in all islet cells in small islets and predominantly in B-cells in large islets. Acetone-fixed frozen sections of human pancreas (5–6 μm) were immunostained with a rabbit antiserum (1700) to human GAD 65 (A–C, E), or with guinea pig antiserum to insulin (D). The magnification is the same for A, B, C, and D. A is enlarged to form E. Two small islets from different sections are illustrated in A, B, and E. All islet cells contain immunoreactive GAD 65. The intensely stained peripheral cells (arrows in E) have the morphology of δ-cells. In the same large islet in sections 20 μm apart, immunostained for GAD 65 (C) and insulin (D), most but not all islet cells contain immunoreactive GAD 65; fewer contain insulin. The characteristic vascular connective tissue separating the cords and trabeculae of endocrine cells is clearly seen in this large islet.

contained GAD 65, but, as in small islets, their staining intensity varied. Immunoreactive GAD 65 is more widely distributed than insulin, suggesting that in the human islet, GAD is not entirely B-cell specific.[34] Two other studies on human[71,84] and one on the pigtail macaque pancreas[60] have also shown that GAD 65 protein is islet restricted but not B-cell specific within the primate pancreas. In this regard, cells positive for GAD, but not for insulin or C-peptide, were found in the postmortem pancreata of two patients dying 10 days and 2 years, respectively, after the diagnosis of IDDM.[84]

Islet cells arise sequentially from a common ductal cell progenitor.[65] In rodents, the separate islet cell types appear to arise in the order A-, B-, and then D cells, and intermediate A/B and B/D phenotypes occur. A similar developmental sequence is likely for human islet cells. Some proteins that are expressed by islet cells early in development may continue to be expressed into the postnatal period.[65] It is possible that one or both GADs may also be expressed developmentally. One explanation for the expression of GAD 65 in at least three types of islet cell in the adult human is that expression is induced in islet progenitor cells at the appropriate stages and that expression persists into adulthood.

Subcellular localization. Synaptic-like microvesicles (SLMVs) have been demonstrated in a number of endocrine cells, including B-cells and B-cell lines.[50,70] In the CNS, GABA is stored in and released from synaptic vesicles. B-cell SLMVs contain at least two of the integral membrane components of synaptic vesicles, namely, synaptophysin and p29. By immunogold histochemistry and electron microscopy, GAD was found to surround the SLMVs, and GABA was concentrated in areas rich in SLMVs. In the β-TC3 cell line, SLMVs were found to have a GABA-transport activity dependent on an electrochemical gradient generated by a vacuolar proton ATPase.[70] Thus, B-cells have mechanisms for the synthesis, uptake, storage, and controlled release of GABA that are very similar to those used in the CNS.

Summary. The study of different animal species, the use of either cultured or fresh unmanipulated tissues, and the adoption of a wide variety of techniques to detect GAD activity, protein, or mRNA, have all contributed to inconsistencies in the reported patterns of GAD expression in the pancreas. It is clear that under physiological conditions, both GAD genes are expressed in the pancreas of rats, mice, and humans. Nevertheless, qualitative and quantitative species differences exist and are relevant to understanding the role of GAD as an autoantigen in IDDM. In the pancreas of adult rats and mice, GAD 67 mRNA is the physiologically dominant form and is expressed in a B-cell restricted pattern. GAD 65 mRNA is expressed at a lower level in the adult rat, and its expression is similarly B-cell restricted. In the adult mouse, however, GAD 65 mRNA is expressed at very low levels, if at all, and its site of expression remains to be determined. The levels of GAD mRNA and protein expression appear to be parallel. As described later, GAD mRNA and protein are expressed widely outside the pancreas, at least in rodents, and GAD 67 again appears to be the dominant form. In the human pancreas, although both GAD mRNAs are expressed, the relative levels are reversed with respect to the rodent: GAD 65 mRNA is the physiologically dominant form. The levels of GAD 65 mRNA and protein appear to be parallel, and their locations within the islet correspond. Both mRNA and protein are islet specific, but not B-cell restricted. Both the site of GAD 67 mRNA expression and whether GAD 67 mRNA is actually translated into protein have to be determined.

GAD expression outside the pancreas

Other endocrine tissues. GAD and GABA are expressed in the endocrine component of the pancreas. Other endocrine tissues in which GAD and GABA are expressed are the testis, ovary, adrenal gland, and endocrine cells in the gastrointestinal tract. Both forms of GAD, and GABA, have been detected in the human, rat, and mouse testis[31,38] and in the rat oviduct.[46] In the testis, GAD 67 and GAD 65 mRNAs are expressed in the same cells within the seminiferous tubules of the mouse and rat testis, although their relative levels vary. GAD immunoreactive protein is found in the cells that also express GAD mRNAs.

Immunoreactive GAD is found in endocrine cells in the rat oxyntic and pyloric mucosa,[63] and GAD and/or GABA have been found in the duodenum, jejunum, ileum, and colon of the rat by other workers.[85] The chromaffin cells of the adrenal medulla and anterior pituitary gland also contain GAD.[86] It is of note that the rat pituitary gland preferentially expresses the smaller GAD form, that is, GAD 65, in the neurointermediate lobe.[87] When screened for expression of the B-cell–specific 64 kDa antigen, pituitary gland was found to be negative.[88]

Autonomic nervous system. GABAergic neurons have been found in both parasympathetic and sympathetic nervous systems and in the myenteric plexus.[86] However, several biochemical pathways are known to be capable of generating GABA[85] independent of GAD activity. Thus, the presence of GABA within a cell does not necessarily indicate that it also expresses GAD. GAD immunoreactivity has been found in GABAergic neurons in rat sympathetic ganglia[89,90] and in preganglionic nerve fibers.[91] The mRNA and protein for both GAD 65 and GAD 67 are found in the ileum of the rat, mouse, and guinea pig.[92] GAD enzymatic assays have shown that the majority of the GAD activity in the ileum is found in the myenteric plexus.[92–94]

Nonendocrine peripheral tissues. The data on the expression of the two GAD forms in nonendocrine peripheral tissues are far from complete. GAD enzymatic activity has been reported in homogenates of kidney, liver, lung, spleen,[46] and gallbladder.[95] GAD 67 but not GAD 65 mRNA has been found in mouse and rat spleen, and in mouse liver,[31] and also in thymus (M.B. French, unpublished data). However, GAD mRNA could not be detected in rodent kidney or lung by RNase protection assay. The GAD activity reported in homogenates of kidney may thus result from translation of very small amounts of mRNA that cannot be detected by this assay. Alternatively, GAD activity may result from another form of GAD. The site of GAD mRNA or protein expression in these peripheral organs has still to be determined. In the spleen and thymus, lymphoid cells are known to express neuroendocrine proteins.[96] It is possible that GAD 67 may be a further example. Data on expression of the two GAD forms are summarized in Table 2.

GAD and IDDM

The two distinct forms of GAD are expressed in both the central nervous system and a wide variety of peripheral tissues. An islet-specific GAD form or variant has not yet been identified. In addition, GAD expression in human islets is not restricted to B-cells, the target of autoimmune attack. The suggested pathogenic role of GAD in rodents and humans must be considered against this backdrop. Historically, GAD was first identified as the 64 kDa antigen in humans. However, we review first the autoimmunity to GAD in two rodent models of IDDM, the NOD mouse and the BB rat, and then in humans.

GAD in rodent IDDM

Both the NOD mouse[97] and the BB rat[98] develop antibodies to a 64 kDa antigen. These antibodies rapidly disappear in NOD mice and decline in a portion of BB rats, after the onset of IDDM. Thus, as in humans, this 64 kDa antigen is also implicated in the pathogenesis of IDDM. The identification of the 64 kDa

Table 2. Expression of GAD 67 and 65 in human, rat, and mouse

	Species		
Tissue	Human	Rat	Mouse
Brain	67√ 65√ Relative levels not determined	67√ 65√ 67 > 65 mRNA	67√ 65√ 67 > 65 mRNA
Testis	67√ 65 ?	67√ 65√ 67 > 65 mRNA GAD mRNA localized to spermatocytes	67√ 65√ Relative levels of GAD mRNA variable
Pancreas	67√ 65√ 65 >>> 67 mRNA 65 protein only 65 mRNA and protein Islet cell but not B-cell restricted	67√ 65√ 67 > 65 mRNA Both GAD mRNAs B-cell restricted	67√ 65√ 67 >>> 65 mRNA GAD 67 mRNA B-cell restricted
Spleen	—	67 mRNA√ GAD activity +	67 mRNA
Liver	—	Negative	67 mRNA√
Thymus	—	—	67 mRNA√

antigen in humans as GAD has prompted investigation of its role in IDDM in NOD mice and BB rats.

GAD antibodies. GAD antibodies (GADAb) and GAD-reactive T-cells have been reported in both animal models of IDDM.[9,10,99–101] However, not all investigators have been able to detect GADAb,[99,102–104] and most have had difficulty in detecting GAD-reactive T-cells. Although the exact timing varies between NOD mouse colonies, the insulitis lesion begins to develop at 3–4 weeks of age.[105] In a small study, antibodies to both GAD 67 and GAD 65 were found to be present in 4-wk-old mice, and their antibody titers increased with age.[10] A larger study found GADAb in 50% of female mice and 40% of male mice by 160 days of age. Interestingly, GADAb developed in younger animals and rose to higher levels in NOD-WEHI than in NOD-Leiter mice.[100] The NOD-WEHI subline has a lower incidence of spontaneous IDDM,[106] suggesting an inverse correlation between the development of GADAbs and IDDM.

Not all workers have detected GADAb in NOD mice, using a wide variety of assays, including RIA immunoprecipitation, Western blot, and ELISA.[99,102–104] This probably reflects the generally low levels of GADAb in NOD mice. In BB rats, the prevalence of GADAb has been reported to be as high as 88% at 120 days of age.[101]

GAD-reactive T-cells. Several groups have now demonstrated that NOD mouse T-cells react to GAD.[9,10,99,104] Like GADAb, the first T-cell responses to GAD are found at 4 wk of age.[9,10] The appearance of GAD-targeted autoimmunity is thus synchronous with the initial development of the insulitis lesion and precedes the development of diabetes. Two early studies reported that T-cells from 4-wk-old female NOD mice responded to recombinant GAD 65[9,10] and GAD 67.[10] The responses were maximal in 8-wk-old mice[9,10] but persisted at least until 20 wk of age.[10] In the second of these two studies, the responding T-cells were further characterized as $CD4^+$.[10] Interestingly, T-cell responses were first directed to the C-terminus of GAD and then spread to include other regions of the protein.[9] It is likely that the frequency of GAD-reactive T-cells is low in vivo. Some workers have been unable to detect GAD-reactive T-cells in either the NOD mouse or the BB rat, unless the animals were primed with GAD or GAD peptides in vivo (our unpublished observations).[107–109]

GAD tolerization. Probably the strongest evidence in support of a pathogenic role for GAD in IDDM has come from tolerization studies.[9,10] GAD 67 and GAD 65, when injected as the native or recombinant protein or as small peptides, can reduce the severity of insulitis and delay or prevent the onset of diabetes (our unpublished data).[9,10,103,110] Intrathymic,[10,110] i.v.,[9] and i.p.,[103] injections of GAD or GAD peptides have all reduced the severity of insulitis and have delayed the onset of diabetes. However, not all have altered the eventual incidence of diabetes.[110] An interesting point is that GAD-injected mice developed similar, or even higher, levels of GADAb compared with control mice, indicating that they were not tolerized at the B-cell level.[10,103] Finally, recombinant proteins expressed in *E. coli* contain endotoxin, a potent inducer of T-cell proliferation. The interpretation of T-cell or tolerance studies using *E. coli*–expressed recombinant GAD is only possible if endotoxin contamination has been controlled for.

GAD as a pathogen. In the NOD mouse, B-cell destruction appears to be T-cell mediated and highly selective. The insulitis lesion persists until B-cells are destroyed, leaving residual islet mantle cells.[1,2] One reason for this degree of specificity is B-cell–specific expression of a target antigen. Both GAD 67 and GAD 65 are widely expressed outside the pancreas in rodents. To date, there is no known islet-specific variant of either GAD form. Within the pancreas, GAD expression is restricted to B-cells. However, the predominant physiological form is GAD 67. Indeed, in the normal adult mouse, it is debatable whether GAD 65 is always expressed. As GAD 65 is presumed to be synonymous with the 64 kDa antigen, and therefore to be the "pathogenic" form of GAD, perhaps predominantly the "wrong" form of GAD is expressed in the "right" anatomical site in the rodent. It is conceivable that under appropriate physiological or pathological conditions, GAD 65 expression could be induced or upregulated in mouse B-cells.

If T-cell immunity to GAD at least drives an ongoing insulitis lesion, why is this lesion restricted to B-cell destruction within islets if GAD is a ubiquitous protein? We could postulate that despite the fact that GAD is expressed elsewhere, the presentation of GAD peptides on B-cells or by antigen-presenting cells within the islets in IDD must, nevertheless, be unique. In other words, the presentation of GAD (auto)epitopes recognized by pathogenic autoreactive T-cells in IDD is islet B-cell specific. The inability to present GAD peptides in an immunogenic form may explain why some organs that express GAD at moderately high levels, such as the ovaries and testes,[46,31] rarely become infiltrated by immune cells.[1,2] It is also worth noting that NOD mice regularly develop autoimmune infiltrates in organs such as the salivary, lachrymal, and thyroid glands,[1,2] which have not been reported to express GAD at any appreciable level.

Autoimmunity to GAD clearly occurs in rodents, and its development parallels that of the insulitis lesion. Although immune responses to GAD precede those to some specific B-cell antigens,[9,10] they lag behind others. T-cell responses to as yet unidentified B-cell components have been reported in NOD mice as young as 8 days old.[111] This suggests, therefore, that GAD may not be the primary autoantigen in the immune attack. The strongest evidence for GAD as a pathogenic autoantigen is that its administration can reduce the incidence of, or prevent, the onset of diabetes in NOD mice. Against this view, GAD-reactive T-cells have still not been shown to be able to transfer diabetes to an appropriate host animal, whereas B-cell–specific T-cell clones have.[112–114] Finally, there may not be a correlation between GAD autoimmunity and the progression to diabetes. GAD-reactive T-cells can be found in 24-wk-old nondiabetic NOD mice.[10] In addition, there is no relationship between the presence of GADAb and the development of diabetes.[10,100,101] Thus, in rodents at least, GAD is a target for autoimmunity, and modification of the immune response to GAD may alter the clinical course of the insulitis lesion. GAD, however, is unlikely to initiate islet pathology.

GAD in human IDDM

GAD antibodies. In addition to GAD, autoantigens in human IDDM include the islet antigen(s) in frozen sections of human pancreas recognized by islet cell antibodies (ICAs)[115] and (pro)insulin.[116] GAD antibodies (GADAb) were first identified in the stiff man syndrome, a rare neurological condition characterized by generalized muscle stiffness and autoimmune endocrine disease, including IDDM.[117] This association led to the identification of the 64 kDa autoantigen as GAD in 1990.[4] With early reports indicating a prevalence of GADAb in IDDM >70%,[59] there has been much interest in the use of GADAb as a screening test for IDDM. GAD 65, which is the main form expressed in human islets,[81] is the main target of autoantibodies. Thus, in two comparisons of antibody frequencies, GAD 65 antibodies occurred in 70% of recent-onset IDDM patient sera and GAD 67 antibodies were detected in <20%.[59,118] Preincubation of IDDM sera with native[119] or recombinant GAD 65[120] has been shown to block ICA in some sera, indicating that GAD comprises a portion of ICA. This is supported by several studies in which a strong association between GADAb and ICA has been noted.[121,122] GADAb are present before the onset of IDDM,[5,123] and levels are stable for several years in at-risk first-degree relatives positive for ICA.[122] However, the additional risk conferred by GADAb in these "at-risk" subjects is controversial.[124–127]

GADAb are most commonly measured by radiobinding assay with recombinant or native GAD, by ELISA or immunoprecipitation of GAD enzymatic activity. A large-scale comparison of these assay formats has revealed that the radiobinding assay has the highest overall sensitivity.[128] However, no large comparison of GADAb and ICA has yet been done to compare their sensitivity and specificity as screening tests for "preclinical" IDDM. Available evidence suggests that GADAb do not, by themselves, cause IDDM. Thus, transfer of peripheral blood mononuclear cells from patients with IDDM to mice with severe combined immunodeficiency (SCID) led to formation of GADAb in some mice, but impairment of B-cell function or evidence of islet cell damage did not occur.[129] Furthermore, GADAb are found in ICA-positive subjects who have normal B-cell insulin release.[130]

GAD-reactive T-cells. The cellular composition of the islet lesion in recent-onset IDDM patients is consistent with direct evidence from the NOD mouse model that B-cell destruction is mediated by T-cells, but the evidence in humans remains circumstantial. When fetal pig pro-islets, which contain GAD as well as other potential autoantigens such as (pro)insulin, were used to stimulate peripheral blood T-cells from IDDM subjects and their ICA-positive at-risk relatives, proliferative responses of CD4 T-cells were significantly higher than in HLA-matched healthy controls or autoimmune disease controls.[131] When GAD was depleted from a pro-islet extract by adsorption with GAD-1 monoclonal Ab, T-cell proliferative responses in at-risk relatives decreased by >50% in the majority of responders.[132] Thus, a significant proportion of the response to whole islet extract appeared to be accounted for by GAD-reactive T-cells.

At-risk relatives and recent-onset IDDM subjects both had higher T-cell proliferative responses to recombinant GAD 65 than ICA negative relatives or healthy controls.[133] However, there was no control for endotoxin in this study, which would have been present in the *E. coli*–produced GAD, and the control subjects were not HLA matched. Recombinant human brain and islet GAD 67 were expressed in *E. coli* in three sections as fusion proteins with glutathione-S-transferase (GST), and GST alone was similarly expressed as a control protein. Proliferation of peripheral blood T-cells to the central region of recombinant islet or brain GAD 67 (amino acids 208–404) minus proliferation to GST alone, was significantly higher in both at-risk relatives and recent-onset IDDM subjects than in other autoimmune disease subjects and HLA-matched healthy controls.[134] The probability of endotoxin in both the GAD-GST and GST proteins also makes these antigen preparations less than ideal, but the difference in responses is evidence for GAD specificity. Interestingly, there was an inverse correlation between the levels of antibodies to GAD and GAD-reactive T-cells in at-risk relatives,[135] raising the possibility that the balance between humoral and cellular immunity to GAD may be critical in the pathogenesis of IDDM. To investigate this possibility, the reactivity of T-cells to GAD 67 (amino acids 208–404) and denatured recombinant human insulin has been measured in at-risk relatives over a period of 2.5 years. Of four relatives who developed IDDM during this time, all had very high levels (stimulation index >12) of GAD-reactive T-cells, but not of insulin-reactive T cells, up to 15 months prior to clinical diagnosis.[136] Only a minority of those who did not progress to IDDM had high levels of GAD-reactive T-cells (M. Honeyman et al., unpublished). Thus, sequences from both GAD 65 and 67 are antigenic for T-cells in human IDDM, and GAD-reactive T-cells may be more useful than GADAb levels as a predictive marker for IDDM in at-risk individuals.

More recently, peripheral blood mononuclear cells from both IDDM and healthy control subjects were shown to proliferate in response to peptides derived from the GAD 65 or GAD 67.[137] In particular, one region of GAD 65 (aa 473–555) was associated with responses in IDDM but not control subjects, raising the possibility that this region contains an IDDM-specific T-cell epitope that could be used in the prediction of IDDM. However, this result was not reproduced in a further study, in which T-cells from both at-risk relatives and IDDM subjects reacted most strongly to a peptide from the center of GAD 65 (aa 247–279), NMYAMMIARFKMF<u>PEVKEK</u>G, compared with non–HLA-matched control subjects. This region contains a small area of high sequence similarity to a peptide from the P2-C protein of Coxsackie B4 virus (underlined).[138] The results from both studies need to be confirmed using larger numbers and HLA-matched controls, because the homologous peptide (aa 253–273) in GAD 67 AISNMYSIMAA<u>RYKYFPEVKT</u> has been shown to induce proliferation in T-cells not only from at-risk relatives and recent-onset IDDM subjects, but also from HLA-matched control subjects.[139]

These apparently conflicting results could, however, be consistent with a scenario in which a healthy individual bearing any or all of the HLA susceptibility alleles for IDDM (DR3, DR4, DQ2, DQ8) is infected by a virus (e.g., Coxsackie) and sets up an immune response to a viral peptide that mimics the central region of GAD 65 and 67. If B-cells were damaged during the infection so that the mimicked GAD peptides became available for presentation by the same HLA alleles, autoimmunity could be established. Responses to other peptides of GAD capable of being presented would then be expected to occur.

GAD fulfills many of the criteria for being a pathogenic autoantigen in IDDM. It is a target of both autoantibodies and T-cells in humans with IDDM and those at risk for IDDM, and GAD autoantibodies at least are a highly specific and sensitive marker of islet autoimmunity in humans. The finding of T-cell reactivity to GAD in the NOD mouse and the ability of GAD to tolerize at the T-cell level and to reduce both insulitis and diabetes in the NOD mouse argues strongly for the pathogenic role of GAD. Further studies are required to optimize schedules and to elucidate the mechanisms of islet antigen-specific tolerance induction in the NOD mouse model, as a guide toward the development of specific immunotherapy for the prevention of IDDM in humans. There are reasons, however, for exercising caution when considering GAD as a tolerogen in humans. Unlike the other major B-cell antigen, (pro)insulin, GAD is not B-cell specific. Without further animal studies and knowledge of the epitopes within GAD that elicit T-cell reactivity to human B-cells, it would seem unwise to attempt to manipulate immunity to this key enzyme, which is also present in brain and peripheral tissues other than islets. Further understanding of the mechanisms by which autoimmunity to GAD in IDDM results in selective destruction of B-cells within the islets may one day allow GAD or GAD peptides to be used therapeutically to prevent IDDM in humans.

Acknowledgments

We thank Margaret Thompson for secretarial assistance.

References

1. Kikutani H, Makino S. The murine autoimmune diabetes model: NOD and related strains. Adv Immunol 51:285–322, 1992.
2. Mordes JP, Desemone J, Rossini AA. The BB rat. Diabetes Metab Rev 3:725–750, 1987.
3. Honeyman MC, Harrison LC. The immunologic insult in type 1 diabetes. Springer Semin Immunopathol 14:253–274, 1993.
4. Baekkeskov S, Aanstoot HJ, Christgau S, et al. Identification of the 64K autoantigen in insulin-dependent diabetes as the GABA-synthesizing enzyme glutamic acid decarboxylase. Nature 347:151–6, 1990.
5. DeAizpurua H, Wilson YM, Harrison LC. Glutamic acid decarboxylase autoantibodies in preclinical insulin-dependent diabetes. Proc Natl Acad Sci USA 89:9841–45, 1992.
6. DeAizpurua HJ, Harrison LC. Glutamic acid decarboxylase in insulin-dependent diabetes mellitus Diabetes Metab Rev 8:133–147, 1992.
7. Baekkeskov S, Nielsen JH, Marner B, et al. Autoantibodies in newly diagnosed diabetic children immunoprecipitate human pancreatic islet cell proteins. Nature 298:167–9, 1982.

8. Harrison LC. Islet cell antigens in insulin-dependent diabetes: Pandora's box revisited. Immunol Today 13:348–52, 1992.
9. Kaufman DL, Clare-Salzer M, Tian J, et al. Spontaneous loss of T-cell tolerance to glutamic acid decarboxylase in murine insulin-independent diabetes. Nature 366:69–72, 1993.
10. Tisch R, Yang X-D, Singer SM, et al. Immune response to glutamic acid decarboxylase correlates with insulitis in non-obese diabetic mice. Nature 366:72–5, 1993.
11. Roberts E, Frankel S. γ-aminobutyric acid in brain: Its formation from glutamic acid. J Biol Chem 187:55–63, 1950.
12. Kaufman DL, Houser CR, Tobin AJ. Two forms of the gamma-aminobutyric acid synthetic enzyme glutamate decarboxylase have distinct intraneuronal distributions and cofactor interactions. J Neurochem 56:720–3, 1991.
13. Jackson FR, Prokaryotic and eukaryotic pyidoxal-dependent decarboxylases are homologous. J Mol Evol 31:326–9, 1990.
14. Maras B, Sweeney G, Barra D, et al. The amino acid sequence of glutamate decarboxylase from *Escherichi coli*. Evolutionary relationship between mammalian and bacterial enzymes. Eur J Biochem 204:93–8, 1992.
15. Smith DK, Kassam T, Singh B, Elliott JF. *Escherichia coli* has two homologous glutamate decarboxylase genes that map to distinct loci. J Bacteriol 174:5820–26, 1992.
16. Hao R, Schmit JC. Cloning of the gene for glutamate decarboxylase and its expression during conidiation in *Neurospora crassa*. Biochem J 293:735–8, 1993.
18. Baum G, Chen Y, Arazi T, et al. A plant glutamate decarboxylase containing a calmodulin binding domain. Cloning, sequence, and functional analysis. J Biol Chem 268:19610–17, 1993.
19. Phillips AM, Salkoff LB, Kelly LE. A neural gene from *Drosophila melanogaster* with homology to vertebrate and invertebrate glutamate decarboxylases. J Neurochem 61:291–301, 1993.
19. Jackson FR, Newby LM, Kulkarni SJ. Drosophila GABAergic systems: Sequence and expression of glutamic acid decarboxylase. J Neurochem 54:1068–78, 1990.
20. Martin DL, Martin SB, Wu SJ, Espina N. Regulatory properties of brain glutamate decarboxylase (GAD): The apoenzyme of GAD is present principally as the smaller of two molecular forms of GAD in brain. J Neurosci 11:2725–31, 1991.
21. Martin DL, Rimvall K. Regulation of gamma-aminobutyric acid synthesis in the brain. J Neurochem 60:395–407, 1993.
22. Kobayashi Y, Kaufman DL, Tobin AJ. Glutamic acid decarboxylase cDNA: Nucleotide sequence encoding an enzymatically active fusion protein. J Neurosci 7:2768–72, 1987.
23. Kaufman DL, McGinnis JF, Krieger NR, Tobin AJ. Brian glutamate decarboxylase cloned in lambda gt-11: Fusion protein produces gamma-aminobutyric acid. Science 232:1138–40. 1986.
24. Cram DS, Barnett LD, Joseph JL, Harrison LC. Cloning and partial nucleotide sequence of human glutamic acid decarboxylase cDNA from brain and pancreatic islets. Biochem Biophys Res Commun 716:1239–44, 1991.
25. Kelly CD, Edwards Y, Johnstone AP, et al. Nucleotide sequence and chromosomal assignment of a cDNA encoding the large isoform of human glutamate decarboxylase. Ann Hum Genet 56:255–65, 1992.
26. Bu DF, Erlander MG, Hitz BC, et al. Two human glutamate decarboxylases, 65-kDa GAD and 67-kDa GAD, are each encoded by a single gene. Proc Natl Acad Sci USA 89:2115–19, 1992.
27. Suzuki R, Asami I, Amann E, Wagatsuma M. Sequences of two porcine glutamic acid decarboxylases (65- and 67-kDa GAD). Gene 152:257–60, 1995.
28. Wyborski RJ, Bond RW, Gottlieb DI. Characterization of a cDNA coding for rat glutamic acid decarboxylase. Mol Brain Res 8:193–8, 1990.

29. Julien JF, Samama P, Mallet J. Rat brain glutamic acid decarboxylase sequence deduced from a cloned cDNA. J Neurochem 54:703–5, 1990.
30. Erlander MG, Tillakaratne NJ, Feldblum S, et al. Two genes encode distinct glutamate decarboxylases. Neuron 7:91–100, 1991.
31. Faulkner-Jones BE, Cram DS, Kun J, Harrison LC. Localisation and quantitation of expression of two glutamate decarboxylase genes in pancreatic β-cells and other peripheral tissues of mouse and rat. Endocrinology 133:2962–72, 1993.
32. Katarova Z, Szabo G, Mugnaini E, Greenspan RJ. Molecular identification of the 62 kd form of glutamic acid decarboxylase from the mouse. Eur. Neurosci. 2:190–202, 1990.
33. Michelsen BK, Petersen JS, Boel E, et al. Cloning, characterization, and autoimmune recognition of rat islet glutamic acid decarboxylase in insulin-dependent diabetes mellitus. Proc Natl Acad Sci USA 88:8754–58, 1991.
34. Cram DS, Faulkner-Jones B, Kun J, Harrison LC. Glutamic acid decarboxylase-67 (GAD 67): Expression relative to GAD 65 in human islets and mapping of autoantibody epitopes. Endocrinology 136:1111–19, 1995.
35. Kawasaki E, Moriuchi R, Watanabe M, et al. Cloning and expression of large isoform of glutamic acid decarboxylase from human pancreatic islet. Biochem Biophys Res Commun 192:1353–59, 1993.
36. Yamashita K, Cram DS, Harrison LC. Molecular cloning of full-length glutamic acid decarboxylase 67 from human pancreas and islets. Biochem Biophys Res Commun 192:1347–52, 1993.
37. Christgau S, Schierbeck H, Aanstoot HJ, et al. Pancreatic beta cells express two autoantigenic forms of glutamic acid decarboxylase, a 65-kDa hydrophilic form and a 64-kDa amphiphilic form which can be both membrane-bound and soluble. J Biol Chem 266:21257–64, 1991.
38. Persson H, Pelto HM, Metsis M, et al. Expression of the neurotransmitter-synthesizing enzyme glutamic acid decarboxylase in male germ cells. Mol Cell Biol 10:4701–11, 1990.
39. Lee DS, Tian J, Phan T, Kaufman DL. Cloning and sequence analysis of a murine cDNA encoding glutamate decarboxylase (GAD). Biochim Biophys Acta 1216:157–60, 1993.
40. Karlsen AE, Hagopian WA, Grubin CE, et al. Cloning and primary structure of a human islet isoform of glutamic acid decarboxylase from chromosome 10. Proc Natl Acad Sci USA 88:8337–41, 1991.
41. Huang WM, Reed FL, Wu E, Wu JY. Molecular cloning and amino acid sequence of brain L-glutamate decarboxylase. Proc Natl Acad Sci USA 87:8491–95, 1990.
42. Brilliant MH, Szabo G, Katarova Z, et al. Sequences homologous to glutamic acid decarboxylase cDNA are present on mouse chromosomes 2 and 10. Genomics 6:115–22, 1990.
43. Vassort C, Riviere M, Bruneau G, et al. Assignment of the rat genes coding for dopa decarboxylase (DDC) and glutamic acid decarboxylases (GAD1 and GAD2). Mamm Genome 4:202–6, 1993.
44. Bond RW, Jansen KR, Gottlieb DI. Pattern of expression of glutamic acid decarboxylase mRNA in the developing rat brain. Proc Natl Acad Sci USA 85:32312–14, 1988.
45. Szabo G, Katarova Z, Greenspan R. Distinct protein forms are produced from alternatively spliced bicistronic glutamic acid decarboxylase mRNAs during development. Mol Cell Biol 14:7535–45, 1994.
46. Tillakaratne NJ, Erlander MG, Collard MW, et al. Glutamate decarboxylases in nonneural cells of rat testis and oviduct: Differential expression of GAD 65 and GAD 67. J Neurochem 58:618–27, 1992.
47. Bond RW, Wyborski RJ, Gottlieb DI. Developmentally regulated expression of an exon containing a stop codon in the gene for glutamic acid decarboxylase. Proc Natl Acad Sci USA 87:8771–75, 1990.

48. Behar T, Schaffner A, Laing P, et al. Many spinal cord cells transiently express low molecular weight forms of glutamic acid decarboxylase during embryonic development. Brain Res 72:203–18, 1993.
49. Behar T, Ma W, Hudson L, Barker JL. Analysis of the anatomical distribution of GAD 67 mRNA encoding truncated glutamic acid decarboxylase proteins in the embronic rat brain. Brain Res 77:77–87, 1994.
50. Reetz A, Solimena M, Matteoli M, et al. GABA and pancreatic beta-cells: Colocalization of gluamic acid decarboxylase (GAD) and GABA with synaptic-like microvesicles suggests their role in GABA storage and secretion. EMBO J 10:1275–84, 1991.
51. Christgau S, Aanstoot HJ, Schierbeck H, et al. Membrane anchoring of the autoantigen GAD 65 to microvesicles in pancreatic beta-cells by palmitoylation in the NH2-terminal domain. J Cell Biol 118:309–20, 1992.
52. Solimena M, Aggujaro D, Muntzel C, et al. Association of GAD-65, but not of GAD-67, with the Golgi complex of transfected Chinese hamster ovary cells mediated by the N-terminal region. Proc Natl Acad Sci USA 90:3073–77, 1993.
53. Soimena M, Dirkx R. Jr, Radzynski M, et al. A signal located within amino acids 1-27 of GAD 65 is required for its targeting to the Golgi complex region. J Cell Biol 126:331–41, 1994.
54. Shi Y, Veit B, Baekkeskov S. Amino acid residues 24-31 but not palmitoylation of cysteines 30 and 45 are required for membrane anchoring of glutamic acid decarboxylase, GAD. J Cell Biol 124:927–34, 1994.
55. Dirkx R, Thomas A, Linsong L, et al. Targeting of the 67 kDa isoform of glutamic acid decarboxylase to intracellular organelles is mediated by its interaction with the NH_2-terminal region of the 65-kDa isoform of glutamic acid decarboxylase. J Biol Chem 270:2241–46, 1995.
56. Oertel WH, Schmechel DE, Tappaz ML, Kopin IJ. Production of a specific antiserum to rat brain glutamic acid decarboxylase by injection of an antigen-antibody complex. Neuroscience 6:2689–700, 1981.
57. Li L, Jiang J, Hagopian WA, et al. Differential detection of rat islet and brain glutamic acid decarboxylase (GAD) isoforms with sequence-specific peptide antibodies. J Histochem Cytochem 43:53–9, 1995.
58. Chang YC, Gottlieb DI. Characterization of the proteins purified with monoclonal antibodies to glutamic acid decarboxylase. J Neurosci 8:2123–30, 1988.
59. Hagopian WA, Michelsen B, Karlsen AE, et al. Autoantibodies in IDDM primarily recognize the 65,000-M(r) rather than the 67,000-M(r) isoform of glutamic acid decarboxylase. Diabetes 42:631–6, 1993.
60. Hagopian WA, Karlsen AE, Petersen JS, et al. Regulation of glutamic acid decarboxylase diabetes autoantigen expression in highly purified isolated islets from *Macaca nemestrina*. Endocrinology 132:674–81, 1993.
61. Greif KF, Erlander MG, Tillakaratne NJ, Tobin AJ. Postnatal expression of glutamate decarboxylases in developing rat cerebellum. Neurochem Res 16:235–42, 1991.
62. Velloso LA, Kampe O, Eizirik DL, et al. Human autoantibodies react with glutamic acid decarboxylase antigen in human and rat but not in mouse islets. Diabetologia 36:39–46, 1993.
63. Gilon P, Tappaz M, Remacle C. Localization of GAD-like immunoreactivity in the pancreas and stomach of the rat and mouse. Histochemistry 96:355–65, 1991.
64. Okada Y, Taniguchi H, Schimada C. High concentration of GABA and high glutamate decarboxylase activity in rat pancreatic islets and human insulinoma. Science 194:620–2, 1976.
65. Alpert S, Hanahan D, Teitelman G. Hybrid insulin genes reveal a developmental lineage for pancreatic endocrine cells and imply a relationship with neuron. Cell 53:295–308, 1988.
66. Rorsman P, Berggren PO, Bokvist K, et al. Glucose-inhibition of glucagon secre-

tion involves activation of GABAA-receptor chloride channels. Nature 341:233–6, 1989.
67. Weir GC, Bonner-Weir S. Islets of Langerhans: The puzzle of intraislet interactions and their relevance to diabetes. J Clin Invest 85:983–7, 1990.
68. Gilon P, Bertrand G, Loubatieres MM, et al. The influence of gamma-aminobutyric acid on hormone release by the mouse and rat endocrine pancreas. Endocrinology 129:2521–9, 1991.
69. Robbins MS, Grouse LH, Sorenson RL, Elde RP. Effect of muscimol on glucose-stimulated somatostatin and insulin release from the isolated, perfused rat pancreas. Diabetes 30:168–71, 1981.
70. Thomas-Reetz A, Hell JW, During MJ, et al. A γ-aminobutyric acid transporter driven by a proton pump is present in synaptic-like microvesicles of pancreatic b-cells. Proc Natl Acad Sci USA 90:5317–21, 1993.
71. Petersen JS, Russel S, Marshall MO, et al. Differential expression of glutamic acid decarboxylase in rat and human islets. Diabetes 42:484–95, 1993.
72. Kim J, Richter W, Aanstoot HJ, et al. Differential expression of GAD 65 and GAD 67 in human, rat and mouse pancreatic islets. Diabetes 42:1799–808, 1993.
73. Bjork E, Kampe O, Karlsson FA, et al. Glucose regulation of the autoantigen GAD 65 in human pancreatic islets. J Clin Endocrinol Metab 75:1574–76, 1992.
74. Bjork E, Kampe O, Andersson A, Karlsson FA. Expression of the 64 kDa/glutamic acid decarboxylase rat islet cell autoantigen is influenced by the rate of insulin secretion. Diabetologia 35:490–3, 1992.
75. Vincent SR, Hokfelt T, Wu JY, et al. Immunohistochemical studies of the GABA system in the pancreas. Neuroendocrinology 36:197–204, 1983.
76. Garry DJ, Garry MG, Sorenson RL. Ultrastructural immunocytochemical localization of L-glutamate decarboxylase and GABA in rat pancreatic zymogen granules. Cell Tissue Res 252:191–7, 1988.
77. Sorenson RL, Garry DG, Brelje TC. Structural and functional considerations of GABA in islets of Langerhans. Beta-cells and nerves. Diabetes 40:365–74, 1991.
78. Krieg PA, Melton DA. In vitro RNA synthesis with SP6 RNA polymerase. Methods Enzymol 155:397–415, 1987.
79. Efrat S, Linde S, Kofod H, et al. Beta-cell lines derived from transgenic mice expressing a hybrid insulin gene-oncogene. Proc Natl Acad Sci USA 85:9037–41, 1988.
80. Hamaguchi K, Gaskins HR, Leiter EH. NIT-1, a pancreatic beta-cell line established from a transgenic NOD/Lt mouse. Diabetes 40:842–9, 1991.
81. Karlsen AE, Hagopian WA, Petersen JS, et al. Recombinant glutamic acid decarboxylase (representing the single isoform expressed in human islets) detects IDDM-associated 64,000-M(r) autoantibodies. Diabetes 41:1355–9, 1992.
82. Gepts W, LeCompte PM. The pancreatic islets in diabetes. Am J Med 70:105–115, 1981.
83. Orci L. Macro- and micro-domains in the endocrine pancreas. Diabetes 31:538–65, 1982.
84. Vives PM, Somoza N, Vargas F, et al. Expression of glutamic acid decarboxylase (GAD) in the alpha, beta and delta cells of normal and diabetic pancreas: Implications for the pathogenesis of type I diabetes. Clin Exp Immunol 92:391–6, 1993.
85. Gilon P, Remacle C. GABA and endocrine cells of the gastro-entero-pancreatic system: A review. In: GABA outside the CNS. Erdo SL, eds. Springer-Verlag, Heidelberg pp 111–32, 1990.
86. Erdo SL, Wolff JR. Gamma-aminobutyric acid outside the mammalian brain. J Neurochem 54:363–72, 1990.
87. Henry S, Tappaz M. Relative enrichment of the lighter 59 kDa form of glutamic acid decarboxylase in nerve endings: An immunoblotting study in pituitary neurointermediate lobe. Neurosci Lett 131:253–6, 1991.
88. Christie MR, Pipeleers DG, Lernmark A, Baekkeskov S. Cellular and subcellular

localization of an Mr 64,000 protein autoantigen in insulin-dependent diabetes. J Biol Chem 265:376–81, 1990.
89. Happola O, Paivarinta H, Soinila S, et al. Localisation of L-glutamate decarboxylase and GABA transaminase immunoreactivity in the sympathetic ganglia of the rat. Neuroscience 21:271–81, 1987.
90. Kenny SL, Ariano MA. The immunofluorescence localization of glutamate decarboxylase in the rat superior cervical ganglion. J Auton Nerv Syst 17:211–15, 1986.
91. Dobo E, Joo F, Wolff JR. Distinct subsets of neuropeptide Y-negative principle neurons receive basket-like innervation from enkephalinergic and GABAergic axons in the superior cervical ganglion of adult rats. Neuroscience 75:833–44, 1993.
92. Williamson S, Faulkner-Jones BE, Cram DS, et al. Transcription and translation of two glutamate decarboxylase genes in the ileum of rat, mouse and guinea pig. J Auton Nerv Syst. 55:18–28, 1995.
93. Miki Y, Taniyama K, Tanake C, Tobe T. GABA, glutamic acid decarboxylase and GABA transaminase levels in the myenteric plexus in the intestine of humans and other mammals. J Neurochem 40:861–6, 1983.
94. Tanaka C. GABA in peripheral tissues. Life Sci 37:2221–35, 1985.
95. Saito N, Taniyama K, Tanaka C. Uptake and release of GABA in guinea pig gall bladder. Am J Physiol 249:G129–96, 1985.
96. Ericsson A, Geenen V, Robert F, et al. Expression of preprotachykinin-A and neuropeptide-Y messenger RNA in the thymus. Mol Endocrinol 4:1211–18, 1990.
97. Atkinson MA, McLaren NK. Autoantibodies in non obese diabetic mice immunopreciptate 64000-Mr islet antigen. Diabetes 37:1587–90, 1988.
98. Baekkeskov S, Dyrberg T, Lernmark A. Autoantibodies to a 64-kilodalton islet cell protein precede the onset of spontaneous diabetes in the BB rat. Science 224:1348–50, 1984.
99. Elliott JF, Qin HY, Bhatti S, et al. Immunization with the larger isoform of mouse glutamic acid decarboxylase (GAD 67) prevents autoimmune diabetes in NOD mice. Diabetes 43:1494–99, 1994.
100. DeAizpurua HJ, French MB, Chosich N, Harrison LC. Natural history of humoral immunity to glutamic acid decarboxylase in non-obese diabetic (NOD) mice. J Autoimmun 7:643–53, 1994.
101. Zeigler M, Schlosser M, Hamann J, et al. Autoantibodies to glutamate decarboxylase detected in diabetes-prone BB/OK rats do not distinguish onset of diabetes. Exp Clin Endocrinol 102:98–103, 1994.
102. Velloso LA, Eizirik DL, Karlsson FA, Kämpe O. Absence of autoantibodies against glutamate decarboxylase (GAD) in the non-obese diabetic (NOD) mouse and low expression of the enzyme in mouse islets. Clin Exp Immunol 96:129–37, 1994.
103. Petersen JS, Karlsen AE, Markholst H, et al. Neonatal tolerisation with glutamic acid decarboxylase but not with bovine serum albumin delays the onset of diabetes in NOD mice. Diabetes 44:1478–84, 1994.
104. Bieg S, Seissler J, Herberg L, et al. GAD 65 is recognized by T-cells, but not by antibodies from NOD-mice. Autoimmunity 17:189–94, 1994.
105. Pozzilli P, Signore A, Williams AJ, Beale PE. NOD mouse colonies around the world—recent facts and figures Immunol Today 23:193–6, 1993.
106. Baxter AG, Koulmanda M, Mandel TE. High and low diabetes incidence non obese diabetic (NOD) mice: Origins and characterisation. Autoimmunity 9:61–7, 1991.
107. Diaz JL, Ways J, Hammonds P. T-lymphocyte lines specific for glutamic acid decarboxylase (GAD) the 64K beta-cell antigen of IDDM. Diabetes 41:118–21, 1992.
108. Matsuba H, Jitsukawa T, Yamagata N, et al. Establishment of rat glutamic acid decarboxyalse (GAD)-reactive T-cell clones from NOD mice. Immunol Lett 42:101–3, 1994.
109. Chen S-L, Whiteley PJ, Freed DC, et al. Responses of NOD congenic mice to a glutamic acid decarboxylase-derived peptide. J Autoimmun 7:635–41, 1994.

110. Gerling IC, Atkinson MA, Leiter EH. The thymus as a site for evaluting the potency of candidate β-cell autoantigens in NOD mice. J Autoimmun 7:851–8, 1994.
111. Gelber C, Paborsky L, Singer S, et al. Isolation of nonobese diabetic mouse T-cells that recognize novel autoantigens involved in the early events of diabetes. Diabetes 43:33–9, 1994.
112. Wegmann DR, Shehadah N, Lafferty KJ, et al. Establishment of islet-specific T-cell lines and clones from islet isografts placed in spontaneously diabetic NOD mice. J Autoimmun 6:517–27, 1993.
113. Wegmann DR, Gill RG, Norbury-Glaser M, et al. Analysis of spontaneous T cell response to insulin in NOD mice. J Autoimmun 7:833–43, 1994.
114. Haskins K, McDuffie M. Acceleration of diabetes in young NOD mice with a CD4+ islet-specific T cell clone. Science 249:1433–36, 1990.
115. Bottazzo GF, Florin-Christensen A, Doniach D. Islet cell antibodies in diabetes mellitus with autoimmune polyendocrine deficiencies. Lancet 2:1279–83, 1974.
116. Palmer JP, Asplin CM, Clemons P, et al. Insulin antibodies in insulin-dependent diabetics before insulin treatment. Science 222:1337–39, 1983.
117. Solimena M, Folli F, Denis-Donini S, et al. Autoantibodies to glutamic acid decarboxylase in a patient with stiff-man syndrome, epilepsy and type 1 diabetes mellitus. N Engl J Med 318:1012–20, 1988.
118. Seissler J, Amann J, Mauch L, et al. Prevalence of autoantibodies to the 65- and 76-kD isoforms of glutamate decarboxylase in insulin-dependent diabetes mellitus. J Clin Invest 92:1394–99, 1993.
119. Genovese S, Bonifacio E, McNally JM, et al. Distinct cytoplasmic islet cell antibodies with different risks for type 1 (insulin-dependent) diabetes mellitus. Diabetologia 35:385–8, 1992.
120. Atkinson MA, Kaufman DL, Newman D, Tobin AJ, Maclaren NK. Islet cell cytoplasmic autoantibody reactivity to glutamate decarboxylase in insulin-dependent diabetes. J Clin Invest 91:350–6, 1993.
121. Martino GV, Tappaz ML, Braghi S, et al. Autoantibodies to glutamic acid decarboxylase (GAD) detected by an immuno-trapping enzyme activity assay: Relation to insulin-dependent diabetes mellitus and islet cell antibodies. J Autoimmun 4:915–23, 1991.
122. Schmidli RS, DeAizpurua HJ, Harrison LC, Colman PG. Antibodies to glutamic acid decarboxylase in at-risk and clinical insulin-dependent diabetic subjects: Relationship to age, sex and islet cell antibody status, and temporal profile. J Autoimmun 7:55–66, 1994.
123. Tuomilehto J, Zimmet P, Mackay IR, et al. Antibodies to glutamic acid decarboxylase as predictors of insulin-dependent diabetes mellitus before clinical onset of disease. Lancet 343:1383–85, 1994.
124. Schmidli RS, Colman PG, Harrison LC. Do glutamic acid decarboxylase antibodies improve the prediction of IDDM in first-degree relatives at-risk for IDDM? J Autoimmun 7:873–9, 1994.
125. Bingley PJ, Christie MR, Bonifacio E, et al. Combined analysis of autoantibodies improves prediction of IDDM in islet cell antibody-positive relatives. Diabetes 43:1304–10, 1994.
126. Thivolet CH, Tappaz M, Durand A, et al. Glutamic acid decarboxylase (GAD) autoantibodies are additional predictive markers of type 1 (insulin-dependent) diabetes mellitus in high risk individuals. Diabetologia 35:570–6, 1992.
127. Aanstoot HJ, Sigurdsson E, Jaffe M, et al. Value of antibodies to GAD 65 combined with islet cell cytoplasmic antibodies for predicting IDDM in a childhood population. Diabetologia 37:917–24, 1994.
128. Schmidli RA, Colman PG, Bonifacio E, and participating labs. Disease sensitivity and specificity of 52 assay for glutamic acid decarboxylase antibodies. Diabetes 44:636–40, 1995.

129. Petersen JS, Marshall MO, Baekkeskov S, Hejnaes KR, Hoier-Madsen M, Dyrberg T. Transfer of type 1 (insulin-dependent) diabetes mellitus associated autoimmunity to mice with severe combined immunodeficiency (SCID). Diabetologia 36:510–15, 1993.
130. Wagner R, Genovese S, Bosi E, et al. Slow metabolic deterioration towards diabetes in islet cell antibody positive patients with autoimmune polyendocrine disease. Diabetologia 37:365–71, 1994.
131. Harrison LC, Chu SX, DeAizpurua HJ, et al. Islet-reactive T cells are a marker of preclinical insulin-dependent diabetes. J Clin Invest 89:1161–65, 1992.
132. DeAizpurua HJ, Honeyman MC, Harrison LC. A 65 kDa antigen/glutamic acid decarboxylase (GAD) in fetal pig pro-islets: Co-precipitation with a 38 kDa protein and recognition by T cells in humans at risk for insulin-dependent diabetes. J Autoimmun 5:759–70, 1992.
133. Atkinson MA, Kaufman DL, Campbell L, et al. Response of peripheral-blood mononuclear cells to glutamate decarboxylase in insulin-dependent diabetes. Lancet 339:458–9, 1992.
134. Honeyman MC, Cram DS, Harrison LC. Glutamic acid decarboxylase 67-reactive T cells: A marker of insulin-dependent diabetes. J Exp Med 177:535–40, 1993.
135. Harrison LC, Honeyman MC, DeAizpurua HJ, et al. Inverse relation between humoral and cellular immunity to glutamic acid decarboxylase in subjects at risk of insulin-dependent diabetes. Lancet 341:1365–69, 1993.
136. Honeyman MC, Stone N, Dempsey-Collier M, et al. Islet antigen-specific T cells and HLA alleles in subjects at risk for IDD. Proc 13th Intl Immunol Diab Wrkshp, Paris, p 108, 1994.
137. Lohmann T, Leslie RD, Hawa M, et al. Immunodominant epitopes of glutamic acid decarboxylase 65 and 67 in insulin-dependent diabetes mellitus. Lancet 343:1607–8, 1994.
138. Atkinson MA, Bowman MA, Campbell L, et al. Cellular immunity to a determinant common to a glutamate decarboxylase and coxsackie virus in insulin-dependent diabetes. J Clin Invest 94:2125–9, 1994.
139. Honeyman MC, Stone N, Brusic V, Harrison LC. Identification of T-cell epitopes containing DR binding motifs in human glutamic acid decarboxylase (GAD) 67. Proc 13th Intl. Immunol Diab Wrkshp, Paris 101, 1994.

2

Molecular genetics of insulin-dependent diabetes mellitus: Analysis of congenic strains

HIROSHI IKEGAMI,[1] SUSUMU MAKINO,[2] AND TOSHIO OGIHARA[1]
[1]*Department of Geriatric Medicine, Osaka University Medical School, Osaka;*
[2]*Aburahi Laboratories, Shionogi & Co. Ltd., Osaka, Japan.*

Key words: insulin-dependent diabetes mellitus, IDDM, genetics, NOD mouse, congenic mice, major histocompatibility complex, MHC, autoimmune disease, ancestral haplotype.

Insulin-dependent diabetes mellitus (IDDM) is caused by autoimmune destruction of insulin-producing B-cells of the pancreas in genetically susceptible individuals. Both in animal models and humans, major genetic susceptibility to IDDM has been mapped to the major histocompatibility complex (MHC). In addition, recent genome screening has mapped non-MHC susceptibility genes to chromosomal segments in the mouse and human genomes. For most loci, however, only chromosomal locations are known, with no information on the responsible genes and their functions.

Diabetogenic genes in the NOD mouse

Breeding studies in crosses of NOD with nondiabetic control strains have suggested that IDDM in NOD mice is controlled by multiple genes, and that at least one of them is located in the MHC.[1–3] The number of estimated diabetogenic genes varies depending on the inbred strain to which the NOD mouse is outcrossed. This is because most diabetogenic alleles at susceptibility loci are not NOD-specific mutations, but are variants that are also found in some control strains, as in the case of susceptibility genes for other common, complex diseases. This is true even for MHC-linked genes: An NOD allele of the Ab gene, which was initially reported to be unique to the NOD mouse,[4] is now known to be present in several NOD-related[5–7] and unrelated strains.[8]

Correspondence: Dr. Hiroshi Ikegami, Department of Geriatric Medicine, Osaka University Medical School, 2-2 Yamadaoka, Suita, Osaka 565, Japan.

Lessons from Animal Diabetes VI. Ed. E. Shafrir.

In breeding studies, only loci that differ between NOD and control strains can be analyzed for their contribution to IDDM susceptibility. When two strains are identical for a susceptibility locus, that is, a control strain has the same susceptible allele as the NOD mouse, then linkage of this locus with IDDM cannot be detected in breeding studies because in such studies all mice are homozygotes for the susceptible allele at this locus, regardless of the origin of the allele (i.e., derived from NOD or control strains). Thus, different crosses detect different sets of diabetogenic genes, some of which are common among several breeding studies and some are detected only in a certain breeding cross. An example of the former is the MHC-linked gene(s) on chromosome 17, which was detected in all breeding studies reported so far.[1,2,5,9–12] An example of the latter is *Idd13* on chromosome 2, which was detected only when the NOD mouse was outcrossed to the NOR mouse.[13] Thus, as the number of strains to which NOD is outcrossed increases, the number of loci whose contribution to disease susceptibility are detected increases.

In general, the more remote the genetic background of a strain from that of the NOD mouse, the greater the number of detectable susceptibility loci. An extreme case is that only one strain (e.g., strain A) has the resistant allele for IDDM at a locus and all other strains, including NOD, have the same susceptible allele at the locus. In this case, only breeding studies between NOD and the strain A will detect linkage of this locus with IDDM, with all other breeding studies failing to detect linkage. With respect to population genetics, such a locus is not so important because most individuals have the same susceptible allele at the locus and, therefore, polymorphism at this locus cannot be used as a marker to detect high risk individuals in a population. This locus, however, is important for analyzing disease mechanisms and for developing a preventive strategy, because a gene at this locus does contribute to disease susceptibility and molecules encoded by this gene contribute to disease mechanisms. Therefore, identification of any loci that contribute to disease susceptibility is important.

To date, >15 diabetogenic loci have been mapped to chromosomal segments in the mouse genome (Table 1).[1,10,12–18] As expected, NOD alleles at most loci are associated with increased susceptibility to IDDM, whereas alleles of control strains are associated with resistance to the disease. In some cases, however, an NOD allele is associated with resistance in comparison with a control allele. In the case of *Idd7* and *Idd8* loci, which were detected in crosses of NOD with B10.H-2^{g7} mice, homozygosity of NOD alleles was associated with resistance to IDDM, whereas B10 alleles were associated with disease susceptibility.[14] Similarly, homozygosity for NON alleles at marker loci on chromosome 6 was reported to be associated with susceptibility to IDDM in crosses of NOD with NON.H-2^{g7}.[16] This is again explained by the fact that IDDM is a common, complex disease and susceptible alleles at most, if not all, loci are not NOD-specific mutations, but relatively common variants found in some control strains. Since the NOD mouse is derived from a diabetic mouse found by chance during the process of inbreeding of a strain from an outbred colony,[19] the NOD mouse can be considered as a strain resulting from an unfortunate combination of variants at several susceptibility loci. Therefore, although the NOD mouse has sufficient

Table 1. Susceptibility loci for IDDM of the NOD mouse

	Chromosome	Candidate	Diabetes	Insulitis	Cross	Refs.
Idd1	17	Class II MHC	+	+	C3H, B6, B10, NON	1,2,9,10
Idd2	9		+	−	NON,	10
					B10.H-2^{g7}	14
					NON.H-2^{g7}	16
Idd3	3	*Il2*	+	+	B10.H-2^{g7}	14
					NON.H-2^{g7}	16
Idd4	11		+	−	B10.H-2^{g7}	14
Idd5	1	*Nramp, Il1r*	+	+	B10.H-2^{g7}, B6	52,53
Idd6	6		+	−	B10.H-2^{g7}	14
					NON.H-2^{g7}	16
Idd7	7	*Tgfb1*	+	−	B10.H-2^{g7}	14
					NON.H-2^{g7}	16
Idd8	14		+	−	B10.H-2^{g7}	14
Idd9	4		+	−	B10.H-2^{g7}	14,55
					B6.PL-Thy1^a/Cy	55
					NON.H-2^{g7}	16
Idd10	3	*Cd1,Cd2, Csfm,Cd53*	+	+	B10.H-2^{g7}	39
					NON.H-2^{g7}	16
Idd11	4	*Nhe1*	+	?	B6, SJL	15
Idd12	14		+	?	B6, SJL	15
Idd13	2	*Il1*	+	−	NOR	13
Idd14	13		+	?	NON.H-2^{g7}	16
Idd15	5		+	?	NON.H-2^{g7}	16
Idd16	17	Class I MHC(?)	+	−(?)	CTS	18

susceptibility genes for the development of IDDM, the combination of NOD alleles is not necessarily the most efficient one. More efficient combinations of susceptibility alleles than NOD may thus be possible. The above-mentioned data suggest that replacement of NOD alleles with B10 alleles at *Idd7* and *Idd8* loci may increase penetrance of the disease and that a better animal model of IDDM than the NOD mouse may be established.

Congenic strains

One of the best ways to directly demonstrate that chromosomal regions where susceptibility genes were mapped in breeding studies contribute to disease susceptibility is to make NOD strains that are congenic for the regions encoding diabetes resistance alleles and to demonstrate that the strains are less susceptible to the disease. To date, only a limited number of groups have used congenic NOD strains to study susceptibility loci for IDDM.

Chromosome 17 (Idd1). Several NOD strains that are congenic for MHC on chromosome 17 were established to study the contribution of *Idd1* to disease susceptibility.[17,20–23] Reciprocal congenic strains with NOD MHC on the genetic background of control strains were also established.[3] All but one of the MHC congenic NOD strains reported to date were completely resistant to IDDM when NOD MHC was replaced by control MHC in the homozygous state (Table 2). Among these congenic strains, NOD.H-2^b, NOD-H-2^{i7}, NOD-H-2^{h4}, and NOD.NCT-H-2 do not express I-E molecules, as in the case of the NOD mouse, and NOD.H-2^b and NOD-H-2^{h4} have the same D^b molecules as the NOD. The complete resistance to IDDM in these strains indicates that lack of I-E expression and/or D^b is not sufficient for disease susceptibility. These data, together with a rare NOD allele of the Ab gene, suggest that the Ab gene alone or in combination with other NOD MHC may be responsible for susceptibility encoded by *Idd1*. This, however, has not previously been proven due to lack of the NOD allele of Ab gene in MHC haplotypes other than NOD.

To prove that *Idd1* is encoded by the Ab gene (alone or in combination with other NOD MHC genes) and to fine-map *Idd1*, it is essential to find intra-MHC recombinant strains that possess a segment of the NOD MHC encoding a candidate Ab gene and to demonstrate that NOD mouse strains that are congenic for the recombinant MHC develop IDDM. The recombination frequency within the MHC, however, is too low to generate intra-MHC recombinant haplotypes by standard breeding between NOD and other inbred strains. Recombination events, however, have occurred during many historical meioses, giving rise to intra-MHC recombinant haplotypes in NOD-related strains. We previously found such a recombinant haplotype in a sister strain of the NOD mouse. The CTS mouse has the same class II MHC, including candidate A and E genes, as the NOD mouse, but different class I MHC from the NOD mouse.[5,6]

Identification of an intra-MHC recombinant haplotype, which possesses a segment of NOD MHC encoding candidate A and E genes, gave us an opportunity to test whether *Idd1* is encoded by the class II MHC of the NOD mouse. To test this, NOD.CTS-H-2 congenic mice were established by repetitive backcrossing of CTS to NOD with selection of the CTS MHC.[3,24] In contrast to the complete resistance to IDDM in previously reported MHC congenic NOD strains, the NOD.CTS-H-2 congenic mice, which have CTS MHC in the homozygous state,

Table 2. NOD MHC congenic strains[3,18,23]

		MHC		K	I-A	I-E	D	IDDM
NOD		NOD	(g7)	d	g7	(−)	b	82%
NOD.H-2^b	N6F1	B10	(b)	b	b	(−)	b	0%
NOD.H-2^{i7}	N6F1	B10.D2	(i7)	b	b	(−)	d	0%
NOD.H-2^{i5}	N6F1	B10.A(5R)	(i5)	b	b	k	d	0%
NOD.H-2^{h4}	N6F1	B10.A(4R)	(h4)	k	k	(−)	b	0%
NOD.H-2^{h2}	N6F1	B10.A(2R)	(h2)	k	k	k	b	0%
NOD.H-2^k	N6F1	B10.BR	(k)	k	k	k	k	0%
NOD.NON-H-2	N6F1	NON	(nbl)	non	non	non	b	0%
NOD.NCT-H-2	N3F1	NCT	(nct)	s	s	(−)	nct	0%
NOD.CTS-H-2	N13-19F1	CTS	(cts)	cts	g7	(−)	cts	44%

non: unique allele for the NON mouse
not: unique allel for the NCT mouse
cts: unique allele for the CTS mouse

developed IDDM.[3] Analysis of the CTS MHC in comparison with the NOD MHC revealed that the 0.5 cM region encompassing class II Oa to class III Hsp 70 loci was identical in the two strains (Fig. 1).[18,20] These data indicate that *Idd1* is encoded in the 0.5 cM segment of the CTS MHC, which is probably derived from the same ancestral chromosome as the NOD MHC. The most likely candidates are A and E genes. This is further supported by the fact that transgenic NOD mice that express I-A or I-E molecules of the control strains are resistant to IDDM,[25–27] although conflicting data were also reported (McDevitt et al., unpublished results).

Although the congenic NOD.CTS-H-2 developed IDDM, the cumulative incidence of IDDM in CTS MHC homozygotes was much lower than that in the NOD parental strain (95% reduction in males and 56% reduction in females at N13-N19F1 generations).[18] Since the IDDM incidence in NOD MHC homozygotes was the same as that in the NOD parental strain, the difference appears to be due to the difference in chromosome 17. This was further supported by the fingerprinting of the genome of the congenic strain. Fingerprinting with more than 100 microsatellite markers indicated that only a <11 cM segment of chromosome 17 was of CTS origin and the rest of the genome was completely replaced by the NOD genome.[18]

These data indicate that the reduction of IDDM incidence is caused by a gene or genes located in the <11 cM segment of chromosome 17. Since the congenic mice contain the same A and E genes as the NOD mouse, the lower incidence of IDDM in NOD. CTS-H-2 congenic mice indicates that the A and E genes are not sufficient to confer full susceptibility to IDDM. These data suggest that *Idd1* consists of two or more components, one of which is located adjacent to, but distinct from, previously known *Idd1* candidates, the A and E genes. We propose the designation of the MHC-linked gene located in the 0.5 cM region as *Idd1* and a second component newly mapped to outside of this segment as *Idd16* (Table 1). Recently, Hattori et al.[28] reported that a second MHC-linked gene(s) mapped to the region between K and Lmp2 loci in the MHC in breeding studies in crosses of NOD with the B10.A(R209), which has a hot spot that cause intra-MHC recombinations between MHC class I-K and class II I-A region. The gene(s) mapped by Hattori et al. may be identical to *Idd16*.

Since the *Tnfa* gene is a major candidate gene for IDDM and is located within the region where a second MHC-linked gene was mapped, we determined the nucleotide sequences of all the exons and exon–intron junctions of the NOD and CTS alleles of the *Tnfa* gene.[18] A 3 bp deletion at nt1502–1504 and several nucleotide substitutions (G to A transition at nt1718 and 2514, and insertion of C at nt2088) were found in the NOD allele as compared with the BALB/c allele.[29] The same sequence, however, has recently been reported for the C57BL/6 allele.[30] The CTS allele was different from the NOD allele, as well as previously reported alleles, in that it had no deletion at nt1502–1504, but the rest of the sequence was identical to the NOD and C57BL/6 alleles. These data, together with difference in *Tnfa* microsatellite alleles between NOD and CTS mice,[20] indicate that *Tnfa* regions of NOD and CTS mice were derived from different ancestral chromosomes. Although differences in DNA sequences were found between the two strains, all were in noncoding sequences, and the coding sequences and splic-

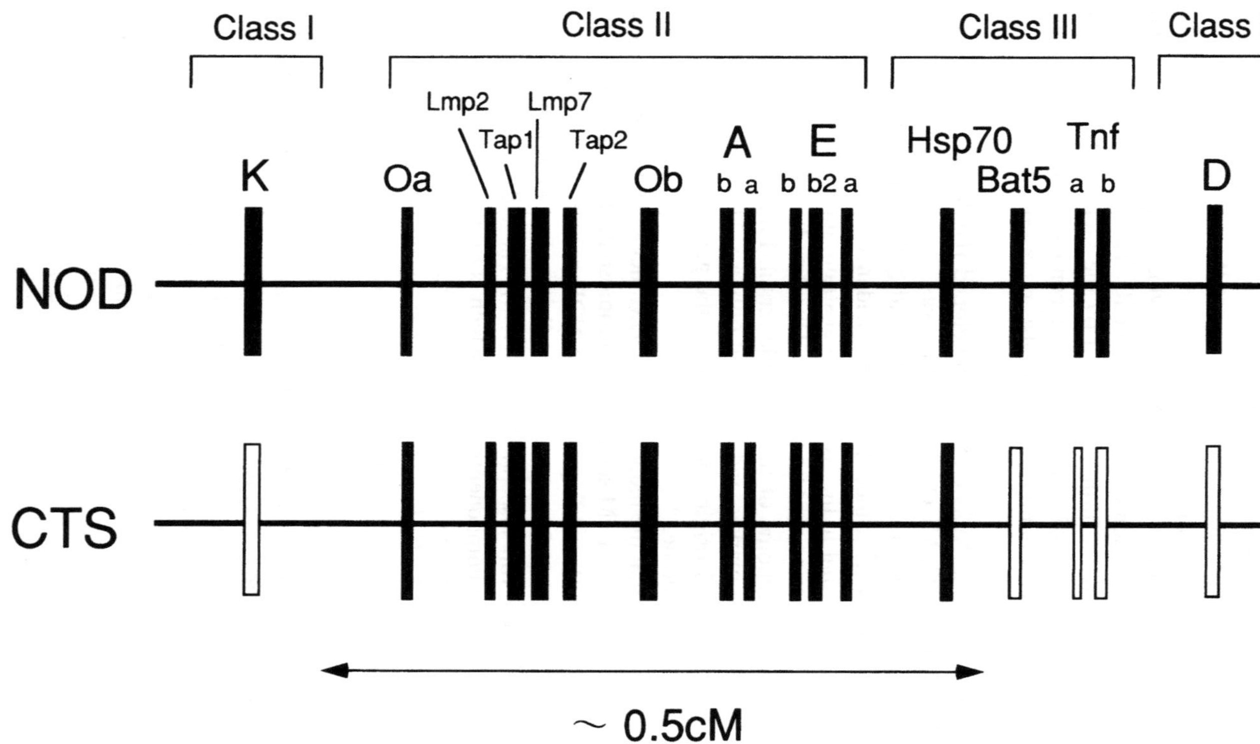

Figure 1. MHC of the NOD and CTS mouse. Shaded box, NOD allele; open box, non-NOD allele.

ing donor and acceptor sequences of NOD and CTS alleles were identical to the published sequences of the BALB/c allele. These data indicate that the primary structure of the TNF molecule in NOD mice is identical to that in CTS as well as control mice, and that amino acid changes in the TNF molecule are unlikely to be a determinant of IDDM susceptibility. Although the coding sequence of the *Tnfa* gene is identical in NOD and CTS mice, there might be a difference in the regulatory region, as suggested by different microsatellite alleles and intron sequences in the two strains. The difference in the regulatory region affects the production of TNF[31,32] and may thereby modulate the incidence of IDDM.

Other candidates are class I MHC genes. Our data suggest that allelic polymorphisms of class I *K* and/or *D* genes may be one of the mechanisms in the reduction of the incidence of IDDM in NOD.CTS-H-2 congenic mice. Consistent with this, suppression of insulitis was reported in transgenic NOD mice expressing the L^d gene.[33] Transgenic expression of NOD alleles of class I molecules (K^d and/or D^b) in NOD.CTS-H-2 congenic mice might help clarify the contribution of class I gene variants to IDDM susceptibility.

Ancestral haplotype congenic mapping. The same class II MHC as that in the NOD mouse was found not only in the CTS mouse, but also in one mouse in the original colony, Jcl:ICR mice.[6] The same class II MHC was also found in other NOD-related strains derived from the same outbred Jcl:ICR colony[7] (Ikegami et al., to be published). Identification of the same class II MHC in both the sister and the original strains indicates that class II MHC of the NOD mouse, which was originally reported to be unique to the NOD mouse, is not specific for NOD but was derived from an ancestral haplotype that was contained in the original closed colony. Thus, MHC of NOD, CTS, and other related strains, such as ILI, are the product of recombination among ancestral haplotypes contained in the original outbred colony. Overall, we found one recombinant in 10 Jcl:ICR mice[6] and three recombinants in 25 NOD-related strains (Ikegami et al., manuscript in preparation). In contrast, >4000 mice are reported to be needed to generate one recombinant between the class I K and class II Ab genes,[34] and many more mice are necessary to generate a double recombinant, as in the case of the CTS MHC. These data indicate the power of our method to identify intra-MHC recombinants. Identification of intra-MHC recombinant haplotypes and the establishment of a congenic NOD strain with the recombinant MHC enabled us to detect a second component of *Idd1*,[18] which could not be detected by usual breeding studies or transgenic techniques.

The method, termed *ancestral haplotype congenic mapping*, can be applied to other *Idd* loci, as well as other complex traits. For example, the NOD allele of the *Il2* gene, a candidate for *Idd3*, was found in NOD-related strains (Ikegami et al., to be published). Since these strains have different microsatellite markers adjacent to the *Il2* locus, a segment of chromosome 3 flanking the *Il2* locus in these strains can be distinguished from the NOD chromosome. Introgression of these chromosomal segments in congenic NOD strains will enable molecular dissection and fine-structure mapping of *Idd3*. More generally, the method will provide a powerful tool for genetic dissection of common, complex traits in inbred animal models that are established from outbred colonies.

Chromosome 3 (Idd3 and Idd10). Initial breeding studies mapped a non-MHC susceptibility gene to the large interval on chromosome 3 linked to a microsatellite marker, *D3Nds1*.[35] To further localize this locus, Wicker et al.[36] established several NOD strains that are congenic for chromosome 3 (Table 3). Their data indicated that the strong linkage of chromosome 3 with IDDM was due to summation of at least two *Idd* loci, now termed *Idd3* and *Idd10*.

Introgression of the proximal segment of chromosome 3 from the diabetes-resistant B6.PL-*Thy1*^{a} strain onto NOD mice significantly reduced the cumulative incidence of IDDM: Only 28% of females and 1% of males developed IDDM by 7 months of age compared with 78% and 38% in females and males, respectively, in the NOD parental strain (Table 3).[36] By using NOD strains that are congenic for different segments of chromosome 3, *Idd3* has been localized to proximal chromosome 3 in a region of <10 cM between *D3Nds12* and *D3Nds1*.[36] More recently, Lord et al.[37] localized *Idd3* to a region of less than 4.1 cM on chromosome 3.[37] A strong candidate gene in this segment is *Il2*, which encodes interleukin-2, the major growth hormone of T-lymphocytes. DNA sequences of NOD, B6, and B10 alleles of *Il2* were determined, and the NOD allele was found to be different from B6[38] or B10 alleles.[14] In particular, these strains differ in the number of trinucleotide repeats that encode glutamine residues in the polyglutamine stretch. Variation in the number of trinucleotide repeats has been shown to cause several genetic diseases. In addition, a four amino acid insertion/deletion towards the end-terminal region was also reported. The introgression of smaller segments of chromosome 3 around the *Il2* locus in congenic NOD strains and the functional analysis of NOD and B10 alleles of the *Il2* gene will clarify whether or not *Idd3* is encoded by the *Il2* gene.

By using the same set of congenic NOD strains as that used for *Idd3* mapping, *Idd10* has been localized to distal chromosome 3 in a region of <34 cM between *D3Mit51* and *D3Nds9*.[36] A 7 cM interval in this segment of chromosome 3 showed strong linkage to diabetes and insulitis in breeding studies in crosses of NOD with B10.H-2^{g7} mice.[39] A fully penetrant immunological defect, the IgG2a$^+$ phenotype, also mapped to this interval, which contains a strong candi-

Table 3. NOD mice congenic for chromosome 3[36]

		Chromosome 3		Incidence of IDDM	
		Idd3	*Idd10*	Females	Males
NOD		NOD	NOD	78%	38%
NOD.B6$^{Il2\text{-}Tshb}$	N6F3–5	**B6**	**B6**	1%	0%
NOD.B10$^{Il2\text{-}Tshb}$	N7F4–5	**B10**	**B10**	3%	0%
NOD.B6^{Il2}	N7F2–3	**B6**	NOD	28%	1%
NOD.B6^{Tshb}	N7F3–5	NOD	**B6**	33%	14%
NOD.B10^{Tshb}	N7F4–5	NOD	**B10**	39%	5%

date gene, *Fcgr1*, which encodes the high affinity Fc receptor for IgG.[39] The NOD allele of *Fcgr1* was different from B10 allele.[39] In particular, a deletion of 4 bp created a premature stop codon, which resulted in ~ 70% deletion in the cytoplasmic tail. This mutation caused a 73% reduction in the turnover of cell surface receptor–antibody complexes, which appears to be related to $IgG2a^+$ phenotype. These data suggest that *Idd10* may be encoded by the *Fcgr1* locus. Wicker et al.,[17] however, recently reported that *Fcgr1* was eliminated as a candidate gene for *Idd10* in their studies on NOD strains congenic for smaller segment of chromosome 3.

Chromosome 2 (Idd13). In crosses of NOD with NOR mice, a recombinant congenic strain in which limited regions of the NOD genome have been replaced by the genome from C57BL/KsJ strain, a susceptibility gene, *Idd13*, has been mapped to chromosome 2 linked to the *Il1a* gene.[13] This was further confirmed by construction of a congenic NOD strain that contains a segment of chromosome 2 from NOR strain: At the N7F1 generation, the cumulative incidence of IDDM in NOR homozygotes (4/11 = 36%) was much lower than that in the NOD homozygotes (12/12 = 100%) by 1 year of age.[13] Since only the <4 cM segment of distal chromosome 2 around the *Il1* locus in the NOR mouse is of C57BL/KsJ origin, with the rest of distal chromosome 2 derived from the NOD mouse, *Idd13* has been localized to the <4 cM segment of distal chromosome 2.[13] The NOR allele of *Idd13* provides resistance to IDDM, whereas the NOD allele confers susceptibility to the disease. Since F2 mice either homozygous or heterozygous for NOD in this segment were reported to be equally susceptible to IDDM, the susceptibility allele of the NOD mouse appears to be dominant over the NOR-derived resistance allele. Histological examination of the pancreas indicated that NOR homozygosity at the *Idd13* locus did not affect the expression of insulitis.[13]

Candidate genes in this region are *Il1a* and *Il1b* genes, which encode cytokines interleukin-1 (IL1) α and IL1β, respectively, although there has been no evidence that suggest primary defects in the coding sequences of either *Il1* structure gene is the cause of IDDM.

Application to human IDDM

Among *Idd* loci mapped in the NOD mouse, *Idd1* is the only locus (loci) whose association with IDDM has also been demonstrated in humans.[40–42] The CD3 gene region on human chromosome 11q23, which is syntenic to the *Idd2* region on mouse chromosome 9, was previously studied for its contribution to susceptibility to IDDM, but no evidence of association or linkage was found between IDDM and the chromosome 11q23 region, including candidate CD3E and CD3D genes.[43,44]

Recently, genome-wide affected sib-pair analysis using microsatellite markers was applied to human IDDM, and several IDDM loci were mapped to the human genome (Table 4).[45–51] Among these was IDDM7,[48] which is located in a region of human chromosome 2q that shows conserved synteny with the region

Table 4. Susceptibility loci for human IDDM

	Chromosome	Candidate	Refs.
IDDM1	6p21	HLA	40,41
IDDM2	11p15	INS	56–58
IDDM3	15q		47,51
IDDM4	11q13		45,46
IDDM5	6q25		45
IDDM6	18q		50
IDDM7	2q31		48,49,51
IDDM8	6q25–27		51
IDDM9	3q21–q25		49
IDDM10	10p11.2–q11.2		49
	7p	GCK	59
	Xp21-p11		49,60

of mouse chromosome 1 containing *Idd5*.[52] A microsatellite marker, *D2S152*, in this region was shown to be linked to, and associated with, IDDM in families from U.S., Italian, and Sardinian populations. A similar tendency was also observed in Japanese population (Y. Nakagawa et al., unpublished observation). Thus, *IDDM7* might be the human homologue of *Idd5*. There are several candidate genes for IDDM susceptibility on chromosome 2q, such as interleukin-1α (*IL1A*), interleukin-1β (*IL1B*), interleukin receptor antagonist (*IL1PN*) genes, and the human homologue of the mouse natural resistance–associated macrophage protein (*Nramp*) gene. These genes, however, are located several tens of cM distal to *D2S152*, and therefore are unlikely to be in linkage disequilibrium with *IDDM7*. In fact, no evidence of linkage to or association with IDDM was observed for polymorphisms in these genes.[48]

Conclusions

Congenic strategy is a powerful approach to prove that any of the *Idd* loci encode significant effects, although it takes time and is laborious to establish congenic strains. Recent progress in complete genetic linkage maps, however, have made it possible to construct *speed congenics* by typing background genes of backcross progeny and breeding only those that have lost a larger portion of the undesired genome to create the next generation.[54] In fact, microsatellite markers have recently been successfully used for fingerprinting of background genes and introgression of a small interval of a chromosomal segment in the construction of congenic strains.[18,37]

For fine-structure mapping of *Idd* loci, ancestral haplotype congenic mapping may prove to be one of the most powerful strategies, as illustrated by the recent success in fine-structure mapping of *Idd1*.[18] The idea is to find recombinant haplotypes, including candidate variants of an inbred animal model, such as the NOD mouse, among mice in the original closed colony and/or related inbred strains derived from the same colony, and to make congenic strains with the recombinant haplotypes. Even in a region where strong linkage disequilibrium has been established between a genetic marker and a disease gene, there is a high possibility of finding recombinant haplotypes in the region of interest among the animals of the original colony and/or related inbred strains derived from the same colony because the high frequency of heterozygosity in the outbred colony provides a high chance of recombination in each meiosis and such a process started long before the breeding of an inbred strain of interest was initiated. Fine-structure mapping is an essential step for physical mapping and positional cloning of *Idd* genes. Progress in fine-structure mapping may reveal that each *Idd* effect is encoded by a number of genes, as in the case of *Idd1*. Thus the number of *Idd* genes will become much larger than currently known (Table 1). Identification and positional cloning of all of the *Idd* genes will lead to a better understanding of the pathogenesis of IDDM and will facilitate the establishment of strategies for effective prevention of the disease.

References

1. Hattori M, Buse JB, Jackson RA, et al. The NOD mouse: Recessive diabetogenic gene in the major histocompatibility complex. Science 231:733–5, 1986.
2. Ikegami H, Yano N, Sato T, et al. In: Immunotherapy of diabetes and selected autoimmune diseases. Eisenbarth GS, ed. CRC Press, Boca Raton, FL, pp. 23–33, 1989.
3. Ikegami H, Makino S. In: Lessons from Animal Diabetes IV. Shafrir E, ed. Smith-Gordon, London, 4:39–50, 1993.
4. Acha-Orbea H, McDevitt H. The first external domain of the non-obese diabetic mouse class II I-Aβ chain is unique. Proc Natl Acad Sci USA 84:2435–39, 1987.
5. Ikegami H, Makino S, Harada M, et al. The cataract Shionogi mouse, a sister strain of the non-obese diabetic mouse: Similar class II but different class I gene products. Diabetologia 31:254–8, 1988.
6. Ikegami H, Eisenbarth G, Hattori M. Major histocompatibility complex-linked diabetogenic gene of the nonobese diabetic mouse: Analysis of genomic DNA amplified by the polymerase chain reaction. J Clin Invest 85:18–24, 1990.
7. Koide Y, Yoshida T. The unique nucleotide sequence of the Ab in the NOD mouse is shared with its nondiabetic sister strains, the IL1 and the CTS mouse. Int Immunol 2:189–92, 1989.
8. Liu GY, Baker D, Fairchild S, et al. Complete characterization of the expressed immune response genes in Biozzi AB/H mice: Structural and functional identity between AB/H and NOD A region molecules. Immunogenetics 37:296–300, 1993.
9. Wicker LS, Miller BJ, Cocker LZ, et al. Genetic control of diabetes and insulitis in the nonobese diabetic (NOD) mouse. J Exp Med 165:1639–54, 1987.
10. Prochazka M, Leiter E, Serreze D, et al. Three recessive loci required for insulin-dependent diabetes in nonobese diabetic mice. Science 237:286–9, 1987.
11. Livingstone A, Edwards CT, Shizuru JA, et al. Genetic analysis of diabetes in the nonobese diabetic mouse. I. MHC and T cell receptor beta gene expression. J Immunol 146:529–34, 1991.
12. de Gouyon B, Melanitou E, Richard M, et al. Genetic analysis of diabetes and insulitis in an interspecific cross of the nonobese diabetic mouse with *Mus spretus*. Proc Natl Acad Sci USA 90:1877–81, 1993.
13. Serreze DV, Prochazka M, Reifsnyder PC, et al. Use of recombinant congenic and congenic strains of NOD mice to identify a new insulin-dependent diabetes resistance gene. J Exp Med 180:1553–58, 1994.
14. Ghosh S, Palmer S, Rodrigues N, et al. Polygenic control of autoimmune diabetes in nonobese diabetic mice. Nature Genet 4:404–9, 1993.
15. Morahan G, McClive P, Huang D, et al. Genetic and physiological association of

diabetes susceptibility with raised Na^+/H^+ exchange activity. Proc Natl Acad Sci USA 91:5898–902, 1994.
16. McAleer MA, Reifsnyder P, Palmer SM, et al. Cross of NOD mice with the related NON strain: A polygenic model for IDDM. Diabetes 44:1186–95, 1995.
17. Wicker L, Todd J, Peterson L. Genetic control of autoimmune diabetes in the NOD mouse. Annu Rev Immunol 13:179–200, 1995.
18. Ikegami H, Makino S, Yamato E, et al. Identification of a new susceptibility locus for insulin-dependent diabetes mellitus by ancestral haplotype congenic mapping. J Clin Invest 96:1936–1942, 1995.
19. Makino S, Kunimoto K, Muraoka Y, et al. Breeding of a non-obese, diabetic strain of mice. Exp Anim 29:1–13, 1980.
20. Ikegami H, Kawaguchi Y, Ueda H, et al. MHC-linked diabetogenic gene of the NOD mouse: Molecular mapping of the 3′ boundary of the diabetogenic region. Biochem Biophys Res Commun 192:677–82, 1993.
21. Prochazka M, Serreze D, Worthen S, et al. Genetic control of diabetogenesis in NOD/Lt mice: Development and analysis of congenic stocks. Diabetes 38:1446–55, 1989.
22. Wicker L, Appel M, Dotta F, et al. Autoimmune syndromes in major histocompatibility complex (MHC) congenic strains of nonobese diabetic (NOD) mice. The NOD MHC is dominant for insulitis and cyclophosphamide-induced diabetes. J Exp Med 176:67–77, 1992.
23. Podolin P, Pressey A, DeLarato N, et al. I-E^+ nonobese diabetic mice develop insulitis and diabetes. J Exp Med 178:793–803, 1993.
24. Ikegami H, Makino S, Hattori M. In: Frontiers of mucosal immunology. Tsuchiya M, ed. 2:429–432, 1991.
25. Miyazaki T, Uno M, Uehira M, et al. Direct evidence for the contribution of the unique I-A^{NOD} to the development of insulitis in non-obese diabetic mouse. Nature 345:722–4, 1990.
26. Lund T, O'Reilly L, Hutchings P, et al. Prevention of insulin-dependent diabetes mellitus in non-obese diabetic mice by transgenes encoding modified I-Ab chain or normal I-Ea chain. Nature 345:727–9, 1990.
27. Slattery R, Kjer-Nielsen L, Allison J, et al. Prevention of diabetes in non-obese diabetic I-A^k transgenic mice. Nature 345:724–6, 1990.
28. Yamato E, Itoh N, Matsumoto E, et al. Second MHC-linked diabetogenic gene located in the proximal side to Lmp-2 in the MHC class I-K region in NOD mice. Diabetes 44(Suppl 1):18A, 1995.
29. Shirai T, Shimizu N, Shiojiri S, et al. Cloning and expression in *Escherichia coli* of the gene for mouse tumor necrosis factor. DNA 7:193–201, 1988.
30. Madema J, Streilein J, Graser R, et al. Nucleotide sequence of the tumor necrosis factor: A gene in seven different inbred strains. Immunogenetics 40:243–4, 1994.
31. Han J, Huez G, Beutler B. Interactive effects of the tumor necrosis factor promotor and 3′-untranslated regions. J Immunol 146:1843–48, 1991.
32. Pociot F, Briant L, Jongeneel CV, et al. Association of tumor necrosis factor (TNF) and class II major histocompatibility complex alleles with the secretion of TNF-α and TNF-β by human mononuclear cells: A possible link to insulin-dependent diabetes mellitus. Eur J Immunol 23:224–31, 1993.
33. Miyazaki T, Matsuda Y, Toyonaga T, et al. Prevention of autoimmune insulitis in nonobese diabetic mice by expression of major histocompatibility complex class I L^d molecules. Proc Natl Acad Sci USA 89:9519–23, 1992.
34. Klein J. Biology of the mouse major histocompatibility complex. Springer-Verlag, Berlin, pp. 192–230, 1975.
35. Todd J, Aitman T, Cornall R, et al. Genetic analysis of autoimmune type 1 diabetes mellitus in mice. Nature 351:542–7, 1991.
36. Wicker L, Todd J, Prins J-B, et al. Resistance alleles at two non-major histocompatibility complex-linked insulin-dependent diabetes loci on chromosome 3, Idd3

and Idd10, protect nonobese diabetic mice from diabetes. J Exp Med 180:1705–13, 1994.
37. Lord C, Bohlander S, Hopes E, et al. Mapping of the diabetes polygene Idd3 on mouse chromosome 3 using novel congenic strains. Mamm Genome 6:563–570, 1995.
38. Chesnut K, She J-X, Cheng I, et al. Characterization of candidate genes for IDD susceptibility from the diabetes-prone NOD mouse strain. Mamm Genome 4:549–54, 1993.
39. Prins J-B, Todd J, Rodrigues N, et al. Linkage on chromosome 3 of autoimmune diabetes and defective Fc receptor for IgG in NOD mice. Science 260:695–8, 1993.
40. Todd J, Bell J, McDevitt H. HLA-DQβ gene contributes to susceptibility and resistance to insulin-dependent diabetes mellitus. Nature 329:599–604, 1987.
41. Todd J, Mijovic C, Fletcher J, et al. Identification of susceptibility loci for insulin-dependent diabetes mellitus by trans-racial gene mapping. Nature 338:587–9, 1989.
42. Ikegami H, Kawaguchi Y, Yamato E, et al. Analysis by the polymerase chain reaction of histocompatibility leucocyte antigen-DR9-linked susceptibility to insulin-dependent diabetes mellitus. J Clin Endocrinol Metab 75:1381–85, 1992.
43. Hyer RN, Julier C, Buckley JD, et al. High-resolution linkage mapping for susceptibility genes in human polygenic disease: Insulin-dependent diabetes mellitus and chromosome 11q. Am J Hum Genet 48:243–57, 1991.
44. Pritchard LE, Kawaguchi Y, Reed PW, et al. Analysis of the CD3 gene region and type 1 diabetes: Application of fluorescence-based technology to linkage disequilibrium mapping. Hum Mol Genet 4:197–202, 1995.
45. Davies J, Kawaguchi Y, Bennett S, et al. A genome-wide search for human type 1 diabetes susceptibility genes. Nature 371:130–6, 1994.
46. Hashimoto L, Habita C, Beressi J, et al. Genetic mapping of a susceptibility locus for insulin-dependent diabetes mellitus on chromosome 11q. Nature 371:161–4, 1994.
47. Field L, Tobias R, Magnus T. A locus on chromosome 15q26 (IDDM3) produces susceptibility to insulin-dependent diabetes mellitus. Nature Gneet 8:189–94, 1994.
48. Copeman J, Cucca F, Hearne C, et al. Linkage disequilibrium mapping of a type 1 diabetes susceptibility gene (IDDM7) to chromosome 2q31-q33. Nature Genet 9:80–5, 1995.
49. Todd JA. Genetic analysis of type 1 diabetes using whole genome approaches. Proc Natl Acad Sci USA 92:8560–8565, 1995.
50. Owerbach D, Gabbay KH. The HOXD8 locus (2q31) is linked to Type I diabetes: Interaction with chromosome 6 and 11 disease susceptibility genes. Diabetes 44:132–6, 1995.
51. Luo, Bui MM, Muir A et al. Affected-sibpair mapping of a novel susceptibility gene for insulin-dependent diabetes mellitus (IDDM8) on chromosome 6q25-27. Am J Hum Genet 57:911–9, 1995.
52. Cornall R, Prins J-B, Todd J, et al. Type 1 diabetes in mice is linked to the interleukin-1 receptor and Lsh/Ity/Bcg genes on chromosome 1. Nature 353:262–5, 1991.
53. Garchon H-J, Bedossa P, Eloy L, et al. Identification and mapping to chromosome 1 of a susceptibility locus for periinsulitis in non-obese diabetic mouse. Nature 353:260–2, 1991.
54. Lander ES, Schork NJ. Genetic dissection of complex traits. Science 265:2037–48, 1994.
55. Rodrigues N, Cornall R, Chandler P, et al. Mapping of an insulin-dependent diabetes locus, Idd-9, in NOD mice to chromosome 4. Mamm Genome 5:167–70, 1994.
56. Bell GI, Horita S, Karam JH. A polymorphic locus near the human insulin gene is associated with insulin-dependent diabetes mellitus. Diabetes 33:176–83, 1984.
57. Julier C, Hyer R, Davies J, et al. Insulin-IGF2 region on chromosome 11p encodes a gene implicated in HLA-DR4-dependent diabetes susceptibility. Nature 354: 155–9, 1991.

58. Bennett ST, Lucassen AM, Gough SCL, et al. Susceptibility to human type 1 diabetes at IDDM2 is determined by tandem repeat variation at the insulin gene minisatellite locus. Nature Genet 9:284–92, 1995.
59. Rowe RE, Wapelhorst B, Bell GI et al. Linkage and association between insulin-dependent diabetes mellitus (IDDM) susceptibility and markers near the glucokinase gene on chromosome 7. Nature Genet 10:240–2, 1995.
60. Cordell HJ, Kawaguchi Y, Todd JA, Farrall M. An extension of the Maximum Lod Score method to X-linked loci. Ann Hum Genet 59:435–449, 1995.

3

Role of macrophages and cytokines in B-cell death

ANTJE FAUST, RAINER KLEEMANN, HELGA ROTHE, AND HUBERT KOLB
Diabetes Research Institute, Heinrich-Heine-University of Dusseldorf, D-40225 Dusseldorf, Germany.

Key words: BB rats, NOD mice, immunopathogenesis, macrophages, inducible NO synthase, intervention.

Involvement of macrophages in the pathogenesis of IDDM

In recent years data have accumulated suggesting an important contribution of macrophages in the pathogenesis of IDDM, participating as primary inductor cells of autoimmunity, or as effector cells of B-cell destruction. In the spontaneously diabetic NOD mouse, macrophages and dendritic cells are the first cells infiltrating at perivascular and peri-insular regions in early insulitis beginning at the age of 3 wk.[1,2] These cells start to invade the inner islet tissue at 30 days of age.[3] Comparable features were found earlier in prediabetic insulitis of BB rats, with macrophage infiltration preceding and being essential for later invasion of T-lymphocytes.[4] The relevance of macrophages for the initiation of the disease could be shown by prevention of insulitis and diabetes development after modulation of macrophages in vivo by silica administration in both animal models[4–7] or by interference with macrophage adhesion receptors.[8] The prevention of islet infiltration in NOD mice by silica treatment was dependent on the young age of the recipient.[9] These studies indicate the involvement of macrophages in initial steps of the pathogenesis in animal models. Although silica specifically targets macrophages, this does not lead to their elimination. Rather, redirection of monocyte homing to sites of inflammation, silica deposits, and functional impairment of macrophages are seen.

An additional role of macrophages is their function as effector cells in B-cell destruction. In vitro activated rat macrophages could be shown to lyse specifi-

Correspondence: Hubert Kolb, Diabetes Research Institute, Auf'm Hennekamp 65, D-40225 Düsseldorf, Germany

Lessons from Animal Diabetes VI. Ed. E. Shafrir.

cally B-cells or insulinoma cells in coculture experiments.[10] The highly reactive free radical nitric oxide (NO) was identified,[11] which is also involved in IL-1–mediated suppression of islet function[12] and islet cell lysis,[13] to be the main toxic agent in these experiments.

Investigation of macrophages in the rodent animal models revealed aberrant functional states in NOD mouse and BB rat. Defective proliferation and maturation of precursor macrophages to Mac3+ cells was shown at the level of bone marrow cells in NOD. The proliferative defect was correlated with aberrant stimulation of mature macrophages by interferon-γ, resulting in decreased expression of MHC class I molecules.[14] Defects in antigen presentation, seen as a reduced ability to induce regulatory cells, was related to the unique MHC class II allele in NOD mice.[14] The finding that NOD and BB peritoneal macrophages secreted lower amounts of IL-1 and TNFα after stimulation with LPS[15,16] led to experiments with substitution of TNFα in vivo. Chronic administration of this cytokine to late prediabetic NOD mice and BB rats prevented the development of diabetes.[15,17] However, the role of TNF in disease development is not unequivocal, since high levels of TNFα could be detected in sera of prediabetic NOD mice and diabetic BB rats, and in supernatants of activated macrophages from prediabetic BB rats.[18–20] In agreement with this finding, short-term administration of TNFα from birth resulted in accelerated disease development in NOD mice.[21]

Several susceptibility genes have been identified in the NOD mouse.[22] Some of these genes outside the MHC locus on chromosome 17 (*Idd1*) were mapped to regions containing loci with relevance to macrophage functions. One of these genes is located in the proximal region of chromosome 1, which includes the IL-1 receptor (*Idd5*).[23] On chromosome 3 the gene underlying the maturation defect of bone marrow macrophages was mapped to the region comprising the loci for two other susceptibility loci (*Idd3* and *Idd10*),[24] which show a strong linkage to diabetes and insulitis. *Idd10* is linked to the gene coding for the high affinity IgG Fc-receptor (FcγRI).[25] Investigation of this receptor in NOD mice, in fact, revealed a mutation of the cytoplasmic tail, resulting in defective endocytosis.[25] *Idd4* was found to be located near the gene for inducible NO-synthase (iNOS) on chromosome 11,[26] which is of considerable interest with regard to B-cell destruction via nitric oxide. In the BB rat at least three susceptibility loci were found.[27] Segregation studies in our own laboratory found the genetic trait of abnormal TNF production by activated macrophages to be linked to the gene(s) responsible for lymphopenia and defective CD45R expression.[28]

Activation of macrophages in the development of diabetes

In the BB rat, macrophages during early islet infiltration appear to be activated as judged from their phenotypes $ED1^+$ and $ED2^-$.[4] Another activation marker discovered in recent years is iNOS, which catalyzes the production of NO by macrophages in large amounts upon stimulation with IFN-γ and LPS in vitro.[29,30] In prediabetic BB rats, islets without immunohistochemically detectable infiltration were found to lack the expression of iNOS mRNA or iNOS protein.[31] In contrast, islets with advanced infiltration in late prediabetes showed strong expression of iNOS transcripts and protein (Table 1). This expression seems to be disease associated, since islets of age-matched diabetes-resistant (DR-BB) animals and normal Wistar rats were devoid of iNOS. In the NOD mouse, islets of 70-day-old prediabetic female animals were negative for iNOS expression of protein, with little amounts of mRNA compared with age-matched Balb/c mice.[32] After acceleration of diabetes development with cyclophosphamide, which might work in vivo as a macrophage activator,[33,34] prediabetic islets showed strong expression of iNOS mRNA and protein.[32] In this study animals were sacrificed 7–10 days after administration of cyclophosphamide, at which time the animals are in the late prediabetic stage, with enhanced intraislet infiltration.[35]

Confinement of iNOS expression to infiltrating macrophages

Within the inflamed islet there are several potential sources of NO release: infiltrating macrophages, endothelium, and endocrine islet cells. All these cell types have been shown to produce NO upon stimulation by IL-1β.[29,36,37] Activation of adjacent endothelial cells by cytokines released from macrophages (e.g., IL-1 or TNFα) might occur during diabetes induction, which in fact could be shown in vitro by release of NO from islet endothelium after cytokine treatment.[37] In our immunohistochemical experiments we found iNOS expression exclusively at areas of macrophage infiltration (Fig. 1). Endocrine islet cell areas showed only minimal staining. These results indicate a predominant contribution of macrophages and possibly endothelial cells to iNOS expression and NO release during insulitis. The relevance of other sources of NO production (endocrine cells) remains to be clarified with regard to the amount of NO production and contribution to B-cell destruction in vivo.

Table 1. iNOS expression in islets of prediabetic animals

	Islets with expression of iNOS (%)[a]	
Animals	Without insulitis[b]	With insulitis[b,c]
DP-BB rat (70–80 days of age)	0	>90
NOD mouse (70 days of age)	0	<10
NOD mouse (8–10 days after CY)[d]	0	>80

[a] Adapted from Rothe et al.,[32] Kleemann et al.,[31] and Faust et al., unpublished, with permission.
[b] At least 20 islets were analyzed.
[c] Only islets with massive peri-insular infiltration (≥10 mononuclear cells) or with intrainsular infiltration (≥5 mononuclear cells) were counted.
[d] Mice treated with 250 mg/kg cyclophosphamide i.p. on day 70.

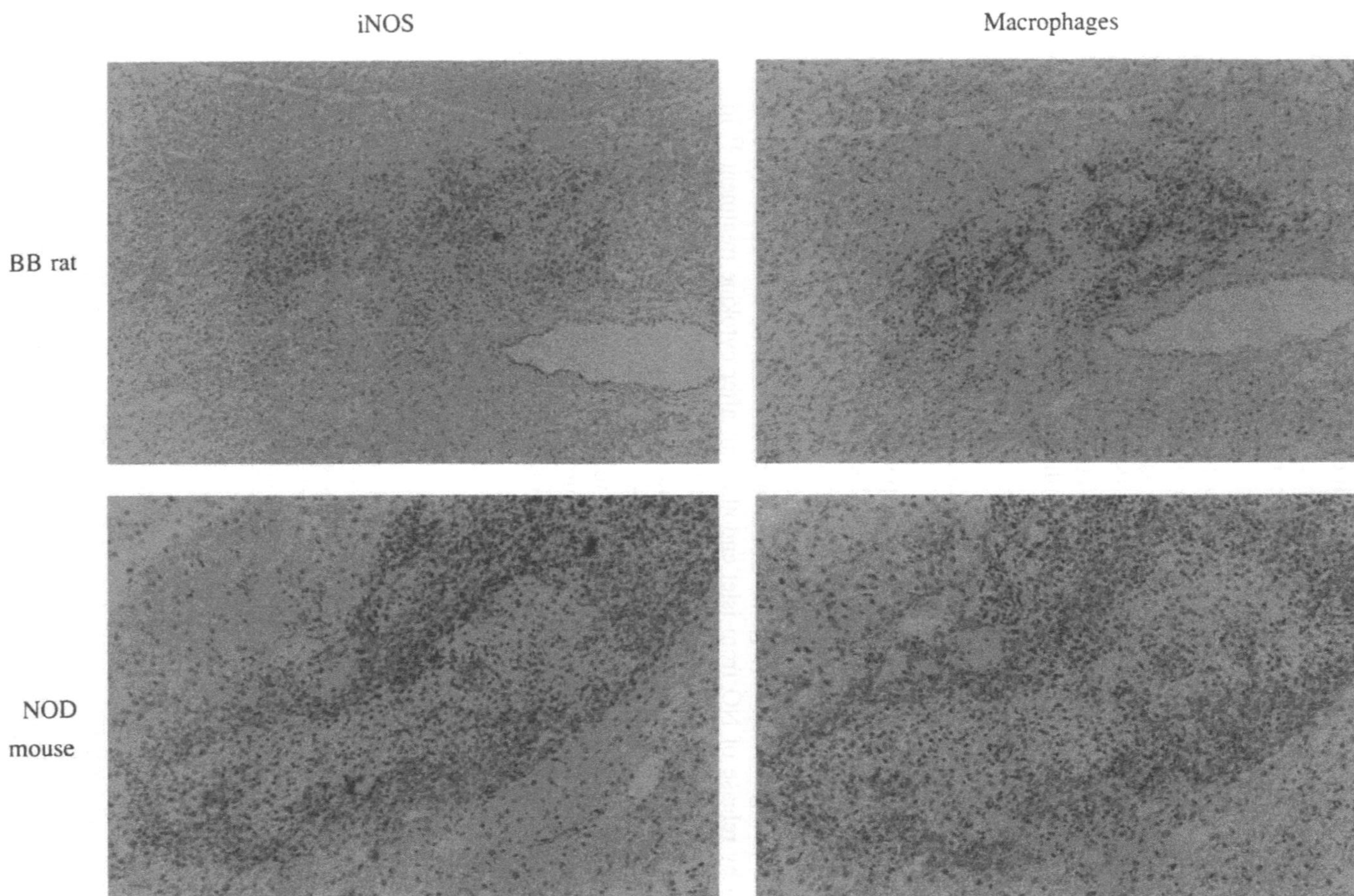

Figure 1. Consecutive sections of infiltrated islets from prediabetic BB rats and from NOD mice 7–10 days after cyclophosphamide treatment were stained by indirect immunohistochemistry with polyclonal anti-iNOS rabbit serum or with monoclonal antibody directed to macrophage antigens ED1 (rat) or F4/80 (mouse), respectively.[31,32]

Relevance of iNOS expression in vivo for the development of diabetes

The presence of iNOS mRNA and protein in inflamed islets does not necessarily demonstrate a direct contribution to diabetes induction. Therefore, specific inhibitors of iNOS were employed to suppress diabetes development in vivo. In the model of immunologically mediated low-dose streptozotocin diabetes (LDSTZ) treatment with the derivatives of L-arginine, NMA and NAME partially suppressed the development of hyperglycemia without major effects on the degree of insulitis (Table 2).[38,39] In contrast, aminoguanidine, a competitive inhibitor of NO formation in islet cells,[40] did not show any effect on hyperglycemia and insulitis in the same model.[41] These discordant results may be seen in the context of some variability of the LDSTZ model. Whereas in several mouse strains LDSTZ diabetes development is clearly T-cell dependent, this is not true for some athymic or SCID mouse strains.[42,43] In diabetes-prone (DP-BB) rats, the administration of NAME significantly reduced diabetes incidence.[44] In NOD mice, treatment with aminoguanidine caused a delay in the onset of diabetes by 3–8 days in the model of adoptive transfer of diabetes without an effect on insulitis.[45] In the model of spontaneously developing diabetes in NOD mice, however, no effect of aminoguanidine treatment on incidence and insulitis scores could be observed, despite sufficient uptake of the drug into the circulation.[46] Our own observations confirm this finding in that cyclophosphamide-induced diabetes in NOD mice could not be reversed by daily treatment for 21 days after the onset of diabetes with aminoguanidine (Faust et al., unpublished).

In our laboratory we established a model of isologous islet transplantation in the NOD mouse after diabetes acceleration with cyclophosphamide. In this model we transplanted islets of young NOD mice within 6 days after diabetes manifestation, which results in primary nonfunction of the islets in >80% of the recipients (no achievement of normoglycemia within 3 days after transplantation).

Table 2. Intervention studies with inhibitors of NO synthase in animal models

NOS inhibitor	Animal model	Effects	Reference
NMA	LDSTZ	Hyperglycemia ↓, insulitis ↓, islet destruction ↓	Lukic et al.[38]
NAME	LDSTZ	Hyperglycemia ↓, insulitis no effect	Kolb et al.[39]
Aminoguanidine	LDSTZ	No effect on hyperglycemia and insulitis; increased mortality	Holstad & Sandler[41]
NAME	BB rat	Diabetes incidence ↓	Rossiter et al.[42]
Aminoguanidine	Adoptive transfer NOD	Delay of diabetes onset 3–8 days, no effect on insulitis	Corbett et al.[43]
Aminoguanidine	NOD mouse	No effect on diabetes incidence	Bowman et al.[44]

In this model, treatment of recipients with aminoguanidine did not show any effect on the survival of the transplanted islets (Table 3). The drug was administered i.p. at a daily dose of 2 × 25 mg/kg, beginning from the day of diabetes manifestation in the later graft recipients.

Two different functional states of macrophages in the development of diabetes

In a model of cyclophosphamide-accelerated NOD mouse diabetes, we investigated the correlation of macrophage activation (iNOS induction) with the occurrence of T-helper type 1 (Th1) or T-helper type 2 (Th2) cytokines. Expression of iNOS and the cytokines was evaluated at the levels of transcription (mRNA in total pancreas detected by RT-PCR) and translation (protein detected by immunohistochemistry of islet sections).[32] In 70-day-old prediabetic NOD mice, a slightly higher level of pancreas iNOS mRNA could be detected in comparison with Balb/c mice. This was not accompanied by recognizable iNOS protein in islet sections, despite the presence of large numbers of infiltrating macrophages, which apparently were iNOS negative (Table 4). Seven to 10 days after a single injection of cyclophosphamide, iNOS mRNA was significantly (sevenfold) elevated. The iNOS protein was localized by immunohistochemical analysis to regions of infiltrating macrophages in the islets (see Fig. 1). This state of macrophage activation was confirmed by elevation in NO_2^- released from isolated islets of cyclophosphamide-treated animals.[32] Concomitant with iNOS induction, a significant fivefold increase in IFN-γ mRNA was found after treatment with cyclophosphamide. This was again confirmed at the level of gene transcription. In contrast to these findings with IFN-γ mRNA, the amount of IL-4 mRNA did not change after diabetes induction with cyclophosphamide. We conclude from these data that in early prediabetic NOD mice a state of Th2 activation is predominant without detectable translation of iNOS enzyme in macrophages and little IFN-γ mRNA in lymphocytes. This is shifted to a Th1

Table 3. Effect of aminoguanidine treatment on primary nonfunction in isologous islet transplantation in NOD mice

Treatment of recipient mice	Number of recipients with primary nonfunction[a]	Number of recipients with primary function and later rejection or long-term survival
Aminoguanidine	4	0
NaCl	7	1

[a] Primary nonfunction was defined as no achievement of normoglycemia (<200 mg/dl) within 3 days after transplantation of 500–600 isologous islets from 3- to 5-week-old NOD mice. Recipients were female NOD mice, rendered diabetic by a single injection of cyclophosphamide. Transplantation was performed within 6 days after manifestation; treatment with 2 × 25 mg aminoguanidine daily was started on the day of manifestation of diabetes (>300 mg/dl).

Table 4. Two different functional states of macrophages in prediabetes[a]

Expression of	Balb/c	70-day-old NOD	NOD 7–10 days after cyclophosphamide
iNOS mRNA[b]	–	(+)	+++
iNOS protein	n.d.[c]	–	+++
IFN-γ mRNA[b]	–	–	+++
IFN-γ protein	n.d.	–	+++
IL-4 mRNA[b]	++	++	++
IL-4 protein	n.d.	+++	+++

[a] Adapted from Rothe et al.,[32] with permission.

[b] RNA was extracted from the total pancreas. For detection of the respective mRNA, the RNA was transcribed by reverse transcriptase and was amplified with specific primers in the PCR reaction. The product was transferred to nitrocellulose and hybridized with radiolabeled probe binding between primer sequences. The strength of the signal was normalized to the measured G3PDH product by phosphorimager.[32] Protein was detected by indirect immunohistochemical staining of serial cryostat pancreatic sections (see legend to Fig. 1).[32]

[c] n.d., not done.

dominance with activation of NO production as a consequence of iNOS transcription and translation (see Table 4).

Conclusions

The data presented show two different functional states of macrophages during the progression of insulitis. Early mostly peri-insular infiltration is associated with macrophages lacking iNOS gene expression, while late-stage destructive intrainsulitis is characterized by strong iNOS expression in macrophages. These findings are suggestive of, but do not prove, a contribution of NO to B-cell destruction. In fact, attempts to prevent diabetes development by administration of iNOS inhibitors remain inconclusive. This may not be surprising because NO is an important mediator of immunosuppressive signals among immune cells. Hence, iNOS inhibition, in part, has immunostimulatory consequences. Finally, our data indicate that macrophage function during insulitis is under cytokine control, with iNOS expression only seen in conjunction with IFN-γ. We conclude that the cytotoxic attack of macrophages against B-cells can be suppressed by immune intervention targeting macrophage stimulatory cytokines.

References

1. Jarpe AJ, Hickman J, Anderson JT et al. Flow cytometric enumeration of mononuclear cell populations infiltrating the islets of Langerhans in prediabetic NOD mice: Development of a model of autoimmune insulitis for type I diabetes. Reg Immunol 3:305–17, 1991.
2. Jansen A, Homo-Delarche F, Hooikaas H, et al. Immunohistochemical characterization of monocytes-macrophages and dendritic cells involved in the initiation of the insulitis and β-cell destruction in NOD mice. Diabetes 43:667–75, 1994.

3. Reddy S, Liu W, Elliott RB. Distribution of pancreatic macrophages preceding and during early insulitis in young NOD mice. Pancreas 8:602–8, 1993.
4. Hanenberg H, Kolb-Bachofen V, Kantwerk-Funke G, Kolb H. Macrophage infiltration precedes and is a prerequisite for lymphocytic insulitis in pancreatic islets of pre-diabetic BB rats. Diabetologia 32:126–34, 1989.
5. Oschilewski U, Kiesel U, Kolb H. Administration of silica prevents diabetes in BB rats. Diabetes 34:197–9, 1985.
6. Lee K-U, Amano K, Yoon J-W. Evidence for initial involvement of macrophage in development of insulitis in NOD mice. Diabetes 37:989–91, 1988.
7. Charlton B, Bacelj A, Mandel TE. Administration of silica particles or anti-Lyt2 antibody prevents β-cell destruction in NOD mice given cyclophosphamide. Diabetes 37:930–5, 1988.
8. Hutchings P, Rosen H, O'Reilly L, et al. Transfer of diabetes in mice prevented by blockade of adhesion-promoting receptor on macrophages. Nature 348:639–42, 1990.
9. Ihm S-H, Yoon J-W. Studies on autoimmunity for initiation of β-cell destruction. VI. Macrophages essential for development of β-cell-specific cytotoxic effectors and insulitis in NOD mice. Diabetes 39:1273–78, 1990.
10. Appels B, Burkart V, Kantwerk-Funke G, et al. Spontaneous cytotoxicity of macrophages against pancreatic islet cells. J Immunol 142:3803–8, 1989.
11. Kröncke K-D, Kolb-Bachofen V, Berschick B, et al. Activated macrophages kill pancreatic syngeneic cells via arginine-dependent nitric oxide generation. Biochem Biophys Res Commun 173:752–8, 1991.
12. Southern C, Schulster D, Green IC. Inhibition of insulin secretion by interleukin-1β and tumor necrosis factorα via an L-arginine-dependent nitric oxide generating mechanism. FEBS Lett 276:42–4, 1990.
13. Bergmann L, Kröncke K-D, Suschek C, et al. Cytotoxic action of IL-1β against pancreatic islets is mediated via nitric oxide formation and is inhibited by N^G-monomethyl-L-arginine. FEBS Lett 299:103–6, 1992.
14. Serreze DV, Gaskins HR, Leiter EH. Defects in the differentiation and function of antigen presenting cells in NOD/Lt mice. J Immunol 150:2534–43, 1993.
15. Jacob CO, Aiso S, Michie SA, et al. Prevention of diabetes in nonobese diabetic mice by tumor necrosis factor (TNF): Similarities between TNF-α and interleukin 1. Immunology 87:968–72, 1990.
16. Lapchak PH, Guilbert LJ, Rabinovitch A. Tumor necrosis factor production is deficient in diabetes-prone BB rats and can be corrected by complete Freund's adjuvant: A possible immunoregulatory role of tumor necrosis factor in the prevention of diabetes. Clin Immunol Immunopathol 65:129–34, 1992.
17. Satoh J, Seino H, Shintani S, et al. Inhibition of type I diabetes in BB rats with recombinant human tumor necrosis factor-α. J Immunol 145:1395–99, 1990.
18. Barel D, Brennan DC, Jevnikar AM, et al. Enhanced tumor necrosis factor in anti-CD3 antibody stimulated diabetic NOD mice: Modulation by PGE_1 and dietary lipid. Autoimmunity 13:141–9, 1992.
19. Rothe H, Fehsel K, Kolb H. Tumor necrosis factor alpha production is upregulated in diabetes prone BB rats. Diabetologia 33:573–5, 1990.
20. Tschöpe D, Schwippert B, Schettler B, et al. Increased GPIIB/IIIA expression and altered DNA-ploidy pattern in megakaryocytes of diabetic BB-rats. Eur J Clin Invest 22:591–8, 1992.
21. Yang X-D, Tisch R, Singer SM, et al. Effect of tumor necrosis factor α on insulin-dependent diabetes mellitus in NOD mice. I. The early development of autoimmunity and the diabetogenic process. J Exp Med 180:995–1004, 1994.
22. Ghosh S, Palmer SM, Rodrigues NR, et al. Polygenic control of autoimmune diabetes in nonobese diabetic mice. Nature Genet 4:404–9, 1993.
23. Cornall RJ, Prins J-B, Todd JA, et al. Type 1 diabetes in mice is linked to the interleukin-1 receptor and Lsh/Ity/Bcg genes on chromosome 1. Nature 353:262–5, 1991.

24. Serreze DV, Leiter EH, Wicker LS, Peterson LB. The *IDD10* region on NOD chromosome (Chr)3 inhibits the T-cell co-stimulatory function of IL-1β produced by macrophages. Autoimmunity 21:53, 1995 (Abstract).
25. Prins J-B, Todd JA, Rodrigues NR, et al. Linkage on chromosome 3 of autoimmune diabetes and defective Fc receptor for IgG in NOD mice. Science 260:695–8, 1993.
26. Jenkins NA, Rothe H, Gilbert DJ, et al. Mapping of the gene for inducible nitric oxide (NO) synthase of mouse macrophages to chromosome 11, close to Evi-2, nu, and Idd-4. Genomics 19:402–4, 1994.
27. Jacob HJ, Pettersson A, Wilson D, et al. Genetic dissection of autoimmune type I diabetes in the BB rat. Nature Genet 2:56–60, 1992.
28. Rothe H, Schuller I, Richter G, et al. Abnormal TNF production in prediabetic BB rats is linked to defective CD45R expression. Immunology 77:1–6, 1992.
29. Stuehr DJ, Gross SS, Sakuma I, et al. Activated murine macrophages secrete a metabolite of arginine with the bioactivity of endothelium-derived relaxing factor and the chemical reactivity of nitric oxide. J Exp Med 169:1011–25, 1989.
30. Xie Q-W, Cho HJ, Calaycay J, et al. Cloning and characterization of inducible nitric oxide synthase from mouse macrophages. Science 256:225–8, 1992.
31. Kleemann R, Rothe H, Kolb-Bachofen V, et al. Transcription and translation of inducible nitric oxide synthase in the pancreas of prediabetic BB rats. FEBS Lett 328:9–12, 1993.
32. Rothe H, Faust A, Schade U, et al. Cyclophosphamide treatment of female nonobese diabetic mice causes enhanced expression of inducible nitric oxide synthase and interferon-gamma, but not of interleukin-4. Diabetologia 37:1154–58, 1994.
33. Szcepanik M, Bryniarski K, Pryjma J, Ptak W. Distinct populations of antigen-presenting macrophages are required for induction of effector and regulatory cells in contact sensitivity response in mice. J Leukocyte Biol 53:320–6, 1993.
34. Kasuga A, Maruyama T, Takei I, et al. The role of cytotoxic macrophages in nonobese diabetic mice: Cytotoxicity against murine mastocytoma and beta-cell lines. Diabetologia 36:1252–57, 1993.
35. Faust A, Burkart V, Ulrich H, et al. Effect of lipoic acid on cyclophosphamide-induced diabetes and insulitis in non-obese diabetic mice. Int J Immunopharmacol 16:61–6, 1994.
36. Corbett JA, Wang JL, Sweetland MA, et al. Interleukin 1*β* induces the formation of nitric oxide by *β*-cells purified from rodent islets of Langerhans. J Clin Invest 90:2384–91, 1992.
37. Suschek C, Fehsel K, Kröncke K-D, et al. Primary cell cultures of rat islet capillary endothelial cells. Am J Pathol 145:685–95, 1994.
38. Lukic ML, Stosic-Grujicic S, Ostojic N, et al. Inhibition of nitric oxide generation affects the induction of diabetes by streptozotocin in mice. Biochem Biophys Res Commun 178:913–20, 1991.
39. Kolb H, Kiesel U, Kröncke K-D, Kolb-Bachofen V. Suppression of low dose streptozotocin induced diabetes in mice by administration of a nitric oxide synthase inhibitor. Life Sci 49:213–7, 1991.
40. Corbett JA, Tilton RG, Chang K, et al. Aminoguanidine, a novel inhibitor of nitric oxide formation, prevents diabetes vascular dysfunction. Diabetes 41:552–6, 1992.
41. Holstad M, Sandler S. Aminoguanidine, an inhibitor of nitric oxide formation, fails to protect against insulitis and hyperglycemia induced by multiple low dose streptozotocin injections in mice. Autoimmunity 15:311–4, 1993.
42. Kolb H, Kröncke K-D. IDDM, Lessons from the low-dose streptozotocin model in mice. Diabetes Rev 1:116–26, 1993.
43. Gerling IC, Friedman H, Greiner DL, et al. Multiple low-dose streptozotocin-induced diabetes in NOD-scid/scid mice in the absence of functional lymphocytes. Diabetes 43:433–40, 1994.
44. Lindsay RM, Smith W, Rossiter SP, et al. Nw-nitro-L-arginine methyl ester reduces the incidence of IDDM in BB/E rats. Diabetes 44:365, 1995.

45. Corbett JA, Mikhael A, Shimizu J, et al. Nitric oxide production in islets from nonobese diabetic mice: Aminoguanidine-sensitive and -resistant stages in the immunological diabetic process. Proc Natl Acad Sci USA 90:8992–95, 1993.
46. Bowman M, Simelli O, Look Z, et al. Pharmacokinetic and therapeutic analysis of aminoguanidine (AG) in non-obese diabetic (NOD) mice. Diabetes 43 (Suppl 1):235A, 1994.

4

Immunotherapy of insulin-dependent diabetes mellitus: Tolerance after short-term anti–IL-2 receptor/ cyclosporine therapy in BB/OK rats

BEATE KUTTLER, SILKE LUCKE, LUTZ VOGT, INGRID KLÖTING, TIBOR DIAMANSTEIN, AND HANS-JÜRGEN HAHN

Institute of Diabetes "G. Katsch", Karlsburg, Medical School, University of Greifswald and Institute of Immunology, Medical School, Free University Berlin, Federal Republic of Germany.

Key words: syngeneic islet transplantation, diabetic BB/OK rats, immunocytochemistry of insulitis, lymphocyte phenotypes, lymphocyte transfusion, monoclonal antibody treatment, induction of tolerance.

There is convincing evidence that the destruction of pancreatic B-cells leading to insulin-dependent diabetes mellitus (IDDM) is immunologically mediated[1–3] and occurs in all mammals.[4] Investigations of animal strains predominantly used in diabetes research, the NOD mouse and the BB rat, demonstrated that the acute onset of hyperglycemia is the final result of a chronic preclinical phase (prediabetes). Already during prediabetes and definitely at the onset of the disease, the number of pancreatic B-cells is reduced.[5–11] In addition to the diminished number of B-cells, the pancreatic islets of prediabetic or newly diagnosed diabetic animals or patients are characterized by an invasion of immunocytes, so-called insulitis.[5,6,9–14] The immunocytochemistry of invading mononuclear cells (MNC) reveals the presence of all major subsets, such as T-lymphocytes (CD4 and CD8 T-cells), B-lymphocytes, macrophages, and NK-cells.[8,10–15] Electron microscopic investigations of inflamed pancreatic islets demonstrated a close contact between immune cells and pancreatic B-cells. As a result of B-cell destruction, B-cell granules have been observed in the extracellular space.[15–17] Although the molecular mechanism of B-cell killing is still not established, cytotoxic reactivities of CD4 T-cells,[18] CD8 T-cells,[19] NK cells,[20] and macrophages[21] against pancreatic B-cells have been demonstrated. In addition, it has been shown that the secretory products of immunocytes, the cytokines, may directly disrupt B-cells,[22] or

Correspondence: Prof. Dr. H. J. Hahn, Nepziner Weg 14m, D-17495 Karlsburg, Germany.

Lessons from Animal Diabetes VI. Ed. E. Shafrir.

may induce surface proteins as MHC-antigens or adhesion molecules, which allow or facilitate cell-mediated killing of B-cells.[23] Consequently, there are a vast amount of experimental trials aimed at interrupting B-cell killing at different levels of the assumed pathogenic processes.[24–26] In this review we summarize our experience with a T-lymphocyte targeted immune therapy in BB/OK rats using a monoclonal antibody directed against the α-chain of the interleukin-2 (IL-2) receptor.

Background for an IL-2 receptor targeted immunotherapy of IDDM

Interleukin-2 plays a central role in stimulating the immune system.[27] Antigen-specific clonal proliferation of T-lymphocytes is initiated through a process of signal transduction, in which the interaction of antigen/major histocompatibility complex (MHC) molecules and the CD3/T-cell receptor complex (TCR) plus the CD4 molecule triggers the production of IL-2 and the expression of its homologous receptor. Whereas IL-2 production is mainly restricted to activated $CD4^+$ T-lymphocytes and some cells of the inflammatory phenotype, the IL-2 receptor (IL-2R) is expressed on different immunologically relevant cells, including T- and B-lymphocytes, NK cells, monocytes/macrophages, intrathymic T-cell precursors, and some other cells, such as mast cells or lymph-borne dendritic cells. One feature of the IL-2R is the existence of three receptor components, designated as the IL-2R β- and γ-chain, which are constitutively expressed on the plasma membrane, and the IL-2R α-chain, which is expressed after stimulation to form the mature receptor.[28,29]

In autoimmune diseases, IL-2 producing and IL-2R expressing cells are detectable during the initial effector phase at the site of immunological inflammation, for example, in autoimmune thyroiditis within the thyroid, in multiple sclerosis within the brain, and in IDDM within the pancreatic islets.[15] Furthermore, IL-2 dependent autoreactive T-cell lines or clones have been isolated in experimentally induced or spontaneously developed autoimmune diseases, such as encephalomyelitis, thyroiditis, uveoretinitis, adjuvant- and collagen-induced arthritis, and IDDM.[15] Such cells secrete IL-2 upon contact with autoantigens, and many of them are capable of mediating autoimmune diseases when injected into normal histocompatible recipients.[15] Therefore, IL-2 secretory cells are implicated in the pathogenesis of autoimmune diseases in an obligatory fashion. Consequently, the abrogation of IL-2 secretion (e.g., by cyclosporin A or by an anti–CD4-targeted therapy) or the prevention of IL-2 action (e.g., by an anti–IL-2 receptor targeted therapy) have been successfully used to intervene in cell-mediated autoimmune reactions.[15,30]

Monoclonal antibodies directed against the inducible IL-2R α-chain have the advantage of targeting only activated immunocytes and are expected to exert a more selective immunosuppression (modulation) than antibodies directed against constitutive membrane determinants. We selected the monoclonal antibody ART18 (mouse anti–rat IgG1),[29] which is bound in vivo to $IL\text{-}2R^+$ cells at the site of inflammation.[15] The antibody is capable of interrupting the rejection of allogeneic grafted organs and tissues as long for as it is applied, and of prolonging the survival time of allogeneic grafted organs after a time-limited therapy.[30–33] In contrast, in minor MHC-different donor-recipient combinations, the IL-2R targeted immunotherapy leads to indefinite graft survival.[33]

When pancreatic islets of BB rats, NOD mice, or IDDM patients are investigated immunohistochemically before or at diabetes manifestation, a predominant part of the insulitis cells express the IL-2R.[8,13,14,17] In newly diagnosed diabetic BB/OK rats, morphometric analysis reveals the presence of the IL-2R on ~30% of immunocytes invaded (Table 1). Furthermore, the number and distribution of TCR^+ and $IL\text{-}2R^+$ cells is rather similar, demonstrating the activated state of the majority of T-lymphocytes within inflamed islets. Despite this marked accumulation at the site of inflammation, we do not observe any differences in $IL\text{-}2R^+$ cells in the lymphatic organs (peripheral blood, spleen, lymph nodes) of newly diagnosed diabetic

Table 1. Lymphocyte phenotypes within pancreatic islets of newly diagnosed diabetic and age-matched normoglycemic BB/OK rats

		Diabetic BB/OK Rats	Normoglycemic BB/OK Rats	*P*
N		20	7	
Sex		12m, 8f	2m, 5f	
Age	days	108 ± 12	90 ± 5	n.s.
Plasma glucose	mmol/l	18.3 ± 1.7	7.2 ± 0.3	<0.01
Pancreatic insulin content	pmol/mg	1.60 ± 0.05	9.51 ± 2.70	<0.01
Islet volume density	%	0.30 ± 0.05	0.79 ± 0.10	<0.01
B-cell volume density	%	0.17 ± 0.04	0.59 ± 0.12	<0.01
Phenotypes of insulitis cells				
R73	cells/mm²	1465 ± 162	293 ± 172	<0.01
W3/25	cells/mm²	2242 ± 258	619 ± 202	<0.01
OX8	cels/mm²	1085 ± 220	195 ± 115	<0.05
Ki-M2R	cells/mm²	901 ± 135	191 ± 52	<0.01
ASGM-1	cells/mm²	299 ± 98	257 ± 140	n.s.
OX6	cells/mm²	3685 ± 709	1191 ± 505	<0.05
ART18	cells/mm²	1211 ± 323	59 ± 33	<0.05

Tail blood was used to determine plasma glucose using a glucose analyzer. Pancreatic insulin content was measured after acid extraction by RIA.[95,96] For morphological investigations, pancreata were either quickly frozen in liquid nitrogen and stored at −70°C until use or fixed in Bouin's solution. For light microscopy, the fixed tissue was embedded in paraffin, cut into 7 μm serial sections, and stained either with hematoxylin-eosin to measure relative islet volume density or with an anti-insulin monoclonal antibody (K36aC10) to estimate relative β-cell volume density.[97] The frozen material was used to prepare cryostat sections (6 μm), which were fixed with aceton and stained with monoclonal antibodies recognizing T-cells [R73 (αβTCR)], NK cells [ASGM-1], macrophages [Ki-M2R], T_H-cells [W3/25 (CD4)], $T_{S/C}$-cells [OX8 (CD8αβ)], and activated cells [OX6 (MHC class II), ART18 (CD25)]. The bound antibodies were visualized using the alkaline phosphatase anti-alkaline phosphatase (APAAP) technique.[98] The number of islet infiltrating cells with different phenotypes was determined morphometrically and given in cells/mm², as described elsewhere.[99]
m, male; f, female.

BB/OK rats in comparison with age-matched normoglycemic BB/OK rats,[34] even if a part of these cells is capable of lysing pancreatic B-cells in vitro.[35,36]

Accumulation of IL-2R$^+$ cells within the pancreas is tempting to treat newly diagnosed diabetic BB/OK rats with ART18 (1 mg/kg body wt),[31,32,37] in combination with a subtherapeutical dose of cyclosporin A (CsA, 1.5 mg/kg), which exerts synergistic effects by blocking IL-2 secretion.[37–39] A 10-day treatment period results in normalization of plasma glucose in ~70% of the animals, which is maintained for at least 120 days and is accompanied by normalized glucose tolerance.[40] In newly diagnosed diabetic animals, the metabolic effect of any immunotherapy can only be observed if the animals have a substantial number of functionally active pancreatic B-cells (>10%) at the time of beginning the therapy. This implies that the therapy should be started in the prediabetic state as was done in NOD mice,[41] since the percentage of newly diagnosed diabetic animals with a sufficient amount of B-cells is rather low.[10]

To exclude these difficulties, we developed a model of an experimentally induced prediabetes, the syngeneically islet-grafted diabetic BB/OK rat. This model has been used to investigate the efficacy of an IL-2R targeted immune therapy in the BB/OK rat, a well-characterized BB rat subline bred at our institute in Karlsburg since 1982 for >35 generations.[42–44] BB/OK rats are characterized by overt hyperglycemia in 55% of animals, with an age at onset between 96 and 140 days. In diabetic animals, the number of B-cells is drastically reduced and the islets are characterized by marked insulitis[10,13] accompanied by numerous immunological abnormalities.[45–49] The breeding system, the genetic and phenotypic characterization, as well as the metabolic features have been described in detail.[50–56]

Efficacy of ART18/CsA therapy in diabetic BB/OK rats with a syngeneic islet graft

Strong support for an autoimmune cell-mediated mechanism of B-cell destruction is the observation that transplantation of a segmental pancreas obtained from a donor twin is not able to cure IDDM of the recipient twin, provided that the recipient is not on immunosuppressive drugs.[57] Similarly, when diabetic animals (NOD mice or BB rats) are grafted with syngeneic pancreatic islets, nonimmunosuppressed animals relapse into hyperglycemia after a normoglycemic period of about 2 wk.[58–61] The graft is invaded with lymphocytes at relapse into hyperglycemia (graft insulitis), and the immunocytochemistry of the invading cells is rather similar, as observed in islets of newly diagnosed diabetic animals. It has been suggested that the mechanism of B-cell destruction due to autoimmune destruction by recall of antigenic memory is comparable with the final stage of mechanisms that are responsible for the development of IDDM. The time between normalization of plasma glucose after syngeneic islet transplantation and relapse into hyperglycemia is a limited "prediabetic" period, which offers various possibilities of investigating the autoaggression before recurrence of diabetes.

Surprisingly, some diabetic BB/OK rats that are grafted with collagenase-isolated BB/OK rat islets[62] beneath the kidney capsule maintain normoglycemia for

>60 days without any immunosuppression of the recipients.[58] We observe this spontaneous acceptance in ~15% of BB/OK rats, whereas in other BB sublines nearly all animals reestablished tolerance after syngeneic islet transplantation,[63,64] again indicating the heterogeneity of the immune reactions in different BB rat sublines. More than 85% of the grafted animals relapse into hyperglycemia within 11–17 days. This relapse can be observed regardless of whether the islets are obtained from 8- to 12-day-old or from adult donors.

When the recipients are treated for only 10 days with 1 mg/kg ART18 in combination with 1.5 mg/kg CsA, 74% of animals (71 of 96) maintain permanent normoglycemia (Table 2). In 25 of 96 (26%) BB/OK rats, the therapy is not able to prevent islet destruction. In these animals the grafted islets are destroyed within 13–19 days, which is reflected by the significant decrease in graft insulin content (Table 2). In successfully treated BB/OK rats, we observe not only a permanent (>240 day) normoglycemia (Fig. 1; see Table 2) accompanied by a nearly normal glucose tolerance,[58] but also maintenance of pancreatic insulin content (and B-cells), and an increase of graft insulin content (see Table 2), indicating sufficient interruption of B-cell destruction in those animals. At present the results can be explained by the different individual sensitivities of animals to the immune therapy applied. Analysis of unsuccessfully treated BB/OK rats reveals no differences in the efficacy of the ART18 batches used, the age at diabetes onset of the recipients, the period of insulin treatment before islet transplantation, the amount of the remaining pancreatic insulin, or the amount of graft insulin (Table 2). However, as Table 3 illustrates, there is an effect of the families from

Table 2. Metabolic outcome after transplantation of 2000 isolated BB/OK rat islets beneath the kidney capsule of diabetic BB/OK rats that were temporarily treated with an anti-CD25 monoclonal antibody (ART18, 1.0 mg/kg bw, i.p.) and low doses of Cyclosporin A (1.5 mg/kg bw, i.p.)

Metabolic outcome	Units of measure	Positive (acceptors)	Negative (failures)	*P*
N		71/96 = 74%	25/96 = 26%	
Sex		32 m, 39 f	11 m, 14 f	
MST	days	>60	16 ± 3	<0.01
Age at diabetes diagnosis	days	116 ± 4	120 ± 7	n. s.
Diabetes Duration prior to Tx	days	32 ± 3	33 ± 5	n. s.
Pancreatic insulin content				
at Tx	pmol/mg	0.57 ± 0.08	0.90 ± 0.15	n. s.
at termination of the experiment	pmol/mg	0.73 ± 0.15	0.54 ± 0.22	n. s.
Graft insulin content				
at Tx	pmol	12017 ± 515*	12448 ± 792*	n. s.
at termination of the experiment	pmol	20640 ± 2356	1274 ± 460	<0.01

Prior to the transplantation (Tx), a pancreatic biopsy was taken[100] to determine pancreatic insulin content by RIA[95] after acid extraction.[96] Islet insulin content was measured after acid ultrasonication by RIA[95] to calculate graft insulin content. At termination of the experiment, the pancreas and graft were removed for determination of insulin content.[95,96]

* $P < 0.01$ vs. termination of the experiment.

MST, mean survival time; m, male; f, female; Tx, transplantation.

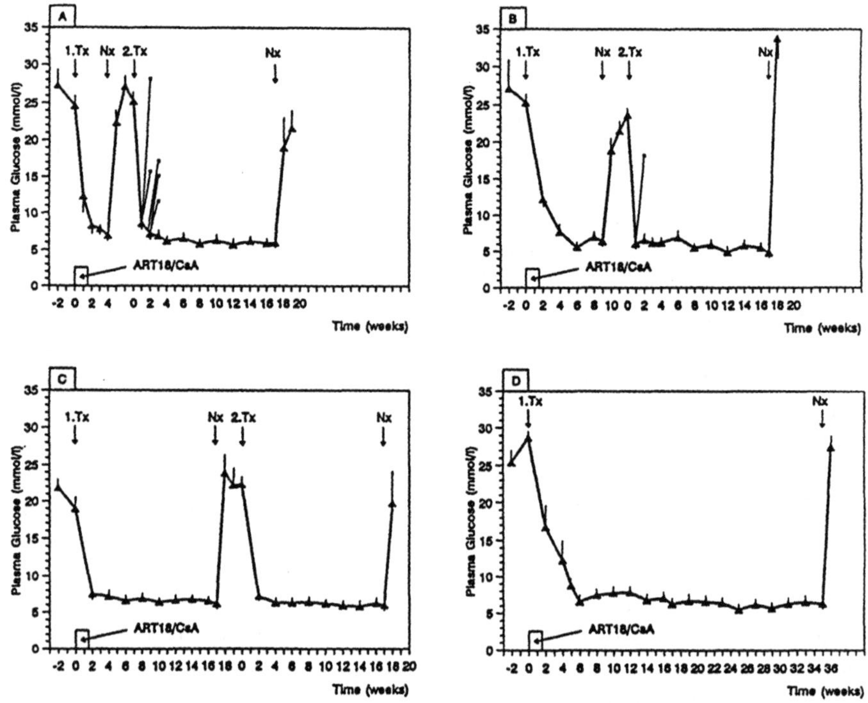

Figure 1. Plasma glucose course of diabetic BB/OK rats grafted with BB/OK rat islets beneath the kidney capsule and treated for 10 days with 1.0 mg/kg body wt ART18 and 1.5 mg/kg body wt cyclosporin A [1st (transplantation (Tx), A–D]. The short-term treatment resulted in a permanent normoglycemia for >240 days (D, n = 6), and graft removal (Nx) caused immediate hyperglycemia (A–D). To prove tolerance development, the first grafts were removed (Nx) after different time periods (A, after 30 days; B, after 60 days; C, after 120 days) and the rats were grafted again with BB/OK rat islets (2nd Tx) without any further immunotherapy (A–C). After 30 days 62% of the rats (8 of 12) accepted the second islet graft (A). The acceptance of the second graft increased to 86% (B: 6 of 7) after 60 days and 100% after 120 days (C: 6 of 6).

which the recipients were derived. Whereas in family 1, >30% of the recipients cannot maintain normoglycemia, family 2 is characterized by an acceptance of >90% (Table 3).

Consequences of ART18/CsA therapy in successfully treated BB/OK rats

Successful immunotherapy results in a long-term normalization of plasma glucose of diabetic BB/OK rats, which are grafted with syngeneic pancreatic islets.[58,65] Graft removal by nephrectomy causes an immediate relapse into hyperglycemia (>20 mmol/l glucose). In addition, the immune therapy reduces the number of immunocytes within native pancreatic islets (Table 4). The W3/25$^+$ cells representing macrophages *and* CD4 T-lymphocytes are not altered by the therapy, especially since the number of macrophages, as demonstrated by the

MoAb KiM-2R, is not affected. On the other hand, the number of $OX8^+$ (CD8 T-lymphocytes *and* NK cells) and $ASGM1^+$ (NK cells) cells is already significantly reduced 30 days after initiation of the therapy. This effect could be observed up to 120 days after transplantation. Even more pronounced is the effect of the therapy on the expression of early (IL-2R) and late (MHC class II antigens) activation markers on immunocytes, as measured by the MoAb ART18 and OX17 (see Table 4). Thus, the number of $IL\text{-}2R^+$ cells decreases continuously, being negligible on day 120. The amount of MHC class II antigen-expressing cells is not that drastically pronounced but is significantly ($p < 0.01$) reduced within the islets. In this connection, we have to take into consideration that not only immunocytes express MHC antigens. The time-dependent decrease of MHC class II antigen-expressing cells in treated BB/OK rats is not age related, as suggested in other animals,[66] since the cell number in the periphery (blood, spleen) is comparable in 100-day-old and >200-day-old BB/OK rats (e.g., blood: 36.7 ± 2.6% in 100-day-old BB/OK rats vs. 39.3 ± 4.3% in >200-day-old BB/OK rats, n = 14). At present there is no evidence that the pancreatic islets become free of insulitis due to the therapy. However, as the morphometric data demonstrate, the numbers of the most important cytotoxic subpopulations within native pancreatic islets, as CD4-, CD8 T-lymphocytes and NK cells, are markedly reduced and the cells that are still present are in a nonactivated state (see Table 4).

In successfully treated normoglycemic BB/OK rats, morphometric analysis of the graft reveals a fulminant invasion of immunocytes on day 30. Such a marked graft insulitis, which is composed of all immune subpopulations, is normally not present in syngeneically grafted islets. The degree of lymphocytic infiltration with T- and NK cells is similar to that observed in syngeneic grafts of untreated BB/OK rats at the time of relapse into hyperglycemia (S. Lucke, personal communication). These data show that the therapy employed does not prevent graft insulitis; however, with time the number of cytotoxic cells is reduced and, again, as in the native pancreatic islets, the majority of immunocytes are in a nonactivated state (see Table 4).

Table 3. Relation of diabetic BB/OK rats that are grafted with BB/OK rat islets and treated with ART18 and Cyclosporin A to their families by a computer program[55,56,101] to evaluate the influence of the family origin on the number of therapy failures

		Family 1	Family 2	*P*
N		24	21	
Diabetes incidence	%	49.5 ± 5.2	58.4 ± 4.6	n. s.
Age at diabetes diagnosis	days	107.4 ± 5.6	94.4 ± 3.3	n. s.
Diabetes duration prior to Tx	days	46.1 ± 6.6	37.6 ± 5.9	n. s.
Pancreatic insulin content	pmol/mg	0.65 ± 0.14	0.44 ± 0.03	n. s.
Graft insulin content	pmol	11,400 ± 580	12,940 ± 960	n. s.
Therapy failures	N (%)	9 (37.5)	2 (9.5)	<0.05*

* Statistical significance was evaluated by the chi^2 test.

Tx, transplantation.

Table 4. Effect of temporary ART18/CsA therapy on peripheral blood lymphocyte phenotypes, pancreatic and graft insulitis of diabetic BB/OK rats that were grafted with BB/OK rat islets after different time periods

Days after Tx	Units of measure	0 day	30 days	60 days	120 days	240 days
N		39	15	7	6	11
Sex		20 m, 19 f	9 m, 6 f	5 m, 2 f	3 m, 3 f	3 m, 8 f
Plasma glucose	mmol/l	26.1 ± 1.4	7.1 ± 0.8**	6.4 ± 0.3**	6.2 ± 0.3**	6.3 ± 0.3**
Pancreatic insulin content	pmol/mg	0.50 ± 0.08	0.62 ± 0.21	0.26 ± 0.11	0.52 ± 0.29	1.67 ± 0.47**
Graft insulin content	pmol	13,115 ± 718	13,002 ± 3139	n. d.	** 18,212 ± 2957	13,474 ± 1811
Peripheral blood						
N		9	6	7	6	7
OX19	%	7.2 ± 1.0	13.3 ± 3.5	14.3 ± 1.0**	11.7 ± 2.7	11.9 ± 2.8
W3/25	%	19.1 ± 3.1	9.2 ± 0.6*	28.1 ± 4.5	18.8 ± 3.0	22.8 ± 5.7
OX8	%	9.9 ± 1.9	6.3 ± 1.7	20.9 ± 2.6**	10.9 ± 2.6	11.1 ± 2.6
Ki-BrI	%	n. d.	27.7 ± 3.5	47.2 ± 2.7	15.6 ± 2.5	43.0 ± 5.3
ASGM-1	%	14.8 ± 3.6	n. d.	16.8 ± 2.2	19.6 ± 4.5	9.9 ± 1.1
ART18	%	1.6 ± 0.7	0.5 ± 0.3	5.5 ± 1.1**	0.7 ± 0.3	5.3 ± 1.3*

Pancreatic insulitis						
N		10	5		7	
OX19	cells/mm²	334 ± 64	394 ± 223	n. d.	209 ± 94	n. d.
W3/25	cells/mm²	749 ± 73	511 ± 83	n. d.	569 ± 244	n. d.
OX8	cells/mm²	1388 ± 342	264 ± 135*	n. d.	171 ± 127*	n. d.
Ki-M2R	cells/mm²	537 ± 67	284 ± 107	n. d.	435 ± 198	n. d.
ASGM-1	cells/mm²	1048 ± 132	267 ± 154**	n. d.	117 ± 34**	n. d.
OX17	cells/mm²	1227 ± 114	1559 ± 436	n. d.	577 ± 137**	n. d.
ART18	cells/mm²	1461 ± 268	206 ± 134**	n. d.	32 ± 17**	n. d.
Graft insulitis						
N			7	8	5	3
OX19	cells/mm²	—	820 ± 183	221 ± 72°°	33 ± 10°°	31 ± 11°
W3/25	cells/mm²	—	851 ± 191	775 ± 106	404 ± 6	196 ± 106
OX8	cells/mm²	—	576 ± 142	102 ± 31°°	30 ± 11°°	27 ± 14°
Ki-M2R	cells/mm²	—	111 ± 39	106 ± 13	38 ± 15	51 ± 32
ASGM-1	cels/mm²	—	543 ± 131	230 ± 60°	20 ± 4°°	44 ± 11°
OX17	cells/mm²	—	1929 ± 320	874 ± 165°°	419 ± 44°°	301 ± 19°
ART18	cells/mm²	—	171 ± 53	49 ± 15°	53 ± 6	18 ± 3

Lymphocytes were prepared by Ficoll/Visotrast gradient centrifugation.[34] The cell phenotypes were determined by means of flow cytometry, as described elsewhere.[99] Pancreata and kidneys with islet grafts were quickly frozen in liquid nitrogen and stored at −70°C until use. The frozen material was used to prepare cryostat sections (6 μm), which were stained with monoclonal antibodies.[98] The number of stained cells was determined morphometrically and given in cells/mm², as described elsewhere.[97] Following primary monoclonal antibodies recognizing T-cells [OX19 (CD5)], B-lymphocytes [Ki-BrI], NK cells [ASGM-1], macrophages [Ki-M2R], T_H-cells [W3/25 (CD4)], $T_{S/C}$-cells [OX8 (CD8$\alpha\beta$)], and activated cells [OX17 (MHC class II), ART18 (CD25)] were used.

Tx, transplantation; m, male; f, female.

* $P < 0.05$, ** $P < 0.01$ vs. 0 day, ° $P < 0.05$, °° $P < 0.01$ vs. 30 days.

In contrast to the pronounced effects of the therapy on pancreatic and graft insulitis, peripheral blood lymphocytes are nearly not altered. It is only remarkable that at the time when a substantial number of immunocytes are accumulated within the graft (day 30), the number of potential cytotoxic immunocytes ($W3/25^+$ plus $OX8^+$ cells) is reduced in the periphery (see Table 4). Whether this result really indicates transfer of immunocytes into the graft is presently still an open question. However, as Table 4 also illustrates, later on these immunocytes recover in the peripheral blood.

Indications for the (re)development of tolerance against pancreatic B-cells in ART18/CsA treated BB/OK rats

When BB/OK rats, which escaped diabetes manifestation spontaneously (e.g., >200 days old), are grafted with syngeneic islets after a toxic B-cell destruction (e.g., with streptozotocin), these animals accept the islets permanently, indicating that the autoaggression cannot be restimulated by application of the antigen.[67,68] Additionally, lymphocytes obtained from >200-day-old normoglycemic BB/OK rats are able to prevent diabetes onset when applied into prediabetic animals.[69] They are also able to prevent autoimmune destruction of B-cells after transfusion into diabetic BB/OK rats that are grafted with syngeneic islets.[69] Taking these observations into consideration, we hypothesize that the majority of BB/OK rats escape diabetes manifestation due to spontaneous redevelopment of tolerance to their own B-cells,[69] although it is not clear if this cell-mediated process is due to induction or transfer of suppressor cells. In other BB rat sublines, it was possible to transfer diabetes with lymphocytes obtained from old normoglycemic BB/Mol rats,[70] which reflects the opposite behavior and again emphasizes the variability between different BB rat sublines.

Long-term normoglycemia in short-term treated animals indicates a marked change of autoaggressive immune reactions in the direction of the development of unresponsiveness and/or tolerance to the syngeneically grafted BB/OK rat islets. To prove the latter hypothesis, the first graft of successfully treated recipients is removed by nephrectomy at 30, 60, and 120 days after transplantation. After the reappearance of hyperglycemia (<2 days), the BB/OK rats are provided with a second syngeneic islet graft but, in contrast to the first transplantation, without further immunomodulation. When ART18/CsA therapy has induced a stable tolerance, the second graft should normalize plasma glucose and maintain normoglycemia despite antigeneic restimulation. Indeed, as Figure 1 illustrates, the second islet grafts are accepted without immunotherapy, when performed 60 and 120 days after the primary transplantation. The normoglycemia is caused by the graft, since its removal results in immediate relapse into hyperglycemia (see Fig. 1B and 1C). However, when the second transplantations are performed on day 30, only 8 of 13 recipients accept the graft without immunotherapy. More than one third of the animals relapse into hyperglycemia within 3 wk (see Fig. 1A), indicating the incomplete development of unresponsiveness within this period (Fig. 2). The latter observation is in agreement with the marked graft insulitis observed at this time (see Table 4). The results support

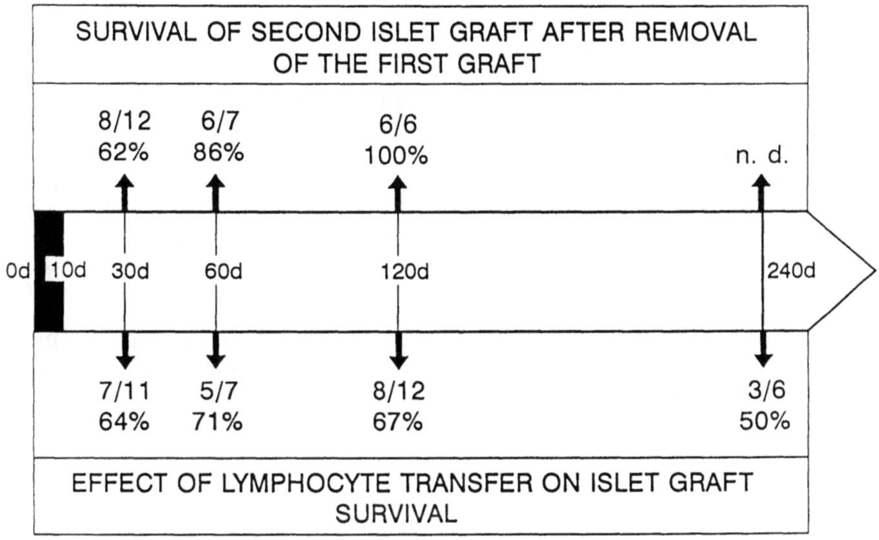

Figure 2. Summary of the results obtained after investigation of tolerance development in BB/OK rats following syngeneic islet transplantation and short-term ART18/CsA therapy. Two criteria served for the assessment of tolerance development: the survival of second islet grafts after removal of the first grafts (top) and the ability of lymphocytes obtained from successfully treated rats to prevent graft destruction in naive BB/OK rats (bottom).

the impact of antigen presence (syngeneic pancreatic islets) for tolerance development. The development of tolerance is the result of an active time-requiring process (>30 days) in which the immune system has to deal with the antigen. However, if LEW.1A rat islets are provided as the second islet graft, the islets are destroyed within 20–30 days (Table 5). The rejection of these allogeneically grafted islets demonstrates that the short-term ART18/CsA therapy does not lead to a general immunosuppression and shows the specificity of the tolerance development to BB/OK rat islets.

The spontaneous development of tolerance in BB/OK rats[69] prompted us to investigate if lymphocytes of successfully treated BB/OK rats are involved in maintaining the stability of tolerance. For that reason, isolated MNC (lymph node cells and spleen cells) obtained 30, 60, 120, and 240 days after the primary transplantation are transfused into naive diabetic BB/OK rats, which have been grafted with syngeneic islets. If tolerance is developed, the immunocytes transferred should be able to prevent autoimmune B-cell destruction. The amount of lymphocytes obtained from the different experimental groups is not significantly different (Table 6). Due to the lymphopenia of BB/OK rats, the cell number isolated is rather low. Despite the different phenotypical composition of the cells transfused (Table 6), two thirds of animals maintain normoglycemia for >120 days in each group investigated. Surprisingly, the graft acceptance has been observed in response to either $29.5 \pm 4.8 \times 10^5$ or only $8.2 \pm 1.3 \times 10^5$ $OX8^+$ cells. An individual phenotypical analysis of the lymphocytes that were trans-

Table 5. Specificity of tolerance development in BB/OK rats

First islet donor Second islet donor	Units of measure	BB/OK rats BB/OK rats	BB/OK rats LEW.1A rats	*P*
N		6	5	
Plasma glucose at Tx	mmol/l	22.4 ± 1.1	19.3 ± 1.0	n. s.
Pancreatic insulin content at Tx	pmol/mg	0.23 ± 0.13	0.48 ± 0.22	n. s.
Graft insulin content at Tx	pmol	14850 ± 1660	12,000 ± 1169	n. s.
Graft insulin content at t_{120}/t_{25}	pmol	19142 ± 2505	1304 ± 639	<0.01
MST	days	>120 ± 0	25 ± 5	<0.01

At 120 days after a syngeneic islet transplantation beneath the kidney capsule of diabetic BB/OK rats and the 10-day treatment with ART18 and Cyclosporin A, the first grafts were removed by nephrectomy. After redevelopment of hyperglycaemia the rats were grafted with either syngeneic (BB/OK) or allogeneic (LEW.1A) islets (= second grafts) without any further immunomodulation. Tx, transplantation; t_{120}/t_{25}, termination of the experiments (MST > 120 days or 25 ± 5 days); MST, mean survival time.

fused into the naive BB/OK rats does not result in any significant correlation to the metabolic outcome. Nevertheless, our results demonstrate the capability of lymphocytes obtained from treated normoglycemic BB/OK rats to prevent B-cell destruction or to induce prevention of B-cell destruction, for example, to trans-

Table 6. Characterization of lymphocytes transfused into naive diabetic BB/OK rats simultaneously grafted with BB/OK rat islets.

Time after Tx and ART18/CsA of Lymphocyte Preparation (days)		30	60	120	240
N		11	7	12	6
Sex		1m, 10f	5m, 2f	5m, 7f	1m, 5f
Prevention of islet destruction		7/11 = 63.6%	5/7 = 71.4%	8/12 = 66.7%	3/6 = 50.0%
Total lymphocyte number	$\times 10^6$	16.4 ± 1.2	14.9 ± 1.5	15.6 ± 1.5	13.5 ± 2.3
OX19	$\times 10^5$	12.9 ± 1.3	20.8 ± 1.9†	20.2 ± 2.4*	18.1 ± 4.9
W3/25	$\times 10^5$	16.8 ± 1.4	18.8 ± 3.5	25.6 ± 2.8*	24.7 ± 6.6
OX8	$\times 10^5$	8.2 ± 1.3	29.5 ± 4.8†	17.6 ± 2.6†,‡	9.5 ± 2.2§
Ki-BrI	$\times 10^5$	55.7 ± 5.8	95.5 ± 10.2†	51.3 ± 10.8‡	80.4 ± 9.7*
ASGM-1	$\times 10^5$	15.0 ± 2.4	28.8 ± 4.6*	23.5 ± 2.6*	16.6 ± 4.1
ART18	$\times 10^5$	0.8 ± 0.7	4.7 ± 1.1†	3.3 ± 0.9*	3.1 ± 0.8

Lymphocyte donors were syngeneically islet grafted and ART18/CsA-treated BB/OK rats. Lymphocytes were prepared by Ficoll/Visotrast gradient centrifugation.[34] The cell phenotypes were analyzed by means of flow cytometry[99] using monoclonal antibodies recognizing T-cells [OX19 (CD5)], B-lymphocytes [Ki-BrI], NK cells [ASGM-1], TH-cells [W3/25 (CD4)], TS/C-cells [OX8 (CD8$\alpha\beta$)], and activated cells [ART18 (CD25)].

Tx, transplantation; m, male; f, female.

* $P < 0.05$; † $P < 0.01$ vs. 30 days; ‡ $P < 0.05$, § $P < 0.01$ vs. 60 days.

fer tolerance. The latter observation is also supported by an immediate relapse into hyperglycemia after graft removal.

Discussion and conclusions

While still incomplete, the current models of IDDM pathogenesis strongly indicate autoimmune destruction of the pancreatic B-cells as the underlying cause of insulinopenia. A number of immunotherapies, which intervene at different levels of the pathogenic process,[24–26] have been suggested. The modulation of autoaggression has been performed before the initial (inductive) event of B-cell destruction, as well as after the inductive event, when the immunocytes start to invade pancreatic islets, or when the account of B-cells has already been diminished with or without an impaired glucose tolerance, at overt diabetes or at termination of the initial remission phase. It is still a matter of debate if the different stages of immune inflammation may or may not represent a uniform pathogenic process and if one or several inductive events cause this process. Analysis of the cells invading pancreatic islets at different "levels" of B-cell destruction in prediabetic animals reveals variable participation of the immunocytes.[10,11] Consequently, the immunotherapies should be adapted to the stage of inflammation to obtain the best therapeutic results.

This strategy is definitely not necessary in the case of permanent immunosuppression, which inhibits or suppresses immune responses completely, with autoaggressive reactions included. The future goal of immunotherapy is immunomodulation, which includes temporary treatment of prediabetic probands with the aim of reinducing immunological tolerance to their own islet B-cells. Animal experiments designed to achieve this aim may help to introduce new immunomodulatory approaches. Those investigations that demonstrate complete prevention of IDDM by postnatal modulation of the immune system or by permanent immunosuppression[2,3,24–26] have minor therapeutic implications, due to severe side effects, including the pancreatic B-cells.[71] Trials in NOD mice that started in young adult "prediabetic" animals and in which either a temporary or an intermittent therapy was applied, gave variable results with respect to diabetes prevention, since the level of islet inflammation in individual animals was not taken into consideration.[41,72,73] Those experiments in which newly diagnosed diabetic animals were treated were of limited success, since the amount of the remaining B-cells was often too low for obtaining normoglycemia.[40]

Study of diabetic animals has, however, the advantage of investigating animals with an expressed disease, and not only with a genetic predisposition. Normoglycemia in diabetic animals, however, can only be obtained when the number and the function of B-cells are capable of maintaining normal insulinemia. Therefore, we grafted syngeneic pancreatic islets into diabetic BB/OK rats, which has the advantage of investigating the therapeutic outcome under restimulation of immune-mediated B-cell destruction,[57–61,67] probably the most important mechanism in B-cell loss during prediabetes. Consequently, this model is insufficient for the investigation of immunomodulation of primary and/or early events in the pathogenesis of IDDM, but it is useful in animals with ongoing B-cell destruction, as shown by a permanent normoglycemia in temporarily treated BB rats.[58,65] However, the selected experimental situation (transplantation of rats with a fully expressed disease = experimental "prediabetes") is sufficient to obtain metabolic and immunological results in defined time periods with a relatively small number of animals (for control and experiment, n = 16; when investigating diabetes-prone rats at least 40 animals are necessary).

The results also emphasize that this therapy is not able to prevent islet destruction in all treated animals. The family accumulation of metabolic failures can be taken as indirect evidence for either a heterogeneic pathogenesis of IDDM or for a different sensitivity of the targeted cells to the therapy used. Further research to improve the therapeutic regimen may ameliorate the efficacy of immune intervention, especially in combination with other inhibitors of IL-2 secretion.[39,74] With a prolonged treatment period, we and others failed to improve the therapeutic outcome.[75,76] The application of a single dose of ART18 binds about 70% of the IL-2R$^+$ insulitis cells within 4 h and nearly all IL-2R$^+$ cells in 24 h.[15] At the latter time, we also observed a decrease in these cells. On the other hand, we did not observe a marked peripheral depletion of IL-2$^+$ cells, even after a 10-day treatment period, since the number of activated peripheral cells was rather low. The consequence of antibody treatment is either a depletion[77] or a modulation[38] of IL-2R$^+$ cells. Independent of the antibodies used (e.g., in the rat system: NDS62, OX39, ART18) and independent of the experimental model employed (rejection, autoaggression),[33,74,77] the invasion of immunocytes is not prevented, but at the site of inflammation the binding of IL-2R MoAbs to the IL-2 receptor prevents clonal expansion of activated T-cells and reduces the number of IL-2R expressing cells, both within the graft and the pancreas. The inhibition of clonal expansion prevents the development of cell-mediated cytotoxicity against the target cells,[78,79] indicating the functional incompetence of invading cells.

The acceptance of a second identical graft without any further immune therapy after removal of the first graft is strong experimental evidence for a stable development of tolerance to the putative antigen.[80] Our results summarized in Figure 2 demonstrate the time dependency of this process, which we interpret as active participation of the immune system. Furthermore, the process is specific for the syngeneic antigen, since MHC-incompatible tissue is rejected. There is increasing evidence for the possibility of inducing peripheral tolerance in adult animals,[81–85] even though the underlying mechanism is still a matter of discussion. In contrast to transplantation tolerance, in which the immune system is guided to accept foreign antigens, the reestablishment of tolerance in autoimmune diseases is the reconstitution of normal conditions (reestablishment of self-tolerance), which can be achieved by immunomodulation instead of permanent immunosuppression. This may happen 1) during the prediabetic period in animals that escape diabetes manifestation[86,87]; 2) spontaneously after transplantation of syngeneic islets[55,61,65]; or 3) after immunotherapy, which selectively prevents the functional maturation of immunocytes, as demonstrated after application of ART18/CsA. If the latter mechanism alone is responsible for maintaining long-term normoglycemia, other immune modulators also targeting the function of invading or invaded immunocytes should give similar or identical results. Indeed, when grafted BB/OK rats are treated for 10 days with an IL-2 immunotoxin[88] or with higher doses of FK 506,[89] a substantial number of animals develop permanent normoglycemia.

The majority of antibody/CsA–treated BB/OK rats develops tolerance, which can be transferred by lymphocytes. The number of lymphocytes prepared from the lymphopenic BB/OK rats and thereafter transfused into naive BB/OK rats is markedly lower than that which is transfused to demonstrate the stability of a developed allogeneic tolerance.[90] Although the anti–IL-2R targeted therapy spares suppressor cells,[91] it is questionable to explain the transfer-mediated prevention of B-cell destruction simply as a consequence of the application of suppressors, even if a suppressor cell deficiency has been suggested as a possible cause of autoaggression. We assume there are one or more cell populations with a regulatory effect on the immune system that have not yet been cytologically and/or functionally characterized.[92] When, for example, B-lymphocyte–depleted thoracic duct lymphocytes obtained from ART18/CsA treated BB/OK rats with an accepted MHC-identical allogeneic islet graft are phenotypically analyzed, >90% of the cells are $CD4^+$ T-lymphocytes, and if CD45RC is coexpressed, it is only found with a low antigen density (T_{H2} cells?).[93] These cells are probably regulatory T-cells in the periphery, which are capable of preventing autoimmune diabetes.[93,94]

In summary, our results demonstrate that short-term immunotherapy in BB/OK rats induces a permanent normoglycemia due to reestablishment of self-tolerance in the majority of animals that were grafted with syngeneic islets. As far as B-cell destruction can be diagnosed in the prediabetic state, such an immunomodulatory therapy should be employed to prevent diabetes. Since anti–IL-2R MoAbs are already being used clinically,[79] it is reasonable to hope that further development of immunomodulation will help in finding a pathogenesis-based "ideal" therapy for IDDM.

Acknowledgments

The original work quoted was kindly supported by the Juvenile Diabetes Foundation (Di 189744), by KAI e.V. (018297/M), by the Ministry of Health of former GDR, the Ministry of Research and Technology (FKZ 0702484), by the German Research Council (Ku 799/1-3) and by the German Diabetes Association (to S.L.).

References

1. Hahn HJ. Tierexperimenteller Diabetes. In: Diabetes mellitus. Bibergeil H, ed. Fischer Verlag, Jena, pp 108–17, 1989.
2. Leiter EH, Prochazka M, Coleman DL. Animal model of human disease. The non-obese diabetes (NOD) mouse. Am J Pathol 128:380–3, 1987.
3. Like AA, Rossini AA. Spontaneous autoimmune diabetes mellitus in the Bio-Breeding/Worcester rat. Surv Synth Pathol 3:131–8, 1984.
4. Buschard K. The thymus-dependent immune system in the pathogenesis of type 1

(insulin-dependent) diabetes mellitus. Animal model and human studies. Dan Med Bull 32:139–51, 1985.
5. Foulis AK, Farquharson MA. Aberrant expression of HLA-DR antigens by insulin-containing beta-cells in recent onset type I diabetes mellitus. Diabetes 35:1215–24, 1986.
6. Gepts W, De Mey J. Islet cell survival determined by morphology: An immunocytochemical study of the islets of Langerhans in juvenile diabetes mellitus. Diabetes 27(Suppl 1):251–61, 1978.
7. Gepts W. Pathologic anatomy of the pancreas in juvenile diabetes mellitus. Diabetes 14:619–33, 1965.
8. Bottazzo GF, Path MRC, Dean BM, et al. In situ characterization of autoimmune phenomena and expression of HLA molecules in the pancreas in diabetic insulitis. N Engl J Med 313:353–60, 1985.
9. Hanafusa T, Miyazaki A, Miyagawa J, et al. Examination of islets in the pancreas biopsy specimens from newly diagnosed type I (insulin-dependent) diabetic patients. Diabetologia 33:105–11, 1990.
10. Lucke S, Besch W, Kauert C, Hahn HJ. The endocrine pancreas of BB/OK-rats before and at diagnosis of hyperglycemia. Exp Clin Endocrinol 91:161–70, 1988.
11. Hanenberg H, Kolb-Bachofen V, Kantwerk-Funke G, Kolb H. Macrophage infiltration precedes and is a prerequisite for lymphocytic insulitis in pancreas of prediabetic BB rats. Diabetologia 32:126–34, 1989.
12. Dean BM, Walker R, Bone AJ, et al. Pre-diabetes in the spontaneously diabetic BB/E rat: Lymphocyte subpopulations in the pancreatic infiltrate and expression of rat MHC class II molecules in endocrine cells. Diabetologia 28:464–6, 1985.
13. Hahn HJ, Gerdes J, Lucke S, et al. Phenotypical characterization of the cells invading pancreatic islets of diabetic BB/OK rats: Effect of interleukin 2 receptor-targeted immunotherapy. Eur J Immunol 18:2037–42, 1988.
14. Lucke S, Diamantstein T, Hahn HJ. Different lymphocyte subset distribution within "insulitis" islets of normoglycaemic and prediabetic BB/OK rats of similar age. Exp Clin Endocrinol 95:57–63, 1990.
15. Diamantstein T, Hahn HJ. Interleukin-2 receptor-targeted immune therapy. In: Monoclonal antibodies and peptide therapy in autoimmune diseases. Bach JF, ed. Marcel Dekker, New York, pp 295–318, 1993.
16. Maruyama T, Ogawa M, Kobayashi F, et al. Electron microscopic studies on the interaction of pancreatic islet cells and splenic lymphocytes in non-obese diabetic (NOD) mice. Biomed Res 9:67–73, 1988.
17. Miyagawa J, Hanafusa T, Itoh N, et al. Ultrastructural pathology of insulitis in the pancreas biopsy specimens from newly diagnosed type I diabetic patients. Abstract 15th IDF Congress, 1994, Kobe, Japan, p 434.
18. Haskins K, McDuffie M. Acceleration of diabetes in young NOD mice with a $CD4^+$ islet-specific T-cell clone. Science 249:1433–36, 1990.
19. Nagata M, Yoon JW. Studies on autoimmunity for T-cell mediated β-cell destruction. Distinct differences in β-cell destruction between CD4+ and CD8+ T-cell clones erived from lymphocytes infiltrating the islets of NOD mice. Diabetes 41:998–1008, 1992.
20. Hosszufalus N, Chan E, Teruya M, et al. Quantitative pheotypic and functional analysis of islet immune cells before and after diabetes onset in the BB rat. Diabetologia 36:1146–54, 1993.
21. Appels B, Burkhart V, Kantwerk-Funke G, et al. Spontaneous cytotoxicity of macrophages against pancreatic islet cells. J Immunol 142:3803–8, 1989.
22. Nerup J, Mandrup-Poulsen T, Molvig J. The HLA-IDDM association for etiology and pathogenesis of IDDM. Diabetes Metab Rev 3:779–802, 1987.
23. Kuttler B, Wanka H, Dunger A, Hahn HJ. Expression of MHC antigens on pancreatic islet cells. In: Frontiers of insulin secretion and pancreatic B-cell research. Flatt PR, Lenzen S, eds. Smith-Gordon, London, pp 509–15, 1994.

24. Skyler JS. Immune intervention studies in insulin-dependent diabetes mellitus. Diabetes Metab Rev 3:1017–35, 1987.
25. Vialettes B, Atlan-Gepner C, Hermitte L. Specific immuno-therapy for the prevention of diabetes: Theoretical and practical aspects. Diabetes Metab Rev 9:251–5, 1993.
26. Muir A, Schatz DA, Pozzilli P, MacLaren NK. Intervention therapies for insulin-dependent diabetes. Autoimmunity 16:301–10, 1993.
27. Smith KA. Interleukin-2: Inception, impact and implications. Science 240:1169–76, 1988.
28. Waldmann TA. Multichain interleukin-2 receptor: A target for immunotherapy in lymphoma. Immunol Rev 81:914–23, 1989.
29. Diamantstein T, Osawa H. The interleukin-2 receptor, its physiology and a new approach to a selective immuosuppressive therapy by anti-interleukin-2 receptor monoclonal antibodies. Immuol Rev 92:5–27, 1986.
30. Diamantstein T, Tilney NL, Strom TB, et al. Interleukin-2 receptor targeted immunosuppressive therapy with anti-interleukin-2 receptor monoclonal antibodies directed against the L-chain of the interleukin 2 receptor: Studies on selectivity and mode of action. In: Advances in immuopharmacology 4. Hadden JW, Spreafico F, Yamamura Y, et al., eds. Pergamon Press, UK, pp 95–104, 1988.
31. Hahn HJ, Kuttler B, Dunger A, et al. Prolongation of rat pancreatic islet allograft survival by teatment of recipient rats with monoclonal anti-interleukin-2 receptor antibody and cyclosporine. Diabetologia 30:44–6, 1987.
32. Kupiec-Weglinski JW, Hahn HJ, et al. Cyclosporin potentiates the immunosuppressive effects of anti-interleukin 2 receptor monoclonal antibody therapy. Transplant Proc 2:207–16, 1988.
33. Kupiec-Weglinski JW, Diamantstein T, Tilney NL. Interleukin 2 receptor-targeted therapy—rationale and applications in organ transplantations. Transplantation 46:785–92, 1988.
34. Kuttler B, Heym S, Volk HD, et al. No correlation between changes in lymphocyte subsets and pancreatic B-cell destruction. Horm Metab Res 22(Suppl 25):182–7, 1990.
35. Wanka H, Kuttler B, Hahn HJ. Detection of β-cell autoreactive cells from BB rats after sensibilization of lymphocytes by syngeneic pancreatic islets. Diabetologia 35(Suppl 1):A213, 1992.
36. Wanka H, Kuttler B, Hahn HJ: Is the in vitro destruction of pancreatic islet cells MHC-dependent? Diabetologia 36(Suppl 1):A9, 1993.
37. Hahn HJ, Kuttler B, Volk HD, et al. Interleukin 2 receptor-targeted therapy in islet transplantation research. Horm Metab Res 25(Suppl 1):177–80, 1989.
38. Tellides G, Dallman MJ, Morris PJ: Synergistic interaction of cyclosporine A with interleukin 2 receptor monoclonal antibody therapy. Transplant Proc 20(Suppl 2):202, 1988.
39. Granelli-Piperno A, Keane M, Steinman RM: Evidence that cyclosporine inhibits cell-mediated immunity primarily at the level of the T lymphocytes rather than the accessory cell. Transplantation 46:53S–60S, 1988.
40. Hahn HJ, Lucke S, Klöting I, et al. Curing BB rats of freshly manifested diabetes by short-term treatment with a combination of a monoclonal anti-interleukin 2 receptor antibody and a subtherapeutic dose of cyclosporin A. Eur J Immunol 17:1075–78, 1987.
41. Kelley VE, Gailton GN, Hattori M, et al. Anti-interleukin 2 receptor antibody suppresses murine diabetic insulitis and lupus nephritis. J Immunol 140:59–61, 1988.
42. Stark O, Klöting I, Reiher K, Kohnert KD: The major histocompatibility complex and insulin-dependent diabetes in BB rats. Acta Biol Med Germ 41:1129–33, 1982.
43. Klöting I, Stark O, Hahn HJ. Animal model of the insulin-dependent diabetes mellitus in BB rats: Their $RT1^u$ homogeneity and prolonged survival of allogeneic skin grafts. Folia Biol 30:24–32, 1984.

44. Klöting I, Stark O, Brdicka R. Incidence of insulin-dependent mellitus in BB rats: Their genetic heterogeneity and susceptibility to infection. Folia Biol 30:33–42, 1984.
45. Klöting I, Stark O. Occurrence of antibodies against gene products of the major histocompatibility complex in normal sera of diabetes-prone BB rats. Folia Biol 31:34–7, 1985.
46. Knospe S, Köhler E, Klöting I. Cell-mediated immune reactions against islets of Langerhans in diabetes-prone BB rats. Exp Clin Endocrinol 89:290–6, 1987.
47. Ziegler B, Klöting I, Besch W, Ziegler M, Hahn HJ. Cytotoxic activity of sera from diabetic BB rats against BB rat islet—a functional study. Diabetes Res 4:67–72, 1987.
48. Schröder D, Hehmke B, Klöting I, Besch W. Effect of sera from diabetes-prone BB/OK rats on neonatal rat pancreatic islets and islet cell suspensions. Exp Clin Endocrinol 93:187–92, 1989.
49. Hehmke B, Lucke S, Schröder D, Klöting I, Kohnert KD. Complement-dependent antibody-mediated cytotoxicity in the spontaneously diabetic BB/OK rat: Association with beta-cell volume density. Eur J Immunol 20:1091–96, 1990.
50. Klöting I, Reiher K. Einige Aspekte zur Haltung, Zucht und Reproduktion spontandiabetischer BB-Ratten. Z Versuchstierkd 27:5–12, 1985.
51. Klöting I, Vogt L. Breeding problems in diabetes-prone BB rats after "wet hysterectomy." Z Versuchstierkd 31:19–22, 1988.
52. Klöting I, Vogt L, Stark O, Fischer U. Genetic heterogeneity in different BB rat subpopulations. Diabetes Res 6:145–9, 1987.
53. Klöting I, Vogt L. BB/O(ttawa)K(arlsburg) rats: Features of a subline of diabetes-prone BB rats. Diabetes Res Clin Exp 18:79–87, 1991.
54. Klöting I, Vogt L. On the influence of metabolic state of parents at conception and pregnancy on incidence and age at onset of diabetes in the progeny of BB/OK rats. Diabetes Res 19:37–40, 1992.
55. Vogt L, Klöting I. Model-based prediction of diabetes incidence in BB/OK rats. Diabetes Metab 19:183–7, 1993.
56. Vogt L, Klöting I. BB-RADABA: A computer program as a tool for breeding and management of laboratory animals. Comp Method Progr Biomed 38:73–4, 1992.
57. Sibley RK, Sutherland DER, Goetz F, Michael AF. Recurrent diabetes mellitus in the pancreas iso- and allograft. A light and electron microscopic and immunohistochemical analysis of four cases. Lab Invest 53:132–44, 1985.
58. Hahn HJ, Kauert C, Dunger A, et al. Interleukin 2 receptor targeted immunotherapy in the prevention of autoimmune beta cell destruction as observed in diabetic BB/OK rats grafted with BB/OK rat islets. Autoimmunity 5:139–42, 1989.
59. Gottlieb PA, Berrios JP, Mariani G, et al. Autoimmune destruction of islets transplanted into RT6-depleted resistant BB/Wor rats. Diabetes 39:643–5, 1990.
60. Mathieu C. Immune intervention in animal models of type I diabetes: Pathogenetical and therapeutical aspects. Thesis, Katholieke Universiteit Leuven, Faculteit Geneeskunde, Leuven, 1994.
61. Wang Y, Pontesilli O, Gill RG, et al. The role of $CD4^+$ and $CD8^+$ T cells in the destruction of islet grafts by spontaneously diabetic mice. Proc Natl Acad Sci USA 88:527–31, 1991.
62. Hahn HJ. Die isolierte Langerhanssche Insel, ein Modell zur Untersuchung der Insulinsekretion in vitro. Endokrinologie 71:308–24, 1978.
63. Kuttler B, Mathieu C, Waer M, et al. Lack of disease recurrence in diabetic BB/Pfd rats after syngeneic islet transplantation. Autoimmunity 15:107–12, 1993.
64. Mathieu C, Kuttler B, Waer M, et al. Spontaneous reestablishment of self-tolerance in BB/Pfd rats. Transplantation 58:349–54, 1994.
65. Woehrle M, Pullmann J, Stuenkel KG, et al. Effect of anti-interleukin-2 receptor antibody treatment on recurrence autoimmune insulitis in the BB rat. Transplant Proc 26:734–5, 1994.

66. Holowachuk EW, Greer MK, Martin DR. The complete sequence of the MHC class II chain RT.1D alpha u of the diabetic BB rat: mRNA levels of RT1.D alpha in lymphocytes. Nucleic Acids Res 15:10551–567, 1987.
67. Hahn HJ, Ziegler B, Lucke S. Streptozocin diabetic BB/OK rats accept permanently BB/OK-islet grafts without immunosuppression. Z Versuchstierkd 32:135–40, 1989.
68. Woehrle M, Pullmann J, Bretzel RG, Federlin K. Prevention of recurrent autoimmune diabetes in the BB rat by islet transplantation under the renal capsule. Transplantation 53:1099–102, 1992.
69. Kuttler B, Dunger A, Volk, et al. Prevention and suppression of autoimmune pancreatic beta-cell destruction in BB rats by syngeneic lymphocytes obtained from long-term normoglycaemic donors. Diabetologia 34:74–7, 1991.
70. Mac Kay P. Adoptive transfer of diabetes to and from old normoglycaemic BB rats. Diabetologia 38:145–52, 1995.
71. Hahn HJ, Kuttler B. Influence on the pancreatic β-cell of manipulations with therapeutic potential for prevention of insulin-dependent diabetes. In: Frontiers of insulin secretion and pancreatic B-cell research. Flatt PR, Lenzen S, eds. Smith-Gordon, London, pp 597–604, 1994.
72. Hutchings P, O'Reilly L, Parish NM, et al. The use of a non-depleting anti-CD4 monoclonal antibody to re-establish tolerance to β-cells in NOD mice. Eur J Immunol 22:1913–18, 1992.
73. Charlton B, Mandel TE. Recurrence of insulitis in the NOD mouse after early prolonged anti-CD4 monoclonal antibody treatment. Autoimmunity 4:1–7, 1989.
74. Steinbrüchel DA, Larsen S, Kristensen T, et al. Survival, function, morphology and serological aspects of rat renal allografts. Effect of short-term treatment with cyclosporin A, anti-CD4 and anti-interleukin-2 receptor monoclonal antibodies. APMIS 100:682–94, 1992.
75. Kuttler B, Kauert C, Diamantstein T, Hahn HJ. Long-term immunosuppression inhibits the development of tolerance in BB rats. Symposium Intern Diabetes Immunother Group (IDIG) Istanbul, 1993.
76. Vialettes B, Bardet S, Hirn D, et al. Pilot study of combination anti IL2R mc antibody-low dose of Cyclosporin A in recently diagnosed type 1 diabetes. Diabetologia 34(Suppl 2):A180, 1991.
77. Ueda H, Hancock WW, Cheung YC, et al. The mechanism of synergistic interaction between anti-interleukin 2 receptor monoclonal antibody and cyclosporine therapy in rat recipients of organ allografts. Transplantation 50:545–50, 1990.
78. Sido B, Morris PJ, Dallman M. Phenotypic and functional analysis of graft-infiltrating cells following therapy with anti-interleukin-2 receptor monoclonal antibody and cyclosporine A in the rat. Transplant Proc 24:2549–50, 1992.
79. Soulillou JP, Cantarovich D, Dantal J, et al. Anti-IL-2 receptor and anti-LFA-1 monoclonal antibodies in transplantation: Rationale and Nantes clinical experience. In: A critical analysis of monoclonal antibody therapy in transplantation. Burlingham WJ, ed. CRC Press, Boca Raton, FL, pp 47–66, 1992.
80. Wood MJA, Sloan DJ, Dallman MJ, Charlton HM. Specific tolerance to neural allografts induced with an antibody to the interleukin 2 receptor. J Exp Med 177:597–603, 1993.
81. Waldmann H, Cobbold S. Monoclonal antibodies for the induction of transplantation tolerance. Curr Opin Immunol 5:753–8, 1993.
82. Leong LYW, Qin S, Cobbold SP, Waldmann H. Classical transplantation tolerance in the adult: The interaction between myeloablation and immunosuppression. Eur J Immunol 22:2825–30, 1992.
83. Qin S, Cobbold SP, Pope H, et al. Infectious transplantation tolerance. Science 259:974–77, 1993.
84. Tomonari K. Tolerance in vivo and in vitro. Immunol Rev 116:139–57, 1990.
85. Alters SE, Grossman D, Fathmann CG. Anti-CD4-mediated transplantation toler-

ance: Mechanisms and memory. In: A critical analysis of monoclonal antibody therapy in transplantation. Burlingham WJ, ed. CRC Press, Boca Raton, FL, pp 79–99, 1992.

86. Hahn HJ, Lucke S, Kuttler B, et al. Investigation of diabetes-prone normoglycaemic BB rats. In: Lessons from animal diabetes. Shafrir E, ed. Smith-Gordon, London, 3:30–3, 1990.
87. Hahn HJ, Lucke S, Klöting I, Besch W. Prospective investigations of long-term normoglycaemic BB/OK rats: Serial determination of glucose tolerance, insulitis, B-cell volume density and pancreatic insulin content. Exp Clin Endocrinol 98:54–61, 1991.
88. Hahn HJ, Diamantstein T. Monoclonals and/or immunotoxims as new therapeutic agents in the treatment of autoimmune diseases. In: Diabetes mellitus. Wojcikowski C, ed. Verlag Normex, Gdansk, pp 1–10, 1992.
89. Hahn HJ, Kauert C, Kuttler B. Efficacy of FK506 in the prevention of autoaggression in syngeneicly grafted diabetic BB rats. Horm Metab Res 25:56(A24), 1993.
90. Chen Z, Cobbold S, Waldmann H, Metcalf S. Stability of tolerance in mice generated by CD4 and CD8 monoclonal antibody treatment: Cell transfer experiments. Transplant Proc 25:790–1, 1993.
91. Kupiec-Weglinski JW, Diamanstein T, Tilney NL, Strom TB. Therapy with monoclonal antibody to interleukin 2 receptor spares suppressor T cells and prevents or reverses acute allograft rejection in rats. Proc Natl Acad Sci USA 83:2624–27, 1986.
92. Rashba Ej, Reich EP, Janeway CA, Sherwin RS. Type 1 diabetes mellitus: An imbalance between effector and regulatory T cells? Acta Diabetol 30:61–9, 1993.
93. Kuttler B, Kauert C, Wanka H, et al. Tolerance to MHC-identical allogeneic rat islets is mediated by CD4+ helper T-lymphocytes. Diabetologia 38(Suppl 1):A84, 1995.
94. Fowell D, Mason D. Evidence that the T cell repertoire of normal rats contains cells with the potential to cure diabetes. Characterization of the $CD4^+$ T cell subset that inhibits this autoimune potential. J Exp Med 177:627–36, 1993.
95. Hahn HJ, Hellman B, Lernmark A, et al. The pancreatic B-cell recognition of insulin secretagogues. X. Influence of neuraminidase treatment on the release of insulin, and the islet content of insulin, sialic acid, and adenosine 3′5′-cyclic monophosphate. J Biol Chem 249:5275–84, 1974.
96. Ziegler B, Hahn HJ, Ziegler M. Insulin recovery in pancrease and host organs of islet grafts. Exp Clin Endocrinol 85:53–60, 1985.
97. Lucke S, Ziegler B, Diaz-Alonso JM, Hahn HJ. Eignung spezifischer Färbemethoden für die Bestimmung des β-Zellvolumens im Rattenpankreas mit normalem und reduziertem Insulingehalt. Acta Histochem 45:107–16, 1985.
98. Cordell UL, Falini B, Erber WN, et al. Immunoenzymatic labeling of monoclonal antibodies using immune complexes of alkaline phosphatase anti-alkaline phosphatase (APAAP complexes). J Histochem 32:219–29, 1984.
99. Kuttler B, Lehmann M, Lacha J, et al. Anti-CD4 therapy induces a donor-specific, organ-unspecific tolerance in allogeneic kidney graft recipients. Transplant Proc 26:728–9, 1994.
100. Logothetopoulus J, Valiquette N, Madura E, Cvet D. Onset and progression of pancreatic insulitis in the overt, spontaneously diabetic, young adult BB rat studied by pancreatic biopsy. Diabetes 33:33–6, 1984.
101. Vogt L, Klöting I. Genetic analysis of frequencies of phenotypes in the spontaneously diabetic BB rat—a main animal model of autoimmune type-I-diabetes. Diabetes Res 22:105–13, 1993.

5

Role of retroviruses in the pathogenesis of diabetes with emphasis on the NOD mouse

TOSHIAKI HANAFUSA,[1] KOJI TOMITA,[1] HIROMU NAKAJIMA,[1] CHISA NAKAGAWA,[2] JUN-ICHIRO MIYAGAWA,[1] MASAMICHI KUWAJIMA,[1,3] SEIICHIRO TARUI,[2] AND YUJI MATSUZAWA[1]

[1]Second Department of Internal Medicine, Osaka University Medical School, Suita 565; [2]Otemae Hospital, Osaka 540, Japan; [3]Present address: Department of Laboratory Medicine, School of Medicine, Tokushima University, Tokushima 770, Japan

Key words: NOD mouse, IDDM, retrovirus, pathogenesis, insulitis, virus cloning, electron microscopy.

Viruses have long been suggested to play a role, in the pathogenesis of IDDM. However, it is very difficult to demonstrate the pathogenic role of viruses in human IDDM. On the other hand, many diabetogenic viruses have been investigated in animal models in which diabetes was induced by exogenous inoculation of cytopathic viruses. One of the rare exceptions is the NOD mouse, in which diabetes develops spontaneously and retroviruses have been found in pancreatic B-cells. Other examples are the multiple low-dose streptozotocin (STZ)-induced diabetes mouse and the C3H-*db*/*db* mouse. In this chapter, we first review the studies in the latter two mouse models, and then describe our results in the NOD mouse.

Multiple low-dose streptozotocin-induced diabetes mouse and C3H-*db*/*db* mouse

One injection of a large dose of STZ induces diabetes in animals by directly destroying pancreatic B-cells. On the other hand, multiple subdiabetogenic doses of STZ are considered to induce diabetes mainly through autoimmune attack by lymphocytes against B-cells. Morphologically, insulitis is observed in the pancreas of diabetic male CD-1 mice injected with multiple low STZ doses.

Electron microscopic examination of the pancreas of mice sacrificed 5–6 days

Correspondence: T. Hanafusa, Second Department of Internal Medicine, Osaka University Medical School, Suita 565, Japan.

Lessons from Animal Diabetes VI. Ed. E. Shafrir.

after the completion of STZ injections revealed the presence of numerous type C viruses in B-cells as well as various features of B-cell disruption.[1] Filamentous and circular profiles were frequent, as were forms budding from these membrane surfaces. Virus particles were not observed budding from the plasma membranes and were not identified within macrophages or lymphocytes. Most virus-containing B-cells were structurally intact, albeit relatively degranulated. Interestingly, these features are very similar to those of retrovirus particles seen in B-cells of the NOD mouse. The islets of mice that are sacrificed 1, 6, and 12 months after the completion of STZ injections showed significantly reduced insulitis and variable numbers of type C viruses.

The significance of induction of type C virus remains uncertain in the STZ-induced insulitis-diabetes model. There is a difference between mouse strains in the induction of type C virus in B-cells. STZ injections induce an insulitis-diabetes syndrome in C57BKsJ mice in the absence of type C virus induction; type A virus induction, however, is observed in these inbred mice. The striking morphological similarity between the retrovirus in the NOD mouse and type C virus in the B-cells of multiple low-dose STZ-diabetic mice suggests that retroviruses are likely to play a significant role in the pathogenesis of diabetes in these mouse models.

Mouse pancreas appears to be a very conducive site for the expression of endogenous retroviral genes, and this expression has been enhanced in pancreatic disease states, including duct cell adenocarcinoma, B-cell adenoma, hereditary diabetes, and spontaneous necrosis of exocrine pancreatic cells.[2] When the effect of an autosomal recessive obesity mutation, diabetes (*db*), was studied, the severity of diabetes was associated with type C retrovirus induction in B-cells and not with the H-2 haplotype of C3H-*db*/*db* mice.[3] This result indicates the importance of type C retrovirus expression in beta cells in the progression of diabetes in these mice. The data obtained in the above-mentioned two mouse models suggest that type C retrovirus expression is closely associated with diabetes in mice, at least in some situations in certain strains. This might be quite relevant to the following mouse diabetes model, the NOD mouse.

NOD mouse

The NOD mouse exhibits insulitis at 4–5 wk of age, and overt diabetes starts to develop at 14–15 wk of age. A genetic predisposition plays a very important role in the pathogenesis of insulitis and diabetes, but the precise sequence of events leading to the development of diabetes still needs to be clarified. Recent research on the NOD mouse revealed that the development of insulitis and diabetes is controlled by several genes, and that the most important gene is within the major histocompatibility complex (MHC).[4] Insulitis and the development of diabetes is probably mediated by cellular autoimmune reactions involving both $CD4^+$ and $CD8^+$ cells. However, the factors triggering the autoimmune reactions remain to be demonstrated. One possible triggering factor could be virus. Fujita and coworkers[5,6] have revealed the presence of retrovirus-like particles in pancreatic B-cells of the NOD mouse. Therefore, the NOD mouse could be an excellent model for analyzing the role of viruses in the development of IDDM.

Retrovirus particles in B-cells of the NOD mouse. Figure 1 shows an islet B-cell of an 8-wk-old female NOD mouse. Virus-like particles are seen in the cisternae of the rough endoplasmic reticulum. Many of the particles are circular, but some of them have a cylindrical shape. The circular shape is probably a cross-sectional view, and the cylindrical shape a sagittal view of virus particles. The morphological characteristics of these particles very much resemble those of retroviruses.[7,8] Therefore, from these morphological characteristics, it can be assumed that these viruses are retroviruses. The particle is enclosed by an envelope that consists of two layers of unit membrane structure. Inside the envelope, ring-shaped and membrane-like structure, a nucleocapsid, can be recognized. The nucleocapsid is packed with amorphous or filamentous materials.

In some cases, virus particles were seen budding from a limiting membrane of rough endoplasmic reticulum into the cisterna. However, this budding phenomenon was restricted to the membrane of rough endoplasmic reticulum and was not seen in the plasma membrane of B-cells. Based on these morphological characteristics, these retroviruses were concluded to be immature type C retroviruses.

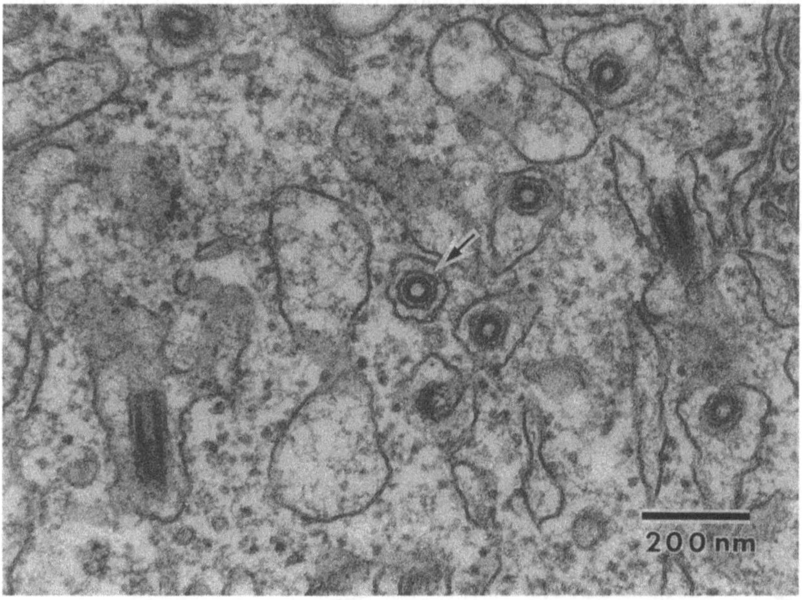

Figure 1. Electron micrograph of a pancreatic B-cell of an 8-wk-old female NOD mouse (×71,000). Virus-like particles (arrow) are seen in the cisternae of the rough endoplasmic reticulum (RER). Many of the particles are circular, but some of them are cylindrical. The particle is enclosed by an envelope consisting of two layers of unit membrane structure. In some cases, virus particles are seen budding from a limiting membrane of the RER into the cisterna. Based on these morphological characteristics, these retroviruses are considered to be immature type C retroviruses.

It is important that these retroviruses were seen only in B-cells and not in A-cells, suggesting that the expression of these retroviruses is B-cell specific and that these retroviruses may play a role in the pathogenesis of diabetes in the NOD mouse. Figure 2 schematically illustrates the relative abundance of the retrovirus particles revealed by electron microscopy. Interestingly, the particles are detectable in the fetus by 18 days. This indicates that the retroviruses are vertically transmitted and that the expression of these viruses in B-cells precedes the start of insulitis in the NOD mouse.

Expression of* gag *p30 in the islet of the NOD mouse. Figure 3 shows the schematic structure of mouse retrovirus. Retrovirus is composed of three regions, namely, *gag*, for group specific antigen; *pol,* for polymerase; and *env*, for envelope protein. These three regions lie between two long terminal repeats, LTRs, which have a promoter activity. The *gag* region is subdivided into four regions, p15, p12, p30, and p10.

The expression of retrovirus-associated protein in the NOD mouse pancreas was supported by Western blot analysis of the NOD mouse pancreas homogenate, using anti-retrovirus *gag* p30 antiserum.[9] The presence of a 30 kDa band was detected in the islet homogenate of NOD mice. The 30 kDa band was not detectable in control ICR mouse islets. The specificity of the reaction was examined by an absorption experiment. After preabsorption of the antiserum with Moloney murine leukemia virus (MMLV), the 30 kDa band became undetectable in the NOD mouse islets, confirming that the positive 30 kDa band seen in the NOD mouse islet homogenate was derived from the retrovirus. Following these observations, we cloned the retrovirus expressed specifically in B-cells of the NOD mouse.[10]

Molecular cloning of the retrovirus in the NOD mouse. For cloning the NOD-mouse-specific retrovirus we started the search of the retrovirus by performing

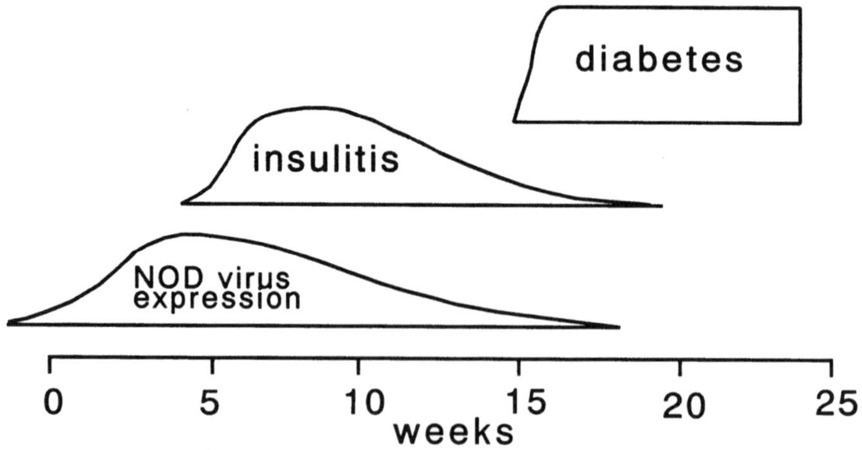

Figure 2. Schematic illustration of the relative abundance of the retrovirus particles revealed by electron microscopy. The particles could be detected in the fetus by 18 days.

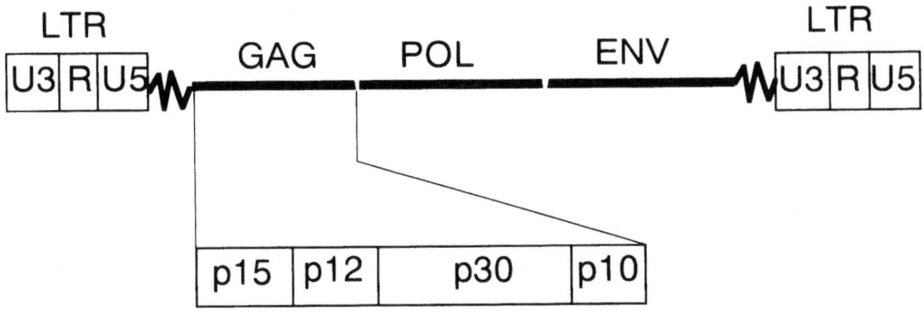

Figure 3. Schematic structure of mouse retrovirus composed of three regions, *gag*, *pol*, and *env*. These three regions lie between two long terminal repeats (LTRs), which have promoter activity. The *gag* region is subdivided into four regions, p15, p12, p30, and p10.

reverse transcription-polymerase chain reaction (RT-PCR) in the pancreas of the NOD mouse. RT-PCR analysis with retroviral gag-specific primers and total RNA from the pancreas of NOD and control mice (NON, B10.G.D, and ICR mouse) yielded two types of retrovirus-associated sequence. Using degenerative primers we identified the retrovirus-like sequence from the cDNA derived from the pancreas of the NOD mouse. This sequence was considered to be NOD-mouse-specific. A candidate for the NOD mouse specific retroviral cDNA was amplified. The level of expression as estimated by RT-PCR tended to correlate with the severity of insulitis.

From the genomic DNA of the NOD mouse, we further isolated four overlapping clones. The sequence of these clones represented the full-viral genome of this retrovirus. Thus, we finally cloned a novel retroviral. The sequence of this retrovirus possessed a strikingly unique feature, compared to that of a replication-defective virus, such as mouse AIDS virus.

Conclusions

Type C retrovirus-like particles were seen in B-cells of the NOD mouse and 30 kDa protein crossreactivity with antibodies to the retrovirus gag protein p30 was detected in islets of the NOD mouse. The full sequence of proviral NOD virus was cloned and determined from the genomic DNA of the NOD mouse.

Acknowledgments

We express our sincere thanks to Dr. E. H. Leiter, Jackson Laboratories, Bar Harbor, ME, USA, for helpful discussions. This work was supported in part by the Scientific Research Fund from the Ministry of Education, Science and Culture of Japan, and a grant from the Ministry of Health and Welfare of Japan (Pathogenesis Section), a grant for Diabetes Research from Otsuka Pharmaceutical Co., Ltd., and grants from the Mochida Memorial Foundation for Medical and Pharmaceutical Research, the Japan Diabetes Foundation, the Yamanouchi

Foundation for Research on Metabolic Disorders, and the Cell Science Research Foundation.

References

1. Like AA, Appel MC, Williams RM, Rossini AA. Streptozotocin-induced pancreatic insulitis in mice. Lab Invest 38:470–86, 1978.
2. Leiter EH, Wilson GL. Viral interactions with pancreatic B-cells. In: The pathology of the endocrine pancreas in diabetes. Lefebvre PJ, Pipeleers DG, eds. Springer-Verlag, Berlin, pp 85–105, 1988.
3. Leiter EH. Type C retrovirus production by pancreatic beta cells: Association with accelerated pathogenesis in C3H-db/db ("diabetes") mice. Am J Pathol 119:22–32, 1985.
4. Ghosh S, Palmer SM, Rodrigues NR, et al. Polygenic control of autoimmune diabetes in nonobese diabetic mice. Nature Genet 4:404–9, 1993.
5. Fujita H, Fujino H, Nonaka K, et al. Retrovirus-like particles in pancreatic B-cells of NOD (non-obese diabetic) mice. Biomed Res 5:67–70, 1984.
6. Fujino-Kurihara H, Fujita H, Hakura A, et al. Morphological aspects on pancreatic islets of non-obese diabetic (NOD) mice. Virchows Arch B [Cell Pathol] 49:107–20,1985.
7. Bernhard W. Type A and C virus particles in murine and other mammalian leukemias and sarcomas. In: Ultrastructure of animal viruses and bacteriophages: An atlas. Dalton AJ, Haguenau F, eds. Academic Press, New York, pp 283–305, 1973.
8. Leiter EH, Wilson GL. Viral interactions with pancreatic B-cells. In: The pathology of the endocrine pancreas in diabetes. Lefèbvre PJ, Pipeleers DG, eds. Springer-Verlag, Berlin, pp 85–105, 1988.
9. Nakagawa C, Hanafusa T, Miyagawa J, et al. Retrovirus *gag* protein p30 in the islets of non-obese diabetic mice: Relevance for pathogenesis of diabetes mellitus. Diabetologia 35:614–8, 1992.
10. Tomita K, Nakajima H, Nakagawa C, et al. Cloning of type C retrovirus in the pancreas of NOD mouse. Abstracts, 12th Int Immunol Diabetes Workshop, Orlando, FL, 1993.

6

Occurrence of pretype I diabetes (pre-IDDM) and type II diabetes (NIDDM) in BC1 [(NOD × Mus spretus) F1 × NOD] mice

MASAKAZU HATTORI,[1] EIJI YAMATO,[1] EISAKU MATSUMOTO,[1] NAOTO ITOH,[1] TETSUSHI TOYONAGA,[1] MARIA PETRUZZELLI,[1] MASAHIRO FUKUDA,[1] MASASHI KOBAYASHI,[2] AND VERNE M. CHAPMAN[3]

[1]Section on Immunology and Immunogenetics, Joslin Diabetes Center, Department of Medicine, Harvard Medical School, Boston, Massachusetts 02215, USA; [2]First Department of Medicine, Toyama Medical and Pharmaceutical University, Toyama 930-01, Japan; [3]Department of Molecular and Cellular Biology, Roswell Cancer Research Institute, Buffalo, New York 14263-0001, USA.

Key words: insulin dependent diabetes, non-insulin dependent diabetes, NOD mice, wild mice, crossbreeding, insulitis, insulin resistance.

Insulitis and diabetes in the NOD mouse

Type I human diabetes (IDDM), the BB rat, and the NOD mouse appear to result from autoimmune B-cell destruction in a setting of genetic predisposition.[1,2] The NOD/Shi mouse was established in 1980 at the Shionogi Aburahi Laboratories.[3] The NOD mouse develops IDDM secondary to B-cell destruction by infiltrating immune cells (insulitis) (Fig. 1; Table 1). Insulitis appears as early as 5 wk of age. To develop overt diabetes, at least 90% of the total islets need to be damaged by infiltrating immune cells. The immune cells invade islets and are often seen adjacent to the ducts and blood vessels. The incidence of insulitis is 90% in NOD/Shi females and 70% in NOD/Shi males at 9 wk of age, and 100% in both sexes at 5 months of age. The incidence and degree of insulitis increase with age, resulting in a lack of intrinsic insulin secretion, hyperglycemia, loss of body weight, and ketosis, as seen in human IDDM. The NOD/Shi mouse develops diabetes usually by 7 months of age.

Correspondence: Masakazu Hattori, Joslin Diabetes Center, One Joslin Place, Boston, MA 02215, USA. E-mail: hattorim@joslab.harvard.edu

Lessons from Animal Diabetes VI. Ed. E. Shafrir.

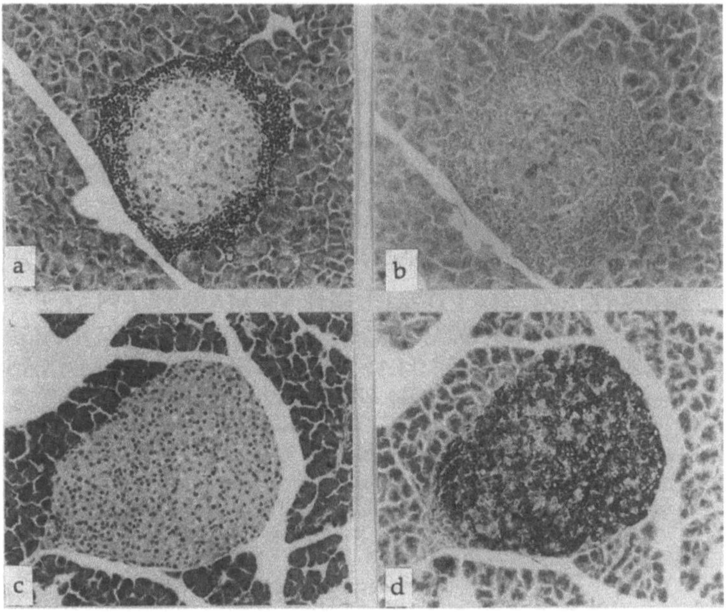

Figure 1. (a) Insulitis in a diabetic NOD female, H-E. (b) Loss of insulin granules in the islet cells from a diabetic NOD female, aldehyde-fuchsin-Masson staining. (c) Islet from a control ICR mouse, H-E staining. (d) Islet from a control ICR mouse aldehyde-fuchsin-Masson staining. Note the insulin granules.

Mus spretus. The mice that provided the bulk of the genome of most laboratory mouse strains are descendants of the western European house mouse, *Mus domesticus.* They were separated from other wild mice approximately 1–2.5 million years ago. Since then, the ancestors of laboratory mice have been bred within the same species. From the evolutionary point of view, the genetic backgrounds of laboratory mouse strains, including NOD, NON, ILI, BALB/c, C3H/He, and C57BL/6, are close to each other. This close genetic background causes diffi-

Table 1. Criteria and degree of insulitis

Sections of pancreas from parental and backcross animals were stained with hematoxylin and eosin, and subjected to microscopic observation. Insulitis is defined as destruction of islet B-cells by infiltrating lymphocytes and macrophages. Each islet was graded according to the following criteria (0–4):

Grade 0:	No infiltrating immune cells
Grade 1:	Peri-insulitis; immune cells adjacent to islets but do not enter the islets (pre-stage leading insulitis).
Grade 2:	Infiltrating immune cells occupy <25% of the islet area.
Grade 3:	Infiltrating cells occupy 25–50% of the islet area.
Grade 4:	Infiltrating cells occupy >50% of the islet

culty in finding polymorphisms between inbred laboratory mouse strains. Wild mice, *Mus spretus,* represent an interesting source of polymorphisms at many loci, compared with the inbred laboratory mouse strains.[4] Our preliminary studies show that *Mus spretus* did not develop obesity, diabetes, insulin resistance, hypertrophy of islets, and insulitis by the age of 12 months.[5]

Influential factors in the development of IDDM (multifactorial disease)

IDDM is a multifactorial disease, and its development appears to be influenced by genetic and environmental factors, including viral infection, diets, and chronic stress. Viruses have been implicated in the pathogenesis of B-cell destruction. In early studies[6–8] on the association of viruses with diabetes, viruses were regarded as agents causing diabetes. However, the appearance of animal models for spontaneous IDDM, the BB rat and the NOD mouse, raised under viral-free conditions has changed the concept of the role of viruses in the development of diabetes. Viruses are divided into two groups: viruses accelerating (or inducing) and those suppressing the development of IDDM. Environmental viral pathogens enhance or inhibit the process of islet cell destruction, depending on the immunologic reactions of the host.[2,9,10] Mechanisms by which viruses may induce diabetes include modification of B-cell antigens, molecular mimicry, direct lysis of B-cells, and virus-induced functional changes in effector or regulatory lymphocytes in the host. Environmental organisms may induce or suppress diabetes in susceptible animals and humans by stimulating genetically programmed effector cells or by disrupting a balanced network of autoreactive and regulatory cells.

A semipurified AIN-76 diet also prevents the development of diabetes in the BB rat[11] and the NOD mouse.[12] Pregestimil is a low or non-allergic infant formula, based on enzymatically hydrolyzed cow's milk casein, and further treated with charcoal to remove antigenic peptides. Pregestimil was shown to reduce islet damage and to completely prevent the development of diabetes (0%) up to 40 wk of age.[13] When the chloroform-methanol extract of the OG 96 mouse diet was added to Pregestimil, NOD mice showed a high incidence of diabetes (75%).[12] Environmental stress, especially chronic moderate stress, accelerates the development of diabetes in the BB rat.[14]

Insulin resistance

Insulin resistance is present if insulin is not able to normally stimulate cellular glucose uptake or to suppress hepatic gluconeogenesis. Insulin resistance can originate in a number of different ways, including 1) a reduction in the affinity or number of insulin receptors, 2) failure of insulin to stimulate the tyrosine kinase activity of the receptor, 3) an impairment in the movement of glucose transporters from the interior of the cell membrane, or 4) defects in the metabolic pathways required to oxidize glucose or convert it to glycogen.[15] Whichever is responsible, the consequence of insulin resistance is that glucose utilization by tissues is impaired and excess glucose accumulates in the circulation (hyperglycemia). The hyperglycemia stimulates the pancreas to produce more insulin in an effort to overcome insulin resistance. Thus, simultaneous elevation of both glucose and insulin is suggestive evidence that insulin resistance is present.[16]

Type II diabetes mellitus (NIDDM)

NIDDM manifests three major pathophysiologic defects: 1) the uptake of glucose into insulin-sensitive tissues (predominantly muscle and adipose tissue) is impaired, 2) excessive production of hepatic glucose is not reduced in response to insulin, and 3) a low rate of insulin response respective to the degree of hyperglycemia. Chronic hyperglycemia itself can exacerbate and sustain the defect in insulin release (glucose toxicity) whenever NIDDM is established.[16]

Criteria for pretype I, type I, and type II diabetes

It is important to have solid phenotypes for genetic analysis of susceptibility to diseases. Therefore, the criteria for IDDM and NIDDM were prepared from our preliminary data in BC1 [(NOD × *Mus spretus*)F1 × NOD] mice (Table 2). To distinguish IDDM and NIDDM, each animal was examined for levels of insulin and blood glucose; insulin resistance at 3, 6, and 12 months of age; and histological changes in the pancreas. NOD mice and hybrid animals of NOD with *Mus spretus* were examined for glucose in urine (glucosuria) using Tes-Tape (Eli Lilly, Indianapolis, IN). The animals were screened weekly for 12 months. When the animals showed glucosuria, blood glucose levels were measured using Accuchek II or a glucometer. Diabetic animals received an insulin injection twice a day (a mixture of Regular Iletin I insulin and NPH Iletin I insulin at 1:1 proportion, Eli Lilly) and were maintained at least 1 wk prior to pathological examination.

Table 2. Criteria for pre-IDDM, IDDM, and NIDDM

	Pre-IDDM	IDDM	NIDDM
Blood glucose	<300 mg/dl	>300 mg/dl Persistent hyperglycemia	>300 mg/dl Persistent or intermittent hyperglycemia, depending on conditions of food intake
Insulin	<150 μU/ml	<40 μU/ml	≥150 μU/ml
Insulin resistance	Absent (sensitive)	Absent (sensitive)	Present (resistant)
Insulitis	Present, >grade II	Present, grade II–IV	Absent
Hypertrophy of islets	Absent	Absent	Present

Insulin tolerance test for insulin resistance and collection of blood samples for blood glucose and insulin levels were performed between 1 and 5 p.m. This is because mice usually eat during the night. Otherwise, the animals were fasted for 15 h prior to the tests.

Breeding studies of NOD with *Mus spretus*

NOD mouse and* Mus spretus *in the present breeding studies. NOD/Shi//Jos mice (registered by the National Research Council, Institute of Laboratory Animal Resources, Washington, DC, USA) were originally obtained from the colony of Dr. Susumu Makino at the Aburahi Laboratories, Shionogi Ltd., Shiga, Japan). NOD/Shi//Jos females develop diabetes at 92% incidence and NOD/Shi//Jos males (hereafter NOD) at <60% incidence by 30 wk of age. Inbred *Mus spretus* used in the present breeding studies were obtained along with I.D. numbers from the Jackson Laboratory (Bar Harbor, ME, USA). *Mus spretus* females and males did not develop obesity, insulin resistance, diabetes, and insulitis by the age of 12 months. The animals in our breeding studies were fed a commercial mouse chow Prolab 3000 Formula (protein 22%, fat 5%; Agway, Syracuse, NY, USA) and were housed in cages with filter tops under conventional conditions (12 h light period, 12 h dark period).

Incidence of diabetes and insulitis (pretype I diabetes) in BC1 [(NOD × *Mus spretus*) F1 × NOD] mice. To examine the genetic mode of inheritance of diabetes and insulitis, NOD mice were mated with *Mus spretus* to produce F1 animals. Then, F1 animals were backcrossed to NOD mice reciprocally. F1 (NOD × *Mus spretus*) males were infertile. Therefore, only female F1 of NOD with *Mus spretus* were backcrossed to NOD males. The F1 and backcross (BC) animals were screened for the development of diabetes weekly for 1 year, and then were subjected to histological examination for insulitis (see Table 1 and Fig. 1). The penetrance of diabetes in our NOD/Shi//Jos colony is ~90% in female NOD mice and 50% in male NOD mice. The incidence of diabetes and insulitis in the F1 animals was 0%. This result suggests that the development of diabetes and insulitis is of a "recessive" nature.

The incidence of diabetes in BC1 [(NOD × *Mus spretus*) F1 × NOD] mice was 3% in females and 12% in males (Table 3). The development of diabetes was predominant in the male BC1 [(NOD × *Mus spretus*) F1 × NOD] mice. It is important to clearly distinguish IDDM and NIDDM diabetes for genotyping. Therefore, we prepared the diabetes criteria based on the data from our experiments (Table 2). In the present experiment, none of the diabetic BC1 males and females showed insulitis (suggesting no IDDM). The diabetic BC1 males and females showed hyperinsulinemia and insulin resistance in the absence of insulitis, suggesting NIDDM as described in the section on NIDDM in BC1 [(NOD × *Mus spretus*) F1 × NOD] mice. Approximately half of the diabetic BC1 males and females showed hypertrophied islets (Fig. 2). The BC1 mice had insulitis (>grade II) at an incidence of 16% in females and 5% in males (see Table 3). In the previous breeding studies by Avner's group,[17] cyclophosphamide-treated BC1 [(NOD × *Mus spretus*) F1 × NOD] mice did not develop IDDM, but had insulitis at a similar incidence as in our present study.

In contrast, the incidence of diabetes in BC1 mice of NOD with C3H/He, NON, and BALB/c from our own breeding studies at the Joslin Diabetes Center was 11–16% in females and 2–3% in males, suggesting at least two or more genes

Table 3. Incidence of diabetes and insulitis in the first backcross (BC1) animals of NOD with *Mus spretus*, C3H/He, NON, and BALB/c at 1 year of age (backcrossed to NOD)

	NOD *M. spretus* (%)	NOD × C3H/He (%)	NOD × NON (%)	NOD × BALB/c (%)
Diabetes in BC1				
Female	5/170 (3)	12/77 (16)	12/99 (12)	11/99 (11)
Male	16/130 (12)	1/58 (2)	2/58 (3)	1/50 (2)
Insulitis in diabetic BC1				
Female	0/3 (0)	12/12 (100)	12/12 (100)	11/11 (100)
Male	0/14 (0)	1/1 (100)	2/2 (100)	1/1 (100)
Insulitis with grade >2[a]				
Female	7/43 (16)[b]	14/28 (50)	12/24 (50)	8/19 (42)
Male	2/42 (5)	ND	ND	ND

[a] Grade 2 of insulitis indicates that 25% of the islet area is damaged by infiltrating immune cells.
[b] In addition to the seven mice with grade >2, six mice showed peri-insulitis (lymphocytic infiltration adjacent to the islets). Peri-insulitis is a pre-stage leading to insulitis. If the six mice were included in the animal group with insulitis, the incidence would be 30% (13 of 43).

Each islet was graded according to the criteria (0–4) listed in Table 1. After grading 15–30 islets, BCl animals were diagnosed for insulitis when islets showed lymphocytic infiltration (grade >2).

The breeding study of NOD with *Mus spretus* was performed after the breeding studies of NOD with BALB/c; C3H/He and NON were performed in parallel at the Joslin Diabetes Center.
ND; not done.

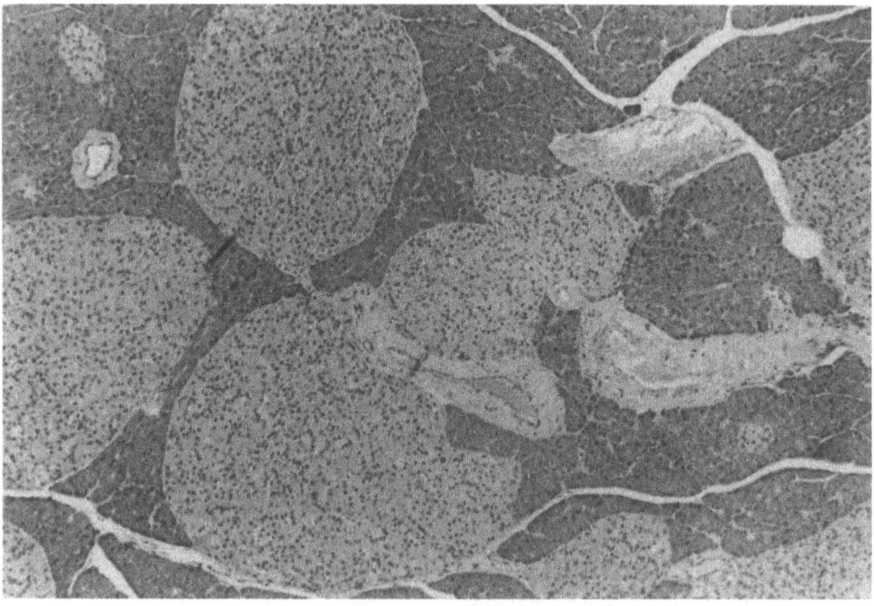

Figure 2. Hypertrophied islets without insulitis from a diabetic BC1 [(NOD × *Mus spretus*) F1 × NOD]. H-E staining (×80).

contributing to the development of diabetes (see Table 3). All the diabetic BC1 animals had insulitis (IDDM). The data from our breeding studies are consistent with the previous reports.[18,19]

IDDM in BC2 (BC1 × NOD). In the present breeding study, female and male BC1 [(NOD × *Mus spretus*) F1 × NOD] mice developed pretype I diabetes (insulitis only). To develop overt diabetes, at least 90% of the total islets need to be damaged by infiltrating immune cells. When the BC1 animals are repeatedly backcrossed to NOD mice, the backcross animals should develop more severe insulitis and IDDM. BC1 [(NOD × *Mus spretus*) F1 × NOD] mice, homozygous for NOD-MHC, were backcrossed to NOD (BC2) and were screened for the development of diabetes without cyclophosphamide treatment. The incidence of diabetes in BC2 was 22% in females and 15% in males. The incidence of insulitis in BC2 was 60% in females and 15% in males (Hattori, to be published). All the diabetic BC2 mice developed insulitis. In contrast to the high incidence of NIDDM in BC1 males, female BC2 mice showed a higher incidence of IDDM than male BC2 mice.

The data from the BC1 and BC2 mice suggest three recessive major susceptibility genes to IDDM and two genes to insulitis in the crossed animals of NOD with *Mus spretus* (Hattori, to be published). Our present results are consistent with the previous report by Avner's group[17] that, in addition to the MHC-linked diabetogenic gene (*Idd1*), there are three recessive major susceptibility genes (markers: *D3Nds2, D6Nds1, D15Nds6*) to IDDM in cyclophosphamide-treated BC1 [(NOD × *Mus spretus*) F1 × NOD] and BC2 (BC1 × NOD).

Occurrence of pre-IDDM and NIDDM in BC1 [(NOD × *Mus spretus*) F1 × NOD] mice

Pre-IDDM in nondiabetic BC1 [(NOD ×* Mus spretus*) F1 × NOD]. Nondiabetic BC1 [(NOD × *Mus spretus*) F1 × NOD] mice were examined for the development of insulitis at 1 year of age. The incidence of insulitis was 16% (7 of 43) in nondiabetic BC1 females and 5% (2 of 42) in nondiabetic BC1 males (pre-IDDM) (Table 4). BC1 [(NOD × *Mus spretus*) F1 × NOD] mice did not develop overt IDDM in the present breeding studies. Of the BC1 females, six mice showed peri-insulitis (grade 1: immune cells adjacent, but not invading the islets, prestage leading to insulitis). If the six mice with peri-insulitis were included with the animals with insulitis (grade >2), the incidence of peri-insulitis and insulitis would be 30% (13 of 43) in BC1 females.

NIDDM in BC1 [(NOD ×* Mus spretus*) F1 × NOD] mice. The incidence of NIDDM in BC1 [(NOD × *Mus spretus*) F1 × NOD] was 2% (3 of 168) in females and 11% (14 of 128) in males at 1 year of age (Fig. 3 and Table 4). The occurrence of NIDDM was observed only in BC1 [(NOD × *Mus spretus*) F1 × NOD] but not in BC1 [(NOD × C3H/He) F1 × NOD], BC1 [(NOD × NON) F1 × NOD], and BC1 [(NOD × BALB/c) F1 × NOD] mice.

The parental NOD, *Mus spretus*, and F1 (NOD × *Mus spretus*) strains did not

Table 4. Incidence of pre-IDDM and NIDDM in BCl [(NOD × *Mus spretus*) F1 × NOD]

	Diabetes	Pre-IDDM (insulitis only, no diabetes)	IDDM (with insulitis)	NIDDM with insulin resistance (no insulitis)
Female	5/170 (3%)[a]	7/43 (16%)	0/168 (0%)	3/168 (2%)
Male	16/130 (12%)[a]	2/42 (5%)	0/128 (0%)	14/128 (11%)

[a]Two female and two male diabetic BCl mice were not identified for type of diabetes because they were not subjected to histological examination.

develop NIDDM. The occurrence of pretype I and type II diabetes was a unique phenomenon observed in the BC1 [(NOD × *Mus spretus*) F1 × NOD] mice. Our linkage analysis for NIDDM in BC1 [(NOD × *Mus spretus*) F1 × NOD] mice suggests that at least one locus is homozygous for NOD and another locus is heterozygous for NOD/*Mus spretus* (Hattori, to be published). These observations suggest an interaction of genetic factors from the NOD mouse and *Mus spretus* contributing to the development of NIDDM. In previous breeding studies,[17] cyclophosphamide-treated BC1 [(NOD × *Mus spretus*) F1 × NOD] mice developed insulitis at an incidence similar to our present study, but did not develop any type of diabetes by the age of 9 months. Cyclophosphamide was not used to accelerate the development of insulitis and diabetes in our study. The discrepancy between the two independent breeding studies may be due to the length of the observation period, diet, and environment.

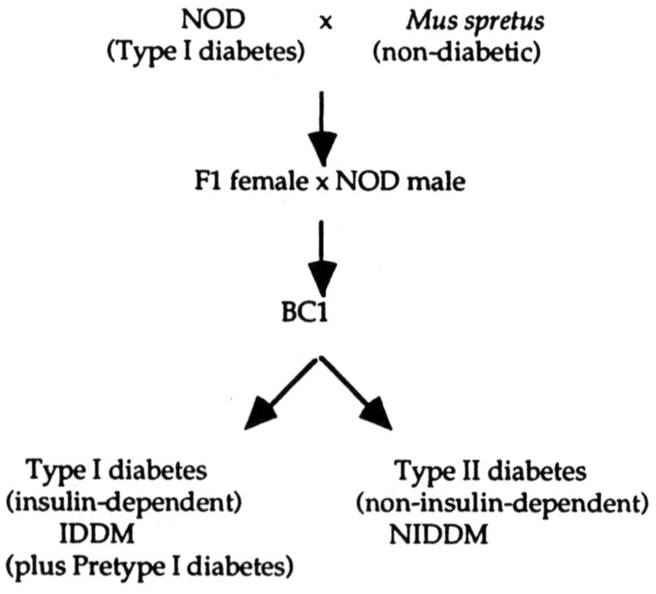

Figure 3. Occurrence of NIDDM and pre-IDDM.

Characteristics of NIDDM in BC1 [(NOD × *Mus spretus*) F1 × NOD] mice

The incidence of diabetes in BC1 [(NOD × *Mus spretus*) F1 × NOD] was 3% (5 of 170) in females and 12% (16 of 130) in males (Table 4). Of the 5 female and 16 male diabetic BC1 mice, 3 females and 14 males were subjected to histologic examination for insulitis. None of the BC1 animals developed insulitis. The islets from the diabetic BC1 mice were rather hypertrophied, resembling the initial features of NIDDM, with insulin resistance, hyperglycemia, and hyperinsulinemia. Approximately half of the BC1 [(NOD × *Mus spretus*) F1 × NOD] animals developed insulin resistance (Table 5). Diabetes was not correlated with obesity in BC1 [(NOD × *Mus spretus*) F1 × NOD] mice.

Hypertrophied islets from diabetic BC1 [(NOD ×* Mus spretus*) F1 × NOD] mice. Approximately half of BC1 [(NOD × *Mus spretus*) F1 × NOD] animals (62% in males, 60% in females) with NIDDM showed hypertrophied islets without insulitis (Fig. 2). NIDDM animals showed hyperinsulinemia and insulin resistance. *Mus spretus* did not develop obesity and hypertrophied islets.

Hyperinsulinemia in BC1 [(NOD ×* Mus spretus*) F1 × NOD] mice. The concentrations of insulin in the peripheral blood were measured in diabetic and nondiabetic BC1 [(NOD × *Mus spretus*) F1 × NOD], NOD, *Mus spretus,* and BALB/c mice by RIA (Table 6). Rat insulin was obtained from Novo BioLabs (Bagsvaerd, Denmark) and was used as a standard for the RIA. Male nondiabetic BC1 [(NOD × *Mus spretus*) F1 × NOD] mice, as well as male and female diabetic BC1 mice, had high levels of plasma insulin in comparison with female BC1 and NOD, *Mus spretus,* and BALB/c.

Insulin resistance precedes the development of NIDDM

Regular Iletin insulin (rapid action type) was obtained from Eli Lilly (Indianapolis, IN, USA). Regular Iletin insulin is beef-pork zinc insulin crystals. The regular insulin (0.25 U/kg body wt) wa sinjected i.p. into diabetic and nondiabetic BC1 [(NOD × *Mus spretus*) F1 × NOD] females and males, and blood glucose was measured for insulin resistance at 15-min intervals for 1 h between 1

Table 5. Incidence of NIDDM with insulin resistance and no insulitis in BCl [(NOD × *Mus spretus*) F1 × NOD] at 1 year of age

Sex	NIDDM in BCl (%)	Diabetes with insulin resistance, no insulitis (%)	Insulin resistance (%)
Male	14/128 (11)	14/14 (100)[a]	65/111 (59)
Female	3/168 (2)	3/3 (100)	24/75 (32)

[a] Insulin resistance preceded the development of overt diabetes.

Table 6. Insulin and glucose concentrations in the peripheral blood from *Mus spretus*, NOD, BALB/c, and BCl [(NOD × *Mus spretus*) F1 × NOD] mice under nonfasting conditions

Strain	Sex	No. mice	Age	Insulin (μU/ml)	Glucose (mg/ml)
Mus spretus	M	2	1.5 yr	17	159
NOD	M + F	8	6–9 mo	27 ± 6	195 ± 5
BALB/c	M + F	7	7–12 mo	25 ± 8	123 ± 10
Nondiabetic BCl	M	59	1–1.5 yr	155 ± 8	179 ± 13
Nondiabetic BCl	F	86	1–1.5 yr	65 ± 6	146 ± 7
Diabetic BCl	M	6	1 yr	1,050 ± 417	>300
Diabetic BCl	F	1	9 mo	906	458

[a] $P = 2.12 \times 10^{-13}$ vs. nondiabetic BCl females (Wilcoxon's rank-sum test). Values are mean ± SE.

and 5 p.m. (insulin tolerance test, ITT). This is because mice usually eat during the night (Fig. 4).

The following criteria were used for the assessment of insulin resistance. Insulin-sensitive, >20% reduction of the initial blood glucose level; insulin-resistant, <10% reduction; borderline, 10–20% reduction (in this case, the test was repeated under fasting conditions). The NOD mouse, *Mus spretus,* and F1 (NOD × *Mus spretus*) were insulin sensitive. Approximately half of the BC1 mice were insulin resistant. Male BC1 mice showed insulin resistance more frequently (59%, 65 of 111 mice) than female BC1 mice (32%, 24 of 75 mice). Insulin resistance was observed in BC1 mice as early as 34 days prior to the development of NIDDM. Insulin resistance and hyperinsulinemia preceded the development of NIDDM in BC1 mice. For example, mice BC1 no. 289 was found to have insulin resistance at 107 days of age (blood glucose, 144 mg/dl), an insulin level of 408 μU/ml, and a blood glucose of 190 mg/dl at 134 days of age, and developed overt diabetes (blood glucose, 326 mg/dl) at 236 days of age.

Conclusions

The NOD mouse develops IDDM, secondary to islet B-cell destruction by infiltrating lymphocytic and monocytic cells. The clinical and pathological features of diabetes in the NOD mouse, a multifactorial disease in a setting of genetic susceptibility, are similar to those in human IDDM. When the NOD/Shi//Jos mice were mated with inbred *Mus spretus,* a wild mouse strain, BC1 animals developed pre-IDDM, insulin resistance, and NIDDM. The NIDDM in hybrids was characterized by preceding insulin resistance, hyperglycemia, hyperinsulinemia, large islets, absence of insulitis, late onset, and male predominance. *Mus spretus* did not develop diabetes and insulitis. Interactions of genetic factors from the NOD mouse and *Mus spretus* resulted in the occurrence of both pre-IDDM and NIDDM with insulin resistance.

IDDM was not found in the present breeding studies but has been found in our second breeding studies (Hattori, to be published). In the present study, 3 of 5 diabetic BC1 females and 14 of 16 diabetic BC1 males were subjected to his-

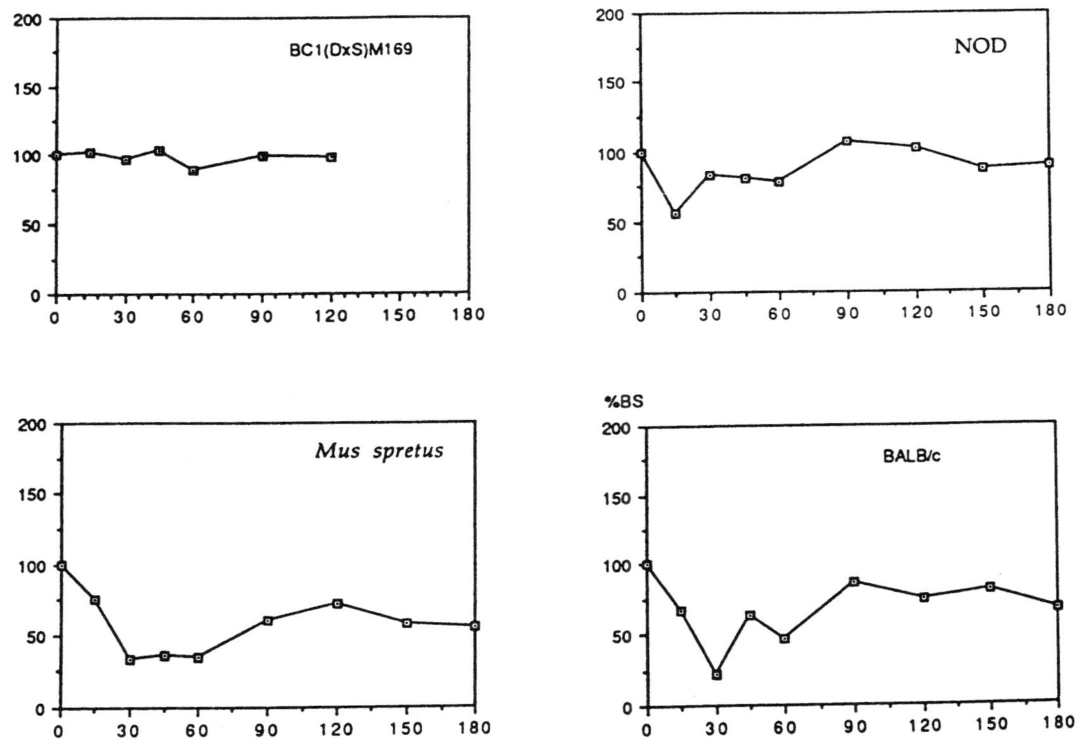

Figure 4. Insulin-resistance test in diabetic BC1 [(NOD × *Mus spretus*) F1 × NOD], NOD, *Mus spretus*, and BALB/c. The vertical axis represents blood glucose levels (mg/dl) and the horizontal axis represents the time course (min) after insulin injection (0.25 U/kg body wt).

tological examination, confirming the absence of insulitis. Two female and male diabetic BC1 mice were not subjected to histological examination. Of 300 BC1 mice in the present study, four diabetic BC1 mice were not identified as to type of diabetes (Table 4). We repeated the breeding studies in inbred NOD/Jos mice and inbred *Mus spretus* using the same breeding scheme. In our second breeding studies BC1 animals developed pre-IDDM, IDDM, and NIDDM (Hattori, to be published). These animals should provide a unique opportunity to study common mechanisms of pre-IDDM (insulitis), insulin resistance, and NIDDM.

Acknowledgments

We thank Dr. Eric S. Lander for his unlimited support, constructive criticism, and encouragement for the current project on hybrids of NOD mice with *Mus spretus,* and Dr. David Houseman for his advice for the use of *Mus spretus*. This work was supported in part by grants from the National Institutes of Health and by a grant from Sumitomo Pharmaceuticals Co., Ltd. EM was a recipient of the Sankyo/Joslin Fellowship. EY and MF were recipients of the Juvenile Diabetes Foundation International Fellowship. M. Hattori is especially grateful to Professor Eleazar Shafrir and late Professor Albert Renold for their encouragement of animal diabetes studies. We regret the sudden passing of Dr. Verne M. Chapman on August 30, 1995.

References

1. Keller RJ, Eisenbarth GS. Immunopathogenesis of type 1 diabetes mellitus. In: Eisenbarth GS, ed. Immunotherapy of diabetes and selected autoimmune diseases. CRC Press, Boca Raton, FL, pp 1–22, 1989.
2. Rossini AA, Greiner DL, Friedman HP, Mordes JP. Immunopathogenesis of diabetes mellitus. Diabetes Rev 1:43–75, 1993.
3. Makino S, Kunimoto K, Muraoka Y, et al. Breeding of a non-obese diabetic strain of mice. Exp Anim 29:1–13, 1980.
4. Guenet JL, Montagutelli X. The contribution of wild specimens to be establishment of the mouse genetic map. In: Genetics in wild mice. Moriwaki K, Shiroishi T, Yonekawa H, eds. Japan Sci Soc Press/S. Karger, Tokyo/Basel, pp 285–98, 1994.
5. Hattori M, Yamato E, Hirokawa KJ, et al. Male backcross mice of NOD with *Mus spretus Spain* predominantly develop diabetes regardless of MHC homozygosity and heterozygosity. Diabetes 41(Suppl 1):93A, 1992.
6. Craighead JE. Viral diabetes mellitus in man and experimental animals. Am J Med 70:127–34, 1981.
7. Notkins AL, Yoon J-W. Virus-induced diabetes mellitus. In: Concepts in viral pathogenesis. Notkins AL, Oldstone MBA, eds. Springer-Verlag, New York, pp 241–7, 1984.
8. Yoon J-W. The role of viruses and environmental factors in the induction of diabetes. Curr Top Microbiol Immunol 164:95–123, 1990.
9. Thomas VA, Woda BA, Handler ES, et al. Altered expression of diabetes in BB/Wor rats by exposure to viral pathogens. Diabetes 40:255–8, 1991.
10. Like AA, Guberski DL, Butler L. Influence of environmental viral agents on frequency and tempo of diabetes mellitus in BB/Wor rats. Diabetes 40:259–62, 1991.
11. Scott FW, Mongeau R, Kardish M, et al. Diet can prevent diabetes in the BB rat. Diabetes 34:1059–62, 1985.

12. Coleman DL, Kuzava JE, Leiter EH. Effect of diet on incidence of diabetes in nonobese diabetic mice. Diabetes 39:432–6, 1990.
13. Elliott RB, Reddy SN, Bibby NJ, Kida K. Dietary prevention of diabetes in the non-obese diabetic mouse. Diabetologia 31:62–4, 1988.
14. Lehman CD, Rodin J, McEwen B, Brinton R. Impact of environmental stress on the expression of insulin-dependent diabetes mellitus. Behav Neurosci 105:241–5, 1991.
15. Goldstein BJ, Kahn CR. Insulin receptors in normal and diseases states. In: International textbook of diabetes mellitus. Alberti KGMM, DeFronzo RA, Zimmet O, eds. Wiley, New York, pp 357–83, 1992.
16. Simonson DC. Insulin resistance, non-insulin-dependent diabetes mellitus. In: The insulin resistance syndrome. The Upjohn Company, Kalamazoo, MI, pp 6–12, 1992.
17. De Gouyon B, Melanitou E, Richard MF, et al. Genetic analysis of diabetes and insulitis in an interspecific cross of the nonobese diabetic mouse with *Mus spretus*. Proc Natl Acad Sci USA 90:1877–81, 1993.
18. Prochazka M, Leiter EH, Serreze DV, Coleman DL. Three recessive loci required for insulin-dependent diabetes in NOD mice. Science 237:286–9, 1987.
19. Wicker LS, Miller BJ, Cocker LZ, et al. Genetic control of diabetes and insulitis in the nonobese diabetic (NOD) mouse. J Exp Med 165:1639–54, 1987.

7

Okamoto model for B-cell damage: Recent advances

HIROSHI OKAMOTO
Department of Biochemistry, Tohoku University School of Medicine, Sendai 980-77, Miyagi, Japan.

Key words: ADP ribosyl cyclase, CD38, cyclic ADP ribose, cyclic ADP ribose hydrolase, NAD^+, nicotinamide, nitric oxide, poly (ADP ribose).

Before describing some of our more recent experiments on pancreatic B-cell damage and its prevention, the historical background of this work and the physiological and pathological significance of NAD^+ and poly (ADP-ribose) in pancreatic B-cells are discussed. In the second part, a novel metabolite of NAD^+, cyclic ADP-ribose, a second messenger for insulin secretion in B-cells, is described. Then, the synthesis and degradation of cyclic ADP-ribose, in which a human leukocyte antigen, CD38, plays a central role, is described. Finally, cytokine-induced B-cell damage and its prevention by nicotinamide are discussed.

Suicidal response of B-cells to damaged DNA

In 1981, we proposed a unifying model for the action of alloxan and streptozotocin on pancreatic B-cells.[1–4] According to our model, breaks in the nuclear DNA of B-cells result from either an accumulation of oxygen radicals or from alkylation of DNA, inducing DNA repair that involves the activation of poly (ADP-ribose) polymerase, using cellular NAD^+ as a substrate. Consequently, intracellular levels of NAD^+ fall dramatically, precluding cellular activities such as insulin synthesis and secretion. The B-cell then ultimately dies. Thus, the attempt of B-cells to repair the damaged DNA seems to be a "suicidal" response.

It is important to consider the difference between alloxan and streptozotocin (STZ) with respect to the initial mechanism of their toxic actions on B-cell DNA. It has been suggested that alloxan acts through the formation of a hydroxyl radical that is produced by the interaction between superoxide and peroxide.[5,6] Superoxide dismutase and catalase remove superoxide and peroxide, respectively,

Correspondence: Hiroshi Okamoto, Department of Biology, Tohoku University School of Medicine, 2-1 Seiryo-machi, Aoba-ku, Sendai 980-77, Miyagi, Japan.

Lessons from Animal Diabetes VI. Ed. E. Shafrir.

and hence can inhibit the formation of the hydroxyl radical. Uchigata et al.[3] showed that the combined administration of superoxide dismutase and catalase more effectively protects against alloxan-induced islet DNA breaks, as well as against proinsulin synthesis inhibition, than the administration of either of these scavenging enzymes alone. Takasu et al.[7] reported that alloxan generates hydrogen peroxide and induces DNA fragmentation in pancreatic islets. On the other hand, STZ-induced islet DNA strand breaks were unaffected by radical scavengers, such as superoxide dismutase and catalase.[3] Therefore, in contrast to alloxan, it is unlikely that STZ acts on islet DNA through the formation of oxygen radicals. The STZ-induced breakage of DNA is thought to be associated with the alkylating activity[8] or nitric oxide (NO) generating activity[9] of the agent. Furthermore, the cytotoxic effect of STZ is related to specific recognition of the drug in B-cells as a transported substrate by GLUT2.[10]

The model for the mechanism of action of alloxan and STZ may be extended to the effects of viruses and inflammation,[1,11,12] especially immune-mediated variants.[11,12] Recently, it has been suggested that cytokines and NO are also involved in this process, as described later.

Therefore, although B-cell damage can be caused by many different processes, such as immunologic abnormalities, inflammatory tissue damage, and B-cytotoxic substances, the final pathway for the toxic agents will be the same (Fig. 1), involving DNA damage, poly (ADP-ribose) polymerase activation, and NAD^+ depletion. Because it is the fall in cellular NAD^+ that nullifies cellular activities, insulin-dependent (type I) diabetes may, in theory, be preventable by suppress-

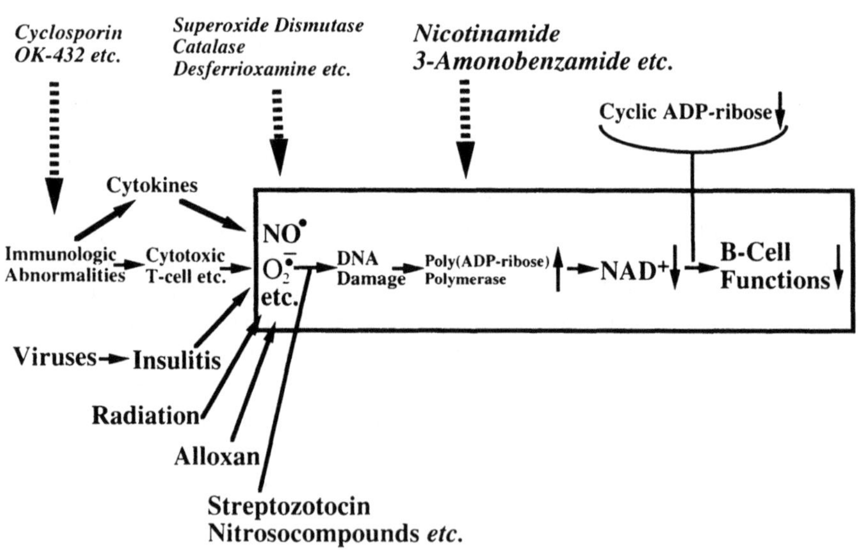

Figure 1. The Okamoto model for B-cell damage and its prevention in toxin- or virus-induced and immune diabetes. NAD^+ depletion results in a decrease in the cyclic ADP-ribose level, leading to the inhibition of B-cell functions such as insulin secretion. (Adapted from refs. 11, 21, and 63, with permission.)

ing immune reactions, scavenging free radicals, and inhibiting poly (ADP-ribose) polymerase by nicotinamide and 3-aminobenzamide.

A novel second messenger in B-cells affecting Ca^{2+} mobilization, cyclic ADP-ribose

As described earlier, for B-cells to secrete and synthesize insulin, maintenance of the cellular NAD^+ level is essential. The reason why maintenance of the NAD^+ level is necessary for B-cells to perform these functions is as follows. Recently, we found that a metabolite of NAD^+, cyclic ADP-ribose, is a second messenger for insulin secretion in B-cells.[13,14] An increase in the intracellular Ca^{2+} concentration plays a major role in biochemical events that couple glucose stimulation to insulin secretion. Mobilization of Ca^{2+} from intracellular stores in the endoplasmic reticulum is an important signaling pathway in these processes.[15] Inositol trisphosphate (IP_3) is generally considered to be a mediator for Ca^{2+} mobilization from intracellular stores.[16,17] Recently, Lee and colleagues[18–20] have reported that in sea urchin eggs cyclic ADP-ribose is as potent and powerful a release of intracellular Ca^{2+} as IP_3.

More recently, we found that cyclic ADP-ribose induces Ca^{2+} release from pancreatic islet microsomes, whereas IP_3 does not;[13,14] we used rat pancreatic islet microsomes as a cell-free system to study Ca^{2+} release, and cyclic ADP-ribose induced Ca^{2+} release from islet microsomes, as indicated by the observed prompt increase in fluorescence. Because of the depletion of the responsive Ca^{2+} pool in microsomes, repeated additions of cyclic ADP-ribose resulted in attenuation of Ca^{2+} release in response to cyclic ADP-ribose. IP_3 did not release Ca^{2+} from islet microsomes, but at this point islet microsomes were still responsive to cyclic ADP-ribose. In rat cerebellum microsomes, IP_3 induced Ca^{2+} release from cerebellum microsomes, and repeated additions of IP_3 resulted in attenuation of the Ca^{2+}-release response to IP_3. Cyclic ADP-ribose also induced Ca^{2+} release from cerebellum microsomes, and repeated additions of cyclic ADP-ribose resulted in attenuation of the Ca^{2+} release response to cyclic ADP-ribose. But at this point, cerebellum microsomes were still responsive to the addition of IP_3. Moreover, heparin, an inhibitor of IP_3 binding to its receptor, blocked IP_3-induced Ca^{2+} release from cerebellum microsomes, but did not block cyclic ADP-ribose–induced Ca^{2+} release, indicating that cerebellum microsomes respond to both cyclic ADP-ribose and IP_3 via different mechanisms. The cyclic ADP-ribose–induced Ca^{2+} release showed a steep concentration-response relationship, with a clear effect at concentrations as low as 0.1 μM and a near-maximum release at 0.5 μM.[14]

Using digitonin-permeabilized pancreatic islets, we next examined the effects of cyclic ADP-ribose and IP_3 on insulin secretion.[14] Cyclic ADP-ribose (1 μM) as well as Ca^{2+} (0.2 μM) induced insulin secretion but IP_3 did not. The combined addition of cyclic ADP ribose and Ca^{2+} did not induce significantly more insulin secretion than the addition of cyclic ADP-ribose or Ca^{2+} alone. The cyclic ADP-ribose–induced insulin secretion was inhibited by the addition of EGTA. These results suggest that Ca^{2+} mobilization from islet microsomes mediated the cyclic ADP-ribose–induced insulin secretion. Cyclic ADP-ribose formation in islets by glucose stimulation attained a near-maximal level by 3 min.[14] These results suggest that glucose stimuli induce cyclic ADP-ribose formation from NAD^+ in B-cells and cyclic ADP-ribose, serving as a second messenger, then mobilizes Ca^{2+} from the endoplasmic reticulum (Fig. 2).

As explained earlier, diabetogenic agents such as STZ deplete intracellular NAD^+ in islets, and this depletion can be prevented by inhibiting the poly (ADP-ribose) polymerase with nicotinamide and 3-aminobenzamide. When we incubated islets with STZ, prepared the islet extract, and assessed its cyclic ADP-ribose content from the Ca^{2+}-mobilizing activity, we found that STZ caused a great reduction in the cyclic ADP-ribose content and that poly (ADP-ribose) polymerase inhibitors such as nicotinamide and 3-aminobenzamide reversed this reduction.[14] As summarized in Figure 2, B-cytotoxins such as STZ and alloxan cause DNA strand breaks that activate nuclear poly (ADP-ribose) polymerase, forming poly (ADP-ribose) and leading to NAD^+ depletion. Therefore, the NAD^+ depletion results in the inhibition of the cyclic ADP-ribose–mediated Ca^{2+}

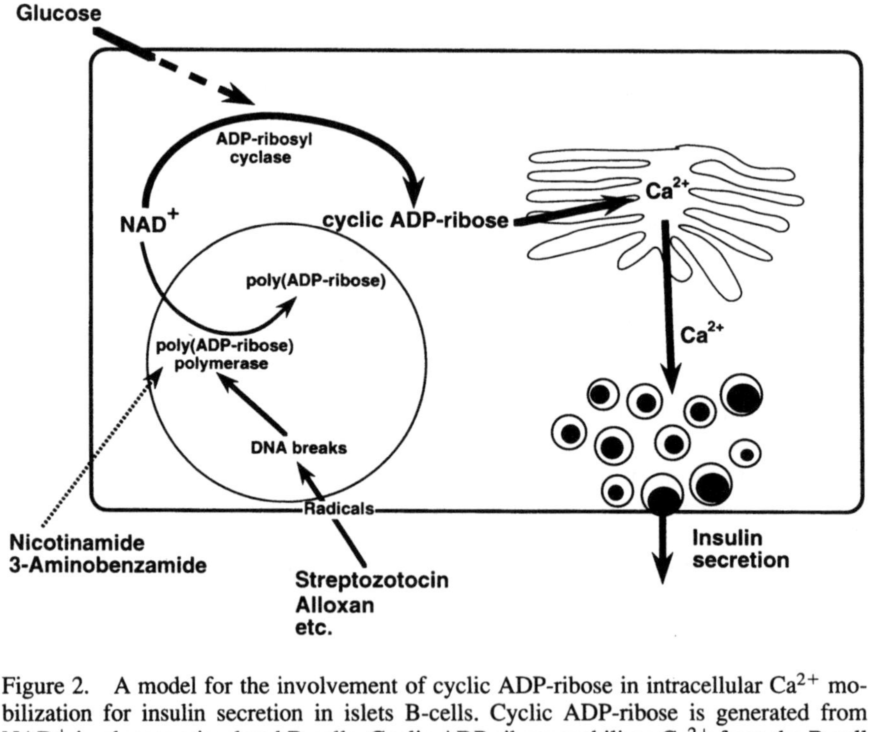

Figure 2. A model for the involvement of cyclic ADP-ribose in intracellular Ca^{2+} mobilization for insulin secretion in islets B-cells. Cyclic ADP-ribose is generated from NAD^+ in glucose-stimulated B-cells. Cyclic ADP-ribose mobilizes Ca^{2+} from the B-cell endoplasmic reticulum to secrete insulin. Streptozotocin and other B-cytotoxins cause DNA strand breaks that activate nuclear poly (ADP-ribose) polymerase, leading to NAD^+ depletion, which results in inhibition of cyclic ADP-ribose–mediated Ca^{2+} mobilization. Poly (ADP-ribose) polymerase inhibitors such as nicotinamide and 3-aminobenzamide prevent NAD^+ depletion, and, therefore, the glucose-induced insulin secreting process continues to function. (Adapted from refs. 21, 42, and 44, with permission.)

mobilization. Poly (ADP-ribose) polymerase inhibitors, such as nicotinamide and 3-aminobenzamide, prevent this depletion and therefore the glucose-induced insulin secreting machinery continues to function, even in the presence of streptozotocin and alloxan.

Enzymes affecting the metabolism of cyclic ADP ribose

Synthesis and degradation of cyclic ADP-ribose. ADP-ribosyl cyclase, which catalyzes the formation of cyclic ADP-ribose from NAD^+, was first purified from the soluble fraction of *Aplysia* ovotestes by sequential column chromatography.[22–24] The amino acid sequence of *Aplysia kurodai* cyclase shows a significant homology (31% identity) with that of a human lymphocyte surface antigen, CD38; the 10 cysteine residues in both proteins are especially conserved (see Fig. 4).[25–28] Therefore, *Aplysia* cyclase and CD38 appear to have similar overall conformation and the same catalytic property. When we expressed human CD38 cDNA in COS-7 cells,[26] we found that CD38 was mainly expressed in the plasma and microsomal membrane fractions. We then incubated $^{32}P\text{-}NAD^+$ with the plasma membrane fraction for 20 min and analyzed the reaction product by HPLC. The formation of a small amount of cyclic ADP-ribose was observed during the first 10 min, and the cyclic ADP-ribose level decreased thereafter. In the

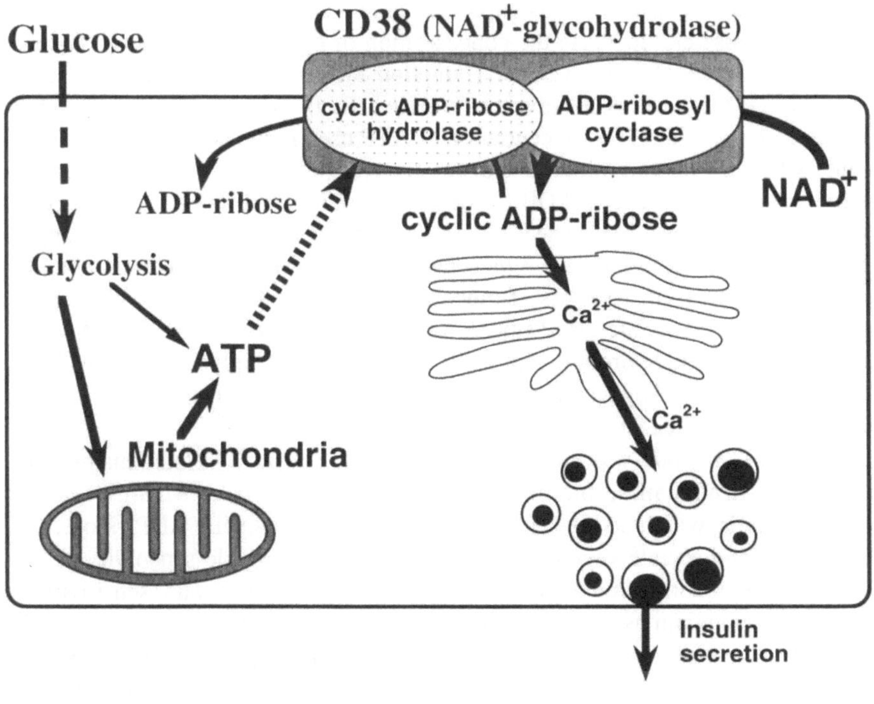

Figure 3. Diagram showing the role of CD38 in glucose-induced insulin secretion in B-cells. (Adapted from refs. 21, 42, and 63, with permission.)

overall reaction, the decrease in NAD^+ and the concomitant stoichiometric formation of ADP-ribose suggested that the membrane-bound CD38 has both ADP-ribosyl cyclase activity and cyclic ADP-ribose hydrolase activity to produce ADP-ribose.[26] To confirm this, we prepared ^{32}P-labeled cyclic ADP-ribose and incubated it with the membrane fraction, and observed that cyclic ADP-ribose was stoichiometrically converted to ADP-ribose.[26] Both the ADP-ribosyl cyclase and the cyclic ADP-ribose hydrolase activities were inhibited by the addition of an anti-CD38 antibody, suggesting that CD38 has both cyclase and hydrolase activities.[26]

The physiological functions of CD38, originally identified as a surface antigen of lymphocyte differentiation,[29,30] have not yet been identified. We analyzed RNAs of various human tissues by blot hybridization. Different levels of CD38 mRNA were detected in all human tissues and cell lines examined, but relatively high levels of expression were detected in high glucose-utilizing tissues, such as insulinomas, pancreas, liver, kidney, and brain.[26] Glucose is the primary stimulus of insulin secretion and synthesis in pancreatic B-cells.[1,31,32] ATP, produced by the metabolism of glucose, has been suggested to be involved in glucose-stimulated insulin secretion by the inactivation of an ATP-sensitive K^+ channel in pancreatic B-cells.[33–35] Our results indicate that ATP inhibits the cyclic ADP-ribose hydrolyzing activity (cyclic ADP-ribose hydrolase) of CD38, thereby increasing the level of cyclic ADP-ribose (see Fig. 3).[26] In addition, we isolated the rat homologue to human CD38 cDNA from an islet cDNA library and also found that the rat homologue is expressed in the islet.[27] This raises the question of how glucose can increase the concentration of cyclic ADP-ribose in B-cells. Glucose increases B-cell oxygen consumption and evokes rises in ATP.[36,37] ATP then inhibits the cyclic ADP-ribose hydrolase activity to increase the cyclic ADP-ribose concentration. The concentration of ATP used in the experiment was 1–10 mM, within the range estimated to occur in islets.[38–42] Therefore, glucose causes an elevation of the cyclic ADP-ribose concentration in B-cells, and the cyclic ADP-ribose induces Ca^{2+} release from the endoplasmic reticulum, resulting in the secretion of insulin. Figure 3 presents our model for insulin secretion by glucose in B-cells. In the process of glucose metabolism, ATP is generated. Then, ATP induces the cyclic ADP-ribose formation from NAD^+ by inhibiting the cyclic ADP-ribose hydrolase activity of CD38 or NAD^+ glycohydrolase. Thus, cyclic ADP-ribose is generated, serving as a second messenger for Ca^{2+} mobilization for insulin secretion.

Rat CD38 mRNA was expressed in the islets and in STZ/nicotinamide–induced rat insulinomas, but not in rat insulinoma-derived RINm5F cells.[27] RINm5F cells, which do not release Ca^{2+} in response to cyclic ADP-ribose,[43–45] synthesize and secrete very little insulin and show negligible sensitivity to glucose.[46] Streptozotocin/nicotinamide–induced insulinomas retain sensitivity to glucose and contain as much insulin mRNA as normal islets.[47–50] These results are consistent with our finding that cyclic ADP-ribose synthesized by CD38 acts as a second messenger for glucose-induced insulin production in B-cells.

More recently, we produced transgenic mice overexpressing human CD38 in B-cells.[42,67] In islets of transgenic mice, in which the enzymic activity of CD38 and the Ca^{2+} mobilizing activity from the cyclic ADP-ribose–sensitive Ca^{2+} pool were increased, the glucose-induced insulin secretion was significantly higher than in control islets. In the glucose tolerance test, the transgenic serum insulin level was also significantly higher than control and the transgenic blood glucose level was reduced. These results provide in vivo evidence that CD38 has a regulatory role in insulin secretion and that cyclic ADP-ribose acts as a second messenger.

Molecular and catalytic properties of CD38 and its mutants. As described earlier, CD38 has both ADP-ribosyl cyclase and cyclic ADP-ribose hydrolase activities. On the other hand, the *Aplysia* enzyme shows only ADP-ribosyl cyclase but not hydrolase. In Figure 4, the amino acid sequences of the human,[30] rat,[27] and mouse[51] CD38s and *Aplysia* ADP-ribosyl cyclases[52,53] are compared. The human, rat, and mouse CD38s have conserved the 13 cysteine residues (amino acid residues 16, 67, 82, 99, 119, 160, 173, 180, 201, 254, 275, 287, and 296 in human CD38). Three of the 13 cysteines (amino acid residues 16, 119, and 201 in human CD38) are not conserved in *Aplysia* ADP-ribosyl cyclases. Neither of the two *Aplysia* proteins have the region corresponding to Cys-16 of the human CD38. The corresponding amino acid residues of Cys-119 and Cys-201 in the

```
                                                                                                ▽
Human CD38               MANCEFSPVS GDKPCCRLSR RAQLCLGVSI LVLI-LVVVLA VVV--PRWRQTW SGPGTTKRFP ETVLARCVKY  70
Rat CD38                 •••Y•••Q•• E•R•G•••T• K••IG•••GL •L•VA••••VV I•LW-••SPLV• K•KP•••H•A DII•G••LI•  72
Mouse CD38               •••Y•••Q•• ••R•G••••• K••IG•••GL ••••A•••GIV •ILLR••SLLV• T•EP•••H•S DIF•G••LI•  73
A.californica cyclase                                    M•P VAI•-AC•C•• •TL-TS-ISPS- EAIVP•RELE NVF•G••KD•  42
A.kurodai cyclase                                        M•P VAIV-AC•C•• •TL-TR-ISPS- EAIFP•PELQ NVF•G••KD•  42

                                                      119
                ▽                 ▽                    ⇓
TEI-H--PEMRHV --DCQSVWDAFK GAFISKHPCN ITEEDYQPLM KLGTQTVPCN KILLWSRIKD LAHQFTQVQR DMFTLEDTLL  150
•Q•LR--••••DQ --••KKILST•• RG••••N••• ••N•••A••V ••V•••I••• •T•F••KS•H ••••Y•WI•G K•••••••••  153
•Q•LR--••••DQ --N••EILST•• •••V••N••• ••R•••A••V ••V•••I••• •T•F••KS•H ••••Y•WI•G K•••••••••  154
EIT-RYLDILPR• RS••SAL•KD•F K••SF•N••D LDLGS•KDFF TSAQ•QL•K• •VMF••GVY• E••DYANTG• KYI••••••P  126
EIT-RYLTILPR• KS••RAL•TN•F K••SF•A••• LOLGS•KDFF QRAQ••L•K• •VMF••GVY• E••D•ADDG• KYI••••••P  126

                                                     201
         ▽              ▽       ▽                     ⇓
GYLADDLTWC GEFNTSKINY QSCPDWRKDC SNNPVSVFWK TVSRRFAEAA CDVVHVMLNG SRSKI--FDKNS TFGSVEVHNL  230
••I••••R•• •DPS••DM•• D•••H•SEN• P••••A•••N VI•QK•••D• •G••Q••••• •L•EP--•YR•• •••••••F••  233
••V••••R•• •DPS••DM•• V•••H•SEN• P•••ITM••• VI•QK•••D• •G••Q••••• •LREP--•Y••• ••••L••FS•  234
••MLNS•V•• •QRANPGF•E KV•••F-•T• PVQARES••G MA•SSY•HS• EGE•TY•VD• •NP•VPAYRPD• F••KY•LP••  207
••MLNS•V•• •QRDKPGF•Q KV•••F-••• PVQARES••G •A•SSY•HS• EGD•TY•VD• •NP•VPAYRPD• F••KY•LP••  207

                          ▽                       ▽             ▽         ▽
QPEKVQTLEA WVIHGGREDS RDLCQDPTIK ELESIISKRN IQFSCKNIYR PDKFLQCVKN PEDSSCTSEI             300
D•N••HK•Q• ••M•DIKGT• SNA•SS•S•N ••K••VN••• MI•A•QDN•• •VR••••••• ••HP••RLNV             303
D•N••HK•Q• ••M•DIEGA• SNA•SSSSLN ••KM•VQ••• MI•A•VDN•• •AR••••••• ••HP••RLNT             304
-TN••TRVKV I•L•RLG•KI IEK•GAGSLL D••KLVKAKH FA•D•VENP• AVL••L•SD• •NARE•RLAK RFYRIA      282
-TN••TKVKV I•L•QLGQKI IER•GAGSLL D••MVVKAKK FG•D•VENPK SVL••L•AD• •NARE•QLAK RYYRIA      282
```

Figure 4. Alignment of amino acid sequences of mammalian CD38s (ADP-ribosyl cyclase/cyclic ADP-ribose hydrolase) and *Aplysia* ADP-ribose cyclases. Dots indicate amino acid residues that are identical to human CD38, and dashes indicate sequence gaps resulting from optimization of alignment. Cysteine residues conserved among all the proteins are marked with reverse triangles, and the two cysteine residues (C-119 and C-201 in human CD38) found in mammalian proteins but not in those of *Aplysia* are marked with arrows. (Adapted from refs. 21 and 28, with permission.)

mammalian protein are lysine and glutamic acid, respectively, in both *Aplysia* proteins.

This finding suggests that the amino acid residues 119 and 201 in CD38 and the corresponding residues in *Aplysia* proteins (residues 95 and 176) may account for the important differences in their enzyme activities. We therefore introduced site-directed mutations into the human CD38 cDNA by replacing the two cysteine residues (Cys-119 and Cys-201) with lysine and glutamic acids, as was found in the *Aplysia* enzymes.[28] The resulting cDNA was then expressed transiently in COS-7 cells. ADP-ribosyl cyclase and cyclic ADP-ribose hydrolase activities were measured using the homogenate of COS-7 cells into which the mutants had been introduced as an enzyme source. As shown in Figure 5, ADP-ribosyl cyclase activity was detected not only in the mutant with two residue replacements (C119K and C201E) but also in the mutant having one cysteine replacement for the same residue as *Aplysia* enzymes (C119K or C201E). We could not detect ADP-ribosyl cyclase activity in the other mutants, even when there was a conservative replacement (C119R or C201D). In contrast with ADP-ribosyl cyclase, cyclic ADP-ribose hydrolase activity was not detected in any of the mutants.

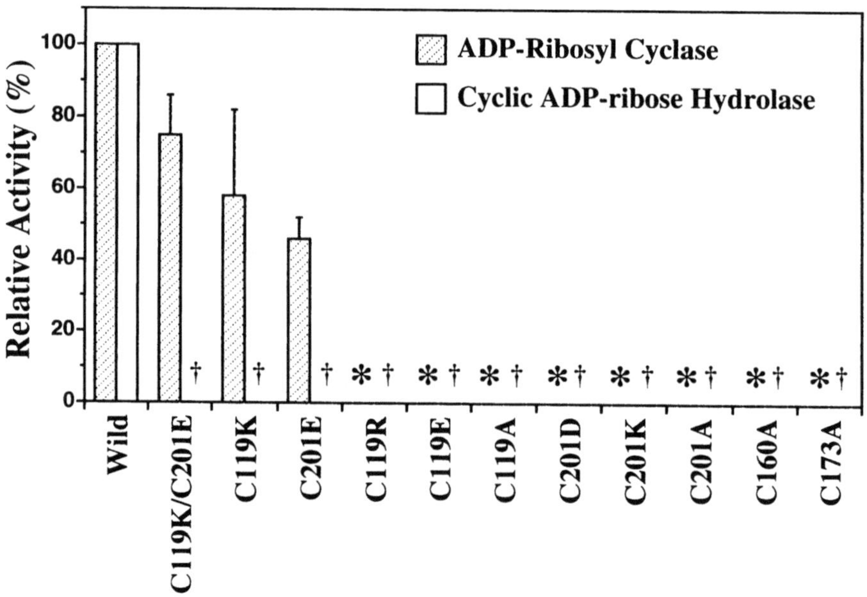

Figure 5. ADP-ribosyl cyclase and cyclic ADP-ribose hydrolase activities of the site-directed mutants of CD38. The bar graph shows the average results from three experiments. Error bars represent the standard errors of the mean. The average ADP-ribosyl cyclase activity of the wild-type CD38 in the three experiments was 1.6 ± 0.3 nmol/min/mg protein. * ADP-ribosyl cyclase activity below detection. The average of cyclic ADP-ribose hydrolase activity of the wild-type CD38 in the three experiments was 8.7 ± 1.9 nmol/min·mg protein. † cyclic ADP-ribose hydrolase activity below detection. (Adapted from refs. 21 and 28, with permission.)

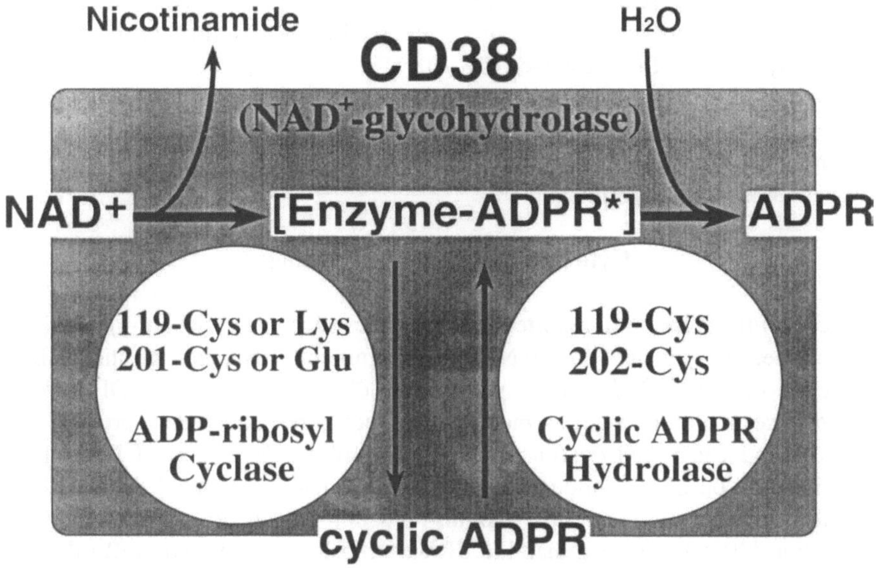

Figure 6. Role of cysteine residues 119 and 201 of CD38 in the synthesis and hydrolysis of cyclic ADP-ribose. ADPR, ADP-ribose. [Enzyme-ADPR*] is proposed as an enzyme-stabilized ADP-ribosyl oxocarbonium ion intermediate.[64–66] (Adapted from ref. 21, with permission.)

To confirm the significance of both Cys-119 and Cys-201 of CD38, we constructed *Aplysia kurodai* ADP-ribosyl cyclase expression plasmids and introduced them into COS-7 cells. The COS-7 cells into which the *Aplysia* expression vector had been introduced exhibited ADP-ribosyl cyclase but not cyclic ADP-ribose hydrolase. We then introduced mutations of K95C and/or E176C of the ADP-ribosyl cyclase, which correspond to the residues 119 and 201 of CD38, and measured the expressed activity. Only the mutant with both K95C and E176C exhibited cyclic ADP-ribose hydrolase activity, in addition to ADP-ribosyl cyclase activity.[28]

These results indicate that the amino acid residues 119 and 201 of CD38 play an essential role in both ADP-ribosyl cyclase and cyclic ADP-ribose hydrolase activities; with cysteine (wild-type) or lysine (*Aplysia* type) in position 119, together with cysteine (wild-type) or glutamic acid (*Aplysia* type) in position 201, the ADP-ribosyl cyclase activity was expressed, but only the wild-type amino acids, cysteines, in these two positions caused the protein to have cyclic ADP-ribose hydrolyzing activity as well (Fig. 6). CD38 and its mammalian homologues have been demonstrated to have both ADP-ribosyl cyclase and cADPR hydrolase activities.[26,27,44,54,55] The overall reaction is classified as an NAD^+-glycohydrolase (NAD^+ase) reaction (EC 3.2.2.5). None of the mutants that lost ADP-ribosyl cyclase and/or cADPR hydrolase activities exhibited ADPR formation from NAD^+ (NAD^+-glycohydrolase activity). This suggests that the NAD^+-glycohydrolase activity in CD38 results from the existence of ADP-ribosyl cyclase/cADPR hydrolase activity and that an enzyme-stabilized ADP-ribosyl oxocarbonium ion intermediate is involved in the reaction for the synthesis and hydrolysis of cADPR (see Fig. 6).

Recently, we determined the nucleotide sequences of *Aplysia kurodai* ADP-ribosyl cyclase[53] and human CD38 genes.[56] Both genes contain eight exons and seven introns. We also mapped the CD38 gene to human chromosome 4p15 by fluorescence in situ hybridization.[57]

Cytokine-induced B-cell damage

More recently, it has been reported that cytokines such as interleukin-1β (IL-1β) induce islet NO synthase and that NO causes significant DNA damage and NAD^+ depletion in islets.[58,59] However, we recently observed that poly (ADP-ribose) polymerase inhibitors, such as nicotinamide, could not completely prevent the cytokine-induced B-cell dysfunction.[60] We examined the effects of IL-1β and nicotinamide on glucose-induced insulin secretion in rat isolated islets. IL-1β greatly reduced insulin secretion, and this reduction was prevented by a high concentration (20 mM) of nicotinamide. However, a low concentration (2 mM) of nicotinamide, which completely inhibited islet poly (ADP-ribose) polymerase[3] and other poly (ADP-ribose) polymerase inhibitors, did not completely prevent the reduction of insulin secretion.[60] Low-concentration (2 mM) nicotinamide and other poly (ADP-ribose) polymerase inhibitors did not block the IL-1β–induced NO formation in rat isolated islets, and only a high concentration (20 mM) of nicotinamide inhibited NO formation.

To elucidate the mechanism of NO formation and its prevention by a high dose of nicotinamide, we determined the level of NO synthase mRNA in rat isolated islets by RT-PCR. The NO synthase mRNA was induced by IL-1β, and the induction was inhibited by the high concentration of nicotinamide but not by low concentrations of nicotinamide.[60] In order to understand the mechanisms of the IL-1β–induced decrease in insulin secretion, we measured the islet ATP content. The ATP content was greatly decreased by IL-1β, and the decrease in ATP content was completely prevented by the addition of the high concentration of nicotinamide but not by low concentrations of the agent.[60]

From the results shown in Figure 7, it is reasonable to assume that cytokines such as IL-1β act on the inducible type of NO synthase gene and express the mRNA and then the NO synthase, resulting in the overproduction of NO. In addition to the B-cell DNA damage, NO can also affect mitochondrial functions to decrease the ATP production. Therefore, cyclic ADP-ribose hydrolase is still active and cyclic ADP-ribose–mediated insulin secretion cannot occur. It is also suggested that a depletion of ATP impairs NAD^+ synthesis because NAD^+ is synthesized by an ATP-requiring two-step reaction utilizing nicotinamide as a precursor.[8] Therefore, the fall of ATP may accelerate the inhibition of cyclic ADP-ribose–mediated Ca^{2+} mobilization by inhibition of the ATP-dependent NAD^+ synthesis. The high concentration of nicotinamide prevents the induction of nitric oxide synthase gene expression and, therefore, the glucose-induced insulin secreting machinery continues to function, even in the presence of cytokines.

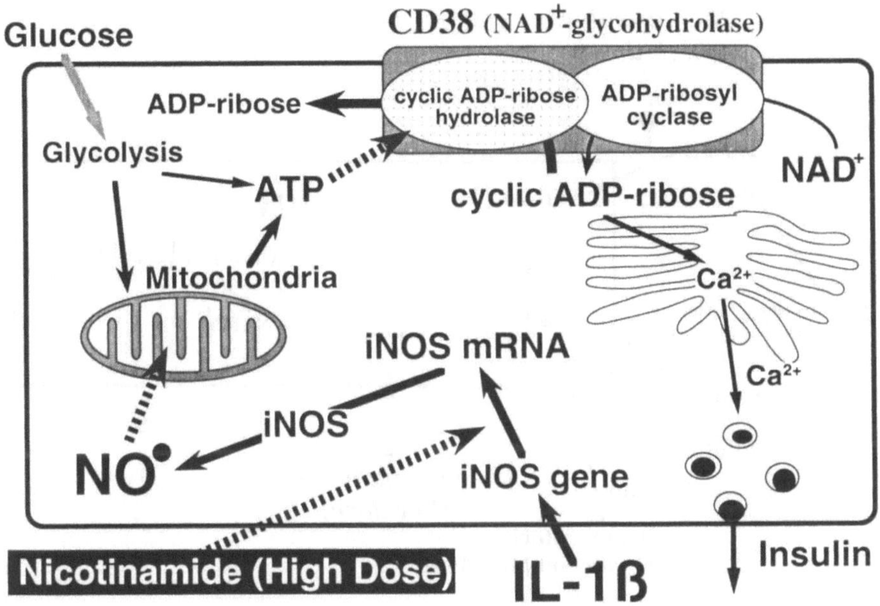

Figure 7. Cytokine-induced B-cell damage and its prevention. IL-1β, interleukin-1β; iNOS, inducible type of nitric oxide synthase.

Thus, to completely prevent IL-1β–induced B-cell damage, the administration of a high dose of nicotinamide is necessary, in the presence of which B-cell function as well as the intracellular level of cyclic ADP-ribose are preserved (Fig. 8). In fact, using poly (ADP-ribose) polymerase disrupted mice, Kolb and his colleagues[61] recently demonstrated that poly (ADP-ribose) polymerase activation and subsequent NAD^+ depletion play an essential role in radical-induced B-cell damage. They also suggested the existence of an alternative pathway of B-cell damage independent of poly (ADP-ribose) polymerase activation and NAD^+ depletion.[61] At higher concentrations of alloxan and STZ, excessive production of hydroxyl radicals[62] and reactive carbonium ions[8] have been reported to attack mitochondria as well as DNA in B-cells. It is quite possible that the alternative pathway involves mitochondrial damage and ATP depletion in B-cells, as described earlier and as shown in Figure 8.

Conclusions

The evidence presented in this paper indicates that a novel metabolite of NAD^+, cyclic ADP-ribose, acts as a second messenger for insulin secretion in B-cells, suggesting that NAD^+ and cyclic ADP-ribose and their derivatives activate B-cell functions, such as insulin secretion, and can be used for the treatment of type I diabetes. Our recent experiments with transgenic mice indicate that CD38 (ADP-ribosyl cyclase/cyclic ADP-ribose hydrolase)–overexpressing mice exhibit in-

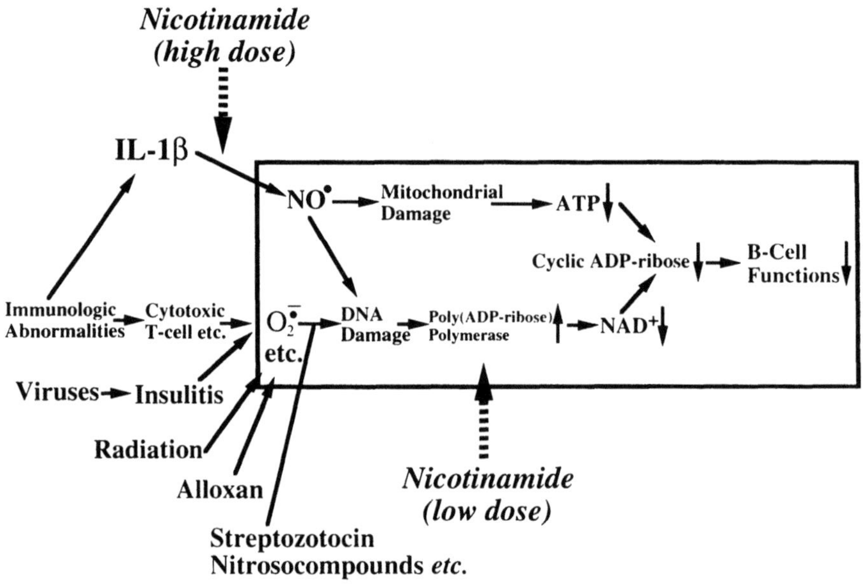

Figure 8. The Okamoto model for the mechanisms of action of low and high doses of nicotinamide on toxin- and cytokine-induced and immune-mediated B-cell damage.

sulin secretion by glucose at a level higher than normal control mice. Low and high doses of nicotinamide can prevent immune-mediated or cytokine-induced B-cell damage, as well as toxin-induced B-cell damage.

References

1. Okamoto H. Regulation of proinsulin synthesis in pancreatic islets and a new aspect to insulin-dependent diabetes. Mol Cell Biochem 37:43–61, 1981.
2. Yamamoto Y, Uchigata Y, Okamoto H. Streptozotocin and alloxan induce DNA strand breaks and poly (ADP-ribose) synthetase in pancreatic islets. Nature 294:284–6, 1981.
3. Uchigata Y, Yamamoto H, Kawamura A, Okamoto H. Protection by superoxide dismutase, catalase, and poly (ADP-ribose) synthetase inhibitors against alloxan- and streptozotocin-induced islet DNA strand breaks and against the inhibition of proinsulin synthesis. J Biol Chem 257:6084–88, 1982.
4. Uchigata Y, Yamamoto H, Nagai A, Okamoto H. Effect of poly (ADP-ribose) synthetase inhibitor administration to rats before and after injection of alloxan and streptozotocin on islet proinsulin synthesis. Diabetes 32:316–18, 1983.
5. Haber F, Weiss J. The catalytic decomposition of hydrogen peroxide by iron salts. Proc R Soc Lond A 147:332–51, 1934.
6. McCord JM, Day ED Jr. Superoxide-dependent production of hydroxyl radical catalyzed by iron-EDTA complex. FEBS Lett 86:139–42, 1978.
7. Takasu N, Asawa T, Komiya I et al. Alloxan-induced DNA strand breaks in pancreatic islets: Evidence for H_2O_2 as an intermediate. J Biol Chem 266:2112–14, 1991.
8. Wilson GL, Hartig PC, Patton NJ, LeDoux SP. Mechanisms of nitrosourea-induced β-cell damage: Activation of poly (ADP-ribose) synthetase and cellular distribution. Diabetes 37:213–16, 1988.

9. Turk J, Corbett JA, Ramanadham S, et al. Biochemical evidence for nitric oxide formation from streptozotocin in isolated pancreatic islets. Biochem Biophys Res Commun 197:1458–64, 1993.
10. Schnedl WJ, Ferber S, Johnson JH, Newgard BC. STZ transport and cytotoxicity: Specific enhancement in GLUT2-expressing cells. Diabetes 43:1326–33, 1994.
11. Okamoto H. Molecular basis of experimental diabetes: Degeneration, oncogenesis, and regeneration of pancreatic B-cells of islets of Langerhans. BioEssays 2:15–21, 1985.
12. Okamoto H. The molecular basis of experimental diabetes. In: Molecular biology of the islets of Langerhans. Okamoto, H. ed. Cambridge University Press, Cambridge, pp 209–31, 1990.
13. Okamoto H, Takasawa S, Nata K, Yonekura H. Cyclic ADP-ribose, a novel second messenger for intracellular Ca^{++} mobilization in pancreatic islets. Abstracts, First conference of IUBMB. Biochemistry of diseases, Nagoya, Japan. Bridge Ltd., Nagoya, p 218, 1992.
14. Takasawa S, Nata K, Yonekura H, Okamoto H. Cyclic ADP-ribose in insulin secretion from pancreatic β cells. Science 259:370–3, 1993.
15. Prentki M, Matschinsky FM. Ca^{2+}, cAMP, and phospholipid-derived messenger in coupling mechanisms of insulin secretion. Physiol Rev 67:1185–248, 1987.
16. Berridge MJ, Irvine RF. Inositol trisphosphate, a novel second messenger in cellular signal transduction. Nature 312:315–21, 1984.
17. Berridge MJ, Irvine RF. Inositol phosphates and cell signaling. Nature 341:197–205, 1989.
18. Clapper DL, Walseth TF, Dargie PJ, Lee HC. Pyridine nucleotide metabolites stimulate calcium release from sea urchin egg microsomes desensitized to inositol trisphosphate. J Biol Chem 262:9561–68, 1987.
19. Lee HC, Walseth TF, Bratt GT, et al. Structural determination of a cyclic metabolite of NAD^+ with intracellular Ca^{2+}-mobilizing activity. J Biol Chem 264:1608–15, 1989.
20. Dargie PJ, Agre MC, Lee HC. Comparison of Ca^{2+} mobilizing activities of cyclic ADP-ribose and inositol trisphosphate. Cell Regul 1:279–90, 1990.
21. Okamoto H, Takasawa S, Tohgo A. New aspects of the physiological significance of NAD, poly ADP-ribose and cyclic ADP-ribose. Biochimie 77:356–63, 1995.
22. Hellmich MR, Strumwassert F. Purification and characterization of a molluscan egg-specific NADase, a second-messenger enzyme. Cell Regul 2:193–202, 1991.
23. Lee HC, Aarhus R. ADP-ribosyl cyclase: An enzyme that cyclizes NAD^+ into a calcium-mobilizing metabolite. Cell Regul 2:203–9, 1991.
24. Okamoto H, Takasawa S, Nata K, et al. ADP-ribosyl cyclase that synthesizes cyclic ADP-ribose, a second messenger for Ca^{++} mobilization in β-cells. Diabetes 4(Suppl 1):76A, 1993.
25. States DJ, Walseth TF, Lee HC. Similarities in amino acid seqeunces of *Aplysia* ADP-ribosyl cyclase and human lymphocyte antigen CD38. Trends Biochem Sci 17:495, 1992.
26. Takasawa S, Tohgo A, Noguchi N, et al. Synthesis and hydrolysis of cyclic ADP-ribose by human leukocyte antigen CD38 and inhibition of the hydrolysis by ATP. J Biol Chem 268:26052–54, 1993.
27. Koguma T, Takasawa S, Tohgo A, et al. Cloning and characterization of cDNA encoding rat ADP-ribosyl cyclase/cyclic ADP-ribose hydrolase (homologue to human CD38) from islets of Langerhans. Biochim Biophys Acta 1223:160–2, 1994.
28. Tohgo A, Takasawa S, Noguchi N, et al. Essential cystine residues for cyclic ADP-ribose synthesis and hydrolysis by CD38. J Biol Chem 269:28555–557, 1994.
29. Reinherz FL, Kung PC, Goldstein G, et al. Discrete stages of human intrathymic differentiation: Analysis of normal thymocytes and leukemic lymphoblasts of T-cell lineage. Proc Natl Acad Sci USA 77:1588–92, 1980.
30. Jackson DG, Bell JI. Isolation of a cDNA encoding the human CD38 (T10) mole-

cule, a cell surface glycoprotein with an unusual discontinuous pattern of expression during lymphocyte differentiation. J Immunol 144:2811–15, 1990.
31. Hedeskov CJ. Mechanism of glucose-induced insulin secretion. Physiol Rev 60:442–509, 1980.
32. Itoh N, Okamoto H. Translational control of proinsulin synthesis by glucose. Nature 283:100–2, 1980.
33. Cook DL, Hales N. Intracellular ATP directly blocks K^+ channels in pancreatic B-cells. Nature 311:271–3, 1984.
34. Ashcroft FM, Harrison DE, Ashcroft SJH. Glucose induces closure of single potassium channels in isolated rat pancreatic β-cells. Nature 312:446–8, 1984.
35. Rajan AS, Aguilar-Bryan L, Nelson DA et al. Ion channels and insulin secretion. Diabetes Care 13:340–63, 1990.
36. Hutton JC, Malaisse WJ. Dynamics of O_2 consumption in rat pancreatic islets. Diabetologia 18:395–405, 1980.
37. Matschinsky FM, Ghosh AK, Meglasson MD, et al. Metabolic concomitants in pure, pancreatic beta cells during glucose-stimulated insulin secretion. J Biol Chem 261:14057–61, 1986.
38. Ashcroft SJH, Weerasinghe LCC, Randle PJ. Interrelationship of islet metabolism, adenosine triphosphate content and insulin release. Biochem J 132:223–31, 1973.
39. Ashcroft SJH, Nino S. Effects of phloretin and dextran-linked phloretin on pancreatic islet metabolism and insulin release. Biochim Biophys Acta 538:334–42, 1978.
40. Bukowiecki L, Trus M, Matschinsky FM, Freinkel N. Alterations in pancreatic islet phosphate content during secretory stimulation with glucose. Biochim Biophys Acta 583:370–7, 1979.
41. Kawazu S, Sener A, Couturier E, Malaisse WJ. Metabolic, cationic and secretory effects of hypoglycemic sulfonylureas in pancreatic islets. Naunyn-Schmiedebergs Arch Pharmacol 312:277–83, 1980.
42. Okamoto H, Takasawa S, Kato I, et al. The role of cyclic ADP-ribose in stimulus-secretion coupling in pancreatic β-cells. In: Diabetes 1994. Baba S, ed. Elsevier, Amsterdam, pp 244–46, 1995.
43. Islam MDS, Larsson O, Berggren P-O. Cyclic ADP-ribose in β cells. Science 262:584–5, 1993.
44. Takasawa S, Nata K, Yonekura H, Okamoto H. Cyclic ADP-ribose in β cells. Science 262:585, 1993.
45. Galione A. Cyclic ADP-ribose in β cells. Science 262:585–6, 1993.
46. Praz GA, Halban PA, Wollheim CB, et al. Regulation of immunoreactive-insulin release from a rat cell line (RINm5F). Biochem. J 210:345–52, 1983.
47. Itoh N, Okamoto H. Synthesis of preproinsulin with RNA preparation of streptozotocin-nicotinamide induced B-cell tumor. FEBS Lett 80:111–14, 1977.
48. Itoh N, Nose K, Okamoto H. Purification and characterization of proinsulin mRNA from rat B-cell tumor. Eur J Biochem 97:1–9, 1979.
49. Masiello P, Wollheim CB, Gori Z, et al. Streptozotocin-induced functioning islet cell tumor in the rat: High frequency of induction and biological properties of the tumor cells. Toxicol Pathol 12:274–80, 1984.
50. Yamagami T, Miwa A, Takasawa S, et al. Induction of rat pancreatic B-cell tumors by the combined administration of streptozotocin or alloxan and poly (adenosine diphosphate ribose) synthetase inhibitors. Cancer Res 45:1845–49, 1985.
51. Harada N, Santos-Argumedo L, Chang R, et al. Expression cloning of a cDNA encoding a novel murine B cell activation marker: Homology to human CD38. J Immunol 151:3111–18, 1993.
52. Glick DL, Hellmich MR, Beushausen S, et al. Primary structure of a molluscan egg-specific NADase, a second-messenger enzyme. Cell Regul 2:211–18, 1991.
53. Nata K, Sugimoto T, Tohgo A, et al. The structure of the *Aplysia kurodai* gene encoding ADP-ribosyl cyclase, a second-messenger enzyme. Gene 158:213–18, 1995.
54. Zocchi E, Franco L, Guida L, et al. A single protein immunologically identified as

CD38 displays NAD^+ glycohydrolase, ADP-ribosyl cyclase and cyclic ADP-ribose hydrolase activities at the outer surface of human erythrocytes. Biochem Biophys Res Commun 196:1459–65, 1993.

55. Howard M, Grimaldi JC, Bazan JF, et al. Formation and hydrolysis of cyclic ADP-ribose catalyzed by lymphocyte antigen CD38. Science 262:1056–59, 1993.
56. Takamura T, Nata K, Karasawa T, et al. The genomic structure of human ADP-ribosyl cyclase/cyclic ADP-ribose hydrolase (CD38) gene. Seikagaku 66:835, 1994.
57. Nakagawara K, Mori M, Takasawa S, et al. Assignment of CD38, the gene encoding human leukocyte antigen CD38 (ADP-ribosyl cyclase/cyclic ADP-ribose hydrolase), to chromosome 4p15. Cytogenet Cell Genet 69:38–9, 1995.
58. Kallmann B, Burkart V, Kröncke KD, et al. Toxicity of chemically generated nitric oxide towards pancreatic islet cells can be prevented by nicotinamide. Life Sci 51:671–8, 1992.
59. Radons J, Heller B, Bürkle A, et al. Nitric oxide toxicity in islet cells involves poly (ADP-ribose) polymerase activation and concomitant NAD^+ depletion. Biochem Biophys Res Commun 199:1270–77, 1994.
60. Akabane A, Kato I, Takasawa S et al. Nicotinamide inhibits IRF-1 mRNA induction and prevents IL-1β-induced nitric oxide synthase expression in pancreatic β cells. Biochem Biophys Res Commun 215:524–30, 1995.
61. Heller B, Wang Z-Q, Wagner EF, et al. Inactivation of poly(ADP-ribose) polymerase gene affects oxygen radical and nitric oxide toxicity in islet cells. J Biol Chem 270:11176–80, 1995.
62. Malaisse WJ: Alloxan toxicity to the pancreatic B-cell: A new hypothesis. Biochem Pharmacol 31:3527–34, 1982.
63. Okamoto H: New aspects to the functioning and regeneration of pancreatic β-cells. Cyclic ADP-ribose and Reg gene. In: Physiology and pathophysiology of the islets of Langerhans. Soria B, ed. Plenum Press, New York, 1996, in press.
64. Kim H, Jacobson EL, Jacobson MK: Synthesis and degradation of cyclic ADP-ribose by NAD glycohydrolases. Science 261:1330–33, 1993.
65. Schuber F, Travo P, Pascal M: On the mechanism of action of calf spleen NAD^+ glycohydrolase. Bioorg Chem 8:83–90, 1979.
66. Tarnus C, Muller HM, Schuber F. Chemical evidence in favor of a stabilized oxocarbonium-ion intermediate in the NAD^+ glycohydrolase-catalyzed reaction. Bioorg Chem 16:38–51, 1988.
67. Kato I, Takasawa S, Akabane A, et al. Regulatory role of CD38 (ADP-ribosyl cyclase/cyclic ADP-ribose hydrolase) in insulin secretion by glucose in pancreatic β cells: Enhanced insulin secretion in CD38-expressing transgenic mice. J Biol Chem 270:30045–50, 1995.

8

Alloxan derivatives as a tool for the elucidation of the mechanism of the diabetogenic action of alloxan

SIGURD LENZEN,[1] MARKUS TIEDGE,[1] ANNE JÖRNS,[2] AND REX MUNDAY[3]

[1]Institute of Clinical Biochemistry, Hannover Medical School, D-30623 Hannover, Germany; [2]Department of Anatomy I, Hannover Medical School, D-30623 Hannover, Germany; [3]AgResearch, Ruakura Agricultural Research Centre, Hamilton, New Zealand.

Key words: alloxan diabetes, alloxan, alloxan derivatives, diabetogenicity, physicochemical properties.

Historical aspects

In 1818, the Italian Brugnatelli obtained a substance from the oxidation of uric acid that he named *ossieritrico*. This name, from the Greek "to make red," referred to the property of the substance to stain the skin a characteristic red color.[1] Twenty years later, the German chemists, Wöhler and Liebig, studied uric acid oxidation in detail and obtained the same compound, which they named *alloxan*, apparently from conflation of the words *allantoin* and *oxalsäure*.[2] The early history of alloxan has been reviewed in detail elsewhere.[3] *Alloxan* became generally accepted as the common name for this substance. The compound as synthesized via aqueous workup is a so-called hydrate, actually a *gem*-diol. The systematic name of this substance is 5,5,-dihydroxy-2,4,6 (1H, 3H, 5H)-pyrimidinetrione.

Alloxan remained solely in the domain of chemists until 1943, when Dunn et al.[4] reported that this substance causes necrosis of pancreatic islets when administered to rabbits or rats. Alloxan is selectively toxic to the B-cells of the pancreas, thereby causing diabetes in experimental animals.[3,5–7] This substance then became of great interest to diabetologists, becoming a standard tool for the induction of diabetes in animals. The mechanism of this process has also been of

Correspondence: Prof. Dr. S. Lenzen, Institute of Clinical Biochemistry, Hannover Medical School, D-30623 Hannover, Germany.

Lessons from Animal Diabetes VI. Ed. E. Shafrir.

constant interest, since very few compounds are known that show such selective toxicity to pancreatic B-cells. Current theories[7] indicate a role for "active oxygen" species, formed during redox cycling of alloxan, in the initiation of the changes that lead to B-cell death.

Shortly after the initial report of alloxan diabetes, a number of scientists became interested in the relationship between the chemical structure and diabetogenicity of alloxan. A number of alloxan derivatives had been described in the early German literature. Biltz in 1912 had reported the synthesis of N-methylalloxan and N,N′-dimethylalloxan by oxidation of uric acid derivatives in a manner analogous to the synthesis of alloxan itself.[8] The former substance was obtained by oxidation of theobromine and the latter by oxidation of caffeine. In 1930, Biilmann and Berg[9] described a general synthesis of N-substituted alloxans from the corresponding N-substituted ureas.

Leaders in the study of alloxan derivatives were Gerhard Brückmann (1909–1948), Ernst Chaim Wertheimer (1893–1978), and Emil Abderhalden (1877–1950). All three scientists had detailed knowledge of the chemistry of alloxan and were familiar with the early work[8,9] on the synthesis of alloxan and its derivatives. Following the procedures of the early workers, a number of alloxan derivatives were synthesized and tested for diabetogenic activity. It soon became clear that the selective toxicity of alloxan to the pancreatic B-cell was critically dependent upon structure, and comparatively small changes to the alloxan molecule had a major impact upon diabetogenicity (Table 1). Thus, while N-methylalloxan was of comparable activity to the parent compound alloxan (ED_{50} 0.30 mmol/kg vs. 0.28 mmol/kg), N-ethylalloxan was somewhat less active (ED_{50} 0.36 mmol/kg), while N-propylalloxan was, at most, weakly diabetogenic. Substitution with larger alkyl groups, as in N-butylalloxan, N-isobutylalloxan, N-benzylalloxan, and N-phenylalloxan, or substitution at both nitrogen atoms, as in N,N′-dimethylalloxan, completely abolished diabetogenicity.[10–13] Such substitution did not, however, abolish toxicity to other organs and tissues, and the derivatives generally showed higher acute toxicity than the parent compound. Like alloxan, they were also toxic to the kidney, an effect that was particularly pronounced with the N,N′-dimethyl derivative.

The scientific careers and courses of the lives of Brückmann, Wertheimer, and Abderhalden mirror in a very typical way the tragic events in the Europe of the 1930s and 1940s. Brückmann was born in Berlin in 1909, obtained his Ph.D. in chemistry from the University of Berlin in 1933, and in the same year emigrated to Palestine. He worked in the Chemical Laboratories of the Municipal Hospital in Tel Aviv and later in the Pharmacology Department of the Hebrew University in Jerusalem. The head of this department was Wertheimer, also an emigré from Nazi Germany. He had moved to Palestine in 1934, having previously worked with Emil Abderhalden in Halle. Abderhalden had been involved in many biological and chemical studies on alloxan and ninhydrin, and Wertheimer was well aware of the early work in this field when he teamed up with Brückmann for a detailed study of the diabetogenicity of alloxan derivatives. The syntheses were conducted by Brückmann in collaboration with Isaacs at the Hebrew Uni-

Table 1. Chemical formulae of alloxan derivatives and their diabetogenic activity in rats.

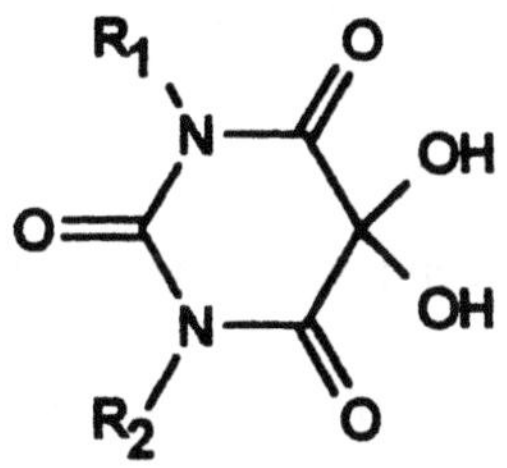

Compound	R_1	R_2	Diabetogenic activity
Alloxan	H	H	High (ED_{50}, 0.28 mmol/kg)
N-Methylalloxan	CH_3	H	High (ED_{50}, 0.30 mmol/kg)
N-Ethylalloxan	C_2H_5	H	Moderate (ED_{50}, 0.36 mmol/kg)
N-Propylalloxan	C_3H_7	H	Low (ED_{50} not determined)
N-Butylalloxan	n-C_4H_9	H	None
N-Isobutylalloxan	i-C_4H_9	H	None
N-Benzylalloxan	$C_6H_5CH_2$	H	None
N-Phenylalloxan	C_6H_5	H	None
N,N′-Dimethylalloxan	CH_3	CH_3	None

versity,[14] and results of the toxicological studies were published in two papers coauthored by Brückmann and Wertheimer.[10,13]

Emil Abderhalden had to leave his chair at the University of Halle when Prussian Saxony became part of the former East Germany. As a Swiss citizen, he returned to Zürich, where he taught physiological chemistry and physiology for a short time at the university.[15] Alloxan studies continued at the University of Zürich for a while, with Albert Renold studying alloxan diabetes and the effects of the two alloxan derivatives, N-methylalloxan and N,N′-dimethylalloxan, for his medical thesis work[16] under the supervision of Karl Bernard.

The studies on alloxan derivatives were extremely valuable, and they would certainly have had a major impact on the progress of research on alloxan diabetes had they continued. However, the tragic death of Brückmann, who fell in Jerusalem during the Israel War of Independence in 1948 (Nomi Bruckmann, personal communication, 1992), put an end to the study of alloxan derivatives in Israel. Abderhalden died in 1950 and the studies in Switzerland were also discontinued.

Physicochemical properties and biological effects of alloxan derivatives

During recent years we have tried to define the chemical properties of alloxan responsible for its selective B-cell toxicity and its diabetogenicity.[17] For this purpose we repeated the syntheses of the N-alkyl substituted alloxan derivatives described by the earlier workers and attempted to relate their physicochemical properties to the selective toxicity.[17,18] We were able to define four factors that may be of importance in determining the toxicity of alloxan, hydrophilicity, stability, reactivity toward thiol groups, and generation of free radicals during redox cycling with glutathione[17,18]:

1. Alloxan is a very hydrophilic compound. The log value of the partition coefficient for alloxan is -1.8.

2. Alloxan is an unstable substance at neutral pH with a half-life of 4.7 min at 25°C and of 1.4 min at 37°C.

3. Alloxan reacts with thiols, oxidizing protein-bound thiol groups, as reflected by inhibition of the thiol enzymes glucokinase and hexokinase. Alloxan is reduced to dialuric acid by thiols such as glutathione and dithiothreitol. With glutathione, subsequent reactions lead to the formation of a substance of unknown structure ("Compound 305") with a characteristic absorbance at 305 nm.

4. Due to the ready autoxidation of dialuric acid, reaction of alloxan with glutathione initiates a redox cycle, generating superoxide radical, hydrogen peroxide, and, in the presence of an iron catalyst, hydroxyl radical.

In many respects, the reactions of N-methylalloxan (MA), N-ethylalloxan (EA), N-propylalloxan (PA), N-butylalloxan (BA), N-isobutylalloxan (IBA), N-benzylalloxan (BZA), and N,N′-dimethylalloxan (DMA) were found to be closely similar to those of the parent compound.[17,18]

1. All alloxan derivatives formed dialuric acids by reduction with dithiothreitol. These substances were also formed by reaction with glutathione, and subsequent reactions gave substances analogous to "Compound 305."

2. All derivatives were found to undergo redox cycling in the presence of glutathione, indicating that substituted dialuric acids undergo autoxidation in the same way as the parent compound. With the exception of N-benzylalloxan, the rate of redox cycling (and hence the rate of "active oxygen" formation) was higher with derivatives than with alloxan itself.

3. All derivatives were unstable at neutral pH, with half-lives comparable with or even shorter than that of the parent compound.

4. Major differences were identified among derivatives, however, in their relative solubilities in aqueous and lipid milieus. With increase of the chain length of the substituent the hydrophilicity of the N-substituted compounds decreased. Although all derivatives tested were less hydrophilic than the parent compound, the diabetogenic N-methylalloxan and N-ethylalloxan can still be classified as hydrophilic, and the weakly diabetogenic N-propylalloxan as only slightly lipophilic. In contrast, the nondiabetogenic N-butylalloxan, N-isobutylalloxan, and N-benzylalloxan are strongly lipophilic compounds and a significant positive correlation can be recorded between diabetogenicity in vivo and hydrophilicity. The only exception is N,N′-dimethylalloxan, which, although nondiabetogenic, is a hydrophilic compound.

These studies indicate that stability and reactivity toward thiol groups are not correlated with diabetogenicity. Furthermore, although "active oxygen" species are suggested to play a role in the mechanism of alloxan-induced B-cell death, the ability to generate such species is not a sufficient condition for diabetogenicity. The observation with regard to hydrophilicity may be important, however, and suggests that the crucial site of action of alloxan and its diabetogenic derivatives may be located at a site in or on the pancreatic B-cell which is in an aqueous rather than a lipid domain.

Although alloxan derivatives had been extensively studied in vivo, their toxicity to isolated pancreatic islets had never been investigated. It was considered possible that the difference between diabetogenic and nondiabetogenic derivatives could reside in an intrinsic difference in B-cell susceptibility, that is, there would be a correlation between the ability of the compounds to cause damage to B-cells in vitro and their diabetogenicity in vivo. In order to investigate this possibility, we have recently performed an electron microscopy study of alloxan and derivatives in collagenase-isolated *ob/ob* mouse islets of Langerhans using an in vitro incubation system.[19] Batches of 70–100 islets were incubated for 5 min in 40 μl Hepes-buffered Krebs-Ringer bicarbonate medium (pH 7.4) with one of the test agents, but without glucose. Thereafter the medium was replaced by the basal medium without test agent and the incubation was continued for 4 h. Islets were ultrastructurally well preserved after this incubation period (Fig. 1). Both

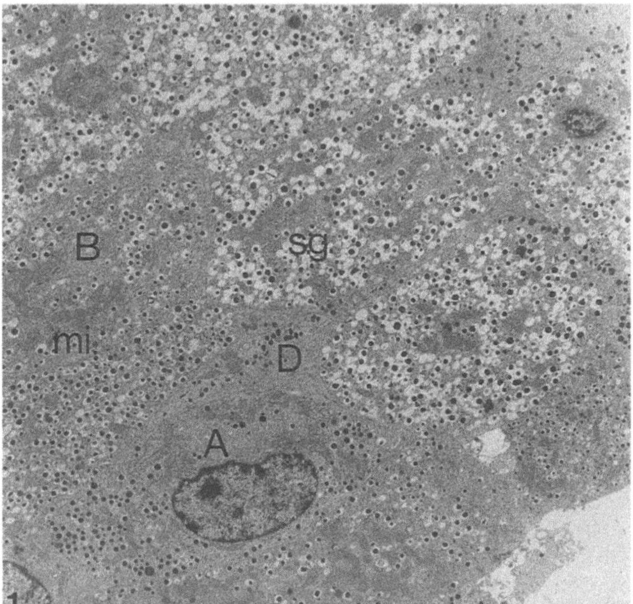

Figure 1. Ultrastructure of endocrine cells in an isolated *ob/ob* mouse islet in a control situation after a 4 h incubation. The endocrine cell types can be identified by the special secretory granule morphology as B-cells (B), A-cells (A), and D-cells (D). Cell organelles such as secretory granules (sg) and mitochondria (mi) are well preserved in all B-cells. ×4500.

B-cells and the other pancreatic islet cell types exhibited no degenerative changes (Fig. 1).

Nearly 50% of the pancreatic B-cells incubated with alloxan at a concentration of 2.5 mmol/l were necrotic (Fig. 2). In the necrotic B-cells the membranes of the cells and of the cell organelles, such as the secretory granules and the cisternae of the rough endoplasmic reticulum and Golgi complex, were ruptured. In addition, the inner membranes of mitochondria lost their structure (Fig. 2). No degenerative lesions were recorded in the islet A-cells, D-cells or exocrine parenchymal cells, even at the highest concentrations of alloxan, at which all B-cells were destroyed.

All alloxan derivatives caused a concentration-dependent selective necrosis of the pancreatic B-cells, with degenerative changes identical to those recorded with the parent compound, as shown in Figure 3 for N-butylalloxan (0.25 mM). Far from being a correlation between toxicity in vitro and diabetogenicity in vivo, however, the nondiabetogenic compounds were more toxic to pancreatic B-cells in vitro than the diabetogenic ones. For comparison, we have examined the ultrastructural changes induced by STZ, another selectively B-cell toxic agent, on incubated *ob/ob* mouse on pancreatic islets. It is interesting that streptozotocin causes the same kind of morphological damage as alloxan and the alloxan derivatives, although the mechanism of action is believed to differ from that of alloxan (Fig. 4).

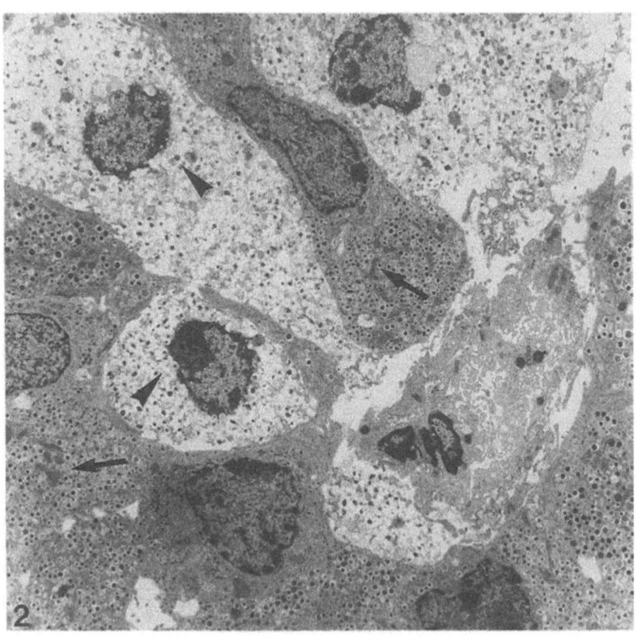

Figure 2. Ultrastructure of pancreatic B-cells in an isolated *ob/ob* mouse islet after preincubation with alloxan (2.5 mmol/l) for 5 min and a subsequent incubation period of 4 h in the absence of the test compound. At this concentration about half of the B-cells show signs of degeneration, such as swollen mitochondria, pale cytoplasm, or rupture of membranes of the cell or its organelles (arrowheads), while other B-cells are still intact (arrows). ×4500.

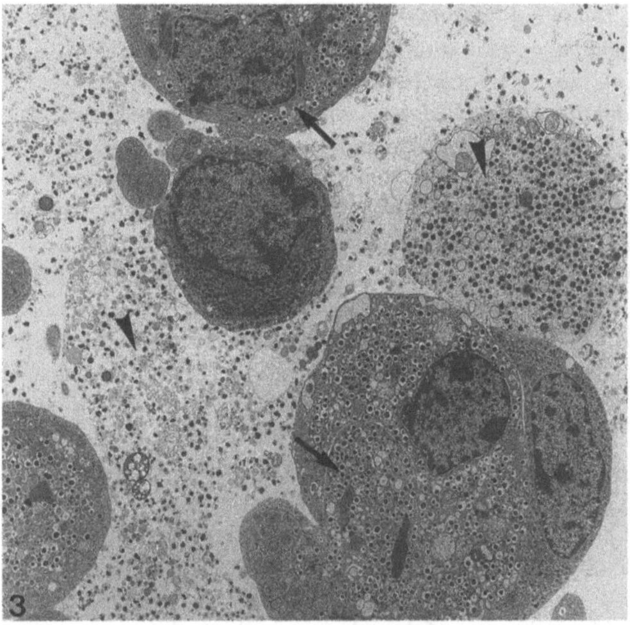

Figure 3. Ultrastructure of pancreatic B-cells in an isolated *ob/ob* mouse islet after preincubation with N-butylalloxan (0.25 mmol/l) for 5 min and subsequent incubation period of 4 h in the absence of the test compound. The extent of pancreatic B-cell destruction in response to this low concentration of the alloxan derivative is comparable to the destructive effects caused by 2.5 mmol/l alloxan. At this concentration some B-cells show signs of degeneration, such as swollen mitochondria, pale cytoplasm, or rupture of membranes of the cell or its organelles (arrowheads), while other B-cells are still intact (arrows). ×4500.

The toxicity to pancreatic B-cells in vitro increased with the increase of the chain length of the substituent (Fig. 5) and thus in parallel to the increase of the lipophilic character of the derivative.[18] As shown in Figure 5, the EC_{50}, that is the concentration at which 50% of the B-cells were destroyed, decreased in the following order: alloxan >> N-methylalloxan (MA) > N-ethylalloxan (EA) > N-propylalloxan (PA) > N-butylalloxan (BA) = N-isobutylalloxan (IBA). N,N′-dimethylalloxan (DMA) (1 mmol/l) also destroyed all B-cells. The most toxic derivatives, N-butylalloxan and N-isobutylalloxan, that is, derivatives with the longest chain substituents, which are also the most lipophilic compounds, were >10 times more toxic than the parent compound alloxan (see Fig. 5).

It is interesting to note that there is a relationship between lipophilicity and toxicity to pancreatic B-cells in vitro, which is the converse of that seen for diabetogenicity in vivo. This is an observation of crucial importance in understanding the mechanism of alloxan diabetes. It is clear that all alloxan derivatives, if presented to islets at sufficient concentration, will selectively destroy B-cells and are therefore potentially diabetogenic. The fact that they are not indicates that following administration they do not reach the islets in such concentration, and suggests that differences in pharmacokinetics may be the key to understanding the differences between alloxan and its derivatives. It is possible that the lipophilic derivatives are taken up by tissue compartments not accessible to the hydrophilic derivatives, and are thus more evenly distributed throughout the body. In this way, they may not reach concentrations in the pancreatic islets high enough to cause diabetes, but may be sufficiently concentrated in the kidney during excretion to cause damage to this organ.

Over the years, many chemical and biochemical properties of alloxan have been discovered, and at one time or another almost every such property has been suggested to be involved in the mechanism whereby alloxan selectively destroys the B-cells of the pancreas. The trouble with this approach is that only a single compound has ever been investigated, and there is therefore no way of knowing whether or not any particular property is truly involved in the toxic mechanism. By the use of a series of derivatives, we have eliminated a number of properties and hope to be able in the future to pinpoint the property or properties that distinguish the alloxan derivatives that are diabetogenic from those that are not and therefore identify the true mechanism of alloxan diabetogenicity. We owe a debt of gratitude to the early chemists and, particularly, to Brückmann, Wertheimer, and Abderhalden, whose careful work may facilitate the answer to a problem that

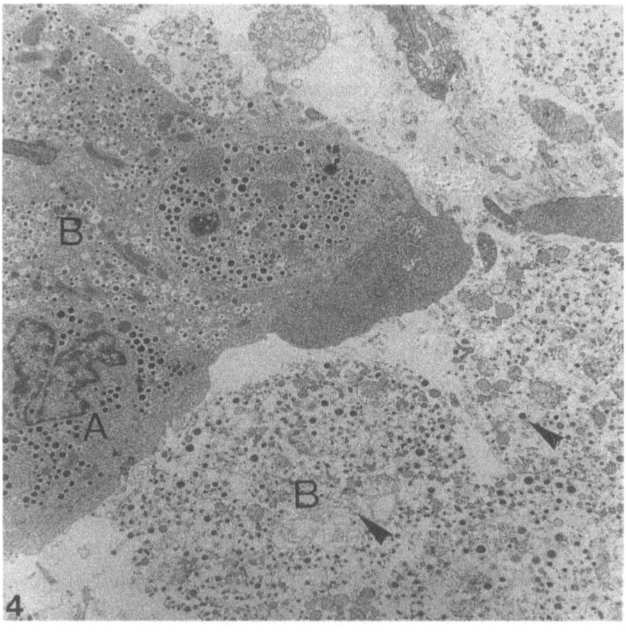

Figure 4. Ultrastructure of endocrine cells in an isolated *ob/ob* mouse islet after preincubation with streptozotocin (1 mmol/l) for 5 min and a subsequent incubation period of 4 h in the absence of the test compound. Pancreatic A-cells (A) are well preserved, whereas most B-cells are necrotic (arrowheads). ×4500.

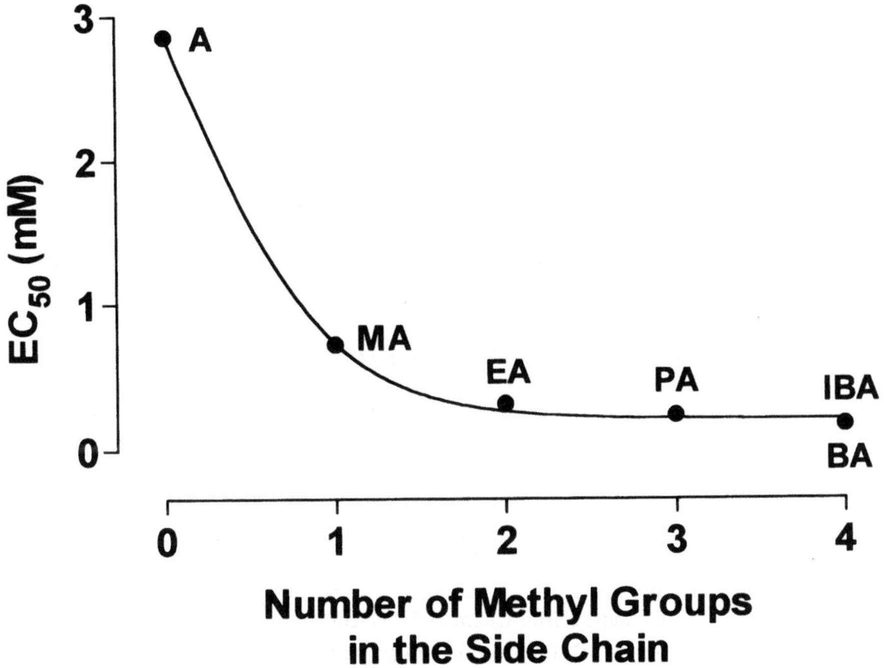

Figure 5. Relation between the EC_{50} (mM) for the pancreatic B-cell necrotizing effect of alloxan (A) and of alloxan derivatives N-methylalloxan (MA), N-ethylalloxan (EA), N-propylalloxan (PA), N-butylalloxan (BA), and N-isobutylalloxan (IBA), and the length of the side chain of the derivative. The EC_{50} decreased significantly ($P < 0.05$; ANOVA) in parallel with the increase of length of the side chain of the derivative, and thus in parallel with the increase of the lipophilicity of the derivative. EC_{50} is defined as the concentration at which the respective test compound has destroyed 50% of pancreatic B-cells. The EC_{50} values for the pancreatic B-cells necrotizing effect have been calculated for alloxan and its derivatives from concentration dependencies (0; 0.25, 0.5; 1.0; 2.5; 5.0 mmol/l).

has taxed the minds of many scientists throughout the world over the past 50 years.

Acknowledgments

The authors are grateful to Ms. Nomi Bruckmann (Jerusalem) for providing bibliographical information on her father, Dr. Gerhard Brückmann.

References

1. Brugnatelli G. Sopra i cangiamenti che avvengono nell'ossiurico (ac. urico) trattato coll'ossisettonoso (ac. nitroso). G Fis Chim 117–29, 1818.
2. Wöhler F, Liebig J. Untersuchungen über die Natur der Harnsäure. Ann Pharm 26:241–340, 1838.

3. Lenzen S, Panten U. Alloxan: History and mechanism of action. Diabetologia 31:337–42, 1988.
4. Dunn JS, Sheehan HL, McLetchie NGB. Necrosis of islets of Langerhans produced experimentally. Lancet 244:484–7, 1943.
5. Webb JL. Alloxan. In: Enzyme and metabolic inhibitors. Webb JL, ed. Academic Press, New York, pp 367–419, 1966.
6. Cooperstein SJ, Watkins D. Action of toxic drugs on islet cells. In: The islet of Langerhans. Cooperstein SJ, Watkins D, eds. Academic Press, New York, pp 387–425, 1981.
7. Oberley LW. Free radicals and diabetes. Free Radic Biol Med 5:113–24, 1988.
8. Biltz H. Alloxan-anhydrid und seine Methylderivate. Ber Dtsch Chem Ges 45:3659–75, 1912.
9. Biilmann E, Berg N. Über die Reduktionspotentiale der Alloxantine und über die Darstellung der Alloxane und Alloxantine. Chem Ber 63:2188–204, 1930.
10. Brückmann G, Wertheimer E. Diabetogenic action of alloxan derivatives. Nature 155:267–8, 1945.
11. Hidy PH. Structural requirements for diabetogenic action in alloxan and related compounds. J Biol Chem 163:307–11, 1946.
12. Bernhard K, Favarger M, Renold A, Spühler O. Beiträge zur Alloxan-Glucosurie. Helv Chim Acta 30:1666–72, 1947.
13. Brückmann G, Wertheimer E. Alloxan studies: The action of alloxan homologues and related compounds. J Biol Chem 168:241–56, 1947.
14. Brückmann G, Isaacs SD. Preparation and properties of new derivatives of alloxan. J Am Chem Soc 71:390–2, 1949.
15. Gabathuler J. Emil Abderhalden. Sein Leben und Werk. Buchhandlung Ribaux AG, St. Gallen, pp 1–362, 1991.
16. Renold AE. Der Alloxan-Diabetes. Zürich, Dissertation, pp 1–45,1948.
17. Lenzen S, Munday R. Thiol-group reactivity, hydrophilicity, and stability of alloxan, its reduction products and its N-methyl derivatives and a comparison with ninhydrin. Biochem Pharmacol 42:1385–91, 1991.
18. Munday R, Ludwig K, Lenzen S. The relationship between physicochemical properties and the biological effects of alloxan and several N-alkyl substituted alloxan derivatives. J Endocrinol 139:153–63, 1993.
19. Jörns A, Munday R, Tiedge M, Lenzen S. Comparative toxicity of alloxan, N-alkylalloxans and ninhydrin to ob/ob mouse pancreatic islets in vitro. unpublished.

9

Role of glucokinase in regulation of insulin secretion: Lessons from transgenic mice

SHIMON EFRAT
Department of Molecular Pharmacology, Albert Einstein College of Medicine, Bronx, NY 10461, USA.

Key words: antisense RNA, glucose phosphorylation, glucose sensing, MODY, ribozyme.

Impairments in the insulin secretory response to glucose in both type II and early type I diabetes have prompted studies into the mechanisms that allow pancreatic B-cells to sense and respond to physiological changes in blood glucose. Glucose-induced insulin secretion requires the metabolism of glucose in B-cells,[1] and the phosphorylation of glucose to glucose-6-phosphate, which determines the rate of glycolysis, is considered the major glucose-sensing step for regulating insulin secretion.[2] B-cells and hepatocytes express a high-K_m member of the hexokinase family, glucokinase (GK), which is responsible for the majority of glucose phosphorylation activity in B-cells.[2] While in the liver transcription of the GK gene is induced by insulin,[3,4] GK expression in B-cells is primarily regulated by glucose at the translational and post-translational levels.[3,5]

Recent DNA polymorphism studies have established a correlation between mutations in the GK locus and diabetes in patients with maturity-onset diabetes of the young (MODY).[6,7] This is a form of NIDDM characterized by an early age of onset and an autosomal dominant inheritance. GK genes from MODY patients contain a number of nonsense and missense mutations that are associated with regions of the protein involved in glucose and ATP binding.[6,7] The molecular mechanism that renders these mutations dominant remains unknown. The inheritance pattern of the disease suggests that the patients' B-cells contain normal enzyme molecules encoded by the wild-type allele. The mutant proteins manifest drastically reduced activities.[8,9] Since GK expression in B-cells is not transcriptionally regulated, the wild-type allele cannot compensate for this reduction. The decreased GK activity may be sufficient to shift the threshold for glucose

Correspondence: Shimon Efrat, Department of Molecular Pharmacology, Albert Einstein College of Medicine, Bronx, NY 10461, USA.

Lessons from Animal Diabetes VI. Ed. E. Shafrir.

sensing, thereby resulting in impaired insulin secretion at physiological glucose levels. However, it remains unclear whether abnormal glucose metabolism in the liver contributes to the disease.

The understanding of the human disease can benefit from an animal model, which would facilitate detailed biochemical and physiological characterization of the abnormalities induced by the GK gene mutations in the context of the entire organism. Such an animal model can also provide a system for investigating the influence of additional genetic and environmental factors on the onset and severity of diabetes, and allow evaluation of therapeutic approaches. Cell-specific gene expression in transgenic mice, as well as targeted gene disruption by homologous recombination in embryonic stem cells, represent powerful approaches for generation of such animal models. To evaluate the relative contribution of impaired GK function in B-cells and hepatocytes to the development of MODY, we initially aimed at creating a mouse model in which GK activity is specifically reduced in B-cells. To this end, a glucokinase ribozyme was expressed in B-cells in transgenic mice under control of the insulin promoter. Ribozymes are RNA molecules that possess RNA catalytic activity.[10] By flanking the ribozyme catalytic element with two gene-specific fragments in antisense orientation, the ribozyme activity can be targeted against a unique RNA sequence, thus reducing target mRNA levels and activity.

Materials and methods

Plasmid constructs. Two oligonucleotides were synthesized containing two 12-base fragments derived from mouse glucokinase gene exon 3 sequence[11] that flank a hammerhead ribozyme catalytic element.[10] Annealing of the sense oligonucleotide 5′GATCCTCTCCCACTTTCTGATGAGTCCGTGAGGACGAAACCAGCATCACCGGTAC3′ and the antisense oligonucleotide 5′CGGTGATGCTGGTTTCGTCCTCACGGACTCATCAGAAAGTGGGAGAG3′ created a fragment (GKRZ) with *Bam*HI and *Kpn*I protruding ends, which was ligated downstream of a hybrid intron element[12] and upstream of the SV40 virus late polyadenylation site in pMLSIS.CAT.[12] The combined 675-bp fragment was cloned into the *Xba*I and *Sal*I sites of pRIP-Tag[13] downstream of the rat insulin II gene promoter to form pRIP-GKRZ.

Generation of transgenic mice. Linearized pRIP-GKRZ DNA was microinjected into C3HeB/FeJ mouse embryos, and transgenic mice were generated and bred according to established procedures.[14]

Glucose phosphorylation. Islets were isolated from the pancreas by collagenase infusion through the bile duct.[15] Islet homogenates from 1- to 3-month-old mice were incubated with different glucose concentrations in the presence of ATP, glucose-6-phosphate dehydrogenase, and NAD, and the formation of fluorescent NADH was monitored as described.[5]

Pancreas perfusion. Anesthetized 4- to 7-month-old mice were canulated through the aorta and portal vein, and were perfused with oxygenated Krebs-Ringer buffer containing a gradient of increasing glucose concentrations. Samples were collected and assayed for glucose by a glucose analyzer, and for insulin by RIA.

Dexamethasone treatment. Mice were implanted subcutaneously with a pellet containing 5 mg dexamethasone, designed for slow release over 21 days (Innovative Research, Toledo, Ohio).

Results and discussion

Generation of mice with reduced islet GK activity. We chose exon 3 of the GK transcript as a target because it encodes the putative ATP binding site of the protein.[16] The ribozyme element was introduced into the construct to increase the efficiency of the antisense RNA molecules by providing them with the ability to cleave the target mRNA. The RIP-GKRZ construct (Fig. 1) was microinjected

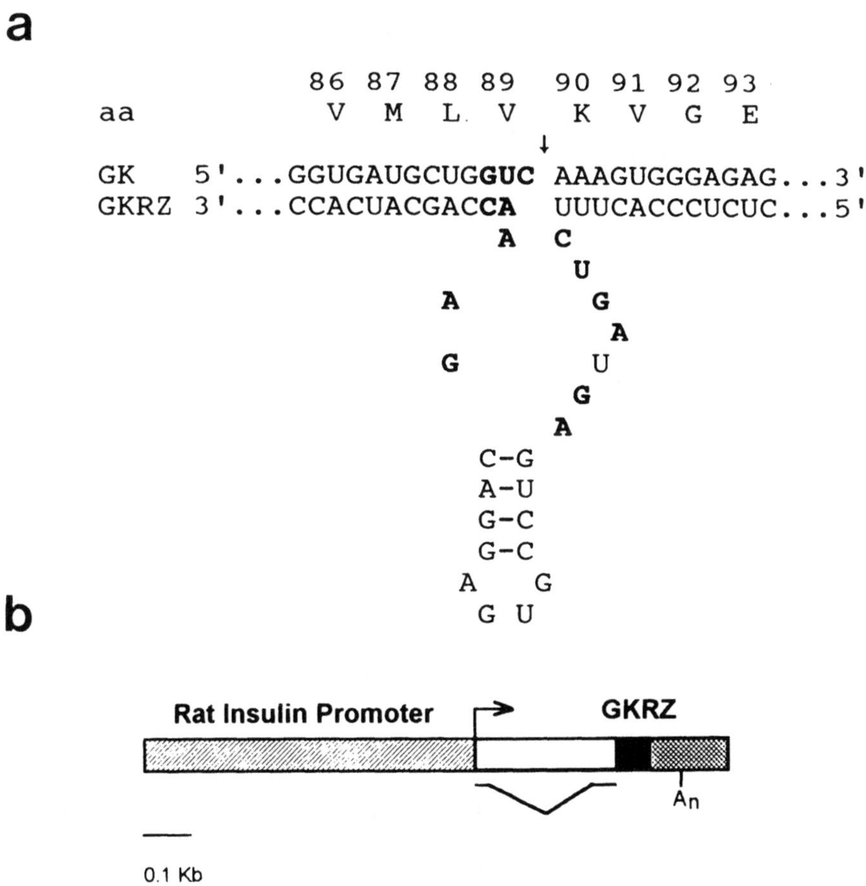

Figure 1. Design of the RIP-GKRZ construct. (a) The GKRZ transcript hybridized with the target GK mRNA. The conserved ribozyme nucleotides are shown in bold letters. The arrow marks the putative cleavage site. (b) The RIP-GKRZ hybrid gene consists of a synthetic DNA fragment encoding the ribozyme flanked by GK antisense sequences (GKRZ), an upstream intron element, and a downstream polyadenylation site. (Reprinted with permission from ref. 17.)

into mouse embryos, and seven transgenic mouse lineages were generated, two of which expressed the transgene, as judged by PCR analysis of islet cDNA.[17] Immunohistochemical analysis of pancreas sections with a GK antiserum revealed reduced staining intensity in transgenic islets, compared with normal controls.[17] Glucose phosphorylation activity at various glucose concentrations was assayed by a fluorimetric method in islets isolated from the transgenic mice. GK activity in RIP-GKRZ islets was reduced by 70%, compared with normal islets, while the activity of the ubiquitous low-K_m hexokinase remained essentially unaffected[17] (Fig. 2). These results demonstrate the target specificity of this approach. RNA quantitation suggested that the reduction in GK activity was due only in part to a decrease in GK mRNA levels, with the rest likely resulting from inhibition of GK mRNA translation. GK expression in B-cells may be more sensitive to inhibition at these levels than in hepatocytes, since it is mainly regulated at the translational and post-translational levels, rather than at the transcriptional level.[3,5] The incomplete inhibition of expression obtained with the antisense approach may represent an advantage for studying its consequences in vivo, since it mimics the situation in MODY. In addition, total shut-off of expression may be lethal.

Phenotype of the RIP-GKRZ mice. Although no data on islet GK activity are available from MODY patients, it is assumed that the wild-type GK allele pro-

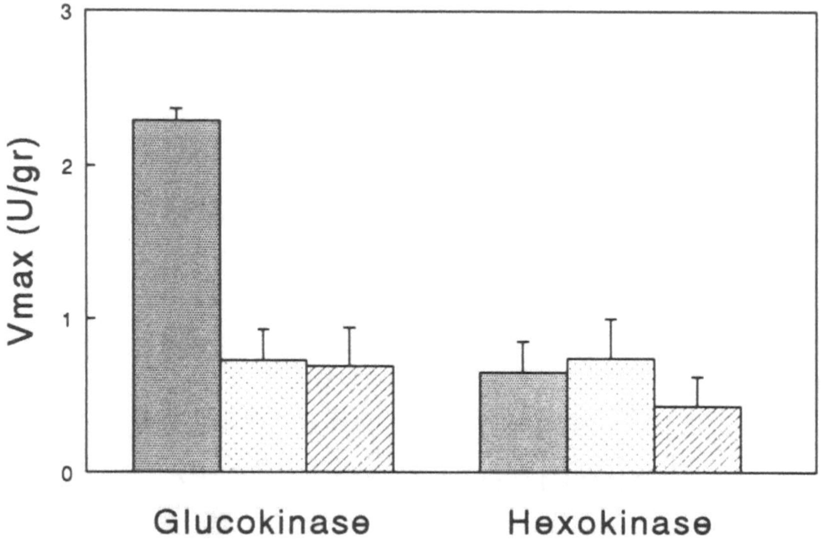

Figure 2. Glucose phosphorylation activity in RIP-GKRZ islets. V_{max} of glucokinase and hexokinase was calculated from Eadie–Hofstee plots. The great difference in K_m, 0.05 and 8 mM for hexokinase and glucokinase, respectively, allows the two enzymatic activities to be distinguished. Solid bar, normal islets; dotted bar, RIP-GKRZ2 islets; hatched bar, RIP-GKRZ4 islets. Values are expressed in units per gram of islet protein. 1 U = 1 μmol product per minute. Values are mean ± SE (n = 3). Transgenic and normal GK activity differences are statistically significant ($P < 0.0001$ by t test). (Reprinted with permission from ref. 17.)

Table 1. Plasma glucose and insulin levels in RIP-GKRZ mice

Mice	Glucose (mg/dl)	Insulin (ng/ml)
A. Fasted		
Normal	122 ± 22 (13)	0.64 ± 0.22 (5)
RIP-GKRZ	124 ± 8 (18)	0.65 ± 0.12 (12)
B. Dexamethasone-treated, fed		
Normal	147 ± 5 (10)	20 ± 3 (6)
RIP-GKRZ	150 ± 5 (10)	28 ± 8 (6)

(A) Mice (3–5 months old) were fasted overnight. The plasma glucose and insulin levels were determined.

(B) Three-month-old mice were treated for 3 wk with 8 μg/g per day dexamethasone from a s.c. implanted slow-release pellet. They were then bled, and the plasma glucose and insulin levels were determined.

Values are means ± SE (n is indicated in parentheses).

duces half of the normal activity. Therefore, islet GK activity in the RIP-GKRZ mice is likely to be as low or lower than that of MODY patients. However, unlike MODY patients, the RIP-GKRZ mice maintained normal fasting plasma glucose levels (Table 1) and manifested normal glucose tolerance. Fasting plasma insulin levels were similar to those of normal controls (Table 1). In contrast, analysis of glucose-induced insulin secretion from in situ–perfused pancreas[17] revealed a markedly reduced response of the transgenic pancreas, compared with that of normal controls, in the glucose concentration range of 75–200 mg/dl (Fig. 3). These findings are similar to those observed in MODY patients.[18] The discrepancy between the impaired response of the perfused pancreas to glucose and the normal plasma insulin levels indicates that unknown compensatory mechanisms may act in B-cells of intact RIP-GKRZ mice to correct insulin output, as has been noted in MODY.[19] Moreover, these results indicate that an impaired liver function, in addition to that of B-cells, may be required for the induction of overt diabetes by GK deficiency. Such liver impairments have been documented in MODY patients.[20] The finding that partial attenuation of expression of the normal GK protein is sufficient to impair the sensitivity of B-cells to glucose supports the interpretation of the dominance of the GK mutations in MODY as a gene-dosage effect, rather than a gain-of-function negative dominant effect of the mutant protein.

Effects of insulin resistance on RIP-GKRZ mice. To determine whether the reduced islet glucose sensitivity would interact with insulin resistance and obesity, as occurs in type II diabetes, in generating an overt diabetic phenotype, the RIP-GKRZ mice were crossed with a diabetes-prone mouse mutant denoted A^{vy} (agouti viable yellow). Normally, the agouti gene is expressed only in the skin. However, the A^{vy} mutation results in ubiquitous expression, which is probably due to a change in the promoter.[21] This is a dominant mutation, which affects hair pigment synthesis and induces insulin resistance and obesity, probably by

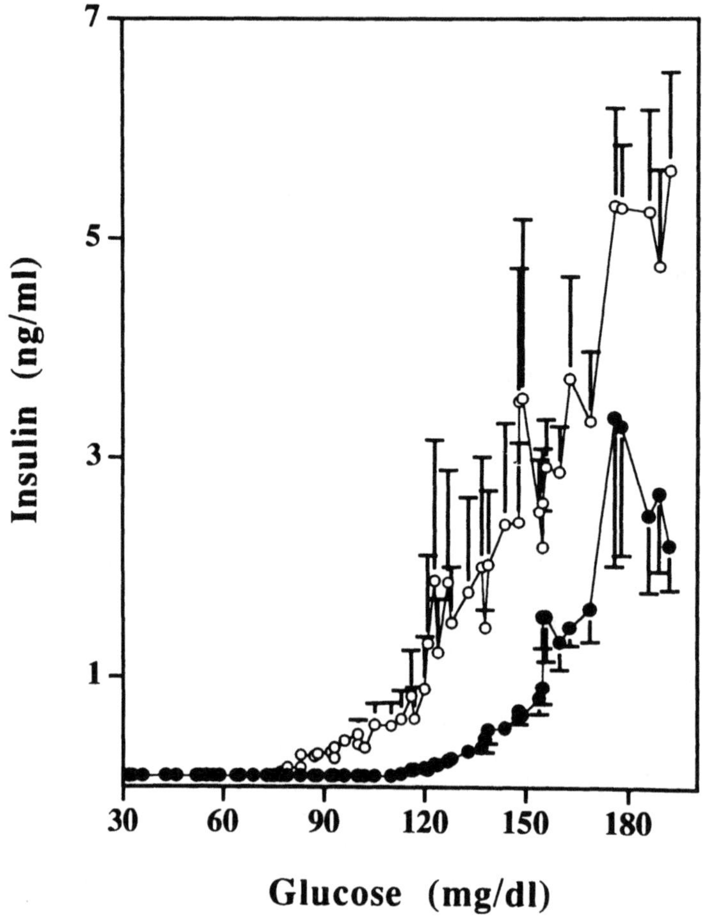

Figure 3. Insulin secretion from in situ–perfused control (open circle) and transgenic (closed circle) mouse pancreas. Values are mean ± SE (n = 5). (Reprinted with permission from ref. 17.)

raising Ca^{2+} levels in target cells. Males develop hyperglycemia and hyperinsulinemia by 4–5 months of age, while females are hyperinsulinemic but euglycemic.[22] The dominance of the mutation allows analysis of F_1 mice from crosses with RIP-GKRZ, which eliminates the need for repeated backcrosses that are required with other diabetogenic recessive mutations. A^{vy} males were mated with RIP-GKRZ females. The inheritance of the A^{vy} mutation was monitored by the characteristic yellow coat color. The control group consisted of mice generated by breeding A^{vy} males with normal C3H females. Mice were monitored for fed, fasting, and glucose tolerance plasma glucose and insulin levels. The reduced GK activity did not induce a detectable exacerbation in the A^{vy} phenotype in either males or females.

In addition to the A^{vy} crosses, we attempted to induce insulin resistance in the RIP-GKRZ mice by placing the mice on a high-fat diet (57% of energy from fat) and by treatment with dexamethasone. Both transgenic and control mice on a high fat diet gained weight at twice the rate of the respective mice that were fed a regular diet during a 3-month treatment. No effect on blood glucose or insulin was noted in any of the groups. In contrast, a 3-week dexamethasone treatment resulted in a 30- to 40-fold increase in plasma insulin levels in both transgenic and control mice; however, the glucose levels remained normal in both groups (Table 1).

Taken together, these results indicate that the RIP-GKRZ mice do not manifest an increased susceptibility to diabetes induced by obesity. In addition, the euglycemia maintained in the dexamethasone-treated mice demonstrates a remarkable ability of the RIP-GKRZ B-cells to upregulate their insulin production and secretion. With only 30% of the normal B-cell GK activity, these mice are able to cope with insulin resistance in a manner indistinguishable from normal controls. The results indicate that a further reduction of GK activity in B-cells may be required to impair their ability to respond to glucose. We are employing gene targeting approaches to address this possibility, as well as to impair GK activity in both liver and B-cells of the same animals.

Conclusions

These results demonstrate that a significant decrease in islet GK activity results in an increased threshold of the glucose concentration required for stimulating insulin secretion. However, this decrease in GK activity is not sufficient to induce hyperglycemia in vivo. It is possible that unknown compensatory mechanisms act in B-cells to generate a normal insulin output. The findings also suggest that GK deficiency in the liver, in addition to that in the islets, may play a role in the induction of diabetes in MODY patients. Gene targeting experiments in mice that will affect GK function in both B-cells and hepatocytes may provide insights into this possibility.

Acknowledgments

This research was supported by the Juvenile Diabetes Foundation International.

References

1. Meglasson MD, Matschinsky FM. New perspectives in pancreatic islet glucokinase. Am J Physiol 246:E1–E13, 1984.
2. Meglasson MD, Matschinsky FM. Pancreatic islet glucose metabolism and regulation of insulin secretion. Diabetes Metab Rev 2:163–214, 1986.
3. Iynedjian PB, Pilot PR, Nouspikel T, et al. Differential expression and regulation of the glucokinase gene in liver and islets of Langerhans. Proc Natl Acad Sci USA 86:7838–42, 1989.
4. Magnuson MA, Shelton KD. An alternate promoter in the glucokinase gene is active in the pancreatic beta cell. J Biol Chem 264:15936–942, 1989.
5. Liang Y, Najafi H, Maschinsky FM. Glucose regulates glycokinase activity in cultured islets from rat pancreas. J Biol Chem 265:16863–866, 1990.
6. Vionnet N, Stoffel M, Takeda J, et al. Nonsense mutation in the glucokinase gene causes early-onset non-insulin-dependent diabetes mellitus. Nature 356:721–2, 1992.

7. Stoffel M, Froguel P, Takeda J, et al. Human glucokinase gene: Isolation, characterization, and identification of two missense mutations linked to early-onset non-insulin-dependent (type 2) diabetes mellitus. Proc Natl Acad Sci USA 89:7698–7702, 1992.
8. Gidh-Jain M, Takeda J, Xu LZ, et al. Glucokinase mutations associated with non-insulin-dependent (type 2) diabetes mellitus have decreased enzymatic activity: Implications for structure/function relationships. Proc Natl Acad Sci USA 90:1932–6, 1993.
9. Takeda J, Gidh-Jain M, Xu LZ, et al. Structure/function studies of human B-cell glucokinase. J Biol Chem 268:15200–204, 1993.
10. Hasellof J, Gerlach WL. Simple RNA enzymes with new and highly specific endoribonuclease activities. Nature 334:585–91, 1988.
11. Hughes SD, Quaade C, Milburn JL, et al. Expression of normal and novel glucokinaes mRNAs in anterior pituitary and islet cells. J Biol Chem 266:4521–30, 1991.
12. Huang MTF, Gorman CM. Intervening sequences increase efficiency of RNA 3′ processing and accumulation of cytoplasmic RNA. Nucleic Acids Res 18:937–47, 1990.
13. Hanahan D. Heritable formation of pancreatic B-cell tumours in transgenic mice expressing recombinant insulin simian virus 40 oncogenes. Nature 315:115–22, 1985.
14. Hogan B, Costantini F, Lacy E. Manipulating the mouse embryo. Cold Spring Harbor Laboratory Press, Cold Spring Harbor, NY, 1986.
15. Gotoh M, Maki T, Kiyoizumi T, et al. An improved method for isolation of mouse pancreatic cells. Transplantation 40:437–8, 1985.
16. Andreone TL, Printz RL, Pilkis SJ, et al. The amino acid sequence of rat liver glucokinase deduced from cloned cDNA. J Biol Chem 264:363–9, 1989.
17. Efrat S, Leiser M, Wu Y-J, et al. Ribozyme-mediated attenuation of pancreatic β-cell glucokinase expression in transgenic mice results in impaired glucose-induced insulin secretion. Proc Natl Acad Sci USA 91:2051–55, 1994.
18. Byrne M, Sturis J, Clement K, et al. Insulin secretory abnormalities in subjects with hyperglycemia due to glucokinase mutations. J Clin Invest 93:1120–30, 1994.
19. Sturis J, Kurland IJ, Byrne MM, et al. Compensation in pancreatic β-cell function in subjects with glucokinase mutations. Diabetes 43:718–23, 1994.
20. Sakura H, Kawamori R, Kubota M, et al. Glucokinase gene mutations and impaired glucose uptake by liver. Lancet 341:1532–33, 1993.
21. Bultman SJ, Michaud EJ, Woychik RP. Molecular characterization of the mouse agouti locus. Cell 71:1195–204, 1992.
22. Yen TT, Bue JM, Gill AM. The obese-diabetic syndrome of the viable yellow mouse and pharmacological interventions. In: Lessons from animals diabetes. Shafrir E, ed. Smith-Gordon, London, 3:294–9, 1990.

10

Consequences of human IAPP expression in transgenic mice

C. BRUCE VERCHERE, DAVID A. D'ALESSIO, AND STEVEN E. KAHN

Division of Metabolism, Endocrinology and Nutrition, Department of Medicine, University of Washington, Seattle, and Veterans Affairs Medical Center, Seattle, Washington 98108, USA.

Key words: islet amyloid, islet amyloid polypeptide, amylin, insulin, transgenic mice, non-insulin dependent diabetes, NIDDM.

Amyloid deposits occur in pancreatic islets in non–insulin-dependent diabetes mellitus (NIDDM) and insulinomas.[1–3] Islet amyloid may contribute to the progressive worsening of B-cell function that is observed in NIDDM,[4] since its presence is associated with decreased B-cell function in monkeys[5] and amyloid was recently demonstrated to cause death of pancreatic islet cells in vitro.[6] While it is known that the major constituent of islet amyloid is a 37 amino acid peptide termed *islet amyloid polypeptide* (IAPP or amylin[7,8]), it is not known why IAPP forms amyloid in NIDDM. The inherent amyloidogenicity of the peptide is thought to reside in the midsection of the molecule (amino acids 20–29), which is predicted to form beta-pleated sheets and does so in vitro.[9] Several species, including primates and cats, produce amyloidogenic forms of IAPP and are prone to islet amyloid development in NIDDM-like syndromes. In species in which non-amyloidogenic IAPP is produced, such as rats and mice, islet amyloid deposits are not observed. However, it appears that production of amyloidogenic IAPP is in itself insufficient for islet amyloid formation to occur, since dog IAPP contains the necessary amyloidogenic sequence, yet amyloid deposits have not been observed in hyperglycemic dogs.[10] This finding suggests that some fundamental alteration in B-cell function, perhaps one common to NIDDM and insulinomas,[11] must be present for IAPP to deposit as islet amyloid. Alternatively, it has been suggested that overproduction of amyloidogenic IAPP may lead to islet amyloid formation.[12]

Correspondence: C. Bruce Verchere, Ph.D., Veterans Affairs Medical Center (151), 1660 S. Columbian Way, Seattle, WA 98108, USA.

Lessons from Animal Diabetes VI. Ed. E. Shafrir.

The biological function of IAPP is also still unclear, although several studies have suggested that IAPP has effects on both insulin secretion and insulin sensitivity.[13] However, most studies have examined the effects of acute administration of IAPP, and there has been considerable controversy over whether the concentrations of IAPP administered were physiologically relevant.[13,14] In rodents, for example, it was found that fasting plasma IAPP levels vary widely, from 2 to 100 pM.[13] Using these concentrations as a guide, it is possible that a physiological role for IAPP exists based on studies that showed that acute administration of 75 pM IAPP was sufficient to inhibit insulin release from the perfused rat pancreas,[15] while 30 pM IAPP inhibited glucose uptake in perfused rat hindlimb muscle.[16] Because these studies suggest that IAPP might impair both insulin secretion and insulin sensitivity, and since both B-cell dysfunction and insulin resistance are characteristic of NIDDM, it has been proposed that excessive production and secretion of IAPP could contribute to the pathogenesis of this disease. Still, there have been no studies examining the effects of long-term administration of IAPP on carbohydrate metabolism.

In order to study the potential role of IAPP in the pathophysiology of NIDDM, we have developed a line of transgenic mice that express a human IAPP transgene in their pancreatic B-cells. Since human IAPP, unlike rodent IAPP, is amyloidogenic, these mice may provide a useful animal model for studying the mechanism of islet amyloid formation. In addition, since these mice chronically overproduce IAPP, they should also provide a model for examining the long-term metabolic effects of elevated plasma IAPP levels. With this in mind, we have used these mice to attempt to answer the following two questions: 1) Is (over)production of amyloidogenic IAPP sufficient for islet amyloid formation to occur?, and 2) Does chronic excess IAPP production lead to alterations in B-cell function?

Methods

Transgenic mice. A line of transgenic mice (RIP hIAPP 25-3) with B-cell expression of human IAPP (hIAPP) was generated at the Howard Hughes Medical Institute, University of Washington.[17] The transgene consists of a 600 bp fragment of the rat insulin II gene promoter (RIP II), a 676 bp *SmaI/BglI* fragment of hIAPP cDNA that encodes the entire human proIAPP sequence (kindly provided by Dr. D. Steiner, University of Chicago), and a fragment containing the 5′ untranslated region of the mouse MHC Eα class II gene encoding the polyadenylation sequence. This line of transgenic mice, which has been maintained by backcrossing of heterozygous transgenic mice to C57 BL/6 × DBA/2 mates (Simonsen Labs, Gilroy, CA), was used in all analyses reported here. All comparisons between transgenic and nontransgenic mice were made using sex- and age-matched pairs at 3–6 months of age, unless otherwise indicated.

Analysis of pancreatic RNA. RNA was extracted from frozen samples of mouse, human, and rat pancreas as described by Han et al.[18] ^{32}P-labeled riboprobes complementary to hIAPP, rat IAPP (rIAPP), and rat insulin I mRNA, and a 106-bp fragment of the mammalian 18S ribosomal RNA (rRNA) were synthesized (IAPP and insulin cDNA templates kindly provided by Dr. D. Steiner, University of Chicago; 18S ribosomal template purchased from Ambion, Austin, TX) and were used for Northern blot analysis of the pancreatic RNA.[17] IAPP and insulin riboprobes were labeled to a specific activity of $\sim 1 \times 10^9$ cpm/μg and 18S rRNA riboprobes to 1×10^6 cpm/μg with ^{32}[P]UTP (Amersham, Arlington Heights, IL). The amount of riboprobe added to each hybridization was estimated to be in excess of the immobilized RNA. The resulting autoradiographs were quantified by densitometric imaging (Image V.1.44 software; NIH, Bethesda, MD), and the signal intensity of IAPP and insulin mRNA for each sample was normalized for that of 18S rRNA.

Analysis of pancreatic and plasma peptide levels. Frozen pancreas samples from transgenic and nontransgenic mice were homogenized in 2 M acetic acid, boiled for 15 min, centrifuged ($3500 \times g$ for 15 min), and stored at −20°C until analysis by radioimmunoassay (RIA) or HPLC. Serum samples were drawn from the orbital sinus of fasted (4–6h) mice anesthetized with sodium pentobarbital. Prior to RIA for the presence of IAPP-like immunoreactivity (IAPP-LI), serum samples were extracted on SepPak C18 cartridges (Waters, Milford, MA) and dried by vacuum centrifugation in a SpeedVac Plus (Savant, Farmingdale, NY).

HPLC analysis of pancreatic plasma extracts. HPLC analysis of pancreatic extracts for the presence of hIAPP-LI and mouse IAPP-LI (mIAPP-LI) was performed on a C18 column (Phenomenex, Torrance, CA) using a Hewlett-Packard (Waldbronn, Germany) 1090 LC system. Prior to injection the extracts were purified on SepPak C18 cartridges, dried by vacuum centrifugation, and reconstituted in 0.1% trifluoroacetic acid (TFA). Human and mouse IAPP were separated on a linear gradient of 28–38% acetonitrile (plus 0.1% TFA) over 20 min. Synthetic rIAPP (which is identical to mIAPP[13]) and hIAPP (both from Bachem, Torrance, CA) were used as standards. Fractions were collected every 0.5 min (0.5 ml/fraction) into polystyrene tubes containing 100 μl Triton X-100 (0.5%). Each fraction was dried by vacuum centrifugation and stored at −20°C for subsequent IAPP-LI RIA.

Perfused pancreas. In vitro secretion of IAPP-LI and insulin immunoreactivity (IRI) was assessed using the isolated perfused pancreas. The technique for surgical isolation and vascular perfusion of the mouse pancreas was identical to that described for the rat,[19] with minor modifications for the mouse.[20] The perfusate was a Krebs-bicarbonate buffer containing 3% dextran (Sigma, St. Louis, MO), 0.1% bovine serum albumin (RIA grade; Sigma), and either 4.4 or 16.7 mM glucose. It was gassed with 95% O_2/5% CO_2 to maintain a pH of 7.4, heated to 37°C, and pumped through the pancreatic vasculature at a flow rate of 1 ml/min. Fractions of venous effluent were collected from the portal vein at 1 min intervals in polystyrene tubes containing 25 μl acetic acid (10 M). Pooled fractions containing 0.75–3.0 ml effluent were extracted on Bond Elut C18/OH columns (Varian, Harbor City, CA), dried by vacuum centrifugation, and stored at −20°C for RIA for IRI or IAPP-LI.

Radioimmunoassays. The RIAs for measurement of IAPP-LI, IRI and somatostatin-like immunoreactivity (SLI) have been previously described.[21,22] The antiserum used in the IAPP RIA recognizes mIAPP and hIAPP on an equimolar basis. Dried extracts of pancreas, plasma, or perfusion effluent were reconstituted in assay buffer prior to RIA.

Histochemical analysis. The histochemical techniques used to analyze mouse pancreas tissue for the presence of hIAPP-LI and islet amyloid have been described elsewhere.[17] Antiserum #5436, which recognizes hIAPP but not mIAPP, was used for immunostaining of human and transgenic mouse pancreas. Sections of pancreas from transgenic mice and nontransgenic mice were examined for the presence of islet amyloid by staining with thioflavin S, a classic histochemical technique for determining the presence of amyloid[23] that has been previously used to detect amyloid deposits in the brains of transgenic mice expressing the β-amyloid precursor protein (β-APP) of Alzheimer's disease.[24] As a positive control for thioflavin S staining of amyloid, we also stained sections of pancreas from diabetic monkeys (*Macaca nemestrina*), which have been previously shown to have islet amyloid deposits.[5,17]

Data analysis. All data are presented as mean $\pm$ SE. Statistical comparisons between transgenic and nontransgenic groups were performed using either the Mann-Whitney U test or Student's t-test. Pearson correlation coefficients were calculated for pairs of continuous variables. $P < 0.05$ was considered statistically significant.

Characterization of transgenic mice

To first establish that these mice expressed the hIAPP transgene in their pancreatic B-cells, extracts of pancreatic RNA from transgenic and nontransgenic mice were analyzed for the presence of hIAPP and mIAPP mRNA by Northern blot analysis. As shown in Figure 1A, a ^{32}P-labeled riboprobe complementary to hIAPP mRNA hybridized to RNA extracted from the pancreas of transgenic mice, producing a single band of $\sim$1.4 kb. The hIAPP mRNA probe also hybridized to extracts of human pancreas RNA, producing a band of 1.6 kb. However, the probe did not hybridize to pancreatic RNA obtained from nontransgenic mice or rats. Thus, the pancreas of these transgenic mice produce a transcript of the hIAPP cDNA that is similar in size to the hIAPP message found in human pancrease. In contrast, a riboprobe complementary to rIAPP mRNA did not hybridize to human pancreatic RNA, but did hybridize to pancreatic RNA from both transgenic and nontransgenic mice as well as from rats, producing a band of approximately 1.0 kb (Fig. 1B). This finding demonstrates that, as expected, the native (mouse) IAPP gene is also expressed by both transgenic and nontransgenic mice.

To determine whether the transgene is expressed in tissues other than pancreas, extracts of RNA from various tissues of transgenic mice were hybridized to the hIAPP riboprobe (Fig. 2). Surprisingly, marked hybridization of the hIAPP probe was observed to kidney RNA, suggesting that the transgene is also

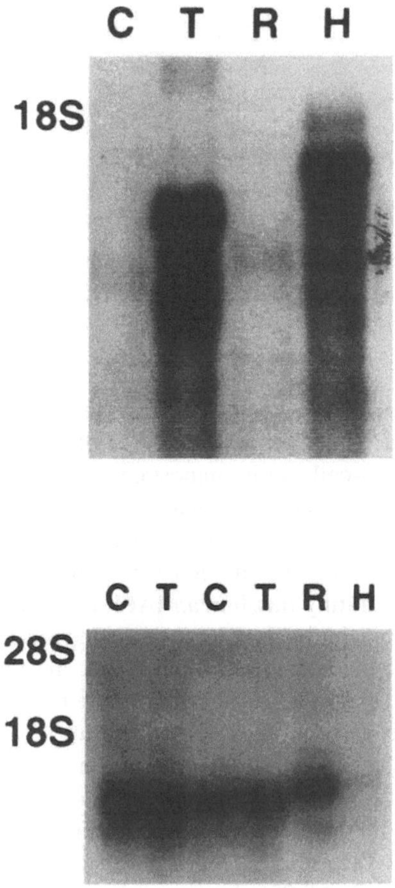

Figure 1. Northern blots of total RNA extracted from pancreata of control mouse (C), transgenic mouse (T), human (H), and rat (R) and hybridized to ^{32}P-labeled riboprobes complementary to (A) hIAPP and (B) rIAPP mRNA. Hybridization to the hIAPP riboprobe is seen only in transgenic mouse and human pancreas, whereas hybridization to the rIAPP riboprobe is seen only in rat, transgenic mouse, and nontransgenic mouse pancreas. (Reprinted with permission from ref. 17.)

being expressed in this tissue. However, the presence of hIAPP mRNA was not detectable in any of the other tissues tested in transgenic mice, including the lung, liver, stomach, and intestine (as shown in Fig. 2); and spleen, brain, heart, and skeletal muscle (data not shown). Since the transgene promoter used (rat insulin II promoter; RIP II) was intended to direct expression of the transgene to the pancreatic B-cell only, it is unclear why hIAPP mRNA is also present in the kidney. However, it is of interest that kidney expression of a similar transgene (hIAPP linked to the RIP II) has also been observed in other transgenic mouse lines.[25] Interestingly, Fox et al.[26] found expression of the hIAPP transgene in brain and pituitary, but not in the kidney. Although we have not specifically examined the pituitaries of our mice for hIAPP expression, our mice do not appear to express the transgene in brain. Such differences in transgene expression likely arise from differences in the construct promoter regions, since the rat insulin I gene promoter (used by Fox and colleagues) has been previously shown to direct transgene expression to brain as well as pancreas,[27] whereas RIP II has been found to direct transgene expression to the kidney.[28,29]

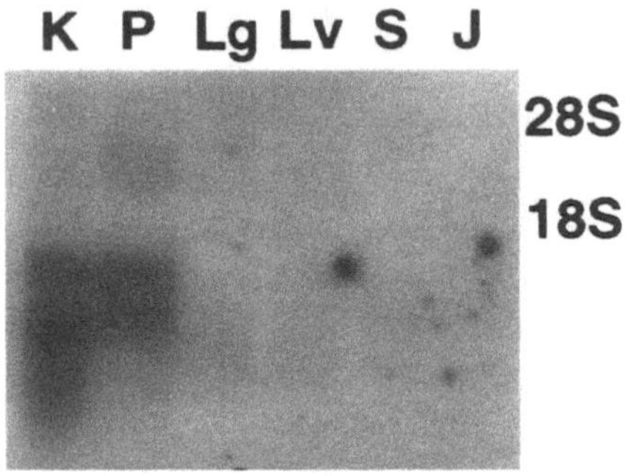

Figure 2. Northern blot of total RNA extracted from various tissues of a transgenic mouse, including kidney (K), pancreas (P), lung (Lg), liver (Lv), stomach (S), and jejunem (J). Hybridization to the hIAPP riboprobe is seen only in transgenic mouse pancreas and kidney.

To establish whether the transgenic message is translated into peptide in these animals and whether the source of this transgenic message is, in fact, the pancreatic B-cell, we immunostained sections of pancreas from transgenic and nontransgenic mice using an antiserum (#5436) that we have shown specifically recognizes hIAPP-LI.[17] As shown in Figure 3, the pancreatic islets of transgenic, but not nontransgenic, mice immunofluoresced when exposed to this antiserum, demonstrating that human IAPP is indeed being synthesized by the islets of the transgenic animals. Moreover, since this immunoreactivity appeared to be localized to the islet core, which is known to be rich in B-cells,[30] it is likely that the hIAPP expression is specific for the B-cell, although the possibility that the transgene is also being expressed in islet A- or D-cells has not yet been determined.

We next sought to determine whether increased levels of IAPP-LI were being produced by the islets of these transgenic mice by measuring IAPP-LI in extracts of whole pancreas. As shown in Figure 4A, pancreatic content of IAPP-LI was two- to threefold higher in the transgenic mice (6.6 ± 1.3 vs. 2.9 ± 0.3 fmol/μg protein; $P < 0.05$), demonstrating that the transgenic animals overproduce IAPP-LI. When pancreatic extracts of transgenic mice were analyzed by HPLC, a major peak of IAPP-LI was observed that co-eluted with synthetic hIAPP (Fig. 4B). In contrast, no hIAPP-LI was observed in pancreatic extracts of the nontransgenic animals. In addition, extracts from both groups of mice contained peaks of IAPP-LI that co-eluted with synthetic rodent IAPP on HPLC. Thus, the pancreatic islets of these transgenic mice produce both human and mouse IAPP, whereas those of the nontransgenic animals produce the native peptide only. Since transgenic and nontransgenic animals appear to produce similar amounts of mouse IAPP-LI (see magnitude of mouse IAPP-LI peaks in Fig. 4B), the two- to threefold excess pancreatic IAPP-LI seen in the transgenic mice appears to be comprised mainly of hIAPP-LI. This observation also suggests that the expression of the native gene is not significantly altered in the transgenic mice, an idea that is supported by our finding using RNA-dot hybridization that levels of mIAPP mRNA were similar in RNA extracts of pancreas from transgenic and nontransgenic animals.[17] Finally, our finding that a large peak of IAPP-LI co-eluted with synthetic hIAPP suggests that the transgenic propeptide (human proIAPP) is efficiently processed by the mouse B-cell to the 37 amino acid human peptide.

IAPP-LI was also measured in plasma of transgenic and nontransgenic mice.

Similar to what was observed in the pancreatic extracts of these mice, circulating IAPP-LI levels were elevated in 2.5-fold in the plasma of the transgenic animals (Fig. 5). This observation suggests that the B-cells of these mice not only produce excessive amounts of hIAPP, but that they also secrete excess amounts of the peptide, a finding further supported by our pancreatic perfusion data (see later).[20]

Characterization of these transgenic mice has thus shown that they produce excessive quantities of an amyloidogenic form of IAPP (hIAPP) in their pancreatic B-cells. Moreover, we have shown that their circulating levels of IAPP-LI are two- to threefold elevated over that of the nontransgenic mice. Therefore, these mice appear suitable for studying two divergent aspects of the possible role of IAPP in the pathogenesis of NIDDM: its role in islet amyloid formation, and its long-term effects on insulin sensitivity and secretion.

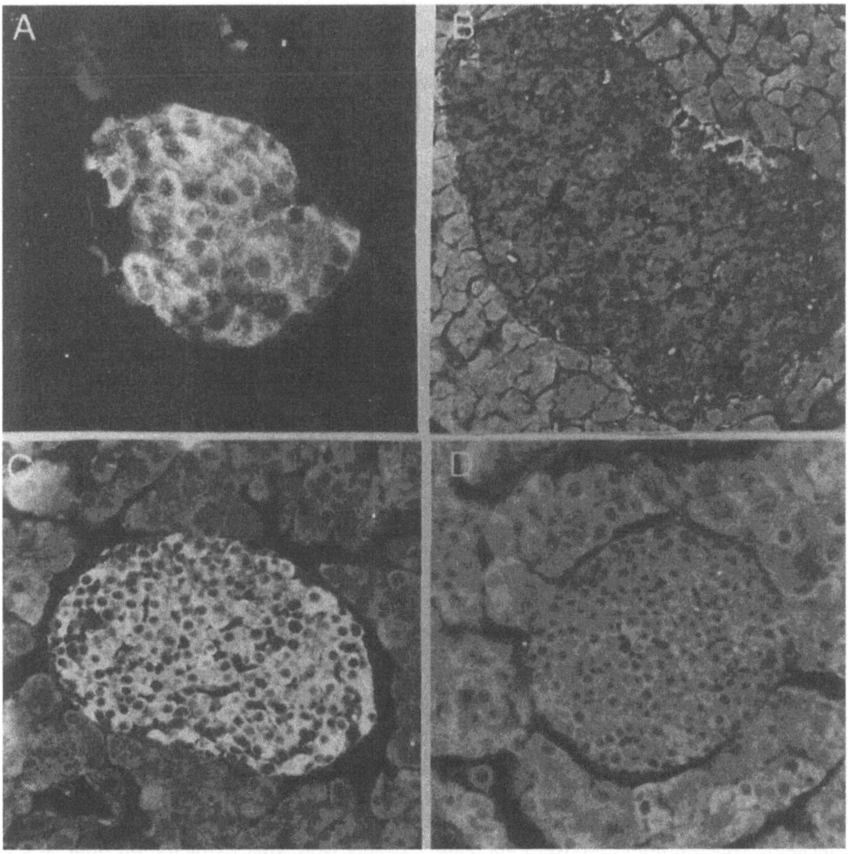

Figure 3. Immunohistochemical analysis of sections of pancreas from human (A), rat (B), transgenic (C), and nontransgenic (D) mouse using an antiserum (#5436) specific for hIAPP. Immunofluorescence is observed in pancreatic islet cells in human and transgenic mouse pancreas, but not rat or nontransgenic mouse pancreas (approximate magnification 100×). (Reprinted with permission from ref. 17.)

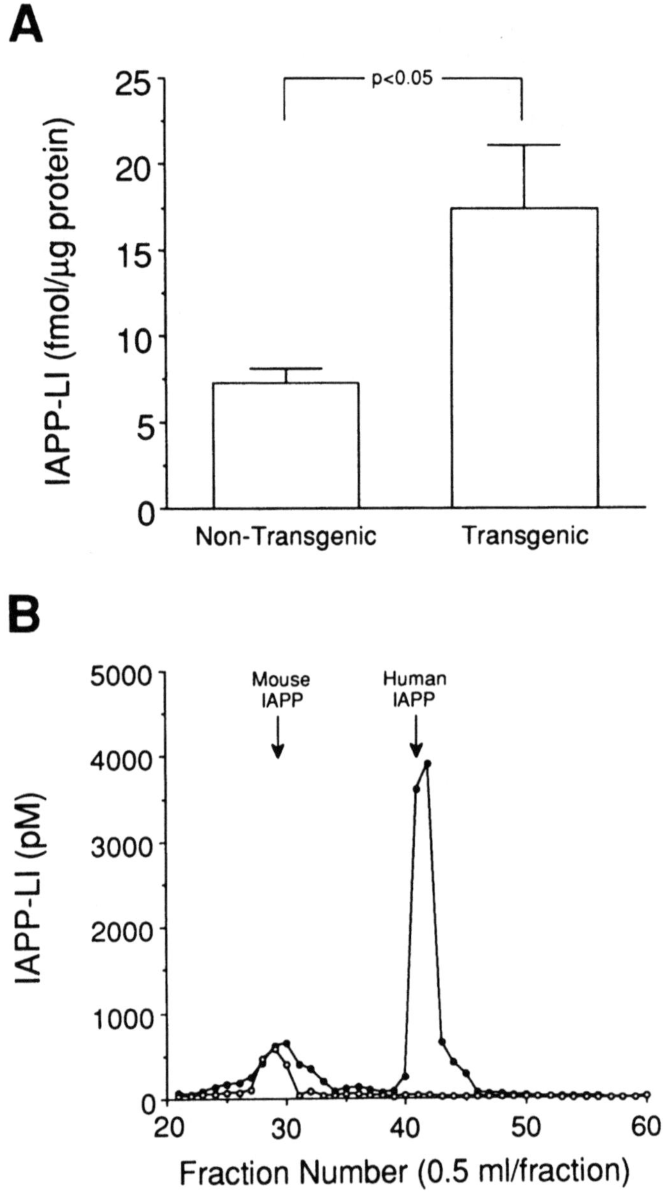

Figure 4. (A) Pancreatic content of IAPP-LI from 12 pairs of transgenic and nontransgenic mice. Data are mean ± SE. (Data adapted from ref. 17.) (B) Representative chromatographs of IAPP-LI in pancreatic extracts from a transgenic (solid circles) and a nontransgenic (open circles) mouse. Pancreatic extracts were first extracted on SepPak C18 cartridges, dried, and reconstituted in 0.1% TFA. Equivalent amounts of pancreatic protein (400 μg) were injected from the transgenic and nontransgenic pancreatic extracts. Peptides were eluted by reverse-phase HPLC on a gradient of 28–38% acetronitrile (0.1% TFA) over minutes 5–25; fractions were collected every 0.5 min. hIAPP and mIAPP were detected by an RIA that detects hIAPP and mIAPP on an equimolar basis.

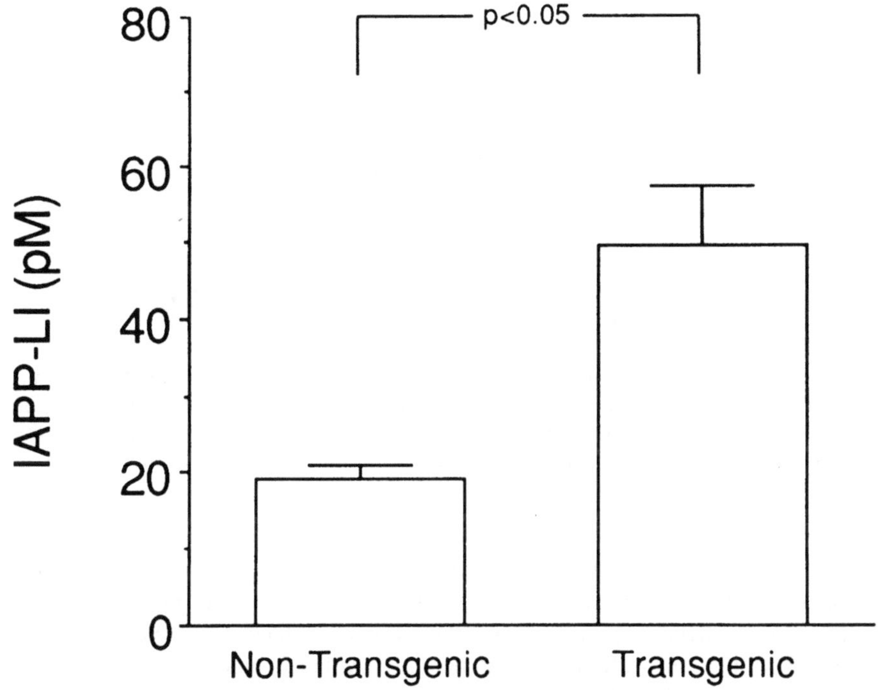

Figure 5. Plasma concentrations of IAPP-LI in transgenic (n = 9) and nontransgenic mice (n = 8) following a 4–6 h fast. Data are means ± SE.

Overproduction of amyloidogenic IAPP and islet amyloid formation

To test the hypothesis that overproduction of amyloidogenic IAPP is sufficient for islet amyloid formation, pairs of heterozygous transgenic mice and age- and sex-matched nontransgenic animals were sacrificed at various ages (6–19 months) and their pancreata were examined for the presence of islet amyloid by thioflavin S staining. At least four sections from each mouse pancreas were obtained, and at least 10 islets were examined from each section. No islet amyloid deposits were observed in any samples of mouse pancreas, including those of three transgenic mice aged 19 months (Fig. 6A). In contrast, when sections of pancreas from a diabetic monkey (*Macaca nemestrina*) were stained with thioflavin S, numerous islets were found to have extensive amyloid deposits (Fig. 6B). This finding is consistent with that from other groups also studying lines of transgenic mice with B-cell expression of a hIAPP transgene[25,26] and with data from humans. Although some amyloid deposits are seen in samples of pancreas obtained from older, nondiabetic humans, extensive islet amyloidosis is only observed in individuals with NIDDM or insulinomas.[1] Together, these observations suggest that simple production of an amyloidogenic form of IAPP is insufficient for islet amyloid formation to occur during the life span of a mouse.

To further increase hIAPP production in transgenic animals, we bred pairs of

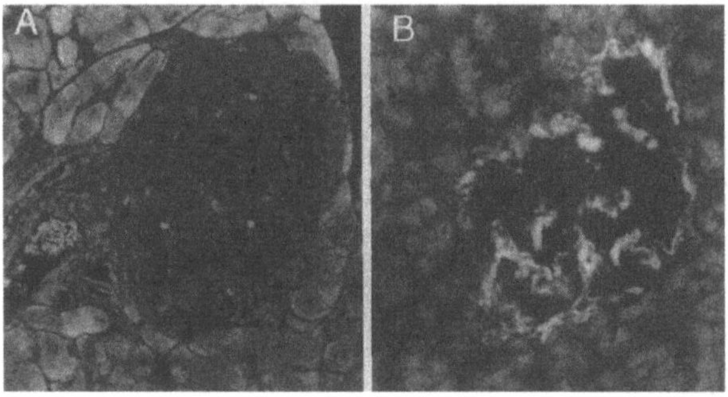

Figure 6. Thioflavin S stain of sections of pancreas from a 12-month-old transgenic mouse (A) and a diabetic monkey (*Macaca nemestrina;* B). Note extensive staining of material in the monkey islet, particularly near blood vessels and in the periphery of the islet. No staining is visible in the mouse pancreas (approximate magnification 100×).

heterozygous transgenic mice to produce mice homozygous for the transgene. These homozygous animals would be predicted to have twice as many copies of the transgene as heterozygous mice and thus should produce even greater amounts of hIAPP. Consistent with this idea, pancreatic IAPP-LI content in these animals was two-fold that of the heterozygous mice (24.5 ± 7.3 vs. 12.9 ± 1.9 fmol/μg protein; $P = 0.06$; Fig. 7). To date, pancreata of six homozygous mice 12–16 months of age have been examined following thioflavin S staining, and none have been found to have islet amyloid. In a second experiment, we administered nicotinic acid (50 μg/day) via the drinking water to three heterozygous transgenic mice to produce experimental insulin resistance.[31] Since the transgene in these mice is driven by the insulin promoter and since insulin synthesis is known to increase as a compensatory mechanism in the face of insulin resistance,[32] it would be predicted that experimental insulin resistance would increase B-cell synthesis of hIAPP. Despite a marked increase in pancreatic IAPP-LI content over that of untreated transgenic mice (46.9 ± 19.6 vs. 12.9 ± 1.9 fmol/μg protein; $P < 0.005$; Fig. 7), after 11 months of treatment no islet amyloid deposits were clearly detectable in the pancreata of the nicotinic acid–treated mice. Taken together, these findings suggest that overproduction of the amyloidogenic peptide will not lead to islet amyloid formation in vivo in a mouse lifetime.

Similar observations have been obtained in other laboratories studying different lines of mice expressing a hIAPP transgene.[25,26] In some lines of hIAPP-transgenic mice, Hoppener and colleagues[25,33] reported plasma levels of IAPP-LI more than 10-fold higher than those of the nontransgenic animals, yet no amyloid was found in sections of pancreas from these mice, further supporting the concept that excessive hIAPP production is not in itself sufficient for the development of islet amyloid within 1 year. Although electron microscopic examination of pancreatic tissue from these mice failed to detect the presence of amyloid fibrils, accumulation of nonfibrillar IAPP-immunoreactive material was found between pancreatic B-cells and capillaries in hIAPP-transgenic mice, but not in nontransgenic mice or mice expressing a rat IAPP transgene.[33] The authors suggested that this material may represent the very early stages of islet amyloid formation. Thus, it remains possible that the relatively short life span of a mouse (~2 years) compared with humans does not allow sufficient time for amyloidosis to occur. However, it should be noted that extensive islet amyloidosis has been observed in diabetic monkeys and cats at only 3–5 years of age.[34,35] Moreover, in vitro formation of amyloid has been shown to occur rapidly (minutes to hours) when human IAPP (or other amyloidogenic proteins, such as the β-amyloid protein of Alzheimer's disease) are incubated at high concentrations in solution.[6,36] Further, de Koning et al.[37] recently showed that amyloid formation occurs in isolated islets from hIAPP-transgenic mice after only 7 days of culture in high glucose, suggesting that under some experimental conditions, islet amyloid formation may occur very rapidly. However, it must be remembered that islets are likely to be exposed to very different conditions in vivo where mechanisms (e.g., macrophages[33]) that protect against amyloid formation may also be operating.

Since the present data suggest that overproduction of hIAPP is not a short-term factor in islet amyloid deposition, it is possible that a defect in the intra-

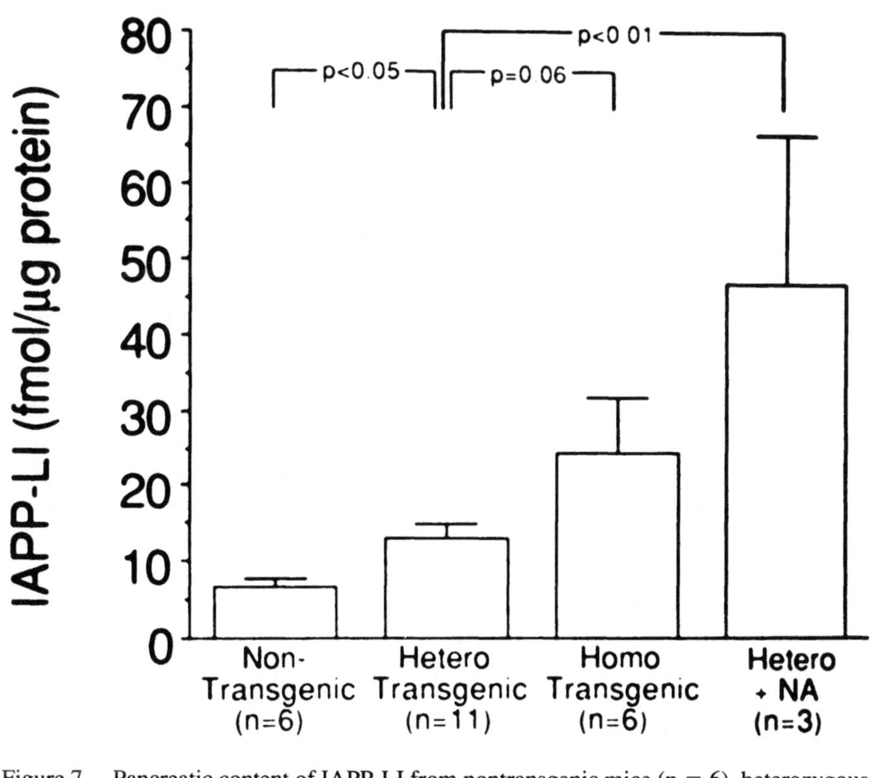

Figure 7. Pancreatic content of IAPP-LI from nontransgenic mice (n = 6), heterozygous transgenic mice (n = 11), homozygous transgenic mice (n = 11), and heterozygous transgenic mice treated with nicotinic acid (NA; n = 3). Data are means ± SE.

cellular handling of the peptide by the diabetic B-cell is a contributor to amyloid formation. Such a defect might occur at one or more of several levels of peptide handling by the B-cell, including sorting,[38] processing,[11,39] storage, secretion,[38] or degradation.[12] Since amyloid deposits are also seen in insulinomas, which share characteristics with the B-cell in NIDDM such as decreased responsiveness to glucose and increased secretion of proinsulin, we have suggested that B-cells in NIDDM and insulinomas share a common defect in peptide handling that leads to amyloid deposition.[11] However, it is also possible that hyperglycemia per se may contribute to amyloid formation, since glucose appears to increase the rate of amyloid formation in hIAPP-transgenic mouse islets in vitro.[37] Further studies aimed at inducing islet amyloid formation in hIAPP-transgenic mice both in vitro and in vivo should help us gain a better understanding of the mechanism(s) underlying amyloid deposition in NIDDM.

Metabolic consequences of chronic overproduction of IAPP in transgenic mice

Since acute administration of IAPP has been suggested to impair both insulin sensitivity and secretion, we hypothesized that the long-term excess circulating IAPP-LI observed in our transgenic mice might lead to changes in B-cell synthesis and secretion of insulin. As shown in Fig. 8A, pancreatic IRI content was 1.5-fold elevated in the transgenic mice compared with control animals. This increased IRI content appears to be linked to the increased IAPP production by these mice, since there was a highly significant correlation between the pancreatic IRI content and that of IAPP-LI in the transgenic animals ($r = 0.864$; $P < 0.001$; Fig. 8B). Further, this change was specific for the B-cell, since pancreatic SLI content was not significantly elevated in the transgenic mice (3.6 ± 0.5 vs. 3.2 ± 0.4 fmol/μg protein). To determine whether the increase in pancreatic insulin content arose from an increase in insulin synthesis by the B-cell, we examined levels of proinsulin mRNA in the pancreata of transgenic and nontransgenic animals. Figure 9A shows a Northern blot of pancreatic RNA from five pairs of mice from each group probed for proinsulin mRNA and 18S rRNA. When the intensity of the proinsulin mRNA band from these mice was analyzed by densitometric imaging and normalized to their respective 18S bands, a significantly higher signal intensity was observed in the transgenic animals (Fig. 9B). Thus, the transgenic mice have increased pancreatic content of both proinsulin mRNA and insulin, suggesting that insulin synthesis is elevated in mice overproducing IAPP.

To determine whether IAPP-LI and IRI secretion increased in parallel with the increase in B-cell synthesis of these two peptides, we measured their secretion from the isolated, perfused pancreas preparation. Similar to the increases in pancreatic IAPP-LI and IRI content, the secretion of both these peptides was significantly elevated in the transgenic mice. Thus, both basal (193 ± 26 vs. 72 ± 12 fmol/15 min; $P < 0.05$; Fig. 10A inset) and 16.7 mM glucose-stimulated (1234 ± 83 vs. 767 ± 215 fmol/40 min; $P < 0.05$; Fig. 10A) IAPP-LI release were greater in the transgenic animals. Similarly, both basal (17.3 ± 4.9 vs. $6.2 \pm$

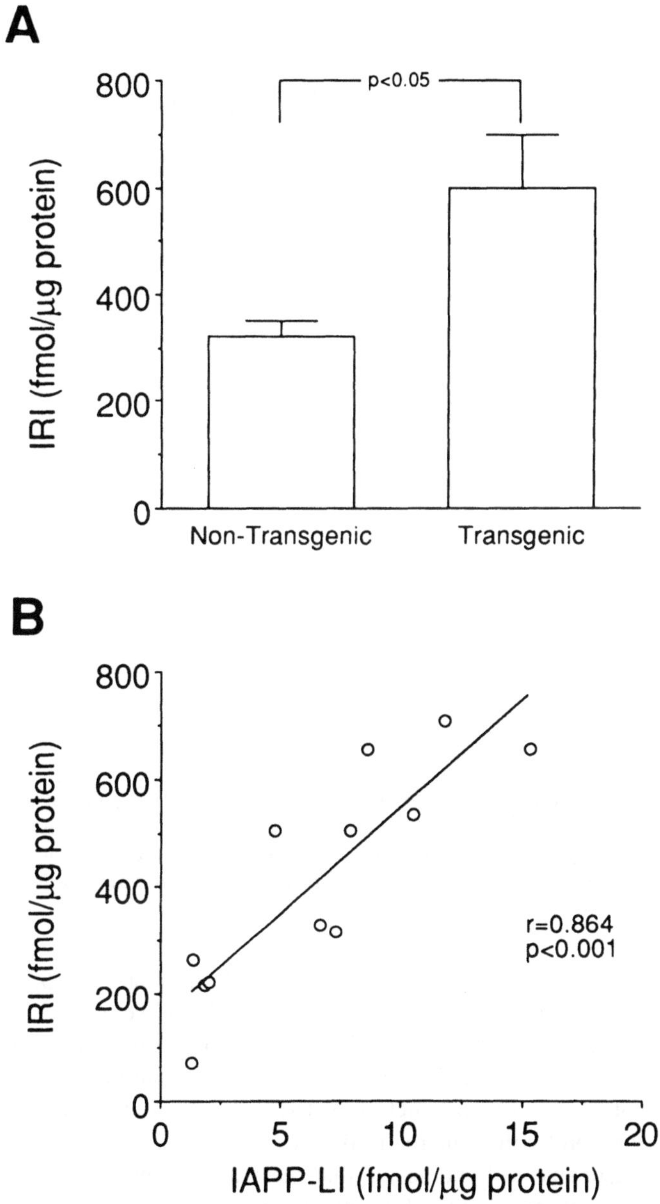

Figure 8. (A) Pancreatic content of IRI from 12 pairs of transgenic and nontransgenic mice. Data are mean ± SE. (Data adapted from ref. 17.) (B) Linear relationship between pancreatic IRI content and pancreatic IAPP-LI content in 12 transgenic mice ($r = 0.864$; $P < 0.001$).

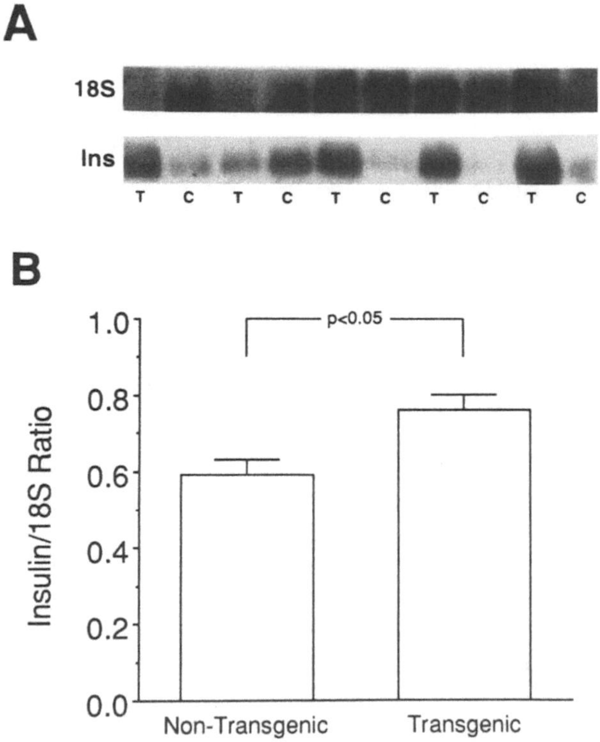

Figure 9. (A) Northern blot of total RNA extracted from five pairs of age and sex-matched transgenic (T) and control (C) mice and hybridized to a ^{32}P-labeled riboprobe for rat insulin. (B) Quantification of insulin mRNA in transgenic and control mice from Northern blot in (A). The signal intensity of insulin mRNA in each mouse pancreas is normalized to that of 18S rRNA.

1.4 pmol/15 min; $P < 0.05$; Fig. 10B inset) and stimulated (1152 ± 330 vs. 416 ± 93 pmol/40 min; $P < 0.05$; Fig. 10B) IRI release were also significantly elevated compared with control mice. These data indicate that hIAPP-transgenic mice not only synthesize but also secrete more insulin than nontransgenic mice, suggesting that chronic overproduction of IAPP results in increased insulin synthesis and secretion by the pancreatic B-cell.

It seems reasonable to hypothesize that these alterations in insulin synthesis and secretion result from some effect of the high circulating levels of IAPP-LI on muscle and/or the pancreatic B-cell, two tissues that have been suggested to be sensitive to the actions of IAPP.[13] For example, IAPP may be acting directly on the B-cell to stimulate B-cell growth and/or insulin synthesis. However, most studies suggest that the direct effect of IAPP on the B-cell, if any, is inhibitory, not stimulatory.[13,15] A more likely explanation is that insulin synthesis and secretion are increased as an adaptation to IAPP-induced insulin resistance, since IAPP has been proposed to be a noncompetitive antagonist of insulin on muscle.[16,40] Increased insulin secretion has commonly been observed as an adapta-

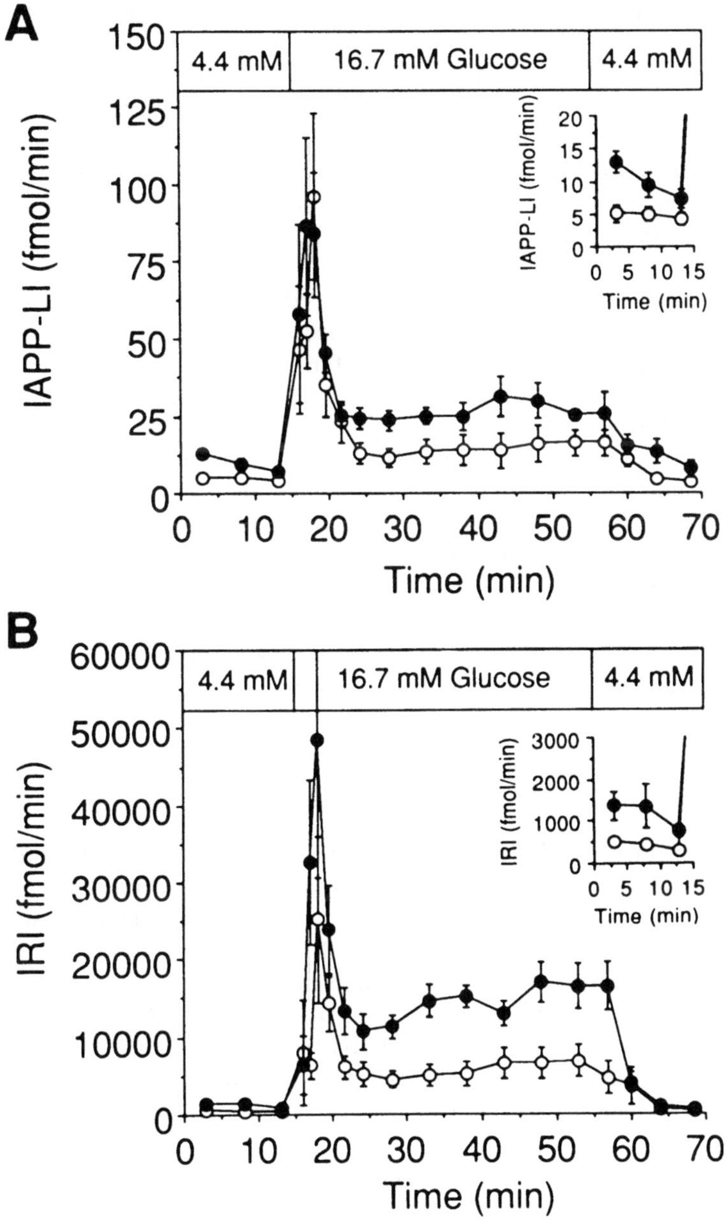

Figure 10. Secretion of (A) IAPP-LI and (B) IRI from the isolated, perfused pancreas of five pairs of transgenic (solid circles) and nontransgenic (open circles) mice in response to 16.7 mM glucose. Insets: (A) IAPP-LI and (B) IRI secretion during basal (4.4 mM) glucose perfusion. Each point represents the mean ± SE. (Reprinted with permission from ref. 20.)

tion to insulin resistance in humans[32] as well as in animal models of insulin resistance such as the obese Zucker rat,[41] or the desert gerbil *Psammomys obesus*.[42] However, although elevated fasting plasma insulin levels are a commonly observed manifestation of insulin resistance,[32] we have found no significant difference between these transgenic mice and controls in fasting plasma insulin and glucose concentrations,[17] an observation consistent with reports by other groups studying similar lines of hIAPP-transgenic mice.[25,26] Further studies will likely reveal the cause of the increased insulin synthesis and secretion we observed in our mice.

In summary, we and others have developed lines of transgenic mice with B-cell expression of hIAPP. The data obtained from these mice suggest that simple overproduction of an amyloidogenic form of IAPP by the pancreatic B-cell is insufficient for islet amyloid formation in the short term, suggesting that some other B-cell defect is likely to be present in human NIDDM. Additional studies have demonstrated that chronic IAPP overproduction is associated with increased insulin synthesis and secretion, compatible with the idea that IAPP may play a physiological role in the regulation of carbohydrate metabolism. Thus, hIAPP-transgenic mice appear to be promising models for the study of the potential role of IAPP in the pathogenesis of NIDDM.

Note added in proof: Since submission of this manuscript, we have observed the development of islet amyloid deposits in 81% of our male hIAPP-transgenic mice greater than 1.5 years.[43] The formation of these deposits was associated with an increarse in dietary fat and an increased frequency of hyperglycemia.

Acknowledgments

The authors thank Dr. Daniel Porte, Jr. for helpful suggestions and critical reading of this manuscript and Dr. Richard Palmiter for generation of the transgenic mice. The authors also wish to acknowledge the expert technical assistance of Maggie Abrahamson, Vicki Hoagland, Richard Chan, Robin Vogel, Ruth Hollingsworth, Ellen Laschansky, Hong Nguyen, and Stein Wang. Care of the mice was supervised by Cheryl Echlepp, D.V.M. These studies were supported by National Institute of Health grants DK-12829 and DK-35816, and by the Medical Research Service of the Department of Veterans Affairs. C.B.V. was supported by a Medical Research Council of Canada Post-Doctoral Fellowship.

References

1. Westermark P, Wilander E, Westermark GT, Johnson KH. Islet amyloid polypeptide-like immunoreactivity in the islet B cells of type 2 (non-insulin-dependent) diabetic and non-diabetic individuals. Diabetologia 30:887–92, 1987.
2. Clark A, Saad MF, Nezzer T, et al. Islet amyloid polypeptide in diabetic and non-diabetic Pima Indians. Diabetologia 33:285–9, 1990.
3. Westermark P, Grimelius L, Polak JM, et al. Amyloid in polypeptide hormone producing tumors. Lab Invest 37:212–15, 1977.
4. Porte D Jr. β-cells in type II diabetes mellitus. Diabetes 40:166–80, 1991.

5. Howard CF Jr. Longitudinal studies on the development of diabetes in individual *Macaca nigra*. Diabetologia 29:301–6, 1986.
6. Lorenzo A, Razzaboni B, Weir GC, Yankner BA: Pancreatic islet cell toxicity of amylin associated with type 2 diabetes mellitus. Nature 368:756–60, 1994.
7. Westermark P, Wernstedt C, Wilander E, et al. Amyloid fibrils in human insulinoma and islets of Langerhans of the diabetic cat are derived from a neuropeptide-like protein also present in normal islets. Proc Natl Acad Sci USA 84:3881–85, 1987.
8. Cooper GJS, Willis AC, Clark A, et al. Purification and characterization of a peptide from amyloid-rich pancreases of type 2 diabetic patients. Proc Natl Acad Sci USA 84:8628–32, 1987.
9. Westermark P, Engstrom U, Johnson KH, et al. Islet amyloid polypeptide: Pinpointing amino acid residues linked to amyloid fibril formation. Proc Natl Acad Sci USA 87:5036–40, 1990.
10. Jordan K, Murtaugh MP, O'Brien TD, et al. Canine IAPP cDNA sequence provides important clues regarding diabetogenesis and amyloidogenesis in type 2 diabetes. Biochem Biophys Res Commun 169:502–8, 1990.
11. Porte D Jr, Kahn SE. Hyperproinsulinemia and amyloid in NIDDM: Clues to etiology of islet β-cell dysfunction? Diabetes 38:1333–36, 1989.
12. Steiner DF, Ohagi S, Nagamatsu S, et al. Is islet amyloid polypeptide a significant factor in pathogenesis or pathophysiology of diabetes? Diabetes 40:305–9, 1991.
13. Cooper GJS. Amylin compared with calcitonin gene-related peptide: Structure, biology, and relevance to metabolic disease. Endocr Rev 15:163–99, 1994.
14. Westermark P, Johnson KH, O'Brien TD, Betsholtz C: Islet amyloid polypeptide—a novel controversy in diabetes research. Diabetologia 35:297–303, 1992.
15. Degano P, Silvestre RA, Salas M, et al. Amylin inhibits glucose-induced insulin secretion in a dose-dependent manner. Study in the perfused rat pancreas. Regul Pept 43:91–6, 1993.
16. Young AA, Gedulin B, Wolfe-Lopez D, et al. Amylin and insulin in rat soleus muscle: Dose responses for cosecreted noncompetitive antagonists. Am J Physiol 263:E274–81, 1992.
17. D'Alessio DA, Verchere CB, Kahn SE, et al. Pancreatic expression and secretion of human IAPP in a transgenic mouse. Diabetes 43:1457–61, 1994.
18. Han J, Stratowa C, Rutter W. Isolation of full-length putative lysophospholipase cDNA using improved methods for mRNA isolation. Biochemistry 26:1617–25, 1987.
19. Verchere CB, Kwok YN, Brown JC. Inhibitory actions of a somatostatin analog, SMS 201-995, on pancreatic and gastric endocrine secretion. Pharmacology 46:54–60, 1993.
20. Verchere CB, D'Alessio DA, Palmiter RD, Kahn SE. Transgenic mice overproducing human islet amyloid polypeptide have increased insulin storage and secretion in vitro. Diabetologia 37:725–8,1994.
21. Ensinck JW, Laschansky EC, Vogel RE, et al. Circulating prosomatostatin-derived peptides. Differential responses to food ingestion. J Clin Invest 83:1580–89, 1989.
22. Kahn SE, Fujimoto WY, D'Alessio DA, et al. Glucose stimulates and potentiates islet amyloid polypeptide secretion by the B-cell. Horm Metab Res 23:577–80, 1991.
23. Schwartz P: Amyloidosis: Cause and manifestation of senile dementia. Charles C. Thomas, Springfield, IL, 1970.
24. Games D, Adams D, Alessandrini R, et al. Alzheimer-type neuropathology in transgenic mice overexpressing V717F β-amyloid precursor protein. Nature 373:523–7, 1995.
25. Hoppener JWM, Verbeek JS, de Koning EJP, et al. Chronic overproduction of islet amyloid polypeptide/amylin in transgenic mice: Lysosomal localization of human islet amyloid polypeptide and lack of marked hyperglcaemia or hyperinsulinaemia. Diabetologia 36:1258–65, 1993.
26. Fox N, Schrementi J, Nishi M, et al. Human islet amyloid polypeptide transgenic

mice as a model of non-insulin dependent diabetes mellitus (NIDDM). FEBS Lett 323:40–4, 1993.
27. Dandoy-Dron F, Monthioux E, Jami J, Bucchini D. Regulatory regions of rat insulin I gene necessary for expression in transgenic mice. Nucleic Acids Res 19:4925–30, 1991.
28. Lo D, Burkly LC, Widera G, et al. Diabetes and tolerance in transgenic mice expressing class II MHC molecules in pancreatic beta cells. Cell 53:159–68, 1988.
29. Roman LM, Simons LF, Hammer RE, et al. The expression of influenza virus hemagglutinin in the pancreatic beta cells of transgenic mice results in autoimmune diabetes. Cell 61:383–96, 1990.
30. Orci L. Macro- and micro-domains in the endocrine pancreas. Diabetes 31:538–65, 1982.
31. Kahn SE, Beard JC, Schwartz MW, et al. Increased β-cell secretory capacity as mechanism for islet adaptation to nicotinic acid-induced insulin resistance. Diabetes 38:562–8,1989.
32. Kahn SE, Prigeon RL, McCulloch DK, et al. Quantification of the relationship between insulin sensitivity and β-cell function in human subjects. Evidence for a hyperbolic function. Diabetes 42:1663–72, 1993.
33. de Koning EJP, Hoppener JWM, Verbeek JS, et al. Human islet amyloid polypeptide accumulates at similar sites in islets of transgenic mice and humans. Diabetes 43:640–44, 1994.
34. Johnson KH, Stevens JB: Light and electron microscopic studies of islet amyloid in diabetic cats. Diabetes 22:81–90, 1973.
35. Howard CF Jr. Insular amyloidosis and diabetes mellitus in *Macaca nigra*. Diabetes 27:357–64, 1978.
36. Lorenzo A, Yankner BA. Beta-amyloid neurotoxicity requires fibril formation and is inhibited by congo red. Proc Natl Acad Sci USA 91:12243–247, 1994.
37. de Koning EJP, Morris ER, Hofhuis FMA, et al. Intra- and extracellular amyloid fibrils are formed in cultured pancreatic islets of transgenic mice expressing human islet amyloid polypeptide. Proc Natl Acad Sci USA 91:8467–71, 1994.
38. Kahn SE, Verchere CB, D'Alessio DA, et al. Evidence for selective release of rodent islet amyloid polypeptide through the constitutive secretory pathway. Diabetologia 36:570–3, 1993.
39. Westermark P, Engstrom U, Westermark GT, et al. Islet amyloid polypeptide (IAPP) and pro-IAPP immunoreactivity in human islets of Langerhans. Diabetes Res Clin Pract 7:219–26, 1989.
40. Leighton B, Cooper GJS. Pancreatic amylin and calcitonin gene-related peptide cause resistance to insulin in skeletal muscle in vitro. Nature 335:632–5, 1988.
41. Kuffert A, Stern JS, Curry DL. Pancreatic hypersensitivity to glucose by young obese Zucker rats (fa/fa). Metabolism 37:952–7, 1988.
42. Shafrir E. Diabetes in Animals. In: Ellenberg and Rifkin's Diabetes Mellitus, 5th ed. Porte D Jr, Sherwin R, eds. Appleton & Lange, Norwalk, CT, 1996, in press.
43. Verchere CB, D'Alessio DA, Palmiter RD, et al. Islet amyloid formation associated with hyperglycemia in transgenic mice with beta cell expression of human islet amyloid polypeptide. Proc Natl Acad Sci USA 93:3492–6, 1996.

11

Immunostimulatory versus immunosuppressive roles of IL-10 in IDDM: Analysis with IL-10–producing transgenic NOD mice

MITSUO ITAKURA,[1] MAKI MORITANI,[1] KATSUHIKO YOSHIMOTO,[1] CHIKARA HASHIMOTO,[2] SATOSHI OTSUKA,[1] TOSHIAKI SANO,[3] AND JUN-ICHI MIYAZAKI[2]

[1]*Otsuka Department of Clinical and Molecular Nutrition, School of Medicine, The University of Tokushima, Tokushima, 770;* [2]*Department of Molecular Embryology Institute of Development, Aging and Cancer, Tohoku University, Sendai, 980-77 Japan;* [3]*Department of Pathology, School of Medicine, The University of Tokushima, Tokushima, 770 Japan.*

Key words: interleukin 10, NOD mice, IDDM, islet A-cells, islet B-cells, glucagon promoter, transgenic mice, reverse transcription-polymerase chain reaction.

Insulin-dependent diabetes mellitus (IDDM, type I diabetes) is an autoimmune disease resulting from autoimmune destruction of pancreatic islet B-cells. The models of human IDDM, non-obese diabetic (NOD) mice and Bio Breeding (BB) rats, have provided a large amount of information on the pathogenesis of IDDM. In these animal models, immune cells, which include helper T ($CD4^+$) and cytotoxic T ($CD8^+$) lymphocytes, B-lymphocytes, and macrophages, infiltrate pancreatic islets (insulitis) preceding islet B-cell destruction, with consequent overt diabetes and ketosis. Insulitis and diabetes in NOD mice or BB rats can be induced by adoptive transfer of lymphocytes and can be prevented by the immunosuppressive agents. Thus, insulitis and the consequent diabetes in NOD mice are caused by the cell-mediated organ-specific autoimmune mechanism. Although it is clear that the destructive process of islet B-cells requires both $CD4^+$ and $CD8^+$ T-cell subsets, the relative contribution of each subset in provoking diabetes is uncertain.

In addition to the role of immune cells, cytokines and polypeptide products produced by immune cells recognizing autoantigens of islet B-cells and by other surrounding cells are also recognized as major regulators of immune and inflam-

Correspondence: Mitsuo Itakura, MD, PhD, Otsuka Department of Clinical and Molecular Nutrition, School of Medicine, The University of Tokushima, Kuramoto, Tokushima-city, 770 Japan.

Lessons from Animal Diabetes VI. Ed. E. Shafrir.

matory responses. Because different Th cell subsets secrete different cytokines, and because the systemic cytokine administration and local in situ transgenic expression of certain cytokines change the natural course of autoimmune insulitis and diabetes in NOD mice, induction or progression of diabetes mediated by $CD4^+$ and $CD8^+$ T-cells may well be regulated through induction of cytokines in various cells. The immunological signals of self-reacting $CD4^+$ T-cells are thus partly, if not totally, regulated by various cytokines in autocrine or paracrine fashions.

In the following sections, the immunopathogenic roles of cytokines, especially interleukin (IL)-10, in IDDM in various animal models, and the results of our study using transgenic NOD mice expressing IL-10 in islet A-cells (NOD-IL-10) are presented.

General aspects of immune regulation by Th1 and Th2 cells

Antigen-activated $CD4^+$ T-cells are termed *Th cells.* At least two distinct cell types, Th1 and Th2, have been described, both in murine[1] and human[2] Th cells. This classification is based on the pattern of secreted cytokines in response to antigen stimulation. Although these subsets were first identified in the in vitro analysis of murine T-cells, similar subsets were reported to exist in vivo in mice, rats, or humans. Th1 cells produce and secrete IL-2, IFN-γ, and TNFβ (lymphotoxin, LT),[1,3–5] whereas Th2 cells produce and secrete IL-4, -5, -6, -10, and -13 (P-600).[3,6] Both subsets of Th cells produce cytokines, such as tumor necrosis factor (TNF)α, IL-3, and granulocyte-macrophage colony stimulating factor (GM-CSF). The third category of Th cells (Th0) is characterized by production of cytokines of both Th1 and Th2 types, including IL-2, IL-3, IFN-γ, IL-4, IL-5, IL-10, and GM-CSF. Th0 cells are thought to be obligatory precursors of Th1 and Th2 cells.[2] The different patterns of cytokine production in Th1 and Th2 cells are paralleled by their markedly different functions.

Two immune responses, cell-mediated immunity, which induces delayed-type hypersensitivity (DTH), and antibody-mediated immunity, are mutually exclusive.[7,8] Th1 cells are associated with cell-mediated immunity in vivo.[4,9] Th1 cells destroy the infected cells with intracellular pathogens. IL-2 or IFN-γ that are produced by Th1 cells activate macrophages. Activated macrophages and $CD8^+$ cells that recognize major histocompatibility complex (MHC) molecule–associated antigens on APC eliminate the antigen-bearing cells. In contrast, Th2 cells stimulate humoral immune responses, mainly through producing such cytokines as IL-4, IL-5, and IL-10. Th2 cells thus stimulate immunoglobulin (Ig) production and inhibit cell-mediated DTH responses.

Because responses of Th1/Th2 cells appear to be mutually inhibitory, especially in strong immune responses, the cross-regulation of Th1/Th2 cells is believed to determine the consequent immune responses. Cross-inhibitory regulators of the Th1/Th2 balance include several cytokines: IFN-γ inhibits the proliferation of Th2 cells,[10,11] whereas IL-4 and IL-10 inhibit IFN-γ production or proliferation of Th1 cells.[12,13] The patterns of cytokines secreted from Th1 and Th2 cells determine activation or inactivation of certain immune responses, and the ultimate direction of immune responses of either cell-mediated or antibody-mediated type. Thus, the Th1/Th2 balance is supported to influence the type of immune responses, as reported in infections of various pathogens.[14,15] The relation of the Th1/Th2 balance with IDDM and factors affecting destruction of islet B-cells and appearance of diabetes in NOD mice are schematically illustrated in Figures 1 and 2. These schemes supply a simplified central dogma or a current working hypothesis on the pathogenesis in NOD mice, which are no doubt subject to future revision as the detailed molecular mechanisms become elucidated.

Multifaceted actions of IL-10

Production and expression of IL-10. Murine IL-10 (mIL-10) is produced by Th2 and Th0 cells, some types of $CD8^+$ T-cells, but not by Th1 cells.[16–18] On the other hand, human IL-10 (hIL-10) is produced by Th0, Th1, and Th2 cells.[19,20] Thus, in humans the production of IL-10 may not be precisely limited to T-cell subsets. In addition to T-cells, a number of other cell types produce IL-10. Activated macrophages and mast cells appear to be a major source of IL-10.[17,21] Ly-1 B-lymphocytes,[22] B-lymphomas,[23] and normal mouse B-lymphocytes produced IL-10 after stimulation, but the major B-lymphocytes with IL-10 production are Ly-1 B-lymphocytes. IL-10 is also produced by keratinocytes and keratinocyte cell lines.[24]

Central dogma. Since Th1 and Th2 pathways are often mutually exclusive, it was suggested that Th1 and Th2 cells may have cross-regulatory effects. Indeed, IFN-γ produced by Th1 cells was initially shown to inhibit proliferation of Th2 cells. A cytokine synthesis inhibitory factor (CSIF) produced by Th2 cells that inhibited cytokine synthesis, especially IFN-γ production by Th1 cells in the presence of APC, was then discovered.[16] This activity was initially described as CSIF and later was designated as IL-10.[25] This CSIF activity was originally detected in the supernatant of activated Th2 cells, but was not found in supernatant of activated Th1 cells. An enormous amount of information on biological activities of IL-10 in 5 years since its discovery substantiated that it is a powerful immunosuppressive factor, and that the original activity IL-10 is to inhibit the production of IFN-γ by Th1 cells, as discussed later. Thus, IL-10 is assumed to exhibit a pivotal role in controlling the immunological responses toward Th2 type by inhibiting cytokine production, especially of IFN-γ from Th1 cells.

Immunosuppressive properties of IL-10. EFFECT ON MACROPHAGES. IL-10 inhibits the synthesis of several cytokines that are normally secreted by human and mouse monocytes/macrophages in response to activation by lipopolysaccharide (LPS). These cytokines include IL-1α, GM-CSF, TNFα, IL-6, IL-8, IL-10, and IL-12.[21,26] In particular, synthesis of IL-1α, GM-CSF, and TNFα was strongly inhibited, whereas the synthesis of IL-1β, IL-6, IL-8, or G-CSF was partially suppressed. This variable inhibition can probably be attributed to the relatively early production of the latter cytokines following monocyte activation. The production of IL-10 by macrophages can be inhibited by IL-10 itself[21]; thus, the secretion of IL-10 by macrophages appears to be self-limited.

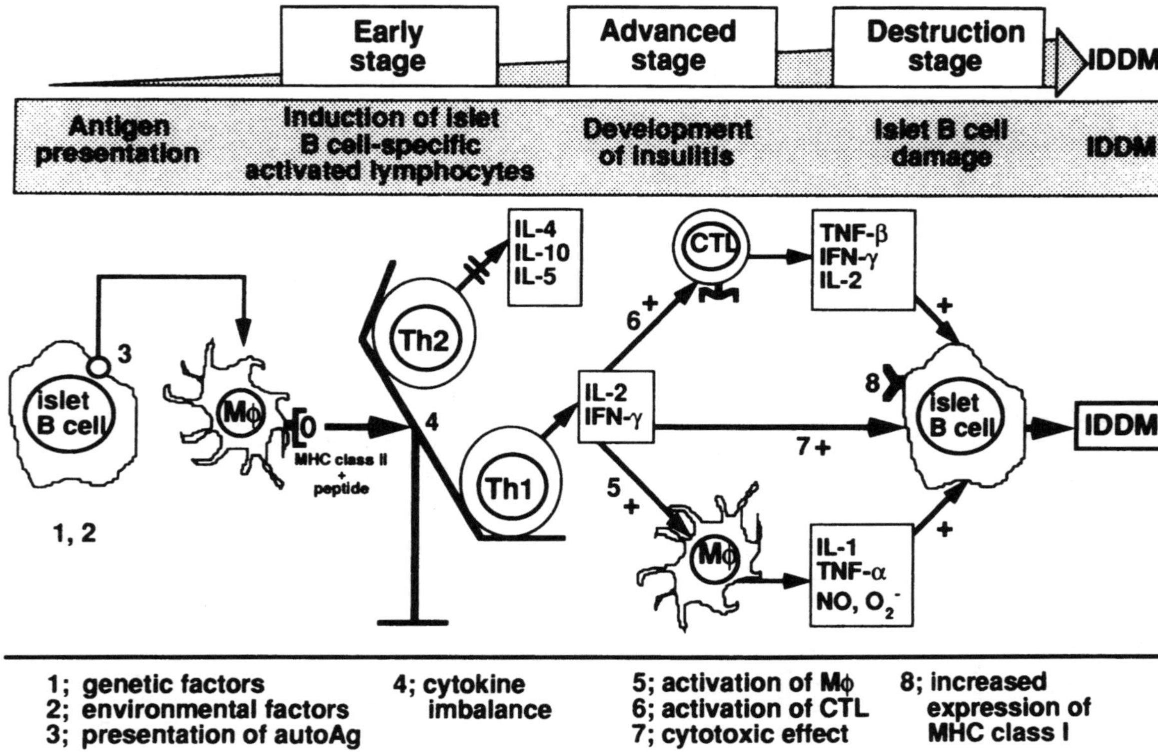

Figure 1. Schematic presentation on the relation of the Th1/Th2 balance to IDDM.

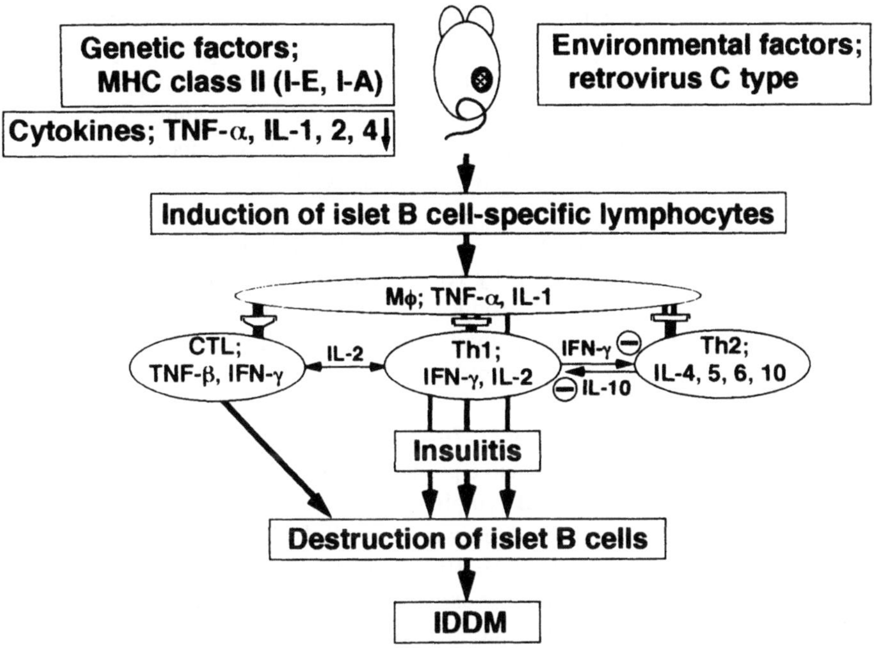

Figure 2. Schematic presentation on factors affecting the destruction of islet B-cells and the appearance of diabetes in NOD mice.

Inhibitory effects of IL-10 on macrophage cytokine synthesis are associated with secondary inhibitory effects on macrophage functions. For example, IL-10 inhibits the production of reactive nitric oxide (NO) and reactive oxygen intermediates (H_2O_2) by macrophages following activation by IFN-γ.[27] Production of NO by IFN-γ in activated macrophages is dependent on endogenous TNFα, and inhibition of NO production by IL-10 is mainly related to the inhibition of TNFα production.[28] The production of NO is implicated in the killing of intracellular parasites. IL-4 and TGFβ have similar effects.[28] Thus, IL-10 as well as IL-4 and TGFβ, which are produced during parasite infections, inhibit IFN-γ production by Th1 cells, and contribute to the suppression of IFN-γ–induced and macrophage-mediated immune destruction of parasites. In addition to inhibition of cytokine production by activated macrophages, IL-10 strongly suppresses the APC function of macrophages by downregulating the expression of MHC class II-associated antigens on these cells.[29]

EFFECT ON T-CELLS. IL-10 inhibits cytokine production at both the mRNA and protein levels by Th1 cells in response to macrophage-presented antigens or stimulation with soluble anti-CD3. This inhibition was not effective when splenic B lymphocytes were used as APC.[16,30] The inhibitory effects of IL-10 were predominant on IFN-γ, which was produced relatively late after T-cell activation, whereas, the levels of cytokines like GM-CSF, TNFβ, and TNFα, which are produced in the early stage, are not or only slightly affected.[16,30] In contrast, cytokine synthesis by Th2 cells was not significantly affected. IL-10 can also indirectly suppress the proliferation of T-cells by inhibiting the production of IL-2.[31] Although IL-10 strongly inhibits the effector function of mature Th1 cells, IL-10 appears to weakly suppress the differentiation of T-cells from precursor cells (Th0).[32]

EFFECT ON NATURAL KILLER (NK) CELLS. IL-10 strongly inhibits the production of IFN-γ by NK cells stimulated by IL-2.[33] Thus, IL-10 can inhibit the synthesis of IFN-γ by all three major producers of IFN-γ: Th1, CD8$^+$, and NK cells. This depends on the stimulation condition, because Th1 cells stimulated by B lymphocytes as APC are not susceptible to IL-10.

Immunostimulatory properties of IL-10. EFFECT ON B-LYMPHOCYTES. While IL-10 holds a potent immunosuppressive properties on macrophages, it exerts a wide immunostimulatory effects on B-lymphocytes. IL-10 is a B-lymphocyte–stimulating factor, exerting at least two effects on small dense splenic B-lymphocytes collected from unstimulated mice. IL-10 upregulates the expression of MHC class II molecule without upregulating the expression of MHC class I molecule in these cells and enhances their in vitro viability. IL-4 also mediates the same effect on resting B-lymphocytes.[34] Neutralizing monoclonal antibodies that specifically recognize IL-4 or IL-10 suggest that IL-4 or IL-10 stimulated B-lymphocytes independently.

In terms of B-lymphocyte activation, IL-10 stimulates both B-lymphocyte proliferation and differentiation into antibody-secreting cells. Activated B-lymphocytes secrete a large amount of IgG, IgA, and IgM, and IL-10 also induces their differentiation into antibody-producing cells and morphological differentiation to those resembling plasma cells.[35] The observed production of a substantial amount of IL-10 by mouse Ly-1 B-lymphocytes and B-lymphomas suggested a possible role for IL-10 as a growth and differentiation factor for mouse B-lymphocytes.

EFFECT ON CYTOTOXIC T CELLS. IL-10 also functions as a differentiation factor for cytotoxic T-cells (CTL, CD8$^+$ cells). IL-10 promotes IL-2–activated CTL precursors to proliferate and differentiate into effector CTL.[36] Furthermore, IL-10 significantly increases cytotoxic effector functions. In single-cell cultures of CTL containing IL-2 and IL-10, IL-10 markedly expanded the clone size and enhanced the lytic potential of these cells with the increased production of IFN-γ, when assayed using appropriate target cells.[36] In contrast, IL-10 did not significantly enhance the lymphokine-activated killer activity of CTL.

EFFECT ON MAST CELLS AND OTHER CELLS. In addition to the effect of IL-10 on B-lymphocytes, it stimulates numerous other cell types. IL-10 stimulates the growth of murine mast cells in the presence of IL-3 and IL-4.[37] IL-10 increases the in vitro proliferative response of murine thymus T-cells that have been activated with IL-2 and IL-4. Thus, IL-10 is a thymus T-cell co-growth factor in the presence of IL-2.

Autoimmunity and IL-10. Human autoimmune diseases are either systemic or organ specific. The prototypes of systemic autoimmunity are the connective tissue disorders, including lupus erythematosus and rheumatoid arthritis. Organ-specific human autoimmune disease commonly affects the thyroid gland and islet B-cells. Another classification of autoimmunity relates to whether the pathogenesis of the disease involves primarily a humoral or a cell-mediated effector mechanism. The humoral immune response predominates in Graves' disease and Hashimoto's thyroiditis. In contrast, autoimmune diseases such as IDDM are considered to be included in the cell-mediated category.

The idea that the autoimmune response in IDDM is mediated by Th1 cells has been supported by studies on NOD mice (Fig. 1).[38,39] IL-10 is assumed to exhibit a pivotal role in controlling immunological responses by deactivating macrophages and downregulating cytokine production by Th1 cells. Thus, IL-10 inhibits IFN-γ, TNFα, TNFβ, and IL-1 synthesis in Th1 cells in vitro. Because IFN-γ is presumed to be etiologic for autoimmune diabetes,[40,41] downregulation of its production by IL10 is expected to suppress cell-mediated autoimmune destruction of islet B-cells and diabetes. In contrast, simultaneous increase of IL-10 and IFN-γ in Th1-dominant responses to pathogens has been reported to persist as late as 10 days after infection.[42,43] These results suggest that the role of IL-10 in the regulation of immune responses is much more complex than expected and that it may generate immunostimulatory effects. The adoptive transfer experiment of the mixture of self-reacting Th1 and Th2 cells with the same T-cell receptor (TCR) showed the nonsuppressive role of Th2 cells on Th1-induced insulitis and diabetes in NOD mice.[44] Thus, the simple Th1/Th2 cell ratio may not be the determinant of the consequent immune responses.

Mechanisms of pancreatic islet B-cell destruction in NOD mice

Immune cells and cytokines. NOD mice are a good model of human IDDM. Insulitis is observed at high incidence both in female and male NOD mice after 5 wk of age. The cumulative incidence of insulitis reaches almost 100% by 30 wk of age. Insulitis is characterized by infiltration of mononuclear cells (MNC) consisting predominantly of $CD4^+$ lymphocytes, a small number of $CD8^+$ lymphocytes, and macrophages or B-lymphocytes to the lesser degree (Fig. 2).

The evidence that T-cells play a major role in the progression of diabetes in NOD mice is now extensive: 1) Immunohistological analysis has revealed that the majority of cells in the islet lesion are T-cells[45]; 2) disease does not appear in athymic NOD mice or neonatally thymectomized mice[46,47]; 3) preventive effects aginst diabetes by treatment with a variety of anti–T-cell mAb[45,48]; 4) diabetes can be prematurely induced in irradiated young or newborn NOD mice by the transfer of diabetogenic T-cells.[49,50]

Analysis of cytokine production in situ has proved that the production of IL-1,[51] TNFα,[52] IL-2,[53] and IL-4[54] in NOD mice is significantly reduced when compared with other strains. Recently, using in situ mRNA hybridization techniques or the reverse transcription–polymerase chain reaction (RT-PCR) method, cytokine production could be detected and localized. Expression of IL-1α, TNFα, and IFN-γ in islet-infiltrating lymphocytes was detected in the pancreas during the development of insulitis and diabetes in NOD mice.[55,56] In another study, cytokine mRNA expression was examined by RT-PCR in a specific cell population obtained by fluorescence-activated cell sorting of the insulitis lesion in NOD mice.[57] According to these studies, MHC class II positive macrophages expressed IL-1β mRNA; a small number of CD8$^+$ T-cells expressed IFN-γ, TNFβ, and perforin; and a large number of CD4$^+$ T-cells have been shown to express IL-4, IL-5, IL-6, and IL-10. These results suggested possible roles of cytokines in islet B-cell destruction. The destruction process can now be mainly attributed to macrophages, T-cells, and their cytokine products in NOD mice.

Apparent paradoxes of virus, autoantigen, and cytokine (TNFα, IL-2, IL-10, etc.) action in NOD mice

A viral etiology of IDDM has been proposed in animal models. A common rat parvovirus, Kilham rat virus, has been identified as a viral trigger of IDDM in BB rats.[58] In NOD mice, retrovirus-like particles have been identified in islet B-cells,[59] suggesting a pathogenic role of the virus in islet B-cell destruction. However, infection of NOD mice with some viruses, including lymphocytic choriomeningitis (LCM) virus, mouse hepatitis virus, encephalomyocarditis virus, and lactate dehydrogenase virus, were reported to paradoxically prevent the development of IDDM.[60] NOD mice bred under strictly pathogen-free conditions were reported to be associated with the increased incidence of diabetes, and deliberate viral infection in this case decreased the incidence of diabetes.[61]

The administration of glutamic acid decarboxylase (GAD) 65, one of islet B-cell autoantigens, can significantly reduce the development of diabetes in NOD mice.[62] This preventive effect is associated with the induction of GAD 65 tolerance, suggesting the suppression of GAD 65– responsive Th1 clones.[62] The same preventive effect has been reported for insulin immunization.[63]

To determine whether a certain cytokine is cytotoxic or protective toward islet B-cells in vivo as assessed in vitro, cytokines have been systemically administered to NOD mice. IL-1α, TNFα, and TNFβ are cytotoxic to islet B-cells in vitro. IL-1 at low concentration or in long-term exposure inhibits insulin secretion. At high concentration or longer exposure IL-1 leads to cell death of isolated islets.[64,65] Combination of IL-1 and TNFα, TNFβ, or IFN-γ produces islet B-cell destruction in vitro.[66] However, systemic administration of IL-1α, TNFα, or TNFβ to NOD mice paradoxically suppresses the incidence of diabetes.[67,68]

Treatment of NOD mice with IL-2, IL-4, or IL-10 in vivo significantly reduced the incidence of insulitis and diabetes.[54,69,70] NOD transgenics (Tg) expressing IL-2 in islet B-cells, however, showed dramatically accelerated diabetes.[71] Also, Tg expressing IL-10 in either islet B- or A-cells exhibited enhanced cell-mediated autoimmune destruction of islet B-cells,[72,73] which is discussed in more detail later.

In addition, TNFα is a cytokine that has potent immunoregulatory functions. TNFα administered to NOD mice in the neonatal period before 2 wk of age causes an early onset and an increased incidence of diabetes, while administration of TNFα to NOD mice >4 wk of age suppressed the incidence of diabetes.[74] This demonstrated one example of age-dependent effects of one cytokine on IDDM. These data strongly suggest that TNFα has a critical role in the early development of autoimmunity toward islet B-cells, and that TNFα may exhibit specific functions at specific time points during the development of an immune system. Because in vivo or in vitro cytokine administration systems exhibited these apparently paradoxical effects, the authentic in situ property of cytokines of interest cannot be assessed by systemic administration.

Analysis of the role of cytokines by transgenic mice technology

Transgenic mice technology has been used to examine the role of various cytokines and MHC protein in the immunopathogenesis of insulitis and diabetes. Cytokines, such as IL-1α, IL-1β, and TNFα, mainly secreted from macrophages, and IFN-γ, IL-2, IL-4, IL-6, and IL-10, mainly secreted from lymphocytes, are now regarded as major modulatory factors of the autoimmune process in IDDM. Targeted expression of several cytokines to islet B-cells can potentially modify IDDM. Tg that express IFN-γ in islet B-cells under the control of a rat insulin promoter developed severe insulitis and diabetes.[75] It appears that IFN-γ induced a series of autoimmune processes, because 1) these IFN-γ Tg, when backcrossed to severe combined immunodeficiency (SCID) mice, did not develop overt diabetes and 2) these IFN-γ Tg rejected transplanted histocompatible islet grafts.[41] Similarly, Tg expressing IFN-α in islet B-cells induced insulitis, selective islet B-cell destruction, and diabetes. Both insulitis and diabetes in IFN-α Tg were prevented with a neutralizing antibody to IFN-α.[76]

Tg expressing TNFα or TNFβ in islet B-cells showed cellular infiltration composed of $CD4^+$ and $CD8^+$ T-cells and B-lymphocytes, and macrophages to a lesser extent, but diabetes never developed, even though several potentially aggravating conditions were superimposed.[77] Similarly, Tg expressing TNFα and TNFβ were reported to develop massive insulitis but not diabetes by another group.[78] In triple-Tg mice ($H\text{-}2K^b$ in islet B-cells, $H\text{-}2K^b$ specific TCR, and IL-2 production from islet B-cells) insulitis and diabetes were induced.[79]

Tg expressing IL-10 in islet B-cells in the BALB/c background induced inflammatory changes in pancreas without insulitis or diabetes, but was associated with the expression of MHC class II molecules and adhesion molecules on vascular endothelium.[80] In addition, transgenic expression of IL-10 in islet B-cells in the N2 and N3 backcrosses toward NOD mice accelerated the onset of diabetes.[72]

In summary, transgenic programming of cytokine production by islet B-cells in nondiabetic mice revealed that local production of various cytokines can induce lymphocytic infiltration but that diabetes results only in the case of IFN-γ, IFNα, or IL-2 overproduction. However, the expression of introduced exogenous genes in islet B-cells was inevitably associated with nonspecific disturbing effects on the function of highly differentiated islet B-cells, as exemplified by MHC class I, II, β_2-microglobulin, Ras, or calmodulin.[75,81–85] Thus, the induced diabetes in IFN-γ, IFNα, or IL-2 Tg is theoretically not free from non-immune destruction of islet B-cells. To test local effects of cytokines on islet B-cells, it is essential to develop a paracrine delivery system of cytokines without directly manipulating islet B-cells.

Our study on transgenic mice expressing IL-10 in pancreatic islet A-cells

To study the paracrine effects of IL-10 on autoimmune insulitis and diabetes, we produced IL-10 transgenic NOD mice (NOD-IL-10) in which murine IL-10 was expressed in pancreatic islet A-cells under the control of a rat glucagon promoter without directly manipulating pancreatic islet B-cells (Fig. 3). NOD IL-10 Tg was generated as described previously.[73] To determine chimerical sequences between each of flanking region of rabbit β-globin exon 3 and the inserted mIL-10 cDNA, 50 ng of genomic DNA from tail biopsy were introduced into a PCR analysis.

Detection of IL-10 mRNA in pancreatic islets. Of 57 progenies screened by PCR, 11 founder mice were obtained. Four were males and seven were females, as presented in Table 1. The transgenic expression of IL-10 was confirmed by RT-PCR of the total RNA sample obtained from pancreatic islets of a transgenic mouse in the B-N3 generation (Fig. 4), as described previously.[73] The DNA fragments of 137 or 167 bp amplified with sets of primers, respectively, were consistent with the amplified products of spliced transgenic IL-10 mRNA. In situ hybridization confirmed the local expression of IL-10 mRNA in islet A-cells (data not shown).

Blood glucose concentrations. Nonfasting blood glucose concentration was monitored once a week using a glucose analyzer (Dri-Chem 100, Fuji Film, Kanagawa, Japan). Glucosuria was monitored weekly with glucose testing tape (Uristix, Miles-Sankyo, Tokyo, Japan). Blood was drawn from incised tails of mice.

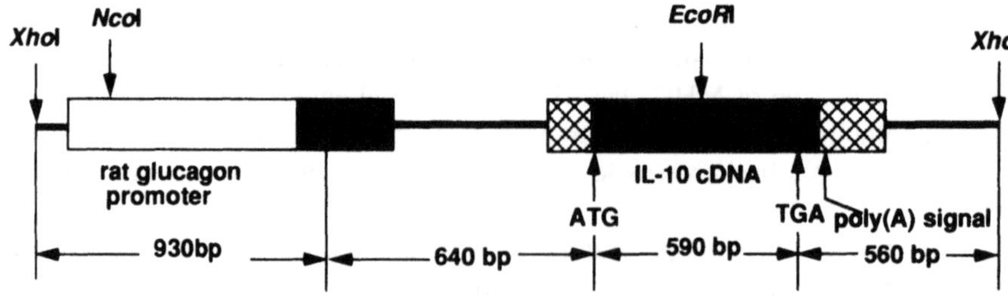

Figure 3. Schematic presentation of an IL-10 transgene construct. The linearized *Xho* I fragment contains a rat glucagon promoter (930 bp), exon 2, intron 2, and a part of exon 3 of the rabbit β-globin gene (640 bp), IL-10 cDNA (590 bp); and a rabbit β-globin exon 3 and poly(A) addition signal sequence (560 bp).

Table 1. List of NOD IL-10 founder mice

Mouse no. (sex)	Diabetes and its onset (wk)[a]	Age of histology (wk)
1 (F)	−	nd[b]
5 (F)	+ (10)	13
17 (M)	+ (10)	nd[b]
20 (M)	+ (6)	7
21 (M)	+ (8)	18[c]
23 (F)	+ (8)	12
25 (F)	−	nd[b]
36 (F)	−	11
39 (F)	+ (9)	10
47 (M)	+ (7)	8
51 (F)	−	10

[a]A nonfasting blood glucose concentration of >250 mg/dl was considered as diabetic.
[b]nd, not determined. The founder mice were kept to produce transgenic lines.
[c]Mouse no. 21 died at the age of 18 wk of an unknown cause, and thus the histology was not available.

The time course of changes in blood glucose concentration of 11 NOD IL-10 founder mice is shown in Figure 6. Surprisingly, all 4 male NOD IL-10 out of 11 founder mice developed diabetes at the age of 6, 7, 8, and 10 wk. All glucosuric male mice had blood glucose concentration >300 mg/dl, with a mean of 489 mg/dl (n = 4). These male NOD IL-10 appeared markedly weak, even before the onset of diabetes. Three of seven female NOD IL-10 became diabetic at the age of 8, 9, and 10 wk. Female NOD IL-10 were classified into two groups with regard to glucose concentration. The first group consisted of three hyper-

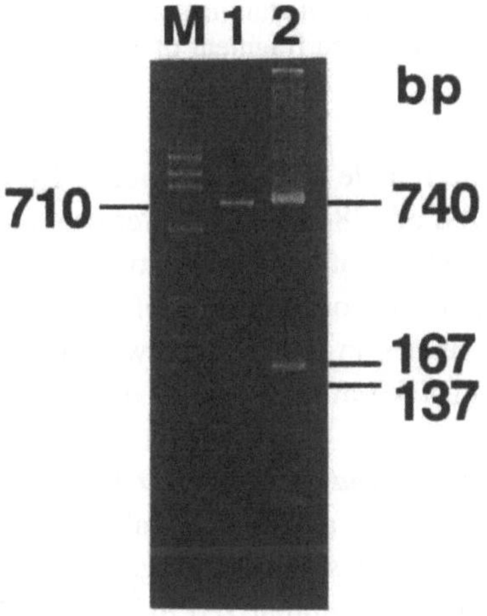

Figure 4. PCR detection of transgenic expression of IL-10 in pancreatic islets of NOD IL-10. Ethidium bromide–stained polyacrylamide gels showing PCR products for each pair of primers. A total of 1 μg of total RNA isolated from pancreatic islets from a mouse in the B-N3 generation was reverse-transcribed using 777 as a primer. Nonquantitative PCR amplification of 35 cycles was performed. The arrows indicate the position of amplified fragments. Lane 1 shows the 137 bp PCR product, and lane 2 shows the 167 bp PCR product. The 710 or 740 bp PCR product is derived from the integrated transgene.

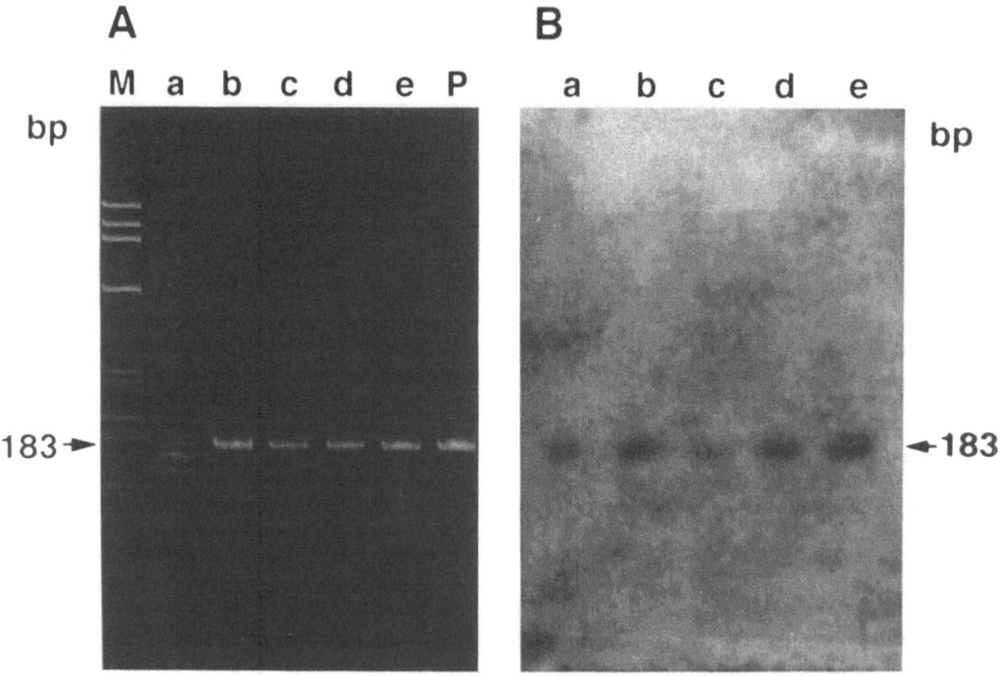

Figure 5. PCR detection of endogenous IL-10 expression in pancreatic islets of nontransgenic NOD mice. (A) Ethidium bromide–stained polyacrylamide gels showing PCR products for each pair of primers. A total of 2 μg of total RNA isolated from pancreatic islets of nontransgenic NOD mice, 5–25 wk old was reverse transcribed using 777 as a primer. Arrows indicate the position of amplified fragments; the 183 bp PCR product is faintly visible in islets of 5 wk old NOD mice (a), but is clearly visible in NOD mice at the age of 8 (b), 10 (c), 15 (d), and 25 wk (e), and in a murine IL-10 cDNA plasmid (P) used as a positive control. (B) PCR Southern blot analysis with a digoxigenin-labeled oligonucleotide probe. PCR products were electrophoresed on a 1.7% agarose gel and transferred to nylon membrane. The blot was hybridized with a DIG-labeled oligonucleotide probe and the luminescent signal was detected for 5 min. Arrows indicate the position of hybridized fragment of 183 bp.

glycemic female mice (mice no. 5, 23, and 39), with a mean blood glucose concentration of 478 mg/dl. Two female mice that were consistently normoglycemic, even at 34 wk of age (mice no. 1 and 25), form the second group, with a mean blood glucose concentration of 116 mg/dl. At the age of 10 wk, the incidence of overt diabetes in NOD IL-10 with a blood glucose concentration >250 mg/dl is 4 of 4 (100%) in male, and 3 of 7 (43%) in female, NOD IL-10.

Histological analysis of NOD IL-10 founder mice. Eight NOD IL-10, including three males and five females, were histologically examined in the nondiabetic or diabetic stage at the age of 7–13 wk, with eight nontransgenic NOD mice as controls, as summarized in Table 2. Mice in the B-N2 generation were sacrificed at the age of 39 wk. Histological and immunohistochemical analysis was performed with hematoxylin-eosin and guinea pig Ab to porcine insulin (Dako). All NOD IL-10 exhibited insulitis of variable degrees, as shown in Figure 7A and 7B. In one female NOD IL-10 (no. 5), an abscess was found in several organs, including the liver, brain, and spleen. Lymphocytes predominated the islet-infiltrating cells, although plasma cells were occasionally recognized (data not shown). The liver, lung (Fig. 7C and 7D), eye, salivary gland, and kidney, in this order, showed lesser degrees of lymphcytic infiltration. Lymphocytic infiltration in the liver and lung, was not, or was rarely, observed in age- and sex-matched nontransgenic NOD mice, as shown in the right-hand column in Table 2. Nontransgenic NOD mice with matched age and sex exhibited no remarkable histology, except for insulitis and rare and mild lymphocytic infiltration to the lung and salivary gland.

Of eight histologically examined NOD IL-10, six (three males and three females) were diabetic and two others were nondiabetic. The diabetic pancreas revealed severe insulitis with the substantial destruction of not only islets themselves, but also acinar cells around islets, exhibiting ductal proliferation and fibrosis (Fig. 7B), which were not observed in control nontransgenic NOD mice.

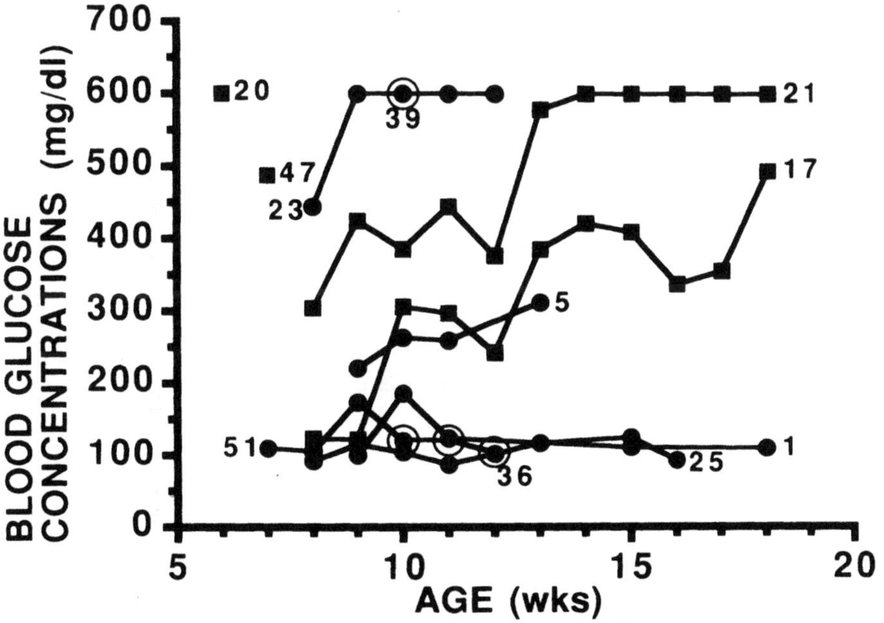

Figure 6. Time course of changes in blood glucose concentrations. The time course of changes in blood glucose concentrations is shown for 11 founder mice, including 4 males and 7 females. Squares and circles represent male and female founder mice, respectively. Large circles denote blood glucose concentrations of two female NOD IL-10 (no. 36 and 39) to delineate the overlapping data. The upper limit of the glucose level is 600 mg/dl with the monitor used. In our colony, diabetes starts at 15 wk of age, and the cumulative incidence of diabetes in female and male NOD mice is 67% at 27 wk of age and 25% at 39 wk of age.

Table 2. Summary of histological analysis of NOD IL-10 founder mice

Mouse no. (sex)	5 (F)	20 (M)	21 (M)	23 (F)	36 (F)	39 (F)	47 (M)	51 (F)	Control NOD mice[c]
Age of histology (wk)	13	7	18	12	11	10	8	10	
Diabetes	+	+	+	+	−	+	+	−	−
Pancreas infiltration	++	++	++	++	+	++	++	+	+ ~ ++ (6/8)
Others[a]	++	++	++	+	+	++	+	+	−
Liver	+[b]	nd	+	+	+	+	+	+	−
Lung	−	nd	nd	+	++	+	+	+	+ (1/8)
Eye	−	nd	−	+	+	−	−	+	−
Salivary gland	−	nd	nd	−	+	−	−	+	+ (1/8)
Kidney	−[b]	nd	−	−	−	−	−	+	−
Thyroid	nd	nd	nd	+	nd	−	−	−	−
Heart	−[b]	nd	nd	−	−	−	−	−	−
Thymus	−	nd	nd	−	−	−	−	−	−
Spleen	−	nd	nd	−	−	−	−	−	−
Brain	−	nd	nd	−	−	−	+	−	−
Pituitary	nd	nd	nd	nd	nd	nd	nd	−	nd
Stomach	−	nd	nd	−	−	−	−	−	−
Intestine	−	nd	nd	−	−	−	−	−	−
Adrenal	−	nd	nd	−	−	−	nd	−	−
Bone marrow	−	nd	nd	nd	−	nd	−	nd	nd

Histology of eight transgenic founder mice, including three males and five females, and of eight control nontransgenic NOD mice, including three males and five females.

[a]"Others" represents the presence of ductal proliferation and fibrosis. Severity was classified as −, +, and ++, respectively, corresponding to absence, clear but mild degree, and severe massive degree of lymphocytic infiltration.

[b]Presence of abscess.

[c]The right-hand column shows the histology of age- and sex-matched nontransgenic NOD mice, with the incidence shown in parentheses.

nd, not determined.

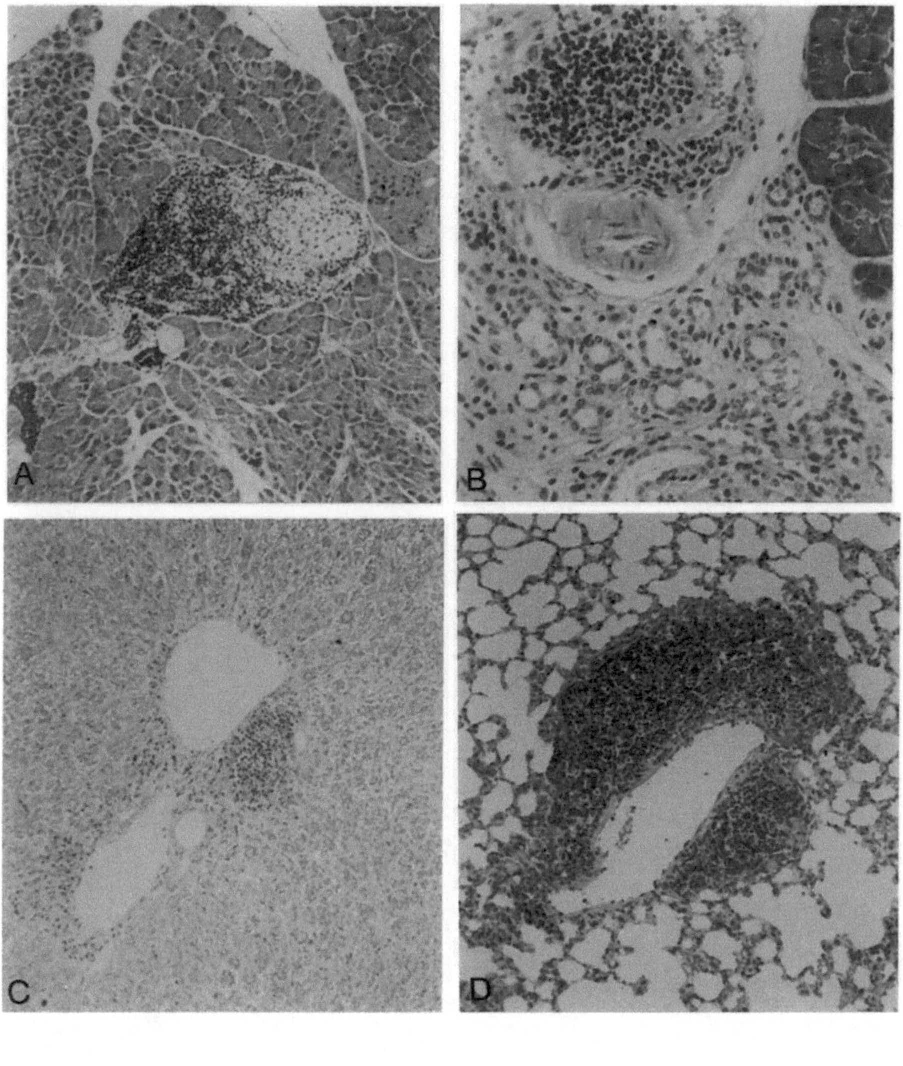

Figure 7. Histology of pancreas, lung, and liver. (A) Section of the pancreas from no. 23 female NOD IL-10 at diabetic stage 12 wk, showing severe insulitis. Magnification ×120. (B) Section of the pancreas from no. 51 female NOD IL-10 at nondiabetic stage 10 wk. The pancreas shows the presence of lymphocytic infiltration and prominent ductal proliferation. Magnification ×240. (C) Section of the liver from no. 36 female NOD IL1-0 at nondiabetic stage 11 wk. The liver shows the focal lymphocytic infiltration. Magnification ×120. (D) Section of the lung from no. 51 female NOD IL-10 at nondiabetic stage of 10 wk. The lung shows localized lymphocytic infiltration. Magnification ×120.

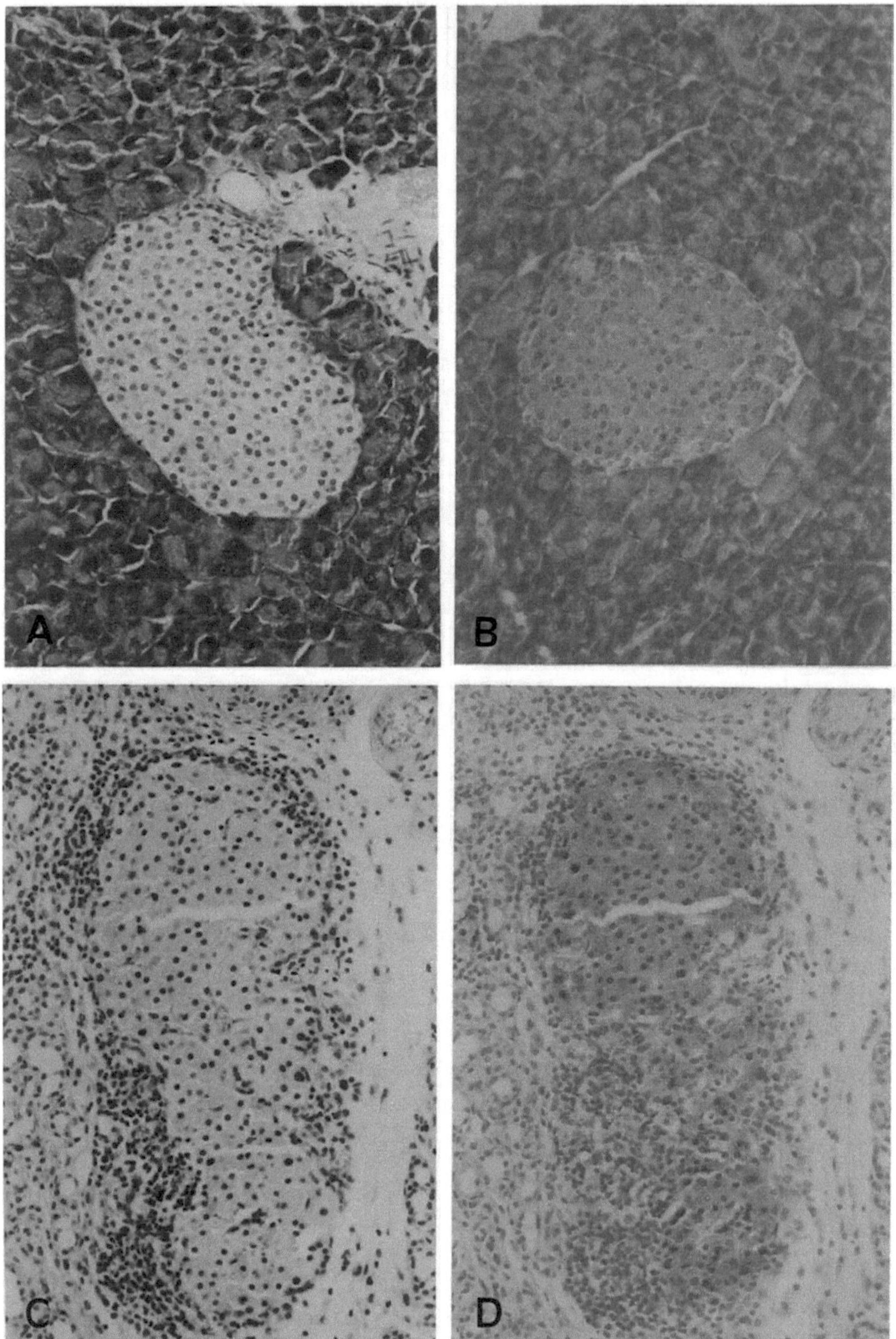

Figure 9. Histology of the pancreas in the B-N2 generation at 39 wk of age. (A) and (C) show the absence and presence, respectively, of lymphocytic peri-insulitis and ductal proliferation in a nontransgenic littermate and a transgenic mouse in the B-N2 generation at 39 wk of age. (B) and (D) show immunohistochemical staining of insulin in a non-transgenic littermate and a transgenic mouse in the B-N2 generation of 39 wk of age, respectively.

The nondiabetic NOD IL-10 showed some intact islets without infiltration, but most islets showed severe insulitis. No acinar change was observed.

Genetic backcross. One diabetic NOD IL-10 founder mouse (no. 17, male) was crossed to a female C57BL/6 mouse, and the resulting F1 was then backcrossed toward C57BL/6 mice to obtain the B-N2 and B-N3 generation, which carries 75% and 87.5% of the C57BL/6 genes, respectively. This F1 was also backcrossed once toward NOD mice to obtain the N-N2 generation, which carries 75% of the NOD gene. The time course of changes in blood glucose concentrations was monitored in these backcrosses and the histology of the pancreas was examined. The presence or absence of the *H-2K^b^* gene was determined for four mice in the N-N2 generation by PCR analysis of tail DNA. Changes in blood glucose concentration in F1 (n = 1), B-N2 (n = 6), and B-N3 (n = 10) were examined. As shown in Figure 8, none of these mice developed diabetes (mean

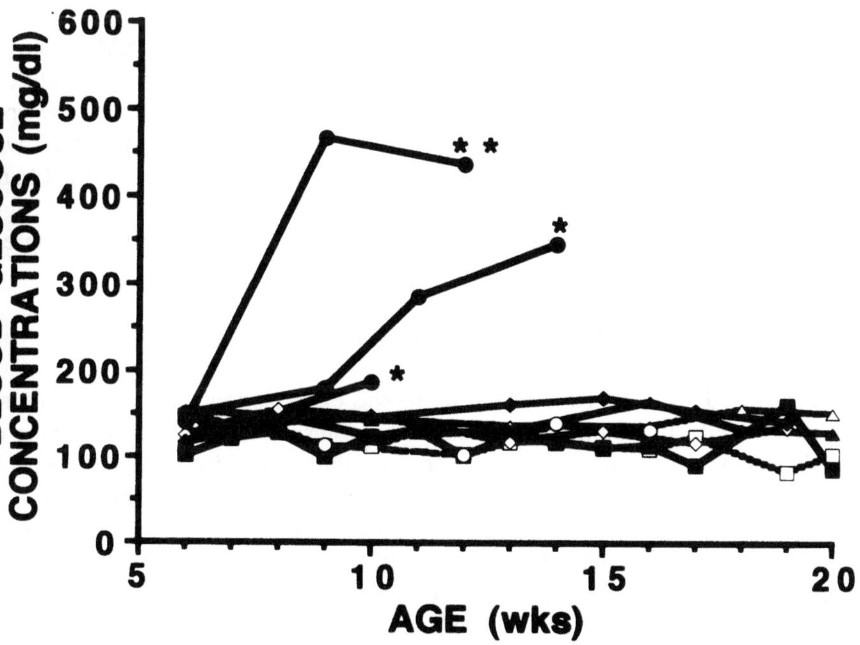

Figure 8. Change of blood glucose concentrations in F1 (C57BL/6 × NOD), B-N2, and B-N3 toward C57BL/6 mice, and N-N2 toward NOD mice. One NOD IL-10 founder mouse (no. 17, male) was crossed to a female C57BL/6 mouse, and the resulting F1 was then backcrossed two times to C57BL/6 mice to obtain the B-N2 and B-N3 generations, which carry 75% and 87.5% of the C57BL/6 genes, respectively. The F1 was also backcrossed once to NOD mice to obtain the N-N2 generation, which carries 75% of the NOD genes. The time course of changes in blood glucose concentrations of F1 (□), the mean of B-N2 (△), the mean of B-N3 (◇), and the mean of normoglycemic N-N2 (○) are shown, respectively. Two N-N2 mice developed early-onset diabetes (●). Solid lines (——) denote data for transgenic mice, while broken lines (- - - -) denote those for nontransgenic littermates. ** and * denote homozygous and heterozygous, respectively, for the *H-2K^b^*.

blood glucose concentration 116, 124, and 154 mg/dl in F1, B-N2, and B-N3, respectively). On the other hand, diabetes developed in 50% of backcrossed progenies toward NOD mice in the N-N2 generation (2 of 4), which carry 75% of the NOD genes, before 10 wk of age. The mean blood glucose concentration of diabetic N-N2 mice was 300 mg/dl at 10 wk of age. One diabetic N-N2 mouse showed a blood glucose concentration of 470 mg/dl at 10 wk of age and died of diabetes at 11 wk of age. The MHC genotype of the N-N2 mice was determined by PCR using $H\text{-}2K^b$. One diabetic mouse in the N-N2 generation was homozygous for the *H-2*, while others, including one diabetic and two nondiabetic mice in the N-N2 generation, were heterozygous for the *H-2*.

Although mice of the B-N2 generation did not develop diabetes, peri-insulitis and prominent ductal proliferation were observed in these mice at 39 wk of age (Fig. 9). The immunohistochemical staining of insulin showed the preserved insulin both in a Tg mouse and in its littermate at 39 wk of age.

Detection of IL-10 in islets of nontransgenic NOD mice. The expression of IL-10 mRNA was detected in pancreatic islets from nontransgenic and nondiabetic NOD mice of 5 wk of age and later, as shown by the PCR-amplified fragment of 183 (Fig. 5A) or 203 bp (data not shown). PCR Southern blotting showed the single major hybridization of a 183 bp fragment with the IL-10–specific probe (Fig. 5B). This confirmed that the amplified 183 bp fragment was actually derived from the IL-10 sequence. The intensity of the amplified band was relatively weak at 5 wk of age, and relatively strong at 8–25 wk of age, suggesting that IL-10 expression in islets may be increased with the progression of insulitis.

Discussion

We produced NOD IL-10 in which IL-10 was expressed in islet A-cells, but not in islet B-cells. Our study demonstrated that transgenic paracrine expression of IL-10 in pancreatic islet A-cells accelerates autoimmune insulitis and diabetes in NOD mice. The roles of various cytokines for IDDM have been studied by two approaches. Firstly, effects of systemic administration in vivo were comparatively studied with the direct effects on cultured islet B-cells in vitro.[52,66,67] Secondly, transgenic expression of various cytokines was directed to islet B-cells.[41,75,76,78,79] Some resulted in diabetes due to islet B-cell dysfunction, but without insulitis. Some, for example, IFN-γ Tg, resulted in established autoimmunity with the rejection of transplanted pancreatic islets. However, the induced diabetes in IFN-γ Tg or α Tg is theoretically not free from non-immune destruction of islet B-cells. Paracrine production of IL-10 without directly manipulating islet B-cells should enable the determination of whether IL-10 is immunostimulatory or immunosuppressive in the destruction process of islet B-cells.

The expression of IL-10 in islet A-cells induced severe insulitis and diabetes in NOD mice, even before the age of 10 wk. In our male and female NOD IL-10, the incidence of overt diabetes was 100% and 43%, respectively, at 10 wk of age. This clearly showed that the natural process of insulitis and diabetes can be significantly modified by the local paracrine supply of IL-10 from islet A-cells. As for two female founder NOD IL-10 of no. 1 and 25 with persistent normoglycemia, the possibility exists that the level of IL-10 expression was low, because two other normoglycemic founder NOD IL-10 of no. 36 and 51 showed similar ductal proliferation in histology at 11 and 10 wk of age, respectively. Alternatively, those persistently normoglycemic female founder NOD IL-10 may well have been mosaic for the integration of the IL-10 transgene.

Because of the high incidence of diabetes in multiple independent founder NOD IL-10, it is concluded that early-onset diabetes observed in our NOD IL-10 was due to the locally and excessively produced IL-10. This is partly consistent with the inflammatory nature of IL-10 disclosed in other IL-10 Tg mice by an insulin promoter.[80,72] They showed a 92% incidence of diabetes in the N2 and N3 generations between IL-10 Tg BALB/c mice and NOD mice.

The importance of the NOD background was demonstrated. Transgenic expression of IL-10 in the non-NOD genetic background, that is, C57BL/6 or BALB/c, caused peri-insulitis without diabetes. In our study, the absence of diabetes in the F1, B-N2, or B-N3 generation, and a 50% incidence of diabetes in the N-N2 generation (2 of 4) was shown. Another IL-10 Tg toward NOD mice[72] accelerated insulitis and diabetes, suggesting that the NOD genetic background is required for local IL-10 to induce infiltration insulitis other than peri-insulitis. Also, the role of MHC types in IL-10–induced diabetes, and heterozygosity or homozygosity in the type of MHC for an NOD type, is not crucial for IL-10–induced diabetes, in contrast to spontaneous diabetes in NOD mice, and it is the subject of a future study.

Histological examination of pancreas of NOD IL-10 revealed severe insulitis. In both female and male control nontransgenic NOD mice, insulitis started to be observed at the age of 5 wk,[38] became prominent at the age of 9–10 wk, but male NOD mice usually did not develop overt diabetes. Most islets in our NOD IL-10 exhibited severe insulitis associated with substantial atrophy of islet B-cells and also acinar cells, showing ductal proliferation. Lymphocytic infiltration in the liver and lung suggests a systemic effect of IL-10.

Based on these, it is concluded that transgenic expression of IL-10 functions as an inflammatory cytokine in situ at the very locus of cell-mediated autoimmune destruction of islet B-cells in NOD mice. This inflammatory nature of IL-10 is in line with the appearance of an IL-10 message from the early phase of the Th1 response in the infection of various pathogens[14,43] or during accelerated rejection of transplanted IL-10–producing islets.[86]

The results of early-onset diabetes associated with severe insulitis suggested that IL-10 secreted from islet A-cells in NOD IL-10 enhanced cell-mediated autoimmune destruction of islet B-cells. In addition, the PCR detected endogenous IL-10 mRNA expression in nontransgenic NOD mice and immunohistochemical staining with anti–IL-10 IgM mAb, has shown that IL-10 is not only expressed in islet A-cells, but also in islet-infiltrating lymphocytes in NOD IL-10 (data not shown). This suggests that IL-10 in situ plays a similar accelerating rather than decelerating role in the cell-mediated autoimmune natural course of destruction of islet B-cells in NOD mice. This result is in a sharp contrast with the established "macrophage-deactivating," "anti-Th1," presumed immunosuppressive activities of IL-10 against Th1-dependent cell-mediated autoimmunity.

The discrepant immunostimulatory effect in our transgenic model vs. the established immunosuppressive activities of IL-10 may be due to the following possibilities: 1) Persistent transgenic expression of IL-10 might have modified organ development. In other words, the effect of IL-10 is not limited to immune-related cells but it might have strongly affected other cells surrounding pancreatic islets, including endothelial cells. The overwhelming effects of IL-10 on these cells might have concealed its immunosuppressive activity on macrophages and lymphocytes. 2) The disproportionately high level expression around the islets might have exhibited chemotactic activity toward $CD8^+$ T-cells and might have induced cytotoxic effector functions, as discussed by Chen and Zlotnic.[36] 3) The role of local and paracrine effects of IL-10 at high concentration is presumably different from the immunosuppressive effect of IL-10 observed in systemic administration to NOD mice.[70] The explanation of the discrepant results is expected to be provided by regulating expression of the foreign gene in vivo, not only in a site-specific paracrine fashion as in our study, but also in a time-specific and intensity-regulated manner.

Our study shows the validity of using transgenic paracrine expression of cytokines in islet A-cells to assess the role of cytokines in islet biology. Although it demonstrates the unexpected inflammatory nature of IL-10 in situ, it presents a theoretical basis for modifying the autoimmune process by changing local cytokine concentrations through a paracrine mechanism.

Future directions to intervene with the islet B-cell destruction in IDDM

The extensive evidence discussed earlier supports an important role of the immune system for islet B-cell destruction. The methods to intervene in the sequence of events leading to autoimmune diabetes in NOD mice are expected to be clinically useful in human IDDM subjects. Immunostimulatory agents affecting macrophages and T-cells, such as treatment with complete Freund's adjuvant, GAD, Bacillus Calmette-Guerin, some viruses, and oral insulin, were reported to reduce the incidence of diabetes in NOD mice.[87,88] Administration of antibody against adhesion molecules (LFA-1, ICAM-1, and L-selectin) was reported to reduce diabetes in NOD mice.[89] Somatic gene therapy trials in an animal model for IDDM by transplantation of proinsulin-producing fibroblasts with a safety system were developed.[90,91]

Immunosuppressive cytokines, if delivered around islet B-cells at significant concentrations, are expected to show therapeutic and protective effects against autoimmune destruction of islet B-cells. Although our NOD IL-10 Tg accelerated insulitis and diabetes, constitutive transgenic expression of IL-10 during the development of islet cells, even before birth, is expected to induce many cell type- and time-specific functions, which may well be different from the effect of IL-10 expressed several weeks after birth. As a matter of fact, the study on the administration of TNFα in NOD mice by Yang et al.[74] demonstrated one example of reverse effects of one cytokine on IDDM depending upon the timing of administration. These data strongly suggest that the cytokines exhibit specific functions in a site- and time-specific manner. Because cytokines exhibit their functions locally at the target organs, the local effects of IL-10 on the immunological destruction of islet B-cells must also be tested by adoptively transferring retrovirally transduced islet-specific $CD4^+$ T-cells.

In our ongoing study, IL-10 cDNA was transduced to $CD4^+$ T-cell clones that were isolated from islet infiltrates of NOD mice. IL-10–transduced $CD4^+$, T-cells were adoptively transferred to young NOD mice in conjunction with wild-type $CD4^+$ T-cells. Adoptive transfer of IL-10–transduced $CD4^+$ T-cells suppressed insulitis and the incidence of diabetes induced by the adoptive transfer of wild-type $CD4^+$ T cells. These results suggest 1) local delivery of immunosuppressive cytokines to islet B-cells using adoptive transfer of genetically modified lymphocytes can potentially prevent a natural destructive course of autoimmune diabetes and 2) the results of this study support the concept of a new type of somatic gene therapy for IDDM in human subjects. Our approach using $CD4^+$ T-cells with IL-10 gene transduction is a subject for future research leading to the treatment of IDDM in humans.

Acknowledgments

This study was supported in part by a grant from Otsuka Pharmaceutical Factory, Inc. to the Otsuka Department of Clinical and Molecular Nutrition, School of Medicine, The University of Tokushima; a Grant-in-Aid for Co-operative Research (A) #06304034; and Juvenile Diabetes Foundation Research grant #193141. We thank Dr. K.W. Moore for providing us with murine IL-10 cDNA, rIL-10, and hybridomas of SXC1 and SXC2; Dr. S. Efrat for providing us with a glucagon promoter; and S. Hamaguchi, R. Katashima, and M. Kondo for expert technical assistance.

References

1. Mosmann TR, Cherwinski H, Bond MW, et al. Two types of murine helper T cell clone. I. Definition according to profiles of lymphokine activities and secreted proteins. J Immunol 136:2348–57, 1986.
2. Mosmann TR, Moore KW. The role of IL-10 in crossregulation of TH1 and TH2 responses. Immunol Today 12:A49–53, 1991.
3. Cherwinski HM, Schumacher JH, Brown KD, Mosmann TR. Two types of mouse helper T cell clone. III. Further differences in lymphokine synthesis between Th1 and Th2 clones revealed by RNA hybridization, functionally monospecific bioassays, and monoclonal antibodies. J Exp Med 166:1229–44, 1987.
4. Mosmann TR, Coffman RL. TH1 and TH2 cells; Different patterns of lymphokine secretion lead to different functional properties. Ann Rev Immunol 7:145–73, 1989.
5. Mosmann TR, Coffman RL. Heterogeneity of cytokine secretion patterns and functions of helper T cells. Adv Immunol 46:111–47, 1989.
6. Brown KD, Zurawski SM, Mosmann TR, Zurawski G. A family of small inducible proteins secreted by leukocytes are members of a new superfamily that includes leukocyte and fibroblast-derived inflammatory agents, growth factors, and indicators of various activation processes. J Immunol 142:679–87, 1989.

7. Parish CR. The relationship between humoral and cell-mediated immunity. Transplant Rev 13:35–66, 1972.
8. Katsura Y. Cell mediated and humoral immune responses in mice. III. Dynamic balance between delayed-type hypersensitivity and antibody response. Immunology 32:227–35, 1977.
9. Cher DJ, Mosmann TR. Two types of murine helper T cell clone. II. Delayed-type hypersensitivity is mediated by Th1 clones. J Immunol 138:3688–94, 1987.
10. Gajewski TF, Fitch FW. Anti-proliferative effect of IFN-γ in immune regulation. I. IFN-γ inhibits the proliferation of Th2 but not Th1 murine helper T lymphocytes clones. J Immunol 140:4245–52, 1988.
11. Fernandez-Botran R, Sanders VM, Mosmann TR, Vitetta ES. Lymphokine-mediated regulation of the proliferative response of clones of T helper 1 and T helper 2 cells. J Exp Med 168:543–58, 1988.
12. Swain SL, Weinberg AD, English M, Hutson G. IL-4 directs the development of Th2-like helper effectors. J Immunol 145:3796–806, 1990.
13. Mosmann TR. Properties and functions of interleukin-10. Adv Immunol 56:1–26, 1994.
14. Svetic A, Jian YC, Lu P, et al. *Brucella abortus* induces a novel cytokine gene expression pattern characterized by elevated IL-10 and IFN-γ in $CD4^+$ T cells. Int Immunol 5:877–83, 1993.
15. Yamamura M, Uyemura K, Deans RJ, et al. Defining protective responses to pathogens; cytokine profiles in leprosy lesions. Science 254:277–9, 1991.
16. Fiorentino DF, Bond MW, Mosmann TR. Two types of mouse T helper cell. IV. Th2 clones secrete a factor that inhibits cytokine production by Th1 clones. J Exp Med 170:2081–95, 1989.
17. Moore KW, O'Garra A, de Waal Malefyt R, et al. Interleukin-10. Ann Rev Immunol 11:165–90, 1993.
18. Mosmann TR, Schumacher JH, Fiorentino DF, et al. Isolation of monoclonal antibodies specific for IL-4, IL-5, IL-6, and a new Th2-specific cytokine (IL-10), cytokine synthesis inhibitory factor, by using a solid phase radioimmunoadsorbent assay. J Immunol 145:2938–45, 1990.
19. Yssel H, de Waal Malefyt R, Roncarolo MG, et al. IL-10 is produced by subsets of human $CD4^+$ T cell clones and peripheral blood T cells. J Immunol 149:2378–84, 1992.
20. Barnes PF, Abrams JS, Lu S, et al. Patterns of cytokine production by mycobacterium-reactive human T cell clones. Infect Immun 61:197–203, 1993.
21. de Waal Malefyt R, Abrams J, Bennett B, et al. Interleukin 10 (IL-10) inhibits cytokine synthesis by human monocytes: An autoregulatory role of IL-10 produced by monocytes. J Exp Med 174:1209–20, 1991.
22. O'Garra A, Change R, Go N, et al. Ly-1 B (B-1) cells are the main source of B cell-derived interleukin 10. Eur J Immunol 22:711–17, 1992.
23. O'Garra A, Stapleton G, Dhar V, et al. Production of cytokines by mouse B cells: B lymphomas and normal B cells produce interleukin 10. Int Immunol 2:821–32, 1990.
24. Rivas JM, Ullrich SE. Systemic suppression of delayed-type hypersensitivity by supernatants from UV-irradiated keratinocytes: An essential role for keratinocyte-derived IL-10. J Immunol 149:3865–71, 1992.
25. Moore KW, Vieira P, Fiorentino DF, et al. Homology of cytokines synthesis inhibitory factor (IL-10) to the Epstein-Barr virus gene BCRFI. Science 248:1230–34, 1990.
26. Fiorentino DF, Zlotnik A, Mosmann TR, et al. IL-10 inhibits cytokine production by activated macrophages. J Immunol 147:3815–22, 1991.
27. Bogdan C, Vodovotz Y, Nathan C. Macrophage deactivation by interleukin 10. J Exp Med 174:1549–55, 1991.
28. Oswald IP, Gazzinelli RT, Sher A, James SL. IL-10 synergizes with IL-4 and TGF-β to inhibit macrophage cytotoxic activity. J Immunol 148:3578–82, 1992.

29. de Waal Malefyt R, Haanen J, Spits H, et al. Interleukin 10 (IL-10) and viral IL-10 strongly reduce antigen-specific human T cell proliferation by diminishing the antigen-presenting capacity of monocytes via downregulation of class II major histocompatibility complex expression. J Exp Med 174:915–24, 1991.
30. Fiorentino DF, Zlotnik A, Vieira P, et al. IL-10 acts on the antigen-presenting cell to inhibit cytokine production by Th1 cells. J Immunol 146:3444–51, 1991.
31. Taga K, Tosato G. IL-10 inhibits human T cell proliferation and IL-2 production. J Immunol 148:1143–48, 1992.
32. Swain SL, Weinberg AD, English M. $CD4^+$ T cell subsets: Lymphokine secretion of memory cells and effector cells that develop from precursors *in vitro*. J Immunol 144:1788–99, 1990.
33. Hsu DH, Moore KW, Spits H. Differential effects of IL-4 and IL-10 on IL-2-induced IFN-γ synthesis and lymphokine-activated killer activity. Int Immunol 4:563–9, 1992.
34. Go NF, Castle BE, Barrett R, et al. Interleukin 10, a novel B cell stimulatory factor: Unresponsiveness of X chromosome-linked immunodeficiency B cells. J Exp Med 172:1625–31, 1990.
35. Rousset F, Garcia E, Defrance T, et al. Interleukin 10 is a potent growth and differentiation factor for activated human B lymphocytes. Proc Natl Acad Sci USA 89:1890–93, 1992.
36. Chen WF, Zlotnic A. IL-10: A novel cytotoxic T cell differentiation factor. J Immunol 147:528–34, 1991.
37. Thompson-Snipes L, Dhar V, Bond MW, et al. Interleukin 10: A novel stimulatory factor for mast cells and their progenitors. J Exp Med 173:507–10, 1991.
38. Kikutani H, Makino S. The murine autoimmune diabetes model: NOD and related strains. Adv Immunol 51:285–322, 1992.
39. Makino S, Kunimoto K, Muroka Y, et al. Breeding of a non-obese, diabetic strain of mice. Exp Anim 29:1–13, 1980.
40. Bottazzo GF, Pujol-Borrell R, Hanafusa T. Hypothesis: Role of aberrant HLA-DR expression and antigen presentation in induction of endocrine autoimmunity. Lancet 2:1115–18, 1983.
41. Sarvetnick N, Shizuru J, Liggitt D, et al. Loss of pancreatic islet tolerance induced by β-cell expression of interferon-γ. Nature 346:844–7, 1990.
42. Svetic A, Jian YC, Lu P, et al. *Brucella abortus* induces a novel cytokine gene expression pattern characterized by elevated IL-10 and IFN-γ in $CD4^+$ T cells. Int Immunol 5:877–83, 1993.
43. Flesch IEA, Kaufmann SHE. Role of macrophages and $\alpha\beta$ T lymphocytes in early interleukin 10 production during *Listeria monocytogenes* infection. Int Immunol 6:463–8, 1994.
44. Katz JD, Benoist C, Mathis D. T helper cell subsets in insulin-dependent diabetes. Science 268:1185–88, 1995.
45. Koike T, Itoh Y, Ishi T, et al. Preventive effect of monoclonal anti-L3T4 antibody on development of diabetes in NOD mice. Diabetes 36:539–41, 1987.
46. Makino S, Harada M, Kishimoto Y. Absence of insulitis and overt diabetes in athymic nude mice with NOD genetic background. Exp Anim 35:495–8, 1986.
47. Ogawa M, Maruyama T, Hasegawa T, et al. The inhibitory effect of neonatal thymectomy on the incidence of insulitis in non-obese diabetes (NOD) mice. Biomed Res 6:103–5, 1985.
48. Shizuru JA, Taylor-Edwards C, Banks BA, et al. Immunotherapy of the nonobese diabetic mouse: Treatment with an antibody to T-helper lymphocytes. Science 240:659–62, 1988.
49. Haskin K, Mcduffie M. Acceleration of diabetes in young NOD mice with a $CD4^+$ islet-specific T cell clones. Science 249:1433–36, 1990.
50. Nakano N, Kikutani H, Nishimoto H, Kishimoto T. T cell receptor V gene usage of islet B cell-reactive T cells is not restricted in non-obese diabetic mice. J Exp Med 173:1091–97, 1991.

51. Rabinovitch A. Roles of cytokines in IDDM pathogenesis and islet β-cell destruction. Diab Rev 1:215–40, 1993.
52. Satoh J, Seino H, Abo T, et al. Recombinant human tumor necrosis factor α suppresses autoimmune diabetes in nonobese diabetic mice. J Clin Invest 84:1345–48, 1989.
53. Yokono K, Kawase Y, Nagata M, et al. Suppression of concanavalin A-induced responses in splenic lymphocytes by activated macrophages in the non-obese diabetic mice. Diabetologia 32:67–73, 1989.
54. Rapoport MJ, Jaramillo A, Zipris D, et al. Interleukin 4 reverses T cell proliferative unresponsiveness and prevents the onset of diabetes in nonobese diabetic mice. J Exp Med 178:87–99, 1993.
55. Held W, Macdonald HR, Weissman IL, et al. Genes encoding tumor necrosis factor α and granzyme A are expressed during development of autoimmune diabetes. Proc Natl Acad Sci USA 87:2239–43, 1990.
56. Anderson JT, Cornelius JG, Jarpe AJ, et al. Insulin-dependent diabetes in the NOD mouse model. II. β cell destruction in autoimmune diabetes is a TH2 and not a TH1 mediated event. Autoimmunity 15:113–22, 1993.
57. Jarpe AJ, Hickman MR, Anderson JT, et al. Flow cytometric enumeration of mononuclear cell populations infiltrating the islets of Langerhans in prediabetic NOD mice: Development of a model of autoimmune insulitis for type I diabetes. Reg Immuol 3:305–17, 1991.
58. Guberski DL, Thomas VA, Shek WR, et al. Induction of type I diabetes by Kilham's rat virus in diabetes-resistant BB/Wor rats. Science 254:1010–13, 1991.
59. Fujita H, Fujino H, Nonaka K, Tarui S. Retrovirus-like particles in pancreatic B-cells of NOD mice. Biomed Res 5:67–70, 1984.
60. Rabinovitch A. Immunoregulatory and cytokine imbalances in the pathogenesis of IDDM: Therapeutic intervention by immunostimulation? Diabetes 43:613–21, 1994.
61. Leiter EH. The role of environmental factors in modulating insulin-dependent diabetes. In: Current topics in immunology and microbiology: The role of microorganisms in non-infectious disease. de Vries R, Cohen I, Van Rood JJ, eds. Springer-Verlag, Berlin, pp 39–55, 1990.
62. Tisch R, Yang X-D, Singer SM, et al. Immune response to glutamic acid decarboxylase correlates with insulitis in non-obese diabetic mice. Nature 366:72–5, 1993.
63. Muir A, Peck A, Clara-Salzler M, et al. Insulin immunization of nonobese diabetic mice induces a protective insulitis characterized by diminished intraislet interferon-γ transcription. J Clin Invest 95:628–34, 1995.
64. Mcdaniel ML, Hughes JH, Wolf BA, et al. Descriptive and mechanistic considerations of interleukin 1 and insulin secretion. Diabetes 37:1311–15, 1988.
65. Mandrup-Poulsen T, Bendtzen K, Nielsen JH, et al. Cytokines cause functional and structural damage to isolated islets of Langerhans. Allergy 40:424–9, 1985.
66. Pukel C, Baquerizo H, Rabinovitch A, et al. Destruction of rat islet cell monolayers by cytokines: Synergistic interaction of interferon-γ, tumor necrosis factor, lymphotoxin, and interleukin 1. Diabetes 37:133–6, 1988.
67. Jacob CO, Aiso S, Michie SA, McDevitt HO. Prevention of diabetes in nonobese diabetic mice by tumor necrosis factor (TNF): Similarities between TNF-α and interleukin 1. Proc Natl Acad Sci USA 87:968–72, 1990.
68. Seino H, Takahashi K, Satoh J, et al. Prevention of autoimmune diabetes with lymphotoxin in NOD mice. Diabetes 42:398–404, 1993.
69. Serreze DV, Hamaguchi K, Leiter EH. Immunostimulation circumvents diabetes in NOD/Lt mice. J Autoimmun 2:759–76, 1989.
70. Pennline KJ, Roque-Gaffney E, Monahan M. Recombinant human IL-10 prevents the onset of diabetes in the nonobese diabetic mouse. Clin Immunol Immunopathol 71:169–75, 1994.
71. Allison J, McClive P, Oxbrow L, et al. Genetic requirements for acceleration of diabetes in non-obese diabetic mice expressing interleukin-2 in islet β-cells. Eur J Immunol 24:2535–41, 1994.

72. Wogensen L, Lee MS, Sarvetnick N. Production of interleukin 10 by islet cells accelerates immune-mediated destruction of β cells in nonobese diabetic mice. J Exp Med 179:1379–84, 1994.
73. Moritani M, Yoshimoto K, Tashiro F, et al. Transgenic expression of IL-10 in pancreatic islet A cells accelerates autoimmune insulitis and diabetes in non-obese diabetic mice. Int Immunol 6:1927–36, 1994.
74. Yang XD, Tisch R, Singer ST, et al. Effect of tumor necrosis factor α on insulin-dependent diabetes mellitus in NOD mice. I. The early development of autoimmunity and the diabetogenic process. J Exp Med 180:995–1004, 1994.
75. Sarvetnick N, Liggitt D, Pitts SL, et al. Insulin-dependent diabetes mellitus induced in transgenic mice by ectopic expression of class II MHC and interferon-gamma. Cell 52:773–82, 1988.
76. Stewart TA, Hultgren B, Huang X, et al. Induction of type I diabetes by interferon-α in transgenic mice. Science 260:1942–46, 1993.
77. Higuchi Y, Herreta P, Munies P, et al. Expression of a tumor necrosis factor α transgene in murine panceratic B cells results in severe and permanent insulitis without evolution towards diabetes. J Exp Med 176:1719–31, 1992.
78. Picarella D, Kratz A, Li C, et al. Transgenic tumor necrosis fcator (TNF)-α production in pancreatic islets leads to insulitis, not diabetes; Distinct patterns of inflammation in TNF-α and TNF-β transgenic mice. J Immunol 150:4136–50, 1993.
79. Heath WR, Allison J, Hoffmann MW, et al. Autoimmune diabetes as a consequence of locally produced interleukin-2. Nature 359:547–9, 1992.
80. Wogensen L, Huang X, Sarvetnick N. Leukocyte extravasation into the pancreatic tissue in transgenic mice expressing interleukin 10 in the islets of Langerhans. J Exp Med 178:175–185, 1993.
81. Allison J, Campbell IL, Morahan G, et al. Diabetes in transgenic mice resulting from over-expression of class I histocompatibility molecules in pancreatic B cells. Nature 333:529–33, 1988.
82. Lo D, Burkly LC, Widera G, et al. Diabetes and tolerance in transgenic mice expressing class II MHC molecules in pancreatic beta cells. Cell 53:159–68, 1988.
83. Allison J, Malcolm L, Culvenor J, et al. Overexpression of β_2-microglobulin in transgenic mouse islet β cells results in defective insulin secretion. Proc Natl Acad Sci USA 88:2070–74, 1991.
84. Efrat S, Fleisher N, Hanahan D. Diabetes induces in male transgenic mice by expression of human H-ras oncoprotein in pancreatic β cells. Mol Cell Biol 10:1779–83, 1990.
85. Epstein PN, Overbeek PA, Means AR. Calmodulin-induces early-onset diabetes in transgenic mice. Cell 58:1067–73, 1989.
86. Lee MS, Wogensen L, Shizuru J, et al. Pancreatic islet production of murine interleukin-10 does not inhibit immune-mediated tissue destruction. J Clin Invest 93:1332–38, 1994.
87. Sedelain MWJ, Qin HY, Lauzon J, Singh B. Prevention of type I diabetes in NOD mice by adjuvant immunotherapy. Diabetes 39:583–9, 1990.
88. Bach JF. Insulin-dependent diabetes mellitus as an autoimmune disease. Endocr Rev 15:516–42, 1994.
89. Hasegawa Y, Yokono K, Taki T, et al. Prevention of autoimmune insulin-dependent diabetes in non-obese diabetic mice by anti-LFA-1 and anti-ICAM-1 mAb. Int Immunol 6:831–8, 1994.
90. Kawakami Y, Yamaoka T, Hirochica R, et al. Somatic gene therapy for diabetes with an immunological safety system for complete removal of transplanted cells. Diabetes 41:956–61, 1992.
91. Yoshimoto K, Murakami R, Moritani M, et al. Loss of ganciclovir sensitivity by exclusion by thymidine kinase gene from transplanted proinsulin-producing fibroblasts as a gene therapy model for diabetes. Gene Therapy, in press.

12

Polyol accumulation versus diabetic retinopathy and nephropathy in transgenic mice expressing human aldose reductase

MITSUO ITAKURA, SETSUKO II, MIKA OHTA, AND TAKASHI YAMAOKA
Otsuka Department of Clinical and Molecular Nutrition, School of Medicine, The University of Tokushima, Tokushima, 770 Japan.

Key words: transgenic mice, aldose reductase, diabetic complications, nephropathy, retinopathy, galactose, aldose reductase inhibitor.

Diabetes mellitus is caused by a relative or absolute insulin deficiency, and it affects >5% of the world's population. Diabetic retinopathy is a leading cause of acquired blindness in adults, and diabetic nephropathy is a major cause of morbidity and mortality in diabetes.[1,2] In a population study of 973 subjects, diabetic nephropathy was frequently associated with retinopathy.[3] Of patients with diabetes, 18% (66 of 365) had proteinuria and 59% of those patients (39 of 66) had diabetic retinopathy. Of patients with proliferative diabetic retinopathy, 40% (19 of 47) had proteinuria. Of patients with preproliferative retinopathy, 16% (20 of 124) had proteinuria.[4]

Chronic hyperglycemia has been recognized as the cause of diabetic complications, and the biochemical basis of diabetic complications includes nonenzymatic glycation and excessive flux through the polyol pathway. Frank[5] has suggested nonenzymatic protein glycation as an etiology for diabetic retinopathy. Glycation of critical proteins within the cell and in the extracellular matrix was supposed to be responsible for diabetic complications. In addition to glycation, fructosylation may be important, because the rate of nonenzymatic condensation of fructose with protein is faster than that of glucose[6] ($k_1 = 4.5$ and 0.6×10^{-3}/mM/h, respectively).

Diabetic hyperglycemia rapidly results in high glucose levels in lens, retina, nerve, and kidney tissues, where glucose uptake is independent of insulin. In these tissues, a portion of excessive glucose that is not metabolized by hexoki-

Correspondence: Mitsuo Itakura, M.D., Ph.D., Otsuka Department of Clinical and Molecular Nutrition, School of Medicine, The University of Tokushima, Kuramoto-cho, Tokushima, 770 Japan.

Lessons from Animal Diabetes VI. Ed. E. Shafrir.

nase is metabolized via the polyol pathway (Fig. 1). In this pathway, glucose is first converted to sorbitol by NADPH-dependent aldose reductase, AR; EC 1.1.1.21. Sorbitol is then converted to fructose by NAD-dependent polyol dehydrogenase (sorbitol dehydrogenase, SDH). Sorbitol accumulates in several tissues in diabetes due to 1) accelerated synthesis of sorbitol from glucose, 2) low SDH activity relative to AR, and 3) the relative impermeability of sorbitol through the cell membrane. Sorbitol accumulation was shown to contribute to diabetic neuropathy and cataract.[7,8] In the following sections an overview of diabetic retinopathy and nephropathy, and the results of our study using human aldose reductase (hAR) transgenic mice (hAR-Tg), are presented, with a focus on the polyol pathway.

Aldose reductase

Aldose reductase (AR) catalyzes the reduction of several aldohexoses, including glucose and galactose, to the corresponding sugar alcohols by an NADPH-dependent reaction. The published crystal structure of hAR contains a parallel $\beta 8/\alpha 8$-barrel motif within the active site pocket facing the nicotinamide ring of NADPH and other key active site residues (Tyr48, His110, and Cys298).[9]

Using hAR cDNA, AR has been mapped to chromosome 7q31-q35, and the sequences on other seven chromosomes (1, 3, 9, 11, 13, 14, and 18) may represent other active non-AR genes, homologous sequences, or pseudogenes.[10] In the promoter region of the hAR gene, a consensus sequence of an androgen response element is present at -396 to -382 base pairs (bp). The hAR promoter, thus, has important regulatory response elements for expression during puberty.[11]

AR is a putative osmoregulatory protein, at least in the renal medulla. Only renal medullary cells are routinely exposed to a hyperosmotic milieu in mammals. A principal osmolyte accumulated in these cells during adaptation to hy-

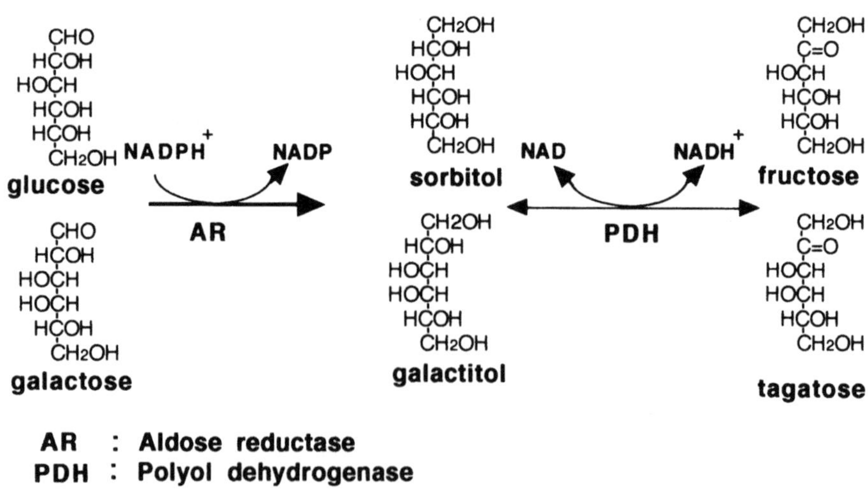

Figure 1. Diagram of polyol pathway.

perosmotic stress is sorbitol.[12] In cell lines derived from rabbit renal papilla or rat renal tubular cells, extracellular hyperosmolarity induces transcription of the AR gene,[13,14] resulting in an increase in AR mRNA, the rate of AR protein synthesis, and ultimately sorbitol accumulation. The upstream fragment (at −3429 to −192 bp) of the rabbit AR gene contains an osmotic response element(s).[15]

AR mRNA abundance, AR activity, and sorbitol content in the rat inner medulla were related to water balance, which is mediated through either vasopressin V1 or V2 receptors, because combined administration of the V1 and V2 antagonists significantly suppressed the increase in these parameters.[16] Streptozocin (STZ) diabetes induced a 4-fold increase in the inner medullary AR activity as determined in vitro, but increased the sorbitol levels in intact inner medullary collecting duct cells only by 1.3-fold.[17] Although AR mRNA was detected in renal medulla, but not in renal cortex,[18] the presence of AR mRNA and AR enzyme protein was recently demonstrated in rat mesangial cells in normal osmotic culture medium.[19] Sorbitol formation catalyzed by AR is important for physiological cellular osmoregulation,[14] but the role of AR in the kidney apart from the medulla is unclear.

In the in vivo system, the changing pattern of AR mRNA and AR activity in the kidney was not in synchrony with the temporal changes in the concentration of the specific tissue metabolites, galactitol, galactose, or myoinositol.[20] Changes in tissue AR activities observed as a result of galactose feeding were not related to the alterations in tissue AR mRNA or galactitol levels. Xylose feeding, despite being associated with intracellular xylitol accumulation, did not result in alterations in AR mRNA in the kidney.

Homologous proteins and genes of AR. Mouse vas deferens protein (MVDP), a member of the aldo-keto reductase superfamily, is produced exclusively in the epithelial cells of the deferent duct under androgenic regulation. A minimal region (−121 to +41 bp) in the androgen response elements in the 5′-flanking DNA of MVDP was identified as being necessary for androgen-regulated gene expression,[21] but hypertonicity-related regulation is not known.

A differential display approach to identify fibroblast growth factor (FGF)-1–inducible genes in NIH/3T3 cells led to the discovery of a gene, termed FGF-regulated (FR)-1, which encodes a protein with 70% similarity to hAR. FR-1 mRNA is detectable in a number of mouse tissues but is most abundant in newborn liver and in adult intestine, ovary, and testis.[22] These results raise the possibility that AR-related proteins may play a role in FGF-1–stimulated mitogenesis.

Aldehyde reductase. Aldehyde reductase, another NADPH-dependent reductase, is present in retinal pigment epithelium. The tissue content of this enzyme is much smaller than that of AR and appeared to be too low to contribute to polyol production in retinal pigment epithelial cells of diabetic patients.[23] However, when AR and aldehyde reductase were purified from human kidney cortex and medulla, the levels of aldehyde reductase exceeded those of AR in both cortex and medulla.[24] AR in the human kidney has a pI of 5.7 and a molecular mass of 37 kDa, while aldehyde reductase has a pI of 5.2 and a molecular mass of 39 kDa. Both enzymes reduce D-glucose and D-galactose to their respective sugar alcohols in vitro, and also are inhibited in vitro by some aldose reductase inhibitors (ARI).[24] Although aldehyde reductase has a 20- to 100-fold lower Km to glucose than AR, it might contribute to polyol production in the kidney, in addition to AR, due to its high content in both the cortex and medulla of human kidney.

SDH. SDH is a zinc-containing enzyme lacking activity toward aliphatic alcohols. SDH belongs to a family of proteins, including ζ-crystallin and zinc-containing alcohol dehydrogenases, which are grouped into five classes according to structural properties.[25] Methylglyoxal, 1,2-propanediol, and glycerol were considerably inefficient substrates, compared with sorbitol, for sheep liver SDH.[26] Although the polyol pathway involves both AR and SDH, the potential contribution of SDH to the abnormal polyol metabolism in diabetes has been virtually ignored. It is pertinent that red blood cell SDH deficiency in a family has been linked with diabetic pathological conditions, such as cataract formation.[27] SDH mRNA was detected in many organs and tissues, such as the liver, kidney, cornea, and seminal vesicles, but not in the small intestine.[28]

Sorbitol accumulation and myoinositol depletion. An excess of sorbitol was first demonstrated in the lens of diabetic rats by van Heyningen[8] and in the lens of diabetic patients with cataract by Kinoshita.[29] A pathogenic role of AR for nerve damage was then proposed.[7] Intracellular concentrations of nonionic organic osmolytes, such as sorbitol, myoinositol, taurine, betaine, and glycerophosphorylcholine, increase in response to external hyperosmolarity, thereby maintaining the intracellular ionic milieu. Thus, osmoregulation may be the primary physiological function of AR. Glucose-induced sorbitol accumulation in the iso-osmotic hyperglycemic state is associated with depletion of myoinositol and taurine. Because myoinositol is the precursor of phosphoinositide synthesis, its depletion may impair cell membrane biosynthesis[30] and (Na/K) ATPase activity.

In a human retinal pigment epithelial cell line, the in vitro response pattern of myoinositol to a high glucose concentration or hyperosmolarity of mannitol was related to the basal level of the AR gene expression.[31] In cultured rabbit aortic smooth muscle cells and rat glomerular mesangial cells, intracellular myoinositol content was reduced by exposing the cells to high glucose concentrations. Aldose reductase inhibitor (ARI) blocked the high glucose concentration-induced decrease in intracellular myoinosoitol content.[32,33]

In cultured neuroblastoma cells, polyol accumulation is linked with myoinositol metabolism.[34] The culture medium containing 30 mM glucose, galactose, or mannose increased the levels in cells of galactitol > sorbitol = mannitol and significantly decreased (Na/K)-ATPase activity, myoinositol content, and bradykinin-stimulated ^{32}P incorporation into phosphatidylinositol. When the ARI sorbinil was added to the culture medium, these cellular functions, as well as myoinositol content and polyol levels, were suppressed close to control values.

Redox state and protein kinase C (PKC). The change in redox state, reflected by the NADPH/NADP$^+$ and/or NADH/NAD$^+$ ratio, is another disease mechanism. NADPH is an obligate cofactor for both AR and nitric oxide synthetase. Increased flux through the polyol pathway leads to depletion of nitric oxide by cofactor competition, causing vasoconstriction and oxidative stress to diabetic endothelial cells.[35]

A recent revision and expansion of the redox hypothesis specifically emphasizes the coupling of increase in the cytoplasmic NADH/NAD$^+$ redox with the SDH reaction. In isolated glomeruli, endothelial and mesangial cells cultured at a high glucose concentration increased diacylglycerol (DAG) through de novo synthesis from glycolytic intermediates.[36] Activation of PKC by an increased cellular level of DAG increases the permeability of endothelial cells to albumin and stimulates matrix protein synthesis in mesangial cells. An increase in the cytosolic NADH/NAD$^+$ ratio induced by increased conversion of sorbitol to fructose may favor increased de novo synthesis of DAG from glucose.

On the other hand, in cultured retinal pigment epithelial cells, glucose-induced CDP-diglyceride accumulation and ^{32}P-incorporation into phosphatidylinositol paralleled myoinositol depletion, but not the change in the cytoplasmic redox state.[17] According to these results, myoinositol depletion, but not the altered redox, is supposed to be the primary cause of glucose-induced damage of cultured retinal pigment epithelial cells.

Cataract. The biochemical mechanisms of cataract formation, associated with diabetes and galactosemia, were demonstrated using various animal models and ARI.[38] Mice are known to be more resistant than rats to cataract formation during hyperglycemia. In the mouse lens, the amount of AR mRNA was shown to be lower than that of hexokinase. On the other hand, the amount of AR mRNA was four times larger than hexokinase mRNA in the rat lens. SDH mRNA was at the same level in the mouse and rat lens.[39]

The expression of hAR in the lens epithelial cells of transgenic mice was driven by a mouse α crystalline promoter.[40] On a 50% galactose diet these transgenic mice in five lines developed cataracts at a rate proportional to the hAR levels in their lens, although they did not develop a diabetic cataract. Galactitol accumulated up to 40 μmol/g wet wt in lenses of galactose-fed transgenic mice, but sorbitol accumulated only up to 3 μmol/g wet wt in lenses of diabetic mice. The total polyol accumulation in the lens was thus regarded as responsible for the formation of the cataract.

Aldose reductase inhibitors

Although interruption of the polyol pathway by inhibiting AR holds considerable promise as a therapeutic measure to prevent or delay the onset and severity of diabetic complications, the effects of ARI on diabetic complications are controversial. The effects of ARI on AR differ not only among tissues but also among species. The extent of participation of the polyol pathway is assumed to vary in different diabetic complications.

A mechanism of AR inhibition by alrestatin using the wild-type hAR and a variety of active site mutants was reported.[41] Alrestatin preferentially binds to the enzyme/NADP complex, but its binding and enzyme inhibition were abolished in some active site mutants. Enzymatic degradation of ARI is dependent not only on pH and temperature, but also on species and the type of tissues or body fluid.[42] Chemical structures of ARI are divided into four types. The first is a carboxylic acid type, which includes epalrestat, statil, tolrestat, AD-5476, and FK-366. The second is a hydantoin type, which includes sorbinil, M-79175, alconil, and SNK-860. The third is a flavonoid type. The fourth type comprises other chemicals, such as ADN-138.

ARI and neuropathy. Several studies of ARI treatment in experimental diabetic neuropathy have demonstrated beneficial effects on metabolic, functional, and structural abnormalities.[43] In diabetic BB rats, the effects of ponalrestat on the decrease of motor nerve conduction velocity (MNCV) and the increase of resistance to hypoxic conduction block (RHCV) were examined. Ponalrestat treatment prevented the decrease of MNCV in diabetic BB rats. The effect on RHCV in the ponalrestat-treated diabetic group was intermediate between the untreated diabetic and nondiabetic groups.[44] Significant improvement in myelinated nerve fiber size and fiber occupancy, and partial improvement in MNCV, have been reported for diabetic rats treated with ponalrestat for 28 wk.[45] The reduced myoinositol content and reduced axonal transport of choline acetyltransferase activity in the sciatic nerves of diabetic animals were prevented by the ARI statil, but the decrease of MNCV in diabetic rats was prevented only by statil and insulin in combination.[46]

IDDM patients, aged 18–56 years, were treated with sorbinil or a placebo by the Sorbinil Retinopathy Trial Research Group.[47] Clinical assessment of distal symmetric polyneuropathy in regard to median motor and sensory nerve conduction velocity showed very little difference between the two groups. In addition, there were no benefits in maximum amplitudes nor changes in conduction velocities. Although no evidence was obtained to support that the early clinical signs and symptoms of diabetic neuropathy were altered by sorbinil, at 30-months and at maximum follow-up visits, the distribution of changes in MNCV for the peroneal nerve showed an overall improvement in the sorbinil group. Patients with advanced diabetic neuropathy were treated for 12 months with tolrestat or with placebo-controlled drug.[48] Although no statistically significant differences in nerve morphometry emerged between the two groups, the treatment group exhibited increased nerve fiber regeneration, and normalization of axoglial dysfunction and segmental demyelination.

ARI and retinopathy. In regard to diabetic retinopathy, sorbinil prevented retinal vessel changes in age- and sex-matched beagle dogs fed a 30% galactose diet for 3 years.[49] Vascular changes, which include selective degeneration of pericytes, formation of microaneurysms and acellular capillaries, and vessel dilation, were investigated in this model. The onset and progression of these retinal changes were retarded by ARI in a dose-dependent manner. On the other hand, sorbinil failed to prevent retinopathy in diabetic or galactosemic dogs for 5 years in other experiments (see Chapter 22).[50] The development of retinopathy assessed by trypsin digests of retinal vessels and the width of retinal capillary basement membrane were not significantly influenced by ARI.

Cultured bovine pericytes were used to investigate the effect of tolrestat.[51] A high glucose concentration reduced the replication of pericytes in a dose-dependent manner. Tolrestat completely inhibited the production of sorbitol in cells exposed to a high glucose concentration but failed to protect cells from glucose toxicity reflected by cell replication. These results suggest that sorbitol accumulation in these cells is probably not the major mechanism for certain glucose toxicities.

Capillary basement membrane in retina and glomerulus became thickened as a result of either galactosemia in dogs for 5 years or alloxan-induced diabetes in dogs for 5 years. ARI did not affect capillary basement membrane thickening in the retina or glomerulus of these diabetic or galactosemic dogs, despite substantial inhibition of polyol accumulation.[52] On the other hand, the ARI ponalrestat completely prevented basement membrane thickening in the deep capillary bed but not in the superficial capillary bed of the retina of diabetic rats.[53]

Extravasation of albumin in retina, which is a marker of blood retinal barrier breakdown, as observed in human diabetic retinopathy, was demonstrated at 2.5 months in galactosemic rats.[54] Sorbinil prevented diffuse cytoplasmic positivity for albumin of retinal vascular endothelial cells in galactocemic rats. AR was immunocytochemically demonstrated in some retinal vascular endothelial cells in galactosemic rats but not in rats fed a normal chow.

A total of 497 patients with IDDM for 1–15 years were randomly assigned to take 250 mg daily sorbinil or placebo, and were followed up for a median of 41 months. Although it is unlikely that sorbinil administered for 3 years has a clinically important effect on the course of retinopathy in adults, the number of microaneurysms increased at a slightly lower rate in the sorbinil group than in the placebo group.[55]

ARI and nephropathy. Albuminuria is believed to be due to various mechanisms, including increased filtration of albumin through the glomerular basement membrane (GBM) and decreased tubular reabsorption. With regard to diabetic increase in urinary albumin excretion (UAE) ARI did[56–61] or did not[62,63] suppress increased UAE in rats, in spite of unanimously reported lowering of renal sorbitol content by ARI.[57–60,62]

Heparan sulfate is the predominant glycosaminoglycan in GBM. It confers a negative charge on GBM, and its loss has been related to the appearance of albuminuria. Diabetic rats synthesized less glomerular heparan sulfate and excreted more albumin than normal rats. Diabetic rats treated with statil did not impove heparan sulfate synthesis nor albuminuria despite normalization of glomerular sorbitol content.[64]

The effect of high glucose on mesangial extracellular matrix (ECM) production was investigated using rat glomerular mesangial cells in culture.[65] Fibronectin, laminin, and collagen type IV and VI accumulation were increased by 30 mM glucose, but not by iso-osmolar mannitol. In cells grown under high glucose concentrations, polyol concentration was increased, and that of myoinositol was reduced. The lactate/pyruvate ratio, an index of the cytosolic redox state, progressively increased under this condition. ECM changes induced by high galactose concentration were not prevented by ARI. Type IV collagen content is increased in mesangial cells cultured under a high glucose concentration, which is corrected by adding ARI into the medium.[66]

Ponalrestat was administered to IDDM patients at 600 mg/day for 3 months. The renal function in IDDM with incipient nephropathy (urinary albumin excretion >20 mg/min) in this study did not significantly improve, including glomerular filtration rate.[67]

New aldose reductase inhibitors. The lenses incubated at a high glucose concentration showed a decrease level of ATP. The lenses treated with a new ARI, AD-5467 (Takeda) showed 70–80% protection of the decrease in ATP.[68] SPR-210 (2-[4-(4,5,7-trifluolobenzothiazol-2-yl) methyl-3-oxo-3,4-dihydro-2h-1,4-benzothiazin-2-yl] acetic acid) highly inhibited partially purified AR from porcine lens ($IC_{50} = 9.5 \times 10^{-9}$ M) or human placenta ($IC_{50} = 1.0 \times 10^{-8}$ M).[69] Spiro [isoquinoline-4(1H), 3′-pyrrolidine]-1,2′,3,5′-(2H)-tetrones] inhibited partially purified AR from bovine lens.[70]

(2S,4S)-6-fluoro-2′,5′ dioxospiro-[chroman-4,4′-imidazolidine]-2-carboxamide (SNK-860) improved functional, biochemical, and structural changes in the peripheral nerve of STZ-diabetic rats. SNK-860 prevented the delayed MNCV and the reduction in myoinositol levels. Morphometric analysis of myelinated fibers also disclosed the prevention of axonal atrophy, distorted axonal circularity, and preservation of large-sized fibers following SNK-860 treatment.[71]

Epalrestat (5-[(1Z,2E)-2-methyl-3-phenylpropenyliden]-4-oxo-2-thioxo-3-thiazolidine acetic acid) is a carboxylic acid derivative. Epalrestat at 150 mg/day administered to patients with diabetic neuropathy for 12 wk improved motor and sensory nerve conduction velocity, and vibration threshold compared with placebo administration. Subjective symptoms, including pain, numbness, coldness in the extremities, muscular weakness, dizziness, and orthostatic fainting, also improved.[72] Epalrestat, 300 mg/day for 1 or 3 years, was also significantly superior to placebo in improving electroretinographic parameters, photo-stress recovery time, signs by funduscopy, and signs by fluorescent angiography in patients with diabetic retinopathy. Epalrestat appeared most effective in patients with less severe diabetes and more recent onset of diabetic complications. Data on its use in other diabetic complications such as nephropathy are lacking. Available data suggest that epalrestat produces some improvement in the late-onset diabetic neuropathy and retinopathy, and that epalrestat may be a useful drug in an area where there is a need for effective therapy.[72] Overall, the role of the polyol pathway in the evolution of diabetic nephropathy and the effectiveness of ARI are not well clarified yet.

Sorbitol dehydrogenase (SDH) inhibitors

Several ARI were shown to inhibit SDH, but only at a concentration unlikely to be reached clinically.[73] Pyrophosphate, tripolyphosphate, some bisphosphates, and several other metal chelators are shown to inactivate SDH, causing the loss of the catalytic zinc atom.[74] Aliphatic thiols are generally the most potent inhibitors of SDH activity.[73] For example, DL-2-bromo-3-(5-imidazolyl) propionic acid was shown to be a competitive inhibitor of SDH.

2-Methyl-4-[N,N-dimethylsulfamoyl-piperazino]-pyrimidine (S-0773), given in drinking water to diabetic rats for 5 wk, decreased diabetes-associated increases in fructose levels and lactate/pyruvate ratios in retina and sciatic nerve. S-0773 also decreased, but did not completely normalize, impaired caudal nerve conduction and vascular dysfunction in ocular tissues, sciatic nerve, and aorta in diabetic rats.[75]

Galactose feeding

Dietary galactose is mainly derived from lactose, the disaccharide of mammalian milk. Hydrolysis of lactose by the galactosidase or lactase of the intestinal microvilli results in the release of glucose and galactose. Most dietary galactose is converted to glucose through four enzymatic steps, including galactokinase, galactose-1-phosphate uridyltransferase, UDP-galactose-4-epimerase, and UDP-glucose pyrophosphorylase in the Leloir pathway. Galactose is also a component of glycolipids, proteoglycans, and glycoproteins in the body. Inability to metabolize galactose occurs in galactosemia, which is caused by inherited defects in any of the first three of four metabolizing enzymes.[76]

Galactose feeding is a commonly used method to examine the effect of polyol accumulation, first because AR has higher affinity to galactose than glucose, second because galactitol is barely metabolized by SDH, and third because it hardly permeates through the cell membrane. Experimental galactosemic dogs develop a retinopathy that is morphologically indistinguishable from that seen in diabetic patients and diabetic dogs.[49] Galactitol accumulates in the kidney of galactose-fed rats, especially in the renal papillae[77] and galactosemia in rats is associated with increased albumin excretion.[49,59]

On the other hand, rats fed galactose for 8 months or dogs fed galactose for 5 years did not develop significant proteinuria, mesangial expansion, glomerular obliteration, or nephromegaly. The width of GBM in galactose-fed dogs and rats was larger than that in those fed a normal diet.[78–80] The abnormal endothelium-dependent vessel relaxation in rats fed a 40% galactose diet was completely prevented by treating the rats with ponalrestat.[81] Galactose-fed rats serve as a useful model to study the mechanisms of diabetic albuminuria, especially with regard to the role of polyol accumulation in the etiology of diabetic complications.[82]

Study of transgenic mice expressing hAR

We studied the role of the polyol pathway in simple retinopathy and diabetic albuminuria in transgenic mice (Tg), in which expression of hAR is driven by a murine MHC class I ($H\text{-}2K^d$) promoter (hAR-Tg). hAR-Tg was generated using a murine $H\text{-}2K^d$ promoter fused to an hAR coding sequence, as described previously.[83] The founder mouse of BDF1 was crossed to BDF1 mice, and integration of the transgene was detected by PCR. At 4 wk of age, DNA was extracted from a punched ear lobe according to Qi et al.[84] To detect chimerical sequences between each of flanking region of rabbit β-globin exon 3 and the inserted hAR cDNA, oligonucleotide primers were designed (Fig. 2). PCR conditions have been described previously.[85]

Assay of hAR protein by ELISA. In hAR-Tg, hAR was expressed in the brain, lung, heart, liver, sciatic nerve, and kidney by Northern and Western blot analyses, while no hAR protein was detected in these organs of littermates.[83] hAR proteins in the retina, lens, and kidney of hAR-Tg and the littermates at 16 wk of age were assayed by ELISA, as reported previously.[83] In the retina and kidney of hAR-Tg, the expression of hAR was confirmed. The amount of hAR protein in μg/g tissue was 84 ± 20 in retina (mean ± SD, n = 5), 1.8 ± 0.5 in lens

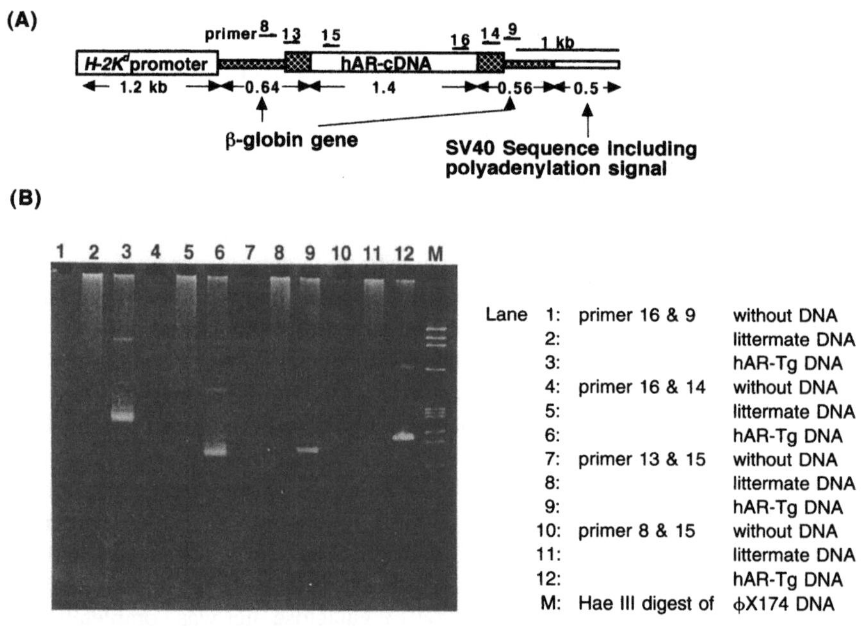

Figure 2. PCR detection of the transgene. (A) Schematic presentation of the PCR primers; their relative locations within the DNA construct are indicated by arrows. The sequence includes a murine MHC class I (*H-2K^d*) promoter; a part of exon 2, intron 2, and a part of exon 3 of the rabbit β-globin gene; hAR cDNA (1,5 kb); a part of rabbit β-globin exon 3; and a poly A addition sequence. (B) PCR products amplified by each pair of primers, which were subjected to polyacrylamide gel electrophoresis and were visualized by ultraviolet illumination in the presence of ethidium bromide. Lane M: *Hae*III digest of ϕX174 DNA. The PCR primer pair of 9 and 16 was used for lanes 1, 2, and 3; that of 14 and 16 was used for lanes 4, 5, and 6; that for 13 and 15 was used for lanes 6, 7, and 8; and that of 8 and 15 was used for lanes 10, 11, and 12. Lanes 1, 4, 7, and 10 are the control reaction without DNA. Lanes 3, 6, 9, and 12 showed the expected 200, 150, 130, and 230 bp fragment of hAR-Tg, respectively. Lanes 2, 5, 8, and 11 showed no amplified band from DNA of the littermates.

(n = 5), and 481 ± 111 in kidney (n = 8) of hAR-Tg. In contrast, hAR protein was not detected in these organs of littermates.

Grouping of experiment animals. At 6 wk of age mice, including both hAR-Tg and littermates, were divided into the following five groups: Group 1 (Control) was nondiabetic mice fed a standard chow. Groups 2 (DM) and 3 (DM + ARI) were diabetic mice not treated and treated with ARI, respectively. Diabetes was induced by a single i.p. administration of 200 mg/kg STZ to male mice and 220 mg/kg to female mice (Wako Pure Chemicals, Osaka) in 0.05 M citrate buffer, pH 4.5. Mice with blood glucose >250 mg/dl were considered diabetic. The ARI epalrestat (Ono Pharmaceutical Co., Osaka), was administered as a mixture, 0.08% by weight, in the standard chow (Oriental Bio-Service, Inc. Kyoto). This treatment diet was started on the day of STZ administration of 6 wk of age. Groups 4 (Gal) and 5 (Gal + ARI) were mice fed a chow containing 30% (wt/wt) Gal (Bio-Serve, Frenchtown, NJ) and chow mixed with ARI, respectively. The number of mice in each group ranged from 4 to 11, as shown in Tables 1 and 2. Mice were maintained on a 12-h light/12-h dark cycle, and were fed ad libitum.

Ophthalmoscopic examination of optic fundi. To examine the retinopathy, ophthalmoscopic examination, and fluorescent retinal angiography were performed. Optic fundi of diabetic- or galactose-fed mice with or without ARI treatment were examined once every 3 or 4 wk by indirect ophthalmoscopy, and the results were recorded with a hand-held fundus camera (Genesis, Kowa Ltd., Nagoya, Japan). Fluorescent retinal angiography was performed with a fluorescein dye (Alcon, Fort Worth, TX) on some mice with retinal red spots.

As shown in Figure 3 (left), retinal red spots, which were suggestive of diabetic microaneurysms, were detected by indirect ophthalmoscopy on the small retinal vein of hAR-Tg within 4 wk after starting galactose feeding, but not in diabetic hAR-Tg or littermates. Some retinal red spots were reversible. As seen in Figure 3 (right), fluorescent retinal angiography showed the saccular staining and weak leakage of fluorescent dye observed in diabetic microaneurysms. The incidence of retinal red spots, such as those shown in Figure 3, was assessed at each time point by determining their frequency relative to the total number of examined eyes. As shown in Figure 4, the frequency of eyes with these spots in male hAR-Tg and littermates increased with time. By analysis of variance (ANOVA), the frequency in male hAR-Tg, including the time points of 10, 12, 13, 17, 20, 21, and 26 wk of age, were significantly higher than that in the littermates ($P < 0.001$). ARI significantly suppressed this frequency in hAR-Tg and the littermates, including data on the above-mentioned time points ($P < 0.001$ and $P < 0.005$, respectively).

In galactose-fed animals, cataract was reported to arise from osmotic effects associated with polyol accumulation.[86] In our experiment, cataract was not observed either in galactose-fed hAR-Tg at 28 wk of age or in diabetic hAR-Tg at 14 wk of age. Although the MHC-class I promoter was supposed to direct expression of its downstream gene in various organs, hAR activity in the lens was at the same level as that in littermates, and it was lower than in retina of hAR-

Table 1. Parameters of kidney and body wt at 10 wk of age

Items	Group 1 control		Group 2 DM		Group 3 DM + ARI		Group 4 Gal		Group 5 Gal + ARI	
examined	Tg	LM	Tg	LM	Tg	LM	Tg	LM	Tg	LM
Kidney	134	125	128	151	162	171	100	98	106	116
Weight[a]	± 10	± 23	± 19	± 30	± 4	± 23	± 14	± 13	± 5	± 18
Protein	154	131	148	143	146	162	198†	169†	170	139
concent.[b]	± 36	± 20	± 22	± 25	± 14	± 27	± 12	± 16	± 45	± 35
Total	20.6	16.0	18.7	21.0	23.6‡	27.3‡	19.7	16.7	18.0	16.6
protein[c]	± 3.4	± 6.0	± 2.8	± 2.6	± 2.1	± 2.5	± 1.7	± 3.3	± 4.1	± 6.3
n[d]	(5)	(5)	(9)	(12)	(5)	(6)	(4)	(5)	(5)	(5)
Body wt[e]	21.8	20.2	13.3*	15.0*	17.0§	18.2§	19.2	18.0	18.1	18.5
	± 1.1	± 1.0	± 2.6	± 3.5	± 2.0	± 2.5	± 1.2	± 1.5	± 0.5	± 1.1
Kidney wt/	6.2	6.4	9.1	8.3†	9.6†	9.5†	5.2†	5.5	5.8	6.2
body wt[f]	± 0.5	± 0.9	± 2.5	± 1.4	± 1.0	± 1.3	± 0.6	± 0.5	± 0.3	± 0.6
n[g]	(4)	(7)	(5)	(4)	(5)	(6)	(4)	(5)	(5)	(5)

[a]Kidney wt in mg.

[b]Protein concentration per kidney wt in mg/g wet kidney wt.

[c]Total protein content per kidney in mg.

[d]Number of animals used for the parameters presented above.

[e]Body wt in g.

[f]Kidney wt/body wt ratio.

[g]Number of animals used for determining the body to kidney wt ratio.

Each value represents the mean ± SD. Significant differences from the control Tg or LM by the unpaired *t*-test.
$*P < 0.005$; $†P < 0.05$.

Significant differences from each DM-Tg or DM-LM by an unpaired *t*-test: $‡P < 0.01$; $§P < 0.05$.

Tg, transgenic mice overexpressing aldose reductase; LM, littermates; ARI, aldose reductase inhibitor.

Table 2. Changes in 24 h urinary volume

Age (wk)	Group 1 Control		Group 2 DM		Group 3 DM + ARI		Group 4 Gal		Group 5 Gal + ARI	
	Tg	LM	Tg	LM	Tg	LM	Tg	LM	Tg	LM
6	2.5 ± 0.6 (3)	2.8 ± 0.5 (3)	2.5 ± 0.6 (3)	2.8 ± 0.5 (3)	2.5 ± 0.6 (3)	2.8 ± 0.5 (3)	2.5 ± 0.6 (3)	2.8 ± 0.5 (3)	2.5 ± 0.6 (3)	2.8 ± 0.5 (3)
7	1.7 ± 0.5 (5)	1.7 ± 0.3 (5)	13.0 ± 1.0 (4)	11.1 ± 2.5 (5)	12.6 ± 1.0 (4)	12.4 ± 1.4 (4)	8.8 ± 1.5 (5)	6.5 ± 0.8 (5)	8.3 ± 2.0 (5)	6.6 ± 2.6 (5)
8	2.0 ± 0.5 (8)	2.1 ± 0.5 (8)	13.4 ± 1.9 (6)	15.1 ± 0.8 (6)	12.2 ± 2.4 (6)	12.7 ± 1.3 (7)	7.4 ± 0.7 (5)	6.3 ± 1.5 (5)	6.9 ± 1.4 (4)	8.1 ± 1.2 (5)
10	2.0 ± 0.4 (5)	1.7 ± 0.6 (5)	15.5 ± 2.1 (8)	14.5 ± 1.4 (10)	11.7 ± 1.0 (9)	12.7 ± 1.4 (8)	9.5 ± 1.1 (5)	7.7 ± 1.3 (5)	7.0 ± 1.2 (4)	8.9 ± 1.6 (4)
14	1.7 ± 0.6 (11)	1.1 ± 0.5 (9)	15.3 ± 1.2 (7)	15.3 ± 0.3 (5)	12.4 ± 2.5 (4)	12.2 ± 0.8 (4)	5.0 ± 0.3 (5)	4.6 ± 0.9 (5)	2.8 ± 0.6 (4)	2.6 ± 0.9 (4)

Each value represents the mean ± SD in ml/24 h. The number of animals is shown in parentheses. The groups are as in Table 1.

Tg and littermates. The negligible level of transgenic expression of hAR in the lens is supposedly responsible for the absence of cataract in this animal model.

Assay of polyol. Sorbitol and galactitol in the kidney of diabetic and galactose-fed mice were assayed by HPLC using a modification of the method of Dethy.[87] Detection of polyol was based on the formation of a urethane bond between phenylisocyanate and the hydroxyl groups of polyols, as described before.[85] Briefly, kidney was homogenized in 4 vol deionized water containing xylitol as

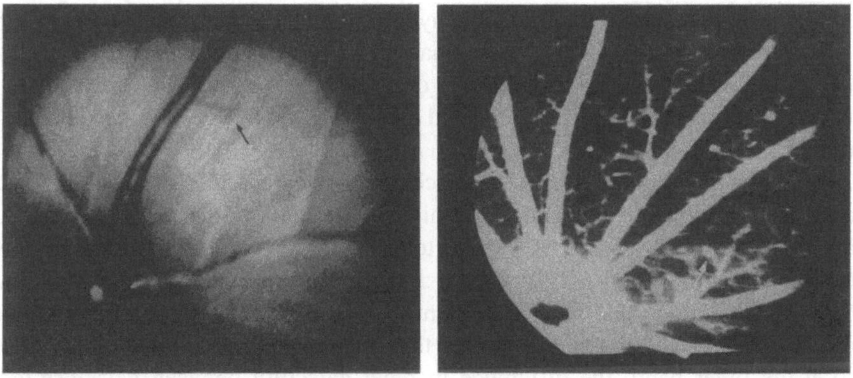

Figure 3. Retinal red spots in hAR-Tg (left) and a fluorescent retinal angiography of hAR-Tg (right). Retinal red spots in hAR-Tg were observed by indirect ophthalmoscopy.

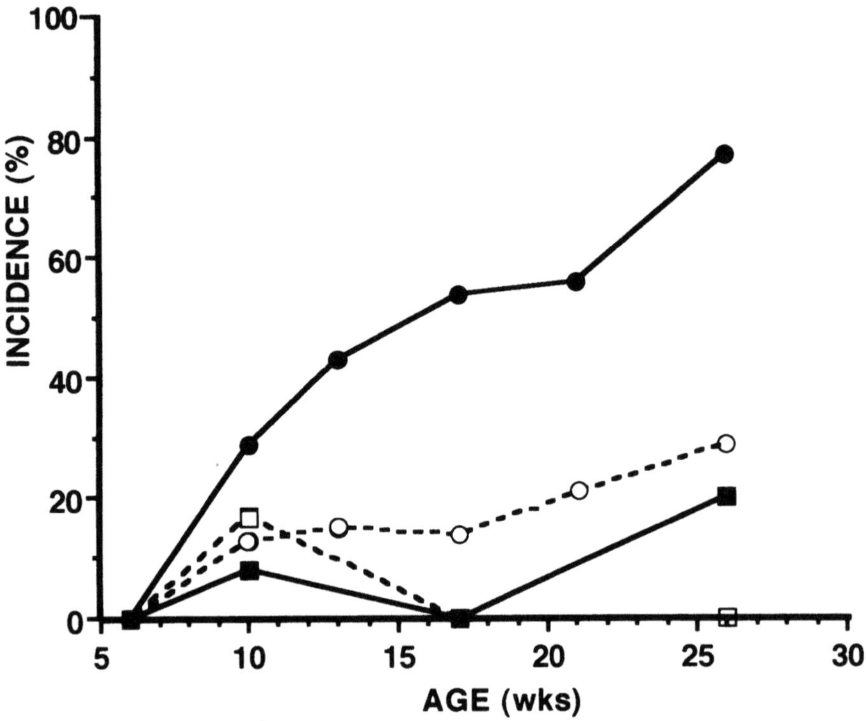

Figure 4. Incidence of retinal red spots in group 4, including Gal-fed hAR-Tg (●) and Gal-fed littermates (○), and those in group 5, including Gal-fed hAR-Tg with ARI (■) and Gal-fed littermates with ARI (□).

an internal standard. After precipitating the protein, the supernatant was lyophilized. The lyophilized sample was heated after adding pyridine. Then, 2 vol phenylisocyanate were added and the mixture was incubated at 55°C for 30 min. After cooling, methanol was added to react with excess phenylisocyanate and the mixture was filtered through a hydrophilic PTFE membrane. Samples of 5 μL were applied to a C-18 reverse phase column with a gradient of 55–100% acetonitrile with 0.05% TFA in water, as described previously.[85] Statistical analysis was performed by Student's unpaired *t* test. Two-way ANOVA was then performed.

As shown in Figure 5, renal sorbitol content in diabetic hAR-Tg of 1.3 $\pm$ 0.2 mg/g protein showed a tendency to be higher than that in littermates (1.1 $\pm$ 0.3, $P < 0.08$) at 10 wk of age. A similar tendency was oberved at 14 wk of age (0.86 $\pm$ 0.3 in diabetic hAR-Tg vs. 0.6 $\pm$ 0.1 in diabetic littermates, $P < 0.08$). With ANOVA, the renal sorbitol content in hAR-Tg, including the time points of 7, 10, and 14 wk of age, was higher than that in littermates with $P < 0.001$.

On the other hand, the difference in renal galactitol between galactose-fed hAR-Tg and littermates was not significant at any of the four time points examined (Fig. 6). Although renal sorbitol of diabetic hAR-Tg and littermates increased from the first week after STZ administration at 6 wk of age, and then decreased gradually up to 14 wk of age, renal galactitol in galactose-fed hAR-Tg and littermates increased linearly. ARI suppressed renal sorbitol in diabetic hAR-Tg and littermates at the examined time points of 7, 10, and 14 wk of age to the nondiabetic level. Renal galactitol in hAR-Tg was suppressed to 58% ($P < 0.01$) and that in littermates to 50% ($P < 0.01$) only at 14 wk of age.

Assay of body weight and renal parameters. The kidney wt, renal protein concentration, total protein content per kidney, body wt, and kidney wt/body wt were determined for hAR-Tg and littermates in 5 groups at 10 wk of age. Results are summarized in Table 1. Renal protein concentrations were assayed according to the method of Bradford.[88] Body wt of the hAR-Tg in group 3 (DM + ARI) was higher than that of hAR-Tg in group 2 (DM) ($P < 0.05$). These data suggest that diabetes decreased body wt, which recovered toward normal by concomitant administration of ARI in hAR-Tg. The kidney wt/body wt ratio of hAR-Tg and lit-

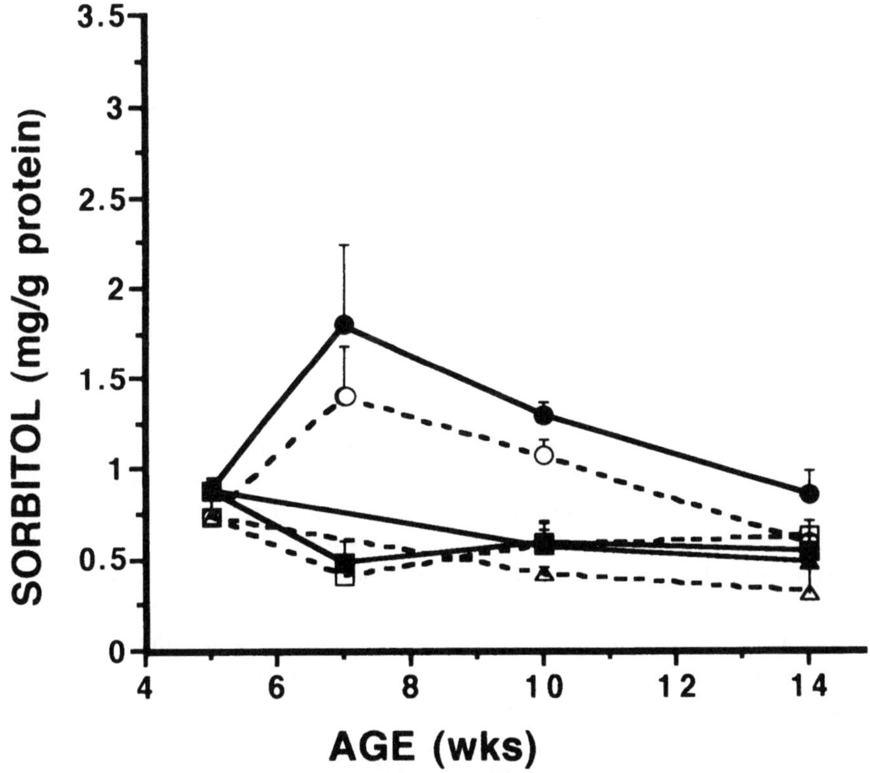

Figure 5. Changes in renal sorbitol content of hAR-Tg and the littermates in groups 1 (Control), 2 (DM), and 3 (DM + ARI). Marks for different groups are as follows: Group 1 includes nondiabetic hAR-Tg (▲) and nondiabetic littermates (△), group 2 includes diabetic hAR-Tg (●) and diabetic littermates (○), and group 3 includes diabetic hAR-Tg with ARI (■) and diabetic littermate with ARI (□). Values are means ± SE in all figures.

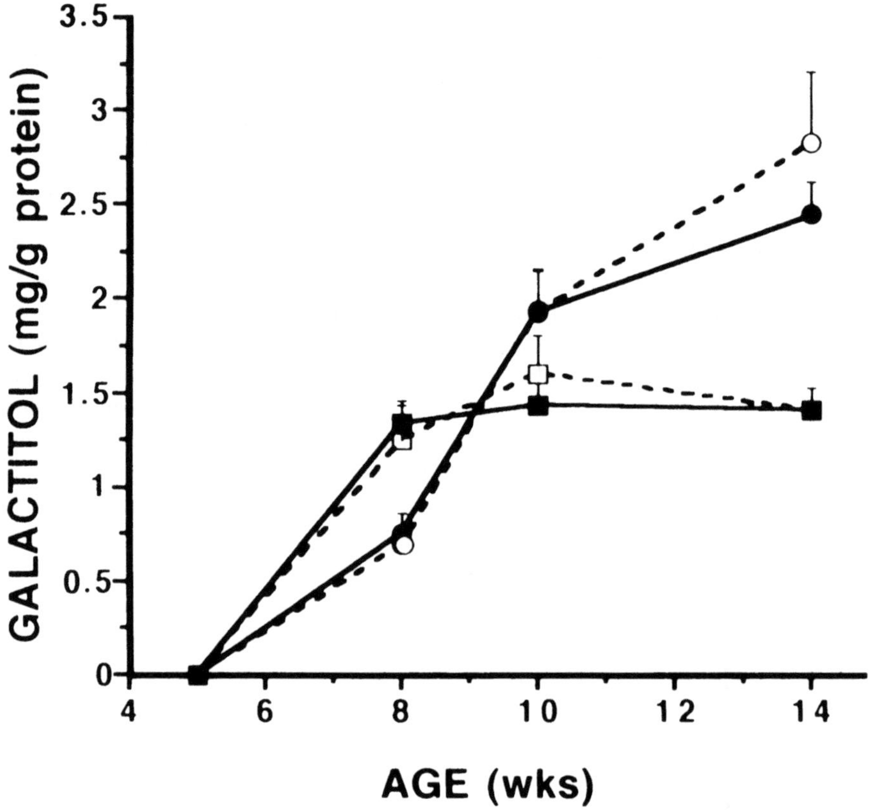

Figure 6. Changes in renal galactitol content in groups 4 (Gal) and 5 (Gal + ARI). The symbols for the different groups are as follows: group 4 includes Gal-fed hAR-Tg (●) and Gal-fed littermate (○), and group 5 includes Gal-fed hAR-Tg with ARI (■) and Gal-fed littermate with ARI (□).

termates in group 2 (DM) was significantly higher than of those in group 1 (control). In galactose-fed hAR-Tg and littermates, body wt and protein/kidney did not show any changes, but the kidney wt/body wt ratio decreased in comparison with control mice in group 1. Although on ARI treatment body wt recovered in diabetic hAR-Tg, ARI did not change the kidney wt/body wt ratio in diabetic or galactose-fed mice. Although total kidney protein in diabetic mice treated with ARI was higher than that in diabetic hAR-Tg and littermates without ARI ($P <$ 0.01 for both), that in galactose-fed mice treated with ARI did not show any difference compared with that in galactose-fed mice not treated with ARI.

Changes in urinary albumin excretion (UAE) in diabetic or galactose-fed mice. Microalbuminuria (albumin excretion of 20–200 μg/min) is a key abnormality in the evolution of clinically relevant diabetic nephropathy. It has been documented in several studies that microalbuminuria predicts overt renal disease and renal failure in IDDM and NIDDM. Thus, the earliest clinically detectable renal involvement in diabetes is the appearance of microalbuminuria.

At the end of wk 0, 1, 4, and 8 after STZ administration or initiation of galactose feeding, at 6, 7, 10, and 14 wk of age, respectively, urine was collected for 3 consecutive days. Urinary volumes (UV) and urinary albumin concentrations at 24 h were assayed by an immuno-nephelometric method[89] using the Beckman Array Protein System (Brea, CA) with anti-mouse albumin rabbit IgG antibody (Organo Teknika Cooperation, Durham, NC). The mean of three values of 24 h UAE in each mouse was used as the assay result.

As shown in Figure 7, UAE in units of μg/24 h in both diabetic hAR-Tg and the diabetic littermates in group 2 increased remarkably from the first week after STZ administration compared with those in nondiabetic hAR-Tg and littermates in group 1. Although UAE in galactose-fed hAR-Tg and in the littermates was lower than in diabetic mice, it increased from the first week after galactose feeding in comparison with control mice in group 1 (Fig. 8). The differences in UAE between hAR-Tg and littermates were not significant at any of the four time points examined in diabetic group 2 or Gal-fed group 4. ARI tended to suppress UAE down to 56% ($0.05 < P < 0.10$) in diabetic hAR-Tg and to 64% ($P < 0.10$) in littermates, to 53% ($P < 0.05$) in galactose-fed hAR-Tg at 10 wk of age but the UAE levels were not lowered to those in control mice of group 1. The effects of ARI by two-way ANOVA on the overall changes, including data at 6, 7, 10, and 14 wk of age, showed suppression of UAE in diabetic hAR-Tg ($0.05 < P < 0.10$) as well as in galactose-fed AR-Tg and littermates ($P < 0.05$ for both).

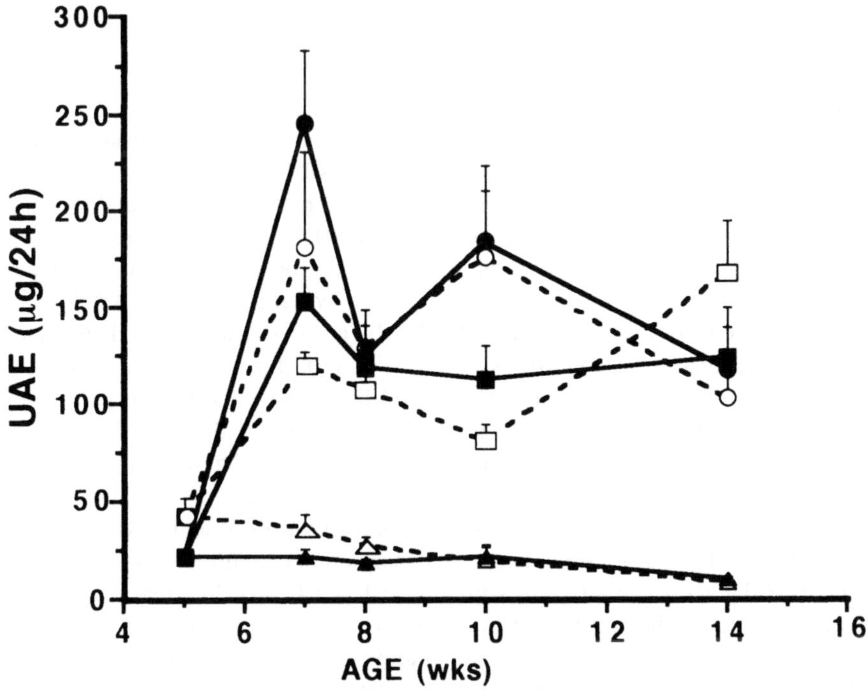

Figure 7. Changes in UAE in hAR-Tg and the littermates in groups 1 (Control), 2 (DM), and 3 (DM + ARI). The symbols are the same as in Figure 6.

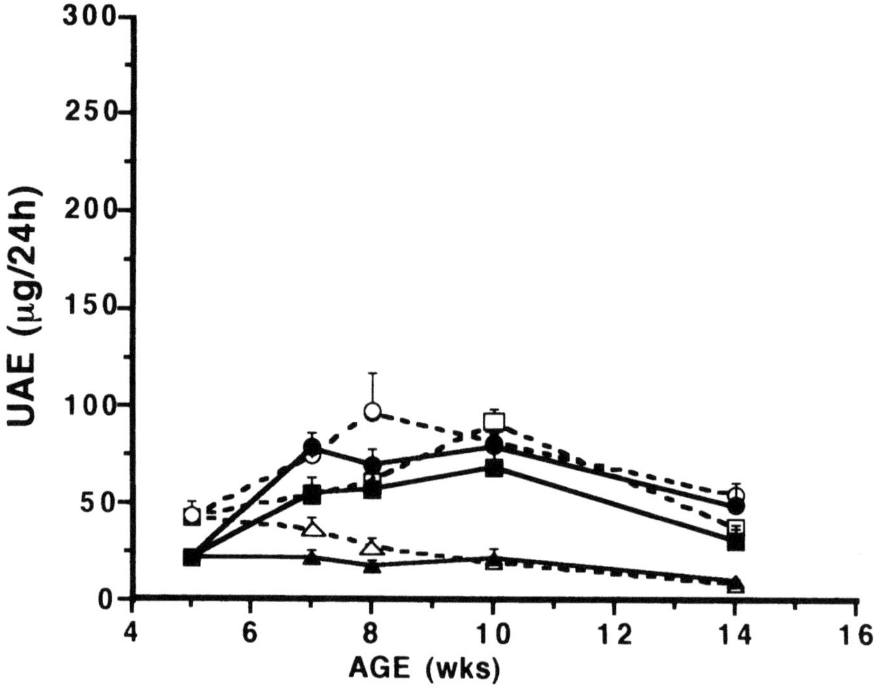

Figure 8. Changes in UAE in hAR-Tg and the littermates in groups 1 (Control), groups 4 (Gal), and 5 (Gal + ARI). The symbols are the same as in Figure 6.

Changes of urinary volume. As summarized in Table 2, all diabetic and galactose-fed mice had frank polyuria compared with control mice in Group 1. ARI suppressed the urinary volume (UV) of diabetic hAR-Tg at 10 and 14 wk of age ($P < 0.0005$ and $P < 0.05$, respectively); that of diabetic littermates at 7, 10, and 14 wk of age ($P < 0.005$, $P < 0.02$, and $P < 0.0001$, respectively); that of galactose-fed hAR-Tg at 10 and 14 wk of age ($P < 0.05$ for both); and that of galactose-fed littermates at 14 wk of age ($P < 0.05$). When the UV of hAR-Tg and littermates were compared, galactose-fed hAR-Tg showed larger UV than galactose-fed littermates at 7 and 10 wk of age ($P < 0.05$ for both). With ANOVA, ARI suppressed UV in diabetic hAR-Tg and littermates, and that in galactose-fed hAR-Tg at all time points examined ($P < 0.002$, $P < 0.005$, and $P < 0.005$, respectively). With ANOVA, UV in galactose-fed hAR-Tg were significantly higher than those in galactose-fed littermates ($P < 0.001$).

The significant positive correlation observed between UV and UAE in galactose-fed hAR-Tg and littermates was in a sharp contrast with its absence in diabetic hAR-Tg and the littermates with $r = 0.01$ and $P = 0.28$. Urea-induced osmotic diuresis has been reported to cause glomerular hyperfiltration and renal hypertrophy.[90] Thus, the diuresis with albuminuria in galactose-fed mice in our study may be due, in part, to the osmotic diuretic effect of galactose. Because UV in galactose-fed hAR-Tg was significantly higher than that in galactose-fed littermates ($P < 0.001$) and because ARI suppressed UV in galactose-fed hAR-Tg with ANOVA of the data at 7, 8, 10, and 14 wk of age ($P < 0.005$), galactitol accumulation is supposed to increase UV. Increased transglomerular pressure[56] and/or decreased reabsorption of albumin[61] might be responsible for the increased UAE in galactose-fed mice in our study. Because UAE correlates with UV in the time course after the start of galactose feeding, the increased UAE is considered to be regulated by the increased UV. The absence of a correlation between UV and UAE in STZ-diabetic mice, however, suggests that the increased UV is not the major determinant of the increased UAE in diabetes.

Discussion

Our study was undertaken to examine the role of AR in the etiology of diabetic simple retinopathy and albuminuria by periodically examining the frequency of retinal red spots, UAE, and renal polyol content in the kidney of diabetic or galactose-fed hAR-Tg and littermates.

Retinopathy. With regard to diabetic retinopathy, no opthalmoscopic retinal changes were observed in diabetic hAR-Tg during the time period examined until 14 wk of age. Examination of ocular fundi detected retinal red spots (see Fig. 3) in galactose-fed mice at 4 wk after starting galactose feeding (see Fig. 4). Since fluorescent retinal angiography showed saccular staining and weak leakage of fluorescent dye, as observed in diabetic microaneurysms (see Fig. 3), these retinal red spots were considered to represent diabetic microaneurysms. Microaneurysms in ocular fundi are regarded as a marker of diabetic retinopathy. The development of microaneurysms after 4 wk of galactose feeding is the most rapid development of simple diabetic retinopathy reported in any animal model. This is in contrast to STZ-diabetic rats fed 72% fructose, in which retinal changes were detected after 3–5 months, and 30% galactose-fed dogs, in which retinal changes were detected only after 2–3 years.[86] The incidence of these spots was higher in hAR-Tg than in the littermates up to 26 wk of age, and was significantly suppressed by ARI (see Fig. 4). The higher incidence of retinal red spots in hAR-Tg than in littermates, and the reversal of galactose-induced retinopathy with ARI both in hAR-Tg and littermates suggest the importance of the polyol pathway in the development of diabetic retinopathy. However, the absence of retinopathy in diabetic hAR-Tg suggests that additional, hitherto unknown factors may operate in triggering its development.

Albuminuria. In addition to frank polyuria in diabetic and galactose-fed mice, UV in galactose-hAR-Tg was significantly increased compared with galactose-fed littermates. ARI suppressed UV in diabetic hAR-Tg, diabetic littermates, and galactose-fed hAR-Tg. These observations suggest a possible effect of polyol accumulation on UV in diabetes and galactose feeding. The observed diuresis (see Table 2) and increased UAE (see Fig. 8) in galactose-fed mice were about half of those in diabetic mice (see Fig. 7). Galactose feeding simulates the diuresis, albuminuria, and polyol accumulation of the diabetic condition, but differs from diabetes with regard to renal hypertrophy (see Table 1) and the time course of polyol accumulation (see Fig. 6). Furthermore, the magnitude of changes in UAE and UV differed between the two models, suggesting that galactose-fed mice are a useful model for the study of some aspects of diabetic albuminuria.

Although the reversal of decreased body wt in diabetic hAR-Tg suggests a beneficial effect of ARI, the increased kidney wt/body wt ratio was not reversed by ARI (see Table 2). In addition, total kidney protein increased with ARI treatment in diabetic animals, but did not change in galactose-fed mice (see Table 2). This seems to rule out a nonspecific renal side effect of ARI, but suggests differences in the action of the polyol pathway in diabetes compared with galactose feeding. The etiologic role of polyol accumulation in diabetes or galactose-induced nephropathy was substantiated by the suppression of UAE by ARI.[56–58] In our study, ARI significantly suppressed renal polyols and tended to suppress UAE in diabetic or galactose-fed hAR-Tg. The modest suppression of UAE in diabetic animals following ARI treatment is compatible with the reported partial suppression to 66% in diabetic rats 6 wk after STZ administration.[91] Although the dose of ARI as a mixture of 0.08% by wt in the chow roughly corresponds to a dose 40 times higher than used for diabetic patients, the larger dose may be necessary to completely suppress galactitol accumulation and UAE induced by a 30% galactose diet.

The high hAR in hAR-Tg was not reflected by obvious increases in UAE compared with the littermates in nondiabetic, diabetic, or galactose-fed animals (see Figs. 5–8). On the other hand, high hAR in hAR-Tg was reflected by the following observations: first, the renal sorbitol concentration in diabetic hAR-Tg was higher than that in diabetic littermates; second, 24 h UV in galactose-fed hAR-Tg was higher than that in galactose-fed littermates (see Table 2); third, UAE in diabetic hAR-Tg tended to be suppressed with ARI treatment, although UAE in diabetic littermates treated with ARI was not significantly suppressed (see Fig. 7); lastly, UV in galactose-fed hAR-Tg was suppressed with ARI treatment, although UV in galactose-fed littermates was not suppressed with ARI (see Table 2). The absence of clear-cut effects of transgenic expression of hAR on galactose- or diabetes-induced increases in UAE, in contrast to the effects on galactose-induced retinopathy, as discussed earlier, and the high hAR protein level in the kidney of hAR-Tg suggest that increased flux through the polyol pathway in the kidney may not be a major determinant for the increased UAE.

Conclusions

Our results suggest that polyol accumulation is essential for diabetic retinopathy, and is modestly important for galactose- or diabetes-induced increases in UAE. This hAR-Tg model may be suitable to selective study of the role of hAR in the pathogenesis of diabetic complications, demonstrating a remarkably early onset and increased incidence of galactose-induced simple retinopathy. In addition, hAR-Tg demonstrated that ARI is highly effective in preventing galactose-induced retinopathy and is modestly effective in preventing nephropathy.

Future directions. Periodical assays of AR and SDH enzyme activity in the kidney and retina of diabetic or galactose-fed mice will provide useful information on the regulation of polyol accumulation by intrinsic mouse AR and SDH. In our study, renal sorbitol in diabetic mice increased within 1 wk after STZ administration, decreased thereafter, but still remained higher than that in nondiabetic mice at 14 wk of age. This was strikingly different from the constant linear increase in renal galactitol in galactose-fed mice (see Figs. 5 and 6). The possibility of increased leakage of sorbitol as a consequence of chronic cellular polyol accumulation[92] is unlikely, because galactitol continued to accumulate in galactose-fed mice. The decreased renal sorbitol content may instead be due to increased SDH activity and/or an increased NAD^+/NADH ratio in the diabetic kidney. A decreased NADPH/$NADP^+$ ratio, as well as changes in the redox state in diabetic rats, is believed to play a significant role in the control of sorbitol synthesis.[17]

Examination of the effect of SDH inhibitors on diabetic or galactose-fed nephropathy, and retinopathy in hAR-Tg and littermates, may clarify which step in polyol metabolism is responsible for cell damage, that is, reduction of glucose to sorbitol with a high oxidized cytosolic ratio of NADPH/$NADP^+$, or oxidation of sorbitol to fructose with a reduced cytosolic ratio of NADH/NAD^+. Diabetic vascular dysfunction might be more closely linked to increased oxidation of sorbitol to fructose than to reduction of glucose to sorbitol. Administration of an SDH inhibitor for 5 wk to diabetic rats markedly attenuated diabetes-induced vascular dysfunction in the sciatic nerve.[93] Because glomerular mesangial cells and retinal pericytes play a central role in the pathogenesis of diabetic nephropathy and retinopathy, cultured mesangial cells and pericytes obtained from hAR-Tg are expected to clarify the roles of polyol accumulation and cellular redox states, in addition to whole organ study in hAR-Tg.

Acknowledgments

This study was supported in part by a grant from Otsuka Pharmaceutical Factory to the Otsuka Department of Clinical and Molecular Nutrition, School of Medicine, the University of Tokushima. We thank Drs. Y. Kokai and J. Fujimoto for the initial production of hAR-Tg, Drs. Y. Takahashi and Y. Tano for examining ocular fungi of mice, Dr. T. Nakayama for his suggestions on the assay of polyols, Dr. C. Nisimura for ELISA of hAR, and E. Adachi and Y. Mizuguchi for expert technical assistance.

References

1. Blom ML, Green WR, Schachat AP. Diabetic retinopathy: A review. Del Med J 66:379–88, 1994.
2. American Diabetes Association. Diabetes 1993 vital statistics. American Diabetes Association, Alexandria, VA, 1993.
3. West KM, Erdreich LJ, Stober JA. A detailed study of risk factors for retinopathy and nephropathy in diabetes. Diabetes 29:501–8, 1980.

4. Jerneld B, Algvere P. Proteinuria and blood glucose levels in a population with diabetic retinopathy. Am J Ophthalmol 104:283–9, 1987.
5. Frank RN. On the pathogenesis of diabetic retinopathy. A 1990 update. Ophthalmology 98:586–93, 1991.
6. Bunn HF, Higgins PJ. Reaction of monosaccharides with proteins: Possible evolutionary significance. Science 213:222–4, 1981.
7. Greene DA, Lattimer S, Ulbrecht J. Glucose-induced alterations in nerve metabolism: Current perspective on the pathogenesis of diabetic neuropathy and future directions for research and therapy. Diabetes Care 8:290–9, 1985.
8. van Heyningen R. Formation of polyols by the lens of the rat with sugar cataract. Nature 184:194–5, 1959.
9. Wilson DK, Bohren KM, Gabbay KH, Quiocho FA. An unlikely sugar substrate site in the 1.65 Å structure of the human aldose reductase holoenzyme implicated in diabetic complications. Science 257:81–4, 1992.
10. Bateman JB, Kojis T, Heinzman C, et al. Mapping of aldose reductase gene sequences to human chromosomes 1, 3, 7, 9, 11, and 13. Genomics 17:560–5, 1993.
11. Wang K, Bohren KM, Gabbay KH. Characterization of the human aldose reductase gene promoter. J Biol Chem 268:16052–8, 1993.
12. Garcia-Perez A, Burg MB. Renal medullary organic osmolytes. Physiol Rev 71:1081–115, 1991.
13. Smardo FL Jr, Burg MB, Garcia-Perez A. Kidney aldose reductase gene transcription is osmotically regulated. Am J Physiol 262:C776–82, 1992.
14. Burg MB. Molecular basis for osmoregulation of organic osmolytes in renal medullary cells. J Exp Zool 268:171–5, 1994.
15. Ferraris JD, Williams CK, Martin BM, et al. Cloning, genomic organization, and osmotic response of the aldose reductase gene. Proc Natl Acad Sci USA 91:10742–6, 1994.
16. Nakanishi T, Yamauchi A, Nakahama H, et al. Organic osmolytes in rat renal inner medulla are modulated by vasopressin V1 and/or V2 antagonist. Am J Physiol 267:F146–52, 1994.
17. Grunewald RW, Weber II, Kinne-Saffran E, Kinne RK. Control of sorbitol metabolism in renal inner medulla of diabetic rats: Regulation by substrate, cosubstrate and products of the aldose reductase reaction. Biochim Biophys Acta 1225:39–47, 1993.
18. Bondy C, Cowley BD Jr, Lightman SL, Kador PF. Feedback inhibition of aldose reductase gene expression in rat renal medulla. J Clin Invest 86:1103–8, 1990.
19. Kikkawa R, Umemura K, Haneda M, et al. Identification and characterization of aldose reductase in cultured rat mesangial cells. Diabetes 41:1165–71, 1992.
20. Wu RR, Lyons PA, Wang A, et al. Effects of galactose feeding on aldose reductase gene expression. J Clin Invest 92:155–9, 1993.
21. Fabre S, Manin M, Pailhoux E, et al. Identification of a functional androgen response element in the promoter of the gene for the androgen-regulated aldose reductase-like protein specific to the mouse vas deferens. J Biol Chem 269:5857–64, 1994.
22. Donohue PJ, Alberts GF, Hampton BS, Winkles JA. A delayed-early gene activated by fibroblast growth factor-1 encodes a protein related to aldose reductase. J Biol Chem 269:8604–9, 1994.
23. Sato S, Lin LR, Reddy VN, Kado PF. Aldose reductase in human retinal pigment epithelial cells. Exp Eye Res 57:235–41, 1993.
24. Sato S, Kador PF. Human kidney aldose and aldehyde reductases. J Diabetes Complications 7:179–87, 1993.
25. Jörnvall H, Persson B, Krook M, Hempel J. Alcohol and aldehyde dehydrogenases. In: The molecular pathology of alcoholism. Palmer N, ed. Oxford University Press, Oxford, pp 130–56, 1991.
26. Lindstad RI, McKinley-McKee JS. Methylglyoxal and the polyol pathway. Three-

carbon compounds are substrate for sheep liver sorbitol dehydrogenase. FEBS Lett 330:31–5, 1993.
27. Vaca G, Ibarra B, Bracamontes M, et al. Red blood cell sorbito dehydrogenase deficiency in a family with cataracts. Hum Genet 61:338–41, 1982.
28. Estonius M, Danielsson O, Karlsson C, et al. Distribution of alcohol and sorbitol dehydrogenases: Assessment of mRNA species in mammalian tissues. Eur J Biochem 215:497–503, 1993.
29. Kinoshita JH. Mechanisms initiating cataract formation. Invest Ophthalmol 13:713–24, 1974.
30. Stevens MJ, Lattimer SA, Kamijo M, et al. Osmotically-induced nerve taurine depletion and the compatible osmolyte hypothesis in experimental diabetic neuropathy in the rat. Diabetologia 36:608–14, 1993.
31. Stevens MJ, Henry DN, Thomas TP, et al. Aldose reductase gene expression and osmotic dysregulation in cultured human retinal pigment epithelial cells. Am J Physiol 265:E428–38, 1993.
32. Sakakibara F, Hotta N, Koh N, Sakamoto N. Effects of high glucose concentrations and epalrestat on sorbitol and myoinositol metabolism in cultured rabbit aortic smooth muscle cells. Diabetes 42:1594–600, 1993.
33. Haneda M, Kikkawa R, Arimura T, et al. Glucose inhibits myo-inositol uptake and reduces myo-inositol content in cultured rat glomerular mesangial cells. Metabolism 39:40–5, 1990.
34. Yorek MA, Dunlap JA, Stefani MR, Davidson EP. Reduced Na^+/K^+ ATPase transport activity, resting membrane potential, and bradykinin-stimulated phosphatidylinositol synthesis by polyol accumulation in cultured neuroblastoma cells. Neurochem Res 19:321–30, 1994.
35. Tesfamariam B. Free radicals in diabetic endothelial cell dysfunction. Free Radic Biol Med 16:383–91, 1994.
36. Derubertis FR, Craven PA. Activation of protein kinase C in glomerular cells in diabetes. Diabetes 43:1–8, 1994.
37. Thomas TP, Porcellati F, Kato K, et al. Effects of glucose on sorbitol pathway activation, cellular redox, and metabolism of myo-inositol, phosphoinositide, and diacylglycerol in cultured human retinal pigment epithelial cells. J Clin Invest 93:2718–24, 1994.
38. Kador PF, Akagi Y, Kinoshita JH. The effect of aldose reductase and its inhibition on sugar cataract formation. Metabolism 35:15–19, 1986.
39. Wen Y, Bekhor I. Levels of expression of hexokinase, aldose reductase and sorbitol dehydrogenase genes in lens of mouse and rat. Curr Eye Res 12:323–32, 1993.
40. Lee AYW, Chung SK, Chung SSM. Demonstration that polyol accumulation is responsible for diabetic cataract by the use of transgenic mice expressing the aldose reductase gene in the lens. Proc Natl Acad Sci USA 92:2780–4, 1995.
41. Ehrig T, Bohren KM, Prendergast FG, Gabbay KH. Mechanism of aldose reductase inhibition: Binding of $NADP^+$/NADPH and alrestatin-like inhibitors. Biochemistry 33:7157–65, 1994.
42. Tanaka Y, Suzuki A. Enzymatic hydrolysis of zenarestat 1-O-acylglucuronide. J Pharm Pharmacol 46:235–9, 1993.
43. Sima AA. Structure-function interactions in the therapeutic response of diabetic neuropathy. J Diabetes Complications 6:64–8, 1992.
44. Carrington AL, Ettlinger CB, Tomlinson DR. Increased resistance to hypoxic conduction block in sciatic nerves of diabetic rats: Effects of extracellular glucose concentration and of aldose reductase inhibition. J Diabetes Complications 8:33–9, 1994.
45. Yagihashi S, Kamijo M, Ido Y, Mirrlees DJ. Effects of long-term aldose reductase inhibition on development of sural nerve in streptozocin-induced diabetic rats. Diabetes 39:690–6, 1990.
46. Willars GB, Townsend J, Tomlinson DR, et al. Studies on peripheral nerve and lens

in long-term experimental diabetes: Effects of the aldose reductase inhibitor statil. Metabolism 37:442–9, 1988.
47. Sorbinil Retinopathy Trial Research Group. The sorbitol retinopathy trial: Neuropathy results. Neurology 43:1141–9, 1993.
48. Sima AA, Greene DA, Brown MB, et al. Effect of hyperglycemia and the aldose reductase inhibitor tolrestat on polyneuropathy. The tolrestat study group. J Diabetes Complications 7:157–69, 1993.
49. Kador PF, Akagi Y, Takahashi Y, et al. Prevention of retinal vessel changes associated with diabetic retinopathy in galactose-fed dogs by aldose reductase inhibitors. Arch Ophthalmol 108:1301–9, 1990.
50. Engerman RL, Kern T. Aldose reductase inhibition fails to prevent retinopathy in diabetic and galactosemic dogs. Diabetes 42:820–5, 1993.
51. Chibber R, Molinatti PA, Wong JSK. The effect of aminoguanidine and tolrestat on glucose toxicity in bovine retinal capillary pericytes. Diabetes 43:758–63, 1994.
52. Engerman RL, Kern TS, Garment MB. Capillary basement membrane in retina, kidney, and muscle of diabetic dogs and galactosemic dogs and its response to 5 years aldose reductase inhibition. J Diabetes Complications 7:241–5, 1993.
53. Chakrabarti S, Sima AA. Effect of aldose reductase inhibition and insulin treatment on retinal capillary basement membrane thickening in BB rats. Diabetes 38:1181–6, 1989.
54. Vinores SA, Van-Niel E, Swerdloff JL, et al. Electron microscopic immunocytochemical evidence for the mechanism of blood-retinal barrier breakdown in galactosemic rats and its association with aldose reductase expression and inhibition. Exp Eye Res 57:723–35, 1993.
55. Sorbitol Retinopathy Trial Research Group. A randomized trial of sorbinil, an aldose reductase inhibitor, in diabetic retinopathy. Arch Ophthalmol 108:1234–44, 1990.
56. Tilton RG, Chang K, Pugliese G, et al. Prevention of hemodynamic and vascular albumin filtration changes in diabetic rats by aldose reductase inhibitors. Diabetes 38:1258–70, 1989.
57. Watanabe T, Kawazu S, Katayama S, et al. A possible role of polyol pathway in diabetic nephropathy. Sakamoto N, Kinoshita JH, Kador PF, Hotta N, eds. Elsevier, Amsterdam, 1988, pp. 360–4.
58. McCaleb ML, Sredy J, Millen J, et al. Prevention of urinary albumin excretion in 6 months streptozocin-diabetic rats with the aldose reductase inhibitor tolrestat. J Diabetes Complications 2:16–18, 1988.
59. Pugliese G, Tilton RG, Speedy A, et al. Vascular filtration function in galactose-fed versus diabetic rats: The role of polyol pathway activity. Metabolism 39:690–7, 1990.
60. Chang WP, Dimitriadis E, Allen T, et al. The effect of aldose reductase inhibitors on glomerular prostaglandin production and urinary albumin excretion in experimental diabetes mellitus. Diabetologia 34:225–31, 1991.
61. Tucker BJ, Rasch R, Blantz RC. Glomerular filtration and tubular reabsorption of albumin in preproteinuric and proteinuric diabetic rats. J Clin Invest 92:686–94, 1993.
62. Daniel BS, Hostetter TH. Aldose reductase inhibition and glomerular abnormalities in diabetic rats. Diabetes 38:981–6, 1989.
63. Mauer SM, Steffes MW, Azar S, et al. Effects of sorbinil on glomerular structure and function in long-term-diabetic rats. Diabetes 38:839–46, 1989.
64. Reddi AS, Jyothirmayi GN. Aldose reductase inhibition by ponalrestat (statil) does not prevent proteinuria in long-term diabetic rats. J Diabetes Complications 7:233–40, 1993.
65. Pugliese G, Pricci F, Pugliese F, et al. Mechanisms of glucose-enhanced extracellular matrix accumulation in rat glomerular mesangial cells. Diabetes 43:478–90, 1994.

66. Kikkawa R. Aldose reductase inhibitor. Int Symp Prog Glomerulopathy (abs), 11, 1994.
67. Ranganathan S, Krempf M, Feraille E, Charbonnel B. Short term effect of an aldose reductase inhibitor on urinary albumin excretion rate and glomerular filtration rate in type 1 diabetic patients with incipient nephropathy. Diabetes Metab 19:257–61, 1993.
68. Win T, Asano G, Shimizu Y. The efficacy of aldose reductase inhibitor, antioxidant butylated hydroxytoluene and cysteine protease inhibitor E64 on hyperglycemia-induced metabolic changes in the organ-cultured bovine lens. J Nippon Med Sci Ika 61:180–9, 1994.
69. Matsui T, Nakamura Y, Ishikawa H, et al. Pharmacological profiles of a novel aldose reductase inhibitors, SPR-210, and its effects on streptozotocin-induced diabetic rats. Jpn J Pharmacol 64:115–24, 1994.
70. Malamas MS, Hohman TC, Millen J. Novel spirosuccinimide aldose reductase inhibitors derived from isoquinoline-1,3-diones: 2-[(4-bromo-2-fluorophenyl)methyl]-6-fluorospiro[isoquinoline-4(1H),3′-pyrrolidine]-1,2′,3,5′(2H)-tetrone and congeners. J Med Chem 37:2043–58, 1994.
71. Kato N, Mizuno K, Matsubara A, et al. Effect of long-term treatment with a new aldose reductase inhibitor, (2S, 4S)-6-fluoro-2′,5′-dioxospiro-[chroman-4,4′-imidazolidine]-2-carboxamide (SNK-860), on peripheral neuropathy in streptozotocin-induced diabetic rats. J Diabetes Complications 8:27–32, 1994.
72. Steele JW, Faulds D, Goa KL. Epalrestat a review of its pharmacology, and therapeutic potential in late-onset complications of diabetes mellitus. Drugs Aging 3:532–55, 1993.
73. Lindstad RI, Hermansen LF, McKinley-McKee JS. Inhibition and activation studies on sheep liver sorbitol dehydrogenase. Eur J Biochem 221:847–54, 1994.
74. Reiersen H, Lindstad RI, McKinley-McKee JS. The inactivation of sheep liver sorbitol dehydrogenase by pyrophosphate and some analogous metal chelators. Arch Biochem Biophys 311:450–6, 1994.
75. Tilton RG, Chang K, Nyengaard JR, et al. Inhibition of sorbitol dehydrogenase effect on vascular and neural dysfunction in streptozocin-induced diabetic rats. Diabetes 44:234–42, 1995.
76. Cohn RM, Segal S. Galactose metabolism and its regulation. Metabolism 22:627–42, 1973.
77. Gabbay K, O'sullivan J. The sorbitol pathway in diabetes and galactosemia: Enzyme and substrate localization and changes in kidney. Diabetes 17:239-43, 1968.
78. Tilton RG, Pugliese G, LaRose LS, et al. Discordant effects of the aldose reductase inhibitor, sorbinil, on vascular structure and function in chronically diabetic and galactosemic rats. J Diabetes Complications 5:230–7, 1991.
79. Engerman RL, Kern TS. Hyperglycemia and development of glomerular pathology: Diabetes compared with galactosemia. Kidney Int 36:41–5, 1989.
80. Kern TS, Engerman RL. Kidney morphology in experimental hyperglycemia. Diabetes 36:244–9, 1987.
81. Cameron NE, Cotter MA. Contraction and relaxation of aortas from galactosaemic rats and the effects of aldose reductase inhibition. Eur J Pharmacol 243:47–53, 1993.
82. Lorentz WB Jr, Shihabi ZK, Weidner N. Galactosemic nephropathy in the rat. Clin Physiol Biochem 5:261–7, 1987.
83. Yamaoka T, Nishimura C, Yamashita K, et al. Acute onset of diabetic pathological changes in transgenic mice with human aldose reductase cDNA. Diabetologia 38:255–61, 1995.
84. Qi S, Akagi K, Araki K, et al. Rapid identification of transgenic mice with PCR amplification of DNA from ear punching. Methods Mol Cell Biol 2:119–22, 1991.
85. Ohta M, Ii S, Yamaoka T, et al. Galactose-induced albuminuria in transgenic mice expressing human aldose reductase. Nutr Res, 15:1343–53, 1995.

86. Kador PF, Kinoshita JH. Diabetes and galactosemic cataracts. Ciba Found Symp 106:110–23, 1984.
87. Dethy JM, Callaert-Deveen C, Janssens M, Lenaers A. Determination of sorbitol and galactitol at the nanogram level in biological samples by high-performance liquid chromatography. Anal Biochem 143:119–24, 1984.
88. Bradford MM. A rapid and sensitive method for the quantitation of microgram quantities of protein utilizing the principle of protein-dye binding. Anal Biochem 72:248–54, 1976.
89. Elving LD, Bakkeren JAJM, Jansen MJH, et al. Screening for microalbuminuria in patients with diabetes mellitus: Frozen storage of urine samples decreases their albumin content. Clin Chem 35:308–10, 1989.
90. Ogino Y, Okada S, Ota Z. Effects of chronic, urea-induced osmotic diuresis on kidney weight and function in rats. Diabetologia 37:225–31, 1994.
91. Tilton RG, Chang K, Pugliese G, et al. Inhibitors of aldose reductase block diabetes-induced increases in blood flow and GFR and markedly reduce 24 h urinary albumin excretion in the rat. Diabetes 36(Suppl 1):43A, 1987.
92. Beyer-Mears A, Ku L, Cohen MP. Glomerular polyol accumulation in diabetes and its prevention by oral sorbinil. Diabetes 33:604–7, 1984.
93. Oates P, Williamson J, Chang K, et al. Attenuation of diabetes-induced vascular dysfunction by an inhibitor of sorbitol dehydrogenase (abs). Diabetes 43(Suppl):3, 1994.

13

Transgenic hyperinsulinemia: A mouse model of insulin resistance and glucose intolerance without obesity

S. LEE MARBÁN[1] AND JESSE ROTH[2]

[1]*Division of Neonatology, Department of Pediatrics, Sinai Hospital of Baltimore;* [2]*Division of Geriatric Medicine and Gerontology, Department of Medicine, Johns Hopkins University School of Medicine, Baltimore, Maryland, USA.*

Key words: transgenic, hyperinsulinemia, insulin resistance, type II diabetes mellitus, human insulin gene.

Type II diabetes represents the most common form of diabetes in humans and is a major cause of morbidity and mortality.[1] In any individual patient, the primary metabolic abnormality initiating this disease process remains elusive, in spite of extensive study of the human condition and multiple animal models.[2-5] Most of these models share several features with human type II diabetes, including glucose intolerance associated with hyperinsulinemia, insulin resistance, and obesity. We describe here transgenic mice that represent a novel model of early type II diabetes. They share many physiologic characteristics with other rodent models of type II diabetes, but they are not obese. They are not the result of a poorly defined mutation that may cause extensive abnormalities beyond those seen in glucose homeostasis, and they are not the result of surgical or pharmacologic manipulation. The metabolic abnormalities seen in these transgenics result from the introduction of multiple copies of the normal human insulin gene into their genome.

The production of these transgenics has been previously described.[6] Briefly, an 8.8 kilobase (kb) human genomic DNA fragment was introduced by pronuclear microinjection. This DNA fragment includes the entire human preproinsulin gene plus ~2 kb of 5′ flanking sequences and >5 kb of 3′ genomic sequence. This 5′ region contains all the known *cis*-acting elements that regulate the insulin gene.[7-11] The two homozygous transgenic lines analyzed in our study were derived from two independent founder animals, and carry 8 and 32 copies of the human insulin gene, respectively. These copies of the human gene are sta-

Correspondence: Dr. S. Lee Marbán, Department of Pediatrics, Sinai Hospital of Baltimore, 2401 W. Belvedere Ave., Baltimore, MD 21215, USA.

Lessons from Animal Diabetes VI. Ed. E. Shafrir.

bly integrated and transmitted to offspring in an autosomal dominant fashion, implying that insertion occurred in each founder at a single site, with the multiple copies of the transgene inherited as a single genetic unit. Expression of the human gene was confirmed at both the RNA and protein levels, and correct tissue specificity of expression has been maintained.[6]

These transgenics were found to be chronically hyperinsulinemic, with total plasma insulin levels roughly correlating with insulin gene copy number, suggesting a gene dosage effect. In these animals, hyperinsulinemia is the primary event, resulting from the introduction of additional insulin genes. The data presented here demonstrate that hyperinsulinemia itself can result in physiologic abnormalities typical of early stages of type II diabetes.

Methods

Animals. Two lines of transgenic mice, carrying multiple copies of the intact human insulin gene and its endogenous 5′ flanking sequences, were produced by pronuclear microinjection as previously described.[6] The original hemizygous animals were designated INS 8 and INS 11, indicating the birth order of the founder mice.[6] These animals were bred independently to homozygosity for the transgene sequences. The resulting homozygous lines were renamed hINS 8 and hINS 32, respectively, to reflect their human gene copy number of 8 and 32 per diploid cell. A line of mice established from nontransgenic littermates was used as control in all studies. Mice were fed ad libitum a standard rodent chow (Rodent Laboratory Chow 5001, Purina, St. Louis, MO). All physiologic studies were performed on mature adult mice, age matched at 10–12 months.

Measurements of plasma, insulin, C-peptide, and glucose levels. Blood samples for all determinations were collected by bleeding from the orbital sinus into tubes containing potassium oxalate crystals as an anticoagulant. Samples were kept on ice and centrifuged immediately; plasma was stored at −20°C for assay at a later time. Fasting samples were collected after a 15–18 h overnight fast (food removed at 5–6 p.m. and samples collected at 9–11 a.m.). Random fed samples were collected after food was removed for 1–2 h to avoid immediate postprandial effects. Total insulin, mouse and human, was measured by radioimmunoassay (RIA) using a commercially available kit (Binax Inc., South Portland, ME), with modifications to allow the use of small plasma volumes, as described previously.[6] Both human (Velosulin Human, Nordisk-USA, Rockville, MD) and rat insulins (Novo BioLabs, Bagsvaerd, Denmark) were used as standards in all assays. Human C-peptide was measured by a commercially available RIA (Medgenix, High Wycombe, UK), using conditions recommended by the manufacturer; rodent C-peptides show no crossreactivity in this assay. Therefore, C-peptide measurements reflect plasma insulin derived solely from the transgenes, while insulin measurements reflect plasma insulin from both the transgenes and the two endogenous mouse insulin genes. Plasma glucose was measured by the glucose oxidase method (Glucose Trinder, Sigma, St. Louis MO).

Fructosamine assay. Glycated serum proteins were measured using a fructosamine assay (RoTAG, Roche Diagnostic Systems, Nutley, NJ), based on the ability of ketoamine bonds between glucose and free amino groups of peptides to reduce nitroblue tetrazolium, with a resulting color change.[12,13] Plasma was used for the assay; plasma and serum samples yield equivalent results in fructosamine determinations.[13]

Glucose tolerance testing. Animals that had been fasted overnight were anesthetized with i.p. Avertin (tribromoethanol, tertamyl alcohol, Aldrich Chemical, Milwaukee, WI). After drawing a fasting blood sample for glucose, insulin, and C-peptide determinations, each animal was injected i.p. with glucose, 2 mg/g of body wt.[14] Subsequent samples were drawn at 20, 40, 60, 90, and 120 min after glucose injection. No individual animal was used for more than four time points to avoid excessive blood loss and hypovolemia.

Insulin tolerance testing. Animals were anesthetized as described earlier for glucose tolerance testing. After drawing a blood sample for glucose at time zero, each animal was injected i.p. with pork insulin (Regular Iletin, Eli Lilly, Indianapolis, IN) 0.5 mU/g of body wt.[14] Blood samples were drawn again at 30 and 60 min post-injection for glucose determination.

Detection of anti-insulin and anti–C-peptide antibodies. Immunoglobulin (Ig) concentrates from control and transgenic animals were prepared from whole plasma by ammonium sulfate precipitation.[15] Ammonium sulfate crystals were added slowly to plasma to a final concentration of 40% (w/v), and samples were allowed to precipitate overnight at 4°C. The precipitated protein fractions were sedimented at 1300 $\times$ g for 15 min at 4°C; the supernatant was discarded and the precipitate was resuspended in phosphate buffered saline (PBS), desalted, and concentrated with two PBS washes using Centricon 10 microconcentrators (Amicon, Danvers, MA). The final concentrates were resuspended in PBS at half the original plasma volume, thus concentrating the Ig fractions twofold over whole plasma. The presence of anti-insulin or anti–C-peptide antibody (Ab) was determined by the ability of the Ig fractions to bind [^{125}I]insulin or [^{125}I]C-peptide. Radioactive tracers and standard anti-insulin and anti–C-peptide Abs from the RIAs (see earlier) were used to determine maximal [^{125}I] binding. These standard Ab preparations were handled identically to plasma samples to demonstrate that Ab was not affected by processing. These Ab were resuspended in their original volume to maintain standard concentrations used in the RIAs. In all samples, [^{125}I]-labeled insulin or C-peptide bound to Ab was precipitated with ethanol[16] and counted.

Insulin extraction from pancreas. Insulin was extracted from pancreas of control and transgenic animals by an acid-ethanol method.[17] Animals were sacrificed by cervical dislocation and the pancreata were dissected, quick-frozen in liquid nitrogen, and stored at −70°C prior to extraction. Each individual pancreas was weighed frozen and immediately dropped into 2 ml of 95% ethanol:water:H_3PO_4 (volume ratio 90:9.9:0.5), pH 3.0, and homogenized using a Brinkman Polytron (Westbury, NY). Cell debris and noninsulin peptides were pelleted by centrifugation and several washes with decreasing ethanol concentration, as previously described.[18] The final supernatant was concentrated to a final volume of <100 μl in a Speed-Vac concentrator (Savant Instruments, Hicksville, NY). For HPLC analysis, 100 μl of water (HPLC grade, Fisher Scientific, Fairlawn, NJ) with 0.1% trifluoroacetic acid (TFA; HPLC grade, Pierce Chemical, Rockford, IL) was added to the pancreatic extract and incubated on ice for 1 h. Particulate material was removed by centrifugation ($1300 \times g$, 10 min at 4°C), followed by filtration of the resulting supernatant through a 0.2 μm filter (Centrex, Schleicher and Schuell, Keene, NH). A 10 μl aliquot of filtrate was removed for total protein determination (Bio-Rad protein assay, Bio-Rad Laboratories, Richmond, CA). The remainder was used for HPLC.

Reverse-phase high performance liquid chromatography. The Waters HPLC system (Milford, MA) consisted of two model 510 HPLC pumps, a model 720 system controller, a model 441 absorbance detector, and a Micro-Bondapak analytical C-18 reverse-phase column (4.6×250 mm). HPLC grade water and acetonitrile were purchased from Fisher Scientific. All solvents were degassed before use. The column was equilibrated with 29.6% acetonitrile/0.1% TFA. Samples of ~150 μl were injected and eluted with a linear gradient of 29.6–34.4% acetonitrile over 50 min at room temperature with a flow rate of 1 ml/min. Fractions of 0.33 ml were collected into polypropylene tubes containing 50 μl of 0.2% insulin-free bovine serum albumin (BSA; Reheis Chemical, Kankakee, IL). Fractions were lyophilized to remove HPLC solvents, dissolved in 40 μl 0.01 N HCl, and diluted with RIA buffer to an appropriate volume for insulin RIA.

DNA sequencing. The 2 kb of 5′ sequences flanking the human insulin gene and included in the DNA fragment used for microinjection were directly sequenced, as described in detail elsewhere.[19]

Liver membrane insulin receptor preparations. Animals were sacrificed by cervical dislocation. The liver was removed quickly, frozen in liquid nitrogen, and stored at −70°C until use. Each liver was weighed frozen and immersed in 20 volumes of ice-cold 1 mM $NaHCO_3$ containing the following protease inhibitors (Sigma): 2 mM phenylmethylsulfonylfluoride (PMSF), 10 μg/ml leupeptin, and 1 TIU/ml aprotinin. Samples were homogenized as described by Shemer et al.[20] Homogenates were centrifuged at $600 \times g$ for 10 min at 4°C; the pellet was discarded, the supernatant was centrifuged at $20{,}000 \times g$ for 30 min at 4°C, and >90% of [^{125}I]insulin binding activity was recovered in the precipitate.[21] The membranes were washed once with $NaHCO_3$ buffer and resuspended in Ca^{2+}-free Krebs-Ringer phosphate (KRP) buffer, pH 7.8, at a protein concentration of 1.2 mg/ml (Bio-Rad protein assay) and stored at −70°C until use.

[^{125}I]Insulin binding studies. [^{125}I]Insulin binding was assayed using crude liver membrane preparations as described by Havrankova et al.[22] with minor modifications.[20] The final membrane protein concentration was 0.4 mg/ml. The assay mixture included 10 mg/ml insulin-free BSA, 1 mg/ml bacitracin (Sigma), 0.35 ng/ml [^{125}I]insulin (human recombinant, SA 280–370 mCi/mg, Amersham, Arlington Heights, IL), and unlabeled human recombinant insulin (a gift from Eli Lilly and Co.) at a final concentration from 0–1.33 × 10^4 ng/ml. The [^{125}I]insulin that bound to membranes in the presence of 1.33 × 10^4 ng/ml unlabeled insulin was designated nonspecific binding, and was subtracted from total binding to obtain specific binding for each sample. All data were normalized as the percent of total ^{125}I counts/60 μg protein.

Triglyceride assay. Plasma triglycerides were measured using a colorimetric assay (Triglyceride [GPO-Trinder], Sigma).

Statistical analyses. Comparisons between experimental groups were made using Student's t-test of statistical significance. Differences were accepted as significant at the $P < 0.05$ level.

Results

Hyperinsulinemia and glycemic control in transgenic mice. Fasting insulin levels were significantly higher in the two transgenic lines than in controls (Fig. 1A). The RIA does not differentiate mouse and human insulins; the results, therefore, represent total plasma insulin levels. Animals of the hINS 8 line, with 8 copies of the human insulin gene, had fasting insulin levels in plasma that were almost twice those of control. Animals of the hINS 32 line, with 32 copies of the human gene, showed fasting insulin values that were almost twice those of the hINS 8 line. Therefore, total plasma insulin in the three lines correlated roughly with insulin gene copy number.

In fed animals, both transgenic lines again had significantly higher insulin levels than control, with hINS 32 demonstrating higher levels than hINS 8. However, fed insulin levels were not significantly different from fasting levels for any of the three lines. Fed samples were collected in late morning, after food had been removed for 1–2 h to avoid immediate postprandial effects. This period may have allowed insulin levels to return to near-fasting levels, or the actual period without food intake may have been longer than anticipated during daylight hours. An alternative conclusion, that the level of insulin is fixed and unresponsive to glycemic stimuli, is unlikely based on the plasma insulin responses to i.p. glucose, discussed later.

Because the amino acid sequences of the A and B chains of human and mouse insulin differ in only a few positions,[23] the insulin RIA used for these studies detected both the mouse and human hormones. However, the sequences of the C-peptides are much more divergent.[24] An RIA specific for human C-peptide allowed the determination of the level of expression of the human insulin transgene. The two transgenic lines, in both fasted and fed states, showed levels of human C-peptide >1.5 ng/ml (Fig. 1B). As with total insulin levels, the level of human C-peptide roughly correlated with gene copy number. Also, human C-peptide

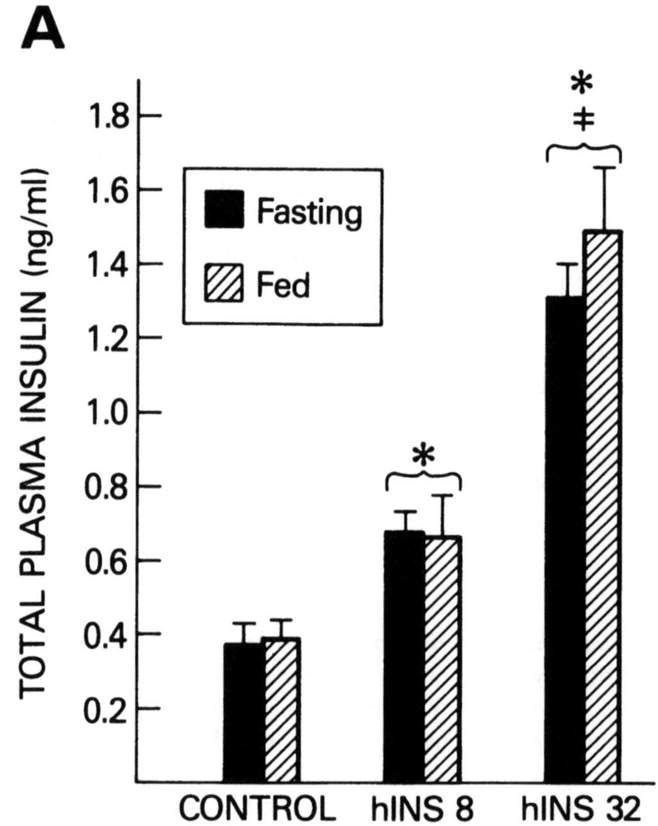

Figure 1. Total plasma insulin (A) and human C-peptide (B) levels in fasting and fed states. Total insulin and human-specific C-peptide were measured in plasma by RIA. Each value represents the mean ± SE of 14–25 individual measurements. Solid bars represent fasting samples; crosshatched bars indicate fed samples. In each of the three lines, there was no significant difference between fasting and fed values for insulin or human C-peptide. In control animals, both fasting and fed human C-peptide levels were below the limit of detection of the assay (0.15 ng/ml), shown by the broken line. Significant differences among the three groups in both assays are as indicated. *$P < 0.001$ vs. control. ‡$P \leq 0.001$ hINS 32 vs. HINS 8.

levels in the fed state were not significantly different from fasting levels in either transgenic line, probably indicating a return to fasting levels within the period of food removal. Under both fasting and fed conditions, human C-peptide levels were undetectable in the plasma of control animals, demonstrating there was no crossreactivity of mouse C-peptide in this assay.

As an initial indicator of glycemic control, plasma glucose levels were measured in fasting and fed animals. Normal fasting glucose levels were maintained in both transgenic lines in spite of their hyperinsulinemia compared with control (Fig. 2). This implies a degree of insulin resistance in the transgenics. In control and both transgenic lines, fed glucose levels were significantly higher than fast-

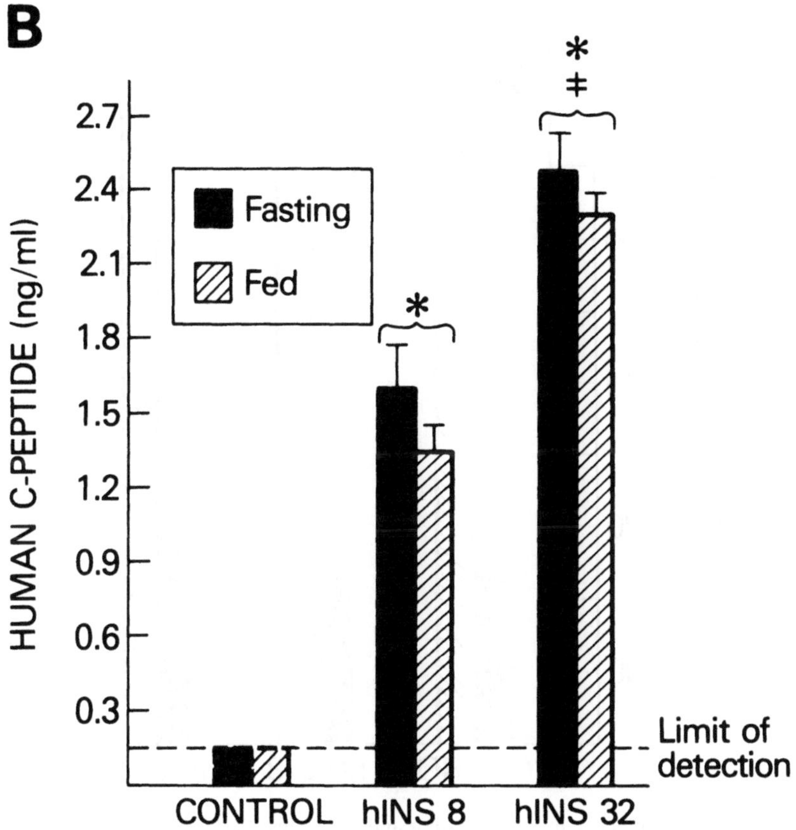

Figure 1 (continued).

ing values. In contrast to the fasting glucose data, however, both transgenic lines demonstrated fed glucose levels significantly above control fed values. This suggests that the transgenics are glucose intolerant in the fed state, although they are able to maintain normal glucose homeostasis in the fasting state.

The long-term glycemic control was further assessed using measurements of glycated plasma proteins. The transgenics did not show elevated levels of fructosamine (Table 1); fructosamine levels of control and transgenic lines fell within the normal range for nondiabetic humans. We are unable to explain why the fructosamine value of hINS 8 animals, while within this range, was significantly lower than the values of control and hINS 32 animals. Hyperinsulinemia may accelerate plasma protein turnover; a modest acceleration in turnover could lower fructosamine levels. The levels of glycated plasma proteins provide an indication of average glycemic control over a 3–6 wk period[14,25] and correlate well with hemoglobin A_{1C} levels.[25-27] Fructosamine levels also correlate with fasting glucose concentrations,[14,26,27] and, therefore, normal fructosamine levels would be predicted in the transgenics since they maintain fasting euglycemia.

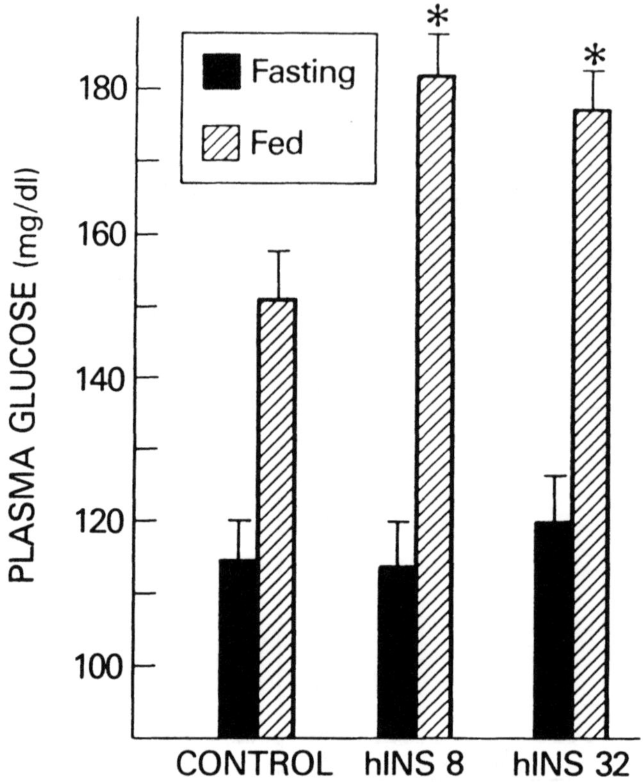

Figure 2. Plasma glucose levels in fasting and fed states. Each value is mean ± SE for 13–25 individual determinations. Fasting glucose values were not significantly different between control and either transgenic line. However, in each of the three lines, the fed glucose level was significantly above fasting ($P < 0.001$). Both transgenic lines had fed glucose levels significantly above control (*$P \leq 0.004$ vs fed control).

Glucose tolerance testing. In response to a 2 mg/g i.p. glucose challenge, animals of all three lines showed a rapid rise in plasma glucose (Fig. 3A). In control animals, by 60 min glucose had begun to fall toward baseline levels, and plasma glucose was near fasting levels at 120 min. In contrast, both transgenic

Table 1. Fructosamine levels in control and transgenic animals

	Fructosamine (mmol/l)
Control	2.10 ± 0.07
hINS 8	1.70 ± 0.14*
hINS 32	1.81 ± 0.20

Results are mean ± SE of 13 individual measurements for each group of mice. The normal range for nondiabetic human subjects is 1.5–2.7 mmol/l.
Comparisons among the three lines showed statistically significant differences only as indicated. *$P < 0.03$ hINS 8 vs. control.

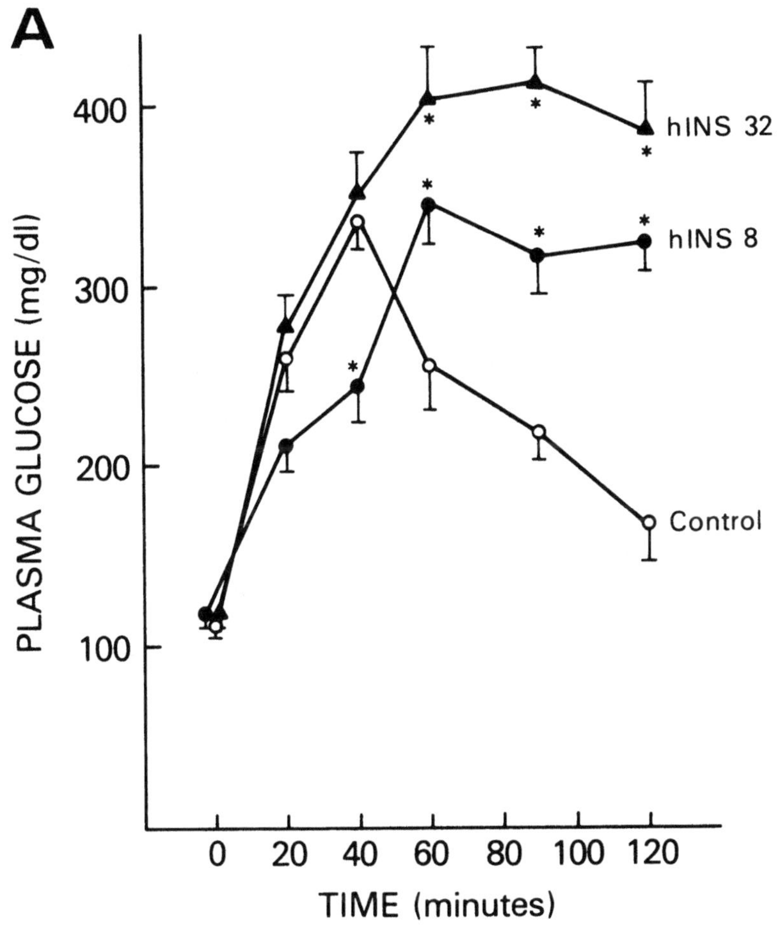

Figure 3. Intraperitoneal glucose tolerance tests. After an overnight fast, animals were injected i.p. with glucose 2 mg/g body wt. Blood samples were taken at time points up to 120 min post-injection for glucose, total insulin, and human C-peptide determinations. (A) Plasma glucose; each point represents the mean ± SE of nine or more individual measurements.

lines showed a prolonged hyperglycemia and delay in glucose disposal, with sustained plasma glucose levels of >300 mg/dl at 120 min after the glucose load.

Control animals demonstrated a biphasic insulin response to the glucose challenge, with a significant rise in plasma insulin levels by 20 min, a fall to near baseline levels at 60 min, and a second large response by 120 min (Fig. 3B). The basis for this pattern of response is obscure. The pattern of insulin response by the transgenics, however, was markedly different. Starting from their higher fasting levels, both transgenic lines showed no significant rise in insulin until 60–90 min after the glucose load; insulin levels then rose sharply at 120 min. The transgenic animals, therefore, can respond to a glycemic challenge with increased insulin release, but this response is very delayed with apparent loss of the first phase response.

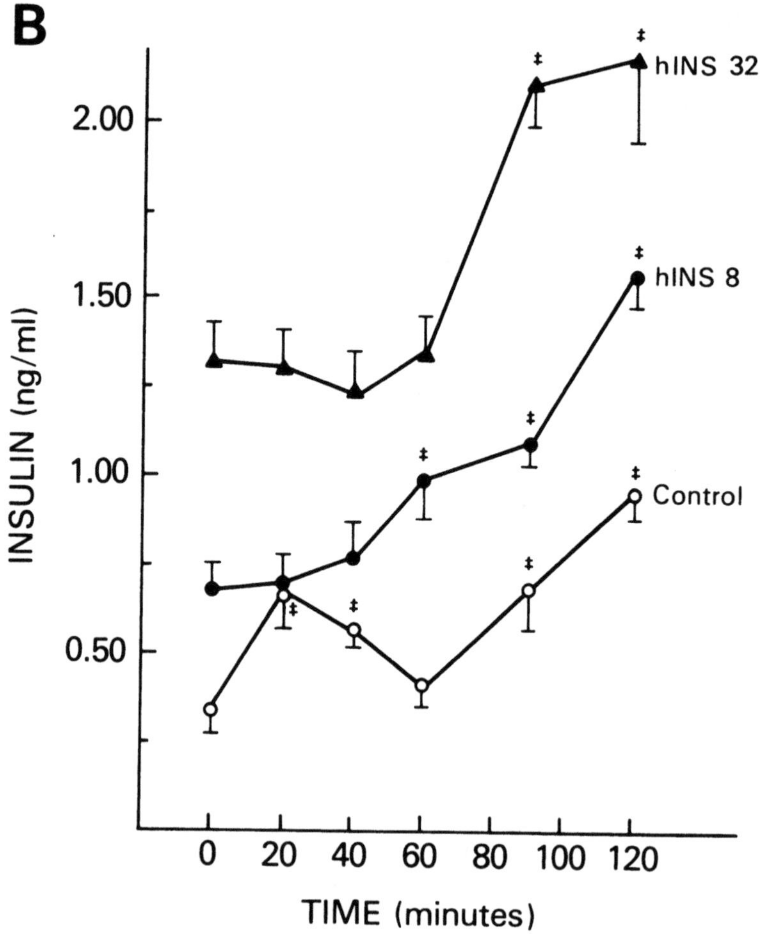

Figure 3 (continued). (B) total plasma insulin measurements; each point represents the mean ± SE of nine or more individual measurements.

The patterns of human C-peptide in plasma after the glucose load were quite similar to those of total insulin (Fig. 3C). The transgenics had no early increase in human C-peptide levels; a significant rise above fasting levels did not occur until 120 min after the glucose load. The level of human gene expression, therefore, is not fixed, but response to a glucose challenge is delayed. The delay in output of insulin and C-peptide in the transgenic animals, with a normal or increased late response, is typical of early stages of type II diabetes in humans and in animals.

Evidence of insulin resistance in transgenic mice. Hyperinsulinemia with normal fasting blood glucose levels implies that the transgenic animals are insulin resistant. This was further documented by insulin tolerance testing. When animals were injected with pork insulin, 0.5 mU/g body wt, mice of both transgenic

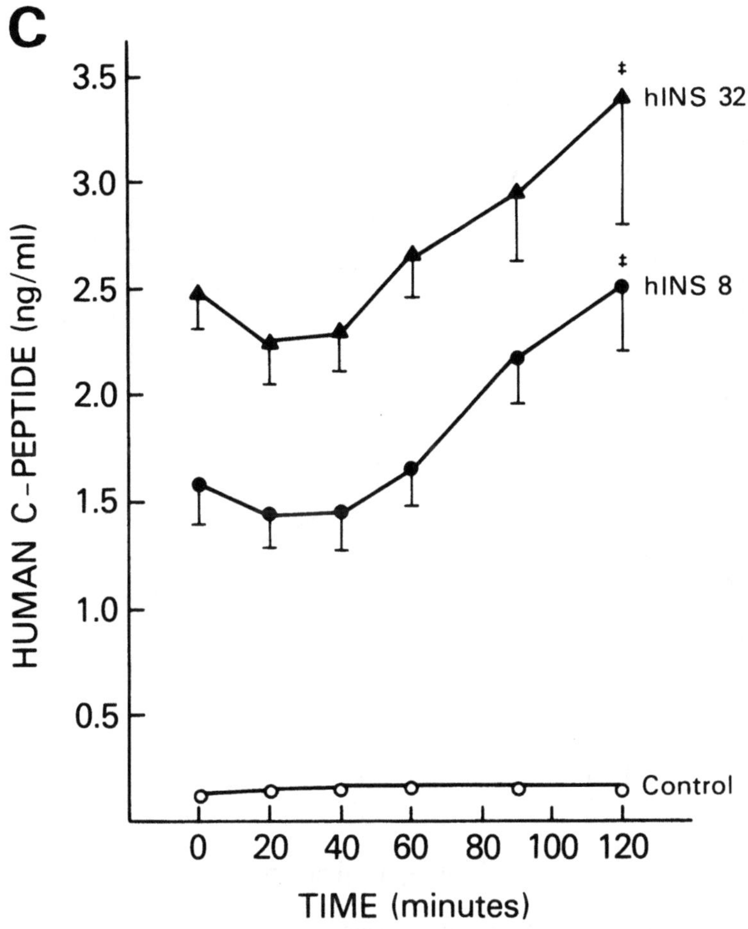

Figure 3 (continued). (C) Human C-peptide. Each time point represents the mean ± SE of a mininum of five individual determinations. *Statistically significant difference from control at that time point. ‡Statistically significant difference from T_0 value.

lines showed a significantly smaller drop in blood glucose than controls (Fig. 4), confirming their insulin resistance. The impaired response to injected insulin was similar to the impaired response to endogenous insulin.

Tests for anti-insulin and anti–C-peptide antibodies. One possible cause of insulin resistance is inactivation of the hormone by Ab directed at the mature hormone or its precursor, proinsulin. The human insulin produced in transgenic animals is a foreign protein that could elicit an Ab response. The protein products of most transgenes have not been found to elicit an immune response, presumably because the protein is expressed early in development, a condition that favors immune tolerance. However, since there have been reports of Ab directed against a transgene product,[28] the insulin transgenics were tested for the pres-

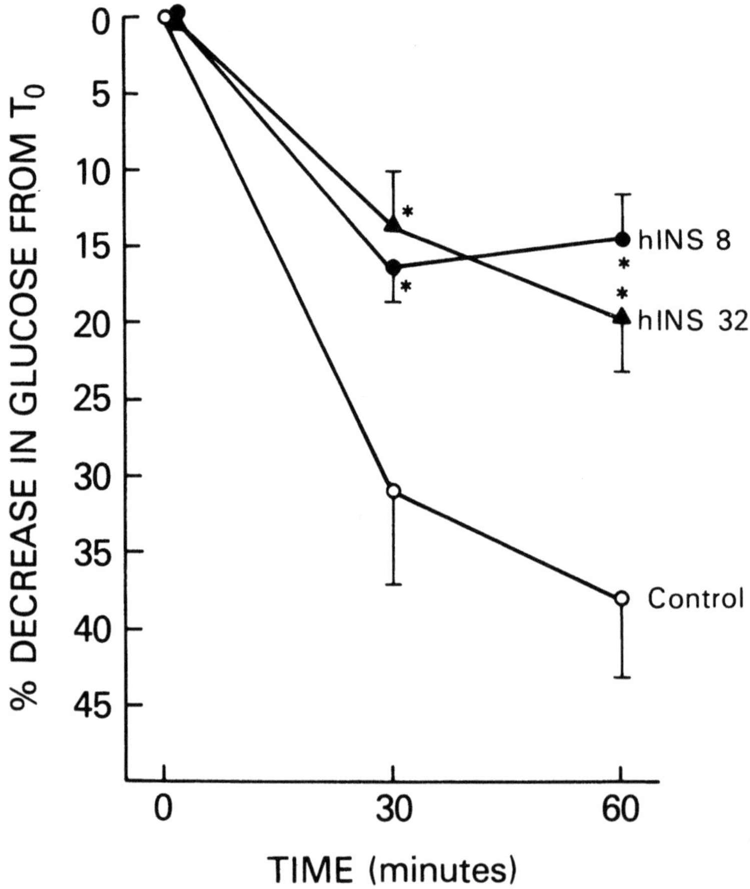

Figure 4. Insulin tolerance tests. Animals were injected i.p. with insulin 0.5 mU/g body wt. Blood samples for glucose determination were drawn prior to and at 30 and 60 min after insulin injection. Each point represents the mean ± SE of 6–10 individual glucose determinations. $*P < 0.02$ vs. control.

ence of Ab against insulin or the C-peptide portion of proinsulin (Table 2). Ig concentrates from both transgenic lines showed no evidence of anti-insulin or anti–C-peptide reaction by precipitation of [^{125}I]-labeled insulin or C-peptide. Therefore, immunologic inactivation of the hormone cannot explain the observed insulin resistance. Also, histologic evaluation of the transgenic pancreas revealed normal islet architecture and no evidence of lymphocytic infiltration (data not shown).

HPLC analysis of pancreatic extracts. Another possible cause of insulin resistance is the production of an abnormal and biologically inactive hormone. After extensive handling, a mutation may have been introduced into the microinjected human insulin gene. To examine the product of this human gene, pancreatic extracts from transgenic animals were analyzed by a sensitive reverse-phase HPLC system that clearly separated the mouse and human insulins. In all individual extracts analyzed, the human insulin peak eluted at the exact position of the human insulin standard (Fig. 5), excluding all but the most subtle mutations.

Extracts from both transgenic lines were kindly studied by Dr. Nina Raben on another highly sensitive HPLC system (data not shown). This system, which utilized an isocratic gradient of 29.7% acetonitrile with triethylammonium phosphate buffer,[29] has been shown to distinguish single amino acid changes in human insulin mutants.[30] Again, the human insulin produced by both transgenic lines eluted at a position identical to the human standard. Therefore, by two separate and sensitive HPLC methods, the insulin produced by the human transgene was shown to be normal; hyperinsulinemia and impaired response to insulin cannot be explained by the presence of a mutant insulin.

Total pancreatic insulin content was not significantly different among the three lines, but in both transgenic lines more than 30% of the insulin was human (Table 3). These HPLC results document that the endogenous mouse insulin genes continue to be expressed in spite of a high level of human insulin production. The hyperinsulinemia in these animals is, therefore, not the result of overproduction of human insulin alone, but rather the result of the combined expression of both mouse and human insulin genes.

DNA sequence of human insulin gene 5′ flanking region. The 2 kb of DNA 5′ to the human insulin gene included in the DNA fragment used for microinjection contains all the known *cis*-acting regulatory elements of the human in-

Table 2. Test for anti-insulin and anti-C–peptide antibodies

	^{125}I-insulin binding		^{125}I-C–peptide binding	
	cpm	% of total counts	cpm	% of total counts
Total counts	3029 ± 12	100	5561 ± 33	100
Maximal binding (B_o)	1117 ± 18	36.9	1994 ± 111	35.9
Nonspecific binding	89 ± 7	2.9	190 ± 43	3.4
Control Ig concentrate	88 ± 7	2.9	172 ± 13	3.1
hINS 8 Ig concentrate	69 ± 6	2.3	189 ± 23	3.4
hINS 32 Ig concentrate	79 ± 7	2.6	191 ± 7	3.4

Immunoglobulin (Ig) concentrates were prepared from plasma of control and transgenic mice and were assayed for their ability to precipitate [^{125}I]-labeled insulin or C-peptide. Total counts represent the standard amount of ^{125}I-labeled insulin or C-peptide added to each sample. All other values are reported as raw counts/min (cpm) and as the percent of total counts. Maximal binding (B_o) represents [^{125}I]-labeled insulin or C-peptide precipitated by a standard amount of known anti-insulin or anti–C-peptide antibody. Nonspecific binding represents [^{125}I] counts precipitated with no antibody present. Values for total counts, maximal binding, and nonspecific binding are mean ± SE of triplicate samples. Values for control and transgenic samples are mean ± SE of five individual Ig preparations for each group.

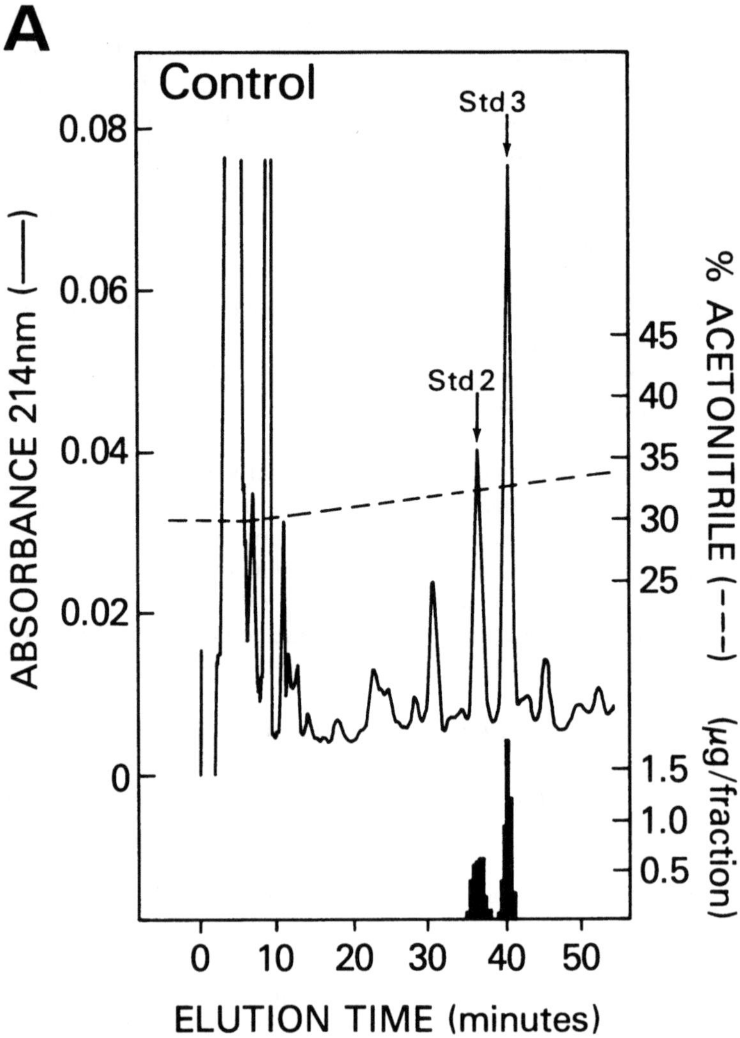

Figure 5. HPLC analysis of pancreatic extracts. Tracings are representative HPLC profiles for pancreatic extracts from (A) control.

sulin gene.[7-11] A mutation introduced into this region during cloning or handling of the DNA fragment prior to microinjection, which could result in altered regulation of human insulin gene expression in both transgenic lines, was ruled out. The sequence of the entire 2 kb region was found to be normal (data not shown).

Down regulation of insulin receptors. Binding of labeled insulin to specific receptors in preparations of liver membrane was measured to determine whether the observed insulin resistance could be attributed to altered receptor affinity for the hormone or decreased receptor number. Competitive insulin binding curves were parallel for all three mouse lines (Fig. 6). The concentration of unlabeled

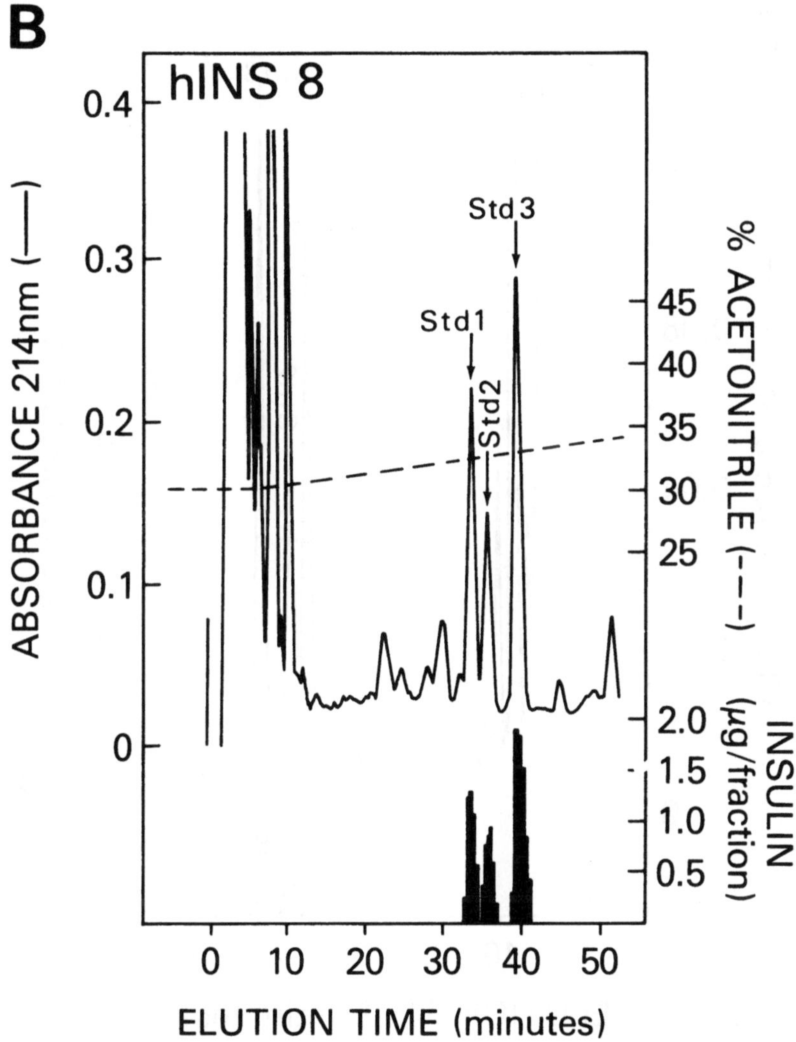

Figure 5 (continued). (B) hINS 8.

insulin required to inhibit labeled insulin binding to receptor by 50% (B_{50}) was not significantly different between control and transgenics; therefore, insulin binding affinity was not altered in the transgenics. The maximal binding of a tracer of labeled insulin, however, was reduced by >30% in both transgenic lines compared with control, consistent with receptor down regulation, that is, a decrease in receptor number, in the transgenics. This reduction in receptor number observed with crude membrane preparations is probably an underestimate of the true decrease in the concentration of plasma membrane receptors. Previous studies[31] have shown that progressive purification of plasma membranes results in higher estimates of the receptor deficiency as other contaminating membrane

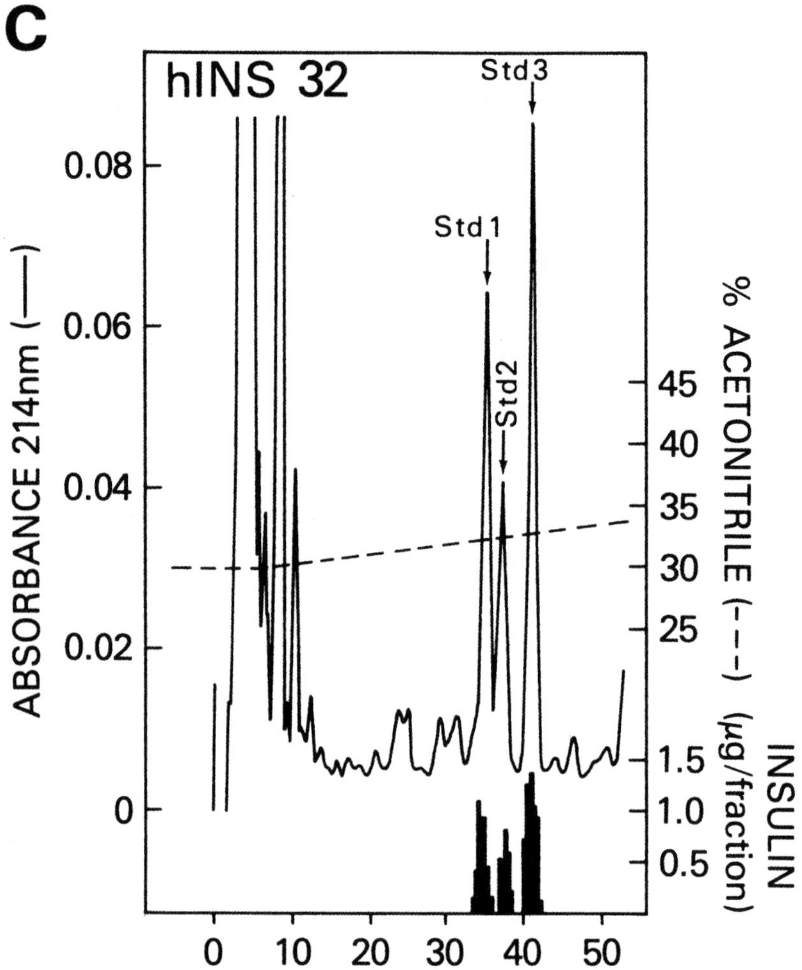

Figure 5 (continued). (C) hINS 32 mice. Solid lines represent UV absorbance at 214 nm. The broken line shows the gradient of acetonitrile. Bar graphs below represent insulin content per fraction, determined by RIA. Arrows indicate the positions where standard insulins eluted in separate runs. Std 1 = human insulin; Std 2 = rat insulin I; Std 3 = rat insulin II.

fractions are removed. Overall, these insulin binding data at least partially explain the observed insulin resistance in the transgenics. Counter-regulatory hormones could also play a role; levels of glucagon and adrenal hormones have not been determined in the transgenics.

Plasma triglycerides and body weights. Because of the strong association of hyperinsulinemia and insulin resistance with lipid abnormalities and obesity, we measured plasma triglyceride levels and body wt in transgenic and control ani-

Table 3. Insulin content in control and transgenic pancreatic extracts

	Pancreatic insulin content (μg/100 μg protein)				
	Human	Mouse I	Mouse II	Total	% Human
Control	N.D.	1.23 ± 0.35	1.89 ± 0.49	3.11 ± 0.83	N.D.
hINS 8	1.13 ± 0.39	0.79 ± 0.16	1.53 ± 0.37	3.46 ± 0.88	31.8 ± 3.9
hINS 32	1.87 ± 0.45	1.08 ± 0.24	2.20 ± 0.54	5.15 ± 1.21	35.5 ± 1.8

Each individual insulin peak separated by HPLC was identified by its elution time relative to known standards and was quantified by RIA. All insulin values were normalized per 100 μg protein content in the extract. Each value is the mean ± SE of results from six to nine individual extracts per line.
N.D., not detectable.

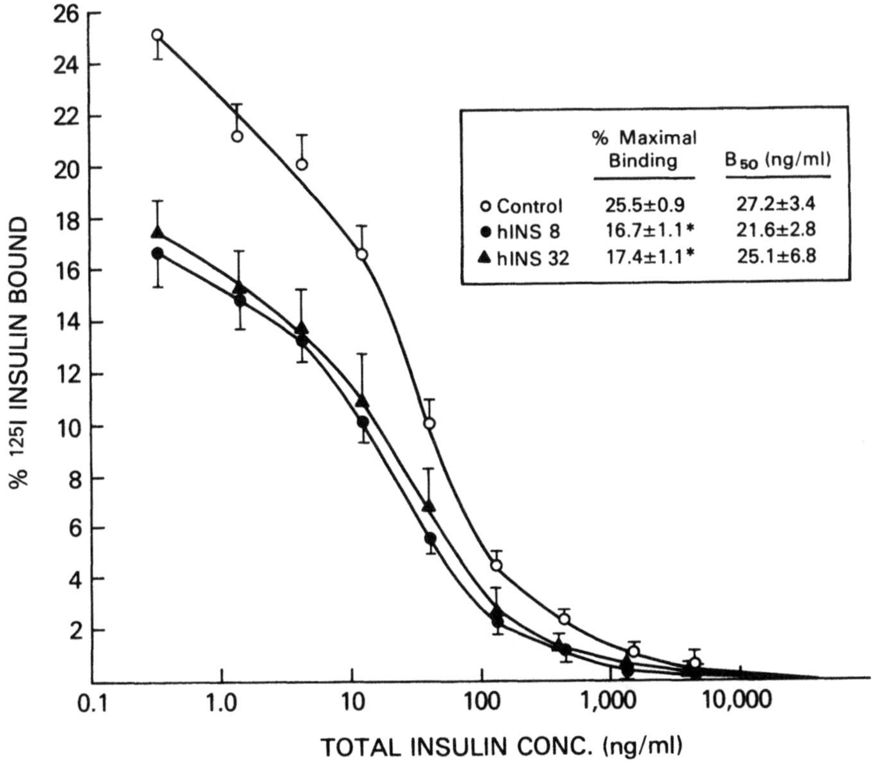

Figure 6. Competitive binding of [^{125}I]insulin to crude liver membrane preparations. [^{125}I]insulin binding to liver membrane preparations was measured in the presence of unlabeled insulin at 0–1.33 × 10^4 ng/ml. Nonspecific binding, defined as the labeled insulin bound in the presence of 1.33 × 10^4 ng/ml cold insulin, was <5% of maximal binding in all cases, and was subtracted from total binding to obtain specific binding for each sample. All data were normalized as [^{125}I] bound per 60 μg of membrane protein. Each value is the mean ± SE of eight membrane preparations per group. B_{50} represents the concentration of unlabeled insulin at which 50% maximal binding occurred. There were no significant differences in B_{50} values among the three lines. Differences in maximal binding values were significant at $P < 0.001$ vs. percent maximal binding for control.

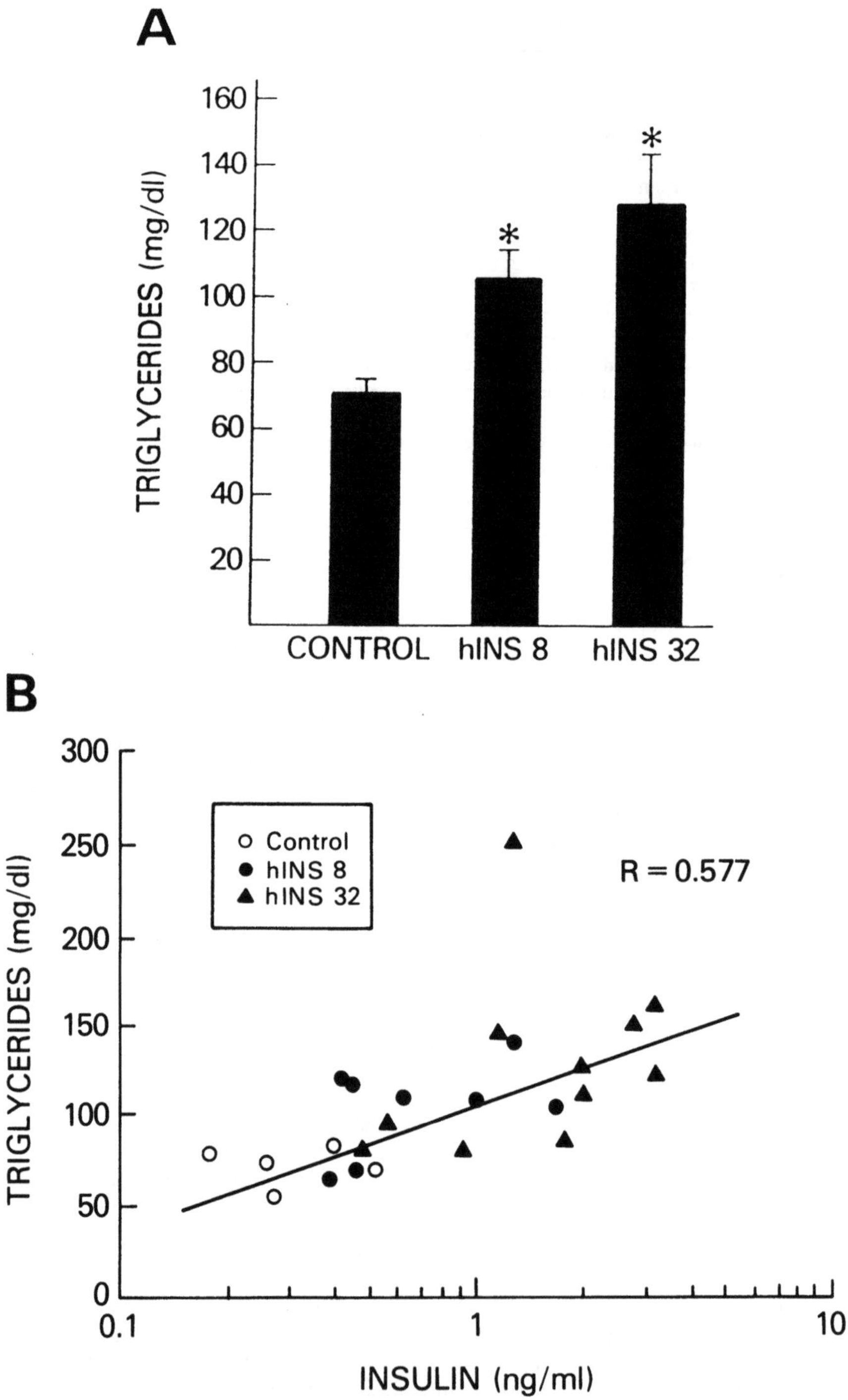

Figure 7. Plasma triglyceride concentrations and their relationship to plasma insulin. (A) Plasma triglyceride levels. Values are the mean ± SE of measurements made on blood from 9–14 individual mice/line. (B) Relationship between plasma triglyceride and insulin levels. Points represent single determinations from individual animals. *$P < 0.001$ vs. control.

mals. Both transgenic lines had a significant elevation in triglycerides compared with control (Fig. 7A). Although the difference between hINS 8 and hINS 32 animals did not reach statistical significance, there was a strong positive correlation between triglyceride and insulin levels in the three groups (Fig. 7B). Comparing weights for male or female mice, there were no significant differences among the three lines (Table 4). Even when fed a high fat diet (9 vs. 4.5% fat in standard rodent chow), no wt difference was seen between controls and transgenics, although wt did increase in all three groups on the high fat intake (data not shown). Therefore, unlike certain human conditions and several other animal models, the hyperinsulinemia, glucose intolerance, and insulin resistance seen in these transgenics are not associated with obesity.

Discussion

We have examined the consequences of hyperinsulinemia in two lines of transgenic mice. Compared with control, the transgenic lines, with 8 and 32 copies of the human insulin gene per cell, had elevated fasting insulin levels. Expression of the human transgenes was confirmed by finding significant fasting levels of human C-peptide in the circulation. HPLC analysis of pancreatic extracts confirmed that the normal human form of insulin was produced, as well as the endogenous mouse insulins, also contributing to the hyperinsulinemia. Insulin resistance was evident by normal fasting blood glucose in the face of hyperinsulinemia, and the decreased response of the transgenics to exogenous insulin. Fed glucose levels, however, were elevated in transgenics. This glucose intolerance was confirmed by i.p. glucose tolerance testing in which the transgenics had prolonged hyperglycemia and delayed insulin response.

There was no evidence of immunologic inactivation of the hormone or cellular immune attack on the pancreas. Insulin binding studies revealed a significant decrease in insulin receptor number, with no change in binding affinity. Triglycerides were also elevated in transgenics and were correlated with fasting insulin levels. Unlike many other animal models of hyperinsulinemia and type II diabetes,[2-5] these transgenics did not become obese, even when placed on a high fat diet. In the transgenics, no abnormalities other than those in insulin production and glucose homeostasis have been detected. We have not excluded the possibility that integration of the transgenes disrupted some untested function.

Table 4. Body weights of control and transgenic animals

	Weight (g)	
	Male	Female
Control	43.5 ± 1.0	39.2 ± 0.8
hINS 8	42.7 ± 0.9	38.5 ± 1.1
hINS 32	42.4 ± 1.0	39.4 ± 0.9

Numbers are mean ± SE of body wt for 24 male and 24 female mice in each group. Comparisons within each sex showed no statistically significant differences among the three groups.

Transgenic technology has made possible selective perturbation of complex physiologic processes.[32,33] Expression of a normal gene can be redirected to a different tissue, or the level of expression can be altered. A foreign or mutated gene can be introduced. More recently, expression of a selected gene can be prevented,[34] or an existing gene can be replaced by a specifically altered one.[35,36] Using these approaches, selected physiologic defects can be produced and models of human disease created.

Several transgenic models of type I or insulin-dependent diabetes have been described.[37,38] Recently, the application of transgenic methods to the study of type II diabetes has gained momentum.[39] Questions about insulin processing[40] and insulin secretion[41] have been approached with transgenic constructs. Such studies may help clarify the B-cell dysfunction in type II diabetes. Similarly, transgenic models have helped to define the role of B-cell glucose sensing through glucokinase in maintaining glucose sensitivity.[42-44] Other groups have achieved expression of normal or mutant insulin receptor genes in muscle of transgenic animals.[45,46] Recently the consequences of chronic hyperglycemia were studied in a transgenic model.[47] A soluble form of the human insulin receptor was expressed and acted as a high-affinity insulin-binding protein in the plasma. These transgenic animals were chronically hyperglycemic, showing increased hepatic glucose production and hyperinsulinemia.

The possible roles of amylin and the GLUT2 glucose transporter in the pathophysiology of type II diabetes have been studied in transgenic models. While initial reports suggested that amylin overexpression in transgenic mice led to impaired glucose tolerance,[48] other transgenic lines have failed to show altered glucose homeostasis in the presence of elevated islet amylin content.[49,50] The potential role of GLUT2 downregulation in type II diabetes[51] has been questioned based on results in a transgenic model. Expression of H-*ras* in transgenic pancreas resulted in decreased GLUT2 levels, but no change in glucose homeostasis or glucose-stimulated insulin release.[52]

The enzyme PEPCK plays a crucial role in hepatic gluconeogenesis. Failure of insulin to suppress hepatic glucose production appears to be an important factor in the pathophysiology of type II diabetes. Regulatory elements controlling PEPCK gene expression have been defined in transgenic mice expressing PEPCK promoter–growth hormone fusion genes.[53] Transgenic mice have been made that carry multiple copies of a PEPCK minigene and have increased levels of the enzyme.[54] These animals developed glucose intolerance and insulin resistance, suggesting that an increase in hepatic glucose production as the primary event can result in a type II diabetic phenotype.

The human insulin gene, under control of its own promoter, has also been used by two other groups (in addition to the work described here) to construct transgenic lines.[55,56] In both cases, the human insulin gene was expressed, but the transgenics maintained normal insulin levels and normal glucose homeostasis. In one line, primer extension analysis of pancreatic mRNA revealed a twofold increase in insulin message in transgenic animals.[57] The increased insulin mRNA was attributed solely to expression of the human insulin transgenes. The fact that pancreatic insulin content was not increased suggested post-transcriptional regulation.

The reason for the striking differences seen in other insulin transgenics from findings in our work is unclear. In all three insulin transgenic studies, similar genomic constructs were used for microinjection. The human gene copy number integrated in the transgenics was not significantly different, ranging in the three reports from 1 to 50 copies per cell.[6,55,56] Factors known to enhance expression, such as the presence of introns and the removal of plasmid sequences[58,59] were similar. In general, the site of integration is thought to be the most important determinant of the level of transgene expression,[58,59] and the differences may have resulted from chance integration events.

In the construction of these transgenic lines, the only manipulation was the introduction of multiple copies of a normal insulin gene. Any resulting physiologic abnormalities, therefore, would logically result from changes in insulin production. These results suggest that primary hyperinsulinemia, independent of obesity or other genetic abnormalities, can lead to glucose intolerance, insulin resistance, receptor downregulation, and lipid abnormalities typical of type II diabetes.

The underlying causes of type II diabetes have been a matter of long-standing investigation.[60-62] Many patients are hyperinsulinemic before they develop overt glucose intolerance.[60-62] Populations at high risk of developing type II diabetes, such as Pima Indians[63,64] and Mexican Americans,[65] have similarly been found to have hyperinsulinemia before the onset of hyperglycemia. Likewise, offspring of diabetic parents are hyperinsulinemic compared with a control population.[66-68] This hyperinsulinemia is often assumed to be indicative of insulin resistance. It is assumed that the primary defect is one of insulin resistance, and that insulin levels rise as a compensatory mechanism. The insulin transgenics described here offer an additional possible sequence. Hyperinsulinemia would be the initiating event, leading to a compensatory resistance to the effects of insulin, and subsequent impaired glucose metabolism. Further insulin secretion may then follow to improve glucose homeostasis, with further insulin resistance and glucose intolerance.

Conclusions

We have described a unique transgenic model of type II diabetes. The animals carry and express multiple copies of the human insulin gene. They are hyperinsulinemic, insulin resistant, and glucose intolerant. Unlike most rodent models of type II diabetes, however, they are not obese. They suggest a scheme for the pathogenesis of type II diabetes in which hyperinsulinemia is the primary event.

Acknowledgments

Our studies were conducted at the Diabetes Branch, National Institute of Diabetes, Digestive and Kidney Diseases, National Institutes of Health, Bethesda, MD. Additional grant support was provided by the Diabetes Research and Education Foundation (SLM). We thank Dr. Nina Raben for assistance with HPLC analyses and DNA sequencing, Dr. Paul Lacy for review of histology slides, and Valerie Barr and Laurie Scott for technical assistance.

References

1. Shafrir E, Bergman M, Felig P. The endocrine pancreas: Diabetes mellitus. In: Endocrinology and metabolism. Felig P, Baxter JD, Broadus AE, Frohman LA, eds. McGraw-Hill, New York, 1987, pp 1043–1178.
2. Herberg L, Coleman DL. Laboratory animals exhibiting obesity and diabetes syndromes. Metabolism 26:59–99, 1977.
3. Mordes JP, Rossini AA. Animal models of diabetes. In: Joslin's diabetes mellitus. Marble A, Krall LP, Bradley RF, Christlieb AR, Soeldner JS, eds. Lea & Febiger, Philadelphia, 1985. pp. 110–137.
4. Coleman DL. Obesity and diabetes: Two mutant genes causing diabetes-obesity syndromes in mice. Diabetologia 14:141–8, 1978.
5. Coleman DL. Diabetes-obesity syndromes in mice. Diabetes 31:1–6, 1982.
6. Marbán SL, DeLoia JA, Gearhart JD: Hyperinsulinemia in transgenic mice carrying multiple copies of the human insulin gene. Dev Genet 10:356–64, 1989.
7. Walker MD, Edlund T, Boulet AM, Rutter WJ: Cell-specific expression controlled by the 5′ flanking region of insulin and chymotrypsin genes. Nature 306:557–61, 1983.
8. Hanahan D. Heritable formation of pancreatic β-cell tumours in transgenic mice expressing recombinant insulin/simian virus 40 oncogenes. Nature 315:115–22, 1985.
9. Karlsson O, Edlund T, Moss JB, et al. A mutational analysis of the insulin gene transcriptional control region: Expression in beta cells is dependent on two related sequences within the enhancer. Proc Natl Acad Sci USA 84:8819–23, 1987.
10. Crowe DT, Tsai MJ: Mutagenesis of the rat insulin II 5′-flanking region defines sequences important for expression in HIT cells. Mol Cell Biol 9:1784–89, 1989.
11. Fromont-Racine M, Bucchini D, Madsen O, et al. Effect of 5′-flanking sequence deletions on expression of the human insulin gene in transgenic mice. Mol Endocrinol 4:669–77, 1990.
12. Johnson RN, Metcalf PA, Baker JR: Fructosamine: A new approach to the estimation of serum glycosylated protein. Clin Chim Acta 127:87–95, 1982.
13. Glinkmanas G, Sarmini H, Bigorie B, et al. Evaluation of Roche fructosamine test: Use for diabetic patient monitoring. Clin Biochem 21:319–21, 1988.
14. Larkins RG: Defective insulin secretory response to glucose in the New Zealand obese mouse. Diabetes 22:251–5, 1973.
15. Green AA, Hughes WL. Protein fractionation on the basis of solubility in aqueous solutions of salts and organic solvents. Methods Enzymol 1:67–90, 1955.
16. Heding LG. Radioimmunological determination of human C-peptide in serum. Diabetologia 11:541–8, 1975.
17. Kimmel JR, Hayden LJ, Pollock HC: Isolation and characterization of a new pancreatic polypeptide hormone. J Biol Chem 250:9369–76, 1975.
18. Shuldiner AR, Bennet C, Robinson EA, Roth J: Isolation and characterization of two different insulins from an amphibian *Xenopus laevis*. Endocrinology 125:469–77, 1989.
19. Raben N, Barbetti F, Cama A, et al. Normal coding sequence of insulin gene in Pima Indians and Nauruans, two groups with highest prevelance of type II diabetes. Diabetes 40:118–22, 1991.
20. Shemer J, Penhos JC, LeRoith D: Insulin receptors in lizard brain and liver: Structural and functional studies of α and β subunits demonstrate evolutionary conservation. Diabetologia 29:321–9, 1986.
21. Shemer J, Ota A, Adamo M, LeRoith D: Insulin-sensitive tyrosine kinase is increased in livers of adult obese Zucker rats: Correlation with prolonged fasting. Endocrinology 123:140–8, 1988.
22. Havrankova J, Roth J, Brownstein M. Insulin receptors are widely distributed in the central nervous system of the rat. Nature 272:827–9, 1978.

23. Dayhoff MO. Insulin amino acid sequences. Atlas Prot Sequence Structure 5:127–30, 1976.
24. Dayhoff MO. C-peptide amino acid sequences. Atlas Prot Sequence Structure 5:150–1, 1978.
25. Cefalu WT, Parker TB, Johnson CR: Validity of serum fructosamine as index of short-term glycemic control in diabetic outpatients. Diabetes Care 11:662–4, 1988.
26. Baker JR, O'Connor JP, Metcalf PA, et al. Clinical usefulness of estimation of serum fructosamine concentration as a screening test for diabetes mellitus. Br Med J 287:863–7, 1983.
27. Baker JR, Metcalf PA, Johnson RN, et al. Use of protein-based standards in automated colorimetric determinations of fructosamine in serum. Clin Chem 31:1550–54, 1985.
28. Adams TE, Alpert S, Hanahan D: Non-tolerance and autoantibodies to a transgenic self-antigen expressed in pancreatic β cells. Nature 325:223–8, 1987.
29. Robbins DC, Shoelson SE, Rubenstein AH, Tager HS. Familial hyperinsulinemia. J Clin Invest 73:714–19, 1984.
30. Shoelson S, Haneda M, Blix P, et al. Three mutant insulins in man. Nature 302:540–3, 1983.
31. Kahn CR, Neville DM, Roth J: Insulin-receptor interaction in the obese hyperinsulinemic mouse. J Biol Chem 248:244–50, 1973.
32. Saffer JD: Transgenic mice in biomedical research. Lab Anim: 30–8, 1992.
33. Merlino GT: Transgenic animals in biomedical research. FASEB J 5:2996–3001, 1991.
34. Munir MI, Rossiter BJF, Caskey CT: Antisense RNA production in transgenic mice. Somat Cell Mol Genet 16:383–94, 1990.
35. Capecchi MR: Altering the genome by homologous recombination. Science 244:1288–92, 1989.
36. Barinaga M: Knockout mice: Round two. Science 265:26–8, 1994.
37. Lipes MA, Eisenbarth GS: Transgenic mouse models of type I diabetes. Diabetes 39:879–84, 1990.
38. Harrison LC. Transgenic approaches to understanding the role of MHC genes in insulin-dependent diabetes mellitus. Clin Endocrinol Metab 5:439–47, 1991.
39. Moller DE: Transgenic approaches to the pathogenesis of NIDDM. Diabetes 3:1394–1401, 1994.
40. Carrol RJ, Hammer RE, Chan SJ, et al. A mutant human proinsulin is secreted from islets of Langerhans in increased amount via an unregulated pathway. Proc Natl Acad Sci USA 85:8943–47, 1988.
41. Welsh M, Hammer RE, Brinster RL, et al. Stimulation of growth hormone synthesis by glucose in islets of Langerhans isolated from transgenic mice. J Biol Chem 261:12915–917, 1986.
42. Epstein PN, Boschero AC, Atwater I, et al. Expression of yeast hexokinase in pancreatic beta cells of transgenic mice reduces blood glucose, enhances insulin secretion, and decreases diabetes. Proc Natl Acad Sci USA 89:12038–42, 1992.
43. Voss-McCowan ME, Xu B, Epstein PN: Insulin synthesis, secretory competence and glucose utilization are sensitized by transgenic yeast hexokinase. J Biol Chem 269:15814–818, 1994.
44. Efrat S, Leiser M, Wu Y-J, et al. Ribozyme-mediated attenuation of pancreatic B-cell glucokinase expression in transgenic mice results in impaired glucose-induced insulin secretion. Proc Natl Acad Sci USA 91:2051–5, 1994.
45. Benecke H, Flier JS, Rosenthal N, et al. Muscle-specific expression of human insulin receptor in transgenic mice. Diabetes 42:206–12, 1993.
46. Chang P-Y, Benecke H, Le Marchand-Brustel Y, et al. Expression of a dominant-negative mutant human insulin receptor in the muscle of transgenic mice. J Biol Chem 269:16034–40, 1994.
47. Schaefer EM, Viard V, Morin J, et al. A new transgenic mouse model of chronic hyperglycemia. Diabetes 43:143–53, 1994.

48. Hoppener JWM, Oosterwijk C, Verbeek SJ, et al. IAPP/amylin transgenic mice as an in vivo model system for type 2 diabetes mellitus? Biochem Soc Trans 21:285, 1992.
49. Fox N, Schrement J, Nishi M, et al. Human islet amyloid polypeptide transgenic mice as a model of non-insulin-dependent diabetes mellitus FEBS Lett 323:40–4, 1993.
50. D'Alessio DA, Verchere CB, Kahn SE, et al. Pancreatic expression and secretion of human islet amyloid polypeptide in a transgenic mouse. Diabetes 43:1457–61, 1994.
51. Unger RH: Diabetic hyperglycemia: Link to impaired glucose transport in pancreatic beta cells. Science 251:1200–5, 1991.
52. Tal M, Wu Y-J, Lweiser M, et al. [Val 12]Hras downregulates GLUT2 in beta cells of transgenic mice without affecting glucose homeostasis. Proc Natl Acad Sci USA 89:5744–8, 1992.
53. Patel YM, Yun JS, Liu J, et al. An analysis of regulatory elements in the phosphoenolpyruvate carboxykinase (GTP) gene which are responsible for its tissue-specific expression and metabolic control in transgenic mice. J Biol Chem 269:5619–28, 1994.
54. Valera A, Pujol A, Pelegrin M, Bosch F. Transgenic mice overexpressing phosphoenolpyruvate carboxykinase develop noninsulin-dependent diabetes mellitus. Proc Nat Acad Sci USA 91:9151–4, 1994.
55. Bucchini D, Ripoche MA, Stinnakre M-G, et al. Pancreatic expression of human insulin gene in transgenic mice. Proc Natl Acad Sci USA 83:2511–15, 1986.
56. Selden RF, Skoskiewicz MJ, Howie KB, et al. Regulation of human insulin gene expression in transgenic mice. Nature 321:525–8, 1986.
57. Schnetzler B, Murakawa G, Abalos D, et al. Adaptation to supraphysiologic levels of insulin gene expression in transgenic mice: Evidence for the importance of post-transcriptional regulation. J Clin Invest 92:272–86, 1993.
58. Brinster RL, Chan HJ, Trumbauer ME, et al. Factors affecting the efficiency of introducing foreign DNA into mice by microinjecting eggs. Proc Natl Acad Sci USA 82:4438–42, 1985.
59. Bonnerot C, Grimber G, Briand P, et al. Patterns of expression of position-dependent integrated transgenes in mouse embryo. Proc Natl Acad Sci USA 87:6331–35, 1990.
60. Reaven GM. Role of insulin resistance in human disease. Diabetes 37:1595–607, 1988.
61. DeFronzo RA: Pathogenesis of type II (non-insulin-dependent) diabetes mellitus: A balanced overview. Diabetologia 35:389–97, 1992.
62. Taylor SI, Accili D, Imai Y: Insulin resistance or insulin deficiency: which is the primary cause of NIDDM? Diabetes 43:735–40, 1994.
63. Bogardus C: Insulin resistance in the pathogenesis of NIDDM in Pima Indians. Diabetes Care 16:228–31, 1993.
64. Lillioja S, Mott DM, Spraul M, et al. Insulin resistance and insulin secretory dysfunction as precursors of non-insulin-dependent diabetes mellitus: Prospective studies of Pima Indians. N Engl J Med 329:1988–92, 1993.
65. Haffner SM, Stern MP, Hazuda HP, et al. Hyperinsulinemia in a population at high risk for non-insulin-dependent diabetes mellitus. N Engl J Med 315:220–4, 1986.
66. Johnston C, Ward WK, Beard JC, et al. Islet function and insulin sensitivity in the non-diabetic offspring of conjugal type 2 diabetic patients. Diabet Med 7:119–25, 1990.
67. Warram JH, Martin BC, Krolewski AS, et al. Slow glucose removal rate and hyperinsulinemia precede the development of type II diabetes in the offspring of diabetic parents. Ann Intern Med 113:909–15, 1990.
68. Martin BC, Warren JH, Krolewski AS, et al. Role of glucose and insulin resistance in development of type 2 diabetes mellitus: Results of a 25-year follow-up study. Lancet 340:925–9, 1992.

14

Spontaneously diabetic rat "OLETF" as a model for NIDDM in humans

KAZUYA KAWANO, TSUKASA HIRASHIMA, SHIGEHITO MORI, AND TAKASHI NATORI

Tokushima Research Institute, Otsuka Pharmaceutical Co., Ltd., Tokushima 771-01, Japan.

Key words: spontaneous, genetic, rat, diabetes, diabetic nephropathy, obesity, hyperglycemia, hyperinsulinemia.

A spontaneous polyuria, polydipsia, and mild obesity was discovered in 1984 in an outbred colony of Long Evans (LE) rats, that had been purchased from Charles River Canada (St. Constant, Quebec, Canada) in 1982. A strain of rats was developed by selective breeding from this rat and designated OLETF (Otsuka Long-Evans Tokushima Fatty). The OLETF line has been maintained since then by brother–sister mating at the Tokushima Research Institute, Otsuka Pharmaceutical Co., Ltd.[1,2] From the same LE colony stock, a line developing insulin-dependent diabetes mellitus (IDDM) has been isolated (LETL)[3-4] and a control line, LETO, has also been established.

General and basic kinetic studies on OLETF rats have been described in the previous report[2] and are summarized as follows: 1) After 24 wk of age, elevation of plasma glucose and immunoreactive insulin, and the oral glucose tolerance test (OGTT), became marked, indicating diabetes mellitus. These increases were more prominent at 55 wk. At 65 wk of age, plasma glucose levels rose to >22.4 mmol/l and the insulin values became <40 pmol/l. 2) The cumulative incidences of diabetes and impaired glucose tolerance (IGT) were 87.8% (341 of 388) and 7.2% (28 of 388), respectively, in male OLETF rats after the 20th generation. Female OLETF rats displayed diabetes in very few individuals at 30 wk of age, though approximately 33.3% (5 of 15) of females showed diabetes when 60 wk old. On the contrary, no diabetic rat was found among LETO rats. 3) Male OLETF rats began to gain weight faster than LETO rats beginning at 5 wk, and the difference gradually increased with age, being about 200 g at 40 wk of age.

Correspondence: Kazuya Kawano, Otsuka Pharmaceutical Co., Ltd., 463-10, Kagasuno Kawauchi-cho, Tokushima 771-01, Japan.

Lessons from Animal Diabetes VI. Ed. E. Shafrir.

Then in the early stage of diabetes with glucosuria, the body wt decreased slightly, with the decrease becoming more prominent with the increase in diabetes severity. 4) At 40 wk of age, the plasma cholesterol level was 4.73 ± 0.68 mmol/l in OLETF rats and 2.71 ± 0.15 in LETO rats. The plasma triglyceride level was 6.45 ± 2.26 mmol/l in OLETF rats and 1.31 ± 0.35 in LETO rats, with both differences being highly significant ($P < 0.001$). 5) Histopathologic changes in the pancreas were classified into three stages: an early stage (<9 wk), a hyperplastic stage (at 10–40 wk), and a final stage >40 wk.

The pathogenesis of diabetes is not yet fully understood. The fact that in identical twins the rate of onset of NIDDM is >90%, whereas that of IDDM is ~50%, supports a genetic basis of the disease.[5] Some candidate genes responsible for developing IDDM were reported in NOD mice[6,7] and BB rats.[8] However, there are few reports on genes responsible for NIDDM, except some gene abnormalities related to insulin,[9] glucokinase,[10-12] glucose transporter,[13] and mitochondria.[14] Recently, we discovered that one locus responsible for the onset of NIDDM in the OLETF rats mapped to the X-chromosome.[15]

As the stage of diabetes progressed, renal complications became inevitable in all animals.[2] Furthermore, our recent data revealed that restriction of feeding suppressed the development and progression of both diabetes and renal complications.[16] We examined the histopathological nature and the prevalence of renal lesions in OLETF rats, evaluating urinary findings concomitantly, and we attempted to determine whether the OLETF strain might be a useful model for the study of human NIDDM and diabetic nephropathy.

Genetic segregation study

F1, F2, and backcross progenies were obtained by reciprocal mating of OLETF and an unrelated strain, F344, or a related control strain, LETO. We used only male progeny, since female OLETF rats did not develop diabetes at 30 wk. Table 1 shows the results of segregation studies. No diabetic rats were found in reciprocal F1 progenies mated with F344 rats. In F2 progenies, 11.8% (diabetes type 5.0%, IGT type 6.8%) of the progenies were diabetic when female OLETF rats were used for mating. No diabetic animal was found in the reverse mating combination.

We next performed a segregation study using OLETF rats and the related control strain, LETO. Of the (OLETF × LETO) F1 progenies, ~46% of rats were diabetic (diabetes type 3.2%, IGT type 42.9%), whereas in the reverse combination only 2 of 63 F1 progenies (3.2%) were diabetic (IGT type). In (OLETF × LETO) F2 progenies, the frequencies of diabetes and IGT types were 13.1 and 22.2%, respectively. Similar results were observed in F2 progenies with the reverse mating combination (diabetes type 7.1%, IGT type 22%). Backcross progenies, (OLETF × LETO) F1 × OLETF showed a higher incidence of diabetes (diabetes type 44.4%, IGT type 22.2%), whereas those obtained by (OLETF × LETO) F1 × LETO showed a low incidence of diabetes (diabetes type 3.6%, IGT type 17.9%). The backcross progenies obtained from OLETF × (LETO × OLETF) F1 matings showed the highest frequencies of diabetes, (diabetes type 37.8%, IGT type 40.0%).

Table 1. Incidence of diabetes mellitus (DM) in various generations of crosses of OLETF and F344/Ducrj or LETO rats

Female Male	n	Diabetes	IGT	Normal
(OLETF × F344/Ducrj) F1	22	0(0%)	0(0%)	22(100%)
(F344/Ducrj × OLETF) F1	22	0(0%)	0(0%)	22(100%)
(OLETF × F344/Ducrj) F2	161	8(5.0%)	11(6.8%)	142(88.2%)
(F344/Ducrj × OLETF) F2	44	0(0%)	0(0%)	44(100%)
(F344/Ducrj × OLETF) F1 × OLETF	18	12(66.7%)	2(11.1%)	4(22.2%)
(F344/Ducrj × OLETF) F1 × F344/Ducrj	18	0(0%)	0(0%)	18(100%)
(OLETF × LETO) F1	63	2(3.2%)	27(42.9%)	34(54.0%)
(LETO × OLETF) F1	63	0(0%)	2(3.2%)	61(96.8%)
(OLETF × LETO) F2	99	13(13.1%)	22(22.2%)	64(64.6%)
(LETO × OLETF) F2	127	9(7.1%)	28(22.0%)	90(70.9%)
(OLETF × LETO) F1 × OLETF	36	16(44.4%)	8(22.2%)	12(33.3%)
(OLETF × LETO) F1 × LETO	28	1(3.6%)	5(17.9%)	22(78.6%)
OLETF × (LETO × OLETF) F1	45	17(37.8%)	18(40.0%)	10(22.2%)
LETO × (LETO × OLETF) F1	32	0(0%)	0(0%)	32(100%)

From Kawano K, Hirashima T, Mori S, Natori T. Diabetes Res Clin Pract 24 Suppl., 1994, with permission.
IGT, impaired glucose tolerance.

Plasma glucose levels of 160 F2 individuals were quantitatively measured. The AR and PRPSII polymorphisms were screened in the F2 progeny as X-chromosome markers. The linkage analysis with the Mapmaker program has demonstrated that the gene controlling the 120 min plasma glucose level is located on the X chromosome 14 cM distant from the *AR* gene (LOD = 2.598) (Fig. 1A). We have designated the gene as *ODB-1*. As shown in Figure 1B, the mapping order is *AR-ODB-1-PRPSII*. The relative distance between *AR* and *PRPSII* is 33.4 cM.

The association of the development of diabetes with the rat MHC (RT1) haplotype was examined using (OLETF($RT1^u$) × F344($RT1^l$)) F2 animals. The RT1 haplotype is clearly segregated at a mendelian ratio of 1:2:1. However, diabetes was evenly distributed in animals with these genotypes, showing no association with the $RT1^u$ haplotype.

Effects of castration and administration of testosterone

We examined the effects of castration with or without testosterone replacement on plasma glucose levels during OGTT at 30 wk after birth. As shown in Figure 2, in both orchidectomized male and ovariectomized female OLETF rats, testosterone converted rats with a normal OGTT response to a diabetes type response. This effect of testosterone was confirmed by determining the incidence of diabetic animals among castrated rats. The effect was more apparent in orchidectomized males, which showed an increase in diabetes incidence to 87.5%, than in ovariectomized females, which showed an increase in incidence to 71.4%.

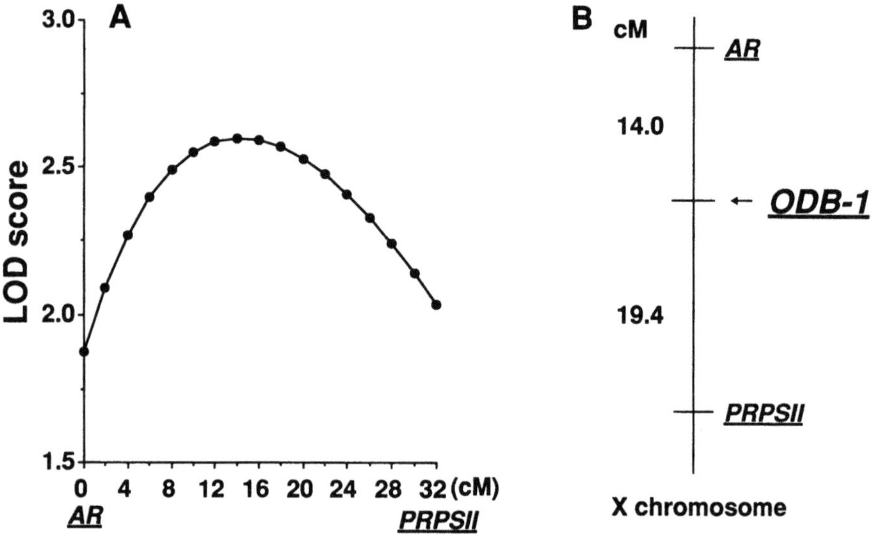

Figure 1. Genetic mapping of a diabetogenic gene *ODB-1*. Linkage analysis was performed using a total of 160 (OLETF ♀ × F344 ♂) F2 offsprings. Values for the 120 min plasma glucose level during the OGTT and phenotypes of the AR or PRPSII microsatellite region were analyzed by the Mapmaker/QTL program. The results show that the most possible site of *ODB-1* is 14 cM from the *AR* gene on the X-chromosome (LOD = 2.598). The relative order (and distance) of *AR, PRPSII,* and *ODB-1* is *AR*-(14.0 cM)-*ODB-1*-(19.4 cM)-*PRPSII*. (From Hirashima T, Kawano K, Mori S, et al. Diab Res Clin Prac, 27:91–96, 1995, with permission.)

Diabetic nephropathy

OLETF rats began to gain weight faster than LETO rats from 5 wk on, and the difference gradually increased with age. After 70 wk of age, body wt began to decrease and at 90 wk of age, body wt was lower than that of the control LETO rats (Fig. 3A). Kidney wt in OLETF rats was usually 1.5 times higher than that of LETO rats at all ages (Fig. 3B). Urinary protein levels at various ages are shown in Figure 4. Urinary protein began to increase more than that of LETO rats from ~30 wk and drastically increased thereafter, ultimately reaching >800 mg/dl.

Histopathologically, early visual changes of glomeruli were focally and segmentally observed as the slight mesangial matrix widening at 23 wk of age before the elevation of urinary protein level. At 29 wk of age, besides mesangial lesion, a few glomeruli showed the segmental lesion with PAS-positive deposits in the mesangium or capillary that seemed to be identical to the fibrin(hyalin) cap in the exudative lesion commonly observed in human diabetic glomerulopathy. On the other hand, at an early stage before 30 wk of age, a number of glomeruli without mesangial and exudative lesions often showed numerous eosinophilic granular deposits entrapped in epithelial cells (podocytes) that were darkly stained by PAM stain. These granules were thought to be identical to

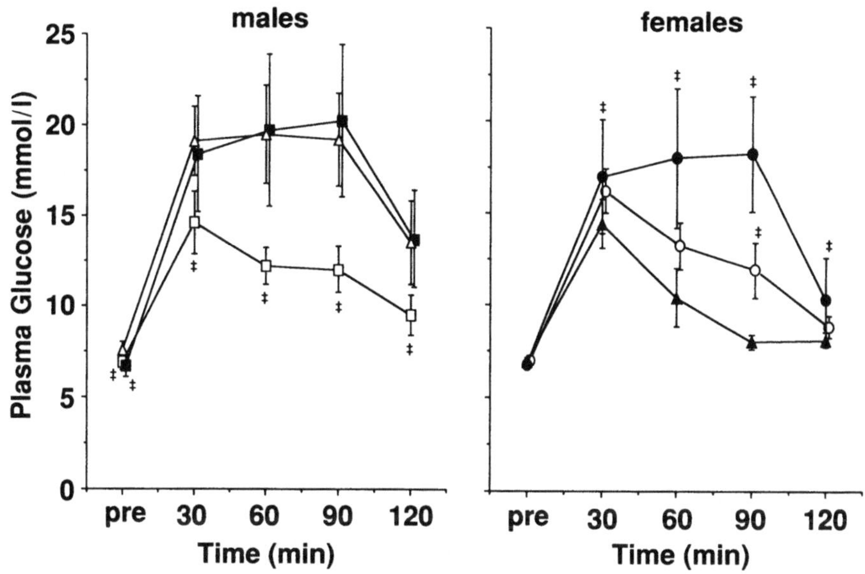

Figure 2. Effects of castration and testosterone treatment on OGTT plasma glucose responses of 30-wk-old OLETF rats (2 g/kg). Rats were fasted for 16 h before the test. Untreated males (△, n = 8); orchidectomized males (□, n = 8); orchidectomized and testosterone-treated males (■, n = 8); untreated females (▲, n = 8); ovariectomized females (○, n = 8); ovariectomized and testosterone-treated females (●, n = 7). Values are means ± SD. ‡Significance of differences from corresponding values of the untreated group $P < 0.01$, Dunnett's test.

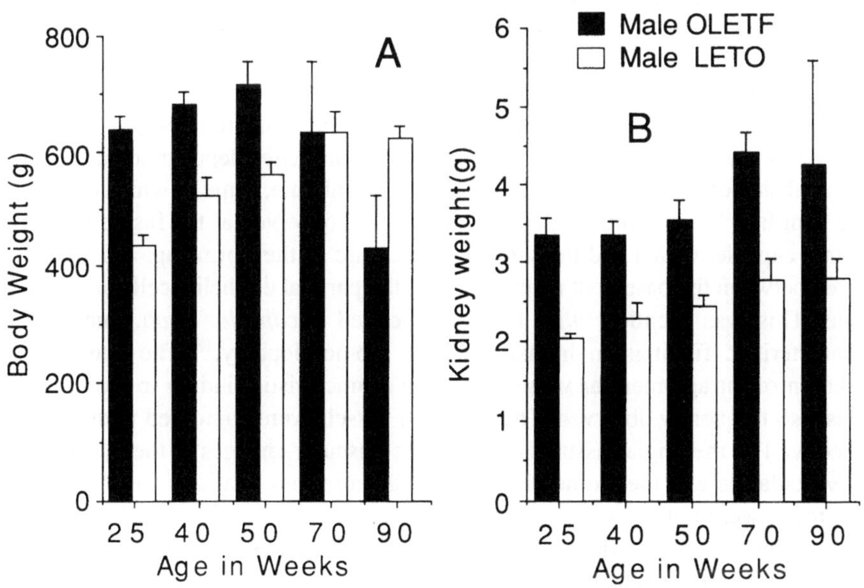

Figure 3. Body wt (A) and kidney wt (B) changes in male OLETF (filled bars, n = 12) and LETO (open bar, n = 12) rats. Columns and bars show mean ± SD.

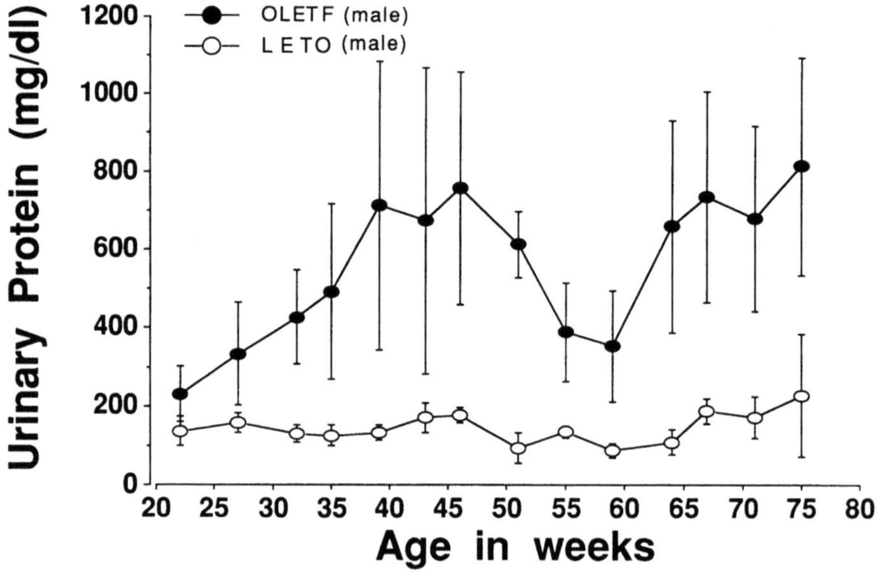

Figure 4. Urinary protein levels in male OLETF (●, n = 12) and LETO (○, n = 12) rats. Columns and bars show mean ± SD.

hyalin droplets regarded to be the reabsorbed protein that is often seen in various human nephrotic syndromes.[17] After 50 wk of age, the glomeruli showing the exudative lesion with prominent fibrin caps became numerous and more dominant than the mesangial lesion. Most of these changes were characteristic in that they arose as focal and segmental lesions in the early stage and exhibited segmental obsolescence in the later stage. However, a few lesions advanced so much as to become global exudative lesions.

In terms of fine structure, the fibrin caps in the exudative lesions were observed as the accumulation of amorphous electron-dense deposits located at the luminal side of a slightly thickened basement membrane, some of which occluded the capillary lumen. Similar deposits were often observed at the inside of Bowman's capsule, which had the same fine structure as the fibrin cap, and were located between the basement membrane and the parietal epithelial cells of the capsule. This was identical with the lesion called a *capsular drop,* one of the characteristic findings in human diabetic glomerulopathy.[18] The aneurysmal dilation of intraglomerular vessels, another characteristic finding in glomerulus, was also frequently observed. These dilated vessels were contoured with a thickened wall of mesangial tissue. However, the vascular changes in the interstitium or vascular pole of the glomerulus, such as the arteriolo-hyalinosis and arteriolosclerosis, were not seen.

Through all stages, the severity of mesangial lesions increased with aging, some of which showed nodular expansion of the mesangial matrix. However, they were not identical with the typical nodular Kimmelstiel–Wilson lesion[19] because they were at most a minor proportion and, if present, lacked the consolidated appearance with an acellular core. After 55 wk of age, segmental or global obsolescence (sclerosis) of glomeruli appeared and involved numerous glomeruli, ultimately developing into an end-stage kidney by 96 wk of age. The glomerular changes of LETO rats were limited to a few mesangial lesions of minimal degree at an early age, which progressed slightly at a later age to show focal mesangial lesions of mild degree and segmental exudative lesions that were very low in number.

The degree and time course of glomerular injury in OLETF rat kidneys were quantitatively analyzed. The three types of glomerular injury were discerned optically: a glomerulus showing only a mesangial lesion with a widened matrix and proliferated cells, a glomerulus showing an exudative lesion, and a glomerulus showing obsolescence. The numbers of each type of glomerular injuries within a cross-sectioned kidney were counted, respectively, for each of five cases at 23, 30, 55, 80, and 100 wk of age. Before 30 wk the mesangial lesion was the predominant type of injury and amounted to 16.5% at 23 wk and 26.2% at 30 wk of age. The proportion of exudative lesions rose to 16.7% at 55 wk and reached a maximum of 52.7% at 80 wk of age. The obsolescent glomeruli appeared at 55 wk and increased to 18.4% at 80 wk of age. At 100 wk of age, the proportion of damaged glomeruli, including both exudative lesions and obsolescence, was >70%.

Discussion and concluding remarks

Male-dependent development of diabetes is a characteristic clinical feature of OLETF rats, similar to that occurring in Wistar Fatty rats or ZDF fatty rats.[20,21] However, our experiments showed that females are also potentially diabetic, because early ovariectomy converted some of them to the diabetic state. Moreover, females might carry one of the genes responsible for inducing diabetes because some (OLETF × LETO) F1 hybrids developed diabetes.

A genetic segregation study using F1, F2, and backcross progenies obtained by matings with an unrelated strain, F344, or a related control strain, LETO, showed that no diabetic syndrome was observed in F1 progenies mated with F344 rats, suggesting that the diabetogenic gene(s) are recessive. As the cumulative incidence of diabetes in (OLETF × F344) F2 progeny was less than the expected value of 25% (11.8%), more than one recessive gene may be involved in the development of diabetes. Some diabetic rats were found among F1 progenies with the related control strain, LETO, and in (OLETF × LETO) F1 × LETO backcross progenies, suggesting that LETO rats share some (or most, but not all) diabetogenic genes with OLETF rats. Since the LETO strain is derived from the same original colony,[2] it is very likely that the LETO shares most of genetic background with the OLETF but lacks at least one diabetogenic gene, preventing the onset of diabetes in this strain.

As shown in Table 1, the X-linked gene(s) of OLETF rats augmented the rate of onset in F1 and backcross progenies compared with LETO rats. These results strongly suggest that one of the diabetogenic genes is located on the X-chromosome of OLETF rats. In fact, linkage analysis has confirmed the association between the gene determining 120 min plasma glucose levels and the *AR* gene (LOD = 2.598) on the X chromosome. Since one of the criteria for diabetes required by WHO[22] is that the 120 min glucose level in the 75 g OGTT is >11.1 mmol/l in human NIDDM, it is reasonable to assume that this parameter reflects the function of one of the diabetogenic genes. Thus, we have designated the gene as *ODB-1*. It would be interesting to determine if a similar gene is responsible for human NIDDM. It is not likely that the mitochondrial genome of OLETF rats is involved in the development of diabetes, because >70% of (F344 × OLETF) F1 × OLETF backcross progenies developed diabetes. In our study, the mean body wt of diabetic rats was significantly higher than those of normal individuals. This indicates an association of diabetes with an increase in body wt. However, the obesity gene[23] or fatty gene[24] was reported to be autosomal recessive, which is clearly different from our X-linked diabetogenic gene (*ODB-1*).

A genetic factor(s) is known to be involved in the development of human NIDDM,[5] but the only genetic studies reported so far are on a nonsense mutation of the glucokinase gene[10-12] and glucose transporter gene polymorphisms.[13] Since there is no report suggesting the involvement of an X-linked gene(s) in humans or experimental animals,[25,26] to the best of our knowledge this is the first report of a gene responsible for NIDDM linked with the X chromosome.

Interestingly, RT1 has no linkage with the diabetogenic gene(s) in OLETF rats. In clear contrast, a MHC gene, namely a certain class II gene, was found to be involved in human IDDM[27] and in experimental animals.[28,29] Diabetic LETL rats of the IDDM type, which we have established from the same breeding colony of Long-Evans rats, carry the $RT1^u$ haplotype. This $RT1^u$ haplotype is necessary for the development of IDDM in LETL rats,[3,4] but in OLETF rats it is not necessary, although they do carry the haplotype. Thus, the difference in the involvement of the RT1 gene(s) between the LETL and OLETF strains may account for the distinct pathogenic origins of the diabetic syndromes in these two strains. The mouse MHC (H-2) was also reported not to be involved in developing streptozotocin-induced NIDDM in C57BL mice.[30]

In our experiments, phenotypically all males and ~30% of females develop diabetes at >60 wk of age. The difference in these incidences seems to be due to sex hormones. This possibility was confirmed by the following experiments. As shown in Figure 2, orchidectomy suppressed the development of diabetes in males, whereas ovariectomy increased the development of diabetes in females. Furthermore, administration of testosterone increased the incidence of diabetes in both castrated male and female rats. Administration of testosterone to castrated F344 rats had no effect on diabetes development (S. Mori, unpublished data).

Plasma levels of testosterone in male rats (OLETF, 3.1 nmol/l (n = 7, 40 wk of age); LETO, 5.2 ± 2.8 nmol/l (n = 12, 40 wk of age); F344, 3.5 ± 1.0 nmol/l (n = 7, 22 wk of age) were not significantly different. These results clearly indicate that testosterone is an important hormonal factor involved in promoting the development of diabetes, even in females. The results indicate that female rats (phenotypically nondevelopers or late developers) also carry a set of diabetogenic genes. In diabetic animals such as WBN/kob rats,[31] KK mice,[32] and *db/db* mice,[33] testosterone was reported to promote the development of diabetes.

Lewis et al.[34] found that administration of testosterone to castrated rats with 95% resection of the pancreas increased the rate of diabetes onset, whereas administration of estrogen suppressed or delayed diabetes development. Taken together, this indicates that for some unknown reason estrogen prevents OLETF females from developing diabetes, even though they carry diabetogenic gene(s).

We consider that in humans too sex hormones may have strong regulatory effects on clinical development of the disease; that is, imbalance of sex hormones (relative excess of testosterone to estrogen) may promote the development of NIDDM in genetically prepared individuals. It is also likely that the influence of estrogen on the development of clinical NIDDM may mask underlying genetic factor(s), even if an X-linked gene is involved. Further study of the gene should be very useful for prediagnosis and prevention of the disease.

The renal dysfunction of OLETF rats began to appear at around 30 wk of age, judging from the analysis of urinary protein, and continued to progress chronically. The early morphologic change in the kidney was focal mesangial matrix widening at 23 wk of age before clinical diabetic nephropathy. The glomerular lesions with fibrin (hyalin) caps observed in a few glomeruli at 29 wk were thought to be identical to the exudative lesion described in human diabetic glomerulopathy,[35] although they differed from the human exudative lesion in their focal and segmental pattern of involvement. Mesangium of OLETF rats was focally and segmentally involved in an early stage and later showed mild to moderate matrix widening in a nonspecific pattern. Mesangial lesions showing a nodular accent were rare and, if present, seemed not to be identical with the typical nodular Kimmelstiel–Wilson lesion.[19]

The structural pathogenesis of nodular lesions remains incompletely understood. Previous studies have supported the suggestion that mesangiolysis and subsequent capillary microaneurysms may evoke Kimmelstiel–Wilson nodules.[36-38] Mesangiolytic change was often found in the mesangium of OLETF rats but was obscured by the superimposed exudative hyalinous lesion. Pure mesangiolytic change was rarely seen. It is evident from our observation that an exudative lesion bearing mesangiolytic changes in OLETF rats fails to evolve into a typical nodular lesion. Based on suggestions of the above-mentioned authors, the probable reason why typical nodular lesions did not develop in OLETF rats is that mesangial expansion was not preceded by pure mesangiolytic change. The cause of the lack in mesangiolysis in OLETF rats remains to be clarified.

The preventive effect of food restriction on diabetes mellitus and its complications has not been sufficiently studied in animal models. In our study,[16] OLETF rats on a 30% restricted diet at 6–80 wk of age showed reduced levels of urinary protein and much less glomerular injury than those without food restriction. These results indicate that the 30% restricted feeding from the early phase is effective in preventing or improving NIDDM and its renal complications. Therefore, glomerular injury of OLETF rats is postulated to be caused exclusively by the prolonged diabetic condition, and the onset of diabetes and of renal complications of OLETF rats have been demonstrated to have a close relation to food intake.

The characteristic features of OLETF rats are as follows: 1) their mode of inheritance involves more than two recessive genes in the development of diabetes, as indicated by (a) both male and female OLETF rats carry a set of diabetogenic genes; (b) LETO rats appear to share some diabetogenic gene with the OLETF strain; (c) a new diabetogenic gene associated with the X-chromosome, named as *ODB-1*. 2) Although the incidence of diabetes in OLETF females is one-third that in males, female OLETF rats also carry diabetogenic genes, since castrated, testosterone-treated females developed diabetes equally as well as males, thus implying that testosterone is an important promoting factor in the clinical development of diabetes. 3) Urinary protein levels increase in male OLETF rats exceed those in LETO rats starting at 30 wk and drastically increase with age. In males 55 wk old, urinary protein levels were >800 mg/dl. The glomerular lesions in OLETF rats, similar to human diabetic nephropathy, were mostly exudative lesions characterized by fibrin (hyalin) caps, capsular drops, and aneurysmal dilatations of intraglomerular vessels.

Our study suggests that OLETF rats should be a useful model for analysis of the pathogenesis and complications of NIDDM and for studies on the development of drugs to regulate the hyperglycemia of diabetes.

References

1. Kawano K, Hirashima T, Mori S, et al. A new rat strain with non-insulin dependent diabetes mellitus, "OLETF." Rat News Lett 25:24–6, 1991.
2. Kawano K, Hirashima T, Mori S, et al. Spontaneous long-term hyperglycemic rat with diabetic complications, Otsuka Long-Evans Tokushima Fatty (OLETF) strain. Diabetes 41:1422–28, 1992.
3. Kawano K, Hirashima T, Mori S, et al. A new strain of rat with insulin-dependent diabetes mellitus "LETL." Rat News Lett 22:14–15, 1989.
4. Kawano K, Hirashima T, Mori S, et al. New inbred strain of Long-Evans Tokushima Lean Rats with IDDM without lymphopenia. Diabetes 40:1375–81, 1991.
5. Barnett AH, Eff C, Leslie RDG, et al. Diabetes in identical twins. A study of 200 pairs. Diabetologia 20:87–93, 1981.
6. Todd JA, Aitman TJ, Cornall RJ, et al. Genetic analysis of autoimmune type 1 diabetes mellitus in mice. Nature 351:542–7, 1991.
7. Prins JB, Todd JA, Rodrigues NR, et al. Linkage on chromosome 3 of autoimmune diabetes and defective Fc receptor for IgG in NOD mice. Science 260:695–8, 1993.
8. Jacob HJ, Pettersson A, Wilson D, et al. Genetic discussion of autoimmune type I diabetes in the BB rat. Nature Genet 2:56–60, 1992.
9. Haneda M, Polonsky KS, Bergenstal RM, et al. Familial hyperinsulinemia due to a structurally abnormal insulin: Definition of an emerging new clinical syndrome. N Engl J Med, 310:1288–94, 1984.
10. Stoffel M, Froguel P, Takeda J, et al. Human glucokinase gene: Isolation, characterization, and identification of two missense mutations linked to early-onset non-insulin-independent (type 2) diabetes mellitus. Proc Natl Acad Sci USA 89:7698–702, 1992.
11. Froguel P, Zouali H, Vionnet N, et al. Familial hyperglycemia due to mutations in glucokinase—definition of a subtype of diabetes mellitus. N Engl J Med 328:697–702, 1993.
12. Stoffel M, Patel P, Lo Y-MD, et al. Missense glucokinase mutation in maturity-onset diabetes of the young and mutation screening in late-onset diabetes. Nature Genet 2:153–6, 1992.

13. Li SR, Baroni MG, Oelbaum RS, et al. Association of genetic variant of the glucose transporter with non-insulin-dependent diabetes mellitus. Lancet 2:368–70, 1988.
14. Ballinger SW, Shoffner JM, Hedaya EV, et al. Maternally transmitted diabetes and deafness associated with a 10.4kb mitochondrial DNA deletion, Nature Genet 1:11–5, 1992.
15. Hirashima T, Kawano K, Mori S, et al. A diabetogenic gene (ODB-1) assigned to the X-chromosome in OLETF rats. Diabetes Res Clin Prac 27:91–96, 1995.
16. Mori S, Hirashima T, Kawano K, et al. Prevention of diabetes mellitus and diabetic nephropathy by diet control in the OLETF rats. J Jpn Diabetes Soc 38:593–599, 1995.
17. Oliver J, MacDowell M, Lee YC. Cellular mechanisms of protein metabolism in the nephron: The structural aspects of proteinuria; tubular absorption, droplet formation, and the disposal of proteins. J Exp Med 99:589–604, 1954.
18. Barrie HJ, Askanazy CL, Smith GW. More glomerular changes in diabetes. Can Med Assoc J 66:428–35, 1952.
19. Kimmelstiel P, Wilson C. Intercapillary lesions in glomeruli of kidney. Am J Pathol 12:83–97, 1936.
20. Ikeda H, Shino A, Matsuo T, et al. A new genetically obese-hyperglycemic rat (Wistar Fatty). Diabetes 30:1045–50, 1981.
21. Peterson RG, Shaw WN, Neel M-A, et al. Zucker diabetic fatty rat as model for non-insulin-dependent diabetes mellitus. ILAR News 32:16–19, 1990.
22. Diabetes mellitus, report of a WHO study group: Technical report series 727, WHO, Geneva, 1985.
23. Coleman DL. Diabetes-obesity syndromes in mice. Diabetes 31:1–6, 1982.
24. Truett GE, Baharry N, Friedman JM, et al. Rat obesity gene fatty (*fa*) maps to chromosome 5: Evidence for homology with mouse gene diabetes. Proc Natl Acad Sci USA 88:7806–9, 1991.
25. Gauguier D, Nelson I, Bernard C, et al. Higher maternal than paternal inheritance of diabetes in GK rats. Diabetes 43:220–4, 1994.
26. Abdel-Halim SM, Guenifi A, Luthman H, et al. Impact of diabetic inheritance on glucose tolerance and insulin secretion in spontaneously diabetic GK-Wistar rats. Diabetes 43:281–8, 1994.
27. Segall M. Perspectives in diabetes. HLA and genetics of IDDM: Holism vs. reductionism? Diabetes 37:1005–8, 1988.
28. Hattori M, Buse JB, Jackson RA, et al. The NOD mouse: Recessive diabetogenic gene in the major histocompatibility complex. Science 231:733–5, 1986.
29. Colle E, Guttmann RD, Seemayer T. Spontaneous diabetes mellitus syndrome in the rat I. Association with the major histocompatibility complex. J Exp Med 154:1237–42, 1981.
30. Leiter EH, Le PH, Coleman DL. Susceptibility to *db* gene and streptozotocin-induced diabetes in C57BL mice: Control by gender-associated, MHC-unlinked traits. Immunogenetics 26:6–13, 1987.
31. Saegusa T, Kon H, Tsuchitani M, et al. Effect of sex hormones on the onset of diabetic syndrome in WBN/Kob rats. Exp Anim 41:481–9, 1992.
32. Nonaka T, Higuchi N, Arai T, et al. Effects of castration and androgen on glycosuria appearance in the diabetic KK mice induced by monosodium glutamate administration. Jpn J Vet Sci 50:1121–23, 1988.
33. Leiter EH, Beamer WG, Coleman DL, et al. Androgenic and estrogenic metabolites in serum of mice fed dehydroepiandrosterone: Relationship to antihyperglycemic effects. Metabolism 36:863–9, 1987.
34. Lewis JT, Foglia VG, Rodrigues RR. The effects of steroids on the incidence of diabetes in rats after subtotal pancreatectomy. Endocrinology 46:111–21, 1950.

35. Olson JL. Diabetes Mellitus. In: Pathology of the kidney, 4th ed. Heptinstall RH, ed. Little, Brown, Boston, pp 1715–63, 1992.
36. Saito Y, Kida H, Takeda S, et al. Mesangiolysis in diabetic glomeruli: Its role in the formation of nodular lesions. Kidney Int 34:389–96, 1988.
37. Stout LC, Kumar S, Whorton EB. Focal mesangiolysis and the pathogenesis of the Kimmelstiel-Wilson nodule. Hum Pathol 24:77–89, 1993.
38. Yajima G. A histopathological study of diabetic nephropathy-light and electron microscopic observations. Acta Pathol Jpn 26:47–62, 1976.

15

Glucose and lipid metabolism in the gold thioglucose injected mouse model of diabesity

STEPHEN C. BLAIR, IAN D. CATERSON, AND GREGORY J. COONEY
Department of Endocrinology, Royal Prince Alfred Hospital, Camperdown, NSW 2050, Australia.

Key words: gold thioglucose (GTG), obese mouse, non–insulin dependent diabetes mellitus (NIDDM), hyperinsulinemia, glucocorticoids, adrenalectomy (ADX), insulin resistance.

Gold thioglucose (GTG) ($C_6H_{11}O_5SAu$) was initially developed and marketed as a therapeutic agent for the treatment of arthritis and rheumatism (Fig. 1). During toxicology studies, it was noted by Brecher and Waxler[1] that a single i.p. or s.c. injection of GTG induced a syndrome of hyperphagia and obesity in mice. This observation was subsequently confirmed by Marshall and colleagues,[2] who reported that GTG induced bilateral necrosis of the ventromedial hypothalamus (VMH) region of the brain and caused damage to the supraoptic nuclei, ventromedial nuclei, arcuate nuclei, and median eminence. These GTG-induced lesions in the hypothalamus of mice did not appear to impair central functions other than the regulation of food intake and body wt.[3] The degree of obesity induced by GTG in mice is dependent on the dose of GTG administered and the particular strain of mice used.[2,4] In general, the larger the dose of GTG administered, the greater the gain in body wt.[5] Administration of a range of doses of GTG has been shown to induce variable wt gain and mortality in the C58, RIII, DBA, and BALB/C strains of mice compared with the CBA strain. CBA mice appear to be unique in that they exhibit a uniform response with respect to these two parameters.[6]

In rats, GTG produces a hypothalamic lesion that is similar to that observed in mice.[7] However, administering GTG to rats at a dose sufficient to induce obesity is associated with a high mortality rate. The increased mortality has been attributed to the effective and lethal doses of GTG being similar,[8] but several groups

Correspondence: Dr. Stephen C. Blair, Department of Endocrinology, Royal Prince Alfred Hospital, Camperdown, NSW 2050, Australia. E-mail: sblair@earth.endocrin.su.oz.au

Lessons from Animal Diabetes VI. Ed. E. Shafrir.

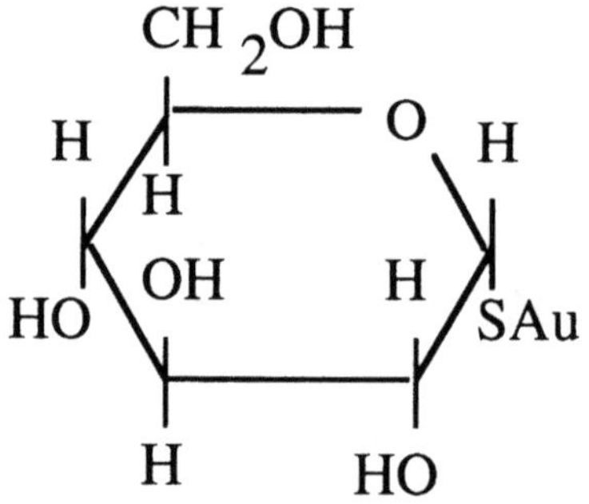

Figure 1. Chemical structure of gold-thioglucose.

have demonstrated that the toxicity in rats can be overcome by administering GTG 1) i.v. in conjunction with sodium hexobarbital, an ultrashort-acting barbiturate used to induce surgical anesthesia[9]; or 2) in small quantities directly into the hypothalamus.[10,11]

Mechanism of GTG action

In mice, administration of GTG produces lesions primarily in the ventromedial region of the hypothalamus.[2] As a result of this lesion, the animals become hyperphagic and overt obesity ensues.[1] A number of studies have shown that glucose analogues, such as 2-deoxyglucose, sodium thioglucose,[7,12] beta thioglucose,[13] and phlorizin,[14] induce acute hyperphagia and prevent GTG-induced lesioning of the hypothalamus. Glick and Mayer[15] proposed that these glucose analogues prevent the GTG-induced necrosis via their inhibition of cellular glucose transport in the hypothalamus. Therefore, it was postulated that 1) GTG enters the cells of the VMH in a similar manner to that of glucose; 2) VMH contains glucoreceptive cells that are involved in the regulation of food intake. These glucoreceptive cells were subsequently identified as oligodendrocytes, occurring in the VMH area.[16,17]

Debons et al.[18] reported the uptake of GTG by the VMH to be dependent on the hormonal status of the animal. In insulin-deficient states, such as IDDM, the VMH was insensitive to GTG. Mice made diabetic with alloxan or anti-insulin serum, prior to GTG injection, failed to become obese. However, on treatment of alloxan-diabetic mice with exogenous insulin (i.v. or intrahypothalamically), the sensitivity of the VMH to GTG was restored and obesity developed. Also, cessation of administration of anti-insulin serum to mice was followed by a return of VMH sensitivity to GTG. Therefore, it was hypothesized that an insulin-deficient state renders the VMH resistant to GTG injury.[19,20] It appears that VMH neurons may be insulin dependent for glucose uptake, unlike neurons outside the VMH. This hypothesis was supported by Oomura,[21] who showed that the firing rate of VMH neurons did not change in response to the electrophoretic application of glucose. However, the firing rate of these neurons was enhanced when both glucose and insulin were applied. In further work designed to examine the effect of insulin and glucocorticoids on the susceptibility of the hypothalamus to lesioning by GTG, Debons et al.[22] examined alloxan-diabetic mice that had been subjected to adrenalectomy (ADX). It was found that glucocorticoid deprivation, and an associated hypophagia, increased the sensitivity of the VMH to GTG, despite the fact that the mice were in an insulin-deficient state. Glucocorticoid administration to the alloxan-diabetic ADX mice decreased the sensitivity of the VMH to lesioning by GTG. However, glucocorticoids did not appear to alter the response of normal nondiabetic mice to GTG.[22] Debons et al.[23] concluded that glucocorticoids were able to influence the susceptibility of the VMH to GTG-induced necrosis by acting as anti-insulin, rather than as anti-inflammatory, agents. It seems that, unlike most cerebral tissue, the uptake of glucose and GTG into the VMH, in particular the glucoreceptive cells (oligodendrocytes), is dependent on the stimulatory action of insulin. Also, glucocorticoids appear to exert an antagonistic effect on the action of insulin at the level of the hypothalamus.[17,23]

It has also been demonstrated that the ability to GTG to lesion the brain is species specific. The VMH of rats and mice is sensitive to lesioning by GTG,[7] whilst chickens,[24] frogs,[25] and gerbils[26] are resistant to GTG-induced obesity. One possible explanation for these observations is that the VMH of GTG-resistant animals differs significantly in structure and/or function from that of rats and mice, by the absence of glucoreceptive, GTG-sensitive cells. In contrast, the VMH of hamsters is susceptible to GTG injury, but this does not result in hyperphagia and obesity.[27]

Specificity of GTG

The effect of GTG on the brain of mice is specific. Other gold thiol compounds, such as gold thiogalactose, gold thiosorbitol, gold thiomalate, gold thiocaproate, gold thioglycoanilide, and gold thiosulphate, do not induce the brain damage that results from the administration of GTG.[7] Neither hyperphagia nor obesity result from such treatments, even though these compounds all possess a similar level of toxicity.[3] On the basis of these observations, Mayer[3] proposed that the toxic gold moiety of accumulated GTG destroys the ventromedial neurons specifically because of the affinity of these cells for the glucose component of the molecule. Studies employing neutron activation analysis and autoradiography to assess the accumulation of radiolabeled gold within the rostral, middle and caudal portions of the brain support this concept.[28,29] In GTG-treated mice, localization of Au^{198} activity was primarily observed in four regions: the hypothalamus, the midline ventral to the anterior commisure, the caudal portion of the septum and ventral hippocampal commisure, and the hindbrain in the midline at the level of the vestibular nuclei in the floor of the fourth ventricle.[28,29] The fact that the administration of GTG leads to such localization suggests that the glucose moiety of GTG is responsible for the focal accumulation of sufficient gold in the hypothalamus to produce a destructive lesion, resulting in hyperphagia and eventually obesity.[28] Administration of GTG also leads to necrosis and gold accumulation in a variety of non-neural tissues, including kidney, liver, heart, and adipose tissue, but this is not thought to play a causal role in the obesity that arises as a result of GTG injection.[30]

Histology of the GTG lesion

Several longitudinal studies were conducted to examine the morphological changes occurring in the hypothalamus of mice as a result of administration of GTG. In one such study,[16] the histological changes occurring during the first 24 h after GTG injection were examined. The first structural abnormality observable under the light microscope appeared at ~6 h after GTG injection. This consisted of a breakdown of the boundary between the oligodendrocytes and the surrounding neuropil in the VMH region. By 7–9 h following GTG injection, a progressive breakdown of the neuropil surrounding the affected oligodendrocytes was observed. Where the area of neuropil dissolution surrounding an oligodendrocyte included a capillary, structural changes were also seen in the wall of the capillary. Neuropil dissolution and capillary breakdown continued to spread until 14–16 h after GTG injection, when a large bilateral area of necrosis adjacent to the third ventricle was evident. Within this necrotic region, the neuropil was represented only by fragmented filaments and the capillaries were in various stages of disintegration; the number of cell nuclei decreased and all remaining nuclei were pyknotic. The necrosis included all types of neurons and glia of the area. Apart from further dissolution of nuclei, the histology of the necrotic area did not undergo any further changes between 16 and 24 h following GTG injection.[16] An earlier study by Brecher et al.[31] had shown that 24 h after GTG injection, the necrotic lesion encompassed the VMH nuclei, paraventricular nuclei, and often the supraoptic nuclei of the hypothalamus. Seven days after injection, the hypothalamic lesion was reduced in size and was present as a narrow band on each side of the ventricle, and by 14 days all that remained of the original necrotic lesion was a narrow scar.[31] It was the sequence of morphological changes occurring in the hypothalamus after GTG injection, in conjunction with the selective cellular distribution of gold in this area, that led Debons et al.[16] to conclude that the oligodendrocytes of the ventromedial area were the glucoreceptor cells. It is this particular cell type that is the primary site of action of the neurotoxin GTG. However, GTG-induced lesions are also known to occur in extrahypothalamic sites.[32]

Development of the diabesity syndrome in GTG-injected mice

Many effects of GTG injection on major parameters of glucose and lipid metabolism have been reported in the literature. These are summarized in Tables 1 and 2 with appropriate references. Some of the major areas of metabolism that have been studied in GTG-injected mice are dealt with in more detail later.

Hyperphagia. GTG-injected mice display acute hypophagia and wt loss during the first few days following GTG injection. Hyperphagia develops within 1–2 wk of injection with GTG.[33] Total body wt is increased 3 wk after GTG injection, but epididymal white adipose tissue (WAT) and interscapular brown adipose tissue (BAT) wt increase occurs as early as 2 wk.[34,35] By 5–6 wk, GTG mice are overtly obese and body wt is 25–40% higher compared with age-matched

Table 1. Metabolic parameters of glucose and lipid metabolism in GTG-obese mice

Metabolic parameter	Tissue change and ref. number				
	Heart	Muscle	Liver	WAT	BAT
Insulin receptor number		↓ 87, 89	↓ 34, 86		↓ 88
Insulin receptor affinity		↔ 90	↔ 34, 86		↔ 88
Insulin receptor tyrosine kinase activity		↓ 89			↓ 88
Insulin receptor substrate 1		↔ 137		↓ 137	
Phosphatidylinositol 3-kinase		↓ 136, 137		↓ 137	
Lipid synthesis			↑ 35, 67	↑ 35, 67	↑ 35
Glucose-6-phosphate dehydrogenase			↑ 56	↑ 56	
Malic enzyme			↑ 56	↑ 56	
Acetyl CoA carboxylase	↔ 55		↑ 55	↔ 55	
Fatty acid content			↑ 117	↑ 117	↑ 117
Triglyceride content		↑ 131	↑ 131		
Glucose transporter numbers (GLUT 4)	↓ 84	↔ 83 **or** ↓ 84		↓ 83–85	↓ 83, 84
Glucose uptake	↓ 66, 74	↓ 66, 87		↑ 66, 105	↔ 105 **or** ↓ 66
Hepatic glucose production			↑ 98, 105		
Pyruvate dehydrogenase	↓ 94, 96, 97	↓ 55, 141	↑ 34, 55	↓ 55, 97	↑ 95, 97
Pyruvate dehydrogenase kinase	↔ 132				↔ 95
Glycogen synthase	↔ 133 **or** ↓ 133	↓ 133 **or** ↑ 104	↔ 102 **or** ↓ 103	↓ 134	
Glycogen phosphorylase	↔ 133	↔ 104 **or** ↓ 133	↑ 102, 103	↓ 134	
Glycogen content	↔ 133	↔ 104	↑ 103, 135	↔ 133	
Leptin mRNA				↑ 146, 149	

Table 2. Blood parameters in GTG-obese mice

Constituent	Change and ref. number
Glucose	↑ 35, 60
Insulin	↑ 68, 134
Glucagon	↔ 33
Glucose tolerance	↓ 60, 68
Glycated hemoglobin	↑ 86
Triglycerides	↑ 34, 93
Free fatty acids	↑ 62
Cholesterol	↑ 61
Corticosterone	↑ 116, 117
Adrenocorticotrophic hormone	↔32
Prolactin	↑ 109
Growth hormone	↓ 109
Thyroxine	↔ 39
Leptin	↑ 146

saline-injected control mice.[34,35] GTG mice continue to gain wt until 12–16 wk post-injection, at which time the animals are no longer consistently hyperphagic and the rate of wt gain is similar to that of control mice.[36-38] Whilst there is little doubt that hyperphagia contributes to the increased wt gain in GTG mice, increased food intake is not the only factor responsible for the obesity that develops in this animal model. Pair feeding did not stop the increased wt gain in GTG mice, and the food intake of GTG mice had to be restricted to 60–80% of control mice to achieve the same rate of wt gain as that observed in control mice.[39,40] Similarly, in other hypothalamic models of obesity, increased wt gain remains evident in the absence of hyperphagia.[41,42]

Hypothalamic administration of neuropeptide Y (NPY) has been shown to promote wt gain by both decreasing energy expenditure and increasing insulin secretion and food intake.[43,44] In two recent studies the amount of NPY and the mRNA level of this neuropeptide were found to be increased in the VMH of GTG mice, suggesting that alterations in NPY may be the cause of the hyperphagia in this animal model.[45,46] However in another study the hypothalamic NPY mRNA levels did not change in response to fasting of GTG-injected mice, but doubled in control mice with fasting.[140] Irrespective of whether alterations in NPY cause the hyperphagia that develops in GTG mice, hyperphagia *per se* is not the only metabolic abnormality responsible for promoting obesity in this animal model. It is likely that other factors, such as energy expenditure, hyperinsulinemia, and hyperglycemia, are important stimuli for the increased wt gain in GTG-injected mice.

Energy balance and thermoregulation. In contrast to genetically obese (*ob/ob*) mice, GTG mice are relatively tolerant of the cold.[47] Exposure to cold invokes a near-normal thermogenic response, involving hypertrophy and hyperplasia of BAT following an increase in sympathetic nervous system (SNS) activity and activation of thermogenesis.[37,39,48] Thus, nonshivering thermogenesis appears normal in GTG-injected mice. However, GTG-injected mice are unable to activate thermogenesis in BAT when fed a cafeteria diet.[37] This defect in diet-induced thermogenesis is thought to be due to the inability of the diet to stimulate the SNS.[47,49] It is thought that the lesion induced by GTG disrupts the neural elements responsible for linking dietary, but not environmental, signals with the sympathetic nerves in BAT.[47] Consequently, the energy that would normally be dissipated in diet-induced thermogenesis in BAT of GTG-injected mice is conserved. Undoubtedly the reduction in diet-induced thermogenesis (and the consequent increase in metabolic efficiency), in conjunction with hyperphagia, leads to an acceleration of the development of obesity in GTG-injected mice.

Lipid metabolism. Obesity is characterized by increased synthesis and deposition of lipid, and therefore the control points of the lipogenic pathway in different tissues represent possible sites of abnormalities that could contribute to the diabesity observed in the different animal models. In previous studies of lipogenesis,[50-54] an increased or decreased lipid synthesis in the same tissues has been reported, but many of these anomalies can be attributed to differences in the stage of obesity and/or the nutritional state of the animals when lipogenesis was measured. In GTG mice this problem has been overcome by measuring the lipogenic response to feeding in liver, WAT, and BAT during the development of obesity (Fig. 2). In the liver of control mice, the lipogenic response to feeding was consistent over the 12 wk experimental period. However, in GTG mice lipid synthesis in liver was much higher following feeding and continued to increase as obesity developed in parallel with the hyperinsulinemia[35] (Fig. 2A and 2B). Hepatic hyperlipogenesis in GTG mice is accompanied by increases in lipogenic enzyme activities (see Table 1).[55,56] Based on findings obtained in other animals with obesity, such as the *ob/ob* mouse, it would appear that the increased activities of these enzymes probably result from a rise in enzyme protein synthesis, more than from activation of existing enzymes.[57,58] In order to store the excess lipid elaborated in GTG-obese mice, the hepatocytes and adipocytes become hypertrophic but not hyperplastic.[59,60] Consistent with the increase in liver lipogenesis are the elevated circulating levels of cholesterol,[61] triglycerides,[34] and free fatty acids (FFA)[62] in GTG-obese mice (see Table 2), as has been shown in VMH-lesioned rats.[63-65] Lipid synthesis in WAT is greatly increased in fed mice for the first 5 wk after GTG injection but after this period, when the fat pads are 6 times larger than those of control animals, the lipogenic rate in WAT of GTG-obese mice decreases (Fig. 2C). This decrease occurs at the same time period that insulin levels are increasing dramatically and thereby represents an increasing insulin resistance in WAT after a period of apparent hypersensitivity to insulin. The period of increased fat synthesis coincides with increased insulin WAT sensitivity to 2-deoxyglucose uptake, observed at 4 wk post-GTG injection.[66] BAT lipogenesis follows a similar pattern to WAT in fed GTG mice, except that the development of insulin resistance with respect to lipid synthesis appears earlier in BAT than in WAT (Fig. 2D). Defective thermogenesis in GTG mice could provide some of the explanation of the rapid accumulation of lipid and development of insulin resistance in BAT. Feeding would normally elicit diet-induced

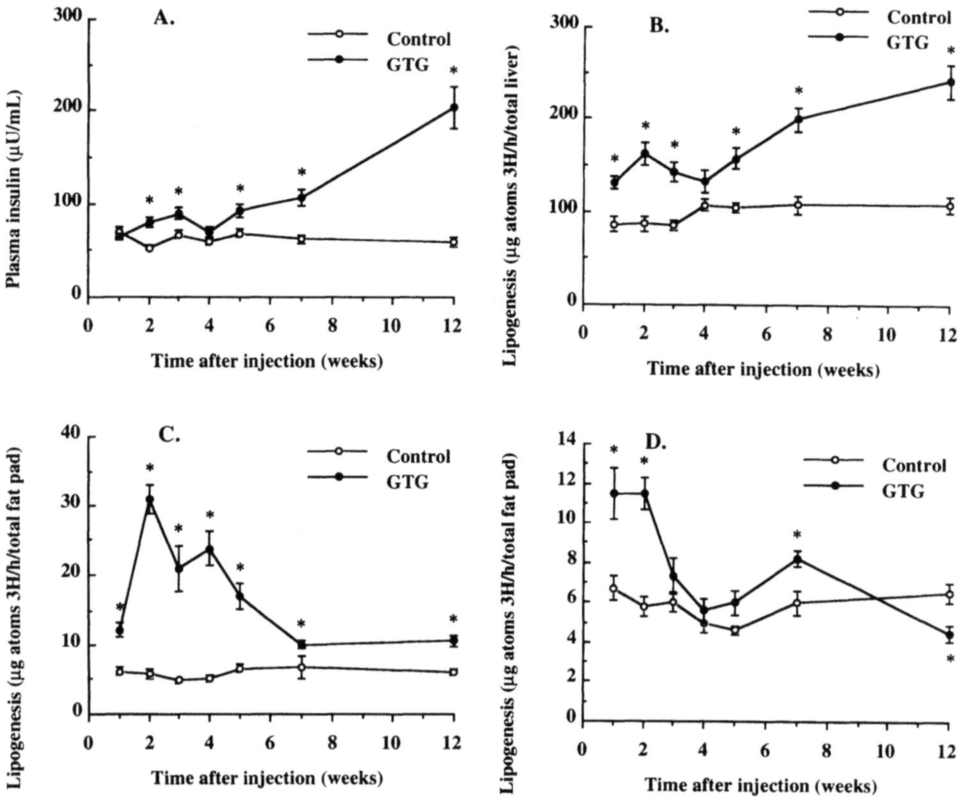

Figure 2. Changes in insulin (A) and rates of fatty acid synthesis in liver (B), epididymal WAT (C), and BAT (D) of fed GTG-injected mice during the development of obesity. Each value is the mean ± SE for 10–15 animals. Statistical analysis was performed by Student's *t*-test. $P \leq 0.05$ for differences between GTG-injected and age-matched control mice. Adapted from [35] with permission.

thermogenesis in BAT, which would use lipid as a substrate and thereby prevent the buildup of lipid in this tissue.

In summary, lipid synthesis is accelerated in liver, WAT, and BAT of mice after GTG injection, but by 5 wk BAT and WAT become insulin resistant and lipid synthesis decreases. Liver, however, remains insulin sensitive with regard to lipogenesis and produces large amounts of lipid, even when the whole animal is overtly insulin resistant, with insulin levels four-fold higher than in fed control animals. It is likely that the early hypersensitivity to insulin observed in the major lipogenic tissues of GTG-injected mice is responsible for the rapid wt gain that characterizes the dynamic phase of obesity in GTG-injected mice.

Although it is clear that there is increased lipogenesis in GTG mouse tissues after ad libitum feeding, it is also well documented that GTG-injected mice are hyperphagic. Therefore, it is possible that the increased lipogenesis in these tissues is simply due to excessive intake of food and not to inherent change in the metabolism of these tissues. To assess whether hyperphagia could explain the initial hyperlipogenesis of GTG mice, we carried out lipogenesis experiments with 2 and 5 wk GTG injected mice that had been starved overnight and then fed 500 mg of standard chow.[67] After overnight starvation, both control and GTG-injected mice began eating the 500 mg chow pellet on presentation and finished consuming the entire meal within 10 min. The rate of lipid synthesis in liver and WAT was shown to be increased in GTG mice compared with age-matched controls despite consuming the same amount of food.[67] Therefore, whilst hyperphagia undoubtedly contributes to the hyperlipogenesis in the tissues of GTG-injected mice, it is not solely responsible for the increased lipid synthesis occurring in these animals.

Insulin secretion. Oversecretion of insulin is evident from an early stage in the development of GTG-induced obesity. Insulin levels in both the overnight fasted state and in response to an i.v. glucose load are increased from 2 wk after the administration of GTG. Moreover, the incremental area under the 0–10 min portion of the insulin curve is increased in 2 wk GTG-injected mice compared with lean control mice (Table 3). At this stage, there is no significant difference between GTG and control mice in terms of body wt or glucose tolerance.[34,35,68] However, by 5–6 wk the GTG mice are significantly heavier than their lean controls and possess a marked glucose intolerance. This is accompanied by overt hyperinsulinemia in both the basal state and also in response to an i.v. glucose load, as indicated by an increase in the incremental area under the curve of insulin (see Table 3).[62,68] At 10 wk after GTG administration, body wt is further increased, as is the degree of glucose intolerance. Whilst 10 wk GTG-injected mice display basal hyperinsulinemia, an i.v. glucose load elicits a much smaller insulin secretory response than that observed in age-matched control animals.[68] Indicative of the reduced insulin secretion in 10 wk GTG-injected mice is the decrease in their insulinogenic index (see Table 3). Thus, it is apparent that during the early stages of GTG-induced obesity, a marked increase in insulin release in response to an i.v. glucose load occurs, but as the animals become chronically hyperinsulinemic and hyperglycemic, the insulin secretory response to i.v. glucose becomes impaired. This sequence of events has similarly been demonstrated in another hypothalamic model of obesity, the VMH-lesioned rat.[69,70]

The refractoriness of insulin secretion in response to i.v. glucose may indicate that the pancreatic B-cells of GTG-injected mice become either insensitive to glucose or depleted of insulin. B-cell exhaustion is, however, unlikely to account for the observed lack of sensitivity to i.v. glucose. Measurement of insulin and glucose levels after ingestion of a standard meal of chow showed a progressive increase in the insulin secretion associated with the development of obesity and glucose intolerance and a marked hyperinsulinemic response to a meal even at 10 wk after GTG injection, when the secretory response to i.v. glucose was essentially absent (Fig. 3; Table 4). This implies that the pancreas of GTG-injected mice remains responsive to stimuli associated with the ingestion of food (perhaps neural signals or other secretagogues in the mixed meal), but becomes insensitive to the prevailing hyperglycemia that develops with obesity. Sufficient

Table 3. Incremental area under the curves of plasma glucose and plasma insulin following administration of an i.v. glucose bolus to GTG mice with varying degrees of obesity

Time post-injection (weeks)	Glucose area (min.mmol/l)		Insulin area (min.μU/ml)		Insulinogenic index (μU/μmol)	
	Control	GTG	Control	GTG	Control	GTG
2	119 ± 12	124 ± 11	142 ± 9	190 ± 19[a]	1.19 ± 0.11	1.53 ± 0.16
5	118 ± 12	192 ± 17[a,b]	136 ± 19	229 ± 26[a]	1.15 ± 0.18	1.19 ± 0.07
10	117 ± 15	230 ± 16[a,b]	112 ± 9	31 ± 7[a,b]	0.96 ± 0.11	0.14 ± 0.01[a,b]

Incremental area under plasma glucose and plasma insulin curves from 0 to 30 min and 0 to 10 min respectively, after i.v. administration of a 0.5 g/kg glucose bolus. By this period of time (30 min for glucose and 10 min for insulin), glucose/insulin levels of control mice had returned to basal values. The insulinogenic index (insulin/glucose ratio) was calculated by dividing the incremental area under the plasma insulin curve by the incremental area under the plasma glucose curve for individual animals. Each value is mean ± SE of the incremental area under the curve for 8–12 GTG or control mice at 2, 5, and 10 wk post-injection.

[a] $P < 0.05$ for differences between GTG and control mice of the same age.

[b] $P < 0.05$ for differences from 2 wk GTG-injected mice.

Adapted from [68] with permission.

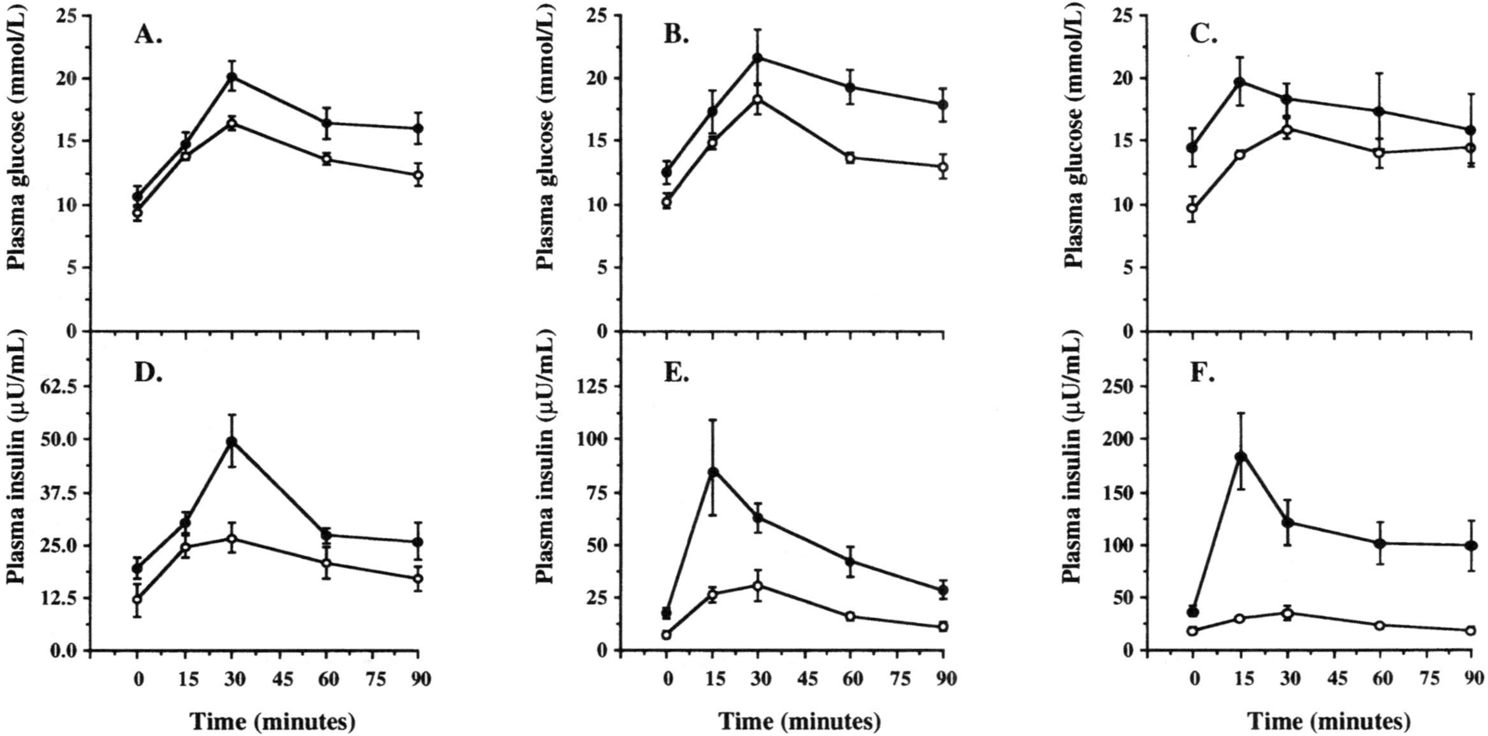

Figure 3. Changes in glycemia and insulinemia in response to ingestion of a standard meal of chow by 2 wk (A, D), 5 wk (B, E), and 10 wk (C, F) GTG-injected and age-matched saline injected control mice. The meal, comprising a 0.25 g pellet of chow, was presented at time 0. Each time point is the mean ± SE of values from five to eight animals. Reproduced from [138] with permission.

Table 4. Incremental area under the curves of plasma glucose and insulin following spontaneous ingestion of a standard meal by GTG mice with varying degrees of obesity

Time post-injection (weeks)	Glucose area (min.mmol/l)		Insulin area (min.μU/ml)		Insulinogenic index (μU/μmol)	
	Control	GTG	Control	GTG	Control	GTG
2	440 ± 68	488 ± 55	857 ± 235	1557 ± 151[a]	1.95 ± 0.18	3.19 ± 0.25[a]
5	395 ± 54	562 ± 110	931 ± 269	2336 ± 253[a]	2.36 ± 0.22	4.16 ± 0.42[a,b]
10	420 ± 96	268 ± 83	1073 ± 289	4042 ± 702[a,b]	2.55 ± 0.29	15.08 ± 1.2[a,b]

Incremental area under plasma glucose and plasma insulin curves from 0 to 90 min and 0 to 60 min, respectively, after presentation of a 0.25 g chow meal. By this period of time (90 min for glucose and 60 min for insulin), glucose/insulin levels of control mice had returned to basal values. The insulogenic index (insulin/glucose ratio) was calculated by dividing the incremental area under the plasma insulin curve by the incremental area under the plasma glucose curve for individual animals. Each value is mean ± SE of the incremental area under the curve for 8–12 GTG or control mice at 2, 5, and 10 wk post-injection.

[a] $P < 0.05$ for differences between GTG and control mice of the same age.

[b] $P < 0.05$ for differences from 2 wk GTG-injected mice.

Adapted from [138] with permission.

insulin secretory capacity throughout the development of GTG-induced obesity is also suggested by 1) increased size of the islet tissue mass, resulting predominantly from hypertrophy and hyperplasia of B-cells,[71] 2) elevated pancreatic content of insulin,[72] and 3) in vitro studies showing enhanced insulin synthesis and secretion in isolated islets.[73,74] Although insulin secretion is increased after administration of GTG, there is no evidence that GTG itself has other than a transient potentiating effect on insulin secretion in isolated islets.[75] Whilst the pancreatic content of insulin is increased in GTG mice, the pancreatic levels of somatostatin and glucagon are not (see Table 2).[72]

The VMH, which is the major area of the brain lesioned by GTG injection, has a role in mediating insulin secretion. Destruction of the VMH has been shown to lead to alterations to the autonomic nervous system with subsequent hyperinsulinemia.[76] Although it has not been directly demonstrated, the increased insulin release in GTG mice is probably caused by the lesioning of the VMH, altering the outflow of the autonomic nervous system, particularly the vagal branches innervating the endocrine pancreas.[63,77] This hypothesis is supported by studies in VMH-lesioned and *fa/fa* rats in which vagotomy and atropine treatment have been shown to attenuate the development of hyperinsulinemia and obesity.[78-81] It has also been shown in the VMH-lesioned rat that an increase in insulin secretion in response to an i.v. glucose load is evident well before the development of hyperphagia, insulin resistance, and obesity.[82] The increase in insulin secretion observed in 2 wk GTG-injected animals is undoubtedly an important abnormality in the initiation of obesity, contributing to the onset of the hyperlipogenesis and insulin resistance in this animal model.

Insulin resistance. The hyperinsulinemia of GTG mice is accompanied by a developing insulin resistance, as indicated by 1) inability of exogenous insulin to decrease circulating glucose levels,[60] 2) loss of sensitivity of insulin-stimulated glucose uptake in BAT and muscle,[66] 3) decrease in adipose and muscle tissue insulin regulatable glucose transporter (GLUT 4) numbers,[83-85] and 4) reduction in insulin binding to receptors prepared from membranes of liver,[34,60,86] muscle,[87] and BAT.[88] The reduction in insulin binding of each of these insulin-sensitive tissues is thought to be due to a decrease in receptor number (downregulation) without any concomitant changes in receptor affinity.[60,87,88] In addition, insulin receptors from skeletal muscle[89] and BAT[88] of GTG-obese mice exhibit a decreased tyrosine kinase activity for phosphorylation of the receptor and of exogenous substrates. This defect in tyrosine kinase activity in tissues of GTG-obese mice is independent of the downregulation of insulin receptor numbers and, as such, represents yet another site of insulin resistance.[89,90] Furthermore, tissue-specific alterations in the guanine-nucleotide–binding protein α-subunits have been documented in GTG-injected mice, and these changes may also contribute to the impaired insulin action.[91]

An additional postbinding site of intracellular insulin resistance is the mitochondrial pyruvate dehydrogenase (PDH) complex, an insulin-sensitive enzyme that is important in the regulation of glucose entry, oxidation and channeling to lipid synthesis. This enzyme complex is inactivated by phosphorylation (PDH kinase) and is activated by dephosphorylation (PDH phosphatase).[92] In GTG-obese mice, the amount of PDH in the active form is decreased in heart,[62,93,94] white adipose tissue,[55] and skeletal muscle,[55] but not in liver[34] and BAT.[95] Acute administration of a fatty acid oxidation inhibitor, 2-tetradecylglycidic acid, to GTG-obese mice has been shown to restore heart PDH complex activity to lean control levels, indicating that the inactivation of the PDH complex in cardiac muscle of GTG-obese mice is most likely a consequence of increased fatty acid oxidation.[96] The decrease in PDH activity, despite high circulating insulin levels,[34,55,97] may be a major factor mediating the inhibitory effects of obesity on glucose oxidation in muscle (cardiac and skeletal), and this may be important in the development of the insulin-resistant state in this animal model of diabesity.

Carbohydrate metabolism. Hyperglycemia and glucose intolerance evolves with the development of obesity in GTG mice. At 2 wk after the administration of GTG, basal glycemia or body wt is not increased compared with the saline injected control mice. In response to an i.v. glucose load or an orally ingested chow meal, the 2 wk GTG-injected mice exhibit no significant increase in the incremental area under the curve (iAUC) of glucose[68] (Fig. 3A; Tables 3 and 4). By 5 wk post-GTG, the mean body wt of the GTG mice is significantly greater than that of their age-matched controls. Basal hyperglycemia is clearly evident by this stage. The excursion in plasma glucose concentrations following an i.v. glucose load[68] or an orally ingested chow meal (see Fig. 3B) is also increased at 5 wk after administration of GTG, clearly demonstrating that these animals are glucose intolerant. The glucose intolerance was reflected in the iAUC of glucose, which was also increased in 5 wk GTG-obese mice, irrespective of whether the animals received an i.v. bolus of glucose or a chow meal (see Tables 3 and 4). At 10 wk after the initiation of obesity, GTG mice are markedly heavier than their lean counterparts. Accompanying this increase in body wt is a further rise in plasma glucose concentrations, basally and in response to an i.v. glucose load[68] or a chow meal (see Fig. 3C). In keeping with the hyperglycemia that develops in this animal model is the glycated hemoglobin level of GTG-obese mice, which is also elevated.[86]

Several mechanisms have been shown to be responsible for the hyperglycemia that develops in GTG mice. Gluconeogenesis is increased in freshly isolated hepatocytes from GTG-obese mice, despite the prevailing hyperinsulinemia, and is not suppressible by supraphysiological doses of insulin.[98] Increased flux through the gluconeogenic pathway is not, however, associated with increased phosphoenolpyruvate carboxykinase (PEPCK) activity (T.M. Greenaway, personal communication), as in KK, *db/db* and *ob/ob* mice,[63] suggesting that gluconeogenesis is regulated by substrate level rather than enzyme activity.[99] Conversely, PEPCK activity is increased in liver of VMH-lesioned rats[100] and is likely to contribute to the higher hepatic glucose production in this animal model.[101] Serum glucagon levels are also not different in GTG-obese mice compared with lean controls,[33] but glucagon binding to hepatic plasma membranes is decreased.[60] Glucose derived from hepatic glycogen stores also contributes to the observed hyperglycemia of GTG-obese mice.[60] Glycogen turnover is increased in the liver of GTG mice, as intimated by the increased glycogen content and phosphorylase activity. Paradoxically, glycogen synthase activity is not increased, at least in response to fasting and refeeding.[102] Muscle glycogen content, glycogen phosphorylase, and glycogen synthase activities are also not different in GTG and lean control mice.[103,104] Together, these findings imply that increased hepatic glucose output resulting from alterations in gluconeogenesis and glycogen turnover, is a significant factor in the onset and maintenance of the hyperglycemia of GTG-obese mice. In support of this assumption is our recent study demonstrating increased hepatic glucose production in GTG mice.[105]

Protein metabolism. Studies in vitro have shown that basal and insulin-stimulated amino acid uptake by skeletal muscles of GTG mice is lower than in lean controls.[106] However, basal and insulin-stimulated rates of protein synthesis and degradation are not changed in the skeletal muscles of GTG-obese mice, implying that protein turnover remains insulin sensitive in this tissue.[107] Therefore, it is apparent that the muscles of GTG-obese mice possess abnormalities that are a consequence of the prevailing hyperinsulinemia, rather than an alteration in protein synthesis and degradation *per se*. In general, GTG-obese mice are not stunted in growth (naso-anal length)[37] and have a greater bone size[108] than their lean counterparts, despite reduced serum levels of growth hormone.[109]

Possible etiology of the diabesity syndrome in GTG mice. The available information presented earlier suggests that lesioning the VMH with GTG produces time- and tissue-dependent alterations in glucose and lipid metabolism in the mouse that result in the development of diabesity. One possible sequence of events that could explain the etiology of diabesity is that the lesion in the VMH leads to inappropriately high insulin release in response to the ingestion of food. This would explain the hyperlipogenesis observed in liver, WAT, and BAT in the early stages of obesity, which must contribute to increased wt gain in GTG-injected animals. At the same time, increased insulin release could also lead to a downregulation of insulin receptor numbers in muscle, which, when coupled with the increased availability of lipid fuels competitively inhibiting glucose utilization, would lead to peripheral insulin resistance. The hyperphagia and decreased energy expenditure due to the loss of diet-induced thermogenesis in BAT would only exacerbate the development of obesity and insulin resistance. A similar mechanism has been proposed to account for the obesity developing in VMH-lesioned rats in which an initial period of increased insulin sensitivity with regard to whole body glucose turnover is followed by progression to whole-body insulin resistance.[101,110] It has also been reported that VMH lesion in rats leads to a vagally mediated increase in insulin release, which has been proposed as a mechanism for the development of insulin resistance by downregulation of insulin receptors and increased availability of lipids for oxidation, which reduces glucose utilization and may lead to hyperglycemia.[82] The early onset of hyperinsulinemia and the subsequent development of a myriad of metabolic changes in the tissues of the hypothalamic models have similarly been documented in animals that are genetically predisposed to develop obesity and NIDDM.[63,111]

Role of adrenal glucocorticoids in the diabesity of GTG-injected mice

One aspect of the obesity in animal models that has not been discussed is the role of glucocorticoids in the development and maintenance of obesity. There is increasing evidence within the literature suggesting that glucocorticoids may play an important role in animal and human obesity. In many animal models of obesity, elevated corticosterone in the circulation, indicating increased activity of the hypothalamo-pituitary-adrenal (HPA) axis, has been reported.[112-116] Mice injected with GTG exhibit elevated corticosterone concentrations and a loss of the diurnal rhythm of this glucocorticoid.[116] Moreover, it appears that the apparent changes in the HPA axis of GTG mice are more evident when the animal is subjected to stress. Exposure of GTG-obese and control mice to ether vapor increases corticosterone concentrations to a greater extent in GTG-obese than lean control mice.[117] In accord with the reported hypercorticosteronemia of GTG-obese mice is the observation that these animals possess significantly larger adrenal glands.[118]

ADX of GTG-obese mice leads to a reduction in food intake and wt loss (Fig. 4). ADX also influences the metabolism of GTG-obese mice, as demonstrated

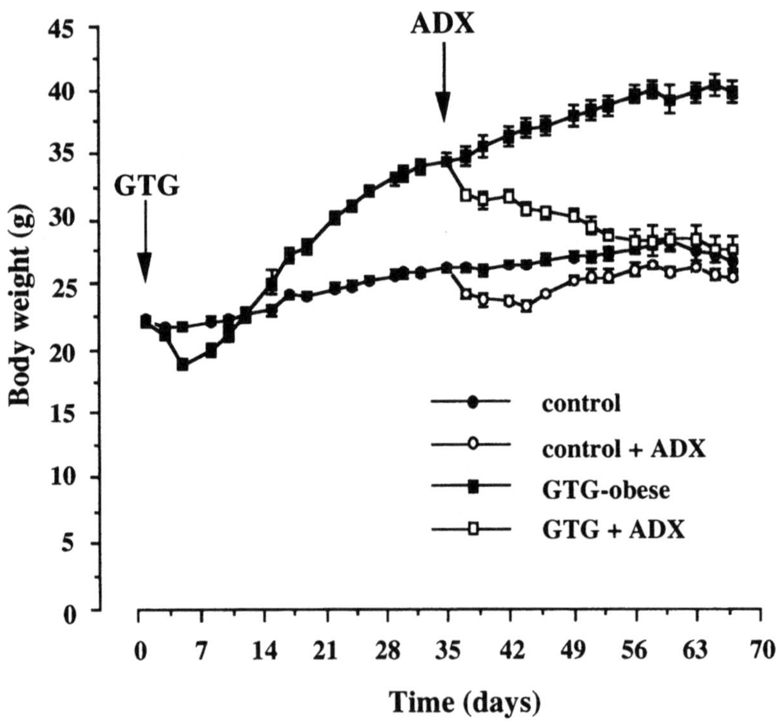

Figure 4. Body wt of GTG and age-matched control mice during the 5 wk prior to and following ADX or sham ADX. Each point is the mean ± SE for 10–12 animals. Two-way ANOVA with repeated measures indicated significant effects of animal type ($P < 0.05$), surgery ($P < 0.05$), and the interaction between animal type and surgery ($P < 0.05$). Reproduced from [117] with permission.

by the decrease in circulating glucose and insulin levels and the increase in FFA levels following ADX of GTG-obese mice.[33] Moreover, carbohydrate and lipid stores are greatly reduced by ADX.[33,117,119] These effects of ADX are predominantly due to a lack of circulating glucocorticoids as i.p. or intracerebroventricular administration of adrenal glucocorticoids, but not mineralocorticoids, is capable of restoring hyperphagia and wt gain in these mice.[120,121] It has been postulated that the wt loss resulting from ADX of GTG-obese mice is mediated solely via a reduction in food intake, and that the reversal by ADX of the metabolic abnormalities of GTG-induced obesity is a consequence of the hypophagia.[33,120] Although there are some reports of ADX causing anorexia, leading to wt loss and death,[121,122] we have previously shown that this wt loss is not due to a reduction in food intake alone.[117] GTG-injected mice lose a substantial amount of wt after ADX, despite consuming the same amount of food as the sham ADX control mice, which gain wt.[117] Thus, it is apparent that the ability of ADX to ameliorate obesity in this animal model cannot be explained solely by a reduction in food intake.

It is apparent from the studies outlined earlier that the obesity and accompanying metabolic abnormalities are dependent upon the presence of adequate levels of corticosterone in the circulation of these animals. Recent studies in rats with experimentally[123,124] or genetically[125] induced obesity suggest that the effects of corticosterone that are responsible for the development and maintenance of obesity are mediated through type II (glucocorticoid) rather than type I (mineralocorticoid) receptors in the central nervous system (CNS). Whether this is also the case in GTG-obese mice remains to be established.

Effects of ADX on metabolism in GTG mice. ADX has been shown to reactivate BAT thermogenesis in GTG-obese mice and other models of obesity.[112,114] It is thought that this ADX-induced increase in BAT thermogenesis is a consequence of the removal of the inhibitory effect of glucocorticoids on sympathetic outflow to the tissue.[126] However, glucocorticoids may also be acting directly on the tissue to reduce (guanosine 5′-diphosphate) GDP binding[127] as well as to inhibit uncoupling protein gene transcription[128] and the amount of the protein in the tissue,[127] thereby reducing BAT thermogenesis. Therefore, it is possible that the reactivation of thermogenesis in BAT of ADX GTG mice is also due to the removal of the suppressive effects of glucocorticoids at the level of the tissue. Increased fatty acid turnover was also observed in BAT of ADX GTG-obese mice, where the fatty acid content of the tissue was decreased despite the rate of fatty acid synthesis being increased (Table 5).[117] Since BAT has a large capacity for thermogenesis, it appears that the increased fatty acid turnover may result from the increased thermogenesis in this tissue, which oxidizes fatty acids to dispose of excess energy.

ADX also reduced the fasting glucose levels and improved the glucose tolerance of GTG-obese mice. Such a reduction in circulating glucose levels may lead to reduction of circulating insulin levels of ADX GTG mice. However, it is clear that the decrease in glycemia that accompanies ADX is not solely responsible for the amelioration of the hyperinsulinemia, as demonstrated by the fact that the

Table 5. Effect of ADX on fatty acid synthesis in liver, WAT, and BAT of GTG mice

Tissue	Animals	Surgery	Tissue wt (g)	Fatty acid synthesis (μg atoms ^{3}H/h/g)
Liver	Control	Sham ADX	1.22 ± 0.03	40.7 ± 3.7
	Control	ADX	1.35 ± 0.05	65.2 ± 6.6[a]
	GTG	Sham ADX	2.20 ± 0.12[b]	121.7 ± 11.2[b]
	GTG	ADX	1.56 ± 0.05[a,b]	66.4 ± 6.9[a,b]
WAT	Control	Sham ADX	0.29 ± 0.04	10.4 ± 1.2
	Control	ADX	0.16 ± 0.03[a]	49.4 ± 5.8[a]
	GTG	Sham ADX	1.59 ± 0.06[b]	9.8 ± 1.1
	GTG	ADX	0.50 ± 0.08[a,b]	18.2 ± 1.5[a,b]
BAT	Control	Sham ADX	0.077 ± 0.005	109.4 ± 19.8
	Control	ADX	0.065 ± 0.006	277.9 ± 46.3[a]
	GTG	Sham ADX	0.256 ± 0.011[b]	40.6 ± 2.8[b]
	GTG	ADX	0.089 ± 0.005[a]	195.2 ± 29.3[a,b]

Fatty acid synthesis in vivo in liver, epididymal white adipose tissue (WAT), and interscapular brown adipose tissue (BAT) of fed GTG and age-matched control mice subjected to ADX or sham ADX, 5 wk previously. Lipogenesis was expressed as μg atoms ^{3}H incorporated into fatty acid/h/g tissue. Each result is the mean ± SE for 10–12 animals.
Statistical analysis was performed by two-way ANOVA. Differences between individual means were assessed by Student's *t*-test: [a]$P < 0.05$ for differences between ADX and sham ADX mice of the same group.
[b]$P < 0.05$ for differences from sham ADX control mice.
Adapted from ref. 117, with permission.

insulin secretory response of ADX GTG-obese mice was returned to normal, despite a significant degree of glucose intolerance still being evident in the same animals (Table 6; Fig. 5). Although it is not known whether the hyperinsulinemia of GTG-obese mice is due to an increase in parasympathetic drive to the pancreas and if such a defect is dependent on the presence of glucocorticoids, the available evidence in another model of obesity, the obese *fa/fa* rat, does, however, suggest that this is a possibility.[129,130] It therefore seems reasonable to hypothesize that removal of glucocorticoids from the circulation of GTG-obese mice leads to a decrease in parasympathetic outflow to the pancreas, which contributes to the decrease in insulin levels following ADX of GTG-obese mice. Support for this effect of ADX on insulin secretion being centrally mediated is also provided by the finding that the insulin levels of ADX GTG-obese mice were restored to normal prior to major wt loss occurring in these animals. To date, the site within the CNS at which corticosterone is acting to modulate autonomic balance remains undefined. In fact, it is possible that a single site of glucocorticoid action does not exist at all and glucocorticoids are acting at numerous sites within the CNS.

ADX reduced the liver wt of GTG-obese mice, primarily due to a decrease in the fatty acid content of the tissue. This increased utilization of fat by ADX GTG-injected mice was accompanied by a decrease in the rate of fatty acid synthesis in the liver (Table 5).[117] ADX similarly reduced fatty synthesis and lowered the fatty acid content of WAT of GTG-obese mice (Table 5). Together, these changes suggest that the turnover of fatty acids is increased in WAT of ADX GTG mice and could account for the wt loss observed in these animals. The effects of ADX on lipid metabolism were not only evident in GTG mice. ADX increased lipid synthesis in WAT of control mice, but despite this increase in fatty acid synthesis (see Table 5), the fatty acid content of WAT was decreased,[117] suggesting an increase in fatty acid oxidation in ADX control mice. This increase in fatty acid turnover, which leads to a loss of adipose tissue in both GTG-obese and control mice, is more obvious in GTG-obese mice owing to their larger fatty acid stores. Whether the ADX-induced decrease in lipid stores is a direct effect on WAT or secondary to the lack of glucocorticoid effects in other tissues has not been elucidated.

Following ADX, GTG-obese mice exhibit decreased food intake, wt loss, enhanced FFA and glucagon levels, and reduced glucose and insulin concentrations.[33,117,121] Abdominal adipose tissue and liver glycogen reserves are also greatly reduced by ADX.[33,119] Since glycogen stores and gluconeogenesis are used to maintain glycemia, it is possible that depleted hepatic glycogen stores, in conjunction with a decreased capacity for gluconeogenesis,[98] may decrease the ability of ADX GTG mice to maintain hyperglycemia. Consistent with such an hypothesis are the findings from our recent in vivo study[105] in which the ability of ADX to reduce the hyperglycemia of GTG-obese mice was shown to be due to a decrease in hepatic glucose production, rather than an increase in glucose uptake by the peripheral tissues. In fact, glucose uptake was shown to be reduced in all the major glucose utilising tissues except BAT following ADX of GTG-obese mice.[105] Additionally, the capacity of the peripheral tissues to oxidise glucose was decreased following ADX of GTG-obese mice as indicated by

Table 6. Incremental area under the curve of plasma glucose and plasma insulin following administration of an i.v. glucose bolus to ADX GTG mice.

Animals	Surgery	Glucose area (min.mmol/l)	Insulin area (min.μU/ml)	Insulinogenic index (μU/μmol)
Control	Sham ADX	297 ± 12	192 ± 5	0.64 ± 0.05
Control	ADX	325 ± 9	159 ± 13[a]	0.49 ± 0.05
GTG	Sham ADX	422 ± 9[b]	230 ± 11[b]	0.54 ± 0.04
GTG	ADX	355 ± 18[a,b]	196 ± 10[a]	0.55 ± 0.04

Incremental area under plasma glucose and insulin curves from 0 to 30 min and 0 to 10 min, respectively, after i.v. administration of a 1 g/kg glucose load. By this period of time (30 min for glucose and 10 min for insulin), glucose/insulin levels of control mice had returned to basal values. The insulinogenic index (insulin/glucose ratio) was calculated by dividing the incremental area under the plasma insulin curve by the incremental area under the plasma glucose curve for individual animals. Each value is the mean ± SE of the iAUC for 5–8 GTG or control mice that had been subjected to ADX or sham ADX, 1 wk previously.

Statistical analysis was performed by two-way ANOVA. Differences between individual means were assessed using Student's *t*-test.

[a]$P < 0.05$ for differences between ADX and sham ADX mice of the same group.

[b]$P < 0.05$ for differences from sham ADX control mice.

Adapted from [139] with permission.

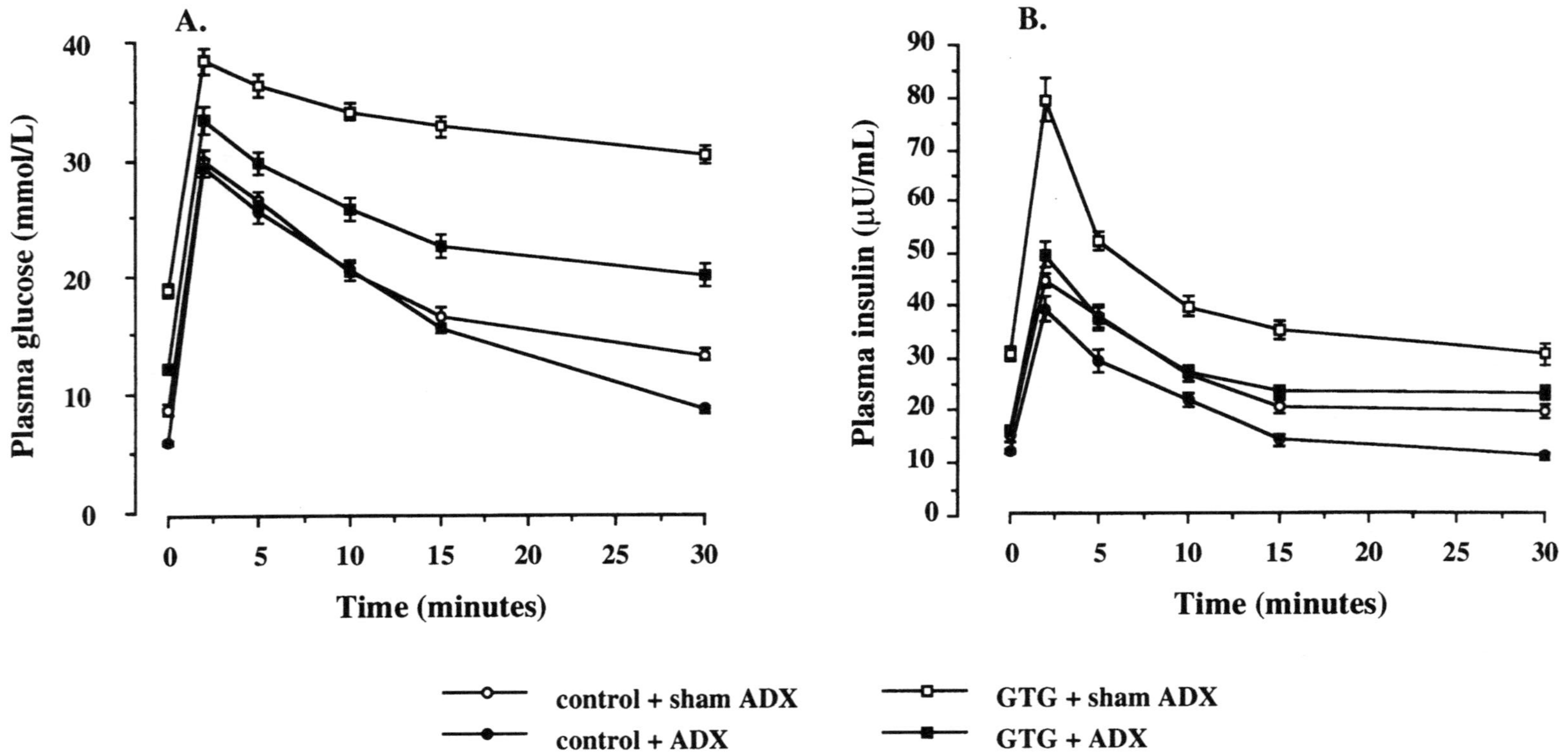

Figure 5. Changes in glycemia and insulinemia in response to i.v. administration of glucose to 5 week GTG and age-matched control mice that had been subjected to ADX or sham-ADX 1 wk previously. Glucose (1g/kg) was administered at time 0. Each curve represents the mean ± SE for five to eight animals. Reproduced from [139] with permission.

the PDH activity which was reduced by varying degrees in all tissues, except quadriceps muscle.[141] Thus, it appears that the ability of ADX to reduce the hyperglycaemia of GTG-obese mice is not due to an increase in the uptake and oxidation of glucose by the peripheral tissues. Instead, it is more likely that the reduction in blood glucose levels observed following ADX of GTG-obese mice is the result of decreased glucose production by the liver.

It would appear that the removal of adrenal hormones, corticosterone in particular, alleviates obesity and reduces hyperglycemia and hyperinsulinemia in GTG-obese mice by a number of mechanisms. The removal of glucocorticoid effects at the level of the VMH may increase sympathetic outflow to BAT and restore diet-induced thermogenesis. A decrease in parasympathetic activity could also account for the decrease in insulin release observed after a glucose challenge in ADX obese mice, which reduces the major stimuli for lipid synthesis in these animals. It also seems that the removal of the counter-regulatory effects of corticosterone at the liver level may decrease glycogen stores, and therefore the ability of the animal to maintain glycemia would be decreased. This inability to maintain glycemia would indirectly lead to an increased use of lipid as an energy source to help preserve the limited carbohydrate stores that are available to the glucocorticoid-deprived animal. The way in which glucocorticoid hormones and insulin interact to influence glucose and lipid metabolism in normal and insulin-resistant obese animal models is still relatively poorly understood and represents an expanding area of current research.

Conclusions

A single injection of GTG in mice lesions the VMH and produces a metabolic syndrome with many of the characteristics of human obesity and NIDDM, that is, diabesity. The GTG brain lesion induces hyperphagia, hyperglycemia, hyperinsulinemia, insulin resistance, triglyceride accumulation, and a range of tissue-specific changes in regulatory enzyme activities of glucose and lipid metabolic pathways. Because GTG lesions only the VMH region of the brain, it follows that all the changes in peripheral tissue metabolism, such as muscle insulin resistance, increased liver lipogenesis, and increased liver glycogen storage, must be secondary to changes in some factor or factors altered by the GTG-induced VMH damage. The VMH lesion of GTG mice increases metabolic efficiency by reducing sympathetic outflow to BAT and decreasing diet-induced thermogenesis. A reciprocal increase in parasympathetic activity after the GTG lesion may well be the cause of the inappropriately high insulin release that occurs early after the lesion and that contributes to the increased lipogenesis, triglyceride accumulation, and insulin resistance that occurs after the GTG lesion. Similar patterns for the development of diabesity have been observed in other experimental, as well as genetic, models.[111] Although the role of glucocorticoids in the development of obesity in GTG-injected mice is not clear, it is apparent that removal of glucocorticoids by ADX has a marked effect on alleviating the metabolic abnormalities and wt gain associated with GTG injection. The effects of ADX are unlikely to be due to a specific reversal of the cause of the diabesity syndrome, because effects of ADX are also observed in control mice. Removal of glucocorticoids will have effects on the CNS to increase sympathetic and decrease parasympathetic outflow, but removal of tissue-specific effects of glucocorticoids on glucose metabolism, glycogen synthesis, and lipid metabolism may be of considerable importance in the wt loss that occurs after ADX. The central interaction of the autonomic nervous system and the HPA axis may well hold the key to understanding the development of diabesity in the GTG mouse and other animal models of this syndrome.

Note added in proof

After this review article was written the mouse obese (*ob*) gene product[142] and its receptor[143] were isolated and cloned. The *ob* gene product, leptin, is a circulating protein which is synthesised and secreted exclusively by adipocytes. Leptin is thought to regulate adipose tissue mass and thus body weight by acting, directly or indirectly, on the hypothalamus to modulate nutrient intake.[142] Additionally, leptin has been shown to be capable of stimulating energy expenditure.[144,145]

In adipocytes of genetically obese (*ob/ob*) mice it appears that the mRNA coding for *ob* is either lacking or flawed and this results in a lack of biologically active leptin in the circulation.[142] The absence of functional leptin in *ob/ob* mice leads to hyperphagia and this probably represents one of the primary metabolic defects responsible for the obesity in this animal model.[144] By contrast, in many of the other rodent models of obesity (*fa/fa* and VMH-lesioned rats; *db/db*, monosodium glutamate-(MSG) and GTG-treated mice) expression of *ob* mRNA in WAT is increased[146–149] and circulating leptin levels are elevated,[146,148] and yet this does not lead to reduced food intake and protect against the development of obesity. In fact, hyperphagia exists in the presence of high circulating leptin levels in these animal models. This resistance to the action of leptin is probably due to the receptors, specifically those located in the hypothalamus, being defective[143] as a result of a genetic mutation (*fa/fa* rat, *db/db* mice) or lesioning of the hypothalamus by electrolytic (VMH-lesioned rat) or chemical (MSG- and GTG-treated mice) means.

Acknowledgments

The preparation of this review and much of the experimental work reported herein would not have been possible without the financial assistance provided by the National Health and Medical Research Council (NH & MRC) of Australia and the Department of Endocrinology, Royal Prince Alfred Hospital, Camperdown NSW 2050, Australia.

References

1. Brecher G, Waxler SH. Obesity in albino mice due to a single injection of gold thioglucose. Proc Soc Exp Biol Med 70:498–501, 1949.

2. Marshall NB, Barnett RJ, Mayer J. Hypothalamic lesions in gold thioglucose injected mice. Proc Soc Exp Biol Med 90:240–4, 1955.
3. Mayer J. The hypothalamic control of gastric hunger contractions as a component of the mechanism of regulation of food intake. Am J Clin Nutr 8:547–61, 1960.
4. Matsuo T, Shino A. Induction of diabetic alterations by gold thioglucose obesity in KK, ICR, and C57BL mice. Diabetologia 8:391–7, 1972.
5. Liebelt RA, Perry JH. Hypothalamic lesions associated with gold-thioglucose injected obesity. Proc Soc Exp Biol Med 95:774–7, 1957.
6. Liebelt RA, Sekiba K, Liebelt AG, Perry JH. Genetic susceptibility to gold thioglucose induced obesity in mice. Proc Soc Exp Biol Med 104:689–2, 1960.
7. Mayer J, Marshall NB. Specificity of gold thioglucose for ventromedial hypothalamic lesions and obesity. Nature 178:1399–1400, 1956.
8. Young JK, Nance DM, Gorski RA. Effects of diet upon hypothalamic vulnerability to gold thioglucose in male mice. Brain Res Bull 4:69–74, 1979.
9. Blech W, Bierwolf B, Weiss I, Ziegler M. *In vitro-* und *in vivo* wirkung von goldthioglucose auf die insulin- und glucagonsekretion des isoliert perfundierten rattenpandreas. Biomed Biochim Acta 45:507–22, 1986.
10. Rietveld WJ, Ten Hoor F, Kooij M, Flory W. Maintenance of 24 hour eating rhythmicity during gold thioglucose induced hypothalamic hyperphagia in rats. Physiol Behav 22:549–53, 1979.
11. Smith CJV, Britt DL. Obesity in the rat induced by hypothalamic implants of gold thioglucose. Physiol Behav 9:391–6, 1971.
12. Mayer J, Marshall NB. Mode d'action de l'aurothioglucose et regulation glucocstatique de la nutrition. C R Acad Sci 242:169–71, 1956.
13. Brown DF, Mc Quirk JP, Larsen SP, Minter SD. Beta-thioglucose inhibits gold thioglucose lesions in the ventromedial hypothalamus. Physiol Behav 46:369–72, 1989.
14. Debons AF, Krimsky I, From A, Pattitian H. Phlorizin inhibition of hypothalamic necrosis by gold thioglucose. Am J Physiol 226:574–8, 1974.
15. Glick Z, Mayer J. Hyperphagia caused by cerebral ventricular infusion of phloridzin. Nature 219:1374, 1968.
16. Debons AF, Krimsky I, From A, Cloutier RJ. Site of action of gold thioglucose in the hypothalamic satiety center. Am J Physiol 219:1397–402, 1970.
17. Debons AF, Krimsky I, Maayan ML, Fani K, Jimenez FA. Gold thioglucose obesity syndrome. Fed Proc 36:143–7, 1977.
18. Debons AF, Krimsky I, Likuski HJ et al. Gold thioglucose damage to the satiety center: Inhibition in diabetes. Am J Physiol 214:652–8, 1968.
19. Debons AF, Krimsky I, From A, Cloutier RJ. Rapid effects of insulin on the hypothalamic satiety center. Am J Physiol 217:1114–18, 1969.
20. Debons AF, Krimsky I, From A. Modification of alloxan induced diabetes: Correlated changes in hypothalamic satiety center. Am J Physiol 224:862–9, 1973.
21. Oomura Y. Glucose as a regulator of neuronal activity. In: Advances in metabolic disorders: CNS regulation of carbohydrate metabolism, Vol 10. Szabo A, ed. Academic Press, New York, pp 31–65, 1983.
22. Debons AF, Krimsky I, From A, Pattitian H. Diabetes-induced resistance of ventromedial hypothalamus to damage by gold thioglucose: Reversal by adrenalectomy. Endocrinology 95:1636–41, 1974.
23. Debons AF, Krimsky I, From A, et al. Action of gold thioglucose on pericapillary structures in the ventromedial hypothalamus. J Pathol 129:73–81, 1979.
24. Karasawa Y, Sugawa-Katayama Y, Danbara H. The effect of intraperitoneally administered gold thioglucose on growth, food consumption and accumulation of gold in various organs of the chicken (*Gallus domesticus*). Comp Biochem Physiol 90:461–4, 1988.
25. Brown DF, Farrar ES. Absence of lesions in the hypothalamus of frogs treated with gold thioglucose. Lab Anim 14:203–4, 1980.

26. Brown DF, Viles JM. Resistance of the gerbil hypothalamus to gold thioglucose lesion formation. Physiol Behav 31:203–8, 1983.
27. Brown DF, Floody OR. Gold thioglucose-induced brain lesions in hamsters. Physiol Behav 39:315–20, 1987.
28. Debons AF, Silver L, Cronkite EP, et al. Localization of gold in mouse brain in relation to gold thioglucose obesity. Am J Physiol 202:743–50, 1962.
29. Kataoka K, Danbara H, Sunayashiki K, et al. Regional and subcellular distribution of gold in brain of gold thioglucose obese mouse. Brain Res Bull 3:257–63, 1978.
30. Sugawa-Katayama Y, Koishi H, Danbara H. Accumulation of gold in various organs of mice injected with gold thioglucose. J Nutr 105:957–62, 1975.
31. Brecher G, Laqueur GL, Cronkite EP, et al. The brain lesions of gold thioglucose induced obesity. J Exp Med 121:395–401, 1965.
32. Perry JH, Liebelt RA. Extra-hypothalamic lesions associated with gold-thioglucose induced obesity. Proc Soc Exp Biol Med 106:55–7, 1961.
33. Debons AF, Tse CS, Zurek LD, et al. Adrenalectomy induced anorexia in gold thioglucose-treated obese mice: Metabolic and hormonal changes. Physiol Behav 38:111–17, 1986.
34. Caterson ID, Astbury LD, Williams PF et al. The activity of the pyruvate dehydrogenase complex in heart and liver from mice during the development of obesity and insulin resistance. Biochem J 243:549–53, 1987.
35. Cooney GJ, Vanner MA, Nicks JL, et al. Changes in the lipogenic response to feeding of liver, white adipose tissue and brown adipose tissue during the development of obesity in the gold-thioglucose-injected mouse. Biochem J 259:651–7, 1989.
36. Chen C, Williams PF, Cooney GJ, Caterson ID. Changes in glycogen metabolism in liver of gold thioglucose injected mice during the development of obesity. Int J Obes 18:627–31, 1994.
37. Hogan S, Himms-Hagen J. Brown adipose tissue of mice with gold thioglucose-induced obesity: Effect of cold and diet. Am J Physiol 244:E581–8, 1983.
38. Likuski HJ, Debons AF, Cloutier RJ. Inhibition of gold thioglucose-induced hypothalamic obesity by glucose analogues. Am J Physiol 212:669–76, 1967.
39. Eley J, Himms-Hagen J. Attenuated response to cold of brown adipose tissue of mice made obese with gold thioglucose. Can J Physiol Pharmacol 67:116–21, 1989.
40. Steinbeck K, Caterson ID, Turtle JR. The activity of the pyruvate dehydrogenase complex in heart muscle in the previously obese mouse model. Biosci Rep 6:1071–75, 1986.
41. Balkan B, Steffens AB, Bruggink JE, Strubbe JH. Hyperinsulinemia and glucose tolerance in obese rats with lesions of the ventromedial hypothalamus: Dependence on food intake and route of administration. Metabolism 40:1092–1100, 1991.
42. Cox JE, Powley TL. Intragastric pair feeding fails to prevent VMH obesity or hyperinsulinemia. Am J Physiol 240:E566–72, 1981.
43. Billington CJ, Briggs JE, Grace M, Levine AS. Effects of intracerebroventricular injection of neuropeptide Y on energy metabolism. Am J Physiol 260:R321–27, 1991.
44. Zarjevski N, Cusin I, Vettor R, et al. Chronic intracerebroventricular neuropeptide-Y administration to normal rats mimics hormonal and metabolic changes of obesity. Endocrinology 133:1753–58, 1993.
45. Young JK. Hypothalamic lesions increase neuronal immunoreactivity for neuropeptide Y. Brain Res Bull 29:375–80, 1992.
46. Young JK, McKenzie JC, Brady LS, Herkenham M. Hypothalamic lesions increase levels of neuropeptide Y mRNA in the arcuate nucleus of mice. Neurosci Lett 165:13–17, 1994.
47. Himms-Hagen J. Brown adipose tissue thermogenesis and obesity. Prog Lipid Res 28:67–115, 1989.
48. Young JB, Landsberg L. Impaired suppression of sympathetic activity during fasting in the gold thioglucose-treated mouse. J Clin Invest 65:1086–94, 1980.

49. Dulloo AG, Young JB. Effects of monosodium glutamate and gold thioglucose on dietary regulation of sympathetic nervous system activity in rodents. Metabolism 40:113–21, 1991.
50. Godbole V, York DA. Lipogenesis in situ in the genetically obese Zucker fatty rat (*fa/fa*): Role of hyperphagia and hyperinsulinaemia. Diabetologia 14:191–7, 1978.
51. Lavau M, Bazin R, Karaoghlanian Z, Guichard C. Evidence for a high fatty acid synthesis activity in interscapular brown adipose tissue of genetically obese Zucker rats. Biochem J 204:503–7, 1982.
52. Mercer SW, Trayhurn P. Developmental changes in fatty acid synthesis in interscapular brown adipose tissue of lean and genetically obese (*ob/ob*) mice. Biochem J 212:393–8, 1983.
53. Saito M, Shimazu T. Decreased rate of fatty acid synthesis in brown adipose tissue of hypothalamic obese rats. FEBS Lett 166:151–4, 1984.
54. Hems DA, Rath EA, Verrinder TR. Fatty acid synthesis in liver and adipose tissue of normal and genetically obese (*ob/ob*) mice during the 24-hour cycle. Biochem J 150:167–73, 1975.
55. Caterson ID, Cooney GJ, Vanner MA, et al. The activities of the pyruvate dehydrogenase complex and of acetyl-CoA carboxylase in various tissues in experimental obesity: Tissue differences and insulin resistance. Diab Nutr Metab 1:65–70, 1988.
56. Martin RJ, Lamprey P. Changes in liver and adipose tissue enzymes and lipogenic activities during the onset of hypothalamic obesity in mice. Life Sci 14:1121–31, 1974.
57. Maragoudakis M, Harkin H, Kalinsky H. Adaptive response of hepatic acetyl CoA carboxylase to dietary alterations in genetically obese mice and their lean littermates. Biochim Biophys Acta 343:590–7, 1974.
58. Volpe JJ, Marasa JC. Regulation of hepatic fatty acid synthetase in the obese-hyperglycemic mutant mouse. Biochim Biophys Acta 409:235–48, 1975.
59. Ashwell M, Meade CJ. Obesity: Do fat cells from genetically obese mice (C57BL/6J *ob/ob*) have an innate capacity for increased fat storage. Diabetologia 15:465–70, 1978.
60. Le Marchand Y, Freychet P, Jeanrenaud B. Longitudinal study on the establishment of insulin resistance in hypothalamic obese mice. Endocrinology 102:74–85, 1978.
61. Katsuki S, Hirata Y, Horino M, et al. Obesity and hyperglycemia induced in mice by gold thioglucose. Diabetes 11:209–15, 1962.
62. Steinbeck K, Caterson ID, Astbury LD, Turtle JR. The effect of diet composition on weight gain and pyruvate dehydrogenase activity in heart muscle in the gold thioglucose obese mouse. Int J Obes 11:507–8, 1987.
63. Bray GA, York DA. Hypothalamic and genetic obesity in experimental animals: An autonomic and endocrine hypothesis. Physiol Rev 59:719–809, 1979.
64. Balkan B, Van Dijk G, Strubbe JH, et al. Exercise-induced sympathetic FFA mobilization in VMH-lesioned rats is normalized by fasting. Am J Physiol 262:R981–5, 1992.
65. Inoue S, Satoh S, Tanaka K, Takamura Y, Murase T. Determinants of fasting hypertriglyceridemia in ventromedial hypothalamic obesity in rats. Am J Physiol 265:R786–91, 1993.
66. Cooney GJ, Astbury LD, Williams PF, Caterson ID. Insulin response in individual tissues of control and gold thioglucose-obese mice *in vivo* with [1-^{14}C] 2-deoxyglucose. Diabetes 36:152–8, 1987.
67. Blair SC, Cooney GJ, Denyer GS, Williams PF, Caterson ID. Differences in lipogenesis in tissues of control and gold-thioglucose obese mice after an isocaloric meal. Biochem Biophys Acta 1085:385–8, 1991.
68. Blair SC, Caterson ID, Cooney GJ. Insulin response to an intravenous glucose load

during development of obesity in gold thioglucose-injected mice. Diabetes 42:1153–58, 1993.
69. Axen KV, Li X, Kung K, Sclafani A. The VMH-dietary obese rat: A new model of non-insulin-dependent diabetes mellitus. Am J Physiol 266:R921–8, 1994.
70. Rohner-Jeanrenaud F, Jeanrenaud B. Consequences of ventromedial hypothalamic lesions upon insulin and glucagon secretion by subsequently isolated perfused pancreases in the rat. J Clin Invest 65:902–10, 1980.
71. Caterson ID. Islet cell metabolism in experimental obesity. PhD Thesis, University of Sydney, 1980.
72. Dolais-Kitabgi J, Le Marchand Y, Freychet P. Somatostatin in the pancreas and hypothalamus of obese mice. Diabetologia 17:257–61, 1979.
73. Caterson ID, Taylor KW. Islet cell function in gold thioglucose-induced obesity in mice. Diabetologia 23:119–23, 1982.
74. Steinbeck KS, Caterson ID, Cooney GJ. Tissue [1-^{14}C] 2-deoxyglucose metabolism in the gold thioglucose obese mouse: Effects of 'dieting.' Int J Obes 18:167–72, 1994.
75. Caterson ID, Taylor KW. Direct effects of gold thioglucose on insulin secretion by isolated rat islets of Langerhans. FEBS Lett 98:351–4, 1979.
76. Storlien LH. The ventromedial hypothalamic area and the vagus are neural substrates for anticipatory insulin release. J Auton Nerv Syst 13:303–10, 1985.
77. Bray GA. Obesity—a state of reduced sympathetic activity and normal or high adrenal activity (the autonomic and adrenal hypothesis revisited). Int J Obes 14(Suppl 3):77–92, 1990.
78. Rohner-Jeanrenaud F, Hochstrasser AC, Jeanrenaud B. Hyperinsulinaemia of preobese and obese *fa/fa* rats is partly vagus nerve mediated. Am J Physiol 244:E317–22, 1983.
79. Rohner-Jeanrenaud F, Jeanrenaud B. Involvement of the cholinergic system in insulin and glucagon oversecretion of genetic pre-obesity. Endocrinology 116:830–4, 1985.
80. Berthoud HR, Jeanrenaud B. Acute hyperinsulinaemia and its reversal by vagotomy after lesions of the ventromedial hypothalamus in anesthetized rats. Endocrinology 105:146–51, 1979.
81. Cox JE, Powley TL. Prior vagotomy blocks VMH obesity in pair fed rats. Am J Physiol 240:E573–83, 1981.
82. Rohner F, Dufour AC, Karakash C, et al. Immediate effect of lesion of the ventromedial hypothalamic area upon glucose-induced insulin secretion in anaesthetised rats. Diabetologia 13:239–42, 1977.
83. Le Marchand-Brustel Y, Olichon-Berthe C, Gremeaux T, et al. Glucose transporter in insulin sensitive tissues of lean and obese mice. Effect of the thermogenic agent BRL 26830A. Endocrinology 127:2687–95, 1990.
84. Machado UF, Shimizu Y, Saito M. Decreased glucose transporter (GLUT 4) content in insulin-sensitive tissues of obese aurothioglucose- and monosodium glutamate-treated mice. Horm Metab Res 25:462–5, 1993.
85. Machado UF, Shimizu Y, Saito M. Reduced content and preserved translocation of glucose transporter (Glut 4) in white adipose tissue of obese mice. Physiol Behav 55:621–5, 1994.
86. Chen C, Williams PF, Heffernan SJ, et al. The effects of fish oil on hepatic insulin receptors in streptozotocin diabetic rats and gold-thioglucose induced obese mice. Horm Metab Res 23:446–8, 1991.
87. Le Marchand-Brustel Y, Jeanrenaud B, Freychet P. Insulin binding and effects in isolated soleus muscle of lean and obese mice. Am J Physiol 234:E348–58, 1978.
88. Tanti JF, Gremeaux T, Brandenburg D, et al. Brown adipose tissue in lean and obese mice. Insulin receptor binding and tyrosine kinase activity. Diabetes 35:1243–48, 1986.
89. Le Marchand-Brustel Y, Gremeaux T, Ballotti R, Van Obberghen E. Insulin re-

ceptor tyrosine kinase is defective in skeletal muscle of insulin resistant obese mice. Nature 315:676–9, 1985.
90. Gremeaux T, Tanti J-F, Van Obberghen E, Le Marchand-Brustel Y. Alteration of insulin receptor kinase in obese, insulin-resistant mice. Biochimie 69:387–93, 1987.
91. Cormont M, Tanti J-F, Van Obberghen E, Le Marchand-Brustel Y. Expression of guanine-nucleotide-binding proteins in lean and obese insulin-resistant mice. Mol Cell Endocrinol 99:169–76, 1994.
92. Randle PJ, Kerbey AL, Espinal J. Mechanisms decreasing glucose oxidation in diabetes and starvation: Role of lipid fuels and hormones. Diab Metab Rev 4:623–38, 1988.
93. Bryson JM, Cooney GJ, Wensley VR, et al. Diurnal patterns of cardiac and hepatic pyruvate dehydrogenase complex activity in gold-thioglucose-obese mice. Biochem J 295:731–4, 1993.
94. Kerbey AL, Caterson ID, Williams PF, Turtle JR. Proportion of active dephosphorylated pyruvate dehydrogenase complex in heart and isolated heart mitochondria is decreased in obese hyperinsulinaemic mice. Biochem J 217:117–21, 1984.
95. Cooney GJ, Denyer GS, Kerbey AL, et al. Pyruvate dehydrogenase-complex activity in brown adipose tissue of gold thioglucose-obese mice. Biochem J 270:257–9, 1990.
96. Caterson ID, Williams PF, Kerbey AL, et al. The effect of body weight and the fatty acid oxidation inhibitor 2-tetradecylglycidic acid on pyruvate dehydrogenase complex activity in mouse heart. Biochem J 224:787–91, 1984.
97. Bryson JM, Cooney GJ, Wensley VR, et al. Tissue differences in the response of pyruvate dehydrogenase complex to a glucose load during the development of obesity in gold thioglucose-obese mice. Biochem J 305:811–16, 1995.
98. Greenaway TM, Cooney GJ, Blair SC, Caterson ID. Increased gluconeogenesis in hepatocytes from GTG-obese mice is insensitive to inhibition by insulin. Int J Obes 16:985–90, 1992.
99. York DA. Animal models for the study of obesity. In: Biochemical pharmacology of obesity. Curtis-Prior, PB ed. Elsevier, Amsterdam, pp 67–104, 1983.
100. Shimazu T, Ogasawara S. Effects of hypothalamic stimulation on gluconeogenesis and glycolysis in rat liver. Am J Physiol 228:1787–93, 1975.
101. Penicaud L, Rohner-Jeanrenaud F, Jeanrenaud, B. *In vivo* metabolic changes as studied longitudinally after ventromedial hypothalamic lesions. Am J Physiol 250:E662–8, 1986.
102. Chen C, Williams PF, Cooney GJ, et al. The effects of fasting and refeeding on liver glycogen synthase and phosphorylase in obese and lean mice. Horm Metab Res 24:161–4, 1992.
103. Chen C, Williams PF, Cooney GJ, Caterson ID. Diurnal rhythms of glycogen metabolism in the liver and skeletal muscle in gold thioglucose induced-obese mice with developing insulin resistance. Int J Obes 16:913–21, 1992.
104. Le Marchand-Brustel Y, Freychet P. Alteration of glycogen synthase activation by insulin in soleus muscles of obese mice. FEBS Lett 120:205–8, 1980.
105. Blair SC, Caterson ID, Cooney GJ. Glucocorticoid deprivation alters in vivo glucose uptake by muscle and adipose tissues of GTG-obese mice. Am J Physiol 269:E927–E933, 1995.
106. Le Marchand-Brustel Y, Moutard N, Freychet P. Aminoisobutyric acid transport in soleus muscles of lean and gold thioglucose-obese mice. Am J Physiol 243:E74–9, 1982.
107. Monier S, Le Cam A, Le Marchand-Brustel Y. Insulin and insulin-like growth factor I. Effects on protein synthesis in isolated muscles from lean and gold thioglucose-obese mice. Diabetes 32:392–7, 1983.
108. De Leeuw I, Van Rooy P, D'Hollander M. Bone mass in obese, gold thioglucose-treated, hyperglycaemic mice. Diabetologia 20:145–7, 1981.
109. Sinha YN, Salocks CB, Vanderlaan WP. Prolactin and growth hormone secretion

in chemically induced and genetically obese mice. Endocrinology 97:1386–93, 1975.
110. Penicaud L, Kinebanyan MK, Ferre P, et al. Development of VMH obesity: *In vivo* insulin secretion and tissue insulin sensitivity. Am J Physiol 257:E255–60, 1989.
111. Shafrir E. Diabetes in animals. In: Diabetes mellitus theory and practice, 4th ed. Rifkin H, Porte D, eds. Elsevier, New York, pp 299–340, 1990.
112. Dallman MF. Viewing the ventromedial hypothalamus from the adrenal gland. Am J Physiol 246:R1–12, 1984.
113. Guillaume-Gentil C, Rohner-Jeanrenaud F, Abramo F, et al. Abnormal regulation of the hypothalamo-pituitary-adrenal axis in the genetically obese *fa/fa* rat. Endocrinology 126:1873–79, 1990.
114. King BM. Glucocorticoids and hypothalamic obesity. Neurosci Biobehav Rev 12:29–37, 1988.
115. Plotsky PM, Thrivikraman KV, Watts AG, Hauger RL. Hypothalamo-pituitary-adrenal axis function in the Zucker obese rat. Endocrinology 130:1931–41, 1992.
116. Saito M, Bray GA. Diurnal rhythm for corticosterone in obese (*ob/ob*) diabetes (*db/db*) and gold-thioglucose-induced obesity in mice. Endocrinology 113:2181–85, 1983.
117. Blair SC, Caterson ID, Cooney GJ: Effect of adrenalectomy on glucose tolerance and lipid metabolism in gold thioglucose obese mice. Am J Physiol 266:E993–1000, 1994.
118. Siclari E, Debons AF. Enhanced adrenal activity in hypothalamic obesity. Fed Proc 40:906, 1981.
119. Tse CS, Debons A. Anorexia after adrenalectomy in gold thioglucose treated obese mice: Role of adipose tissue mass. Physiol Behav 38:105–10, 1986.
120. Debons AF, Zurek LD, Tse CS, Abrahamsen S. Central nervous system control of hyperphagia in hypothalamic obesity: Dependence on adrenal glucocorticoids. Endocrinology 118:1678–81, 1986.
121. Debons AF, Siclari E, Das KC, Fuhr B. Gold thioglucose-induced hypothalamic damage, hyperphagia, and obesity: Dependence on the adrenal gland. Endocrinology 110:2024–29, 1982.
122. Debons AF, Das KC, Fuhr B, Siclari E. Anorexia after adrenalectomy in gold thioglucose-treated obese mice. Endocrinology 112:1847–51, 1983.
123. Okada S, York DA, Bray GA. Mifepristone (RU 486), a blocker of type II glucocorticoid and progestin receptors, reverses a dietary form of obesity. Am J Physiol 262:R1106–10, 1992.
124. Thomas TL, Devenport LD, Stith RD. Relative contribution of type I and II corticosterone receptors in VMH lesion-induced obesity and hyperinsulinemia. Am J Physiol 266:R1623–29, 1994.
125. Langley SC, York DA. Effects of antiglucocorticoid RU 486 on development of obesity in obese *fa/fa* Zucker rats. Am J Physiol 259:R539–44, 1990.
126. Walker HC, Romsos DR. Glucocorticoids in the CNS regulate BAT metabolism and plasma insulin in *ob/ob* mice. Am J Physiol 262:E110–17, 1992.
127. Strack AM, Bradbury MJ, Dallman MF. Corticosterone decreases nonshivering thermogenesis and increases lipid storage in brown adipose tissue. Am J Physiol 268:R183–91, 1995.
128. Moriscot A, Rabelo R, Bianco AC. Corticosterone inhibits uncoupling protein gene expression in brown adipose tissue. Am J Physiol 265:E81–7, 1993.
129. Fletcher JM, McKenzie, N. The parasympathetic nervous system and glucocorticoid-mediated hyperinsulinaemia in the genetically obese (*fa/fa*) Zucker rat. J Endocrinol 118:87–92, 1988.
130. Stubbs M, York DA. Central glucocorticoid regulation of parasympathetic drive to pancreatic β-cells in the obese *fa/fa* rat. Int J Obes 15:547–53, 1991.
131. Benzo CA, Stearns SB. Effects of gold thioglucose treatment on enzymes of glycogen metabolism in liver and skeletal muscle in mice. Biochem Med 26:395–402, 1981.

132. Caterson ID, Kerbey AL, Cooney GJ, et al. Inactivation of pyruvate dehydrogenase complex in heart muscle mitochondria of gold thioglucose induced obese mice is not due to a stable increase in activity of pyruvate dehydrogenase kinase. Biochem J 253:291–4, 1988.
133. Chen C, Williams PF, Caterson ID. Liver and peripheral tissue glycogen metabolism in obese mice: Effect of a mixed meal. Am J Physiol 265:E743–51, 1993.
134. Chen C, Williams PF, Cooney GJ, Caterson ID. Liver and peripheral tissue glycogen metabolism in obese mice with developed insulin resistance: Response to an oral glucose load. Diab Nutr Metab 6:1–9, 1993.
135. Shull KH, Mayer J. The turnover of glycogen in obese hyperglycemic mice. J Biol Chem 218:885–97, 1956.
136. Heydrick SJ, Jullien D, Gautier N, et al. Defect in skeletal muscle phosphotidylinositol-3-kinase in obese insulin-resistant mice. J Clin Invest 91:1358–66, 1993.
137. Heydrick SJ, Gautier N, Olichon-Berthe C, et al. Early alteration of insulin stimulation of PI 3-kinase in muscle and adipocyte from gold thioglucose obese mice. Am J Physiol 268:E604–12, 1995.
138. Blair SC, Caterson ID, Cooney GJ. Insulin response to a spontaneously ingested standard meal during the development of obesity in GTG-injected mice. Int J Obesity (in press) 1996.
139. Blair SC, Caterson ID, Cooney GJ. Glucose tolerance and insulin secretion in gold thioglucose-obese mice following adrenalectomy. J Endocr 148:391–8, 1996.
140. Marks JL, Waite K, Cameron-Smith D, et al. Effects of gold thioglucose on neuropeptide Y messenger RNA levels in the mouse hypothalamus. Am J Physiol (in press) 1996.
141. Blair SC, Greenaway TM, Bryson JM, et al. Hepatic gluconeogenesis and the activity of PDH in individual tissues of GTG-obese mice following adrenalectomy. Obes Res (in press) 1996.
142. Zhang Y, Proenca R, Maffei M, et al. Postional cloning of the mouse obese gene and its human homologue. Nature 372:425–432, 1994.
143. Tartaglia LA, Dembski M, Weng X, et al. Identification and expression cloning of a leptin receptor, OB-R. Cell 83:1263–1271, 1995.
144. Halaas JL, Gajiwala KS, Maffei M, et al. Weight reducing effects of the plasma protein encoded by the obese gene. Science 269:543–546, 1995.
145. Trayhurn P, Duncan JS, Rayner DV. Acute cold-induced suppression of *ob* (obese) gene expression in white adipose tissue of mice: mediation by the sympathetic system. Biochem J 311:729–733, 1995.
146. Maffei M, Halaas J, Ravussin E, et al. Leptin levels in human and rodent: measurement of plasma leptin and *ob* RNA in obese and weight-reduced subjects. Nature Med 1:1155–1161, 1995.
147. Funahashi T, Shimomura I, Hiraoka H, et al. Enhanced expression of rat obese (*ob*) gene in adipose tissues of ventromedial hypothalamus (VMH)-lesioned rats. Biochem Biophys Res Commun. 211:469–475, 1995.
148. Frederich RC, Lollmann B, Hamann A, et al. Expression of *ob* mRNA and its encoded protein in rodents. Impact of nutrition and obesity. J Clin Invest 96:1658–1663, 1995.
149. Maffei M, Fei H, Gwo-Hwa L, et al. Increased expression in adipocytes of *ob* RNA in mice with lesions of the hypothalamus and with mutations at the *db* locus. Proc Natl Acad Sci 92:6957–6960, 1995.

16

Diabetes in the Chinese hamster

BARBARA J. FRANKEL
Department of Histology and Cell Biology, Umeå University, S-901 87 Umeå, Sweden.

Key words: islet B-cell, cytoplasmic calcium, etiology, gluconeogenesis, insulin, insulin resistance, islet of Langerhans, nutrition, complications of diabetes.

How did the Chinese hamster become a laboratory animal?

According to Yerganian,[1] the first recorded experimental use of the Chinese hamster, *Cricetulus griseus,* was for the identification of pneumococcal types prior to treatment of pneumonia at the Peking Medical College Hospital in 1919 because laboratory mice were not available.[2] The Chinese or "striped-back" hamster is recognized by the black stripe that runs down the middle of the otherwise yellowish-brown fur on its back. The hamster is also known for its low chromosome number (2N = 22 as compared with the human 46 or golden hamster 44), thin cheek-pouch membrane, proportionally large and fully descended testes (10% of body wt!),[1] and spontaneous, hereditary diabetes mellitus. The Chinese hamster is referred to as a "dwarf" hamster because young adults weigh only 28–40 g[3] compared with the much larger, 120 g Syrian or golden hamster, *Mesocricetus auratus,* which is used in the laboratory and as a house pet. Due to the unusually small number of chromosomes, Chinese hamster cells are convenient for chromosomal studies. The development of immortal cell lines has made this even easier. During the last decade in several hundred studies cell lines developed from nondiabetic Chinese hamster lung fibroblasts, ovary, and other tissues have been used. Thus, the Chinese hamster has become a significant contributor to medical research.

Correspondence: Barbara J. Frankel, Department of Histology and Cell Biology, University of Umeå, S-901 87 Umeå, Sweden.

Lessons from Animal Diabetes VI. Ed. E. Shafrir.

Breeding techniques for Chinese hamsters

Sexually mature female Chinese hamsters tend to be aggressive,[4-6] presumably because only the most aggressive females in the wild are able to defend the large territories necessary to find enough food for their young. This pugnacity and a relatively low fertility make it more difficult to breed Chinese hamsters than other rodents. Keeping brother–sister pairs together from birth partially reduces the aggressiveness that develops at sexual maturity. Observing each pair of animals after they are placed in a cage together is referred to as *hand breeding* and is very labor intensive and therefore expensive. However, it allows for selection for more docile animals.[4] The *collar seclusion method* involves placing a small, lightweight plastic collar around the neck of each female hamster and having the individual cages interconnected by narrow tubes.[5] An uncollared male is free to enter or leave at will, but the female cannot leave her individual cage.[5,6] Hopefully, selective removal of the most aggressive female hamsters from breeding stocks[6] will reduce this problem over a period of time.

Discovery of genetic diabetes in the Chinese hamster

Discovery of diabetes. Inbreeding of five families of apparently normal, wild-type Chinese hamsters began in 1953 at the Yerganian colony in Boston, Massachusetts (USA).[3] Polydipsia and polyuria[7] arose spontaneously during inbreeding around the fourth generation[3] in 1954 in four families and appeared in another family after 13 generations[3] in 1957. The diabetes was recognized in 1957[8] and was first reported 2 years later.[9,10] The progressive inbreeding of apparently normal Chinese hamsters led to the appearance of hereditary diabetes, presumably by concentrating certain diabetogenic genes that had been present in low levels in the wild population.

An early-onset, "juvenile form" of diabetes was seen with onset prior to weaning. Those animals failed to reach sexual maturity and to reproduce themselves, and that subline was lost from the genetic pool during the first few years of inbreeding.[8] However, a "maturity onset" form was salvaged.[3] Diabetes can be induced in Chinese hamsters with various pharmacological agents, but these forms of diabetes are not discussed in this review.

What is a "prediabetic"? Prior to the onset of glucosuria or hyperglycemia, the animal is considered to be a prediabetic. By following the development of symptoms from the apparently healthy, prediabetic stage to the manifest diabetic stage, one hopes to better understand the cause of the disease. Initially, there were no sublines with a 100% incidence of diabetes, and in each litter of animals it was not clear in advance which animals would become diabetic. The offspring of two ketotic parents show a 100% incidence of diabetes[11] and are therefore all prediabetics. Hamsters with intermittent glucosuria or with newly diagnosed diabetes have been studied when true prediabetics have not been available.

Previous reviews. Many reviews on the spontaneously diabetic Chinese hamster have been published previously, several with extensive references lists as follows: 1960, 42 refs.[12]; 1964, 18 refs.[7]; 1965, 38 refs.[3]; 1967, 15 refs.[13]; 1976, 65 refs.[14]; 1981, 41 refs.[15]; 1982, 131 refs.[16]; 1984, 29 refs.[17]; 1984, 11 refs.[18]

Clinical symptoms: body weight, urine, and blood

Body weight. During the early 1960s one subline of diabetic animals with obesity was isolated,[7] but this subline disappeared relatively quickly. There is no association between diabetes and overt obesity in the Chinese hamster today. Adult diabetic hamsters are reported to have reduced,[19] normal,[19-22] or increased[23] body wt. Prediabetics are also reported to have reduced,[17,20] normal,[17,24] or increased[25] body wt. Despite these conflicting reports, the diabetic hamster today is considered to be a model for non-obese[21,26] diabetes.

Urine. Diabetes in the inbred Chinese hamster was first detected as polyuria,[7,10,27] 10–75 ml/24 h.[16] The onset of symptoms varies among colonies, sublines, and years, for example, onset between 18 and 250 days of age[8] or between 14 and 112 days.[16] Extreme daily urinary output of 50–70 ml in 24 h,[10] roughly twice the body wt of an adult Chinese hamster, or up to 120 ml/day, four times the adult body wt!, have been reported.[7] However, urine volumes in the Umeå colony are now only 1–2 ml/day in normal vs. 5–20 ml/day in diabetic hamsters.[28] Aluminum (Al) levels are normal in the diabetic hamster's urine, but due to the polyuria, daily loss of Al via the urine is excessive (L Juntti-Berggren, P-O Berggren, B Frankel, unpublished data) and may explain why Al levels are decreased in both islet B-cells and exocrine pancreatic cells of the diabetic[29] but not in the (non-polyuric) prediabetic.[30]

Diabetic Chinese hamsters show albuminuria,[18] increased urinary urea nitrogen,[27] glucosuria,[31] and polydipsia. Ketonuria is seen in severely diabetic animals,[3,10] but has become less frequent over the years due to the loss of the most diabetic animals. Gerritsen[16] reported that ketonuria usually occurs several months after the onset of glucosuria or between 6 and 12 months of age, but that it is not predictable. Just as stress tends to raise blood glucose levels in humans with diabetes, when diabetic hamsters fight other hamsters, the incidence and severity of their glucosuria increases.[32]

Blood. Blood glucose levels in diabetic hamsters vary greatly, with normal animals showing 5.7–6.4 mmol/l, diabetics 11.1–33.3 mmol/l, with one individual >45.9 mmol/l according to Meier.[10] The age at onset of hyperglycemia also varies from <1 month to >3 months.[16] Glycohemoglobin is elevated in the blood of the diabetic hamsters.[33] Diabetics show both fed and fasting hyperglycemia,[16-19] and the degree of hyperglycemia varies among different sublines.[16,34] Ketonemia[19,31] is seen in severely diabetic animals.

Some diabetic animals show high levels of total cholesterol, triglycerides, free fatty acids,[19,31,35,36] phospholipids, HDL cholesterol,[36] and lipid peroxide, a precursor to atherosclerosis.[36] Plasma insulin levels are quite variable: They can be subnormal,[16,19,37-39] normal,[16,38,40] or excessive.[16,41] The most severely hyperglycemic animals almost always have subnormal plasma insulin levels. Theophylline given by i.p. injection increases serum insulin levels in diabetic hamsters.[35] Genetically diabetic, but not yet hyperglycemic, animals are referred to as *prediabetic* hamsters and may go through a period of hyperinsulinemia.[17,20,28,42-44] Alternatively, the period of hyperinsulinemia may occur soon after the onset of hyperglycemia.[41] Hyperinsulinemia in the presence of normal glucose levels suggests that the peripheral tissues are resistant to insulin. Originally it was believed that diabetes in the Chinese hamster was primarily due to a B-cell defect, but it is becoming increasingly clear that both pancreatic and extrapancreatic defects are responsible.

After an i.p. glucose load, plasma glucose levels are excessive[31,45] in diabetic animals. Plasma glucose levels are also excessive[35,38,42,44,46] after an *oral* glucose load, and the insulin response is subnormal.[35,38]

Plasma glucagon levels in diabetic animals are normal[47] or excessive,[16,38] depending upon the subline.[16] In prediabetics, plasma glucagon is initially normal (age 14 days) but excessive at 20–23 days before the animals are significantly hyperglycemic.[17] After an i.p. arginine load, plasma glucagon levels are normal or excessive.[38]

Initially it was believed that a certain fraction of α-2 serum proteins was elevated in genetically diabetic hamster sublines and could be used as a marker prior to the onset of diabetes,[48,49] but the fraction was later shown not to be specific for the diabetic or prediabetic animals.[50,51] Blood pressure, as measured with a tail pressure cuff,[27] is normal in diabetic hamsters.

Treatment of the diabetic hamster

Mildly diabetic Chinese hamsters have been given sulfonylureas or biguanides in their drinking water, by stomach tube, or by s.c. or i.p. injection. Severely diabetic hamsters have required insulin injection.[50,52] Initially, many animals were dependent upon hypoglycemic therapy for their continued existence and for the maintenance of fertility.[7,8,10] Diabetic hamsters have also been treated with continuous insulin infusions via s.c.-implanted "minipumps".[53,54] Daily insulin injections or s.c. pumps are labor intensive and thus expensive. Therefore, the most severely diabetic (ketotic) animals have not reproduced well and have more or less disappeared from all colonies.[16] Today, the majority of diabetic hamsters are hyperglycemic but not ketonemic, and they survive without hypoglycemic treatment. Normal hamsters have a life span of 3–4 years, while the (untreated) mildly diabetic animals live 2–3 years, and the severely diabetic animals <2 years.[16]

Initially, it was hoped that the genetically diabetic Chinese hamster would become an important tool for the screening and study of potentially hypoglycemic agents.[10] A new oral hypoglycemic agent, ciglitazone, is believed to reduce insulin resistance and has been shown to reduce plasma glycerol, β-OH-butyrate, and triglycerides in the diabetic hamster.[55] A new adenosine-receptor agonist, dilazep, has been shown to reduce proteinuria in the diabetic hamster.[27] Treatment of diabetic hamsters by limiting food intake or by modifying the diet is discussed in the section on adipose tissue, obesity, food consumption, and effect of diet (p. 284).

Negative selection is necessary when breeding diabetic Chinese hamsters

Constant elimination (*counterselection* or *negative selection*) of the milder diabetics from a breeding stock increases the frequency of diabetes.[16,56] However, even in well-inbred sublines of diabetic hamsters, some of the offspring can show little glucosuria, that is, they do not "breed true" for the diabetic trait.[13,14] To select for sufficiently diabetic animals at the Upjohn colony, blood glucose levels of all potential breeding animals were measured at age 3 months, when all animals that would be diabetic had presumably become glucosuric.[56] Many generations of inbreeding and the genetic bottlenecks in which only a few animals have provided the genetic material for a whole colony or a subline should make the animals very homogeneous. However, the disease does not seem to be phenotypically fixed. If negative selection is not performed, the disease remains in a very mild form or apparently disappears.[56] Selection is reported to have produced diabetic sublines with close to a 100% incidence of diabetes.[14,16,43,56] In an inbred subline the animals all start showing glucosuria at about the same age,[14,16] and therefore normoglycemic animals prior to that age can be considered as prediabetic. The range of blood glucose concentrations within a given subline is quite broad. In one Upjohn subline, it was 6.4–30.5 mmol/l, suggesting that in utero, preweaning, nutritional, or other environmental factors have significant effects on the blood glucose level.[56]

The present colony in Umeå

Acquisition. Yerganian provided animals to the Upjohn Company (Kalamazoo, MI, USA) in 1962, and the colony at the Department of Histology and Cell Biology in Umeå began with several litters of animals from Upjohn in 1982. The 100% normal, subline-M hamsters[16] had been brother–sister inbred for 29–32 generations, and the >90% to 100% diabetic subline-L hamsters[16,56] had been brother–sister inbred for 41–44 generations at Upjohn.

Initial breeding attempts. In Umeå, during the first 1–2 years, the animals were almost entirely brother–sister inbred. However, many litters failed to reproduce because they did not contain both males and females, the brother–sister pairs did not produce young, or the young died before weaning. To increase the production of animals, the hamsters are now (1995) outbred within the descendants of one normal and one diabetic litter that were received from Upjohn in 1982.

During the first 16 months in which hamsters were born in this colony, the animals were "hand bred." Twenty-five percent of the liveborn normal pups and 16% of the diabetic pups died and/or were cannibalized before weaning at 21–23 days of age. Chinese hamsters are quite sensitive to noise, and the dams get upset if exposed to too much noise or other disturbances during the first week after delivery. This nervousness is seen throughout life and is much more apparent in the normal animals, presumably because the diabetics are too sick to be as quarrelsome. With improved handling and breeding routines, the loss of liveborn young is now much lower, but no accurate statistics are available. Selection against excessively aggressive females does not seem to have markedly reduced the frequency of territoriality and irritability in the females.

Present breeding technique. The hamsters are now bred in a collar seclusion type of breeding system, with each of five individual cages (for one male and four females) being connected by a long, broad, clear plastic tube ("boulevard") by way of a short, narrow (30 mm internal diameter) clear tube. The male hamster wears no collar and can easily run through the 30 mm tubes at will. Each female wears a 2-mm thick plastic collar with a 13 mm diameter neck hole and four broad "spokes" that stick out to the sides to a diameter of 33 mm. The collar prohibits the female from leaving her own cage but allows her to wash her face with her paws normally. When the collar gets worn out or chewed down to a smaller outer diameter, it is replaced. Daily measurement of urinary glucose/ketones or treatment with insulin or sulfonylureas/biguanides is not performed.

Negative selection. A diabetic is excluded from the breeding stock at approximately 3 months of age if its orbital plasma glucose level after local anesthesia is too close to the normal level when measured 3 h after "lights out." Such measurements represent fed plasma glucose levels because Chinese hamsters consume most of their food during the first few hours of darkness. Although the diabetics in the Umeå colony are derived from a highly inbred, essentially 100% glucosuria-positive, but rarely ketonuric, L subline from Upjohn,[16] some of the Umeå "diabetics" have normal or near-normal plasma glucose levels at age 3 months. Glucose measurements at later time points may show more diabetic levels, and genetically diabetic animals that have shown near-normal plasma glucose levels may produce hyperglycemic offspring. Therefore, the genetic trait has not necessarily been lost, but environmental factors affect the phenotypic expression, and negative selection is necessary to prevent accumulation of the more mildly affected animals. The nature of those environmental factors is not clear.

Diet. Originally the Chinese hamsters were fed a high-fat pelleted feed designed for mouse breeding (Ralston Purina Mouse Chow no. 5015), which contained ≥11.0% mostly animal fat, ≥17.0% protein, ≤3.0% crude fiber, ≤6.5% ash, and 52.5% carbohydrate. Because that pelleted feed is no longer produced, a similar feed is made by AnalyCen AB (Lidköping, Sweden) and contains 11.0% fat (67% animal, 33% vegetable), 20.2% protein, 2.6% plant fiber, 4.8% ash, and 51.4% carbohydrate. Changes in the diet can markedly affect the severity of diabetic symptoms in the genetically diabetic hamster; such changes are discussed in the sections on adipose tissue, obesity, food consumption, and effect of diet (p. 284), and on the effect of diet in prediabetic hamsters (new data) (p. 286).

Assay of Chinese hamster insulin

Chinese hamster insulin has not been isolated in pure form, and the primary structure is unknown. Serial dilutions of pancreatic extracts from Chinese hamsters were compared with rat, beef, and pork insulin standards by radioimmunoassay.[57] Proportional crossreactivity was obtained with pork insulin, nearly proportional crossreactivity was obtained with beef, but little proportionality was obtained with rat insulin standards.[57] This suggests that Chinese hamster insulin is immunologically or structurally closer to pork and human insulin, and is farther from rat insulin. Plasma or pancreatic insulin is now measured by radioimmunoassay using pork[27,40] insulin standards.

Histology of the diabetic Chinese hamster islet

Islet organization and size. Total pancreatic wet wt is normal in the diabetic hamster.[58] As in the rat and mouse, the Chinese hamster's islets are interspersed throughout the exocrine pancreas. The normal islet consists of a large number of insulin-containing B-cells in the middle of the islet,[33,59-61] while a much smaller number of somatostatin-containing D-cells form a thin, incomplete ring around the B-cells, and the glucagon-containing A-cells form a thicker, incomplete ring farthest out in the periphery or interspersed among the D-cells.[33,59,60]

Amylin-like immunoreactivity is localized in the B-cell granules and decreases in parallel with the decrease in insulin immunoreactivity seen with increasing severity of diabetes.[44] In other rodents pancreatic polypeptide is primarily found in islets of the posterior part of the head of the pancreas—the part that develops from the ventral pancreatic bud. Pancreatic polypeptide has not been studied in the Chinese hamster.

In the diabetic Chinese hamster pancreas, the number,[10,58] size,[10] and volume[58,62] of islets is decreased. Infiltration of lymphocytes is occasionally seen and is discussed later in the section on autoimmunity (p. 288). In the prediabetic Chinese hamster, islet size and volume are normal[44,63] or increased.[8] The light[61,62,64,65] and electron microscopic[59,64,66-69] appearances of the islet cells in the Chinese hamster have been described previously.

B-cells. In the diabetic islet, the B-cell area is smaller,[59,62] and there is B-cell degranulation,[41,59,62,64,67,70,71] decreased electron density in the insulin granules,[69] increased glycogen deposits,[10,59,62,69] increased necrosis with phagolysosomes,[59,67] abnormal or "hydropic cytoplasm,"[10,59] and karyolytic nuclei.[10] The individual cells from collagenase-isolated normal hamster islets show normal motility and spread out on Petri dishes in culture over 3–7 days, while diabetic islet cells remain in a oval clump during culture.[17]

Islets from prediabetic hamsters show normal B-cell– and amylin-containing areas[44] or decreased B-cell granulation, decreased B-granule volume, and possibly decreased mitochondrial volume.[63] Islets from hamsters with spontaneous remission or with erratic diabetic syndrome show increased numbers of hyperfunctioning B-cells.[59] Islets from the nondiabetic siblings of diabetic hamsters show reduced B-cell granulation and glycogen infiltration,[62] compared with genetically normal hamsters. Limitation of food intake in prediabetic hamsters reduces diabetic symptoms and partially prevents B-cell degranulation.[42] Prediabetic islets in culture show subnormal migration or motility of individual cells.[17]

B-cells from diabetic hamsters contain decreased levels of aluminum (−61%), and increased levels of copper (+92%), magnesium (+6%), and rubidium (+13%),[29] but these changes appear to be the result of diabetes, rather than the cause, because similar changes are not found in prediabetic hamsters.[30] However, acinar cells in the exocrine pancreas of the prediabetic hamster contain subnormal levels of iron (−10%)[30]; the cause or significance of this change is unknown.

A-cells. In diabetic islets, the A-cell area is decreased[58] or the number, staining, and area are increased.[10,33,59,60] Lysosomal digestion of secretory granules occurs,[68,69] A-cells occupy both the periphery and more central regions of the islet,[60] and dilated rough endoplasmic reticulum containing amorphous material is found in long-term ketonuric hamsters.[69] In the prediabetic islet, the A-cell area is normal.[44]

D cells. In diabetic islets, D-cell area is decreased[58] or increased,[33,60] and D-cells are observed both in the periphery and in the more central parts of the islet.[33,60] D-cells are also found in the diabetic hypothalamus and stomach, where the somatostatin concentrations are lower than normal.[58] Thus, the histological appearance varies somewhat among different studies, but in general the histology parallels the hormone release: Decreased insulin release and excessive glucagon, with milder diabetics and prediabetics showing less marked differences. The role of the D-cells and what causes the changes in the B- and A-cells is unclear.

Islet mitosis frequency in the prediabetic Chinese hamster

Meier and Yerganian[8] published morphological studies that compared the islets from the offspring of diabetic and normal hamsters from the day of birth onward. Among other differences, more mitotic figures were seen in islets from the presumably prediabetic pups at 2 h of age, and at 1 day of age the prediabetic islets were more numerous, larger, and more B-cell rich.[8] Those authors suggested that maternal hyperglycemia in utero may have contributed to the B-cell hyperplasia.[8] The morphological differences between normal and prediabetic young had markedly decreased by 10–14 days of age.[8] Using ^{3}H-thymidine incorporation into nuclei as a measure of cell replication, we found a significantly *lower* mitotic index in the islets of normoglycemic, prediabetic hamsters at age 17–20 days.[28] The mitotic index was not significantly different from normal at 6 wk of age.[28] Thus, hyperglycemia in utero does appear to stimulate B-cell replication in the newborn prediabetic animal, but the ability to replicate islet B-cells seems to be lost when the animal is no longer under the influence of maternal hyperglycemia or the maternal hormones of pregnancy. Like et al.[65] found an increased ^{3}H-thymidine incorporation in B-cells of two hyperglycemic, hyperinsulinemic, recent-onset diabetic hamsters but not in animals that had been diabetic for a longer period of time. Thus, hyperglycemia that is not associated with the intrauterine environment probably also stimulates B-cell mitosis for a short period of time. The reduced B-cell replication at 3 wk of age in the prediabetic hamster, when normal islet cells replicate very rapidly,[28] might explain why islets of adult diabetic hamsters are smaller than normal. It is not clear what causes the decreased mitosis in the normoglycemic, prediabetic animal. Since decreased cell replication is also seen in the prediabetics' adrenal glands,[28] the mitotic defect may be present in other tissues as well.

Hormone synthesis and content of the diabetic Chinese hamster pancreas

Insulin synthesis. After incubation of hamster islets with 20.8 mmol/l glucose and ^{14}C-leucine, insulin can be extracted in acid ethanol and separated on a Sephadex G50 column.[72] The major fraction elutes in a peak that overlaps bovine insulin and has an estimated mol wt of ~7700.[72] During a 4 h incubation, normal islets first show release of ^{14}C-labeled insulin between 60 and 120 min, whereas diabetic islets do not show release until 120–180 min. Total release is severalfold greater from normal islets than from the same number of diabetic islets.[72] At the end of the 4 h incubation, total ^{14}C-labeled insulin is markedly lower in diabetic islets than in normal islets.[72] Thus, insulin synthesis and release occur more slowly in the diabetic islet.

Insulin content in whole pancreas. The pancreatic insulin content is normal[17] or decreased[17,44] in the prediabetic hamster, normal[73] or decreased[41,44,73] in the recent-onset diabetic animal, and decreased[18,31,38,44,47,57,58] in the older diabetic hamster. As the decrease in pancreatic insulin content appears to occur after the onset of hyperglycemia,[73] the islet B-cell might become exhausted as a result of an exaggerated response to extrapancreatic insulin resistance.

Insulin content in isolated islets. In morphological studies diabetic islets are generally smaller than normal islets (see section on islet histology p. 273). Collagenase-isolated diabetic islets tend to be smaller than normal islets when observed with the microscope. However, for batch incubation studies, the researcher tries to choose "normal-sized" diabetic islets for comparison with normal ones. Differences in isolation and culture conditions can affect islet size. Islet wt is reported to be subnormal by dry wt,[40] but not significantly below normal by protein content[39] or calcium content.[39] In collagenase-isolated islets from the diabetic[22,39,72,74] and prediabetic[72] hamster, the insulin content is lower when reported as amount of insulin/islet, but normal when reported as amount of insulin/μg dry islet wt.[40]

Other pancreatic hormones. In the whole diabetic pancreas, glucagon content is normal[38,58,73] or increased,[38,44,45,47] with the greatest increases seen in the most severely glucosuric animals.[38] The glucagon content is greater in isolated islets[60] from diabetic hamsters and in their stomachs.[47] It is normal[17] or increased[17] in the prediabetic pancreas. The somatostatin content is decreased.[58] The amylin content is normal in the prediabetic, but decreased in the older and more severely diabetic pancreas.[44]

B-cell function in the diabetic Chinese hamster

Insulin release from the in vitro perfused pancreas. The insulin release in response to glucose is biphasic in the normal hamster pancreas.[44,57] Both phases are subnormal in the diabetic hamster.[44,57] The diabetic pancreas also shows subnormal amylin[44] release. The more severely diabetic the hamster, the lower the insulin and amylin release.[44]

Glucose + arginine together, or K+ alone, cause subnormal insulin release from the diabetic pancreas.[15,34] The responses vary somewhat from subline to subline. Theophylline alone causes normal[34,57] or excessive insulin release from the diabetic pancreas,[34] depending upon the subline. Glucose + theophylline together cause subnormal insulin release from the diabetic pancreas.[75] Insulin release from the perfused pancreas in response to glucose, tolbutamide, mannoheptulose, and puromycin in the normal hamster (and in an unspecified number of diabetic hamsters) has also been studied.[76]

Insulin release from young diabetic hamsters. Hormone release from the in vitro perfused pancreas of young diabetic hamsters has been studied in two different colonies with slightly different results. We found normal insulin, glucagon, and somatostatin release from the pancreases of 6–12-wk-old hyperglycemic, diabetic animals, but subnormal insulin release from the even more hyperglycemic, 17 to 21-wk-old diabetics.[73] Nakajima et al.[44] found subnormal insulin release but not significantly suppressed amylin release from normoglycemic 9- to 17-wk-old, genetically diabetic pancreata. Thus, the age at which pancreatic insulin release starts to fail may vary slightly between different colonies, but islet failure occurs much later (at 2–4 months of age) than when excessive gluconeogenesis is seen at 2[22] to 7[43] wk of age.

Insulin release from pieces of whole pancreas. Glucose-stimulated insulin release from ~8-mg pieces of whole pancreas was measured.[70] Overtly diabetic hamsters showed subnormal insulin release and intermittently glucosuric animals showed excessive insulin release.[70]

Insulin release from isolated islets. Glucose-stimulated insulin release is subnormal from diabetic islets during batch incubation[22,37,38,40,74] and perifusion.[39,60,72,74] The insulin release is potentiated, but not normalized, by the calcium channel agonist, perchlorate[40] and the phosphodiesterase inhibitor, 3-isobutyl-1-methylxanthine (IBMX).[39,74] As from the whole, perfused pancreas, insulin release is biphasic from normal perifused hamster islets, and both phases of insulin release are suppressed in the diabetic perifused islets.[39,60,72] Diabetic hamster islets require higher glucose levels than normal to stimulate half-maximal insulin release.[74] Insulin release from normal islets is suppressed by somatostatin in a dose-dependent manner;[31] no somatostatin suppression data are available for diabetic islets.

Effect of treatment on pancreatic function. When diabetic hamsters receive continuous insulin infusions during a 4-wk period via the s.c. minipump, hyperglycemia and glucosuria progressively decrease.[53] First-phase insulin release from their in vitro perfused pancreata is improved after blood glucose normalization.[53] However, suppression of blood glucose to *subnormal* levels (by the same dose of insulin given to less severely diabetic animals) is associated with suppression of pancreatic insulin release.[53] Thus, pancreatic B-cells, like brain and other cells, do not function well during hypoglycemia. Treatment of prediabetic hamsters for 4 wk with a continuous, low-dose insulin infusion does not prevent the onset of diabetes.[54] In fact, it seems to make the diabetes slightly worse, possibly by promoting a small wt gain.[54]

Glucose utilization in isolated islets. The production of ^3H_2O from ^3H-labeled glucose was determined in isolated islets from normal and diabetic Chinese hamsters as a measure of the combined fluxes through glycolysis and the pentose phosphate shunt. Glucose utilization in diabetic islets, measured as glucose equivalents produced per dry islet wt, was 84–85% of the normal level at both 3 and 20 mmol/l glucose[1]. This suppression could be accounted for by the decreased proportion of B-cells in the diabetic islets, ~75% of normal.[62] This suggests that glucose utilization is not significantly suppressed in the individual diabetic B-cell.

Stimulation-secretion coupling, glucose-stimulated insulin release, and ion movements. The process[77] by which glucose is believed to cause insulin release from the islet B-cell includes the following: glucose metabolism with production of NAD(P)H and ATP, closure of ATP-sensitive K^+ channels, opening of voltage-gated Ca^{2+} channels, influx of extracellular Ca^{2+}, Ca^{2+} accumulation in the cytoplasm, and insulin granule exocytosis. Voltage differences across the plasma membrane and reduced nucleotides in the islet B-cell have not been studied in the Chinese hamster. Glucose utilization is probably not markedly different from normal in the diabetic islet (see earlier), K^+ uptake, as approximated by $^{86}Rb^+$ uptake, is normal.[37] However, closure of K^+ channels, as determined by following $^{86}Rb^+$ efflux, is abnormal in the diabetic islet.[37] Glucose-stimulated $^{45}Ca^{2+}$ uptake is not different from normal[39] or is subnormal,[37,40] even with the help of the Ca-channel agonist, perchlorate.[40] Ca^{2+} efflux is subnormal in response to glucose, even in the presence of IBMX.[39]

Islet cAMP. The cAMP-raising methyl xanthine, IBMX, increases glucose-stimulated insulin release from batch-incubated normal islets fivefold and increases the diabetics' subnormal insulin release by threefold.[74] During batch incubations with IBMX, normal, but not diabetic, islets show a glucose-induced increase of cAMP levels in the islets.[74] IBMX increases cytosolic Ca^{2+} in the presence of glucose in both normal and diabetic islet B-cells, but it is not yet clear how great the effect is in normal vs. diabetic B-cells (see later). Theophylline given by i.p. injection increases serum insulin levels in diabetic hamsters.[35] These data together suggest that glucose fails to raise cAMP levels normally in the diabetic islet, and that cAMP is necessary for raising intracellular Ca^{2+} levels prior to in-

[1]Lindström P, Sehlin J, Frankel BJ: Glucose-stimulated elevation of cytoplasmic calcium is defective in the diabetic Chinese hamster islet B cell. Europ J Endocrinol 134: in press, 1996.

sulin granule exocytosis. Activation of the G-protein–linked A_1-adenosine receptor[78,79] is believed to excessively inhibit adenylate cyclase in *ob/ob* mouse fat cells and in some obese humans. Could a similar defect be present in the diabetic Chinese hamster?

Cytoplasmic Ca^{2+} levels are subnormal in the isolated islet B-cells from the diabetic Chinese hamster (new data)

Normal (subline N) and diabetic (subline L) animals' islets were isolated using collagenase, separated into individual cells by incubation with trypsin and DNAse I, and then allowed to settle on glass coverslips during culture for 1 or 2 days. Then, the coverslips with attached cells were incubated with 1 μmol/l Fura-2/AM (Molecular Probes Inc., Eugene, OR, USA) for 45 min at 37°C. After rinsing, the cells were perifused on the object stage of an inverted microscope (Zeiss IM) with an attached XBO 75W/2 xenon arc lamp equipped for dual wave-length epifluorescence microfluorometry. The coverslip was superfused at a flow rate of 0.6 ml/min; medium in the chamber was kept at 37°C by warming the pump tubing and the object stage.

The cytoplasmic Ca^{2+} activity in a single cell was calculated from the ratio of emitted light at 510 nm, measured with a photomultiplier, when the cell was illuminated with 340-nm or 380-nm excitation light. Measurements at the two wavelengths were done every 5 sec during flashes of 125 msec at each wavelength with 200 msec between flashes. Calibration of the Fura-2 signal was done using droplets of "intracellular-like" medium and non-esterified Fura-2. Such calibration showed that a 340/380 nm ratio of 1 corresponded to a cytoplasmic calcium concentration of 47 nmol/l, and a ratio of 2 corresponded to 229 nmol/l. Immunostaining of the cells showed that 77–90% of cells from normal hamster islets were positive for insulin and 11–16% were positive for glucagon. The corresponding figures for the diabetic cells were 88–94% insulin- and 23% glucagon-positive cells.[1]

As seen in Figure 1, glucose caused a transient, slight suppression, and then a marked increase, in cytoplasmic Ca^{2+} in the normal hamster B-cell (filled squares, solid line). The marked rise disappeared after a few minutes, despite continued glucose perifusion. In the diabetic B-cell (open triangles, dotted line), the slight suppression of cytoplasmic Ca^{2+} occurred later than normal and was less "deep"; the "spike" was many min delayed, less "sharp," and smaller than in the normal cell. (In diabetic B-cells the initial suppression is often decreased or missing, and the large spike is often decreased and delayed (unpublished data). As seen in Figure 2, the normal B-cell responded to a combination of glucose + IBMX with a transient suppression, a "spike," and a relatively constant Ca^{2+} elevation throughout the rest of the perifusion period. In the diabetic B-cell, the suppression was delayed but clearly present, the spike was somewhat delayed but sharp, and the post-spike Ca^{2+} plateau was not different from normal.

The glucose-induced changes in cytoplasmic Ca^{2+} in the normal B-cell are believed to correspond to Ca^{2+} uptake into intracellular organelles and the opening of Ca^{2+} channels through the plasma membrane. IBMX partially normalized

[1]See footnote p. 277.

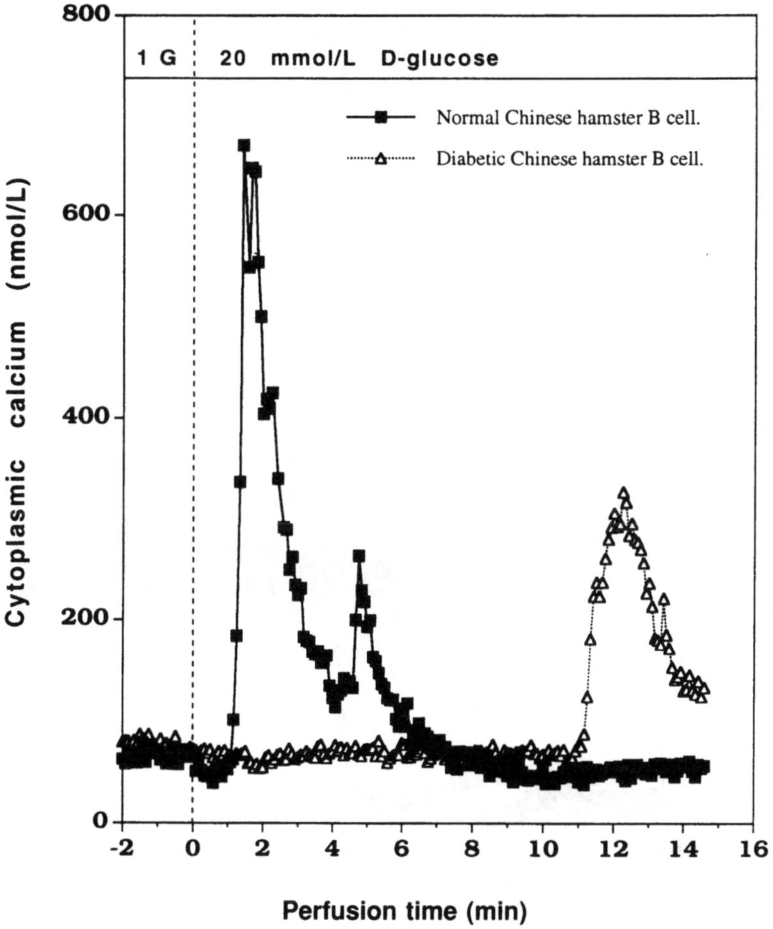

Figure 1. Effect of 20 mmol/l D-glucose on cytoplasmic Ca^{2+} in one isolated normal (filled squares, solid line) and one diabetic (open triangles, dotted line) Chinese hamster islet B-cell. Glucose caused a 1-min suppression followd by a 6-min elevation in the cytoplasmic Ca^{2+} level in the normal B-cell. In the diabetic B-cell the initial suppression was delayed and less "deep"; the rise was delayed and blunted. (P. Lindström, J. Sehlin, and B. Frankel, unpublished data.)

the transient suppression, made the spike occur sooner, and increased the cytoplasmic Ca^{2+} level after the spike in the diabetic B-cell. It appears as if the diabetic B-cell fails to increase cAMP levels sufficiently during glucose stimulation and that cAMP participates in several different reactions: Ca^{2+} uptake into intracellular organelles (transient Ca^{2+} depression), the later increases in cytosolic Ca^{2+} (Ca^{2+} influx), and the accompanying insulin release. Perhaps the A_1-adenosine receptor,[78,79] which is believed to be excessively active in suppressing cAMP levels in the *ob/ob* mouse fat cell, could be overly active in the diabetic Chinese hamster islet B-cell.

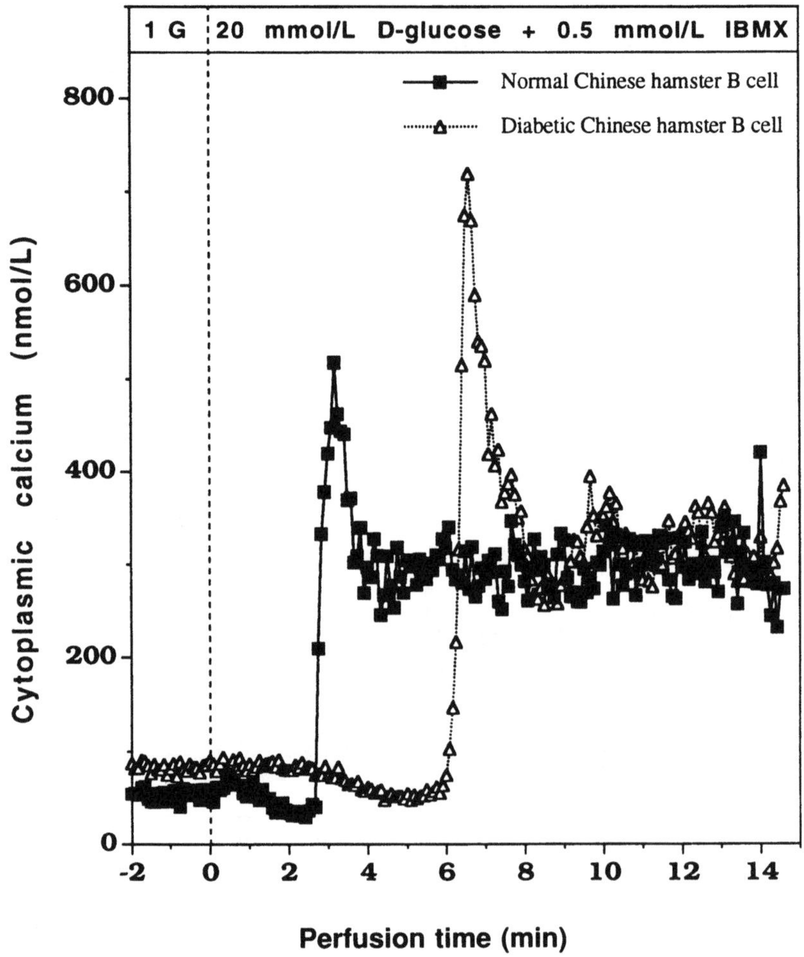

Figure 2. Effect of 20 mmol/l D-glucose + 0.5 mmol/l IBMX on cytoplasmic Ca^{2+} in one isolated normal and one diabetic Chinese hamster islet B-cell. Glucose + IBMX caused a suppression, spike, and elevated plateau in the cytosolic Ca^{2+} level in the normal B-cell. In the diabetic B-cell the initial suppression was delayed, the spike was delayed, but the plateau was not different from normal. (P. Lindström, J. Sehlin, and B. Frankel, unpublished data.)

A-cell function in the diabetic Chinese hamster

Pancreas perfusion. Basal glucagon release from the diabetic pancreas perfused in vitro is not different from normal.[57] However, glucagon release is not suppressed as quickly or as completely during glucose perfusion as for the normal pancreas.[57] Arginine alone causes a biphasic glucagon release at 20 mml/l, but no insulin release, from both the normal and the diabetic hamster pancreas when perfused in vitro.[75] However, that pharmacological dose of arginine in the absence of any glucose may be toxic to the pancreas, because pancreatic contents of both insulin and glucagon after perfusion are abnormally low.[75] A combination of arginine + glucose stimulates glucagon release from the hamster pancreas perfused in vitro. Arginine + glucose–stimulated glucagon release is normal from the perfused pancreas of the young diabetic.[73] It becomes excessive in some, but not all, diabetic sublines, when the diabetics get older.[15,34] K+ alone, at 18 mmol/l, also causes glucagon release, and the response is normal[15,34] or excessive[34] in the diabetic pancreas, depending on the subline. Theophylline alone also causes glucagon release, and the response is normal[57,75] or excessive[34] in the diabetic pancreas, depending upon the subline. Glucose + theophylline cause some glucagon release, which is not significantly different from normal.[57,75] In the presence of theophylline + glucose + arginine, glucagon release is excessive.[75]

Perifused and batch-incubated islets. Basal glucagon release from diabetic islets perifused in vitro is excessive,[60] and this release is not at all suppressed by glucose.[60] Basal glucagon release from isolated, batch-incubated islets is excessive from diabetic islets, and glucose fails to suppress glucagon release in all but the most mildly diabetic animals.[38]

D-cell function in the diabetic Chinese hamster

Somatostatin-containing D-cells occupy a greater area in the islets of diabetic hamsters,[33] but somatostatin concentrations (as ng/g wet wt of pancreas) are normal in prediabetic and diabetic hamster pancreas.[33] Somatostatin release from the in vitro perfused pancreas from the young diabetic hamster[73] is normal in response to glucose + arginine.

Kidney and bladder

Kidney and uriniferous tubule. Diabetic Chinese hamsters show polyuria and polydipsia,[10,27] albuminuria,[18,27] increased urinary urea nitrogen,[27] and normal to increased kidney size.[80-82]. Hydronephrosis was seen in some animals in the 1950s and 1960s.[10] Tubular damage and collapse,[10,80] and, in a few cases, glycogen in the collecting tubules, were also seen.[69,80] The diabetic kidney contains excessive levels of glucose, sorbitol, and fructose, substances involved in development of nephropathy,[83] not affected by treatment with an aldose reductase inhibitor, ONO-2235.[83]

Glomerulus. Diabetic hamsters may show increased[27,84] or normal[80] glomerulus size, thickening of Bowman's capsule,[80] and glycogen in Bowman's space.[69] In the capillary tuft one sees fusion of podocyte foot processes,[84] capillary dilation and coalescence,[80,84] nuclear pyknosis, and loss of podocyte, endothelial, and mesangial cell nuclei.[10,80,84] The glomerular basement membrane (GBM) is diffusely thickened, but seldom nodularly thickened.[27,80,84] There is periodic acid-Schiff (PAS)–positive material within capillary tufts[10] and a decreased number of anionic sites in the GBM.[27] GBM abnormalities can be partially corrected by treatment with dilazep, an adenosine potentiator and platelet aggregation inhibitor.[27]

Enzyme activities. In the diabetic kidney gluconeogenic enzyme activities are increased.[85] Acid glycohydrolases, which are involved in glycoprotein metabolism, vary among different sublines but are generally decreased in the kidneys of most diabetic sublines.[82] These changes may be related to the GBM dysfunction seen in the diabetic glomerulus.

Bladder. The wet wt of the diabetic urinary bladder is increased.[86] In the isolated bladder perfused in situ the threshold volume (pressure) that causes spontaneous contractions is increased in the diabetic animals.[86] In the diabetic bladder, the levels of the neuropeptide, vasoactive intestinal peptide (VIP), are elevated,[20] and parasympathetic innervation is reduced.[87] Thus many, but not all, of the changes in the diabetic hamster's urinary tract are similar to those found in the human with diabetic nephropathy.

Retinopathy and the eyes

Diabetic Chinese hamsters show increased levels of glucose, sorbitol, and fructose in their lenses,[83,88] which are associated with an increased incidence of cataracts,[89] although cataracts are not seen in all diabetic hamster colonies.[83] Diabetic hamsters also show increased levels of glucose, sorbitol, and fructose in their retinas.[83] Diabetic hamsters do not show retinal hemorrhages or aneurysms similar to those seen in human diabetics,[89] but do show irregular thickening of the capillary basement membranes, proliferation or degeneration of the intramural pericytes, and reduced numbers of pericytes as compared with endothelial cells.[89-91] Diabetic hamsters show lesions in the retina, and various staining procedures suggest either glycoprotein or glycogen accumulation.[92] Some of the lesions are associated with the capillaries; some are not.[92]

Both diabetic and prediabetic hamsters show decreased wet wt of the eyes.[20] VIP, a neuropeptide found in certain nerves, is found in excessive quantities in the eyes of diabetic and prediabetic hamsters, and the changes seem to involve VIP-immunoreactive nerve fibers associated with the blood vessels in the choroid layer.[20] Thus, diabetes clearly affects the diabetic hamster's eyes. Some of the effects are similar to those seen in human diabetics, while others are not.

Nervous system

Brain. In diabetic hamsters the brain wt is normal,[93] but the cerebral hemispheres show structural abnormalities in nerve cells, myelin, and blood vessels.[94] Somatotrophs are increased in the adenohypophysis of the diabetic hamster,[95] and several different types of cells in or near the median eminence are morphologically abnormal.[23] In the hypothalamus of severely diabetic female hamsters, several nuclei (paraventricular, ventromedial, dorsomedial, arcuate, and premammillary) show decreased size, cell number, or cell density.[93] Glucose uptake by the brain is normal in prediabetic hamsters.[43]

Peripheral nerve structure and function. The peripheral nerves of diabetic hamsters show segmental, axonal degeneration[87,96] and decreased conduction velocities.[97] The degree of structural abnormalities is correlated with the duration of diabetes.[96] (In humans, decreased conduction is believed to be due to hyperglycemia-induced suppression of myoinositol and related phospholipids,[98] and diabetic neuropathy is believed to be due to hyperglycemia-induced sorbitol accumulation with subsequent Schwann cell loss and demyelinization.[98]) The sciatic nerve contains subnormal levels of inositol in the diabetic hamster.[83] This nerve (but not the spinal cord[88]) contains excessive levels of glucose, sorbitol, and fructose.[83,88]

Norepinephrine. Norepinephrine levels are increased in diabetic hamster kidney, liver, cerebral cortex,[81] heart muscle,[19] and 24-h collected urine.[28] The excessive norepinephrine levels in the kidney, liver, and cerebral cortex are at least partially reversible upon insulin treatment,[81] suggesting that they are a result of hyperglycemia. However, the lack of insulin may not be the cause of excessive norepinephrine levels, because we found excessive norepinephrine levels in hyperglycemic but normoinsulinemic diabetic hamsters.[28] Excessive norepinephrine levels are suggested to be associated with insulin resistance in humans.[99] Thus, the diabetic Chinese hamster clearly suffers from a peripheral neuropathy that is similar to that seen in human diabetics.

Liver

Morphology and histology. In the diabetic Chinese hamster, the liver wt is normal[47] or excessive.[22,100] The liver contains subnormal,[101] normal,[22] or excessive amounts of glycogen,[10] and contains excessive PAS-positive material[10] and excessive lipid peroxide, a precursor to atherosclerosis.[36] In an effort to discover genetic markers for diabetes and prediabetes, liver proteins were separated by two-dimensional gel electrophoresis.[17,102] The positions of two specific proteins were believed to indicate if an animal is genetically diabetic or not. However, these proteins did not differentiate between normal and diabetic animals.

Enzymes in the adult liver. The glycolytic and pentose phosphate shunt enzyme activities are decreased in the livers of some diabetic hamsters,[18,22,47] and the gluconeogenic enzyme activities are increased in vitro.[18,22,35,43,47,85,101] Cholesterol synthesis is normal.[100]

Enzymes in the prediabetic liver. In prediabetic animals, gluconeogenesis is normal in vivo[43] but is increased in vitro.[22] Glycolysis is decreased.[17,22] The gluconeogenic and glycogenolytic enzymes respond abnormally to enzyme inhibitors.[103] Liver levels of glycogen,[31] triglycerides,[43] and lipid peroxide[36] are increased.

Insulin receptor. ^{125}I-Insulin binding to its receptor on the diabetic liver is increased, mainly due to an increased receptor number.[104,105] Insulin receptor kinase activity and autophosphorylation are decreased in the diabetic liver.[106] The prediabetic insulin receptor has not been studied.

Heart muscle and blood vessels

Heart. Ketotic diabetic Chinese hamsters, with plasma insulin levels only one fifth of the normal level, show increased heart wt, ventricular septum, and ventricular wall thickness.[19] There is an increase in glycogen, triglyceride, cholesterol, and norepinephrine content; a decreased ATP and glycolytic enzyme activity;[19] and a decreased glucose oxidation to CO_2.[101] Diabetic hearts contain normal amounts of lipid peroxide.[36] Myocardial degeneration is seen by light microscopy in some but not all diabetic animals.[19] Prediabetic hamsters with hyperinsulinemia show normal heart uptake of 2-deoxyglucose.[43]

Vessels. Uptake of ^{14}C-cholesterol by preparations of intimal lining from the abdominal and thoracic aorta of diabetic hamsters is increased.[7] The aorta shows an increased lipid peroxide content and morphological changes in the intima that resemble the early phase of atherosclerosis.[36] Abnormal capillaries, arterioles, and thickened capillary basement membranes are seen in the cerebral cortex of diabetic hamsters[94] and in the kidney glomerulus.[80,84]

Skeletal muscle and skeleton

In the diabetic hamster diaphragm, glucose incorporation into glycogen is normal, and the response to added insulin is normal, but oxidation of glucose to CO_2 is reduced.[101] Sorbitol and fructose levels are not elevated in the diabetic hamster's gastrocnemius.[88] In prediabetic hamsters, glucose uptake into certain skeletal muscles (adductor longus, extensor digitorium longus, triceps brachii) is decreased,[43] but not in others (tibialis anterior,[43] diaphragm[31,43]). Spinal cord and joint changes that normally occur with advanced age in normal hamsters are seen at younger ages in diabetic Chinese hamsters.[107,108] Thus, both diabetic and prediabetic hamsters show abnormal metabolism in skeletal muscle, but the defects do not occur simultaneously or in all muscles.[43]

Adrenal morphology and function

Adrenal histology is normal in diabetic Chinese hamsters.[28] However, adrenal wet wt and volume are decreased (cortex volume -12%, medulla -37%).[28] Cortisol, aldosterone, and DHEA sulfate levels in plasma are not different from normal.[28] The decreased adrenal wt may be explained by a decreased mitotic rate (as measured by ^3H-thymidine incorporation into premitotic nuclei) at 17–20 days of age,[28] but the cause of the decreased mitotic rate is unknown. The mitotic rate in tissues other than the adrenal and the islet has not been studied.

Adipose tissue, obesity, food consumption, and effect of diet

Adipose tissue. A variety of enzymes has been studied in the epididymal adipose tissue of diabetic hamsters, but no significant differences are found as compared with normal hamsters.[47] Glucose conversion to CO_2 or to fatty acids is essentially normal in the epididymal fat pad from diabetic hamsters,[21,31,101] and the tissue shows a normal response to insulin in vitro.[31] However, the uptake of 2-deoxyglucose into interscapular brown adipose tissues and white adipose tissue from the inguinal, dorsal, and epididymal sites is subnormal in prediabetic hamsters.[43] It was suggested that the defective glucose uptake by adipose tissue might prevent the diabetic hamster from becoming grossly obese,[43] as in some other rodent models of diabetes. Large quantities of PAS-positive material are found in the pericardial adipose tissue of diabetic hamsters,[10] but the significance of that finding with regard to carbohydrate metabolism is unclear. It could simply be a result of the hamsters' hyperglycemia.

Obesity. Except for the single, isolated subline of hamsters with obesity in the early 1960s,[7] the diabetic Chinese hamster is not considered to be obese.[26] However, at least some prediabetic hamsters have increased body wt and an increased percentage of body fat prior to becoming diabetic.[25] Therefore, the genetic defect or defects in the genetically diabetic Chinese hamster appear to affect many tissues, including some metabolic functions in the adipocyte. The molecular nature of the defects is not known.

Food consumption. Diabetic Chinese hamsters consume more food than normal hamsters,[8,21,46] and much of this excess is lost as glucose and protein in the urine.[31,46] Prediabetic hamsters also consume more food than normal hamsters,[24,25] but the prediabetics' hyperphagia cannot be a result of losses via urine and feces. The genetic or metabolic cause of the hyperphagia is not known. Deslex and Rossi[95] suggest that the prediabetic hamster might suffer from a defective hypothalamic-pituitary function that could steer the appetite. Gerritsen[16] suggests that the hyperphagia results from the diabetic milieu in utero.

Effect of diet. Prediabetic hamsters may go through a period of hyperphagia, increased body wt, and increased total body fat prior to becoming diabetic.[17,25] Limitation of food intake starting in the prediabetic stage reduces the severity of diabetes.[14,25,42] Prediabetics restricted to a normal food intake for 30 months remain essentially clinically normal thereafter.[17,25] The cause of the hyperphagic phase is unclear.

Both the quantity and the quality of the diet markedly affect the severity of diabetic symptoms in the adult diabetic hamster. For instance, maintenance of adult diabetic animals on certain diets (e.g., low-fat, vegetable rather than animal fat) reduces the severity of their symptoms.[16,45] Guar gum, 8% in the diet also significantly reduces polydipsia, polyuria, and glucosuria.[109] Exactly which components in the diet are diabetogenic is not known. Because a high-fat diet does not cause glucosuria in normal hamsters, only genetically (pre)diabetic animals seem to be particularly sensitive to the (unidentified) component(s). Because insulin and glucagon release from the in vitro perfused pancreas of adult diabetic hamsters is not altered in diabetic hamsters after a switch to a more diabetogenic diet (B Frankel, GM Grodsky, unpublished data), the diabetogenic factor probably affects the extrapancreatic tissues and not the pancreas directly.

Effect of diet in prediabetic hamsters (new data)

To determine the effect of quantity and type of dietary fat on diabetic symptoms, prediabetic hamsters were fed ad libitum one of two different diets, which differed among other things, in fat content and type of fat (animal or vegetable) starting upon weaning at the age of 3 wk. The "4% diet" was the Berkeley Diet—standard rat and mouse food (Feedstuffs Processing Co., San Francisco, CA, USA); it contained 3.4 kcal/g as 4% fat, 23% protein, and 52% carbohydrate. The only source of fat was vegetable oil. The "11% diet" was Purina mouse chow; it contained 3.8 kcal/g as 11% fat, 17% protein, and 54% carbohydrate. The only source of fat (when this study was performed) was beef and pork fat. The percent contents are based on the manufacturers' analyses, assuming 10% moisture, if not stated by the manufacturer. These data were produced in collaboration with GC Gerritsen, MC Blanks of the Upjohn Company, and GM Grodsky of the University of California, San Francisco, CA, USA.

Despite an occasionally greater total caloric intake on the 4%-fat diet, the animals fed the 4%-fat diet consumed less calories as fat, consumed less saturated fat, and had lower plasma glucose levels (Table 1). This suggests that the less-saturated vegetable oil or the lower percentage of total calories from fat significantly reduced the hyperglycemia in genetically diabetic hamsters. The diets were obtained from two different companies and therefore could have had other differences.

Intestinal motility

In diabetic hamsters, barium X-ray pattern of the intestinal tract show prolonged transit times,[110] abnormal dilation,[27,110] and decreased motility in some areas.[110] Small and large intestine weights are increased.[100] Electron microscopy of the unmyelinated axons in Auerbach's myenteric plexus show glycogen deposition, filament accumulation, and dense residual bodies in abnormally swollen axons.[110] Cholesterol synthesis is increased in the small and large intestine of the diabetic.[100]

Reproduction

In the severely diabetic female hamster, ovarian wt is normal, serum estradiol is normal, but serum progesterone is reduced, the number of secondary follicles in the ovary is reduced, and the number of atretic follicles is increased.[93] Diabetic hamster mothers show an increased rate of abortion, incomplete pregnancy, fetal death, malformations such as omphalocele, and underdevelopment of the young.[8,111] In the severely diabetic male hamster, testis and germinal epithelium volumes are decreased, sperm/spermatid number and Leydig cell number are reduced, and spermatogenesis is completely arrested at the stage of the primary spermatocytes.[112]

Periodontal disease

Severely diabetic hamsters show an increased frequency of calculus-like deposits on the teeth, gingival inflammation, migration of the epithelial attachment and

Table 1. Effect of diet on plasma glucose in (pre)diabetic Chinese hamsters.

Diet	Age (wk)						
	4	6	8	10	12	14	16
4%	16 ± 1	18 ± 1	20 ± 1	22 ± 1	23 ± 1	24 ± 1	24 ± 1
11%	15 ± 1	22 ± 1	23 ± 1	25 ± 1	27 ± 1	27 ± 1	28 ± 1
P	NS	<0.001	<0.01	NS	<0.05	NS	<0.025
Plasma glucose (mmol/l)							
4%	6.5 ± 0.1	5.4 ± 0.3	6.1 ± 0.2	5.8 ± 0.5	5.5 ± 0.3	6.6 ± 0.2	7.7 ± 0.4
11%	6.5 ± 0.1	8.7 ± 0.6	10.0 ± 0.7	8.3 ± 0.8	10.4 ± 1.2	11.0 ± 0.5	13.4 ± 2.0
P	NS	<0.001	<0.001	<0.02	<0.005	<0.001	<0.02
Calorie consumption (kcal/g body wt/day)							
4%	—	0.74 ± 0.02	0.65 ± 0.03	0.58 ± 0.02	0.63 ± 0.01	0.60 ± 0.02	0.62 ± 0.02
11%	—	0.68 ± 0.01	0.56 ± 0.02	0.58 ± 0.03	0.51 ± 0.02	0.51 ± 0.03	0.55 ± 0.03
P	—	NS	<0.02	NS	<0.001	<0.05	NS
Fat consumption (kcal/g body wt/day)							
4%	—	0.08 ± 0.00	0.07 ± 0.00	0.06 ± 0.00	0.07 ± 0.00	0.06 ± 0.00	0.07 ± 0.00
11%	—	0.18 ± 0.00	0.15 ± 0.01	0.15 ± 0.01	0.13 ± 0.01	0.13 ± 0.01	0.14 ± 0.01
P	—	<0.001	<0.001	<0.001	<0.001	<0.001	<0.001

Data are given as mean ± SE, statistical differences were determined by two-tailed Student's *t*-test; NS = not significant.
Calorie and fat consumption were based on food intake during each 2-week period.
These data were produced in collaboration with M.C. Blanks and G.C. Gerritsen at Diabetes and Atherosclerosis Research; The Upjohn Company, Kalamazoo, MI, USA; and G.M. Grodsky at the Department of Biochemistry and Biophysics, University of California, San Francisco, CA, USA.

pocket formation, alveolar resorption, and generalized degeneration of pulpal fibroblasts and odontoblasts.[113,114]

Immune system

Diabetic hamsters show abnormal wound healing: delayed infiltration of fibroblasts and polymorphonuclear leukocytes, delayed growth of new capillaries, and excessive edema.[115] Both initial total-antibody and later IgG-antibody responses of diabetic and prediabetic Chinese hamsters to sheep erythrocytes are subnormal and delayed.[116] Because even prediabetic hamsters respond abnormally, the immune defect seems to be primary and not secondary to hyperglycemia.[116] The genetic cause of this immune defect is not known. Theoretically, such an inborn defect could make the genetically diabetic hamster susceptible to otherwise mild environmental insults such as viruses. However, antibody titers to Coxsackie B and several other viruses are normal in diabetic hamsters.[18]

Autoimmunity is unlikely to cause diabetes in the Chinese hamster

Boquist[117] occasionally saw lymphocytic infiltration close to capillaries in islets of newly diagnosed diabetic hamsters. Gerritsen occasionally saw lymphocytic infiltration near islets in both normal and diabetic animals (personal communication). Malaisse et al.[70] saw interlobular and perivascular lymphocytic infiltration in the *exocrine tissues* in about one third of the normal and diabetic animals they studied. In the Asahikawa colony, insulitis was previously seen in nondiabetic and mildly diabetic hamsters, while the severe diabetics showed frequent vacuolation but no insulitis.[18] However, more recent studies have not shown insulitis during the prediabetic stage in that colony.[44]

The hamster colony at the Asahikawa Medical College, Japan[111] can be traced back to apparently normal animals from the Yerganian colony in 1969. Diabetes was first observed in the Asahikawa colony in 1979 (A Morikawa, personal communication). Because the Asahikawa and Umeå animals can be traced back to Yerganian, the colonies should be relatively similar now, except for any genetic drift that has taken place since they separated.

Insulitis in normal animals might represent a viral infection in the whole colony that is not severe enough to cause diabetes in the genetically nonsusceptible, normal animals. In the Asahikawa colony no differences in antibody titers to Coxsackie B and several other viruses were found in normal vs. diabetic hamsters.[18]

Prediabetic hamsters from the Upjohn colony were treated by us with the immunosuppressant cyclophosphamide but we were not able to significantly postpone the onset or reduce the severity of the hyperglycemia or glucosuria.[118] Antilymphocyte serum and neonatal thymectomy also did not prevent the diabetes.[119] Therefore, it seems unlikely that autoimmune destruction of the islets is the precipitating cause of diabetes in the Chinese hamster. However, overtaxed islet cells might fall apart and induce an immune reaction (insulitis).

What causes diabetes in the Chinese hamster?

A genetic cause. The cause of diabetes in the Chinese hamster has not been pinpointed. The disease is inherited, and the allele or alleles were probably present in non-inbred or wild animals. The allele(s) may have improved survival during periods of famine (thrifty genotype[120]). The same mutations that had a selective advantage in the hamster may also have been selected for in humans long ago. Despite the hypotheses of one,[3] four,[121] or two[56] recessive genes in the diabetic Chinese hamster, the number of mutations is unknown. Published studies sometimes give widely conflicting results, which can be explained on the basis of differences in experimental techniques, and possibly even genetic differences. An inbred subline can change if it has been separated from its sibling subline for more than just a few generations (*genetic drift*).[122] A single genetic defect in a membrane receptor, a cytoplasmic second messenger, or an essential building block that is present in many different tissues could cause the many phenotypically different effects, depending upon the tissue involved.

What is the earliest observable change? At 2 h after birth, excessive mitotic frequency and B-cell hypertrophy are seen in the pancreatic islets,[8] but this could be the result of maternal hyperglycemia.[8] At 15 days of age, when the young are still at least partly suckling, the animals show hyperphagia.[24,25] Could this be caused by a defect in some satiety factor as suggested in the *ob/ob* mouse,[123,124] or some defect in the hypothalamus-adenohypophysis area[23,95]? Excessive hepatic gluconeogenic activity is seen at 15 days of age.[22] Too few B-cells are undergoing mitosis at 17–20 days.[28] Glucose uptake into certain skeletal muscles is decreased in normoglycemic prediabetic hamsters at ~50 days of age.[43] However, suppression of insulin release does not occur until 2–4 months of age.[44,73] Note, however, that the ages at which hyperglycemia occurs and pancreatic insulin release starts to fail may vary between colonies. A period of hyperinsulinemia is sometimes seen soon after the onset of hyperglycemia. Could that represent transient B-cell overwork that later results in "exhaustion"? Alternatively, the B-cell failure may be due to the same hypothetical, inborn, intracellular error that affects many other tissues as well. Some symptoms, for example, retinopathy,[88,89,92] neuropathy,[20,97,98] and joint degeneration,[107,108] are most likely the result of hyperglycemia, while others, for example, abnormal gluconeogenesis, glucose uptake, or glycolysis,[17,22,43] may be more closely related to insulin resistance or to the inborn error(s).

Insulin resistance. The inappropriate overproduction of glucose in the presence of functioning insulin is due to hepatic insulin resistance. In the diabetic Chinese hamster, the insulin receptor[104-106] on hepatocytes functions abnormally. In the desert gerbil, *Psammomys obesus*, a high-calorie diet induces obesity, insulin resistance, and diabetes, and the defect is believed to lie at or near the hepatic insulin receptor.[125,126] Glucose oxidation by diabetic Chinese hamsters in vivo does not appear to be improved after 5 days of insulin treatment.[127] However, epididymal adipocytes appear to respond normally to one dose (250 μU/ml = 1500 pmol/l) of exogenous insulin in vitro.[31]

Several different tissues appear to contribute to the hyperglycemia in the diabetic Chinese hamster, and the primary defect(s) may lie at the cytosolic, mitochondrial, or cell surface sites and not at the insulin receptor. Once hyperglycemia is firmly established, hepatic overproduction of glucose seems to aggravate the diabetes.[43] Takahashi et al.[43] suggest that the decreased glucose uptake into adipose tissue might be the reason that these insulin-resistant animals do not become grossly obese. Thus, it is clear that the genetically diabetic Chinese hamster exhibits insulin resistance, but the nature of the primary defect remains to be determined.

Diet and diabetic symptoms. Some dietary factors, such as a high percentage of fat in the diet, saturated fat in the diet, and access to diet ad libitum aggravate the diabetes in the genetically diabetic Chinese hamster. The same diet does not affect genetically normal Chinese hamsters. The exact nature of the factors and how they exacerbate the diabetes are unclear. Some fat-related molecule, or fatty acids themselves, may increase the already excessive gluconeogenesis.

Calcium channels. Total B-cell calcium appears normal,[29] but glucose-induced $^{45}Ca^{2+}$ uptake,[37,40] $^{45}Ca^{2+}$ efflux,[39] and cytosolic Ca^{2+} levels are abnormal. Could the voltage-gated Ca^{2+} channels be insensitive to the ATP, NAD(P)H, or cAMP produced by the metabolism of glucose?

cAMP production. A sluggish cAMP-producing adenylate cyclase or an overly active cAMP-suppressing A1-adenosine receptor,[78,79] both plasma membrane-bound proteins, could explain the ability of theophylline[35,57,75] and IBMX[39,74] to increase B-cell cytosolic Ca^{2+}, islet cAMP, islet insulin release, and plasma insulin levels.[35] However, these methyl xanthines do not completely normalize the Ca^{2+}, cAMP and insulin levels. Therefore, an additional defect must still be identified in the B-cell. Identification of a specific enzymatic defect may lead to a better treatment than is now available for human (and hamster) diabetics. The same defect could be causing both the subnormal insulin release and the insulin resistance. The daily administration of the adenosine-potentiating drug, dilazep, reduces proteinuria but does *not* affect fasting plasma glucose or glycohemoglobin levels in young diabetic hamsters.[27] Perhaps a modification that lengthens the drug's half-life would be more effective.

Conclusions

Apparently normal, wild-type Chinese hamsters were initially inbred to create a homogeneous, inbred line of animals that naturally have a relatively small number of chromosomes. Glucosuria and polyuria appeared after a few generations of inbreeding. The "diabetic allele" or "alleles" had probably been present in the wild, and may have provided a selective advantage by helping the animals to survive periods of famine with increased gluconeogenesis, but diabetes has not been reported to occur in the wild.

Originally, diabetic Chinese hamsters were considered to be a model for non-obese, IDDM because some animals required daily insulin injections for survival. However, over the years the most severely diabetic animals have died out, and immunosuppression does not stop the development of the disease. The diabetic Chinese hamster is now considered to be an animal model for non-obese, NIDDM.

Chinese hamsters are touchy and produce fewer young than mice. The feistiness has not been lost despite attempts at selection for docile animals. Perhaps territoriality also has a selective advantage in nature.

Chinese hamsters suffer from both pancreatic and extrapancreatic defects. The islet defects include insufficient B-cell mitosis, reduced rate of insulin synthesis, subnormal insulin release, and excessive glucagon release in response to glucose. The extrapancreatic defects include excessive gluconeogenesis; decreased glycolysis; decreased glucose uptake in liver, skeletal muscle, kidney, and possibly adipose tissue; and abnormal function of hepatocyte insulin receptors. Dietary factors (saturated fats?) markedly exacerbate the diabetic symptoms, possibly by aggravating the already unbalanced glucose utilization.

Many complications plague the diabetic hamster: retinopathy, nephropathy, peripheral neuropathy, neurogenic bladder, intestinal motility defects, infertility, abnormal wound healing, abnormal immune responses, atherosclerosis, hypertriglyceridemia, hypercholesterolemia, and early joint degeneration. These sequelae are similar, but not always identical, to the diabetic sequelae of human diabetics. Many symptoms appear prior to the onset of hyperglycemia: excessive appetite, decreased islet and adrenal cell division, abnormal glucose metabolism in liver and skeletal muscle, increased triglyceride content in hepatocytes, and subnormal antibody production.

The chromosome, gene, and subcellular locations for the mutation or mutations that cause all of these defects are not known. Could all of the observed defects be the result of one single defective enzyme, cofactor, second messenger, or protein that is present in many different tissues? Several processes in the diabetic islet B-cell (pancreatic insulin release, cytosolic Ca^{2+} levels, plasma insulin) are improved by cAMP stimulation. Therefore, one defect may lie in cAMP production or in calcium channels in the B-cell.

The genetically diabetic Chinese hamster remains a useful model for continued investigation of the molecular defects (and possible treatments) that cause impaired insulin release, abnormal glucose metabolism, and diabetic sequelae.

Acknowledgments

I thank Ms. Eva Lundgren-Hallin, Gunilla Forsgren, Kristina Forsgren, and Berith Lundström for excellent technical assistance. Some previously unpublished data are reported here from collaborations with Margaret C. Blanks, George C. Gerritsen, Gerold M. Grodsky, Per Lindström, and Janove Sehlin. Some personal communications are reported here from George C. Gerritsen and Akizuki Morikawa. Parts of this work were supported in part by the Swedish Medical Research Council (project no. 12X-4756).

References

1. Yerganian G. The striped-back or Chinese hamster, *Cricetulus griseus*. J Natl Cancer Inst 20:705–21, 1958.
2. Hsieh ET. A new laboratory animal (*Cricetulus griseus*). Natl Med J China 5:20–4, 1919.
3. Yerganian G. Spontaneous diabetes mellitus in the Chinese hamster *Cricetulus griseus*. Current trends and projected views. In: On the nature and treatment of diabetes. Leibel BS, Wrenshall GA, eds. Excerpta Medica, Amsterdam, pp 612–26, 1965.
4. Yerganian G. The Chinese hamster (*Cricetulus griseus*). In: The UFAW handbook on the care and management of laboratory animals, 3rd ed. UFAW Staff, Lane-Petter W, Worden AN, Hill FB, et al., Williams & Wilkins, Baltimore, MD, pp 340–52, 1967.
5. Belčić VI. Zucht des Chinesischen Hamsters nach der Kragenmethode. Z Versuchstierk 12:90–2, 1970.
6. Calland CJ, Wightman SR, Neal SB. Establishment of a Chinese hamster breeding colony. Lab Animal Sci 36:183–5, 1986.
7. Yerganian G. Spontaneous diabetes mellitus in the Chinese hamster, *Cricetulus griseus*. IV. Genetic aspects. In: Aetiology of diabetes mellitus and its complications. Ciba Foundation Colloquia on Endocrinology, Vol. 15. Cameron MP, O'Connor M, eds. Churchill, London, pp 25–41, 1964.
8. Meier H, Yerganian G. Spontaneous diabetes mellitus in the Chinese hamster (*Cricetulus griseus*). II. Findings in the offspring of diabetic parents. Diabetes 10:12–8, 1961.
9. Yerganian G, Meier H. Spontaneous hereditary diabetes mellitus in the Chinese hamster (*Cricetulus griseus*). A preliminary report on clinico-pathological findings, genetic aspects, breeding and responses to hypoglycemic drugs. Fed Proc 18:514, 1959.
10. Meier H, Yerganian GA. Spontaneous hereditary diabetes mellitus in Chinese hamster (*Cricetulus griseus*). 1. Pathological findings. Proc Soc Exp Biol Med 100:810–5, 1959.
11. Gerritsen GC, Needham LB, Schmidt FL, Dulin WE. Studies on the prediction and development of diabetes in offspring of diabetic Chinese hamsters. Diabetologia 6:158–62, 1970.
12. Meier H. Diabetes mellitus in animals. A review. Diabetes 9:485–9, 1960.
13. Sirek OV, Sirek A. The colony of Chinese hamsters of the C.H. Best Institute. A review of experimental work. Diabetologia 3:65–73, 1967.
14. Gerritsen GC, Blanks MC, Schmidt FL, Dulin WE. Environmental influences on the manifestation of diabetes mellitus in Chinese hamsters. In: The genetics of diabetes mellitus. Creutzfeldt W, Köbberling J, Neel JV, eds. Springer-Verlag, Berlin, pp 165–87, 1976.
15. Grodsky GM, Frankel BJ. Diabetes mellitus in the Chinese hamster. In: Etiology and pathogenesis of insulin-dependent diabetes mellitus. Martin JM, Ehrlich RM, Holland FJ, eds. Raven Press, New York, pp 239–249, 1981.
16. Gerritsen GC. The Chinese hamster as a model for the study of diabetes mellitus. Diabetes 31(Suppl 1):14–21, 1982.
17. Dulin WE, Gerritsen GC, Chang AY, et al. Characteristics of the prediabetic state in the Chinese hamster and evaluation of environmental factors that affect expression of the diabetic genes. In: Lessons from animal diabetes. Shafrir E, Renold AE, eds. J Libbey, London, pp 199–209, 1984.
18. Watanabe K, Morikawa A, Etoh M, et al. Characterization of diabetic Chinese hamsters in the Asahikawa colony. In: Lessons from animal diabetes. Shafrir E, Renold AE, eds. J Libbey, London, pp 99–105, 1984.
19. Eto M, Watanabe K, Sekiguchi M, et al. Metabolic and morphological changes of

the heart in Chinese hamsters (CHAD strain) with spontaneous long-term diabetes. Diabetes Res Clin Pract 3:297–305, 1987.
20. Diani AR, Peterson T, Sawada GA, et al. Elevated levels of vasoactive intestinal peptide in the eye and urinary bladder of diabetic and prediabetic Chinese hamsters. Diabetologia 28:302–7, 1985.
21. Dulin WE, Chang AY, Gerritsen GC. Comparison of diabetes in the Chinese hamster, KK mouse and db mouse. In: Diabetes. Proc 7th IDF Congress, Buenos Aires, 1970. Rodriguez RR, Vallance-Owen J, eds. Excerpta Medica, Amsterdam, pp 868–80, 1971.
22. Chang AY, Schneider DI. Metabolic abnormalities in the pancreatic islets and livers of the diabetic Chinese hamster. Diabetologia 6:180–5, 1970.
23. Bestetti G, Rossi GL. Hypothalamic changes in diabetic Chinese hamsters. A semiquantitative, light and electron microscopic study. Lab Invest 47:516–22, 1982.
24. Gerritsen GC, Blanks MC. Preliminary studies on food and water consumption of prediabetic Chinese hamsters. Diabetologia 6:177–9, 1970.
25. Gerritsen GC, Blanks MC, Miller RL, Dulin WE. Effect of diet limitation on the development of diabetes in prediabetic Chinese hamsters. Diabetologia 10:559–65, 1974.
26. Hunt CE, Lindsey JR, Walkley SU. Animal models of diabetes and obesity, including the PBB/Ld mouse. Fed Proc 35:1206–17, 1976.
27. Morikawa A, Eto M, Makino I. The effect of dilazep on urinary protein excretion in spontaneous diabetic Chinese hamster. Nephron 65:125–32, 1993.
28. Frankel BJ, Korsgren O, Andersson A. DNA replication in pancreatic islets and adrenals in diabetic Chinese hamsters. Horm Res 39:67–72, 1993.
29. Juntti-Berggren L, Lindh U, Berggren P-O, Frankel BJ. Proton microprobe analysis of 15 elements in pancreatic B cells and exocrine pancreas in diabetic Chinese hamsters. Biosci Rep 7:33–41, 1987.
30. Juntti-Berggren L, Lindh U, Berggren P-O, Frankel BJ. Elemental composition in the pancreatic B cell is normal in the prediabetic Chinese hamster. Pancreas 7:11–14, 1992.
31. Gerritsen GC, Dulin WE. Characterization of diabetes in the Chinese hamster. Diabetologia 3:74–84, 1967.
32. Ehrentheil OF, Reyna LJ, Yerganian G, Chen ET. Studies in stress glycosuria. Prolonged glycosuria in Chinese hamsters after repeated stress. Diabetes 13:83–6, 1964.
33. Iwashima Y, Watanabe K, Morikawa I. Changes in the pancreatic A-, B- and D-cell populations during development of diabetes in spontaneously diabetic Chinese hamsters of the Asahikawa colony (CHAD). Diabetes Res Clin Pract 8:201–14, 1990.
34. Frankel BJ, Heldt AM, Grodsky GM. Insulin and glucagon release in the diabetic Chinese hamster: Differences among inbred sublines. Diabetologia 22:292–5, 1982.
35. Loge O, Losert W, Jahn P. Insulinfreisetzende Wirkung von Glymidin-natrium (Glykodiazin), Isopropylnoradrenalin, Glucagon und Theophyllin an chinesischen Streifenhamstern mit spontanem oder streptozotocininduziertem Diabetes Mellitus. Res Exp Med 160:292–306, 1973.
36. Eto M, Watanabe K, Iwashima Y, et al. Plasma and tissue lipid peroxides in spontaneously diabetic Chinese hamsters. J Clin Biochem Nutr 8:33–9, 1990.
37. Frankel BJ, Sehlin J. Abnormalities in glucose-stimulated insulin release, ^{45}Ca uptake, and ^{86}Rb efflux in diabetic Chinese hamster islets. Diabetes 36:648–53, 1987.
38. Iwashima Y, Watanabe K, Eto M, et al. Insulin and glucagon response of the diabetic Chinese hamster in the Asahikawa colony. Diabetes Res Clin Pract 1:87–94, 1985.
39. Siegel EG, Wollheim CB, Sharp GWG, et al. Defective calcium handling and insulin release in islets from diabetic Chinese hamsters. Biochem J 180:233–6, 1979.
40. Frankel BJ, Sehlin J. Effect of perchlorate on glucose-stimulated insulin release

and $^{45}Ca^{2+}$ uptake in pancreatic islets from diabetic Chinese hamsters. Pancreas 9:550–7, 1994.
41. Frankel BJ, Cajander S, Boquist L. Islet morphology in young, genetically diabetic Chinese hamsters during the hyperinsulinemic phase. Pancreas 2:625–31, 1987.
42. Gerritsen G, Blanks MC. Effect of diet limitation on development of diabetes in nonhyperphagic prediabetic Chinese hamsters. Adv Exp Med Biol 119:237–42, 1978.
43. Takahashi M, Eto M, Makino I. Peripheral insulin resistance precedes the onset of hyperglycemia in spontaneously diabetic Chinese hamsters of Asahikawa colony. Diabetes Res Clin Pract 20:101–9, 1993.
44. Nakajima K, Morikawa A, Makino I. Natural history of B-cell dysfunction in spontaneously diabetic Chinese hamsters. Diabetes Res Clin Pract 24:131–42, 1994.
45. Grodsky GM, Frankel BJ, Gerich JE, Gerritsen GC. The diabetic Chinese hamster: In vitro insulin and glucagon release; the "chemical diabetic"; and the effect of diet on ketonuria. Diabetologia 10:521–8, 1974.
46. Gerritsen GC, Blanks MC. Characterization of Chinese hamsters by metabolic balance, glucose tolerance and insulin secretion. Diabetologia 10:493–9, 1974.
47. Chang AY, Noble RE, Wyse BM. Comparison of highly inbred diabetic and nondiabetic lines in the Upjohn colony of Chinese hamsters. Diabetes 26:1063–71, 1977.
48. Green MN, Yerganian G, Meier H. Elevated α-2-serum proteins as a possible genetic marker in spontaneous hereditary diabetes mellitus of the Chinese hamster (*Cricetulus griseus*). Experientia 16:503–4, 1960.
49. Green MN, Yerganian G, Gagnon HJ. Prediction of spontaneous hereditary diabetes mellitus in Chinese hamsters by means of elevated α-2 serum-levels. Nature 197:396, 1963.
50. Gerritsen GC, Dulin WE. Serum proteins of Chinese hamsters and response of diabetics to tolbutamide and insulin. Diabetes 15:331–5, 1966.
51. Yerganian G. Discussion. In: Aetiology of diabetes mellitus and its complications. Ciba Foundation Colloquia on Endocrinology, Vol. 15. Cameron MP, O'Connor M, eds. Churchill, London, pp 41–8, 1964.
52. Meier H, Yerganian G. Spontaneous diabetes mellitus in the Chinese hamster (*Cricetulus griseus*). III. Maintenance of a diabetic hamster colony with the aid of hypoglycemic therapy. Diabetes 10:19–21, 1961.
53. Frankel BJ, Schmid FG, Grodsky GM. Effect of continuous insulin infusion with an implantable seven-day minipump in the diabetic Chinese hamster. Endocrinology 104:1532–39, 1979.
54. Frankel BJ, Grodsky GM. Effect of continuous low-dose insulin treatment on subsequent incidence of diabetes in genetically prediabetic Chinese hamsters. Diabetes 28:544–7, 1979.
55. Chang AY, Wyse BM, Gilchrist BJ, et al. Ciglitazone, a new hypoglycemic agent. I. Studies in ob/ob and db/db mice, diabetic Chinese hamsters, and normal and streptozotocin-diabetic rats. Diabetes 32:830–8, 1983.
56. Gerritsen GC, Dulin WE, Connell MA, et al. Epidemiology and genetics of the spontaneously diabetic Chinese hamster. In: Lessons from animal diabetes. Shafrir E, Renold AE, eds. J Libbey, London, pp 81–92, 1984.
57. Frankel BJ, Gerich JE, Hagura R, et al. Abnormal secretion of insulin and glucagon by the in vitro perfused pancreas of the genetically diabetic Chinese hamster. J Clin Invest 53:1637–46, 1974.
58. Petersson B, Elde R, Efendić S, et al. Somatostatin in the pancreas, stomach and hypothalamus of the diabetic Chinese hamster. Diabetologia 13:463–6, 1977.
59. Like AA, Gerritsen GC, Dulin WE, Gaudreau P. Studies in the diabetic Chinese hamster: Electron microscopy of pancreatic islets. Diabetologia 10:509–20, 1974.
60. Iwashima Y, Watanabe K, Eto M, et al. Paradoxical glucagon response in perifused islets of the diabetic Chinese hamster. Diabetes Res Clin Pract 1:273–80, 1986.

61. Boquist L. Morphology of the pancreatic islets of the nondiabetic adult Chinese hamster, *Cricetulus griseus*. Light microscopical findings. Acta Soc Med Upsal 72:331–44, 1967.
62. Carpenter AM, Gerritsen GC, Dulin WE, Lazarow A. Islet and beta cell volumes in diabetic Chinese hamsters and their non-diabetic siblings. Diabetologia 3:92–6, 1967.
63. Carpenter AM, Gerritsen GC, Dulin WE, Lazarow A. Islet and beta cell volumes in offspring of severely diabetic (ketotic) Chinese hamsters. Diabetologia 6:168–76, 1970.
64. Boquist L. Pancreatic islet morphology in diabetic Chinese hamsters. A light and electron microscopic study. Acta Pathol Microbiol Scand 75:399–414, 1969.
65. Like AA, Gerritsen GC, Dulin WE, Gaudreau P. Studies in the diabetic Chinese hamster: Light microscopy and autoradiography of pancreatic islets. Diabetologia 10:501–8, 1974.
66. Boquist L. Morphology of the pancreatic islets of the nondiabetic adult Chinese hamster, *Cricetulus griseus*. Ultrastructural findings. Acta Soc Med Upsal 72:345–57, 1967.
67. Luse SA, Caramia F, Gerritsen G, Dulin WE. Spontaneous diabetes mellitus in the Chinese hamster: An electron microscopic study of the islets of Langerhans. Diabetologia 3:97–108, 1967.
68. Orci L, Stauffacher W, Dulin WE, et al. Ultrastructural changes in A-cells exposed to diabetic hyperglycaemia. Observations made on pancreas of Chinese hamsters. Diabetologia 6:199–206, 1970.
69. Soret MG, Dulin WE, Mathews J, Gerritsen GC. Morphologic abnormalities observed in retina, pancreas and kidney of diabetic Chinese hamsters. Diabetologia 10:567–79, 1974.
70. Malaisse W, Malaisse-Lagae F, Gerritsen GC, et al. Insulin secretion in vitro by the pancreas of the Chinese hamster. Diabetologia 3:109–14, 1967.
71. Williamson JR. Electron microscopy of glycogenic changes in beta cells in experimental diabetes. Diabetes 9:471–80, 1960.
72. Chang AY. Insulin synthesis and secretion by isolated islets of spontaneously diabetic Chinese hamsters. In: The structure and metabolism of the pancreatic islets. Wenner-Gren Symposium no. 16. Falkmer S, Hellman B, Täljedal IB, eds. Pergamon Press, Oxford, pp 515–26, 1970.
73. Frankel BJ, Heldt AM, Gerritsen GC, Grodsky GM. Insulin, glucagon, and somatostatin release from the prediabetic Chinese hamster. Diabetologia 27:387–91, 1984.
74. Rabinovitch A, Renold AE, Cerasi E. Decreased cyclic AMP and insulin responses to glucose in pancreatic islets of diabetic Chinese hamsters. Diabetologia 12:581–7, 1976.
75. Frankel BJ, Gerich JE, Fanska RE, et al. Responses to arginine of the perfused pancreas of the genetically diabetic Chinese hamster. Diabetes 24:272–9, 1975.
76. Gerhards E, Rühl M. Insulin secretion of the isolated perfused pancreas of the Chinese hamster (*Cricetulus griseus*). Acta Endocrinol 76:302–18, 1974.
77. Pralong WF, Bartley C, Wollheim CB. Single islet β-cell stimulation by nutrients: Relationship between pyridine nucleotides, cytosolic Ca^{2+} and secretion. EMBO J 9:53–60, 1990.
78. Libert F, Passage E, Parmenter M, et al. Chromosomal mapping of A1 and A2 adenosine receptors, VIP receptor, and a new subtype of serotonin receptor. Genomics 11:225–7, 1991.
79. LaNoue KF, Martin LF. Abnormal adenosine receptor function in genetic obesity. FASEB J 8:72–80, 1994.
80. Lawe JE. Renal changes in hamster with hereditary diabetes mellitus. Arch Pathol 73:166–174, 1962.
81. Feldman JM, Gerritsen GC. Tissue norepinephrine concentration in spontaneously diabetic Chinese hamsters. Horm Res 29:168–75, 1988.

82. Chang AY, Greenberg HS. Acid glycohydrolase in Chinese hamster with spontaneous diabetes. III. Line-dependent variance. Biochim Biophys Acta 525:134–41, 1978.
83. Sekiguchi M, Watanabe K, Eto M, et al. Polyol pathway in tissues of spontaneously diabetic Chinese hamsters (*Cricetulus griseus*) and the effect of an aldose reductase inhibitor, ONO-2235. Comp Biochem Physiol 98B:637–40, 1991.
84. Shirai T, Welsh GW 3rd, Sims EAH. Diabetes mellitus in the Chinese hamster. II. The evolution of renal glomerulopathy. Diabetologia 3:266–86, 1967.
85. Chang AY, Schneider DI. Rate of gluconeogenesis and levels of gluconeogenic enzymes in liver and kidney of diabetic and normal Chinese hamsters. Biochim Biophys Acta 222:587–92, 1970.
86. Itoh H, Morikawa A, Makino I. Urinary bladder dysfunction in spontaneously diabetic Chinese hamsters. Diabetes Res Clin Pract 22:163–70, 1994.
87. Dail WG, Evan AE, Gerritsen GC, Dulin WE. Abnormalities in pelvic visceral nerves: A basis for neurogenic bladder in the diabetic Chinese hamster. Invest Urol 15:161–6, 1977.
88. Holcomb GN, Klemm LA, Dulin WE. The polyol pathway for glucose metabolism in tissues from normal, diabetic, and ketotic Chinese hamsters. Diabetologia 10:549–53, 1974.
89. Federman JL, Gerritsen GC. The retinal vasculature of the Chinese hamster: A preliminary study. Diabetologia 6:186–91, 1970.
90. Ghosh M, Hausler HR, Basu PK, Stachowska B. An electron microscopic study of retinal capillaries of normal and spontaneously diabetic Chinese hamsters. Can J Ophthalmol 5:187–93, 1970.
91. Rossi GL, Bestetti GE, Kunz E. The retinal capillary microangiopathy of spontaneously diabetic Chinese hamsters is evident in all vascular wall components. Exp Clin Endocrinol (Life Sci Adv) 6:303–8, 1987.
92. Soret MG, Dulin WE, Gerritsen GC. Microangiopathy in animals with spontaneous diabetes. In: Early diabetes. Advances in metabolic disorders. Suppl. 2, Vascular and neurological changes in early diabetes. Camerini-Dávalos RA, Cole HS, eds. Academic Press, New York, pp 291–8, 1973.
93. Garris DR, Smith C, Davis D, et al. Morphometric evaluation of the hypothalamic-ovarian axis of the ketonuric, diabetic Chinese hamster: Relationship to the reproductive cycle. Diabetologia 23:275–9, 1982.
94. Luse SA, Gerritsen GC, Dulin WE. Cerebral abnormalities in diabetes mellitus: An ultrastructural study of the brain in early onset diabetes mellitus in the Chinese hamster. Diabetologia 6:192–8, 1970.
95. Deslex P, Rossi GL. Quantitative evaluation of somatotrophic cells (SC) in the adenohypophysis of normal and diabetic male Chinese hamsters. Diabetologia 12:489–93, 1976.
96. Schlaepfer WW, Gerritsen GC, Dulin WE. Segmental demyelination in the distal peripheral nerves of chronically diabetic Chinese hamsters. Diabetologia 10:541–8, 1974.
97. Kennedy WR, Quick DC, Miyoshi T, Gerritsen GC. Peripheral neurology of the Chinese hamster. Diabetologia 23:445–51, 1982.
98. Brown MR, Dyck PJ, McClearn GE, et al. Central and peripheral nervous system complications. Diabetes 31(Suppl 1):65–70, 1982.
99. Daly PA, Landsberg L. Hypertension in obesity and NIDDM. Role of insulin and sympathetic nervous system. Diabetes Care 14:240–8, 1991.
100. Feingold KR, Lear SR, Moser AH. De novo cholesterol synthesis in three different animal models of diabetes. Diabetologia 26:234–9, 1984.
101. Sims EAH, Landau BR. Diabetes mellitus in the Chinese hamster. I. Metabolic and morphologic studies. Diabetologia 3:115–23, 1967.
102. Sammons DW, Adams LD, Tollaksen SL, Anderson NG. A genetic marker in spontaneously diabetic Chinese hamsters. Diabetes 29(Suppl 2):37A, 1980.

103. Wapnir RA, Stiel L. Regulation of hepatic gluconeogenesis and glycogenolysis by phosphorylated glycerol and glycolytic intermediates in diabetic and control Chinese hamsters. Biochem Med Metab Biol 37:228–34, 1987.
104. Oshima E, Watanabe K, Ishii K. Insulin receptors on hepatocytes from spontaneously diabetic Chinese hamsters. Endocrinol Jpn 35:1–9, 1988.
105. Hepp KD, Langley J, von Funcke HJ, et al. Increased insulin binding capacity of liver membranes from diabetic Chinese hamsters. Nature 258:154, 1975.
106. Oshima E, Watanabe K, Makino I. Decreased autophosphorylation and kinase activity of insulin receptors in diabetic Chinese hamsters. Diabetes Res Clin Prac 10:35–43, 1990.
107. Silberberg R, Gerritsen G. Aging changes in intervertebral discs and spondylosis in Chinese hamsters. Diabetes 25:477–83, 1976.
108. Silberberg R, Gerritsen G, Hasler M. Articular cartilage of diabetic Chinese hamsters. Arch Pathol Lab Med 100:50–4, 1976.
109. Chin BC, Track NS. Diabetic Chinese hamsters. Metabolic effects of long term guar gum consumption. Diabetes Res Clin Pract 1:197–202, 1985.
110. Diani AR, Grogan DM, Yates ME, et al. Radiologic abnormalities and autonomic neuropathology in the digestive tract of the ketonuric diabetic Chinese hamster. Diabetologia 17:33–40, 1979.
111. Funaki K, Mikamo K. Developmental-stage-dependent teratogenic effects of maternal spontaneous diabetes in the Chinese hamster. Diabetes 32:637–43, 1983.
112. Schöffling K, Federlin K, Schmitt W, Pfeiffer EF. Histometric investigations on the testicular tissue of rats with alloxan diabetes and Chinese hamsters with spontaneous diabetes. Acta Endocrinol 54:335–46, 1967.
113. Shklar G, Cohen MM, Yerganian G. A histopathologic study of periodontal disease in the Chinese hamster with hereditary diabetes. J Periodontol 33:14–21, 1962.
114. Cohen MM, Shklar G, Yerganian G. Pulpal and periodontal disease in a strain of Chinese hamsters with hereditary diabetes mellitus. Oral Surg Oral Med Oral Pathol 16:104–12, 1963.
115. Weringer EJ, Arquilla ER. Wound healing in normal and diabetic Chinese hamsters. Diabetologia 21:394–401, 1981.
116. Fletcher-McGruder BL, Gerritsen GC. Deficient humoral antibody response of the spontaneously diabetic Chinese hamster. Proc Soc Exp Biol Med 175:74–8, 1984.
117. Boquist L. Histochemistry and electron microscopy of islets. In: The diabetic pancreas. Volk BW, Wellman KF, eds. Plenum Press, New York, pp 129–69, 1977.
118. Gishizky ML, Frankel BJ, Grodsky GM. Cyclophosphamide treatment of prediabetic Chinese hamsters. Acta Physiol Scand 121:81–4, 1984.
119. Fletcher-McGruder BL, Gerritsen GC. Effect of antilymphocyte serum and neonatal thymectomy upon onset of diabetes in the Chinese hamster. Proc Soc Exp Biol Med 180:92–7, 1985.
120. Neel JV. Diabetes mellitus: A "thrifty" genotype rendered detrimental by "progress"? Am J Hum Genet 14:353–62, 1962.
121. Butler L, Gerritsen GC. A comparison of the modes of inheritance of diabetes in the Chinese hamster and the KK mouse. Diabetologia 6:163–7, 1970.
122. Falconer DS. Genetic aspects of breeding methods. In: The UFAW handbook on the care and management of laboratory animals, 3rd ed. UFAW staff, Lane-Petter W, Worden AN, Berton Hill, et al., eds. Williams & Wilkins, Baltimore, MD, pp 72–96, 1966.
123. Zhang Y, Proenca R, Maffei M, et al. Positional cloning of the mouse *obese* gene and its human homologue. Nature 372:425–32, 1994.
124. Marx J. Obesity gene discovery may help solve weighty problem. Science 266:1477–78, 1994.
125. Kanety H, Moshe S, Shafrir E, et al. Hyperinsulinemia induces a reversible impairment in insulin receptor function leading to diabetes in the sand rat model of

non-insulin-dependent diabetes mellitus. Proc Natl Acad Sci USA 91:1853–57, 1994.
126. Shafrir E, Gutman A. *Psammomys obesus* of the Jerusalem colony: A model for nutritionally induced, non-insulin-dependent diabetes. J Basic Clin Physiol Pharmacol 4:83–99, 1993.
127. Gundersen K, Yerganian G, Lin BJ, et al. Diabetes in the Chinese hamster. Some clinical and metabolic aspects. Diabetologia 3:85–91, 1967.

17

Studies on the pathogenesis of NIDDM in the GK (Goto-Kakizaki) rat

CLAES-GÖRAN ÖSTENSON,[1] SAMY M ABDEL-HALIM,[1] ARNE ANDERSSON,[2] AND SUAD EFENDIC[1]

Department of [1]Molecular Medicine, Endocrine and Diabetes Unit, Karolinska Institute and Hospital, Stockholm; [2]Department of Medical Cell Biology, Uppsala University, Uppsala, Sweden.

Key words: non-insulin-dependent diabetes mellitus, inheritance, insulin secretion, insulin resistance, glucose metabolism, stimulus-secretion coupling, glucotoxicity.

NIDDM has a hereditary background and mainly develops on the basis of defective insulin secretion and insulin resistance in liver and extrahepatic tissues.[1,2] It is still a matter of debate as to which of these defects is primary. However, impaired insulin secretion seems to be a prerequisite for the development of NIDDM.[3,4] When the B-cells fail to secrete enough insulin to compensate for insulin resistance, hyperglycemia develops. In contrast, subjects with intact B-cell secretory function can handle major insulin resistance and maintain normoglycemia by increasing their insulin secretion.

The cellular mechanisms underlying the impaired insulin secretion in NIDDM have not been clarified. Studies in human pancreatic islets are hampered by ethical considerations and practical reasons. However, animal models of diabetes have offered great possibilities for pathophysiological studies at the B-cell level. Several rodent models of spontaneous NIDDM have been described, in most of which obesity is coexisting with hyperinsulinemia and insulin resistance.[5] The GK (Goto-Kakizaki) rat is a normal weight rodent with the NIDDM-like syndrome that became available in the 1970s,[6] but detailed studies on islet function in this animal were performed only during the past 5 years.

Correspondence: Dr. Claes-Göran Östenson, Department of Molecular Medicine, The Endocrine and Diabetes Unit, Karolinska Hospital, S-171 76 Stockholm, Sweden.

Lessons from Animal Diabetes VI. Ed. E. Shafrir.

Glucose intolerance in GK rats

The GK rat strain was established from normoglycemic Wistar rats by repeated breeding, in each successive generation, of animals with the highest blood glucose levels during an oral glucose tolerance test (GTT).[6,7] Inbreeding, that is, brother–sister mating, was systematically used from the ninth generation.[8] Glucose intolerance was evident in all rats from the sixth generation in which the glucose-induced insulin response, as tested in the perfused pancreas, was significantly reduced.[9] After 40–50 generations, GK rats demonstrated a worsened GTT compared with the early generations[10] however, they were still only mildly to moderately diabetic with fasting blood glucose levels of 7–10 mmol/l and without ketosis.[8,10-13] Decreased glucose tolerance develops early in GK rats, since significant hyperglycemia was present already in 1-wk-old pups compared with Wistar rats with the same age and body wt.[13]

Genetic considerations

The hereditary nature of diabetes in the GK rat is obvious from the breeding procedure used to produce the strain.[6-8] It has been shown that F1 hybrid rats of GK and Wistar rats were intermediate between parents with regard to both GTT and glucose-induced insulin responses.[13,14] We furthermore found that F1 hybrids with a GK father were as glucose intolerant as those with a GK mother, suggesting that maternal hyperglycemia played no or a minimal role in the glucose tolerance of the offspring.[13] Interestingly, F1 hybrid rats of a diabetic mother had a better insulin response to glucose in the perfused pancreas, a finding supported by another study by Gauguier et al.[15] on F1 hybrids of GK/Wistar rats in which insulin responses to glucose in vivo were measured. In the latter study, however, glucose tolerance was worse, and in parity to that in GK rats, in F1 hybrids with maternal compared with F1 rats with paternal inheritance. The reason for the different results of maternal impact on glucose tolerance in offspring between the two studies is not obvious. However, differences in methodological procedures may be important in this context; for instance, we did use an i.p. GTT and transferred all studied litters to lactating, healthy foster mothers, whereas the French group injected glucose i.v. and kept the litters with the proper mother, GK or Wistar, during the lactation period. In this context, it is also of interest that no deletion in the mitochondrial DNA, which may be implicated in maternal inheritance, was detected in GK rats.[15]

Repeated backcrossing of F1 rats with healthy Wistar rats successively improved glucose tolerance.[13] This result suggests that several additive genes are responsible for glucose intolerance in the GK rat and is in agreement with the original observation made by Goto et al.[6] and also with the general proposition of a polygenic mode of inheritance in most cases of human NIDDM. As is the case in humans, it is still unclear which genes are causing diabetes in the GK rat. However, recent studies employing linkage analysis have elucidated this issue.[16,17]

Pancreatic hormone release

Glucose-stimulated insulin secretion is markedly impaired in the GK rat. This defect was present both in vivo[10,15,17,18] and in the isolated perfused pancreas,[9-13] as well as in isolated pancreatic islets.[11,19-21] Also, the insulin response to glyceraldehyde, a glycolytic intermediate, is markedly decreased in GK rat islets.[21,22] In this context it is of interest that the insulin response to glucose was somewhat better preserved when tested in batch-incubated islets than in the intact perfused pancreas or in vivo. Similar results have been obtained previously when comparing insulin responses to glucose in isolated islets and perfused pancreas of neonatally streptozotocin (STZ)-injected rats, another animal model of NIDDM.[23] Although this discrepancy between different test systems suggests that either a suppressive factor of insulin release, for example, somatostatin or catecholamines,[24,25] is overacting in the intact pancreas, or a B-cell stimulatory agent, such as glucagon,[23,26] is accumulating during batch incubations, the importance of such a factor has so far not been demonstrated.[25,26]

In the isolated, perfused pancreas, the insulin response to glucose was minute in female and entirely absent in male GK rats (Fig. 1).[10-13] In fact, an enhanced basal release of insulin at 3.3 mmol/l glucose tended to be suppressed when the glucose level of the perfusate was increased to 16.7 mmol/l.[11,12] The reason for this tendency, as well as for an "off response," that is, an increased insulin release when the perfusate glucose was changed back to a low 3.3 mmol/l level, is not known but may be related to a defective stimulus-secretion coupling for glucose in the GK rat B-cell.[11,12]

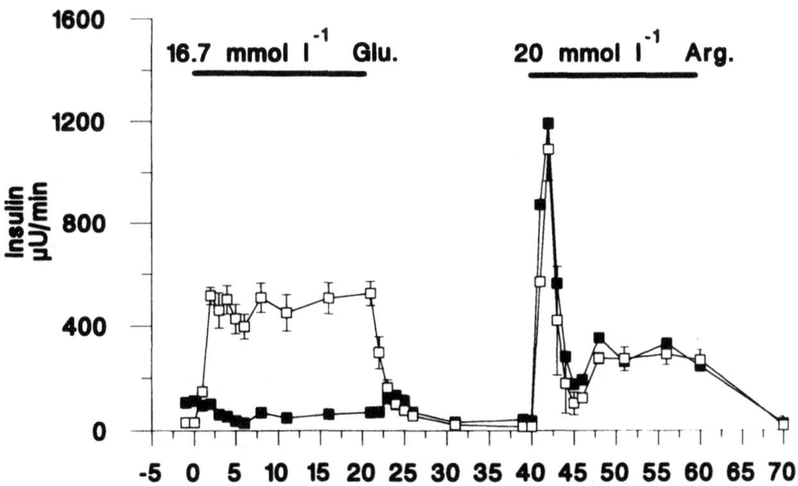

Figure 1. Insulin responses to 16.7 mmol/l glucose (Glu.) and 20 mmol/l arginine (Arg.) in the perfused pancreas of male nondiabetic Wistar rats (□, n = 14) and GK (■, n = 8) rats. The perfusion medium contained 3.3 mmol/l glucose during the arginine stimulation and when not otherwise indicated. (Reprinted with permission from ref. 12.)

It is still a matter of debate whether impaired glucose-induced insulin release is a primary feature of the GK rat, or is present only in older animals. Recently, Hughes and coworkers[21] reported normal insulin responses to various glucose concentrations in isolated islets of 8-wk-old hyperglycemic GK rats of the London colony, whereas the responses were impaired by at least 50% in islets from 14-wk-old animals as compared with islets from age-matched Wistar rats.[21] These data were interpreted to indicate that the defect in the glucose stimulus-secretion coupling in the B-cells is not a primary cause of diabetes in the GK rat, and that the deterioration of insulin response is rather secondary to some factor associated with the diabetic state. In the Stockholm colony, however, we demonstrated impairment of glucose-stimulated insulin release in the perfused pancreas as well as in batch-incubated islets, from 2- to 3-month-old GK rats.[11-13] In support of an early appearance of an insulin secretory defect, we also found almost abolished insulin response in vivo in 4-wk-old GK rats after an i.v. glucose challenge (Lundqvist, Östenson, unpublished observations). Furthermore, glucose-induced insulin release from batch-incubated islets of 1-wk-old F1 hybrid rats of GK and Wistar parents was impaired compared with the release from islets of age-matched Wistar rats.[27] It is pertinent that Serradas and coworkers,[28] studying the Paris GK rat colony, found decreased insulin responses both at low and high glucose concentrations in fetal GK rat islets. Since the fold increase in hormonal response between high and low glucose was similar in fetal GK and control islets, these authors, however, concluded that there was a normal responsiveness to glucose in fetal islets of GK rats. Although the issue of whether an insulin secretory defect is present early in the GK rat is not fully clarified, a number of observations related earlier suggest the presence of a primary defect at the level of the B-cell.

Among nonglucose insulinogogues, arginine-induced insulin release from the GK rat pancreas has been shown to be normal or even exaggerated as compared with control release (see Fig. 1).[9,10,12,13,21] Since preperfusion for 50–90 min with a buffer containing no glucose markedly diminished the insulin response to arginine in GK, but not control pancreata,[10] it is possible that previous exposure to glucose, in vivo and/or during the perfusion (see Fig. 1), potentiates the secretion of insulin during subsequent arginine stimulation. Hence, arginine-induced insulin release may be impaired in GK rats as in NIDDM patients, as was demonstrated after normalization of glycemia.[29,30] Further studies are needed to elucidate the potentiating role of glucose on the insulin response to arginine and other insulinogogues.

Substrates that are oxidized in the Krebs cycle, without preceding metabolism via glycolysis, such as α-ketoisocaproate (KIC) and octanoate, have been shown to elicit normal insulin responses in both GK and F1 hybrid rats.[21,31,32] However, the potentiating effect of glucose on KIC- and octanoate-induced insulin release was impaired in F1 rat islets.[31] At variance with the data obtained with KIC, the amino acid leucin stimulated insulin secretion less from isolated GK rat than from control rat islets,[19] despite the fact that leucin is metabolized to KIC and then to acetyl-CoA in the Krebs cycle. Inhibitors of phosphodiesterase, isobutyl-methylxanthin (IBMX) and theophylline, are known to enhance cAMP levels in B-cells. In studies on isolated islets from GK and F1 hybrid rats, these compounds stimulated insulin secretion to the same extent as in control islets.[19,31]

As to glucagon, arginine was shown to stimulate release of this hormone normally in the perfused pancreas of GK rats[9] and F1 hybrids of GK/Wistar rats.[13] In contrast, the somatostatin response to arginine in these organ systems was exaggerated, while glucose-stimulated somatostatin release was minute.[12,13] A similar pattern of D-cell responsiveness has previously been described in the perfused pancreas of the neonatally STZ-injected rat.[33]

Islet morphology and hormone production in GK rats

Since the insulin response, at least to glucose, is impaired in the GK rat, it is of potential interest to consider the properties of the "insulin factory" in detail. Thus, questions arise whether altered islet morphology pertaining to B- and/or non–B-cells, reduced replicatory activity of B-cells, or defective proinsulin/insulin biosynthesis may contribute to the poor secretory performance of B-cells in GK rats.

Early studies of islet morphology in GK rats revealed that, from the age of 3 months, an increasing proportion of islets become irregular in shape, so-called starfish-like islets,[7] with fibrous tissue intermingled among the endocrine cells. No signs of insulitis have been detected throughout the life of GK rats, but in older animals degranulation and decrease in number of B-cells were noted.[8] We recently studied islet morphology in 2- to 3-month-old GK rats and accordingly found only occasional starfish-like islets.[34] The reason for this change in islet shape is unknown. Since the alterations in islet structure, including fibrosis, seem to develop only after several months, it is likely that these changes are secondary to hyperglycemia or some other factor pertaining to the diabetic state. In these 2- to 3-month-old GK rats, we and others, furthermore, found a normal pancreatic endocrine cell and islet B-cell density,[34-36] as well as a normal pancreatic insulin content.[12] In contrast to these findings, however, Serradas et al.[28] reported decreased B-cell mass as well as insulin content in the fetal pancreas of GK rats.

Since data from the Japanese group[8] suggested degeneration and final paucity of B-cells with aging in the GK rat, it was of interest to study B-cell replicatory activity at different ages of the animals. For this purpose, male GK and control Wistar rats aged 1, 3, and 6 months were given an i.p. injection of ^{3}H-thymidine (1 μCi/g; Amersham, UK) 1 h before killing. After fixation of a piece of the pancreas in Bouin's solution and embedding in paraffin, 7 μm sections were stained for insulin by immunohistochemistry, using the PAP technique with an anti-bovine insulin guinea-pig serum (ICN Immuno Biologicals, Il, USA) as the primary antibody. Thereafter, autoradiographic processing was performed as described.[37] The labeling index of B-cells was determined by counting radioactively labeled nuclei of insulin-positive cells at a high magnification ($\times$1000), expressing them as a percentage of the total number of nuclei scored. Insulin-positive cells were considered labeled with ^{3}H-thymidine if $>$20 grains were detected over their nuclei. Most labeled cells indeed had $>$20 grains, and the background count was $\leq$2 grains per nucleus.

As seen in Figure 2, body wt was slightly but significantly ($P < 0.001$) lower in 3- and 6-month-old GK rats as compared with age-matched control rats. The B-cell labeling index was not significantly different between GK and control animals at any age studied. As expected, the high labeling index of almost 2% in 1-month-old rats decreased to levels of 0.5% or less in older animals. In the light of these findings, it seems unlikely that the described changes in islet morphology of GK rats with aging are due to altered B-cell replicatory activity.

Islet (pro)insulin biosynthesis was studied in 4-month-old GK and control rats by immunoprecipitation.[19] In islets incubated at 2.8 but not at 16.7 mmol/l glucose, both (pro)insulin and total protein biosynthesis were significantly higher in GK than in controls. However, the fractional (pro)insulin biosynthesis (as percent of total protein biosynthesis), as well as the incremental rate of (pro)insulin biosynthesis induced by raising the glucose concentration from 2.8 to 16.7 mmol/l, were similar in islets of GK and control rats. This was confirmed in a preliminary study, in which largely normal processing of proinsulin as well as of the proinsulin-converting enzymes PC1 and PC2 was also demonstrated.[38]

To conclude, in young GK rats (2–3 months old) islet morphology is grossly normal, while it seems that islet shape irregularities, due to increasing fibrosis, and a decrease in B-cell density occur in the aging GK rat. Furthermore, B-cell replicatory capacity as well as proinsulin/insulin biosynthesis appear normal, although decreased pancreatic insulin content has been reported in older GK rats.[7,8,10] Biosynthesis of glucagon and somatostatin in GK rat islets has not been investigated, but the content of these hormones in GK pancreata has been shown to be normal or slightly increased, respectively, compared with control pancreata.[12]

Islet microcirculation

Similar to other diabetic rodent models, both whole pancreatic and islet blood flow, measured with the microsphere technique, were shown to be increased in GK rats[39,40] as well as in the more mildly diabetic F1 hybrids of GK/Wistar rats.[41] Increased islet perfusion has also been found in other conditions with increased functional demand on the B-cells, for example, pregnancy, obesity, and hyperglycemia after glucose infusion.[42] Vagotomy decreased pancreatic and islet blood flow, to a greater extent in diabetic than in control rats, so that the flow rates were similar in both types of vagotomized animals.[41] Thus, the increased islet perfusion in GK rats seems to depend on a mechanism involving the vagal nerve. This mechanism could well be mediated by nitric oxide (NO), since administration of the NO synthase inhibitor N^G-nitro-L-arginine to GK rats before microcirculation measurements reduced the increased pancreatic and islet blood flow rates to levels seen after vagotomy.[39] It is possible that increased cholinergic stimulation by the vagal nerve is a compensatory mechanism in the diabetic rat with poor B-cell function, since it has been shown in another rat model of NIDDM, the neonatally STZ-injected rat, that cholinergic binding to B-cells, as well as carbachol-induced insulin release, are enhanced.[43] The relationships between islet blood flow and B-cell secretory function, however, appear to be com-

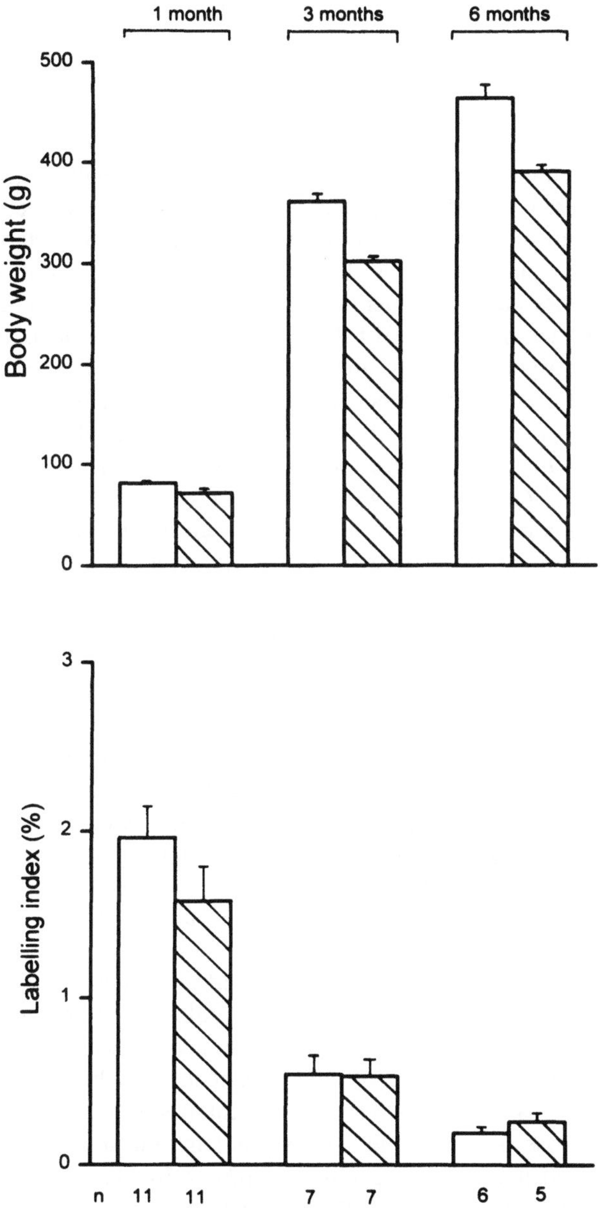

Figure 2. Body wt and labeling index of B-cell replicatory activity in 1-, 3-, and 6-month-old nondiabetic Wistar (unfilled bars) and GK (striped bars) rats. The number of animals (n) in each group is given in the bottom of the figure. See the text for other details.

plex, because the insulinogenic peptide GLP-1 (7-36) amide also was a potent suppressor of the increased islet perfusion in GK rats.[44]

Stimulus-secretion coupling in B-cells

Stimulus-secretion coupling in the normal B-cell is complexely regulated by substrates and a series of endocrine, neurocrine, paracrine, and autocrine factors.[45] However, in case of glucose-induced insulin secretion, the metabolism of the hexose seems to play a key role. Under normal conditions, glucose is rapidly transported across the B-cell membrane, and this step is governed by specific glucose transporter molecules, GLUT-2, localized to the cell membrane.[46,47] Glucose is then phosphorylated by gluco/hexokinase to glucose 6-phosphate.[48] A rise in extracellular glucose concentration thus immediately increases the metabolism of glucose in the B-cell, leading to enhanced oxidation of the hexose and thereby an increased intracellular ATP/ADP ratio. The latter state closes ATP-sensitive K^+ channels and, when most of these channels (>99%) are closed, the B-cell membrane depolarizes and voltage-dependent L-type Ca^{2+} channels open, giving rise to action potentials. The resulting increase in cytosolic calcium leads to exocytosis of insulin from the B-cell.[49]

In our and other studies on glucose metabolism in islets of GK rats, a number of alterations have been demonstrated (Fig. 3). Ohneda et al.[36] reported underexpression of GLUT-2 (step 1 in Fig. 3) in GK rat islets. They concluded, however, that neither the magnitude of this underexpression of the glucose transporter nor the reduction in glucose transport function in GK rat islets is sufficient to explain the profound impairment of glucose-induced insulin secretion. This view is supported by the finding of normal activities of glucokinase/hexokinase isoenzymes in islets isolated from GK rats.[32,50] Furthermore, glucose utilization in GK relative to control rat islets was found to be increased (see later), which strongly speaks against a limiting role of either glucose transport or glucose phosphorylation in the metabolism of the hexose.

The glucose cycle activity was two- to three-fold increased, to about 16%, in GK vs. control rat islets (Fig. 4).[11] This is a futile cycle that consists of phosphorylation of glucose to glucose-6-phosphate, and its dephosphorylation back to glucose. For every glucose molecule participating in the glucose cycle, one molecule of ATP is consumed. The cycle is catalyzed by glucokinase and glucose-6-phosphatase (step 2 in Fig. 3), and the latter enzyme also present in the B-cells.[51] An enhanced glucose cycling in diabetic B-cells may possibly lead to a reduction in the cytosolic ATP/ADP ratio, and thereby impaired insulin secretion. The rate of glucose cycling in B-cells of the GK rat is modest, however, in comparison with that of islets from obese hyperglycemic (*ob/ob*) mice, in which ≥30% of glucose-6-phosphate is directly dephosphorylated.[52,53] In islets of neonatally STZ-injected rats, the rate of glucose cycling (15%) was close to that of GK rat islets.[54]

It has been suggested that glycogen accumulation in B-cells during sustained hyperglycemia may contribute, by interference of glycogenolysis with glycolysis, to so-called B-cell glucotoxicity.[55,56] In collaboration with Malaisse's group,

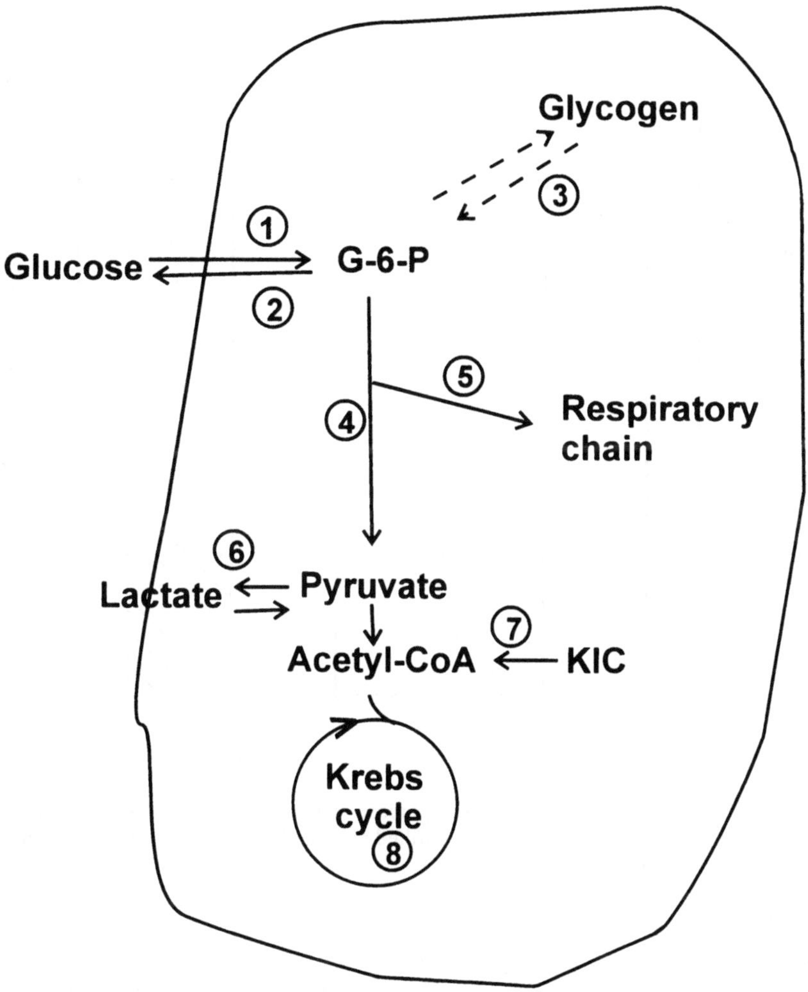

Figure 3. Metabolism of glucose in the B-cell of GK rats. The circled numbers refer to pathways discussed in the text.

we investigated glycogen metabolism (step 3 in Fig. 3) in GK rat islets. Unexpectedly, a small amount of glycogen was demonstrated in B-cells, and consequently no evidence was found of interference of glycogen-derived substrates with glycolysis and insulin release.[35] In addition, glycogen synthase, glycogen phosphorylase, and α-amylase activities were not significantly different in homogenates of GK and control islets.[57]

Determination of islet glucose utilization, reflected as the production of ^3H_2O from 5-^{3}H-glucose, showed a two- to threefold increased rate of glycolysis (step 4 in Fig. 3) in islets of 2- to 3-month-old GK rats compared with islets from control rats of identical age.[11] Similar results were obtained also in 8-wk-old GK rats from the London colony,[21] as well as in 17-wk-old rats from the Paris

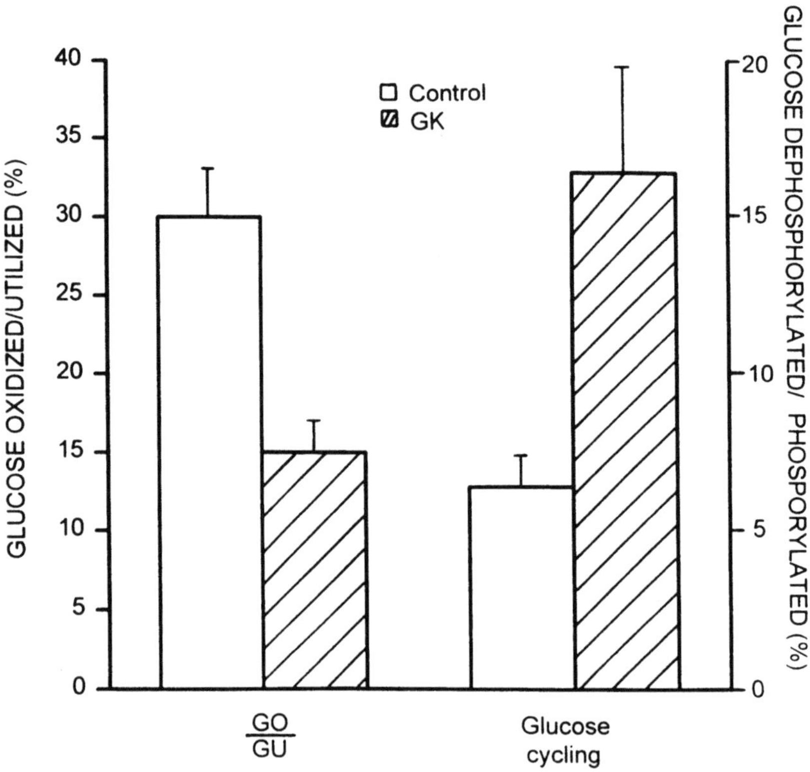

Figure 4. Ratio of oxidized to utilized glucose (left panel) and glucose cycling (right panel) in islets from control Wistar rats (unfilled bars) and GK rats (striped bars). (Adapted from ref. 11, with permission.)

colony.[20] However, the latter group, in another study, found lower rates of glucose utilization also in 17-wk-old GK rats that were close to those in control rats.[19]

The further fate of the utilized glucose in GK rat islets, and it contribution to production of energy-rich compounds, will be considered in relation to the glycerol phosphate shuttle, the production of lactate, the insulinogenic effect of KIC and other metabolizable substrates, and eventually the Krebs cycle (steps 5–8 in Fig. 3). The glycerol phosphate shuttle is a mechanism operating in B-cells by which NADH is oxidized to NAD, associated with oxidative phosphorylation and thus ATP production.[58] In collaboration with Malaisse and his group,[50] we found in GK rat islets a 60–70% decrease in the activity of mitochondrial FAD-linked glycerophosphate dehydrogenase, the regulatory enzyme of the glycerol phosphate shuttle, whereas other islet mitochondrial enzymes, such as glutamate dehydrogenase, 2-ketoglutarate dehydrogenase, as well as glutamic acid decarboxylase, exhibited normal activities. Similar impairment of the shuttle has been shown also in islets of *db/db* mice and neonatally STZ-injected rats.[59]

Further support for the importance of the glycerol phosphate shuttle as a site responsible for defective glucose metabolism in GK rat B-cells was given by the demonstration of both impaired closure of the ATP-sensitive K^+ channels and decreased insulin response to dihydroxyacetone, which is preferentially metabolized via the shuttle.[60] It is likely that in islets of GK rats the oxidation of NADH occurs at an increased rate of lactate production (step 6 in Fig. 3), yielding the NAD required for further augmentation of the anaerobic glycolysis. Hypothetically, an impaired glycerol phosphate shuttle, resulting in decreased oxidation of NADH, may be an event that triggers increased anaerobic glycolysis. In preliminary studies, lactate production in GK islets was found to be increased about twofold (unpublished data), which verifies previous data from radioisotopic studies of metabolic fluxes in GK rat islets.[61] As with increased glucose cycling, an impaired glycerol phosphate shuttle activity could contribute to a reduced ATP/ADP ratio. Although it is not clear whether this occurs to such an extent that it leads to defective insulin secretion, Giroix et al.[20] indeed demonstrated a lower ATP/ADP ratio in GK rat islets than in control islets incubated at a high (16.7 mmol/l) glucose concentration. Indirect evidence also supporting a role for a decreased ATP/ADP ratio in this context was provided by our experiments demonstrating that short-term treatment of *ob/ob* mice with dexamethasone, known to augment glucose-6-phosphatase activity, increased glucose cycling in the islets to comprise 56% of the phosphorylated glucose. In parallel, glucose-induced insulin release decreased by 60%.[62]

Oxidation of glucose in Krebs cycle, estimated as the production of $^{14}CO_2$ from ^{14}C-labeled glucose has been reported to be decreased,[19] similar, or slightly increased[11,20,21] in GK rat islets in comparison with control islets. In relation to glucose utilization rates in each of the studies, however, the ratio of oxidized to utilized glucose was decreased in GK rat islets vs. control islets (12–15% vs. 25–30%, respectively; see Fig. 4). Giroix and coworkers[61] have proposed, based on results from incubations of islets with both 6-^{14}C- and 3,4^{14}C-glucose, that the Krebs cycle itself is not malfunctioning in GK rat B-cells. Instead, a low metabolic flux through a step proximal to the Krebs cycle, such as the generation of acetyl residues in the reaction catalyzed by pyruvate dehydrogenase (PDH), may be responsible for the decreased ratio of oxidized to utilized glucose in GK rat islets. Indeed, a decreased activity of PDH, likely to be due to enhanced PDH kinase activity, has been demonstrated recently in GK rat islets.[63]

The normal function of the Krebs cycle is further emphasized by the above-mentioned fact that KIC, in the absence of glucose, induced similar insulin responses in GK and control rat islets.[21,31,32] In concert with this observation, Tsuura et al.[32] demonstrated with the patch-clamp technique that KIC, in contrast to glucose, closed the ATP-sensitive K^+ channels normally in GK rat islets. In addition, these channels were closed by exogenous ATP; hence, an intact channel activity is present in GK rat B-cells. So far, the function of steps distal to the ATP-sensitive K^+ channels, that is, the L-type Ca^{2+} channels, the exocytotic process, and their modulating factors, have not been sufficiently clarified.

Taken together, a number of alterations have been demonstrated in the glucose metabolism of GK rat islets, such as increased glucose cycling, reduced glycerol phosphate shuttle activity, and a decreased ratio of oxidized to utilized glucose. Each of these abnormalities may contribute to defective insulin release by interfering with the ATP/ADP ratio in the B-cells. At present, it is unclear whether each of these abnormalities is a primary phenomenon or not.

Role of glucotoxicity in the GK rat

B-cell secretion may be impaired by prolonged exposure to higher than normal concentrations of glucose, so-called glucotoxicity.[64] Conversely, long-term normalization of glycemia by insulin or phlorizin treatment of diabetic rats has at least partly restored B-cell responsiveness to glucose.[65,66] The mechanisms behind the glucotoxic effect are largely unknown. Hypothetically, hyperglycemia in GK rats could evolve gradually due to decreased insulin sensitivity, and then subsequently impair insulin secretion. Indeed, decreased insulin sensitivity in the liver has been demonstrated in adult GK rats.[8,67] It is not known, however, whether the hepatic insulin resistance is a primary defect, or in itself rather a "glucotoxic" effect by hyperglycemia due to B-cell dysfunction. Moderate insulin resistance has also been shown in extrahepatic tissues, for example, muscle and fat cells, of GK rats.[67] In this context it is of interest that insulin-stimulated glycogen synthase activity was reduced in skeletal muscle of GK rats,[68] suggesting that defective activation of glucose conversion into glycogen by this tissue may contribute to the hyperglycemia. Glycogen synthase in adipocytes from GK rats was, however, activated normally by insulin.[68]

To investigate to what extent a glucotoxic effect in GK rats might have contributed to the severely reduced insulin response to glucose, six male GK rats (2–3 months old) were treated for 1 wk with a daily dose, at 4 p.m., of 8–10 U of heat-treated Ultralente insulin (Novo-Nordisk A/S, Denmark) prior to perfusion of the pancreas. Six control GK rats were injected with saline for 1 wk. Before treatment, blood glucose levels (mean $\pm$ SE) at 9 a.m. and 4 p.m. were 13.8 ± 2.3 and 11.1 ± 2.0 mmol/l, respectively, in the insulin-treated group and 11.4 ± 3.2 and 9.5 ± 2.2 mmol/l, respectively, in the saline-treated group of rats. During insulin treatment, blood glucose concentrations at 9 a.m. and 4 p.m. decreased to 3.3 ± 0.3 and 4.7 ± 0.9 mmol/l, respectively ($P < 0.02$ vs. pretreatment). In the saline-treated group, blood glucose levels remained high at 9.4 ± 2.8 and 9.6 ± 1.1 mmol/l, respectively.

As seen in Figure 5, the insulin response to 16.7 mmol/l glucose in the perfused pancreas was not improved by the insulin treatment. The high basal insulin release (at 3.3 mmol/l glucose), however, tended to be decreased after insulin treatment. The slightly lower-than-normal blood glucose levels induced by insulin could potentially explain the lack of recovery of insulin response to glucose through activation of the sympathetic nervous system. However, in the perfused pancreas of Wistar rats, subjected to mild insulin treatment and with similar glycemia as in the insulin-injected GK rats, no marked impairment of glucose-stimulated insulin release was noted (data not shown).

It could be argued that 1 wk of normalization of glycemia was not sufficient to restore the insulin response to glucose. However, when islets from mildly diabetic F1 hybrid rats were transplanted under the kidney capsule of normo-

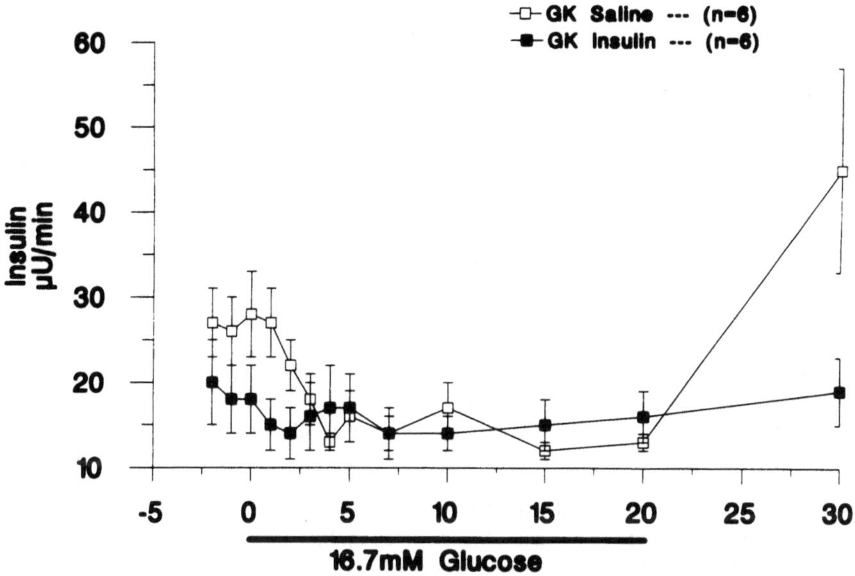

Figure 5. Glucose-stimulated insulin secretion in the isolated perfused pancreas of GK rats after 1 wk of normalization of glycemia by insulin treatment (■). Control GK rats (□) received daily saline injections over 1 wk. See the text for details.

glycemic, immunotolerant nude mice, the insulin responses to glucose from the perfused islet grafts after 4 wk in the normoglycemic milieu in vivo were still lower in F1 hybrid than in control islets.[27] In this context, it is of interest that F1 hybrid rats, which present glucose intolerance and impaired insulin response to glucose as early as at 1 wk of age, show signs of decreased extrahepatic insulin sensitivity only at adult age.[69] Thus, insulin-mediated glucose transport in the epitrochlearis muscle of 1-month-old F1 hybrid rats was normal, while that of 2-month-old hybrid rats was moderately decreased compared with age-matched healthy rats. These observations suggest the primacy of a B-cell defect in the F1 hybrid, and most likely also in the GK rat, and that decreased insulin sensitivity occurs secondary to hyperglycemia.

Conclusions

The GK rat is a non-obese animal model of NIDDM in which the diabetic state appears to be conveyed by several additive genes. Glucose intolerance is not due to major derangement of islet morphology or B-cell replicatory rate, nor to decreased (pro)insulin biosynthesis and B-cell stores of insulin. An impaired insulin secretion seems to be the early and, probably most important, factor in the development of glucose intolerance. Several abnormalities in glucose metabolism of the B-cell, such as increased glucose cycling, decreased glycerol phosphate shuttle activity, and decreased ratio of oxidized to utilized glucose, could contribute to a defective stimulus-secretion coupling for glucose in the GK rat.

Acknowledgements

Studies performed on GK rats of the Stockholm colony were supported by the Swedish Medical Research Council (00034, 109, 8982), the Swedish Diabetes Association, the Novo-Nordisk Foundation, the Hagbergs Foundation, the Magn. Bergvalls Foundation, and funds from the Karolinska Institute.

References

1. Efendic S, Luft R, Wajngot A. Aspects of the pathogenesis of type 2 diabetes. Endocr Rev 5:395–409, 1984.
2. DeFronzo RA, Bonadonna RC, Ferrannini E. Pathogenesis of NIDDM. A balanced overview. Diabetes Care 15:317–68, 1992.
3. Cerasi E, Luft R. The plasma insulin response to glucose infusion in healthy subjects and in diabetes mellitus. Acta Endocrinol 55:278–304, 1967.
4. Efendic S, Grill V, Luft R, Wajngot A. Low insulin response: A marker of prediabetes. Adv Exp Med Biol 246:167–74, 1988.
5. Shafrir E: Diabetes in animals. In: Diabetes mellitus. Theory and practice, 4th ed. Rifkin H, Porte D, eds. Elsevier, New York, pp 299–340, 1990.
6. Goto Y, Kakizaki M, Masaki N. Spontaneous diabetes produced by selective breeding of normal Wistar rats. Proc Jpn Acad 51:80–5, 1975.
7. Goto Y, Suzuki K-I, Sasaki M, et al. GK rat as a model of nonobese, noninsulin dependent diabetes: Selective breeding over 35 generations. In: Lessons from animal diabetes. Shafrir E, Renold AE, eds. J Libbey, London, 2:301–3, 1988.
8. Suzuki K-I, Goto Y, Toyota T: Spontaneously diabetic GK (Goto-Kakizaki) rats. In: Lessons from animal diabetes. Shafrir E, ed. Smith-Gordon, London, 4:107–16, 1992.
9. Kimura K, Toyota T, Kakizaki M, et al. Impaired insulin secretion in the spontaneous diabetes rats. Tohoku J Exp Med 137:453–9, 1982.
10. Portha B, Serradas P, Bailbe D, et al. β-cell insensitivity to glucose in the GK rat, a spontaneous nonobese model for type II diabetes. Diabetes 40:486–91, 1991.
11. Östenson C-G, Khan A, Abdel-Halim SM, et al. Abnormal insulin secretion and glucose metabolism in pancreatic islets from the spontaneously diabetic GK rat. Diabetologia 36:3–8, 1993.
12. Abdel-Halim SM, Guenifi A, Efendic S, Östenson C-G. Both somatostatin and insulin responses are impaired in the perfused pancreas of spontaneously non-insulin-dependent diabetic GK rats. Acta Physiol Scand 148:219–26, 1993.
13. Abdel-Halim SM, Guenifi A, Luthman H, et al. Impact of diabetic inheritance on glucose tolerance and insulin secretion in spontaneously diabetic GK-Wistar rats. Diabetes 43:281–8, 1994.
14. Goto Y, Kakizaki M, Yagihashi S: Neurological findings in spontaneously diabetic rats. Excerpta Med Int Congr Ser 581:26–38, 1982.
15. Gauguier D, Nelson I, Bernard C, et al. Higher maternal than paternal inheritance of diabetes in GK rats. Diabetes 43:220–4, 1994.
16. Galli J, Li L-S, Östenson C-G, et al. Genetic analysis of non-insulin dependent diabetes mellitus in the GK rat. Nature Genetics 12:31–7, 1996.
17. Gauguier D, Froguel P, Parent V, et al. Chromosomal mapping of genetic loci associated with non-insulin dependent diabetes in the GU rat. Nature Genetics 12:38–43, 1996.
18. Goto Y, Kakizaki M, Toyota T, et al. Spontaneous diabetes produced by repeated selective breeding of normal Wistar rats. In: Diabetes. Bajaj JS, ed. Excerpta Medica, Amsterdam, pp 703–10, 1977.

19. Giroix M-H, Vesco L, Portha B: Functional and metabolic perturbations in isolated pancreatic islets from the GK rat, a genetic model of noninsulin-dependent diabetes. Endocrinology 132:815–22, 1993.
20. Giroix M-H, Sener A, Bailbe D, et al. Metabolic, ionic, and secretory response to D-glucose in islets from rats with acquired or inherited non-insulin-dependent diabetes. Biochem Med Metab Biol 50:301–21, 1993.
21. Hughes SJ, Suzuki K, Goto Y. The role of islet secretory function in the development of diabetes in the GK Wistar rat. Diabetologia 37:863–70, 1994.
22. Östenson C-G, Khan A, Efendic S, et al. Abnormal insulin secretion and glucose metabolism in pancreatic islets from the spontaneously diabetic GK-rat. Diabetes 40(Suppl 1):293A, 1991.
23. Halban PA, Bonner-Weir S, Weir GC. Elevated proinsulin biosynthesis in vitro from a rat model of non-insulin-dependent diabetes mellitus. Diabetes 32:277–81, 1983.
24. Östenson C-G, Cattaneo AG, Doxey JC, Efendic S: α_2-Adrenoceptors and insulin release in pancreatic islets of normal and diabetic rats. Am J Physiol 257:E439–43, 1989.
25. Östenscn C-G, Hjemdahl P, Efendic S: Release of catecholamines is increased but does not contribute to the impaired insulin secretion in the perfused pancreata of diabetic rats. Pancreas 8:34–8, 1993.
26. Grill V, Östenson C-G: The influence of a diabetic state on insulin secretion: Studies in animal models of non-insulin dependent diabetes. In: Pathogenesis of non-insulin-dependent diabetes mellitus. Grill V, Efendic S, eds. Raven Press, New York, pp 93–106, 1988.
27. Abdel-Halim SM, Östenson C-G, Andersson A, et al. A defective stimulus-secretion coupling rather than glucotoxicity mediates the impaired insulin secretion in the mildly diabetic F_1 hybrids of GK-Wistar rats. Diabetes, 44:1280–84, 1995.
28. Serradas P, Saulnier C, Giroix MH, et al. Characterization of islet architecture and B-cell function in the fetal pancreas of the GK rat, a genetic model of non-insulin dependent diabetes. Diabetologia 37(Suppl 1):A42, 1994.
29. Porte D Jr. β-cells in type II diabetes mellitus. Diabetes 40:166–80, 1991.
30. Pigon J, Waldelöf K, Östenson C-G. Arginine-induced insulin secretion is impaired in type 2 diabetic patients compared with control with identical glycaemia and insulinaemia. Diabetologia 34(Suppl 2):A108, 1991.
31. Östenson C-G, Sundén M, Abdel-Halim SM, Efendic S. Krebs cycle substrates but not glucose elicit normal insulin release in islets of rats with decreased glucose tolerance. Diabetologia 36(Suppl. 1):A28, 1993.
32. Tsuura Y, Ishida H, Okamoto Y, et al. Glucose sensitivity of ATP-sensitive K^+ channels is impaired in β-cells of the GK rat. Diabetes 42:1446–53, 1993.
33. Östenson C-G, Efendic S, Grill V: Abnormal regulation by glucose of somatostatin secretion in the perfused pancreas of NIDDM rats. Pancreas 5:347–53, 1990.
34. Guenifi A, Abdel-Halim SM, Höög A, et al. Preserved β-cell density in the endocrine pancreas of young, spontaneously diabetic Goto-Kakizaki (GK) rats. Pancreas 10:148–53, 1995.
35. Sener A, Malaisse-Lagae F, Östenson C-G, Malaisse WJ: Metabolism of endogenous nutrients in islets of Goto-Kakizaki (GK) rats. Biochem J 296:329–34, 1993.
36. Ohneda M, Johnson JH, Inman LR, et al. GLUT2 expression and function in β-cells of GK rats with NIDDM. Diabetes 42:1065–72, 1993.
37. Korsgren O, Jansson L, Eizirik DL, Andersson A: Functional and morphological differentiation of fetal porcine islet-like cell clusters after transplantation into nude mice. Diabetologia 34:379–86, 1991.
38. Hutton JC, Guest PC, Clark A, et al. Proinsulin biosynthesis and processing in the GK rat model of type 2 diabetes. Diabetologia 37(Suppl 1):A42, 1994.
39. Svensson AM, Östenson C-G, Sandler S, et al. Inhibition of nitric oxide synthase by N^G-nitro-L-arginine causes a preferential decrease in pancreatic islet blood flow

in normal rats and spontaneously diabetic GK rats. Endocrinology 135:849–53, 1994.
40. Atef N, Portha B, Pénicaud L. Changes in islet blood flow in rats with NIDDM. Diabetologia 37:677–80, 1994.
41. Svensson AM, Abdel-Halim SM, Efendic S, et al. Pancreatic and islet blood flow in F1-hybrids of the non-insulin-dependent diabetic GK-Wistar rat. Eur J Endocrinol 130:612–16, 1994.
42. Svensson AM. Pancreatic islet blood flow in the rat. Thesis, Acta Universitatis Upsaliensis, Vol. 483, Uppsala 1994.
43. Östenson C-G, Grill V: Evidence that hyperglycemia increases muscarinic binding in pancreatic islets of the rat. Endocrinology 121:1705–10, 1987.
44. Svensson AM, Östenson C-G, Efendic S, Jansson L: Effect of GLP-1 (7-36) amide on pancreatic and intestinal blood flow in the spontaneously diabetic GK rat. Diabetologia 37(Suppl 1):A72, 1994.
45. Berggren P-O, Rorsman P, Efendic S, et al. Mechanisms of action of entero-insular hormones, islet peptides and neural input on the insulin secretory process. In: Nutrient Regulation of insulin secretion. Flatt P, ed. Portland Press, London, pp 289–318, 1992.
46. Orci L, Thorens B, Ravazzola M, Lodish HF: Localization of the pancreatic β-cell glucose transporter to specific plasma membrane domains. Science 245:295–7, 1990.
47. Johnson JH, Newgard CB, Milburn JL, et al. The high K_m glucose transporter of islets of Langerhans is functionally similar to the low affinity transporter of liver and has an identical primary sequence. J Biol Chem 265:6548–51, 1990.
48. Meglasson MD, Matschinsky FM: New perspectives on pancreatic islet glucokinase. Am J Physiol 246:E1–13, 1984.
49. Efendic S, Kindmark H, Berggren P-O: Mechanisms involved in the regulation of the insulin secretory process. J Intern Med 229(Suppl 2):9–22, 1991.
50. Östenson C-G, Abdel-Halim SM, Rasschaert J, et al. Deficient activity of FAD-linked glycerophosphate dehydrogenase in islets of GK rats. Diabetologia 36:722–6, 1993.
51. Täljedal I-B. Presence, induction and possible role of glucose-6-phosphatase in mammalian pancreatic islets. Biochem J 114:387–94, 1969.
52. Khan A, Chandramouli V, Östenson C-G, et al. Evidence for the presence of glucose cycling in pancreatic islets of the ob/ob mouse. J Biol Chem 264:9732–33, 1989.
53. Khan A, Chandramouli V, Östenson C-G, et al. Glucose cycling is markedly enhanced in pancreatic islets of obese hyperglycemic mice. Endocrinology 126:2413–16, 1990.
54. Khan A, Chandramouli V, Östenson C-G, et al. Glucose cycling in islets from healthy and diabetic rats. Diabetes 39:456–9, 1990.
55. Malaisse WJ, Marynissen G, Sener A: Possible role of glycogen accumulation in B-cell glucotoxicity. Metabolism 41:814–19, 1992.
56. Malaisse WJ, Maggetto C, Leclerque-Meyer V, Sener A: Interference of glycogenolysis with glycolysis in pancreatic islets from glucose-infused rats. J Clin Invest 91:432–6, 1993.
57. Zhang T-M, Östenson C-G, Malaisse WJ: Glycogen synthase, glycogen phosphorylase and α-amylase activity in homogenates of islets of GK rats: Comparison with hepatic and pancreatic extracts. Cell Biochem Funct 12:185–9, 1994.
58. Rasschaert J, Malaisse WJ: Hexose metabolism in pancreatic islets. Glucose-induced and Ca^{2+}-dependent activation of FAD-linked glycerophosphate dehydrogenase. Biochem J 278:335–40, 1991.
59. Giroix M-H, Rasschaert J, Bailbe D, et al. Impairment of the glycerol phosphate shuttle in islets from rats with diabetes induced by neonatal streptozotocin. Diabetes 40:227–32, 1991.
60. Tsuura Y, Ishida H, Okamoto Y, et al. Reduced sensitivity of dihydroxyacetone on

ATP-sensitive K^+ channels of pancreatic beta cells in GK rats. Diabetologia 37:1082–87, 1994.
61. Giroix M-H, Sener A, Portha B, Malaisse WJ. Preferential alteration of oxidative relative to total glycolysis in pancreatic islets of two rat models of inherited or acquired Type 2 (non-insulin-dependent) diabetes mellitus. Diabetologia 36:305–9, 1993.
62. Khan A, Östenson C-G, Berggren P-O, Efendic S. Glucocorticoid increases glucose cycling and inhibits insulin release in pancreatic islets of ob/ob mice. Am J Physiol 263:E663–6, 1992.
63. Zhou Y-P, Östenson C-G, Ling Z-C, Grill V: Deficiency of pyruvate dehydrogenase activity in pancreatic islets of diabetic GK rats. Endocrinology 136:3546–51, 1995.
64. Rossetti L, Giaccari A, DeFronzo RA: Glucose toxicity. Diabetes Care 13:610–30, 1990.
65. Kergoat M, Bailbe D, Portha B: Insulin treatment improves glucose-induced insulin release in rats with NIDDM induced by streptozotocin. Diabetes 36:971–7, 1987.
66. Rosetti L, Smith D, Shulman GI, et al. Correction of hyperglycaemia with phlorizin normalizes tissue sensitivity to insulin in diabetic rats. J Clin Invest 79:1510–15, 1987.
67. Bisbis S, Bailbe D, Tormo M-A, et al. Insulin resistance in the GK rat: Decreased receptor number but normal kinase activity in liver. Am J Physiol 265:E807–13, 1993.
68. Villar-Palasi C, Farese RV: Impaired skeletal muscle glycogen synthase activation by insulin in the Goto-Kakizaki (G/K) rat. Diabetologia 37:885–8, 1994.
69. Nolte LA, Abdel-Halim SM, Martin IK, et al. Development of decreased insulin-induced glucose transport in skeletal muscle of glucose intolerant hybrids of diabetic GK rats. Clin Sci 88:301–6, 1995.

18

Role of immune and non-immune aspects of genetics in different expressions of diabetes in db/db *mice*

KONSTANTIN BLOCH AND PNINA VARDI
The National Center for Childhood Diabetes, Schneider's Children Medical Center, Petah-Tikvah, Israel.

Key words: C57 BKS db/db mice mutation, genetics, NOD mice, lupus-like syndrome, immunodeficiency, retrovirus, radiation, islet B-cell, streptozotocin diabetes.

The availability of several animal species with diabetes contributed significantly to the understanding of the pathogenesis of the disease and to the development of novel treatment and prevention strategies. Additional contribution derived from investigation of spontaneous and induced forms of diabetes resembling human IDDM and NIDDM, two entities distinguished by their pathogenic background and their clinical presentation. Among the animal models the *db/db* mouse, with its autosomal recessive *db* (diabetes) mutated gene on chromosome 4, is of particular importance for the understanding of the complexity of diabetes, since it displays characteristics of NIDDM during the initial stage of the syndrome and of IDDM during the final stage as a result of B-cell necrosis and islet atrophy. Homozygosity for the *db* gene in mice is characterized by hyperglycemia and hyperinsulinemia early in the process of diabetes, followed by a stage of insulin deficiency and overt IDDM, although the clinical expression differs according to the different genetic background of the mouse. The increased susceptibility of the *db/db* mouse to the diabetogenic action of viruses and chemical compounds before overt diabetes appears makes this animal an ideal model to study the mechanisms by which diabetogenic or antidiabetogenic genes interact with other genes and/or environmental factors. Previous reviews described different aspects of diabetes in the *db/db* mouse,[1–3] whereas we focus on the role of genetically determined immune and non-immune mechanisms and describe the effect of environmental factors in inducing different expressions of diabetes in *db/db* mice.

Correspondence: Dr. K. Bloch, National Center for Childhood Diabetes, Schneider's Children Medical Center, Kaplan 14, Petah-Tikvah, Israel.

Lessons from Animal Diabetes VI. Ed. E. Shafrir.

Influence of the genetic background on the expression of the *db* mutation

History of the db/db mouse. In 1965 a new autosomal-recessive mutation with complete penetrance, named *db* (diabetes), was discovered in the C57BLKS/J (BKS) mouse strain at the Jackson Laboratory (Bar Harbor, ME, USA).[4] Five separate spontaneous mutations allelic with the *db* gene have been described since the initial report.[5] Molecular mapping of the *db* mutation showed that it maps to the medial region of chromosome 4 in linkage with several genes, such as *jun, GLUT1*, and *Lck*, that are represented by syntenic grouping on human chromosome 1p.[6] Probably, the autosomal recessive fatty (*fa*) mutation in rats on chromosome 5, causing the obesity syndrome with hyperinsulinemia, hyperlipidemia, and multiple endocrine abnormalities, is the apparent homologue of the mouse *db* mutation based upon linkage to the α-interferon and *GLUT1* genes.[7] Positional cloning of the mouse obese gene suggests that the *db* gene may be the central nervous system receptor for the "lipostatin"-like hormone, presumably secreted by the adipocyte.[8]

Heterozygous mice with the *db* gene display glucose intolerance but do not progress to overt diabetes. In vitro stimulation of isolated pancreatic islets of such rodents show an abnormal insulin secretory response to glucose stimulation.[9,10] In the homozygous state (*db/db*), when such mutation is coexpressed with the genetic BKS background, hyperphagia, progressive obesity, and severe diabetes with insulin resistance appear, with hyperglycemia and transient elevation of the plasma insulin concentration. The histological findings at the initial stage of the disease are characterized by islet hyperplasia, which subsequently evolves into typical islet degeneration. At the later stage of the disease, the clinical picture is that of weight loss and a shortened life span.[1,11]

When the *db* gene is coexpressed with different genetic backgrounds of other inbred mice strains, such as the C57BL/6J (B6) and the 129/J, some of the *db*-dependent diabetogenic effects lose their strength. It should be noted that diabetes resistance in homozygous B6 *db/db* mice is characterized by milder hyperinsulinemia, associated with pronounced islet cell hyperplasia but no evidence of islet destruction.[11,12]

To assess the number of genes involved in the development of diabetes and their influence on *db*-related diabetes expression, genetic analysis was done by several investigators.[11–14] Earlier studies performed at the Jackson Laboratory using (B6 × BKS) F1 *db/db* hybrid mice, backcrossed with each of the parental strains, showed that modification of diabetes expression is under multigenic control, and that modifiers of the B6 genome responsible for such a milder diabetic form are associated with the appearance of islet hypertrophy. This characteristic seemed to be dominant over that determined by the BKS genome, which induces severe diabetes with islet atrophy.[11,14] Interestingly, controversial results were reported by Leiter et al.,[13] showing that *db/db* mice of either F1 or F2 generations exhibit severe hyperglycemia accompanied by a histologic picture of pancreatic islet atrophy. These findings suggest that the influence of the BKS mouse genes responsible for the severe diabetic syndrome dominated over the milder form determined by the B6 genome. Such inconsistency could be speculated to happen as a result of a genetic drift occurring in these mice strains throughout the long period of time between the two studies. Conclusive results were obtained in later experiments showing that diabetes susceptibility was induced by the BKS genes and that their expression dominates over that of the diabetes-resistant 129/J mice. Moreover, findings suggested that at least two genetic loci are responsible for the different degrees of diabetes severity in the BKS and the 129/J strains.[12]

Role of H-2 and non–H-2 genes in db-dependent diabetes. Human and animal genetic studies have shown that the role of the MHC in the genetic control of IDDM is predominant.[15] A recent report suggested that this chromosomal region is probably involved in a subclass of patients with NIDDM.[16] To assess the impact of the H-2 haplotype on the expression of diabetes and obesity syndromes associated with the mutated *db* gene, eight inbred strains with different H-2 haplotypes carrying the *db* gene were evaluated. The B6 and 129/J mice, carrying the H-2b haplotype, were found to be resistant to *db*-dependent diabetes, and the H-2d of both BKS and DBA/2 was found to confer susceptibility, while mice carrying the H-2k haplotype clearly exhibited both diabetes resistance and susceptibility.[17] Such results should be interpreted with caution because inbred mice strains into which the *db* gene was introduced could differ in their response not only as a result of pure H-2 expression, but also due to the influence of genetic loci other than H-2. To minimize such genetic polymorphism between different inbred animal strains, the degree of genetically induced diabetes of congenic-resistant BKS.B6 H-2b mice was compared with that induced by the BKS H-2d, showing a lack of association between diabetes severity and a single H-2 Haplotype.[13] These findings are consistent with other reports that did not find an H-2b protective effect by comparing diabetes severity expressed in H-2k vs. H-2b segregants in C3H *db/db* mice.[18] Thus, neither resistance nor susceptibility to *db*-dependent diabetes is restricted to a single H-2 haplotype.

The lack of a significant H-2 haplotype influence on the expression and severity of diabetes in *db/db* mice points out the involvement of non–H-2–related factors. In some inbred strains, it was evident that gender has a major effect on diabetes susceptibility, as shown by *db* gene expression in CBA/Lt *db/db* (H-2k) mice. Males of this strain had a severe form of diabetes with B-cell necrosis, a variable degree of leukocytic infiltration, and a high mortality rate (100% by 6 months), compared with only transient episodes of hyperglycemia in females. These data suggest that non-MHC genes controlling sex steroid metabolism could interfere with the determination of *db*-gene–dependent diabetes expression.[2,17] A high level of testosterone seems to accelerate diabetogenesis, whereas estrogen seems to exert a protective effect against *db*-gene–dependent diabetes.[13] Estrone-treated BKS *db/db* mice did not develop hyperinsulinemia, hyperglycemia, or islet atrophy, but they were characterized by hyperphagia and marked obesity,[19] suggesting that the balance between active estrogens and androgens is critical to glucose homeostasis. Enzymes responsible for the sulfation/desulfation cycle may play an important role in regulating the level of active androgens and estrogens within a tissue. Sex steroid sulfotransferases remove steroids from the active pool by sulfation (sulfated forms are receptor inactive). Leiter[2] suggested that expression of the diabetic syndrome is dependent on the strength of interaction between the *db* gene and the sulfotransferase enzyme genes. Recently, it was found that mRNA transcripts level from the *Std* locus (chromosome 7), encoding DHEA sulfotransferase, and from the *Ste* locus (chromosome 5), encoding estrogen sulfotransferase, are aberrantly regulated by the *db* mutation in the BKS strain. These *db* gene–induced effects were specific for the sex steroid sulfotransferases, since other potential sex steroid metabolizing enzymes (phenol sulfotransferase, sex steroid sulfohydrolase, and UDP-glucuronyltransferase) were unaffected.[20] Diabetogenic susceptibility in BKS mutant mice of both sexes was associated with marked estrone sulfotransferase activity and depression of preandrogen/androgen sulfation, producing a hyperandrogenized hepatic tissue state. These shifts in steroid sulfation were not observed in the diabetes-resistant C3HeB *db/db* females. Thus, genes controlling steroid sulfotransferase were qualified as genomic modifiers of gender-dependent diabetes susceptibility in mice.[21]

It is probable that one of the non–H-2–linked genes determining *db* gene expression is the *m* (misty) autosomal recessive gene, which is located on chromosome 4. This gene in the homozygous state induces a change of color in BKS mice from black to misty. In BKS mice with double homozygosity for both the *db* and *m* (*dbm/dbm*), these two genes are closely situated on the same chromosome. Such double homozygosity produces syndromes comparable with those induced in the B6 *db+/db+* mice: lower blood glucose concentration, no signs of ketosis, longer life span, and no evidence of islet degeneration in histology until the age of 12 months.[22] Moreover, homozygosity for the *m* gene when coexpressed with the genetic background of BKS mice confers diabetes protection from multiple low doses of streptozotocin (STZ).[23] Yet, up to now the mechanisms by which the *m* gene confers resistance to diabetes are not clear. Some other genes seem to play a role in both the regulation of the coat pigmentation and in processes that are unrelated to pigmentation. For example, the viable yellow (A^{vy}) mutation at the agouti locus exerts a pleiotropic effect, including obesity and diabetes.[24]

db *gene and non-obese diabetic (NOD) mouse genetic background.* Several investigators reported on the immunopathology associated with the *db* gene–dependent syndrome in mice. In some parameters it is similar to that found in NOD mice and BB rats.[2,25] Most of the studies described the BKS inbred strain in which *db/db* homozygous animals exhibit early thymic involution with a decreased level of thymic hormone,[26] T-cell peripheral lymphopenia, and T-cell imbalance.[27] Such mice spontaneously develop antibodies to islet cells and to retroviruses.[28,29] Anti-islet complement-dependent antibody and antibody-dependent cell cytotoxicity were detectable in sera of BKS *db/db* mice on the 10th day of life; hyperglycemia was not detected until the first month, and lasted throughout the entire life span of these animals.[30] Interestingly, in contrast to the susceptible BKS mouse, the diabetes-resistant B6 mouse does not produce antibodies to the B-cell, suggesting that the *db* susceptibility is associated with an increased level of autoreactivity to B-cell antigen found in the BKS strain.[29] When the mutant *db* gene is expressed in NOD mice, the NOD characteristics become intensified with 100% diabetes incidence in both males and females, compared with only 50–80% IDDM in females and 10–20% in males in most NOD colonies. In homozygous NOD *db/db* mice, severe insulitis with evidence of temporary islet hyperplasia can be detected at the initial stages of the disease, followed by islet destruction and hypoinsulinemia.[31] When NOD mice were crossed with obese diabetes-prone BKS or diabetes-resistant B6 mice, only about 1% of the F1 developed diabetes, indicating that the susceptibility genes inherited from the NOD strain are recessive.[2,32]

db *gene expression and the lupus-like syndrome.* In humans, an association between systemic autoimmunity and the development of diabetes is reflected by the higher prevalence of IDDM among relatives of patients with rheumatoid arthritis,[33] and the high incidence of detection of nonspecific organ immunity in patients with IDDM.[34] In addition, insulin resistance was reported in patients with systemic lupus erythematosus (SLE).[35] The genetic predisposition of several mutant inbred mice to develop systemic autoimmunity may serve as a model to evaluate the impact of such an inherited tendency on the form of *db*-related diabetes. For this purpose, the *Y*-linked accelerator gene (*Yaa*), which induces a lupus-like syndrome,[36] was introduced into (C3HeB/FeJ × C3H/HeJ) F2 *db/db* mice. At 2 and 3 months of age, F2 *db/db Yaa/Yaa* males developed a higher blood glucose level than their grandparent diabetic (*db/db*) males. Beyond this age, no difference in the level of blood glucose and mortality rate was observed between the two groups, suggesting that the *Yaa* gene accelerating effect is restricted only to earlier stages of the *db*-related diabetes process.[37]

A second model suitable for evaluation of the impact of autoimmunity on the degree of diabetes induced by the *db* gene is the NZB and the (NZB × NZW) F1 mouse, which develops a spontaneous lupus-like disease, in contrast to the inbred NZW strain. Females of this strain develop autoimmune phenomena more easily than males.[38] Interestingly, (NZB × NZW) F1 and NZB of both sexes display impaired islet function with evidence of autoimmune insulitis.[39] A severe form of lupus-like syndrome is found in (BKS × NZB) F2 *db/db* mice when compared with (BKS × NZW) F2 *db/db* mice, and the severity of the "lupus" is directly correlated with the degree of diabetes of these animals.[40,41] F2 males of (BKS × NZB) *db/db* mice have a higher blood glucose level with greater impairment of glucose tolerance at the prediabetic stage than (BKS × NZW) F2 *db/db* males (Fig. 1.I). In F2 *db/db* male mice carrying the NZB genome, hyperinsulinemia was detected early in the course of the disease, alternating with a significant reduction of serum insulin level (Fig. 1.II) and weight loss later on.

In addition, mortality rate reaches 75% among F2 hybrid *db/db* males carrying NZB genes at the age of 5 months, whereas it reaches only 20% in the F2 of hybrid *db/db* males carrying NZW genes (Fig. 1.III). As in other animal models, gender seems also to play a role here in the expression of the disease; males of both genotypes suffer from a more serious form of the disease than females.[41] In contrast to (BKS × NZW) F2 *db/db* mice, the coexistence of the *db* mutation in (BKS × NZB) F2 mice with a "lupus" background induces the appearance of insulitis, the production of islet cell surface antibodies, and complement-depen-

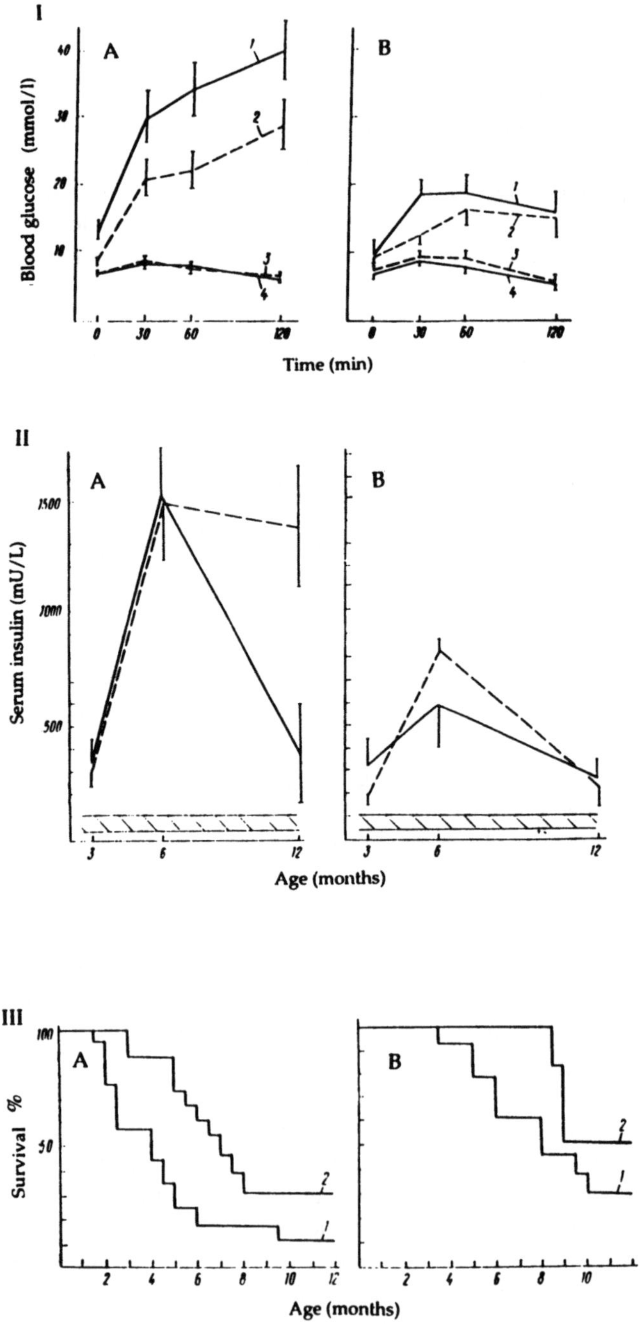

Figure 1. Expression of the *db* mutation in (BKs × NZB) F2 and (BKS × NZW) F2 mice. I: Glucose tolerance test in male (——) and female (– – –) *db*/*db* mice (1,2) and control +/? mice (3,4). (A) (BKS × NZB) F2 mice. (B) (BKS × NZW) F2 mice. II: Age-related changes of serum insulin level in male (——) and female (– – –) (BKS × NZB) F2 *db*/*db* mice (A) and (BKS × NZW) F2 *db*/*db* mice (B). The shaded area represents the insulin level in control +/? mice. III: Survival of male (A) and female (B) (BKS × NZB) F2 *db*/*db* mice (1) and (BKS × NZW) F2 *db*/*db* mice (2). (Adapted from ref. 41, with permission.)

dent antibody-mediated cytotoxicity against the islets (Bloch et al., unpublished data). Thus it is probable that inherited generalized immunopathology intensifies the expression of the *db* mutation in different genetic backgrounds.

db *gene and the immunodeficiency state.* To evaluate the role of the immune system in *db* diabetes, BKS *db/db* mice were rendered T-lymphocyte deficient, either by performing adolescent thymectomy, followed by lethal irradiation and bone marrow reconstruction, or by crossing strains congenic for the *db* mutation with inbred mice carrying the recessive mutations, such as combined immunodeficiency (*scid*), x-linked immunodeficiency (*xid*), and nude.[37] Diabetes was studied in double homozygous animals obtained in the F2 generation. None of these manipulations prevented the animals from progression to diabetes, as expressed by severe hyperglycemia, massive B-cell necrosis, and a high mortality rate. All these findings were observed in the absence both of normal T- and B-cell functions, as assessed by lack of a T-cell response to mitogens, and of B-cell antibodies to insulin and to retrovirus p73. Thus, it can be assumed that B-cell autoimmunity is not the primary event in the etiopathogenesis of diabetes induced in the *db/db* mouse, but it is likely to be a secondary phenomenon.

db *gene and the viral hypothesis.* The facilitated tendency to develop diabetes in certain inbred strains of mice carrying the *db* mutation is believed by several authors to depend on gene expression of endogenous retroviruses incorporated in the B-cell genome. When evaluated by electron microscopy, the ultrastructural appearance of the B-cells obtained from severely diabetic C3H.SW *db/db* mice show expression of endogenous type C retrovirus.[42] The 73.000-dalton core protein antigen (p73) of intracisternal type A particles (IAP), which are retrovirus-like structures, were detected in inbred strains of mice predisposed to develop severe diabetes (BKS, *DB*A/2J, CBA/LtJ, C3HeB/FeJ). In contrast, *db*-resistant strains (B6,129/J, MA/J) did not display any B-cell morphologic changes and glucose-inducible IAP expression in the islets,[43] suggesting that p73 may represent a hidden B-cell neoantigen that, once expressed (stimulated by glucose), induces anti-islet immunity.[44] In addition, early retroviral gene expression has been detected in pancreatic B-cells of mice susceptible to multiple low dose STZ diabetes[45] and in B-cells of the NOD mouse.[46]

db *gene and B-cell proliferation.* One of the most important factors in diabetes development is the limited ability of the pancreatic B-cells to proliferate.[47,48] Several reports showed that diabetes-prone (BKS) and -resistant (B6) inbred strains display marked differences of their B-cell proliferative capacity. In vitro and in vivo studies indicate that the islet proliferative rate of B6 mice is about twice as high as that of BKS mice, irrespective of age and glucose concentration.[49] Intrasplenic transplantation of syngeneic pancreatic islets using insufficient islet number to reverse hyperglycemia (150 islets) into alloxan-induced diabetic BKS mice led to considerable shrinkage of the transplanted islet volume, whereas syngeneic islets implanted into B6 remained intact in most cases. In parallel with the reversal of diabetes and the achievement of normoglycemia in some inbred B6 mice following islet transplantation, an increase in the transplanted islet volume was observed.[50] Such findings suggest that insufficient islet capacity to proliferate in response to metabolic demands may be a crucial factor in the development of human IDDM.[47] The distinct fate of BKS vs. B6 islets grafted into the spleen of syngeneic, diabetic recipients, is perhaps due to differential inducibility of B-cell "junk" genes. Such difference is represented by the glucose-inducible and strongly immunogenic IAP genomes expressed in BKS, but not in B-cells of B6 mice.[51] C3H.SW/SnJ mice are another example that supports the importance of B-cell retrovirus expression. Despite a non-obese C3H.SW/SnJ phenotype, male mice have the incredible genetic ability to expand B-cell mass by hyperplasia, and the presence of the *db* mutation in the homozygous state gives a severe BKS-*db*/*db*–like pathology, which presumably results from expression of the endogenous B-cell retroviruses (both type C and IAP) in response to chronic hyperglycemia.[52] In addition to the insufficient B-cell proliferative response observed in diabetes-prone BKS mice carrying the *db* mutation, an increased insulin resistance and low insulin response to glucose were detected.[53] Overexpression of the human insulin-regulatable glucose transporter (GLUT4) overcomes the glucose transporter translocation defect and alleviates insulin resistance in transgenic BKS *db*/*db* mice, thus resulting in markedly improved glycemic control.[54]

In summary, the expression of the diabetogenic effects induced by the mutant *db* gene seems to strongly depend upon expression of additional modifying genes inside or outside of the MHC (H-2) locus. Such genes may determine the rate of islet cell replication and insulin synthesis, insulin resistance, expression of endogenous retroviruses, and active islet autoimmunity, resulting in different expression of the diabetes syndrome.

Role of the *db* mutation in susceptibility to various diabetogenic factors

Streptozotocin-induced diabetes and the* db *mutation. STZ-induced diabetes is a most attractive model of chemically induced disease because its development and pathogenic mechanisms resemble human IDDM. The B-cell–selective action of STZ may relate to the preferential islet uptake of the drug due to the glucose moiety of the molecule. STZ induces DNA breakage and activation of the poly (ADP-ribose) polymerase, with acute depletion of NAD in B-cells in the process of DNA repair.[55] The diabetogenic action of STZ strongly depends on its mode of administration: a high single dose rapidly induces overt diabetes due to acute chemical B-cell destruction,[56] while multiple low-dose STZ leads to delayed onset of diabetes associated with the appearance of cellular and humoral B-cell autoimmunity.[57] In addition to STZ, genetic factors influence the outcome of this treatment. At the prediabetes stage, young *db*/*db* mice are highly susceptible to either single high dose or multiple low doses of STZ compared with their *db*/*m* and *m*/*m* littermates. A single dose of STZ results in almost complete disappearance of pancreatic islets in the *db*/*db* mouse 48 h following injection, while the *db*/*m* and *m*/*m* mice have lower levels of blood glucose with less pronounced islet destruction (Fig. 2A).[58] *db*/*db* mice show the highest sensitivity to the action of low-dose STZ injection, particularly when given at normoglycemia. A lower level of STZ sensitivity is found in heterozygous *db*/*m* mice, and in mice

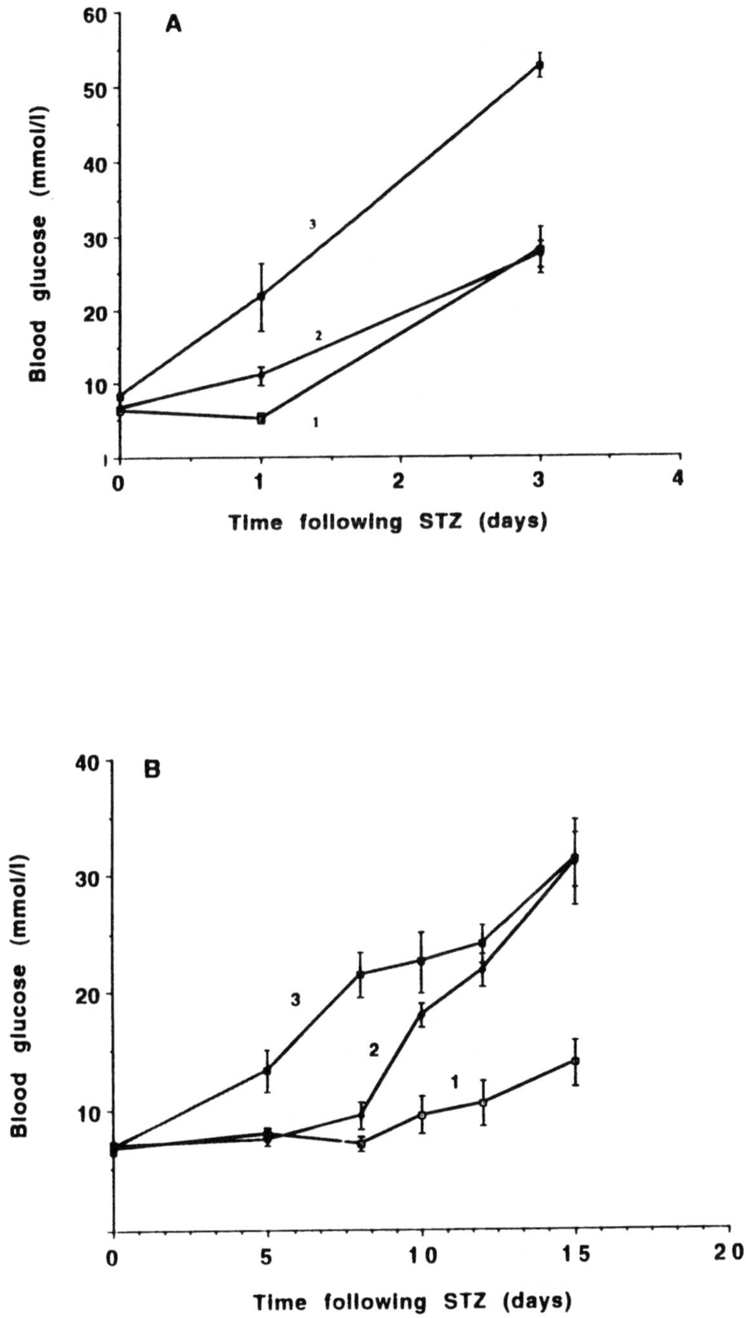

Figure 2. Streptozotocin-induced hyperglycemia in males of mutant BKS mice. 1. *m/m mice*; 2, *db/m* mice; 3, *db/db* mice. (A) High-dose STZ. (B) Multiple low-dose STZ. (Adapted from refs. 23 and 58, with permission.)

homozygous for the *m* gene (Fig. 2B). The level of hyperglycemia detected in these experiments directly correlated with the degree of the dystrophic islet changes and with the appearance of insulitis.[23] Unlike the post-STZ diabetes of the *db/m* heterozygous mice, diabetes still develops in young *db/db* mice following radiation, which inhibits post-STZ insulitis (Fig. 3), suggesting that STZ-induced insulitis does not directly cause B-cell destruction when low doses of STZ are given to young *db/db* mice at the prediabetic stage.[59] These data suggest that at an early stage of life, the *db/db* mouse is highly susceptible to the diabetogenic action of STZ, but, unlike most mice strains, this form of induced diabetes does not result from an immune attack.

db ***gene and radiation effects.*** A high dose of radiation was reported to protect against the development of IDDM, probably by eliminating anti-islet immunity. In contrast, low doses of radiation induce IDDM in rats with a defect of the immune system (thymectomy). Clinical features of the radiation-induced syndrome include hyperglycemia, insulinopenia, ketosis and corresponding islet atrophy, insulitis, and the appearance of autoantibodies to islet cells.[60] Such autoimmune

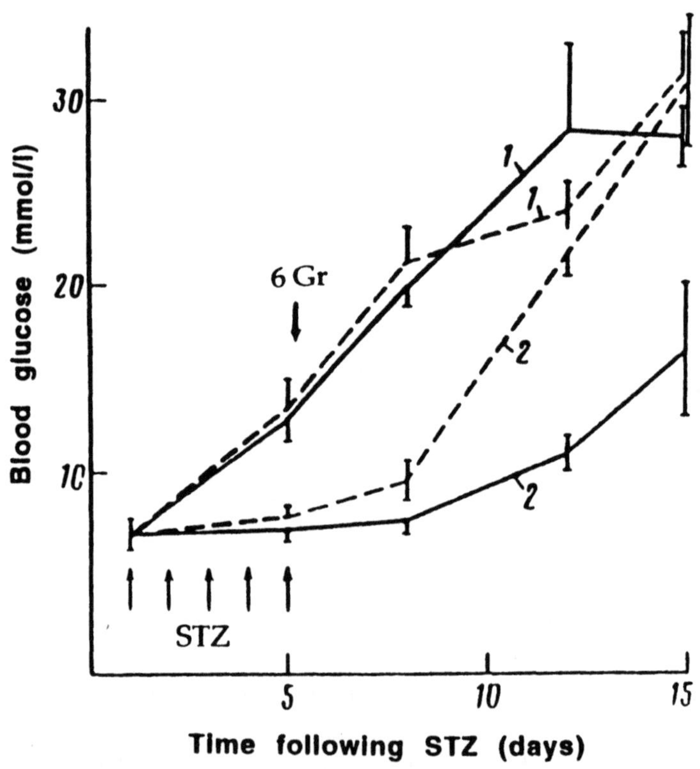

Figure 3. Level of blood glucose in *db/db* mice (1) and *db/m* mice (2) after STZ treatment (– – –) and STZ + irradiation [6 Gr] (———). (Adapted from ref. 59, with permission.)

phenomena could result from a selective depletion of a regulatory T-cell population. Interestingly, BKS *db/db* mice at the prediabetic period were significantly more sensitive to high and low doses of γ-radiation with a 100% rate of mortality, as compared with 5% in *db/m* mice (death occurred within 30 days after whole body irradiation using 6.0 Gr). In addition, a more profound postradiation (0.5 Gr) DNA degradation was found in radiosensitive organs, such as the thymus and the spleen, compared with *db/m* and *m/m* mice.[59,61] These results suggest that the increased radio sensitivity of the young *db/db* mouse could result from a transient immunodeficiency state at the prediabetic stage.

db/db *B-cell response to secretagogues.* Long-term sulfanilamide therapy has been shown to result in reduced islet function in normal rodents,[62] while it potentiates the effect of STZ-induced diabetes.[63] Recently it was found that long-term administration of glibenclamide to BKS mice coexpressing *db* genes enhanced B-cell damage with the development of insulitis; this effect was not obtained in normal littermates. Again, the *db* gene seems to interfere with various diabetes-related processes, including the B-cell cytotrophic effect obtained by long-term sulfanilamide therapy, and with B-cell immune recognition.[64]

Dietary effects in* db/db *mice. Mutant *db/db* mice are hyperphagic and, when allowed food ad libitum, impressive obesity and diabetes result.[4] Restriction of overeating results in delay of the development of diabetes and its complications.[65] B-cell necrosis and islet atrophy in the final stages of disease appear to be due to a dietary component, especially high carbohydrate intake. Total substitution of carbohydrate for casein markedly reduces the severity of the diabetic syndrome in *db/db* mice.[66,67] Such a diet extends the life span, slows down the decrease in islet insulin content,[68] and delays B-cell necrosis.[66] However, in parallel with the beneficial effects, restriction of food intake in the BKS *db/db* mice may induce unfavorable changes, such as immunodeficiency (absolute and relative deficiency in spleen lymphocyte subsets and total number of spleen cells; inability to produce agglutinating antibody to sheep erythrocytes and specific antibody to noninfectious CB4), resulting in high susceptibility to diabetogenic viruses.[69] Of importance is the finding that heterozygous *db* mice are more resistant to starvation than normal mice, as reflected by a longer life span.[70] Analogously, *db*-like genes in humans could be preserved during evolution due to their adaptive advantages in conditions of chronic food deficiency.

Susceptibility of mutant* db/db *mice to the action of diabetogenic viruses. The role of viral agents in triggering the development of IDDM was suggested by many investigators, but no convincing evidence was produced to sustain this hypothesis. To evaluate the role of the *db* gene in the development of B-cell susceptibility, BKS mice carrying different dose of this gene were infected with the diabetogenic TC-83 vaccine containing Venezuelan encephalitis virus (TC-83VE). Following inoculation, the mortality rate varied between 5% and 10% and did not differ among the three groups of mice (+/+, *db/* + , *db/db*) in the study. However, in *db/db* mice, a profound decrease in B-cell granulation with marked mitochondrial degeneration was observed in the islets, while such changes were absent in the other groups. In addition, the most striking effect of the vaccine was seen in homozygous *db/db* mice that had a higher level of glucose intolerance, hypoinsulinemia, and reduced pancreatic insulin content.[71] Moreover, a significant correlation was found between the dose of *db* gene in the BKS mouse genome and the degree of susceptibility to the human diabetogenic CB4 viruses.[72] Life-table analysis for the mortality rate and the severity of islet pathology in *db/db* mice inoculated with the CB4 vaccine showed that homozygous *db/db* mice were more sensitive than their heterozygote (*db*/+) and control (+ /+) littermates. The increased susceptibility in *db/db* mice is virus induced but is mediated by the host genotype and not by the degree of virus replication.[73] The host response to the CB4 virus differed between *db/db* mice with and without overt diabetes and, unlike ad libitum–feeding hyperglycemic animals, diet-restricted normoglycemic *db/db* mice developed immunodeficiency and were unable to produce specific antibody to noninfections CB4. Because of their immunodeficient state, such mice were more susceptible to viral infection than those with phenotypic expression of the *db* mutation.[69] Thus, the prediabetic immunodeficiency in BKS *db/db* mice seems to support the idea that viral infection could directly elicit B-cell damage with diabetes onset.

Concluding remarks

Human diabetes is now more profoundly understood, not only due to dynamic research conducted on humans but also due to information generated in animal models of different diabetic syndromes. In this context, the autosomal recessive *db* mutation offers both an optimal approach to the understanding of diabetogenic gene expression, when acting together with modifying genes of different genomic backgrounds, and offers a model to study the influence of the *db* gene on susceptibility to various environmental factors, such as chemical substances, viruses, drugs, and radiation. Particularities of diabetogenesis in the *db/db* mouse allow the study not only of mechanisms leading to the development of inherited NIDDM, but also complex processes leading to insulin deficiency, and in this context the latest research data demonstrate that the *db* gene not only determines NIDDM development, but also increases susceptibility to different diabetogenic environmental factors, resulting in B-cell necrosis and progression to IDDM in early stages of the disease.

The heterogeneity of human diabetes, which is grossly divided in NIDDM and IDDM, still does not explain the complex clinical presentation. Secondary failure to oral hypoglycemic agents in adult NIDDM patients and their transfer to insulin dependency indicate that the B-cell may undergo dynamic changes from a hypersecreting to a hyposecreting organ in the same individual. Whether this group of patients constitutes a new "intermediate" group between IDDM and NIDDM with a different genetic background, or their clinical syndrome merely reflects the result of environmental interference on the same genetic background, is not yet clear. The *db* model suggests that NIDDM and IDDM may develop in the same animal but in different stages of their life, and nongenetic factors are of major importance in determining the type of diabetes.

Such findings are supported by recent reports on human diabetes showing that the risk of developing IDDM in offspring of parents with NIDDM is increased. Moreover, a significant cumulative risk for IDDM probands with NIDDM parents and increased frequency of HLA-DR4 was found in a study of subjects with NIDDM.[16,74] Recently it has been shown that a subgroup of patients with NIDDM presents features characteristic of IDDM, such as various circulating autoantibodies to pancreatic islets and HLA haplotypes, which determine the susceptibility to develop IDDM.[75] The previously mentioned transformation of the clinical picture in adult diabetic subjects from NIDDM to insulin dependency may suggest that the rate of B-cell destruction is either initiated relatively late in life or that the rate of destruction is very slow. In addition, accumulated data on the *db/db* mouse suggest that this animal reflects much of the same etiopathogenic factors that come into play in women with polycystic ovarian syndrome that progresses to obesity, insulin resistance, and NIDDM.[76] Thus, similar to many other cases in which animal models made major contributions to the understanding of human pathology, the *db* mutation offers a useful model for the study of human diabetes without prejudiced ideas.

References

1. Coleman DL, Hummel KP. Studies with the mutation, diabetes, in mouse. Diabetologia 3:328–48, 1967.
2. Leiter EH. The genetics of diabetes susceptibility in mice. FASEB J 3:2231–41, 1989.
3. Shafrir E. Animal models of non-insulin-dependent diabetes. Diabetes Metab Rev 8:179–208, 1992.
4. Hummel KP, Dickie MM, Coleman DL. Diabetes, a new mutation in the mouse. Science 153:1127–28, 1966.
5. Lyon MG, Searl AG. Genetic strains and variants of the laboratory mouse. Oxford University Press, Oxford, pp 265–6, 1989.
6. Bahary N, Leibel RL, Joseph L, et al. Molecular mapping of the mouse *db* mutation. Proc Natl Acad Sci USA 87:8642–46, 1990.
7. Truett G, Bahary N, Friedman J, et al. Rat obesity gene fatty (fa) maps to chromosome 5: Evidence for homology with the mouse gene diabetes (*db*). Proc Natl Acad Sci USA 88:7806–9, 1991.
8. Zhang Y, Proenca R, Maffei M, et al. Positional cloning of the mouse obese gene and its human homologue. Nature 372:425–32, 1994.
9. Molina JM, Premdas FH, Klenck RE, et al. The dynamic insulin secretory response of isolated pancreatic islets of the diabetic mouse. Evidence for a gene dosage effect on insulin secretion. Diabetes 33:1120–23, 1984.
10. Chick WL, Lavine RL, Like AA. Studies in the diabetic mutant mouse. 5. Glucose tolerance in mice homozygous and heterozygous for the diabetes (*db*) gene. Diabetologia 6:527, 1970.
11. Hummel KP, Coleman DL, Lane PW. The influence of genetic background on expression of mutations at the diabetes locus in the mouse. 1. C57B1/KsJ and C57BL/6J strains. Biochem Genet 7:1–13, 1972.
12. Kaku K, Province M, Permutt MA. Genetic analysis of obesity-induced diabetes associated with a limited capacity to synthesize insulin in C57BL/KS mice: Evidence for polygenic control. Diabetologia 32:636–43, 1989.

13. Leiter EH, Le PH, Coleman DL. Susceptibility to *db* gene and streptozotocin-induced diabetes in C57BL mice: Control by gender-associated, MHC-unlinked traits. Immunogenetics 26:6–13, 1987.
14. Coleman DL, Hummel KP. Influence of genetic background on expression of mutations at the diabetes locus in the mouse. 2. Studies on background modifiers. Isr J Med Sci 11:708–13, 1975.
15. Bach JF. Insulin-dependent diabetes mellitus as an autoimmune disease. Endocr Rev 15:516–42, 1994.
16. Rich SS, French LR, Sprafka JM, et al. HLA-associated susceptibility to type 2 (non-insulin-dependent) diabetes mellitus: The Wadena City Health Study. Diabetologia 36:234–8, 1993.
17. Leiter EH, Coleman DL, Hummel KP. The influence of genetic background on the expression of mutation at the diabetes locus in the mouse. 3. Effect of H-2 haplotype and sex. Diabetes 30:1029–34, 1981.
18. Leiter EH. Genetic control of pathogenesis of diabetes in C3H mice. Influence of the major histocompatibility complex. Diabetes 33:1068–71, 1984.
19. Prochazka M, Premdas F, Leiter E, et al. Estrone treatment dissociates primary versus secondary consequences of "diabetes" gene expression in mice. Diabetes 35:725–8, 1986.
20. Leiter EH, Chapman H. Obesity-induced diabetes (diabesity) in C57BLKS/J mice produces aberant trans-regulation of sex steroid sulfotransferase genes. J Clin Invest 93:2007–13, 1994.
21. Leiter EH, Chapman HD, Coleman DL. The influence of genetic background on the expression of mutations at the diabetes locus in the mouse. V. Interaction between the *db* gene and hepatic sex steroid sulfotransferases correlates with gender-dependent susceptibility to hyperglycemia. Endocrinology 124:912–22, 1989.
22. Gunnarson R. Function of the pancreatic B-cell during the development of hyperglycaemia in mice homozygous for the mutation "diabetes" (*db*) and "misty" (*m*). Diabetologia 11:431–8, 1975.
23. Bloch KO, Poltorak VV, Brindak OI, et al. Importance of *db* gene for the development of low dose streptozotocin diabetes in C57BL/KsJ mice. Probl Endokrinol Mosk 34:73–5, 1988 (in Russian).
24. Gill AM, Leiter EH, Powell JG, et al. Dexamethasone-induced hyperglycemia in obese A^{vy}/a (viable yellow) female mice entails preferential induction of a hepatic estrogen sulfotransferase. Diabetes 43:999–1004, 1994.
25. Jackson R, Rassi N, Crump T, et al. The BB diabetic rat: Profound T-cell lymphocytopenia. Diabetes 30:887–9, 1981.
26. Dardenne M, Savino W, Gastinel LN, et al. Thymic dysfunction in the mutant diabetec (*db/db*) mouse. J Immunol 130:1195–99, 1983.
27. Boillot D, Assan R, Dardenne M, et al. T-lymphopenia and T-cell imbalance in diabetic *db/db* mice. Diabetes 35:198–203, 1986.
28. Sai P, Kremer M, Maurel C. Antibodies spontaneously bound to islet cells in diabetic C57BL/KsJ *db/db* mice. Diabetologia 27:139–42, 1984.
29. Leiter EH, Yoon JW, Kuff EL. Spontaneous autoantibody production against insulin and retrovirus distinguishes C57B1/KsJ from C57B1/6J mice. Diabetes 34(Suppl 1):319A, 1985.
30. Debray-Sachs M, Dardenne M, Sai P, et al. Anti-islet immunity and thymic disfuction in the mutant diabetic C57BLKS/J *db/db* mouse. Diabetes 32:1048–54, 1983.
31. Nishimura M, Miyamoto H. Immunological influence of A^{vy}, *db*, *ob* and *nu* genes placed on the inbred NOD background as murine models for human type 1 diabetes. J Immunogen 14:127–30, 1984.
32. Prochazka M, Leiter EH, Serreze DV, et al. Three recessive loci required for insulin-dependent diabetes in nonobese diabetic mice. Science 237:286–9, 1987.
33. Pancrel P, Meretey K, Illyes Y. Immunogenetic study of the familial accumulation

of type 1 (IDDM) diabetes mellitus and rheumatoid arthritis. Diabetologia 30:566A, 1987.

34. Huang SW, Maclaren NK. Antibodies to nucleic acids in juvenile onset diabetes. Diabetes 27:1105–11, 1978.
35. Petri M, Yoo SS. Insulin resistance in systemic lupus erythematosus. Arthritis Rheumat 37:S323, 1994.
36. Murphy ED, Roth JB. A Y chromosome associated factor in strain BxSB producing accelerated autoimmunity and lymphoproliferation. Arthritis Rheumat 22:1188, 1979.
37. Leiter EH, Prochazka M, Shultz LD. Effect of immunodeficiency on diabetogenesis in genetically diabetic (*db/db*) mice. J Immunol 138:3224–29, 1987.
38. Steinberg AD, Krieg AM, Gourley MF, et al. Theoretical and experimental approaches to generalized autoimmunity. Immunol Rev 118:129–63, 1990.
39. Kolb H, Freytag G, Kiesel U, et al. Spontaneous autoimmune reaction against pancreatic islets in mouse strains with generalized autoimmune disease. Diabetologia 19:216–21, 1980.
40. Bloch KO, Poltorak VV. Diabetogenic effects of the *db* (diabetes) gene in mice with inherited immunopathology. In: Hormonal regulation in health and pathology. Kharkov, USSR, p 11, 1989 (in Russian).
41. Bloch KO, Poltorak VV. Expression of the *db* (diabetes) gene in mice with hereditary load by generalized autoimmune pathology. Tsitol Genet 25:40–5, 1991 (in Russian).
42. Leiter EH. Type C retrovirus production by pancreatic beta cells. Association with accelerated pathogenesis in C3H-*db/db* ("diabetes") mice. Am J Pathol 119:22–32, 1985.
43. Leiter EH, Kuff EL. Intracisternal Type A particles in murine pancreatic B cells. Immunocytochemical demonstration of increased antigen (p73) in genetically diabetic mice. Am J Pathol 114:46–55, 1984.
44. Leiter EH, Fewell JW, Kuff EL. Glucose induces intracisternal type A retroviral gene transcription and translation in pancreatic beta cells. J Exp Med 163:87–100, 1986.
45. Appel MC, Rossini AA, Williams RM, et al. Viral studies in streptozotocin-induced pancreatic insulitis. Diabetologia 15:327–36, 1978.
46. Fujita H, Fujino H, Nonaka K, et al. Retrovirus-like particles in pancreatic beta-cells of NOD (non-obese diabetic) mice. Biomed Res 59:67–70, 1984.
47. Swenne I. Pancreatic beta-cell growth and diabetes mellitus. Diabetologia 35:193–201, 1992.
48. Eizirik DL, Sandler S, Palmer JP. Repair of pancreatic beta-cells. A relevant phenomenon in early IDDM? Diabetes 42:1383–91, 1993.
49. Sweene I, Andersson A. Effect of genetic background on the capacity for islet cell replication in mice. Diabetologia 27:464–7, 1984.
50. Andersson A. The influence of hyperglycaemia, hyperinsulinaemia and genetic background on the fate of intrasplenically implanted mouse islets. Diabetologia 25:269–72, 1983.
51. Leiter EH. Analysis of differential survival of syngeneic islet transplanted into hyperglycemic C57BL/6J versus C57BLKS/J mice. Transplantation 44:401–6, 1987.
52. Leiter EH. Control of spontaneous glucose intolerance, hyperinsulinemia, and islet hyperplasia in nonobese C3H.SW male mice by Y-linked locus and adrenal gland. Metabolism 37:689–96, 1988.
53. Orland MJ, Permutt MA. Genetic susceptibility to diabetes in inbred strains of mice: Measurements of proinsulin mRNA and response to dexamethasone. Diabetologia 30:934–9, 1987.
54. Gibbs EM, Stock JL, Mccoid SC, et al. Glycemic improvement in diabetic *db/db* mice by overexpression of the human insulin-regulatable glucose transporter (GLUT4). J Clin Invest 4:1512–18, 1995.

55. Okamoto H. Regulation of proinsulin synthesis in pancreatic islets and new aspect to insulin-dependent diabetes. Mol Cell Biochem 37:43–61, 1981.
56. Takieten N, Rakieten M, Nadkarni M. Studies on the diabetogenic action of streptozotocin. Cancer Chemother Rep 29:91, 1963.
57. Kolb H, Kroncke KD. IDDM. Lessons from the low-dose streptozotocin model in mice. Diabetes Rev 1:116–26, 1993.
58. Bloch KO, Poltorak VV, Brindak OI. Effect of mutant genes *db*(diabetes) and m(misty) on the severity of high-dose streptozotocin diabetes in C57b1/KsJY mice. Tsitol Genet 25:58–61, 1991 (in Russian).
59. Bloch KO, Bloch OV, Poltorak VV, et al. The effect of irradiation on the development of streptozotocin diabetes in C57BL/KsJY mice heterozygous and homozygous for gene *db* (diabetes). Radiobiologiia 31:43–7, 1991 (in Russian).
60. Stumbles PA, Penhale WJ. IDDM in rats induced by thymectomy and irradiation. Diabetes 42:571–8, 1993.
61. Bloch KO, Bloch OV. Influence of mutant gene *db* on radiosensibility of C57BL/KsJY mice. First All-Union Radiobiology Congress, Moscow, 2:43–4, 1989.
62. Ikeda T, Takeuchi T, Ito Y, et al. Potentiating effect of glibenclamide on the occurrence of streptozotocin diabetes. Horm Metab Res 18:352, 1986.
63. Sodoyez JC, Sodoyez-Goffiaux F, Dunbar JC, et al. Reduction in the activity of the pancreatic islets induced in normal rodents by prolonged treatment with derivatives of sulfonylurea. Diabetes 19:603–9, 1970.
64. Gladkikh AI, Poltorak VV, Brindak OI, et al. Effect of gene *db* dose on the development of beta-cytotropic and immunotropic effects of long-term glibenclamide administration in C57BL/KsJY mice. Tsitol Genet 26:20–6, 1992 (in Russian).
65. Wyse BM, Dulin WE. The influence of age and dietary conditions on diabetes in the *db* mouse. Diabetologia 6:267–73, 1977.
66. Leiter EH, Coleman DL, Eisenstein AB, et al. Dietary control of pathogenesis in C57BL/KsJ *db/db* diabetes mice. Metabolism 30:554–62, 1981.
67. Leiter EH, Coleman DL, Ingram DK, et al. Influence of dietary carbohydrate on the induction of diabetes in C57BL/KsJ *db/db* diabetes mice. J Nutr 113:184–95, 1983.
68. Shafrir E. Nonrecognition of insulin as gluconeogenesis suppressant—a manifestation of selective hepatic insulin resistance in several animal species with type II diabetes: Sand rats, spiny mice and *db/db* mice. In: Lessons from animal diabetes. Shafrir E, Renold AE, eds. Libbey, London, 2:304–15, 1988.
69. Montgomery LB, Loria RM, Chinchilli VM. Immunodeficiency as primary phenotype of diabetes mutation *db*. Studies with coxsachievirus B4. Diabetes 39:675–82, 1990.
70. Coleman DL. Lessons from studies with genetic forms of diabetes in the mouse. Metabolism 32:162–4, 1983.
71. Rayfield EJ, Seto Y, Goldberg SL, et al. Venezuelan encephalitis virus-induced alterations in carbohydrate metabolism in genetically diabetic mice. Diabetes 28:799–803, 1979.
72. Webb SR, Loria RM, Madge GE, et al. Susceptibility of mice to group B Coxsackie virus is influenced by the diabetic gene. J Exp Med 143:1239–48, 1976.
73. Loria RM, Montgomery LB, Corey LA, et al. Influence of diabetes mellitus heredity on susceptibility to coxsackievirus B4. Arch Virol 81:251–62, 1984.
74. Quatraro A, Consoli G, Ceriello A, et al. Heterogeneity of diabetes mellitus [letter]. Diabetologia 30:678–9, 1987.
75. Groop L, Miettinen A, Groop PH, et al. Organ-specific autoimmunity and HLA-DR antigens as markers for beta-cell destruction in patients with type II diabetes. Diabetes 37:99–103, 1988.
76. Dunaif A. Insulin resistance in polycystic ovarian syndrome. Ann NY Acad Sci 687:60–4, 1993.

19

Inositols—Potential roles in insulin action and in diabetes: Evidence from insulin-resistant nonhuman primates

BARBARA C. HANSEN AND HEIDI K. ORTMEYER
Obesity and Diabetes Research Center, Department of Physiology, School of Medicine, University of Maryland at Baltimore, Maryland, USA

Key words: insulin action, insulin mediators, inositol, diabetes, monkey, NIDDM, diabetic complications

Myoinositol is an essential component of all animal cells and is found in plants, fungi, and some bacteria.[1] According to Billington,[2] myoinositol was first identified in cardiac muscle by Scherer in 1850. It is a component of membrane phospholipids, and the function of these membrane phosphoinositides may prove to be important to the action of insulin, as discussed later. Myoinositol, a cyclitol, is the most common of the nine isomers of hexahydroxycyclohexane. Two stereoisomers of myoinositol are D-chiroinositol and L-chiroinositol. In humans myoinositol is found mainly either in its free form or covalently bound to phospholipid, with three principal forms: phosphatidylinositol (PI), phosphatidylinositol-4-P (PIP or PI-4-P), or phosphatidylinositol 4,5-P_2 (PI-4,5-P_2 or PIP_2). The structures of myoinositol, of D-chiroinositol, and of several phosphatidylinositols are shown in Figure 1.

Hauser and Finelli[3] showed that mammalian tissues actively synthesize both free myoinositol and phosphatidylinositol, and later Clements and Diethelm[4] estimated the magnitude of the synthesis in the human kidneys alone to be about 4 g/day. Biosynthesis in the kidney as well as in other tissues greatly exceeds the amounts ingested (~1 g/day). PI biosynthesis has been localized to the endoplasmic reticulum and is probably transported by intracellular transfer proteins, which may play a role in PI function at the membrane.[5] D-glucose 6-phosphate is converted to inositol-1-phosphate and then dephosphorylated. Labeled inositol has been shown to be taken up by an active transport or carrier-mediated process and incorporated into phosphoinositides, and within several hours it was

Correspondence: Dr. Heidi K. Ortmeyer, Obesity and Diabetes Research Center, University of Maryland at Baltimore, 10 S. Pine St., Rm 6-00, Baltimore, MD 21201, USA. E-mail: hortmeye@umabnet.ab.umd.edu

Lessons from Animal Diabetes VI. Ed. E. Shafrir.

Free Myo-inositol D-chiroinositol

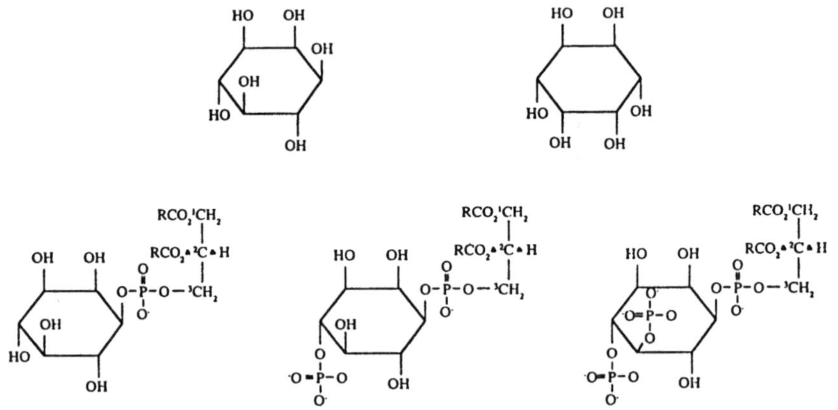

Phosphatidylinositol (PI)(Ptdlns) Phosphatidylinositol 4-P (PI4-P)(PIP) Phosphatidylinositol 4, 5-P2 (PI4, 5-P2)(PIP2)

Figure 1. Structures of inositols.

found to be concentrated in spleen, liver, pituitary gland, kidney, and thyroid gland,[6] but little was concentrated by muscle. The fatty acid compositions of PI, PIP, and PIP_2 are high in stearic and arachidonic acids.

The normal plasma concentration of free myoinositol is ~30 mM,[7] and smaller amounts circulate with lipoproteins. The myoinositol concentration is 20-fold higher in breast milk.[8] The normal plasma concentrations of D- and L-chiroinositol are ~0.10 and 0.03 μM, respectively.[9]

The dietary forms of inositol include free myoinositol, chiroinositol, inositol-containing phospholipids (phosphatidylinositol, PI), and inositol hexaphosphate (phytic acid). Americans are estimated to ingest 1 g/day of myoinositol,[10] and milk contains 30–80 mg/l.[11] Free inositol is actively transported across the small intestine.[12] The feeding of myoinositol significantly increased the levels of free myoinositol in the plasma, liver, kidney, and intestine of rats,[13] and in the plasma of human subjects.[14] Intestinal bacteria can also synthesize myoinositol.[15] Dietary deprivation of myoinositol can produce a significant increase in liver triacylglycerols due to decreased secretion rates[16] and increased fatty acid mobilization from adipose tissue.[17]

Myoinositol was found in urine from subjects with Bright's disease in 1856, and in 1858 in was discovered to be in large amounts in the urine of subjects with diabetes.[18] The kidney is the major organ of regulation of plasma levels of myoinositol and of the catabolism of myoinositol, with its breakdown leading to the pentose phosphate cycle. Free myoinositol is also excreted intact in small amounts.[4] In diabetes, myoinositoluria may be the result of glucose inhibition on renal tubular reabsorption, as well as reduced general cellular uptake, and it is restored to normal by insulin treatment.[14]

Inositol-containing phopholipids are found in the liver microsomal and Golgi vesicle membranes,[19] on the internal surface of the plasma membrane of platelets,[20] and in the inner monolayer of the sarcoplasmic reticulum lipid bilayer.[21] Phosphoinositols appear to play an important role in regulating enzyme activity and transport (e.g., activation of the (Na/K)ATPase in microsomes from kidney,[22] association with alkaline phosphatase and 5′-nucleotidase in plasma membrane[23]). PI has been identified as an essential constituent of acetyl-CoA carboxylase in rat liver,[24] and as being involved in anchoring acetylcholinesterase to the plasma membrane.[25] It is important in the formation of the eicosanoids, such as prostaglandins, thromboxanes, and leukotrienes. Free arachidonic acid release from the turnover of membrane PI occurs directly via phospholipase C cleavage (by phosphodiesterase activity) or indirectly by conversion to phosphoinositides hydrolyzed to diacylglycerols.[26] Phospholipase C (phosphodiesterase) activity in the degradation of membrane PI may or may not be involved in the regulation of CA^+ fluxes in response to hormones.[27] Further, Kirk et al.[28] have related enhanced PI metabolism to activation of glycogen phosphorylase. The possible involvement of PIs in receptor-mediated cellular responses (specifically in the activation of glycogen synthase) is illustrated in Figure 2. This figure is a composite of the work of many individuals, including J. Larner, G. Romero, C.R. Kahn, M. White, and P. Cohen.[29,30]

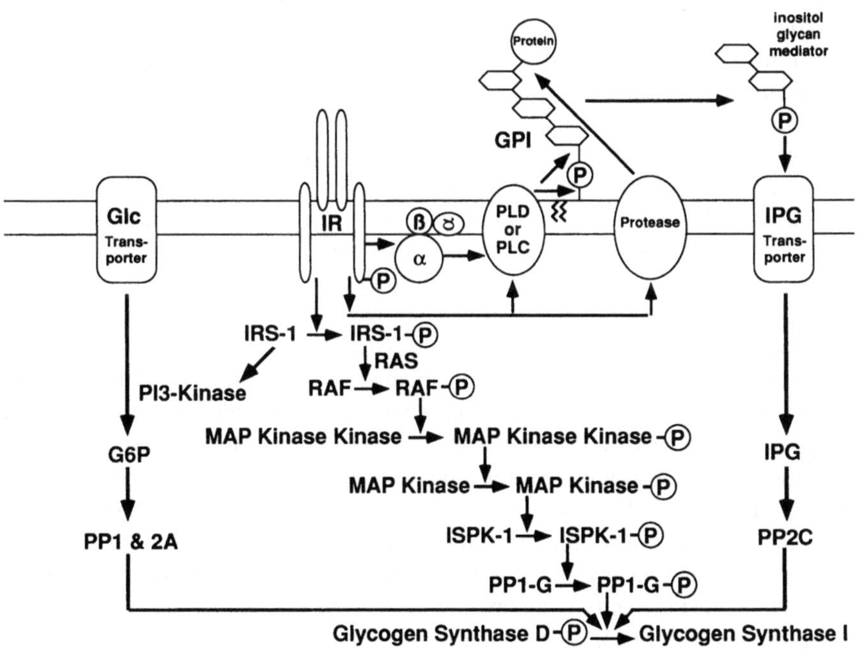

Figure 2. Model of insulin action on skeletal muscle glycogen synthase. IR, insulin receptor; P, phosphate; IRS-1, insulin receptor substrate-1; MAP, mitogen-activated protein; ISPK-1, insulin-stimulated protein kinase-1; PP1-G, glycogen-associated protein phosphatase; PLC, phospholipase C; PLD, phospholipase D; GPI, glycosyl-phosphatidylinositol; PP2C, protein phosphatase-2C.

One stereoisomer of myoinositol, chiroinositol, was first identified in an animal by Hipps et al.,[31] who mentioned its prior identification in plants as early as 1907. Chiroinositol was subsequently quantified in the urine of normal and uremic subjects, and was found to be elevated in uremia.[32]

Myoinositol and peripheral neuropathy

Three independent lines of evidence link possible defects in inositol metabolism to diabetes. A role for myoinositol depletion has been hypothesized in diabetic neuropathy and in diabetic nephropathy, and a role for defects in the inisotol-containing putative mediators of insulin action has been suggested to be involved in the insulin resistance of non–insulin-dependent diabetes mellitus (NIDDM).

Inositols were first found to be involved in neural transmission when polyphosphoinositides were shown to be rapidly metabolized in nerve trunks.[33] In experimental hyperglycemia in rodents, concentrations of myoinositol in peripheral nerve were reduced,[34] despite normal circulating levels of free myoinositol, and structural changes were also observed.[35] Myoinositol levels in nerves of diabetic humans were also shown to be reduced.[36] Motor nerve conduction velocity (MNCV) was slowed in experimentally hyperglycemic rodents,[37] monkeys with spontaneous NIDDM,[38] and diabetic humans.[39] Hyperglycemia probably inhibits both inositol transport into cells and renal tubular reabsorption of inositol.

Both insulin treatment and dietary supplementation with myoinositol improved nerve function. Clements and Reynertson[14] obtained elevated plasma levels of myoinositol following a 3 g oral load and suggested that such supplementation could be useful in diabetic patients. Salway et al.[40] showed improvement in the neurophysiology of diabetic patients with 0.5 g oral inositol daily for 2 wk. However, in one study high levels of dietary supplementation (3%) were associated with decreased MNCV in both normal and experimentally diabetic rodents.[34]

Dietary supplementation with myoinositol, with the goal of increasing myoinositol content in nerves and of improving hyperglycemia-associated neuropathy, was attempted by Greene et al.[34] They were the first to demonstrate that a diet containing 1% myoinositol improved MNCV in streptozotocin (STZ) diabetic rats. MNCV in these myoinositol-treated rats was still lower than in control rats, but was significantly higher compared with untreated diabetic rats. One group of rats showed a positive relationship between MNCV and nerve myoinositol content. Although several rodent studies have supported these original findings of improved MNCV with myoinositol supplementation,[41–44] others have not found myoinositol to improve MNCV in rats[45,46] or in humans.[47–49]

Other alterations in myoinositol and phosphoinositide metabolism lead to reduced (Na/K)ATPase activity by unknown mechanism, possibly via reduced availability of diacylglycerol, as hypothesized by Kim et al.[50] The (Na/K)ATPase defect in nerve associated with myoinositol depletion in STZ-diabetic rats was improved by myoinositol supplementation in vivo.[50] A similar reduction in (Na/K) ATPase activity has been shown in the pancreatic B-cells, but, at least in cell cultures, myoinositol supplementation did not alter glucose-induced islet desensitization.[50]

Uptake of myoinositol is via a sodium- and energy-dependent transporter, and the gene has been cloned.[52] Simmons et al.[53] provided evidence that the myoinositol transporter, which is inhibited by glucose, was also affected by intracellular lipid metabolism. This glucose-induced reduction in myoinositol uptake was reversed in cultured neuroblastoma cells by their return to normal glucose levels.[54] This decrease in myoinositol transport caused a decrease in (Na/K)ATPase activity, which was reversed by myoinositol supplementation of STZ-diabetic rats.[55] In addition, the reduced Mg^{2+} levels in diabetes could be a factor in reduced inositol transport.[56] These reports suggest a key role for myoinositol deficiency in diabetic neuropathy. Oral administration of inositol (12 g/day) to eight humans increased CSF inositol levels by ~70%,[57] suggesting the potential of other neural effects of systemic increases in myoinositol that should be further considered.

Relation of early renal changes of diabetes to inositol

The development of diabetes is accompanied by an increased glomerular filtration rate (GFR) and by increased activity of (Na/K)ATPase. The increase in renal medullary (Na/K)ATPase in STZ-diabetic rats was completely prevented by 1–2 wk of oral administration of myoinositol,[58] while the increased GFR was unaffected. It seems clear that extracellular glucose competitively inhibits the sodium-dependent uptake of myoinositol[59]; however, Whiteside and Thompson[60] demonstrated an upregulation of myoinositol transport that might overcome this inhibition. Guzman and Crews[61] found that the upregulation of inositol transport in cultured glomerular mesangial cells, induced by hyperglycemia, involved activation of protein kinase C, possibly representing an adaptation to hyperglycemia. Thus, intracellular myoinositol levels in diabetic tissues have been reported to be increased, decreased, or unaffected, suggesting the possibility that long-term regulation, perhaps by intracellular myoinositol itself, counters impaired uptake.[62]

Inositol-containing insulin mediators

Inositol has been found to be a component of substances thought to be involved in some of the actions of insulin. In recent years, within the context of broad efforts to understand the molecular basis of insulin action, research has been concentrated on two types of intracellular events that follow insulin binding to the insulin receptor. One is the examination of changes in the phosphorylation state of the receptor, specifically autophosphorylation of the β-subunit,[63] resulting in a receptor with tyrosine kinase activity further phosphorylating intracellular substrates such as IRS-1, and subsequent activation of serine/threonine kinases.[64] The other type of intracellular event believed to be involved in insulin action is the generation/release of intracellular mediators/modulators that regulate the activity of protein kinases and phosphatases. Among the candidates is a family of insulin-dependent enzyme modulators first described by Saltiel and Cuatrecacas[65] that contain inositol. Larner et al.[66] had previously partially purified some insulin-stimulated glycopeptides, suggesting they contained amino acids and sugars, and Saltiel and Cuatrecacas[65] identified their components as an inositol and an amino sugar, and probably neutral sugars.

Myoinositol had earlier been determined to be a component of some phospholipids (phosphatidylinositol), and to be a constituent of membrane lipids in animal cells, and was subsequently shown to be a constituent of anchors of many membrane proteins.[67] Myoinositol was further shown to be a component of specific insulin-sensitive glycophospholipids,[65,68] or phospho-oligosaccharides,[69] which are structurally related to the glycosyl-phosphatidylinositol (GPI) anchors of membrane proteins.[67] These compounds are also referred to as *inositol glycans* or *inositol phosphoglycan mediators*.[67] Efforts to determine the composition and structure of these proposed mediators of insulin action are as yet incomplete, although evidence has been developed in the past 8 years concerning components of such mediators. By various techniques, a number of different possible mediators, but not their detailed structures and functions, have been suggested.

Chiroinositol has also been identified in the hydrophobic anchoring domain of acetylcholinesterase[25,70] and of human placental alkaline phosphatase.[70] Mato et al. identified a chiroinositol-containing insulin-sensitive glycophospholipid,[68] and this was confirmed to be the D-isomer of chioroinositol by Larner et al.[71] Larner et al.[72] separated two putative mediators based on their elution at pH 2.0 (containing chiroinositol and galactosamine) and at pH 1.3 (containing myoinositol and glucosamine). The former activates pyruvate dehydrogenase (PDH) phosphatase, and the latter inhibits cAMP-dependent protein kinase.

The mechanism of release of these insulin-sensitive inositol phosphate glycans (IPG) was hypothesized by Saltiel et al.[73] to be as follows: 1) Insulin activates the insulin receptor tyrosine kinase, 2) the activated receptor stimulates (possibly via a G protein) a GPI-specific phospholipase C (PI-PLC), 3) the phospholipase C catalyzes the hydrolysis of a free GPI on the cytoplasmic side of the plasma membrane, and 4) this results in the intracellular generation/release of the IPG and the generation/release of diacylglycerol from a specific lipid precursor. Romero and Larner[30] have hypothesized that the GPI lipid is located at the outer leaflet of the plasma membrane so that the PLC hydrolysis of the GPI results in the IPG being released to the extracellular medium. The IPG must then be transported into the cell and exert its effects via direct interactions with specific kinases and phosphatases (see Fig. 2).

The inositol phosphoglycan insulin mediators have been shown to have many biological effects, including activation of glycogen synthase phosphatase,[74] PDH,[75] PDH phosphatase,[76] and cAMP phosphodiesterase,[77] and inhibition of cAMP-dependent protein kinase.[74] Note that some of the actions of the putative insulin mediators were demonstrated before they were known to be inositol phosphoglycans. The myoinositol- and chiroinositol-containing insulin mediators were shown to stimulate glucose incorporation into glycogen and to lower plasma glucose concentrations in STZ-diabetic rats.[78] The D-chiroinositol–containing insulin mediator was also shown to activate glycerol-3-phosphate acyltransferase in adipocytes of the Goto-Kakizakì (GK) rat (a lean animal model of NIDDM).[79] Recently, it has been hypothesized that the D-chiroinositol–containing mediator activates glycogen synthase by allosterically activating protein phosphatase 2C.[80]

Urinary excretion of inositol and insulin resistance

As noted earlier, a high urinary excretion rate of myoinositol in diabetic subjects has been found, and could be hypothesized to reflect defective metabolism of inositol and of inositol-containing mediators of insulin action. The mechanism of this high urinary excretion rate of myoinositol in diabetes was proposed by Daughaday and Larner[81] to be glucose inhibition of the renal reabsorption of myoinositol. Elevated myoinositol content in urine has recently been shown to be accompanied by low D-chioroinositol excretion in both humans and nonhuman primates with NIDDM.[82] The obese insulin-resistant rhesus monkeys also had a tendency toward lower urine chiroinositol content compared with normal monkeys. This observation, along with the finding that in two insulin-deficient hyperglycemic experimental models, STZ-treated rats, and partially pancreatectomized dogs, no increase in urine chiroinositol content was found, suggested that urinary excretion of chiroinositol might be related to insulin resistance. The inbred lean diabetic GK rat was also found to have significantly lower urine chiroinositol excretion compared with Wistar rats.[83] Recently, urinary chiroinositol excretion was found to be significantly lower in non-obese NIDDM Japanese subjects compared with impaired glucose-tolerant (IGT) and normal glucose-tolerant subjects.[84]

In studies in three groups of monkeys — normal, insulin-resistant, and NIDDM — we examined whether the altered urinary glucose content was a direct consequence of elevated plasma glucose concentrations, or whether it might be more linearly related to insulin resistance and the progressive pathophysiology of diabetes.[85] Euglycemic hyperinsulinemic clamps were performed and compared with the urinary excretion rates of chiroinositol in 27 rhesus monkeys.[85] Maximal whole-body insulin-mediated glucose uptake rate, as an indicator of insulin sensitivity, was found across this range of normal, insulin-resistant, and diabetic monkeys to be correlated with urinary chiroinositol excretion rate, with the highest levels in normals and the lowest in diabetics ($r = 0.47$, $P < 0.02$). When only those monkeys with low urinary chiroinositol excretion rates were considered, this relationship became even stronger ($r = 0.65$, $P < 0.01$), suggesting that there is a wide range of normal urinary excretion of chiroinositol, but when a reduced level of excretion is observed, insulin resistance is much more likely to be present. Urinary chiroinositol excretion was also correlated with the insulin sensitivity index and glucose effectiveness in the normal glucose-tolerant, IGT, and non-obese diabetic Japanese subjects.[84]

Our experiments also included evaluation of the ability of insulin to activate glycogen synthase.[85] In muscle biopsies obtained prior to and during the euglycemic hyperinsulinemic clamp, insulin activation of skeletal muscle glycogen synthase activity was also found to be correlated with urinary chiroinositol excretion rate ($r = 0.38$, $P < 0.05$). As with whole-body insulin resistance, when only the group of monkeys with low urinary chiroinositol excretion rates was considered, the relationship was even stronger ($r = 0.73$, $P < 0.001$). The urinary chiroinositol excretion rates, whole-body insulin-mediated glucose disposal rates, and skeletal muscle glycogen synthase activities in the normal, insulin-resistant, and NIDDM monkeys are shown in Figure 3.

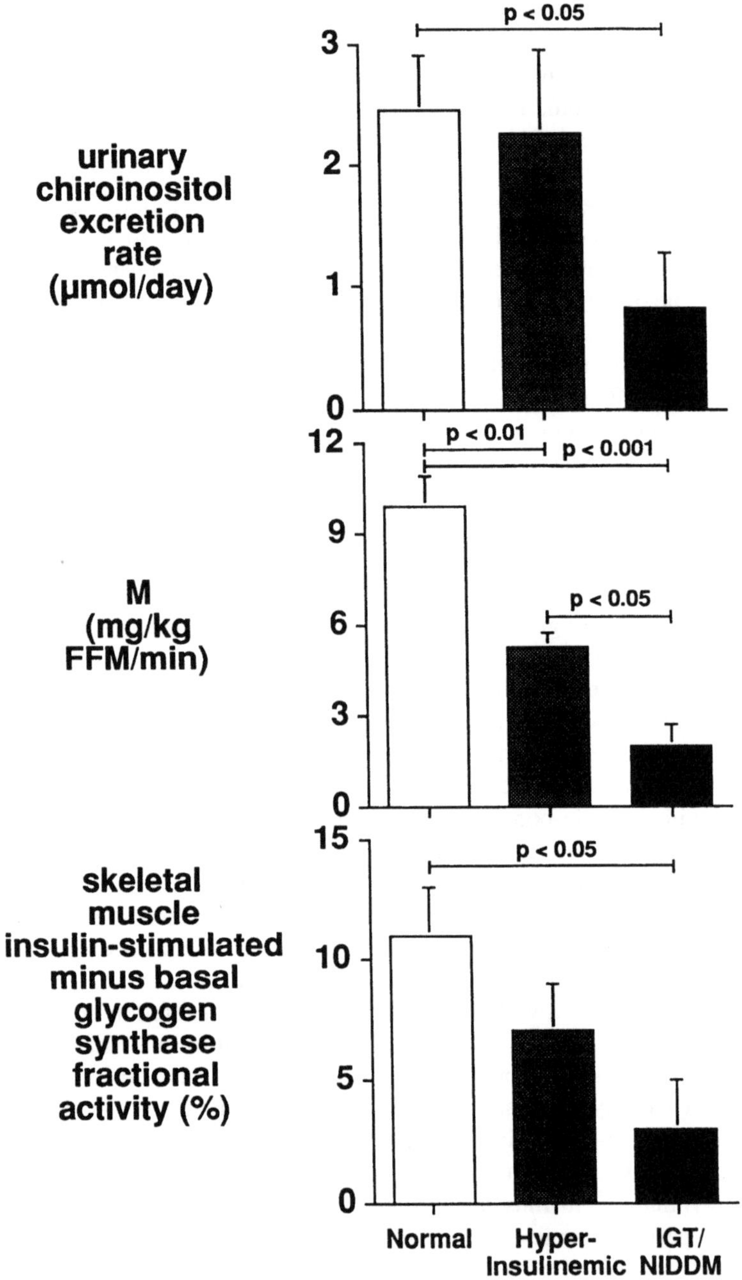

Figure 3. Urinary chiroinositol excretion rate, whole-body insulin-mediated glucose disposal rate (M), and skeletal muscle glycogen synthase activity in normal, hyperinsulinemic, and impaired glucose-tolerant/diabetic monkeys (IGT/NIDDM). (Redrawn, with permission, from ref. 85.)

Inositol administration and glucose metabolism

The previous studies of myoinositol supplementation to augment deficient myoinositol levels in the nerves of diabetics, together with the reduced urinary chiroinositol excretion rate in insulin-resistant subjects, and the identification of possible inositol-containing mediators of insulin action, led to examination of the possible therapeutic benefit of augmenting inositol levels. Numerous studies have examined the effects of myoinositol administration on glucose metabolism in STZ-diabetic rats[44,45,86–90] and in IDDM humans.[47–49] The primary goal of several of these studies was to examine the effects of myoinositol on peripheral neuropathy, as noted previously. The doses used in STZ-diabetic rats ranged up to 2% of diet by weight for 9 months, and up to 1 g/kg per day for 8 wk. The doses used in the studies of human subjects ranged up to 280 mg/kg per day for 2 wk and in another study ~40 mg/kg per day for 24 wk. In none of these studies was any effect of myoinositol on lowering hyperglycemia found, despite changes sometimes observed in plasma and muscle myoinositol content and improvements in MNCV.

In the first study that resulted in a positive effect of inositol on lowering plasma glucose concentrations, Narayan et al.[91] administered pinitol [D-chiro (+)-o-methyl inositol] orally to alloxan-diabetic animals. This compound was obtained by alcohol extraction of the leaves of *Bougainvillea spectabilis*. A dose of 0.01 g/kg was given to normal mice and to alloxan-diabetic mice at 24, 28, 48, 52, and 72 h post-alloxan to examine blood glucose–lowering effects. They found significant lowering of glucose in normal fasted albino mice, and in alloxan-diabetic mice, and improved glucose tolerance in the latter group.

Several years later, the effect of D-chiroinositol was examined in normal and in STZ-diabetic rats.[92] Free D-chiroinositol was administered intragastrically in a single dose (10 mg/kg) to normal and STZ-diabetic rats to examine the acute effects. In normal rats a single intragastric dose of D-chiroinositol (2–25 mg/kg) produced a 30–50% decrease in plasma glucose when a glucose load was given i.p. 2 h later. Also, the glucose disappearance rate was increased by 50% ($P < 0.05$). In STZ-diabetic rats, a single dose of intragastric D-chiroinositol produced a 30–40% decrease in plasma glucose ($P < 0.05$). A dose of 10 mg/kg myoinositol had no effect on the glucose disappearance rate.[92]

In an associated study, spontaneously insulin-resistant and diabetic rhesus monkeys were given i.v. D-chiroinositol in a single dose of 100 mg/kg. Both the glucose and insulin disappearance rates in the 30 min following an i.v. glucose load were increased, indicating a mean improvement in glucose tolerance by 30%, without increased insulin output.[92]

In another study of the effects of i.v. D-chiroinositol in monkeys, a bolus i.v. infusion of 1 g/kg D-chiroinositol was superimposed on a steady-state euglycemic hyperinsulinemic clamp to determine whether it might alter insulin action in insulin-resistant monkeys.[93] Both the independent activity ($P < 0.05$) and the fractional velocity ($P < 0.05$) of glycogen synthase were increased, and the independent activity ($P < 0.05$) and activity ratio ($P < 0.05$) of glycogen phosphorylase were decreased by D-chiroinositol, while total activities were unaf-

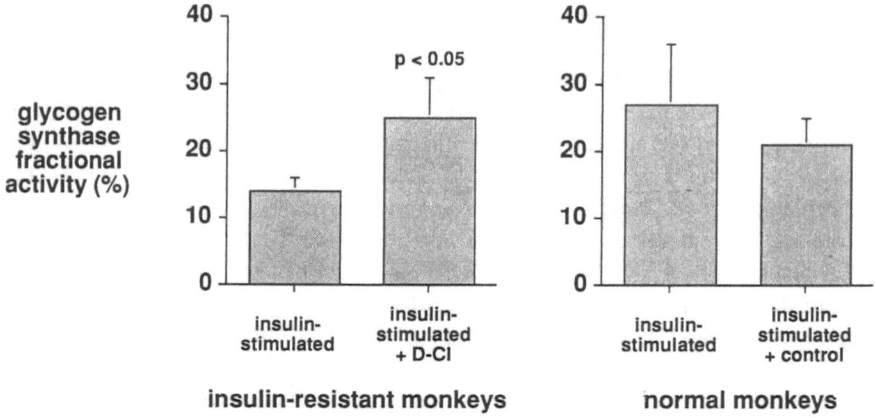

Figure 4. Effect of i.v. D-chiroinositol (D-CI) on glycogen synthase activity during a euglycemic hyperinsulinemic clamp in insulin-resistant monkeys and the effect of vehicle on glycogen synthase activity during a euglycemic hyperinsulinemic clamp in normal monkeys. (Redrawn, with permission, from ref. 93.)

fected. Figure 4 shows the skeletal muscle glycogen synthase activity before and after D-chiroinositol administration in insulin-resistant monkeys compared with vehicle in normal monkeys. Figure 5 shows the skeletal muscle glycogen phosphorylase activity before and after D-chiroinositol administration in insulin-resistant monkeys compared with vehicle in normal monkeys.

As noted earlier, intragastric administration of chiroinositol into normal or STZ-diabetic rats was shown to lower postprandial glucose in both groups.[93] This finding served as the basis for further detailed studies of oral administration of chiroinositol to insulin-resistant rhesus monkeys.

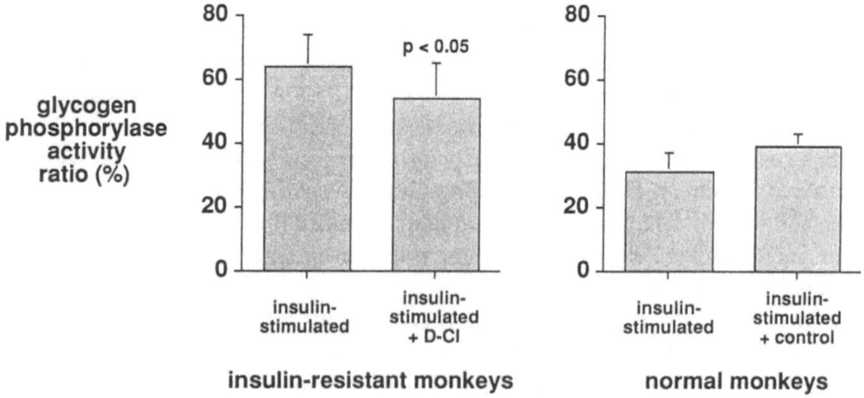

Figure 5. Effect of i.v. D-chiroinositol (D-CI) on glycogen phosphorylase activity during a euglycemic hyperinsulinemic clamp in insulin-resistant monkeys and the effect of vehicle on glycogen phosphorylase activity during a euglycemic hyperinsulinemic clamp in normal monkeys. (Redrawn, with permission, from ref. 93.)

To examine the acute effects of orally administered D-chiroinositol, meal tolerance tests were carried out with and without inclusion of chiroinositol in the meal. A dose of 0.5 g/kg incorporated into a liquid diet meal resulted in significantly lower postprandial plasma glucose concentrations in six insulin-resistant rhesus monkeys during the period 120–300 min after meal ingestion ($P < 0.05$).[94] This glucose lowering effect was accomplished without any increase in plasma insulin levels above those of the meal alone.

The effects of oral administration of another inositol, myoinositol, were examined using doses higher than those previously used in rodent or human studies of myoinositol supplementation for neuropathy-related purposes. In a group of six obese insulin-resistant monkeys, myoinositol was administered by incorporation into the diet of the animals.[95] Five days of dietary supplementation of 1.5 g/kg per day significantly lowered the urine glucose excretion rate ($P < 0.01$) and postprandial plasma glucose concentrations ($P < 0.05$) in these monkeys. There were no increases in plasma insulin concentrations during the myoinositol-supplemented period. It is likely that at least some of the myoinositol was converted to chiroinositol, and that the apparent glucose-lowering effect of myoinositol was accomplished by chiroinositol. The in vivo conversion of labeled myoinositol to labeled chiroinositol in rat urine, liver, muscle, blood, intestine, brain, kidney, and heart has been described by Pak et al.[96]

Thus, supplementation with both D-chiroinositol and myoinositol has provided evidence for a possible positive therapeutic effect in diabetic subjects through the apparent effects of these inositols to chronically lower postprandial glucose levels. Large doses of myoinositol (4% w/v of diet) may be required for this effect, and such a dose level is generally higher than those previously studied in attempts to mitigate the complications of diabetes. Thus, the possibility of toxicity of chronic dosing will require further consideration.

Conclusions

Abundant evidence supports the hypothesis that inositol metabolism is abnormal in experimental forms of diabetes in rodents and in spontaneous diabetes in nonhuman primates and in humans. Myoinositol supplementation has been shown by some investigators to improve abnormal nerve function associated with depleted nerve inositol, and has been demonstrated in recent studies to lower plasma glucose concentrations in monkeys with IGT and insulin resistance. Chiroinositol has been shown to be related to insulin sensitivity and insulin action on glycogen synthase in monkeys. D-chiroinositol supplementation lowers plasma glucose concentrations in impaired glucose-tolerant and insulin-resistant monkeys, and improves insulin action on skeletal muscle glycogen synthase activity in monkeys. The generation of the D-chiroinositol–containing mediator may activate a protein phosphatase (protein phosphatase 2C), which in turn activates glycogen synthase by dephosphorylation. Also, the phosphoinositide-derived diacylglycerol may activate protein kinase C, thus simultaneously altering the protein phosphorylation/dephosphorylation ratio.

The therapeutic potential of myoinositol and D-chiroinositol for both lowering insulin resistance and improving postprandial and urinary glucose profiles deserves further consideration for possible application to humans.

References

1. Holub BJ. Metabolism and function of myo-inositol and inositol phospholipids. Ann Rev Nutr 6:563–97, 1986.
2. Billington DC. The Inositols. In: The inositol phosphates: Chemical synthesis and biological significance. pp 1–8, 1993. VCH (pub) Weinheim, New York
3. Hauser G, Finelli VN. The biosynthesis of free and phosphatide myoinositol from glucose by mammalian tissue slices. J Biol Chem 238:3224–28, 1963.
4. Clements RSJ, Diethelm AG. The metabolism of myo-inositol by the human kidney. J Lab Clin Med 93:210–19, 1979.
5. Helmkamp GMJ: Phosphatidylinositol transfer proteins: Structure, catalytic activity, and physiological function. Chem Phys Lipids 38:3–16, 1985.
6. Lewin LM, Yannai Y, Sulimovici S, Kraicer PF: Studies on the metabolic role of myo-inositol. Biochem J 156:375–80, 1976.
7. Holub BJ: Nutritional, biochemical, and clinical aspects of inositol and phosphatidylinositol metabolism. Can J Physiol Pharmacol 62:1–8, 1984.
8. Burton LE, Wells WW: Studies on the developmental pattern of the enzymes converting glucose 6-phosphate to myo-inositol in the rat. Dev Biol 37:35–42, 1974.
9. Ostlund REJ, McGill JB, Herskowitz I, et al. D-*chiro*-inositol metabolism in diabetes mellitus. Proc Natl Acad Sci USA 90:9988–92, 1993.
10. Goodhart RS. Bioflavonoids. In: Modern nutrition in health and disease. Goodhart RS, Shils ME, eds. Lea & Febiger, Philadelphia, pp 259–67, 1973.
11. Machlin LJ. Handbook of vitamins: Nutritional, biochemical, and clinical aspects. 1984. M. Dekker (Pub) New York
12. Caspary WF, Crane RK. Active transport of myo-inositol and its relation to the sugar transport system in hamster small intestine. Biochim Biophys Acta 203:308–16, 1970.
13. Burton LE, Ray RE, Bradford JR, et al. Myo-inositol metabolism in the neonatal and developing rat fed a myo-inositol-free diet. J Nutr 106:1610–16, 1976.
14. Clements RS, Reynertson R. Myoinositol metabolism in diabetes mellitus: Effect of insulin treatment. Diabetes 26:215–21, 1977.
15. Hayashi E, Maeda T, Hasegawa R, Tomita T: The effect of myo-inositol deficiency on lipid metabolism in rats. I. The alteration of lipid metabolism in myo-inositol-deficient rats. Biochim Biophys Acta 360:134–45, 1974.
16. Hoover GA, Nicolosi RJ, Corey JE, et al. Inositol deficiency in the gerbil: Altered hepatic lipid metabolism and triglyceride secretion. J Nutr 108:1588–94, 1978.
17. Hayashi E, Maeda T, Tomita T: The effect of myo-inositol deficiency on lipid metabolism in rats. II. The mechanism of triacylgycerol accumulation in the liver of myo-inositol-deficient rats. Biochim Biophys Acta 360:146–55, 1974.
18. Vohl H: Ueber das Auftreten des Inosits im Harn bei Nierenkrankheiten und die Verwandlung des Diabetes inositus. Arch Physiol 2:410–12, 1858.
19. Sundler R, Sarcione SL, Alberts AW, Vagelos PR: Evidence against phospholipid asymmetry in intracellular membranes from liver. Proc Natl Acad Sci USA 74:3350–54, 1977.
20. Perret B, Chap HJ, Douste-Blazy L. Asymetric distribution of arachidonic acid in the plasma membrane of human platelets: A determination using purified phospholipases and a rapid method for membrane isolation. Biochim Biophys Acta 556:434–46, 1979.
21. Herbette L, Blasie JK, Defoor P, et al. Phospholipid asymmetry in the isolated sarcoplasmic reticulum membrane. Arch Biochem Biophys 234:235–42, 1984.
22. Mandersloot JG, Roelofsen B, DeGier J. Phosphatidylinositol as the endogenous ac-

tivator of the (NA+/K+)-ATPase in microsomes of rabbit kidney. Biochim Biophys Acta 508:478–85, 1978.

23. Low MG, Finean JB: Specific release of plasma membrane enzymes by a phosphatidylinositol-specific phospholipase C. Biochim Biophys Acta 508:565–70, 1978.
24. Heger HW, Peter HW: Effects of phospholipids in the action of acetyl-CoA carboxylase from rat liver. Zeitschrift fur Naturforschung-Section C- Biosciences. 32:97–100, 1977.
25. Futerman AH, Low MG, Ackermann KE, et al. Identification of covalently bound inositol in the hydrophobic membrane-anchoring domain of torpedo acetylcholinesterase. Biochem Biophys Res Commun 129:312–17, 1985.
26. Keough KMW, Thompson W: Soluble and particulate forms of phosphoinositide phosphodiesterase in ox brain. Biochim Biophys Acta 270:324–36, 1972.
27. Bleasdale JE, Eichberg J, Hauser G: Inositol and phosphoinositides: Metabolism and regulation. 1985. Human Press (Pub) Clifton, NJ
28. Kirk CJ, Rodrigues LM, Hems DA: The influence of vasopressin and related peptides on glycogen phosphorylase activity and phosphatidylinositol metabolism in hepatocytes. Biochem J 178:493–6, 1979.
29. Kahn CR, Goldfine AB. Molecular determinants of insulin action. J Diabetes Complications 7:92–105, 1993.
30. Romero G, Larner J. Insulin mediators and the mechanism of insulin action. Adv Pharmacol 24:21–50, 1993.
31. Hipps PP, Holland WH, Sherman WR. Identification and measurement of chiro-inositol in the American cockroach, *Periplanta americana* L. Biochem Biophys Res Commun 46:1903–8, 1972.
32. Niwa T, Yamamoto N, Maeda K, et al. Gas chromatographic-mass spectrometric analysis of polyols in urine and serum of uremic patients: Identification of new deoxyalditols and inositol isomers. J Chromatogr 277:25–39, 1983.
33. Yagihara Y, Salway JG, Hawthorne JN: Incorporation of ^{32}P *in vitro* into triphosphoinositide and related lipids of rat superior cervical ganglia and vagus nerves. J Neurochem 16:1133–39, 1969.
34. Greene DA, Jesus PVD, Winegrad AI: Effects of insulin and dietary myoinositol on impaired peripheral motor nerve conduction velocity in acute streptozotocin diabetes. J Clin Invest 55:1326–36, 1975.
35. Fukuma M, Carpentier J-L, Orci L, et al. An alteration in internodal myelin membrane structure in large sciatic nerve fibres in rats with acute streptozotocin diabetes and impaired nerve conduction velocity. Diabetologia 15:65–72, 1977.
36. Mayhew JA, Gillon KRW, Hawthorne JN: Free and lipid inositol, sorbitol and sugars in sciatic nerve obtained post-mortem from diabetic patients and control subjects. Diabetologia 24:13–5, 1983.
37. Sharma AK, Thomas PK. Animal models: Pathology and pathophysiology. In: Diabetic neuropathy. Dyck PJ, Thomas PK, Asbury AK, et al., eds. Saunders, Philadelphia, pp 237–52, 1987.
38. Cornblath DR, Hillman MA, Striffler JS, et al. Peripheral neuropathy in diabetic monkeys. Diabetes 38:1365–70, 1989.
39. Gregersen G. Diabetic neuropathy: Influence of age, metabolic control and duration of diabetes on motor conduction velocity. Neurology 17:972–80, 1967.
40. Salway JG, Finnegan JA, Barnett D, et al. Effect of *myo*-inositol on peripheral-nerve function in diabetes. Lancet 2:1282–84, 1978.
41. Palmono KP, Whiting PH, Hawthorne JN: Free and lipid myo-inositiol in tissues from rats with acute and less severe streptozotocin-induced diabetes. Biochem J 167:229–35, 1977.
42. Greene DA, Latimer SA: Sodium- and energy-dependent uptake of *myo*-inositol by rabbit peripheral nerve: Competitive inhibition by glucose and lack of an insulin effect. J Clin Invest 70:1009–18, 1982.

43. Mayer JH, Tomlinson DR: Prevention of defects of axonal transport and nerve conduction velocity by oral administration of *myo*-inositol or an aldose reductase inhibitor in streptozotocin-diabetic rats. Diabetologia 25:433–8, 1983.
44. Carrington AL, Calcutt NA, Ettlinger CB, et al. Effects of treatment with myo-inositol or its 1,2,6-trisphosphate (PP56) on nerve conduction in streptozotocin-diabetes. Eur J Pharmacol 237:257–63, 1993.
45. Jefferys JGR, Palmano KP, Sharma AK, Thomas PK: Influence of dietary myoinositol on nerve conduction and inositol phospholipids in normal and diabetic rats. J Neurol Neurosurg Psychiat 41:333–9, 1978.
46. Thomas PK, Jefferys JGR, Sharma AK, Bajada S. Nerve conduction velocity in experimental diabetes in the rat and rabbit. J Neurol Neurosurg Psychiat 44:233–8, 1981.
47. Gregersen G, Børsting H, Theil P, Servo C. Myoinositol and function of peripheral nerves in human diabetics: A controlled clinical trial. Acta Neurol Scand 58:241–8, 1978.
48. Gregersen G, Bertelsen B, Harbo H, et al. Oral supplementation of myoinositol: Effects on peripheral nerve function in human diabetics and on the concentration in plasma, erythrocytes, urine and muscle tissue in human diabetics and normals. Acta Neurol Scand 67:164–72, 1983.
49. Arendrup K, Gregersen G, Hawley J, Hawthorne JN: High-dose dietary *myo*-inositol does not alter the ischaemia phenomenon in human diabetes. Acta Neurol Scand 80:99–102, 1989.
50. Kim J, Kyriazi H, Greene DA: Normalization of Na+/K+-ATPase activity in isolated membrane fraction from sciatic nerves of streptozocin-induced diabetic rats by dietary myo-inositol supplementation in vivo or protein kinase C agonists in vitro. Diabetes 40:558–67, 1991.
51. Xia M, Laychock SG: Insulin secretion, myo-inositol transport, and Na+/K+-ATPase in glucose-desensitized rat islets. Diabetes 42:1392–400, 1993.
52. Kwon HM, Yamauchi A, Uchida S, et al. Cloning of the cDNA for a Na+/myo-inositol cotransporter, a hypertonicity stress protein. J Biol Chem 267:6297-301, 1992.
53. Simmons D, Ng LL, Bomford J: Relationship between myoinositol influx and lipids in diabetic neuropathy. Acta Diabetol 30:233–7, 1993.
54. Yorek MA, Dunlap JA, Stefani MR, Davidson EP. Reversal of hyperglycemic-induced defects in myo-inositol metabolism and Na^+/K^+ pump activity in cultured neuroblastoma cells by normalizing glucose levels. Metabolism 42:1180–89, 1993.
55. Yorek MA, Wiese TJ, Davidson EP, et al. Reduced motor nerve conduction velocity and Na^+/K^+-ATPase activity in rats maintained on L-fucose diet. Diabetes 42:1401–6, 1993.
56. Grafton G, Bunce CM, Sheppard MC, et al. Effect of Mg^{2+} on Na^+-dependent inositol transport: Role for Mg^{2+} in etiology of diabetic complications. Diabetes 41:35–9, 1992.
57. Levine J, Rapaport A, Lev L, et al. Inositol treatment raises CSF inositol levels. Brain Res 627:168–70, 1993.
58. Cohen RA, MacGregor LC, Spokes KC, et al. Effect of myo-inositol on renal Na-K-ATPase in experimental diabetes. Metabolism 39:1026–32, 1990.
59. Haneda M, Kikkawa R, Arimura T, et al. Glucose inhibits myo-inositol uptake and reduces myo-inositol content in cultured rat glomerular mesangial cells. Metabolism 39:40–5, 1990.
60. Whiteside CI, Thompson JC: Upregulation of D-*myo*-inositol transport in diabetic rat glomerular cells. Am J Physiol 262:E301–6, 1992.
61. Guzman NJ, Crews FT: Regulation of inositol transport by glucose and protein kinase C in mesangial cells. Kidney Int 42:33–40, 1992.
62. Olgemöller B, Schwaabe S, Schleicher ED, Gerbitz KD: Upregulation of *myo*-

inositol transport compensates for competitive inhibition by glucose. An explanation for the inositol paradox? Diabetes 42:1119–25, 1993.
63. Kasuga M, Karlsson FA, Kahn CR. Insulin stimulates the phosphorylation of the 95,000-dalton subunit of its own receptor. Science 215:185–7, 1982.
64. Chou CK, Dull TJ, Rusell DS, et al. Human insulin receptors mutated at the ATP-binding site lack protein kinase activity and fail to mediate postreceptor effects of insulin. J Biol Chem 262:1842–7, 1987.
65. Saltiel AR, Cuatrecasas P. Insulin stimulates the generation from hepatic plasma membranes of modulators derived from an inositol glycolipid. Proc Natl Acad Sci USA 83:5793–7, 1986.
66. Larner J, Cheng K, Schwartz C, et al. A proteolytic mechanism for the action of insulin via oligopeptide mediator formation. Fed Proc 41:2724–29, 1982.
67. Low MG, Saltiel AR: Structural and functional roles of glycosyl-phosphatidylinositol in membranes. Science 239:268–75, 1988.
68. Mato JM, Kelly KL, Abler A, et al. Partial structure of an insulin-sensitive glycophospholipid. Biochem Biophys Res Commun 146:764–70, 1987.
69. Alemany S, Mato JM, Stralfors P: Phospho-dephospho-control by insulin is mimicked by a phospho-oligosaccharide in adipocytes. Nature 330:77–9, 1987.
70. Low MG, Futerman AH, Ackermann KE, et al. Removal of covalently bound inositol from *Torpedo* acetylcholinesterase and mammalian alkaline phosphatases by deamination with nitrous acid: Evidence for a common membrane-anchoring structure. Biochem J 241:615–19, 1987.
71. Larner J, Huang LC, Schwartz CRW, et al. Rat liver insulin mediator which stimulates pyruvate dehydrogenase phosphatase contains galactosamine and d-chiroinositol. Biochem Biophys Res Commun 151:1416–26, 1988.
72. Larner J, Huang LC, Tang G, et al. Insulin mediators: Structure and function. Cold Spring Harb Symp Quant Biol 53:965–71, 1988.
73. Saltiel AR, Osterman DG, Darnell JC, et al. The function of glycosyl phosphoinositides in hormone action. Phil Trans R Soc Lond 320:345–58, 1988.
74. Larner J, Galasko G, Cheng K, et al. Generation by insulin of a chemical mediator that controls protein phosphorylation and dephosphorylation. Science 206:1408–10, 1979.
75. Jarett L, Seals JR: Pyruvate dehydrogenase activation in adipocyte mitochondria by an insulin-generated mediator from muscle. Science 206:1407–8, 1979.
76. Larner J, Huang LC, Suzuki S, et al. Insulin mediators and the control of pyruvate dehydrogenase complex. Ann NY Acad Sci 573:297–305, 1989.
77. Saltiel AR: Insulin generates an enzyme modulator from hepatic plasma membranes: Regulation of adenosine 3′,5′-monophosphate phosphodiesterase, pyruvate dehydrogenase, and adenylate cyclase. Endocrinology 120:967–72, 1987.
78. Huang LC, Fonteles MC, Houston DB, et al. Chiroinositol deficiency and insulin resistance. III. Acute glycogenic and hypoglycemic effects of two inositol phosphoglycan insulin mediators in normal and streptozotocin-diabetic rats *in vivo*. Endocrinology 132:652–7, 1993.
79. Farese RV, Standaert ML, Yamada K, et al. Insulin-induced activation of glycerol-3-phosphate acyltransferase by a *chiro*-inositol-containing insulin mediator is defective in adipocytes of insulin-resistant, type II diabetic, Goto-Kakizaki rat. Proc Natl Acad Sci USA 91:11040–44, 1994.
80. Abe S, Huang LC, Larner J. Dephosphorylation of PDH by phosphoprotein phosphatases and its allosteric regulation by inositol glycans. In: Alpha-keto acid dehydrogenase complexes. In press, 1996. M.S. Patel, T.E. Roche & R.A. Harris (eds) Berhauser Verlag A.G. In press
81. Daughaday WH, Larner J. The renal excretion of inositol in normal and diabetic human beings. J Clin Invest 33:326–32, 1954.
82. Kennington AS, Hill CR, Craig J, et al. Low urinary *chiro*-inositol excretion in non-insulin-dependent diabetes mellitus. N Engl J Med 323:373–8, 1990.

83. Suzuki S, Taneda Y, Hirai S, et al. Molecular mechanism of insulin resistance in spontaneous diabetic GK (Goto-Kakizaki) rats. In: New directions in research and clinical works for obesity and diabetes mellitus. Sakamoto N, Angel A, Hotta N, eds. Elsevier, New York, pp 197–203, 1991.
84. Suzuki S, Kawasaki H, Satoh Y, et al. Urinary chiro-inositol excretion is an index marker of insulin sensitivity in Japanese type II diabetes. Diabetes Care 17:1465–8, 1994.
85. Ortmeyer HK, Bodkin NL, Lilley K, et al. Chiroinositol deficiency and insulin resistance. I. Urinary excretion rate of chiroinositol is directly associated with insulin resistance in spontaneously diabetic rhesus monkeys. Endocrinology 132:640–5, 1993.
86. Kofo-Abayomi A, Lucas PD: A comparison between atria from control and streptozotocin-diabetic rats: The effects of dietary myo-inositol. Br J Pharmacol 93:3–8, 1988.
87. Pugliese G, Tilton RG, Speedy A, et al. Modulation of hemodynamic and vascular filtration changes in diabetic rats by dietary *myo*-inositol. Diabetes 39:312–22, 1990.
88. Gousse A, Yoshida M, Weiss RM, Latifpour J. Beta adrenergic receptor alterations in diabetic rat prostate: Effects of insulin and dietary myoinositol. Prostate 19:121–31, 1991.
89. Latifpour J, Gousse A, Yoshida M, Weiss RM. Effect of insulin and dietary myoinositol on muscarinic receptor alterations in diabetic rat bladder. J Urol 147:760–3, 1992.
90. Tilton RG, Faller AM, LaRose LS, et al. Dietary *myo*-inositol supplementation does not prevent retinal and glomerular vascular structural changes in chronically diabetic rats. J Diabetes Complications 7:188–98, 1993.
91. Narayanan CR, Joshi DD, Mujumdar AM, Dhekne VV: Pinitol—a new antidiabetic compound from the leaves of *Bougainvillea spectabilis*. Curr Sci (Bangalore, India) 56:139–41, 1987.
92. Ortmeyer HK, Huang LC, Zhang L, et al. Chiroinositol deficiency and insulin resistance. II. Acute effects of D-chiroinositol administration in streptozotocin-diabetic rats, normal rats given a glucose load, and spontaneously insulin-resistant rhesus monkeys. Endocrinology 132:646–51, 1993.
93. Ortmeyer HK, Bodkin NL, Hansen BC, Larner J: In vivo D-chiroinositol activates skeletal muscle glycogen synthase and inactivates glycogen phosphorylase in rhesus monkeys. J Nutr Biochem 6:499–503, 1995.
94. Ortmeyer HK, Larner J, Hansen BC. Effects of D-chiroinositol added to a meal on plasma glucose and insulin in hyperinsulinemic rhesus monkeys. Obesity Res, 3:605S–608S, 1995.
95. Ortmeyer HK, Hansen BC. Myoinositol added to the diet results in lower urine glucose and in lower postprandial plasma glucose in obese rhesus monkeys. Diabetes Care 18:A12, 1995.
96. Pak YB, Huang LC, Lilley K, Larner J. In vivo conversion of [^{3}H]myoinositol to [^{3}H] chiroinositol in rat tissues. J Biol Chem 267:16904–910, 1992.

20

Antidiabetic characteristics of dehydroepiandrosterone in animal and human studies

MARGOT P. CLEARY
The Hormel Institute, University of Minnesota, Austin, MN 55912, USA.

Key words: obesity, insulin, glucose tolerance, insulin resistance, mice, rats, humans.

Interest in dehydroepiandrosterone (DHEA) as a metabolic factor involved with diabetes in relation to obesity dates back to a study published by Šonka and coworkers[1] 30 years ago. They reported that no DHEA was detected in the urine of 27 of 32 elderly obese diabetic subjects. This and other studies led Šonka[2] to the speculation that DHEA, a 17-ketosteroid produced in the adrenal gland, may have important regulatory functions in the body. Since these early observations were made, ameliorative effects resulting from the administration of DHEA were obtained in numerous studies. Most studies have used mice and rats, but additional work has been presented in dogs, monkeys, rabbits, and humans. An array of effects was attributed to DHEA treatment, including prevention of cancer, aging, obesity, viral diseases, and atherosclerosis.[3–6] Following in Šonka's path there have also been further investigations into the role of DHEA as an antidiabetic agent. My overview of the literature is pertinent to this potential characteristic of DHEA.

DHEA treatment in mice

Yen and coworkers[7] reported that administration of DHEA at a dose of 500 mg/kg 3× weekly by intubation to yellow obese A^{vy}/a mice resulted in lower weight (wt) gain than that found in untreated mice. This was seen in both male and female obese mice, and also in lean mice. Of particular interest was that the change in body wt was independent of an effect on food intake. In the obese DHEA-treated mice, substantial decreases in body fat were reported. Other investigators

Correspondence: Margot Cleary, The Hormel Institute, University of Minnesota, Austin, MN 55912, USA.

Lessons from Animal Diabetes VI. Ed. E. Shafrir.

have confirmed that DHEA administration, either by intubation, injection, or as a food ingredient, results in body wt changes in a number of mice strains.[8–12]

Coleman et al.[13] presented the first report of a potential antidiabetic effect of DHEA treatment in mice. Diabetic (*db/db*) BKS mice that progress from mild to severe diabetes during the first few months of life received DHEA at a level of 0.4% in their diet beginning at either 4 (females) or 2 (males) months of age. Blood glucose concentrations of the DHEA-treated *db/db* mice were rapidly lowered into the range of normal mice (Fig. 1). The blood glucose levels of DHEA-treated mice were maintained within this normal range over the 18 wk experimental period. However, initiating DHEA treatment at 4 months of age did not consistently result in normal blood glucose levels in diabetic mice. In contrast to the results described earlier for other strains of mice in which DHEA treatment resulted in reduced body wt gain, BKS *db/db* DHEA-treated mice had similar body wt compared with nontreated mice. This finding was probably due to the fact that food intake of the *db/db* DHEA-treated mice was 50% higher than that of nontreated mice. In response to oral glucose tolerance tests (OGTT), *db/db* DHEA-treated mice were found to be less glucose intolerant than nontreated mice. The expected age-related decreases in plasma insulin concentration and pancreatic insulin content that are characteristic of this syndrome were not found in these DHEA-treated mice. In another aspect of this investigation, normal lean BKS mice were made diabetic by streptozotocin (STZ) administration and were treated with DHEA, but this did not alter the course of the chemically induced diabetes.

The response of mice that have obesity associated with diabetes was further investigated by evaluating the interaction of genetic background and DHEA treatment.[14] The BKS mice described earlier, which are obese with life-shortening diabetes, were compared with BL6 mice that are severely obese but have a tran-

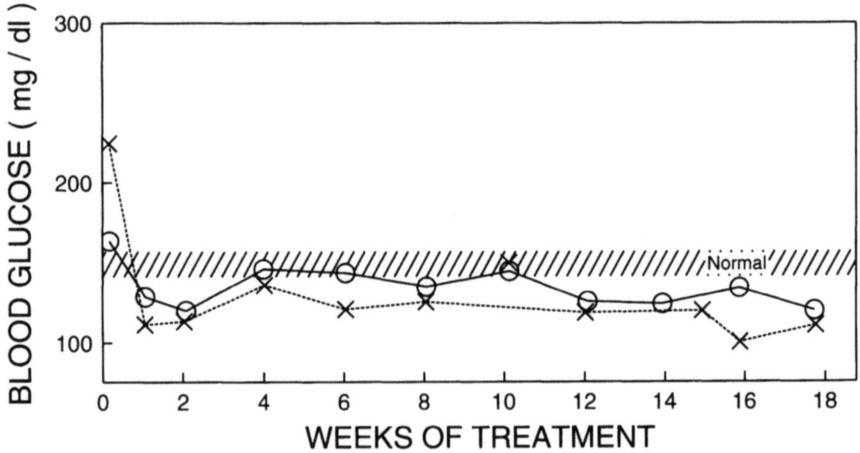

Figure 1. Effect of dehydroepiandrosterone treatment on blood glucose levels in male (X– – –X) and female (O——O) BKS (*db/db*) mice. (Adapted from ref. 13, with permission.)

sient, well-compensated diabetes. The mice were fed diets containing 0.4% DHEA from weaning. A summary of some of the results for the *db/db* mice from these two different backgrounds is shown in Table 1. In general, BL6 mice were found to be more sensitive to DHEA treatment than BKS mice. In fact, some effects of DHEA on body wt and food intake in the BL6 *db/db* mice could be seen at a treatment level of 0.1% DHEA.

Nondiabetic BL6 2-yr-old female mice were also treated with 0.4% DHEA in their diet.[14] After 4 wk an improved response to an OGTT was seen in the DHEA-treated mice. After 8 wk DHEA-treated mice had plasma insulin levels significantly lower than mice fed chow alone. Plasma glucose levels were similar in the two groups. These results were indicative of an improvement in tissue insulin sensitivity in DHEA-treated mice.

Additional studies[15] were conducted to determine if some metabolite of DHEA was the active compound. Three major metabolites, DHEA-sulfate (DHEA-S), α-hydroxyetiocholanolone, and β-hydroxyetiocholanolone, were fed at 0.1% in the diet and another metabolite, 17β-estradiol, was fed at 0.005%. Treatment with these metabolites was found to prevent the development of the severe diabetes characteristic of BKS mice, and there was little effect on either body wt or food intake. These results were comparable with those of 0.4% DHEA treatment.

Coleman[16] further investigated the use of etiocholanolones as potential antidiabetic agents. Both α- and β-etiocholanolone treatment at 0.1% in the diet of obese diabetic (*db/db*) mice on the BL6 background resulted in lower body wt gains, as well as improvements in serum insulin and glucose levels. Similar responses were reported for A^{vy}/A mice. In addition, the etiocholanolones did not produce an estrogen effect, as occurred with DHEA treatment. In evaluating the overall picture, Coleman speculated that administration of either DHEA or etiocholanolone was most successful in mice with mild diabetes and severe obesity, such as in the BL6 mice.

To further evaluate the antidiabetic role of DHEA, Leiter et al.[17] measured serum metabolites following DHEA treatment. Normal and *db/db* BKS male mice had large increases in serum DHEA, androstenedione (A), testosterone (T), and dihydrotestosterone (DHT) concentrations after 2 wk of 0.4% DHEA treatment. However, only diabetic mice had elevations in serum estrogen levels. It was noted that the metabolism of DHEA did not require the presence of gonads or adrenal glands. In SWXJ mice DHEA treatment initiated shortly after weaning also resulted in elevations of serum DHEA, T, A, and DHT concentrations.[18,19] Elevations in serum estrone and 17β-estradiol were only found in those mice that developed ovarian granulosa cell tumors.[18] Body wt was lower as a consequence of DHEA treatment in this mice strain.[18]

These findings indicate that DHEA's potential usefulness as a therapeutic agent in mice might be limited because of its conversion to androgens, and in some cases to estrogen. Development of analogues with limited potential to be converted to these metabolites may overcome these problems.[20] The response of BKS *db/db* mice to a DHEA analogue, 16α-fluoro-5-androsten-17-one, was recently evaluated.[21] This compound at a level of 0.2–0.3% in the diet markedly reduced plasma glucose levels relative to nontreated control mice. Body wt and

Table 1. Effects of DHEA treatment on nondiabetic and diabetic *(db/db)* BKS and BL6 mice

Strain genotype	Food consumption (g/day)	Blood glucose (mg/dl)	Plasma insulin (μU/l)	Pancreatic insulin (U/g)	Pancreatic granulated β-cells (%)
BKS	4.04 ± 0.20	131 ± 2	42 ± 2	3.84 ± 0.63	68.8 ± 7.1
Nondiabetic control (n = 6)					
Nondiabetic DHEA (n = 3)	3.95 ± 0.18	101 ± 1	51 ± 3	3.89 ± 0.13	79.4 ± 5.8
db/db control (n = 11)	4.19 ± 0.16	>450	205 ± 29	1.06 ± 0.25	4.1 ± 1.1
db/db DHEA (n = 11)	6.21 ± 0.11*	156 ± 19	2374 ± 1883*	10.40 ± 0.08*	44.2 ± 6.2*
BL6					
Nondiabetic control (n = 6)	3.82 ± 0.15	140 ± 4	37 ± 5	2.67 ± 0.17	76.3 ± 2.4
Nondiabetic DHEA (n = 5)	3.03 ± 0.13	118 ± 4	37 ± 9	2.29 ± 0.12	70.3 ± 3.4
db/db control (n = 16)	4.82 ± 0.17	140 ± 15	2302 ± 454	8.50 ± 1.89	14.4 ± 4.4
db/db DHEA (n = 10)	2.78 ± 0.18	111 ± 9	1414 ± 345	8.97 ± 1.48	49.9 ± 6.8*

Data are means ± SE for male mice treated from weaning until 18 wk of age.
*Significant difference from *db/db* control mice.
Adapted from ref. 14, with permission.

food intake were not affected by treatment with 16α-fluoro-5-androsten-17-one, and no steroid hormone side effects were reported.

Effects of DHEA treatment in rats

In rats direct antidiabetic effects of DHEA administration have not been investigated in an IDDM model. However, aspects of glucose metabolism and insulin sensitivity have been reported for several different rat strains treated with DHEA. In the majority of these studies the obese (*fa/fa*) Zucker rat has been used. This model of juvenile-onset obesity develops hyperinsulinemia and insulin resistance with age. These rats share many characteristics associated with early stages of NIDDM. Initial studies indicated that DHEA treatment at a level of 0.6% in the diet of young growing Zucker rats from 5 wk until 20 wk of age resulted in lower body wt and a lower rate of wt gain, as well as decreased body fat and fat cell size compared with nontreated obese rats.[22] It was also noted that serum insulin levels of obese DHEA-treated rats were significantly lower than those of control obese rats.[23] In normoinsulinemic lean rats, there was no effect of DHEA on serum insulin concentrations.[23] This insulin-lowering effect after long-term DHEA treatment was confirmed in a number of additional studies in either fed or fasted obese Zucker rats of various ages.[24–26] There was no effect of DHEA on serum glucose levels. Recently Svec et al.[27] confirmed these findings for serum insulin and glucose measurements. It was also reported that DHEA treatment for 1 month did not affect C-peptide levels but did result in a significant increase in the C-peptide to insulin ratio in both lean and obese Zucker rats.

The finding of both lower body fat content and smaller fat cell size in conjunction with lower serum insulin concentrations in the obese DHEA-treated rats led to the hypothesis that tissue insulin sensitivity was improved by DHEA administration. This was investigated using isolated adipocytes obtained from rats after 22 wk of DHEA treatment from 6 to 29 wk of age.[26] To account for the lower food intake of the obese DHEA-treated rats resulting from their smaller body size, an additional nontreated control group, pair-fed to the treated rats, was included. The obese pair-fed rats had body wt, fat pad wt, and a serum insulin concentration intermediate between the obese DHEA-treated and obese ad libitum–fed groups. Insulin resistance was evaluated by determining the fate of radiolabeled glucose (incorporation into CO_2, fatty acids, and glyceride-glycerol) in the absence (basal) and presence (stimulated) of insulin in isolated fat cells. In general, DHEA treatment did not improve glucose utilization in fat cells of obese rats in either the basal or insulin-stimulated condition. Lean rats treated with DHEA did have higher levels of glucose metabolized to CO_2 in both the absence and presence of insulin than control lean rats. A study done in parallel indicated that pancreatic insulin levels were lower in obese rats after 22 wk of DHEA treatment.[28] A shift toward islets of smaller size for both lean and obese DHEA-treated rats was found, but only obese rats had significantly lower serum insulin levels. This suggested that DHEA may have a direct effect on the pancreas.

Short-term effects of DHEA treatment on factors associated with insulin sensitivity were also determined using the obese Zucker rat.[29] Rats were treated from

5 wk until 10 wk of age with 0.6% DHEA in their diet. A wt-matched obese group was included to account for the lower body wt of the DHEA-treated obese rats. After 5 wk, neither pancreatic wt nor insulin levels were different among ad libitum–fed, wt-matched, or DHEA-treated obese rats. However, DHEA-treated obese rats had significantly lower serum insulin levels than the other two groups. Adipocyte and muscle glucose metabolism were also measured after short-term DHEA administration, but there were no significant effects attributable to DHEA treatment on insulin sensitivity in these two tissues. In an additional study no differences in the OGTT response were found in DHEA-treated obese vs. ad libitum–fed obese rats.[30]

These findings suggest there may be an effect of DHEA treatment on pancreatic insulin release. This could be mediated through changes in islet size, as it was previously reported that the elevated number of larger islets present in the obese compared with the lean Zucker rats was responsible for the increased insulin secretion in obese rats.[31] Insulin clearance by the liver might also be affected by DHEA, as many hepatic metabolic changes have been reported in DHEA-treated rats.[32] These possibilities were recently investigated (M.P. Cleary, T. Zabel, unpublished observations). Female obese rats, 10 wk old, were treated for 5 wk with 0.6% DHEA in their food. Isolated pancreatic islets were prepared and purified on Ficoll gradients. There was a similar yield of islets from both DHEA-treated and control obese rats. However, mean islet size was significantly lower in the DHEA-treated obese compared with control obese rats (204 ± 13 vs. 231 ± 10 μm, $P < 0.05$ mean ± SD). Portal vs. vena cava measurements of insulin, glucagon, and corticosterone were also made (Table 2). Portal vein and vena cava insulin levels of DHEA obese rats were 33 and 66%, respectively, of those of obese control rats. Glucagon values were significantly lower in the portal vein of obese DHEA-treated rats compared with control rats, but were similar in the vena cava. Corticosterone levels were not significantly altered by DHEA treatment. These findings do not suggest selective hepatic clearance of insulin or other hormones in DHEA-treated obese rats.

In an additional experiment (M.P. Cleary, T. Zabel, unpublished observations), 5-wk-old obese female rats were treated with DHEA for 5 wk. Isolated islets were prepared and divided into three size ranges based on diameter: <125 μm, 125–250 μm, and >250 μm. Groups of five islets were then incubated in 0.3 mg/ml or 3 mg/ml glucose. Total insulin secretion was not different between DHEA-treated and control obese rats at either low or high glucose concentrations or at any of the islet sizes (Table 3). As expected, more insulin was secreted from the larger islets in comparison with the smaller islets. These results, that is a shift toward smaller islet size with no effect on insulin secretion in islets of a specific size, provides a partial explanation for DHEA's effects on serum insulin. How DHEA alters pancreatic islet size is not presently known. Possibly, this could be mediated through the known inhibition by DHEA of mammalian glucose-6-phosphate dehydrogenase activity.[33–35]

Sprague-Dawley rats have been used as a model for age-related changes in the development of insulin resistance.[36–38] In several studies that have potential relevance to diabetes, the effects of DHEA administration were assessed in this rat

Table 2. Serum insulin, glucagon, and corticosterone levels in portal vein and vena cava from obese control and obese DHEA[a]-treated Zucker rats[b]

	Insulin mU/l		Glucagon (ng/l)		Corticosterone (μg/l)	
	Portal vein	Vena cava	Portal vein	Vena cava	Portal vein	Vena cava
Obese control	625 ± 284	225 ± 46	859 ± 114	612 ± 78	344 ± 111	571 ± 138
Obese DHEA treated	204† ± 88	148 ± 38	650* ± 161	571 ± 98	283 ± 118	402 ± 191

[a]DHEA included in the diet at 0.6% from 10 to 15 wk of age.

[b]Unpublished data (MP Cleary, T Zabel).

Results are expressed as mean ± SD for six mice.

*$P < 0.05$, †$P < 0.025$ determined using Student's *t*-test; obese DHEA-treated rats significantly different from obese control rats.

Table 3. Insulin secretion from isolated pancreatic islets from obese control and obese DHEA[a]-treated Zucker rats[b]

	Pancreatic islet size					
	<125 μm		125–250 μm		>250 μm	
	Glucose concentration (mg/ml)					
	0.3	3.0	0.3	3.0	0.3	3.0
Obese control	256* ± 70	308† ± 87	359* ± 30	665† ± 223	475* ± 115	983† ± 326
Obese DHEA-treated	226* ± 54	367* ± 112	326* ± 72	674† ± 132	487* ± 104	914† ± 280

[a]DHEA included in the diet at 0.6% from 5 to 10 wk of age.

[b]Unpublished data (M.P. Cleary, T. Zabel).

Results are as μU insulin (mean ± SD for four mice).

*,†Within an islet size, values with different symbols are significantly different as determined by one-way ANOVA followed by the Neuman-Keul's test.

strain. For example, after 6 wk of DHEA treatment, 23-wk-old Sprague-Dawley rats with diet-induced obesity had lower serum insulin levels than either non-treated chow-fed rats or nontreated diet-induced obese rats.[39] In another aspect of this study, chow-fed rats, dietary-induced obese rats, and diet-resistant rats (that did not become obese on the fat-enriched diet), were treated with 0.6% DHEA for 2 wk. In all three DHEA-treated groups, both serum insulin and glucose concentrations were significantly lower compared with the values in the respective control groups (Fig. 2). This occurred despite the fact that there were no significant effects on either epididymal or retroperitoneal fat pad wt or fat cell size between treated and untreated groups after this short treatment period.

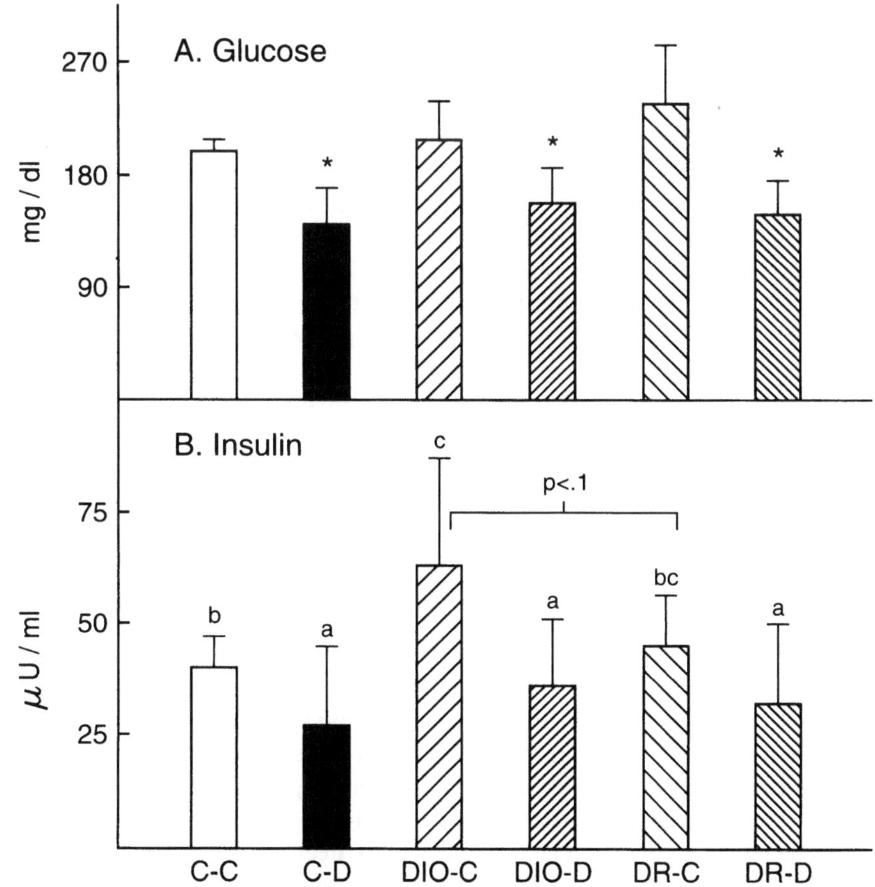

Figure 2. Serum glucose (A) and insulin (B) levels in Sprague-Dawley rats. C-C, chow-control; C-D, chow-DHEA; DIO-C, diet-induced obese-control; DIO-D, diet-induced obese-DHEA; DR-C, diet-resistant-control; DR-D, diet-resistant-DHEA. Values are means ± SD. *DHEA-treated significantly different from control rats by 2 × 3 ANOVA. Bars with different superscripts are significantly different from each other by the F test after 2 × 3 ANOVA with significant interactions. (Adapted from ref. 39, with permission.)

The BHE cdb rat strain has been reported to have a genetic disposition for glucose intolerance, particularly when fed a high carbohydrate diet.[40] Although a number of studies have been reported using these rats to evaluate consequences of DHEA treatment, only one study presented data related to diabetes. McIntosh and Berdanier[41] reported that after 7 wk of DHEA treatment (0.35 mol/kg), BHE rats fed a high glucose diet had similar OGTT response as nontreated rats fed the same diet. This occurred despite the fact that the DHEA-treated BHE rats had significantly lower body wt and body fat than the nontreated animals.

DHEA treatment has been used to induce polycystic ovaries in rats.[42,43] An extensive evaluation of female Sprague-Dawley rats treated with daily injections of DHEA (6 mg/100 g) beginning at 27 days of age indicated that there was a heterogeneous response to this intervention.[44] This was reflected both by ovarian pathology and serum hormone profiles. In general, after 20 days of DHEA administration, estrone, estradiol, A, T, and DHT concentrations were all elevated in treated vs. control rats, although the increased levels varied in relation to ovarian morphology. This elevation in serum androgen measurements following DHEA treatment was similar to the response of mice described earlier.

Thus, a number of changes, such as lower body wt, fat pad wt, fat cell size, and serum insulin level, result from DHEA treatment in rats. These observations suggest that glucose utilization and insulin resistance should be improved in these animals. However, when measurements are made to more directly determine if tissue glucose metabolism and insulin sensitivity are affected by DHEA treatment, the results have not shown that such changes occur.

DHEA treatment in humans

The effects of short-term DHEA administration have been evaluated also in human subjects. Nestler et al.[45] reported that following 28 days of receiving 1600 mg/day DHEA, young men of normal wt responded with a 31% decrease in body fat with no change in body wt. Tissue sensitivity to insulin was also assessed using the hyperinsulinemic-euglycemic clamp technique. Despite the significant change in body fat, there was no effect on insulin sensitivity attributable to DHEA administration. Six healthy obese young men were studied in a similar protocol by the same research group.[46] Body wt and body fat changes were variable among the subjects in response to DHEA treatment, and no significant differences were found in these two parameters between the placebo and DHEA treatment periods. In agreement with the above-mentioned investigation, there were also no changes in tissue insulin sensitivity and insulin-independent fractional glucose disappearance between the placebo and DHEA treatment periods. In an additional study based on a double-blind, placebo-controlled crossover design,[47] six postmenopausal obese women were treated with 400 mg/4× day DHEA for 28 days or were given a placebo. Initial OGTT was performed and repeated after both placebo and DHEA treatment periods. There were no changes in either body wt or body fat, and serum insulin and glucose levels were not altered by DHEA treatment. After the DHEA treatment period, there was a significant 50% increase in insulin response for both peak increments reached and for the integrated insulin area under the curve (AUC) in response to the OGTT. There was no significant effect on serum glucose levels. These findings led to the conclusion that an insulin-resistant state was induced by DHEA treatment in these women. In another experiment, postmenopausal women were given 50 mg/day DHEA for 3 wk in a prospective, placebo-controlled, randomized, blinded, crossover study.[48] There was no effect on body wt or fat, and OGTT results were similar for insulin and glucose AUC for both the DHEA and placebo periods. However, T-lymphocyte insulin binding and degradation was significantly increased after the DHEA treatment. This finding was an indication that there was an improvement of insulin sensitivity attributable to DHEA.

In a study using a longer period of DHEA treatment, Morales et al.[49] used DHEA (50 mg/day) to raise serum DHEA and DHEA-S in older subjects to levels found in young adults. A randomized, double-blind, placebo, crossover design was employed using both men ($n = 13$) and women ($n = 17$). Insulin sensitivity and insulin-dependent fractional glucose disappearance and glucose effectiveness were not affected by a 3-month DHEA treatment period in either men or women. In a small subset (three men and two women), there was also no effect of glucose disposal using the hyperinsulinemic-euglycemic clamp technique.

In another study,[50] obese men and women were treated with etiocholanedione (a naturally occurring metabolite of etiocholanolone, which in turn is a metabolite of DHEA). Subjects were to consume 4 g/day for 10 wk during the experimental period in this randomized, double-blind, crossover study. Actual intake was assessed to be on average 3.63 g/day. Subjects were also placed on energy-restricted diets. During the etiocholanedione treatment period, subjects lost more wt than during the placebo period. Unfortunately, no measurements of insulin or glucose metabolism were made in this study.

In one paper[51] data were presented on DHEA treatment, given to a moderately obese NIDDM female subject who was also hyperandrogenic. The subject was treated for 1 month with 0.25 mg/day dexamethasone followed by an additional month of dexamethasone plus DHEA (150 mg/2× day). There was little change in body wt or body fat due to either treatment. Serum glucose levels during an OGTT after the dexamethasone plus DHEA treatment were lower than with either no or dexamethasone treatment. The fasting serum insulin level was markedly lower after the dexamethasone plus DHEA treatment period, and there was a decrease in the insulin AUC. It was further reported that the glucose disappearance rate in response to i.v. insulin was increased after dexamethasone plus DHEA treatment compared with the two other interventions. These results were interpreted as an amelioration of insulin resistance in response to DHEA administration. However, there was no period of DHEA treatment alone, and the results are for only one subject. Thus, it is difficult to draw conclusions from this study on the effect of DHEA on diabetes treatment.

Thus, few studies have been published on effects of DHEA treatment in humans that include factors associated with glucose tolerance. Although body wt or fat may be altered by DHEA or metabolite administration, in general, these findings do not indicate that DHEA affects glucose metabolism in a manner that would indicate improved glucose utilization. At this time there is little evidence to conclude that DHEA would be an effective antidiabetic agent. It remains to be seen whether further studies with larger and well-defined populations would provide additional information as to whether replacement or pharmacological doses of DHEA might have antidiabetic effects.

Additional aspects of DHEA in relation to diabetes

The focus of the presentation to this point has been on the effects of DHEA administration on metabolic factors associated with diabetes. However, another consideration would be the relation of physiological levels of DHEA to aspects of the diabetes syndrome. DHEA and DHEA-S are the most abundantly circulating steroids in humans, yet these compounds have no known physiological role. Serum concentrations of DHEA and DHEA-S peak in the late twenties and decline with aging.[52–54] This contrasts with serum levels of cortisol and other adrenal steroids, which remain relatively unchanged as humans age.[55] In recent articles the possible role of physiological levels of DHEA/DHEA-S in relation to obesity and insulin resistance, cardiovascular disease, and breast cancer have been proposed.[56–59]

Several studies have evaluated the potential relationship of serum DHEA/DHEA-S to blood glucose levels. In samples from 231 men who participated in the Rancho Bernardo study,[60] fasting blood glucose concentrations >140 mg/dl were associated with somewhat lower DHEA-S levels than those found when blood glucose levels were <140 mg/dl. There was no relation of these findings to diabetes per se. In another report[61] studying 848 nondiabetic women from the same population, it was found that serum DHEA-S levels were positively correlated with blood glucose levels. Subjects with a history of diabetes or with serum glucose levels >140 mg/dl were excluded from this protocol, precluding making any correlation of the results to diabetes or to the results for men cited earlier.

In a third study,[62] also on subjects from the same community, NIDDM men ($n = 44$) were found to have significantly lower serum DHEA-S levels than nondiabetic men ($n = 88$), but there were no differences in serum DHEA levels. Herrington et al.[63] also found lower DHEA-S concentrations in NIDDM subjects. They reported 30% lower age-adjusted DHEA-S levels in 19 diabetic patients (7 men and 12 women) who were undergoing diagnostic coronary angiography compared with 185 nondiabetic subjects undergoing the same procedure, but these differences were not statistically significant.

Nyholm et al.[64] measured serum DHEA-S levels in 20 postmenopausal insulin-treated diabetic women and in 16 postmenopausal control women. The serum DHEA-S levels of the insulin-treated diabetic women were 37% higher than those of the control group, although this difference was not statistically significant. In an additional study,[65] adult-onset diabetic nonhirsute women (56-yr-old) were compared with groups of adult-onset diabetic hirsute women (60-yr-old), young and old hirsute nondiabetic women, and young and old control women. Most subjects were obese. All diabetic subjects were treated with diabetic diets, some received oral hypoglycemic agents, and one half received insulin therapy. Serum DHEA levels of the diabetic group were higher than those of the old control and diabetic hirsute subjects, but were not different from the other three groups. In a potentially related experiment, Remer and Pietrzik[66] have reported that 1 wk of insulin treatment in young male rats resulted in increased serum DHEA-S levels both in the basal condition and following ACTH stimulation. They speculated that chronic insulin treatment leads to increased serum DHEA-S levels. They also hypothesized that there is a different metabolic response to acute and chronic insulin administration that explains why acute insulin administration leads to lower DHEA-S levels (see later). However, in a 16-yr-old female subject with acanthosis nigricans, infusion of insulin over time resulted in hyperandrogenism but did not alter DHEA or DHEA-S concentrations.[67]

Several of the above-cited studies,[64,67] in which subjects had other hormonal alterations in addition to their diabetic status, call attention to the fact that individuals can have combined metabolic disturbances. Such subjects may provide opportunities for identifying mechanisms through which metabolic factors can interact. Women with polycystic ovarian syndrome (PCOS), who are usually insulin resistant and/or hyperinsulinemic, have provided such an opportunity to determine the potential relationship of elevated serum insulin to serum DHEA-S (and/or DHEA). There have been reports of elevated serum DHEA/DHEA-S accompanying elevated serum insulin in PCOS subjects.[68–71] However, subjects with PCOS have also been reported to have either elevated serum insulin but normal DHEA-S levels,[72,73] or to have normal serum insulin and DHEA-S concentrations.[74] Azziz et al.[75] divided hyperandrogenic women into hyperinsulinemic and normoinsulinemic groups. Basal serum DHEA-S but not DHEA concentrations were higher only in the normoinsulinemic group compared with control values. In some studies evaluations of the correlation between insulin and DHEA levels have been made but, as in the above-mentioned cases, the results have been varied.[69–73,75–78.]

Results of these studies are difficult to interpret and thus draw conclusions from because of differences in study populations (age, body wt, body fat, hormonal profiles, and control or comparison groups used). In one study six different groups were used.[76] It was found that lean and obese hirsute women, regardless of whether they had PCOS, had higher serum DHEA-S concentrations than did control lean and obese subjects. However, the obese hirsute women had lower DHEA-S than did lean hirsute women. Serum insulin levels were highest in the obese hirsute women with PCOS, and this group had the highest insulin AUC following OGTT compared with the other hirsute group. Thus, in the hirsute (hyperandrogenic) women, those with the highest serum insulin had lower serum DHEA-S levels. These results are summarized in Table 4. There was no significant correlation of DHEA-S with either baseline insulin or insulin AUC when all groups were combined.

A series of studies have been presented using a number of different experimental and control groups. Initially, Schriock et al.[79] studied three groups of women in whom OGTT was performed. The first group was composed of normal menstruating women, the second group of women with elevated DHEA-S

Table 4. Insulin, insulin area under the curve (AUC), androstenedione (A), testosterone (T), and DHEA-S levels in hirsute patients (obese and lean) and in control subjects (obese and lean)

	Obese			Lean		
		Hirsute			Hirsute	
	Controls	PCOS	Non-PCOS	Controls	PCOS	Non-PCOS
Baseline insulin (pmol/l)	149 ± 78[a]	239 ± 121[b,f,h]	129 ± 49[a]	92 ± 37	129 ± 27[a,e]	86 ± 50
Insulin AUC	10802 ± 3708[c]	15594 ± 791[d,e,g]	9059 ± 5217[c]	3198 ± 1220	5410 ± 751[a,f]	3322 ± 351
A (nmol/l)	7.4 ± 2.6[b]	7.3 ± 2.6[b]	7.1 ± 2.9[b]	4.9 ± 2.5	10 ± 3[c]	5.9 ± 1.9
T (nmol/l)	1.7 ± 0.6	1.9 ± 0.6[a]	1.6 ± 0.7	1.4 ± 0.4	2.3 ± 1.4[a]	1.5 ± 0.2
DHEA-S (μmol/l)	5.9 ± 4.5	8.1 ± 3.1[b,j]	8.4 ± 3.6[b,i]	5.1 ± 1.9	11.7 ± 4.6[c]	12.7 ± 4.7[c]

Adapted from ref. 76, with permission.

Data are means ± SD. [a]$P < 0.05$ vs. lean controls; [b]$P < 0.02$ vs. lean controls; [c]$P < 0.01$ vs. lean controls; [d]$P < 0.005$ vs. lean controls; [e]$P < 0.05$ vs. obese controls; [f]$P < 0.01$ vs. obese controls; [g]$P < 0.05$ vs. obese non-PCOS; [h]$P < 0.02$ vs. obese non-PCOS; [i]$P < 0.05$ vs. lean non-PCOS; [j]$P < 0.05$ vs. lean PCOS.

levels, and the third group of insulin-resistant women with PCOS and acanthosis nigricans. No body wt was given for the normal menstruating women, but the insulin-resistant group was reported to have a mean wt of 123 kg and the high DHEA-S group had a mean wt of 85.5 kg. As expected, insulin AUC was found to be highest in the insulin-resistant group. These results are summarized in Table 5. A negative correlation between insulin AUC and basal serum DHEA-S was found for all three groups. In contrast, T levels were positively correlated with serum insulin concentrations in these women. In this same study,[79] in vitro insulin binding to erythrocytes was found to be increased in the high DHEA-S group relative to binding for the normal subjects, while the insulin-resistant group had much lower binding compared with the other two groups. In a follow-up investigation, Buffington et al.[80] divided hyperandrogenic women into two groups: 1) obese PCOS subjects who were hyperinsulinemic with acanthosis nigricans and elevated serum T but normal serum DHEA levels, and 2) obese adrenal hyperandrogenic subjects with normal serum insulin levels and elevated androgen levels. Two control groups were used, one of normal wt and an obese group with elevated serum insulin concentrations. Serum DHEA levels were positively correlated with insulin binding and were inversely correlated with basal insulin and insulin AUC after OGTT among the four groups. In fact, the best correlation was obtained using DHEA/T ratios. In an earlier study,[77] these researchers had shown that baseline levels of DHEA were higher in PCOS subjects with NIDDM than in either obese control or PCOS subjects with normal glucose tolerance. The two PCOS groups had similar serum insulin levels that were threefold higher than in obese control subjects.

Subjects with PCOS were also treated with interventions to perturb their hormonal milieu. In one experimental protocol, obese PCOS and obese control subjects were studied before and after a 17 h infusion with either 1 mg/h DHEA or saline alone.[81] The two groups had similar basal DHEA and DHEA-S levels, but the PCOS women had higher serum insulin levels, higher insulin AUC, and lower insulin binding to lymphocytes. DHEA infusions had little effect on OGTT in either group. However, insulin binding was increased in the PCOS subjects, and

Table 5. Dehydroepiandrosterone-sulfate (DHEA-S), testosterone (T), insulin, insulin area under the curve (AUC), and insulin binding in three groups of women

	DHEA-S (μg/dl)	T (ng/dl)	Insulin (μU/ml)	AUC (μU insulin/ml·min)	Insulin binding to RBC (%)
Normal women (n = 11)	190 ± 33	41 ± 5	12 ± 5	7497 ± 1293	11.6 ± 1.5
High DHEA-S women (n = 9)	614 ± 50	84 ± 12	14 ± 5	10,454 ± 2006	14.6 ± 0.4
Insulin-resistant women (n = 6)	180 ± 27	124 ± 16	77 ± 41	43,562 ± 6511	5.5 ± 1.0

Adapted from ref. 79, with permission.
Data are means ± SD.
RBC, red blood cell.

both groups had increased pyruvate dehydrogenase activity in lymphocytes, indicating improved insulin sensitivity. Another approach employed by several researchers has been to use drug therapy to ameliorate high androgen levels. Diamanti-Kandarakis et al.[82] treated lean and obese PCOS subjects with flutamide for 3 months. There was little effect on insulin sensitivity over the treatment period, but serum DHEA-S levels were lowered. In addition, a significant positive correlation of change in glucose utilization rate and change in DHEA-S was reported. Another drug, ketoconazole, was used in a 10 month trial.[83] Serum DHEA-S concentration was lowered by the treatment, but no effect on either basal or peak insulin after OGTT was reported.

Another approach to evaluate the potential role of DHEA/DHEA-S levels in diabetes would be to examine individuals with either prediabetic or hyperinsulinemic status. Pima Indians as a group have an increased risk of developing NIDDM. Weiss et al.[84] divided young Pima Indian women into two groups based on high vs. low serum insulin levels. Those with hyperinsulinemia had significantly lower serum DHEA-S concentrations than those with low serum insulin levels. There were no differences in either serum T or A concentrations between the two groups. However, in another study premenopausal obese hyperinsulinemic women and lean normoinsulinemic women were found to have similar DHEA and DHEA-S levels.[85] There was an inverse correlation between DHEA and body mass index (BMI) in the obese women.

In general, the interrelation of serum insulin and DHEA levels in men has not been studied to the extent it has been in women. However, in the past few years there have been a number of publications pertaining to this topic. In a recent study Haffner et al.[86] reported that for normal wt men there was a negative correlation with DHEA-S and fasting serum insulin, as well as with insulin 2 h after the initiation of OGTT. Phillips[87] reported no correlation of insulin and DHEA-S levels for obese men. Pasquali et al.[88] found that obese and normal wt men had similar plasma DHEA-S concentrations. The obese subjects as a group had higher insulin levels than the normal wt men. But when the obese men were divided into two groups based on waist-to-hip ratio, only the obese subjects with a waist-to-hip ratio >0.97 had higher serum insulin levels than normal wt men. DHEA-S was negatively correlated with BMI. No correlation of insulin and DHEA-S was made by the authors. In a recent report,[89] obese men with normal OGTT results were reported to have a positive correlation of the subscapular-to-triceps skinfold ratio (another indicator of central obesity) and a negative correlation of DHEA-S with insulin AUC. Haffner et al.[90] had previously reported a significant negative correlation of waist-to-hip ratio with DHEA-S levels in a group of men with a mean BMI of 27, which was substantially lower than the BMI of 35 reported for the men in the study of Herranz et al.[89] Neither study found correlations of DHEA-S with BMI. However, Tchernof et al.[91] reported that lean men with a mean BMI of 22.7 and waist-to-hip ratio of 0.87 had higher serum DHEA levels than did obese men with a BMI of 29.9 and a waist-to-hip ratio of 0.98. DHEA levels were negatively correlated with total body fatness, BMI, and waist-to-hip ratio. Although there is a suggestion that in obese men, particularly with central obesity, high serum insulin may be associated with lower

DHEA-S, the results are not conclusive. The results of another recent study by Tchernof et al.[92] may provide some explanation for these reported differences. They found that although there was a negative correlation of DHEA with insulin AUC for men with a BMI over a range from 19.4 to 34.2, when the results were corrected for fat mass, BMI, or visceral adipose tissue area there was no longer a correlation.

There are a number of studies suggesting sex-related differences in serum DHEA or DHEA-S levels. Women tend to have lower levels than men.[52,53,93,94] Nestler and Kahwash[95] reported that there are sex-specific differences in the effects of insulin on the metabolic clearance rate (MCR) of DHEA in humans. Using the hyperinsulinemic-euglycemic clamp technique, they found that in men, the MCR_{DHEA} rose during the insulin infusion but was unchanged when saline was infused. In women, however, there was no difference in MCR_{DHEA} during the insulin vs. the saline infusion. The subjects were nonobese and ranged in age from 20 to 30 years. In a different investigation, Farah et al.[73] found that MCR_{DHEA} was higher in obese hyperinsulinemic women than in normal wt normoinsulinemic women. MCR_{DHEA} for obese women with PCOS plus acanthosis nigricans was similar to that of the obese women, although the latter group had even higher serum insulin levels than the obese group. In an additional study, Haning et al.[96] reported the $MCR_{DHEA\text{-}S}$ for obese women with elevated serum DHEA-S levels was significantly lower than in normal wt women with normal DHEA-S levels.

In a number of studies, plasma DHEA/DHEA-S levels have been altered by various interventions. Nestler et al.[97,98] found a decline of serum DHEA-S levels in both normal men and women during the course of experimentally induced hyperinsulinemia using the hyperinsulinemic-euglycemic clamp technique. This finding has been corroborated for male subjects[99] and for female subjects following OGTT.[100,101] Falcone et al.[72] reported that following an ivGTT, normal women and insulin-sensitive PCOS women had decreased DHEA levels but insulin-resistant PCOS women did not. However, several other investigators have reported no differences in serum DHEA-S levels in either control obese, control nonobese, PCOS obese, or PCOS nonobese women following OGTT.[74,102]

Nestler et al.[98] have speculated that hyperinsulinemia leads to a reduction in serum DHEA/DHEA-S concentrations by inhibition of steroid production as well as by stimulating clearance. Some of their results support that the effect on steroid production is at least partially mediated by insulin's inhibition of adrenal 17,21 lyase.[103] Additional studies support that high serum insulin levels result in low serum DHEA-S levels. For example, when insulin levels were reduced in obese hypertensive men in response to drug treatment with two different calcium channel blocks, both nitrendipine[104] and amlodipine[105] resulted in increased serum DHEA-S levels. In a small group of obese PCOS women treated with diazoxide, however, reductions in serum insulin levels were not accompanied by changes in DHEA-S concentrations.[106] A third calcium channel blocker, diltiazem, was given to obese hypertensive men and postmenopausal women.[107] In both groups post-treatment insulin levels were lower than placebo values; however, only in the men were DHEA and DHEA-S levels increased significantly. This study, in addition to those cited earlier, further supports the notion that there are apparently differences in DHEA metabolism between men and women.

Several other drug interventions that also are known to lower serum insulin levels have been used to rule out that some other mechanism of action of the calcium channel blockers is responsible for the reported effects on serum DHEA/DHEA-S concentrations. In one study treatment with the biguanide, metformin, was found to significantly decrease the serum insulin level with an increase in serum DHEA-S in both obese hypertensive and non-obese normotensive men.[108] Neither serum DHEA nor BMI were altered. In another study, treatment with benfluorex resulted in lower serum insulin in middle-aged (30–50 yr old) but not elderly (60–80 yr old) men.[109] However, in both groups treatment resulted in significantly lower insulin AUC and significant increases in both serum DHEA and DHEA-S concentrations.

Conclusions

An attempt was made here to evaluate and integrate the results of various studies to determine if DHEA had "antidiabetic" properties. Many intriguing aspects of DHEA, both as a pharmacological agent and as a physiological metabolite related to diabetes, have been reported. As a pharmacological agent, particularly in rodents and possibly in humans, DHEA or metabolite treatment appeared to improve factors usually associated with NIDDM, such as body wt, body fat, and fat cell size. However, there were usually no measured changes in glucose tolerance/insulin resistance, except in some strains of mice. This may be due to a type of insulin resistance resulting from elevated androgen levels induced by DHEA treatment, particularly in female subjects in a manner similar to that described for testosterone.[110] On the other hand, there appears to be the possibility that hyperinsulinemia may result in alterations in serum androgens, as indicated by studies cited earlier as well as others.[111] Also, consideration should be given to differences in androgen metabolism between humans and rodents, since rodent adrenal glands have been reported not to synthesize androgens,[112,113] although not all reports support this finding.[114,115] In addition, differences in sex-related response to DHEA treatment, as well as its metabolism, need to be taken into consideration in designing studies and interpreting results. Several recent studies using insulin binding to lymphocytes indicate that insulin sensitivity may be improved by DHEA treatment. This suggests that changes induced by DHEA may need more sensitive testing procedures. This is supported by a recent report using human fibroblasts.[116] Incubation of cultured fibroblasts from normal wt men and women with DHEA but not DHEA-S resulted in increased 2-deoxyglucose uptake. In addition, transcripts for GLUT1 were also increased, and added insulin increased the sensitivity of glucose uptake to DHEA. This and some of the other results presented indicate that it may be important to distinguish metabolism and the effects of DHEA from DHEA-S. To date, there appears to be little evidence of DHEA having preventive or ameliorative effects for IDDM.

Some data suggest that hyperinsulinemia/insulin resistance may lower serum DHEA levels. Although there has not been a definitive mechanism of action for

DHEA/DHEA-S identified to date, the fact that it circulates at relatively high concentrations in blood has suggested to many that it plays some as yet undefined, but important, biological role. If hyperinsulinemia leads to lower DHEA levels and/or alterations in its metabolism, this could potentially have long-term consequences for the individuals involved.

Establishment of the biological role for DHEA would, in turn, make it easier to determine whether continued investigations of antidiabetic actions of DHEA are, in fact, justified and/or worthwhile. At the present time the different study designs and measurements made do not allow an easy interpretation of results, particularly when trying to compare work from different research groups. Whether analogues of DHEA, such as 16α-fluoro-5-androsten-17-one,[21] would be effective as antidiabetic agents, particularly in humans, remains to be determined. Recently, Lardy et al.[117] have identified 7-oxo derivatives of DHEA that have antiobesity action in rats but have not been evaluated as antidiabetic agents. This provides another approach to understanding DHEA by identifying metabolites of DHEA that may be the active physiological compounds. In addition, dosages for comparison of DHEA and its analogues and metabolites between animals and human studies should be established. Experimental groups should be carefully defined so that meaningful comparisons can be made within a study and in relation to other studies. Clearly, sex-related differences in DHEA metabolism between men and women need to be delineated. The prospect of a naturally occurring compound or metabolite that may be able to ameliorate some of the negative consequences of diabetes is an exciting area for continued investigation.

References

1. Šonka J, Gregorová I, Páv J, Skrha F. Dehydroepiandrosterone in obese diabetics. Lancet 2:44–5, 1964.
2. Šonka J. Dehydroepiandrosterone metabolic effects. Acta Univ Carol Med Monogr 71:1–171, 1976.
3. Kalimi M, Regelson W. The biologic role of dehydroepiandrosterone (DHEA). Walter de Gruyter, New York, 1990.
4. Gordon GB, Schantz LM, Talalay P. Modulation of growth, differentiation and carcinogenesis by dehydroepiandrosterone. Adv Enzyme Regul 26:355–82, 1987.
5. Cleary MP. The antiobesity effect of dehydroepiandrosterone in rats. Proc Soc Exp Biol Med 196:8–16, 1991.
6. Berdanier CD, Parente JA Jr, McIntosh MK. Is dehydroepiandrosterone an antiobesity agent? FASEB J 7:414–9, 1993.
7. Yen TT, Allan JA, Pearson DV, Acton JM. Prevention of obesity in $A^{vy/a}$ mice by dehydroepiandrosterone. Lipids 12:409–13, 1977.
8. Schwartz AG. Inhibition of spontaneous breast cancer formation in female C3H ($A^{vy/a}$) mice by long term treatment with dehydroepiandrosterone. Cancer Res 39:1129–32, 1979.
9. Weindruch R, McFeetus G, Walford RL. Food intake reduction and immunologic alterations in mice fed dehydroepiandrosterone. Exp Gerontol 19:297–304, 1984.
10. Cleary MP, Seidenstat R, Tannen RH, Schwartz AG. The effect of dehydroepiandrosterone on adipose tissue cellularity in mice. Proc Soc Exp Biol Med 171:276–84, 1982.

11. Granholm NH, Staber LD, Wilken PJ. Effects of dehydroepiandrosterone on obesity and glucose-6-phosphate dehydrogenase activity in the lethal yellow mouse (strain 129/Sv-A^y/A^w). J Exp Zool 242:67–74, 1987.
12. Pashko LL, Fairman DK, Schwartz AG. Inhibition of proteinuria development in aging Sprague-Dawley rats and C57BL/6 mice by long-term treatment with dehydroepiandrosterone. J Gerontol 41:433–8, 1986.
13. Coleman DL, Leiter EH, Schwizer RW. Therapeutic effects of dehydroepiandrosterone (DHEA) in diabetic mice. Diabetes 31:830–3, 1982.
14. Coleman DL, Schwizer RW, Leiter EH. Effect of genetic background on the therapeutic effects of dehydroepiandrosterone (DHEA) in diabetes-obesity mutants and in aged normal mice. Diabetes 33:26–32, 1984.
15. Coleman DL, Leiter EH, Applezweig N. Therapeutic effects of dehydroepiandrosterone metabolites in diabetes mutant mice (C56BL/KsJ-*db/db*). Endocrinology 115:239–43, 1984.
16. Coleman DL. Antiobesity effects of etiocholanolones in diabetes (*db*), viable yellow (A^{vy}), and normal mice. Endocrinology 117:2279–83, 1985.
17. Leiter EH, Beamer WG, Coleman DL, Longcope C. Androgenic and estrogenic metabolites in serum of mice fed dehydroepiandrosterone: Relationship to antihyperglycemic effects. Metabolism 36:863–9, 1987.
18. Beamer WG, Shultz KL, Tennent BJ. Induction of ovarian granulosa cell tumors in SWXJ-9 mice with dehydroepiandrosterone. Cancer Res 48:2788–92, 1988.
19. Tennent BJ, Shultz KL, Beamer WG. Genetic susceptibility for C19 androgen induction of ovarian granulosa cell tumorigenesis in SWXJ strains of mice. Cancer Res 53:1059–63, 1993.
20. Schwartz AG, Lewbart ML, Pashko LL. Novel dehydroepiandrosterone analogues with enhanced biological activity and reduced side effects in mice and rats. Cancer Res 48:4817–22, 1988.
21. Pashko LL, Schwarz AG. Antihyperglycemic effect of dehydroepiandrosterone analogue 16α-fluoro-5-androsten-17-one in diabetic mice. Diabetes 42:1105–8, 1993.
22. Cleary MP, Shepherd A, Jenks B. Effect of dehydroepiandrosterone on growth in lean and obese Zucker rats. J Nutr 114:1242–51, 1984.
23. Shepherd A, Cleary MP. Metabolic alterations due to dehydroepiandrosterone in the obese Zucker rat. Am J Physiol 246:E123–8, 1984.
24. Cleary MP, Billheimer J, Finan A, et al. Metabolic consequences of dehydroepiandrosterone in lean and obese adult Zucker rats. Horm Metab Res 16(Suppl 1):43–6, 1984.
25. Cleary MP, Zisk JR. Antiobesity effect of two different levels of dehydroepiandrosterone treatment in lean and obese middle-aged female Zucker rats. Int J Obes 10:193–204, 1986.
26. Muller S, Cleary MP. Glucose metabolism in isolated adipocytes from lean and obese Zucker rats following treatment with dehydroepiandrosterone. Metabolism 34:278–84, 1985.
27. Svec F, Hilton CW, Wright B, et al. The effect of DHEA given chronically to Zucker rats. Proc Soc Exp Biol Med 209:92–7, 1995.
28. Gansler TS, Muller S, Cleary MP. Chronic administration of dehydroepiandrosterone (DHEA) reduces pancreatic β-cell hyperplasia and hyperinsulinemia in genetically obese rats. Proc Soc Exp Biol Med 180:155–62, 1985.
29. Cleary MP, Zabel T, Sartin JL. Effects of short-term dehydroepiandrosterone treatment on serum and pancreatic insulin in Zucker rats. J Nutr 118:382–7, 1988.
30. Mohan PF, Cleary MP. Comparisons of dehydroepiandrosterone and clofibric acid treatments in obese Zucker rats. J Nutr 119:496–501, 1989.
31. Hayek A, Woodside W. Correlation between morphology and function in isolated islets of the Zucker rat. Diabetes 28:565–9, 1979.
32. Cleary MP. Effect of dehydroepiandrosterone treatment on liver metabolism in rats. Int J Biochem 22:305–10, 1990.

33. Marks PA, Banks J. Inhibition of mammalian glucose-6-phosphate dehydrogenase by steroid hormones. Proc Natl Acad Sci USA 46:447–52, 1960.
34. Oertel GW, Benes P. The effects of steroids on glucose-6-phosphate dehydrogenase. J Steroid Biochem 3:493–6, 1972.
35. Beitner R, Naor Z. The effect of adenine nucleotides and dehydroepiandrosterone on the isoenzymes of $NADP^+$- and NAD^+-glucose 6-phosphate dehydrogenase and 6-phosphogluconate dehydrogenase from rat adipose tissue. Biochim Biophys Acta 286:437–40, 1972.
36. Hissin PJ, Foley JE, Wardzala LJ, et al. Mechanism of insulin-resistant glucose transport activity in the enlarged adipose cell of the aged, obese rat. Relative depletion of intracellular glucose transport systems. J Clin Invest 70:780–90, 1982.
37. Richardson DK, Czech MP. Diminished activities of fatty acid synthesis enzymes in insulin-resistant adipocytes from spontaneously obese rats. Horm Metab Res 11:427–31, 1979.
38. Olefsky JM. The effect of spontaneous obesity on insulin binding, glucose transport and glucose oxidation of isolated rat adipocytes. J Clin Invest 57:842–51, 1976.
39. Mohan PF, Ihnen JS, Levin BE, Cleary MP. Effects of dehydroepiandrosterone treatment in rats with diet-induced obesity. J Nutr 120:1103–14, 1990.
40. Berdanier CD. The BHE rat: An animal model for the study of noninsulin-dependent diabetes mellitus. FASEB J 5:2139–44, 1991.
41. McIntosh MK, Berdanier CD. Antiobesity effects of dehydroepiandrosterone are mediated be futile substrate cycling in hepatocytes of BHE/cdb rats. J Nutr 121:2037–43, 1991.
42. Roy S, Mahesh VB, Greenblatt RB. Effect of dehydroepiandrosterone and Δ^4-androstenedione on the reproductive organ of female rats; production of cystic changes in the ovary. Nature 196:2–5, 1962.
43. Parker CR, Mahesh VB. Interrelationship between excessive levels of circulatory androgens in the blood and ovulatory failure. J Reprod Med 17:75–90, 1976.
44. Lee M-T, Anderson E, Lee GY. Changes in ovarian morphology and serum hormones in the rat after treatment with dehydroepiandrosterone. Anat Rec 231:185–92, 1991.
45. Nestler JE, Barlascini CO, Clore JN, Blackard WG. Dehydroepiandrosterone reduces serum low density lipoprotein levels and body fat but does not alter insulin sensitivity in normal men. J Clin Endocr Metab 66:57–61, 1988.
46. Usiskin KB, Butterworth S, Clore JN, et al. Lack of effect of dehydroepiandrosterone in obese men. Int J Obes 14:457–63, 1990.
47. Mortola JF, Yen SSC. The effects of oral dehydroepiandrosterone on endocrine-metabolic parameters in postmenopausal women. J Clin Endocr Metab 71:696–704, 1990.
48. Casson PR, Faquin CC, Stentz FB, et al. Replacement of dehydroepiandrosterone enhances T-lymphocyte insulin binding in postmenopausal women. Fertil Steril 63:1027–31, 1995.
49. Morales AJ, Nolan JJ, Nelson JC, Yen SSC. Effects of replacement dose of dehydroepiandrosterone in men and women of advancing age. J Clin Endocr Metab 78:1360–67, 1994.
50. Zumoff B, Strain GW, Heymsfield SB, Lichtman S. A randomized double-blind crossover study of the antiobesity effects of etiocholanedione. Obes Res 2:13–8, 1994.
51. Buffington CK, Pourmotabbed G, Kitabchi AE. Case report: Amelioration of insulin resistance in diabetes with dehydroepiandrosterone. Am J Med Sci 306:320–4, 1993.
52. Orentreich N, Brind JL, Rizer RL, Vogelman JH. Age changes and sex differences in serum dehydroepiandrosterone sulfate concentrations throughout adulthood. J Clin Endocr Metab 59:551–5, 1984.
53. Carlstrom K, Brody S, Lunell NO, et al. Dehydroepiandrosterone sulphate and de-

hydroepiandrosterone in serum: Differences related to age and sex. Maturitas 10:297–306, 1988.
54. Orentreich N, Brind JL, Vogelnab JH, et al. Long-term longitudinal measurements of plasma dehydroepiandrosterone sulfate in normal men. J Clin Endocr Metab 75:1002–4, 1992.
55. Parker LN. Adrenal androgens in clinical medicine. Academic Press, San Diego pp 118–34, 1989.
56. Ebeling P, Koivisto VA. Physiological importance of dehydroepiandrosterone. Lancet 343:1479–81, 1994.
57. Nafziger AN, Herrington DM, Bush TL. Dehydroepiandrosterone and dehydroepiandrosterone sulfate: Their relationship to cardiovascular disease. Epidem Rev 13:267–93, 1991.
58. Nestler JE, McClanahan MA. Diabetes and adrenal disease. Balliere's Clin Endocr Metab 6:829–47, 1992.
59. Nestler JE, Clore JN, Blackard WG. Dehydroepiandrosterone: The "missing link" between hyperinsulinemia and atherosclerosis. FASEB J 6:3073–75, 1992.
60. Barrett-Connor E, Khaw KT, Yen SSC. A prospective study of dehydroepiandrosterone sulfate, mortality, and cardiovascular disease. N Engl J Med 315:1519–24, 1986.
61. Khaw KT, Barrett-Connor E. Fasting plasma glucose and endogenous androgens in non-diabetic postmenopausal women. Clin Sci 80:199–203, 1991.
62. Barrett-Connor E. Lower endogenous androgen levels and dyslipidemia in men with non-insulin-dependent diabetes mellitus. Ann Intern Med 117:807–11, 1992.
63. Herrington DM, Gordon GB, Achuff SC, et al. Plasma dehydroepiandrosterone sulfate in patients undergoing diagnostic coronary angiography. J Am Coll Cardiol 16:862–70, 1990.
64. Nyholm H, Djursing H, Hagen C, et al. Androgens and estrogens in postmenopausal insulin-treated diabetic women. J Clin Endocr Metab 69:946–9, 1989.
65. Szpunar WE, Blair AJ, McCann DS. Plasma androgen concentrations in diabetic women. Diabetes 26:1125–29, 1977.
66. Remer T, Pietrzik K. Chronic insulin treatment, but not chronic ACTH administration increases plasma dehydroepiandrosterone sulfate levels in adolescent male rats. Exp Clin Endocr 101:222–9, 1993.
67. DeClue TJ, Shah SC, Marchese M, Malone JI. Insulin resistance and hyperinsulinemia induce hyperandrogenism in a young type B insulin-resistant female. J Clin Endocr Metab 72:1308–11, 1991.
68. Chang RJ, Nakamura RM, Judd HL, Kaplan SA. Insulin resistance in nonobese patients with polycystic ovarian disease. J Clin Endocr Metab 57:356–9, 1983.
69. Shoupe D, Kumar DD, Lobo RA. Insulin resistance in polycystic ovarian syndrome. Am J Obstet Gynecol 147:588–92, 1983.
70. Smith S, Ravnikar VA, Barbieri RL. Androgen and insulin response to an oral glucose challenge in hyperandrogenic women. Fertil Steril 48:72–7, 1987.
71. Vidal-Puig A, Muñoz-Torres M, Jodar-Gimeno E, et al. Hyperinsulinemia in polycystic ovary syndrome: Relationship to clinical and hormonal factors. Clin Invest 72:853–7, 1994.
72. Falcone T, Finegood DT, Fantu G, Morris D. Androgen response to endogenous insulin secretion during the frequently sampled intravenous glucose test in normal and hyperandrogenic women. J Clin Endocr Metab 71:1653–57, 1990.
73. Farah MJ, Givens JR, Kitabchi AE. Bimodal correlation between circulating insulin level and production of rate of dehydroepiandrosterone: Positive correlation in controls and negative correlation in polycystic ovary syndrome with acanthosis nigricans. J Clin Endocr Metab 70:1075–81, 1990.
74. Tropeano G, Lucisano A, Liberale I, et al. Insulin, C-peptide, androgens, and β-endorphin response to oral glucose in patients with polycystic ovary syndrome. J Clin Endocr Metab 78:305–9, 1994.

75. Azziz R, Bradley EL, Potter HD, et al. Chronic hyperinsulinemia and the adrenal androgen response to acute corticotropin-(1-24) stimulation in hyperandrogenic women. Am J Obstet Gynecol 172:1251–56, 1995.
76. Toscano V, Bianchi P, Balducci R, et al. Lack of linear relationship between hyperinsulinemia and hyperandrogenism. Clin Endocr 36:197–207, 1992.
77. Buffington CK, Givens JR, Kitabchi AE. Opposing actions of dehydroepiandrosterone and testosterone on insulin sensitivity. In vivo and in vitro studies in hyperandrogenic females. Diabetes 40:693–700, 1991.
78. Dunaif A, Graf M, Mandeli J, et al. Characterization of groups of hyperandrogenic women with acanthosis nigricans, impaired glucose intolerance, and/or hyperinsulinemia. J Clin Endocr Metab 65:499–507, 1987.
79. Schriock ED, Buffington CK, Hubert GD, et al. Divergent correlations of circulating dehydroepiandrosterone and insulin receptor sulfate and testosterone with insulin levels binding. J Clin Endocr Metab 66:1329–31, 1988.
80. Buffington CK, Givens JR, Kitabchi AE. Enhanced adrenocortical activity as a contributing factor to diabetes in hyperandrogenic women. Metabolism 43:584–90, 1994.
81. Schriock ED, Buffington CK, Givens JR, Buster JE. Enhanced post-receptor insulin effects following dehydroepiandrosterone infusion. J Soc Gynecol Invest 1:74–8, 1994.
82. Diamanti-Kandarakis E, Mitrakou A, Hennes MMI, et al. Insulin sensitivity and antiandrogenic therapy in women with polycystic ovary syndrome. Metabolism 44:525–31, 1995.
83. Vidal-Puig A, Muñoz-Torres M, Garcia-Calvente C, et al. Reduction of endogenous, ovarian and adrenal androgens with ketoconazole does not alter insulin response in the polycystic ovary syndrome. J Endocr Invet 17:647–52, 1994.
84. Weiss DJ, Charles MA, Dunaif A, et al. Hyperinsulinemia is associated with menstrual irregularity and altered serum androgens in Pima Indian women. Metabolism 43:803–7, 1994.
85. DePergola G, Giagulli VA, Garruti G. Low dehydroepiandrosterone circulating levels in premenopausal obese women with very high body mass index. Metabolism 40:187–90, 1991.
86. Haffner SM, Valdez RA, Mykkanen L, et al. Decreased testosterone and dehydroepiandrosterone sulfate concentrations are associated with increased insulin and glucose concentrations in diabetic men. Metabolism 43:599–603, 1994.
87. Phillips GB. Relationship between serum sex hormones and the glucose-insulin-lipid defect in men with obesity. Metabolism 42:116–20, 1993.
88. Pasquali R, Casimirri F, Cantobelli S, et al. Effect of obesity and body fat distribution on sex hormones and insulin in men. Metabolism 40:101–4, 1991.
89. Herranz L, Megia A, Grand C, et al. Dehydroepiandrosterone sulphate, body fat distribution and insulin in obese men. Int J Obes 19:57–60, 1995.
90. Haffner SM, Valdez RA, Stern MP, Katz MS. Obesity, body fat distribution and sex hormones in men. Int J Obes 17:643–9, 1993.
91. Tchernof A, Deprés J-P, Bélanger A, et al. Reduced testosterone and adrenal C_{19} steroid levels in obese men. Metabolism 44:513–19, 1995.
92. Tchernof A, Després J-P, Dupont A, et al. Relation of steroid hormones to glucose tolerance and plasma insulin levels in men. Diabetes Care 18:292–9, 1995.
93. Parker LN. Adrenal androgens in clinical medicine. Academic Press, San Diego p 615, 1989.
94. Zumoff P, Rosenfeld GW, Strain GW. Sex difference in twenty-four hour mean plasma concentrations of dehydroepiandrosterone (DHA) and dehydroepiandrosterone sulfate (DHAS) and the DHA to DHAS ratio in normal adults. J Clin Endocr Metab 51:330–3, 1980.
95. Nestler JE, Kahwash Z. Sex-specific action of insulin to acutely increase the metabolic clearance rate of dehydroepiandrosterone in humans. J Clin Invest 94: 1484–89, 1994.

96. Haning RV Jr, Carlson IH, Flood CA, et al. Metabolism of dehydroepiandrosterone sulfate (DS) in normal women and women with high DS concentrations. J Clin Endocr Metab 73:1210–15, 1991.
97. Nestler JE, Clore JN, Strauss III JF, Blackard WG. The effects of hyperinsulinemia on serum testosterone, progesterone, dehydroepiandrosterone sulfate, and cortisol in normal women and in a woman with hyperandrogenism, insulin resistance, and acanthosis nigricans. J Clin Endocr Metab 64:180–4, 1987.
98. Nestler JE, Usiskin KS, Barlascini CO, et al. Suppression of serum dehydroepiandrosterone sulfate levels by insulin: An evaluation of possible mechanisms. J Clin Endocr Metab 69:1040–46, 1989.
99. Ebeling P, Stenman U-H, Seppälä M, Koivisto VA. Acute hyperinsulinemia, androgen homeostasis and insulin sensitivity in healthy men. J Endocrinol 146:63–9, 1995.
100. Diamond MP, Grainger DA, Laudano AJ, et al. Effect of acute physiological elevation of insulin on circulating androgen levels in nonobese women. J Clin Endocr Metab 72:883–7, 1991.
101. Hubert GD, Schriock ED, Givens JR, Buster JE. Suppression of circulating Δ4-androstenedione and dehydroepiandrosterone sulfate during oral glucose tolerance test in normal females. J Clin Endocr Metab 73:781–4, 1991.
102. Buyalos RP, Bradley EL, Judd HL, et al. No acute effect of physiological insulin increase on dehydroepiandrosterone sulfate in women with obesity and/or polycystic ovarian disease. Fertil Steril 56:1179–82, 1991.
103. Nestler JE, McClanahan MA, Clore JN, et al. Insulin inhibits 17, 20-lyase activity in man. J Clin Endocr Metab 74:362–7, 1992.
104. Beer NA, Jakubowicz DJ, Beer RM, et al. Effects of nitrendipine on glucose tolerance and serum insulin and dehydroepiandrosterone sulfate levels in insulin-resistant obese and hypertensive men. J Clin Endocr Metab 76:178–83, 1993.
105. Beer NA, Jakubowicz DJ, Beer RM, Nestler JE. The calcium blocker amlodipine raises serum dehydroepiandrosterone sulfate and androstenedione, but lowers serum cortisol, in insulin-resistant obese and hypertensive men. J Clin Endocr Metab 76:1464–69, 1993.
106. Nester JE, Barlascini CO, Matt DW, et al. Suppression of serum insulin by diazoxide reduces serum testosterone level in obese women with polycystic ovary syndrome. J Clin Endocr Metab 68:1027–31, 1989.
107. Beer NA, Jakubowicz DJ, Beer RM, Nestler JE. Disparate effects on insulin reduction with diltiazem on serum dehydroepiandrosterone sulfate in obese hypertensive men and women. J Clin Endocr Metab 79:1077–81, 1994.
108. Nestler JE, Beer NA, Jakubowicz DJ, Beer RM. Effects of a reduction in circulating insulin by metformin on serum dehydroepiandrosterone sulfate in nondiabetic men. J Clin Endocr Metab 78:549–54, 1994.
109. Nestler JE, Beer NA, Jakubowicz DJ, et al. Effects of insulin reduction with benfluorex on serum dehydroepiandrosterone (DHEA), DHEA sulfate and blood pressure in hypertensive middle-aged and elderly men. J Clin Endocr Metab 80:700–6, 1995.
110. Björntorp P. Hyperandrogenicity in women—a prediabetic condition. J Intern Med 234:579–83, 1993.
111. Stuart CA, Prince MJ, Peters EJ, Meyer WJ. Hyperinsulinemia and hyperandrogenemia: In vivo androgen response to insulin infusion. Obstet Gynecol 69:921–5, 1987.
112. van Weerden WM, Biering HG, van Steenbrugge GJ, et al. Adrenal glands of mouse and rat do not synthesize androgens. Life Sci 50:857–61, 1992.
113. Bélanger B, Bélanger A, Labrie F, et al. Comparison of residual C-19 steroids in plasma and prostatic tissue of human, rat and guinea pig after castration: Unique importance of extratesticular androgens in men. J Steroid Biochem 32:695–8, 1989.
114. Edery M, Carreau S, Simon M-J, Drowdowsky MA. *In vitro* pregnenolone me-

tabolism by mouse adrenal gland: II—biosynthesis of androgens. Steroids 39:191–200, 1982.

115. Ando S, Aquila S, Beraldi E, et al. Physiological changes in androgen plasma levels with elapsing of time from castration in adult male rats. Horm Metab Res 20:96–9, 1988.
116. Nakashima N, Haji M, Sakai Y, et al. Effecy of dehydroepiandrosterone on glucose uptake in cultured human fibroblasts. Metabolism 44:543–8, 1995.
117. Lardy H, Partridge B, Kneer N, Wei Y. Ergosteroids: Induction of thermogenic enzymes in liver of rats treated with steroids derived from dehydroepiandrosterone. Proc Natl Acad Sci USA 92:6617–19, 1995.

21

Endoneurial vessel abnormalities in diabetic animal models

SOROKU YAGIHASHI, KAZUHIRO SUGIMOTO, AND RYU-ICHI WADA
Department of Pathology, Hirosaki University, School of Medicine, Hirosaki, 036 Japan.

Key words: neuropathy, microangiopathy, animal model, structure, blood flow, pathogenesis.

Neuropathy is a common and serious complication in diabetic patients. Both metabolic and vascular factors are considered to play a crucial role in its pathogenesis. Among the metabolic factors, the enhanced polyol pathway mediated by the enzyme aldose reductase are suggested to induce acutely functional impairments of peripheral nerve fibers and endoneurial microvessels, followed by irreversible structural changes.[1–3] Increased nonenzymatic glycation of neural and microvascular proteins may also elicit functional and structural alterations of the peripheral nerve, contributing to the development of neuropathy.[4] Thus, the alterations of endoneurial vessel changes may primarily or secondarily facilitate neuropathic changes in the diabetic condition.[5] Evaluation of endoneurial microvessels is therefore important not only for determining the role of microvascular lesions in the development of neuropathy, but also for elucidation of the progression of systemic microangiopathy in diabetes.

Classic pathological studies depicted angiopathic changes in the central and peripheral nervous systems in diabetic patients. Fagerberg[6] summarized typical features of microangiopathy of endoneurial vessels as well as sclerotic changes of epineurial arterioles, which he ascribed to be the major reason for the systematic neurological complications in diabetic patients. Subsequent sporadic reports also pointed out a variety of microangiopathic changes in human diabetic neuropathy.[7,8] However, these studies failed to demonstrate clear evidence supporting the notion that vascular lesions are related to the degree of nerve fiber pathology. In addition to the microangiopathic changes, the formation of fibrin

Correspondence: Dr. Soroku Yagihashi, Department of Pathology, Hirosaki University School of Medicine, 5 Zaifu-cho, Hirosaki, 036 Japan.

Lessons from Animal Diabetes VI. Ed. E. Shafrir.

thrombi, as well as endothelial closure of vessels are encountered in diabetic patients.[9,10] These studies though, were not performed in a systematic manner, and no clear relationship between nerve fiber degeneration and vascular lesions emerged.

Based on data from extensive morphometric analysis, Dyck et al.[11] demonstrated a significant relationship between vascular changes and nerve fiber pathology. Systematic studies indicated that the focal fiber loss was a characteristic feature in the proximal nerve, whereas fiber loss was diffuse in the distal nerve, probably due to the summation of multifocal fiber loss of the proximal region.[12,13] In their earlier reports, capillary closure with thickening of the vascular wall and endothelial swelling was ascribed to cause the endoneurial ischemia,[11–13] although the later study could not reproduce the closure of capillary lumen, which was similarly detected in nondiabetic control patients or other neuropathic patients.[14–16] Johnson et al.[17] also reported focal myelinated fiber loss in the biopsied sural nerves from patients with diabetes. On the other hand, others claimed that the focality of fiber loss was not specific for diabetes but could also be demonstrated in other neuropathies of nonvascular origin.[18] It is thus shown that vascular lesions occur in diabetic nerve, but we still lack evidence that nerve fiber pathology occurs as a consequence of vascular lesions.

In the absence of microangiopathic structural changes, functional derangements of endoneurial vessels should be taken into account as a potentially important contributor to the development of neuropathy. Alterations of the blood–nerve barrier may affect the microenvironment of the peripheral nerve. In experimental animal models of diabetes, perturbation of the blood–nerve barrier was demonstrated even in early stages of diabetes.[19,20] Disruption of the blood–nerve barrier with disintegration of the endothelial cell junction was detected ultrastructurally in the endoneurial microvessels in human diabetic patients.[16]

These endoneurial vascular abnormalities are associated with increased permeation of plasma albumin or immunoglobulins in the peripheral nerves of human diabetic patients.[21,22] Such an enhanced vascular permeability is relevant to increased nerve water content or interstitial edema in diabetic nerve.[23] These changes further elicit microcirculatory disturbances by physical compression of the perineurial and transperineurial vessels and widening of intercapillary distances, resulting in reduced tissue oxygen tension.[24,25] In hyperglycemic or galactosemic conditions, an increase in tissue osmolality is an important factor for the alterations of endoneurial blood flow and oxygen tension.[26–28] It is thus likely that structural and functional changes of endoneurial vessels may contribute in a major way to the pathological process resulting in diabetic neuropathy. In this article, we address the changes of endoneurial vessels in diabetic animal models and discuss the significance of these vascular changes in the pathogenesis of diabetic neuropathy.

Vascular systems in the peripheral nerve

Extrinsic and intrinsic (nutritive) vascular supply of peripheral nerve. The peripheral nerve has two functionally independent circulations, an extrinsic and an intrinsic vascular bed.[29–32] These two systems have extensive anastomoses between them (Fig. 1). The extrinsic system consists of regional arteries that originate from nearby large arteries, as well as from adjacent muscular and periostal vessels. They reach the epineurial and perineurial regions, and transform into small arterioles. These nutrient arterioles form an anastomotic plexus within the epi-perineurial layers and finally penetrate the perineurium to form capillary networks in the endoneurium, constituting the intrinsic vascular system.[29,30] The intrafascicular capillaries run mostly longitudinally with nerve fibers.

Distribution of the intrinsic vascular supply and capillary density. Longitudinally running capillaries distribute diffusely within the endoneurium. The density of capillaries and distances between them appear to be critical determinants for adequate perfusion of the nerve. The density of capillaries in the endoneurium is extremely low, <100 vessels/mm^2, compared with other organs.[31] Bell and Weddell[30] found that intercapillary distances are greater in nerve than in other tissues, such as muscle. Experimentally induced ischemia has revealed focal areas of low perfusion, so-called watershed zones within the endoneurium.[33] The areas of fiber damage coincided with the sites of ischemia. By using microspheres, Nukada and Dyck[34] artificially produced impaired perfusion of the intrinsic en-

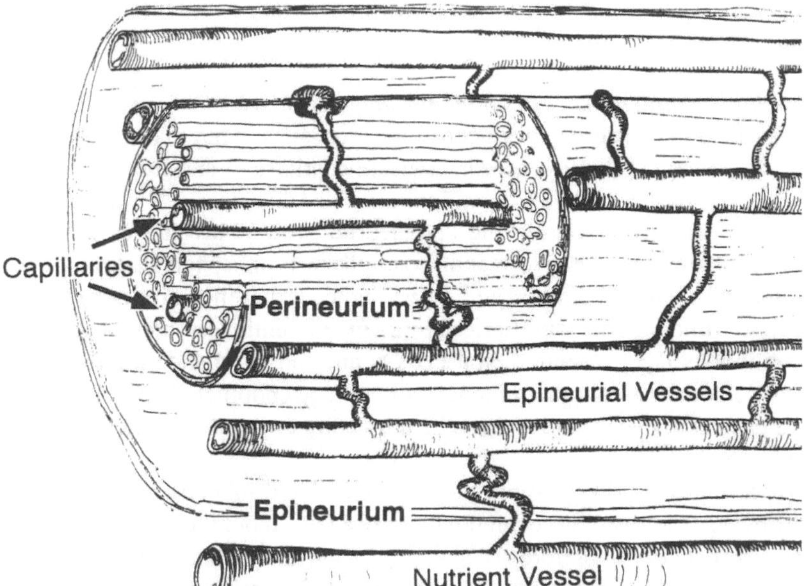

Figure 1. Vascular supply of the peripheral nervous system. The peripheral nerve is supplied with an extrinsic arterial-arteriolar system, consisting of nutrient vessels (arteries) interconnected with networks of epineurial vessels (arterioles). The epineurial vessels penetrate the perineurium and transform into endoneurial capillaries, which constitute the intrinsic blood supply system for the peripheral nerve.

doneurial vascular system, secondary to occlusion of the outer epineurial plexus and the subperineurial regions, which resulted in centrifascicular infarction in the sciatic nerve, sparing the subperineurial area. This focality has been ascribed to the structural findings that capillary density is greater in the subperineurial region in the sciatic nerve.[29] By contrast, single arterial occlusion by ligation (femoral artery) induced nerve fiber degeneration in the subperineurial area, suggesting a different topography of the watershed area between intrinsic and extrinsic vascular systems.[35]

In contrast to the results from morphological studies, physiological studies failed to document any differences in nerve blood flow between the center and the periphery of rat peripheral nerves at rest or even during severe hypoxic conditions, following ligation of nutritive arterioles.[36,37] These findings suggest that sufficient perfusion can be preserved by abundant radial vascular anastomoses of peripheral nerve. On the other hand, along the longitudinal axis of the peripheral nerve, regional blood flow is reduced by the occlusion of nutrient epineurial arterioles.[37] It may therefore be assumed that longitudinal anastomoses are not as efficient as regional (radial) anastomoses. Blunt and Stratton[33] cautioned that such effects of intrafascicular anastomoses may be true only in distal segments, probably compensating for the relative scarcity and the lower pressure of nutrient arterioles as compared with those supplying the proximal nerve. The reason for this discrepancy in the effects of ischemia between nerve blood flow and anatomical vascular arrangements is unknown. It also remains open whether the same variability in capillary distribution applies to human nerves.

Structural characteristics of endoneurial microvessels. Endoneurial capillaries are structurally different from those in other sites. They have larger calibers than a typical capillary bed (e.g., muscle capillaries) and are widely spaced, with large luminal areas and long intercapillary distances.[29–31] The appearance of these large capillaries in the endoneurium resembles that of postcapillary venules. However, positive reactions to alkaline phosphatase, lack of permeability and their susceptibility to injury indicate that they are indeed true capillaries.[23] Endoneurial capilllaries are a type of continuous capillaries, lined by a continuous layer of endothelial cells connected by abundant tight junctions. They do not show endothelial fenestrations. The endothelial cells are supported by a basement membrane and are incompletely covered by pericytes at the abluminal side.[29]

Adjacent to the perineurium, transperineurial and subperineurial arterioles are distributed. Their wall is covered with only a few smooth muscle cell layers. The arteriolar caliber is therefore not likely to vary with humoral and autonomic stimuli. The structural characteristics of endoneurial microvessels make the peripheral nerve susceptible to small changes in blood volume and perfusion pressure. The long intercapillary distances make perfusion of the endoneurium inefficient and renders the nerve particularly vulnerable to endoneurial edema. The arterioles and capillaries undergo collapse, with further compromise of blood flow secondary to edema and increased endoneurial pressure.[37,38]

Regulation of nerve blood flow by epineurial and perineurial arteriole

The peripheral nerve axon is far from the parent cell body, and its energy utilization is dependent on the glucose uptake and tissue oxygen along its course.[39] Nerve environment is therefore crucial to the maintenance of the energy supply and oxygenation from blood flow. To preserve function and structure, an autoregulation mechanism operates in the vascular systems. Mural contraction of smooth muscles is responsible for the autoregulatory mechanism in most tissues. Because perineurial and endoneurial arterioles are covered with poorly developed smooth muscle layers, it is anticipated that endoneurial blood flow autoregulates poorly, which has been confirmed under experimental conditions.[40,41]

In contrast to the poor autoregulation of peripheral nerve blood flow, it may be regulated via nervous control mechanisms. This is supported by the findings of an abundant autonomic nerve supply to the walls of epineurial and perineurial arterioles. Rechtland et al.[42] demonstrated adrenergic nerve plexa around epineurial and perineurial vessels but not accompanying endoneurial vessels. Cholinergic and nonadrenergic peptidergic fibers have also been demonstrated.[43] In aged rats or rats with experimental diabetes, adrenergic fibers of vasa nervorum were demonstrated to be denervated, whereas sympathetic nervous activity of the paravertebral area was enhanced.[43–45]

Recent electron microscopic studies of human sural nerve have demonstrated the presence of dense autonomic nerve endings (adrenergic, cholinergic, and non-adrenergic-noncholinergic) on the adventitia of perineurial and transperineurial arterioles (Fig. 2).[46] It may therefore be expected that arteries and large, and possibly small, arterioles are under neural control for regulation of nerve blood flow. However, this is not likely at the level of endoneurial microvessels.

Vascular changes in human diabetic nerves

Epineurial arteriole. Arterio-arteriolosclerosis is commonly found in diabetic patients. Epineurial arterioles are not exemted. Fagerberg[6] reported that the ratio of wall thickness to lumen diameter in arterioles of sural nerves was increased in diabetic patients compared with control subjects, and in diabetic patients with neuropathy compared with those without neuropathy. The method of measurements in that study using thick paraffin sections was not accurate enough to differentiate arterioles from venules and capillaries, and to discriminate the vascular layers. Korthals and associates[47] reported that the intimal area and the number of intimal nuclei in epineurial arterioles were increased in peripheral nerves from patients with diabetic polyneuropathy. Unfortunately, these findings were not correlated with the severity of nerve fiber degeneration. It is unlikely that these changes are the immediate cause for diabetic neuropathy, since the changes described here are not specific for diabetes and are commonly encountered in patients with arteriosclerosis or vasculopathic neuropathy of other causes.

As already described, epineurial and perineurial arterioles are supplied with abundant sympathetic nerve endings.[42–45] Such innervation suggests that it has an important role in the regulation of endoneurial blood flow, which may be di-

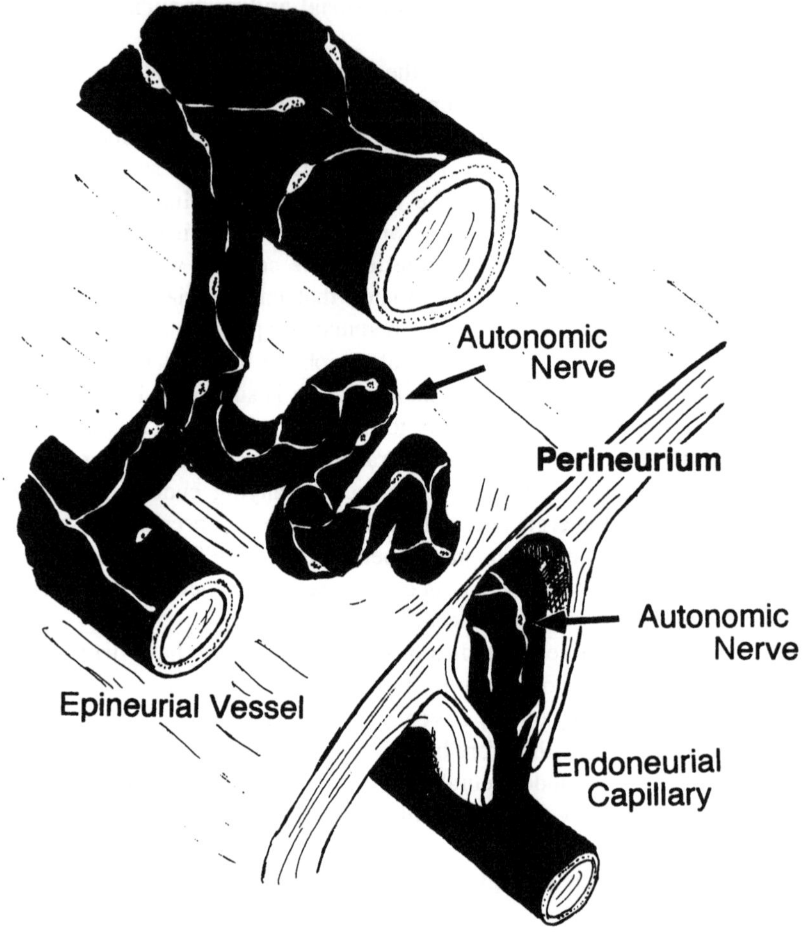

Figure 2. Autonomic nerve control of epineurial and transperineurial vessels (arterioles) Autonomic (sympathetic) nerves are abundantly distributed on the wall of the epineurial and transperineurial vessels, and regulate nerve blood flow.

rectly related to the nerve function. Hence, impaired autonomic nerve function in diabetes may therefore influence the manifesations of somatic neuropathy. Grover-Johnson et al.[48] demonstrated degenerative changes of sympathetic nerve endings of the vasa nervorum in patients with diabetes even shortly after the onset of the disease. It is therefore likely that alterations in nerve blood flow may be an early manifestation in diabetic patients, although not yet proven. By contrast, in experimental animal models alterations of nerve blood flow were repeatedly reported, even in acutely diabetic animals,[38,49,50] although there are no conclusive data showing degeneration or loss of sympathetic nerve endings in epineurial and perineurial arterioles.

Nitric oxide (NO) release from endothelial cells may be relevant to regulation of nerve blood flow at the arteriolar level. Knowledge of the role of NO system in vasoregulation has expanded greatly in the last few years.[51] Regulation by angiotensin II or endothelins may also play important roles in arteriolar regulation of nerve blood flow (see later).[52–55]

Endoneurial microvessels. At the light microscopic levels, it has been known for a long time that endoneurial vessels undergo characteristic microangiopathic changes, exemplified by hyalinous thickening with PAS-positive staining.[6] Occasional occlusion of lumen has also been encountered.[6] These findings led initially to the theory of a vascular basis for human diabetic neuropathy. With the emergence of electron microscopic studies of the biopsied samples of diabetic sural nerves, microangiopathic changes were substantiated by findings such as thickening of the basement membranes and endothelial hyperplasia.[7,8]

No relationships between microvascular changes and nerve fiber pathology were described, however, until Dyck et al.[11,14] revitalized the vascular theory based on their findings of focal myelinated fiber loss and a significant relationship between vascular changes and nerve fiber pathology. In their earlier studies,[11] the thickening of vascular walls and closure of vascular lumina were the most characteristic changes that correlated with the severity of nerve fiber pathology,[11,12] although other investigators argued against their conclusion by findings of frequent closure of endoneurial vessels in nondiabetic patients.[15,16] In their subsequent studies, they could not reproduce the findings of closure of endoneurial vessels and revised their original assumption to the conclusion that overall functional and structural derangements including ischemia, as well as increased permeation of plasma components, not necessarily relating to the occlusion of vessels, may contribute to the development of neuropathy.

Recent ultrastructural morphometric studies on the nerve biopsy samples from diabetic patients confirmed the universal changes of endoneurial microvessels, characterized by reduplicated, homogeneously thickened basement membranes, and proliferative and swollen endothelial cells, as depicted by Dyck's observation.[56–59] The severity of microangiopathic changes correlated with the severity of clinical neuropathy and nerve fiber pathology, reinforcing the importance of vascular factors in the etiology of neuropathy.[57,59] In our own studies on Japanese NIDDM patients, we found similar changes of endoneurial microvessels and found that the thickening of basement membrane correlated with the degree of reduced myelinated fiber density, but capillary closure was not contributory.[60]

A controversy has arisen over the significance of basement membrane thickening of endoneurial microvessels in human diabetic nerves. Bradley et al.[15] raised an argument against the vascular basis for diabetic neuropathy by their observations that both capillary closure and thickening of endoneurial vascular walls were similarly altered in diabetic nerves and in nerves from nondiabetic patients with hereditary neuropathic patients. It is widely known that capillary basement membranes thicken with increasing age.[16,61,62] Sima et al.[16] found a significant thickening of basement membranes of endoneurial microvessels in both IDDM and NIDDM patients, which is not related to aging but is associated with duration of diabetes. By contrast, Giannini and Dyck[63] could not find an age-associated increase in the basement membrane area of endoneurial microvessels, although they found an age-associated decrease in microvessel density. In their series of studies on endoneurial microvessels, they reconfirmed that thickening and reduplication of basement membrane were conspicuous in human diabetic nerves.[64] It has thus been shown that microangiopathic changes composed by thickened basement membranes are the most characteristic feature of human diabetic nerves.

Diabetic microangiopathic changes are often associated with ultrastructural changes indicating degeneration and loss of pericytes.[61,62,64,65] Cellular debris or fragmented cytoplasmic organelles supposed to be derived from destroyed pericytes are often demonstrated in the wall of endoneurial vessels in human diabetic nerves.[64] Vracko and Benditt[65] considered that the thickened and reduplicated basement membrane in diabetic capillaries may be a consequence of repeated cell death of endothelial cells and pericytes. Other mechanisms related to the increased activity of the polyol pathway and increased glycation of structural proteins have also been implicated in basement membrane thickening by inducing increased collagen synthesis and reduced degradation.[1,2,4,62]

Vascular changes in diabetic animal models

Changes in STZ-diabetic rats. Alterations of peripheral nerve blood flow have been constantly demonstrated repeatedly in STZ-diabetic rats, by the early stages of diabetes.[37,38,49,50,54,55] Acute changes of endoneurial blood flow correlate with functional impairments of peripheral nerve in the early stage of the diabetic state. As early as 1 week after the induction of diabetes, sciatic nerve blood flow was reduced by about 60% of normal levels[37,50] when measured by the hydrogen clearance method and >50% when measured by laser Doppler flowmetry.[49] The reduction in peripheral nerve blood flow induces nerve ischemia, reduced oxygen tension, and subsequent tissue hypoxia.[37,38] Such a hypoxic state correlates well with the reduction of motor nerve conduction velocity (MNCV).[38,49,50]

Improvement in nerve blood flow significantly improves the reduced MNCV.[49,50,54,55] Prostaglandin analogues,[49] angiotensin-converting enzyme inhibitor,[54] angiotensin receptor blocker,[55] nictonic acid derivative,[67] as well as aldose reductase inhibitors (ARI)[49] all significantly improve peripheral nerve blood flow and MNCV. In contrast to these reports, Williamson and his group, by using radiotracers, demonstrated an increase in sciatic nerve blood flow in STZ-diabetic rats.[68,69] Administration of ARI or acetyl-L-carnitine significantly inhibited the increased blood flow and prevented the nerve conduction delay.[68,69]

In recent experiments using the hydrogen clearance method, Zochodne and Ho[70] could not find a significant decrease in sciatic nerve blood flow in STZ-diabetic rats. Thus, the alterations of nerve blood flow in experimental diabetes are still controversial, and less invasive methods for the measurement of blood flow should be explored. The mechanisms relating altered nerve blood flow to impaired nerve function are not known. Ischemia may directly cause reduced oxygen tension and tissue hypoxia, which eventually leads to poor energy production and utilization.[28,37] Derangement of intracellular phosphoinositide metabolism has been claimed to be a major sequel of hyperglycemia-induced activation of the polyol pathway.[2,3,69] Diminished (Na,K)ATPase activity also appears to play a central role in explaining the functional deficits of the peripheral nerve under ischemic or metabolically induced hypoxic conditions.[2,3,39,67,69]

In relation to reduced nerve blood flow, Cameron et al.[54,71,72] reported that capillary density is an important determinant of nerve blood volume. Improvement in nerve blood flow significantly correlated with increased density of capillaries, believed to be due to angiogenesis, in particular after treatment with angiotensin II receptor blocker[54] or essential fatty acid supplementation.[71] Their identification of endoneurial capillaries was based on positive alkaline phosphatase staining. In contrast, Yasuda et al.[49] reported, by using semithin sections of the transverse peripheral nerve, that capillary density of diabetic rats was not different from control rats in spite of the reduced nerve blood flow. Paradoxically, they found an increase in the population of enlarged vessels.

On the other hand, our ultrastructural studies disclosed that the appearances of endoneurial microvessels in STZ-diabetic rats were different from those of human diabetic nerves. In our studies on STZ-diabetic rats after 4–10 months of diabetes duration, we could not demonstrate a decrease in capillary density, narrowing, or closure of endoneurial vessels, but found increased vascular areas and widened vascular lumina (Fig. 3).[73,74] The endothelial cells did not show swelling nor any particular abnormal structural changes. There was no apparent increase in the vascular wall area, and the basement membrane area was decreased in diabetic rats.[73] From these findings we concluded that capillary size may be a more important determinant than the capillary density for blood flow in the peripheral nerve. The widened capillary lumina detected structurally may be consistent with an increase in total blood volume at a given time, although the nerve blood flow may be stagnated.

Recent morphometric studies demonstrated a trend toward enlarged capillary lumina in peripheral nerves of mildly diabetic patients compared with those of nondiabetic patients.[59] With progression of disease, closure or occlusion of vessels became apparent in some cases. It may therefore be postulated that the enlargement of capillary lumina is an early structural sign of microangiopathy in diabetic nerves, in particular of IDDM. It should be of note, however, that such widened vessels may be a secondary response to tissue anoxia, because tissue anoxia induces collapse of endothelial cells with secondary dilatation of capillaries in the hypoxic brain.[75]

With much longer duration of diabetes, of >2 years, Powell et al.[76] described endoneurial microvessels with reduplicated basement membranes and deposition of paracrystalline materials in the peripheral nerve of alloxan-diabetic rats. Microinfarctions secondary to platelet thrombi were found in ventral roots of insulin-treated diabetic BB rats with 8–11 months diabetes duration, simulating the proximal motor neuropathy in human diabetic patients.[77] Although these findings appear to be probably uncommon in diabetic animals, there are commmonalities between structural changes of microvessels in human and diabetic animals.

In our morphometric analyses on endoneurial vessels, the pericyte area was significantly reduced in the tibial nerve of STZ-diabetic rats with a 10 month diabetes duration as compared with age-matched control rats (Fig. 3).[73] The reduction of pericyte area appears to be a consistent finding in STZ-diabetic rats. In recent studies on diabetic rats with shorter diabetes duration, the sural nerve exhibited similar and significant reductions in pericyte areas (Sugimoto et al.,

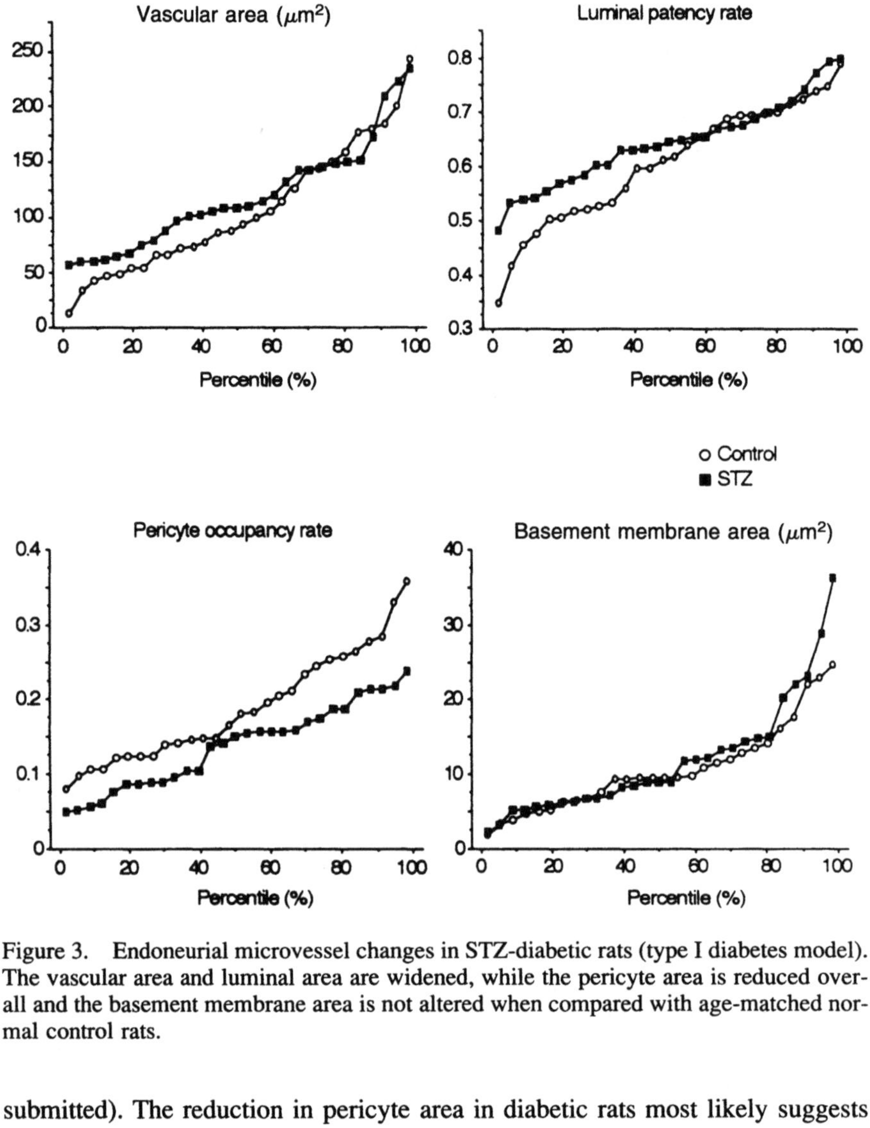

Figure 3. Endoneurial microvessel changes in STZ-diabetic rats (type I diabetes model). The vascular area and luminal area are widened, while the pericyte area is reduced overall and the basement membrane area is not altered when compared with age-matched normal control rats.

submitted). The reduction in pericyte area in diabetic rats most likely suggests degeneration and loss of pericytes.

Degeneration and loss of microvascular pericytes are well documented in diabetes. It occurs in other capillary beds, such as cardiac[61] or skeletal muscles,[62] and most importantly in retinal microvessels, which undergo typical microaneurysmal formation and hemorrhage.[78,79] Recent ultrastructural studies have confirmed the presence of active degeneration of pericytes in endoneurial microvessels in the nerves of human diabetics.[64] For the cellular injury of pericytes, activation of the polyol pathway under hyperglycemic conditions is considered to be important, since pericytes are rich in aldose reductase.[80] Inhibition of the polyol pathway by ARI was found to significantly prevent the loss of pericytes in galactose-fed dogs and also in diabetic rats.[81] Loss of pericytes and microaneurysm formation were similarly prevented by long-term administration of aminoguanidine, a potent inhibitor of advanced non-enzymatic glycation in chronically STZ-diabetic rats.[79] These results seem to suggest that the pathogenesis of pericyte injury in diabetes is multifactorial.

To summarize the findings of endoneurial microvascular changes in STZ-diabetic animals, they are characterized by an enlarged vascular lumen and a reduced area of pericytes. These changes are associated with functional and structural changes of peripheral nerves—slowed MNCV and distal myelinated fiber atrophy.[82] The findings of endoneurial vessels in STZ-diabetic rats are in keeping with the vascular changes found in mild human diabetic patients, except for the thickening of the basement membrane.[58] Functional and structural improvements elicited by vasoactive compounds described earlier in diabetic neuropathy of STZ-diabetic rats[49,50,54,71,72] may suggest a pathogenic role for vascular changes in the development of neuropathy. It still remains unclear, however, what role basement membrane thickening may play in the genesis of diabetic neuropathy.

Neuropathic changes in GK-rats. The GK (Goto-Kakizaki) rat is a model of type II diabetes. It is a non-obese, hyperglycemic, but non–insulin-dependent animal. These rats were produced by repetitive selective breeding of glucose-intolerant rats who originated from a normal rat population.[83,84] Recent studies on islet function in this model have demonstrated impaired signal transduction of phosphatidylinositol metabolism following stimulation with glucose. The model also shows peripheral insulin resistance. Early glomerular basement membrane thickening as a sign of glomerulopathy and neuropathic changes, characterized by slow MNCV and myelinated fiber atrophy, have been demonstrated in GK rats.[85,86]

Sima et al.[87] reported in a detailed morphometric analysis of biopsied sural nerve samples that the structural expression of neuropathy is different in IDDM and NIDDM patients. The main feature in diabetic nerves from IDDM patients was axonal atrophy, as repeatedly demonstrated in IDDM animal models, while it was not apparent in NIDDM patients. Our experimental studies suggest a different neuropathic pattern in STZ-diabetic rats and GK rats, which are IDDM and NIDDM models, respectively. The former are characterized by severe conduction slowing and distal axonal atropy, whereas the latter show mild slowing of nerve conduction and fiber atrophy without axonal atrophy (Table 1).[88] This occurs despite the fact that both models show similar biochemical abnormalities, such as elevated sorbitol and reduced myoinositol levels in the sciatic nerve.[84]

In contrast to the changes demonstrated in STZ-diabetic rats, endoneurial microvessels in GK rats exhibited narrowing of the vascular lumina, and marked thickening of the vascular wall secondary to increased pericyte areas and basement membrane thickening (Fig. 4).[74] These findings are consistent with the data of NIDDM patients showing narrowing of the vascular lumen and basement membrane thickening.[87] Such differences in the vascular pathology may contribute to the heterogeneous expression of the pathology of diabetic neuropathy in type I and type II diabetes.

Table 1. Neuropathic patterns of type I and II diabetic animal models

	Type I (STZ rat)	Type II (GK rat)
Glucose levels	markedly elevated	mildly elevated
Nerve conduction	markedly slowed	slightly reduced
Nerve sorbitol	elevated	slightly elevated
myo-inositol	reduced	slightly reduced
Na-K-ATPase	reduced	?
Myelinated fibers	atrophic	atrophic
Axonal atrophy	present	absent
Axonal neurofilaments	depleted	?
Endoneurial Vessels		
Vascular lumen	dilated	narrowed
Basal Lamina	not thickened	thickened
Pericyte area	reduced	increased
Pathogenesis	*Metabolic*	*Genetic (?)*

Changes in other diabetic animal models and hypoglycemic animals. In other diabetic animal models, microvascular abnormalities are not yet well defined. Although the presence of neuropathy has been proposed, systematic examinations of endoneurial vessels have not been performed in C57Bl Ks (*db/db*) mice, diabetic Chinese hamsters, Wistar/fatty rats, and OLETF rats. It is particularly noteworthy that the spontaneously diabetic WBN/Kob rat, with both endocrine and exocrine pancreatic insufficiency, shows marked thickening of the vascular wall and degenerative changes of endoneurial endothelial cells and pericytes.[89] In this species, a predominant motor neuropathy with perivascular demyelination are the characteristic features of neuropathy, which differs from other diabetic animal models. The close relationship between microvascular changes and peripheral nerve fiber pathology suggests the importance of vasculopathy in the progression of the neuropathy in this model. Although the precise mechanisms underlying the diabetic syndrome in the WBN/Kob rat are unclear, systemic excessive iron deposition suggests the possibility of an inborn error in iron metabolism (Wada, submitted).

Marked thickening of the vascular wall and pericyte degeneration with narrowing of the lumen were recently demonstrated in hypoglycemic insulinoma-bearing STZ-treated rats.[90] These rats also showed marked myelinated fiber loss with active degenerative and demyelinative changes. Yasaki and Dyck[91] found the most marked myelinated fiber degeneration in the centrofascicular area of the peripheral nerve fascicle in rats rendered acutely hypoglycemic by excessive insulin injection. They suggested that hypoglycemia induces ischemic nerve injury. Nerve blood flow has recently been found to be reduced in STZ-diabetic rats after excessive insulin injection.[92] Thus, insulin therapy with hypoglycemic events may have deleterious effects on peripheral nerve structure and function in diabetic subjects.

Relationship between metabolic changes and microvascular abnormalities

Results from recent studies testing a variety of vasoactive compounds have strengthened the vascular theory as a pathogenic mechanism in diabetic neuropathy. Prostaglandins, α-blockers, and angiotensin-converting enzyme inhibitors all ameliorate the peripheral nerve blood flow and conduction deficit in STZ-diabetic rats.[49,50,54,55,67] Such vascular changes now appear to be closely interrelated with hyperglycemia-induced metabolic changes. Treatment with ARI improves nerve blood flow in STZ-diabetic rats.[49] The mechanisms of this improvement have recently been suggested to be mediated via the nitric oxide (NO) system.[93] In the peripheral nerve, NO potentiates the blood flow at the arterio-

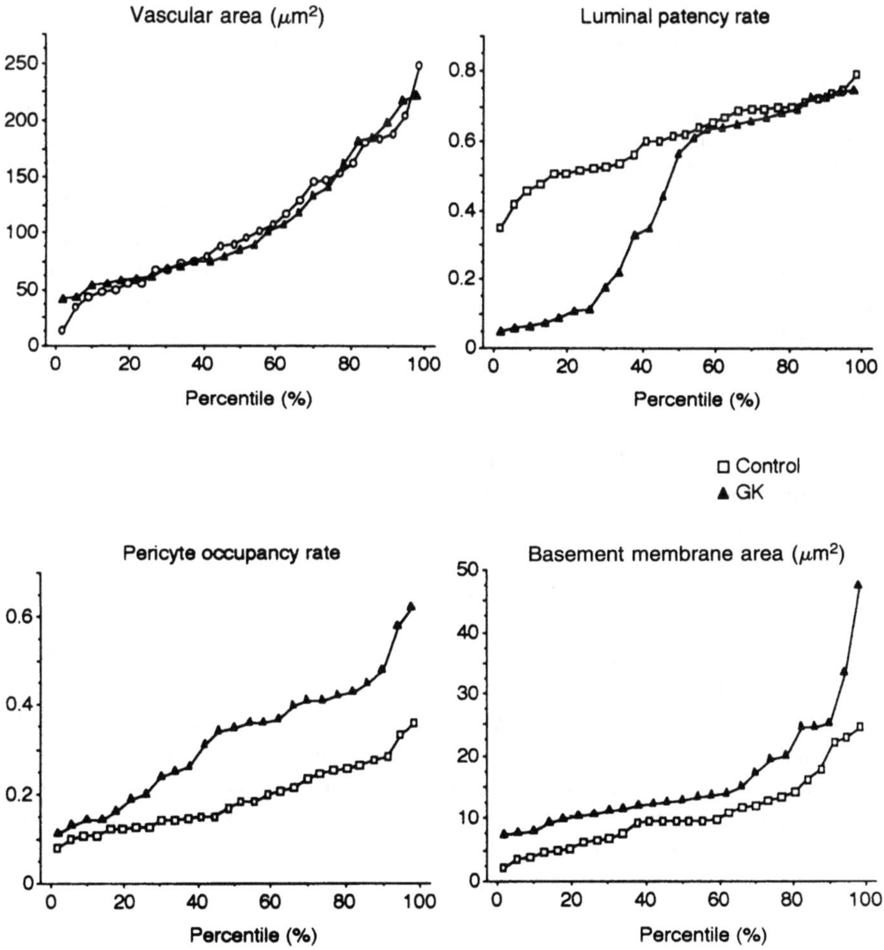

Figure 4. Endoneurial microvessel changes in Goto-Kakizaki rats (type II diabetes model). While the vascular area is not altered, a narrowed vascular lumen and increased pericyte and basement membrane areas are found when compared with those in age-matched normal control rats.

lar level and maintains nerve conduction. When hyperglycemia activates the polyol pathway, NADPH, which has a high affinity to aldose reductase, is utilized for conversion of glucose to sorbitol. For the production of NO, NO synthase requires NADPH for the production of L-citrulline from arginine. This metabolic relationship indicates that activation of the polyol pathway by large consumption of NADPH competitively inhibits the NO synthase, thereby reducing the release of NO from vascular endothelial cells and reducing nerve blood flow. Cameron et al.[94] recently described a significant reduction of nerve blood flow associated with slowed MNCV in rats treated with NO synthase inhibitor. The beneficial effect of ARI on nerve conduction velocity in STZ-diabetic rats is eliminated when they are cotreated with ARI and NO synthase inhibitor.[93]

Increased glycation of nerve structural proteins appears to perturb nerve blood flow and contributes to the development of functional and structural abnormalities of the peripheral nerve.[4] Preliminary data from immunohistochemical studies have localized glycated proteins to endoneurial microvessels (Sugimoto et al., unpublished observation). Advanced glycation end products are reported to diminish the NO activity with reduced blood flow.[95] Aminoguanidine is a potent inhibitor of the advanced glycation of proteins.[4] In our previous studies, continuous treatment with aminoguanidine effectively inhibited the slowing of MNCV and axonal atrophy in STZ-diabetic rats.[96] We also performed a morphometric analysis of the endoneurial microvessels from aminoguanidine-treated diabetic rats. The results demonstrated normalization of widened vascular areas and luminal patency rate.[97] We concluded that aminoguanidine had a protective effect on microvascular changes elicited by the diabetic state and that the improvement in peripheral nerve abnormalities may be mediated by the prevention of microvascular changes. In fact, Kihara et al.[98] found an improvement in MNCV associated with peripheral nerve blood flow in STZ-diabetic rats after treatment with aminoguanidine. The beneficial effect of aminoguanidine was further confirmed by Cameron et al.[99]

There are other reports with respect to the nature of aminoguanidine action. It has been suggested that aminoguanidine competitively inhibits the production of NO because of its chemical structure being similar to arginine.[100] Therefore, treatment with aminoguanidine will reduce NO levels, resulting in a vasoconstrictive action. Williamson et al.[62] documented increased blood flow and vascular permeability in the peripheral nerve, retina, and kidney in STZ-diabetic rats, which were significantly inhibited by treatment with aminoguanidine.[100] It is not known at present whether aminoguanidine serves as an alternate substrate for NO synthase. Thus, the relationship between altered nerve blood flow and the effects of aminoguanidine remains to be clarified.

Concluding remarks

The endoneurial microvessels undergo characteristic structural alterations in diabetic animal models. The expression appears to be different in type I (STZ-rat) and type II (GK-rat) diabetic models. Widening of the vascular lumen and a reduced pericyte area are characteristic changes in STZ rats, whereas GK rats show narrowing of the lumen and increased basement membrane area. The structural alterations of microvessels may be the consequence of hyperglycemia-induced activation of the polyol pathway and increased glycation of structural proteins in type I model, whereas there is a possibility that microvascular changes are relevant to genetic factors in the type II model. Such vascular changes may also reflect the abnormalities of peripheral nerve blood flow and may significantly influence the development of diabetic neuropathy.

Acknowledgments

This study was supported by a special grant for the promotion of diabetes research from the Ministry of Health and Welfare, Japan and a grant for educational research from Hirosaki University.

References

1. Winegrad AI. Does a common mechanism induce diverse complications of diabetes? Diabetes 36:396–406, 1989.
2. Greene DA, Sima AAF, Stevens MJ et al. Complications: Neuropathy, pathogenetic considerations. Diabetes Care 15:1902–25, 1992.
3. Williamson JR, Chang K, Frangos M et al. Hyperglycemic pseudohypoxia and diabetic complications. Diabetes 42:801–13, 1993.
4. Brownlee M. Glycation and diabetic complications. Diabetes 43:836–41, 1994.
5. Thomas PK. Diabetic neuropathy: Models, mechanisms and mayhem. Can J Neurol Sci 19:1–7, 1992.
6. Fagerberg SE. Diabetic neuropathy: A clinical and histological study on the significance of vascular affections. Acta Med Scand 164(Suppl 345):1–97, 1954.
7. Behse F, Buchthal F, Carlsen F. Nerve biopsy and conduction studies in diabetic neuropathy. J Neurol Neurosurg Psychiat 40:1072–82, 1977.
8. Yagihashi S, Matsunaga M. Ultrastructural pathology of human diabetic neuropathy. Tohoku J Exp Med 129:357–66, 1979.
9. Timperly WR, Ward JD, Preston FE et al. Clinical and histological studies in diabetic neuropathy: A reassessment of vascular factors in relation to vascular coagulation. Diabetologia 12:237–43, 1976.
10. Williams E, Timperly WR, Ward JD, Duckworth T. Electron microscopical studies of vessels in diabetic peripheral neuropathy. J Clin Pathol 33:462–70, 1980.
11. Dyck PJ, Hansen S, Karnes J et al. Capillary number and percentage closed in human diabetic sural nerve. Proc Natl Acad Sci USA 82:2513–17, 1985.
12. Dyck PJ, Karnes JL, O'Brien P et al. The spatial distribution of fiber loss in diabetic polyneuropathy suggests ischemia. Ann Neurol 19:440–9, 1986.
13. Dyck PJ, Lais A, Karnes JL et al. Fiber loss is primary and multifocal in sural nerves in diabetic polyneuropathy. Ann Neurol 19:425–39, 1986.
14. Yasuda H, Dyck PJ. Abnormalities of endoneurial microvessels and sural nerve pathology in diabetic neuropathy. Neurology 37:20–8, 1987.
15. Bradley J, Thomas PK, King RHM et al. Morphometry of endoneurial capillaries in diabetic sensory and autonomic neuropathy. Diabetologia 33:611–18, 1990.
16. Sima AAF, Nathaniel V, Prashar A et al. Endoneurial microvessels in human diabetic neuropathy. Endothelial cell dysjunction and lack of treatment effect by aldose reductase inhibitor. Diabetes 40:1090–99, 1991.
17. Johnson PC, Doll SC, Cromey DW. Pathogenesis of diabetic neuropathy. Ann Neurol 19:450–7, 1986.

18. Llewlyn JG, Thomas PK, Gilbey SG et al. Pattern of myelinated fiber loss in the sural nerve in neuropathy related to type I (insulin dependent) diabetes. Diabetologia 31:162–7, 1988.
19. Knudsen GH, Jakobsen J, Juhler M, Paulson OB. Decreased blood-brain barrier permeability to sodium in early experimental diabetes. Diabetes 35:1371–73, 1986.
20. Rechtland E, Smith QR, Latker CH, Rapoport SI. Altered blood-nerve barrier permeability to small molecules in experimental diabetes mellitus. J Neuropathol Exp Neurol 46:302–14, 1987.
21. Ohi T, Poduslo JF, Dyck PJ. Increased endoneurial albumin in diabetic polyneuropathy. Neurology 35:1790–91, 1985.
22. Poduslo JF, Curran GL, Dyck PJ. Increase in albumin, IgG, and IgM blood-nerve barrier indices in human diabetic neuropathy. Proc Natl Acad Sci USA 85:4879–83, 1988.
23. Lundborg G. Structure and function of the intraneural microvessels as related to trauma, edema formation, and nerve function. J Bone Joint Surg Am 57:938–48, 1975.
24. Myers RR, Powell HC, Shapiro HM, et al. Changes in endoneurial fluid pressure, permeability, and peripheral nerve ultrastructure in experimental lead neuropathy. Ann Neurol 8:392–401, 1980.
25. Wadhwani KC, Caspers-Velu LE, Murphy VA et al. Prevention of nerve edema and increased blood-nerve barrier permeability-surface area product in galactosemic rats by aldose reductase or thromboxane synthetase inhibitors. Diabetes 38:1469–77, 1989.
26. Myers RR, Powell HC. Galactose neuropathy: Impact of chronic endoneurial edema on nerve blood flow. Ann Neurol 16:587–94, 1984.
27. Myers RR, Murakami H, Powell HC. Reduced nerve blood flow in edematous neuropathies: A biochemical mechanism. Microvasc Res 32:145–51, 1986.
28. Low PA, Nukada H, Schmelzer JD et al. Endoneurial oxygen tension and radial topography in nerve edema. Brain Res 341:147–54, 1985.
29. Lundborg G, Branemark PI. Microvascular structure and function of peripheral nerves. Adv Microcirc 1:66–88, 1968.
30. Bell MA, Weddell AGM. A descriptive study of the blood vessels of the sciatic nerve in the rat, man and other mammals. Brain 107:871–98, 1984.
31. Bell MA, Weddell AGM. A morphometric study of intrafascicular vessels of mammalian sciatic nerve. Muscle Nerve 7:524–34, 1984.
32. Wadhwani KC, Rapoport SI. Transport properties of vertebrate blood-nerve barrier: Comparison with blood-brain barrier. Prog Neurobiol 43:235–79, 1994.
33. Blunt MJ, Stratton K. The immediate effects of ligature of vasa nervorum. J Anat 90:204–16, 1956.
34. Nukada H, Dyck PJ. Microsphere embolization of nerve capillaries and fiber degeneration. Am J Pathol 115:275–87, 1984.
35. Nukada H, Powell HC, Myers RR. Spatial distribution of nerve injury after occlusion of individual major vessels in rat sciatic nerves. J Neuropathol Exp Neurol 52:452–9, 1993.
36. Sladky JT, Greenberg JH, Brown MJ. Regional perfusion in normal and ischemic rat sciatic nerves. Ann Neurol 17:191–5, 1985.
37. Low PA, Lagerlund TD, McManis PG. Nerve blood flow and oxygen delivery in normal, diabetic, and ischemic neuropathy. Int Rev Neurobiol 31:355–438, 1989.
38. Myers RR, Powell HC. Galactose neuropathy: Impact of chronic endoneurial edema on nerve blood flow. Ann Neurol 16:587–94, 1984.
39. Greene DA, Lattimer SA. Sodium- and energy-dependent uptake of myo-inositol by rabbit peripheral nerve. Competitive inhibition by glucose and lack of an insulin effect. J Clin Invest 70:1009–18, 1982.
40. Smith DR, Kobrine AI, Rizzoli HV. Absence of autoregulation in peripheral nerve blood flow. J Neurol Sci 33:347–52, 1977.

41. Kihara M, Low PA. Regulation of rat nerve blood flow: Role of epineurial a-receptors. J Physiol 422:145–52, 1990.
42. Rechtland E, Hervonen A, Sato S, Rapoport SI. Distribution of adrenergic innervation of blood vessels in peripheral nerve. Brain Res 374:185–9, 1986.
43. Appenzeller O, Dhital KK, Cowen T, Burnstock G. The nerves to blood vessels supplying blood to nerves: The innervation of vasa nervorum. Brain Res 304:383–6, 1984.
44. Dhital K, Lincoln J, Appenzeller O, Burnstock G. Adrenergic innervation of vasa and nervi nervorum of optic, sciatic, vagus and sympathetic nerve trunks in normal and streptozotocin-diabetic rats. Brain Res 367:39–44, 1986.
45. Koistiraho J, Wadhwani KC, Rapoport SI. Adrenergic innervation of the tibial and vagus nerves in rats of different ages. Mech Ageing Dev 52:195–205, 1990.
46. Beggs J, Johnson PC, Olafsen A et al. Transperineurial arterioles in human sural nerve. J Neuropathol Exp Neurol 50:704–18, 1991.
47. Korthals JK, Gieron MA, Dyck PJ. Intima of epineurial arterioles is increased in diabetic polyneuropathy. Neurology 38:1582–86, 1988.
48. Grover-Johnson NM, Baumann FG, Imparato AM et al. Abnormal innervation of lower limb epineurial arterioles in human diabetes. Diabetologia 20:310–18, 1981.
49. Yasuda H, Sonobe M, Yamashita M et al. Effect of prostaglandin E1 analogue TFC 612 on diabetic neuropathy in streptozocin-induced diabetic rats. Comparison with aldose reductase inhibitor ONO 2235. Diabetes 38:832–8, 1989.
50. Cameron NE, Cotter MA, Low PA. Nerve blood flow in early experimental diabetes in rats: Relation to conduction deficits. Am J Physiol 261:E1–8, 1991.
51. Moncada S, Palmer MJ, Higgs EA. Nitric oxide: Physiology, pathophysiology, and pharmacology. Pharmacol Rev 43:109–42, 1991.
52. Yanagisawa M, Masaki T. Molecular biology and biochemistry of the endothelins. Trends Pharmacol Sci 10:374–8, 1989.
53. Sakurai T, Yanagisawa M, Masaki T. Molecular characterization of endothelin receptors. Trends Pharmacol Sci 13:103–8, 1992.
54. Cameron NE, Cotter MA, Robertson S. Angiotensin converting enzyme inhibition prevents development of muscle and nerve dysfunction and stimulates angiogenesis in streptozotocin-diabetic rats. Diabetologia 35:12–18, 1992.
55. Maxfield EK, Cameron NE, Cotter MA, Dines KC. Angiotensin II receptor blockade improves nerve function, modulates nerve blood flow and stimulates endoneurial angiogenesis in streptozotocin-diabetic rats. Diabetologia 36:1230–37, 1993.
56. Powell HC, Rosoff J, Myers RR. Microangiopathy in human diabetic neuropathy. Acta Neuropathol 68:295–305, 1985.
57. Britland ST, Young RJ, Sharma AK, Clarke BF. Relationship of endoneurial capillary abnormalities to type and severity of diabetic polyneuropathy. Diabetes 39:909–13, 1990.
58. Malik RA, Newrick PG, Sharma AK et al. Microangiopathy in human diabetic neuropathy: Relationship between capillary abnormalities and the severity of neuropathy. Diabetologia 32:92–102, 1989.
59. Malik RA, Veves A, Masson EA, et al. Endoneurial capillary abnormalities in mild human diabetic neuropathy. J Neurol Neurosurg Psychiat 55:557–61, 1992.
60. Yagihashi S. Pathogenesis of diabetic neuropathy. Peripheral Nerve 1:43–9, 1990 (in Japanese).
61. Vracko R. A comparison of the microvascular lesions in diabetes mellitus with those of normal aging. J Am Geriat Soc 30:201–5, 1982.
62. Williamson JR, Tilton RG, Chang K, Kilo C. Basement membrane abnormalities in diabetes mellitus: Relationship to clinical microangiopathy. Diabetes Metab Rev 4:339–70, 1988.
63. Giannini C, Dyck PJ. Ultrastructural morphometric features of human sural nerve endoneurial microvessels. J Neuropathol Exp Neurol 52:361–9, 1993.

64. Giannini C, Dyck PJ. Ultrastructural morphometric abnormalities of sural nerve endoneurial microvessels in diabetes mellitus. Ann Neurol 36:408–15, 1994.
65. Vracko R, Benditt EP. Capillary basal lamina thickening. Its relationship to endothelial cell death and replacement. J Cell Biol 47:281–5, 1970.
66. Tilton RG, Faller AM, Burkhardt JK et al. Pericyte degeneration and acellular capillaries are increased in the feet of human diabetic patients. Diabetologia 28:895–900, 1985.
67. Hotta N, Kakuta H, Fukusawa H et al. Effect of niceritrol on streptozocin-induced diabetic neuropathy in rats. Diabetes 41:587–91, 1992.
68. Tilton RG, Chang K, Hasan KS et al. Prevention of diabetic vascular dysfunction by guanidines: Inhibition of nitric oxide synthase versus advanced glycation end product formation. Diabetes 42:221–32, 1993.
69. Ido Y, McHowat J, Chang KC et al. Neural dysfunction and metabolic imbalances in diabetic rats. Prevention by acetyl-L-carnitine. Diabetes 43:1469–77, 1994.
70. Zochodne DW, Ho LT. Diabetes mellitus prevents capsaicin from inducing hyperemia in the rat sciatic nerve. Diabetologia 36:493–6, 1993.
71. Cameron NE, Cotter MA, Robertson S. Essential fatty acid diet supplementation. Effects on peripheral nerve and skeletal muscle function and capillarization in streptozocin-induced diabetic rats. Diabetes 40:532–9, 1991.
72. Cameron NE, Cotter MA, Ferguson K et al. Effects of chronic α-adrenergic receptor blockade on peripheral nerve conduction, hypoxic resistance, polyols, Na^+-K^+-ATPase activity, and vascular supply in STZ-D rats. Diabetes 40:1652–58, 1991.
73. Sugimoto K, Nishida N, Yamagishi S et al. Structural changes in endoneurial microvessels in rats with chronic streptozotocin-induced diabetes-morphometric analysis. J Jpn Diab Soc 36:771–8, 1993.
74. Yagihashi S, Sugimoto K, Wada R. Different neuropathic patterns between type I and type II animal models. In: Pathogenesis and treatment of NIDDM and its related problems. Sakamoto N, Alberti KGMM, Hotta N, eds. Elsevier Science, Amsterdam, pp 401–5, 1994.
75. Doukas J, Cutler AH, Baswell CA et al. Reversible endothelial cell relaxation induced by oxygen and glucose deprivation. A model of ischemia in vitro. Am J Pathol 145:211–19, 1994.
76. Powell HC, Myers RR. Axonopathy and microangiopathy in chronic alloxan diabetes. Acta Neuropathol 65:128–37, 1984.
77. Sima AAF, Thibert P. Proximal motor neuropathy in the BB-Wistar rat. Diabetes 31:784–8, 1982.
78. Cogan DG. Aldose reductase and complications of diabetes. Ann Intern Med 101:527–37, 1984.
79. Hammes HP, Martin S, Federlin K et al. Aminoguanidine treatment inhibits the development of experimental diabetic retinopathy. Proc Natl Acad Sci USA 88:11555–558, 1991.
80. Chakarabarti S, Sima AAF, Nakajima T et al. Aldose reductase in the BB rat: Isolation, immunological identification and localization in the retina and peripheral nerve. Diabetologia 30:244–51, 1987.
81. Kador PF, Akagi Y, Takahashi Y et al. Prevention of pericyte ghost formation in retinal capillaries of galactose-fed dogs by aldose reductase inhibitors. Arch Ophthalmol 106:1099–120, 1988.
82. Yagihashi S, Kamijo M, Watanabe K. Reduced myelinated fiber size correlates with loss of axonal neurofilaments in peripheral nerve of chronically streptozotocin diabetic rats. Am J Pathol 136:1365–73, 1990.
83. Goto Y, Kakizaki M, Masaki N. Spontaneous diabetes produced by selective breeding of normal Wistar rats. Proc Jpn Acad 51:80–5, 1975.
84. Suzuki K-I, Goto Y, Toyota T. Spontaneously diabetic GK (Goto-Kakizaki) rats. In: Lessons from Animal Diabetes. Shafrir E, ed. Smith Gordon, London, 4:107–16, 1993.

85. Yagihashi S, Goto Y, Kakizaki M, Kaseda N. Thickening of glomerular basement membrane in spontaneously diabetic rats. Diabetologia 15:309–12, 1978.
86. Yagihashi S, Tonosaki A, Yamada K-I et al. Peripheral neuropathy in selectively-inbred spontaneously diabetic rats: Electrophysiological, morphometrical and freeze-replica studies. Tohoku J Exp Med 138:39–48, 1982.
87. Sima AAF, Nathaniel V, Bril V et al. Histopathological heterogeneity of neuropathy in insulin-dependent and non-insulin-dependent diabetes, and demonstration of axo-glial dysjunction in human diabetic neuropathy. J Clin Invest 81:349–64, 1988.
88. Yagihashi S, Kamijo M, Nagai K. Peripheral neuropathy in diabetic animals: Heterogeneous expression of neuropathic patterns in different animal models. In: Lessons from animal diabetes. Shafrir E, ed. Smith-Gordon, London, 3:459–63, 1991.
89. Yagihashi S, Wada R, Kamijo M, Nagai K. Peripheral neuropathy in the WBN/Kob rat with chronic pancreatitis and spontaneous diabetes. Lab Invest 68:296–307, 1993.
90. Sugimoto K, Yagihashi S. Peripheral nerve pathology in rats with streptozotocin-induced insulinoma. Acta Neuropathol. 1996 (in press)
91. Yasaki S, Dyck PJ. Spatial distribution of fiber degeneration in acute hypoglycemic neuropathy in rat. J Neuropathol Exp Neurol 50:681–92, 1991.
92. Kihara M, Zollman PJ, Smithson IL et al. Hypoxic effect of exogenous insulin on normal and diabetic peripheral nerve. Am J Physiol 266:E980–5, 1994.
93. Stevens MJ, Dananberg J, Feldman EL et al. The linked roles of nitric oxide, aldose reductase and, (Na^+,K^+)-ATPase in the slowing of nerve conduction in the streptozotocin diabetic rat. J Clin Invest 94:853–9, 1994.
94. Cameron NE, Cotter MA, Dines KC, Maxfield EK. Pharmacological manipulation of vascular endothelium function in non-diabetic and streptozotocin-diabetic rats: Effects on nerve conduction, hypoxic resistance and endoneurial capillarization. Diabetologia 36:516–22, 1993.
95. Bucala R, Tracey KJ, Cerami A. Advanced glycosylation products quench nitric oxide and mediate defective endothelium-dependent vasodilatation in experimental diabetes. J Clin Invest 87:432–8, 1991.
96. Yagihashi S, Kamijo M, Baba M, et al. Effect of aminoguanidine on functional and structural abnormalities in peripheral nerve of STZ-induced diabetic rats. Diabetes 41:47–52, 1992.
97. Sugimoto K, Nishida N, Wada R. Effects of aminoguanidine on endoneurial vascular abnormalities in experimental diabetic neuropathy. Diabetes 43(Suppl 1):16A, 1994.
98. Kihara M, Schmelzer JD, Poduslo JF et al. Aminoguanidine effect on nerve blood flow, vascular permeability, electrophysiology, and oxygen free radicals. Proc Natl Acad Sci USA 88:6107–11, 1991.
99. Cameron NE, Cotter MA, Dines KC, Love A. Effects of aminoguanidine on peripheral nerve function and polyol pathway metabolites in streptozotocin-diabetic rats. Diabetologia 35:946–50, 1992.
100. Corbett JA, Tilton RG, Chang K et al. Aminoguanidine, a novel inhibitor of nitric oxide formation, prevents diabetic vascular dysfunction. Diabetes 41:552–6, 1992.

22

Dog and rat models of diabetic retinopathy

TIMOTHY S. KERN, RENU KOWLURU, AND RONALD L. ENGERMAN
Department of Ophthalmology and Visual Science, University of Wisconsin-Madison, Madison, Wisconsin 53706, USA.

Key words: dog, rat, retinopathy, galactosemic retinopathy, polyol pathway, aldose reductase, oxidative stress.

Diabetic retinopathy is a significant complication of Type 1 and Type 2 diabetes mellitus, being observed in most patients after 15 years of diabetes, and increasing the risk of blindness 25-fold above normal.[1,2] The natural history of clinically demonstrable retinopathy has been carefully documented in patients, and important stages (formation of capillary microaneurysms, excessive vascular permeability, vascular occlusion, proliferation of new vessels and fibrous tissue, and contraction of the fibrovascular proliferations) have been identified.[3] The earliest stages of the retinopathy (before microaneurysms appear), however, are not apparent clinically, and can be studied in patients only by noninvasive means, such as fluorescein angiography, or by relying on eyes collected at autopsy or at surgery.

Animal models of diabetic retinopathy have proved valuable in efforts to unravel the pathogenesis of this retinopathy and to identify therapies to inhibit it. Species used in studies of the effect of diabetes on the retina have included animals with spontaneous and experimentally induced diabetes. Retinas of spontaneously diabetic animals studied have included fish,[4] mice,[5,6] rats,[7,8] cats,[9] dogs,[10–12] and monkeys,[13] but the reports (with the exception of studies of rats) have been mostly descriptive. To ensure an adequate number of diabetic animals for experimentation, and to minimize effects of unanticipated genetic differences among genetically diabetic animals and controls, many investigators have relied on experimentally induced forms of diabetes. A variety of animal species have been examined over the years for their ability to develop diabetic retinopathy, but the best characterized experimental model of diabetic retinopathy is in dogs.

Correspondence: T.S. Kern, PhD, Department of Ophthalmology & Visual Science, University of Wisconsin, 1300 University Ave., Madison, WI 53706, USA.

Lessons from Animal Diabetes VI. Ed. E. Shafrir.

Dogs as models of diabetic retinopathy

Retinopathy was initially identified in a single spontaneously diabetic dog,[14] and soon after in dogs with experimentally induced diabetes.[10–12,15,16] The anatomic features of retinopathy in diabetic dogs have been shown repeatedly to be morphologically indistinguishable from those of background retinopathy seen in diabetic humans.[16–18] The lesions characteristic of background retinopathy in both diabetic humans and dogs involve the microvasculature, and include capillary microaneurysms, acellular (and nonperfused) capillaries, pericyte ghosts, varicose and dilated capillaries, and dot and blot hemorrhages (Fig. 1). Arteriolar smooth muscle cell loss has also been observed in humans and dogs.[19,20] The lesions in diabetic dogs are secondary to insulin deficiency, since they develop irrespective of how the diabetes was induced (alloxan, growth hormone, pancreatectomy), and can be inhibited by strict regulation of glycemia with exogenous insulin.[21] Improved glycemic control has recently been found to significantly inhibit the development and progression of retinopathy in patients as well.[22]

The importance of hyperglycemia in the pathogenesis of diabetic retinopathy was demonstrated a number of years ago by study of normal, nondiabetic dogs fed a galactose-rich diet.[23,24] During the 3–5 years of study, normal dogs fed a diet enriched with 30% galactose were found to develop a retinopathy that was indistinguishable from that of diabetic dogs and patients, including microaneurysms, vaso-obliteration, pericyte ghosts, and hemorrhages.[23–26] The morpho-

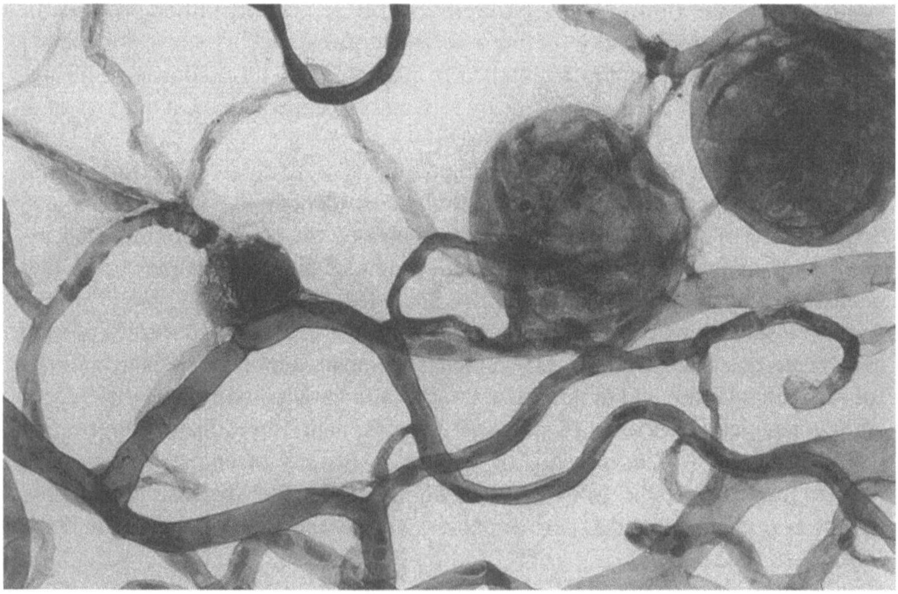

Figure 1. Retinopathy in an alloxan-diabetic dog kept in poor glycemic control for 5 years. Saccular microaneurysms, acellular capillaries, and pericyte ghosts are present. Hematoxylin and PAS.

logic similarity of the retinopathy observed in galactosemic dogs, diabetic dogs, and diabetic patients suggests that the two dog models and patients share a common pathway leading to the retinopathy. The galactose-induced retinopathy, however, develops despite the absence of many of the systemic abnormalities of metabolism that are characteristic of diabetes (such as those involving concentrations of glucose, insulin, fatty acids, etc).[24] The available evidence suggests that excessive blood hexose (either glucose or galactose) is important in the initiation of retinopathy, as is the duration of hyperglycemia.

As is true in diabetic humans, there is a long interval before retinopathy becomes manifest in diabetic or galactosemic dogs, with capillary aneurysms usually not beginning to appear in dogs until about 2–3 years after induction of elevated hexose levels. Likewise, after about 2 years of hyperglycemia, increasing numbers of retinal capillaries are found to possess endothelial cells but few or no pericytes. Within 5 years of insulin-deficient diabetes or galactosemia, all dogs have marked retinopathy. The reason for the prolonged interval before retinopathy becomes manifest is unknown, but any explanation of this latent period might offer valuable insight into the pathogenesis of the retinopathy. Cataract becomes appreciable after 1–2 years of diabetes or galactosemia, thus impairing visualization of the fundus while the dogs are alive.

Neovascularization has been observed to develop in both diabetic and galactose-fed dogs. The new vessels, identified by their lack of basement membrane[27] and their characteristic "chicken-wire" pattern, have developed within the retina in diabetic dogs and experimentally galactosemic dogs, but have not extended into the vitreous during the initial 5 years of study. New vessels extending into the vitreous have been reported in 2 of 9 dogs fed galactose for 6–7 years,[28] thus confirming the evidence that the dog is a useful model to investigate the pathogenesis of diabetic retinopathy.

Rats as models of diabetic retinopathy

Consistent with findings in dogs, rats fed a galactose-rich diet likewise develop diabetic-like microvascular disease in the retina,[29,30] and galactosemic rats have been proposed as a convenient model of diabetic retinopathy. In the past, diabetes in rats has not been found consistently to produce a comparable retinopathy (Table 1). Some investigators reported microvascular lesions, such as microaneurysms, pericyte loss, and acellular capillaries in diabetic rats, whereas others have failed to observe those lesions. The reasons for this lack of reproducibility are not clear, but may have been due in part to a failure to maintain a hyperglycemia severe enough to develop particular lesions, or to study the animals for a sufficiently long duration of diabetes. Studies of diabetic rats commonly have been conducted for about a year or less, with little or no attention to the severity of hyperglycemia throughout the study.

The retinal disease seen in alloxan-diabetic rats has been compared by us to that in rats receiving a galactose-rich diet, with careful attention being paid to the severity of hyperhexosemia.[31] Diabetic rats were kept severely hyperglycemic, being given too little insulin to allow more than a slow weight gain.

Table 1. Retinal lesions in diabetic rats and galactose-fed rats[1]

	Duration (mos)	Micro-aneurysms	Pericyte loss	Acellular capillaries	Basement membrane thickening
Diabetes					
Sharma et al.[79]	6		Yes		
Hammes et al.[44]	6	Yes	Yes	Yes	
Robison et al.[80]	6–8		Yes		
Tilton et al.[81]	6–9		No		Yes
Luenberger[34]	6–12	Yes			Yes
Sima et al.[7]	8–11	No	Yes		No
Chakrabarti et al.[41]	14	No	Yes		
Hammes et al.[37]	6–17	Yes	Yes	Yes	
Kern and Engerman[31]	20	No	Yes	Yes	Yes
Galactosemia					
Kern and Engerman[31,33]	15–26	No	Yes	Yes	Yes
Robison et al.[29]	29	Yes	Yes	Yes	Yes

[1]Literature prior to 1972 is cited elsewhere.[82]

By $1^1/_2$ years of diabetes or galactosemia, both diabetic rats and experimentally galactosemic rats were found always to have developed retinal microvascular disease (Fig. 2).

The most reproducible lesions observed both in diabetic and galactose-fed rats include pericyte loss (assessed by the presence of pericyte ghosts), acellular capillaries, and capillary basement membrane thickening. Some acellular capillaries, pericyte ghosts, and sudanophilic deposits occasionally were observed at shorter durations as well, but since mild changes have also been found occasionally in normal animals and patients, it is uncertain whether or not the lesions seen before the $1^1/_2$ year duration were produced by the hyperglycemia. Even at $1^1/_2$ years of study, the severity of retinal disease varied greatly within experimental groups, as indicated by the large SD shown in Figure 2. Acellular capillaries have been reported by some to become greater than normal after as little as 6 months of diabetes.[32] It is not clear why different investigators have found lesions to develop at different rates, but the differences might be due to differences between rat strains, in criteria to identify acellular capillaries, or in the severity of hyperglycemia.

Some lesions typical of diabetic retinopathy in humans and dogs were found to be less reproducible in diabetic and galactosemic rats, being observed in some but not all animals at durations of up to 22 months. Foci of dilated, hypercellular vessels (IRMA) were found in about three quarters of the rats fed 50% galactose for 20 months or more, and in about one third of rats fed a 30% galactose diet for a similar duration. These lesions were more frequent at 26 or more months of galactosemia.[29,30] Diabetic rats also develop IRMA, but at an apparently slower rate. Saccular microaneurysms, such as those typ-

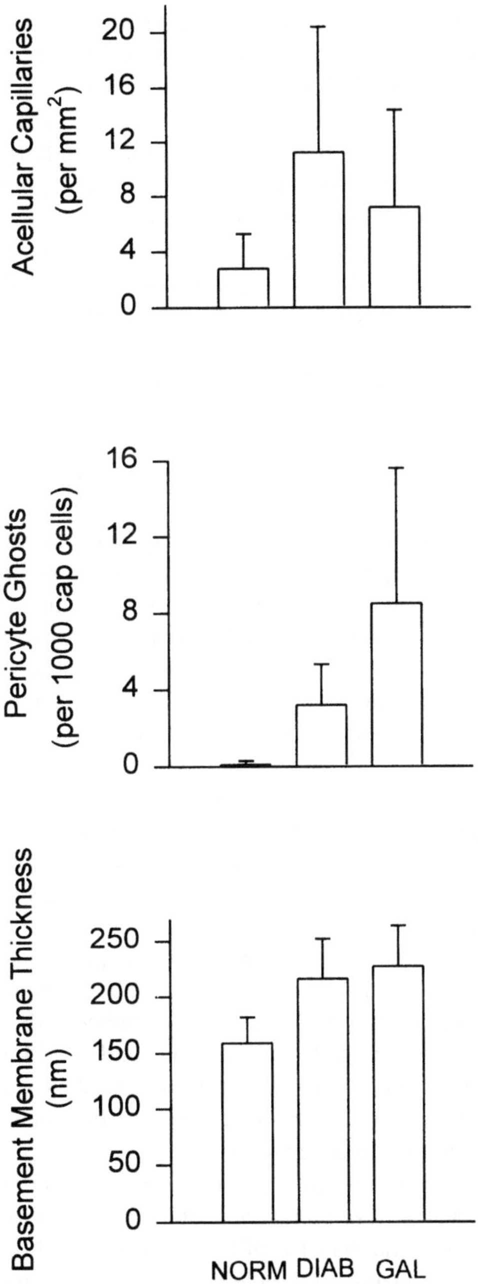

Figure 2. Comparison of the frequency of the most reproducible capillary lesions found in alloxan-diabetic rats and normal rats fed a 30% galactose diet for 20 months. Results are the mean ± SD. (Data from Kern and Engerman[31]).

ical of patients or dogs having diabetes, have not been observed by us in diabetic or galactosemic rats at durations of up to 22 and 26 months, respectively, although unidentified clusters of vessel cells occasionally have been observed.[31,33] The latter, unusual lesions occurred with a frequency similar to IRMA, but bore little resemblance to the saccular capillary microaneurysms seen characteristically in patients and dogs. Microaneurysms have previously been claimed to develop in diabetic or galactosemic rats,[29,34–37] but the lesions described usually have more closely resembled vessel kinks or have been ill-defined masses of cells bearing only a superficial resemblance to the saccular microaneuryms characteristic of diabetic retinopathy in humans and dogs.

Thus, diabetes and experimental galactosemia in rats lead to similar retinal lesions, and the lesions are consistent at least with the early stages of diabetic retinopathy. Lesions that developed in diabetic rats also developed in galactosemic rats, and lesions that failed to develop in galactosemic rats were also absent from diabetic rats. In humans and dogs, pericyte loss and capillary obliteration precede the development of microaneurysms and advanced stages of retinopathy, so the development of pericyte ghosts and acellular capillaries in rats probably represents a significant stage in the development of retinopathy. Rats offer a relatively inexpensive animal model for study of the early stages of retinopathy, and for screening potentially useful but untested therapies for their effects on diabetic retinal disease. Rapidly developing cataracts in diabetic or galactose-fed rats, however, preclude evaluation of the retinopathy while the animals are alive.

Chemically induced diabetes in rats presents a special problem when studies exceed 1 year in duration. Between one third and one half of alloxan-diabetic rats have been observed by us to undergo spontaneous remission of hyperglycemia, even after 1–1½ years of severe insulin deficiency.[31] Evidence of islet regeneration has been observed also in other studies,[38,39] especially in the presence of estrogen.[40] Inasmuch as the severity of hyperglycemia is one of the most important determinants for the development of diabetic retinopathy, systematic and regular monitoring of glycemia is essential. Diabetic BB rats do not experience spontaneous remission of hyperglycemia, and by 14 months of hyperglycemia have been reported to develop lesions consistent with the early stages of diabetic retinopathy, including pericyte loss and basement membrane thickening (and an absence of microaneurysms).[41]

Rats fed very high concentrations of sucrose or fructose (~70% in the diet) also have been reported to develop retinal lesions, including loss of pericytes and endothelial cells, and formation of capillary strands.[42,43] A possibility that this retinopathy is due to a nutritional deficiency secondary to such a high carbohydrate diet seems unlikely because of the relative lack of retinal pathology in rats fed a diet enriched with a similarly high concentration of starch. The mechanisms by which fructose or sucrose induce retinal pathology have been little investigated, but the retinopathy in sucrose-fed spontaneously diabetic rats reportedly can be inhibited by islet transplantation.[44]

Aldose reductase and other sequelae of hyperglycemia in the pathogenesis of diabetic retinopathy

The available evidence suggests that one or more sequelae of hyperglycemia that are shared in common between diabetes and galactosemia initiate the retinopathy. Much attention to date has focused on the potential role of aldose reductase and the polyol pathway in the development of the retinopathy. The findings to date, however, have been controversial (Table 2). Some investigators have reported that aldose reductase inhibitors (ARIs) prevent galactose-induced retinopathy in dogs and rats, or thickening of retinal capillary basement membrane in galactosemic or diabetic rats.[26,29,35,45–47] On the contrary, in a more lengthy study in which polyol accumulation in the retina and other tissues was inhibited by up to 100%, retinopathy was uninfluenced by inhibition of aldose reductase.[48] Despite the absence of any effect on retinal microaneurysms, pericyte loss, acellular capillaries, or capillary basement membrane thickness, ARI therapy in the latter study did effectively inhibit a peripheral nerve dysfunction seen in the same diabetic animals.[49] This evidence suggests that reported effects of any drug on the development of retinopathy are not likely to be mediated by inhibition of retinal polyol *accumulation.*

Possible reasons for the disparate conclusions regarding the ability of ARI to prevent retinal pathology have been discussed elsewhere,[48,50] and may include differences in the degree of inhibition of the polyol pathway, differences in experimental design, differences in the duration of study, and differences in methods of analysis of retinal lesions. Prevention of cataract has been suggested as a biomarker of aldose reductase inhibition. Cataract development in our ARI-treated dogs was significantly inhibited (although not completely prevented), and was also not prevented in galactose-fed dogs given ARI by Kador et al.,[51] in

Table 2. Studies reporting a beneficial effect or no effect of aldose reductase inhibition on the development of retinopathy

	Ref.	% inhibition of retinal polyol
Inhibition of retinopathy		
Dog, galactose fed	Kador et al.[26,47]	Not determined
Rat, galactose fed	Robison et al.[29]	Not determined
Rat, fructose fed diabetic	Kojima et al.[53]	Not determined
No inhibition of retinopathy		
Dog, diabetic	Engerman and Kern[48]	100%
Dog, galactose fed	Engerman and Kern[48]	96%
Rat, galactose fed	Kern and Engerman[33]	62%
Clinical trials		
Sorbinil	Sorbinil Retinopathy Trial Research group[83]	Not determined
Ponalrestat	Kohner et al.[84]	Not determined
Ponalrestat	Arauz-Pacheo et al.[85]	Not determined

which retinopathy was said to have been retarded by the drug. In a latter study by the same group,[47] higher doses of ARI are said to have kept lenses clear in 4 of 13 galactose-fed dogs, but retinopathy data were presented for only 8 animals, with no explanation of how those particular dogs were selected for histologic evaluation of the retina. The ability or inability to inhibit lens aldose reductase provides no information about polyol production in the retina, inasmuch as tissues differ in the extent to which polyol production is inhibited by a given level of drug.[52]

At present, no suitable method is available to monitor flux through the polyol pathway in the retina in vivo, and, accordingly, one may speculate that it is the excessive flux rather than polyol accumulation that is responsible for the retinopathy in diabetes. In experimental galactosemia, the galactitol concentration in each tissue is the best estimate of polyol production at present, since the galactitol produced by aldose reductase is further metabolized little, if at all. Unfortunately, past reports claiming that ARI prevents retinopathy[26,29] contain no estimate of the retinal polyol concentration or pathway flux for comparison with our studies.

Another possible explanation for contradictory conclusions regarding the role of polyol production in the development of retinopathy focuses on the importance of hyperglycemia as the cause of retinopathy. Studies in our laboratory were designed to ensure that the blood hexose elevation (glucose or galactose) was as severe in the ARI-treated animals as in animals not given the drug. In the studies claiming a drug effect on retinopathy, this important variable was inadequately monitored, and the available evidence suggests that glycemia in those studies was in fact less severe in the ARI-treated groups than in controls. In studies of galactose-fed dogs by Kador et al.[47,51] the animals in which retinopathy was supposedly inhibited by ARI showed an elevation of glycohemoglobin that was only half as much as that in control animals not given the drug,[47,51] indicating that the drug-treated animals had lower blood hexose levels than control animals. In other studies, glycemia was not measured, except for a single determination at autopsy,[29] or the ARI-treated group failed to show the diabetes-induced growth inhibition observed in a control diabetic group (the ARI-treated group being apparently less insulin deficient),[53] or the ARI-treated group showed levels of glycohemoglobin lower than controls during the experiment.[8] The evidence that hyperglycemia might have been less severe in these drug-treated groups is not to be overlooked, and might account for the development of less retinopathy, regardless of the ARI tested.

Possible roles of other sequelae of hyperglycemia in the pathogenesis of retinopathy are being investigated. Aminoguanidine, which can block late sequelae of nonenzymatic glycation and inhibit nitric oxide synthesis, has been reported to have a variety of beneficial effects in diabetic animals, including inhibition of diabetes-induced albuminuria and thickening of the capillary basement membrane in glomeruli, normalization of blood flow,[54–58] and, importantly, inhibition of the development of acellular capillaries and other retinal lesions in diabetic rats.[32,37,59] Fluorescence of trypsinized retinal vessels at 370 nm (excitation) and 440 nm (emission) has been attributed to nonenzymatic glycation, and has been found to be increased in diabetes and to be reduced in aminoguanidine-treated animals. A calcium channel blocker was previously claimed to inhibit the development of acellular capillaries in retinas of diabetic rats,[60] but the duration of diabetes (8 wk) was far shorter than we have found necessary to produce acellular capillaries.

Experimental galactosemia as a model of hyperglycemia

Experimental galactosemia is a model of hyperglycemia, and not merely of polyol pathway overactivity. Experimental galactosemia has been found to reproduce many sequelae of hyperglycemia observed in diabetes, including alterations in retinal blood flow, increased nonenzymatic glycation and formation of advanced glycation end products, increased activity of protein kinase C and levels of diacylglycerol (a regulator of protein kinase C), and altered expression of genes.[61–67] Each of these sequelae of hyperglycemia has been shown to cause detrimental effects in vivo and might contribute to the development of diabetic complications. Experimental galactosemia in dogs and rats also reproduces additional abnormalities of retinal metabolism found in diabetes, including decreased levels of glutathione (GSH) and activities of (Na,K)ATPase and calcium ATPase (Fig. 3), and increased levels of diacylglycerol or activity of protein kinase C.[66,68,69] Detrimental effects of hyperglycemia on enzyme activity apparently are somewhat specific; activities of retinal (Na,K)ATPase and calcium ATPase, for example, are subnormal in diabetes and galactosemia, whereas (Mg)ATPase is not affected. The cause of subnormal GSH in retina seems similar in diabetes and experimental galactosemia; activities of γ-glutamyl transpeptidase, glutathione peroxidase, and glutathione reductase are reduced, whereas enzymes involved in the synthesis of GSH seem unaffected.[70,71]

Dissimilar therapies, such as antioxidants and ARI, can correct a variety of hyperglycemia-induced abnormalities of retinal metabolism, suggesting interac-

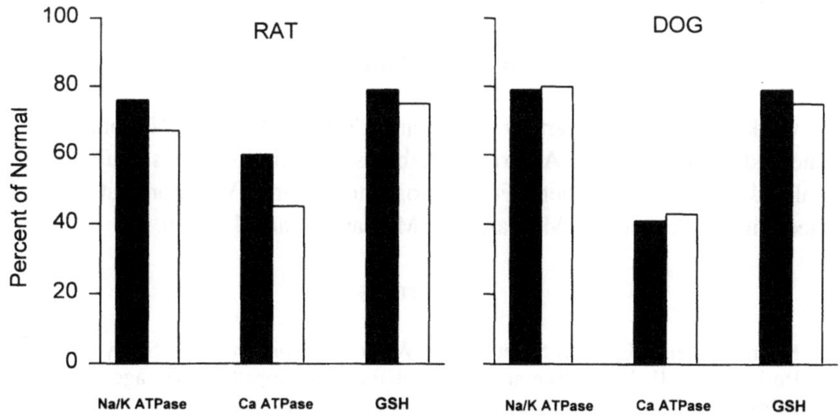

Figure 3. Diabetes (■) and experimental galactosemia (□) cause similar defects in the activity of retinal (Na,K)ATPase and calcium ATPase, and in glutathione levels in two species. Measurements were made after 2 months of diabetes or galactosemia.

tions between oxidative stress and the polyol pathway in the retina.[71–73] Investigations into the relationship between the development of diabetic retinopathy and hyperglycemia-induced abnormalities of retinal metabolism and physiology are now underway.

Comparison of retina to kidney and nerve

The similarity between the effects of diabetes and of galactosemia on the retina is not necessarily reproduced in other tissues. Experimental galactosemia of 5 years duration in dogs does not reproduce the nerve conduction defect characteristic of diabetic dogs and humans, nor does it reproduce most aspects of diabetic renal disease in dogs.[74–76] Renal hypertrophy, glomerular enlargement, mesangial expansion, and glomerulosclerosis are absent from kidneys of dogs in which a galactose-rich diet produces retinopathy. Only thickening of the capillary basement membrane occurs in glomeruli of both the diabetic dog and the experimentally galactosemic dog. The evidence in dogs fed galactose suggests that the renal disease and abnormal nerve function in diabetes apparently develop by mechanisms different from that leading to retinopathy. Findings in rats possibly differ from those in dogs, in that galactose-feeding in rats leads to a defect in nerve conduction velocity[77] and, in one report, to mesangial expansion in renal glomeruli.[78] The basis for the apparent species differences remains unclear.

Conclusions

Studies involving diabetic and experimentally galactosemic animals have clearly shown that hyperhexosemia plays a critical role in the development of diabetic retinopathy. The mechanism by which the hyperglycemia causes retinopathy, however, remains unclear. Comparison of two dissimilar models of retinopathy (diabetes and galactosemia) in multiple animal species provides a unique opportunity to provide insight into the pathogenesis and treatment of diabetic retinopathy.

Acknowledgments

This work was funded in part by PHS grant EY00300 from the National Eye Institute and grants from the American Diabetes Association/Lions Clubs International and the Juvenile Diabetes Foundation International. We appreciate the skillful technical assistance of M. Garment, M. Larson, and J. Heen.

References

1. Klein R, Klein BE, Moss SE et al. The Wisconsin Epidemiologic Study of Diabetic Retinopathy. II. Prevalence and risk of diabetic retinopathy when age at diagnosis is less than 30 years. Arch Ophthalmol 102:520–26, 1984.
2. Klein R, Klein BE, Moss SE et al. The Wisconsin Epidemiologic Study of Diabetic Retinopathy. II. Prevalence and risk of diabetic retinopathy when age at diagnosis is 30 or more years. Arch Ophthalmol 102:527–32, 1984.

3. Davis MD. Diabetic retinopathy. A clinical overview. Diabetes Care 15:1844–74, 1992.
4. Yokote M. Retinal and renal microangiopathy in carp with spontaneous diabetes mellitus. In: Early Diabetes. Adv. Metab. Disord. (Suppl 2) Academic Press 1973:pp 299–304.
5. Kaczurowski M. Angiopathy of retinal vessels in diabetic mice. Arch Ophthalmol 84:316–20, 1970.
6. Duhault J, Lebon F, Boulanger M. KK mice as a model of microangiopathic lesions in diabetes. In: 7th Europ. Conf. Microcirc. Karger, Aberdeen, pp 453–8, 1973.
7. Sima AAF, Garcia-Salinas R, Basu PK. The BB Wistar rat: An experimental model for the study of diabetic retinopathy. Metabolism 32(Suppl 1):136–40, 1983.
8. Chakrabarti S, Sima AAF. Effect of aldose reductase inhibition and insulin treatment on retinal capillary basement membrane thickening in BB rats. Diabetes 38:1181–86, 1989.
9. Toussaint D. Contribution a l'etude anatomique et clinique de la retinopathie diabetique chez l'homme et chez l'animal. Pathol Europ 108–48, 1968.
10. Hausler HR, Sibay TM, Campbell J. Retinopathy in a dog following diabetes induced by growth hormone. Diabetes 13:122–6, 1964.
11. Gepts W, Toussaint D. Spontaneous diabetes in dogs and cats. Diabetologia 3:249–64, 1967.
12. Sibay TM, Hausler HR. Eye findings in two spontaneously diabetic related dogs. Am J Ophthalmol 63:289–94, 1967.
13. Laver N, Robison WG Jr, Hansen BC. Spontaneously diabetic monkeys as a model for diabetic retinopathy. ARVO Abstracts. Invest Ophthalmol Vis Sci 35 (Suppl):1733, 1994.
14. Patz A, Maumenee AE. Studies on diabetic retinopathy. I. Retinopathy in a dog with spontaneous diabetes mellitus. Am J Ophthalmol 532–41, 1962.
15. Engerman RE, Bloodworth JMB Jr. Experimental diabetic retinopathy. Invest Ophthalmol 3:466–7, 1964.
16. Engerman RL, Bloodworth JMB Jr. Experimental diabetic retinopathy in dogs. Arch Ophthalmol 73:205–10, 1965.
17. Engerman RL, Finkelstein D, Aguirre G et al. Appropriate animal models for research on human diabetes mellitus and its complications. Ocular complications. Diabetes 31(Suppl 1):82–8, 1982.
18. Engerman RL. Pathogenesis of diabetic retinopathy. Diabetes 38:1203–6, 1989.
19. Ashton N. Arteriolar involvement in diabetic retinopathy. Br J Ophthalmol 37:282–92, 1953.
20. Gardiner TA, Stitt AW, Anderson HR, Archer DB. Selective loss of vascular smooth muscle cells in the retinal microcirculation of diabetic dogs. Br J Ophthalmol 78:54–60, 1994.
21. Engerman RL, Bloodworth JMB Jr, Nelson S. Relationship of microvascular disease in diabetes to metabolic control. Diabetes 26:760–9, 1977.
22. Diabetes Control and Complications Trial Research Group. The effect of intensive treatment of diabetes on the development of long-term complications in insulin-dependent diabetes mellitus. N Engl J Med 329:977–86, 1993.
23. Engerman RL, Kern TS. Experimental galactosemia produces diabetic-like retinopathy. Diabetes 31(Suppl):26A, 1982.
24. Engerman RL, Kern TS. Experimental galactosemia produces diabetic-like retinopathy. Diabetes 33:97–100, 1984.
25. Kador PF, Akagi Y, Terubayashi H et al. Prevention of pericyte ghost formation in retinal capillaries of galactose-fed dogs by aldose reductase inhibitors. Arch Ophthalmolmol 106:1099–1102, 1988.
26. Kador PF, Akagi Y, Takahashi Y et al. Prevention of retinal vessel changes associated with diabetic retinopathy in galactose-fed dogs by aldose reductase inhibitors. Arch Ophthalmolmol 108:1301–9, 1990.

27. Wallow IH, Engerman RL. Permeability and patency of retinal blood vessels in experimental diabetes. Invest Ophthalmolmol 16:447–61, 1977.
28. Kador PF, Takahashi Y, Wyman M, Ferris F III. Diabeteslike proliferative retinal changes in galactose-fed dogs. Arch Ophthalmolmol 113:352–54, 1995.
29. Robison WG Jr, Nagata M, Laver N et al. Diabetic-like retinopathy in rats prevented with an aldose reductase inhibitor. Invest Ophthalmol Vis Sci 30:2285–92, 1989.
30. Kern TS, Engerman RL. Galactose-induced retinal lesions in rats and effects of aldose reductase inhibition. Diabetes 41(Suppl 1):20A, 1992.
31. Kern TS, Engerman RL. Comparison of retinal lesions in alloxan-diabetic rats and galactose-fed rats. Curr Eye Res 13:863–7, 1994.
32. Hammes H-P, Syed S, Uhlmann M et al. Aminoguanidine does not inhibit the initial phase of experimental diabetic retinopathy in rats. Diabetologia 38:269–73, 1995.
33. Kern TS, Engerman RL. Galactose-induced retinal microangiopathy in rats. Invest Ophthalmol Vis Sci 36:490–6, 1994.
34. Leuenberger P, Cameron D, Stauffacher W et al. Ocular lesions in rats rendered chronically diabetic with streptozotocin. Ophthalmol Res 2:189–204, 1971.
35. Robison WG Jr, Tillis TN, Laver N, Kinoshita JH. Diabetes-related histopathologies of the rat retina prevented with an aldose reductase inhibitor. Exp Eye Res 50:355–66, 1990.
36. Schröder S, Palinski W, Schmid-Schönbein GW. Activated monocyte and granulocytes, capillay nonperfusion and neovascularization in diabetic retinopathy. Am J Pathol 139:81–100, 1991.
37. Hammes H-P, Martin S, Federlin K et al. Aminoguanidine treatment inhibits the development of experimental diabetic retinopathy. Proc Natl Acad Sci USA. 88:11555–58, 1991.
38. von Sallmann L, Grimes P. Histopathologic changes in the eyes and kidneys of rats with long term streptozotocin diabetes. In: Contemporary ophthalmology. Williams & Wilkins, Baltimore, MD, Bellows JG, ed pp 502–11, 1972.
39. Martin JM, Gregor WH, Lacy PE, Evans RG. The effect of hyperglycemia upon islet regeneration in rats. Diabetes 12:538–44, 1963.
40. Rodriguez RR. Alloxan diabetes in the rat: Recovery following estrogen treatment. Endocrinology 55:1–9, 1954.
41. Chakrabarti S, Sima AAF, Tze WJ, Tai J. Prevention of diabetic retinal capillary pericyte degeneration and loss by pancreatic islet allograft. Curr Eye Res 6:649–58, 1987.
42. Yanko L, Michaelson IC, Cohen AM. The retinopathy of sucrose-fed rats. Isr J Med Sci 8:1633–36, 1972.
43. Boot-Handford R, Heath H. Identification of fructose as the retinopathic agent associated with the ingestion of sucrose-rich diets in the rat. Metabolism 29:1247–52, 1980.
44. Hammes H-P, Klinzing I, Wiegand S et al. Islet transplantation inhibits diabetic retinopathy in the sucrose-fed diabetic Cohen diabetic rat. Invest Ophthalmol Vis Sci 34:2092–96, 1993.
45. Robison WG Jr, Kador PF, Kinoshita JH. Retinal capillaries: Basement membrane thickening by galactosemia prevented with aldose reductase inhibitor. Science 221:1177–79, 1983.
46. Vinores SA, Campochiaro PA. Prevention or moderation of some ultrastructural changes in the RPE and retina of galactosemic rats by aldose reductase inhibition. Exp Eye Res 49:494–510, 1989.
47. Kador PF, Takahashi Y, Sato S, Wyman M. Amelioration of diabetes-like retinal changes in galactose-fed dogs. Prev Med 23:717–21, 1994.
48. Engerman RL, Kern TS. Aldose reductase inhibition fails to prevent retinopathy in diabetic and galactosemic dogs. Diabetes 42:820–25, 1993.
49. Engerman RL, Kern TS, Larson ME. Nerve conduction velocity and aldose reduc-

tase inhibition during 5 years diabetes or galactosemia in dogs. Diabetologia 37:141–4, 1994.
50. Engerman RL, Kern TS. Aldose reductase inhibition and retinopathy [letter to editor]. Diabetes 43:338–9, 1994.
51. Kador PF, Takahashi Y, Sato S, Wyman M. Aldose reductase, retinal vessel changes and cataracts in galactose-fed dogs. In: Diabetes 1991. Rifkin H, Colwell JA, Taylor SI, eds Elsevier, Amsterdam, pp 373–8, 1991.
52. Kador PF, Kinoshita JH, Tung WH, Chylack LT Jr. Differences in the susceptibility of various aldose reductases to inhibition. Invest Ophthalmol Vis Sci 19:980–2, 1980.
53. Kojima K, Matsubara H, Harada T et al. Effects of aldose reductase inhibitor on retinal microangiopathy in streptozotocin-diabetic rats. Jpn J Ophthalmol 29:99–109, 1985.
54. Corbett JA, Tilton RG, Chang K et al. Aminoguanidine, a novel inhibitor of nitric oxide formation, prevents diabetic vascular dysfunction. Diabetes 41:552–6, 1992.
55. Kihara M, Schemlzer JD, Poduslo JF et al. Aminoguanidine effects on nerve blood flow, vascular permeability, electrophysiology, and oxygen free radicals. Proc Natl Acad Sci USA 88:6107–11, 1991.
56. Soulis-Liparota T, Cooper M, Papazoglou D et al. Retardation by aminoguandine of development of albuminuria, mesangial expansion, and tissue fluorescence in streptozotocin-induced diabetic rat. Diabetes 40:1328–34, 1991.
57. Tilton RG, Chang K, Hasan KS et al. Prevention of diabetic vascular dysfunction by guanidines. Inhibition of nitric oxide synthase versus advanced glycation end-product formation. Diabetes 42:221–32, 1993.
58. Yagihashi S, Kamijo M, Baba M et al. Effect of aminoguanidine on functional and structural abnormalities in peripheral nerve of STZ-induced diabetic rats. Diabetes 41:47–52, 1992.
59. Hammes H-P, Brownlee M, Edelstein D et al. Aminoguanidine inhibits the development of accelerated diabetic retinopathy in the spontaneous hypertensive rat. Diabetologia 37:32–5, 1994.
60. Pierce GN, Afzal N, Kroeger EA et al. Cataract formation is prevented by administration of verapamil to diabetic rats. Endocrinology 125:730–35, 1989.
61. Urbanowski JC, Cohenford MA, Dain JA. Nonenzymatic galactosylation of human serum albumin. J Biol Chem 257:111–5, 1982.
62. Tilton RG, Chang K, Weigel C et al. Increased ocular blood flow and ^{125}I-albumin permeation in galactose-fed rats: Inhibition with sorbinil. Invest Ophthalmol Vis Sci 29:861–8, 1988.
63. Tilton RG, Chang K, Pugliese G et al. Prevention of hemodynamic and vascular albumin filtration changes in diabetic rats by aldose reductase inhibitors. Diabetes 37:1258–70, 1989.
64. Richard S, Tamas C, Sell DR, Monnier VM. Tissue-specific effects of aldose reductase inhibition on fluorescence and cross-linking of extracellular matrix in chronic galactosemia. Diabetes 40:1049–56, 1991.
65. Nagaraj RH, Fogarty JF, Kern TS et al. Advanced maillard reaction in dura mater collagen of chronic hyperglycemic dogs: Effect of glycemic control and sorbinil treatment. Diabetes 42(Suppl):240A, 1993.
66. Zia P, Inoguchi T, Kern TS et al. Characterization of the mechanism for the chronic activation of DAG-PKC pathway in diabetes and hypergalactosemia. Diabetes 43:1122–29, 1994.
67. Cagliero E, Roth T, Roy S, Lorenzi M. Characteristics and mechanisms of high-glucose-induced overexpression of basement membrane components in cultured human endothelial cells. Diabetes 40:102–10, 1991.
68. Kern TS, Kowluru R, Engerman RL. Abnormalities of retinal metabolism in dia-

betes or galactosemia. ATPases and glutathione. Invest Ophthalmol Vis Sci 35:2962–67, 1994.
69. Shiba T, Inoguchi T, Sportsman JR et al. Correlation of diacylglycerol and protein kinase C in rat retina to retinal circulation. Am J Physiol 265:E783–93, 1993.
70. Kowluru R, Kern TS, Engerman RL. Abnormalities of retinal metabolism in diabetes or galactosemia II. Comparison of gamma-glutamyl transpeptidase in retina and cerebral cortex, and effects of antioxidant therapy. Curr Eye Res 13:891–6, 1994.
71. Kowluru R, Kern TS, Engerman RL. Antioxidants prevent retinal glutathione dysmetabolism in diabetes. ARVO abstracts. Invest Ophthalmol Vis Sci 36:S1066, 1995.
72. Kern TS, Kowluru R, Engerman RL. Correction of retinal dysmetabolism in diabetic rats and galactosemic rats by antioxidants, calcium channel blocker, or aldose reductase inhibitor. Abstracts 15th Int Diabetes Fed Congr Japan p 369, 1994.
73. Kowluru R, Kern T, Engerman R. Antioxidant therapy corrects subnormal free radical defense mechanism in diabetic retina. Diabetes 44(Suppl 1):117A, 1995.
74. Kern TS, Engerman RL. Kidney morphology in experimental hyperglycemia. Diabetes 36:244–9, 1987.
75. Engerman RL, Kern TS. Hyperglycemia and development of glomerular pathology: Diabetes compared with galactosemia. Kidney Int 36:41–5, 1989.
76. Engerman RL, Kern TS, Larson ME. Nerve conduction velocity in dogs is reduced in diabetes and not by galactosemia. Metabolism 39:638–40, 1990.
77. Sharma AK, Thomas PK, Baker RWR. Peripheral nerve abnormalities related to galactose administration in rats. J Neurol Neurosurg Psychiatry 39:794–802, 1976.
78. Tilton RG, Pugliese G, LaRose LS. Discordant effects of the aldose reductase inhibitor sorbinil on vascular structure and function in chronically diabetic and galactosemic rats. J Diabetes Complications 5:230–7, 1991.
79. Sharma AK, Gardiner TA, Archer DB. A morphologic and autoradiographic study of cell death and regeneration in the retinal microvasculature of normal and diabetic rats. Am J Ophthalmol 100:51–60, 1985.
80. Robison WG Jr, McCaleb ML, Feld LG, et al. Degenerated intramural pericytes ('ghost cells') in the retinal capillaries of diabetic rats. Curr Eye Res 10:339–50, 1991.
81. Tilton RG, LaRose LS, Kilo C, Williamson JR. Absence of degenerative changes in retinal and uveal capillaries in diabetic rats. Invest Ophthalmol Vis Sci 27:716–21, 1986.
82. Engerman RL. Animal models of diabetic retinopathy. Trans Am Acad Ophthalmol Otolaryngol 81:710–15, 1976.
83. Sorbinil Retinopathy Trial Research Group. A randomized trial of sorbinil, an aldose reductase inhibitor, in diabetic retinopathy. Arch Ophthalmol 108:1234–44, 1990.
84. Kohner EM, Caldwell G, Plehwe WE, et al. Ponalrestat in early diabetic retinopathy. Diabetes 39 (Suppl 1):62A, 1990.
85. Arauz-Pacheo C, Ramiriz LC, Pruneda L et al. The effect of the aldose reductase inhibitor, ponalrestat, on the progression of diabetic retinopathy. J Diab Compl 6:131–7, 1992.

23

Glucose homeostasis in three interstrains (LA/N-BN/Crl cp/cp; *Zuc13M-BN/Crl* fa/fa; *and Zuc13M-LA/N* fa/cp) *of genetically obese rats*

E. BOWIE KAHLE,[1] KENNETH G. BUTZ,[1] RUDOLPH L. LEIBEL,[2] CARL T. HANSEN,[3] SAM J. BHATHENA,[4] AND OTHO E. MICHAELIS IV[4]*

[1]Departments of Physiology and Biology, Marshall University, Huntington, West Virginia, USA; [2]Rockefeller University, New York, New York, USA; [3]NIH, Bethesda, Maryland, USA; [4]Beltsville Human Nutrition Research Center, Beltsville, Maryland, USA.

Key words: glucose, insulin, diabetes, corticosterone, interstrains, *fa* mutation, *cp* mutation.

Glucose homeostasis in the postabsorptive state is affected by *fa* and *cp* mutations in the rat.[1,2] The fatty (*fa*)[3] and corpulent (*cp*)[4,5] genes are recessive rat mutations that cause obesity. Interbreeding of heterozygous lean Zucker rats, carrying the *fa* mutation, and Koletsky rats, carrying the *cp* mutation, produced obese F_1 progeny at a 3:1 lean:obese ratio.[6] These obese interstrain rats appeared to confirm that *fa* and *cp* are mutations of the same gene, but that conclusion has not been confirmed by molecular genetic analysis. Also, metabolic characteristics of the *fa/cp* obese hybrids have not been reported.

The *cp* mutation arose spontaneously[5] and was interbred into the LA/N rat developed at the National Institutes of Health (NIH), producing the LA/N *cp* rat.[4] The obese LA/N *cp* rats exhibit metabolic characteristics associated with type IV hyperlipoproteinemia (carbohydrate sensitivity) in humans,[7,8] including normoglycemia or mild hyperglycemia, and basal hyperinsulinemia. High circulating levels of glucocorticoids, characteristic of some obesities,[9,10,11] interact with insulin to enhance food intake in LA/N *cp* rats[12] and to facilitate the deposition of energy by insulin.[13]

Correspondence: E. Bowie Kahle, Ph.D., Department of Physiology, Marshall University School of Medicine, Huntington, WV 25755, USA.
*Deceased on May 27, 1995.

Lessons from Animal Diabetes VI. Ed. E. Shafrir.

The *fa* mutation in the Zucker (Zuc13M) rat[3] also arose spontaneously and shows strong similarities to human obesity syndromes. The Zucker fatty rat is hyperinsulinemic with normal blood glucose levels,[1] and develops adipocyte hypertrophy and hyperplasia.[14] The Zuc13M *fa/fa* has been outcrossed to the Brown Norway (BN/Crl) strain[15] in order to permit molecular mapping of *fa*. The Zuc13M-BN/Crl animals segregating for *fa* have not been described in detail.

In preparation to apply the mapping techniques described here to locate the *cp* mutation, we constructed outcrosses to BN/Crl segregating the *cp* and, separately, the *fa* mutation. We report the phenotypes of *cp/cp* on LA/N-BN/Crl, of *fa/fa* on Zuc13M-BN/Crl, and of the compound mutant animals on a Zuc13M-LA/N background.

Methods

Crosses segregating for the *fa* and *cp* mutations were constructed from either LA/N *cp/+* (C.T. Hansen, NIH, Bethesda, MD) or Zuc13M *fa/+* (Vassar College, Poughkeepsie, NY) of both sexes and BN/Crl +/+ rats (Charles River Breeding Laboratories, Wilmington, MA). Half of the F_1 progeny were predicted to be carriers of the mutant gene. Heterozygote (*cp/+* or *fa/+*) animals were identified by random brother–sister test matings. Progeny of F_1 LA/N-BN/Crl and Zuc13M-BN/Crl interstrain crosses were weaned at 28 days of age, and littermates (average 7–9 animals per cage) were visually scored for the presence of obese pups between 28 and 42 days of age. Animals were maintained during the active obesity development phase to 10 wk of age on lab chow (PMI Feeds, St. Louis, MO; proximate analysis: 58% carbohydrate from grains and cane molasses, 6% fiber, 23.5% protein, 4.5% fat, remaining 8% ash plus vitamin and mineral supplements). At this age, fasting (15–20 h) heparinized blood was obtained from anesthetized animals (50 mg/kg body wt pentobarbital sodium) by heart puncture preceding sacrifice.

In a separate experiment, breeding protocols producing apparent *fa/cp* compound mutants were carried out with heterozygous LA/N *cp/+* rats of both sexes (C.T. Hansen, NIH, Bethesda, MD) and Zuc13M *fa/+* rats (Vassar College, Poughkeepsie, NY). Progeny of Zuc × LA/N crosses were visually scored for the presence of obese pups between 28 and 42 days of age. These animals were fed a semipurified diet[2] (54% sucrose, 6% fiber, 20% protein, 16% fat, 3.1% mineral mixture, 1% vitamin mixture) from 9.5 to 13 wk of age preceding analysis of fasting plasma obtained as detailed earlier. These data on apparent *fa/cp* mutants are included to permit comparative evaluation with established information on Zucker *fa*[16] and LA/N *cp* animals[2,12] fed a lipogenic, high sucrose diet.

Breeding experiments took place in the Marshall University Animal Facilities with routine husbandry guidelines recommended by the Marshall University Animal Care Committee. Plasma analyses included insulin (Immunonuclear Corp., Stillwater, MN) using rat insulin standards (gift of Eli Lilly, Indianapolis, IN) and corticosterone (ICN Biochemicals, Inc., Costa Mesa, CA) by RIA. Plasma glucose was analyzed enzymatically (hexokinase) using a Centrifichem Automated System (Trace-America, Miami, FL).

Results and discussion

The F_2 progeny of 14 LA/N-BN/Crl litters approximated the 3:1 ratio typical of an autosomal recessive trait, yielding 81 lean and 31 obese animals. A similar lean:obese ratio, 34 lean and 11 obese rats, was recorded in 7 litters of F_2 Zuc13M-BN/Crl animals. From 6 litters of LA/N-Zuc13M crosses, 32 lean and 18 obese F_1 progeny were produced, a 1.8:1 ratio. Physiological characterization of the progeny of the rat interstrain crosses segregating *cp* and *fa*, as examined in this protocol, illustrates subtle differences between the mutations and defines their homology (Tables 1–3).

Characteristics of glucose homeostasis in F_2 progeny of the LA/N-BN/Crl corpulent (*cp*) and the Zuc-BN/Crl fatty (*fa*) interstrain animals are given in Table 1. Outcross of *cp* on the LA/N strain to BN/Crl rat produced animals with glucose homeostasis characteristics different from those of the parent LA/N *cp* strain. Fasting plasma glucose concentrations were higher in obese animals of both sexes compared with their lean littermates ($P < 0.05$). Glucose levels of obese LA/N-BN/Crl male rats appear to be higher than those of LA/N male animals reported previously.[12] Both sexes of this obese animal also exhibited plasma insulin concentrations highly elevated above their lean counterparts ($P < 0.05$). Together these data suggest that obese LA/N-BN/Crl *cp*/*cp* males are glucose intolerant

Table 1. Fasting glucose, insulin, and corticosterone concentrations in obese and lean LA/N-BN/Crl *cp* and Zuc13M-BN/Crl *fa* rats

		n	Glucose (mmol/l)	Insulin (pmol/l)	Corticosterone (nmol/l)
LA/N-BN/Crl *cp*[c]					
Male	Obese	10	8.0 ± 1.0[a]	2174 ± 861[a]	1189 ± 274[b]
	Lean	41	6.2 ± 0.3	488 ± 86	1068 ± 75[b]
Female	Obese	20	6.7 ± 0.3[a]	2734 ± 524[a]	1484 ± 199
	Lean	38	6.0 ± 0.3	258 ± 50	1564 ± 144
Both sexes	Obese	30	7.2 ± 0.4[a]	2547 ± 438[a]	1385 ± 159
	Lean	79	6.1 ± 0.2	373 ± 50	1313 ± 84
Zuc13M-BN/Crl *fa*[d]					
Male	Obese	3	6.3 ± 1.1	1787 ± 1277	421 ± 170[a]
	Lean	13	5.3 ± 0.5	524 ± 323	231 ± 98
Female	Obese	5	7.1 ± 1.4	1543 ± 926[a]	1521 ± 277
	Lean	9	6.2 ± 1.2	265 ± 108	1192 ± 231
Both sexes	Obese	8	6.8 ± 0.9	1636 ± 775[a]	1111 ± 271
	Lean	22	5.7 ± 0.6	423 ± 194	964 ± 113

Values are means ± SE.

[a]Different phenotypic (obese vs. lean) values within treatment at $P < 0.05$ by the independent *t*-test.

[b]Different genotypic (obese vs. obese; lean vs. lean) values between treatment at $P < 0.05$ by the independent *t*-test.

[c]Kahle et al. (current study): age, 10 wk; diet, 54% sucrose; 16 h fast F_2 generation brother × sister heterozygote LA/N-BN *cp*/+ matings.

[d]Kahle et al. (current study); age, 10 wk; diet, 54% sucrose; 16 h fast F_2 generation brother × sister heterozygote Zuc-BN *fa*/+ matings.

Table 2. Glucose and hormonal profiles of obese and lean Zuc13M-LA/N *fa/cp* rats compared with established Zucker *fa* and LA/N *cp* rat data

		n	Glucose (mmol/l)	Insulin (pmol/l)	Corticosterone (nmol/l)
Zuc13M-LA/N *fa/cp*[b]					
Male	Obese	5	8.5 ± 0.6[a]	1442 ± 323[a]	2205 ± 199[a]
	Lean	8	5.9 ± 0.4	187 ± 93	1403 ± 159
Female	Obese	4	8.6 ± 0.6	1069 ± 179	2442 ± 266
	Lean	2	5.7 ± 0.3	122 ± 72	2364 ± 234
Both sexes	Obese	10	8.6 ± 0.2[a]	1299 ± 273[a]	2681 ± 254[a]
	Lean	17	5.8 ± 0.4	165 ± 14	1676 ± 185
LA/N *cp*[c]					
Male	Obese	6	6.2 ± 0.2	3559 ± 509[a]	1036 ± 75[a]
	Lean	6	5.4 ± 0.4	538 ± 100	774 ± 72
JCR:LA *cp*[d]					
Female	Obese	9	6.9 ± 0.5	1866 ± 1155	502 ± 473
SHR/N *cp*[e]					
Female	Obese		17.5 ± 2.0	12,000 ± 500	1879 ± 113[a]
	Lean		11.0 ± 0.5	7250 ± 200	1031 ± 150
SHHF/Mcc *cp*[f]					
Male	Obese	4	7.3 ± 0.7[a]	2892 ± 237[a]	
	Lean	4	4.4 ± 0.6	387 ± 72	
Female	Obese	4	7.3 ± 0.3[a]	1048 ± 72[a]	
	Lean	4	4.8 ± 0.2	387 ± 129	
Zuc13M *fa*[g,h]					
Female	Obese	6	6.8 ± 0.3[a]	1406 ± 158[a]	
	Lean	6	6.0 ± 0.3	474 ± 86	
Female	Obese	8			1403 ± 78
	Lean	8			1351 ± 115

Values are means ± SE.
[a]Different phenotypic (obese vs. lean) values within treatment at $P < 0.05$ by the independent *t*-test.
[b]Kahle et al. (current study): age, 13 wk; diet, 54% sucrose; 16 h fast.
[c]Ellwood et al., 1985: age, 16 wk; diet, 54% sucrose; 12 h fast.
[d]Brindley et al., 1992: age, 5–6 mo; diet, lab chow; 2 h fast.
[e]Wiesenfeld et al., 1990: age, 16 wk; diet, 54% sucrose; overnight fast.
[f]McCune et al., 1990: age, 8 mo; diet, not specified; 24 h fast.
[g]Bhathena et al., 1987: age, 17 wk; diet, 54% sucrose; overnight fast.
[h]Yukimura et al., 1978: age, 8–10 wk; diet, lab chow; fast not specified.

and mildly diabetic. Outcrossed *fa* on Zuc13M to the BN/Crl rat did not cause changes in glucose homeostasis relative to those reported previously for the Zucker *fa/fa*.[1,17] The obese Zuc-BN/Crl *fa* had fasting insulin concentrations ($P <$ 0.05) higher than their lean counterparts but were not hyperglycemic.

When animals of both sexes were combined, the fasting plasma concentrations of corticosterone did not differ between obese and lean littermates of either the LA/N-BN/Crl *cp* or Zuc-BN/Crl *fa* interstrain crosses. However, when corticosterone levels from male and female animals were analyzed separately, obese male animals in both interstrains had higher fasting corticosterone levels compared with their lean counterparts ($P < 0.05$). Corticosterone levels were similar in both interstrains of female animals. Obese male LA/N-BN/Crl *cp/cp* rats had significantly higher corticosterone levels than did obese ZucBN/Crl *fa/fa* males ($P < 0.05$). Likewise, glucocorticoid levels differed between the lean males of the two interstrains.

In separate experiments we examined fasting hormonal concentrations and basal glucose levels of F_1 progeny of Zucker *fa/+* rats crossed with LA/N *cp/+* animals. The results are presented in Table 2. Obese progeny were presumed to be *fa/cp* compound mutants. These animals were fed a 54% sucrose diet to allow comparison with previous evaluation[2] of this diet in rat obesity models possessing either the *cp* or *fa* mutation. Michaelis et al.[2] and later Ellwood et al.[12] reported more pronounced hyperinsulinemia, but normal fasting glucose concentrations when LA/N *cp* rats were fed a 54% sucrose diet vs. a 54% starch diet. Starch included as the only carbohydrate source putatively would be metabolized with comparable effects on glucoregulatory parameters as lab chow. The lipogenic effect of sucrose-rich replacement diets on obesity development has been studied extensively in rodents[2] and human beings.[8] Our obese *fa/cp* animals exhibited higher fasting plasma glucose ($P < 0.01$), insulin ($P < 0.05$), and corticosterone ($P < 0.01$) concentrations than their lean littermates.

Table 3. Characteristics of corpulent and fatty rat strains

Strain	Insulin resistance	Diabetes	Hyperlipidemia	Hyperphagia	Hypertension
LA/N-BN/Crl *cp*					
Obese	Yes	Mild	NA	Yes	NA
Lean	No	No	NA	No	NA
Zuc13M-BN/Crl *fa*					
Obese	Yes	No	NA	Yes	NA
Lean	No	No	NA	No	NA
Zuc13M-LA/N *fa/cp*					
Obese	Yes	Yes	NA	Yes	NA
Lean	No	No	NA	No	NA
LA/N *cp*					
Obese	Yes	No	Yes	Yes	No
Lean	No	No	No	No	No
JCR:LA *cp*					
Obese	Yes	Mild	Yes	Yes	No
Lean	No	No	No	No	No
SHR/N *cp*					
Obese	Yes	Yes	Yes	Yes	Mild
Lean	No	No	No	No	Yes
SHHF/Mcc *cp*					
Obese	Yes	Yes	Yes	Yes	Mild
Lean	No	No	No	No	Yes
WKY/N *cp*					
Obese	Yes	Severe	Yes	Yes	NA
Zuc *fa*					
Obese	Yes	No	Yes	Yes	No
Lean	No	No	No	No	No

NA, not available.

Mild fasting hyperglycemia has been recorded in some corpulent strains, for example, the SHHF[18] and SHR/N.[19] Fasting levels of glucose usually are not elevated in the LA/N obese animal.[2,20] Fasting glucose concentrations are similar between obese and lean phenotypes of other corpulent rat strains, for example, the starch-fed SHR/N[21] and JCR:LA.[22]

Mild fasting hyperglycemia has been recorded in sucrose-fed but not starch-fed obese Zucker fatty animals.[16] However, most investigators have found the obese Zucker fatty rat is normoglycemic or shows only very slight hyperglycemia.[23,24] According to our data, as well as to all previous research reviewed, the obese phenotypes of corpulent and fatty rat strains are hyperinsulinemic.[1,2,18,25]

Fasting concentrations of corticosterone at the levels recorded for *fa/cp* rats in this study are not usually observed in either corpulent or fatty rat strains, even though obese animals have been found to exhibit higher corticosterone levels than lean littermates.[13,19] We recorded higher corticosterone levels in the LA/N-BN/Crl *cp/cp* compared with the lean littermates, but at concentrations similar to those given by Ellwood et al.[12] for the LA/N *cp* rat, and that appear to be 50% lower than in the *fa/cp* animal.

We know that our compound mutants appear to have different glucose homeostasis characteristics, particularly fasting glucose and corticosterone levels, relative to those reported previously.[12,16,17] However, because proper controls are not available, we can only speculate on factors causing these differences. We do not know if these unique findings are due to 1) the fact that *fa/cp* animals are compound mutants causing a synergistic effect, 2) the admixture of the genetically different LA/N and Zuc13M parent strains, or 3) both. To further define a possible synergism in the *fa/cp* compound mutant, additional control groups and other glucoregulatory hormones, for example, glucagon, somatostatin, and growth hormone, will be analyzed in the future.

A second phenomenon, reported previously, also illustrates the need to better define the genetic nature of *fa* and *cp*. We recently documented the first record of reproduction in a genetically obese female *fa/cp* rat.[26] A cross between this *fa/cp* female and her brother (genotype unknown) produced four lean and six obese F_2 progeny, with all littermates surviving past the age of 6 months. Obese female homozygotes (*fa/fa* or *cp/cp*) rarely conceive or maintain embryos to birth.[27] Again, it is not possible to explain the relationship of this unique reproductive event to the phenotypes of the compound mutant animals on the Zuc13M-LA/N background.

The primary molecular nature of the *fa* and *cp* genes may soon be identified. The work of Truett et al.[15] suggests that the *fa* gene is homologous to the *db* (diabetes) mutation on mouse chromosome 4 based on the location of two flanking genes, *Ifa* and *Glut1*. Rat chromosome 5 and mouse chromosome 4 have been shown to possess synteny homology with a region of human chromosome 1 (1p31-34).[28] Attempts to clone the *fa* and *cp* genes are underway.

Conclusions

Obese *fa/cp* rats are glucose intolerant and diabetic. The factor(s) causing high fasting glucose levels and very high corticosterone levels suggest that either the *fa* and *cp* mutations have a synergistic effect when expressed in the same animal, that these traits are due to an admixture of the genetic backgrounds of the parent strains, or both. The traits also emerge to a lesser degree in male BN/Crl-LA/N *cp/cp* animals. Because the status of glucose homeostasis differs between interstrains expressing the *fa* and/or *cp* mutations, their exact identification must await cloning procedures.

References

1. Zucker LM, Antoniades HN. Insulin and obesity in the Zucker genetically obese rat "fatty." Endocrinology 90:1320–30, 1972.
2. Michaelis OE, Ellwood KC, Hallfrisch J, Hansen CT. Effect of dietary sucrose and genotype on metabolic parameters of a new strain of genetically obese rat: LA/N-corpulent. Nutr Res 3:217–28, 1983.
3. Zucker LM, Zucker TF. Fatty, a new mutation in the rat. J Hered 52:275–8, 1961.
4. Hansen CT. Two new congenic rat strains for nutrition and obesity research (abstract). Fed Proc 42:573, 1983.
5. Koletsky S. Obese spontaneously hypertensive rats: A model for study of atherosclerosis. Exp Mol Pathol 19:53–60, 1973.
6. Yen TT, Shaw WN, Yu PL. Genetics of obesity in Zucker rats and Koletsky rats. Heredity 38:373–7, 1977.
7. Reiser S, Bohn E, Hallfrisch J, et al. Serum insulin and glucose in hyperinsulinemic subjects fed three different levels of sucrose. Am J Clin Nutr 34:2348–58, 1981.
8. Reiser S, Bichard MC, Hallfrisch J, et al. Blood lipids and their distribution in lipoproteins in hyperinsulinemic subjects fed three different levels of sucrose. J Nutr 111:1045–57, 1981.
9. Edwardson JA, Hough CAM. The pituitary-adrenal system of the genetically obese (*ob/ob*) mouse. J Endocrinol 65:999–1073, 1975.
10. Stern JS, Hirsch J. Obesity and pancreatic function. In: Handbook of physiology, endocrinology Vol 1: Endocrine pancreas. Washington DC, American Physiology Society, pp 641–51, 1972.
11. Travis RH. Pathophysiology of syndromes of cortisol excess in man. In: Handbook of physiology, endocrinology Vol 6: Adrenal gland. Washington DC, American Physiology Society pp 271–82, 1975.
12. Ellwood KC, Michaelis OE, Emberland JJ, Bhathena SJ. Hormonal and lipogenic and gluconeogenic enzyme responses in LA/N-corpulent rats. Proc Soc Exp Biol Med 179:163–7, 1985.
13. Brindley DN, Hales P, Al-Sieni AI, Russell JC. Sustained decreases in weight and serum insulin, glucose, triacylglycerol and cholesterol in JCR:LA-corpulent rats treated with D-fenfluramine. Br J Pharmacol 105:679–85, 1992.
14. Johnson PR, Zucker LM, Cruce JA, Hirsch J. Cellularity of adipose depots in the genetically obese Zucker rat. J Lipid Res 12:706–14, 1971.
15. Truett GE, Bahary N, Friedman JM, Leibel RL. Rat obesity gene fatty (*fa*) maps to chromosome 5: Evidence for homology with the mouse gene diabetes (*db*). Proc Natl Acad Sci USA 88:7806–9, 1991.
16. Bhathena SH, Aparicio P, Revett K, et al. Effect of dietary carbohydrates on glucagon and insulin receptors in genetically obese female Zucker rats. J Nutr 117:1291–97, 1987.
17. Yukimura Y, Bray GA, Wolfsen AR. Some effects of adrenalectomy in the fatty rat: Endocrinology 103:1924–28, 1978.
18. McCune SA, Baker PB, Stills HF. SHHF/Mcc-*cp* rat: Model of obesity, non-insulin-dependent diabetes and congestive heart failure. In: Lessons from animal diabetes. Shafrir E, ed. Smith-Gordon, London, 3:397–401, 1990.
19. Wiesenfeld PW, Baldwin J, Chanderbhan R, et al. Comparison of adrenal function

in female SHR/N-*cp* rats on sucrose and starch diets. In: Lessons from animal diabetes. Shafrir E, ed. Smith-Gordon, London, 3:391–6, 1990.
20. Voyles NR, Bhathena SJ, Kennedy B, et al. Tissue somatostatin levels in three models of genetic obesity in rats. Proc Soc Exp Biol Med 185:49–54, 1987.
21. Russell JC, Koeslag DG. JCR:LA-corpulent rat: A strain with spontaneous vascular and myocardial disease. ILAR News 32:27–31, 1990.
22. Bryce GF, Johnson PR, Sullivan AC, Stern JS. Insulin and glucagon: Plasma levels and pancreatic release in the genetically obese Zucker rat. Horm Metab Res 9:366–70, 1977.
23. Ionescu E, Sauter JF, Jeanrenaud B. Abnormal glucose tolerance in genetically obese (*fa/fa*) rats. Am J Physiol 248:E500–6, 1985.
24. Michaelis OE, Carswell N, Hansen CT, et al. A new genetic model of noninsulin-dependent diabetes and hypertension: The spontaneous hypertensive/NIH corpulent rat. In: Lessons from animal diabetes. Shafrir E, Renold AE, eds. Libbey, London, 2:257–64, 1988.
25. Szepes B, Michaelis OE, Hansen CT. Very severe diabetes in a substrain of WKY/N-cp male rats. FASEB J 7:A270, 1993.
26. Kahle EB, Butz KG, Michaelis OE, et al. The first documentation of reproduction in a genetically obese female *fa/cp* rat. ILAR News 36:75–7, 1994.
27. Chelch AM, Edmonds ES. Copulatory behavior and reproductive capacity of the genetically obese female Zucker rat. Physiol Behav 27:331–5, 1981.
28. Searle AG, Peters J, Lyon MF, et al. Chromosome maps of man and mouse. Ann Hum Genet 53:89–140, 1989.

Author Index

Subject Index

MIX
Papier aus verantwortungsvollen Quellen
Paper from responsible sources
FSC® C105338

If you have any concerns about our products,
you can contact us on
ProductSafety@springernature.com

In case Publisher is established outside the EU,
the EU authorized representative is:
Springer Nature Customer Service Center GmbH
Europaplatz 3, 69115 Heidelberg, Germany

Printed by Libri Plureos GmbH
in Hamburg, Germany